A.D.		
1870 (approx)		United States Department of Justice
1870's		American District Telegraph
1878	U.S.	Congress provided for a system of Indian Police
1882		Border Patrol
1882		Office of Intelligence, by Secretary of Navy, Bureau of Navigation
1889		Brinks Incorporated
1890	St. Louis	St. Louis/San Francisco Railroad Police
1891		Immigration and Naturalization Service
1892		Illinois Central Railroad Police
1893		Manchester Dock Police Force—England
1893		National Chiefs of Police Union (International Association of Chiefs of Police)
1900 (approx)		"Hunt Police"—Pawnee (American)
1900		Chicago, Milwaukee, St. Paul, and Pacific Railroad Police
1901		Southern Pacific Railroad Police
1903		Policewoman—Germany
1904 (approx)		Green Crop Society—Ting Hsien, a North China Community
1909		William J. Burns Detective Agency
1914		14,000 Railroad Police working in U.S.
1918		Protective associations (U.S.)
1919		Special Intelligence Unit (I.R.S.)
1921		Protective Section of Association of American Railroads
1924		Federal Bureau of Investigation
1928		Pilferage—Lloyds list reported a strike. Dock laborers demanded a distribution of sugar in addition to wage in Boulogne
1930's		Plant Protection (U.S.)
1930		Narcotics Bureau
1931		Forerunner of National Auto Theft Bureau
1933		Early Study of Private Police (Shalloo)—U.S.
1934		Alcohol Tax Unit (I.R.S.)
1936		Commercial Association on the Water—South China—Canton
1940's		Retail/industrial security expands
1942 (approx)		Kingdom of Nupe in Nigeria—The King, formed of his personal slaves the police force
1947		CIA formed
1948		Office of Special Investigations
1949		Associations for the regulation of cutting lumber. Associations for the protection of fruits and bamboo sprouts—Formosa
1951		Internal Revenue inspections
1953		Fear of public opinion expressed was the chief preventative of mischief and crimes—Kikuyu
1954		Wackenhut Detective Agency
1954		Organized Crime and Racketeering Section (Department of Justice)
1955		American Society for Industrial Security
1963		Guardsmark, Inc.
1970		Rand Corporation Study of the Private Police
1974		American Society for Industrial Security—appoints Professional Certification Board
1975		LEAA—appoints Private Security Task Force

SECURITY ADMINISTRATION
AN INTRODUCTION

Third Edition

Security Administration

An Introduction

By

RICHARD S. POST, Ph.D.
President, Richard S. Post & Associates
Chicago, Illinois
Formerly Executive Vice President
Guardsmark, Inc.
Formerly Chairman, Criminal Justice
University of Wisconsin—Platteville,
Central Intelligence Agency and
U.S. Army Military Police Corp.,
Plant Protection Officer and
Director of Security Services

ARTHUR A. KINGSBURY, Ph.D.
Associate Dean, Business and Public Service Departments
Macomb County Community College
Mt. Clemens, Michigan
Formerly Assistant Director
Department of Criminal Justice
University of Wisconsin, Platteville
United States Treasury Agent
United States Army Counterintelligence Officer

HV
8290
.P6
1977

CHARLES C THOMAS • PUBLISHER
Springfield • Illinois • U.S.A.

Published and Distributed Throughout the World by
CHARLES C THOMAS • PUBLISHER
Bannerstone House
301-327 East Lawrence Avenue, Springfield, Illinois, U.S.A.

This book is protected by copyright. No part of it
may be reproduced in any manner without written
permission from the publisher.

© 1977, by CHARLES C THOMAS • PUBLISHER
ISBN 0-398-03572-5
Library of Congress Catalog Card Number 76-13007

*With THOMAS BOOKS careful attention is given to all details of
manufacturing and design. It is the Publisher's desire to present books that
are satisfactory as to their physical qualities and artistic possibilities and
appropriate for their particular use. THOMAS BOOKS will be true to those
laws of quality that assure a good name and good will.*

Library of Congress Cataloging in Publication Data

Post, Richard S.
   Security administration.

   Bibliography: p.
   Includes index.
   1. Industry—Security measures. 2. Public buildings
—Security measures. I. Kingsbury, Arthur A., joint
author. II. Title.
HV8290.P6 1976   363.2'3   76-13007
ISBN 0-398-03572-5

Printed in the United States of America
C-1

PREFACE

THIS VOLUME REPRESENTS a total revision of the textbook *Security Administration: An Introduction* originally published in 1970. When it was published there were very few sources for the serious student of security and loss prevention. Consequently *Security Administration* attempted to provide a link between what might be considered operational materials and an overview of the entire field of security, loss prevention, crime prevention, and security management. While it was initially successful in accomplishing its original task, other more specialized materials began emerging as the field grew and developed professionally.

The decision to completely revise *Security Administration* and to change its format was not an easy one. Since it had become one of the leading books in its field there was some hesitation by the publisher to "change a winner." It was, however, decided that a book was necessary to focus on the needs of the beginning student in colleges, universities, and in the field. This revised edition is designed to fill that need.

The most relevant portions of the original text have been revised and included in this book along with considerable new and current materials. The presentation is designed to provide a generic overview of each dimension of the security field. Each major concept in prevention, protection, loss control, crime prevention, as well as security administration and management are presented.

The bibliography of *Security Administration* has been retained and considerably expanded in this text. Entries have increased from 2,600 and now exceed 6,000 with a detailed index to assist research and further study into areas of interest.

The purpose of this book is to provide the beginning student of security with a single complete sourcebook for initial study and continued reference throughout his career.

The bibliography is designed to assist the location of supplementary materials. It is presented by subject within each chapter to insure quick and ready reference when needed.

Since the publication of the first edition of *Security Administration* the security field has grown and changed considerably. Thousands of new professionals are working at jobs which did not exist; schools are preparing new students to enter the field in increasingly large numbers; public expectations about the level of service to be provided have grown; and the need for professionally competent, well-prepared and sensitive professionals has never been greater. This volume was prepared in response to this changed environment. It provides a single sourcebook both *about the field* of security and *for it*.

Security Administration: An Introduction is *about the field* since it provides a comprehensive treatment of all its facets. It provides the history, rationale for decisions and organizational structure of current security systems; and the basis for protective programs. *It is for the field* since it presents the various and many procedures, techniques, policies and resources necessary for the successful management and operation of a security program.

Security is at a critical juncture in development. Public law enforcement agencies have come under the close scrutiny of an interested public with demands being made for upgrading personnel and increasing the quality of service. Similarly, a wide range of agencies and individuals that provide security for the many private purposes are also beginning to come under scrutiny. If security is to continue to be a forerunner in protecting the individual in society, an awareness must be created of the implications of the existing organization and activities of security in a democratic society.

Professionalization of security will become a reality only when those engaged in its provisions are well prepared for their tasks and are responsive to the ever-changing needs in their chosen field of interest.

R.S.P.
A.A.K.

ACKNOWLEDGEMENTS

THE TASK OF SUMMARIZING a field of activity such as security is a tremendous undertaking and could not be possible without the assistance and support of many people in the field. Many persons and organizations contributed their time and efforts into the development of the materials, concepts and programs presented in this volume. They include:

James C. Brown
Loss Prevention Consultant
Park Ridge, Illinois

James Brodie
Director of Security
Oak Forest Hospital

Wayne Hanewicz, Coordinator
Criminal Justice Training Center
Macomb Community College

Dr. Wayman Crow
Director, Western Behavioral
Sciences Institute

John W. Kocher
Assistant Professor
Security & Loss Prevention
Lane Community College

Kenneth G. Fauth
Associate Professor
Northern Michigan University

We would also like to acknowledge the support shown by Sangamon State University and Macomb Community College for the clerical and administrative support for this project. Likewise, all the publishers and organizations who cooperated with the undertaking are to be thanked and acknowledged in appropriate places throughout the text.

While the contributions of these individuals and organizations are necessary and most important it is in the last analysis required that someone take responsibility for what is contained in this volume. The materials contained in this book come from and are a part of the security field. It has been our task to collect, evaluate, and present these portions of the body of security knowledge which best expressed the current "State of the Art." The information and knowledge belong to everyone, the omissions and errors are ours.

The Authors

. . . To Penelope and Mary Dell who are still responsible for the completion of this volume.

CONTENTS

	Page
Preface	v
Acknowledgements	vii

PART I
OVERVIEW OF SECURITY

Chapter
1. INTRODUCTION 5
2. HISTORY 27
3. LEGAL BASIS FOR PROTECTIVE SERVICES 76

PART II
SECURITY PROGRAMS

4. GOVERNMENTAL SECURITY PROGRAMS 170
5. BUSINESS SECURITY 220
6. COMMERCIAL PROTECTIVE SERVICES 311
7. INDUSTRIAL SECURITY PROGRAMS 340
8. INSTITUTIONAL PROTECTIVE SERVICES 394
9. INDIVIDUAL PROTECTION PROGRAMS 449
10. GENERIC SECURITY FUNCTIONS 471

PART III
COMPONENTS OF SECURITY

11. PHYSICAL SECURITY 499
12. INFORMATION SECURITY 577
13. PERSONNEL SECURITY 628
14. THE PLANNING PROCESS 665
15. ORGANIZATION FOR SECURITY 735
16. SECURITY TRAINING AND EDUCATION 768

PART IV
PROBLEM AREAS

17. INTERNAL PROBLEMS 787
18. EXTERNAL PROBLEMS 865
19. ISSUES IN SECURITY 891

Index 913

SECURITY ADMINISTRATION
AN INTRODUCTION

PART I
OVERVIEW OF SECURITY

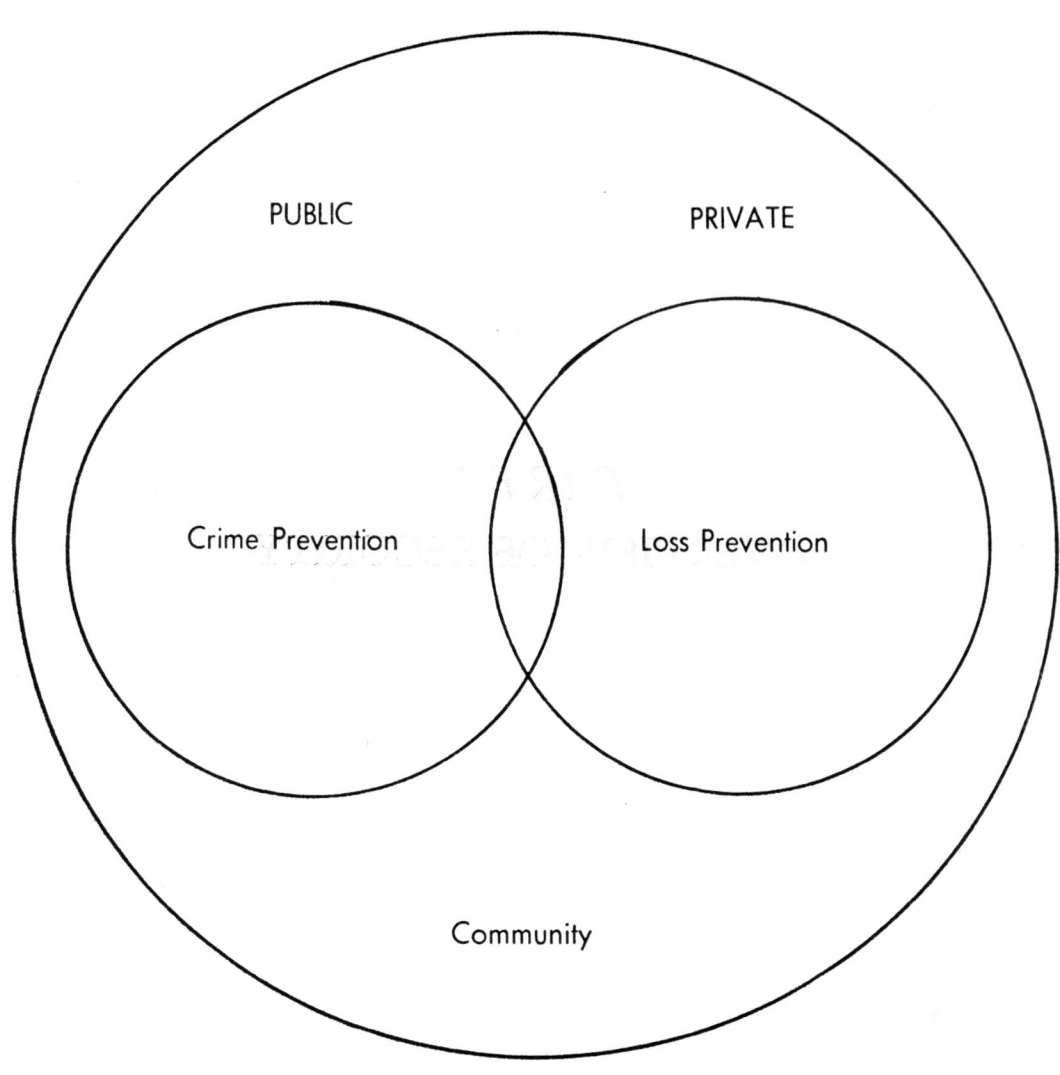

Chapter 1

INTRODUCTION

THE PROTECTION of life, property, and the maintenance of social order are the basic responsibilities of governments in the 20th century. The provision of a secure environment for working and living is viewed as a goal toward which societies, governments, and individuals aspire. Creativity, productivity, and meaningful social progress can be made most viable in an open, dynamic environment in which the individual is free from the fear of unwarranted restrictions on his activities. These restrictions have traditionally come from governmental restrictions on his activities; environmental conditions which precluded him from realizing his potential; or more currently, crime and criminal activities which produce fear, mistrust, social and economic loss, as well as the alienation of persons from each other.

PROTECTIVE SERVICES

In the United States, various levels of government have responded to the need for the protection of the individual, his activities and property through the development and growth of *protective agencies*. Many of these agencies have dual responsibilities for the prevention of crime and disorder as well as the enforcement of established laws governing social conduct. For a variety of reasons, including lack of financial and personnel resources, governmental fragmentation, as well as political, and social reasons, the formally established agencies of protection have not been able to produce the desired level of security for all persons and their property.

The provision of additional measures of security have come to be provided by extragovernmental individuals and organizations who provide the requested level of security on a commercial basis. These individuals and organizations constitute the *private protective services*.

The essential difference between "police" and "security" personnel, the activities they perform, and their roles in providing protection, is the degree and amount of formal sanction or authority given them by the government. The essential difference, in common practice, is that the public security agencies, e.g. police, have focused on active law enforcement with a decreased emphasis on property protection and prevention. The private protective services have, however, focused on prevention and protection and deemphasized their role in enforcement. Thus, it is common to refer to the term "security" as a protection and prevention activity, while "police" is viewed as an enforcement and investigative activity. This distinction is, however, more the result of accommodations historically made between the practitioners in each of these services which have come to be accepted as the true distinction between the functions they perform. To say otherwise would be to indicate that prevention and protection are less a part of public security than law enforcement and order maintenance. To agree with this assertion would be to assent to the assumption that both public "policing" and private "security" are parts of the whole, e.g. public security or protective services.

The legal authority, training, responsibilities, status, and quality of personnel are, at this point in time, documentably superior in the public protective services. Conversely, many of the personnel who currently staff the private agencies have been drawn from the public agencies after retirement. These men are often hired because of the skills or contacts in some aspect of "security" which is required in the private protective services. It thus becomes a moot point that there are significant differences between public and private protective services. The major differences are in the legal authority, focus of attention, quality of operating personnel, lower pay and status, as well as poorly defined career opportunities for the private officers.

The private protective services industry employs approximately 290,000 persons. The public protective services field employs approximately 515,000 (of which 120,000 are guards). The two groups (815,000) account for all the major protective services programs provided in the United States.[1]

A descriptive analysis of the private sector of the protective services industry has been conducted and is reported in the Kakalik and Wildhorn study.[2]

Types of Protective Services

There are three major areas in protective services. These are (1) (Public) government, (2) (Private) proprietary, and (3) commercial programs.

Governmental Programs
(Public Protective Services)

Governmental programs are those provided within each state, local, and federal agency relating to law enforcement, crime prevention, loss control, and property protection aspects of public safety.

Each level of government has a responsibility to provide some type of protective service for itself as well as society. In the case of the federal government, security and enforcement activities are administratively separated from protective functions for federal property. The Federal Protective Service, General Services Administration is responsible for the protection of federal buildings while the Federal Bureau of Investigation is responsible for investigating threats to damage them. Thus, each agency has a specific protective mission which is related to the primary function (either *passive*, e.g. building security, or *active*, e.g. investigation of threats) it is to perform.

In the United States, the concept of criminal justice has emerged as an "umbrella" term to embrace all the activities considered to be part of the formal social control structure. This "umbrella" includes the police, courts, and correctional institutions as well as the prosecutorial function. This focusing on only the formal agencies of social control has overlooked the activities represented by security officers and by organizations for which they work.

The private protective services in the United States represent a total pool of manpower nearly equalling that of all public police agencies. Expenditures for their services are also equal to approximately 60 percent that of expenditures for public law enforcement services.[3] The activites in which they are engaged are identical in many cases to those performed by public law enforcement agencies. Yet, the concept of criminal justice excludes mention of the activities performed by these organizations and systematically denies (by definition) that there is any relationship between nongovernmental protective services and those provided by formal public agencies.

It equally systematically excludes the fact that police services, for the most part, provide only after-the-fact information collection and "emergency response" services. It overlooks the historically demonstrable evidence that patrol services have little positive correlative relationship between crime prevention or deterrence and further overlooks the fact that other social agencies and public agencies as well as private activities relate very directly to lowering the incidence of a wide variety of serious crimes, including homicides, aggravated assault, burglary, and robbery. Studies, the most notable of which is the Crisis Intervention Demonstration Program in New York, show that "untraditional" police activities are yielding much greater results than random roving patrol-type activities. Individuals are free to choose those protective services resources in a community to solve their protective requirements. Figure 1-1 presents the protective choices available in addition to those of the criminal justice system. Nevertheless, the criminal justice concept has permeated the thinking about protective services and has provided a false sense of assumptions about providing protection for members of society.

The freedom from "fear of crime" and the "freedom from fear" itself both interact to provide the environment of effectiveness or ineffectiveness for protective services (police and private) activities in a given community. The public relations efforts of the police historically have been to promote a "sense of security" in the clients they serve. This has been to show that the formal agency will be able to serve their needs and provide the type of protective services required in the community. In fact, the police have been able to do little more than keep measurable criminal activity within *acceptable* limits in a given community. In their role of an "information collection" agency, they report on criminal deviant activities which have occurred in their community which they were unable or powerless to prevent or control.

Private Protective Services

PROPRIETARY PROGRAMS. Proprietary programs are security programs developed and operated by individual companies to protect their assets and operations from loss. According to the Rand Study, this sector comprises a total of 222,000 personnel, plus approximately 10,000 management-level personnel. Proprietary programs are involved in either physical, information, or personnel security programs to prevent loss or damage from occurring to the corporations for which they provide a protective program. They provide functions similar to commercial firms but work only for one specific noncommercial employer.

Commercial Programs

Protective services are sold to clients by security companies on a profit-making basis.

Three basic types of service are offered

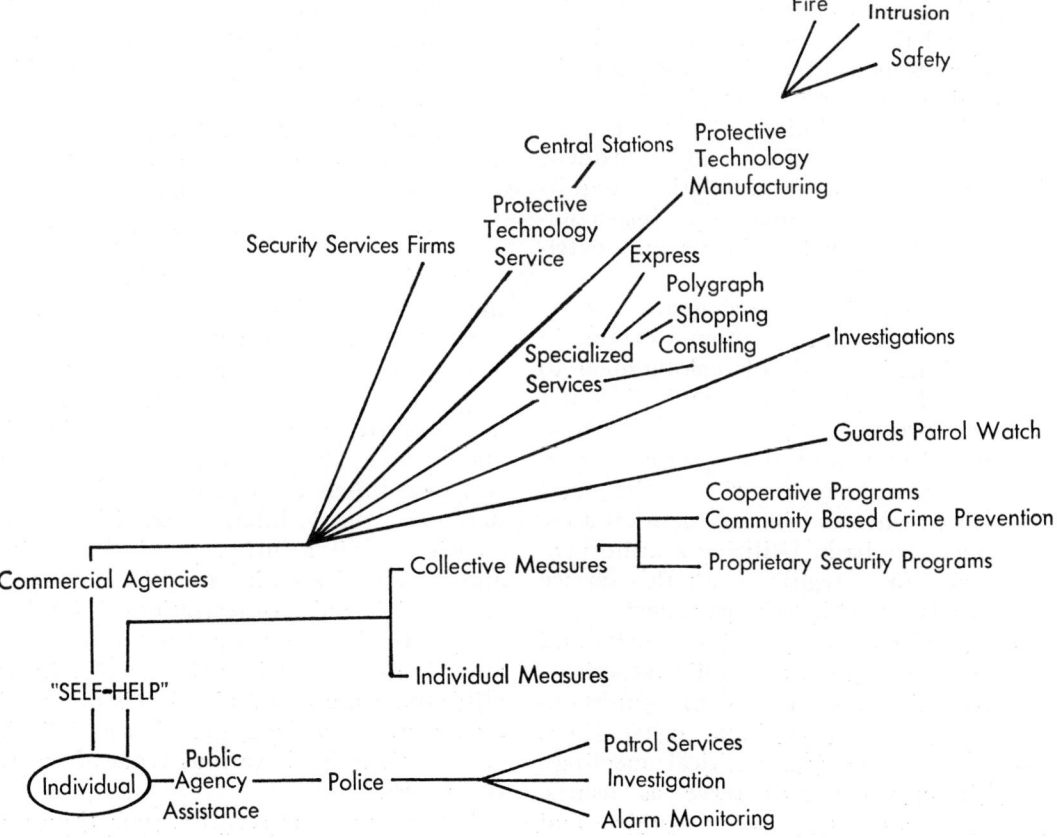

Figure 1-1. Protective choices.

by companies within the commercial industry. These services are:
1. uniformed guard service,
2. central station service, and
3. armored car and courier service.

Some organizations cover more than one category, and those offering guard service provide additional private investigative service.

These services include:[4]
1. Industrial plant security services for building protection, traffic and perimeter control, fire warning, and personnel services. The uniformed guard, basic to plant security, operates effectively as a combined policeman, detective, safety specialist, fire-watch, maintenance man, and public relations representative.
2. Institutional security services for hospitals, schools and colleges, libraries, and other public facilities. Both uniformed and plainclothes personnel help guard against drug traffic and internal theft, protect students and employees, patrol buildings and grounds, and control and direct visitors.
3. High rise building security services, serving office buildings and apartment houses with large populations requiring constant patrol and inspection against internal accidents and external marauders.
4. Uniformed and armed guards for retail establishments, often working in conjunction with plainclothesmen to apprehend shoplifters and dishonest employees.
5. Patrol and inspection services to provide after-hours security to the small store owner and businessman on a cooperative basis. Utilizing a uniformed guard on a regular beat; this service also is available to homeowners.
6. Special events service for security and crowd control at all public gatherings, including fairs, sports and entertainment events, business conventions, trade shows and political meetings. Uniformed guards serve as ushers, ticket sellers/takers and guides, and protect guests against internal disturbances and outside intruders.
7. K-9 patrol services, utilizing the twin advantages of early warning to the guard and a greater deterrent to marauders. A trained German Shepherd alerts the guard to unseen prowlers, protects him, and provides an additional weapon in case of trouble.
8. Investigative services to handle a broad range of criminal and general assignments for business and management, including corporations, legal and financial firms, and private individuals. Work includes internal surveys, surveillance and pre-employment investigations, plus follow-up of fraudulent claims, missing persons, contract violations, trial evidence, and inventory losses. Retailers now spend an estimated $2 billion annually on store security, ranging from guards to closed-circuit TV systems.
9. Personal protection services, to train chauffeurs and staff in methods of protecting company officials from kidnapping, to perform personal bodyguard services, and to escort late-working employees to their autos or to public transportation.

WHAT IS SECURITY?

Security is many things, depending upon the context in which the term is used. Security, viewed in a vacuum, does not exist. Various types and levels of security can and do exist and are meaningful. The term *security* is often used quite loosely and in many different contexts. For example, such things as national security, international security, internal security, private security, retail security, physical security, and industrial security are all enumerated and used in daily conversations. The definitions of these terms are not often clear and are often used interchangeably. In reality, there are two major levels of security: governmental and proprietary. Within each of these levels various types of security do exist.

In governmental security, such things as

international security, national security, and state security are all present. In the broadest sense, the governmental level of security deals with those problems and issues which protect the interests of the government and its dealings with other nations. In its dealings with subnational units of government, or maintenance of power, or the administration of governmental services, demands are made that its operations be free from interruption and that an environment is established which provides citizens with an opportunity to go about their business in relative safety and freedom from inconvenience. This condition is developed through the use of various administrative governmental processes such as the establishment of regulatory, supervisory, and law enforcement agencies. Various agencies charged with responsibility for maintaining internal security or protection must operate within the legal framework established by the judicial system within the country while performing its tasks. The precautions and measures for security which are utilized by protective agencies must be acceptable to the society. The processes and techniques involved likewise must take into consideration the rights of the citizens in the society.

Proprietary security is the other level of security. This includes all measures to be taken by individuals, partnerships, or corporations to protect their private property or interests. In providing security for specific applications, the purpose of security may be described as providing protection of materials, equipment, information, personnel, physical facilities, and preventing influences which are undesirable, unauthorized, or detrimental to the goals of the particular organization being secured. The restrictions which are placed on governmental security likewise should be existent in other types of security application. In a democratic society such as the United States, security at any level should conform to the general standards of governmental operations within the society. In addition to merely meeting regulations, additional emphasis should be given to security operations which stress equity, justice, rationality, compassion, understanding, empathy, and above all, judgment.

Taken in its broadest context, security at any level attempts to do two things: First, it attempts to provide protection against hazards which are man-made, natural, or environmental; second, it attempts to prevent all unlawful events from occurring to nations, states, municipalities, and individuals. Its main goal is to provide protection against all types and kinds of losses.

Obviously then, there are different levels of security which may be discussed. One can discuss the total security of the society, the security of the vast industrial complex or the security of the small privately owned supermarket. The problems and issues involved are quite similar, the variables being the scope of the problem and the specific security goals. The moral and legal issues are, however, identical.

Definitions of Security

The historical development of the term *security* appeared in the adverb, *securus*, which was from the late Latin sense "safe, free from danger" and later passed into the Roman language. It was also adopted in Germany, and hence appears in English as *sicker*. Related translations incompassed the terms "protect, shield from, guard against, render safe, and take effective precautions against."

Original translations of the word, *securus*, describe basic characteristics which are relevant today. The nine areas include:

1. safe,
2. free from danger,
3. feeling no care or apprehension,
4. protected from or not exposed to danger,
5. guardianship,
6. free from risk,
7. to satisfy,
8. to protect, and
9. take effective precautions against.

The problems of defining security and social control are often centered in the variety and changing nature of "society." Coupled with this are semantic and terminological differences. Consequently, the fact that cultures vary from one to another adds new dimensions and variables which have increased the difficulty in defining security.

Since definitions tend to be arbitrary, any attempt to classify them may result in overlap. The definitions of security may, however, be separated into eight categories as shown below:

EIGHT CATEGORIES OF SECURITY DEFINITIONS

1. HISTORICAL—a narrative of past events
2. PSYCHOLOGICAL—the study and interpretation of the human mind
3. SOCIOLOGICAL—the study of human social behavior
4. FUNCTIONAL—the procedural aspects of social control
5. MANAGEMENT—the organizational context of security
6. NORMATIVE—the presenting of norms and standards
7. STRUCTURAL—security viewed in terms of its parts and interrelationships
8. DESCRIPTIVE—a collection of different classifications of elements of security

Historical

History is a narrative of past events. The definitions that identify security as a historical development include:

> (1) Security knowledge and techniques are an accumulation of the human race.
> (2) Private property is interpreted as intrinsic to the person and an attempt to modify this is an attack on the person.
> (3) The philosopher, John Locke, wrote of "being all equal and independent. No one ought to harm another in his life, health, liberty or possessions."

The major characteristics of the historical definition assume security of the person's safety is *intrinsic* to the individual. Yet, a contradiction to the preceding definitions is the historical legalist approach wherein "security derives from the law."[5] If we excuse the fact that taboos, tokens, and customs constituted the social regulation of tribal life in early civilizations, likewise the total protection of the community was probably self-activating and self-executing.[6] In the same way, "whatever notions of jurisdiction are evolved there must be some authority called forth in support of it, procedural devices to execute it, and methods for sustaining it. Thus, the accompanying of a structure for social control is the need for persuasive authority to shore up the framework."[7]

Psychological

The psychological approach is based on the study and interpretations of the human mind. The concept of learning and the sum knowledge of any particular circumstance is all part of its psychological influence on the definitions of security. The major emotional and behavioral characteristics of an individual or group are encompassed within this definition. Three explanations for security include:

> (1) Our position in security is not wholly to catch a thief; we must try to understand *why* people steal and attempt to put up barriers to prevent them from doing so.[8]
> (2) Security functions are essentially protective (not punitive), preventive, and precautionary through the systematic organization of normal operating relationships. Its ultimate reliance is not in power nor the fear of power but an understanding in the ethos of a society and the production of a climate in which functional responsibilities which effect the destined are discharged.[9]
> (3) The security function seeks to pattern operational relationships and the intentions and attitudes of operators, so that its overall result is a danger-free climate in which the functional constituents can freely express themselves.[10]

In this same category, another definition of security may be stated as a collection of different elements of behavior within a group or by individuals: "Security is freedom from exposure to danger; protection, safety or a place of safety; *feeling* of or assurance of safety or certainty; freedom from anxiety; a means of protection."[11]

Sociological

In broad terms, the sociological approach to define and identify security—more specifically, *social control*—incorporates a number of definite characteristics. They are:

>   (1) the study of human social behavior,
>   (2) the study of human society, and
>   (3) the study of organizations and institutions.

Although the term *socialization* is used in an educational framework, the frame of reference is applicable to the definitions of social control:

> Socialization is the process by which one learns to relate socially to the personnel and artifacts of a culture. Enculturation emphasizes the subject matter learned, while socialization emphasizes its effect in the individual in his behavior relating to others.[12]

At the present time, few sociological definitions exist relating to security. Consequently, a "social service" view is summarized relevant to social control.

> A fundamental premise is that private security services fill a perceived need and provide clear social benefits to their consumer, and to some extent, to the general public. Few would argue that, *ceterus paritus*, if private security services were drastically reduced or eliminated, reported crime, fear of crime, and price of retail merchandise would rise.[13]

Thus, the basic consideration for the sociological interpretation of security tends to be oriented to group, organizational, and cultural processes as they relate to social control.

Functional

The term *functional* is used in the context of application. A large portion of the personnel who are associated with the fields of security and have some influence within the total area of social control tend to define security in a rather narrow procedural role. A classic definition was developed by the British Home Office Crime Prevention Centre: "The anticipation, recognition, and appraisal of a crime risk and the initiation of action to remove or reduce it."[14]

Management

The management definition of security has inherent limiting characteristics. The accepted definition of management incorporates negative connotations for the objectivity necessary to define security. This is not to be confused with the procedural concepts of security which promote administrative categorizations within the framework of security, but rather confine or restrain the limited definitions of security in a management context. The definition of "manage" incorporates limiting words such as "to direct, handle, exert control, make submissive to one's authority and discipline." As a result of utilizing the management concept as a basis for defining security *in toto,* a rather provincial and limiting definition is developed. Consequently, security becomes defined in this context as organizational security: "A dynamic and open psycho-social and economic-technical system designed to create a security awareness that fosters mutually acceptable patterns of attitudes, behavior, and relationships within an organization's environments."[15] This organizational security approach is based on the principle that security, organizational, and management theories should recognize and accept contributions from all the disciplines.[16] However, the predominent characteristics of the managerial definitions of security assumes that individual attitudes are easily modified and the ability to integrate the so-called "psycho-social" and economic-technical systems within the framework of organizational and managerial theories is relatively easy.

Normative

The normative definitions of security

TABLE 1-I
CONTEMPORARY SECURITY DEFINITIONS

Function	Industry	Field
to achieve an end	*commercial*	occupational field
an end to be achieved	firms selling services	profession
a state	firms selling product	discipline
a series of activities	*In-house*	
set of procedures	People providing services	
a series of barriers		
process		
function		
a commodity		
a service		

encompass characteristics which include the prescribing of norms and standards. As a result, security under this category is then defined as: "*Protective security,* the art of preserving property for use by the rightful owner by defensive means; *detective security,* the art of detecting crime and which is at its best when it detects crime in the planning stage."[17]

A second normative definition of security prescribes "acceptable" organizational standards: "Security being those measures which are necessary to maintain a state of well-being within a facility and to prevent loss, damage or compromise due to crime; espionage, sabotage, fire, accidents, disasters, strikes, riots."[18]

Structural

A structural definition of security views social control in terms of its parts and their interrelationships.[19] In like manner, the characteristics of the structural definition encompass the notion of arrangement, and in a real sense, the configuration of elements. Two definitions which incorporate the structural concept of security are:

> (1) "Security, the protection of property of all kinds from loss through theft, fraud, fire, and other forms of damage and waste."[20]
>
> (2) "Security in its widest sense is to protect a way of life."[21]

The attempt to embody segments of loss prevention within the total spectrum of social control is generally the hallmark of a structural definition of security. This is graphically seen in the second definition.

Descriptive

The descriptive definition of security tends to characterize by describing a general collection of different classifications of elements of security. This definition of security explains the process by which security may be obtained. "Security provides those means, active or passive, which allow for the activities of the organization to continue in an uninterrupted and orderly manner without disruption."[22]

Definitions of security also appear to be formulated by a number of different influences which include:

1. rational for security,
2. technical applications of security,
3. procedural aspects of security,
4. preventive aspects of social control,
5. motivational influences on security,
6. organizational concepts of needs and their relationships to security, and
7. individual group and cultural needs relevant to social control.

In contemporary usage, the term *security* refers to these things described in Table 1-I.

Thus, depending on the environment or perspective, security is defined and described accordingly.

DEFINITIONAL CHARACTERISTICS. The most often-found characteristics in the eight previously described categories, will eventually dictate the basis for a comprehensive definition of security. They include:

1. Knowledge and techniques of security are an accumulation of the human race.
2. Personal safety is intrinsic to the individual.
3. Social regulations are part of the individual and group well-being.
4. Emotional and behavioral characteristics of an individual or group are important.
5. Security is prevention-oriented.
6. Security is necessary for organizational and cultural processes.
7. Security may be manifested in functional actions.
8. A system or organization must create an awareness that fosters mutually acceptable patterns of attitudes vis-à-vis security.
9. Certain norms or standards are necessary to have security.
10. Security is part of society.
11. Security can be classified by elements.

Assuming that security may be divided into eight separate categories, Figure 1-2 illustrates a definition based on a configurative understanding of each character.

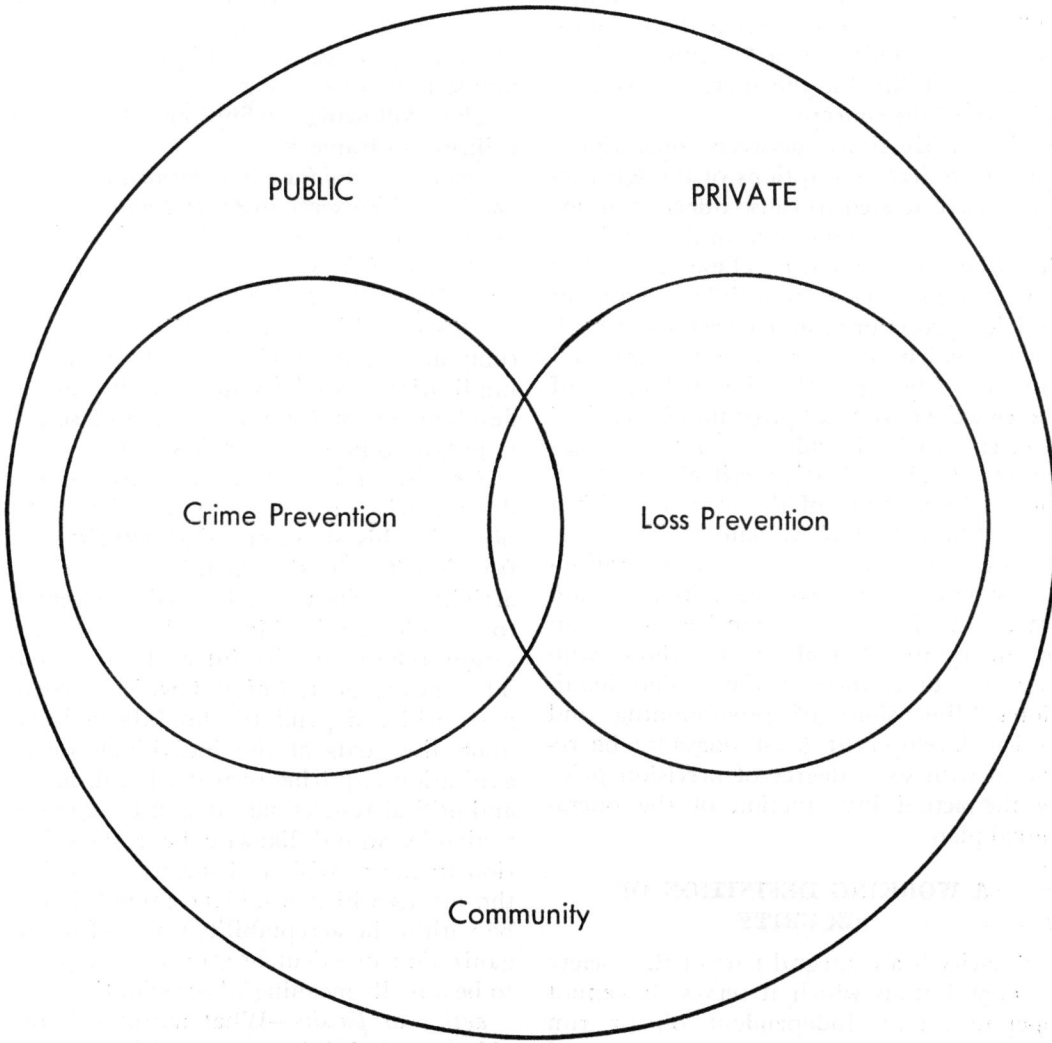

Figure 1-2. Working together to create a secure community and corporate environment.

The individual characteristics are part of each categorical description of security, and as such an understanding of the significance of the total symbolic usage will be necessary prior to defining any given set of principles found in the different categories.

Security has been shown to have a multitude of definitions and meanings based on the context, environment, and purpose. The common element in many of these definitions is that of "protection from" and "prevention of" a variety of events, conditions, and problems. While many of the definitions tend to reflect a negative or defensive image for the activity, process or "state" their main thrust is the development of promotion of a "freedom to" attitude or ability for the user, possessor or initiator of the concept.

The distinctions between operational and theoretical descriptions of the term *security* have tended to cause difficulty in arriving at a common and understandable definition of the term. These differences have made it almost impossible to agree on policies, procedures, and measures of evaluation which are agreeable to user and provider. Consequently, the evaluation of the effectiveness of a "program of security" is often provided under one set of expectations (and evaluative criteria) and reviewed by the user of the program with a much different set of evaluators.

Programs which are based on operational responses to perceived security needs are often not viewed as responsive to human or interpersonal realities by those who must be constrained by them. The definitional dimensions of programming and policy development must therefore be resolved with some degree of precision *prior to* the actual introduction of the operational plan.

A WORKING DEFINITION OF SECURITY

Security is an integral part of the society or organization which it serves. It cannot operate totally independent of, or run parallel to, the power or governmental structure of the organization it serves. Security deals with individuals and their relationships with society or a component part of society which may be business, industrial, or governmental. Since security is *people-oriented,* an operational definition of security must be so structured and framed so that this orientation is taken into consideration. Likewise, societal resources available, such as environmental, cultural, natural, and governmental, would influence the types of security utilized or developed in a given situation. These things being given, a definition of security must follow which encompasses all these factors and universally provides a frame of reference for the establishment of a system, program, process, or technique for the administration of security.

The following definition of security utilizes this framework:

Security provides those means, active or passive, which serve to protect and preserve an environment which allows for the conduct of activities within the organization or society without disruption.

This definition can be divided into six component parts and analyzed so that its implications can be noted relative to the development and administration of security programs in a democratic society:

Security provides those means—How does security provide those means? It is implicit in this statement that security is a function. Obviously, then, as a functional activity, security would be used as a vehicle to provide certain things within an organization relative to its internal protection. The means or techniques which security may utilize depend to the fullest degree upon the goals of the individuals or organization requiring security. Legal, moral, and ethical restrictions, as well as resources available, would likewise be a consideration in the provision of means. Therefore, the means which would be provided must be within the acceptability range of the organization or clientele group to be served to be a really meaningful contribution.

active or passive—What means will provide security? The means which provide for security are one of the most critical areas of security administration. A number of basic questions must be answered before a determination of means could be made. An estimate of the job to be done, goals, objects or things to be protected, and the potential risks must all be determined. The means, techniques, and processes must be developed which most economically, effectively, and efficiently provide the level or degree of protection that the organization requires.

which serve to protect and preserve—These two terms, *protection* and *preservation*, are the hallmarks of any security program. The degree to which preservation and protection must be provided are the direct outgrowth of the necessity for an emphasis upon security programming within an organization.

an environment—This represents the most delicate, sensitive area of a security program. An ethereal effect, or atmosphere, must be developed and maintained by the security program which keeps within the organization goals and philosophies. The activities of the organization, the organizational structure, and the management policies determine the environment within which the organization wishes security to be maintained. The environment to be developed can be anywhere along a continuum . . . from a very secretive, compartmentalized organization or environment, to one which is quite open and relaxed. The overriding consideration in environment establishment and maintenance is the provision of the needed amount of security while providing for the maximum amount of employee efficiency and productiveness within the organization.

which allows for the conduct of activities within the organization or society—This indicates that the processes or techniques utilized in providing for effective security should be integrated into the activities of the organization in such a manner that they are the least obtrusive. Overly restrictive security measures may restrict the employees in the conduct of the primary mission of the organization. Conversely, too liberal security measures may provide employees with too wide a latitude in their performance of duties and may seriously affect the long-run effectiveness of organizational activities.

without disruption—the process or procedures utilized in maintenance of organizational security from any internal or external forces, man-made, or natural, should not be allowed to disrupt the activities of the organization. It is the function of security to prevent and preserve organizational integrity with a minimum of obtrusiveness or interference. Security precautions are designed to give little more than protection from the unforeseeable. With proper planning and proper programming, security can fulfill its organizational function without disrupting the organization's major activities while insuring a reasonably reliable protection factor against disrupting influences.

DIVISIONS OF SECURITY

Security may be categorized according to its function. There are essentially three types of security that may be discussed: *personal, property,* and *governmental* security. In historical development, personal security would be considered the oldest form, followed by property protection, with governmental security being the last in the line of development. These three types of security are present at either of the levels of security previously indicated. Personal and property protection might be either a function at the governmental or the private level. For example, the personal protection of the president of the United States is a governmental security function whereas, conversely, the protection of governmental materials by an industrial contractor would be at the private level of security. Regardless of the level at which these types of security might be performed, each type of security is provided by one or any combination of the three security processes. These processes are physical, personnel, and information security. All security, whether governmental or private, is provided through these three pro-

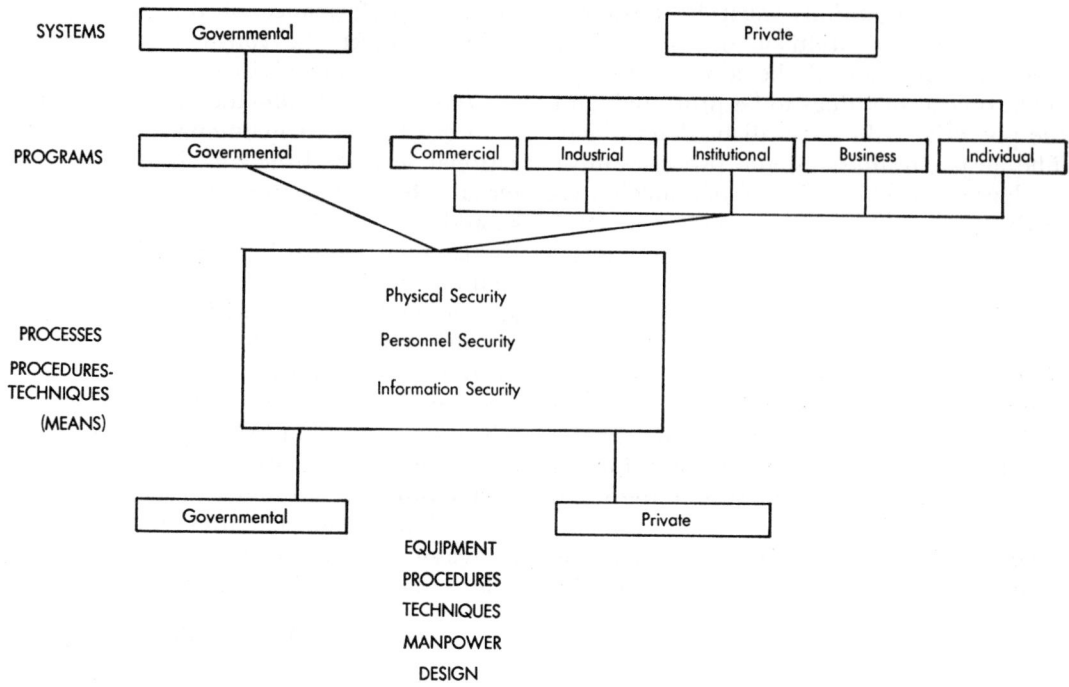

Figure 1-3.

cesses. These processes are incorporated into programs which provide security for a wide range of purposes. These purposes, however, can be categorized into four major program areas: industrial, business, individual, and governmental security. These programs can be divided into two major systems: governmental or proprietary, again based on clientele served.

The final category to be discussed is that of means. This category involves the procedures and techniques used in the actual implementation of security processes within a particular program system. These are the hardware, procedures and techniques. It designates the means which can be utilized in the actual establishment of a security program. This would normally be the operational level of security, involving such things as guard forces, investigative services, safety programs, fire prevention programs, and the like; in other words, the actual *hardware level* of security programming. These means can be employed either as an organic part of the organization internally, or provided by some external source to perform the actual security functions within the programming system.

TOTAL LOSS CONTROL

Over the last several years, security has become synonymous with loss prevention. The concept of *total loss control* or protection from all forms and classes of hazards has thus provided a broad framework for the provision of asset protection. The relatively new concept of security is the generic name which serves to provide a comprehensive systematic method for placing into perspective the component parts from which the individual chooses his protection. It serves to provide an "umbrella" under which all the subsystems of societal and corporate protection are provided. These include insurance, architecture and design, police and fire protection, and safety programs. These subsystems are the components of a "total loss control program."

They work both individually and collectively to provide protection and prevention for vital personnel and property resources.

The term *security* is quickly being replaced in the corporate environment by terminology such as *loss prevention, asset protection,* or *total loss control.* The reliance on a negatively described, defensively oriented managerial subsystem is being replaced by a "systems concept" embracing all aspects of corporate activity. Totally integrated subsystems of protection now include "security" within them rather than viewing it as a separate and distinct operational system.

Total loss control systems attempt to (1) reduce the exposure to risk, (2) deal with breakdowns in risk reduction or operating subsystems within the organizational environment, and (3) take after the fact remedial action to deal with problems relating to loss. Such systems include:

INSURANCE
FIRE
SAFETY   (O.S.H.A.)
SECURITY
DESIGN
HEALTH

These functional subsystems operate to reduce the loss potential for the corporation, its employees, and operations while improving the potential for profitable operations. The integration of these functions enables cost-effective decisions to be made about the allocation of resources for protection. Prior to the development of the total loss control conceptualization, relatively simplistic solutions were often considered adequate to deal with loss problems. Allowable levels of waste or loss were considered as a cost of doing business, as long as the levels were maintained within acceptable limits—those limits normally being how much impact they have on the profitability of the overall operation.

The current realities of industrial and economic loss due to criminal activity have placed a rather grim imperative before the business and industrial community . . . reduce losses or go out of business! Thousands of businesses each year close their doors forever because of uncontrolled losses and thousands more have substantially reduced profits for the same reason. However, many businesses still attempt to lower their losses with partial solutions such as guards, alarm hardware, dogs, or additional insurance coverage. They do not, as yet, feel the imperative which is before them.

In the past several years ecology, the economy, and energy shortages have emerged as the three major problems to be faced by everyone in the world. These three areas of concern have placed an imperative, a rather stern one, on the political decision-makers as well as those within the corporate structure. The solutions to these problems can no longer be a patchwork of short-term palliatives. The actions taken must be of a magnitude equal to that of the problem. In the past, this has not been the case with the current crisis being the result. Narrowly defined, short-term solutions will not solve major problems; particularly those which cut across national boundaries as well as competing economic interests.

Similarly, an imperative exists for the business and industrial community as well as the individual in society. Massive overhauling of the criminal justice system has been recommended since the Wickersham Commission and is currently being advocated by the National Advisory Commission on Criminal Justice Goals and Standards. These recommendations have, at best, been incrementally implemented. The police departments still provide response to breakdowns in social order, criminal activity, and community order maintenance. Except for the use of modern technology, they provide about the same quality and level of service as did their counterparts fifty years ago. It has been estimated that the per capita expenditures for police services (excluding federal funding through the Law Enforcement Assistance Administration, U.S. Department of Justice) has not increased for the past 100 years!

A situation exists wherein the response to the protection of lives and property in the community has not appreciably increased in the face of rising criminal activity and economic losses to citizens. Much the same is true in the business and industrial community.

Policing the plant, requiring employee ID cards, searching lunch boxes, and the myriad of policies and equipment used to reduce or discourage theft or loss has not reduced their occurrence.

Much of the program planning in industrial security has been defensive in nature: building of barriers around the corporation much like the moats which surrounded castles in the Middle Ages. While these barriers tended to keep unwanted persons off company property, they did little else. Internal protection was oriented to insuring that employees observed work rules, safety hazards and equipment malfunctions were reported and that the plant was not carried away at the end of each shift by employees. Most of the activities were of a defensive or after-the-fact action-taking nature; much like the police.

This reliance on after-the-fact action and barrier building was, however, to be expected since most of the managers and directors of security programs have historically been drawn from the ranks of police or investigative agencies. These men attempted to transfer those techniques, attitudes, and programs which were used in their previous employment into the industrial environment. Since there were no other models to compare the value of these programs against, they became the standards against which other "new" approaches might be measured.

A considerable portion of this text deals with the presentation of these standards, how they operate, and their effectiveness. The development of "new" approaches or models will evolve from the field of security as it matures and grows.

SUMMARY

The last several years have seen a dramatic growth in the problems encountered by society in maintaining order and preserving peace in our communities and in our nation. Likewise, the problems of maintaining security for business establishments, retail manufacturing firms, and those services which directly provide for and promote the health and safety of our society have been acute.

The problems of disorder and social disintegration have also become the problems of business and industry. The disruption of business by civil disorders, demonstrations, riots, and disturbances has become common. The problems of shoplifting, pilferage, petty theft, and other allied problems have likewise grown to alarming proportions. Never before in recorded history have the problems of general society impinged more directly upon the private pursuits of its citizens than at the present. The need for providing protection for property and private interests, over and above those which government can provide, have never been more in demand. Into this setting has been placed the security officer who has been made responsible for the protection of individual property interest rights in the face of growing disregard for authority and for the integrity of one's property and possessions. Security officers must be aware of the duties and responsibilities of their professions, and they must be willing to provide the services necessary for the attainment of their goals within the standards of conduct established by the government in a democratic society. In order to do this, the security officer must be aware of the history and development of his profession, of his legal responsibilities, and of the problems and issues confronting him in the conduct of his duties.

The mechanism used by society to protect life and property is a type of public or private agency with safety and/or security responsibilities. Public agencies such as federal, state, and municipal police departments have general responsibility in society for maintenance of order, control and regulation of conduct, protection of the individual and his property, and the enforcement of criminal laws. In its broadest context, private security has the same major goals but only serves a restricted clientele group. Both types of agencies, therefore, have similar general goals. The public agencies have their rightful activities prescribed by statutes or ordinances, whereas the private agencies' specific goals are based on the wishes of the clientele groups. These goals, however, are either in part or totally consistent with those of public agencies. For example, a private agency may be required to prevent and deter shoplifting (larceny) within a large retail store. It accomplishes this goal by utilizing one-way mirrors, closed circuit television, store detectives. Thus, it has an identical part of the goal, property protection, as do the traditional public agencies: "The prevention of crime and disorder, the preservation of the peace, and the protection of life, property, and individual freedom."[23] Obviously then, it can be said that private agencies have specific roles and functions in society which can be defined and circumscribed.

The security function, as performed both by governmental and proprietary agencies, has developed considerably since the first night watchmen were appointed in Jamestown. While the development of the proprietary and governmental activities have not followed a consistent pattern, both varieties have evolved into large scale, effective, protective devices. In tracing the historical development of governmental agencies, we see order maintenance and law enforcement agencies develop, both at the municipal and federal level, in response to specific needs which could not be adequately provided for on an individual basis. For example, the government's desire to protect its currency resulted in the development of the Secret Service. Similarly, the recent need for a coordinating force against organized crime activities, resulted in the establishment of the Organized Crime and Racketeering Section in the Department of Justice.

The development of proprietary security on the other hand, followed a different but parallel pattern. Private security personnel performed law enforcement activities when public law enforcement agencies did not exist or could not perform adequate law enforcement. They currently perform preventative activities because public law enforcement agencies cannot provide the extensive coverage needed by many types of organizations. They likewise supplement law enforcement activities in areas where public agencies cannot adequately perform their functions.

Since the purpose of security agencies is primarily one of prevention, they can be considered to provide true *policing* or *order-maintenance* activities for the organizations they serve. If their activities are viewed in the context of the total society, security can be viewed as providing a *prepolice* function in relation to the formal criminal justice agencies in this country.

We have reached a point in development that proprietary security personnel are involved directly in the criminal justice process. Security agencies are doing prevention work and are also acting as prepolice agencies. Examples are as follows:

1. An unlawful event occurs on private property and the security agency has the option of either doing the investigation themselves, or turning it over to a public law enforcement agency.

2. A private firm has an internal theft of a large amount of money or an embezzlement and may or may not call in a public law enforcement agency. If they do not wish to do so, their investigative staff will conduct an investigation and an internal disposition of the individuals involved. The organization will bring its own sanction within its own *system of justice*.

3. Conversely, if an event such as shoplifting occurs, security officers would act in the same capacity as a law enforcement agency and be called in, and the suspect brought into the formal criminal justice process. In the last example, they are acting as prepolice agents in introducing the person into the formal system.

Similarly, the general preventive and protective activities engaged in by proprietary security agents are of the same type as the original law enforcement officer in this country. The original policeman was a *preserver of the peace* or an *order-maintenance* individual, rather than a law enforcer or detective. The contemporary law enforcement agency is now engaged in two distinct functional activities, one being order-maintenance, and the other, law enforcement. Security agencies through the process of natural selection have chosen to engage in prevention type activities, leaving law enforcement to formal agencies of society.

The issue of role definition and goal orientation for private and public agencies and how their interrelationships affect a fully integrated system of safety-security public protection represent a critical area of concern for security administrators. This problem will be discussed more fully in subsequent chapters.

APPENDIX 1A

INTERNATIONAL SECURITY ASSOCIATIONS

A compilation of the major security associations is herein presented as a guide to the many and varied professional groups.

In a like manner, a comparative view of three international security associations are presented. The three organizations include:

A. American Society for Industrial Security (U.S.A.)
B. Industrial Police and Security Association (U.K.)
C. Industrial and Commercial Security Association of South Africa Ltd. (South Africa)

A compendia of the individual society's goals, membership catagories and special projects are listed.

SUMMARIZED EXTRACTS FROM THE BY-LAWS OF EACH SOCIETY

American Society for Industrial Security (A.S.I.S.)—U.S.A.

A.S.I.S. is generally considered the most prominent professional security society within the continental U.S. This is not to suggest that the many security societies which are in existence within the U.S. are not excellent, but rather most of the related security associations are "trade" oriented, whereas A.S.I.S. encompasses all facets of loss prevention. A.S.I.S. has recently taken major steps in developing chapters throughout the world and also sponsoring seminars on an international basis.

GOALS: The Society's By-laws indicate the objectives of A.S.I.S. will be to:

*1. Promote interchange among members of the Society relevant to information, ideas, methods, relating to security, etc.
*2. Promote a means for maintaining a representative and centralized body or organization to collect, collate, coordinate and distribute information, ideas relevant to security, etc.
*3. Establish ethical and professional standards for its members, etc.
*4. Print, publish, distribute and circulate books, pamphlets, etc., in connection with and in furtherance of the activities and purpose of the society, etc.
*5. Make surveys and studies, hold conferences and forums, and arrange for the presentation of lectures, etc., on matters of interest, etc.
*6. Solicit, collect, raise and obtain money for any of the purposes of the society, etc.

---

* Please refer to the A.S.I.S. By-laws for a comprehensive and detailed description.

MEMBERSHIP:

The membership of A.S.I.S. has five (5) categories consisting of:
1. Active Member
2. Affiliate Member
3. Student Member
4. Life Member
5. Honorary Member

SPECIAL PROJECTS:

The Society publishes a number of technical works, scholarly research papers, magazines *(Security Management)* and other related materials. The A.S.I.S. Foundation, Inc., is a nonprofit entity within the total framework of the society.

Industrial Police and Security Association—United Kingdom

The I.P.S.A. is the largest and most active security association in the United Kingdom. Similarly, the I.P.S.A., like A.S.I.S., is only one of many security societies within the United Kingdom, and likewise the other associations are chiefly "trade-oriented." This society has done an excellent job in taking appropriate steps toward the "professionalization" of the field of security by establishing an *Institution of Industrial Security,* whose membership is based on experience and successful completion of written exams.

GOALS: The Association's constitution and rules indicate the objectives of I.P.S.A. will be to:
* 1. Establish, promote, and encourage the science and professional practice of security, etc.
* 2. Promote and make more effective security measures in industry and community by establishing liaison and exchange of ideas between members and other interested individuals, etc.
* 3. Establish ethical professional standards, etc.
* 4. Print, publish, and distribute materials in connection with security, etc.

---

* Please refer to the I.P.S.A. Constitution and Rules for a comprehensive and detailed description.

* 5. Study, hold conferences and training programs to further security, etc.
* 6. Provide and maintain a centralized body of information for the security professional and members, etc.
* 7. Establish a placement bureau for its members in particular, etc.
* 8. Administer the *Institution of Industrial Security,* etc.

MEMBERSHIP:

The membership of I.P.S.A. includes five (5) categories:
1. Member
2. Corporate Member
3. Associate Member
4. Group Member
5. Honorary Member

SPECIAL PROJECTS:

The I.P.S.A. has an outstanding security magazine, *Security Protection,* and offers training programs and administers the *Institution of Industrial Security,* which confers (upon appropriate completions of specific requirements) the categories of Graduate, Member, and Fellow.

Industrial and Commercial Security Association of South Africa, Ltd. (I.C.S.A.S.A.)

In like manner, the I.C.S.A.S.A. is an active and dynamic security society within South Africa. Many of its programs and duties parallel A.S.I.S. and I.P.S.A.'s responsibilities.

GOALS: The I.C.S.A.S.A. articles promote the following objectives:
† 1. To establish, promote, and encourage the science and professional prevention of security, etc.,
† 2. promote more effective security measures within industry, etc.,
† 3. collect and maintain a centralized body of security knowledge, etc.,
† 4. establish, foster, and encourage ethical and professional standards of work in security, etc.,
† 5. provide liaison between associations

---

† Please refer to the I.C.S.A.S.A. articles for a comprehensive and detailed description.

and other parties with mutual interest, etc.,
†6. establish and promote studies, forums, and training courses in security, etc., and
†7. provide advice in the selection and training of security personnel, etc.

MEMBERSHIP:
There are four categories of membership:
    1. Ordinary Member
    2. Corporate Member
    3. Associate Member
    4. Group Member

SPECIAL PROJECTS:
An active training and seminar program is regularly offered by the I.C.S.A.S.A. and the continued work of the *Institute of Industrial and Commercial Security* confers the categories of Student, Graduate, Member, and Fellow which are based on an annual examination.

APPENDIX 1B

A SELECTED COLLECTION OF SECURITY AND RELATED ASSOCIATIONS

Airport Security Council (for metropolitan NYC airports only)
97 45 Queens Boulevard
Forest Hills, NY 11274

American Polygraph Association
Box 74
Linthicum Heights, MD 21090

American Society for Industrial Security
2000 K Street, N.W., Suite 651
Washington, DC 20006

American Society of Safety Engineers
850 Busse Highway
Park Ridge, IL 60068

Associated Locksmiths of America
11 Elmendorf Street
Kingston, NY 12401

Associated Special Investigators & Police International
P.O. Box 434
Saint John, NB E2L 4L9

Association of British Private Detectives
37A Horse Ferry
Birmingham 1, England

Association of Burglary Insurance Surveyors
130 Fenchurch Street
London, GC 3, England

Association of Federal Investigators
815 15th Street, N.W., Suite 825
Washington, DC 20005

Association of Transportation Security Officers
P.O. Box 92220
Los Angeles, CA 90009

Automatic Fire Alarm Association
119 W. 23rd Street
New York, NY 10010

British Investigators, Association of
293 Kingston Road
Leatherhead Surrey, England

British Security Industry Association
14 Tottenham Street
London, W.1, England

California Association of Licensed Investigators
1700 Pontius Avenue, Room 202
W. Los Angeles, CA 90025

Canadian Association of University Security Directors
Address Unknown

Canadian Society for Industrial Security
926 Connaught Street
Ottawa, Ontario K2B 5MB

Central Station Electrical Protection Association
1000 Vermont Avenue, N.W.
Washington, DC 20005

Computer Security Institute
43 Boston Post Road/West Main Street
Northboro, MA 01532

Council of International Investigators
Box 5646
Baltimore, MD 21210

Credit Card Investigators, Association of
P.O. Box 813
Novato, CA 94947

Electrical & Electronics Engineers, Institute of
345 East 47th Street
New York, NY 10017

Harness Tracks Security
150 East 42nd Street
New York, NY 10017

Independent Armored Car Operator's Association, Inc.
724 York Road
Towson, MD 21204

Industrial and Commercial Security Association of South Africa, Ltd.
210 Lyndhurst Road
Lyndhurst Johannesburg, S. Africa

Industrial Police & Security Association
4 Syon Park Gardens
1 Sieworth, Middlesex, England

Institute of Industrial Security
4 Syon Park Gardens
1 Sieworth, Middlesex, England

Insurance Crime Prevention Bureaux
365 Evans Avenue, Suite 410
Toronto, Ontario M8Z 1K2

Insurance Crime Prevention Institute
21 Charles Street
Westport, CT 06880

Insurance Security Association
Audit Dept.
Aetna Life and Casualty Co.
Hartford, CT 06115

International Association of Arson Investigators
P.O. Box 1208
Springfield, IL 62705

International Association of College and University Security Directors
2600 Dixwell Avenue
Hamden, CT 06514

International Association of School Security Directors
1320 South West Fourth Street
Fort Lauderdale, FL 33310

International Association of Security Services
Box 378
Northfield, IL 60093

International Guards Union of America
1070 S. Knox Court
Denver, CO 80219

International Union, United Plant Guard Workers of America
P.O. Box 435
East Detroit, MI 48021

Jeweler's Security Alliance of the United States
535 Fifth Avenue
New York, NY 10017

Loss Exccutives Association
Hartford Insurance Group
Hartford Plaza
Hartford, CT 06115

National Armored Car Carriers (a division of the Contract Carriers Conference, ATA
1616 P Street, N.W.
Washington, DC 20036

National Burglar and Fire Alarm Association, Inc.
1730 Pennsylvania Ave., N.W.
Washington, DC 20006

National Civil Defense Advisory Council
Office of Emergency Preparedness
Executive Office Building Annex
Washington, DC 20504

National Classification Management Society
Box 7453
Alexandria, VA 22307

National Counter Intelligence Corps Association
P.O. Box 762
Baltimore, MD 21203

National Crime Prevention Association, Inc.
210 Gardiner Ln.
Louisville, KY

National District Attorneys Association
211 East Chicago Avenue, Suite 1204
Chicago, IL 60611

National Fire Protection Association
470 Atlantic Avenue
Boston, MA 02210

National Security Industrial Association
Union Trust Building, Suite 700
Washington, DC 20005

Property Loss Research Bureau
20 North Wacker Drive
Chicago, IL 60606

Research Security Administrators
P.O. Box 358
North Hollywood, CA

Safe Manufacturers National Association (12 companies)
366 Madison Avenue
New York, NY 10017

Security Equipment Industry Association
233 East Erie Street
Chicago, IL 60611

Society of Former Special Agents of the FBI
Statler-Hilton Hotel
Mezzanine Floor
New York, NY 10001

World Association of Detectives
Box 5068
San Mateo, CA 94402

FOOTNOTES

1. Sorrel Wildhorn and James Kakalik, *Nature and Extent of Private Police in the United States* (Santa Monica, Rand Corporation, 1971), p. 13.
2. Wildhorn, *Nature and Extent of Private Police in the United States.*
3. Wildhorn, *Nature and Extent of Private Police in the United States.*
4. Excerpted from *The Cost of Crimes Against Business,* U.S. Department of Commerce, Domestic and International Business Administration, Washington, D.C. Also, *Security Management,* March 1975, p. 9.
5. K. G. Wright, *Cost—Effective Security* (Maidenhead, McGraw-Hill, 1972), p. 3.
6. See Freud, *Totem and Taboo,* Straching Transl. 1952; Kelsen, *General Theory of Law and State,* Wedberg Transl. 1945; Maine, *Ancient Law 67,* Everyman's Library ed. 1954; Maine, *Early History of Institutions* (1878) Summer, *Folkways* (1940).
7. W. L. Clark and W. L. Marshall, *A Treatise of Its Law of Crimes* (Chicago, Callaghan and Company, 7th ed.), pp. 11-17.
8. Ruth M. Owens, "The Human Element of Security," *Industrial Security,* December, 1967, p. 45.
9. Paul Emerson Knight and Alan M. Richardson, *The Scope and Limitation of Industrial Security* (Springfield, Thomas, 1963), p. 98.
10. Emerson and Richardson, *The Scope and Limitation of Industrial Security,* p. 99.
11. Richard Post, *Determining Security Needs* (Madison, Oak Security Publications, 1973), p. 101.
12. August F. Kerber and Wilford R. Smith, *A Cultural Approach to Education* (Dubuque, Kendall/Hunt, 1972), p. 368.
13. James S. Kakalik and Sorrel Wildhorn, *Private Police in the U.S.: Findings and Recommendations,* Vol. I, LEAA (Government Publication Office, 1972), p. VII.
14. Arthur A. Kingsbury, *Introduction to Security and Crime Prevention Surveys* (Springfield, Thomas, 1973), p. 6.
15. Henry Ursic and Leroy Pagano, *Security Management Systems* (Springfield, Thomas, 1973), p. 38.
16. Ursic and Pagano, *Security Management Systems.*
17. Lord Hayter, "Security and Society," *Security—Attitudes and Techniques for Management,* ed. Noel Currer-Briggs (London, Hutchinson, 1968), p. XVII.
18. David Paine, *Basic Principles of Industrial Security* (Madison, Oak Security Publications, 1972), p. 23.
19. Suggested by Kerber and Smith, *A Cultural Approach to Education,* p. 5.

20. Eric Oliver and John Wilson, *Practical Security in Commerce and Industry* (London, Gower, 1972), p. 3.
21. Briggs, *Security Management and Techniques for Management*, p. 123.
22. Richard S. Post and Arthur A. Kingsbury, *Security Administration: An Introduction* (Springfield, Thomas, 1973), p. 5.
23. A. C. Germann, Frank D. Day, and Robert R. J. Gallati, *Introduction to Law Enforcement*, 7th ed. (Springfield, Thomas, 1968), p. 227.

BOOKS

Gray, and Farnsworth, L. W.: *Security in a World of Change*. Duxbury, 1969.

Green, Gion, and Farber, Raymond C.: *Introduction to Security*. Los Angeles, Security World Publishing Company.

Knight, Paul Emerson, and Richardson, Alan: *The Scope and Limitation of Industrial Security*. Springfield, Thomas, 1968.

Lipman, Ira A.: *How to Protect Yourself from Crime*. New York, Atheneum, 1975.

Lipson, Milton: *On Guard*. New York, Quadrangle/The New York Times Book Company, 1975.

Official Yearbook of Security Police and Fire Services—International Security Directory. Security Gazette, Ltd., 1969.

Peel, John D.: *The Story of Private Security*. Springfield, Thomas, 1971.

Wiles, Paul and McClintock, F. H., eds.: *The Security Industry in the United Kingdom*. Cambridge, Institute of Criminology, 1972.

PERIODICALS

Astor, Saul D.: New look in corporate security. *Business Management*, 33-36, March 1969.

Armstrong, Henry G.: Three primary topics affecting police & security officer liaison-panelist. *Industrial Security*, January 1964.

Anonymous and staff-written: The state of the art. *Security World*, 5: (No. 9), October 1968.

Ahern, John J.: Liaison between police and industry remarks, 1963 seminar. *Industrial Security*, October 1963.

Barnett, Frank R.: Security is everybody's business. *Industrial Security*, 2:No. 8, October 1958.

Bopp, William J.: Police science and the concept of prevention. *Security World*, April 1970.

Brose, Patricia: The security professional—take one. *Industrial Security*, December 1971.

Bronstron, Curtis: Liaison between police and industrial security officers. *Police, 8:*No. 6, July-August 1964.

Davis, J. R.: Police relations with industry. *Indiana Police Chief*, June 1955.

Fidel, Edward A. Jr.: Analysis of industrial security membership. *Industrial Security, 10:* No. 6, December 1966.

Gale, R. F.: Security, Britain's new growth industry. *Police, 4:*No. 8, 24, April 1972.

Goddard, Robert J.: Professionalism in security—fact or fiction. *Industrial Security, 9:*No. 10, January, 1965.

Gulinello, Leo: Security personnel. *Proceedings of the Seminar on Urban Design, Security and Crime*. Washington, U.S. Department of Justice, January 1973, p. 17.

Hansen, Paul: The growth of ASIS. *Industrial Security, 2:*No. 18, October 1958.

How useful are our ethical codes? *Chemical Engineering*, 7, Sept. 2, 1963.

Kakalik, James S., and Sorrel Wildhorn: *Private Police in the United States: Findings and Recommendations*. Santa Monica, Rand Corporation, 1971, Vol. 1.

———: *The Private Police Industry: Its Nature and Extent*. Santa Monica, Rand Corporation, 1971, Vol. 2.

———: *Current Regulations of Private Police: Regulatory Agency Experience and Views*, Santa Monica, The Rand Corporation, 1971, Vol. 3.

———: *The Law and Private Police*, Santa Monica, The Rand Corporation, 1971, Vol. 4.

———: *Special-Purpose Public Police*, Santa Monica, The Rand Corporation, 1971, Vol. 5.

Kingsbury, Arthur A.: Arriving at a definition of security. *Security and Protection*, 13-14, May 1975.

———: Guidelines for security education, ethics. *Security Management*, Vol. 18, No. 5, 27, November 1974.

Liebling, Joseph J.: New trends in industrial security. *Industrial Security*, December 1971.

Mart, V. C.: Private police. *Police Journal*, April-June 1975, 122-132.

McAvory, S. K.: Federal protection force takes progressive action. *Industrial Security*, October 1970.

McCartney, Fred C.: Industrial police. *Police Yearbook*, 1963.

McCarthy, Mai B.: Women in security. *Industrial Security,* June 1966.

Norton, O. P.: Industrial security: An essential element of modern society. *Top Security,* 44-46, May 1975.

Pleece, Sydney: The nature and potential of the security industry. *Police Journal, XLIV:* No. 1, 41, January-March 1972.

Post, Richard S.: Creating a secure environment. *Industrial Security,* December 1970.

———: Contemporary protective services. *Security Register,* January-February 1974, 24-30.

———: Relations with private police services. *Police Chief,* March 1971, 54.

Port, A. Tyler: The challenge of the future. *Industrial Security, 3:*No. 4, January 1959.

Private police forces in growing demand. *U.S. News and World Report, 74:*No. 4, 54-56, January 29, 1973.

Professional standards of senior officers. *Security Gazette,* August 1964.

Ranstad, Harold: Changes and trends in security. *Industrial Security,* 4, June 1968.

Scheltemu, Stades, Ottoline: Das stehlen bei kinden und jugendlichen. *Amsterdam,* 1949.

Scholl, Dr. C. E.: Professionalism and . . . you. *Industrial Security,* July 1959.

Scott, Thomas and McPherson, Marlys: The development of the private sector of the criminal justice system. *Law and Society Review,* 1971.

Security gains ground in U.S.A. *Security Gazette, 11:*No. 8, 338, 1969.

State of the art, the. *Security World,* October 1970.

Steinberg, Stanley: Private police practices and problems. *Law and Social Order,* 585, 1972.

Tamm, Quinn: The IACP field service division —a new venture of police security. *Industrial Security,* July 1961.

———: The economics of social immorality. *Industrial Security, 10:*No. 5, October 1966.

Toward regulation of the security industry. *Security Gazette, 12:*270, 1970.

United States committee on education and labor. *Private Police Systems,* 1971.

Crime expense. *U.S. News and World Report,* October 26, 1970.

Walsh, Timothy J.: Some reflections on industrial security as a profession. *Industrial Security,* April 1963.

———: Professionalism in industry security. *Industry Security, 5:*No. 6, April 1961.

———: A glossary of common security terms. *Security World,* 26-28, November 1974.

———: Ethics in the security profession. *Protection of Assets Manual, 2:*1974.

Wright, W. D. Jr.: An industrial security profile. *Industrial Security,* February 1970.

———: An industrial security profile. *Industrial Security,* Part II, December, 1970.

———: ASIS Foundation Inc. *Industrial Security, 12:*No. 28, February, 1968.

Zulligan, Hans: Ueben symbolische diebstable von kindern und jugendlichen. *Biel. Inst. fier Psychology.*

UNPUBLISHED MATERIALS

Fauth, Kenneth G.: Philosophy of security. *Western Illinois University* (unpublished paper), Fall, 1973.

Friend, Bernard D.: Profile of the physical security officer specialization or professionalism. *Unpublished graduate B paper, Michigan State,* 1968.

Garber, Joseph: The history, organization and structure of the New York City Housing Authority Police Department. *B.S. Thesis, Political Science, John Jay College of Criminal Justice,* 1970.

Larkins, Hayes: A survey of experiences, activities and views of the industrial security administration graduates of Michigan State University. *Unpublished Michigan State Thesis,* Fall 1966.

Smiegel-Leibowitz, Arleen: Does crime pay? An economic analysis. *Unpublished M.A. Thesis, Columbia University,* 1965.

Chapter 2

HISTORY

PROTECTION OF THE INDIVIDUAL IN SOCIETY

THE PROTECTION of the individual and his property is one of the oldest tasks in history. It has, for the most part, been neglected until recent years when historians have begun treating the development of law enforcement. It is impossible to separate the growth and development of what is termed private security from that of the development of law enforcement. The patterns of development in private security and of what is commonly known as *law enforcement* do, however, differ to a considerable degree in many respects. The history of law enforcement follows the flow of legal development which has shifted the primary responsibility for protection and prevention of criminal activities from the private individual to the governmental jurisdiction in which he lives. In other words, the individual was no longer legally responsible for protection and prevention, but the public became the aggrieved party and assumed the responsibility for prevention, protection, and apprehension of offenders.

Private security has always been a matter of individual concern and activity. While the legal remedies and the legal responsibilities have shifted in a general sense, practicality has necessitated the perpetuation and increased sophistication for protection of individuals and their property with private nongovernmental agencies or individuals. The inability of government to provide the type of security measures which are often required by individuals has historically produced a need for and use of private security measures. It is this distinction and the development of the distinction which is traced in the history of the development of the protection of the individual in society.

ANCIENT PERIOD

Human beings have always felt the need for providing a degree of safety and security for themselves and their possessions. Since the earliest beginnings of human life, men have sought ways to protect themselves from their environment and other hostile forces in nature. One of the earliest attempts to establish some type of protective device resulted in the use of a variety of physical barriers which were placed between the human and the hostile environment. The use of caves, which were either inaccessible due to height from the ground or the entrance to which could be closed by means of large rocks, served the purpose in many instances. In other cases, the use of fire at the entrance of caves or in the proximity of the dwelling place served this purpose. In other instances, dwellings were built on platforms in lakes or large bodies of water making them inaccessible to other beings. Thus *physical barriers* came to be one of the first things that man used to protect himself and his property.

In a similar manner, the use of *guards* developed at quite an early point in the history of personal protection. The practice of having a person protect property at all hours of the day has been established by many historical documents and biblical references. Night watchmen, sentries, and guards were utilized to insure that property and individuals were kept safe.

The concept of individual protection and the concept of justice among individuals of very primitive groups is synonymous with the maintenance of community equilibrium. If property was stolen from an individual in the community, the economic equilibrium was unjustly disturbed. Such an act constituted a wrong against the individual and equilibrium was reestablished by a settlement between the aggrieved person and the wrong-doer. If the thief restored what had been stolen, the injured party was very often satisfied and no further action was taken. This philosophy has continued with relatively little modification to the present time.

Early codes were developed by such individuals as Hammurabi (approximately 1700 B.C.). Under his code, punishment was taken from private hands and made a function of the government. The stated purpose of his code was to see that "the orphan did not fall prey to the powerful," and that "the man with one coin shall not fall prey to the man with many." Trade, commerce, agriculture, and professions were all rigidly supervised and strict penalties for noncompliance initiated.

After the collapse of these early civilizations, the responsibility for protection again rested with those strong enough to safely maintain their property. The first recorded instance of an established police force was in the Greek city states. Even in Greece at this time the responsibility rested with the citizens, it not being considered a state responsibility. The army was primarily used for traffic control and governmental law enforcement. A *secret police* (people's police) was established at this time to protect the state from subversion.

Caesar Augustus, the emperor of Rome at the time of Christ, created a police force for that city. It has been said that after about 6 A.D. this was the most effective police force up until recent developments in law enforcement. Prior to this time, however, from approximately 400 B.C. to 6 A.D., the law enforcement of Rome was not effective; it was on a private individual basis. The *vigiles* were nonmilitary police detailed for permanent police-fire duty within the city. This was a force of about five thousand men. The *praetorian guard* were established as a personal protective force for the emperor and the palace grounds.

The concept of crime prevention was developed in Rome. The *urban cohorts* became a state police force for the city of Rome as a military organization. The *vigiles* (the vigilantes) were formed as a secret police agency for Rome to protect the state security.

The fall of Rome and the disintegration of the societies which flourished in this early period of history resulted in migrations of people from continental Europe to the British Isles. The development of the protection mechanisms which are currently used can be traced from this early English background directly to contemporary public and private agencies.

ANGLO-SAXON (600 A.D.—1000 A.D.)

The settlers who crossed to the British Isles from the mainland of Europe established themselves for the most part in small groups in separated and widely scattered small localities which were called *tuns*. Each one of these tuns, or towns as they became known, were locales of police jurisdiction. Municipal policing has been said to have originated in these tuns. The process of *hue and cry* as an ordinance in the method of apprehension was established as a means for protecting the community against law violators. An offense against a person was an individual act which was punished by individual means. A death penalty was not utilized during this time, the punishment being branding or voluntary servitude. Since prisons did not exist, when a person who injured a victim was arrested, he was turned over to the family of the person offended and was their slave for a period of time; if a serious crime, for the rest of his life, if for a less serious crime, a portion of his life. This regulation of law enforcement by all citizens was on an individual basis. If the person in servitude escaped, word was sent to the head men of other tuns. When he was caught, he was sent back to his own village to complete the period of servitude. Trial by ordeal to determine guilt or innocence was utilized as was branding and mutilation for punishment of serious crimes. Formal or informal arrangements were kept in effect by the local units of government to protect the individual's rights.

There was, however, the responsibility of the individual to preserve and protect his own property.

NORMAN PERIOD (1066 A.D.—1285 A.D.)

This was another important period in the history of the protection of the individual in society. A revolution of law enforcement and legal concepts developed during this period. Norman William, King of France, invaded and conquered England and established a military regime in England. He was responsible for changing the concepts of crime being an offense against the individual to that of being a crime against the state. After his invasion and conquest, he divided England into fifty-five separate military areas. These areas were called *shires* which are roughly comparable to the current county in the United States. An army officer was placed in charge of each area and was given the title of *rieve*. From *shire-rieve* we get the modern-day derivation of *sheriff*. This is one of the first modern instances of military policing of civilians. Within the shire system, the office called *constabuli* was established. Two of these individuals were assigned to a village to aid the rieve in the conduct of his duties. *Constabuli* originally meant *keeper of the stables*. This was a very important person in those days, since travel and fighting was done on horseback. He was the forerunner of our modern-day constable.

The king decided that the shire-rieve should not try cases of persons arrested. He decided to send judges to hear them. These traveling judges were the forerunner of the circuit judges of modern times. This is the first instance of the division of law enforcement and judicial powers.

In 1116 A.D., in the reign of King Henry I, known as "the law giver," the *leges henrici* were issued. It was stated in them that "there shall be certain offenses against the king's peace; arson, robbery, murder, false coinage, and crimes of violence. These we deem to be felonious." With the issuance of the Laws of Henry, a great division in the history of policing came into existence. Before 1116 A.D., the police acted as a private police; thereafter, they were in public service. A very clear distinction between offenses was made at this time, the division between felonies and misdemeanors was carefully delineated. If a felony (serious crime) had been committed, both the police and the citizens had the broad authority and equal rights to arrest whatever was done in their presence. In the case of misdemeanors (minor offenses), however, only the police had the right to arrest. This is the origin of the present-day laws concerning these matters. Also during this period, grand jury came into existence which was established to make inquiries into the facts of crime and to eliminate the Anglo-Saxon trial by ordeal system of justice. The Magna Carta was also promulgated in 1215 which provided for due process and the restoration of local governmental control, and marked the beginning of separate state and local governments.

During this period, individuals protected their persons and properties using much the same basic techniques which were used in the ancient period. Walled living structures, walled communities, and night watchmen were used to protect individuals and their properties. The establishment of the criminal law during this period and the shift in legal responsibility for enforcement from the individual to the government did not eliminate the necessity for individuals to take precautions in safeguarding their persons and property.

WESTMINSTER PERIOD (1285 A.D.—1500 A.D.)

The Westminster period is so-called because the laws governing policing arose from the capital of England, at that time Westminster. The decrees and laws promulgated during this time effected the entire field of law enforcement. Five major laws were implemented during this period which served to radically change legal law enforcement processes.

Statutes of 1285

Statutes of 1285 (Westminster) originated the *curfew* idea. Gates were closed at sundown to assist in the enforcement of law by keeping undesirables out and keeping those in the city in check. The night watch and the development of the office of *baliff* also came out of these statutes. Checks were made by the baliffs on lodgers at inns in the city and guards were placed at the gates from sunset to sunrise to secure the city.

Statutes of Treason in 1352

The statutes of *treason* in 1352 indicated that whoever would give aid to the enemies of the land should forfeit his life. Likewise, counterfeiting of money was considered to be treason, since the first instance of coinage of money had been placed in the hands of a central government.

Justice of the Peace

The Justice of the Peace in 1361 A.D. established the new county officer, *Justice of the Peace,* because of the inefficiency of the sheriffs. Three or four of these men were established in every county. They had the authority to pursue, arrest, chastise, and imprison with overall authority and no distinction as to type of crime. This first attempt to put the police and judicial functions in one officer failed, however and the two offices were separated for a period of seventy-five years.

Local Government

Local government regulations (1370 A.D.) allowed cities to establish their own ordinances. Attempts were made to control traffic, vice, and other criminal activities under these regulations. These regulations resulted in the establishment of a marching watch or a night patrol for the purpose of safety and security in the cities.

Courts of the Star Chamber

The *Court of the Star Chamber* (1478 A.D.) received its name from the room in which it met. It was a special court established by the king to try those accused of crimes against the state. It was given power to force testimony and resulted in the legalization of third-degree techniques. Opposition against the methods used by this court and its powers led to the *Bill of Rights* in 1688. This doctrine established protection against self-incrimination and provided that one would not be made to say anything which would be self-incriminating. This general principle has continued in existence to the present day.

During this period, the barriers system utilized during the previous periods was also continued. The use of guards, however, became more acceptable for protection of property and also for personal safety (private protection). These individuals were primarily governmental servants providing for collective security as opposed to individual personal and property security.

Thus far, physical security measures had become more and more sophisticated, barriers developing from caves, to fires, to walls, to fences with gates, followed by drawbridges and moats, followed by locking devices such as dead bolts and door bars. Likewise, the use of guards had become more sophisticated with individual watchmen expanding into groups of watchmen patrolling the streets and standing guard at the closed gates of the cities.

PERIOD OF WARD AND WATCH (1500—1800 A.D.)

This was a period of great social and economic upheaval for Great Britain. England was beginning to engage in the world trade. A number of restrictive measures, such as the *Enclosure Act,* placed a considerable hardship on the people and caused a great deal of social unrest and migration within the country. The law enforcement and social control mechanisms within the country were unable to cope with the conditions existing and crime increased rapidly. It was in this period of great turmoil and social upheaval that the first real evidence that the public protective system was unable to cope with the problems of providing a minimal amount of protection for business and industry within the country came dramatically into view.

Beginning in 1500, a number of different types of police agencies were established in attempt to deal effectively with the crime problem. Individual merchants hired private persons to protect their property and business establishments. Banks and merchants hired individuals to protect their interests during business hours. Likewise, night guards and watchmen were hired to watch business establishments when they were closed to prevent thefts and loss of property. The use of the forerunners of private detectives to locate and identify stolen property also developed during this period. When property was stolen, the merchants were more interested in obtaining its return than obtaining a legal conviction of the perpetrator. This led to the practice of using private detectives to locate stolen property and negotiate its return; a practice which was begun prior to this time and continued through this period into the present day. The *parochial* police also came into existence. The people of the cities were divided into religious areas or parishes who banded together and hired their own police to protect them and their property within the parish. These private individuals, in protecting parish property, performed essentially the same service as would a public police officer except that his responsibility was the property in the parish boundaries. At one time, the number of policemen in a parish was so high that there were instances of seven hundred of these policemen being reported in a single parish.

MODERN PERIOD
(1800 to the Present)

Prior to 1800, the industrial revolution began. The development of industry in the cities produced slums and the factory systems which began to increase the blight in the metropolitan areas. Crime became rampant. During this time, the *fence* came into active and well-known existence. Children were used to steal for them. Counterfeiting became so prevalent that for a time there was more bad money than good money in circulation. There were more than fifty false mints documented in London alone. The government, in attempting to combat these conditions, rewrote the penal laws, making punishment severe in an attempt to deter the offenders. There were at one time 150 offenses punishable by death for such things as stealing bread, pickpocketing, etc. This system did not materially reduce the criminal activity; therefore, a system of offering rewards was started. One such award system provided that for catching a highwayman, the reward was 150 pounds; for catching a horse thief, twenty pounds; for catching a common thief, five pounds; and for catching a vagabond, ten shillings was the reward. Likewise, this system did not prove effective. The system of deportation was used to remove the criminals from the countryside. This also proved to be of marginal value.

Also during this period, the concept that "every man's home is his castle" developed. The unlicensed privilege of bearing arms for self-protection was begun. Citizens, in an attempt to protect their personal property, resorted to the use of various protective devices within their homes. Wolf traps were used in an attempt to prevent entry into homes; these traps were set inside doors and windows to catch anyone attempting to break in. Again, even such techniques proved not to be successful. It should be pointed out that these barrier systems utilized during this period attempted to *deter* criminal activity, but not *prohibit* it.

In addition to individual citizens attempting to protect their own property, wide varieties of police came into existence in London. A list of these officers would include (1) special constables, (2) watches and wards, (3) Bow Street Runners, (4) parish police, (5) merchant police, (6) Thames River police, and (7) special detective agencies.

With all these agencies operating in the city, it became very difficult for the unified prevention, detection, and apprehension of criminals. Sir Robert Peel observed the crime situation and concluded that "you cannot have good policing when responsibility is divided." In making his study of the quality of law enforcement agencies in London, Peel said, "They are persons hired to sleep in the open air." He further called the night watch the "shiver and shake watch." He said that these men spend half the night shivering with cold and the other half shaking with fear. Based on his conclusions and recommendations, a major revision of the metropolitan London police was enacted and between 1835 and 1850 there was a complete and total revolution in law enforcement. Modern policing, as it is currently known, came into existence.

While the development of the metropolitan police was begun in England, the type of protection that was needed by various types of individuals in British society was not met by the development of the metropolitan police nor by the *Police Act of 1835* which established *city and borough police forces,* nor the *County Act of 1839* which provided for the establishment of county police, or the *Obligations Act of 1856,* nor the *Local Government Act of 1888.* Private guard systems, private detective agencies were still utilized to locate and recover stolen property as well as provide for the protection of private property which public law enforcement agencies could not adequately provide.

THE AMERICAN DEVELOPMENT

The development of the protection of the individual in the United States can best be characterized by its lack of uniformity and consistency. Because of geographical distances, agencies which provided for protection and social control varied considerably in style and effectiveness in the critical development period of the United States. A number of different agencies, both private and public, were developed within the United States for the enforcement of law and to provide protection and social control for society. The Northeastern United States found the office of constable appropriate for its needs in small communities, whereas in the Southern United States, the office of sheriff was found to be more appropriate. Similarly, in larger cities, the office of city marshal, night watchman, or *rattlewatch* came into existence. As might be expected, these regional differences became part of the formal governmental structures for these areas and many continued up to the present time.

The jurisdiction of all law enforcement agencies in the United States is quite closely circumscribed by constitution, statutes, or ordinances. Municipal police had authority within their cities, county sheriffs within their county, village marshals within their villages, state police agencies within their states, and federal law enforcement officers' responsibilities were limited to specific offenses wherever they might occur. This fragmentation of authority and responsibility resulted in the inability of law enforcement agencies to provide adequate protection in many areas of the United States. Likewise, because of inadequate budgets, poorly selected and trained personnel, graft, corruption, political patronage and interference, many law enforcement agencies have not been able to provide the type and quality of service necessary to adequately protect private individuals and their property. Because of these inabilities, private individuals, business, and industry supported the development of private law enforcement agencies or developed internal protective services.

The development of formal types of public or private protective services for the individual and society are of relatively recent origin in the United States. Chronologically, the development of the various public and private agencies does not follow a pattern. Many private agencies began operations before many of the public agencies. Many of the services performed by private agencies were assumed by public agencies when they were developed. Various activities and techniques developed by the private agencies were incorporated into the activities of public agencies, both municipal and federal.

In 1636, the first formal protective agents were appointed in Boston in the form of watchmen. They were followed in 1658 by the City of New York which also appointed a similar type of person, as did Philadelphia in 1700. This was a time in the development of the United States when there was little need for formal organizations to provide protection and security and the informal watchman, bellman, rattlewatch type of service was considered adequate.

The development of what would be considered the forerunners of municipal police departments began in New York City in 1783. New York was followed in 1801 by Detroit and in 1803 by Cincinnati. However, as late as 1821 there is evidence that a private guard service, called *specials*, provided protection in Boston. In 1828 the first federal investigative service began with the establishment of the Post Office. In 1837, the Chicago Police Department began its operations. It is believed that the first private police in the Midwest were established in St. Louis in 1846, which is the same year that the forerunners of the San Francisco Police Department began full-time operations. Four years later, in 1850, the Pinkerton Detective Agency began operations in Chicago. Also, in this year the Los Angeles Police Department began its operations. It was not until five years later, in 1855, that the Philadelphia Police Department began, and in 1856 the Dallas Police Department began operations as a law enforcement agency.

During the period of the 1850's, the need for security of a very effective type was realized by industrial and business organizations. Because of the lack of development of public law enforcement agencies during this period, detective agencies such as the Pinkerton Agency began protecting private industrial properties, investigating thefts, and arresting criminals which were beyond the scope of local law enforcement agencies (who were unable to investigate or take action due to the limited nature of their jurisdictions).

During the Civil War, the agency established by Allan Pinkerton was retained by the United States Government to supply intelligence gathering for the war effort, since the government did not possess the capability at this time to perform these activities. Also during this time the railroads, which were instrumental in the growth of the United States, were provided with authority in most states to establish their own railway police to enable them to protect their property. As early as 1865, *Railway Police Acts* were established in states to allow the railroads to have their own police with full police powers for the protection of the companies' equipment and rolling stock. The initiation of these acts came only one year after the beginning of a second law enforcement agency at the federal level, that of the Treasury Department in 1864.

In 1870, the United States Department of Justice began its operations. In 1882, the United States Border Patrol began its operations, and in 1889, Brinks Incorporated began protecting payrolls and provided security for private property in the United States.

The period from 1890 to 1914 saw tremendous growth in the scope of activities performed by the private security agencies. The railroads began forming their own protective agencies and in 1890 the St. Louis and San Francisco railroads began their own police service. In 1892, the Illinois Central Railroad followed; in 1900, the Chicago, Milwaukee, St. Paul and Pacific railways also followed. In 1901, the Southern Pacific Railway began. In 1914, there were 14,000 railway police working in the United States. In 1921, the protective section of the Association of American Railways began its operations as a cooperative security organization between all United States rail carriers.

In 1909, the William J. Burns Detective Agency began its operations and became the sole investigating agency for the American Banking Association. It was not until 1924 that the Federal Bureau of Investigation came into existence to provide law enforcement on a nationwide coordinated basis. Until this time, agencies such as the Burns and Pinkerton agencies were the only law enforcement agencies in the United States with the capability to provide national and even international protection of individuals and property. In the 1930's, plant protection and a provision of services for corporate security began to develop. In the 1940's, the programs of industrial plant protection and retail security expanded greatly with many private guard services and corporations establishing guard or internal protective services.

Private Security agents were also used by industry to prevent union organizations and to promote company policies against strikes. Private detective agencies have had a long, bloody, and violent involvement in the U.S. labor movement.[1] The passage of the "Pinkerton Act" by the U.S. Congress in 1894 was in response to the involvement of detectives in violent labor-management confrontations.

The Senate Select Committee on Violations of Free Speech and Rights of Labor[2] issued a report with the following conclusions:

> The operations of private police systems described in this report offer a basis for generalized conclusions concerning the consequences that flow from the use of private armed guards as employers' agents in labor relations. These consequences are particularly grave in their influence upon the civil rights of citizens, the maintenance of public peace and safety, the operations of the economic system, and the functioning of government.
>
> The experiences in Harlan County, Ky., and in the industrial communities indicate clearly that where private police systems are used as instruments of antiunion policy, they (a) abridge and violate the civil liberties of workers and other individuals; (b) violate the rights of labor guaranteed by Federal statutes; (c) result in riots and bloodshed, causing loss of life and injury to persons and property; and (d) endanger the public safety.
>
> On the ecomonic front, the use of private police systems as agents in employers' antiunion policy, causes disorganization of markets and interruptions in the free flow of commerce. The ruthless and brutal activities of armed private guards to prevent union organization (a) give unfair competitive advantage to those employers who oppress labor; (b) create bitterness between labor and management; (c) lead to strikes; and (d) cause interruptions in the flow of commerce.
>
> The use of private deputies in an antiunion campaign is inimical to the maintenance of orderly representative government. It leads to (a) private usurpation of public authority; (b) corruption of public officials; (c) oppression of large groups of citizens under the authority of the State; and (d) perversion of representative government.

This same Senate Committee recommended:[3]

> The functions of private police systems must be restricted to the protection of plant and property. Employers using the channels of interstate commerce should not be permitted to spread and perpetuate a system of repressing the civil rights guaranteed by the Federal Constitution and Federal legislation. When company-controlled police systems cover several states and affect the activities of thousands of workers, it is the place of the Federal Government to intervene. In the interest of industrial peace and the uninterrupted functioning of the national economic system, it is necessary that Federal action confine company police systems to their proper duty of protecting plant and property.
>
> In recommending legislation to correct the evils of company police systems the committee is scrupulous not to interfere with the right of the employer to police his premises or employ persons for the legitimate functions of protection. The committee does not feel that at this time the Congress should undertake detailed regulation of the personnel and conduct of the police systems of employers. Rather it should define those practices which have led to the infringement of civil liberties and industrial disorder and make their commission an offense. In only one respect does it seem advisable to regulate the personnel which employers may hire, and that is to prohibit the employment as armed guards of persons who have previous criminal records showing a tendency toward violence or the dangerous use of deadly weapons.
>
> Legislative remedies, in the opinion of

the committee, should be designed to prohibit labor espionage and the rough shadowing, coercion, and intimidation of workers in ordinary times, and to restrict company police to company property during times of strike. A statutory prohibition of these practices of private police systems, carefully defined, will also cover the similar practices of detective and strikebreaking agencies.

The passage of state and federal legislation halted much of the violence and violations of individual rights.[4]

THE PRESENT

There are currently over 800,000 persons providing either public or private protective services in the United States. One in every 100 persons in the civilian labor force is employed in public or private security work. A total of 515,000 persons are employed in public protective services, 395,000 of which are in public police work and 120,000 employed as government guards or watchmen. An additional 290,000 are employed as private security personnel. If, however, all security-type personnel are considered as performing a similar function, the ratio of security type activities to police type activities is almost one to one. Approximately 36 percent of all security personnel are employed in the private sector and 64 percent in the public sector with the government guards added to the nonpeace officer category. About half of all security personnel have full peace officer powers and half do not. The underlying factors for growth of private security services include the following:

1. An increase in crime and business crime losses.
2. An increased need for security in government activities such as NASA and DOD activities, air hi-jacking, terrorism, bombings, etc.
3. An increase in private and corporate income and more willingness to pay for additional services.
4. Insurance requirements for sophisticated security systems for the ability to purchase insurance.
5. Insurance discounts being provided for the use of private security measures.
6. The availability of electronic detection and deterrent equipment for improved security intrusion detection.
7. The trend toward purchasing specialized services by business and industry.
8. The lack of police ability to provide more than minimal protective services in the community.

The advent of computerized police departments, the National Crime Information Center (NCIC) and the related federal, state and local law enforcement agencies have not decreased the need for private protection. The services provided by commercial agencies in addition to proprietary or company-owned and operated security services are numerous and varied.*
The types of services offered can be categorized into security services, private detective agencies, armored express services, alarm services, specialized patrol services, insurance investigators and polygraph examiners. These services offered by private companies for general public use are regulated by various state licensing and regulating agencies. Their counterparts in the provision of security services for individual corporations are called *proprietary* security forces. These forces, which are part of a business or industrial concern in providing internal security, have developed a private protective force. Many of these forces are of a very significant size, but are not regulated or licensed in most states.

The growth rate of private commercial security services or investigative firms has risen dramatically particularly in agencies such as Pinkerton's Inc., Burns International Security Services, Wackenhut Corporation, Globe Security Systems, and Guardsmark. These firms generated a combined revenue of $488 million in 1973 which was up from $350 million in 1970. These firms account for approximately 50 percent of all expenditures for security services in 1973.[5]

---

*See Chapter 1 for a complete description of these services.

TABLE 2-I

SELECTED MAJOR SECURITY SERVICES
1973 Sales

	1973	Employees (Approximate Numbers)
Wackenhut	$ 90,458.031	18,000
Burns	153,649.	39,000
Pinkerton	174,748.	37,000
Globe	51,376.	9,000
Guardsmark	17,674.506	4,500
	$487,905.537	105,500

The protection and preservation of individual property and persons are the primary objectives of contemporary security organizations. A significant portion of all revenue generated by commercial security service firms and the major portion of the activities provided by proprietary firms is for physical security services. The need for these activities is in excess of those required by a normal individual in society who can accept the minimal degree of security as provided by public law enforcement agencies.[6] These individuals and corporations require additional specialized services and feel that the expenditures they are making for security services provide a greater degree of protection for their investment and minimize the need for additional public resources expenditures for protection of their property or activities.

FOOTNOTES

1. Appendix 2A presents a chronology of private security agents in strikebreaking activities.
2. Report on Private Police Systems, *Violations of Free Speech and Rights of Labor*, 76th Congress, First Session, 1939, p. 214.
3. Report on Private Police Systems, pp. 217-218.
4. See Appendix 2B for a review of legislation and practice concerning strikebreaking.
5. James S. Kakalik and Sorrel Wildhorn, *Private Police in the United States: Findings and Recommendations.* (Rand Corporation, 1971), vol. 1, p. 10.
6. See Appendices 2C and 2D for a more complete discussion of current public-private police relationships.

BOOKS

Aydelotte, F.: *Elizabethan Rogues and Vagabonds.* New York, Barnes and Noble, 1913.

Brown, Lorne, and Brown, Caroline: *An Unauthorized History of the R.C.M.P.,* Toronto, James, Lewis, and Samuel, 1973.

Chambers, Ernest J.: *The Royal North West Mounted Police: A Corps History,* Montreal, The Mortimer Press, 1906.

Colquhoun, Patrick: *A Treatise on the Commerce and Police of the River Thames.* Patterson Smith, Montclair, New Jersey, 1969.

Cooper, William: *Smuggling in Sussex.* 1858.

Dabelow, Christoph Christian Von: *Directariat de Romer,* Halle, Ruff 1804.

Deacon, Richard: *A History of The British Secret Service.* London, Frederick Muller, Ltd., 1969.

Eldridge, Benjamin, and Watis, William B.: *Our Rival the Rascal.* Boston, 1897.

Fitzgerald, Percy: *Chronicles of Bow Street Police-Office.* Montclair, New Jersey, Patterson Smith, 1972.

Graham, Frank: *Smuggling in Cornwall.* Newcastle, 1967.

Horan, James David: *The Pinkertons: The Detective Dynasty that Made History.* New York, Crown Publisher, 1968.

Howard Association: *The Practical Prevention of Violence & Other Crimes.* London, Howard Association. 1875.

Lambarde, William: *Of the Duties of Constables.* London, 1602.

Landsberg, Ernst: *Das Furtum des Bosglaubigen Besitzers,* Bonn, 1888.

Lee, Captain W. L. Melville: *A History of Police in England.* Montclair, New Jersey, Patterson Smith, 1971.

MacDonald, J.: *Crime Is a Business.* Stanford University Press, 1939.

MacMullen, R.: *Enemies of the Roman Order, Treason, Unrest, and Alienation in the Roman Empire,* Harvard University Press, 1966.

McClintock, F. W.: *Crime in England and Wales.* London, 1968.

Morrison, William D.: *Crime and Its Causes.* London, Sunnenschein, 1891.

Pike, M. A. and Owen, Luke: *A History of Crime in England.* Montclair, New Jersey, Patterson Smith, 1968.

Pinkerton, Allan: *Strikers, Communists, Tramps, and Detectives,* Mass Violence in America Series, 1878.

Plum, W. R.: *The Military Telegraph During the Civil War,* 1882.

Reith, Charles: *A Short History of the British Police.* Oxford University Press, 1948.

———: *The Blind Eye of History: A Study of the Origins of the Present Era,* London, Faber and Faber, n.d., 1952.

Richardson, Albert D.: *The Secret Service, the Field, the Dungeon and the Escape,* Hartford, 1865.

Rumbelow, Donald: *I Spy Blue.* London, Macmillan, St. Martin's Press, 1971.

Schisas, P.: *Offenses Against the State in Roman Law,* University of London Press, 1926.

Sutherland, Edwin H.: *The Professional Thief,* Chicago, U. of Chicago, 1937.

Ullmann, W.: *The Individual and Society in the Middle Ages.* Baltimore, Johns Hopkins Press, 1966.

Wood, H. T.: *Industrial England in the Middle of the Eighteenth Century,* 1910.

PERIODICALS

Anonymous and staff-written: A pictorial history of crime. *Security World, 5:*No. 1, January, 1968.

Cross, Michael S.: The shiner's war: social violence in the Ottawa Valley in the 1830's. *Canadian Historical Review, LIV:*No. 1, 1, March 1973.

Ferdinand, Theodore N.: A comparison of crime in Boston and New Haven since 1849. *San Francisco,* 1967.

———: The criminal patterns of Boston since 1849. *Am J Soc, 73:*84-99, 1967.

Jones, P.: The city state in late medieval Italy. *Transactions of the Royal Historical Society, 15:*71-79, 1965.

Kehoe, M.: History of strike-breaking. *Canadian Labour, 17:*No. 10, 2, 1972.

Kingsbury, Arthur: Guidelines for security education, historical aspects of security. *Security Management, 17:*No. 6, 17.

Neuberger, Richard L.: The royal Canadian mounted police. *Harper's Magazine,* July 1934.

Phillips, Peter: Historical locks. *Security Surveyors,* 54, May 1972.

Sellin, Thorsten: Research memorandum on crime in the depression. *Soc Sci Res Council,* Bulletin 27, 1937.

Shalloo, J. P.: Private police. *American Academy of Political & Social Science.* Monograph No. 1, 1933.

Sheehan, Thomas: The special police. *Police, 13:*No. 1, 91, 1968.

Speicher, F. P.: Midnight mystery meeting: A scientific focus on the future. *Industrial Security, 5:*No. 85, October 1961.

UNPUBLISHED MATERIALS

Hay, Doug: Property, authority, and the law in the eighteenth century. *Unpublished paper, Centre for the Study of Social History, Warwick University,* April, 1972.

Haydon, A. L.: The riders of the plains: A record of the Royal Northwest Mounted Police of Canada 1873-1910. 1910.

Hood, James A.: A panoramic view of the history of the British police organization. *Unpublished graduate thesis, Michigan State,* 1972.

Morgan, Edwin Charles: The Northwest Mounted Police, 1873-1883. *University of Saskatchewan,* 1970.

Schooley, Thomas S.: The historical development of the American municipal police. *Unpublished graduate thesis, Michigan State,* 1972.

Sokolove, Bruce: American vigilantism—historical and comparative perspectives. *Unpublished graduate thesis, Michigan State,* 1970.

Privacy in western history from the ages of Pericles to the American republic. *Unpublished Ph.D. Dissertation, Harvard University,* 1965.

APPENDIX 2A

OFFICIAL SOURCES ON THE USE OF STRIKEBREAKERS AND GUARDS IN INDUSTRIAL DISPUTES*

1. Forty-seventh Congress, second session. Senate. Committee on Education and Labor. Hearings on Relations Between Labor and Capital.

Under a resolution of August 7, 1882, the Senate Committee on Education and Labor was directed—

to take into consideration the subject of the relations between labor and capital * * * the subject of labor strikes, and to inquire into the

---

* A report on Strikebreaking, Violations of Free Speech and Rights of Labor, 76th Congress, First Session (1939), pp. 139-148.

causes thereof and the agencies producing same * * *.

The testimony developed by this committee shows that, owing to the primary economic position of capital, strikeguards were not essentials of strikebreaking technique at that time. Spies, blacklists, forms of economic intimidation and control, and imported strikebreakers were sufficiently effective weapons; although sections of evidence relating to attempts at unionization in mining districts show that physical intimidation was also used. The telegraphers' strike of 1883 is developed in considerable detail. The main weapons of the monopolistic telegraph companies were publicity, blacklists, discrimination, and economic intimidation preceding the strike, and an impregnable economic position during the strike.

2. *Forty-eithth Congress, first session. Senate Report No. 820. An Act to Prohibit Importation and Migration of Foreigners.*

The Senate Committee on Education and Labor reported favorably, June 28, 1884—

An act to prohibit the importation and migration of foreigners and aliens under contract or agreement to perform labor in the United States, its Territories, and the District of Columbia.

In its report the committee cited evidence that foreigners were imported under misrepresentation to replace strikers.

3. *Forty-ninth Congress, second session. House Report No. 4174. Investigation of Labor Troubles in Missouri, Arkansas, Kansas, Texas, and Illinois.*

A House resolution of April 12, 1886 authorized a committee of seven, appointed by the Speaker, to investigate the—

cause and extent of the disturbed condition now existing between the railway corporations engaged in carrying interstate commerce and their employees in the States of Illinois, Missouri, Kansas, Arkansas, and Texas

and to make recommendations to the House.

The "disturbed condition" was the great railway strike on the Southwestern or Gould system, called March 1, 1886, by District Assembly 101 of the Knights of Labor. Violence was precipitated when the railroad companies tried to operate trains under the protection of armed guards. A tragic occurrence on April 9 at East St. Louis, related in the majority report, demonstrates the brutality of these employer tactics:

The Louisville and Nashville roads had determined to run its trains in at all hazards, and for that purpose had at its service guards of a considerable force, consisting of deputy sheriffs and marshalls * * * In the afternoon the company resolved to send a posse of deputies to clear the crossing. For this purpose nine men, armed with Winchester rifles and revolvers, were selected.

When resisted by the strikers—

One of the deputies raised his rifle, fired, and a man was seen to fall. Then showers of stones and pistol-shots from all directions began to rain upon the officers, who returned the fire with their guns and pistols, with deadly effect, into the crowd. The firing was kept up until the crossing was clear * * *.

4. *Fiftieth Congress, second session. House Report No. 4147. Labor Troubles in the Anthracite Regions of Pennsylvania, 1887-88.*

A House committee, whose membership was announced by the Speaker on February 9, 1888, was empowered to investigate the strike on the Reading Railroad and the strike in the anthracite regions of Pennsylvania, and to make recommendations. The strike in the Lehigh region began September 10, 1887; the strike in the Schuylkill region, in January 1888; and the strike on the Reading Railroad began in December 1887. The strikers, organized under the Knights of Labor, were fought with economic and overtly terroristic weapons by the railroad and coal monopolies which controlled the region. Not only did they fight the strikers ruthlessly, but there is some evidence to show that the monopolies also engineered the disastrous strike for the double purpose of breaking the union and curtailing production and expense. The majority report found that—

> most of the recent labor troubles in the anthracite regions of Pennsylvania arise from the railroads in that section being permitted to mine as well as transport coal.

The unequal struggle was described in one of the reports as follows:

> On the one side wealth, intelligence, and a masterful organization, with its soldiery under the guise of "company police," carrying Winchester rifles. On the other side poverty, ignorance, and a scattered mass of men, powerless to do any more than patiently exercise the innate right of every man to refrain from doing labor he has not agreed to perform. The former can live and wait until the latter are "starved" into submission.

The committee found that the monopolies had usurped the rights and duties of civil authority and made vigorous recommendations to change the situation.

5. *Fifty-second Congress, second session. House Report No. 2447. Employment of Pinkerton Detectives.*

Pinkerton's National Detective Agency, Inc., was investigated by the House Committee on the Judiciary under resolutions of May 12 and July 7, 1892. The former resolution directed the committee to investigate the employment of Pinkertons by railroad corporations; while the July 7 resolution, growing out of the Homestead conflict of July 6, directed the committee to investigate the causes and conditions of that clash.

Both majority and minority reports agreed in finding that Pinkerton guards were the cause of violence at Homestead and elsewhere. The majority report stated that Federal legislation on the employment of Pinkertons would be unconstitutional and recommended State regulation. Several of the minority, however, found that Congress had power to prevent the use of private detectives by persons engaged in interstate commerce.

6. *Fifty-second Congress, second session. Senate Report No. 1280. Investigation of Labor Troubles.*

Under the resolution of August 2, 1892, a select Senate committee was created to investigate the character and uses of organized armed bodies of men privately employed, to recommend corrective legislation, and to report regarding a—

> More effective organization and employment of the *posse comitatus* in the District of Columbia and the Territories of the United States for the maintenance and execution of laws.

Like the House investigation, the Senate investigation grew out of the Homestead incident. The select committee found that the employment of Pinkerton guards at Homestead was unnecessary and led to violence. On page xv of its report it stated:

> Whether assumedly legal or not, the employment of armed bodies of men for private purposes, either by employers or employees, is to be deprecated and should not be resorted to. Such use of private armed men is an assumption of the State's authority by private citizens.

As to the legislative power of Congress in this matter, the select committee reported:

> The States have undoubted authority to legislate against the employment of armed bodies of men for private purposes, as many of them are doing. As to the power of Congress to legislate, this is not so clear, though it would seem that Congress ought not to be powerless to prevent the movement of bodies of private citizens from one State to another State for the purpose of taking part, with arms in their hands, in the settlement of disputes between employers and their workmen.

7. *Fifty-third Congress, third session. Senate Executive Documents, vol. 2, No. 7. Chicago Strike of June-July 1894.* (by U.S. Strike Commission)

President Cleveland appointed a commission of three on July 26, 1894, to investigate the Pullman strike. This struggle began between the Pullman Palace Car Co. and the American Railway Union on June 26, 1894, and soon affected, through sympathetic strikes, most of the railroads entering Chicago. The employing interests unified their antiunion efforts through the

General Managers' Association. The Commission found—

From June 22 until the practical end of the strike the General Managers' Association directed and controlled the contest on the part of the railroads, using the combined resources of all the roads to support the contentions and insure the protection of each.

Headquarters were established; agencies for hiring men opened; as the men arrived they were cared for and assigned to duty upon the different lines; a bureau was started to furnish information to the press; the lawyers of the different roads were called into conference and combination in legal and criminal proceedings; the general managers met daily to hear reports and to direct proceedings; constant communication was kept up with the civil and military authorities as to the movements and assignments of police, marshals, and troops.

The Commission condemned the dual role of 3,600 United States deputy marshals in this strike. Selected by and appointed at the request of the General Managers' Association, they "exercised unrestricted United States authority" under the sole direction and control of the Association. The Commission made no findings concerning private detective agencies, although there is some evidence of paid agents provocateur and private detectives in the Commission's record.

8. *Fifty-sixth Congress, first session. Senate Document No. 25. Coeur D'Alene Mining Troubles.*

This is a letter from Edward Boyce, president of the Western Federation of Miners, presented by Mr. Pettigrew and printed as a Senate document. It narrates the struggle of the miners in the Coeur D'Alene region from the time that gold was discovered there in 1882, with a detailed account of the 1899 strike. Boyce refers to the use by employers of both strikeguards and strikebreakers.

9. *United States Industrial Commission. Reports and Testimony, 1901-2. (Washington: Government Printing Office.)*

The Industrial Commission was created by an act of Congress, June 18, 1898, with two principal instructions:

SEC. 2. That it shall be the duty of this commission to investigate questions pertaining to immigration, to labor, to agriculture, to manufacturing, and to business, and to report to Congress and to suggest such legislation as it may deem best upon these subjects.

SEC. 3. That it shall furnish such information and suggest such laws as may be made a basis for uniform legislation by the various States of the Union, in order to harmonize conflicting interest and to be equitable to the laborer, the employer, the producer, and the consumer.

The Commission published its findings in 19 volumes, by general subject. Most relevant to our present topic are volumes VIII, XII, XIV, XV, and XIX. In its final report, volume XIX, pages 890-893, the Commission condemned the importation of workers from other localities to take the places of those on strike, commended State legislation prohibiting the importation of armed men for the protection of property. Speaking of such guards or "Pinkerton men" the Commission said:

Being often from other localities or States, they have no understanding of the matters at issue in the dispute, no sympathy with the working men, and are therefore disposed to go as far as the law allows, or even further, in resisting the acts of the men.

10. *Fifty-eighth Congress, special session. Senate Document No. 6. Report to the President on the Anthracite Coal Strike of May-October 1902 (by the Anthracite Coal Strike Commission).*

The Anthracite Coal Strike Commission was appointed by President Theodore Roosevelt, October 16, 1902, at the request of both operators and miners. Its instruction was—

"to inquire into, consider, and pass upon the questions in controversy in connection with the strike in the anthracite region" of Pennsylvania, "and the causes out of which the controversy arose."

Although its report, issued on March 18, 1903, dealt mainly with technical and eco-

---

S. Rept. 6, 76-1—10.

nomic aspects of the strike, the Commission criticized the use of strikeguards by the operators in the following terms:

The resentment expressed by many persons connected with the strike, at the presence of the armed guards and militia of the State, does not argue well for the peaceable character or purposes of such persons.

11. *Fifty-eighth Congress, third session. Senate Document No. 122. Report on Labor Disturbances in Colorado, 1880-1904. (Prepared by Walter B. Palmer of the United States Department of Commerce and Labor.)*

This report gives a "straight forward history" of 13 of the more serious labor disturbances in Colorado in the 25-year period following the Leadville strike of 1880. Repeated mention is made of the use and effect of strikeguards and strikebreakers.

12. *Fifty-ninth Congress, second session. House Document No. 822. Strikes and Lockouts (21st annual report of the Commissioner of Labor).*

Chapter V of this report (pp. 917-960) is a digest of the law relating to strikes, blacklisting, boycotts, etc., and lists the statutes on these practices in force as of 1906.

13. *Minnesota, Bureau of Labor, Industries, and Commerce, Twelfth Biennial Report, 1909-10.*

The use of strikebreakers and strikeguards in the Minnesota switchmen's strike of 1909 is fully treated on pages 29-57. The strikebreakers were so ill suited to their new jobs that many of them were fired, and as destitute and unemployed people they came to the attention of the Minnesota Bureau of Labor.

14. *Sixty-first Congress, second session. Senate Document No. 521. Report on Strike at Bethlehem Steel Works, South Bethlehem, Pa. (by Bureau of Labor).*

The United States Bureau of Labor undertook this investigation in compliance with a Senate resolution of April 19, 1910. It found that the strike, which began on February 4, 1910, was occasioned by long standing and highly unsatisfactory working conditions. When the workmen struck, they were unorganized. The Bethlehem Steel Co., in fighting the strike, placed its main reliance on State police acting in the role of strikeguards. The material on this is in the primary form of statements and affidavits submitted by all parties to the controversy.

15. *United States Department of Commerce and Labor, Immigration Service. Report on an Investigation in Regard to Alleged Infractions of the Alien Contract Labor Law on the Part of a Car Manufacturing Company at McKees Rocks, Pa., September 1909 (by Inspector A. P. Schell).*

The inspector found in his report that the Pressed Steel Car Manufacturing Co. had engaged strikebreakers through the Bergoff Service Bureau in New York City to replace its workers who had gone on strike July 14, 1909. These strikebreakers were virtually imprisoned in a stockade policed by armed guards secured from the same agency. Insight into the nonprofessional character of these strikebreakers is gained from the following excerpt:

About 900 men were engaged in this manner, and sent from New York, Philadelphia and Chicago. Shortly after their arrival they, in turn, became dissatisfied with the treatment accorded them, and about 400 left the plant and joined forces with the original strikers.

16. *Sixty-second Congress, first session. House. Committee on Labor. Peonage in Western Pennsylvania. Hearings pursuant to House Resolution 90.*

These hearings on the 1909 strike at the Pressed Steel Car Co.'s plant at McKees Rocks, Pa., contain much material on the strikeguard and strikebreaker services of the Bergoff Service Bureau of New York City.

17. *Sixty-second Congress, first session. House. Committee on Rules. Conditions Existing in Westmoreland Coal Fields.*

Hearings pursuant to House Resolution 179.

These hearings on the Westmoreland (Pa.) coal field strike, which began March 10, 1910, are replete with evidences of the vicious unionbreaking tactics of the coal operators in that field. Among these practices the use of strikebreakers and deputized armed guards figures prominently.

18. *Sixty-second Congress, second session. House Document No. 847. Report on the Miners' Strike in Bituminous Coal Field in Westmoreland County, Pa., in 1910-11 (by Bureau of Labor).*

In conformity with House Resolution No. 547, the Bureau of Labor undertook an investigation of the 1910 coal miners' strike. The Bureau's report fully supports the charges advanced in the hearing held before the House Committee on Rules (supra).

19. *Sixty-second Congress, second session. House Document No. 671. The Strike at Lawrence, Mass.*

The hearings on the Lawrence textile strike of 1912 contain many first-hand accounts of the effect of the use of imported strikeguards in this labor dispute.

20. *Sixty-second Congress, second session. Senate Document No. 870. Report on Strike of Textile Workers in Lawrence, Mass., in 1912 (by Frederick C. Croxton of the Bureau of Labor).*

Under a Senate resolution of May 7, 1912, the Bureau of Labor undertook an investigation of the strike of textile workers in Lawrence, Mass., which began on January 11, 1912. The report contains many references to the strikeguard activities of imported thugs, Pinkertons, and deputies.

21. *Sixty-third Congress, first session. Senate. Subcommittee of the Committee on Education and Labor. Conditions in the Paint Creek District, W. Va. Hearings pursuant to Senate Resolution 37.*

Abundant testimony establishes that one of the principal causes of violence in the 1912 strike in the Paint Creek district was the wantonly brutal activity of armed guards supplied to the mine companies by the Baldwin-Felts Detective Agency. There is also much material concerning the procurement and use of imported strikebreakers.

22. *Sixty-third Congress, second session. Senate Report No. 321. Investigation of Paint Creek Coal Fields of West Virginia.*

The investigating committee submitted a joint report, its various members being responsible for specific portions of the resolution. In a summary of these reports, Senator Swanson, chairman of the committee, stated that the importation of strikebreakers and the employment of strikeguards by the coal operators greatly aggravated the strife and disturbance occasioned by the strike.

23. *West Virginia Mining Investigation Commission. Report to Governor Glasscock, November 27, 1912 (Charleston, W. Va.).*

The West Virginia Mining Commission was appointed by Governor Glasscock to investigate the coal miners' strike in the Paint and Cabin Creek district. It condemned the guard system of the coal operators, finding it "vicious, strife-promoting and un-American" and recommended legislation which would remove police power from the control of interested individuals.

24. *Sixty-third Congress, second session. House. Committee on Rules. Industrial Disputes in Colorado and Michigan. Hearings pursuant to House Resolutions 290 and 313.*

These hearings were short preliminary hearings, held in Washington, adumbrating the material which was developed in later hearings on the same subject—employer use of strikeguards and strikebreakers to smash the 1913 strikes in the Colorado coal fields and the Michigan copper mines.

25. *Sixty-third Congress, second session. House. Subcommittee of the Committee*

on Mines and Mining. Conditions in the coal mines of Colorado. Hearings pursuant to House Resolution 387.

In a resolution of January 27, 1914, the House Committee on Mines and Mining was directed to investigate—

* * * conditions existing in the coal fields in the counties of Las Animas, Huerfano, Fremont, Grant, Routt, Boulder, Weld, and other counties in the State of Colorado; and in and about the copper mines in the counties of Houghton, Keweenaw, and Ontonagon, in the State of Michigan * * *.

The subcommittees of the Committee on Mines and Mining proceeded to Colorado and Michigan and collected voluminous testimony on the strikes then in progress in both States. The material developed by both subcommittees was similar in that both hearings showed the evil effects of the antiunion policy of large mining corporations which included the use of imported strikebreakers and strikeguards.

26. *Sixty-third Congress, third session. House Document No. 1630. Report on the Colorado Strike Investigation.*

The majority report on the Colorado coal strike, which began about September 23, 1913, found that mine guards had been deputized and had been sworn in as members of the militia, even though they were not, in some cases, citizens of the State. It found that the Baldwin-Felts Detective Agency had been hired by the coal mine operators, that 12 machine guns and large amounts of ammunition had been purchased for the use of guards and deputies. It found that—

during the strike men were brought in from other States by the car and train load, and were delivered to the mining camps under guard of the militia, many of them being foreigners and unacquainted with the work of mining coal.

It failed, however, to make legislative recommendations to correct these practices.

The minority views of Mr. Austin, based on the same findings, recommended—

Laws to prevent the sale and transportation of firearms and ammunition, making it a felony to carry concealed weapons * * * that mine guards shall not be eligible for service in the State militia or the National Guard, prohibiting the employment of aliens to take the place of native or naturalized citizens in the mines of that State; making it unlawful for deputy sheriffs or other officials to serve on grand or trial juries * * *.

27. *Sixty-third Congress, second session. House. Subcommittee of the Committee on Mines and Mining. Conditions in the Copper Mines of Michigan. Hearings pursuant to House Resolution 387.*

The hearings held by this House subcommittee collected a large amount of evidence concerning the use of strikeguards and strikebreakers by the copper mine operators in the Michigan copper strike of 1913.

28. *Sixty-third Congress, second session. Senate Document No. 381. Michigan Copper District Strike (by Walter B. Palmer, of the Bureau of Labor).*

The Bureau of Labor Statistics investigation of a strike of copper ore miners, which began July 23, 1913, reveals that armed strikeguards and strikebreakers were employed by the large ore-mining corporations. One hundred and twelve guards of the Waddell-Mahon Corporation of New York were sent to Calumet, Houghton County, Mich., in July 1913. They were paid by the mine companies or the county; were deputized and armed. The Ascher Detective Agency of New York sent in 150 guards, similarly armed and paid. The guards of both of these agencies were involved in violence. Two strikers were wantonly murdered by the Waddell men. German immigrants were shipped in from New York as strikebreakers, in ignorance of the strike, and forcibly kept at work by the armed guards.

29. *Sixty-fourth Congress, first session. Senate Document No. 415. Final report and testimony of the United States Commission on Industrial Relations.*

The U.S. Commission on Industrial Relations was created by an act of Congress, August 23, 1912, with authority to—

inquire into the general condition of labor in the principal industries of the U.S., including agriculture, and especially those which are carried on in corporate forms; into existing relations between employers and employees. * * *

and to—

seek to discover the underlying causes of dissatisfaction in the industrial situation * * *.

Under such a mandate, the Commission took testimony relating to practically every major labor dispute that had occurred in the two preceding decades in many parts of the United States. Its final report and testimony, printed as Senate Document No. 415 in 1916, is consequently a rich source of case history on employer use of strikebreakers, strikeguards, and detective agencies. Among these strikes, the Commission's developments of the silk workers' strike in Paterson, N.J., in 1913, and the strike on the Harriman Railroad System in 1910 are outstanding. Men hired through detective agencies contributed largely to the violence that characterized both strikes. The material is accessible through an index of testimony by witnesses, and an index of subjects, both in the back of volume XI. The final report of the Commissioners appears in volume I.

30. *Sixty-sixth Congress, first session. Senate Report No. 289. Investigating Strike in Steel Industries.*

This report of the investigation of the steel strike in 1919 inadequately covers, in understanding and scope, the dispute that the committee was called on to examine. During the conflict the strikers complained of the private detectives of the companies, and their killing of four unionists, but the committee made no findings in this regard. It did, however, comment on police conduct prejudicial to the strikers.

31. *Sixty-seventh Congress, first session. Senate. Committee on Education and Labor. West Virginia Coal Fields. Hearings pursuant to Senate Resolution 80.*

The activities of Baldwin-Felts operatives and guards form a large part of the material contained in these hearings on the coal miners' strike of 1920. There is also reference to the importation by the Williamson Coal Operator's Association of non-English speaking strikebreakers.

32. *United States Coal Commission. Report, transmitted pursuant to the act approved September 22, 1922 Public, No. (347. Washington) Government Printing Office, 1925.*

In the section of its report dealing with civil liberties in the coal fields, the United States Commission found that the employment of thugs and gunmen by mine operators controverted governmental authority. In its recommendations on labor relations in bituminous coal, the Commission stated, in part:

We condemn violence, thuggery, and gun work, violation of the law, and disturbance of the peace. * * *

We recommend that such destructive labor policies as the use of spies, the use of deputy sheriffs as paid company guards * * * be abolished.

33. *Seventieth Congress, first session. Senate. Subcommittee of the Committee on Interstate Commerce. Conditions in the coal fields of Pennsylvania, West Virginia, and Ohio. Hearings pursuant to Senate Resolution 105.*

These hearings, growing out of the coal strike which began in August 1925, disclosed the use by coal mine operators of imported strikeguards and strikebreakers attended by the usual violence. The report of the subcommittee, appearing pp. 344-365, stated, in part:

Everywhere your committee made an investigation in the Pittsburgh district we found coal and iron police and deputy sheriffs visible in great numbers. In the Pittsburgh district your committee understands there are employed at the present time between 500 and 600 coal and iron police and deputy sheriffs. They are all very large men; most of them weighing from 200 to 250 pounds. They all are heavily armed and carry clubs usually designated as a "blackjack."

Everywhere your committee visited they found victims of the coal and iron police who had been beaten up and were still carrying scars on their faces and heads from the rough treatment they had received.

34. 2 N.L.R.B. 626. *Decision in the matter of Remington Rand, Inc., and Remington Rand Joint Protective Board of the District Council Office Equipment Workers, Case No. C-145 (also in pt. 18 of this committee's record, exhibit 3861. pp. 7947-8014).*

The National Labor Relations Board found the use of strikeguards and strikebreakers, attended by the usual pernicious effects, in the *Remington Rand Case,* as well as in other less well-known cases.

APPENDIX 2B

RECENT STATE STATUTES AFFECTING THE STRIKEBREAKING BUSINESS*

The growing realization in recent years of a need of comprehensive regulation of the detective agency business has resulted in statutes in certain states that mark an advance, as far as the control of strikebreaking services is concerned, over the earlier type of legislation, enacted in the latter part of the 19th century, which has been discussed above. Today, five states have laws requiring the licensing of detective agencies engaged in the business of furnishing guards for hire. Two of these state statutes include specific references to strikebreaking services. Another state makes it a misdemeanor to furnish strikebreakers.

In California,[1] Illinois,[2] New York,[3] Wisconsin,[4] and Massachusetts,[5] the statutes either require that agencies engaged in supplying guards to employers be licensed or prohibit the employment of armed guards from unlicensed agencies. In the California and Illinois statutes there is no specific reference to the strikebreaking business. The Wisconsin statute, the pioneer in this field, passed in 1925 and amended in 1931, provides that any agency providing guards and all individuals acting or serving as guards must be licensed by the secretary of state. The secretary of state is empowered, after investigation, to issue a license on the basis of application made by such agencies or professional guards. Persons without such licenses may not engage in the business of acting as private detectives or private police. The requirements under this statute that the prospective private guard must be endorsed by five reputable citizens and the fire and police commission in the city in which he plans to work, and that the secretary of state must be satisfied as to the character, competency and integrity of such prospective guard, should, in the hands of competent administrators, effectively prevent the licensing of professional strikeguards or strikebreakers. Public police, railroad police and private watchmen regularly employed are exempt from the operation of this statute. The effect of this Wisconsin act upon undercover operatives and labor spies and the persistent and flagrant violation of it by detective agencies operating in Wisconsin will be fully discussed in another report shortly to be submitted.[6]

Unlike the Wisconsin Act, the California and Illinois statutes do not require the licensing or registration of the individual guards or operatives employed by the detective agency. Only the agency is required to be licensed under these acts. Neither do these acts prohibit or place any restrictions upon the furnishing of strikeguards, strikebreakers, or other strike services. The officers of the detective agency must, in order to secure a license, satisfy the regulatory body of their good character, and must post indemnity bonds, but this form of regulation obviously does not reach to the employment or recruiting of professional "finks" by them.[7] Stringent applica-

---

* A report on Private Police System, Violations of Free Speech and Rights of Labor, 76th Congress, First Session (1939), pp. 130-132.

tion of the statute might, however, result in revocation of the license for employing men of bad character.

The Massachusetts statute is, in form, an amendment to an earlier act requiring the licensing of detective agencies.[8] The purpose of the amendment is to embrace both strikebreaking services and industrial espionage. The amendment includes strikebreakers and labor spies in the definition of private detective. Agencies engaged in furnishing strikebreakers and labor spies are to be licensed under the act. The act does not require the licensing of individual strikeguards. Since it is strikeguards who cause most of the evils of the strikebreaking business, the act cannot be said to be an entirely satisfactory regulation of that business. There is no provision in this act for hearings on applications for a license, and the conditions for the issuance of a license are not particularly stringent.[9] Another Massachusetts statute provides that no employer may hire armed strikeguards from an unlicensed agency during the continuance of a strike or a lockout. Employees of a licensed agency who are hired as armed guards must be citizens of Massachusetts who have not been convicted of felony.[10]

In 1937, Pennsylvania passed a statute[11] which, while it did not require the registration of individual strikebreakers or strikeguards, made it a misdemeanor for any person, firm, or corporation "not directly involved in a labor strike or lockout" to recruit any persons to take the place of employees in an industry where a strike or lockout is in effect. Licensed or public employment agencies are exempted from the provisions of this act. This act does not apply to strikeguards.

The New York statute, enacted in 1938, provides the most effective regulation of the strikebreaking business. By that statute, detective agencies are prohibited from employing any persons who have been convicted of felonies or any offenses involving moral turpitude or of certain specified misdemeanors. In addition, fingerprints of the employees of detective agencies must be filed with the secretary of state, who is required to compare such fingerprints with the fingerprint file in the Bureau of Criminal Identification. Having thus comprehensively required the registration and identification of all employees of detective agencies the New York statute goes on to make it unlawful for a detective agency to furnish strikebreakers or strikeguards. The pertinent sections of the statute read as follows:

> It shall be unlawful for the holder of a license or for any employee of such licensee, knowingly to commit any of the following acts within or without the State of New York * * * to advertise for, recruit, furnish, or replace or offer to furnish or replace for hire or reward, within or without the State of New York any help or labor, skilled or unskilled, or to furnish or offer to furnish armed guards, other than armed guards theretofore regularly employed for the protection of pay rolls, property or premises, for service upon property which is being operated in anticipation of or during the course or existence of a strike, or furnish armed guards upon the highways, for persons involved in labor dispute * * *.

The statute makes it illegal to engage in the business of private detective, or to furnish guards or patrolmen to protect persons or property without having first obtained a license in accordance with the act.

This New York statute also prohibits the practice of industrial espionage by detective agencies, as will be fully pointed out in a report shortly to be issued.

In the opinion of the committee, the New York act of 1938 is the most comprehensive regulation of the strikebreaking and detective agency business that has ever been attempted by any State. It combines the licensing features of the Wisconsin statute with a sweeping prohibition of the strikebreaking business. The licensing features of the California and Illinois statutes, while they offer a degree of supervision over the guard services of detective agencies, have no direct or prohibitory effect upon the strikebreaking business itself. The New York statute is more effective administratively than flat statutory prohibitions upon the provision of strike-guards or strikebreakers like the Pennsylvania statute. These and earlier state statutes demonstrate that the states have ample police power to regulate the detective agency business in all its aspects in the public interest. The virtue of the New York act is that it provides, through a scheme of inclusive regulation, a method of controlling and eliminating the evils of the strike-breaking and espionage phases of the detective agency business. The committee recommends the New York act as an example to other states desirous of eliminating these evils from their industrial life.

It is too early as yet to pass any judgment upon the effect and operation of the New York statute, or its treatment by the courts. What success it will have in putting a stop to the activities of detective agencies engaged in interstate commerce, and operating from adjoining states remains to be seen.*

---

* Subsequent legislation has virtually prohibited continuation of such activities by detective and guard agencies in the United States.

FOOTNOTES

1. Cal. Laws (1927, amended 1933), *Deering's Gen. Laws* (1937) vol. 1, act 2070a, §§1-11.
2. Ill. Laws (1933, amended 1937), *Smith-Hurd Ann. Stats.*, §§608b-608z.
3. N.Y. Laws (1938), *McKinney's Consolidated Laws* (1938) Supp. vol. 19, *Gen. Business*, §§70-90.
4. Wisc. Laws (1925, amended 1931, amended 1935), Wisc. Stats. (1937), §175.07.
5. Mass. Laws (1934) Ann. Law (1937), Supp. vol. 14, c.149, §23A.
6. Further report on Industrial Espionage.
7. The Illinois act makes it unlawful for an agency or its employees to encourage or incite strikers to do unlawful acts. Ill. Laws, 1933, p. 469, §11, *Smith Hurd Illinois Annotated Statutes* §608 1.
8. Mass. Laws (1879, amended 1919, amended 1937), Ann. Laws (1933) v. 4, c. 147, §§22-30; v. 4, 1937 Supp., c. 147, §§25 A-25C, 26, and 30.
9. The effect of these Massachusetts laws on labor espionage will be more fully discussed in this committee's supplementary report on that subject.
10. Mass. Laws (1934) Ann. Laws (1937), Supp. v. 14, c. 149, §23A.
11. Pennsylvania Laws (1937) approved June 21, 1937, No. 391, p. 1982.

APPENDIX 2C

THE DEVELOPMENT OF THE PRIVATE SECTOR OF THE CRIMINAL JUSTICE SYSTEM*

THOMAS M. SCOTT and MARLYS MCPHERSON
University of Minnesota

One of the more interesting questions that political scientists are likely to overlook in their analyses of public policy, the delivery of services, etc., is the continually changing mix of public and private provision of such policies and services. Present concern with the criminal justice system, in general, and the police in particular, is no exception. Political scientists are interested in questions of police organization, public accountability of police activity, implications of greater involvement in local police activity by state and federal agencies, professionalization of local police officers, etc. But we have tended to concentrate on the political-governmental provision of police services and to ignore both the traditional and the contemporary role of private police as a very important part of the total panoply of police activities in American society.

We have overlooked private police in part because the private police themselves have not operated in the full light of publicity and in part because we have assumed,

---

* Scott, Thomas M. and McPherson, Marlys, "The Development of the Private Sector of the Criminal Justice System, *Law and Society Review*, Volume 1, Number 2 (November, 1971), pp. 267-288.

erroneously no doubt, that their functions were somehow different from, and therefore irrelevant to, the kinds of concerns related to the public provision of police services.

If one takes a standard definition of police functions, e.g. O. W. Wilson's *Police Administration,* crime prevention, crime repression, criminal apprehension, and the regulation of noncriminal behavior and social welfare functions (including traffic control, intervention in domestic squabbles, handling of drunks, etc.), it is clear that private police have been heavily involved in the performance of these functions since, at least, the establishment of the Pinkerton Agency in the 1850's. Certainly, private police are involved in significant ways in the performance of these functions today. We do not really know how many of the reported crimes listed as "solved" by police departments may in fact have been solved by private police who have turned evidence and, in some cases, the suspect over to the public police.

Of greater significance, and more difficult to determine, is the effect of private protection firms performing the functions of crime prevention and criminal apprehension. Police patrol is recognized as the most effective method of preventing crimes from occurring. How much crime, therefore, is prevented by private protective agencies and security guards who regularly patrol and guard certain areas? In how many instances do security guards apprehend, and perhaps arrest, criminal suspects?

In short, there is little doubt that private police have played and will continue to play a major role in the performance of most of the traditionally defined public police functions.

How large and extensive is the role of the private police segment of the criminal justice system? Unfortunately, there are no reliable ways to provide an answer without extensive state-by-state and locality-by-locality surveys. In some states, only the private investigative firm is required to hold a license and it may hire as many operatives as it wishes without obtaining additional licenses. In addition many private police functions are performed by employees of private firms who are not licensed as private police. Finally, many individuals are able to function as private police without ever coming within the licensing requirements of local and state governments.

Nevertheless, it seems clear that the private investigation and private protection industry as a whole has experienced considerable growth in recent years. *Forbes* magazine (1970: 22) quotes a member of the security industry as estimating that "two out of every three law enforcement officers in the nation are actually on private payrolls." Pinkerton's alone has over 23,000 employees. It has been estimated that approximately $1.6 billion was spent for services performed by the private protection firms last year, with an additional $400 million spent for protective fire and criminal alarms (*Forbes,* 1970: 22). The President's Commission on Law Enforcement and Administration of Justice reported that public expenditures in 1965 at the federal, state, and local levels on police, criminal courts, and counsel totaled $3.2 billion. This estimate means that approximately one third of the total amount of money expended in the criminal justice system is spent in the private police and protection sector of the system. Certainly, then, no examination of the role of the police in the criminal justice system can be complete without considering the role of the private police. This is especially true when there is evidence that the private police part of the system is rapidly expanding and when there are many questions concerning accountability, rights of the accused, and law and order.

In this article, we will provide some rudimentary description and analysis of the private police sector of the criminal justice system. The study is limited in four major ways. First, the analysis of the legal structure within which the private police system functions is limited to laws and practices in the State of Minnesota. One obvious way to expand the present study would be to undertake a comparative analysis of such laws and practices among the several states, but resources would not permit it in this case.

Second, data were obtained from interviews, using structured questionnaires, with the heads of ten licensed agencies operating in the Minneapolis-St. Paul metropolitan area. No attempt was made, at this stage, to sample systematically from among the entire range of private police agencies or activities in the Minneapolis-St. Paul area, partly because of limited resources and partly because no clear definition of what the entire universe of private police agencies might look like is available. Three of the agencies included could be described as large and regionally or nationally based. Five were middle-sized and more locally oriented, and two were small, one- or two-man operations. In addition, interviews were conducted with representatives of the State Bureau of Criminal Apprehension, the Hennepin County Sheriff's office, the Minneapolis Police Department, and several suburban police departments.

Third, within the total range of activities carried out by private police agencies, the study has concentrated on those functions that overlap significantly with public police activities, so that some efforts at comparison between the two sectors can be made. This means that we have not considered the types of activity where private police undertake civil investigations involving personal and domestic problems, e.g. marital cases, divorce and custody cases, personal surveillance, and insurance investigation. On the other hand, we have not considered the range of functions performed by public police which are not generally performed by the private police, primarily those noncriminal, regulative, and social welfare service functions alluded to earlier. We have focused on the areas of functional overlap between the public and private police, namely crime prevention and repression and criminal apprehension.

Finally, the generic term "private police" covers a broad range of institutions, agencies, persons, and activities, where the general functions include the protection of property and persons (crime prevention) and the detection, investigation, and apprehension of criminal suspects. Included in such a categorization would be private detective and investigative agencies; firms and individuals who provide security guards and watchmen for hire; and firms manufacturing, selling, and installing burglar alarms, closed circuit television, and other electronic devices specifically designed to detect the occurrence of criminal activity. In addition, a growing number of commercial and business establishments, as well as industrial firms and corporations, maintain their own security divisions whose primary responsibility is to provide for the overall security—internal as well as external—for the company, including the hiring and training of security guards.

In this study we have focused on the private agencies whose sole function is providing both investigative and security services for hire and we have excluded the security forces maintained by individual nonpolice companies. Private agencies exclusively engaged in security and investigation are readily identifiable, whereas it is much more difficult to draw an accurate sample of firms with their own security forces. It should be kept in mind, however, that whether a company maintains its own security division or hires guards from a private agency the same function is being performed and the methods used are the same.

In the report of the study we shall consider the legal framework within which the private police function (especially in Minnesota, including licensing requirements and enforcement problems). Second, we shall examine the structure of the private police system, especially as it compares with the public police system. Third, we shall explore the relationships between the private and public police and, finally, we shall consider some of the implications of the findings.

Legal Framework—Authority

Since there is no federal legislation dealing directly with private police, the legal authority granted to such individuals and agencies is defined by state law, varies from state to state, and may vary within states from one local jurisdiction to another. John Peel in his book, *Fundamentals of Training for Security Officers*, summarizes the prevailing grants of authority to private police:

> Watchmen, guards, security officers, special police officers appointed for the purpose of patrolling, policing, watching and guarding the persons, premises, and property of an area shall have the same powers and authority upon the assigned property or premises which they are appointed to protect, and in the period of their duty, as the regular police officers but not otherwise (Peel, 1970: 65).

Peel also indicates that in some localities in the absence of statutory limitations, private police forces have authority virtually equivalent to public law enforcement officers (Peel, 1970: 65).

It is not clear to what extent Minnesota is unique in its policies regarding private police, which generally contradict those reported by Peel. It is clear from the statutes, however, that Minnesota law has restricted the legal authority of private police in various ways, beginning with and probably stemming from an explicit prohibition of the interference of private police license holders in any way in labor strikes.

Under Minnesota law private police have no legal authority beyond that of the ordinary citizen. In addition, Minnesota law makes it a crime for any private person or agency to imitate or attempt to imitate vehicle markings, badges, emblems, or other means of identification used by public law enforcement officers.

A county sheriff may deputize any individual and thus confer upon him the power of arrest and other powers ordinarily assigned to public law officers. In some rural counties this practice is widespread and often includes professional private detectives. In the metropolitan area (Hennepin County), however, the power of arrest is rarely granted to private investigative and security personnel; the sheriff prefers to operate with his own professionally trained staff.

The Hennepin County sheriff does grant arrest power in some cases to security guards employed by some of the large, well known firms or corporations but limited to the premises of those firms and only after a review of the company's security policies.

While the interviews with the representatives of the sheriff's office suggested that most private police agents would prefer the power of arrest, eight of the ten private police agency heads interviewed denied seeking the arrest power and argued that they preferred the existing arrangements. They felt that legally conferred police power carried with it legal responsibilities that would place undesirable burdens on their security personnel and substantially restrict their methods of investigation.

For security personnel, legal authority would make it mandatory for a security guard to act, i.e. undertake apprehension and attempt arrest, if a crime were committed within his jurisdiction (the premises). While the policy of the security company may require that their guards take action, it remains at the discretion of the individual and/or the security company to decide on the particular procedures to follow under various circumstances (when to undertake apprehension and when to make a citizen's arrest, etc.). Acting without legal authority *and* legal responsibility allows the security guard to avoid personally dangerous situations where a police officer would be forced to act.

In the case of investigation, operating without legal authority permits greater latitude in both criminal and civil investigations and allows the private agent to deal with a variety of crimes extralegally, that is, outside of the officially prescribed standards of the criminal justice process.[1] For example, the following statements were made to the interviewer: "We can rough a guy up if we want to," "we can get a confession in cases where the police can't because we don't have to worry so much about a guy's rights," and "we can use every means possible to secure information."[2]

Despite the fact that private agents in Minnesota do not possess police powers, it is probably true that most individuals when confronted by a uniformed guard or a man stating that he is a "detective" or "investigator" naturally assumes he has some kind of legal authority. Public misunderstanding of the law undoubtedly gives private agents an additional advantage.

Legal Authority—Licensing and Regulation

Peel also describes wide variation in the criteria and procedures for defining and licensing private police personnel and agencies. Licenses are issued by such units as chiefs, superintendents or commissioners of police, directors of public safety, mayors, city managers, county sheriffs, and general licensing agencies. In addition, state statutes and local ordinances and practice vary widely in their license requirements, conditions of license revocation, and extent to which such provisions are enforced (Peel, 1970: 30ff). Since no comparative state data are available on these matters, our discussion of the Minnesota case is only illustrative.

Under operative Minnesota law, a person who engages in any of a variety of specified activities for hire is required to obtain a license from the state director of public safety. Engaging in such activities without a license is a gross misdemeanor. Minnesota law distinguishes between private detectives and private protective agents, both of whom are required to hold a license. However, persons in the employ of the private investigator or protective agent may engage in such activities without themselves being licensed.

An examination of the relevant statutes reveals that there is considerable overlap between the activities defined as appropriate for private police and the crime prevention and repression and criminal investigation and apprehension functions normally assigned to the public police. According to the Minnesota Statutes (Section 326.338):

Subdivision 1: "Persons who for fee or reward or any consideration shall engage in the business of investigators, or who for fee, reward or any consideration shall make investigations for the purpose of obtaining information for others with respect to any of the following matters: Crime or wrongs done or threatened against the government of the United States or of any state or municipal subdivision thereof; the identity, habits, conduct, movements, whereabouts, affiliations, transactions, reputation or character of any person or organization; the credibility of witnesses or other persons; the whereabouts of missing persons; the location or recovery of lost or stolen property; the origin of and responsibility for libels, losses, accidents, or damage or injuries to real or personal property; the affiliation, connection or relation of any person, firm, or corporation with any organization, society or association, or with any official, representative or member thereof; the conduct, honesty, efficiency, loyalty or activities of employees or persons seeking employment, agents, contractors and subcontractors; the evidence to be used before any authorized investigating committee, board of award, board of arbitration, administrative body or officer or in the trial of civil or criminal cases; or the identification or apprehension of persons suspected of crimes or misdemeanors shall be deemed engaged in the business of private detective."

Subdivision 2: "Any person who shall furnish, for hire or reward, watchmen or guards or private patrolmen or other persons to protect other persons or their property or to prevent the theft, unlawful taking of goods, merchandise, money, choses in action, or other valuable things, or to procure the return thereof, shall be deemed engaged in the business of protective agent...."

What appears to legally distinguish the private police from the public police is the purpose for which private agents are licensed and the method of compensation. The private police agent performs functions which are virtually identical in many respects to those carried out by public police but he performs them for other private individuals and is paid for his services a sum agreed upon by both parties without statutory limitations as to the amount.

In Minnesota, basic problems associated with the licensing of private police have been raised both by the current licensing agent and the representatives of the private agencies themselves. The licensing agent (the state crime bureau and director of public safety) feels that present statutes, court interpretations, and shortages of manpower for enforcement (one part-time crime bureau staff person has total responsibility for private police licensing and regulation) restricts its discretion in granting or denying a license and in regulating activity once a license is granted.[3] Indeed, the operative attorney general's opinion requires that every applicant fulfilling the minimum requirements must be granted a license.[4]

At the same time, the Minnesota Association of Private Detectives (founded in 1968 and presently including about half of the licensed agencies in the state) was established with the primary objective of improving the general image and reputation of the profession by raising standards of recruitment, training, and practice uniformly to at least the levels achieved by public law enforcement agencies. The Association supports changes in the statutes which would raise the professional requirements for licensed agents and encourages enforcement of the laws requiring licenses of those now practicing illegally without one. To this extent, and for the time being, at least, the licensor and licensee are on the same side of the quality control issue.

As police practices become more visible, and large nationally based private police corporations become more predominant, the pressures for tighter controls on and upgraded standards for the licensing of private police will continue to increase. In many respects the situation is analogous to the present conflicts between more and less professionalized public police agencies with the difference that the nationally based private police corporations which stand to gain from higher standards and tighter controls are in a strong position, politically and economically, to move the political system in those directions.

Structure of the Private Police System

We began by trying to compare accurately the number of private and public police operating in the Minneapolis metropolitan area. This is very difficult to do because the private police operative's license permits him to function in any part of the state, while public police are bound by the local jurisdictional boundaries. Second, the large private agencies make extensive use of part time employees, especially in the security function where demand may fluctuate considerably over short periods of time. Third, a number of the private agencies were unwilling to divulge such information. Indeed, as the interviewing progressed, fewer and fewer agencies were willing to discuss their work. Finally, because of the enforcement problems discussed earlier an undetermined number of persons are operating in this metropolitan area without a license.

Of the ninety licensed private police agencies operating in the state, 39 to 42 percent are located in the Minneapolis part of the metropolitan area. Nine percent are located on the St. Paul side. These agencies range from one-man operations to large nationally known operations employing over 200 personnel.

Appendix Table 2-I provides a very

APPENDIX TABLE 2-I

LICENSED PRIVATE POLICE AGENCIES LOCATED IN THE MINNEAPOLIS PART OF THE METROPOLITAN AREA WITH AGENCY SIZE AND EMPLOYEE DISTRIBUTION

Size of Agency	Number of Agencies	Percent of Agencies	Projected Number of Employees	Projected Percent
Small employees per agency	13	39.2%	39	3.1%
Medium 25 employees per agency	12	44.7%	425	33.6%
Large 100 employees per agency	8	21.1%	800	63.3%
Totals	38	100%	1264	100%

rough estimate of the structure of the private police system located in the Minneapolis metropolitan area. The table indicates the majority of the private police are employed in the large corporations and only a very small proportion of the total (less than 4%) are employed in the small licensed firms.

In addition to these estimates, the State Crime Bureau reckons that there are as many agencies operating without licenses as with them. They are certain that these are primarily one- and two-man operations, exclusively. If we assume a conservative estimate of another fifty unlicensed operatives, it brings the total to more than 1,200 private police working in the Minneapolis area.

By contrast, the Minneapolis police department employed 869 officers in 1970 and the Hennepin County sheriff employed an additional 150 personnel for a total of 1,019 public police functioning in the same general jurisdiction. While these figures do not include suburban Hennepin County police departments, nor security guards employed directly by commercial and industrial firms, one can conservatively conclude that there are at least as many private police as public police operating in the Minneapolis area.

Size is, of course, an important factor in determining the kinds of specialities an agency develops. Between 80 percent and 100 percent of the total work volume of the small firms is investigative, where demand has been growing but at a slow and steady rate. For the most part, the clients of the private investigators include defense attorneys, insurance companies, and individuals and firms who wish to avoid the publicity of a public police investigation or who are dissatisfied with the extent to which the public police can investigate a case. Most of the investigative work of the smaller agencies involves civil cases, in particular personal and domestic investigations. If the smaller agencies are involved in security work, the work usually involves the installation of security devices and/or individual protection such as personal body guards, bank deposit guards, etc. Many of the small agencies are not oriented toward growth of the agency and explicitly limit the volume of business to the level that can be managed with existing personnel.

For those smaller agencies that are oriented toward growth, however, the pattern of development is similar in most cases. The agency begins with limited personnel and concentrates on all types of investigative work where the capital costs can be minimal. When financially secure, the agency moves into the security specialties where growth is rapid and where capital costs are greater. Expansion at this point occurs primarily by adding security personnel. The stabilizing point seems to be between 20 percent and 35 percent investigative work and 65 percent to 80 percent security work. At a certain point, the agency begins to move toward qualitative rather than quantitative development and becomes more selective in the types of cases and clients it will accept. A number of the larger corporative agencies, for example, refuse to take personal and domestic cases and "questionable" or "sticky" cases, e.g. those involving politicians or providing security guards for firms with labor problems. As a consequence, assuming a continued quantity of the less attractive types of cases, there will presumably continue to be a demand for the smaller, less selective private police agencies.

The area where private demand is growing most rapidly is industrial security. In response to this demand, agencies are developing total comprehensive security plans for firms, including consulting and planning overall plant security, preemployment investigations, hiring and training of security guards, installing and monitoring electronic detection devices, and providing investigative agents to deal with internal security problems (pilferage, embezzlement, espionage) as they arise. There also has been an increase in the number of neighborhood groups who cooperatively hire a protective service to regularly patrol their neighborhoods. The clients of security agencies are those who desire and can afford security and protection in addition to what public police can reasonably provide.

Perhaps the most important factor in the development of the larger regionally and nationally based private police agencies is their ability to utilize efficiently highly sophisticated, expensive equipment. In an era when technological sophistication in the law enforcement field has developed very rapidly and expensively, it seems clear that the large private police agencies are much better equipped than their counterparts in public police agencies.

At a minimum, the private agencies utilize the same kind of equipment as public law enforcement agencies. In performing the crime prevention function, uniformed security guards (frequently armed), two-way, radio-equipped patrol cars, walkie-talkies, and riot equipment are commonly used.[5] The larger national corporations have facilities at least comparable to large public law enforcement departments, such as completely equipped crime laboratories, to assist in the performance of the criminal investigation function. One national company advertises that it maintains a central file containing detailed records on over 6,000,000 individuals.

In addition, the private agencies utilize much more sophisticated, scientifically advanced, technical equipment than most local law enforcement agencies can afford. In Minneapolis, for example, all major equipment purchases by the police department must first be approved by the city council; and there have been few cooperative joint purchases or sharing of major pieces of equipment with other police departments. On the other hand, the national corporations, which have offices throughout the country, maintain centralized equipment which can be dispatched where and when needed, e.g. specially trained dogs, electronically equipped surveillance vehicles, special purpose trucks, trailing devices, photographic equipment, helicopters, intricate security alarms.

Additionally, while we were not able to verify the extent of use, it seems clear that the private agencies are less restricted than public agencies in the use of electronic devices, such as telephone bugs and other intrusive equipment, since the private agencies are not publicly accountable and do not for the most part operate in the glare of publicity. Furthermore, because the private police are paid for services rendered they can maintain surveillance as long and as extensively as the client wants, while the public police may be forced for economic reasons to shift resources to other cases.

In general, the competitive market system has decided advantages that favor the large private police agencies over the smaller agencies and the public police in the crime prevention and criminal investigation function. Centralization of equipment, as well as financial ability to maintain expensive technical equipment, and the time and personnel to pursue a case to its conclusion, lead to efficiency and thoroughness which public law enforcement agencies often cannot attain. At the same time, the prerogative of private agencies to hire and dismiss personnel, as well as employing people for temporary and part-time assignments, in response to varying demands permits economic efficiencies not possible in public agencies operating under civil service regulations.

The final section on the structure of the private police system concerns the quality of personnel as reflected in recruitment and training. For the most part such questions are subsumed under the heading "professionalization." Again, some general comparisons with public police agencies are instructive.

The range of opinion and concern for professionalization of personnel is about the same for private police as for public police. The larger private agencies are quite concerned about high educational and experience levels for recruits, extensive training, and professionally determined mobility. By the same token, some police departments, especially suburban departments in the Minneapolis area provide incentives for educational achievement and have begun to facilitate mobility patterns. On the other hand, many of the smaller private police agencies do not appear to share the same concern for higher standards of recruitment and licensing and these attitudes reflect the feelings of some personnel in local police departments.

When personnel practices are examined, however, some differences between the private and public systems emerge. All the private agencies indicated that college work was not significant in hiring security personnel where previous experience and personality factors such as "stable," "personable," and "not afraid of people" were most important. On the other hand, college education was quite important in hiring investigative personnel (second only to previous investigation and law enforcement experience). Indeed, most investigative employees, especially of the larger firms, have some college work and about half have college degrees.

Public police agencies generally require high school diplomas, although many now provide incentives for college work. On the other hand the public agencies in the Minneapolis area are more rigorous in their testing and evaluation of job applicants while the private firms tend to rely heavily on personal interviews.

In the area of personnel training as a component of professionalization, the public agencies have the clear edge. The Minneapolis Department has a sixteen-week rookie school which all recruits attend, and under law all police officers must receive at least 210 hours of training. For private agencies, on-the-job training is most common, although some of the large national firms have their own training programs that parallel the programs offered to public police recruits. In addition, of course, many personnel hired by the private agencies have previous training and experience in public police departments and to that extent the public police systems serve as a kind of farm system for the larger private firms.

A third component of professionalization involves mobility potential. The 1968 Report of the Minnesota Governor's Commission on Law Enforcement and the Administration of Justice and Corrections, in assessing the degree of professionalism in police work, states:

> A professional has a degree of freedom as to his choice of where he wants to practice his profession. As he becomes more competent at his profession, his range of choices of where and how he wants to practice should increase. Such is not the case with police. . . . Until a greater degree of mobility develops within police work, the police drive for professionalization will continue to be

extremely slow (Governor's Commission on Law Enforcement, 1968: LV-E-1).

The traditional type of mobility in police work is almost exclusively vertical, i.e. promotion within the department for which an officer works. Lateral mobility—transfers from one department to another at the same or a higher level—is, with few exceptions, nonexistent. New men are hired only at the patrolman level and promotions occur at regularized intervals. An individual desiring a career as a detective or investigative officer or in a supervisory position must work his way up through the ranks in that department. If he moves to another police agency it would normally mean starting out again as a patrolman.

A somewhat different pattern of mobility is evident in private agencies. The internal organization of the large agencies is in most cases patterned after police departments. There is an investigative division and a security division (corresponding to police patrol division). Within the security section there are ranks similar to those found in police departments—guard (corresponding to patrolman), sergeant, lieutenant, captain, etc. However, in the absence of civil service regulations, the mobility patterns tend to be more fluid and diverse. Vertical mobility is the predominant pattern within the security division and moves from the security division into the investigative division occur infrequently. Lateral mobility exists to the extent that a private agency, when hiring someone with previous security experience, will start him at a level commensurate with his previous experience.

Within the investigative area, lateral mobility is the predominant pattern. Almost all investigative personnel are recruited either from public law enforcement agencies or other private agencies. Previous experience means beginning with a higher salary. The essential separation of investigative personnel and functions from security personnel and functions is, as indicated, different from the public police pattern. The result is that moving from public law enforcement work into private investigation is desirable because the job is, in most instances, more lucrative. On the other hand, few policemen move into security guard positions (unless part-time) because the pay scale for private security personnel is generally lower than what they receive as police officers.

It is clear that the private police system plays a major role in the overall performance of police functions, especially in the urban setting. It is also clear that the general economic and business rules of success are largely responsible for the developing structure of the private police system. To the extent that large, nationally capitalized firms are better able to purchase, maintain, and utilize the most highly sophisticated technical equipment under fewer constraints and with greater impunity than virtually all urban public police systems, they can provide better security and investigative service to those willing and able to pay. Thus, it is in their interest to improve the "image" of the private police system by emphasizing, both within their own ranks and through state law, higher standards for recruitment and hiring, and better training programs. On the other hand, the large firms do not want to eliminate completely the one- and two-man operations since both they (the large firms) and the public police are not willing and/or able to handle some types of cases.

Finally, the taxpayer is disadvantaged in all this to the extent that his taxes support a training and experience system in public police agencies that (especially with 20-year retirement programs) provides a most valuable manpower source for the private police system. This problem will continue as long as public police agencies continue to operate within locally based civil service restrictions and private police systems function within the rules of an unrestrained market price system.

Public-Private Police Relationships

The final part of our investigation attempted to define the relationships that exist between the two parts of this dual police system. Interviews with members of various public police departments, the county sheriff's office, and the state bureau of criminal apprehension reveal two salient facts. First, there are no official laws or policies on the part of either the public or private police systems defining their relationships with each other. Second, there is a considerable range of attitudes on the part of individuals, particularly among public law enforcement officers, regarding the private police system. These attitudes, in turn, affect the nature of the relationships that develop between the two systems.

Some law enforcement officers still hold a view based on the notion that private detectives are "snoopers" whose methods are unprofessional and often dishonest and that illegal and private security men are "gun-happy kids," "old men" (the popular night watchman image), "a cab driver out to make a fast buck in a business where the demand is growing and no particular skills are required," or, most serious of all, "some criminals who figure that the easiest way to rob a house is to be hired to guard it." Police officers who hold these views generally feel that private agents are tolerated "because there is no law against them." They claim to have little or no contact with private agents and do not believe that the private system provides any useful supplementary service to what the public police system can provide.

Most law enforcement officers interviewed indicated that their attitudes toward the private police system are more positive at best, too ambivalent at least, while no one interviewed indicated that he had favorable attitudes toward and cooperative relationships with all private agents. Many did feel that properly qualified and trained private investigative and security personnel can provide a valuable supplement to public police work.

There appear to be two primary reasons why the more negative attitudes of public toward private police are being replaced by a more positive and cooperative point of view. First, a number of the professionally oriented agencies are consciously pursuing policies to dispel the adverse image of private investigators and security personnel and improve the relationships they have with public law enforcement officers. For example, a number of agencies will not begin investigation of a criminal case until the police have completed their investigation. If a client contacts them before calling the police they specifically request that an official report of the crime be made to the police. Most agencies reported that they always notify the relevant law enforcement agency when operating within their jurisdiction, giving them full details of the case they are working on. Security guards of such agencies are given special training in the procedures to follow if an attempted crime is witnessed. Instructions usually include calling in the police and giving them "full cooperation."

A second, and perhaps more important, factor leading to improved relationships and increased communication is the large number of police officers "moonlighting" on a part-time basis for private agencies. In addition, increasing numbers leave public police departments to start their own private agencies, work full time for an established agency, or join the security division of a business or corporation. Nine of the ten agencies with whom interviews were conducted had employees with previous law enforcement training—police, sheriff's office, F.B.I. or military intelligence, and the majority of the agencies considered previous law enforcement experience as the single most important qualification in hiring for investigative positions. One agency reported that it had employed as many as 200 individual police officers on a part-time, case-by-case basis during the past two years. This type of interaction—the same individuals employed by both private agencies and the public police—has had a major impact on increasing cooperative attitudes and relationships between the public and private systems.

Since so many of the interactions between private and public police are informal, *ad hoc,* and not disclosed freely by either party, it is difficult to describe systematic patterns and developments in these relationships. However, the following examples illustrate some of the types of interactions that do occur in the urban setting. Most frequently mentioned is the mutual referral of cases. The public police will recommend the name of a private investigative or security agency in cases where they cannot conduct the kind of investigation or supply the kind of surveillance requested by the complainant.

Similarly, private agents indicated that in particular cases they would suggest to a client that the problem was one which the police could best deal with (in particular where violence is anticipated). The comments regarding the mutual referral of cases suggest that both the police and the private agents recognize some vague line of demarcation, not clearly delineated, between what constitutes the responsibility of public police and in what areas the private agencies might function.

A second way in which the public law enforcement agencies cooperate with private agencies is through the mutual exchange of information. Both the police and the private detectives maintain their own network of informants. When private agents receive information concerning a case the police are working on they will pass it along to the police, and vice versa. One agency even indicated that one of its investigators attended the official monthly intelligence meetings of local law enforcement officers. This type of relationship is primarily true only of those agencies that have a close cooperative relationship with the police.

Security guards regularly come into contact with law enforcement officials in the course of their work. They may cooperate with the police in apprehending a suspect in the act of committing a crime on the premises they are guarding or if they undertake apprehension without police assistance, the police are called to make the arrest. In such instances, they turn over all relevant information and evidence to the arresting officer. Similarly, when private investigative agents are employed by clients on a case on which the police are also working, private agents may cooperate with the police by turning over to them evidence they have collected in their investigation. In the instances where the police and private agents cooperate in criminal apprehension or criminal investigation leading to arrest, private agents are regarded as providing supplementary services for the police.

Unfortunately, police records do not indicate in which case private agents have played a significant role. Police officials indicated the number was "probably not very large." However, one private agency employing ten investigators and twenty security guards indicated that their firm averaged six felony apprehensions per month. Equally important, but less measurable, is the contribution of private agencies in the number of crimes prevented by the presence of private security personnel.

Finally, private police agencies lend investigative and surveillance equipment to public police agencies under some conditions. This is, of course, a most sensitive area of interaction since such equipment may be used for surveillance or property search purposes that are illegal either under the 1968 Omnibus Crime Control and Safe Streets Act or state laws. As Braun and Lee (1971:562) suggest, such activity is difficult to detect and prosecute whether it is carried on by private or public police agencies. Often the borrowing by public agencies is done because they are prohibited by governmental action or lack of funds from purchasing such equipment. The primary benefit to the private agencies of such sharing of resources is access to information normally available only to public police agencies.

In many respects, the relationships between the private and public police represent one of the most interesting and potentially troublesome aspects of the development of a private police system. To the extent that the private police supplement public police work for those willing and able to pay and such supplements are not subsidized by the average taxpayer, the development of a professionally oriented private police system is probably useful and can actually benefit less affluent parts of the community by reducing some of the demands for public police services. However, to the extent that the private police system in its activities and methods provides a means by which the public police are able to bypass, evade, or subvert systems of accountability and rules of procedure, the unregulated development of a closely interacting private and public police system will inevitably create serious problems.

Conclusions and Discussion

Our study suggests several lines for further analysis both by the policy specialist and the general student of the criminal justice system. In the first instance, the rapid development of large, nationally based, private police agencies along with the pressures for increased professionalization emerging from within the public police system will require, sooner or later, changes in the laws involving functions, licensing, and regulation of the private police. Various categories of licenses may be required to distinguish among the smaller one- to three-man operations, the security personnel employed by nonpolice firms, and nationally based private security and investigative corporations.

For the student of the criminal justice system, a multitude of questions have been raised involving rates of development of the private system, the distribution of various types of cases by types of firms, potential conflicts within the private police profession, the implications of efficiencies and economies of scale available to the private system that have not been utilized by any but the very largest public agencies, etc.

These issues aside, however, we may conclude that the private police system has indeed expanded in recent years. What is more significant is that the expansion has been in directions different from those that prevailed in the recent past. The expansion has come by way of the nationally based, heavily capitalized firms that are able to utilize equipment and methods usually not available to the public police agencies. The expansion has also come from increased use of security personnel employed by firms engaged in commercial or industrial activity.

There are several fundamental issues raised by these developments. To the extent that private police are engaged in the investigation and apprehension of persons who have committed crimes, their functions clearly overlap those assigned to the public police. In the case of the public police, however, the act committed is defined as a crime against the society, the police agency undertakes the investigation and apprehension, and questions of guilt or innocence are decided in the courts. The client of the police officer, as it were, is the community. The community determines the rules under which the investigation, apprehension, and disposition of the case take place and the officer, theoretically at least, is accountable to that system of rules and to the community.

The private police agent, however, in dealing with the same acts that the public officer deals with, defined as crimes, is in the employ of a private individual or firm rather than serving the community. He is to a much greater extent not subject to the same rules for investigation and apprehension and he is accountable only to himself, his profession, and his employer.

He is employed, presumably, to investigate and apprehend the wrongdoer because the client has suffered a direct loss by virtue of the acts of the wrongdoer. From the client's standpoint it is the loss and its potential restitution (or the prevention of future such losses) that is important rather than the more general concept of justice for a crime committed against society. His interests, therefore, are in direct action and results, rather than general principles and rules of appropriate investigative and apprehension behavior established by the community. Such incentives, then, become those directing the behavior of the employee, the private police agency.

This can have two important implications for the criminal justice system. First, many persons are investigated and apprehended for allegedly criminal acts but never move into the judicial part of the criminal justice system. Their cases are "resolved" among themselves, the private police agency, and the client who was, presumably, wronged by the act. Certainly not all criminal investigations conducted by private investigators involve such extralegal resolutions of alleged crimes. However, there are certain kinds of cases, in particular robbery and theft, embezzlement, internal theft, and industrial espionage, where this kind of "solution" to crime is likely to occur.

The following example will illustrate the point. A company suffers a loss of $3,000 and reports the loss to the police. The police make a preliminary investigation—interview people, ascertain the facts of the case, make a report, but uncover no conclusive evidence. The case remains dormant, largely because the police cannot actively pursue the investigation due to limitations of time and manpower. At this point the private investigative agent is called in (he actually may have been called before the police, but preferred to wait until their investigation was completed). The private investigator explains to the client that there are two alternative approaches to the investigation depending upon whether the client desires prosecution or restitution. Our interviews suggest that 75 percent of the victims in such cases prefer restitution. Assuming that restitution rather than prosecution is the goal, the investigation proceeds according to the rules and procedures of the private rather than the public police.

This leads directly to the second major implication of the development of the private police system for the criminal justice system: the potential disregard for the rights of suspects and others connected with the investigation. This may result in investigative methods that are not subject to the same kinds of constraints under which public police operate, and it may result in the use of technical equipment that goes beyond that which the courts have permitted for police activity. Furthermore, once apprehended, the accused person is not protected by the procedures and guarantees now afforded persons accused of crime in the public sector. In short, virtually all the difficult and often controversial procedural rights and protections now guaranteed to persons apprehended and accused of crimes by the police can be and are often ignored by investigative and apprehension procedures used in the private police system, largely because the definition of the "crime," the interest of the "victim," and the incentive of the investigating and apprehending personnel are different from those we assume for the public police system.

Quite apart from the private police implications for the criminal justice system, there are two other issues which we shall raise in conclusion. First is the problem of invasion of privacy by private police agents. Questions of the definition and meaning of privacy are difficult and they are not made any easier by the fact that private police are in a sense licensed for purposes of securing information about individuals with very few legal limitations on the extent to which they can eavesdrop; spy; question friends, acquaintances, and employers; use electronic equipment; etc. The problem is aggravated by the lack of standards and enforcement procedures in the profession itself and by the very difficult procedures through which an individual must go if he wishes to complain about the investigative activities of a licensed private police agent. In Minnesota, while informal complaints are received by the director of public safety, the costs in initiating and following through with the procedures of a formal complaint are so high that very few citizens are willing to pay the price. Since the state does not actively enforce the existing rules regulating private police agencies, the effect is to permit them to function largely unhampered in their methods and techniques.

Finally, there is the problem of consumer protection for the purchaser of private police services. Since the state laws and licensing practices do not establish very much by way of minimum standards, and since the industry itself has not moved very far in its self-regulation, the consumer has little to go on when he considers the purchase of private police services.

FOOTNOTES

1. Braun and Lee (1971:561) argue that existing civil and criminal laws probably provide adequate recourse to citizens who have been treated improperly during the course of arrest or search incident to arrest by private police personnel, but that present laws, both federal and state, do not have much power to protect the citizen from improper private police activity involving surveillance and private property search. In the case of interrogation by private police personnel, they point out that procedural requirements such as those deriving from *Miranda v. Arizona* (384 U.S. 436, 1968) generally have not been applied.
2. While a minority of the agents interviewed made such extreme statements, all of them indicated that they could conduct an investigation without many of the restrictions operating in public police investigations.
3. Braun and Lee (1971:559) indicate that since complaints of improper activity by private police personnel to licensing agents are rarely filed and license revocation rare, there is not much effective control of private police activity exercised through the state's licensing power.
4. Opinion of the Attorney General, 828-D, November 7, 1945.
5. Despite the Minnesota law regarding the imitative use of symbolic markings, a number of agencies use vehicles of a make and model similar to police patrol cars, with badge-like emblems on the sides. Also, the uniforms of some guards are very similar to police uniforms.
6. As Braun and Lee (1971:555) indicate, however, existing civil and criminal laws provide more or less effective constraints against improper private police activity except in the areas of surveillance, private property search and interrogation.

REFERENCES

Braun, Michael, and Lee, David J.: Private police forces: Legal powers and limitations. *University of Chicago Law Review*, 38:555-582.
——: Creeping capitalism. *Forbes*, 106:22ff.
Minnesota Governor's Commission on Law Enforcement: Preliminary report of the governor's commission on law enforcement, administration of justice and corrections. St. Paul, State of Minnesota, 1968.
Peel, John Donald: *Fundamentals of Training for Security Officers*. Springfield, Thomas, 1970.
Wilson, O. W.: *Police Administration*. New York, McGraw-Hill Book Company, 1963.

APPENDIX 2D

THE PLACE OF PRIVATE POLICE IN SOCIETY: AN AREA OF RESEARCH FOR THE SOCIAL SCIENCES*

Social scientists have directed little inquiry into the nature or extent of private police activity in the United States. This article undertakes an examination of private police within the general context of policing in order to pinpoint some of the factors that should be considered in creating a research design.

---

* Theodore M. Becker, "The Place of Private Police in Society: An Area of Research for the Social Sciences," *Society of Social Problems*, Mass., 1974, pp. 438-453.

The development, growth, and present-day functions of private police are discussed, as well as the extent to which private police activity is regulated. These components are analyzed within a conceptual framework that focuses upon the interrelationships between the "public" and "private" sectors of law enforcement. Implications for police research and for society are suggested.

The time for social scientists to turn their attention to the operation of private policing agencies is long overdue. According to a *Forbes* (1970) article, two out of every three law enforcement officers in the nation are actually on private payrolls. In dollars, $1.6 billion was spent by private firms for protection in 1969 (*Forbes,* 1970: 22) as compared with $3.2 billion in public (federal, state, and local) expenditures for police, counsel, and criminal courts in 1965 (Scott and McPherson, 1971:268). Such widespread use of private police and private protection agencies points up a need for a broader understanding of their activities and the functions they fulfill. Yet, to date, very little literature has been devoted to examinations of private law enforcement agencies. While a few authors have concerned themselves with historical documentations of Pinkertons, the largest private law enforcement organization (see, e.g. Rowan, 1931; Horan and Swiggett, 1951; Lavine, 1963; Horan, 1967), these efforts for the most part have relied upon anecdotes to demonstrate the persistence and tenacity of private investigators. There appears to have been no widespread criticism of private detective agencies since the Pinkerton involvement in labor disputes, around the turn of the century (Friedman, 1907). With the exception of one study of private police which concentrated on Pennsylvania (Shalloo, 1933), only recently have specific attempts been made at systematic analysis of the law governing private law enforcement (*Southern California Law Review,* 1967; Braun and Lee, 1971). Although from time to time legislative hearings have been conducted which have touched upon the impact of private law enforcement on society,[1] social scientists are only just beginning to do work in this area (Scott and McPherson, 1971). In this analysis, I attempt to set forth a framework in which social scientific examination of private police can proceed.

CLASSIFYING "PUBLIC" AND "PRIVATE" POLICE

Using "Sponsorship" as the Basis of Distinction

The definition of "private police agencies" is potentially very broad. Perhaps the simplest definition is one based on source of sponsorship.[2] Such a definition can be viewed broadly or narrowly, depending on the qualifiers. If one conceives of "sponsorship" as the act of paying the police, much as *Forbes* (1970) does implicitly in the statistics presented above, one will certainly exclude volunteer vigilante-type groups which from time to time serve policing functions, such as the Hassidic Jewish patrols in New York City, and may also be excluding volunteer policemen who, in some communities, serve without pay as deputies cooperating with public police forces. A broader conception of "sponsorship" might satisfactorily include such groups. In most of those cases in which no payment changes hands, it is still possible to determine whether the volunteer works for the "public" sector (as in the case of deputization) or for the "private" sector (volunteer snoopers or stoolpigeons), or for their own unique purposes (personal protection in the case of the Hassidic Jewish patrols, "national security" in the cases of the "Minutemen" and other nativist paramilitary groups [Albares, 1968]).

It may be unsatisfactory to utilize a definition based even on a broad notion of "sponsorship," however, because many instances are foreseeable in which such a definition would give a researcher a good deal of categorical overlapping and resulting confusion. Such instances suggest areas of convergence between the public and private sectors of law enforcement which, taken together, tend to discourage the use of definitions based on sponsorship. For example, often deputies volunteering services in time of need have motives other than public service. They may be bullies-at-heart, enjoy the license to carry a concealed weapon, or, much worse, be "truculent, disorderly, intolerant, and downright vicious . . . with motives of their own and objectives foreign to the maintenance of civil peace" (Smith, 1960). In such cases, it is less than accurate to characterize those deputies as sponsored by the public.

Even in cases where pay changes hands, it can be difficult to discern the nature of the sponsor. For example, there have been instances in which, because of additional service requirements brought about by the nature of some industries, such as the sports industry, cities have actually charged private companies (in addition to taxes), for "public" police labor (*Forbes*, 1970: 24). State government has encouraged the use and actively sought the support of a private detective agency in a war on crime (*Southern California Law Review*, 1967: 540). There has been, over the past few years, a good amount of exchange of public police agency information for private police agency equipment. Because the larger private police organizations are often extraordinarily well equipped (national private police organizations maintain such equipment for ready dispatch as specially trained dogs, electronically equipped surveillance vehicles, special purpose trucks, photographic equipment, helicopters, and intricate security alarms), private police forces have been known to lend sophisticated investigative and surveillance equipment to public police agencies in exchange for access to information available to public police forces (Scott and McPherson, 1971:278).

Another area of cooperation between public and private police has been identified. One Minneapolis private police agency with which Scott and McPherson (1971: 283) conducted an interview reported that it had employed approximately 200 public police officers on a part-time basis over a two-year period. Nine of the ten Minneapolis private police agencies with which they conducted interviews had employees with previous public law enforcement experience. Many public police officers leave the force to "start their own private agencies, work full-time for an established agency, or join the security division of a business or corporation" (Scott and McPherson, 1971:282). This type of cooperation is indicative of the type of convergence between public and private sectors of law enforcement which would render definitions based on sponsorship ambiguous.

Besides such apparent muddling of the public and private police function there may be other, more subtle, factors at work which tend to make definitions based on sponsorship even less clear and more problematic. A couple of passages from a widely used work on police administration brings to light such subtle factors in a chapter entitled "Informing the Public" (Wilson, 1950:412-413). In one such passage, the author states:

> Important groups should receive the first invitations for tours of inspection. It is usually well to start with the leading luncheon clubs and take the entire list in the order of their importance. Church and industrial groups and professional clubs may follow. When it becomes known that groups have been taken on tours of inspection through police headquarters, many others will wish to do the same; from the police viewpoint this is desirable, and the privilege of inspection should not be denied any group sufficiently interested to make the request. Less extensive tours may be instituted for smaller and less important ones.

Another suggestion offered, again aimed at impressing "the public" favorably (Wilson, 1950:413), is to:

> ... invite influential men of the community to ride in police cars and observe the operation of the police officer on duty so that they may know how the officers patrol their beats and dispose of the unusual situations with which they are so frequently confronted.

It seems that, in light of Wilson's emphasis on the importance to the police of certain elements in society, the "sponsors" of public police might in large parts be the same industrialists and members of leading luncheon clubs who most likely employ private police. When one proceeds on the assumption that public police depend on the public for financial support (Hahn, 1970:646), it may become relevant to consider specifically which segments of the public are most important to public police.

In any case, it becomes clear that distinguishing private from public police by use of a measure of "sponsorship" is not a satisfactory approach. Although it initially would seem reasonable to identify public police as those sponsored by the "public sector" of the economy, i.e. tax dollars and private police as those sponsored by private industry and private dollars, the foregoing analysis has demonstrated that such a simple distinction is not possible. Using sponsorship as the definer and thus the basis by which public and private police are distinguished cannot be reconciled with the passages quoted from Wilson (1950), or with the above cited examples of convergence and overlap between private and public police. Therefore, a definition based on sponsorship is unacceptable.

Using "Services Performed" as the Basis of Distinction

I will begin by drawing a general picture of the scope of services performed by private police forces and by public police forces. It will soon become apparent that, in the area of "services performed," reality does not coincide with what might be the more popular conceptions. Rather than devoting most of their time to the traditional investigative "private-eye" work portrayed in histories of private police agencies (Rowan, 1931; Horan and Swiggett, 1951; Lavine, 1963; Horan, 1967), it has been estimated that most private policing firms do about 20-35 percent investigative work and 65-80 percent security work (Scott and McPherson, 1971). A full 500,000 uniformed persons are included in an estimated total 800,000 "private police," i.e. police paid by private sources as opposed to tax dollars (*Forbes*, 1970: 22). These security functions include campus security forces, special patrol agencies, armored car units, plant protection units, industrial guards, and retail security guards.

Despite the large number of persons already employed in such private police security functions, there is every indication that the demand will steadily grow (*Forbes*, 1970). The reader may recall the above mentioned expanded use of private security forces by state government (see also *Southern California Law Review*, 1967). In addition to this, it has been noted that some private police agencies, structured along the lines of city police forces, have begun to provide patrol car and guard services to neighborhood groups and urban corporations (*Forbes*, 1970:22; Braun and Lee, 1971). Furthermore, to the extent that various broad interest groups, such as tenants, demand increased protection, business and property owners may be compelled to furnish such security in the form of private police. Already there has been some suggestion of this in a recent court action. In *Kline v. 1500 Massachusetts Avenue Apartment Corporation*,[3] the United States Court of Appeals for the District of Columbia Circuit held, in a 2-1 decision, that evidence that a landlord had noticed that tenants were being subjected to crimes against their persons and properties, which were taking place in common hallways of the landlord's building, established that the landlord was aware of conditions which created a likelihood that further criminal attacks on tenants would occur. The standard of care in providing such protection was held to be that standard which the landlord himself was employing at the time the tenant became a resident on the premises.[4] Accordingly, since such a standard had not been maintained by the landlord, he was held liable for resulting injuries to a tenant.

The *Kline* court made its decision on the basis of facts which demonstrated an actual decrease in the quality of security services provided by the landlord. But the language of the court's holding does not necessarily preclude a finding of landlord liability where the level of security has remained constant, but the rate of crimes has increased. This, too, would be a situation in which the standard of care would not be equal to the standard employed when a tenant originally moved onto the premises, because the level of protection effectively would be diminished. If courts in the future are willing to extend the *Kline* decision to such situations, the demand for private police protection in the area of personal and property protection will most certainly increase. Indeed, we have a situation in which, as *Forbes* (1970: 23) has observed, "regular police would have to curtail their activities too much to meet the demands of the business enterprise."

At the same time that we are witnessing an increase in the use of and demand for private police in the security function, we are uncovering a somewhat altered public police function. It is by now relatively well known that public police do more than pursue criminals. A great deal of police activity can be classified under the category "order maintenance" (Wilson, 1968). Much public police activity amounts to sheer service functions. Many community dwellers, especially in neighborhoods inhabited by the poor, have come to depend on such police services, and look to the public police as the primary supplier of public service. Wilson (1968) pinpoints some of these as first aid, rescuing cats, and helping ladies. But there are many more important service functions public police can and do fulfill, such as spending time in ghetto neighborhoods working with troubled youngsters (Bordua, 1968).

Public police increasingly have been devoting much of their time to assistance which is noncrime related, in the nature of health care through ambulance service, mediation of family and neighbor disputes which are civil in nature, environmental disturbances, runaway children, and nonfunctioning of gas, phone, electricity, and water utilities (Bercal, 1970; Schulz, 1969). Recent surveys have noted that public police are called in by poor people to such a great extent in noncriminal and emergency situations (Bayley and Mendelsohn, 1969; Black, 1968) that one literature review concluded "the poor and uneducated, it seems, use the police in the same way that middle-class people use family doctors and clergymen" (Johnson and Gregory, 1971). Specifically, it has been estimated that between 80 and 90 percent of the public policeman's function is in activities unrelated to crime control or law enforcement (Misner, 1967; Epstein, 1962). In Chicago during one month in 1966, of a total of 134,369 calls for police, only 17 percent were classified as "Criminal Incident" (Parnas, 1967).

Thus, while the assumption has been that the public policeman's primary function is to control crime, researchers have disclosed much evidence to the contrary. There seems to be a great deal of concurrence recently among researchers that the public policeman's function has shifted toward much increased performance in noncriminal and emergency situations (Bayley and Mendelsohn, 1969; Cumming, Cumming, and Edell, 1965; Black, 1968; Misner, 1967; Epstein, 1962). A recent analysis of public police budgets suggests that any expansion of the noncriminal functions of public police will necessarily concide with a corresponding decrease in the crime control function. If consumer demand increases for noncrime related services, crime control services will decrease because public police resources cannot allow a net increase in overall per capita police services. This is because public police budgets on the whole have not substantially increased in effective spending power during this century. Bordua and Haurek (1970) have demonstrated that, although annual local police expenditures in the United States between 1902 and 1960 increased over 31-fold, the expenditure change is not due to a real increase in resources available to improve the quality of law enforcement, but due to the joint effect of such components which are largely independent of changes in crime rates, such as population growth, inflation, urbanization, and motor vehicle increase.

In light of the foregoing analysis of the services private and public police actually perform, we uncover no clear-cut basis for distinguishing public and private police as to services performed. At best, private and public police functions seem to be turned around from traditional images. Thus, we find private police performing the more aggressive crime control and apprehension functions, while public police perform more prevention, order maintenance, and community service functions. This is not to say that public police do not still participate to a great extent in investigation and apprehension of perpetrators of serious crimes, particularly crimes of violence. The area in which private police seem to be taking over is that of theft prevention, protection of private premises, and the types of more minor crimes which used to be curtailed by the presence of policemen on the beat (Trojanowicz, 1970). But it is evident that the private police are taking over many functions that public police would ordinarily be called upon to perform, while at the same time public police are performing many traditionally nonpolice functions.

The foregoing discussion suggests that, just as a definition based on "sponsorship" is less than satisfactory when attempting to distinguish private from public police, a definition based on "services performed" also is unsatisfactory. As with "sponsorship," we encounter too much overlapping of services performed and shifting of functions, all of which prevents any clearcut conception of the services performed by either public or private police. The inability to settle upon a basis for distinguishing private from public police, thereby failing to create satisfactory definitions of each, points to the necessity of creating some sorts of empirical measures of convergence and bases of distinction between the functions and sponsors of public and private police. Future research can be devoted to creating and testing such measures.

There are, however, certain courses of analysis which can be pursued without the benefit of clear-cut definitions, but simply with notions of the concepts "public" and "private" police. Such analysis may, in itself, suggest certain useful measures with which police researchers can work.

PRIVATE POLICING AND THE SELF-HELP ETHIC

I suggest that the widespread use of formal police organizations, privately sponsored, represents the logical extension of the institutionalization of self-help. Historically, after several mildly successful attempts at organizing police around self-help notions,[5] the system of modern police was created in England and America. Modern police systems were recommended largely to increase policing efficiency, build more uniformity in policing practices, relieve citizens from policing duties, and prevent the "dangerous classes" from rioting or otherwise threatening middle and upper classes (see Hart, 1951:24-28; Silver, 1967). With the advent of modern police, a system had been created in which citizens could utilize police services rather than resort to direct self-help in protecting property of persons (Reiss and Bordua, 1967: 28). As Reiss and Bordua state:

> The existence of police symbolizes not only that the citizen will be protected from the violator, but that the violator will be protected from the citizen. One way the police serve the cause of legality, therefore, is to assure by their presence and performance that a set of rules prevails which make it unnecessary for the citizen to be continually prepared to defend himself or his property.

The modern police organization soon was recognized as insufficient to render self-defense totally unnecessary. Those respectable classes which had welcomed organized public police in some instances have evidently felt the need to supplement public police activity, but have in the last century chosen not to do so through direct self-help (Silver, 1967). Instead the public police institutional model was paralleled, and private police agencies such as Pinkerton's were established. Use of this model is consistent with a "self-help once removed" ethic, by which those who can afford to hire substitute bodies to act for their own self-defense, and prefer this because direct self-help is not only undesirable and unfashionable, but may not be compatible with public police peacekeeping efforts. Instead, it is preferable to create private police agencies which can supplement the work of public police organizations.

This analysis suggests that the advent of modern police did not mark the demise of citizen self-help, but that citizen self-help simply took a different form for those who could afford it. Thus, while modern police may constitute a social control mechanism essential to the maintenance of order and legality in a modern society (Feagin, 1970: 797), it is not necessarily essential that, as Reiss and Bordua (1967) and Feagin (1970:797) have suggested, ". . . citizen dependence on the police for protection against crime and violence, not on self-help, is critical to maintaining order and legality in a society." Certainly order would be disrupted if certain classes of citizens took to self-defense on an *ad hoc* basis, but the structured private police agency is just as much a manifestation of self-help, but one which does not require the citizens who hire private police to participate directly in self-defense or defense of property. It is primarily the society in which each citizen, regardless of social standing, takes it upon himself to defend himself and his property, that Reiss and Bordua (1967:28) refer to as made unnecessary by modern public police. It cannot be otherwise, since the remarkable growth of private police agencies demonstrates the popularity of self-help among certain interest groups in our society, although it takes the form of "self-help once removed."

Social scientists have often thought the term "self-help" to be more or less synonymous with the use of physical force by private "illegitimate" organizations (mainly volunteer), and have not generally included profit-making private police organizations in definitions of self-help. One such limited definition has been offered by Feagin (1970:798):

> . . . the current American situation being a problematic one in which the use of physical force by various private organizations and individuals—assassinations, ghetto revolts, raids by protective organizations, violent crimes, and similar occurrences—actually appears to be increasing. In recent years even groups of policemen have on occasion behaved as though they were vigilante organizations pursuing their own private ends. Moreover, in many of these cases the private use of force has been regarded as legitimate by certain segments of the American population.

In his study of propensities toward home defense, Feagin (1970:799) found that a majority of both blacks and whites sampled stated ". . . that people like themselves had to be prepared to defend their homes against crime and violence, that such protection could not be left to the police." Had he included "legitimate" profit-making private police agencies in his definition of self-help, he may have been able to add an interesting dimension to his analysis of social and economic correlates. For instance, he found that "for the white sample the proportion home-defense oriented generally goes down as one moves up the occupational and educational ladders" (Feagin, 1970:802). It may have been rewarding to consider the additional dimension of how much private police agency protection the person higher on the occupational and educational ladders utilizes in his living and working situation. Perhaps the more higher-status persons can hire their home defense out to private agencies, or depend more on public police for crime prevention because they call on others for noncrime oriented services, the less they need be home-defense oriented in their expressions. In reality, there may be just as much self-help home-defense orientation among all levels of socioeconomic status, but less need for expressions of it and direct doing among higher-status individuals.

REGULATION OF PRIVATE POLICE

Given the widespread use of private police, it is important to briefly inquire into the manners in which they are regulated. Two law review comments (Braun and Lee, 1971; *Southern California Law Review*, 1967) have undertaken fairly thorough inquiries into the law regulating private police agencies. Therefore, I will simply highlight some important considerations in this area.

Thirty-three states and the District of Columbia require licenses for private police activities. These licenses confer no additional powers (Braun and Lee, 1971: 558). Another type of enabling provision is the state or municipal deputization or commissioning of private guards and watchmen. Certain classes of private police, such as railroad detectives, campus security guards, and retail security guards, may be granted special powers concerning arrest and search. But in the majority of states the majority of private police possess no powers beyond those of the ordinary citizen (Braun and Lee, 1971:559).

As Scott and McPherson (1971) have pointed out, private police may actually prefer not possessing powers beyond that of the ordinary citizen. The reason for this is two-fold. First, the powers at common law of the ordinary citizen are actually quite broad;[6] and legislatures in many jurisdictions actually have expanded the common law arrest powers for private police, without conferring upon them the public policeman's arrest power. For instance, citizen's arrest statutes in some states have eliminated the common law breach of the peace requirement in misdemeanant citizen arrest; and shoplifting statutes allow detention of a suspect whenever reasonable cause to investigate exists (Braun and Lee, 1971:561). Second, without special arrest powers the private police retain the privilege but not the obligation to arrest and detain (Scott and McPherson, 1971:272) and also are not obliged to provide the suspect with his full measure of constitutional rights. Scott and McPherson (1971:272) elicited these statements during the course of their Minnesota interviews:

> "We can rough a guy up if we want to," "we can get a confession in cases where the police can't because we don't have to worry so much about a guy's rights," and "we can use every means possible to secure information."

They noted that while a minority of their private police interviewees made extreme statements such as these, all of their interviewees indicated that they had broader latitude in the investigation process than public police.

The United States Supreme Court, in speaking to the problem of private police arrest powers, has ironically given cause for private policemen to avoid additional grants of legal authority and to operate only on the basis of citizen's arrest power. In *Griffin v. Maryland*,[7] the petitioners were ordered to leave a privately owned amusement park, which had a policy of excluding blacks, by a special private policeman who was acting under his authority as a deputy sheriff. The Court enforced the Fourteenth Amendment ban on such racial discrimination only because the person making the arrest was deputized. Had the park security guard not been a deputy, it is doubtful that the Court would have found the requisite "state action" to make this a violation of the United States Constitution. While the Court recognizes that, whether or not the security guard had been deputized, he still had the right to arrest them in his private capacity as an agent of the park, the holding is narrow and centers upon the security guard's status as deputy. Mr. Justice Clark, in his concurring opinion, makes it clear that he joined the majority only because the "State must be recognized as a joint participant in the challenged activity."[8]

If a private police officer is deputized or commissioned, he may be required to adhere to the rules, regulations, and practices of the local police department (Braun and Lee, 1971:559); but short of this, there are really only three ways in which private police are controlled through some legislative regulations, however indirect. These are license revocation, criminal prosecution, and tort liability (Braun and Lee, 1971:559). Braun and Lee noted that the impact of license revocation is minimal because complaints must be initiated and most arrestees are not in a solid enough position to do this. Thus far it seems that little successful criminal prosecution of private police for abuses of power has been undertaken, although successful civil suits have been initiated (Braun and Lee, 1971:564, 565).

But again one might question the ability of many arrestees, who may have to proceed as paupers in their own criminal defenses, to initiate successful civil actions against private police. In addition, provable damages may not be substantial in situations in which the arrestee has not been physically abused, but where more subtle abuse has taken place in the form of denial of various constitutional rights.

We must look to Court-made law to get an idea of the extent to which private police have been allowed to ignore the constitutional safeguards of a suspect. For the purposes of this discussion, I shall concentrate on custodial interrogation and the issuance of constitutional warnings. Public police are required to recite to suspects they have detained certain rights, such as the right to remain silent and to have an attorney present.[9] Private police, by and large, are not subject to this requirement; and yet admissions and statements tending toward self-incrimination made by the detained suspect to a private policeman without the benefit of constitutional warnings will not necessarily be excluded at trial.

Courts have been lenient with private policemen with regards to the issuance of "Miranda" warnings. Very early the United States Supreme Court established in *Bardeau v. McDowell*,[10] that physical evidence obtained illegally by private citizens is admissible in a criminal prosecution, even where the same evidence would not be admissible if obtained in the same manner by persons acting under the authority of a governmental agency. Decisions subsequent to *Miranda v. Arizona* have limited the warning requirement to interrogations by public police, governmental officials, and private police commissioned under state or local law, and have not required uncommissioned private police to give the warnings.[11] Private police interrogations are not protected by the Fifth and Sixth Amendments to the United States Constitution, since no government agencies are involved and no state action exists.[12] Short of extensions of doctrines concerning what constitutes state action,[13] we may expect that the current status of undeputized and uncommissioned private police will obtain.

The absence of obligation to issue Miranda warnings, and the admissibility of statements obtained from detainees in the absence of such warnings, can have some notable effects on the activities of private police. For instance, one retail security guards' manual cited in *Peak v. W. T. Grant Co.*[14] offered this bit of advice: "In stores in which office is within view of sales floor it is advisable to take shoplifter to stockroom, employees' restroom, or other place. . . ." Where there is no express grant of authority to private police, detention even broader in scope than that allowed by common law has been permitted in retail shoplifting situations. The policy justification for such detention and interrogation that has been offered by the Courts is that it is important to allow the security guard to determine whether to let the detainee go because he is not guilty or whether further detention and ultimate arrest is necessary.

Although private police who lack formal grants of authority often are not held to the standards to which public law enforcement officers are held with respect to custodial interrogation, most uniformed security guards look and act just like public police officers, and the impact of this appearance can be substantial. As Scott and McPherson (1971:272) state:

> Despite the fact that private agents in Minnesota do not possess police powers, it is probably true that most individuals when confronted by a uniformed guard or a man stating that he is a "detective" or "investigator" naturally assume he has some kind of legal authority. Public misunderstanding of the law undoubtedly gives private agents an additional advantage.

The 500,000 uniformed private policemen, who often wear, as complements to uniforms, accessories in the nature of shoulder patches, metal badges, and many of whom carry guns,[15] certainly display all the indicia of public police. Even the most educated of our citizens must often look twice to determine whether they are public or private police. The private police are aware of their advantage. One textbook on plant protection states:

> The mere presence of a uniformed individual contributes a psychological condition of great significance to the average mind. Over the period of many years, the wearer of the uniform has represented a leader, designated and recognized by governmental bodies. This association has been attached to the form of distinctive wearing apparel (Davis, 1957:22).

Thus, private police are accorded not only the psychological advantage over those whom they detain and interrogate, but advantage under the law as well.[16]

CONCLUSION: IMPLICATIONS FOR SOCIETY AND SOCIAL RESEARCH

Given the widespread use of private police in security and protection positions, and the fact that regular public police would have to curtail their regular law enforcement activities severely in order to even begin to meet the demands of those who now hire private police to fulfill their need, it seems reasonable to interpret the massive hiring of private police as an indication that those who hire such protection feel that public police protection is in some way inadequate. People who hire private police are "paying twice" for police protection—once through taxes and once out-of-the-pocket. While, because of the diffusion of tax dollars, an individual taxpayer may not feel that he "pays the police," the person hiring a private policeman will want to make sure he gets his money's worth. In connection with this, it is important to note that those who can afford to hire private police may be the same class or group of people in a given community who are influential in the community and therefore are desirable spokesmen for the public police (Wilson, 1950: 412-413). In light of this, state and local government and public police departments might circumvent criticism of their ability to maintain law and order and derive important political benefits by allowing these influential power groups to exercise a great deal of autonomy concerning their privately hired police. This would serve to explain the relative dearth of regulations and restrictions on private police activity. In this way, influential citizens and corporations who hire private police are able to get their money's worth out of them (perhaps at a loss of constitutional protections for those apprehended by private police) and public police departments receive less adverse publicity for inability to satisfactorily police their domain.

Presentation of a few examples of the ways in which consideration of private police in the manner discussed above could alter recent police research might serve to illustrate the usefulness of including private police in a researcher's definition of "police." For instance, when one speaks of police departments' dependence upon "public" approval for financial support (Hahn, 1970:646), it might be fruitful to inquire as to "which public—the public at large or the more influential citizens who spend large amounts of money employing private police?" Likewise, when one examines a citizenry as the maker of police attitudes and police loyalty (Savitz, 1970), it may be profitable to examine the concept of "citizenry" more closely. Perhaps it is not the "unreliable" citizenry at large which has the most impact on public police, but instead the "sponsoring" citizenry. Studies of police professionalism (Walsh, 1970) could consider the effects on professionalization of a public policeman who "moonlights" as a private policeman, and also consider whether the "clients" of public police are only those citizens with whom the police come into contact (Groves and Rossi, 1970). Researchers who have suggested that police take on certain services which whites abandon in black ghettos and leave to them, thereby co-opting the police (Bordua, 1968), could now consider how, in the wake of public police professionalization and unionism and the resulting desire to improve community relations through community services, private police are now the ones being co-opted into the less desirable positions—the "dirty work." These are just a few of the ways in which research on private police can be relevant not only for a better understanding of private police organizations, but for a better understanding of police systems as a whole. A new perspective should be developed among police researchers whereby researchers are "sensitized" to the existence and importance of private police in society.

There is another area of self-help which does not involve the hiring of private police but which can have potential long-range implications for society if private police behavior is allowed to go unchecked. That large segment of society which cannot afford to hire their own police also in recent years have exhibited a strong propensity toward self-help in self-defense (Feagin, 1970; Marx and Archer, 1971). During the past two centuries of American history, there have been at least 326 instances of vigilante movements (Brown, 1969:154). Lately, contemporary self-defense groups have been organized, such as the Monroe, North Carolina, "Self Defense Guard," the Brooklyn, New York, "Maccabees," the "Deacons" in Louisiana, the "Community Alert Patrol" in Watts, Anthony Imperiale's "North Ward Citizen's Committee" in Newark, New Jersey, and the Black Panther Party for Self-Defense (Marx and Archer, 1971). Marx and Archer (1971:70-71) point out that many of these groups have performed critical catalytic functions because of their bargaining power in a given community.

Certainly many factors exist today which tend to drive citizens, both rich and poor, toward self-help. Among these we might list rising crime rates, anomie, racial attitudes, and the like. It is not within the scope of this analysis to document all the reasons for citizen resort to self-help, but I instead have proceeded on documentation that self-help exists. It does, however, seem necessary to suggest one potential cause of community self-help which is related to the use of private police. Just as the extended use of private police may be a response to community disturbances which threaten the person and property of higher-status citizens, it is possible that vigilante groups may be at some future date organized in response to abuses by private police. This is not to say that all private police in the past necessarily have abused their broad powers, but that there have been instances of such abuse recorded which could be interpreted as good cause for more careful regulation of private police and for research directed toward determinations of when private police alleviate community unrest and when they exacerbate it. If private police in a given community are taken to represent forces of oppression, then even one instance of abuse such as those which have been recorded—either the use of excessive force,[17] false imprisonment,[18] brutality in eliciting confessions,[19] or wrongfully wounding and killing bystanders or suspects[20]—could be the catalytic event for a series of unfortunate community disturbances. Private police presently can have an important impact, especially on urban communities, where their presence can be so widespread that over a decade ago they had already been known to have effected over 10,000 apprehensions per year in one city in the retail security area alone (*Southern California Law Review*:541).

If research discloses situations where the progress made by increasing community services conducted by public police is potentially offset by real or perceived oppression by private police, regulation of private police use may be very much in order. It does not seem an optimal use of human or economic resources to alleviate community strife on the one hand and at the same time to allow private money and private police to contribute to increased community strife. Perhaps this is an area in which, rather than spending more tax dollars on poorer communities, the expenditures of various power groups for private police hiring could be restricted. As an alternative, a host of regulations on private police are possible, which range from restricting their use of indicia of authority such as uniforms to actually granting authority to private police and thus forcing them to assume the legal responsibilities that accompany that authority. Research directed toward understanding the effects of widespread use of private police on various interrelationships and attitudes within a community, including the citizen-public police relationship, is needed in order to structure effective regulations concerning private police use.

FOOTNOTES

1. See especially *Reports of the New York State Legislative Committee on Privacy of Communications and Licensure of Private Detectives, 1957-58.*
2. I use "sponsors" rather than "clients" in this section, and have generally avoided any identification of the clients of either public or private police. This is because I subscribe to the view that police are essentially a service without clients (Reiss and Bordua, 1967:30). It is problematic to label the citizens with whom police come into contact (Walsh, 1970: 707) as clients of the police, because police may come into contact with various groups in society while actually performing in the name of other groups as their clients (Groves and Rossi, 1971). I have chosen to be satisfied with two categories, "sponsors" and "services performed," as potential definers of public and private police in an attempt to circumvent the dilemma posed by the use of the term "clients" while not avoiding those persons who would possibly be called clients of police. Thus, for those instances in which police are carrying out services in the name of a group, the term *sponsors* should suffice, while by viewing "services performed," it should always be possible also to consider those persons for whom the particular service is being performed. This approach would appear to be compatible with the view that the central meaning of police authority is its significance as a mechanism for "managing relationships" (Reiss and Bordua, 1967:25-26).
3. 439 F.2d 477 (D.C. Cir. 1970).
4. In *Kline,* the plaintiff had sustained injuries when she was criminally assaulted and robbed by an intruder in the hallway of her apartment building. At the time the plaintiff had first signed a lease, a doorman was on duty at the main entrance twenty-four hours a day, and several other security precautions were in effect. But at the time the assault on the plaintiff took place, ". . . the main entrance had no doorman, the desk in the lobby was left unattended much of the time, the 15th Street entrance was generally unguarded . . . and the 16th Street entrance was often left unlocked all night." (479 F.2d 477 (D.C. Cir. 1970)).
5. Although space precludes any extended discussion of police history here, I will briefly review the historical basis for self-help in policing. Historians have noted that the Anglo-Saxon policing system can be distinguished from those of most of the world in that it is one which comes from the community and historically has depended upon citizen participation and control as opposed to governmental control (Reith, 1948:1; Jeffries, 1952:17-18; Fosdick, 1915:Chapter 1; Smith, 1960:129). Formal police forces, paid by the government, are a relatively new phenomena in Anglo-Saxon society (Pollock and Maitland, 1911:ii, 582), not appearing formally until well into the nineteenth century (Jeffries, 1952:17-18). Much earlier, systems of collective responsibility for protection had evolved, but such systems were found to be unsatisfactory (Brown, 1936; Hart, 1951:22-26; Reith, 1948:3). In their place, a system was created in which a country gentleman was for a period of time appointed by the other community residents to take responsibility for policing the area (Hart, 1951:22; Chapman, 1962:12). This office was unpaid and compulsory. Later, to avoid the burden of holding this office, gentlemen who had been appointed took to paying deputies to perform the duties in their place (Jeffries, 1952:21-22; Reith, 1948:5; Hart, 1951: 24; Chapman, 1962:12). Thus, in both towns and rural areas in England, the police tradition began with communal responsibility for community well-be-

ing assumed by the middle and upper classes of the community, and evolved into a system whereby the upper classes of citizens could absolve themselves from direct responsibility, by directly paying a member of the lower-classes to perform the law enforcement work in the interest of the upper-class appointee. I prefer to term these practices "self-help" policing which evolved into a system of "self-help once-removed" policing.

During the late eighteenth century, in response to a rapid increase in crime and disorder, supplemental citizen self-help groups were organized. Most notable among these was the Bow Street Runners, organized in large part by novelist Henry Fielding, which became the direct precursor of the modern-day police force (Reith, 1948:19). A few years after the establishment of the Bow Street organization, Patrick Colquhoun originated the Thames Police, organized by him at the request of several West India merchants (Reith, 1948:27). This patrol was paid by merchants and took charge of loading and unloading at the docks. It was soon found that this organization of private police was effective in curtailing theft (Reith, 1948:27). These early examples of more formally organized police forces were later absorbed into a New Police Force sponsored by the government (Reith, 1948:29).

American police were, in effect, a transplantation of the English model. The self-help ethic seems also to have been successfully transplanted and vigorously survived. In the 1850's, shortly after the time modern formalized government-sponsored police forces were created in England and during the time governmental police forces were being created in America (Lane, 1967), private police gained a popularity which was to be ever-growing. The Pinkerton's National Detective Agency began operation with ten people in 1850 in Chicago as the protective agent for railroad property (Horan and Swiggett, 1951:6); and it, along with many other private police agencies, has grown steadily.

6. Misdemeanant arrests were permitted at common law to preserve the public peace and an actual breach of the peace committed in the presence of the arresting party was the cause requirement. Arrests for felonies were permitted whenever a bad act had been committed and there was reasonable cause to suspect. Force was allowed to supplement the arrest, and so was detention, to the extent that was necessary and reasonable in order to deliver the arrestee to a public official.

7. 378 U.S. 130 (1963).
8. 378 U.S. 137, 138 (1963).
9. See *Miranda v. Arizona,* 384 U.S. 436 (1966).
10. 256 U.S. 465 (1921).
11. See *State v. Valpredo,* 75 Wash. 2d 368, 450 P.2d 979 (1969).
12. See *Evalt v. United States,* 359 F.2d 534 (9th Cir. 1966), and *People v. Frank,* 275 N.Y.S.2d 570 (Sup. Ct. 1966).
13. See *Marsh v. Alabama,* 326 U.S. 501 (1946) and *Amalgamated Food Employees Union Local 590 v. Logan Valley Plaza Inc.* 391 U.S. 308 (1968) for examples of expansion of the state action element in other areas which affect a broad spectrum of the public.
14. 409 S.W.2d 58 (Mo. 1960).
15. At least most of them have the right to display guns, as citizens in most states do. See *Southern California Law Review* (1967:544).
16. Training and professionalization cannot presently be relied upon to regulate the activities of private police. Scott and McPherson (1971:278-280) concluded that the quality of personnel as reflected in recruitment and training, while higher in large private police agencies than their smaller counterparts, is higher still in public police agencies. Where private police do have any extensive training and experience, they have usually obtained it at the taxpayer's expense in public police departments (1971:281). Such cross-employment might be expected to affect compliance with regulations and professionalism in any number of ways. To the extent that the public force of which an individual might carry over the standards of operation he adheres to on his "public" job into the realm of his "private" job. On the other hand, the individual

law enforcement officer may view his private job as a breather from his public job—a chance to ignore many of the standards required of him in his public job. Furthermore, to the extent that the public force of which an individual is or has been a member is not well regulated. The chances may be good that his behavior on a private police force may not meet very strict standards. The same goes for professionalization. Even the "professionalized" public police officer may not feel that the professional code under which public police operate is applicable to his private job.

17. See *Tomblinson v. Nobile*, 103 Cal. App. 2d 266, 229 P.2d 97 (Dist. Ct. App. 1951); *McChristian v. Poplin*, 75 Cal. App. 2d 249, 171 P.2d 85 (Dist. Ct. App. 1946).
18. *Hanna v. Raphael Weill & Co.*, 90 Cal. App. 2d 461, 203 P.2d 564 (Dist. Ct. App. 1949).
19. *United States v. Williams*, 341 U.S. 70 (1951), *Williams v. United States*, 341 U.S. 97 (1951).
20. See *Myles v. Meineke*, 82 Ohio App. 126, 78 N.E.2d 917 (1948); *St. John v. Reid*, 17 Cal. App. 2d 5, 61 P.2d 363 (Dist. Ct. App. 1936); *Callum v. Hartford Accident & Indemnity Company*, 186 Cal. App. 2d 885, 337 P.2d 259 (Super. Ct. 1959).

REFERENCES

Albares, R.: Nativist Paramilitarism in the United States: The Minutemen Organization. Working Paper No. 109, Center for Social Organization Studies, University of Chicago, 1968.

Bayley, D., and H. Mendelsohn: *Minorities and the Police*. New York: The Free Press, 1969.

Bercal, T.: Calls for police assistance: consumer demands for governmental service. *American Behavioral Scientist, 13:*(May-August 1970):681-692.

Black, A.: *The People and the Police*. New York: McGraw-Hill, 1968.

Bordua, D.: Comments on police-community relations. Pp. 204-221 in S. I. Cohn (ed.), Law Enforcement Science and Technology II. Chicago: Illinois Institute of Technology Research Institute, 1968.

Bordua, D., and E. Haurek: The police budget's lot: components of the increase in local police expenditures, 1902-1960. *American Behavioral Scientist, 13:*(May-August 1970):667-680.

Braun, M., and D. Lee: Private police forces: legal powers and limitations. *University of Chicago Law Review, 38:*(Spring 1971): 558-582.

Brown, R. M.: The American vigilante tradition. Pp. 154-226 in H. D. Graham and T. R. Gurr (eds.), *The History of Violence in America*. New York: Bantam, 1969.

Brown, R. S.: *The Sarjeants of the Peace in Medieval England and Wales*. Manchester: University of Manchester Press, 1936.

Chapman, S.: *The Police Heritage in England and America: A Developmental Survey*. East Lansing, Michigan: Michigan State University Institute for Community Development and Service, 1962.

Cumming, E., I. Cumming, and L. Edell: Policeman as philosopher, guardian, and friend. *Social Problems*, 12: (Winter 1965) : 22.

Davis: *Industrial Plant Protection,* 1957.

Epstein, C.: *Intergroup Relations with Police Officers*. Baltimore, Maryland: Wilkins and Wilkins Co., 1962.

Feagin, J.: Home defense and the police: black and white perspectives. *American Behavioral Scientist, 13:*(May-August 1970):797-814.

———: Creeping capitalism. *Forbes, 106:*(September 1, 1970):22-28.

Fosdick, R.: *European Police Systems*. New York: The Century Company, 1915.

Friedman, M.: *The Pinkerton Labor Spy*. New York: Wilshire Book Company, 1907.

Groves, W., and P. Rossi: Police perceptions of a hostile ghetto: realism or projection. *American Behavioral Scientist, 13:*(May-August 1970):727-743.

Hahn, H.: The police and the public: an overview. *American Behavioral Scientist, 13:* (May-August 1970):645-647.

Hart, J. M.: *The British Police*. London: George Allen & Unwin, Ltd, 1951.

Horan, J.: *The Pinkertons: The Detective Dynasty that Made History*. New York: Crown, 1967.

Horan, J., and H. Swiggett: *The Pinkerton Story*. New York: G. P. Putnam & Sons, 1951.

Jeffries, C.: *The Colonial Police*. London: M. Parrish, 1952.

Johnson, D., and R. Gregory: Police-community relations in the United States: a review of recent literature and projects. *Journal* of Criminal Law, Criminology and Police Science, 62:(March 1971):94-103.

Lane, R.: *Policing the City: Boston 1822-1885.* Cambridge, Massachusetts: Harvard University Press, 1967.

Lavine, S.: *Allan Pinkerton: America's First Private Eye.* London: Hammond, Hammond & Co., 1963.

Marx, G., and D. Archer: Citizen involvement in the law enforcement process: the case of community police patrols. *American Behavorial Science, 15:* (September-October 1971):52-72.

Misner, G.: The urban police mission. *Issues in Criminology, 3:*(Summer):35-46, 1967.

Parnas, R.: The police response to the domestic disturbance. *Wisconsin Law,* 914-960, 1967.

Pollock and Maitland: *History of the English Law,* 1911.

Reiss, A., and D. Bordua: Environment and organization: a perspective on the police. Pp. 25-55 in D. Bordua (ed.), *The Police: Six Sociological Essays.* New York: John Wiley & Sons, 1967.

Reith, C.: *A Short History of the British Police.* London: Oxford University Press, 1948.

Rowan, R.: *The Pinkertons: A Detective Dynasty.* Boston: Little, Brown & Co., 1931.

Savitz, L.: The dimensions of police loyalty. *American Behavioral Scientist, 13:*(May-August 1970):693-704.

Schulz, D.: Some aspects of the policeman's role as it impinges upon family life in a Negro ghetto. *Sociological Focus, 2:*(Spring 1969):63-72.

Scott, T., and M. McPherson: The development of the private sector of the criminal justice system. *Law and Society Review, 6:* (November 1971):267-288.

Silver, A.: The demand for order in civil society: a review of some themes in the history of urban crime, police, and riot. Pp. 1-24 in D. Bordua (ed.), *The Police: Six Sociological Essays.* New York: John Wiley & Sons, 1967.

Shallo, J. P.: *Private Police: With Special Reference to Pennsylvania.* Philadelphia: The American Academy of Political and Social Science, 1933.

Smith, B.: *Police Systems in the United States.* New York: Harper & Bros., 1960.

Regulation of private police. *Southern California Law Review 40:*540, 1967.

Trojanowicz, R.: Police-community relations: problems and process. *Criminology, 9:*(February 1970):401-423.

Walsh, J.: Professionalism and the police: the cop as medical student. *American Behavioral Scientist, 13:*(May-August 1970):705-725.

Wilson, J.: *Varieties of Police Behavior,* Cambridge: Harvard University Press, 1968.

Wilson, O. W.: *Police Administration.* New York: McGraw-Hill, 1950.

Chapter 3

LEGAL BASIS FOR PROTECTIVE SERVICES

PROTECTIVE SERVICES are more than legal statutes and institutions in society. They encompass those dimensions of social control called "law enforcement" as well as those called "security" and "prevention."

The distinction between public and private protective services is one however which has not been totally dealt with by the practitioners in either the public or the private protective services field. These distinctions are often based on legal authority. The practitioners of protective services in the public sector basically derive their power, authority and responsibility to act from various levels of government, either state, federal, or local. They are often called police officers, federal agents, or government security officers. Their authority to act comes from their intimate relationship with the justice process. They are responsible for information collection, investigation or apprehension of those committing acts defined as criminal.

The private protective service practitioners in many respects perform the same tasks that their public counterparts do, however, their authority is derived from the extension of the private citizen's right to protect his own property.* Private security personnel are thereby employed as:

1. an extension of a property owner's right to this protection; or
2. persons granted a quasi-legal status either through licensing or regulation by a unit of government, state or local, to perform protective services within a given jurisdiction.

Their activities are often considered complementary or supplementary to the duly authorized public protective service agencies operating in the same jurisdiction.

---

*See Chapter 3, Appendix 3A for a description of Special Policing.

When protective services are viewed at the systems level, a very clear distinction can be made between the legal basis for governmental and private security (Fig. 3-1). The legal basis for governmental se-

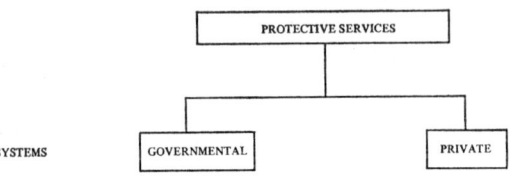

Figure 3-1. Protective Services Systems.

curity programs normally is applicable to security operations in government or in private business or industrial firms which have manufacturing or supply contracts with the United States Government. Governmental security or law enforcement is normally provided in response to specific problems and is usually the result of specific legislative enactments, executive orders, statutes, and directives. Private security, on the other hand, while being established through statute or ordinance is of a much more general nature, often finding its basis in the common law. The legal basis for private security is normally applicable to any business firm, industrial firm, or individual requiring private protection or security.

The wide variety of uses for protective services makes it impossible to point to any one specific law to establish the basis for security operations. It is, however, entirely appropriate to look to specific pieces of legislation or government directives which provide the basis for specific types of security measures.

LEGAL BASIS FOR GOVERNMENTAL SECURITY PROGRAMS

The legal basis for various security programs in the governmental system can be traced to the general requirement for the government to provide for its continued operations. Precisely now the internal integrity of the government is to be maintained is clearly spelled out in various executive orders, acts, directives, memorandums, and statutes. In order to show the development of specific security regulations, their development and refinement will be presented in a chronological manner.

1939—Federal Security Program (Hatch Act)

The Act of August 2, 1939, Section 9A (53 Statute 1148), The Hatch Act, provides that it shall be unlawful for any person employed by the Federal Government whose compensation is paid from funds authorized or appropriated by Congress to have membership in any political party or organization which advocates the overthrow of our constitutional form of government in the United States. This Act prohibited those belonging to a subversive organization from federal employment.

1940—Work Relief Projects

It provided the legal means to remove subversives from government employment. The Act of June 26, 1940 (54 Statute 620, Section 15(f), *Emergency Relief Appropriation Act* stated that aliens, communists, and members of the Nazi (Bund) organization would not be given employment on any work project appropriated under appropriations made by that Act.

The Act of June 28, 1940, Section 6 (54 Statute 679), and the Act of December 17, 1942, Section 3 (56 Statute 1053, ch. 739), was repealed and superseded by 64 Statute 476, ch. 803, post, which authorized the Secretary of War and the Secretary of the Navy during the war to summarily remove any employee in the interest of national security without provision of any law, rules, or regulations governing the removal of employees. An employee so removed was to be given the opportunity to be informed within thirty days of the reasons for such removal and was given the right to submit affidavits within thirty days thereafter to show why he should be retained. This authority was also extended to the Secretary of Defense and Air Force (63 Statute 1023, Section 630 superseded by 64 Statute 476, post).

1941—Civil Service War Regulation

On November 30, 1941, the Civil Service Commission issued War Service Regulation 2, Section 3(g) which provided that a person may be disqualified for appointment or for examination if there existed a reasonable doubt as to his loyalty to the government of the United States. This regulation brought forth one of the first questions dealing with loyalty and suitability for government employment.

1942—Interdepartmental Committee on Investigations

This committee was created by the Attorney General to advise government departments on the handling of loyalty cases. It served primarily to advise governmental agencies on the investigative procedures, nature, and purpose of FBI reports, and the need for sound procedures in evaluating charges of disloyalty.

In April of 1942, the Attorney General created a special interdepartmental committee on investigations composed of officials representing the Interior, Treasury, Commerce, and Justice Departments and the Federal Deposit Insurance Corporation.

1943—Interdepartmental Committee

Executive Order 9300 dated February 5, 1943 (8 F.R. 1701), replaced the Attorney General's interdepartmental committee with a new interdepartmental committee and employee investigations in the Department of Justice. It was composed of officials representing the Treasury and Interior departments, the Federal Reserve Board, the Federal Depositor Insurance Corporation, and Civil Service Commission.

This committee was concerned with permanent employees only and served as an advisory board only, having no authority to enforce its findings in any agency or department. It confined itself to statutory provisions of the applicable laws dealing with the problem and deemed the employee removable on grounds of loyalty only if it was established that the employee was a member of an organization advocating the overthrow of our constitutional form of government; or that the employee advocated the use of force or violence as a means of changing our form of government. The interdepartmental committee was replaced by the Loyalty Review Board in the Civil Service Commission under Executive Order 9835.

1946—Power to Terminate Employment (Summary Suspension)

The Act of July 5, 1946, Title I (60 Statute 458) authorized the Secretary of State in his absolute discretion on or before June 30, 1947, to terminate the employment of any officer or employee of the Department of State or the Foreign Service of the United States whenever he should deem such termination advisable in the interests of the United States. But an employee so terminated was not barred from seeking employment in another agency.

A similar authorization was repeated in the Department of State Appropriations Act 1948 approved July 9, 1947 (61 Statute 288); 1949 approved June 3, 1948 (62 Statute 315, Section 104); and 1950, approved July 20, 1949 (63 Statute 456, Section 4). The latter authorizations repealed (64 Statute 476, Section 4), and superseded by the Act of August 26, 1950 (64 Statute 476, ch. 803, post). However, the provision again appeared in the Department of State Appropriation Act 1951, approved September 6, 1950 (64 Statute 768, Section 1213); 1952, approved October 22, 1951 (65 Statute 581, Section 103); 1953, approved July 10, 1952 (66 Statute 555, Section 103); but was omitted from the Department of State Appropriation Act of 1954, approved August 5, 1953 (Public Law 195, 83rd Congress). A similar provision authorizing the Secretary of Commerce to so terminate the employment of any officer or employee of the Department of Commerce was contained in the Department of Commerce Appropriations Act 1952, approved October 22, 1951 (65 Statute 594, Section 304), and the Department of Commerce Appropriations Act 1953, approved July 10, 1952 (66 Statute 567, Section 304), but was omitted from the Department of Commerce Appropriations Act 1954, approved August 5, 1953 (Public Law 195, 83rd Congress).

1946—Committee on Un-American Activities

The Act of August 2, 1946 (60 Statute 823, 828-9, Section 121(b). The Committee on Un-American Activities, as a whole or by subcommittees, is authorized to make from time to time investigations of (1) the extent, character, and objectives of un-American propaganda activities in the United States; (2) the diffusion within the United States of subversive and un-American propaganda that is instigated from foreign countries or of a domestic origin and attacks the principle of the form of government that is guaranteed by our constitution, and (3) all other questions in relation thereto that would aid Congress in any necessary remedial legislation.

1946—President's Temporary Commission on Loyalty

Executive Order number 9806 dated November 25, 1946 (11 F.R. 13863), established this Commission. The President's authority to establish this Commission stemmed from Revised Statutes Section 1753 which authorized the President to regulate the admission to Civil Service and from Section 214 of the Act of May 3, 1945 (59 Statute 134) which provides that appropriations of the Executive Department and independent establishments of the government shall be available for the expenses of committees, board, or other interagency groups engaged in authorized activities of common interests to such departments and establishments, and composed in whole or in part of representatives thereof who received no additional compensation by virtue of such membership. Its functions were as follows:

1. To inquire into the standards, procedures, and organizational provisions for investigation of federal employees or applicants for federal employment and for the removal or disqualification from employment of any disloyal or subversive person.
2. To make recommendations as to
   a. the adequacy of existing security procedures,
   b. the desirable standards for judging loyalty of employees, and
   c. proper procedures, etc.

1947—Loyalty Review Board

Executive Order 9835* dated March 21, 1947 (12 F.R. 1935), provides for the investigative procedure of applicants for federal employment requiring investigation of persons entering competitive Civil Service be made by the Civil Service Commission and the investigation of persons other than those entering the competitive service be conducted by the employing department or agency. The order lists the following sources of information which shall be referred to:
1. The Federal Bureau of Investigation files,
2. The Civil Service Commission files,
3. Military and Naval Intelligence files,
4. Files of any other appropriate governmental investigative or intelligence agency,
5. House Committee on Un-American Activities files,
6. Local law enforcement files at places of residence and employment of applicant,
7. Schools and colleges,
8. Former employers, or
9. Any other appropriate source.

The order places the responsibility for an effective program of dismissal of disloyal employees directly upon the heads of each agency and department, and provides for the creation of a loyalty board in each agency to hear loyalty cases arising in each agency and to make recommendations with respect to the removal of any officer or employee on grounds relating to loyalty.

The Loyalty Review Board is established in the Civil Service Commission which shall have the authority to review cases involving persons recommended for dismissal by the loyalty boards of the departments or agencies, and to make advisory recommendations thereon to the head of the employing department. The order provides for security measures and investigation. Part Five of the order sets up standards for the refusal of employment or for the removal of employees. This part has been amended by Executive Order 10241 dated April 28, 1951 (16 F.R. 3690), which provides that the standard for refusal of employment or removal from employment in an Executive Department or Agency on grounds relating to loyalty shall be that on all the grounds of the evidence there is a reasonable doubt as to the loyalty of the person involved. The order sets out a list of activities and associations which may be considered in connection with the determination of disloyalty. This order (9835) was revoked by Executive Order 10450 of April 27, 1953.

1947—National Security Act

This act provided for a comprehensive program of security for the United States and the development of procedures to ensure national security. The act of July 26, 1947 (61 Statute 495-510), was amended by the Act of August 10, 1949 (63 Statute 578-592), by reorganization Plan No. 4 of 1949 (63 Statute 1067), and by reorganization Plan No. 3 of 1953 (67 Statute 634), by reorganization Plan No. 7 of 1953 (67 Statute 639), by Act of April 4, 1953 (67 Statute 19, c. 16), and by Public Law 85-599, p. 2; 50 U.S.C. pp. 402-403. . . . Section 2 of the National Security Act of 1947 as amended (50 U.S.C. 401), was amended to read as follows:

> In enacting this legislation, it is the intent of Congress to provide a comprehensive program for the future security of the United States; to provide for the establishment of integrated policies and procedures for the departments, agen-

---

* Revoked by Ex. Order 11785 (June 4, 1974).

cies, and functions of the Government relating to the national security; to provide a Department of Defense, including the three military Departments of the Army, the Navy (including naval aviation and the United States Marine Corps), and the Air Force under the direction, authority, and control of the Secretary of Defense; to provide that each military department shall be separately organized under its own Secretary and shall function under the direction, authority, and control of the Secretary of Defense; to provide for their unified direction under civilian control of the Secretary of Defense but not to merge these departments or services; to provide for the establishment of unified or specified combatant commands, and a clear and direct line of command to such commands; to eliminate unnecessary duplication in the Department of Defense, and particularly in the field of research and engineering by vesting its overall direction and control in the Secretary of Defense; to provide more efficient, effective, and economical administration in the Department of Defense; to provide for the unified strategic direction of the combatant forces, for their operation under unified command, and for their integration into an efficient team of land, naval, and air forces but not to establish a single Chief of Staff over the armed forces nor an overall armed forces general staff.

The act also provided for the establishment of the Central Intelligence Agency and established the National Security Council a fully integrated intelligence and defense mechanism for the United States.

1948—Confidential Status of Employee Loyalty Records

The Presidential Directive of March 13, 1948 (13 F.R. 1359), provides a confidential status for employee loyalty records. This provided that records of hearings would not be made available to unofficial nongovernmental sources.

1949—Establishment of Committees to Coordinate Internal Security

The Presidential Directive of March 27, 1949, as amended stated that permanent committees were established which were responsible for the coordination of internal security. They were established under the National Security Council pursuant to the provisions of Section 101 of the National Security Act. These committees are as follows:

1. The Interdepartmental Intelligence Conference (IIC) is responsible for the coordination of the investigation of all domestic espionage, counterespionage, sabotage, subversion and other related intelligence matters, affecting internal security. It consists of the Director of the Federal Bureau of Investigation, Department of Justice; Assistant Chief of Staff for Intelligence Department of the Army; Director of Naval Intelligence, Department of the Navy; and the Director of Special Investigations, the Inspector General Department of the Air Force.

2. The Interdepartmental Committee on Internal Security (ICIS) is hereby created and shall be responsible for coordinating all phases of the internal security field other than the functions outlined in paragraph above concerning the IIC. It shall be composed of representatives from the Department of State, Treasury, Defense, and Justice.

Both Committees shall invite nonmember agency representatives as *ad hoc* members thereof when matters involving their responsibilities are under consideration.

In accordance with arrangements to be determined in each case, there shall be transferred to the IIC and the ICIS for incorporation as subcommittee or for the absorption of their functions such existing committee as are operating in their respective fields of responsibility. The two committees shall also establish such new subcommittees as will assist them in carrying out their responsibilities.

1950—Protection of Certain Military and Naval Installations

Executive Order 10104 of February 1, 1950 (15 F.R. 597): *Defining Certain Vital Military and Naval Installations and Equipment as Requiring Protection* against the General Dissemination of Information Relative Thereto.

Whenever, in the interests of national defense, the President defines certain vital military and naval installations or equipment as requiring protection against the general dissemination of information relative thereto, it shall be unlawful to make any representation of such vital military and naval installations or equipment without first obtaining permission of the commanding officer of the military or naval post, camp, or stations, or naval vessels, military or naval aircraft, and any separate military or naval command concerned, or higher authority, and promptly submitting the product obtained to such commanding officer or higher authority for censorship or such other action as he may deem necessary.

1950—Summary Suspensions

Employees as Security Risks

The Act of August 26, 1950 (64 Stat. 476, ch. 803), as amended by Section 301 (c) of Public Law 85-568 (72 Stat. 432), provides for the Summary dismissal of certain employees in sensitive agencies as poor security risks. It authorized the heads of certain specified government departments and agencies engaged in sensitive activities to summarily suspend employees considered to be poor security risks and to terminate their service when a subsequent investigation develops facts which support such action. The individual so suspended has a right of employment in other nonsensitive agencies or departments provided the Civil Service Commission determines he is so eligible. However, with respect to the agency from which he has been summarily suspended, the decision of the head of that agency shall be final and conclusive and no appeal from said decision is provided for.

The suspended employee shall be given a written statement of charges against him within thirty days after his suspension and a reasonable opportunity to answer such charges at a hearing through submission of affidavits from officials designated by the head of the agency. The agency head shall review the case before making the final decision of termination.

This Act shall not impair the power invested in the Atomic Energy Commission covering its control over the law of its security of its employees.

The President may extend the coverage of this Act to any department or agency when he deems it to be necessary in the best interest of national security. If the President includes other agencies in the coverage of this Act, he shall submit a report of this action to the Committee on Armed Services of the Congress. The agencies which are covered by this Act are the Departments of State, Commerce, Defense, Justice, Army, Navy, Air Force, Coast Guard, Atomic Energy Commission, National Security Resources Board, and the National Advisory Committee for Aeronautics.

Executive Order 10237 of April 26, 1951 (16 F.R. 3627), extended the provision of this Act to the Panama Canal and the Panama Railroad Company. This coverage was extended to all federal agencies by Executive Order 10450.

1950—The Internal Security Act

(McCarran Internal Security Act)

Act of September 23, 1950 (64 Stat. 987-1031)—The Internal Security Act of 1950—as amended by Act of July 12, 1952 (66 Stat. 590, c. 697); Act of July 29, 1954 (68 Stat. 586, c. 646); by the Communist Control Act of 1954 (68 Stat. 777-780, pp. 6-11); by Act of July 26, 1955 (69 Stat. 375, c. 381); by Act of August 5, 1955 (69 Stat. 539, c. 589)— The Subversive Activities Control Board Tenure Act; by Act of July 31, 1956 (70 Stat. 737, 739, pp. 105 (18), 106 (a, 45); and by Act of August 28, 1958 (Pub. L. 85-791, 72 Stat. 950 pp. 29, 30); 50 U.S.C. pp. 781-844.

The origin of the Internal Security Act of 1950 dates back to 1948 when the House Committee on Un-American Activities developed and submitted to the House, the Broad Communist Control Mundt-Nixon Bill.

This Act is, in fact, a Communist Control Act designed to counteract the influence of Communist front organizations by registration requirements and other disclosures. All organizations determined to be Communist action or Communist front by the Attorney General are referred to the Subversive Activities Control Board which makes a decision based on evidence as to whether the organization is an action or a front group, and whether it is required to register as such with the Attorney General.

Subversive Activities Control Board

Jurisdiction and Authority

The Subversive Activities Control Board is established and its general powers and functions defined in Section 12 and 13 of the Subversive Activities Control Act, first enacted as Title I of the Internal Security Act of 1950, 64 Stat. 987. The Act has been amended in substance by the Communist Control Act of 1954, 68 Stat. 775, and by Public Law 90-237 (Act of January 2, 1968), 81 Stat. 765. Other amendments, in minor details, were made by the Act of July 29, 1954, 68 Stat. 586; the Act of August 28, 1958, 72 Stat. 950; and the Act of May 31, 1962, 76 Stat. 91.

Organization

The Board is an independent executive agency consisting of five members appointed by the President with the advice and consent of the Senate for terms of five years.

Activities

The Board conducts hearings and determines (1) upon petition of the Attorney General under the 1950 Act, whether any organization must register as "Communist-action" or "Communist-front," within the meaning of the Act, and whether any unregistered individual must register as an officer or member of an "action" organization or officer of such "front" organization; (2) upon application by registered organization or individual, whether the registration shall be cancelled; (3) upon petition of the Attorney General under the 1954 amendments, whether any organization is "Communist-infiltrated" and, upon subsequent application by the organization, whether it has ceased to be such.

Following hearings, the Board issues appropriate findings as to the facts and accompanies its findings with appropriate orders. Decisions and orders of the Board may be taken by the party aggrieved to the United States Court of Appeals for the District of Columbia for judicial review and, upon grant of a petition for certiorari, to the Supreme Court of the United States. When an appeal is taken, orders of the Board do not become final unless affirmed on appeal or the appeal is dismissed by the courts. Failure by an organization to comply with an order of the Board is punished pursuant to processes established independently of the Act, and the Board has no functions in such matters.[1]

1950 Safeguarding Vessels and Harbors

Executive Order 10173 of October 18, 1950 (15 F.R. 7005), as amended by Executive Order 10277 and 10352 (16 F.R. 7535, 17 F.R. 4607): "Regulations Relating to the Safeguarding of Vessels, Harbors, Ports and Waterfront Facilities of the United States."

These Executive Orders established a legal basis for the Port Security Programs of the United States.

1953—United Nations and Other International Organizations

Loyalty of United States Citizens

Executive Order 10422 of January 9, 1953 (18 F.R. 239), prescribes procedures for making available to the Secretary General of the United Nations certain information concerning United States citizens employed or being considered for employment at the Secretariat of the United Nations. This order was issued following a recommendation of the Commission of Jurists to the Secretary General of the United Nations. Regarded as of the first importance to refrain from employing or to dismiss from employment on the Secretariat of the United Nations any United States citizen who he has reasonable grounds for believing has been, is, or is likely to be engaged, in espionage or subversive activities against the United States; and that the United States make available to the Secretary General information on which he can make his determination as to whether reasonable grounds exist for believing that a United States citizen employed or being considered for employment on the Secretariat has been, is or is likely to be engaged in espionage or subversive activities against the United States.

International Organizations Employees Loyalty Board

1. There is established in the Civil Service Commission an International Organizations Employee Loyalty Board consisting of not less than three members who shall be officers or employees of the Commission.
2. The Board is authorized, in cases referred to it under this order, to inquire as to the loyalty to the government of the United States of United States citizens employed or considered for employment by international organizations of which the United States is a member. Its major authority is the termination under the standards set forth in Part II of this order for transmission by the Secretary of State to the Executive Head of the International Organizations.

1953—Security Requirements for Government Employment

Executive Order 10450 of April 27, 1953 (18 F.R. 2489) establishes a security program for federal departments and agencies. This order supersedes Executive Order 9835, and establishes a new Federal Employee Program which broadened the scope from loyalty to the security concept. Executive order 10450, April 27, 1953, established security requirements for Government employees and provided the Department of Justice furnish the heads of departments and agencies the name of each foreign or domestic organization, association, movement, group or combinations of persons which the Attorney General, after appropriate investigation and determination, designated as totalitarian, facist, communist, or subversive or having adopted a policy of advocating or approving the commission of Acts of force or violence to deny others their rights under the constitution of the United States or its seeking to alter the form of government of the United States by unconstitutional means.

The principle of Executive Order 10450 is that all persons seeking the privilege of employment or privileged to be employed in the departments and agencies of the government shall be adjudged by minimum standards and by procedures which are mutually consistent among the departments and agencies of the Federal government. Membership in, affiliation with, or sympathetic association with, any organization designated pursuant to Executive Order 10450 is but one of the factors by which a department or agency shall reach its determination and about which information shall be developed as to whether the employment or retention in employment in the Federal service of a person being investigated is clearly consistent with the interests of the national security. The Attorney General's list is designed primarily for use in connection with security determinations concerning Government employees. However, it serves a similar purpose for security determination with respect to military personnel in the Departments of the Army and the Air Force and for industrial contractor personnel.

In implementing the personnel security program, it became necessary to conduct investigations on certain persons to determine their loyalty to the United States. When these investigations were being conducted, it became apparent that several agencies were involved in the conduct of these investigations. The agencies primarily concerned were the FBI, Department of the Army (including the Air Corps), and the Department of the Navy. As the program got under way, it became evident that some type of investigative jurisdiction had to be established. At that time, several agreements were reached, but they proved to lack effectiveness. The investigative agencies concerned still got into one another's way, which resulted in the compromise of some investigations and led to embarrassing situations. In 1949, the heads of the Department of Defense (Army, Navy, Air Force) and the FBI held a meeting to decide what should be done to remedy the situation. This meeting resulted in the *Delimitations Agreement of 1949.* The agreement has been supplemented five times since the basic agreement was reached, and it covers just about all possible situations.

Supplemental Agreement Number 1 provides for close coordination and cooperation among the FBI, NIS, OSI, and Army Intelligence. According to the agreement, representatives of these agencies will maintain close liaison, including a meeting of representatives of these agencies at least twice a month for the purpose of discussing pending and contemplated investigations, and any other subject necessary to insure that there is proper coordination of their investigative work.

Supplemental Agreement Number 2 is concerned with violations of the Atomic Energy Act of 1946 and pertains solely to the FBI.

Supplemental Agreement Number 3 concerns the investigation of persons on vessels of the Military Sea Transportation Service and pertains to the Navy.

Supplemental Agreement Number 4 is concerned with the investigation of private contractors of the Armed Forces and, in brief, provides that the Armed Forces will conduct background investigations on civilian employees or applicants for employment with privately owned plants and firms working under classified contracts for the Armed Forces.

Supplemental Agreement Number 5 gives authority to the Army, Navy, and Air Force to conduct background investigations on members of the inactive Reserve and National Guard who

1. are being called to active duty;
2. are being investigated to determine whether they should be disenrolled;
3. are being selected for positions of trust or positions where they will have access to classified material.

Prior to undertaking such investigations, the FBI will be consulted to determine whether the FBI has any information concerning the individual concerned. Those investigations will not be pursued further than is necessary to make the determination required by other agencies.

1953—Minimum Security Measure to Be Taken by Various Government Agencies

Executive Order 10501 of November 5, 1953 (18 F.R. 7049), effective December 15, 1953, as amended by Executive Order 10816 of May 7, 1959 (24 F.R. 3777), outlines minimum security measures to be taken by agencies of the Government and outlines protection to be taken against unauthorized disclosures of information. The Act essentially establishes guidelines so that certain official information affecting the national defense is protected uniformly against unauthorized disclosures. It provides for the classification categories so that official information, which requires protection in the interest of national defense, should be limited to three categories of classification, which in descending order of importance, shall carry one of the following designations: *Top Secret, Secret,* or *Confidential.* No other designations shall be used to classify defense, and military information as requiring protection in the interest of national defense as expressly provided by statute.

1954—Military Personnel Security Program

Department of Defense Directive Number 5210.9, dated April 7, 1954, as revised by Directive Transmittal Number 54-63, dated June 9, 1954, and by Transmittal Number 54-132, dated December 2, 1954, provided for a Military Personnel Security Program.

1954—Security Clearances

The Mutual Security Act of 1954 (68

Stat. 859; P.L. 665, 83rd Congress) established provisions for security clearances.

> Section 531 Security Clearance. . . . No citizen or resident of the United States may be employed, or if already employed, may be assigned to duties by the Director under this Act for a period to exceed three months unless: (a) such individual has been investigated as to loyalty and security by the Civil Service Commission, or by the Federal Bureau of Investigation in the case of specific positions which have been certified by the director as being of a high degree of importance or sensitivity or in the case that the Civil Service Commission investigation develops data reflecting that the individual is of questionable loyalty. A report thereon has been made to the Director, and until the Director has certified in writing (and filed copies thereof with the Senate Committee on Foreign Relations and the House Committee on Foreign Affairs) that, after full consideration of such reports, he believes such individual is loyal to the United States, its Constitution, and form of government, and is not now and has never knowingly been a member of any organization advocating, contrary views; or (b) such individual has been investigated by a military intelligence agency and the Secretary of Defense has certified in writing that he believes such individual is loyal to the United States and files copies, thereof, with the Senate Committee on Foreign Relations and the House Committee on Foreign Affairs.

1954—Atomic Energy Act

The Atomic Energy Act of 1954 (68 Stat. 919; P.L. 703, 83rd Congress, Section 145, 42 U.S.C.A., Section 2165).

> Security Restrictions. . . . No arrangement shall be made . . . no contract shall be made or continued in effect . . . no license shall be issued . . . unless the person with whom such arrangements is made, the contractor or perspective contractor, or the perspective licensee agrees in writing not to permit any individual to have access to Restricted Data until the Civil Service Commission shall have made an investigation and reported to the Commission on the character, associations, and loyalty of such individuals, and the Commission shall have determined that permitting such a person to have access to Restricted Data will not endanger the common defense and security. (b) except as authorized by the Commission or the General Manager upon the determination by the Commission or by the General Manager that such an action is clearly consistent with the national interest, no individual shall be employed by the Commission nor shall the Commission permit any individual to have access to Restricted Data until the Civil Service Commission shall have made an investigation and reported to the Commission on the character, association, and loyalty of such individual, and the Commission shall have determined that permitting such person to have access to Restricted Data will not endanger the common defense or security.

Atomic Energy Commission

(Legislative and Executive Order Basis)[2]

AEC Personnel Security Program

The personnel security program of the AEC is based on the provisions of Section 145 of the Atomic Energy Act of 1954, as amended; the provisions of Executive Order 10450—Security Requirements for Government Employment, issued April 27, 1953; and the provisions of Executive Order 10865—Safeguarding Classified Information Within Industry, issued February 20, 1960.

The section provides that the Civil Service Commission shall be responsible for conducting security investigations for the AEC. The statutory exceptions to this provision are as follows:

1. If during the investigation the Civil Service Commission develops any data reflecting that the individual who is the subject of the investigation is of questionable loyalty, the Civil Service Commission must refer the case to the Federal Bureau of Investigation for the conduct of a full field investigation.
2. The majority of the members of the

Commission may determine that specific positions in the AEC program are of a high degree of importance or sensitivity and upon such determination the investigation and reports on the individual who is to occupy such a position shall be made by the Federal Bureau of Investigation.
3. The President may cause investigation of any groups or class to be made by the Federal Bureau of Investigation, when he deems it to be in the national interest.
4. Under an amendment to the Atomic Energy Act in 1961, the Commission may accept an investigation and report on the character, associations, and loyalty of an individual made by a Government agency, other than Civil Service Commission or the Federal Bureau of Investigation, provided that a security clearance has been granted to such individual by another Government agency based on such investigation and report.

Statutory authority is provided in Section 145 for the Commission or the General Manager to permit the employment of an individual, by the Commission, or to permit an individual access to Restricted Data prior to completion of an investigation, upon a determination that such action is clearly consistent with the national interest. Section 161 (n) prohibits delegation of this authority.

Section 145 also provides that the Commission may establish standards and specifications as to the scope and extent of investigation to be made by the Civil Service Commission, provided such standards are based on location and class or kind of work to be done and consideration is given to the degree of importance to the common defense and security of the Restricted Data to which access will be permitted. Under this provision, the Commission established the "L" clearance based on a national agency check which permits access to Confidential Restricted Data. A full background investigation is still required for access to Secret and Top Secret Restricted Data.

Provisions of Section 145 (g) adopted in 1958 further permit the Commission during state of war or period of national disaster due to enemy attacks to authorize the employment of individuals and access by individuals to Restricted Data pending receipt of the investigative report and the required statutory determination.

Executive Orders 10450, as amended, and 10865, as amended, established for all departments and agencies of the Government the minimum security standards for direct Government employment and for safe-guarding classified information within industry, respectively.

Executive Order 10450 provides for the maintenance within the Federal Government of an effective program insuring that employment and retention in employment of civilian officers and employees is clearly consistent with the interests of national security. This order requires the designation of sensitivity of positions, the standards of investigation for Government employees occupying such positions, and the authority for suspension of employees. Moreover, Executive Order 10450 contains the security criteria for determining the eligibility for employment or retention in employment.

Executive Order 10865 provides standards and procedures for granting access to classified information within the industrial program of the Federal Government. This order also provides for confrontation of witnesses, with certain well defined exceptions.

Legislation other than that in the Atomic Energy Act which should be mentioned as having a bearing on the personnel security program of the AEC is as follows:

1. Public Law 81-266 of 1949

This Appropriation Act contained a rider which provided that no part of any appropriation contained in the Act for the AEC would be used to confer a fellowship on any individual who advocated or who was a member of an organization or party that advocated the overthrow of the Government of the United States by force or violence or with respect to whom the Commission found, upon investigation and report by the Federal Bureau of Investigation on the character, associations, and loyalty of whom, that reasonable grounds existed for belief that such individuals were disloyal to the Government of the United States. A similar rider has been attached to all subsequent Appropriation Acts except that the Civil Service Commission is the Designated Investigative Agency.

2. Public Law 85-568—National Aeronautics and Space Act of 1958

Provisions contained in Section 304 of the Space Act enable the Commission to authorize its employees, or employees of its contractors or licensees, to permit any member officer or employee of the National Aeronautics and Space Council, or the Administrator of NASA, or any employee of a NASA contractor to have access to Restricted Data relating to aeronautical and space activities which may be required in the performance of duty, if so certified by the Space Council or the Administrator of NASA, and if appropriately cleared under security procedures and standards of NASA.

3. Public Law 87-297—The Arms Control and Disarmament Act of 1961

This Arms Control and Disarmament Act provides that the Commission may authorize the Director of the ACDA, and certain listed classes of people who have been certified by the Director as having a "need-to-know," to have access to Restricted Data but only if the Commission has determined in accordance with Commission security procedures and standards that permitting such persons to have access to Restricted Data will not endanger the common defense and security.

1954—Investigation by the Attorney General of Certain Offenses

The Act of August 31, 1954 (68 Stat. 998; P.L. 725, 83rd Congress) authorized the Attorney General and the Federal Bureau of Investigation to make certain investigations. The Act provides in Part

. . . that notwithstanding any other provision of law and without limiting the authority to investigate any matter which may have been or may, hereafter, be conferred upon them or upon any other department or agency of the Government the Attorney General and the Federal Bureau of Investigation have the authority to investigate any violation of Title 18, United States Code, involving Government officers and employees. Any information, complaint, or allegation received in any department or agency of the Executive branch of Government related to said violations involving Government officers or employees shall be expeditiously reported to the Attorney General by the head of such department or agency unless the responsibility to perform an investigation with respect, thereto, is specifically assigned by another provision of law, or unless the Attorney General, otherwise, directs with respect, as to any department or agency of the Government, to any specified class of information, allegation, or complaint: Provided, that the provision of this section shall not limit, in any way, the existing authority of the military department to investigate persons for offenses over which the Armed Forces have jurisdiction under the Uniform Code of Military Justice: Provides further that the provision of this section shall not limit in any way the primary authority of the Post Master General to investigate postal offenses.

Under Title 18, the United States Code, the Federal Bureau of Investigation is responsible for investigating a number of specific offenses relating to the national security. The following list represents the major sections over which the Federal Bureau of Investigation and other Internal Security Agencies have responsibility.

Conspiracy

18 UNITED STATES CODE, SECTION 371. CONSPIRACY TO COMMIT OFFENSE OR DEFRAUD THE UNITED STATES.

If two or more persons conspire either to commit any offense against the United States, or to defraud the United States, or any agency, thereof, in any manner or for any purpose, and one or more of such persons do any act to effect the object of the conspiracy, each shall be fined not more than $10,000 or imprisoned not more than five years, or both.

Extracts of the Espionage and Sabotage and Other Federal Criminal Statutes

18 UNITED STATES CODE, SECTION 793, GATHERING, TRANSMITTING OR LOSING DEFENSE INFORMATION.

(a) Whoever, for the purpose of obtaining information respecting the national defense with intent or reason to believe that the information is to be used to the injury of the United States, or to the advantage of any foreign nation, goes upon, enters, flies over, or otherwise obtains information concerning any vessel, aircraft, work of defense, navy yard, naval station, submarine base, fueling station, fort, battery, torpedo station, dockyard, canal, railroad, arsenal, camp, factory, mine, telegraph, telephone, wireless, or signal station, building, office, research laboratory or station or other place connected with the national defense owned or constructed, or in progress of construction by the United States or under the control of the United States, or of any of its officers, departments, or agencies, or within the exclusive jurisdiction of the United States, or any place in which any vessel, aircraft, arms, munitions, or other materials or instruments for use in time of war are being made, prepared, repaired, stored, or are the subject of research or development, under any contract or agreement with the United States, or any department or agency thereof, or with any person on behalf of the United States, or otherwise on behalf of the United States, or any prohibited place so designated by the President by proclamation in time of war or in case of national emergency in which anything for the use of the Army, Navy, or Air Force is being prepared or constructed or stored, information as to which prohibited place the President has determined would be prejudicial to the national defense; or

(b) Whoever, for the purpose aforesaid, and with like intent or reason to believe, copies, takes, makes, or obtains, or attempts to copy, take, make, or obtain any sketch, photograph, photographic negative, blueprint, plan, map, model, instrument, appliance, document, writing, or note of anything connected with the national defense; or

(c) Whoever, for the purpose aforesaid, receives or obtains or agrees or attempts to receive or obtain from any person, or from any source whatever, any document, writing, code book, signal book, sketch, photograph, photographic negative, blueprint, plan, map of anything connected with the nation defense, knowing or having reason to believe, at the time he receives or obtains, or agrees or attempts to receive or obtain it, that it has been or will be obtained, taken, made, or disposed of by any person contrary to the provisions of this chapter; or

(d) Whoever, lawfully having possession of, access to, control over, or being entrusted with any document, writing, code book, signal book, sketch, photograph, photographic negative, blueprint, plan, map, model, instrument, appliance, or note relating to the national defense which information the possessor has reason to believe could be used to the injury of the United States or to the advantage of any foreign nation, willfully communicates, delivers, transmits, or causes to be communicated, delivered or transmitted, the same to any person not entitled to receive it, or willfully retains the same and fails to deliver it on demand to the officer or employee of the United States entitled to receive it; or

(e) Whoever, having unauthorized possession of, access to, or control over any document, writing, code book, signal book, sketch, photograph, photographic negative, blue print, plan, map, model, instrument, appliance, or note relating to the national defense or information relating to the national defense which information the possessor has reason to believe could be used to the injury of the United States or to the advantage of any foreign nation willfully communicates, delivers, transmits or causes to be communicated, delivered, or transmitted, or attempts to communicate, deliver, transmit or cause to be communicated, delivered, or transmitted the same to any person not entitled to receive it, or willfully retains the same and fails to deliver it to the officer or employee of the United States entitled to receive it; or

(f) Whoever, being entrusted with or having lawful possession or control of any document, writing, code book, signal book, sketch, photograph, photographic negative, blueprint, plan, map, model instrument, appliance, note, or information, relating to the national defense, (1) through gross negligence permits the same to be removed from its proper place of custody or delivered to anyone in violation of his trust, or to be lost, stolen, abstracted, or destroyed, or (2) having knowledge that the same has been illegally removed from its proper place of custody or delivered to anyone in violation of his trust, or lost, or stolen, abstracted, or destroyed, and fails to make prompt report of such loss, theft, abstraction, or destruction to his superior officer— shall be fined not more than $10,000 or imprisoned not more than ten years, or both.

(g) If two or more persons conspire to violate any of the foregoing provisions of this section, and one or more of such persons do any act to effect the object of the conspiracy, each of the parties to such conspiracy shall be subjected to the punishment provided for the offense which is the object of such conspiracy.

18 UNITED STATES CODE, SECTION 794, GATHERING OR DELIVERING DEFENSE INFORMATION TO FOREIGN GOVERNMENTS.

(a) Whoever, with intent or reason to believe that it is to be used to the injury of the United States or to the advantage of a foreign nation, communicates, delivers, or transmits, or attempts to communicate, deliver, or transmit, to any foreign government, or to any faction or party or military or naval force within a foreign country, whether recognized or unrecognized by the United States, or to any representative, officer, agent, employee, subject, or citizen thereof, either directly or indirectly, any document, writing, code book, signal book, sketch, photograph, photographic negative, blueprint, plan, map, model, note, instrument, appliance, or information relating to the national defense, shall be punished by death or by imprisonment for any term of years or for life.

(b) Whoever, in time of war with intent that the same shall be communicated to the enemy, collects, records, publishes, or communicates or attempts to elicit any information with respect to the movement, numbers, description, condition, or disposition of any of the Armed Forces, ships, aircraft, or war materials of the United States, or with respect to the plans or conduct or supposed plans or conduct of any naval or military operations, or with respect to any works or measures undertaken for or connected with, or intended for the fortification or defense of any place, or any other information relating to the public defense, which might be useful to the enemy, shall be punished by death or imprisonment for any term of years or for life.

(c) If two or more persons conspire to violate this section and one or more of such persons do any act to effect the object of the conspiracy, each of the parties to such conspiracy shall be subject to the punishment provided for the offense which is the object of such conspiracy.

18 UNITED STATES CODE, SECTION 795. PHOTOGRAPHING AND SKETCHING DEFENSE INSTALLATIONS.

(a) Whenever, in the interests of national defense, the President defines certain vital military and naval installations or equipment as requiring protection against the general dissemination of information relative thereto, it shall be unlawful to make any photograph, sketch, picture, drawing, map, or graphical representation of such vital military and naval installations or equipment without first obtaining permission of the commanding officer of the military or naval post, camp, or station, or naval vessels, military and naval aircraft, and any separate military or naval command concerned, or higher authority, and promptly submitting the product obtained to such commanding officer or higher authority for censorship or such other action as he may deem necessary.

(b) Whoever violates this section shall be fined not more than $1,000 or imprisoned not more than one year, or both.

18 UNITED STATES CODE, SECTION 796. USE OF AIRCRAFT FOR PHOTOGRAPHING DEFENSE INSTALLATIONS.

Whoever uses or permits the use of an aircraft or any contrivance used, or designed for navigation or flight in the air, for the purpose of making a photograph, sketch, picture, drawing, map, or graphic representation of vital military or naval installations or equipment, in violation of section 795 of this title, shall be fined not more than $1,000 or imprisoned not more than one year, or both.

18 UNITED STATES CODE, SECTION 797. PUBLICATION AND SALES OR PHOTOGRAPHS OF DEFENSE INSTALLATIONS.

On and after thirty days from the date upon which the President defines any vital military or naval installation or equipment as being within the category contemplated under Section 795 of this title, whoever reproduces, publishes, sells, or gives away any or graphical representation of the vital military or naval installations or equipment so defined, without first obtaining permission of the commanding officer of the military or naval post, camp, or station concerned, or higher authority, unless such photograph, sketch, picture, drawing, map, or graphical representation has clearly indicated thereon that it has been censored by the proper military or naval authority, shall be fined not more than $1,000 or imprisoned not more than one year, or both.

18 UNITED STATES CODE, SECTION 798. DISCLOSURE OF CLASSIFIED INFORMATION.

(a) Whoever knowingly and willfully communicates, furnishes, transmits, or otherwise makes available to an unauthorized person, or publishes, or uses in any manner prejudicial to the safety or interest of the United States or for the benefit of any foreign government to the detriment of the United States any classified information— (1) concerning the nature, preparation or use of any code, cipher, or cryptographic system of the United States or any foreign government; or (2) concerning the design, construction, use, maintenance, or repair of any device, apparatus, or appliance used or prepared or planned for use by the United States or any foreign government of cryptographic or communication intelligence purpose; or (3) concerning the communication intelligence activities of the United States or any foreign government; or (4) obtained by the processes of communication intelligence form the communications of any foreign government, knowing the same to have been obtained by such processes—shall be fined not more than $10,000 or imprisoned not more than ten years, or both.

(b) As used in subsection (a) of this section—the term "Classified information" means information which, at the time of a violation of this section is, for reasons of national security, specifically designated by a United States Government Agency for limited or restricted dissemination or distribution.

The terms "code," "cipher," and "cryptographic system" include in their meanings, in addition to their usual meanings, any method of secret writing and any mechanical or electrical device or method used for the purpose of disguising or concealing the contents, significance, or meanings or communications.

The term "foreign government" includes in its meaning any person or persons acting or purporting to act for or on behalf of any faction, party, department, agency, bureau, or military force of or within a foreign country, or for on behalf of any government or any person or persons purporting to act as a government within a foreign country, whether or not such government is recognized by the United States.

The term "Communication intelligence" means all procedures and methods used in the interception of communications and the obtaining of information from such communications by other than the intended recipients.

The term "unauthorized person" means any person who, or agency which is not authorized to receive information of the categories set forth in subsection (a) of this section, by the President, or by the head of a department or agency of the United States Government which is expressly designated by the President to engage in communication intelligence activities for the United States.

(c) Nothing in this section shall prohibit the furnishing upon lawful demand, of information to any regularly constituted committee of the Senate or House of Representatives of the United States of America, or joint committee thereof.

18 UNITED STATES CODE, SECTION 799. VIOLATION OF REGULATIONS OF NATIONAL AERONAUTICS AND SPACE ADMINISTRATION.

Whoever willfully shall violate, attempt to violate, or conspire to violate any regulation or order promulgated by the Administrator of the National Aeronautics and Space Administration for the protection or security of any laboratory, station, base or other facility, or part thereof, or any aircraft, missile, spacecraft, or similar vehicle, or part thereof, or other property or equipment in the custody of the Administration, or any real or personal property or equipment in the custody of any contractor under any contract with the Administration or any subcontractor of any such contract or, shall be fined not more than $5,000 or imprisoned not more than one year, or both.

SABOTAGE ACT—18 UNITED STATES CODE, SECTION 2153. DESTRUCTION OF WAR MATERIAL, WAR PREMISES, OR WAR UTILITIES.

(a) Whoever, when the United States is at war, or in times of national emergency as declared by the President or by the Congress, with intent to injure, interfere with, or obstruct the United States or any associate nation in preparing for or carrying on the war or defense activities, or, with reason to believe that his act may injure, interfere with or obstruct the United States or any associate nation in preparing for or carrying on the war or defense activities willfully injures, destroys, contaminates or infects, or attempts to so injure, destroy, contaminate or infect any war material, war premises, or war utilities, shall be fined not more than $10,000 or imprisoned not more than thirty years, or both.

(b) If two or more persons conspire to violate this section, and one or more of such persons do any act to effect the object of the conspiracy, each of the parties to such conspiracy shall be punished as provided in subsection (a) of this section.

18 UNITED STATES CODE, SECTION 2154. PRODUCTION OF DEFENSE WAR MATERIAL, WAR PREMISES, OR WAR UTILITIES.

(a) Whoever, when the United States is at war, or in times of national emergency as declared by the President or by the Congress, with intent to injure, interfere with, or obstruct the United States or any associate nation in preparing for or carrying on the war or defense activities, or, with reason to believe that his act may injure, interfere with, or obstruct the United States or any associate nation in preparing for or carrying on the war or defense activities, willfully makes, constructs, or causes to be made or constructed in a defective manner, or attempts to make, construct, or cause to be made or constructed in a defective manner any war material, war premises or war utilities, or any tool, implement, machine, utensil, or receptacle used or employed in making, producing, manufacturing, or preparing any such war materials, war premises or war utilities, shall be fined not more than $10,000 or imprisoned not more than thirty years, or both.

(b) If two or more persons conspire to violate this section, and one or more of such persons do any act to effect the object of the conspiracy, each of the parties to such conspiracy shall be punished as provided in subsection A of this section.

18 UNITED STATES CODE, SECTION 2155. DESTRUCTION OF NATIONAL-DEFENSE MATERIALS, NATIONAL-DEFENSE PREMISES OR NATIONAL-DEFENSE UTILITIES.

(a) Whoever, with intent to injure, interfere with, or obstruct the national defense of the United States, willfully injures, destroys, contaminates or infects, or attempts to so injure, destroy, contaminate or infect any national-defense utilities, shall be fined not more than $10,000 or imprisoned not more than ten years, or both.

(b) If two or more persons conspire to violate this section, and one or more of such persons do any act to affect the object of the conspiracy, each of the parties to such conspiracy shall be punished as provided in subsection (a) of this section.

UNITED STATES CODE, SECTION 2156. PRODUCTION OF DEFECTIVE NATIONAL-DEFENSE MATERIAL, NATIONAL-DEFENSE PREMISES OR NATIONAL-DEFENSE UTILITIES.

(a) Whoever, with intent to injure, interfere with, or obstruct the national defense of the United States, willfully makes, constructs, or attempts to make or construct in a defective manner, any national-defense material, national-defense premises or national-defense utilities, or any tool, implement, machine, utensil, or receptacle used or employed in making, producing, manufacturing, or repairing any such national-defense material, national-defense premises or national-defense utilities, shall be fined not more than $10,000 or imprisoned not more than ten years, or both.

(b) If two or more persons conspire to violate this section, and one or more of such persons do any act to effect the object of the conspiracy, each of the parties to such conspiracy shall be punished as provided in subsection (a) of this section.

1955—Strike or Overthrow of Government

Superseding Section 9A of the Hatch Act of August 9, 1955 (69 Stat. 624, P.L. 330, 84th Congress, 5 United States Code Section 118P-118R) ". . . No person shall accept or hold office in the government of the United States or any agency thereof including wholly owned government corporations who (1) advocates the overthrow of our constitutional form of government in the United States; (2) is a member of an organization that advocates the overthrow of our constitutional form of government in the United States knowing such organization so advocates; (3) participates in any strike or asserts the right to strike against the government of the United States or such agency; or (4) is a member of an organization of government employees that asserts the right to strike against the government of the United States or such agencies knowing that such organization asserts such right.

1958—Federal Aviation Act

The Act of August 23, 1958 (72 Stat. 731; P.L. 85-726) security provisions section 1201. ". . . to establish security provisions which will encourage the maximum use of the navigable air space by civil aircraft consistent with the national security."

Section 1202—Security Control of Air Traffic

The administrator in consultation with the Department of Defense shall establish such zones or areas in the air space of the United States as he may find necessary in the interest of national defense and by rule, regulation, or order restrict or prohibit the flight of civil aircraft which he cannot identify, locate or control with available facilities within such zones or areas.

NATIONAL AERONAUTICS AND SPACE ACT OF 1958 ACT OF JULY 29, 1958 (72 STAT. 426; PUB. L. 85-568) : SECTION 304.

(a) The Administrator shall establish such security requirements, restrictions, and safeguards as he deems necessary in the interest of the national security. The Administrator may arrange with the Civil Service Commission for the conduct of such security or other personnel investigations of the Administration's officers, employees, and consultants, and its contractors and subcontractors and their officers and employees, actual or prospective, as he deems appropriate; and if any such investigation develops any data reflecting that the individual who is the subject thereof is of questionable loyalty the matter shall be referred to the Federal Bureau of Investigation for the conduct of a full field investigation, the results of which shall be furnished to the Administrator.

(b) The Atomic Energy Commission may authorize any of its employees, or employees of any contractor, prospective contractor, licensee, or prospective licensee of the Atomic Energy Commission or any other person authorized to have access to Restricted Data by the Atomic Energy Commission under subsection 145b of the Atomic Energy Act of 1954 (42 U.S.C. 2165b), to permit any member officer, or employee, member of an advisory committee, contractor, subcontractor, or officer or employee of a contractor or subcontractor of the Administration, to have access to Restricted Data relating to aeronautical and space activities which is required in the performance of his duties and so certified by the Council or the Administrator, as the case may be, but only if (1) the Council or Administrator or designee thereof has determined, in accordance with the established personnel security procedures and standards of the Council or Administration, that permitting such individual to have access to such Restricted Data will not endanger the common defense and security, and (2) the Council or Administration or designee thereof finds that the established personnel and other security procedures and standards of the council or Administration are adequate and in reasonable conformity to the standards established by the Atomic Energy Commission under section 145 of the Atomic Energy Act of 1954 (42 U.S.C. 2165). Any individual granted access to such Restricted Data pursuant to this subsection may exchange such Data with any individual who (A) is an officer or employee of the Department of Defense, or any department or agency thereof, or a member of the armed forces, or a contractor or subcontractor of any such department, agency, or armed force or an officer or employee of any such contractor or subcontractor, and (B) has been authorized to have access to Restricted Data under the provisions of section 143 of the Atomic Energy Act.

Statutory and Executive Authority for NASA Security Program[3]

1. Regulations and instructions governing the personnel and industrial security programs of NASA are issued pursuant to the provisions of Section 304 (a) of the National Aeronautics and Space Act of 1958, which authorizes the Administrator, NASA, to ". . . establish such security requirements, restrictions, and safeguards as he deems necessary in the interest of the national security."

2. In addition, the NASA personnel security program is based upon the provisions of Public Law 733, August 26, 1950 (64 Stat. 476) and its implementing Executive Order 10450, as amended through August 5, 1954. Public Law 733, as amended by the National Aeronautics and Space Act of 1958, authorized the Administrator, NASA, to summarily suspend without pay and to terminate employees of the agency when deemed nec-

essary in the interest of national security.
3. The NASA industrial security program is based on the provisions of Executive Order 10865, as amended through January 17, 1961.

Industrial Security Program

1. In June, 1959, the Administrator of NASA and the Acting Secretary of Defense, by an exchange of letters, entered into an agreement which extended the industrial security program of the DOD to include NASA classified contracts. Subsequent to the issuance of Executive Order 10865, the Administrator and the Deputy Secretary of Defense exchanged letters confirming the original agreement, subject to modifications to reflect certain provisions of the executive order.
2. Under the terms of the agreement, NASA accepted the *Armed Forces Industrial Security Regulations* (AFISR) and the *Industrial Security Manual for Safeguarding Classified Information* (ISM) as satisfying NASA requirements for the security of classified information entrusted to NASA contractors. The necessary security services provided for in the AFISR, i.e. inspection and clearance of employees, and monitoring of security practices, are performed for NASA by one of the military services to which DOD assigns security cognizance over the contractor facility engaged in a NASA classified contract. NASA, as the contracting agency, retains the authority and responsibility for those security functions specified in the AFISR and ISM to be performed by a contracting military department. The principal responsibilities are set forth in Section 7, NASA Management Instruction 24-3-1 dated April 21, 1961. This Instruction prescribes the internal NASA policies and procedures for the security processing of NASA contracts where performance will require access to classified information by the contractor or his employees.
3. DOD Directive 5220.6, dated July 28, 1960, issued pursuant to Executive Order 10865, reflects the provisions of the NASA-DOD agreement with respect to the processing of access authorization review cases ("agency cases") arising out of the release of classified information to or within industry by the NASA. In an "agency case" referred for consideration and determination under the access authorization review program, NASA is entitled to representation on the screening and central boards.

Safeguarding Classified Information Within Industry

Executive Order 10865, Feb. 20, 1960 (25 F.R. 1583):

SECTION 1

(a) The Secretary of State, the Secretary of Defense, the Commissioners of the Atomic Energy Commission, the Administrator of the National Aeronautics and Space Administration, and the Administrator of the Federal Aviation Agency, respectively, shall, by regulation, prescribe such specific requirements, restrictions, and other safeguards as they consider necessary to protect (1) releases of classified information to or within United States industry that relate to bidding on, or the negotiation, award, performance, or termination of, contracts with their respective agencies, and (2) other releases of classified information to or within industry that such agencies have responsibility for safeguarding. So far as possible, regulations prescribed by them under this order shall be uniform and provide for full cooperation among the agencies concerned.

(b) Under agreement between the Department of Defense and any other department or agency of the United States, including, but not limited to, those referred to in subsection (c) of this section, regulations prescribed by the Secretary of

Defense under subsection (a) of this section may be extended to apply to protect releases (1) of classified information to or within United States industry that relate to bidding on, or the negotiation, award, performance, or termination of, contracts with such other department or agency, and (2) other releases of classified information to or within industry which such other department or agency has responsibility for safeguarding.

(c) When used in this order, the term "head of a department" means the Secretary of State, the Secretary of Defense, the Commissioners of the Atomic Energy Commission, the Administrator of the National Aeronautics and Space Administration, the Administrator of the Federal Aviation Agency, and, in sections 4 and 8, includes the Attorney General. The term "department" means the Department of State, the Department of Defense, and the Atomic Energy Commission, the National Aeronautics and Space Administration, the Federal Aviation Agency, and, in sections 4 and 8, includes the Department of Justice.

SECTION 2

An authorization for access to classified information may be granted by the head of a department or his designee, including but not limited to, those officials named in section 8 of this order, to an individual hereinafter termed an "applicant," for a specific classification category only upon a finding that it is clearly consistent with the national interest to do so.

SECTION 3

Except as provided in section 9 of this order, an authorization for access to a specific classification category may not be finally denied or revoked by the head of a department or his designee, including, but not limited to, those officials named in section 8 of this order, unless the applicant has been given the following:

(1) A written statement of the reasons why his access authorization may be denied or revoked, which shall be as comprehensive and detailed as the national security permits.

(2) A reasonable opportunity to reply in writing under oath or affirmation to the statement of reasons.

(3) After he has filed under oath or affirmation, a written reply to the statement of reasons, the form and sufficiency of which may be prescribed by regulations issued by the head of the department concerned, an opportunity to appear personally before the head of the department concerned or his designee, including, but not limited to, those officials named in section 8 of this order, for the purpose of supporting his eligibility for access authorization and to present evidence on his behalf.

(4) A reasonable time to prepare for that appearance.

(5) An opportunity to be represented by counsel.

(6) An opportunity to cross-examine persons either orally or through written interrogatories in accordance with section 4 on matters not relating to the characterization in the statement of reasons of any organization or individual other than the applicant.

(7) A written notice of the final decision in his case which, if adverse, shall specify whether the head of the department or his designee, including, but not limited to, those officials named in section 8 of this order, found for or against him with respect to each allegation in the statement of reasons.

The specific provisions of 10865 which form the basis for the governmental industrial security program are delineated in two principle directives issued by the Department of the Defense. These are the Department of Defense Industrial Security Regulation DODISR (DOD 5220-22R) and the Industrial Security Manual for Safeguarding Classified Information (DOD 5220-22M Industrial Security Manual). These documents cover the essential policy and procedure with respect to safeguarding classified information in the hands of industry.

LEGAL BASIS FOR PRIVATE PROTECTIVE SERVICES

As was readily apparent in the legal basis for governmental security, a number of specific legislative enactments exist to give a basis to governmental programs. These executive orders, directives, and statutes are continually updated and revised as conditions and needs for specific types of control develop. The situation for private protective services is essentially the same as governmental security in that statutes and court decisions continually alter the specific circumstances in which an individual may protect his person or property. Nevertheless, private "security" finds its basis in the inalienable rights of citizens of the United States to protect their persons and properties. More specifically, the basis is found in the common law and in state statutory law relative to the protection of persons and their property.

Constitutionally, citizens have the right to defend themselves and their property. This common law right to protect property forms the basis for an individual's hiring a watchman or security officer to protect his property or serves as the basis for large security operations to prevent shoplifting or employee pilferage. This self-defense concept provides that one may protect his property or person against some injury attempted by another.

Many jurisdictions have statutes which give the individual the right to protect himself and property. In Wisconsin, for example. Wisconsin Statute 939.49 (defense of property and protection against shoplifting) indicates the legal jurisdiction of an individual in protecting his property against shoplifting. It states the following:

> A person is privileged to threaten or intentionally use force against another for the purpose of preventing or terminating what he reasonably believes to be an unlawful interference with his property. Only such degree of force or threat thereof, may intentionally be used as the actor reasonably deems it necessary to prevent or terminate the interference. It is not reasonable to intentionally use force intended to likely cause death or great bodily harm for the sole purpose of defense of one's property.

It further states:

> A person is privileged to defend a third person's property from real or apparent unlawful interference by another under the same conditions and by the same means as those under and by which he is privileged to defend his own property from real or apparent unlawful interference, provided that he reasonably believes that the facts are such as would give the third person the privilege to defend his own property. That his intervention is necessary for the protection of the third person's property, and that the third person whose property he is protecting is a member of his immediate family or household or a person whose property he has a legal duty to protect or is a merchant and the actor is the merchant's employee or agent.

This statutory description of a citizen's rights to protect his property or his right to hire an individual to protect his property provides a legal footing for a private security program. The following is also pointed out:

> A general rule that a man is not privileged to use deadly force in the defense of his property and/or person except in the defense of his dwelling place and to prevent a felony attempted by violence or surprise (burglary, robbery, arson, etc.). There is a substantial line of authority which contends that the use of deadly force even in defense of his dwelling, is not privileged unless the actor reasonably believes the envasion of his dwelling threatened him, or privilege of self defense or serious bodily harm.

The basis for private security is then,

state statute, common law, or the constitutional right of citizens to protect their property. Protection is also provided for the whole range of activities covered in the criminal law by private individuals or on a private basis. The primary function of a private security officer is to protect assets. The job is not the same as a public police officer—he does not often investigate a crime nor does he arrest a criminal; but rather protects private property and prevents, if possible, a crime from occurring.

There are precise rules of law to guide the actions of police or security officers in each state. However, the laws, as they now stand, are a maze of privileges and general standards which are interpreted and judged according to the particular circumstances. Specific state statutes will govern each act. Many of the more common areas are presented in this section along with their definitions.

The most important point to remember is that a security officer is *not* a policeman, and he does not have police authority. Authority is generally limited to that of a private citizen in the protection of property and in self-defense. Security officers are only authorized by their clients to act in their behalf in the protection of their property.

One of the primary sources of restrictions on security officer's duties is the *tort law*. Tort law governs the activities of all citizens by allowing an injured person to bring a lawsuit against the person causing the injury and to sue for damages resulting from the injury. The following are the major torts* often encountered by security personnel:

Battery—"Intentionally causing the harmful or offensive touching of another person." No force is necessary; mere touch is considered a battery.

Assault—"Intentionally causing the fear of a harmful or offensive touching." Causing someone worry about being harmed or touched is grounds for the charge of assault.

Infliction of mental distress—"Intentionally causing mental or emotional distress in another person."

False imprisonment—"Intentionally confining or restricting the movement or freedom of another person."

Malicious prosecution—"Groundlessly instituting criminal procedures against another person."

Trespass to land—"Unauthorized entering upon the property of another."

Trespass to personal property—"Unauthorized taking or damaging of another's goods."

Negligence—"Causing injury to persons or property by taking unreasonable risk or failing to use reasonable care"; either too much or too little action.

Defamation (slander or libel)—"Injuring the reputation of another by publicly uttering untrue statements."

Invasion of privacy—"Intruding upon another's physical solitude; disclosing private information about another; publicly placing another in false light."

Security officers are also restricted by the criminal law which imposes fines and jail sentences for improper conduct. Under criminal law, prosecution is possible, for the crimes of manslaughter, negligent homicide, murder, assault, and battery. At times, to avoid criminal liability, an officer must be able to show a legal justification or defense for his actions.*

> Common law crimes, which may be involved in the taking by one person, or property belonging to another are *larceny, embezzlement, and obtaining property by false pretenses*. All three have common elements, but each has distinguishing features.[4]

A common element is the taking and converting of another's property or the taking and carrying away of another person's

---

* Source: *Black's Legal Dictionary*.

* Appendix 3B of this chapter provides a comprehensive set of legal guidelines for security officers. While local court decisions will modify the general rules set forth, it provides a guide to current court decision in major problem areas.

property for your own, an unlawful purpose.

Larceny is the trespass against the possession of another. The person who takes property to which he is not entitled to be in possession of at the time of taking. Embezzlement involves an appropriation of the property of another by one who is lawfully in possession at the time of his act, by virtue of some relationship of trust with the owner of the property. Obtaining property by false pretenses is distinguished on the basis that the criminal has obtained both possession and title of the property by virtue of his wrongdoings.[5]

An addition to these acts is:

Robbery—Felonious taking of personal property in the possession of another, from his person or immediate presence, and against his will, accomplished by means of force or fear.[6]

Burglary—The breaking and entering of the house of another with intent to commit a felony therein, whether the felony be committed or not.[7]

Theft—A popular name for larceny. The fraudulent taking of corporeal personal property belonging to another, from his possession, or the possession of some person holding the same for him, without his consent, with intent to deprive the owner of the value of the same, and to appropriate it to the use or benefit of the person taking, are acts which often requires the individual's protection and action be taken.[8]

LEGAL BASIS FOR COMMERCIAL PROTECTIVE SERVICES

In addition to the requirements for private citizens to provide protective services for themselves, commercial security services must meet additional standards. The provision of "third party" protection for "hire or reward" is regulated by either state or local legislation in many states. In 1968, there were a total of fifteen states which regulated such agencies and companies. At present, there are forty states which require some form of licensing or regulation for the commercial protective services agency. This regulation is often expressed in terms such as:

No person shall act or hold himself out as a private detective, private policeman, or private guard, nor shall any person solicit business or perform any service in this state as a private detective, private police, or private guard, or receive any fees or compensation whatever, for acting as a private detective, private police, or private guard, for any firm, person, or corporation without first having obtained a license and filed a bond provided for. . . .

The amount of licensing and regulation varies considerably between states. There are, for example, presently ten states which do not license agencies or individuals at all. These being:

    Idaho
    Missouri
    Oklahoma
    Oregon
    Rhode Island
    South Dakota
    Utah
    Virginia
    Washington
    Wyoming

There are an additional five states which license for revenue only. These are:

    Alabama
    Alaska
    Louisiana
    Mississippi
    Tennessee

The amount of regulation and control within the forty states which regulate or license private detectives or guard agencies varies considerably. The recent National Advisory Committee on Criminal Justice Goals and Standards—Private Police Task Force summarized the current licensing and regulation requirements. They are presented in Appendix 3C of this chapter.

Legislative Purpose

The primary rationale for the hiring of the commercial firm is to extend the ability of the individual to protect his property. In this respect, the commercial agency is acting as an extension of the private rights of the owner in protecting property. This extension of the private right to protection does not provide a "cloak" of protection or additional authority for the commercial firm. It is often assumed to be present by the general public because of the adopted symbols of "police power" worn by the private protective services officer, e.g. uniforms, badges, and guns. The assumption of power is also at times considered legitimate by the officers or their agencies in carrying out their duties. Many problems, lawsuits, and misunderstandings have occurred because of these assumptions. *The simple fact is that,* unless individual officers are granted "peace officer" powers by some agency of government such as a sheriff's department or a police department, *they act as private citizens.*

While many industrial firms do not wish their commercial security officers to act as policemen there are other firms who would like them to have, or at least act as though they had these powers. The illusion (or deception) of having security personnel present whose legal status is not clearly understood is often viewed as a deterrent asset.*

The powers of arrests are clearly defined both for private and public protective officers. The powers of search and interrogation for the private officer are, however, somewhat more ambiguous with sanctions for improper action dependent upon the nature and extent of abuses, and the deterrent effect of existing sanctions.

The parallels which exist between the public and private sector seem to suggest that similar sanctions and regulations should exist for both sectors. This, however, is not the case currently. The sponsorship of the different sectors as well as the motivation and incentives for operations in the private sector are much different for the public sector. Consequently, to apply public police standards and sanctions to the private, commercial, or proprietary sector would appear unreasonable and unwarranted.

The purpose of private protective services is primarily preventative, *not* enforcement of public laws. The enforcement of public law is more often an adjunct to their preventative function. Therefore, while a common standard of action should exist for the protection of the rights of citizens, it need not apply to private nonenforcement personnel. Specific procedural guidelines should, however, be developed and enforced to insure a thorough understanding of private sector personnel of what enforcement actions are acceptable.

---

* See Chapter 3, Appendix 3D for a discussion of private power, limitation and legal issues.

FOOTNOTES

1. *United States Government Organization Manual, 1966-67.* (Washington, Government Printing Office, 1966), pp. 511-512.
2. United States Personnel Security Practices. *United States Government Organization Manual, 1966-67,* pp. 707-712.
3. United States Personnel Security Practices, pp. 961-966.
4. William A. Rutter: *Law and Procedure,* 5th ed. California Gilbert Law Summaries, 1968, p. 50.
5. Rutter, *Law and Procedure,* p. 50.
6. Henry C. Black: *Black's Law Dictionary,* 4th ed. (St. Paul, West Publishing Co., 1950), p. 1492.
7. *Black's Law Dictionary,* p. 247.
8. *Black's Law Dictionary,* p. 1647.

APPENDIX 3A

THE SPECIAL POLICE*

INTRODUCTION

Of all the legal relationships which characterize the police service, few have received less attention than that of special police officers. Although it is generally correct to categorize police officers as servants[1] of the state and as a municipal officer exercising the state's power within a local jurisdiction, such definition can be misleading. Thus, private and part time special

---

* From Sheehan, Thomas M., *Police.* 13(Sept-Oct):91-96. Courtesy of Charles C Thomas, Publisher.

police officers frequently bring into play relationships not found or emphasized in the traditional police service. This paper will attempt to describe a number of the principles which separates the special officer from his counterpart in the municipal or state system.

Private individuals often request the service of police officers to assist them in maintaining order on private premises. Drive-in movies, restaurants, supermarkets, pubs, and a host of other commercial activities utilize special policemen to keep order and maintain efficient and satisfactory service. Such deployment may be established as a matter of routine, or it may take the nature of an emergency. An officer so assigned, frequently serves in a dual capacity.[2] On the one hand he protects the interest of a private employer in the furtherance of a business. On the other, he may be called upon to enforce the criminal law of the state, and in this respect he resembles, and is considered, a public officer. Where hired privately, he is given the power of a public officer most frequently through deputization. Here the commissioner of police or the county sheriff bestows upon him limited enforcement powers.

As private employment bestows only limited enforcement powers in the hands of the property owner, the principle of agency law and the delegation of governmental power are used to give the owners of property and their police additional, although limited, enforcement powers. It is therefore necessary to briefly preface our discussion of the special police officer on several important facets of the law of agency.

LAW OF AGENCY AND THE SPECIAL POLICE OFFICER

The relationship existing between employer and employee is characterized by a series of duties and rights which may, or may not, be upheld in a court of law. As special policemen frequently hire out their services to an employer, they are best thought of in terms of agency law. This type of relationship, however, becomes strained when public officers are assigned to guard private property at the command of a superior or the police department itself. Such an assignment can create difficult questions pertaining to legal liability for the wrongful acts of the special police officer. This problem can best be brought into focus by first defining three basic definitions that relate to the law of agency and the special police officer:

Civil Agency

S.2 Master; Servant; Independent Contractor.

(1) A Master is a principle who employs an agent to perform service in his affairs and who controls or has the right to control the physical conduct of the other in the performance of the service.

(2) A Servant is an agent employed by a master to perform service in his affairs whose physical conduct in the performance of the service is controlled or is subject to the right to control by the master.

(3) An Independent Contractor is a person who contracts with another to do something for him but who is not controlled by the other nor subject to the other's right to control with respect to his physical conduct in the performance of the undertaking. He may or may not be an agent.[3]

It has been held that a police officer, separate and apart from his official capacity, may undertake to act in a capacity which in law constitutes a civil agency.[4] One court has established, in outline form, four hypotheticals related to agency law and brought into play with the use of the special police officer. They are:

(1) Cases wherein an employee of the party's own selection, with duties which can be legally performed, or more efficiently performed, in whole or in part, by a public officer, and the employer procures the designation and appointment of his employee as a police officer.

(2) Cases in which the law invests employees, such as trainmen, with police powers in the performance of a quasi-judicial service.

(3) Cases where, under statutes of the state, police officers or constables are appointed and commissioned by the Governor for employment by those needing police protection.

(4) Cases wherein a party calls upon the chief of police or sheriff for police protection merely, having no direction or control, often not naming of the personnel, looking to the officer to perform his duty as such, not paying for the time the officer is so engaged.[5]

The legal decisions which have evolved around these hypotheticals have been solved largely in the context of the master-servant, independent contractor relationship mentioned above. It is through these relationships that a special police officer may simultaneously represent public and private interests. Thus, while serving as a special policeman for a railroad company, he may be vested with the public power of arrest when the criminal laws are broken, although serving in what is essentially a commercial capacity. Or, in the alternative, he may be officially designated a public officer while serving as a part time, or full time private officer. The fact of being a police officer does not prevent his being employed, unless local legislation so declares.[6] This duality of power on the part of the special police officer raises several important policy questions relating to the placement of such officers within the hierarchy of law enforcement. In *Ex Parte Newberg* it was thus stated:

> ... that (policing) ... is a public function, and should be offered to all alike, and the expense thereof borne by all alike from the public funds. To allow these expenses to be borne by private parties would, if carried to its final conclusion, result in the proprietor of a large store, where thousands of people might congregate, to be called upon to bear a substantial part of the city police salaries, or a portion of the city firemen's salaries ... to carry such reasoning to its final conclusion, the private citizen might find all the city policemen, and firemen might have their salaries fastened upon those unfortunate enough to be regulated by the city authorities, thus relieving the municipality of one of its reasons for being called into existence as a portion of our political economy.[7]

There is, of course, another very basic underlying reason for judicial suspicion of the special police officer. Private police agencies, financed and controlled through private funds, emanating from a narrow political base, have often been used to support not the public good but the vested interest. Labor disorders during the earlier years of this century prove a case in point. Although public order and the enforcement of the criminal law may, and frequently does, favor the actions taken by the special police officer, the fact that his actions are not necessarily controlled by the commonweal is cause for caution and public control. A number of states and municipalities have passed statutory guidelines for the organization and control of special police. It has been stated that a municipality may impose the cost of the performance of such services (the power to license assumed) upon private owners.[8] Nor are laws requiring employers to have *suitable* persons on duty to maintain order invalid as depriving operators of their right of freedom of contract.[9]

To counteract the rather permissive nature of the law in allowing for divided employment, a number of jurisdictions have passed "moonlighting" regulations restricting the off-duty activity of their officers. The possibility of moonlighting, however, still exists for officers maintained by private employers and those public officers who reside in jurisdictions which do not prohibit such activity as a matter of law or administrative regulation.

PUBLIC OR PRIVATE CAPACITY

When a private person or corporation with the consent of the state employs police officers to represent them or protect their property "and such officers are engaged in the performance of their duties *to their employer* ... they become and are the *servants* and employees of such private persons ..." who are liable in tort for their acts.[10] It has been held, however, that "... a peace officer cannot be adjudged the agent of one who calls him to aid in preventing a breach of the peace (criminal law), even though it may be threatened upon private premises."[11] Nor can a duly appointed special officer deriving his authority from the sovereignty for the purpose of enforcing the law make the city liable for his misfeasance,[12] malfeasance,[13] or nonfeasance.[14]

This distinction between public and private capacity becomes very important in those instances where the police officer is at once a sworn deputy and a special policeman in the employment of a private business. The court, in determining tort liability, must reach a conclusion as to the exact capacity, public or private, in which a particular policeman was performing at the time of the act sued upon. Thus:

> A special police officer, who was a member of the All-State Police, and who was limited in his activities to guarding or protecting buildings and property of landlord, was landlord agent in serving distress warrant on tenant, so as to render landlord liable for injuries sustained by tenant when assaulted by the special officer.[15]

This is clearly an instance where the officer was acting in a private capacity for his employer. However, if the officer while in private employ disregards his orders and acts for a purpose of his own, he, and not his employer, may be held liable for his wrongful acts.[16] In order to be considered without the control of his master and solely responsible for his own actions, a strict standard is drawn:

> If there is no such express direction (from the employer), it may be inferred from the nature of the duties imposed or the services to be rendered, and, if so, the authorization or instigation is established by ways of implication. In such cases the relation of master and servant is made out, and then the question is whether the act done was within the scope or cause of the servants or agents employment.[17]

The test as to whether or not the offending servant-officer has committed the act for which his private employer will be rendered liable may generally be stated to depend upon:

> (1) The act being of the kind the offender is employed to perform;
> (2) It occurs substantially within the authorized time and space limits of the employment; and
> (3) The offender being actuated, at least in part, by a purpose to serve the master.[18]

These three basic tenets assist the court in determining whether a particular servant's activities fall within "the scope of his employment" as a private functionary. If his acts are directly related to the performance of a business duty, injuries arising therefrom may be imputed to the employer. Therefore, a special police officer, equipped and maintained by a private employer, who settles a personal grudge upon finishing his day's work may be said to be acting without "the scope of his employment," and not in the performance of a private enforcement function.

As we have mentioned, the special officer may be serving in a dual capacity. While working in private employment he may have cause to enforce the laws of the state or municipality if he is a sworn peace officer of that jurisdiction:

> ... that special police officers appointed (under statute) derive all their powers from the appointment made by the civil authorities, here by the Governor of the State. The authority to make arrests ... in no sense can be said to arise from the relation of master and servant ... existing between the special officer and the company at whose application he was commissioned, and the fact that he is to serve, at the expense of such company does not affect his status as that of a special police officer. The rule of *respondeat superior* has no application where there is no evidence to show that the Company was instrumental in causing the arrest or subsequent prosecution.[19]

and:

The general rule is that in the absence of statute, a private person or corporation is not responsible for the acts of a special police officer, appointed by public authority, but employed and paid by the private person or corporation, when the acts complained of are performed in the carrying out of his duty as a public officer. The acts of such person are, *prima facie*, those of a public officer, not rendering the private employer liable. It does not follow, however, that because a servant is also a police officer, all his acts are of a public nature; and where he is acting in the performance of the duties for which he is employed or his movements are actively directed by his employer . . . such employer may become liable for his acts.[20]

Whereupon the evidence presented a conclusion that the police officer is acting as the servant of the defendant employer cannot be reasonably supported, it becomes a question of law for the court to decide.[21] Generally, however, the question of whether an officer is acting in an official capacity or within the scope of his employment as a private servant is a question for the jury.[22]

LIABILITY

The controlling factor is whether the duty which the officer was performing was for the pecuniary benefit of his employer, either local or special, or for the benefit of the public generally. Thus, it has been held that a conductor's duty is essentially that of a private and not public nature,[23] that a police officer while off-duty and working as a janitor cannot have a wrong done by him charged to his official character.[24] Furthermore, the difference between liability and the want of it turns upon the proposition as to whether the agent officer carelessly performs what he was employed to do in his private capacity, or as a free moral agent, turns aside, for his own purposes and amusement, and consciously does something not within the scope of his employment.[25] This is not altered because the agent later "superadds" malice or hatefulness to his otherwise wrongful act.[26] The distinction to be drawn between acts properly within the officer's duty, either public or private, and those which are particularly the product of his own intent can be narrow indeed. It is ordinarily a question of fact for the jury as to whether such lawful employment has ceased at the time of the particular event upon which liability is sought to be placed.[27]

Public carriers have been held to a strict standard in controlling special police in their employ.[28] Officers assigned to preserve order in public waiting rooms,[29] stores,[30] amusement parks,[31] drinking establishments,[32] dance halls,[33] and theaters,[34] are usually said to be performing private policing functions for which their employers are responsible. Likewise, a patrolman called in *off the street* into a private establishment to aid in enforcing regulations of a theater becomes the cinema's agent.[35] Not so, however, if in entering the theater he discovers the parties violating the criminal law or an ordinance of the city, for he then arrests as a public officer.[36] Similarly, the policing of a store and the protection of a merchant's goods from shoplifting and burglary involves more than the performance of a public duty. The officer is acting under the control and direction of a private employer.[37] The above holds true even though it has been held that the mere payment of salary, as well as the equipping and maintenance of a force by private money is *not* controlling.[38] These factors do, however, give evidence of control and direction by the employer.

Furthermore, where a special policeman employed to protect private property leaves the premises in trying to apprehend a person suspected of stealing does not exonerate the private employer for assault committed by the officer without the property.[39] In *MacDonald v. Ogan*,[40] however, this factor was thought important by the court: "[the company is not liable] . . . where the acts were committed on property which does not belong to the company, where the acts were not done for the preservation of company property, and where the company did not direct and was in no way connected with the acts of the officer."

The better reasoned opinion should consider the place of the act as only one factor in reaching a decision as to whether the officer was engaged in private employment.

Another important consideration once liability has attached to the private employer involves the nature of the business and environs patrolled. Although mitigation has not found widespread acceptance by the courts in such instances, it is not difficult to sympathize with private employers who must control large crowds, drinking, or dancing. Although these types of establishments exist as a matter of privilege, and not right,[41] and are looked upon suspiciously by the law, their social value permits their existence. In the words of one dissenting judge:

> [Supervising drinking establishments] ... makes the task of these quasi-judicial officers who are hired to maintain peace and order rather hard and difficult. Unconsciously they may exceed the authority which they have under the law. When they do their employer becomes liable for their actions. However, these are all circumstances which should be taken into consideration in mitigation of the damages to be paid.[42]

NEGLIGENCE

The law does not charge a person with all the possible consequences of a negligent act. Negligence carries with it liability for consequences which should reasonably have been anticipated by prudent men.[43] In many states the rule is to effect that the acts of a special policeman are presumed to have been done reasonably and in his capacity as a public officer.[44] Such a presumption shields the private employer from certain legal burdens. Thus, the injured party must come forward with some proof giving rise to an inference that the officer's wrong was committed in the course of the performance of his duty to his private employer, rather than in his official capacity. In *Jacoby v. Southern Pac. Ry. Co.*,[45] involving an assault by a railroad agent, plaintiff typically had the burden to prove, not only the existence of an assault by the agent, but also that the agent was in the employ of the railroad and made the assault while working within the scope of his employment. The foundation for the presumption that a special officer acts in his public capacity is that "one who is invested with authority by the sovereign, commissioned and sworn to faithfully perform the duties pertaining to such commission, must necessarily be supposed to be acting in conformity thereto; anyone who claims that the officer was not so acting must show affirmatively that such was the case."[46] To hold otherwise, to reverse the presumption, so to speak, would give a negative effect to such commission making the bestowal of power by a public agency no better than the master-servant counterpart.[47]

The fact that the special police officer was commissioned by the state is often viewed as protecting his private employer from liability for his acts. This statement of the law must be taken cautiously however. As we have mentioned, much depends on the nature of the act and the propensity of a jury to find either a public or private capacity.[48]

When an act ceases to be one of service to the employer, and becomes one of vindicating the public right or justice, the officer may be said to be acting as a police officer in the public interest.[49] Thus, the enforcement of the criminal law or municipal ordinance becomes a primary test in ascertaining the nature of the officer's conduct. The social utility of his conduct is measured in this term. The fact that he is paid by his employer is, again, no evidence that the officer is not *prima facie* a public officer.[50]

While serving as a public officer the private employer is not exposed to liability for resulting injuries. Thus, special officers serving on a posse,[51] directing traffic for a private party on the officer's day off, but while in uniform and under orders from his superior,[52] have been held to be public functions. In *Kidder v. Whitney*[53] a suit involving a special officer assigned to direct traffic at a drive-in theater, the court remarked:

> [the officer] . . . was a special policeman. . . . He was appointed by the selectmen. . . . He was not under civil service. The town lent him a few pieces of equipment. He produced his own uniform. He was paid for services by the defendant. Whatever instructions he received in directing traffic was given to him by the chief of police, who observed him each week to see how he was performing or had some officer do it. As far as the record goes [the officer] . . . never received any instruction or direction concerning traffic from the defendant. They left that to his judgment and discretion. He sometimes sold tickets but that was not his job. . . .

Although privately employed[54] and equipped, the court held for public employment on the ground that the defendant theater did no more than narrate the material facts to the officer (where to direct traffic and at what hours) [55] leaving him to decide what to do. Discretion precedes responsibility. In this instance it was exercised by the individual officer in conjunction with the police. Similarly, when discretion is lodged exclusively in the hands of public officials and not the private employer or policeman, the courts have usually found the private employer protected from liability:

> It is true that the defendant asked to have an officer appointed . . . and that he paid him. But he did not appoint him, and could not control his official conduct, which was governed by the regulation of the police commissioners, and own sense of duty as a public officer. Over his official acts, the company had no control.[56]

If the power to control rests entirely with the individual (private detective), he is then an independent contractor. Similar to the acts of a private officer performing a public function, the "contractor" frequently insulates the employer from direct liability for his acts. He stands between the employer and the injured party and in so doing himself becomes liable for his negligent acts. Policemen hired to assist in ceremonies celebrating the opening of a new supermarket, and who were given great discretion as to how to keep things "running smoothly" were independent contractors and not servants of the employer.[57] Special police officers, in order to bind their employers to tortious acts, must not act with such discretion as to warrant little or no control over them.[58]

STATUTES

All but a few states have resorted to statutory controls in the area of the special police.[59] The power to create a special police officer exists in the state, which appoints through the office of the governor upon application. As one might expect there is little consensus exhibited over the various provisions in this area. Such factors as the methods of control, licensing, as well as the scope of power vary widely according to jurisdiction. Many states define the activity of the special police in terms of specific industries such as rail, common carriers, and public utilities, while others do not restrict the application of the term to define industries. An example of the latter is the West Virginia code which utilizes a rather broad, comprehensive approach:

> No person, firm, company, partnership or corporation shall engage in the business of private detective or investigator or the business of watch, guard or patrol agency for the purpose of furnishing guards, patrolmen, or other persons to protect persons or property or to prevent the theft or the unlawful taking of goods . . . or in the business of furnishing . . . information as to . . . personal character or activities of any person . . . company . . . for fee, hire, or reward . . . without having first obtained from the office of the secretary of state a license to do so. . . .[60]

This rather broad grant through the licensing power has an interesting, and unusual, proviso at the end of the section:

Provided . . . that this section shall not apply to any person who is a duly qualified and acting police officer under the laws of the State of West Virginia, either while acting in his official capacity or while working for a private employer in his off-duty hours; nor to any person, corporation . . . whose business is the furnishing of information to any employer concerning the business activities of his employees while on the premises of such employer.[61]

Here the legislature has affirmed the basic common law duality of power so prevalent in the court decisions of the first half of the Twentieth Century.

Confusion has been fostered in a number of jurisdictions by the inclusion of the category private detective within the special police statutes (and vice versa). Thus a rather typical statute reads: "The term 'private detective business' shall also mean the furnishing for hire or reward of watchmen or guards or private patrolmen or other persons to protect persons or property, either real or personal, or any other purpose whatsoever. . . ."[62] Legislative bodies should closely scrutinize the actual and potential differences existing between the two entities. Thus, questions may be asked as to whether the status of the private detective differs significantly from the role and legal relations characterized by the special police. Or whether private detectives are, or should be, deputized as public officers. In this regard their secondary role as law enforcers might become a good deal more distinct than is now the case.

Similar to statutory requirements governing "regular police" operations, special policemen are required in nearly all jurisdictions to take an oath, carry a badge, or wear a uniform while on duty. While acting in such capacity the State is absolved from liability.

FOOTNOTES

1. "A servant is a person employed by a master to perform service in his affairs whose physical conduct in the performance of the service is controlled or is subject to the right to control by the master. This court has stated that the right of control and not necessarily the exercise of that right is the test of the relation of master and servant. Basically, it is the distinction between a person who is subject to orders as to how he does his work and one who agrees only to do the work in his own way." *Graalum v. Radisson Ramp*, 245 Minn. 54, 71 N.W.2 904, 908 (1955).
2. "His duties as an officer of the association and as a police officer of the city were not inconsistent, and both duties could be performed simultaneously. . . ." *Waddington v. Stores Mut. Protective Ass'n*, 44 Ga. 826, 163 S.E. 313, 314 (1932).
3. Restatement (Second) of Agency S.2 (1958).
4. *Graalum v. Radisson Ramp*, p. 904.
5. *J. J. Newberry Co. v. Smith*, 227 Ala. 234, 149 So. 669, 671 (1933).
6. *Neallus v. Hutchinson Amusement Co.*, 126 Me. 469, 139 A.2 671 (1927).
7. *Ex Parte Newberg*, 143 Tex. Crim. 211, 143 S.W.2 786 (1940).
8. *Tannenbaum v. Rehm*, 152 Ala. 494, 44 So. 532, 533 (1907): "The ordinance cannot be said to be unreasonable in that the city assumes to designate the man to perform the particular service, or impose the cost of such service upon the manager. The duty of protecting the person or citizen from dangers of fire in the exercise of the police power would seem to carry with it the right to employ the most effective means to that end, and this would include the right of designating competent agents or servants for the performance of such services by firemen on the manager of the theater . . . [has been held]. *City of New Orleans v. Hop Lee*, 104 La. 601, 29 So. 214 (1901), *Harrison v. Baltimore*, 1 Gill (Md.) 264 ( ): "We are of the opinion that the ordinance in question was clearly within the police power of the municipality, and that it is not unreasonable."
9. *County Ballroom v. Bain*, 211 S.W.2 248 (1948).
10. 55 A.L.R. 1198, 1199: Includes a complete listing of citations for this proposition of law.
11. *O'Quin Baptist Mem. Hospital*, 201 S.W.2 694 (1947): "To hold that every citizen

or corporation who calls in the assistance of the police department for quieting disturbances and protecting life and property makes the police officers his or its agent, and liable for the acts of the police officers, would be to seriously impair the usefulness of the police department and to further the cause of violence."

12. " 'Misfeasance' by a public officer is a crime, and is the performance of a discretionary act with an improper or corrupt nature." *State v. Matushefake,* 215 A.2 443, 448 (1965).

13. " 'Malfeasance' such as might constitute misconduct in office is performance of that which an officer has no authority to do and which is positively wrong." *Lowhorn v. Robertson,* 226 P.r 1008, 1015 (1954).

14. " 'Nonfeasance' is substantially failure to perform a required legal duty. . . ." *Schumacker v. State ex rel Finlony,* 78 Nev. 167, 370 p.^{2} 209, 211- (1962). *New York C. & St. Louis Ry. Co. v. Fieback,* 87 Ohio St. 254, 100 N.E. 889, 891 (1912).

15. *Stokes v. Hansberry,* 314 Ill. App. 195, 40 N.E.2 823 (1949).

16. *Hayes v. Sears, Roebuck Co.,* 209 P.2 468, 478 (1949): "If he acts maliciously or in pursuit of some purpose of his own, the defendant is not bound by his conduct, but if, while acting within the general scope of his employment, he simply disregards his master's orders or exceeds his powers, the master will be responsible for his conduct."

17. *McCain v. Baltimore & O. R. Co.,* 65 W. Va. 233, 64 S.E. 18 (1909).

18. *Fournier v. Churchill Downs-Latonia,* 292, Ky. 215, 166 S.W.2 38 (1942).

19. *Maggi v. Pompa,* 287 p. 982 (1930).

20. *Wheatley v. Washington Jockey Club,* 234 P.2 878 (1951), quoting from 22 Am. Jur. 385 (False Imprisonment) s. 45.

21. *Krowka v. Colt Patent Fire Arm Mfg. Co.,* 125 Conn. 705, 8 A.2 5 (1939). See also, *Bradlow v. American Dist. Telegraph Co.,* 131 Conn. 192, 38 A.2 679, 680 (1944): ". . . occasion might arise in which the servant is so clearly without the scope of his authority that the question is one of law."

22. *Sharp v. Erie R. Co.,* 184 N.Y. 100, 76 N.E. 923 (1906): *Jefferson v. Yazoo & M.V.R. R. Co.,* 11 So.2 442, 443 (1943): "The consensus of authority, and the rule in this court, is, the fact that a servant's conduct was unauthorized does not bring it within the scope of his employment, provided it is of the same general nature as that authorized, or incidental to the conduct authorized."

23. *Southern R. Co. v. Grubbs,* 115 Va. 876, 80 S.E. 749 (1914). *Contra, Healey v. Lothrop,* 171 Mass. 263, 50 N.E. 540 (1898).

24. *Dickson v. Waldron,* 135 Ind. 507, 34 N.E. 506 (1893).

25. *Texas Breeders v. Racing Ass'n Blanchard,* 81 F.2 382 (1936).

26. *United States Steel Co. v. Butler,* 69 So.2 685 (1953).

27. *Biniewski v. City of New York,* 267 App. Div. 108, 44 N.Y. S.2 543 (1943).

28. 55 A.L.R. 1210.

29. *Walters v. Stonewall Cotton Mills,* 136 Miss. 361, 101 So. 495 (1924).

30. *Perkins Bros. Co. v. Anderson,* 155 S.W. 556 (1913).

31. *Rice v. Harrington,* 38 R.I. 47, 94 A. 736 (1915).

32. *Moore v. Blanchard,* 35 So.2 667 (1948).

33. *Bounty Ballroom v. Bain,* 211 S.W.2 248 (1948). Here the operator of a dance hall was held liable for the actions of a private officer even though he did not direct, participate or know about the assault until after it had happened.

34. *Dickson v. Waldron,* 35 N.E. 1 (1893). The manager of a theater is responsible for the acts of a special policeman, who was appointed for the theater, at the special request of the manager, by the board of metropolitan police commissioners, and who was employed and paid solely by such manager. *McChristian v. Popkin,* 15 Cal. App.2 249, 171 P.2 85, 89 (1946): "So far as such employees might make arrests for the commission of crimes upon the premises they were acting as police officers, but in checking ticket stubs, restraining unseemly, rude or boisterous conduct, and regulating good order in and about the theater, such employees were acting as agents of the theater owners and not as police officers." *Neallus v. Hutchinson Amusement Co.,* 126 Me. 469, 139 A. 671 (1927).

35. *Dickson v. Waldron,* 135 Ind. 507, 34 N.E. 506 (1893).

36. *Dickson v. Waldron.*
37. *J. J. Newberry Co. v. Smith,* 149 So. 669, 671 (1933).
38. *J. J. Newberry Co. v. Smith.*
39. *Jefferson v. Yazoo & M.V.R. R. Co.,* 194 Miss. 729, 11 So.2 442 (1943).
40. *MacDonald v. Ogan,* 129 P.2 654, 657 (1942).
41. *Bounty Ballroom v. Bain,* 211 S.W.2 248 (1948): "It is the general rule . . . that public dance and dance halls are proper subjects of regulation under the general police power of the state because of the frequency with which they are attended at disorder, disturbances and breaches of the peace and it has been judicially determined that the City of Dallas, operating under a Home Rule charter, has authority to regulate public dance halls."
42. *Moore v. Blanchard,* 35 So.2 667, 671 (1948).
43. *Smith's Adm. R. v. Corder,* 286 S.W.2 512 (1956).
44. *Red River Lumber Co. v. Cardenas,* 95 F.2 157 (1938).
45. 97 S.W.2 515 (1936).
46. *New York C. & St. Louis Ry. Co. v. Fieback,* 87 Ohio St. 254, 100 N.E. 889, 891 (1912).
47. *Norfolk & W. Ry. Co. v. Haun,* 167 Va. 157, 187 S.E. 481, 485 (1936): "To hold that the special police agent was . . . acting as a private servant of the defendant in making the arrest, would render the provision of the statute clothing such officer with police authority nugatory, and make it as perilous for an employer to engage a person duly commissioned a police officer to investigate violations of the criminal law as to employ a private citizen for that purpose."
48. 55 A.L.R. 1197, 1204, 1205.
49. *McKain v. Baltimore & O. R. Co.,* 65 W. Va. 233, 64 S.E. 18 (1909).
50. *Thompson v. Norfolk & W. Ry. Co.,* 116 W. Va. 705, 182 S.E. 880, 883 (1935).
51. *Thompson v. Norfolk & W. Ry. Co.*
52. *Yates v. City of Salem,* 174 N.E.2 368, 369 (1961): "He was injured while in uniform, performing the work of a police officer, by assignment of his superior officer. He was therefore injured in the performance of his duty. In its performance he was not acting in the capacity of employee of the contractor."
53. 336 Mass., 307, 145 N.E.2 684, 685 (1957).
54. *MacDonald v. Ogan,* 129 P.2 654, 657 (1942): ". . . that even where a company procures the appointment and pays the wages of a special . . . officer, it is not liable for his negligent acts, committed in the performance of his public duty. . . ."
55. *St. John v. Reid,* 17 Cal. App.2 5, 61 P.2 363 (1936).
56. *MacDonald v. Ogan,* p.656.
57. *Luz v. Stop & Shop, Inc. of Peabody,* 348 Mass. 198, 202 N.E.2 771 (1964).
58. *Komorowski v. Boston Store of Chicago.* "Under a contract between a department store and a detective agency by the terms of which the agency is to furnish two detectives of its own selection who shall be under the sole control of the agency, and who are authorized to arrest persons upon their own judgment or the judgment of an officer, agent, or employee of the store . . . the detectives are employed by the agency and not by the store. . . ."
59. Statutes governing "private detectives" have been included where they are part of "special police" sections, or, where they represent the only provisions relating to private police. Code of Ala. 1958-55 S.376 et seq.; Ark. Stats. 1947 Anno. S.42-109, 110, 111; West's Anno. Cal. Codes (Gov.) 1400-1413 et. seq.; Colo. Rev. Stats. 1963 99-2-1 et. seq.; Conn. Gen. Stats. Anno. 29-18-to-29-22, 22-124, 7-91, 7-313, 25-44, 26-206, 23-18; Del. Code Anno. 24 S. 1301 et. seq.; Distr. of Col. Cod 1961 6-1006, 4-115, 4-112, 4-133; Fla. Stats. Anno. 354.01 to 354.07 et. seq.; Code of Georgia Anno. S.26-4901-02; Rev. Laws of Hawaii 1955 S.31-6, 165A-; et. seq.; Idaho Code 18-711, 712, 23-804; Smith-Hurd Ill. Anno. Stat. 24S.11-1; Burns Ind. Stat. Anno. 48-6108; Iowa Code Anno. 80A.2, 4, 5, 6, 9, 10, 12, 748.3; Kentucky Rev. Stats. 61360; West's Louisiana Stats. Anno. 3: 2391, 18:376; Maine Rev. Stats. Anno. 5 S.1773, 17 S.3282, 34 S.93; Anno Code of Md. 1957 41 S.61-to-70, 23 S.342-to-348; Anno. Laws of Mass. 147:9-to-10 (E), 149:176; Mich. Stats. Anno. 18.172 et. seq.; Minn. Stats. Anno. 203.42-43, 418.12; Miss. Code 1942 Anno. 3980.5, 4909, 9696-65; Vernon's Ann. Missouri Stats. 73.110, 74.127, 75.110, 84.340, 84.-720, 85.220, 319.010, 319.020, 562.190,

562.200; Rev. Code of Montana 1947 Anno. 94-3920; Rev. Stats. Neb. 1943 28-725-727; Nev. Rev. Stats. 648.010 et. seq.; New Hamp. Rev. Stat. Anno. 1955 (supp.) 4:1, 381:1 et. seq.; N.J. Stats. Anno. 2A:151-43, 15:11-16 et. seq., 40: 47-14, 19, 30:4-14; New Mex. Stats. 1953 Anno. 14-17-2, 14-21-24, 14-42-8, 40A-7-2, 45-20-4, 69-2-20; McKinney's Consol. Laws of New York Anno. (Unconsol. 9107, 9133). (Penal 382). (Gen. Mun. 207 -g), (Towns 158). (Village 189 -a), (Trans. Corp. 30), (Mental Hyg. 34 (4); Gen. Stats. of N.C. 1962 60-83 to 60-87; 66-49.1 to 66-49.8, 74A-; to 74A-6, 81-12, 122-33.34, 98, 129-4, 147.- 15.1; North Dak. Century Code Anno. 43-30-01 et. seq., 49-17-10 et. seq.; Page's Ohio Rev. Code Anno. 3771.01 et. seq., 49-17-10 et. seq.; Page's Ohio Rev. Code Anno, 3771.01 et. seq., 4973.17 et. seq.; Okla. Stats. Anno. 51 S.16, 17, 18, 66 S. 183; Oregon Rev. Stats. 51.260, 142.070, 148.101, 110, 120, 150, 160, 170, 180. 162.570, 180.090, 433.125, 210, 220, 449.-315, 565.240, 565.640; Purdon's Penna. Stats. Anno. 10 S.171, 3, 4, 5, 6, 16 S. 7516, 18 S.4672, 35 S.1201, 2, 3, 53 S. 46121, 125, 127, 53 S.46230, 53 S.56416; Gen. Laws of R.I. 1956 12-2-1 et. seq.; Code of Laws of South Carolina 1962 S53-3, 53-6, 56-641, 56-646; Tenn. Code Anno. 6-2130; Vernon's Tex. Stats. S. 995, PC 484; Utah Code Anno. 1953 67-12-1 et. seq.; Vermont Stats. Anno. S.24-1936-1937; Code of Va. 1950 S.15-562, 15-574.4, 15-570; 19.1-28, 19.1-30, 31, 32, 33-218-20, 52-23, 52-24, 56-330, 56-445-56; Wyo. Stats. S.7-16, 17, 18, 17-134; West's Wisc. Stats. Anno. 134.58, 175.08; W. Va. Code S.30-18-1 to 30-18-8.

60. West Virginia Code 30-18-1.
61. West Virginia Code 30-18-1.
62. Delaware Code Anno. 24 S.1301.

APPENDIX 3B

LEGAL GUIDELINES FOR SECURITY OFFICERS*

This Appendix provides a listing of Case Citations treating the following Protective Services problem areas:

ARREST
ASSAULT
DEFAMATION
EMOTIONAL DISTRESS
EMPLOYMENT, SCOPE OF
PHYSICAL EVIDENCE
FALSE IMPRISONMENT
FORCE, USE OF
MALICIOUS PROSECUTION
PRIVACY
POLYGRAPH
SEARCH
SHOPLIFTING
SPECIAL POLICE
WARNINGS (MIRANDA RULE)

* The materials presented in this Appendix are extracted from *Private Police and the Law*, Richard A. Rifas (Wilmette, Ill., 1973). A complete treatment of these cases and related materials can be found in this publication. This extract is presented courtesy of Richard A. Rifas.

ARREST
What Constitutes an Arrest

The elements in all states of what constitutes an arrest (lawful or unlawful) are:

1. Intent to take the person into the custody of the law. It is intended that the person go to trial for his actions.
2. An arrest is made under authority of law (either as peace officer or as a private citizen). It is here that state arrest for felonies committed in his presence, while in other states a private person can arrest for either misdemeanors or felonies committed in his presence. An arrest made by a private person for a crime for which he has no authority to arrest is an arrest if the other elements are present, but is an unlawful arrest, and subjects the officer to a civil suit for false arrest.
3. Seizure of the person being arrested. Either actual seizure of the person by the use of physical force or constructive seizure—the person

being arrested submits himself to the authority of the officer who intends to take that person into the custody of the law.
4. An understanding by the person who is being arrested that he is being arrested. Understanding may be imparted by words, "You are under arrest." —or by circumstances, "Get into the room and you will stay here until the police arrive." Exception to this element is that a drunk or unconscious person can be arrested even though they do not understand what is going on as long as the other elements are present.

What Constitutes a Detention?

Detention involves investigation of suspicious circumstances or unidentified individuals. It is a stopping to ask questions, "What is your name? What are you doing on this floor?" It is using ones uniform—ones position—ones command presence—to detain a person to find out what is going on. It does not allow the use of physical force to detain a person for investigation (use of physical force in self-defense is obviously allowed, but that is not what is now being considered). If force is used to detain a person, the security officer is subject to possible Civil/Criminal lawsuit for assault and battery. Physical force should not be used except in self-defense or in making an arrest.

Probable Cause to Arrest—How the Courts Define It

Probable cause implies "more than suspicion." Probable cause is judged by facts known at the moment of arrest.

The facts of probable cause vary according to the *modus operandi* of crime. Probable cause must identify the suspect before arrest. Probable cause means the same thing as reasonable grounds.

Sources of Probable Cause to Believe That a Crime Is Being Committed

1. officer's direct observations, that information he receives empirically through one or more of his five senses, including personal corroborative information
2. information from a reporting agent, witness victim, informant, fellow officer, citizen, or other sources such as roll call briefing, police reports, radio communications
3. unusual time involved
4. unusual location involved
5. unusual appearance of person in question
6. unusual activity of person in question such as flight, furtive movements, hiding, attempt to destroy evidence, resistance to officer
7. factors of time and distance involved
8. statements made by the suspected person on his evasive answers or unreasonable explanations
9. hearsay information
10. past record or present conduct of the suspected person
11. type and seriousness of the crime involved
12. experience or training of arresting officer in comparison to that of a lay person
13. emergency nature of the circumstances involved
14. identification of the suspected person by a witness
15. evidence in plain view
16. weapons in plain view
17. fingerprint identification
18. voice identification
19. handwriting identification
20. knowledge that crime has occurred

Assisting Peace Officer

Chapter 38 (Ill. Rev. Statutes) Section 107-8

a. a peace officer making a lawful arrest may command the aid of male person over the age of eighteen

b. a person commanded to aid a peace officer shall have the same authority to arrest as that peace officer

c. a person commanded to aid a peace officer shall not be civilly liable for any reasonable conduct in aid of the officer. A person giving such aid must be duly requested to do so by the officer—a mere volunteer acts at his own peril

Section 31-8

Refusing to Aid an Officer

a. Apprehending a person whom the officer is authorized to apprehend; or

b. preventing the commission by another of any offense, shall be fined not to exceed $100.00.

Section 31-1

Resisting or Obstructing Peace Officer

A person who knowingly resists or obstructs the performance of one known to the person to be a peace officer of any authorized act within his official capacity shall be fined not to exceed $500.00 or imprisoned in a penal institution other than the penitentiary not to exceed one year, or both.

Rights After Arrest

Section 103-1

Rights on Arrest

a. After an arrest on a warrant, the person making the arrest shall inform the person arrested that a warrant has been issued for his arrest and the nature of the offense specified in the warrant.

b. After an arrest without a warrant, the person making the arrest shall inform the person arrested of the nature of the offense on which the arresting is based.

Arrest by Peace Officer

Chapter 38
Section 107-2

A. Peace officer may arrest a person when:

a. He has a warrant commanding that such person be arrested; or

b. he has reasonable grounds to believe that a warrant for the person's arrest has been issued in this state or in another jurisdiction; or

c. he has reasonable grounds to believe that the person is committing or has committed an offense.

Section 107-3

Arrest by Private Person

Any person may arrest another when he has reasonable grounds to believe that an offense other than an ordinance violation is being committed.

Definitions

Chapter 38
Section 107-1

a. A "warrant of arrest" is a written order from a court directed to a peace officer, or some other person specifically named, commanding him to arrest a person.

b. A "summons" is a written order issued by a court which commands a person to appear before a court at a stated time and place.

c. A "notice to appear" is a written request issued by a peace officer that a person appear before a court at a stated time and place.

Method of Arrest

Chapter 38
Sections 107-5

a. An arrest is made by an actual restraint of the person or by his submission to custody.

b. An arrest may be made on any day at any time of the day or night.

c. An arrest may be made anywhere within the jurisdiction of this state.

d. All necessary and reasonable force may be used to effect an entry into any building or property or part thereof to make an authorized arrest.

Issuance of Arrest Warrant Upon Complaint

Chapter 38
Section 109-9

a. When a complaint is presented to a court charging that an offense has been committed it shall examine, upon oath or affirmation, the complaint or any witnesses.

b. The complaint shall be in writing and shall:

1. State the name of the accused if known, and if not known, the accused may be designated by any name or de-

scription by which he can be identified with reasonable certainty;

    2. state the offense with which the accused is charged;

    3. state the time and place of the offense as definitely as can be done by the complaint; and

    4. be subscribed and sworn to by the complaint.

c. A warrant shall be issued by the court for the arrest of the person complained against if it appears from the contents of the complaint and the examination of the complaint or other witnesses, if any, that the person against whom the complaint was made has committed an offense.

d. The warrant of arrest shall:

    1. be in writing;

    2. specify the name of the person to be arrested or if his name is unknown, shall designate such person by any name or description by which he can be identified with reasonable certainty;

    3. set forth the nature of the offense;

    4. state the date when issued and the municipality or county where issued;

    5. be signed by the judge of the court with the title of his office;

    6. command that the person against whom the complaint was made be arrested warrant or if he is absent or most accessible court in the same county; and

    7. specify the amount of bail.

e. The warrant shall be directed to all peace officers in the state. It shall be executed by the peace officer, or by a private person specially named therein, and may be executed in any county in the state.

The following case citations present the major court decisions regarding arrest, particularly as those decisions relate to the private protective services.

Section A: Arrest

Odorizzi v. A.O. Smith Corporation
452 F.2d 229
7th Circuit (Ill/Ind/Wisc)

People v. Kaprelian
6 Ill. App. 3d 1066 (1972)

Potts v. Wright
357 F.Supp. 215 (Pa. 1973)

Tillman v. Holsum Bakeries, Inc.
244 S.2d 681 (La. App. 1971)

Armstead v. Escobedo
488 F.2d 509 (1974)
U.S. Court of Appeals for Texas

United States v. Guana-Sanchez
484 F.2d 590 (1973)
U.S. Court of Appeals for 7th Circuit (Ill/Ind/Wis)

People v. Howlett
Ill. App. 3d 906 (1971)

People v. Mirbell
276 Ill. App. 533 (1934)
Illinois Appellate Court

Green v. No. 35 Check Exchange, Inc.
222 N.E.2d 133
(Ill. App. 1966)

Lindquist v. Friedman's Inc.
8 N.E.2d
(Ill. App. 1937)

People v. Bridges
123 Ill. App. 2d 58 (1970)
Illinois Appellate Court

McWilliams v. Interstate Bakeries
439 F.2d 16 (1971)
U.S. Court of Appeals for Georgia

Monteiro v. Howard
334 F.Supp. 441 (1971)
U.S. District Court for Rhode Island

United States v. Goeden
433 F.2d 430 (1970)
U.S. Court of Appeals for 5th Circuit

United States v. Alexander
415 F.2d 1352 (1969)
U.S. Court of Appeals 7th Circuit (Ill/Ind/Wisc)

People v. Wilkins
104 Cal. Rptr. 89 (1972)
California Appellate Court

People v. Lawson
264 N.W.2d 864
(Ill. App. 1970)

People v. Nelson
260 N.E.2d 251
(Ill. App. 1970)

People v. Lenker
285 N.E.2d 807
(Ill. App. 1970)

United States v. Gonzalez
362 F.Supp. 415 (1973)
U.S. District Court—New York

People v. Duncan
262 N.W.2d 274
(Ill. App. 1970)

Rothschild v. Drake Hotel, Inc.
397 F.2d 419 (1968)
U.S. Court of Appeals
7th Circuit (Ill/Ind/Wisc)

United States v. Strauss
452 F.2d 375 (1971)
U.S. Court of Appeals 7th Circuit (Ill/Ind/Wisc)

Frisbie v. Collins
72 S.Ct. 509 (1952)
United States Supreme Court

People v. Robinson
194 N.W.2d 537
(Mich. App. 1971)
Court of Appeals of Michigan

Section B: Detention

People v. McGarry
294 N.E.2d 718
(Ill. App. 1973)

United States v. Ruffin
389 F.2d 76
7th Circuit (Ill/Ind/Wisc)

ASSAULT

Assault—A person commits an assault when, without lawful authority, he engages in conduct which places another in reasonable apprehension of receiving a battery.

Battery—A person commits battery if he intentionally or knowingly without legal justification and by any means, causes harm to an individual or makes physical conduct of an insulting or provoking nature with an individual.

The following cases set forth the guidelines for officers to follow to prevent liability for assault.

Assault

Levine v. Enlow
462 S.W.2d 50
(Texas 1970)

Silas v. Bowen
277 F.Supp 314 (1967)

Manson v. Wabash Railroad Company
338 S.W.2d 54
(Mo. 1960)

Defamation

Defamation—usually occurs when damage has been caused to a person's reputation in the presence of other individuals.

Defamation may be divided into two classes: slander *per se* and slander *per quod*. In slander *per se*, actual damages need not be proven, whereas, in slander *per quod*, damages and malice must be specifically proven.

A word is *slander per se* if it is obviously injurious; a word is *slander per quod* when an innuendo is required to give the word a slanderous meaning. A word cannot be slander *per se* if the word itself is capable of innocent construction.

In order to render defamation actionable, there must be publication to a third party, other than plaintiff and defendant.

Defamation

Privileged communications involves protection from liability.

Absolute Privilege

Communications absolutely privileged are those where there is no liability even when there is express malice. This situation generally applies to legislative and judicial proceedings.

Additional Privilege

In the absence of malice, falsity of the matter, if presented in good faith, will not subject the person to liability. The existence of express malice will cause the person to lose his conditional privilege.

Criminal Defamation

(Ill. Revised Statutes—Chap. 38)

Section 27.1

Elements of the Offense

a. A person commits criminal defamation when, with intent to defame another, living or dead, he communicates by any means to any person which tends to provoke a breach of the peace.

b. Sentence;
criminal

Section 27.2

Justification

In all prosecutions to criminal defamation, the truth, when communicated with good motives, and for justifiable ends, shall be an affirmative defense.

The following cases describe the hold decisions regarding defamation:

Section I: Defamation

Criminal Defamation

City of Chicago v. Lambert
47 Ill. App. 2d 151 (1964)

Note: Classes of Slander/Libel

Whitby v. Associates Discount Corp.
59 Ill. App. 2d 337 (1965)

Wolfson v. Kirk
273 So.2d 774
(Fla. App. 1973)

Comar v. Greater Niles Twp Pub Corp
13 Ill. App. 2d 267 (1957)

Jones v. Sears, Roebuck and Company
459 F.2d 584 (1972)

Stewart v. Nation-Wide Check Corporation
182 S.E.2d 410
(N.C. 1971)

The People v. Fuller
238 Ill. 116 (1909)

Bradley v. Bakke
306 Ill. App. 569 (194)

Cook v. Safeway Stores, Inc.
511 P.2d 375 (Ore. 1973)
Supreme Court of Oregon

Farnum v. Colbert
293 A.2d 279
(D.C. App. 1972)

McGuire v. Jankiewicz
8 Ill. App. 3d 319 (1972)
Illinois Appellate Court

Smith v. Phoenix Furniture Company
339 F.Supp. 969 (1972)
U.S. District Court for South Carolina

Parker v. Kirkland
298 Ill. App. 340 (1939)

Picard v. Brennan
307 A.2d 833 (Me. 1973)
Supreme Judicial Court of Maine

Shupe v. Rose's Stores, Incorporated
192 S.E.2d 766 (Vir. 1972)
Supreme Court of Virginia

Modla v. Parker
495 P.2d 494 (Ariz. App. 1972)
Court of Appeals of Arizona

Delis v. Sepsis
9 Ill. App. 3d 217 (1971)

Glenn v. Gidel
496 S.W.2d 692
(Tex. Civil 1972)
Court of Civil Appeals of Texas

Great Atlantic and Pacific Tea Co. v. Paul
261 A.2d 731 (Md. 1970)
Court of Appeals of Maryland

Wells v. Shop Rite Foods, Inc.
474 F.2d 838 (1973)
U.S. Court of Appeals for Texas

Rougeau v. Firestone Tire and Rubber Company
274 So.2d 454
(La. App. 1973)
Court of Appeals of Louisiana

Mitchell v. Peoria Journal-Star, Inc.
76 Ill. App. 2d 154 (1966)

Larson v. Doner
32 Ill. App. 2d 471 (1971)

Neece v. Kantu
507 P.2d 447
(N.M. 1973)
Court of Appeals of New Mexico

Cook v. Easy Shore Newspapers, Inc.
327 Ill. App. 559 (1945)

Flannery v. Allyn
47 Ill. App. 2d 308 (1964)

Coursey v. Greater Niles Twp. Pub. Corp.
40 Ill. 2d 257 (1968)

Bridges v. Farmer
483 S.W.2d 939
(Tex. Civ. 1972)
Court of Civil Appeals of Texas

Hooks v. McCall
272 S.2d 925
(Miss. 1973)
Supreme Court of Mississippi

American District Telegraph Co. v. Brink's Incorporated
380 F.2d 131 (1967)

Lotrich v. Life Printing and Pub. Co., Inc.
117 Ill. App. 2d 89 (1969)

Sanders v. Stewart
208 N.E.2d 509
(Ind. 1973)
Court of Appeals of Indiana

Munsell v. Ideal Food Stores
494 P.2d 1063
(Kan. 1972)
Supreme Court of Kansas

Swanson v. Speidel Corporation
293 A.2d 307 (R.I. 1972)
Supreme Court of Rhode Island

Judge v. Rockford Memorial Hospital
17 Ill. App. 2d 365 (1958)

Schlat v. State Farm Mut. Auto Ins. Co.
15 Ill. App. 2d 194 (1954)

Zeinfeld v. Hayes Freight Lines, Inc.
41 Ill. 2d 345 (1968)

Brown v. First National Bank of Mason City
193 N.W.2d 547
(Iowa 1972)

Linn v. United Plant Guard Wkrs. of Amer., Loc 114
86 S.Ct. 657 (1966)

Conrad v. Logan
283 N.E.2d 54 (1972)

John v. Tribune Company
28 Ill. App. 2d 300 (1960)

EMOTIONAL DISTRESS

Any type of disturbance which is actionable and that which causes mental anguish and nervous exhaustion.

Ex: an individual receiving threats on his life, an individual who has been maliciously defamed publicly, an individual being provoked in taking the life of another. In emotional distress, damages are not incurred by physical objects but by mental suffering.

The following court cases describe court ruling on this situation.

Emotional Distress

Knierim v. Izzo
22 Ill. 2d 73 (1961)

Nelson v. Nuccio
131 Ill. App. 2d 261 (1970)

EMPLOYMENT, SCOPE OF

Conduct must be of the same general nature as that authorized or incidental to the conduct authorized.

For scope of employment, the following matters are to be considered:

a. Whether or not the act is one commonly done by such servants;

b. The time, place and purpose of the act;

c. The previous relations between the master and the servant;

d. Whether or not the master has reason to expect that such an act will be done has been furnished by the master to the servant;

e. The extent of departure from the normal method of accomplishing an authorized result.

The question of Scope of Employment is important since the actions of a person may be considered lawful while acting within the scope of his employment but unlawfully if acting outside of it. This is particularly true while acting in a protective capacity for an employer, or as an agent of a security firm.

The following cases outline court decisions on this area of concern.

Employment, Scope of

Gudgel v. Southern Shippers, Inc.
387 F.2d 723
(1967)

Bremen State Bank v. Hartford Accident and Indemnity Co.
427 F.2d 425
(1970)

Ryan v. Associates Investment Co.
297 Ill. App. 544
(1938)

Voytas v. United States
256 F.2d 786 (1959)

Dumas v. Lloyd
286 N.E.2d 566
(Ill. App. 1972)

Gold Mills, Inc. v. Orbit Processing Corporation
297 A.2d 203 (1972)
Superior Court of New Jersey

Nash v. Sears, Roebuck and Company
174 N.W.2d 818
(Mich. 1970)

Gramm v. Armour and Company
132 Ill. App. 2d 1011
(1971)

Pascoe v. Meadowmoor Dairies
41 Ill. App. 2d 52 (1963)

Sixty-Six v. Finley
224 So.2d 381
(Fla. App. 1969)

Weiss v. Furniture in the Raw
306 N.Y.S. 2d 253
(1969)

Herch v. Kentfield Builders, Inc.
189 N.W.2d 286
(Mich. 1972)

Hipp v. Hospital Authority of City of Marietta
121 S.E.2d 273
(Ga. App. 1961)

Stewart Warner Corp v. Burns International Sec. Serv. Inc.
353 F.Supp. 1387
(1973)

Sauer v. Iskowich
80 Ill. App. 2d 202
(1967)

Bonnem v. Harrison
17 Ill. App. 2d 292
(1958)

Daas v. Pearon
319 N.Y.S. 2d 537
(N.Y. 1971)

American Insurance Association v. Smith
439 S.W.2d 418
(Tex. 1969)
Court of Civil Appeals of Texas

Greenbaum v. Brooks
139 S.E.2d 432
(Ga. 1964)
Court of Appeals of Georgia

Kroger Company v. Warren
420 S.W.2d 218
(Tex. App. 1967)

Callaghan v. Harvey
225 Ill. App. 353
(1922)

McMahon v. Chicago City Ry. Co.
239 Ill. 334
(1909)

Carlberg v. Spiegles House Furnishing Co.
178 Ill. App. 424
(1913)

Kovatich v. Ross
230 Ill. App. 330
(1923)

Field v. Kane
99 Ill. App. 1
(1901)

Miller v. Federated Department Stores, Inc.
294 N.E.2d 474
(Mass. App. 1973)

Metzler v. Layton
373 Ill. 88
(1940)

Tuttle v. Forsberg
331 Ill. App. 503
(1947)

Note: Store liable for injury to customer who attempted to stop a theft.

Jacbosma v. Goldberg's Fashion Forum
303 N.E.2d 226
(Ill. App. 1973)

Leach v. Penn-Mar Merchants Association, Inc.
308 A.2d 446
(1973)

Colonial Stores Inc. v. Holt
166 S.E.2d 30
(Ga. App. 1968)
Court of Appeals of Georgia

Nelson v. R. H. Macy and Co.
434 S.W.2d 767
(Mo. 1968)
Missouri Court of Appeals

Drug Fair of Maryland, Inc. v. Smith
283 A.2d 392
(Md. App. 1971)

Gaffney v. William J. Burns Detective Agency Int. Inc.
299 N.E.2d 540
(Ill. App. 1973)

Brien v. 18925 Collins Avenue Corp.
233 So.2d 847
(Fla. App. 1970)

Callum v. Hartford Accident and Indemnity Co.
337 P.2d 259
(Cal. App. 1959)
California Appellate Court

Shannessy v. Walgreen Co.
324 Ill. App. 590
(1945)

Awe v. Striker
263 N.E.2d 345
(Ill. App. 1970)

Buckley v. Edgewater Beach Hotel Co.
247 Ill. App. 239
(1928)

PHYSICAL EVIDENCE

Anything that can be held or seen, e.g. knife, gun, and fingerprints, may be considered to be physical evidence.

The collection of evidence at a crime scene or the preservation of evidence obtained as the result of an arrest must be properly done. The maintenance of a chain of custody for the evidence, proper handling, marking, storing and receipting are all essential for a successful prosecution. The following cases detail the court decisions on the handling of evidence.

Scherber v. State of California
86 S.Ct. 1826
(1966)

United States v. Dionisio
93 S.Ct. 764 (1973)

United States v. Mara
93 S.Ct. 774 (1973)

United States v. Wade
87 S.Ct. 1926 (1967)

Cupp v. Murphy
93 S.Ct. 2000 (1973)

People v. Ardella
276 N.E.2d 302 (Ill. 1971)

United States v. Roberts
481 F.2d 802 (1973)

Pan American World Airlines, Inc., Guided Missiles Range Division, Patrick Air Force Base and International Brotherhood of Teamsters, Chauffeurs, Warehousemen and Helpers of America, Local 172
62-1 ARB 8311 (1962)

Colgate-Palmolive Company, Berkeley Plant and International Longshoremen's and Warehousemen's Union, Warehouse Union Local 6
50 LA 441 (1968)

FALSE IMPRISONMENT

False imprisonment is defined as the unlawful restraint of an individual's liberty or freedom. There must be an intent to restrain the person, but malice is not required. It is the intent, not the malice or absence of malice, that constitutes false imprisonment.

Actual physical force is not required. Detention may be accomplished by the threat of physical force.

There is no false imprisonment based on legal process such as a warrant, even if the judge erred in his judgment in issuing the warrant.

False Imprisonment—False Arrest

False imprisonment based on false arrest —*Ex.:* private police officer makes an arrest for municipal ordinance violation (but only suspicion), that the person is committing an offense and the person is found guilty, etc.

False Imprisonment—Consent

Consent is a defense to false imprisonment—either actual or implied: "Yes, I will go with you to the manager's office." (no objection)

Consent under force, duress, or fraud is *not* a valid consent. The fact that the arrested person is found guilty is not a defense to an action for false imprisonment.

The following cases detail the guidelines under which courts have determined the actions of private officers in specific situations to be legal and acceptable or illegal and culpable.

False Imprisonment

Hanna v. Raphael Weill and Company
203 P.2d 564
District Court of Appeal of California
(Cal. App. 1949)

Gilmer v. Playboy Club of Denver, Inc.
513 P.2d 1965
(Colo. App. 1973)

Martinzef v. Sears, Roebuck and Co.
467 P.2d 37
(N.M. 1970)
Court of Appeals of New Mexico

Webbier v. Thoroughbred Racing Protective Bur., Inc.
254 A.2d 285
(R.I. 1969)

Cox v. Rhodes Avenue Hospital
198 Ill. App. 83
(1916)

Felton v. Coyle
238 N.E.2d 191
(Ill. App. 1968)

Hoffay v. Stanely Department Stores
328 N.Y.S. 2d 798 (1971)

Alvarez v. Reyonold
181 N.E.2d 616
(Ill. App. 1962)

Lucker v. Nelson
341 F.Supp. iii
(1972)

Aldridge v. Fox
348 Ill. App. 96 (1952)

Moon v. Sperry and Hutchinson Co.
465 S.W.2d 330
(Ark. 1971)

False Imprisonment: Malicious Prosecution

Hickox v. J. B. Morin Agency, Inc.
272 A.2d 321
(N.W. 1970)

City of Peoria v. Underwriters at Lloyd's of London, Inc.
290 F.Supp. 890
(1968)

Hughes v. New York Cent System
20 Ill. App. 2d 224
(1959)

Winans v. Congress Hotel Co.
227 Ill. App. 276
(1922)

Pinkerton v. Martin
82 Ill. App. 590 (1898)

People v. Powers
295 F.Supp. 924 (1969)

FORCE, USE OF

Use of force means no more force than is necessary to accomplish the arrest or defense.

Most states allow physical force to be used when:

1. Reasonable physical force allowed to make a lawful arrest. (Most states allow private person to make lawful arrest for felonies.)
2. Reasonable physical force is allowed to effect trespassers without making an arrest (very minimum force).
3. Reasonable physical force is allowed to protect a third party from attacks or injury.
4. Reasonable physical force is allowed in self-defense.

The following decisions delineate the guidelines for the use of force by private protective services officers.

Schnepf v. Grubb
261 N.E.2d 47
(Ill. App. 1970)

People v. Speed
284 N.E.2d 636
(Ill. 1973)

People v. Joyner
278 N.E.2d 756
(Ill. 1972)

People v. Taylor
279 N.E.2d 143
(Ill. App. 1972)

People v. Odum
279 N.E.2d 12
(Ill. App. 1972)

People v. Schwartz
297 N.E.2d 671
(Ill. App. 1973)

People v. Galmore
283 N.E.2d 105
(Ill. App. 1972)

People v. Scott
285 N.E.2d 476
(Ill. App. 1972)

People v. Morgan
252 N.E.2d 730
(Ill. App. 1969)

People v. Smith
88 N.E.2d 444
(Ill. 1949)

People v. Adams
291 N.E.2d 54
(Ill. App. 1972)

People v. Thomas
290 N.E.2d 418
(Ill. App. 1972)

People v. Stombaugh
284 N.E.2d 640
(Ill. 1972)

People v. Dillard
284 N.E.2d 490
(Ill. App. 1972)

MALICIOUS PROSECUTION

Malicious prosecution—A suit for malicious prosecution may be maintained where either a civil or criminal lawsuit was instituted by a person, without probable causes; there was malice in instituting the proceedings; termination of the original lawsuit was in the plaintiff's favor; and the plaintiff suffered injury or damage as a result of that person instituting the initial lawsuit.

Probable cause—Facts as will warrant a reasonable and prudent man in believing that his original action in bringing the lawsuit is just and proper.

A person who, in good faith, submits to an attorney and initiates a civil lawsuit acting on the attorney's advice has a complete defense to an action of the person to win his lawsuit, does not establish a lack of probable cause.

Lack of probable cause is essential for maintaining an action for malicious prosecution. If there was probable cause to institute the original lawsuit—this is a complete and absolute defense for malicious prosecution—the existence of probable cause is a defense no matter how much malice or spite started the original lawsuit. Probable cause cannot be inferred from the showing of malice—it must be proven to have existed at the time of the filing of the original lawsuit.

Civil proceedings—An action for malicious prosecution of a civil proceedings can only be maintained where the action is begun by arrest of the plaintiff or the seizure of his property—there is no cause of action where the lawsuit is started with a summons and there is no arrest or seizure of plaintiff's property.

Criminal proceedings—A conviction is evidence of the existence of probable cause, unless such conviction was obtained by fraud, perjured testimony or false evidence. The fact that the conviction was reversed does not show lack of probable cause in the original lawsuit.

Malice—is not an essential element of malicious prosecution—but standing alone, it will not support an action for malicious prosecution.

Malice is not established by fact that the person was acquitted of the criminal charge. Probable cause may not be inferred from evidence of malice.

The following court decisions treat malicious prosecution in sufficient detail to prevent liability for the private protective services officer while performing his duties.

Section J: Malicious Prosecution

Holiday Magic, Inc. v. Scott
4 Ill. App. 3d 962 (1972)

March v. Cacioppo
37 Ill. 2d 235 (1962)

Barrett v. Baylor
457 F.2d 119 (1972)

Freides v. Sani-Mode Mg. Co.
33 Ill. 2d 291 (1965)

De Correvant v. Lohman
84 Ill. App. 2d 221 (1967)

Hughes v. New York Cent. System
20 Ill. App. 2d 224 (1959)

Ammons v. Jet Credit Sales, Inc.
34 Ill. App. 2d 456 (1926)

Zimbon v. 1400 Lake Shore Drive Corp.
8 Ill. App. 2d 554 (1956)

PRIVACY

Privacy—The right to be left alone, free from unwarranted publicity; the right of a person to withhold himself and his property from outside invasion; the right of the person to be let alone from search and intrusion without probable cause. Some examples of invasion of privacy would be:

a. unwarranted search of an automobile without probable cause,

b. unwarranted search of an individual's home without probable cause, or

c. invasion of private property.

The right of privacy is one major area of concern for the private officer. His actions in dealing with this individual right must be beyond reproach. The following court decisions detail some of the more graphic situations and the court established guidelines for action.

Right of Privacy

Forster v. Manchester
189 A.2d 147 (Pa. 1963)

Nader v. General Motors Corporation
255 N.E.2d 765
 (N.Y. App. 1970)

Galella v. Onassis
353 F.Supp. 196
 (1972)

Pinkerton National Detective Agency Inc. v. Stevens
132 S.E.2d 119
 (Ga. App. 1963)

Solder v. Pendleton Detectives
88 S.2d 761
 (La. App. 1956)

LeCrone v. Ohio Bell Telephone Co.
201 N.E.2d 533
 (Ohio App. 1963)

McDaniel v. Atlanta Coca-Cola Bottling Co.
2 S.E.2d 810
 (Ga. App. 1973)

Fowler v. Southern Bell Telephone and Telegraph Company
343 F.2d 150
 (1965)

Note: FBI had right to investigate bank account—was not invasion of right of privacy.

Fifth Avenue Peace Parade Committee v. Gray
480 F.2d 326 (1973)

Note: Right of Privacy/Liability of Employees

Ellenberg v. Pinkerton's Inc.
188 S.E.2d 921
 (Ga. 1972)
Court of Appeals of Georgia

Markham v. Markham
272 S.2d 813
 (Fla. 1973)
Supreme Court of Florida

Eric v. Perk Dog Food Co.
347 Ill. App. 293 (1952)

Leopold v. Levin
45 Ill. 2d 434 (1970)

Earp v. City of Detroit
167 N.W.2d 841
 (Mich. 1969)

Pearson v. Dodd
410 F.2d 701 (1969)

Thomas v. General Electric Company
207 F.Supp. 792
 (1962)

Gregory v. Bryant Hunt Co.
174 W.2d 510
(Ky. App. 1943)

Yoder v. Smith
112 N.W.2d 862
(Iowa 1962)

Harrison v. Humble Oil and Refining Company
264 F.Supp. 89
(1967)

Norris v. Moskin Stores Inc.
132 S.2d 321
(Ala. 1961)

POLYGRAPH

Polygraph is the test used as an indication to determine whether an individual is lying or telling the truth. This is usually based upon the operator's test results. In many states, the operator is allowed to testify or the company is allowed to use any confession or admission the subject makes while under the psychological compulsion of the machine or an anticipation of taking the polygraph test.

COURT DECISIONS. The courts have heard evidence in cases involving the polygraph test. The courts have questioned the use of this test in terms of the value and reliability of the results of the test.

The evidence includes the following:
1. the basic theory of the polygraph,
2. the reliance on the polygraph by government agencies,
3. the reliance on the polygraph by private industry,
4. the comparative reliability of the polygraph and other scientific evidence such as fingerprint and ballistic evidence, and
5. the opinions of the experts as to whether polygraph evidence would be a valuable aid in connection with the determination of the issue as the one facing the court in this case and in the administration of justice.

The following court decisions have held both for and against the use of the polygraph in a variety of situations and applications.

Polygraph Test

United States v. Ridling
350 F.Supp. 90 (1972)

State v. McDavitt
297 A.2d 849
(N.J. 1972)

A. v. B.
336 N.Y.S. 2d 839 (1972)

Walther v. O'Connell
339 N.Y.S. 2d 386 (1972)

United States v. Hart
344 F.Supp. 522 (1971)

State v. Towns
301 N.E.2d 700
(Ohio App. 1973)
Court of Appeals of Ohio

Polygraph Test/Inadmissable

United States v. Wilson
361 F.Supp. 510 (1973)

People v. Nicholls
42 Ill. 2d 91 (1969)

People v. Hill
302 N.E.2d 373
(Ill. App. 1973)

People v. Zazzetta
189 N.E.2d 260
(Ill. 1963)

People v. Potts
220 N.E.2d 251 (1966)

People v. Boney
192 N.E.2d 920
(Ill. 1962)

People v. Seipel
247 N.E.2d 905
(Ill. App. 1969)

People v. Nimmer
85 N.E.2d 249
(Ill. 1962)

People v. Nelson
210 N.E.2d 212
(Ill. 1965)

People v. Schneemilch
213 N.E.2d 50
(Ill. App. 1965)

People v. McVet
287 N.E.2d 479
(Ill. App. 1972)

People v. Ackerman
269 N.E.2d 737
(Ill. App. 1971)

People v. Brown
269 N.E.2d 735
(Ill. App. 1971)

People v. Taylor
285 N.E.2d 489 (1972)

People v. Flowers
152 N.E.2d 838
(Ill. 1958)

People v. Triplett
226 N.E.2d 30
(Ill. 1967)

People v. Parisie
287 N.E.2d 310
(Ill. App. 1972)

People v. Melquist
185 N.E.2d 825
(Ill. 1962)

Polygraph

People v. Bernett
258 N.E.2d 793
(Ill. 1970)

People v. Stacey
184 N.E.2d 866
(Ill. 1962)

People v. Hill
212 N.E.2d 259
(Ill. App. 1965)

Potheimer v. Poetheimer
180 N.E.2d 356
(Ill. App. 1962)

Austin v. City of East Moline Bd. of Fire and P. Com'rs.
288 N.E.2d 113
(Ill. App. 1972)

Coursey v. Board of Fire and Police Commissioners
234 N.E.2d 339
(Ill. App. 1967)

Seattle Police Officers' Guild v. City of Seattle
494 P.2d 485
(Wash. 1972)

The following are *not* court decisions, but decisions of labor arbitrators in union-management arbitration.

YES:

In re Allen Indus. Inc. and Local 986, UAW, 26 Lab. Arb. 363 (1956)

In re Warwick Electronic, Inc. and International Guards Local 9, 46 Lab. Arb. 95 (1966)

B. F. Goodrich Tire Co., 61-2 ARB 8497

National Electric Coil, 66-2 ARB 8578

NO:

Town and Country Food Co., Inc., and United Packinghouses, Local 753, 39 Lab. Arb. 332 (1962)

In re Lag Drug Co. and Teamsters Local 743, 39 Lab. Arb. 1121 (1962)

In re B. F. Goodrich Tire Co. and Teamsters Local 743, 36 Lab. Arb. 552 (1961)

These are *not* court decisions, but decisions of labor arbitrators in union-management arbitration.

NO:

Lone Star Company, 149, NLRB, No. 67, 1964, 57 LRPM 1365.

Wieboldt Stores, Inc., 65-Brd-Ill

Montgomery Ward and Company, 65-BRd-112

Monogram Models, Inc., 65-BRd-113

United Mills 63-1 ARB 8179

Bowman Transportation 73-2 ARB 8336

Admissibility of the results of polygraph test in arbitration proceedings.

YES:

Illinois Bell Tel. Co. and Int'l of Elec. Workers, 39 Lab. Arb. 470 (1962)

In re Wilkof Steel and Supply Co., and Teamsters Local 92, 39 Lab. Arb. 883 (1962)

Westinghouse Electric Corp. 43 LA 450 (1964)

American Maise Products Co. 45 LA 1155 (1965)

McDonnell Aircraft Corp. 66-1 ARB 8236

Owens-Corning Fiberglass Corp. 67-1 ARB 8278

Koppers Co. 68-1 ARB 8084

NO:

In re Brass-Craft Mfg. Co. and Local 408, UAW, 36, Lab. Arb. 1177 (1961)

In re Continental Air Transp. Co., Inc., and Teamsters Local 727, 38 Lab. Arb. 778 (1962)

In re Marathon Elec. Mfg. Co., and Local 116, UAW, 31 Lab. Arb. 1040 (1959)

In re Spiegel Inc. and Warehouse Employees Local 743, 44 Lab. Arb. 405 (1965)

General American Transportation Corp. 31 LA 355 (1958)

Coronet Phosphate Co. Inc., 31 LA 515 (1958)

Marathon Electric Mfg. Corp., 31 LA 1040 (1959)

Publishers' Assn. of New York City, 32 LA 44 (1959)

Brass-Craft Mfg. Corp., 61-2 ARB 8498

Continental Air Transport Co., 38 LA 778 (1962)

NO:

Town and Country Food Co., 62-3 ARB 9054

Dayton Steel Foundry Co., 63-1 ARB 8003

Lag Drug Co., Inc. 63-1 ARB 8106

United Mills, Inc. 63-1 ARB 8179

Louis Zahn Drug Co. 63-1 ARB 8344

Skaggs-Stone Inc., 40 La 1273 (1963)

Spiegel, Inc. 65-1 ARB 8213

Saveway Inwood Service Station 65-1 ARB 8385

Pann Steel Co. Inc. 66-1 ARB 8310

Kwick Kafeteria, Inc., 66-1 ARB 8359

Neuhoft Packers, 71-1 ARB 8179

Discharge for failure to take the polygraph test.

Allen Industries, Inc.
26 LA 363 (1956)
Arbitrator: Joseph M. Klamon

Warwick Electronics, Inc.
46 LA 95 (1966)
Arbitrator: Carroll P. Daugherty

Lag Drug Co.
39 LA 1121 (1962)
Board of Arbitration: Peter M. Kelliher (impartial arbitrator); Lloyd Yale (company-appointed arbitrator); John Burrzinsk (union-appointed arbitrator)

No discharge for taking the polygraph test.

Pearl Beer Distributing Co.
59 LA 821 (1972)
Arbitrator: Raymond L. Britton
If employee consents to take the polygraph test.

Illinois Bell Telephone Co.
39 LA 470 (1962)
Arbitrator: M. S. Ryder

American Maize Products Co.
56 LA 521 (1971)
Arbitrator: John Day Larkin
Note: Results of polygraph test are not admissible.

Louis Zahn Drug Co.
40 LA 353 (1963)
Arbitrator: John F. Sembower
Note: Employee could not be discharged for failure to take polygraph test.

Bowman Transportation, Inc. and United Steelworkers of America, Local 13600
73-2 ARB 8336 (1973)
Arbitrator: James P. Whyte

In re Bowman Transportation Co. Inc. and United Steelworkers of America, Local 13600
60 LA 837 (1973)
Arbitrator: Paul W. Hardy
Note: Results of polygraph test.

Spiegel, Inc.
44 LA 405 (1965)
Arbitrator: John F. Sembower

Western Harness Packing Assn.
57 LA 373 (1971)
Arbitrator: Irving Helbling

American Hospital Supply Corporation
71-1 ARB 8353 (1971)
Arbitrator: Pearce Davis

Note: National Labor Relation Board

American Oil Company and International Brotherhood of Service Station Operators of America
1971 CCH NLPB
22, 824

National Food Service, Inc. and Teamster Local Union No. 783, International Brotherhood of Teamsters, Chauffeurs, Warehousemen and Helpers of America
1972 CCH NLRB 24059
Note: Threat of lie test concerning union activities is unfair labor practice.

Glazer's Wholesale Drug Co. Inc. and Dallas General Drivers, Warehousemen and Helpers Local Union No. 745 affiliated with the International Brotherhood of Teamsters, Chauffeurs, Warehousemen and Helpers of America
1965 CCH NLRB 9311

National Labor Relations Board, *Petitioner v. the Borden Company*, Respondent
57 LC 12519 (1968)

Falstaff Beer Distributors of Greater Miami and Carroll M. Everett, an individual
1965 CCH NLRB 9430

Lone Star Co. and General Drivers. Warehousemen and Helpers Local Union No. 968
1964 CCH NLRB 13560

Southwire Co. and International Union of Electrical, Radio and Machine Workers AFL-CIO
1966 CCH NLRB 20519

Aladdin Industries, Inc. and United States Steelworkers of America, AFL-CIO and International Brotherhood of Teamsters and Charles I. Mason, Charles N. Lee and Bobby H. Crumlegg. Individuals
1964 CCH NLRB 13279

Peller v. Retail Credit Company
359 F.Supp. 1235
(1973)

SEARCH

A peace officer must follow the constitutional rules of search and seizure in order that evidence is admissible in the criminal trial of the defendant. The exclusionary rule holds that competent, material, and relevant evidence illegally seized—is not admissible in a criminal trial. However, most states do not apply this exclusionary rule to private nondeputized officers and, therefore, the private officer is concerned with the possibility of civil/criminal actions for assault, battery, theft and invasion of privacy, rather than the admissibility of evidence at trial.

The following case citations provide the basic decision about the scope of authority for private services personnel to conduct searches.

People v. Crawford
272 N.E.2d 706 (Ill. 1971)

People v. Santiago
278 N.Y.S. 2d 260 (1967)

United States v. Berger
355 F.Supp. 919 (1937)

People v. Bryant
243 N.E.2d 354 (Ill. App. 1968)

Barnes v. United States
373 F.2d 517
L.A. 5th (1967)

Wright v. United States
224 A.2d 475
(D.C. App. 1966)

People v. Horman
292 N.Y.S. 2d 874 (1968)

Stapleton v. Superior Court of Los Angeles County
73 Cal. Rptr. 575 (1969)

Sackler v. Sackler
203 N.E.2d 481
(N.Y. App. 1964)

Williams v. Williams
221 N.E.2d 622
C.Ct. of C.P. Ohio (1966)

Burdeau v. McDowell
41 S.Ct. 574 (1921)

United States v. Winbush
428 F.2d 357 (1970)
United States Court of Appeals for Ohio

Eisentrager v. Hocker
450 F.2d 490 (1971)
United States Court of Appeals for Nevada

United States v. Harding
475 F.2d 480 (1973)

People v. Lanthier
488 P.2d 625 (Cal. 1971)
Supreme Court of California

People v. Benson
490 P.2d 1287 (Colo. 1971)
Supreme Court of Colorado

People v. Mangiefico
102 Cal. Reptr. 449 (1972)
California Court of Appeals

Consent/Search

People v. Henderson
33 Ill. 2d 225 (1965)

Schneckloth v. Bustamonte
93 S.Ct. 2041 (1973)

People v. Walker
34 Ill. 2d 23 (1966)

Frazier v. Cupp
89 S.Ct. 1420 (1969)
U.S. Supreme Court

People v. Haskell
41 Ill. 2d 25 (1968)

People v. Harvey
199 N.E.2d 236
(Ill. App. 1964)

People v. Rodriguez
79 Ill. App. 2d 26 (1967)

People v. Nunn
288 N.E.2d 88
(Ill. App. 1972)

Chapman v. United States
81 S.Ct. 776 (1961)
U.S. Supreme Court

Stoner v. California
84 S.Ct. 899 (1964)
U.S. Supreme Court

People v. Tidwell
266 N.E.2d 787
(Ill. 1971)

People v. Overton
299 N.E.2d 598
(N.Y. App. 1967)

Piazzola v. Watkins
442 F.2d 285 (1971)

Search Requested by Administrative Agency

Camara v. Municipal Court
87 S.Ct. 1727 (1967)

See v. City of Seattle
87 S.Ct. 1737 (1967)
Fruenhauf Corp.
49 LA 89 (1967)
TRW 71-1 ARB 8411 (1971)

Award Search

Scott Paper Company
69-1 ARB 8471 (1969)

Medical Ancillary Services, Inc.
1972 CCH NLRB 24502 (1972)

Avco Corporation, Avco Electronics Division
1972 CCH NLRB 24660 (1972)

In re McLouth Steel Corporation and United Steelworkers of America Local 2659
57-LA (1972)

The Kroger Company and International Brotherhood of Teamsters, Chauffeurs, Warehousemen and Helpers of America, Local 249
63-1 ARB 8315 (1963)

Chevron Chemical Company, Oronite Additives Division and Oil, Chemical and Atomic Workers International Union, AFL-CIO, Local 4-447
73-2 ARB 8481 (1973)

SHOPLIFTING

Shoplifting involves any person who wrongfully takes actual possession of merchandise from a mercantile establishment.

Mercantile Establishment

A mercantile establishment is a place where one or more persons are employed in which goods, wares, or merchandise are offered for sale, and includes a building, shed, structure, or any part thereof, occupied in connection with such establishment (Ill. Revised Statutes, Chap. 48, Section 8a).

Shoplifting

Probable cause to believe—more than mere suspicion—is necessary in order to be charged. Based on all facts at that time, a reasonable and prudent objective might involve a disinterested non-emotional merchant, agent, or employee who believes that a person has wrongfully taken or has actual possession of wrongfully obtained merchandise.

Illinois Shoplifting Statute—Chapter 38, Section 10-3

People v. Hasty
262 N.E.2d 292 (1970)

Shoplifting—Theft, 38/16-1—Testimony of Security Guard

People v. Thomas
292 N.E.2d 153
(Ill. App. 1972)

People v. Leman
238 N.E.2d 312
(Ill. App. 1968)

Meadow v. F. W. Woolworth Company
254 F.Supp. 907 (1966)

J. S. Dillon and Sons Stores Company v. Carrington
455 P.2d
(Colo. 1969)

Total v. Alexander's
314 N.W.2d 93 (1968)

Doyle v. Douglas
390 P.2d 871
(Okla. 1964)
Supreme Court of Oklahoma

Shaw v. May Department Stores Co.
268 A.2d 607
(D.C. App. 1970)

Davis v. Zion Cooperative Mercantile Institution
509 P.2d 362
(Utah 1973)

Lasseigne v. Walgreen
274 So.2d 480
(La. App. 1937)

Simmons v. J. C. Penney Company
186 So.2d 358
(La. App. 1966)

Delp v. Zapp's Drug and Variety Stores
395 P.2d 137
(Ore. 1964)

Wisner v. S. S. Kresge Company
465 S.W.2d 666
(Mo. App. 1971)

Banks v. Food Town
98 So. 2d 719
(La. App. 1957)

Fleet v. May Department Stores, Inc.
599 P.2d 1054
(Ore. 1972)

J. C. Penney Company v. Cox
148 So. 2d 679
(Miss. 1963)

Steinbaugh v. Payless Drug Store, Inc.
401 P.2d 104
(N.M. 1965)

J. C. Penney Company v. Duran
479 S.W.2d 374
(Tex. 1972)

Clark v. Kroger Company
382 F.2d 562
(1967)

Coblyn v. Kennedy's Inc.
268 N.E.2d 860
(Mass. 1971)

Isaiah v. Great Atlantic and Pacific Tea Co.
174 N.E.2d 128
(Ohio App. 1959)

Washington Country Kennel Club, Inc. v. Edge
216 So.2d 512
(Fla. App. 1968)

Dominguez v. Globe Discount City, Inc.
470 S.W.2d 919
(Tex. App. 1971)

Martinez v. Sears, Roebuck and Co.
467 P.2d 37
(N.M. App. 1970)

Gaffney v. Payless Drug Stores
492 P.2d 474
(Ore. 1972)

Reicheneder v. Skaggs Drug Center
421 F.2d 307
(1970)

Mullins v. Rinks, Inc.
272 N.E.2d 152
(Ohio App. 1971)

King v. Anderson
51 Cal. Rptr. 561
(1966)

Eason v. J. Weingarten, Inc.
219 So.2d 516
(La. App. 1969)

Abner v. W. T. Grant Company
139 S.E.2d 408
(Ga. App. 1964)

Lukas v. J. C. Penney Company
378 P.2d 717
(Ore. 1963)
Supreme Court of Oregon

Zayre of Virginia, Inc. v. Gowdy
147 S.E.2d 710
(Va. 1966)

Cooke v. J. J. Newberry and Co.
232 A.2d 425
(N.J. App. 1967)

Cannon v. Goudehaux's, Inc.
255 So.2d 243
(La. App. 1971)
Court of Appeals of Louisiana

Stewart v. J. C. Penney Co.
267 S.2d 925
(La. App. 1972)
Court of Appeals of Louisiana

Bruce v. Meigers Supermarkets, Inc.
191 N.W.2d 132
(Mich. App. 1971)

Wilde v. Schewgman Bros. Giant Supermarkets, Inc.
160 So.2d 839
(La. App. 1964)

Wolin v. Abraham and Straus
316 N.Y.S. 2d 377
(1970)

Jacques v. Sears, Roebuck and Co.
334 N.Y.S. 2d 632
(1972)

Garvis v. K-Mart Discount Store
461 S.W.2d 317
(Kansas App. 1970)

Moore v. Federal Department Stores, Inc.
190 N.W.2d 262
(Mich. App. 1971)

Note: Employee and store liable for false arrest—question probable cause under Ill. Shopping Statute is for the Jury.

Gas v. Zayre of Illinois, Inc.
305 N.E.2d 704
(Ill. App. 1973)

SPECIAL POLICE

Special Police are private persons who are deputized by either the city or state to have powers of peace officers.

Special Policemen

Any person who, hired for a reward, shall guard or protect any building, structure, premises, person, or property within the city; provided, however, that this shall not apply to regularly appointed police officers of the city or to any sheriff or deputy sheriff of the county.

The following decisions describe the duties, authority and legal responsibilities for Special Police.

Special Police

Hughes v. New York Central System
155 N.E.2d 809
(Ill. App. 1959)

Adler v. White City Construction Co.
147 App. 20
(1909)

Komorowski v. Boston Store of Chicago
263 Ill. App. 88
(1931)

Grimmin v. State of Maryland
84 S.Ct. 1770
(1964)

National Labor Rel. Bd. v. Jones and Laghlin Steel Corp.
67 S.Ct. 1274
(1947)

Martin v. Conlisk
347 F.Supp. 262
(1972)

Adickes v. S. H. Kress and Company
90 S.Ct. 1598
(1970)

Screw v. United States
65 S.Ct. 1031
(1945)

United States v. Price
86 S.Ct. 1152
(1966)

Williams v. United States
71 S.Ct. 576
(1951)

North Carolina Ass'n of Lic. Detectives v. Morgan
195 S.E.2d 357
(N.C. 1973)

Knotts v. State
187 N.E.2d 571
(Ind. 1963)

Neapolitan v. United States Steel Corporation
149 N.E.2d 589
(1956)

Sennett v. Zimmerman
324 P.2d 414
(1957)

Singleton v. United States
225 A.2d 315
(D.C. App. 1967)

Frank v. Wabash Railroad Company
295 S.W.2d 16
(Mo. 1956)
Supreme Court of Missouri Division No. 2

Williams v. Commonwealth
128 S.E. 573
(1925)
Supreme Court of Appeals of Virginia

Rodriguez v. Motor Vehicle Accident Indem. Corp.
282 N.Y.S.2d 625
(1967)

Note: Corporation not liable for negligence of special police officer.

Magg v. Pompoa
287 P. 982
(Cal. App. 1930)

St. John v. Reid
61 P.2d 363
(Cal. App. 1936)

WARNINGS (MIRANDA RULE)

The term *warning* applies to oral statements given by a person as contrasted with physical evidence taken from a person.

Miranda Rule

Voluntary Miranda warnings must have been given.
1. He has a right to remain silent.
2. Anything he says can be used against him in a court of law.
3. He has the right to the presence of an attorney.
4. And, that if he cannot afford an attorney, one will be appointed to him prior to any questioning if he so desires.

The Miranda warning has been held to apply to certain circumstances when an arrest is made by private officers; in other decisions it has not. For a thorough discussion of the applicability, local decisions must be consulted. The general rule should be to provide the warning if the possibility of prosecution exists. The following cases provide the entire spectrum of judicial decisions regarding Miranda.

Miranda Rule/Miranda Warnings

People v. Frank
275 N.Y.S. 2d 570
(N.Y. 1966)

United States v. Castell
476 F.2d 152 (1973)

State v. Peoples
275 N.E.2d 626
(Ohio App. 1971)

United States v. Antonelli
434 2d 335 (1970)

State v. La Rose
174 N.W.2d 247
(Minn. 1970)

People v. Vlek
252 N.E.2d 337
(Ill. App. 1969)

People v. Hawkins
53 Ill. 2d 181 (1972)

Hood v. Commonwealth
448 S.W.2d 388
(Ky. 1969)

State v. Little
439 P.2d 387
(Kansas 1969)

Schaumberg v. State
432 P.2d 500
(Nevada 1967)

State v. Masters
154 N.W.2d 133
(Iowa 1967)

People v. Williams
281 N.Y.S. 2d 251
(N.Y. 1967)

Pratt v. State
263 A.2d 247
(1970)

People v. Jones
301 N.E.2d 85
(Ill. App. 1973)
Ill. App. Ct.

Allen v. Eicher
295 F.Supp. 1184
(1969)
U.S. District Court of Maryland

People v. Haydel
109 Cal. Rpt. 222
(Cal. App. 1973)

In re Lucky, Incorporated and Warehousemen's Union, Local 853, International Brotherhood of Teamsters, Chauffeurs, Warehousemen and Helpers of America
53 LA 1275 (1969)

APPENDIX 3C

SUMMARY OF STATE LICENSING AND REGULATORY REQUIREMENTS FOR SECURITY OFFICERS*

Appendix 3C—Figure 1

*Prepared for presentation to a joint meeting of The Private Security Advisory Council and The Private Security Task Force. Chicago, Illinois, July 10, 1975.

PSAC	Licensee responsible	May employ any number	Employee statement	Temporary need not register	Registration required	Application procedure: Name and address	Aliases, former names	Date and place of birth	Citizenship	Employment: 5 year	3 year	general	Education: 8th	Experience in position	Position duties	Criminal record	Physical description	Fingerprints	Photograph	Qualifications: Never refused a license or revoked	No criminal record	felony	moral turpitude offense	weapon violation	burglars instrument	aiding escape	drugs	Good character	MISCELLANEOUS Identification cards	Badge restrictions	Uniform restrictions	Confidentiality of info	Local regulation allowed	Advertising regulated	Surety required: $25,000	$10,000	$ 5,000	$ 3,000	$ 2,500	$ 2,000	FIREARMS By permit only	Guards may carry
Wyoming																																										
Wisconsin			x		x					x	x	x		x															x						x							
West Virginia																																						x				
Washington																																										
Virginia																																										
Vermont	x	x																											x		x											
Utah																																										
Texas	x			x	x	x	x				x					x	x	x	x	x	x		x	x	x	x	x		x						x							
Tennessee																																										
South Dakota																																										
South Carolina	x			x	x	x		x		x					x	x	x	x	x	x				x									x							x		
Rhode Island																																										
Pennsylvania	x	x	x		x			x		x				x	x			x	x	x	x	x	x	x	x	x			x													x
Oregon																																										
Oklahoma																																										
Ohio				x									x	x	x		x				x	x	x	x																		
North Dakota																									x												x					
North Carolina				x									x	x	x									x		x			x													
New York	x	x	x		x		x	x		x				x			x	x			x	x	x	x	x		x															
New Mexico	x	x	x	x	x	x	x				x						x	x						x	x	x	x	x											x			x
New Jersey																								x	x	x											x	x				
New Hampshire	x	x																		x	x	x						x														
Nevada	x	x			x									x		x					x			x		x		x					x									
Nebraska																												x														
Montana																																										
Missouri																																										
Mississippi																																										
Minnesota	x	x																				x	x	x	x				x						x			x				
Michigan	x	x	x			x					x				x	x					x	x	x	x		x		x	x						x			x				
Massachusetts																						x	x	x											x			x				
Maryland	x	x		x																	x	x	x	x											x	x		x				
Maine		x	x		x		x		x								x		x	x															x							
Louisiana																																										
Kentucky																																										
Kansas																						x	x	x	x		x		x									x				
Iowa																						x	x	x													x					
Indiana			x																			x							x													
Illinois		x	x	x	x				x				x	x	x	x	x					x		x					x													
Idaho																																										
Hawaii	x	x						x								x	x					x	x	x					x													
Georgia			x	x		x		x		x	x		x	x	x		x	x				x			x					x					x				x			
Florida																						x		x						x					x							
Delaware	x	x	x		x	x	x	x					x		x	x	x	x	x	x	x	x	x	x			x		x		x				x	x						
Connecticut		x			x		x	x		x	x				x	x	x	x	x		x	x	x	x			x		x						x				x			
Colorado																																										
California	x			x										x								x	x	x	x	x	x									x						
Arkansas																						x																					x
Arizona	x	x		x	x		x			x	x	x		x	x	x			x	x			x	x								x										
Alaska																																										
Alabama																																										

Appendix 3C—Figure 2

PSAC	Armored car	Central Station alarm	Counter-intelligence service	Couriers	Detection-of-deception examiner	Guards and patrol	Guard-dog service	Private detective or investigator	License for revenue only	No regulation	QUALIFICATIONS FOR LICENSE	Written exam	Residency required	Citizenship: U.S.	Resident Alien	Minimum age: 18	20	21	25	Experience: 1	2	3	4	5	general	Education: High school	No felony conviction	Good character	LICENSING PERIOD	One year	Two years	Five years	APPLICATION PROCEDURES	Photo and fingerprints	Address	References	Employment record	Age, date, and place of birth	Business name	Name and residence of partners, officers, directors	Experience qualifications	Nature of business	Previous residences	Criminal record	Statement of classification seeking	Employer addresses	Branch offices	Info deemed necessary	Letter from sheriff and police	Physical description
Wyoming									x																																									
Wisconsin				x		x											x										x	x									x											x	
West Virginia				x		x								x													x	x							x													x	
Washington						x																																												
Virginia						x																																												
Vermont				x		x		x						x			x										x	x					x	x								x								
Utah						x																																												
Texas	x	x		x	x	x	x		x				x				x				x						x	x	x				x	x			x	x			x	x	x	x			x		x	x	x
Tennessee							x																																											
South Dakota						x																																												
South Carolina				x		x								x				x					x				x	x	x					x	x		x	x							x		x		x	
Rhode Island						x																																												
Pennsylvania				x		x								x				x			x			x			x	x		x				x	x	x	x											x		
Oregon									x																																									
Oklahoma									x																																									
Ohio				x		x			x							x											x	x	x				x	x	x	x	x			x							x	x		x
North Dakota				x	x	x			x		x			x													x	x	x				x																	
North Carolina	x	x	x	x	x	x	x		x				x			x							x				x	x	x	x			x	x	x	x			x	x		x					x	x	x	
New York						x			x				x					x			x						x	x	x	x	x	x	x	x	x				x			x					x	x		
New Mexico						x			x				x	x					x								x	x	x	x			x	x			x	x	x	x		x					x	x		
New Jersey				x		x								x				x					x				x						x																	
New Hampshire				x		x							x														x	x					x	x	x	x			x			x					x	x		
Nevada			x		x	x			x	x						x							x				x	x	x				x	x	x	x	x		x						x	x	x		
Nebraska	x	x				x				x				x				x								x							x																	
Montana				x		x																																											
Missouri								x																																										
Mississippi					x																																													
Minnesota				x		x				x	x	x				x									x		x				x		x	x	x	x	x			x										
Michigan	x	x		x	x	x			x	x				x			x			x			x		x		x	x	x	x			x	x			x	x			x							x		x
Massachusetts				x		x				x				x			x			x			x				x	x	x				x	x																
Maryland				x		x				x				x			x			x			x	x			x	x									x	x	x	x	x	x						x	x	
Maine				x		x					x			x			x			x			x				x	x								x	x		x										
Louisiana							x																																											
Kentucky				x						x				x				x									x						x																	
Kansas				x		x				x				x				x									x	x	x				x	x		x	x	x	x	x			x				x	x		
Iowa				x		x				x				x	x												x	x	x				x																	x
Indiana				x		x				x	x					x			x								x				x		x															x		
Illinois			x			x			x	x				x				x									x	x	x				x	x			x												x	
Idaho							x																																											
Hawaii				x		x				x	x					x											x	x		x			x	x	x	x	x	x				x			x	x			x	
Georgia				x		x					x	x	x	x			x										x	x	x				x	x		x	x	x	x		x	x				x			x	
Florida	x			x	x	x				x				x			x			x		x	x	x		x	x	x	x				x	x	x	x	x	x	x	x	x	x			x	x		x	x	
Delaware				x		x																								x																				
Connecticut				x		x				x					x												x	x				x	x			x	x							x	x		x	x	x	
Colorado					x	x			x				x					x				x					x			x			x	x																
California					x	x	x		x				x					x									x	x		x			x			x			x	x	x	x		x			x	x	x	
Arkansas				x		x				x				x			x				x						x	x	x	x			x	x		x	x	x	x		x	x			x		x	x	
Arizona				x		x				x	x							x									x	x	x	x			x	x			x	x	x	x								x	x	
Alaska						x																																												
Alabama						x	x																																											

Appendix 3C—Figure 3

APPENDIX 3D

PRIVATE POLICE FORCES: LEGAL POWERS AND LIMITATIONS*

Although concern over the conduct and control of public police activities has provoked extensive discussion among legal commentators,[1] the literature has largely ignored the role of private police.[2] However, the rapid growth of the private police industry[3] and the incursion of this industry into areas normally associated with public law enforcement[4] raise important questions concerning the conduct and control of private police activities.

It is difficult to generalize about the private police industry.[5] Diversification exists not only in the degree of training and professionalism of particular private policemen,[6] but in the particular role for which they are employed.[7] Although private police perform numerous functions, including the provision of armored car, patrol, and investigation services, they are used most extensively as uniformed guards in industrial and retail settings.[8] Within these settings private police may serve a variety of purposes, including: the protection of property and persons from theft, fire, and other destructive events; the investigation and surveillance of employees to prevent pilferage and embezzlement; the maintenance of company order and the enforcement of rules and regulations; and the collection of information to influence business, legal, or personal decisions.[9]

These functions indicate that private police are oriented primarily toward prevention and protection, and not toward general law enforcement.[10] The private interests that spawned the use of private police rarely seem to require, or tolerate, a wholesale displacement of public police forces. Thus, to avoid possible adverse public relations and potential tort liability, security guards are typically instructed to utilize public police whenever possible in matters touching the public domain, particularly arrests and searches.[11] In some situations, however, private police must exercise powers analogous to those of their public counterparts.[12] Retail security guards, for example, regularly arrest, search, interrogate, and seek criminal prosecution of suspected shoplifters.[13] This comment explores the extent of these private police powers to arrest, search, and interrogate, and examines the adequacy of existing and potential controls on the exercise of these powers.

GENERAL LEGAL FRAMEWORK

Every citizen possesses certain common law and statutory powers of arrest,[14] search and seizure,[15] and self-defense.[16] The private policeman enjoys these powers no less than any other individual. The question remains, however, whether private police are authorized to exercise powers greater than those granted to citizens generally. Although thirty-three states and the District of Columbia require licenses for certain private police activities,[17] these licenses generally confer no additional powers.[18] At most, the licensing statutes merely regulate the qualifications of licensees[19] and their employees,[20] and in some cases impose restrictions on the conduct of certain private police activities.[21] However, in addition to licensing, some states and municipalities deputize, or "commission," private watchmen and guards.[22] Typically, the commission vests the recipient with some or all of the powers of a public police officer during such hours and upon such premises as the commission may prescribe.[23] In turn, the commission may require the recipient to adhere to the rules,

---

* Michael Braun and David J. Lee, "Private Police Forces: Legal Powers and Limitations." *University of Chicago Law Review*, 38:555-582, 1971. (Reprinted courtesy of The University of Chicago Law School.)

regulations, and practices of the local police department.[24] A commission, unlike a license, is not a prerequisite to engage in private police activities, and comparatively few private police appear to operate under a commission.[25] Finally, some states grant special powers, generally concerning arrest and search, to private police who are employed in connection with the affairs of certain categories of employers. Railroad detectives,[26] campus security officers,[27] and retail security guards[28] are often granted such additional powers. In most states, however, the majority of private police seem to possess no powers beyond those of the ordinary citizen.

Abuses in the exercise of private police power are presently controlled by three factors: license revocation, criminal prosecution, and tort liability. The impact of license revocation appears minimal. In general, outside complainants initiate revocation proceedings.[29] Except perhaps for the overcharged client, a complainant usually has little to gain from a revocation hearing, and where highly personal matters are involved, perhaps much to lose.[30] As a result, complaints are seldom filed and revocation is rare.[31] Thus, the legal burden of controlling private police conduct seems to rest on criminal and civil sanctions. Although it is unlikely that these remedies deter public police misconduct, certain factors suggest that they might be more effective against private police.[32] The remaining sections of this comment examine the efficacy of these and other possible sanctions for controlling abuses of private police power in the areas of arrest, search, and interrogation.

ARREST

The Power to Arrest

An arrest is the apprehension of a criminal suspect for the purpose of bringing him before a public official to answer a criminal charge.[33] Although private investigations and industrial security work may sometimes culminate in a private police arrest, the incidence of arrest is probably highest in the retail security context.[34] Where time does not permit the security guard to summon public police, he must often arrest the suspect himself. Such arrests are necessarily without the protection of a warrant, and where the security guard is not deputized, the power to arrest normally depends on the common law and statutory authorization of citizen's arrest.

Under the common law, citizens and public police officers possessed apparently similar powers to make arrests without a warrant.[35] Misdemeanant arrests were permitted only to preserve the public peace, and the law required an actual breach[36] of the peace committed in the presence[37] of the arresting party. Felony arrests were permitted whenever a felony had actually been committed and there was reasonable cause to believe in the suspect's guilt.[38] In addition, to supplement the arrest power, the arrestor was permitted to employ force if necessary,[39] and to detain the arrestee for such time as was reasonably required to deliver him to an appropriate public official.[40]

Although many states have expanded the arrest powers of public police officers,[41] the scope of permissible citizen's arrest, and the attendant powers of force and detention, have remained relatively constant.[42] The breach of the peace requirement, however, has been eliminated in the context of most citizen misdemeanant arrests,[43] and shoplifting statutes,[44] which allow merchants to "detain" a suspect when reasonable cause exists,[45] have reduced to some extent the impact of the presence requirement.

The limited powers of citizen's arrest seem sufficient to permit private police to fulfill their protective functions. Although the principal limitation on the general power of citizen's arrest—the presence requirement for misdemeanant arrests—might have posed a serious restriction on retail security, since the guard is rarely present during the theft, the shoplifting statutes have largely eliminated this difficulty. In other private police contexts, either the arresting guard is present during the commission of the crime, or the crime constitutes a felony, and hence presence is not required.[46]

Adequacy of Sanctions for Abuse of Arrest Power

Perhaps because the arrest powers of private police are sufficiently broad in those areas where arrests are likely, private police agencies appear to stay well within the legal standards of arrest.[47] Some agencies require adherence to even stricter arrest standards than the law requires.[48] This restraint may be due in part to the sensitivity of these agencies to potentially adverse public relations, since to further its business interests the private police industry must promote an image of restraint and professionalism.[49] However, abuses of arrest power do occur, and in cases involving armed private police[50] injury has sometimes been substantial.[51]

The principal legal restraints on private police abuse of the arrest power are civil and criminal liability for false arrest,[52] false imprisonment,[53] and assault and battery.[54] Although criminal sanctions are rarely invoked to punish such abuses,[55] and the threat of prosecution is therefore unlikely to affect private police arrest conduct, the potential for civil liability appears to deter most illegal arrest practices.

It is generally agreed that civil liability is an ineffective deterrent to public police abuse.[56] This may not be true, however, in the private police context. Although public and private policemen are often judgment proof,[57] liability insurance carried by employers of private police and bonding requirements of most state licensing statutes may create a limited fund from which to satisfy claims against private police.[58] Moreover, the potential prodefendant bias in suits against public policemen[59] may play a lesser role in suits involving private police. Finally, while sovereign immunity often precludes suits against states and municipalities,[60] the doctrine of *respondeat superior* generally enables plaintiffs to proceed against the employer of private police.[61] Those few cases which take a narrow view of *respondeat superior* appear unsound.[62]

In practice, suits against private police are quite common[63] and often successful. And the profit orientation of private police agencies and their employers makes them particularly sensitive to large damage awards.[64] Thus, in view of the moderation generally exercised by private police in arrest situations, existing civil and criminal sanctions seem adequate to deter abuse of the arrest power.

SEARCH

The Power to Search

A private policeman may desire to conduct a search[65] for a variety of reasons. In the retail security context, the private guard may wish to search the person of an apprehended shoplifter to recover stolen property[66] or to collect evidence for use either in a criminal prosecution or in defense of a civil suit for false arrest. Searches would also be useful in other areas of private police activity, particularly private investigation and industrial security. These could include search of a car or dwelling for pilfered goods or the use of electronic surveillance devices[67] to obtain information for use in making legal, business, or personal decisions. In the absence of consent,[68] however, it appears that there is little legal authority for most forms of private search.

Surveillance by private parties, in the form of wiretapping, electronic eavesdropping, or interception of any oral or wire communication without consent, is almost entirely prohibited by Title III of the Omnibus Crime Control and Safe Streets Act.[69] A number of states have also enacted legislation outlawing private wiretapping or electronic eavesdropping.[70] Similarly, a private search of a building or vehicle without consent is generally prohibited,[71] unless incident to an arrest.[72]

Search of a person by a private citizen appears to be limited to search incident to a lawful arrest or detention, though much of the authority for this right is ambiguous. Statutes in at least nine states allow any person making an arrest to seize weapons "about the person" of the arrested party and to deliver them to a magistrate.[73] Two states authorize a personal search and seizure not only of offensive weapons, but of incriminating articles "about the person."[74] Shoplifting statutes generally do not authorize a search of a suspect.[75] The common law authority for a private search is sparse and inconclusive.[76]

In view of the dearth of authority on search incident to a lawful private arrest, some discussion of the issues involved seems appropriate. As a general consideration, since the public police are intended to be society's primary law enforcers, the limitations on public police search should set the upper boundaries of allowable search by private police.[77] The perimeters of permissible public search are, of course, established by the fourth amendment.[78]

The right of public police to conduct a warrantless personal search is governed in part by *Chimel v. California*,[79] which authorizes a postarrest search of a person and the area within his immediate control where the policeman fears use of a dangerous weapon by the arrestee or where he has reasonable cause to believe the search will turn up destructible or easily concealed evidence.[80] Search incident to an investigative "stop" or detention is considered in *Terry v. Ohio*,[81] which authorizes a "frisk" or external pat-down of clothing where the officer fears that he or others may be in danger. If the frisk reveals the presence of weapons, the officer may conduct a personal search to remove them.

When an articulable suspicion of danger exists, granting a private policeman or citizen the authority to search for the purpose of finding and seizing weapons of an arrestee, at least equivalent to the pat-down approved by *Terry*, seems to be a necessary concomitant of the power to arrest.[82] Furthermore, since private police generally have no authority to make an investigative detention or "stop" short of an arrest,[83] it is probably most sensible to grant a right of full personal search when weapon use is feared. The difficult question is whether to allow a nonconsensual search to recover property for return to the owner or for use as evidence. These proprietary and evidentiary concerns can usually be satisfied by simply arresting the suspect and leaving search to the public police. The argument that search of a suspect may enable the parties to settle the matter without further recourse to the public police, and thus avoid stigmatizing the suspect with an arrest record or criminal conviction, ignores the possibility of arranging a valid consent search. If the stolen property is easily destructible or concealable, however, and if there is a substantial likelihood that the thief will destroy or conceal it before the public police arrive, an emergency search by the guard may be justifiable.[84] The right would be analogous to that granted by *Chimel*, but might carefully be hedged by placing a strict burden of proof on the searcher to demonstrate an absolute necessity for the search.

Adequacy of Sanctions for Abuse of Search Power

In discussing both the adequacy of existing sanctions to control private police search and the need for additional controls, it is useful to distinguish search of the person incident to an arrest from surveillance and private property searches. In the area of search incident to an arrest, private police practice does not appear to exceed permissible limits.[85] The retail security guard generally does not search at all, relying instead on detention of a suspect until the arrival of public police, who then handle recovery of the property. Agencies involved in other types of police work—armored car service, for example—usually search only when the presence of dangerous weapons is suspected. Search of a more substantial nature is left to the public police.[86] This lack of abuse is due at least in part to a consideration discussed earlier—the desire to maintain good public relations with customers and clients.[87] Furthermore, deferral to public police is generally a viable alternative to self-help in making a search incident to an arrest. Thus, it appears that the threat of civil liability for trespass or assault and battery[88] may be an adequate deterrent to illegal private police conduct in this area. If particularly flagrant abuses arise, criminal sanctions for assault and battery are available.

In contrast to private police search incident to an arrest, the existing legal prohibitions on surveillance and private property search appear to have little impact on certain forms of private police activity. For example, private investigators, utilized by both domestic and business interests, commonly conduct illegal searches of dwellings[89] and engage in illicit surveillance activities,[90] primarily to obtain information for use in a civil action or to influence some private decision or venture.[91]

The 1968 Crime Control Act prescribes both civil and criminal penalties for violation of its surveillance provisions.[92] Most state statutes which outlaw similar conduct prescribe at least criminal penalties.[93] In addition, general civil and criminal sanctions for trespass are available in cases of illegal property search,[94] and civil sanctions for invasion of privacy exist to control illegal surveillance.[95] Yet the industry practices make it quite clear that these penalties are often inadequate to deal with the problems. This failure may be due to a number of factors. Because of their often surreptitious nature, these forms of search may pass undetected.[96] Moreover, since the search may often involve embarrassing or confidential subject matter, victims may be hesitant to complain.[97] Finally, the information obtained through these forms of search usually cannot be obtained by alternative legal methods or by engaging public police aid.

However, the federal act[98] and some of the state statutes[99] outlawing electronic surveillance provide a further sanction—evidence secured by private parties in violation of the surveillance provisions may not be introduced at judicial proceedings. It would seem that a similar exclusionary rule should apply to the fruits of unauthorized property searches,[100] since the dangers of abuse and the difficulties of control seem identical. Although such a rule would have little impact on invasions of privacy for the purpose of gathering information not to be used in judicial proceedings, the effect on searches used to gather evidence for divorce suits and other types of civil actions might be quite substantial.[101] The deterrent effect of exclusion at criminal trials is more difficult to assess, since private search is apparently rarely intended to gather evidence for use at criminal trials.[102]

While it is unclear, then, whether legislative enactment of a criminal exclusionary rule applicable to surveillance and property searches is necessary,[103] it would seem that, even assuming state action,[104] a judicial extension of the fourth amendment exclusionary rule to the fruits of illegal private search is unwarranted.[105] The fourth amendment rule renders evidence inadmissible in a criminal proceeding if law enforcement officers obtained it as a result of an unconstitutional search.[106] *Burdeau v. McDowell*,[107] decided by the Supreme Court in 1921, held the rule inapplicable to private parties. The Court concluded that the fourth amendment was intended to limit only governmental action. While recent decisions have followed the reasoning of *Burdeau*,[108] some courts have limited the exclusionary rule to public police on the additional ground that the rule would not deter illegal private conduct.[109]

The exclusionary rule represents a constitutional judgment that the need to protect fourth amendment rights outweighs the need to use all available evidence of guilt in criminal prosecutions. The Supreme Court in *Mapp v. Ohio*[110] indicated that such a judgment depends upon two related factual assumptions: (1) that the rule would deter police violation of the fourth amendment,[111] and (2) that the rule was in fact the only effective deterrent.[112] Even with respect to public police the validity of these assumptions is questionable.[113] In addition, some suggested alternatives to the rule might be more effective.[114] Furthermore, the rule deters only searches intended to obtain evidence for use in a criminal prosecution. Thus, one reason assigned for the possible ineffectiveness of the rule in the public police context is that many police activities are not conviction oriented.[115]

Private police agencies appear to be even

less conviction oriented than the public police. They seem to be concerned primarily with protection of property and personnel and with investigation of internal company business. Even the private investigator is rarely concerned with criminal convictions; his activities, when judicial proceedings are involved at all, are directed toward obtaining evidence for civil actions. A civil exclusionary rule, therefore, would be a far more effective deterrent of illegal private police conduct than would an extension of the fourth amendment criminal exclusionary rule.[116] Furthermore, at least in the context of search incident to arrest, the threat of extensive liability in tort and the economic pressure to maintain favorable public relations may well be effective deterrents to private police misconduct.[117] Thus, neither *Mapp* assumption holds true for private police activity,[118] and, even assuming state action, application of a constitutionally based exclusionary rule to private police search appears to be unwarranted as a matter of policy.[119]

INTERROGATION

The right lawfully to detain a suspect is a necessary prerequisite of effective interrogation,[120] without a right to detain, the suspect may simply walk away, just as he might if asked an unpleasant question by a fellow citizen. While there is no express authorization of private police interrogation during detention of a suspect,[121] such interrogation may be justified as a matter of policy. The detention following a shoplifting arrest presents the typical situation in which a private policeman may want to interrogate a suspect.[122] The guard or his supervisor may wish either to persuade the suspect to settle the matter on friendly terms by returning the missing property, or to determine if the situation justifies summoning the public police. The supervisor may also wish to obtain a confession from the shoplifter for use in a criminal prosecution or as a defense to a later action for false arrest. However, none of these justifications, nor the right to detain the suspect, implies any authority to use coercive tactics to compel the suspect to answer questions during an interrogation.[123]

Private police generally do not abuse their right to interrogate. Typically, a shoplifting suspect is taken to the office of the security supervisor[124] and questioned in the presence of the arresting guard.[125] Often the suspect is informed of his right to remain silent and right to counsel.[126] Since postarrest detention is usually permitted for only a brief period,[127] there is less likelihood than in public police interrogations that coercive methods will be effective to extract a confession. Unreasonable extensions of the period of detention and use of force and threats are discouraged by previously discussed economic factors and by tort and criminal sanctions.[128] Nevertheless, the question often arises whether particular confessions should be admissible in a criminal trial of the confessor-suspect.

Exclusion of an improperly obtained confession may be based on purely evidentiary grounds: A coerced or involuntary confession may be unreliable.[129] This rule has shown considerable vitality in the private police context,[130] although the recent trend in evidence law which favors the admission of all probative evidence[131] may point toward a diminishing role for this rule in the future.

In the public police context, a much more pervasive exclusionary rule is applied. Under *Miranda v. Arizona*,[132] any incriminating or exculpatory statement obtained from a criminal suspect during a custodial interrogation[133] is inadmissible at his trial, unless prior to his statement the suspect was effectively advised of and knowingly waived his fifth amendment right to remain silent and his sixth amendment right to counsel.[134]

Post-*Miranda* decisions have limited the rule to interrogations by police, governmental officials, and private police commissioned under state or local law.[135] Interrogations by noncommissioned private police generally have been distinguished on the factual assumption that private-party interrogations are without the inherent "potentiality for compulsion" which the *Miranda* Court found in the public police context.[136] The courts adopting this assumption may have felt that an extension of *Miranda* to private police was unwarranted in view of the rarity of private police abuse and the inability of *Miranda* to deter practices not motivated by a desire to convict.[137] Furthermore, a few courts suggest that private police interrogations are not protected by the fifth and sixth amendments since nongovernmental agencies are involved and no state action exists.[138] This last distinction is perhaps the only persuasive argument in favor of limiting the operation of *Miranda* to public police; the former distinctions may well be unsound.

Prior to *Miranda*, the admissibility of confessions made to governmental officers was controlled by a "voluntariness" doctrine resting on the fourteenth amendment.[139] If it appeared from the facts surrounding a confession that it had been coercively obtained, the confession was excluded at trial as violative of due process. This decisional law represented the response of the Supreme Court to a long history of public police abuse,[140] and the black-letter standards of *Miranda* may be viewed as the culmination of the Court's efforts to control public police interrogation conduct. Moreover, the *Miranda* Court was probably responding to a tendency of the lower courts to condone conduct which the Court felt violated standards of procedural fairness.[141] Thus, to a considerable extent, *Miranda* was designed to deter public police abuse. Given this view of the decision, an extension of *Miranda* to private police interrogations would seem, as a matter of policy, unnecessary. As demonstrated by the apparent rarity of private police abuse in this area, the threat of civil and criminal liability, the fear of declining business, and the possibility of confession exclusion under evidence law seem adequate to deter most private police abuses.

Deterrence of police abuse, however, may not have been the only rationale underlying the *Miranda* decision.[142] The marked shift in procedural standards from a "voluntary" confession to one obtained only after a "knowing and intelligent waiver" of rights[143] and the emphasis on protection of Fifth Amendment rights[144] suggest that the Court in *Miranda* was concerned also with erecting new protections for individual constitutional rights.

This aspect of *Miranda* was emphasized by the Supreme Court of California in *People v. Kelley*,[145] which reversed a state court conviction of a serviceman based, in part, on a confession secured by a military investigator without the proper *Miranda* warnings. Despite the legality of the confession under military law,[146] and even though recognizing that the decision might not deter military police practices, the court held the confession inadmissible:

> One of the important purposes of the [confession] rules . . . was that it was necessary to deter improper police practices . . . that might lead to involuntary confessions [citations omitted]. But it cannot be denied that the whole series of cases—*Escobedo*, . . . *Miranda*, *Johnson*, . . . and their numerous progeny—was fundamentally aimed at protecting the Fifth and Sixth Amendment rights of the accused, that is, to protect against self-incrimination and to protect the right to counsel.[147]

Thus, the Warren Court may have been concerned not only with unconstitutional police abuses encouraged by the privacy and conviction orientation of public police interrogations, but with the inherently coercive capacity of those interrogations, which jeopardized effective exercise of the Fifth and Sixth Amendment rights of an ignorant or fearful suspect.[148] The *Miranda* warnings may therefore have been intended, at least in part, to dispel this fear and ignorance.

The principal factual distinctions between private and public police interrogations seem to be the strictly limited period of permissible private police detention and the unofficial, or informal, character of the questioning.[149] However, numerous post-*Miranda* decisions indicate that neither a short period of detention[150] nor the absence of recognized official character[151] is sufficient to dispel the taint of compulsion in a custodial interrogation. Moreover, these distinctions are largely negated by the similarities between public and private interrogations—both involve detention, privacy, the appearance of authority, and the availability to the interrogator of psychologically coercive methods of questioning.[152] All of these factors were relied upon in *Miranda*.[153] These similarities suggest that a significantly coercive atmosphere exists in the context of many private police interrogations, and that suspects, unwarned of their rights, may waive them through fear or ignorance. Moreover, extension of *Miranda* to private police would probably have a definite impact on their practices, since private police interrogation, in contrast to search or arrest, may be relatively conviction oriented.[154] Thus, as a matter of policy, an extension of both the procedural safeguards and exclusionary sanction[155] of *Miranda* to private police interrogations may well be warranted.

The extension of *Miranda*, however, depends upon one further consideration— whether private police interrogations constitute state action.[156] One possible basis for finding state action concerns the issuance of commissions.[157] This approach was adopted by the Supreme Court in *Williams v. United States*.[158] The Lindsley Lumber Company had suffered numerous thefts and hired petitioner, who held a special police officer card from the City of Miami and operated a detective agency, to ascertain the identity of the thieves. In the course of his investigation, petitioner imprisoned, beat, and brutally coerced confessions from four suspects. The Court upheld petitioner's conviction under the Civil Rights Act of 1866,[159] finding that his acts were performed under "color" of state law.[160] In a 1970 decision, a Maryland court extended the strictures of *Miranda* to commissioned private police.[161] The court held that the commission and its attendant powers made the private guard ". . . a law enforcement officer within the contemplation of *Miranda*."[162] Thus, there appears to be some authority for finding state action in the conduct of commissioned private policemen.

Similarly, there may be authority for finding state action in interrotations by private police who are licensed under state or local law. This rationale has been used already to include within the strictures of the fourteenth amendment the discriminatory practices of licensed commercial establishments which cater to the public.[163] In such cases it was found that the "pervasive" regulation by the state warranted application of the fourteenth amendment.[164] This rationale may be applicable also where the state licenses private individuals to perform specialized police activities.

The scope of state action in the private police context may not, however, be so limited. In *Marsh v. Alabama*,[165] the Supreme Court held that when a state permits a private business to perform all the functions of a municipal corporation, the public activities of the business constitute state action. The routine participation of private police in certain areas of law enforcement may sometimes supplant the public police,[166] and to this extent private police are performing a public function. It is also apparent that private police coordinate their activities with the public police, and while the suggestion of a loose partnership may be too strong, private police are at least aware that the fruits of their investigations may be desired by the public police. Thus, although private police perform only limited public functions, the rationale of *Marsh* suggests that when the state permits private police activities, it may endow these activities with state action.

Aside from the governmental character of these functions, there is the additional factor that private police activities often affect the general public. In *Amalgamated Food Employees Union Local 590 v. Logan Valley Plaza, Inc.*,[167] which extended *Marsh* to protect peaceful picketing of a privately owned shopping center, the Supreme Court considered the state action element in commercial establishments which cater to the general public:

> The more an owner, for his advantage, opens up his property for use by the public in general, the more do his rights become circumscribed by the statutory and constitutional rights of those who use it.[168]

Thus, state action may be present in the activity of a security guard employed by a privately owned shopping center, or by any private organization which deals with the general public. This rationale suggests that whenever private police activity affects a relatively broad spectrum of the public, state action is present and hence the *Miranda* requirements may apply to some private police interrogations.

CONCLUSION

Private police enjoy extensive powers which enable them to perform functions analogous to public police activity. In the area of arrest these powers are clearly defined, and for the most part require little modification. The search and interrogation powers of private police are in some cases more ambiguous, and here legislative delineation may be appropriate. Whether additional sanctions should be placed on the exercise of these powers depends upon the nature and extent of private police abuses and the deterrent effect of existing sanctions. While certain parallels between private and public police at least suggest that similar sanctions should apply to both, this preliminary inquiry indicates that the motivations and incentives of private police differ from their public counterparts, and that blanket application of public police sanctions to private police appears unwarranted. However, courts and commentators alike should be sensitive to the possibility that the existing powers and controls of private police may require alteration.

FOOTNOTES

1. *E.g.*, M. Banton, *The Policeman in the Community* (1964); A. Black, *The People and the Police* (1968); P. Chevigny, *Police Power: Police Abuses in New York City* (1969); J. Skolnick, *Justice Without Trial: Law Enforcement in a Democratic Society* (1966).

2. The only previous treatment of this subject in a legal periodical appears in Note, *Regulation of Private Police*, 40 S. Cal. L. Rev. 540 (1967).

   As used in this comment, the term "private police" refers to individuals or organizations who perform protective or investigative services for profit or in connection with other business, and who are not public employees engaged in the exercise of their official duties. This definition includes such groups as armored car agencies, private detective agencies, special patrol agencies, retail security guards, plant protection units, industrial security units, and campus security guards. The one significant group that does not fit this definition is the citizen volunteer patrol.

3. It has been estimated that $1.6 billion was spent by the private sector for protective services in 1969. Creeping Capitalism, *Forbes*, Sept. 1, 1970, at 22 [hereinafter cited as *Forbes*]. "The security industry itself claims that two out of every three law enforcement officers in the nation are actually on private payrolls." *NLRB v. Jones & Laughlin Steel Corp.*, 331 U.S. 416, 429 (1946).

4. Some private police agencies, structured along the lines of city police forces, have begun to provide patrol car and guard services to neighborhood groups and urban corporations. *Forbes, supra* note 3, at 22.

5. Much of the information set forth in this comment is based on interviews conducted in Chicago, Illinois, from September, 1970, to January, 1971. In many instances interviews were refused, and in other instances the interviewee declined to have his name or firm associated with his remarks. Very few interviews were conducted with small private investigators or private guards. Thus, the reader should note that the generalizations expressed concerning private police activities relate primarily to the larger agencies. For

discussions of the small-scale operators, see *Report of the New York State Legislative Committee on Privacy of Communications and Licensure of Private Detectives* 43-60 (1957) [hereinafter cited as *Report on Private Detectives* (1957)]; *Report of the New York State Legislative Committee on Privacy of Communications and Licensure of Private Detectives* 43-48 (1958) [hereinafter cited as *Report on Private Detectives* (1958)].

6. "To an extent of which the public is perhaps unaware, licensed private detectives often engage men of scant ability and little stability. While generalization is impossible, we have been struck by the sporadic, short term, and poorly paid nature of much of the employment in this occupational group." *Report on Private Detectives* (1957), *supra* note 5, at 53. While this description is perhaps overly broad, the qualifications and training of private police do seem to reflect a relatively low level of responsibility. For example, the starting wage in Chicago, Illinois, for a security guard is $1.95 per hour. A high rate of job turnover also characterizes many private police activities. Most training is on the job; the typical security guard will begin work with, at most, an orientation lecture. In other areas, however, training seems to be quite extensive. *See, e.g.,* R. Seng & J. Gilmour, *Brink's: The Money Movers* 71 (1959). Investigators who have received top security clearance by the federal government are often trained in the full range of police techniques. The supervisory personnel of many private police agencies are well paid and well trained, and often have backgrounds in public law enforcement. Thus, no single label adequately fits the range of expertise prevalent in this field.

Variations in the size and sophistication of private police operations are equally great. There are three corporate giants in the field—Pinkerton's National Detective Agency, Inc., The William J. Burns International Detective Agency, Inc., and The Wackenhut Corporation. The largest, Pinkerton's, Inc., employs upwards of 25,000 field personnel. Interview with Edward Costells, Assistant Manager of the Chicago office, Pinkerton's National Detective Agency, Inc., in Chicago, Ill., Nov. 25, 1970. However, most holders of private detective licenses are either lone operators or small agencies which may employ fewer than fifteen people on a full time basis. *Report on Private Detectives* (1958), *supra* note 5, at 47.

7. "Not only is there variety in the kinds of jobs done but also there is great specialization within well-known categories. One licensee is devoted to investigations in the perfume industry. Another serves exclusively a 'clientele of railroads, steamship lines and common carriers.' One corporation is devoted to observing and reporting 'instances of taxi drivers carrying passengers without recording the fare on their meters.'" *Report on Private Detectives* (1958), *supra* note 5, at 46.

8. Of the estimated 800,000 private police officers, approximately 500,000 are uniformed. *Forbes, supra* note 3, at 22.

9. For a more comprehensive listing, see R. Momboisse, *Industrial Security for Strikes, Riots and Disasters* 52-54 (1968).

10. *See, e.g.,* The William J. Burns Int'l Detective Agency, Inc., *Handbook for Guards* (1962) [hereinafter cited as *Burns Handbook*]. "The protection of a plant by a guard unit is primarily *preventive.*" *Id.* at 8 (emphasis in original).

11. "Whenever possible, have a law enforcement officer make the arrest, but if no law enforcement officer is available and an arrest is necessary, make sure that the law enforcement authorities are contacted immediately." *Id.* at 16.

12. Some agencies provide private police services which parallel or supplant public police forces. *See* note 4 *supra.* The parallel to public police forces is most clearly seen in campus security forces. As indicated in text at note 27 *infra,* many campus security police enjoy full police officer powers while engaged in the performance of their duties. These duties often include full-scale patrolling of both private and public grounds. Where public police

are allocated on the basis of crime rate, and to the extent campus security police are effective, the result may be that the campus security forces supplant the public police within a limited geographical area.

The desire to prevent and deter destructive acts may also invite a direct involvement in law enforcement and the criminal process: "Some advocate criminal prosecution for all thefts, but most [employers of industrial police] prosecute only the big ones. Nevertheless, all agree that occasional prosecution is a good deterrent to future crime." R. Momboisse, *supra* note 9, at 409. The decision to prosecute may also be influenced by other factors. In retail security these include the amount of the theft, the attitude of the shoplifter, and the degree of recidivism. Interview with a Director of Retail Security, in Chicago, Ill., Jan. 6, 1971 (name withheld by request).

13. See note 12 *supra*.
14. See text at notes 34-45 *infra*.
15. See text at notes 69-76 *infra*.
16. See generally *Restatement (Second) of Torts* §§63-75; 1 F. Harper & F. James, *The Law of Torts* §3.11 (1956). The citizen also possesses certain other related rights concerning defense of third parties, defense of unlawful intrusion upon property, and defending dispossession of real and personal property. *Id.* §§3.12-.16.
17. *E.g.*, Cal. Bus. & Prof. Code §§7500-83 (as amended, Supp. 1970); Del. Code Ann. tit. 24, §§1301-21 (Supp. 1968); Ill. Ann. Stat. ch. 38, §§201-51 to 201 51 (Smith-Hurd 1964); N.Y. Gen. Bus. Law §§70 to 89-a.

In addition, some municipalities require adherence to their own licensing provisions. *E.g.*, Los Angeles, Cal. Code §21.117 (1946). In other cases, the state statute authorizes certain classes of municipalities to regulate and license private detectives. *E.g.*, Mo. Ann. Stat. §§73.110 (17), 75.110 (18), 84.340, 84.720 (1952). However, some state licensing statutes prohibit further licensing by any political subdivision of the state. *E.g.*, Mich. Stat. Ann. §18.185 (5) (Supp. 1970).

The coverage and exemptions of these statutes vary considerably. Most of the states require licensing of both private detectives and private patrolmen, although occasionally the statutes are broader, and may include, for example, a "repossessor," Nev. Rev. Stat. §648.015 (1967), or a polygraph operator, Mass. Ann. Laws ch. 147, §22 (1) (1965). Typical exemptions include officers of federal, state, and local governments, deputies and special police while on duty, insurance adjusters, practicing attorneys, and any person employed exclusively in connection with the affairs of his employer. The exemption for private police employed exclusively in connection with the affairs of their employers effectively excludes from coverage a majority of retail and industrial guards. *E.g.*, N.C. Gen. Stat. §66-49.2 (2) (d) (1965); Pa. Stat. Ann. tit. 22, §25 (1955). This exemption has been explained in terms of ". . . [t]he employer's self-advantage in selecting trustworthy employees and in supervising their work. . . ." *Report on Private Detectives* (1957), *supra* note 5, at 46. On the licensing statutes see generally *id.* at 101-17 (statutes as of 1957).

18. *See, e.g.*, License of Niehoff, 9 Pa. D. & C. 2d 410, 48 Berks County L.J. 286 (1956); *Doherty v. Lester*, 4 Misc. 2d 741, 159 N.Y.S.2d 219 (1957). *But ch. Frank v. Wabash R.R.*, 295 S.W.2d 16 (Mo. 1956).

Occasionally the statutes grant some limited power, such as the right to carry a nightstick, Mich. Stat. Ann. §18. 185 (19) (4) (Supp. 1970), or a nonconcealed deadly weapon, N.M. Stat. Ann. §67-33-44 (Supp. 1969). On the other hand, some of the statutes specifically deny certain powers, such as entering on private property without consent, N.M. Stat. Ann. §67-33-25 (F) (Supp. 1969), or inducing a criminal confession by coercion or reward, Colo. Rev. Stat. Ann. §44-1-10 (1963).

19. Among the most common qualifications are general provisions as to "honesty, competency, integrity and trustworthiness," a clean criminal record with respect to felonies and other specified crimes, and some prior law enforcement experience. *E.g.*, N.Y. Gen. Bus.

Law §72 (1); Pa. Stat. Ann. tit. 22, §14 (a) (1955). Sometimes the applicant must pass a written examination. *E.g.,* Ill. Ann. Stat. ch. 38, §201-6A (g) (Smith-Hurd 1964). Character references may also be required. *E.g.,* N.C. Gen. Stat. §66-49.3 (b) (5) (1965). However, training requirements are rare. *But see* Mich. Stat. Ann. §18.185 (31) (Supp. 1970); Vt. Stat. Ann. tit. 32, §9507 (5) (Supp. 1970).

20. Persons employed by a licensee are often required to register with the licensing authority, which usually entails submitting fingerprints and a photograph to the licensing authority. *E.g.,* Ill. Ann. Stat. ch. 38, §201-10b (Smith-Hurd 1964). *See generally Report on Private Detectives* (1957), *supra* note 5, at 52-53.

21. The principal restriction is the furnishing of a bond by the licensee to assure "the faithful and honest conduct of . . . business by the applicant." The bond is usually available to "any person injured by the violation of any of the provisions of [the licensing statute], or by the willful, malicious and wrongful act of the principal or employe." Pa. Stat. Ann. tit. 22, §16 (a) (1955). *See also Myles v. Meinecke,* 82 Ohio App. 126, 78 N.E.2d 917 (1948). The bonds may range anywhere from $1,000 to $25,000, although most average $5,000 or less for individuals and $10,000 or less for agencies.

22. "It is a common practice in this country for private watchmen or guards to be vested with the powers of policemen, sheriffs or peace officers to protect the private property of their private employers." *NLRB v. Jones & Laughlin Steel Corp.,* 331 U.S. 416, 429 (1946). *See also Williams v. United States,* 341 U.S. 97, 98 (1951).

Some statutes permit the governor or some state agency to vest private police with special powers relative to the enforcement of particular state laws. *E.g.,* Me. Rev. Stat. Ann. tit. 32, §§ 3801-02 (1965) (the governor may vest 50 private detectives with the arrest powers of a sheriff for the enforcement of laws relating to shoplifting, gambling, larceny, embezzlement, stolen goods, and certain misfeasance by state officers). Actually such statutes would appear to create a special state police force composed of private detectives. Commissioning, however, is aimed at protecting private interests.

23. *E.g.,* Md. Ann Code art. 41, §§60-70 (1964); Mich. Stat. Ann. §18.185 (30) (supp. 1970); N.C. Gen. Stat. §§74A-1, 74A-2 (1969); Chicago, Ill., Code ch. 173 (1970); New York, N.Y., Administrative Code §434a-7.0 (1963).

24. *E.g.,* Houston, Tex., Code §30-74 (1958) ("Special officers shall at all times be subject to orders and instructions of ranking officers of the police department.") *See also Frank v. Wabash R.R.,* 295 S.W.2d 16, 20, 21 (Mo. 1956).

25. Private police agencies usually have their men commissioned only upon the request of the client. The reasons for this appear to be two-fold. First, the commission is usually limited to a specific address, and cannot be transferred when the guard is reassigned. Second, many agency executives feel that the added powers conferred by a commission are not essential to the limited responsibilities of a security guard.

26. *E.g.,* Okla. Stat. Ann. tit. 66, §183 (1964); W. Va. Code Ann. §61-3-41 (1966).

27. *E.g.,* Ark. Stat. Ann. §7-112 to §7-120 (Supp. 1969) (state institutions); Ill. Ann. Stat. ch. 144, §28 (Smith-Hurd 1964) (University of Illinois); Mass. Ann. Laws ch. 147, §10G (Supp. 1969) (state and private institutions); Tex. Rev. Civ. Stat. Ann. art. 5891A-1 (Supp. 1969) (private institutions).

28. D.C. Code Ann. §4-115 (1966); and *see* notes 44-45 *infra.*

29. *See Report on Private Detectives* (1957), *supra* note 5, at 54.

30. "One must suspect that those who may have been pursued by private detectives are not eager thereafter to prolong the acquaintance by pressing complaints with the licensing authority." *Id.* at 41.

31. "[C]omplaints against licensees [in New York State] average only a half dozen a year, and the disposition of them has been uniformly in favor of the licensee. Moreover, these complaints come from dissatisfied clients, rather

than from harassed subjects of investigation." Id.

Where disciplinary action is administered, the action may amount to little more than a short term suspension of the license. For example, in *Agency for Investigation and Detection, Inc. v. Dep't of State,* 25 App. Div. 2d 738, 739, 169 N.Y.S.2d 168, 169 (1966), the licensee "acted in a shocking and most callous disregard of law and the rights of individuals" when he broke into a private dwelling and "pushed, shoved and committed acts amounting to assaults" upon the occupants. The court affirmed a two-month suspension of his license.

32. *See* text as notes 52-64 *infra*.
33. *See generally* Perkins, *The Law of Arrest,* 25 Iowa L. Rev. 201 (1940).
34. During one twelve-month period the member stores of New York City's Store's Mutual Protective Association apprehended nearly 10,000 suspected shoplifters. S. Curtis, *Modern Retail Security* 779-82 (1960), as cited in Note, *Regulation of Private Police,* 40 S. Cal. L. Rev. 540, 541 n.11 (1967).
35. *See generally* Perkins, *supra* note 33; Wilgus, *Arrest Without a Warrant,* 22 Mich. L. Rev. 541 (1924); Note, *The Law of Citizen's Arrest,* 65 Colum. L. Rev. 502 (1965).
36. *E.g., Commonwealth v. Wright,* 158 Mass. 149, 158-59 (1893) (peace officer); *Radloff v. National Food Stores, Inc.,* 20 Wis. 2d 224, 123 N.W.2d 570 (1963) (citizen). Because not all misdemeanors constitute a breach of the peace—petit larceny, for example—an arrest without a warrant may be unlawful even though the suspect is guilty.
37. *E.g., Lynn v. Weaver,* 251 Mich. 265, 231 N.W. 579 (1930) (peace officer); *Carroll v. United States,* 267 U.S. 132, 157 (1925) (citizen).
38. In some instances, however, a public police officer was permitted to arrest a suspected felon whenever he had reasonable cause to believe a felony had been committed. *See* Perkins, *supra* note 33, at 233-38.
39. *See generally* Note, *Justification for the Use of Force in the Criminal Law,* 13 Stan. L. Rev. 566 (1961). A citizen is generally entitled to use whatever force is necessary to make the arrest. *Id.* at 569. However, deadly force may be employed only to apprehend an escaping felon. *See, e.g., People v. Lathrop,* 49 Cal. App. 63, 192 P. 722 (1920). Some authorities have urged that the power to use deadly force in a felony arrest should be limited to public police officers. *E.g.,* Model Penal Code §3.07 (2) (b) (ii), Comment at 58 (Tent. Draft No. 8, 1958). In any event, a citizen or private policeman is not authorized to use more force than a public police officer. *Berryman v. Int'l Paper Co.,* 139 So. 2d 806 (La. App. 1962).
40. As a rule, the suspect must be taken without unnecessary delay before the appropriate authorities. *E.g., Cline v. Tait,* 116 Mont. 571, 155 P.2d 752 (1945) (peace officer); *Singerman v. William J. Burns Int'l Detective Agency, Inc.,* 219 App. Div. 291, 219 N.Y.S. 724 (1927) (citizen). Moreover, ". . . imprisonment or detention beyond the reasonable time not only renders the imprisonment or detention illegal, but makes the entire transaction (including the arrest) a trespass ab initio." *Great American Indemnity Co. v. Beverly,* 150 F.Supp. 134, 140 (M.D. Ga. 1956).
41. *E.g.,* Cal. Pen. Code §836; Ill. Ann. Stat. ch. 38, §107-2 (c) (Smith-Hurd 1970); Code Ann. §§17-251, -253 (1962). These statutes generally permit public police officers to arrest any suspect without a warrant whenever there is probable cause to believe in the suspect's guilt, or to believe that the suspect committed a crime in the officer's presence.
42. However, in Nebraska and Wyoming, citizens apparently enjoy greater powers of arrest in cases of petit larceny than do the public police officers. Neb. Rev. Stat. §29-402 (1943); Wyo. Stat. Ann. §7-156 (1959).
43. *E.g.,* Cal. Pen. Code §837; Ill. Ann. Stat. ch. 38, §107-3 (Smith-Hurd 1970); N.Y. Code Crim. Proc. §183; S.C. Code Ann. §§17-251, -252 (1962).
44. *See generally* Note, *Survey and Analysis of Criminal and Tort Aspects of*

Shoplifting Statutes, 58 Mich. L. Rev. 429 (1960).

45. The statutes normally provide that the detention shall be effected in a reasonable manner and for a reasonable length of time. They also specify the purpose of the detention. *E.g.,* Ala. Code tit. 14, §334(1) (1959) (to effect recovery of the goods); Ill. Ann. Stat. ch. 38, §10-3 (Smith-Hurd Supp. 1970) (to investigate ownership of the goods); Mass. Ann. Laws ch. 231, §94-B (Supp. 1971) (to question the suspect); Minn. Stat. Ann. §§629.365, 629.366 (Supp. 1971) (to deliver the suspect to a peace officer); Ohio Rev. Code Ann. §2935.041 (Baldwin 1964) (to cause an arrest to be made by a police officer). The statutes normally state that such detention shall not constitute an arrest, although the distinction apparently has no practical significance, other than to limit liability of the merchant. In addition, the statutes sometimes provide special defenses for the merchant or his agents in civil and criminal actions for false arrest, false imprisonment, and assault and battery.

46. In cases where private police investigations culminate in an arrest, some agencies indicate that their usual practice is to delay the arrest until they can apprehend the suspect in the act. Interview with Guido Mattei, Regional Manager of Burns Detective Agency, Inc., in Chicago, Ill., Jan. 8, 1971 [hereinafter cited as Burns Interview].

47. *See* note 46 *supra.*

48. *See, e.g., Burns Handbook, supra* note 10, at 17:

> A guard without police authority has only the power to arrest which the ordinary citizen has. That is, he may arrest a person under the following conditions: (a) The guard *sees* the person commit a felony. (b) The guard has reason to believe a felony was committed and that the person being arrested has committed it. (c) If a person is charged with a felony and the guard knows a warrant has been issued, he can arrest the offender without having the warrant in his possession.
>
> These standards are, in one sense, more restrictive than the law of citizen's arrest, because they would allow only felony arrests. However, the *Handbook* is apparently mistaken when it would allow a felony arrest in the (b) situation above, where there is reason to believe a felony had been committed. Under the common law an actual commission of a felony was required, and only two states seem to have broadened citizen powers to include reasonable belief as to the commission of a felony. *See* Ark. Stat. Ann. §43-404 (1964); Ohio Rev. Code Ann. §2935.04 (Baldwin 1964).

49. One impetus behind enactment of the private police licensing statutes was provided by private police themselves, principally the larger firms, who desired to give their industry a reputable image. Burns Interview, *supra* note 46. And the desire continues:

> This [New York State Committee] has heard suggestions from within the ranks of licensed private detectives for new educational requirements, for qualifying examinations, and for transfer of licensing authority from the Department of State to the Board of Regents—all for the purpose of establishing the private detective as a "professional man."

Report on Private Detectives (1957) *supra* note 5, at 50.

50. It is impossible to estimate what percentage of private police activities are conducted under arms. Agencies appear to arm their men only at the request of the client, or when the assignment is inherently dangerous. Apparently all armored car messengers carry arms. Retail security personnel, other than uniformed patrol guards, are not, in general, armed. The arming of industrial security guards seems to depend on the ". . . type of security interest involved, the number and kind of persons employed at the plant, and the character of the community or area in which the plant is located." R. Momboisse, *supra* note 9, at 43.

51. *E.g., People v. Silver,* 16 Cal. 2d 714, 108 P.2d 4 (1940) (night watchman killed a youth who had trespassed for the purpose of stealing gasoline); *Myles v. Meineke,* 82 Ohio App. 126, 78 N.E. 2d 917 (1948) (plaintiff wrongfully shot by a private guard); *Doherty v. Lester,* 4 Misc. 2d 741, 159 N.Y.S.2d 219

(1957) (private detective shot and seriously wounded a nineteen-year-old suspect).
52. *E.g., Martin v. Castner-Knott Dry Goods Co.,* 27 Tenn. App. 421, 181 S.W.2d 638 (1944); *Peak v. W. T. Grant Co.,* 409 S.W.2d 58 (Mo. 1966).
53. *E.g., Meinecke v. Skaggs,* 123 Mont. 308, 213 P.2d 237 (1949); *Great Atl. & Pac. Tea Co. v. Smith,* 281 Ky. 583, 136 S.W.2d 759 (1940).
54. *E.g., Morgan v. Loyacomo,* 190 Miss. 656, 1 So. 2d 510 (1941); *Greenfield v. Colonial Stores, Inc.,* 110 Ga. App. 572, 139 S.E.2d 403 (1964).
55. Successful criminal prosecutions against public police officers are practically nonexistent: "It is absurd to suggest that any district attorney, or superior officer, is going to take criminal action against one of his subordinates." *White v. Towers,* 37 Cal. 2d 727, 737, 235 P.2d 209, 215-16 (1951) (Carter, J., dissenting). Similarly, it appears that successful criminal prosecutions against private police are rare. Burns Interview, *supra* note 46.
56. "The fact is, however, that there are several million illegal arrests and imprisonments in the United States each year, and that only a handful of damage suits are filed against policemen." Hall, *Police and Law in a Democratic Society,* 28 Ind. L.J. 133, 152 (1953). *See generally* Foote, *Tort Remedies for Police Violations of Individual Rights,* 39 Minn. L. Rev. 493 (1955).
57. *See* Hall, *supra* note 56, at 153-54; Greenstone, *Liability of Police Officers for Misuse of Their Weapons,* 16 Clev.-Mar. L. Rev. 397 (1967). The low salary of an individual guard probably shields him from any substantial judgment. *See* note 6 *supra*.
58. Although these bonds are generally available to injured third parties, their effectiveness is limited both by their relatively small size, and by the numerous exemptions to the bonding requirements. *See* notes 17 & 21 *supra*. It is suggested that all private police who carry arms should furnish a bond, whether or not they are required to be licensed, and that the size of the bond should reflect the use of dangerous weapons. *Cf. Myles v. Meineke,* 82 Ohio App. 126, 78 N.E.2d 917 (1948). Public police are sometimes required to furnish bonds. *See* Greenstone, *supra* note 57, at 397 n.1.
59. "[A] current of sympathy for the police has begun to run through law-abiding segments of all communities in grateful recognition of the dangers and pressures to which these men daily expose themselves." Manos, *Police Liability for False Arrest or Imprisonment,* 16 Clev.-Mar. L. Rev. 415, 427 (1967). Two members of this law-abiding segment have advocated civil immunity for all police activities performed in the course of duty. Jones & Mathes, *Toward a "Scope of Official Duty" Immunity for Police Officers in Damage Actions,* 53 Geo. L.J. 889 (1965).
60. *See generally* Jaffe, *Suits Against Governments and Officers: Sovereign Immunity,* 77 Harv. L. Rev. 1 (1963); Note, *Municipal Immunity in Police Torts,* 16 Clev.-Mar. L. Rev. 448 (1967). Although this doctrine appears to be undergoing steady erosion—*see, e.g., Hargrove v. Town of Cocoa Beach,* 96 So. 2d 130 (Fla. 1957)—it is notable that, under the Federal Tort Claims Act, recovery for false arrest, false imprisonment, assault, and battery is expressly excluded. 28 U.S.C. §2680(h) (1964).
61. *See* cases cited at notes 52-54 *supra*. Moreover, at least one case has suggested that recovery may be permitted against the client of a private police agency, even though the service contract specifically provided that the agency would assume full responsibility for the private police operations. *Komorowski v. Boston Store,* 263 Ill. App. 88, 93-96 (1931).
62. Liability is normally imputed to the employer of a private policeman for those acts which occur within the scope of the employment. In determining this scope, courts sometimes test whether the activity was in furtherance of the employer's interests. *See, e.g., Mackie v. Ambassador Hotel & Inv. Corp.,* 123 Cal. App. 215, 11 P. 2d 3 (1932). In *Mackie* the court held that, although the arrest furthered the employer's interests, the subsequent illegal detention did not, and hence the employer could not be held liable. The test should be expanded to in-

clude all acts performed in connection with the private policeman's employment—that is, all acts performed during duty hours and on duty premises—and should not depend on whether, in fact, an individual act benefited the employer.

63. One agency executive has estimated that in practically every case of unfriendly contact between his operatives and the general public—such as ejecting an unruly patron at a sporting event—a civil suit ensued. Burns Interview, *supra* note 46. *See also* R. Momboisse, *supra* note 9, at 414.

64. Although in many illegal arrests by private police the harm may be little more than dignitary, juries have often returned substantial verdicts. *E.g., National Food Stores, Inc. v. Utley,* 303 F.2d 284 (8th Cir. 1962) ($10,000); *Burke v. New York, N.H. & H.R.R.,* 267 F.2d 894 (2d Cir. 1959) ($2,250); *Montgomery Ward & Co. v. Medline,* 104 F.2d 485 (4th Cir. 1939) ($3,000); *Gibson v. J. C. Penney Co.,* 165 Cal. App. 640, 331 P.2d 1057 (1958) ($5,000).

65. "Search" includes the following: (1) search of the person, including an area within his reach, *Chimel v. California,* 395 U.S. 752 (1969), his clothing and external patting thereof, *Terry v. Ohio,* 392 U.S. 1 (1968), and the surface and orifices of the body, including fingerprinting and taking of blood, *Schmerber v. California,* 384 U.S. 757 (1966); (2) search of private property, including vehicles, *Carroll v. United States,* 267 U.S. 132 (1925); and (3) surveillance, including wiretapping and electronic eavesdropping, *Katz v. United States,* 389 U.S. 347 (1967), and the use of spies or informers, *Osborn v. United States,* 385 U.S. 323 (1966).

The literature on public police search and seizure is voluminous. Particularly useful is *Model Code of Prearraignment Procedure* (Tent. Draft No. 3, 1970) [hereinafter cited as MCPP No. 3]. For a listing of authorities on the subject, see L. Weinreb, *Criminal Process* 764-68 (1969).

66. *E.g., People v. Santiago,* 53 Misc. 2d 264, 278 N.Y.S.2d 260 (Rockland County Ct. 1967).

67. Electronic surveillance or wiretapping is a form of search, and its use, like other forms of search, must conform to the strictures of the fourth amendment when engaged in by governmental authorities. *Katz v. United States,* 389 U.S. 347 (1967); *Berger v. New York,* 388 U.S. 41 (1967). *See generally* Kitch, *Katz v. United States: The Limits of the Fourth Amendment,* 1968 Sup. Ct. Rev. 133; Kitch, *The Supreme Court's Code of Criminal Procedure: 1968-1969 Edition,* 1969 Sup. Ct. Rev. 155, 188-90; Note, *From Private Places to Personal Privacy: A Post-Katz Study of Fourth Amendment Protection,* 43 N.Y.U.L. Rev. 968 (1968).

68. The law on consent search has developed primarily in the public police field. The courts generally require that waiver of fourth amendment rights be made voluntarily, and impose a rather strict burden of proof on the prosecution to show voluntariness. *Johnson v. United States,* 333 U.S. 10 (1948); *Channel v. United States,* 285 F.2d 217 (9th Cir. 1960). *See generally* MCPP No. 3, §§4.01-.03, Comments at 51-57. In the private police context the issue of voluntary consent might arise on a motion to suppress evidence if a statute provides for exclusion, *see* text at notes 98-100 *infra,* or as a defense to a suit for trespass or invasion of privacy. *See generally* W. Prosser, *Torts* §18, §112 at 850-51 (3d ed. 1964); *Restatement (Second) of Torts* §§167-84.

Moreover, a minority of federal courts have held that the principles announced in *Miranda v. Arizona,* 384 U.S. 436 (1966), which established strict standards for waiver of fifth and sixth amendment rights, require advising a suspect of his fourth amendment right to require a search warrant and waiver of that right by the suspect before a valid consent to a search may be obtained. *United States v. Nikrasch,* 367 F.2d 740 (7th Cir. 1966); *United States v. Moderacki,* 280 F.Supp. 633 (D. Del. 1968); *contra, United States ex rel. Harris v. Hendricks,* 423 F.2d 1096 (3d Cir. 1970). *See generally* Note, *Consent Searches: A Reappraisal After, Miran-*

da v. Arizona, 67 Colum. L. Rev. 130 (1967); Note, *Constitutional Law— Miranda v. Arizona and the Fourth Amendment*, 46 N.C.L. Rev. 142 (1967). Analogously, private police might be required (though not on constitutional grounds) to advise a suspect of his right to refuse to allow a search. Failure to do so could be per se evidence of nonconsent, either for the purpose of evidence exclusion or for imposition of tort liability on the private policeman for illegal search.

69. 18. U.S.C. §§2510-11 (1964). Only oral communications "uttered by a person exhibiting an expectation that such communication is not subject to interception under circumstances justifying such expectation" fall within the purview of the Act. 18 U.S.C. §2510(2) (1964). This language was intended to conform to the standards set out in *Katz*. S. Rep. No. 1097, 90th Cong., 2d Sess. 2161-63, 2178 (1968) [hereinafter cited as Senate Report].

70. *See* Subcomm. on Administrative Practice and Procedure of the Senate Comm. on the Judiciary, 89th Cong., 2d Sess., *Laws Relating to Wiretapping and Eavesdropping* (Comm. Print 1966), for a listing of state statutes in effect in 1966. In contrast to the federal statute, some of the state statutes enforce blanket prohibitions on all unconsented-to eavesdropping. *E.g.*, Cal. Pen. Code Ann. §653i (1961); Ill. Ann. Stat. ch. 38, §§14-1-2 (Smith-Hurd 1970). *See also* American Bar Ass'n, Project on Minimum Standards for Criminal Justice, Electronic Surveillance (Tent. Draft, 1968).

71. The only conceivable support for a nonconsensual private search of a building or vehicle is an old common law property right, which granted the owner of personal property or his agent the right to enter upon the land of one who had wrongfully taken such property and return it to his own possession. The right has also seen some use in the shoplifting context. However, reasonable mistake as to the culprit or presence of stolen property was usually no defense to a later suit in trespass brought by the victim of the search. W. Prosser, *supra* note 68, at §22, pp. 119-20; Restatement (Second) of Torts §§100-06, and comment d at 175; *S. H. Kress & Co. v. Musgrove*, 153 Va. 348, 149 S.E. 453 (1929). The right has apparently fallen into desuetude. MCPP No. 3, *supra* note 65, at 27.

72. *See* text at notes 81-84 *infra*. Given the conclusion that private parties should have a limited right to search a person incident to an arrest, particularly for self-defense, there may be occasions when a cursory search of an area within the immediate reach of the arrestee would be equally necessary to fulfill that purpose; for example, when the arrestee is in a car. *Chimel* recognized this in the public police context. 395 U.S. at 763; and *see* MCPP No. 3, *supra* note 65, at 36-37.

73. Ariz. Rev. Stat. Ann. §13-1415 (1956); Cal. Pen. Code Ann. §846 (1961); Hawaii Rev. Laws §708-8 (1968); Idaho Code Ann. §19-613 (1948); Iowa Code Ann. §755.12 (1950); Nev. Rev. Stat. §171.146 (1969); N.D. Cent. Code Ann. §29-06-24 (1960); Okla. Stat. Ann. tit. 22, §206 (1969); S.D. Comp. Laws Ann. §23-22-17 (1967).

74. Mich. Stat. Ann. §28.884 (1954); Del. Code Ann. tit. 11, §2303 (Supp. 1968). Both here and in note 73 *supra*, the statutes speak in terms of "about the person," which might easily be read to allow a search of both the person and the area within his reach. *See* note 72 *supra*. *See also* Tex. Code Crim. Proc. Ann. §18-22 (1966), which has received rather varied interpretation by the Texas courts. It seems fairly clear that search thereunder must be incident to a lawful arrest. *Davis v. State*, 113 Tex. Crim. 421, 21 S.W. 2d 509 (1929).

75. Texas appears to be the only exception. Tex. Pen. Code Ann. art. 1436e(2) (Supp. 1970). *See generally* Comment, *Survey and Analysis of Tort and Criminal Aspects of Shoplifting Statutes*, 58 Mich. L. Rev. 429, 447-49 (1960).

76. One jurisdiction apparently grants the right, *People v. Santiago*, 53 Misc. 2d 264, 278 N.Y.S.2d 260 (Rockland County Ct. 1967); *People v. Williams*, 53 Misc. 2d 1086, 281 N.Y.S.2d 251 (Syracuse City Ct. 1967), while another limits search to that authorized by

its search statute. *People v. Martin*, 225 Cal. App. 2d 91, 36 Cal. Rptr. 924 (1964). One case has totally disallowed such a right. *Application of Fried*, 68 F.Supp. 961 (S.D.N.Y. 1946). *But cf. Agnello v. United States*, 290 F. 671, 684 (2d Cir. 1923), *rev'd in part*, 269 U.S. 20 (1925); *United States v. Viale*, 312 F.2d 595 (2d Cir. 1963).

77. *People v. Williams*, 53 Misc. 2d 1086, 281 N.Y.S.2d 251 (Syracuse City Ct. 1967); see *Williams v. Williams*, 8 Ohio Misc. 156, 221 N.E.2d 622 (C.P. 1966). But *cf.* note 42 *supra*.

78. "The right of the people to be secure in their persons, houses, papers, and effects, against unreasonable searches and seizures, shall not be violated, and no Warrants shall issue, but upon probable cause, supported by Oath or affirmation, and particularly describing the place to be searched, and the persons or things to be seized." U.S. Const. amend. IV.

79. 395 U.S. 752 (1969); see MCPP No. 3, §§3.01-.04, Comments.

80. 395 U.S. at 763.

81. 392 U.S. 1 (1968).

82. The statutory materials directly support this proposition, *supra* note 73. If the private party does arrest, and fears injury from the arrestee, it is doubtful that reasonable civil or criminal penalties would deter search for weapons. On the other hand, imposition of such penalties would force a citizen to arrest at his peril, hardly an incentive for private law enforcement.

83. *See* note 45 *supra*.

84. *People v. Williams*, 53 Misc. 2d 1086, 1090, 281 N.Y.S.2d 251, 256 (Syracuse City Ct. 1967).

85. *See Burns Handbook*, *supra* note 10, at 16, 18: "Always try to avoid physical contact, including body search, when making an arrest. . . . Searches [of the person] should be made only when necessary. . . ." *Cf. People v. Trimarco*, 41 Misc. 2d 775, 245 N.Y.S.2d 795 (Sup. Ct. 1963). This and the generalizations which follow are in part the conclusions reached from the interview-study of private police practice.

86. This is the pattern seen in industrial security, armored car service, and the investigative aspects of retail security. *See* R. Momboisse, *supra* note 9, at 403-15.

87. *See* text at notes 47-49 *supra*.

88. 1 F. Harper & F. James, *supra* note 16, at §§3.1-3.5.

89. *E.g., Sackler v. Sackler*, 15 N.Y.2d 40, 44, 203 N.E.2d 481, 483, 255 N.Y.S.2d 83, 86 (1964) ("Proof of guilt collected in raids by private detectives has been . . . the basis for thousands of divorce decrees in our State."); *Williams v. Williams*, 8 Ohio Misc. 156, 221 N.E. 2d 622 (C.P. 1966); *Report on Private Detectives* (1957), *supra* note 5, at 41; *Agency for Investigation and Detection, Inc. v. Dep't of State*, 25 App. Div. 2d 738, 169 N.Y.S.2d 168 (1966); Bylin, Super Snooper, *Wall Street Journal*, Feb. 17, 1971, at 1, col. 1 [hereinafter cited as *Wall Street Journal*].

90. *See Commonwealth v. Murray*, 423 Pa. 37, 223 A.2d 102 (1966); *Report of the New York State Legislative Committee on Privacy of Communications and Licensure of Private Detectives* 20-22 (1960); *Report of the New York State Legislative Committee on Privacy of Communications and Licensure of Private Detectives* 19-33 (1956); *Wall Street Journal*, *supra* note 89; *Senate Report*, *supra* note 69, at 2154.

91. *See Report on Private Detectives* (1957), *supra* note 5, at 57-58; 2 sources cited note 89 *supra*. Private police agencies concerned with retail and industrial security, who might use an illegal search to uncover evidence of employee misfeasance—see *Wall Street Journal*, *supra* note 89; R. Momboisse, *supra* note 9, at 403-15; *People v. Johnson*, 153 Cal. App. 2d 870, 315 P.2d 468 (1957)—usually denied engaging in such activity. This was not surprising, however, in view of the heavy penalties which often attend such conduct. *See* text at notes 92-95 *infra*.

92. 18 U.S.C. §2511(1) (1964) provides that an offender "shall be fined not more than $10,000 or imprisoned not more than five years, or both." 18 U.S.C. §2520 (1964) provides for civil recovery of actual damages, if more than a prescribed liquidated sum, and both punitive damages and litigation costs.

93. *See* sources at note 70 *supra*. And *see* Ill. Ann. Stat. ch. 38, §14-6 (Smith-Hurd 1964) providing for injunctive relief and both actual and punitive damages.
94. *See* 1 F. Harper & F. James, *supra* note 16, at §§1.1-1.10 (1956).
95. W. Prosser, *supra* note 68, at §112, p. 833.
96. *See Senate Report*, *supra* note 69, at 2156.
97. *See Report on Private Detectives* (1957), *supra* note 5, at 41, 57-58.
98. 18 U.S.C. §2515 (Supp. V, 1965-69) provides for exclusion in "any court, grand jury, department, office, agency, regulatory body, legislative committee, or other authority of the United States, a State, or a political subdivision thereof. . . ." *See Senate Report*, *supra* note 9, at 2184-85.
99. *E.g.*, Ill. Ann. Stat. ch. 38, §14-5 (Smith-Hurd 1964); Nev. Rev. Stat. §200.680 (1969); N.Y. Civ. Prac. Law §4506; Ore. Rev. Stat. §41.910 (1969); R.I. Gen. Laws Ann. §11-35-13 (Supp. 1970).
100. *See* Tex. Code Crim. Proc. Ann. art. 38.23 (1966) (exclusion at any *criminal* trial of evidence against an accused obtained by any person in violation of Texas law or the federal Constitution); *Williams v. Williams*, 8 Ohio Misc. 156, 221 N.E.2d 622 (C.P. 1966); *Del Presto v. Del Presto*, 92 N.J. Super. 305, 223 A.2d 217 (1966) (exclusion of evidence at a *civil* trial when illegally obtained by a private party); *cf. Deiner v. Mid-American Coaches, Inc.*, 378 S.W.2d 509 (Mo. 1964).
101. *Report on Private Detectives* (1957), *supra* note 5, at 57-58.
102. *But cf. Knoll Associates, Inc. v. Dixon*, 232 F.Supp. 283 (S.D.N.Y. 1964), and discussion at note 12 *supra*.
103. *See* text at notes 110-19 *infra*.
104. *See* text at notes 156-68 *infra*.
105. Some comment in the literature has argued that the rule be applied to private police. Comment, *Regulation of Private Police*, 40 S. Cal. L. Rev. 540, 546-49 (1967); Comment, *Seizures by Private Parties: Exclusion in Criminal Cases*, 19 Stan. L. Rev. 608 (1967).
106. *Mapp v. Ohio*, 367 U.S. 643 (1961). The rule also extends to search by private parties where they act at the hire or under the direction of governmental officials. *E.g., Knoll Associates, Inc. v. FTC*, 397 F.2d 530 (7th Cir. 1968); *Corngold v. United States*, 367 F.2d 1 (9th Cir. 1966); *Stapleton v. Superior Court*, 70 Cal. 2d 97, 447 P.2d 967, 73 Cal. Rptr. 575 (1969); *People v. Tarantino*, 45 Cal. 2d 590, 290 P.2d 505 (1955).
107. 256 U.S. 465 (1921). The Court has not since re-examined its position, though certiorari was denied in *People v. Radazzo*, 220 Cal. App. 2d 768, 34 Cal. Rptr. 65 (1963), *cert. denied*, 377 U.S. 1000 (1964).
108. *E.g., Wolf Low v. United States*, 391 F.2d 61 (9th Cir. 1968); *Barnes v. United States*, 373 F.2d 517 (5th Cir. 1967); *United States v. Goldberg*, 330 F.2d 30 (3d Cir. 1964); *People v. Trimarco*, 41 Misc. 2d 775, 245 N.Y.S.2d 795 (Sup. Ct. 1963).
109. *E.g., People v. Radazzo*, 220 Cal. App. 2d 768, 34 Cal. Rptr. 65 (1963), *cert. denied*, 377 U.S. 1000 (1964); *People v. Botts*, 250 Cal. App. 2d 478, 58 Cal. Rptr. 412 (1967). *See also* recent cases involving illegal foreign police search: *Stonehill v. United States*, 405 F.2d 738 (9th Cir. 1968); *Brulay v. United States*, 383 F.2d 345 (9th Cir. 1968); *Commonwealth v. Wallace*, 356 Mass. 92, 248 N.E.2d 246 (1969).
110. 376 U.S. 643 (1961).
111. *Id.* at 656.
112. *Id.* at 651-52; *see Elkins v. United States*, 364 U.S. 206, 220-21 (1960); *People v. Cahan*, 44 Cal. 2d 434, 445, 447, 282 P.2d 905, 911-12, 913 (1955). *See generally* Oaks, *Studying the Exclusionary Rule in Search and Seizure*, 37 U. Chi. L. Rev. 665, 668-72 (1970).
113. *See generally* Oaks, *supra* note 112. Also *see Terry v. Ohio*, 392 U.S. 1, 12-15 (1968).
114. Oaks, *supra* note 112, at 673-74.
115. The Court in *Terry v. Ohio*, 392 U.S. 1, 13 (1968), stated:

> [I]n some contexts the rule is ineffective as a deterrent. Street encounters between citizens and police officers are incredibly rich in diversity. . . . Encounters are initiated by the police for a wide variety of reasons, some of which are wholly unrelated to a desire to prosecute for crime.

See also Lankford v. Gelston, 364 F.2d 197 (4th Cir. 1966).
116. *See* text at notes 101-02 *supra*.
117. *See* text at notes 85-88 *supra*.

118. Note that this conclusion would not follow if *Mapp* were extended as a *civil* exclusionary rule to private police search. This is essentially the mechanism employed by the courts in *Williams v. Williams*, 8 Ohio Misc. 156, 221 N.E.2d 622 (C.P. 1966), and *Del Presto v. Del Presto*, 92 N.J. Super. 305, 223 A.2d 217 (1966), to create judicially a civil exclusionary rule. See note 100 *supra*. The problem with these cases is that they are doctrinally unsound, however correct they may be from a policy standpoint. Not only do they extend *Mapp* to private activity, in direct opposition to the trend of the case law, notes 108-09 *supra*, but they extend *Mapp* to civil proceedings, though that decision has been otherwise extended no farther than to quasi-criminal actions. *See One 1958 Plymouth Sedan v. Pennsylvania*, 380 U.S. 693 (1965); Comment, *The Applicability of the Exclusionary Rule to Civil Cases*, 19 Baylor L. Rev. 263 (1967). Clearly the best solution is legislative enactment of such a civil rule, exemplified by the existing statutes providing a limited civil exclusionary rule in the eavesdropping context. See notes 98 & 99 *supra*.

119. A criminal exclusionary rule in the private police area would be justifiable if carefully drawn to include only those areas of conduct where it might have some deterrent effect, whether on public or private police:

> [A] rigid and unthinking application of the exclusionary rule, in futile protest against practices which it can never be used effectively to control, may exact a high toll in human injury and frustration of efforts to prevent crime.

Terry v. Ohio, 392 U.S. 1, 15 (1968). From this standpoint, the Texas statutory exclusionary rule, Tex. Code Crim. Proc. Ann. art. 38.23 (1965), may be too broadly drawn. *See* note 100 *supra*. And *see* Oaks, *supra* note 112, at 754-57, which concludes that the rule may be unjustifiable even in the public police area.

120. *Miranda v. Arizona*, 384 U.S. 436 (1966), dealt with "the protection which must be given to the privilege against self-incrimination when the individual is first subjected to police interrogation *while in custody at the station or otherwise deprived of his freedom of action in any significant way*." 384 U.S. at 477 (emphasis added). The Court termed such questioning "custodial interrogation." The detention requirement has been emphasized by later decisions. *See, e.g., Orozco v. Texas*, 394 U.S. 324, 326-27 (1969); *Mathis v. United States*, 391 U.S. 1, 4-5 (1968). *People v. Rodney P. (Anon.)*, 21 N.Y. 2d 1, 233 N.E.2d 255, 286 N.Y.S.2d 255 (1967). Note also the shoplifting statutes, which often grant a right to detain "for questioning." *E.g.*, Mass. Ann. Laws ch. 231 §94-B (Supp. 1971) and sources at notes 44 & 45 *supra*.

121. Every citizen has "the liberty . . . to address questions to other persons, for ordinarily the person addressed has an equal right to ignore his interrogator and walk away. . . ." *Terry v. Ohio*, 392 U.S. 1, 32-33 (1968) (Harlan, J., concurring). When questioning is incident to a detention, however, that liberty becomes less clear. In the public context, *Miranda* allows questioning by police only if the suspect has readily assented (assuming counsel is not present).

122. *E.g., State v. Valpredo*, 75 Wash. 2d 368, 450 P.2d 979 (1969); *People v. Williams*, 53 Misc. 2d 1086, 281 N.Y.S.2d 251 (Syracuse City Ct. 1967); *People v. Crabtree*, 239 Cal. App. 2d 789, 49 Cal. Rptr. 285 (1966).

123. This is true in both the public and private police contexts. Compulsion by the public police in interrogating a suspect is strictly prohibited by the fifth amendment as read by *Miranda* (though exclusion at trial is the only penalty). *See also Malloy v. Hogan*, 378 U.S. 1 (1964). Specific criminal penalties are also often imposed on public police. *E.g.*, Tex. Pen. Code art. 1157 (1961). Similarly, private police are subject to criminal penalties as well as liability in tort. *See* note 128 *infra*. Authority to compel testimony exists only where a governmental agency can guarantee immunity from prosecution to the party from whom testimony is sought. *See, e.g.*, 1 K. Davis,

Administrative Law Treatise §3.08 (1958); 8 J. Wigmore, *Evidence* §§ 2281-82 (McNaughton ed. 1961); *Garrity v. New Jersey,* 385 U.S. 493 (1967); cf. *Spevack v. Klein,* 385 U.S. 511 (1967).

124. See the guard manual cited in *Peak v. W. T. Grant Co.,* 409 S.W.2d 58 (Mo. 1966) ("In stores in which office is within view of sales floor it is advisable to take shoplifter to stockroom, employees' restroom, or other place. . . ."); and sources at note 122 *supra.*

125. *State v. Valpredo,* 75 Wash. 2d 368, 450 P.2d 979 (1969), is an example of the typical handling of a shoplifting case. And see *State v. Masters,* 154 N.W.2d 133 (Iowa 1967).

126. *E.g., State v. Valpredo,* 75 Wash. 2d 368, 450 P.2d 979 (1969). Private parties are not constitutionally required to give the *Miranda* warnings. See text at notes 131-34 *infra.* Nevertheless, a substantial sector of the industry requires its operatives to give the *Miranda* warnings at the inception of all arrests or detentions, often despite specific advice of counsel to the contrary. Interview with a Director of Retail Security, in Chicago, Ill., Jan. 6, 1971 (name withheld by request). A variant on this pattern is that the agencies require the warnings to be given in localities where the district attorney has interpreted *Miranda* as applying equally to private and public law enforcement personnel. Interview with a Director of Retail Security, Chicago, Ill., Dec. 6, 1970 (name withheld by request). A leading authority in the field of retail security has written:

I have discussed this problem with District Attorneys in six states . . . and found every one of them supported the belief that the Supreme Court ruling [*Miranda*] is an interpretation of the rights of ALL citizens and therefore is directed just as much at so-called "private police" as at public law enforcement. . . . [A]s a consultant to the National Crime Commission I have had opportunity to discuss this situation with men connected with the commission and they also can see no reason for assuming the Supreme Court guidelines do not apply equally to private and public law enforcement personnel.

Letter from S. J. Curtis, author of *Modern Retail Security* (1968), to Eric T. Lodge, Dec. 16, 1966, on file in the University of Southern California Law Library, excerpt reprinted in Comment, *Regulation of Private Police,* 40 S. Cal. L. Rev. 540, 546 n.36 (1967). In any event, awareness of *Miranda* seems high among supervisors of private police organizations.

127. The rights of a guard or his employer to detain after an arrest or by detention under a shoplifting statute are strictly limited, typically to detention in "a reasonable manner and for a reasonable time." See note 45 *supra.* In retail security practice, the public police are usually called immediately after the detention or arrest. Interviews with Retail Security Directors, Chicago, Ill., Dec. 6, 1970, and Jan. 6, 1971 (names withheld by request).

128. An overlong detention subjects the arrestor to liability for false imprisonment. See note 40 *supra.* Use of threats or force is discouraged not only by standard criminal sanctions for assault, battery, and intimidation, but also by specific provisions in some jurisdictions prohibiting private coercion of confessions. *E.g.,* Nev. Rev. Stat. §199.460 (1967); Colo. Rev. Stat. Ann. §44-1-10 (1963); Ill. Ann. Stat. ch. 38, §12-7 (Smith-Hurd 1964).

129. See 3 J. Wigmore, *Evidence* §§821-30 (1940).

130. *E.g., Mefford v. State,* 235 Md. 497, 201 A.2d 824 (1964); *State v. Christopher,* 10 Ariz. App. 169, 457 P.2d 356 (1969); *People v. Frank,* 52 Misc. 2d 266, 275 N.Y.S.2d 570 (Sup. Ct. 1966). And see Comment, *Confessions Obtained Through Interrogations Conducted by Private Persons, Investigators, and Security Agents,* 4 Willamette L.J. 262, 266-68 (1966). There are also statutory enactments of the rule. *E.g.,* Minn. Stat. Ann. §634.03 (1947); N.Y. Code Crim. Proc. §395; Tex. Code Crim. Proc. §38.21 (1966), §38.22 (Supp. 1970).

131. *See, e.g., On Lee v. United States,* 343 U.S. 747, 757 (1952); *United States v. United Shoe Mach. Corp.,* 89 F.Supp. 349 (D. Mass. 1950); Weinstein, *Probative Force of Hearsay,* 46 Iowa L. Rev. 331 (1961).

132. 384 U.S. 436 (1966). Recent commentary on the decision is extensive. See H. J. Friendly, *Benchmarks* 266-84 (1967); Kamisar, *A Dissent from the Miranda Dissents: Some Comments on the "New" Fifth Amendment and the Old "Voluntariness" Test*, 65 Mich. L. Rev. 59 (1966). See generally the bibliography in L. Weinreb, *Criminal Process* 770-74 (1969).
133. See note 120 *supra*.
134. 384 U.S. at 444-45. Dissatisfaction with *Miranda*, primarily because of its restrictive effect on law enforcement, *see Senate Report, supra* note 69, at 2123-53, has led to a legislative attempt to overrule the decision. In 1968, as part of the Omnibus Crime Control and Safe Streets Act, Congress enacted 18 U.S.C. §3501 (Supp. V, 1965-69) in an attempt to return to pre-*Miranda* admissibility law. See text at notes 138-40 *infra*. The courts have not yet confronted the conflict, but as to its possible outcome, see Burt, *Miranda and Title II: A Morganatic Marriage*, 1969 Sup. Ct. Rev. 81. Moreover, it appears that the Burger Court views the Fifth Amendment in a somewhat less indulgent light than did its predecessor. *See Harris v. New York*, 39 U.S.L.W. 4281 (U.S. Feb. 24, 1971), holding, 5-4, that a confession from a suspect obtained in violation of *Miranda* might be used against him at a later trial to impeach his credibility on the stand.
135. *Pratt v. State*, 9 Md. 220, 263 A.2d 247 (1970) (holding that a citizen acting under a special police commission was subject to the *Miranda* requirements; *see Commonwealth v. Bordner*, 432 Pa. 405, 247 A.2d 612 (1968) (interrogation of a boy by his mother held subject to the *Miranda* requirements when she was acting at the insistence and in the presence of a law enforcement officer). For other "private" parties who may be subject to *Miranda* requirements, see *Procunier v. Atchley*, 39 U.S.L.W. 4125 (U.S. Jan. 19, 1971); *People v. Polk*, 63 Cal. 2d 443, 406 P. 2d 641, 47 Cal. Rptr. 1 (1965); *People v. Frank*, 52 Misc. 2d 266, 275 N.Y.S.2d 570 (Sup. Ct. 1966). *See generally* Annot., *Custodial Interrogation—Miranda Rule*, 31 A.L.R.3d 565 (1970).
136. *E.g., Shaumberg v. State*, 83 Nev. 372, 432 P.2d 500 (1967); *State v. Christopher*, 10 Ariz. App. 169, 457 P.2d 356 (1969); *see* cases compiled in Comment, *Admissibility of Confessions or Admissions of Accused Obtained During Custodial Interrogation by Non-Police Personnel*, 40 Miss. L.J. 139 (1968).
137. Courts have recognized this distinction in other fields. See the cases in note 109 *supra* dealing with the fourth amendment exclusionary rule. One of those cases, *Commonwealth v. Wallace*, 356 Mass. 92, 248 N.E.2d 246 (1969), also recognized this distinction in the *Miranda* context. See discussion at note 142 *infra*.
138. *See Evalt v. United States*, 359 F.2d 534, 542 (9th Cir. 1966); *People v. Frank*, 52 Misc. 2d 266, 275 N.Y.S.2d 570 (Sup. Ct. 1966).
139. *E.g., Greenwald v. Wisconsin*, 390 U.S. 519 (1968); *Rogers v. Richmond*, 365 U.S. 534 (1961); *Haynes v. Washington*, 373 U.S. 503 (1963). *See generally* Paulsen, *The Fourteenth Amendment and the Third Degree*, 6 Stan. L. Rev. 411 (1954).
140. 384 U.S. at 445-49; *see* IV National Comm'n on Law Observance and Enforcement, Report on Lawlessness in Law Enforcement (1931) (Wichersham Comm'n Report); Booth, *Confessions, and Methods Employed in Procuring Them*, 4 S. Cal. L. Rev. 83 (1930).
141. Friendly, *The Fifth Amendment Tomorrow: The Case for Constitutional Change*, 37 U. Cinn. L. Rev. 671, 710-11 (1968). *See e.g., Greenwald v. Wisconsin*, 390 U.S. 519 (1968); *Clewis v. Texas*, 386 U.S. 707 (1967). Moreover, Judge Friendly cites a growing mistrust in the Court of police honesty. Friendly, *supra*, at 711. See also Kitch, *The Supreme Court's Code of Criminal Procedure: 1968-1969 Edition*, 1969 Sup. Ct. Rev. 155, 159-60.
142. *Contra*, Oaks, *supra* note 112, at 671 ("deterrence was the 'single and distinct' purpose" of the *Miranda* decision); and *see Commonwealth v. Wallace*, 356 Mass. 92, 248 N.E.2d 246 (1969). In that case foreign police officials had obtained a statement of guilt from the defendant without properly warn-

ing him of his rights as required by *Miranda*. The confession was subsequently introduced in evidence against the confessor in a stateside criminal proceeding. The Supreme Court of Massachusetts held that the confession was properly admitted since foreign police practice would not be affected or deterred by an extraterritorial extension of *Miranda*. On *Wallace, see United States v. Nagelberg* (2d Cir. Nov. 9, 1970). *But cf. Bram v. United States,* 168 U.S. 532 (1897).
143. 384 U.S. at 475.
144. "The Fifth Amendment privilege is so fundamental to our system of constitutional rule and the expedient of giving an adequate warning . . . so simple, we will not pause to inquire in individual cases whether the defendant was aware of his rights without a warning being given." *Id.* at 468.
145. 66 Cal. 2d 232, 424 P.2d 947, 57 Cal. Rptr. 363 (1967). *See also Bram v. United States,* 168 U.S. 532 (1897); *United States v. Miller,* 261 F. Supp. 442 (D. Del. 1965).
146. The California Supreme Court might have assumed otherwise had the case come up on appeal after the decision in *O'Callahan v. Parker,* 395 U.S. 258 (1969), where the Supreme Court limited military court criminal jurisdiction to cases which were "service connected." *Id.* at 272. In any event, the confession in *Kelley* would be admissible before a military tribunal. Military law does not afford a pretrial right to counsel, nor is that right guaranteed to service personnel by the federal Constitution, 66 Cal. 2d at 248-49, 424 P.2d at 960, 57 Cal. Rptr. at 376.
147. *Id.* at 250, 424 P.2d at 961, 57 Cal. Rptr. at 377. This conclusion finds support in *People v. Varnum,* 66 Cal. 2d 808, 427 P.2d 772, 59 Cal. Rptr. 108 (1967), *appeal dismissed,* 390 U.S. 529 (1968), where the California Supreme Court held that a criminal defendant could not object to use at his trial of a confession obtained from a third party by public police in violation of *Miranda* standards. The court recognized that the defendant would have had standing to object if the evidence sought to be introduced had been secured in violation of the Fourth Amendment, since the Fourth Amendment exclusionary rule was designed primarily to deter illegal police search. But the court rejected the defendant's argument that protection of the *Fifth Amendment* privilege afforded by *Miranda* required analogous treatment:

Noncoercive questioning is not in itself unlawful, however, and the Fifth and Sixth Amendment rights protected by . . . *Miranda* are violated only when evidence obtained without the required warnings and waiver is introduced against the person whose questioning produced the evidence.

66 Cal. 2d at 812, 427 P.2d at 775, 59 Cal. Rptr. at 111.

For other authorities suggesting that *Miranda* was intended to do more than deter police abuse, see *Harris v. New York,* 39 U.S.L.W. 4281, 4294-95 (U.S. Feb. 24, 1971) (Brennan, J., dissenting); letter cited in note 126 *supra;* Note, *Evidence Taken in Violation of the Fourth Amendment and Statements Taken in Violation of the Fifth Amendment Held Admissible in a Domestic Court When Secured by Foreign Police,* 56 Va. L. Rev. 335 (1970).
148. Judge Friendly characterizes this new protection of the Fifth and Sixth Amendments as "a ground bass that resounds throughout the *Miranda* opinion." Friendly, *supra* note 141, at 711. His idea is that the Warren Court sought equal protection of the Sixth Amendment and the Fifth Amendment privilege "by advancing the point at which the privilege became applicable and surrounding the poor man with safeguards in the way of warning and counsel that would put him more nearly on a par with the rich man and the professional criminal." *Id.* at 711.
149. The argument here is that the suspect is aware that he is dealing with private parties and hence feels less compelled to answer questions than if his interrogators were public police. Factors which might dispel this informal atmosphere would be the presence of uniformed guards, a clear statement of intention by the interrogator that

criminal prosecution is likely, or the suspect's awareness that the public police are on the way, typically the case in the shoplifting context. On the uniform question *see People v. Wright,* 249 Cal. App. 2d 692, 694, 57 Cal. Rptr. 781, 782 n.1 (1967).

150. *E.g., Orozco v. Texas,* 394 U.S. 324 (1969); *United States v. Pierce,* 397 F.2d 128 (4th Cir. 1968). *See also Miranda v. Arizona,* 384 U.S. 436, 469 (1966): "The circumstances surrounding in-custody interrogation can operate very quickly to overbear the will of [a suspect]...."
151. *See Mathis v. United States,* 391 U.S. 1 (1968); *Procunier v. Atchley,* 39 U.S. L.W. 4125 (U.S. Jan. 19, 1971); *People v. Arguello,* 13 Cal. 2d 566, 407 P.2d 661, 47 Cal. Rptr. 485 (1965).
152. *See Miranda v. Arizona,* 384 U.S. 436, 448-54 (1966), and sources cited therein. *Id.* at 448 n.8. *See also* F. Inbau & J. Reid, *Criminal Interrogations and Confessions* 24-117 (1967). A point developed earlier, note 6 *supra,* and deserving of further notice here, is that private police often have an extensive background in public law enforcement, particularly at the supervisory level. Furthermore, under present industry practice, field interrogations are generally the province of these supervisors alone. Burns Interview, *supra* note 46; Interviews with Retail Security Directors, Chicago, Ill., Dec. 6, 1970, and Jan. 6, 1971. *Burns Handbook, supra* note 10, at 18: "[Guard] duties do not include investigating persons committing crimes, except to the extent that oral questioning may be necessary. . . ." It would appear, then, that private police interrogation is often conducted with a level of inquisitorial skill and experience analogous to that displayed in the conduct of public police interrogations.
153. 384 U.S. at 449-58.
154. A desire to effect a peaceable return of stolen property without involving the public police would seem to be the only motive for an interrogation not involving a desire to obtain a confession for use in court. This is in marked contrast to a search or arrest, which clearly serve the aims of property protection aside from conviction goals. Furthermore, the industry practice of administering the *Miranda* warnings, *supra* note 126, is highly suggestive of a conviction orientation in private police interrogations. In many localities, the prosecuting attorney will refuse to prosecute a suspect to whom the warnings had not been given at the time of arrest. This is the underlying reason for administration of the *Miranda* warnings by at least one large retail security agency. Interview with a Director of Retail Security, in Chicago, Ill., Jan. 6, 1971 (name withheld by request).
155. It is unlikely that the courts would enforce administration of the *Miranda* warnings by liability in tort. This step has not been taken even in the public police context. *See Allen v. Eicher,* 295 F.Supp. 1184, 1185-86 (D. Md. 1969) (". . . *Miranda* does not per se make an interrogation which violates its precepts into an actionable tort").
156. *See* text and notes at notes 137 & 138 *supra.*
157. *See* text at notes 22-25 *supra.*
158. 341 U.S. 97 (1951).
159. 18 U.S.C. §242 (1964).
160. 341 U.S. at 99.
161. *Pratt v. State,* 9 Md. App. 220, 263 A.2d 247 (1970).
162. *Id.* at 249.
163. *E.g., Irvis v. Scott,* 318 F.Supp. 1246 (M.D. Pa. 1970). And *see Garner v. Louisiana,* 368 U.S. 157, 182-85 (1961) (Douglas, J., concurring); *Lombard v. Louisiana,* 373 U.S. 267, 281-83 (1963) (Douglas, J., concurring).
164. *Seidenberg v. McSorleys' Old Ale House, Inc.,* 317 F.Supp. 593, 599 (S.D.N.Y. 1970).
165. 326 U.S. 501 (1946).
166. *See* notes 4 & 12 *supra.*
167. 391 U.S. 308 (1968).
168. *Id.* at 325, quoting *Marsh v. Alabama,* 326 U.S. 501, 506 (1946).

BOOKS

Barrett, Edward L., Jr.: *The Tenney Committee, Legislative Investigation of Subversive Activities in California.* Ithaca, New York. Cornell University Press, 1951.

Black, M. A., and Campbell, Henry: *Black's Law Dictionary*. The Publisher's Editorial Staff, 4th Edition. St. Paul, Minn., West Publishing Co., 1951.

Calkins, Donald A.: *Cases and Materials on Michigan Criminal Law for the Police*. Research funded by a grant from National Institute of Law Enforcement and Criminal Justice Law Enforcement Assistance Administration, United States Department of Justice.

Fein, Sherman E., J.D., Ed.D. and Maskell, Arthur M., J.D.: *Selected Cases on the Law of Shoplifting*. Springfield, Thomas, 1975.

Chamberlain, Lawrence H.: *Loyalty and Legislative Action: A Survey of Activity by the New York State Legislature*. 1919-1949. Cornell University Press, 1951.

Cushman, Robert E.: *Civil Liberties in the United States: A Guide to Current Problems and Experience*. Cornell University Press, 1956.

George, James, Jr.: *Constitutional Limitations on Evidence in Criminal Cases*. Practising Law Institute, New York 1973.

Hubbard, David G.: *The Skyjacker: His Flights of Fantasy*. New York, Macmillan Company, 1971.

Inbau, Fred E. and Aspen, Marvin E.: *Criminal Law for the Layman*. Philadelphia, Chilton Book Company, 1970.

Kerper, Hazel B., J.D. and Kerper, Janeen, J.D.: *Legal Rights of the Convicted*. Criminal Justice Series, St. Paul, Minn., West Publishing Co., 1974.

Kimball, Spencer L.: *Historical Introduction to the Legal System*. Ann Arbor, Michigan, Overbeck Company, Publishers, 1962.

Klotter, John C., B.A., J.D. and Meier, Carl L., J.D.: *Criminal Evidence for Police*. W. H. Anderson Company, 1971.

Landynski, Jacob W.: *Search and Seizure and the Supreme Court—A Study in Constitutional Interpretation*. Baltimore, Johns Hopkins Press, 1966.

Rifas, Richard A.: *Private Police and the Law*. Private Police Publishers, 2nd Printing, 1974.

Skolnick, J.: *Justice Without Trial*. New York, Wiley, 1966.

Wiretap World Opens a Little. Long Island Press, May 27, 1966.

Taylor, Telford: *Grand Inquest: The Store of Congressional Investigations*. Simon and Schuster, 1955.

PUBLICATIONS OF THE GOVERNMENT, LEARNED SOCIETIES, AND OTHER ORGANIZATIONS

AFL-CIO: Staff Report for the AFL-CIO Executive Council Regarding Charges Against Mr. Dave Beck. Washington, May 9, 1957.

California Senate Judiciary Committee Report on the Interception of Messages by the Use of Electronic and Other Devices. Calif. Legislature, Regular Session, 1957.

California State Senate Committee of the Judiciary. Report on the interception of messages by the use of electronics and other devices and the use of such in the suppression of crime and the use of such by private persons for their own use. California State Printing Office, 1957.

Clarifying and Protecting the Right of the Public to Information and for Other Purposes. Report No. 1210, Sen. Committee on the Judiciary. 88th Cong., 2nd Session, 1964.

Eavesdropping and Wiretapping. Report of the Joint Legislative Committee to Study Illegal Interception of Communications, Legislative Document No. 53, 1956, p. 86.

Eavesdropping, Wiretapping and Licensed Private Detectives. Report of the Joint Legislative Committee to Study Illegal Interception of Communications. March 1957, p. 117.

Electronic Eavesdropping. Report of the Joint Legislative Committee on Privacy of Communications and Licensure of Private Investigators. May 1958, p. 126.

Food and Drug Administration. The Code of Laws of the United States of America. Title 21, Section 221, Washington, Gov. Printing Office.

General Rules and Regulations. SEC Act of 1934. Government Printing Office. Washington, D.C., July, 1963.

Government Electronic Data Processing Systems. Hearings Before the Subcommittee on Census and Statistics of the Committee on Post Office and Civil Service, 89th Cong., 2nd sess., 1966.

Greenwood, Peter W.: An Analysis of the Apprehension Activities of the New York City Police Department, The New York City Rand Institute, Report No. 529. September 1970.

Hearings Before the Senate Subcommittee on Constitutional Rights of the Committee on the Judiciary, 86th Cong. 1st Sess., pt. 3, 1959, p. 679.

Honsberger, John: The Power of Arrest and the Duties and Rights of Citizens and Police. Law Society of Upper Canada Lectures, 1963, p. 1.

Invasion of Privacy (Government Agencies). Hearings Before the Subcommittee on Administrative Practice and Procedure of the Senate Judiciary Committee. 89th Cong., 1st Sess., 1965.

Library of Congress: Survey of Foreign Law of Wiretapping. Hearing on Wiretapping, Eavesdropping, and the Bill of Rights. 85th Cong., 2nd Sess.

McKay, R. B.: Statement for the Committee on the Bill of Rights of the Association of the Bar of the City of New York before the Committee on the Judiciary. United States Senate, 1962, p. 19.

Mehay, Stephen L.: Police and Productivity: Can the Invisible Hand of Competition Extend the Long Arm of the Law? *Federal Reserve Bank of Philadelphia Business Review*, May 1973, p. 3.

Muellar, G. W.: The Law Relating to Police Interrogation Privileges and Limitations. *Police Power and Individual Freedom*, Chicago, 1963.

National Information Center, Hearings Before the Ad Hoc Committee on a National Research Data Processing and Information Retrieval Center of the House Committee on Education and Labor, 88th Cong., 1st Sess., 1963.

National Petroleum Council's Committee on Emergency Preparedness for the Petroleum Industry: Civil Defense and Emergency Planning for the Petroleum and Gas Industries, Vol. 1, *Principles and Procedures*, March 19, 1964.

New York City Bar Association: Monitoring Devices and Lawyers. Record of the Association of the Bar of the City of New York, No. 9, 1965.

Pan American Unions, Dept. of International Law: Strengthening of Internal Security. March 26-April 7, 1951.

Parker, Donn B.: Testimony Before the State of California Assembly Committee on Statewide Information Policy. September 22, 1969.

Report of the California Legislature, Senate Judiciary Committee on the Interception of Messages by the Use of Electronic and Other Devices. Sacramento, 1957, p. 29.

Report of the Joint Legislative Committee on Privacy of Communications and Licensure of Private Investigators. Albany, New York, 1961, p. 86.

Report of the Royal Commission on Security. Queens Printer, Ottawa, 1969.

Report of the President's Commission on Campus Unrest. Washington, D.C., Government Printing Office, 1970.

Report of the New York Legislature Committee on Privacy of Communications and Licensure of Private Detectives. 1957, pp. 43-60, 1958, pp. 43-48.

Revised Statutes of Canada, 1970. Queens Printer, Ottawa, 1971.

Senate Committee on Armed Services, Hearings, National Defense Establishment. 80th Cong., 1st Sess., (3 parts), Washington, D.C., 1947, p. 758.

Senate Committee on the Judiciary, Subcommittee to Investigate the Internal Security Act of 1950, The Wennerstroem Spy Case, How It Touched the U.S. and NATO. 88th Cong., 2nd Sess., Washington, D.C., 1964.

Senate Committee on Naval Affairs, Report to the Secretary of the Navy, Unification of the War and Navy Departments and Postwar Organization for National Security. 79th Cong., 1st Sess., 1945, pp. 12-13, 159-163.

Senate Committee on Rules and Administration, Report, Joint Committee on Central Intelligence Agency, Senate Report No. 1570, 84th Cong., 2nd Sess., Washington, D.C., February 23, 1956.

Senate Judiciary Internal Security Subcommittee, Hearing. Interlocking Subversion in Government Departments. 83rd Cong., 1st Sess., Washington, D.C., part 13, June 25, 1953.

State Statutes on Wiretapping. Subcommittee on Constitutional Rights of the Committee on the Judiciary, United States Senate, 87th Cong., 1st Sess., 1961, p. 110.

U.S. Congress, House: Civil Defense in Western Europe and the Soviet Union. 86th Cong., 1st Sess., No. 300, April 27, 1959.

———: Civil Defense Reorganization. 85th Cong., 2nd Sess., Report No. 26, June 12, 1958.

———: Committee on Government Operations. 84th Cong., 2nd Sess., Report No. 2946, 1956.

———: New Civil Defense Legislation. 85th Cong., 1st Sess., Report No. 2125, 1957.

———: New Civil Defense Program. 87th

Cong., 1st Sess., Report No. 1249, September 21, 1961.

———: Twenty-Fourth Intermediate Report of the Committee on Government Operations. 84th Cong., 2nd Sess., Report No. 2946, 1956.

U.S. Congress House Select Committee on Small Business: Proprietary Rights & Data, 86th Cong., 2nd Sess., Report No. 51, March 29, 30, 31, 1960.

U.S. Congress House, Committee on the Judiciary Counsel Affairs and Security Administration & the Department of State Hearings. 87th Cong., 2nd Sess., Report No. 9904, January 31, 1962 and February 2, 1962.

U.S. Congress, House of Representatives. The Computer and Invasion of Privacy. 50th Cong., 1966.

———. Records of the Invasion of Privacy. *Congressional Record*, May 18, 1965, p. 10821-25.

Wigmore, John H.: *A Treatise on the Anglo-American System of Evidence in Trials at Common Law.* Vol. 8, 3rd ed., Boston, 1940.

Wiretapping. Hearings Before Subcommittee No. 5 of the House Committee on the Judiciary, 84th Cong., 1st Sess., Sept. 2, 1955.

Wiretapping and Eavesdropping Legislation. Hearings Before the Subcommittee on Constitutional Rights of the Committee on the Judiciary. United States Senate, 87th Cong., 1st Sess., 1961.

Wiretapping and The Omnibus Crime Bill and Safe Streets Act of 1968. *Security World,* Vol. 5, No. 8, September, 1968.

Wiretapping for National Security. Hearings Before a Subcommittee of the Senate Judiciary Committee, 83rd Cong., 2nd Sess., 1954.

Working Papers of the National Commission on Reform of Federal Criminal Laws, Vol. II, Part 4, July 1970, p. 916.

PERIODICALS

Adams, E. L.: The right of privacy and its relation to the law of libel, *29 American Law Review,* Vol. 37, 1905.

Adjutant General's Department, State of Ohio: Industrial security, Bulletin No. 6-1 *(Civil Defense Information),* Columbus, Ohio, 1959.

Advisory group to make proposals for private security "model code," *LEAA Newsletter,* 3:No. 1, 14, February-March 1973.

Alvid, Samuel: Planned bankruptcy, *Security World,* IV:No. 7, July-August 1967.

American Bar Association, Section of Criminal Law and Criminology, Report of committee on mercenary crime, 23:94-100, May-June 1932.

American Bar Association commission on campus government and student dissent, *Report,* 1970.

American Civil Liberties Union: Post security program, *New York Times,* May 20, 1963, p. 5.

American Civil Liberties Union: Post security program, *New York Times,* May 22, 1963, p. 18.

American Civil Liberties Union: The wiretapping problem today. New York, 1962.

American Cyanamid: American Cyanamid's action against alleged piracy of its antibiotics secrets, *Chemical and Engineering News,* July 2, 1962.

———: Deciding drug suits, *Chemical Week,* January 18, 1964.

———: Drug company official denies Cyanamid suit. *Chemical and Engineering News,* March 19, 1962.

———: Drug-secrets ring disclosed, *The Record,* June 19, 1962.

———: Theft of Drug Secrets, *Washington Daily News,* June 21, 1962, p. 2.

———: Cyanamid charges Pentagon buys drugs made by Italians with pirated secrets, *Wall Street Journal,* June 26, 1962.

———: Canadian go-between in drugs pirating uncovered in U.S., *The Gazette,* July 8, 1962.

———: 20th century pirates, *Newark Star-Ledger,* July 26, 1962.

———: A federal grand jury has indicted three former employees of American Cyanamid's Lederle Laboratories Divisions, *Chemical and Engineering News,* November 5, 1962, p. 17.

American Cyanamid charges a consultant of Italian firm bought stolen drug data, *Wall Street Journal,* July 9, 1964.

———: Former employee stole Cyanamid's secrets, New York judge rules, *Wall Street Journal,* January 10, 1964, p. 6.

———: Italian firm denies it paid Cyanamid chemist for stolen drug data, *Wall Street Journal,* July 23, 1964, p. 5.

Anonymous and staff written: Wiretapping and the omnibus crime bill & safe streets act of 1968, *Security World,* 5:No. 8, September 1968.

Angell, Ernest: Communism or freedom: The legal profession and the rule of law, *American Bar Association Journal*, October 1957, p. 45.

Barnes, Jack: Fed, state and local gouts tighten laws for security firms, *Security News*, January 30, 1973, pp. 10-15.

Barriers to the flow of technical information: Limitation statements—legal basis, *Research in Education*, January-December 1970.

Bazelon, David: Law and order without justice, *Congressional Record*, March 2, 1966.

Beach, Robert E.: A question of property rights. The government and industrial know-how, *American Bar Association Journal*, 41:Nov. 1955.

Beaverstock, Derek M.: Almost certainly ours, *Security and Protection*, September 1974, pp. 9-10.

————: Nine tenths of the law, *Top Security*, May 1975, pp. 34-35.

Berger, David: Law in the security world: The proposed federal eavesdropping law, *Security World*, 4:No. 4, April 1967.

The Big Ear. Final script of NBC television program, October 31, 1965.

Bilek, Arthur J.: Security regulation in the seventies, *Security Management*, September 1975, pp. 8, 9, 13.

Bishop, J. W., Jr.: The executive's right of privacy: An unresolved constitutional question, *Yale Law Journal*, 66:1957.

Black, Jack: A burglar looks at laws and codes, *Harper's Monthly Magazine*, February 1930, pp. 306-315.

————: A reporter at large: Burglary—II, *New Yorker*, 39:89-91 (B), December 14, 1963.

Black, Susan: A reporter at large: Burglary—I, *New Yorker*, December 7, 1963, pp. 63-64 (A).

Blake, Gene: Law in the security world: The constable blundered, *Security World*, III: No. 7, July-August 1966.

BNDD Bulletin: Eavesdropping with consent, *Security World*, November 1971.

Boten, Bernard, and Gordan A. Murray: Trial of the future, the challenge to the law, *Security World*, 2:No. 5, July-August 1965.

Braun, Michael and David J. Lee: Private police forces: Legal powers and limitations, *University of Chicago Law Review*, 38:555-582, 1971.

Cahill, Thomas: Protection of the company's interest in a possible criminal situation-panelist, *Industrial Security*, October 1963.

Calvert, Arthur: Watch out for wire taps, *Security World*, 2:No. 2, March 1965.

Canada: Report of royal commission to investigate disclosures of secret and confidential information to unauthorized persons, Ottawa: Clouthier, 1946.

Can congress force steel to tell all? *Business Week*, September 15, 1962, p. 30.

Carlin, J. E.: Lawyer's ethics, New York, Russell Sage Foundation, 1966.

Carlson, Alan, and Feeney, Floyd: Handling robbery arrestees: Some sources of fact and policy, The Center on Administration of Criminal Justice, University of California, Davis, 1973.

Chamberlain, J. P.: Anti-fence legislation, *American Bar Association Journal*, 14:1928.

Clark vs. Bunker, 172 U.S.P.Q., 1972.

Collins, Daniel: Security firms get industry bug deals, *Security News*, January 30, 1973, pp. 1, 3, 20.

Commonwealth vs. Fisher, 213 Pa., 1905.

The Constitutional Right to Anonymity, *Yale Law Journal*, 1961.

Consumer Credit Reporting, *Security World*, May 1971.

Courtney, Jeremiah: Section 605 and you, Part I, *Security World*, 1:No. 1, July 1964.

————: Section 605 and you, Part II, *Security World*, 1:No. 2, September 1964.

Creamer, J. Shane: Case Commentaries, *Security World*, VIII, Nos. 1-7, January-August 1970.

Dale, Jerry: Bill of rights violated—by company snoopers, *UAW-Solidarity*, IX:4-5, March 1965.

Daugherty, Fred H.: Shopping center picketing held legal, *Security World*, 5:No. 10, November 1968.

Davis, John R.: Legal aspects of industrial security, *Police*, 4:No. 3, January-February 1960.

Does industrial security suppress civil liberties? *American Society for Industrial Security*, April, 1958, pp. 21-24.

Doherty, Joseph F.: Legal and industrial relations aspects, *Industrial Security*, December 1964.

Don't duck the legal issues, they may bite you, *Industry Week*, September 22, 1975, pp. 35-37.

Drug companies fear secrets might get out as result of proposed law: Moss bill, *Oil, Paint, and Drug Report*, August 1957, pp. 172-73.

Eavesdropping legislation: Down—but not out? *Time,* June 23, 1967, p. 45.

Frantz: The new supreme court decision on the federal civil rights status, *Industrial Personnel Security,* 1951.

Frank: The right to work, *Industrial and Labor Relations Review,* 1952, p. 247.

Fraenkel: The federal civil rights law, *Industrial Personnel Security,* 1947.

Fadiman, Clifton: Please tap my wire, I like it! *Holiday,* July 1964, p. 12.

Fairfield, W. S., and Clift, C.: The wiretappers I, *The Reporter,* 7:8-22, December 23, 1952.

———: The wiretappers II, *The Reporter,* 8:9-20, January 6, 1953.

———: The private eyes, *The Reporter,* 12:14-29, February 10, 1955.

Findley, Robert C., Frank J. Micle and Robert M. Hanlon: Consumer in the marketplace, *Notre Dame Lawyer,* 38:556-613, August 1963.

Foote, C.: Tort remedies for police violations of individual rights, *Minnesota Law Review,* 39:493, 1955.

Gardner, John: Footing the government's new bill, *Security Gazette,* May 1975, pp. 174-175.

Gaseler, John: Insurance for wrongful arrest, *Top Security,* May 1975, p. 48.

Godkin, E. L.: The rights of the citizen to his reputation, *Scribner's,* 58:1890.

Gohr, Phillip R.: Exercising security authority, *Security World,* January 1972, p. 14.

Goldberger, P.: Tony Imperiale stand vigilant for law and order, *New York Times Magazine,* September 29, 1969.

Goodrich, B. F. vs. Wohlgemuth, 192 N.E.2d 99, 117 Ohio Appeals 493.

Gordis, Phillip: Government checks can bounce, *Security World,* 1:No. 2, September-October 1964.

Greenleaf, Simon: *A Treatise on the Law of Evidence,* 4th ed., Boston, 1848.

Goure, Leon: Civil defense in the Soviet Union, Los Angeles, U. of California, 1962.

Gressman: The unhappy history of civil rights legislation, *Industrial Personnel Security,* 1952.

Gross, Phillip J.: Privacy and the law: A legislative summary, *Assets Protection,* Spring, 1975, pp. 61-62.

Hamilton, Peter: Towards regulation in the security industry, *Security Gazette,* July 1970.

———: The police and the security industry. I. The decline of public interest to law and order, *Police Journal,* 41:261-267, 1968.

———: The police and the security industry. II. The rise of the security industry and its role in crime prevention, *Police Journal,* 41:297-303, 1968.

Hamilton, W.: The ancient maxim caveat emptor, *Yale Law Journal,* 40:No. 1133, 1931.

Hays, Robert C.: The California law of unfair competition takes a turn—against the employer, *California Law Review,* 41:1953.

Hinerfeld, Robert E.: Law in security world: Legal aspects of recording telephone conversations, *Security World,* 3:No. 1, December-January, 1966.

Hoebel, E. A.: The law and national survival, *Security World,* December 1970.

Illinois Licensing Law, *Security World,* 1:No. 2, September-October 1964.

Inbau, Fred E., and Reid, John E.: Criminal interrogation and confession, *Security World,* 2:No. 3, May 1965.

Inbau, Fred E.: Interrogation: The law, *Security World,* September 1969.

———: Review of the common law, Part I and II, *Security World,* June-August 1971.

Invasion of privacy, Gallagher Committee, 89th Cong., 1st Sess., 1965.

Jennings, Edward L., Jr.: Loyalty oath rulings, *Industrial Security,* February 1968, p. 6.

———: Effect of a ruling against loyalty laws, *Industrial Security,* June 1967.

Jerome S. Spevack vs. Lewis S. Straus, Supreme Court, October 1957.

Katzell, R. A.: A psychologist examines violations of privacy, *Virginia Law Weekly,* 17:No. 4, 1964.

King, Donald B.: Electronic surveillance and constitutional rights: Some recent developments and observations, *George Washington Law Review,* 33:240-269, 1964.

King, Paul A.: Living with the law, *Dun's Review and Modern Industry,* March 1960.

Landes, William M., and Posner, Richard A.: Private enforcement of law, *Journal of Legal Studies,* 1-46, January 1975.

Law and Private Police, The, *Vol. 4:* The Rand Corporation, 1971.

Law in the security world: The Randazzo decision, *Security World,* 2:No. 2, March, 1965.

Lawrence, David: The lost right of privacy, *American Mercury,* 33:12-18, 1936.

Lear, F. S.: Treason in Roman and Germanic

law, *Collected Papers,* University of Texas Press, 1965.

Legal action on tetracycline piles up, *Chemical and Engineering News,* July 20, 1964, p. 21.

Legislature of Ontario Debates, November 12, 1970, p. 6308; October 23, 1970, p. 5472.

———: April 8, 1965, p. 402; May 7, 1970, p. 2446.

———: May 6, 1965, p. 2651.

———: November 10, 1970, p. 625.

———: June 17, 1971, p. 2851; June 20, 1972, p. 3927.

———: November 24, 1972, p. 4799; November 28, 1972, p. 4891.

———: April 2, 1965, p. 1926; June 12, 1968, p. 4430.

———: December 9, 1960, p. 344.

Lenhoff: The right to work: here and abroad, *Illinois Law Review, 46:*699, 1951.

Lovibond, Oliver: Court procedure, *Security and Protection,* 9-12, November 1974.

Marcus: Civil rights and the anti-trust laws, *University Chicago Law Rev.,* 1951.

Machen, Ernest W., Jr.: Search and seizure, Chapel Hill, Institute of Government.

Mason, W. P., and Marshall, R. N.: A tubular directional microphone, *Acoustical Soc Am, 10:*206-215, January 1939.

McDade, Thomas M.: The Miranda decision and the security officer, *Industrial Security,* December 1967.

McKee, W. F.: Evidentiary problems: Camera surveillance of sex deviates, *Law and Order,* August 1964, p. 72.

Mellin, W. J.: I was a wiretapper, *Saturday Evening Post,* Sept. 10, 1949.

Michael, D. N.: Speculation on the relation of the computer to individual freedom and the right of privacy, *George Washington Law Review,* 1964, pp. 270-286.

Michigan State University: Michigan statutes annotated, *Criminal Law, 24:*Sec. 28.128.-532, 1962.

Miranda vs. the "Private Citizen," Protect Newsletter, September 1973, pp. 1, 2.

Mitchell, Evan: Is moonlighting legal? *Security and Protection,* May 1975, p. 22.

Monsanto vs. Miller, Chemical Engineering, June 2, 1958.

Morse, W.: Remarks on rights of privacy as a constitutional guarantee, *Congressional Record,* February 2, 1967.

Mors, Wallace P.: State regulation of retail installment financing: Progress and problems, *Journal of Business,* October 1950, pp. 199-218, and January 1951, pp. 43-71.

Morton, Desmond: Aid to the civil power: The Canadian militia in support of social order, 1867-1914, *The Canadian Historical Review, 51:*No. 4, 407, December 1970.

Murphy, Walter: *Wiretapping on Trial.* New York, 1965.

Murray: The right to equal opportunity in employment, *California Law Review, 33:*338, 1945.

National security act of 1947, *Public Law, 253:* July 26, 1947.

National Stolen Property Act: Title 18 United States Code, Section 2314.

New Jersey joint legislative committee to study wiretapping and unauthorized recording of speech, *Report,* New Jersey, November 1958.

New security bill—will it be passed or pigeonholed? *Occupational Hazards,* July, 1970, p. 47.

New York City Bar Association: Mental illness, due process, and the criminal defendant, Fordham University Press, 1968.

New York Times: Articles on Mobil oil case, December 10, 11, 12, 1971.

———: High court faces plea on wiretaps, March 13, 1969.

———: Mitchell hopes high court will reverse eavesdropping ruling, March 19, 1969.

———: Bugging of common market offices stirs furor, March 16, 1970.

———: Privacy vs. protection—The bugged society, June 8, 1969.

Nizer, L.: The right of privacy: A half century's development, *Michigan Law Review, 39:*526, 1941.

O'Conner, Daniel: The Green case, *Industrial Security,* July, 1965.

Ohio Treasury Investigation Committee, report of the investigating commission, April 12, 1958.

Osborn, Albert S.: *The problem of proof,* Boyd Printing Company, 1950.

Ostrow, Ronald J.: Wiretap cases bug nation's law officers, *Los Angeles Times,* May 29, 1966, p. 1.

Peterson, Virgil W.: Tipping the scales of justice, *Security World,* July/August, 1969.

Policeman on private premises, The, *Police, 5:* No. 2, 23, October 1972.

Poulantzas, Nicholas M.: The right of hot pursuit in international law. 1969.

Prosser, W.: Privacy, *Columbia Law Review, 48:*383, 1960.

Provost, Nancy Lou: New procedures for in-

dustrial security hearings, *Industrial Security Magazine,* January 1961.
Rauh, Joseph L., Jr.: Non-confrontation in security cases: The Greene decision, *Virginia Law Review,* 1:1190, 1959.
Raymond, Aubrey L.: Legal aspects-panelist, *Industrial Security,* October 1962.
Regulation of private police, *Southern California Law Review,* 40:540, 1967.
Report on national commission review of wiretap law, *Security World,* 30-32, January 1975.
Report of the royal commission inquiry into labour dispute. Queen's Printer, 1968.
Re-proposed new drug patent law in Italy, *New York Times,* September 20, 1963.
Restatement of the law of torts, Committee on Torts, American Law Institute, Section 757, Comment b, 1939.
Rogge: Justice and civil liberties, *Industrial Personnel Security,* 1939.
Royal, Robert F.: Investigation the dishonest act, *Industrial Security,* December 1964.
Rules for the avoidance of organizational conflicts of interest, *Industrial Security,* June 1, 1963.
Rutter, William A.: *Criminal Law and Procedure,* 5th ed., Gilbert Law Summaries, Beverly Hills.
Sachs, J. H.: Criminal law torts—False imprisonment: The shoplifting problem in Wisconsin, *Wisconsin Law Review,* 3:478, 1964.
Saden, George A.: Inquiry into ambulance chasing, *Connecticut Bar Journal,* 34:117-122, June 1960.
Schiff, E. J.: Children have a right to privacy, *The P.T.A. Magazine,* 26, June 1962.
Steefen, Thomas L.: Truth in lending: A viable subject, *George Washington Law Review,* 32:861-892, April 1964.
Sutton, John F., Jr.: Authority of a person not an officer to arrest for a misdemeanor, *Security World,* 2:No. 7, October 1965.
———: How to read a court's opinion, *Security World,* 2:No. 8, November 1965.
———: Law in the security world: Commercial interference and the prima facie tort, *Security World,* 3:No. 3, March 1966.
———: Law in the security world: Escobedo and the right to counsel, *Security World,* 3:No. 4, April 1966.
———: Law in the security world: False imprisonment, *Security World,* 2:No. 6, September, 1965.
Silving, Helen: Testing of the unconscious criminal cases, *Harvard Law Review,* 69: 683-705, 1956.
Sloan: Federal civil rights legislation and the constitution, *Industrial Personnel Security,* 1949.
Smith, Chester H.: What security order and laws must achieve, *Security Management,* 47-51, January 1974, pp. 47-51.
Stanford Law Review, No. 2, 1950, pp. 744, 750.
Unions act on threats to privacy, *Business Week,* March 13, 1965, pp. 87-88.
Van De Kamp, John K.: Search and seizure, part I, *Security World,* 5:No. 1, January 1968.
———: Search and seizure, part II, *Security World,* 5:No. 2, February 1968.
Wagner, S. P.: Records and the invasion of privacy, *Social Science Journal,* January, 1965.
Walsh, Timothy J., and Healy, Richard J.: Security and the civil law, *Protection of Assets Manual,* 2:1974.
Walton, Clarence C., and Cleveland, Frederick W., Jr.: Corporation on trial: The electrical cases, 1964.
Warren: The law and the future, *Fortune,* November 1955.
Weinstein and Farer: State credit card crime act, *Security World,* April 1969.
Westin, A. F.: Bookies and bugs in California, *The Uses of Power,* 1962, p. 138.
———: Wire-tap: The house approves, *New Republic,* April 9, 1954, p. 6.
———: Wiretapping: The sleuth revolution, *Commentary,* April 1960, pp. 333-340.
———: Wiretapping problem, the, *Columbia Law Review,* 52:165, 1952.
Whalen, R. G.: To tap or not to tap, *New York Times Magazine,* December 12, 1948.
Who can tap a wire? *U.S. News & World Report,* April 1, 1955.
Whyte, William H., Jr.: Is anybody listening? New York, S. and S., 1952.
Wire tapped? snoopy operators? scramble, *Printers' Ink,* April 10, 1964, p. 15.
Wiretappers, The, *Reporter,* January 6, 1953, p. 9.
Wire-tapping cases in New York, *Outlook,* May 31, 1916, p. 234.
Wright, J. Skelly: The new role of defense under Escobedo and Miranda, *American Bar Association Journal,* 52:1117, December 1966.
Yeagley, J. Walter: Information for industry on the August 1962 amendment to the in-

ternal security act—panelist, *Industrial Security,* October 1962.

Wolk, L. I.: Some legal aspects of industrial espionage, *Practical Lawyer, 9:*April 1963.

UNPUBLISHED MATERIALS

Johnston, Terence R.: Eavesdropping: The courts, congress, and the government, Unpublished graduate B paper, Mich. State, 1973.

Landever, Arthur Robert: Electronic surveillance and the American constitutional system. Michigan, University Microfilms, Inc., 1969.

Moss, Gorest: The house committee on internal security: An American dilemma. Unpublished graduate thesis, Mich. State, 1970.

Schuster, Joseph Frederick: Electronic eavesdropping and the Fourth Amendment. The Pennsylvania State University, Ph.D., 27/03-A, 1965, p. 809.

PART II

SECURITY PROGRAMS

INTRODUCTION

It is necessary that a distinction be made between security for governmental purposes and security which is proprietary. Governmental security which includes protecting governmental classified information is set up to satisfy the requirements of the *Industrial Security Manual for Safeguarding Classified Information* or to meet the requirements of a particular governmental agency establishing a security program. Regulations detail the requirements for protecting information which is classified by the Department of Defense and other user agencies as well as specific requirements of other agencies requiring security measures. Security needs which do not involve the protection of classified government information have a wide range of sophistication and format.

The industrial, business and private application of security varies from organization to organization. Some measures are intended to protect trade secrets such as the programs utilized by many manufacturers in highly competitive businesses, while others are geared primarily for the protection of merchandise or other production commodities.

There are then two different basic sets of problems involved in providing security, based upon the clientele group to be served. The government has needs for security of a uniform manner, for example, all governmental programs concerning classified information. Private business, industries, or individuals may require a broad scope of varying security needs and measures, quite independent and different in form and function from the unitary governmental systems. It then becomes meaningful to divide security at the systems level into two major subdivisions, these being *governmental security systems* and *proprietary security systems*.

GOVERNMENTAL SECURITY SYSTEMS

The government has a need for security in many of its operations which are unique and peculiar to governmental operations. The federal government is involved in an array of activities both inside the United States and abroad. The need for security of the government can generally be divided into three major areas of concern: security in governmental operations, security in production of materials related to defense operations, and security for the storage of information or material produced for governmental needs. To provide security for these diversified needs, the government's operations are decentralized geographically. In order to provide some uniform basis for security in all phases and locations of governmental operations, various agencies have developed security programs which serve to protect their needs. A wide variety of governmental agencies, to be discussed in a subsequent chapter, provide security and intelligence gathering functions. Various branches of the military services as well as civilian agencies are responsible for internal security, intelligence collection, and law enforcement. These agencies with operational security responsibilities are, however, an integral part of the governmental security system.

Security of production and storage of governmental materials and information, while shared to a considerable degree by related federal agencies, is the major responsibility of the Secretary of Defense. The Secretary of Defense is responsible for the security of all materials produced and stored for the government by either civilian or governmental agencies or industries. The Secretary of Defense is authorized to act in behalf of a number of agencies in administering the production of materials for governmental use. He is responsible for the maintenance of governmental security standards in the production or storage of materials or commodities for the following agencies: The National Aeronautics and Space Administration, Federal Aviation Agency, General Services Administration, National Science Foundation, and Treasury Department. These agencies have been designated as *user* agencies by the government since they utilize the Department of Defense regulations to govern all production of materials for their use.

The Department of Defense has responsibility for the governmental program known as the Government Industrial Security Program. While the name is the official designation for this type of security program, the application of the name Industrial Security Program to the government system for maintaining security of its material production leads to a great deal of confusion. Various types of industrial applications of security could differ from the standards established for governmental needs; therefore, it is more appropriate to speak of the government Industrial Security Program as being a Governmental Production Security Program. This designation appears more meaningful since the various industrial proprietary security needs differ considerably and they perform many functions differing greatly from those established and required for government production security.

Industrial Defense—Industrial Security

In producing and storing materials for governmental use, the government is required to utilize the facilities of proprietary and industrial firms as well as governmental installations. While using industrial and business firms, two terms are utilized by the government when trying to establish different, but closely related types of security programming; these terms are *industrial security* and *industrial defense*.

The industrial security or production security programming involves standards of operations in the production and storage of governmental material. Industrial defense involves programming for making key production and storage facilities less vulnerable to attack and destruction by foreign powers.

Industrial defense measures include the following:

1. dispersion of new facilities and major expansions of existing facilities to locations away from target areas;
2. dispersion of movable supplies and equipment, i.e. spare parts, inventory, vehicles which are not continuously needed in target areas;
3. relocation of key production for critical items to other existing facilities not in target areas;
4. arranging for alternate sources of supply;
5. protective construction;
6. provision for fallout shelters;
7. provision for the evacuation of a facility; and
8. provision for the continuity of management.

It also includes the following:

1. the selection and equipping of alternate headquarters;
2. the establishment of personnel successions lists;
3. the protection and duplication of vital records and documents;
4. the review of all legal documents such as charters, bylaws, etc.;
5. to insure that surviving directors and officers have authority to continue operations;
6. development of emergency financial arrangements;
7. disaster plans for emergency repair and restoration of facilities;
8. organizing and training of employees for self-help;
9. establishing of industrial mutual aid associations for civil defense;
10. the preparation of corporate and plant disaster plan manuals;

11. the preparation of emergency shut-down procedures;

12. preparation of handbooks for employees; and

13. support of an assistance to the communities civil defense efforts.

The difference between industrial defense and industrial security can therefore be clearly distinguished as follows: Industrial security or production security has the goal of providing for day-to-day security of the facilities and its production, whereas industrial defense is concerned with the planning and programming for the protection of strategic production and storage needs for the national defense effort.

PROPRIETARY SECURITY SYSTEMS

We have defined proprietary security as the system involved with security for industrial, business, and private applications. Industrial applications of security included those for the manufacturing and processing industries as well as extracting and supply industries. The business applications of security include retailing and wholesaling as well as service and transportation. Proprietary applications include all individual citizens security needs as well as the needs of industrial or business companies to protect their property and their employees while on their property and enforcement of policies, rules, and regulations as they affect the company's and employees' safety and security. In order to do this, a wide variety of formats are structurally established to provide for the peculiar security needs within an organization. Likewise, the needs of a private individual for security are met in a multiplicity of applications of the security processes through various procedures and techniques.

SYSTEM SIMILARITY

If the government and proprietary systems are analyzed, it readily becomes apparent that certain functional subactivities within the security context are present in both of these systems. The needs for security, while differing in government and proprietary organizations, do show remarkable similarity in the methods which are used to provide for the security. We have seen that various Executive Orders established the requirements for governmental production security in proprietary organizations. When the organizations are involved in the production of materials for governmental use, the two systems are combined and overlap with the proprietary security normally present within the contracting organization. This very often establishes two different types of security systems within a single organization. For example, a defense contractor who manages a diversified production capability very likely has two parallel security systems; one governmental and one proprietary, within the same organization. In such a situation where the two systems exist within the organization, it becomes evident that the major functions in either of the systems are very similar. They can be categorized into administrative, preventive, and investigative functions. The administrative functions are involved with the establishment of system policy and procedure. The preventive functions are involved with the provision of safety and security for the actual ongoing operations through the use of techniques and procedures (such as guards, alarm systems, and personnel security processes). The investigative function makes inquiries into incidents involving breaches of security or violation of policies, rules, and regulations.

Chapter 4

GOVERNMENTAL SECURITY PROGRAMS

THERE ARE NUMEROUS federal agencies, which have either an intelligence or internal security function. It is essential that a general knowledge of the organization, function, and responsibilities of these various agencies and the interrelationships between them be clearly understood. The authority and specific functions of many of these agencies are often shrouded in mystery and consequently misunderstood.

All federal agencies[1] organically have some type of investigative or internal security function. Any agency would therefore provide information of an intelligence or internal security nature. There are, however, a number of agencies with specific intelligence collection responsibilities. Few federal agencies have the singular organizational responsibility for this activity. It most often occurs as a secondary or ancillary production effort.

National internal security requires that many types of information be collected, evaluated, and disseminated in a timely manner. To accomplish this, many federal agencies have been charged with the responsibility for collecting and processing certain types of information and providing it through various coordinating boards to the National Security Council (NSC) and the President of the United States. The development of a coordinated collection and dissemination effort is, however, a recent phenomenon.

HISTORY

Prior to World War II, the President was advised on matters of national security by members of his cabinet. There was very little attempt made to coordinate information obtained from the various governmental agencies, nor was any effort made to adequately check the reliability and consistency of the information provided to the President. Intelligence support efforts were operating relatively independent of each other.

During World War II, in an effort to achieve coordinated national security information, the State-War-Navy-Air Coordinating Committee (SWNACC) was formed. The various departments were represented on the SWNACC by the assistant secretaries of these departments. This committee was superceded by the National Security Council (NSC) as created by the National Security Act of 1947.

One of the first attempts at coordination of intelligence information was through the creation of the office of the Coordinator of Information (OCI) in July of 1941. This office was relatively short-lived and was succeeded in 1942 by the Office of Strategic Services (OSS). At the end of World War II, it was realized that an organized intelligence collection agency was necessary in the United States government. The Office of Strategic Services, however, was disbanded in 1945 and replaced by the Strategic Services Unit (SSU). This organization operated for approximately one year when it was replaced in 1946 by the Central Intelligence Group (CIG).[2] The experiences of operating the CIG were incorporated into the National Security Act of 1947. This act resulted in the establishment of the Central Intelligence Agency (CIA) which was given the responsibility for the collection and dissemination of foreign intelligence for the United States.

Two broad categories of agencies which collect intelligence and security information exist. They are those agencies with:

1. Primary responsibility: Those agencies with a primary organizational purpose of collection of such information, e.g. CIA, DIA.

2. Ancillary responsibility: those agencies which collect this information as part of their normal activities which are unrelated to intelligence collection, e.g. Department of Justice, Department of Treasury.

Each of these categories can further be subdivided by types of agency responsibilities.
  a. Those agencies with direct collection, evaluation, and dissemination responsibilities, e.g. FBI, CIA, Secret Service, Bureau of Customs.
  b. Those agencies with supervisory or coordination responsibilities, e.g. USIB, NSC.

These categories are not mutually exclusive. Agencies like the CIA, for example, may fit into any of these categories and types simultaneously since it has both supervisory and collection responsibilities. The Justice Department on the other hand, has an ancillary responsibility for collection, but its subdivision, the FBI, has a direct collection responsibility. Therefore, a distinction must exist between the organization's primary purpose and goals and its intelligence and security purpose and activities. A discernable difference exists between the various types of intelligence and security agencies. There are administrative-supervisory-coordinating bodies and operational agencies or branches of agencies which have responsibility for intelligence or security functions.

The National Security Council (NSC) and the United States Intelligence Board (USIB) are the two major supervisory bodies; the Central Intelligence Agency, Department of State, Department of Justice, Treasury Department, Civil Service Commission, and Department of Defense being either operational agencies or having units responsible for information-intelligence collection and/or internal security operations. The following agencies comprise what is referred to as the "intelligence community":
1. the National Security Council (NSC);
2. the United States Intelligence Board (USIB);
3. Central Intelligence Agency (CIA);
4. Department of State (State);
5. Department of Justice (Justice);
6. Department of Treasury (Treasury); and
7. Department of Defense (DOD).

A number of other agencies provide internal security intelligence information but are not considered full-time members of the intelligence community; these are (1) the United States Civil Service Commission (CSC), (2) the Atomic Energy Commission (AEC), and (3) the National Security Agency (NSA).

The various intelligence and security agencies each perform specific functions for the Federal Government. Their duties and interrelationships are, therefore, of concern in order to understand the complexity of the entire intelligence and security structure. To clarify their relationships, each major agency will be considered individually as to its purpose, organization, and role in federal intelligence and security operations.

NATIONAL SECURITY COUNCIL

The National Security Council was established by the National Security Act of 1947 (61 Stat 496; 50 U.S.C. 402), amended by the National Security Act Amend-

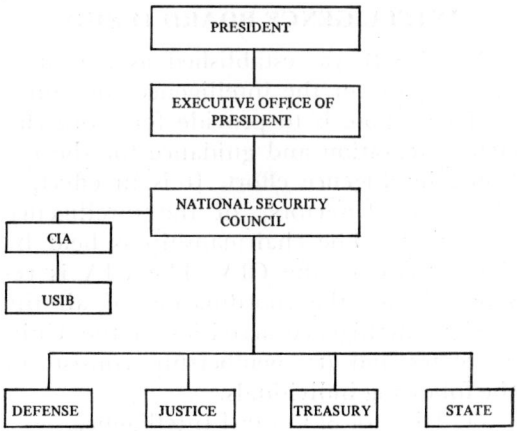

Figure 4-1. National intelligence structure.

ments of 1949 (63 Stat 479; 50 U.S.C. 401 et seq.). Its function is to advise the President with respect to the integration of domestic, foreign, and military policies relating to the national security.

Responsibilities

The specific duties of the NSC are as follows:
1. to appraise and assess objectives, commitments, and risks of the United States in relation to actual or potential military power in the interest of national security;
2. to consider policies in matters of common interest in the departments and agencies of the government concerned with national security and make recommendations to the President; and
3. to perform other such related functions as the President may direct for the purpose of coordinating the policies and functions of the departments and agencies relating to national security. The council is composed of the President, Vice President, Secretary of State, and the Secretary of Defense. The NSC is charged with advising the President with respect to the integration of domestic, foreign, and military policies relating to the national security.

THE UNITED STATES INTELLIGENCE BOARD (USIB)

The USIB was established as a coordinating body for the intelligence community. Its purpose is to provide for more efficient integration and guidance for the national intelligence effort. It is in effect, a "Board of Directors" of the intelligence community. The chairmanship is held by the Director of the CIA. The CIA is responsible for the coordination of all the foreign intelligence activities in the United States and its membership consists of the following individuals:
1. Director of Central Intelligence,
2. Deputy Chairman (Deputy of the CIA),

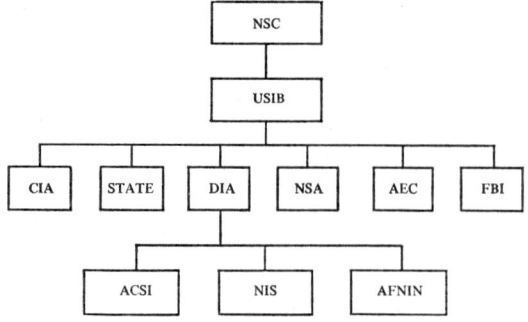

Figure 4-2. Supervisory-coordination responsibilities of USIB.

3. Director of Intelligence and Research Department of State,
4. Director of Defense Intelligence Agency,
5. Assistant Chief of Staff Intelligence, U.S. Army—as an observer,
6. Assistant Chief of Staff for Navel Operations-Intelligence Department of Navy—as an observer,
7. Assistant Chief of Staff Intelligence, Air Force—as an observer,
8. Assistant to the Director, Federal Bureau of Investigation,
9. Assistant General Manager Atomic Energy Commission, and
10. Director of the National Security Agency.

CENTRAL INTELLIGENCE AGENCY (CIA)

Creation and Authority

The Central Intelligence Agency was established under the National Security Council by the National Security Act of 1947 (61 Stat 497; 50 U.S.C. 403).

For the purpose of coordinating the intelligence activities of the several government departments and agencies in the interest of national security, the agency, under the direction of the National Security Council:
1. advises the National Security Council in matters concerning such intelligence activities of the government departments and agencies as relate to the national security;

2. makes recommendations to the National Security Council for the coordination of such intelligence activities of the departments and agencies of the government as related to national security;
3. correlates and evaluates intelligence relating to the national security and provides for the appropriate dissemination of such intelligence within the government using, where appropriate, existing agencies and facilities;
4. performs for the benefit of the existing intelligence agencies, such additional services of common concern as the National Security Council determines can be more efficiently accomplished centrally; and
5. performs such other functions and duties related to intelligence affecting the national security as the National Security Council may from time to time direct.³

During the establishment of the CIA, the need for secrecy of operations was recognized. It was also recognized that danger might arise from an agency not subject to normal governmental inspection. There was also the possibility that the agency might become a threat to internal or international security. History is replete with instances where nations have been taken over by state secret police organizations. To exclude such happenings, a number of significant provisions were included in the National Security Act to prevent this possibility, specifically, "(1) The CIA shall have no police subpoena or law enforcement powers and no internal security functions; (2) The CIA will not exceed most departmental intelligence functions." The Act specifically states that "every department" will continue to collect, correlate, evaluate, and disseminate departmental intelligence; (3) The Director of Central Intelligence, with certain approvals, has a right to inspect the intelligence production of all government security agencies. He must specify to all of

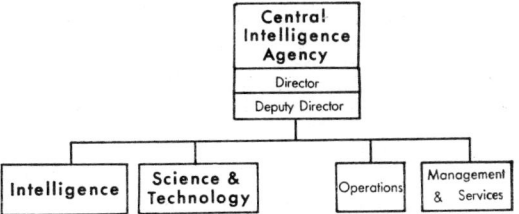

Figure 4-3. Major Divisions of CIA.

these agencies that they must make intelligence available to CIA for correlation, evaluation, and dissemination. This regulation does not apply to the Federal Bureau of Investigation. Their files, however, are made available when necessary through correspondence between the directors of the two agencies. The internal organization of the CIA is not unlike that of other intelligence or internal security operations. The agency is headed by a Director. The Director and Deputy Director are appointed by the President by and with the advice and consent of the Senate. The organization is divided into four functional deputy directorates: Intelligence, Science and Technology, Operations, Management and Services.

The Deputy Directorate of Intelligence is responsible for the assembly, analysis, and evaluation of information and the production of daily and periodic intelligence reports on any country, person, or situation—by the National Security Council and President.

The Deputy Directorate of Science and Technology is responsible for collecting information on techniques being developed in science and weapons including nuclear weapons and unconventional warfare.

The Deputy Directorate for Operations is the official name for the clandestine services division. This is the division which deals with espionage, subversion, sabotage, and paramilitary operations.

The Deputy Directorate for Management and Services is responsible for procuring equipment, security, logistics, communications, records, and so forth.

THE DEPARTMENT OF STATE

Intelligence information must be available to conduct foreign policy effectively. Therefore, a major task of the Department of State is the collection, evaluation, and dissemination of intelligence. Since the Secretary of State is a member of the National Security Council, the information obtained by the State Department can be made available directly to the President. There is a unit within the Department of State whose primary responsibility is the collection of intelligence information.

Bureau of Intelligence and Research

This bureau develops and implements a coordinated program for positive foreign intelligence for the Department of State. Its program includes the production of intelligence studies and intelligence reports pertinent to the formulation and execution of foreign policy. The director of this bureau represents the Department of State on the United States Intelligence Board and all other interdepartmental groups and committees. He is also responsible for developing and maintaining friendly and productive relations with intelligence officials of friendly foreign governments.

Office of External Research

The office is subordinate to the Bureau of Intelligence and Research. Through this office, the bureau maintains liaison with cultural and educational institutions and with other Federal agencies on a wide range of matters relating to government contractual and private foreign affairs research. The bureau contributes to the National Intelligence Estimate and is primarily responsible for political and economic information.

Bureau of Security and Consular Affairs

The second unit within the Department of State responsible for certain types of

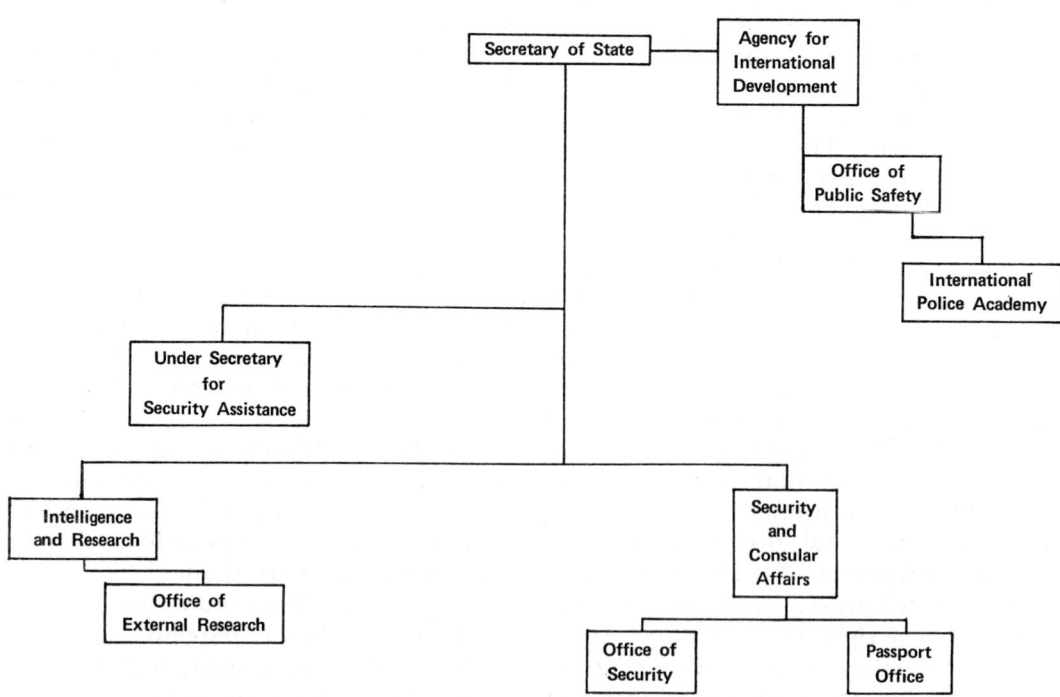

Figure 4-4. Department of State—Intelligence Security Structure.

intelligence information is the Passport Office. It is located within the Bureau of Security and Counselor Affairs (SCA) and develops intelligence information relative to personalities. This office maintains a master index with more than thirty-five million names and intelligence information. Liaison is maintained between the Passport Office and the other federal intelligence agencies.

Office of Security

This final unit is the Office of Security. This unit is responsible for the security of personnel, documents, and the physical security of department installation. This office also has a function of protecting the Secretary of State and visiting heads of state and government.

This office conducts criminal investigations at the request of other elements of the Department of State. The decision for prosecution or referral to the Department of Justice lies with the requesting office. The types of investigations conducted by this Office are as follows:

1. Conducts investigations regarding the violation of criminal statutes relating to U.S. passports. Investigations are conducted at the request of the Passport Office and when completed, the reports and files are returned to the Passport Office.
2. Conducts investigations of violations of visa laws and regulations on request from the Visa Office.
3. At request of the Munitions Control Office, conducts investigations relating to the licensing authority of the Secretary of State.

Office of Public Safety

This office in the State Department's Agency for International Development provided programs of assistance to police in foreign nations. Emphasis was placed on interdicting the flow of narcotics to the United States. Assistance was provided upon request of friendly nations and when it was in the interests of the United States to do so. Assistance in the form of (1) advisers who lived and worked in the country; or (2) training of foreign police officers in the United States, principally at the AID International Police Academy; or (3) provision of selected items of police equipment. It recently was drastically reduced in its operating capabilities.

Cabinet Committee to Combat Terrorism

This committee was established on September 25, 1972, to insure that the varied resources of the Federal Government are fully employed in combating international terrorism. The Cabinet Committee, chaired by the Secretary of State, is composed of the:
1. Secretary of the Treasury
2. Secretary of Defense
3. Attorney General
4. Secretary of Transportation
5. U.S. Ambassador to the U.N.
6. Director of the CIA
7. Assistant to the President for National Security Affairs
8. Assistant to the President for Domestic Affairs
9. Director of the FBI

The committee's responsibilities include (1) coordination of intelligence; (2) tightening up precautionary measures against terrorism; (3) preparation of contingency plans; and (4) intensified efforts to increase international cooperation.

THE DEPARTMENT OF JUSTICE

There are seven major divisions of the Department of Justice which have responsibility for internal security or intelligence collection. They are the Internal Security Division, Federal Bureau of Investigation, the Immigration and Naturalization Service, the Drug Enforcement Administration, the Organized Crime and Racketeering Section of the Criminal Division, U.S. Marshalls, and the Law Enforcement Assistance Administration.

The Internal Security Division

The Internal Security Division (ISD) is responsible for and supervises enforcement of all criminal law related to inter-

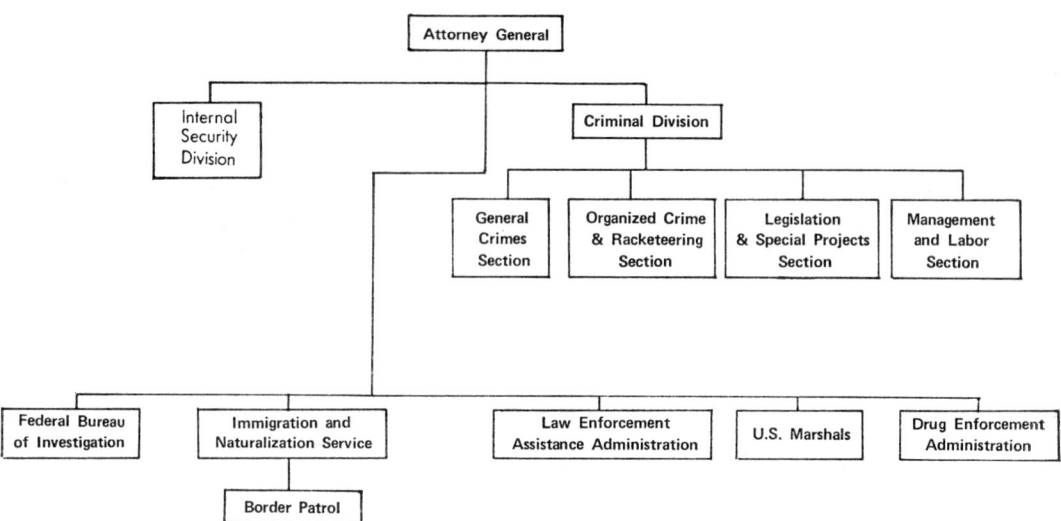

Figure 4-5. Major Divisions of the Justice Department.

nal security. This unit was formed in July, 1954 and the assistant attorney general in charge directs the United States attorneys with respect to prosecution relating to subversive activities including laws relating to espionage, sedition, treason, and criminal prosecutions under the Atomic Energy and Smith Acts.

Interdepartmental Committee on Internal Security

On March 11, 1971, the President appointed the Assistant Attorney General in charge of the Internal Security Division as Chairman of the Interdepartmental Committee on Internal Security. The committee is charged with effecting the coordination of all phases of the internal security field, except those specifically assigned to the Interdepartmental Intelligence Conference.

The Federal Bureau of Investigation

The Federal Bureau of Investigation (FBI) is the principal security agency in the United States government. The FBI was established in 1908 as the Bureau of Investigation. On July 1, 1935, the name was changed to the Federal Bureau of Investigation. As the principal law enforcement arm of the Department of Justice, the FBI is strictly a fact-gathering and fact-reporting agency. It does not issue clearances, draw conclusions, or make recommendations as to the prosecutive or administrative action to the recipients of its reports. The FBI is also charged with protecting the nation from subversion. The domestic intelligence operations in this area are primarily preventative in nature.

It is responsible for investigating violations of all Federal laws with the exception of those specifically assigned to other agencies. There are approximately two hundred such laws within the FBI's jurisdiction.

The FBI has considerable responsibility for the investigation of sabotage, espionage, treason, sedition, and subversive activities. It has primary jurisdiction for gathering intelligence on the activities of subversive organizations in the United States.

Additionally, the FBI has the responsibility for investigating of nonpolitical crimes such as bank robbery, kidnapping,

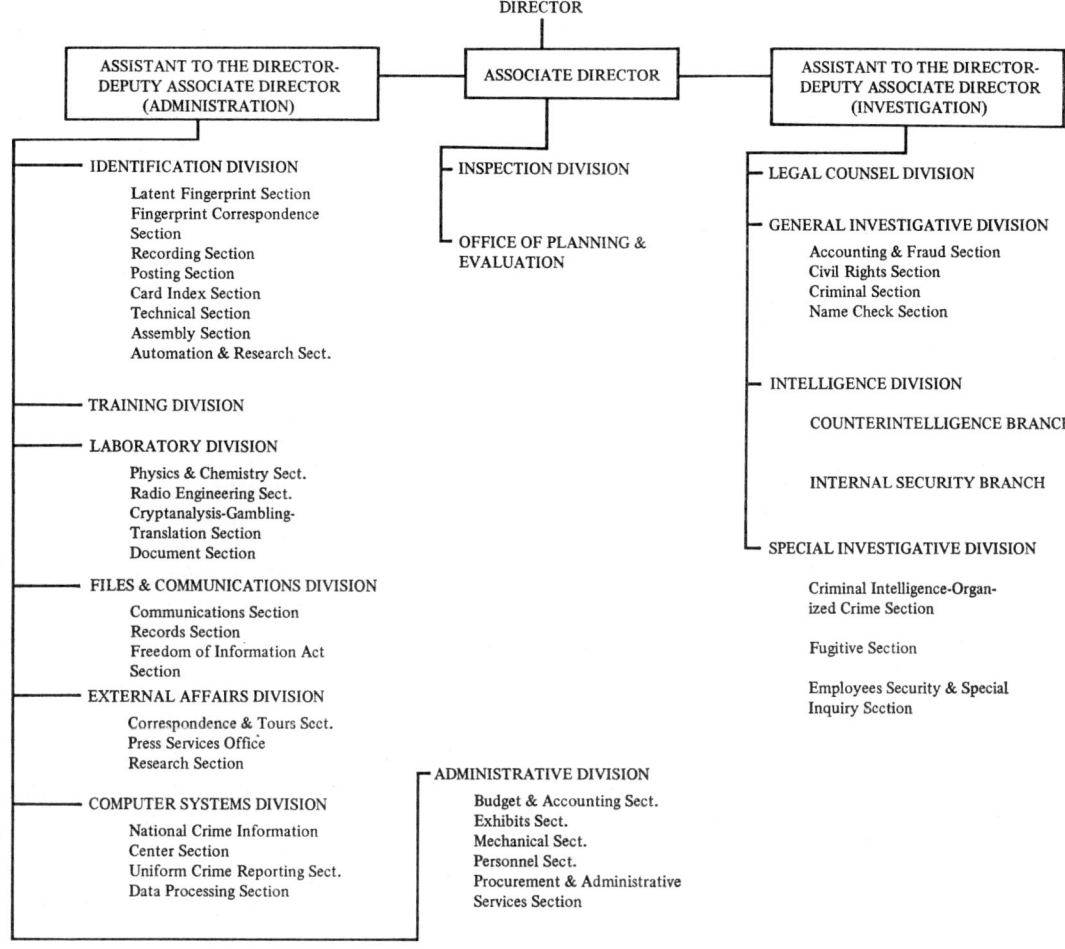

Figure 4-6. Functional organization of the Federal Bureau of Investigation.

interstate transportation, etc. The records division contains a great deal of criminal and intelligence information which is made available to official agencies of the United States government.

The Immigration and Naturalization Service—Border Patrol (INS)

The INS was created March 3, 1891, and placed in the Justice Department in June, 1946, with statutory authority for current activities derived from the Immigration and Nationality Act of June 25, 1952, as amended. The specific law enforcement responsibilities of the INS are:

1. administering and enforcing the immigration and nationalization laws relating to the admission, exclusion, deportation, and naturalization of aliens;
2. investigating alleged violations of these laws;
3. patrolling the borders of the U.S. to prevent illegal entry; and
4. cooperating closely with all law enforcement agencies.

Since abolishment of the Customs Patrol in 1948, the Border Patrol is the only Federal uniformed organization assigned to work regularly between ports of entry on the land borders of the United States. The Patrol is specifically assigned to the protection of the borders of the U.S. against illegal entrants, narcotics traffickers, smugglers, and other violators of Federal law.

INS maintains investigators in some foreign countries and does gather information about individuals and their activities. It also fingerprints and registers aliens in the United States. In performing these functions, it gathers information of intelligence value which it provides to other agencies on request.

Drug Enforcement Administration (DEA)

DEA was established on July 1, 1973, to control narcotic and dangerous drug abuse through enforcement and prevention programs. The Bureau of Narcotics and Dangerous Drugs was the predecessor of the DEA.

The primary responsibility of the DEA is to enforce the laws and statutes relating to narcotic drugs, marihuana, depressants, stimulants, and the hallucinogenic drugs. To fulfill its responsibility, the DEA (1) conducts domestic and international investigations of major drug traffickers concentrating efforts at the illicit supply or diversion; (2) regulates the legal trade in narcotics and dangerous drugs; and (3) provides specialized training in narcotic and dangerous drug control to local, state, and Federal law enforcement officers.

In collecting information relative to its primary function, the DEA collects information of intelligence and security value. This information is transmitted to interested agencies.

Organized Crime and Racketeering (OCRS)

The organized crime and racketeering section of the Department of Justice is responsible for the coordination of all federal enforcement and prosecution efforts involving organized crime activities. This unit is responsible for establishing "strike forces" which are investigative teams supplied by various law enforcement branches of the Federal government to collect information relative to organized criminal activities for prosecution and develop and maintain intelligence files relative to organized crime. Organized crime in its current form presents a definite threat to the internal security of the United States and, therefore, operations of the organized crime and racketeering section are of a security and law enforcement nature.

Criminal Division

This division is responsible for the enforcement of all Federal criminal laws except those specifically assigned to the Antitrust, Civil Rights, and Tax Divisions of the Department of Justice. The specific responsibilities of the division are:

1. supervising and directing U.S. Attorneys in the field in criminal matters and litigation;
2. exercising supervision over international extradition proceedings, and civil as well as criminal litigation arising under the immigration and nationality laws;
3. special responsibility for coordinating enforcement activities against organized crime; and
4. preparing all government legal briefs in its criminal cases before the Supreme Court.

Law Enforcement Assistance Administration (LEAA)

LEAA was established June 19, 1968, by the Omnibus Crime Control and Safe Streets Act of 1968, under general authority of the Attorney General. One general purpose of LEAA is to assist State and local governments to reduce crime.

The Office of Criminal Justice Assistance manages the LEAA operations including (1) action grant operations; (2) technical assistance to the states; and (3) coordinating policy recommendations. Purposes for which action grants may be used are:

1. development of public protection devices;

2. recruiting and training law enforcement personnel;
3. public education;
4. projects to prevent and control civil disorders;
5. improvement of courts and corrections systems; and
6. organizing and training of special units to combat organized crime.

U.S. Marshalls

On May 12, 1969, the Attorney General established the Office of the Director, U.S. Marshalls Service. The Director was charged with supervising and directing the U.S. Marshalls and their deputies in each of the Federal judicial districts. On May 10, 1973, the U.S. Marshalls Service was designated a bureau within the Department of Justice. Among the U.S. Marshalls' responsibilities is the protection of Federal buildings and property, often during periods of civil disorder.

THE DEPARTMENT OF THE TREASURY

The Department of Treasury has many diverse functions relating to law enforcement and internal security-intelligence collection activities. There are three major divisions in the Treasury Department which have these functions. They are the United States Secret Service, Customs Service, Internal Revenue Service, Bureau of Alcohol, Tobacco and Firearms, and the Office of Law Enforcement. These functions are coordinated under the supervision of an assistant.

Secretary (of the Treasury) for Enforcement, Tariff and Trade Affairs, and Operations

Office of the Assistant Secretary (Enforcement, Tariff and Trade Affairs, and Operations)

The Assistant Secretary serves as the United States representative to the International Police Organization. The primary responsibilities of this office are as follows:

1. providing assistance to the Secretary and Deputy Secretary on Treasury law enforcement matters, including the formulation of policies for all Treasury enforcement activities;
2. coordinating law enforcement cooperation with other Federal agencies and with state and local enforcement agencies; and
3. supervising the Consolidated Federal Law Enforcement Training Center, which provides law enforcement training for personnel of participating Federal agencies.

Office of Law Enforcement (OLE)

This office was established by the Assistant Secretary (Enforcement, Tariff and Trade Affairs, and Operations) to carry out all law enforcement functions at the highest level. The Director is responsible for developing and coordinating all Department of the Treasury policy on national and international law enforcement programs, including narcotics and dangerous drugs and organized crime. The OLE provides the focal and contact point for all law enforcement and intelligence initiatives and interactions with the individual bureaus of the Treasury.

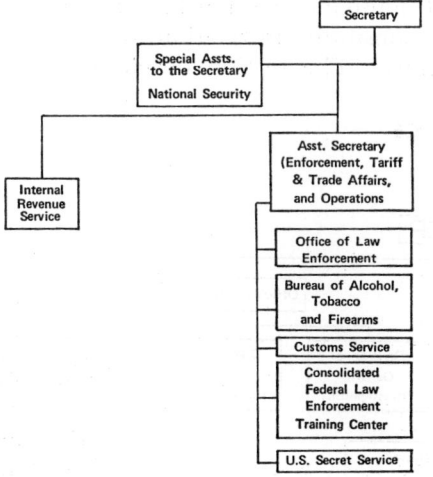

Figure 4-7. Law enforcement and security units of the Treasury Department.

Figure 4-8. Organization of Secret Service.

Secret Service

The U.S. Secret Service was established in 1865 to suppress counterfeiting. Subsequent laws have charged the Service with protecting the President and his family, the Vice President, and other designated officials and dignitaries. Protective procedures established by the Service are:

1. advance planning of the security plan with local agents and other law enforcement officials;
2. establishing the protective patrol activities;
3. briefing all participants in the plan; and
4. collecting, evaluating, steering, and disseminating protective security information. The Office of Protective Intelligence maintains liaison with interested agencies to insure receipt of potential threats. It is also responsible for the detection and arrest of persons violating the United States laws regulating coinage, obligations, and securities. In their activities, any intelligence information which might be obtained would be passed on to the appropriate concerned agency.

U.S. Customs Service

The Treasury Department's Bureau of Customs was renamed the U.S. Customs Service on August 1, 1973. Customs Service was established July 31, 1789, and subsequently placed in the Treasury Department later in that same year. A Division of Customs was created in 1875 and the division became a bureau in 1927. The Service is charged with:

1. enforcing collection of duties on goods entering the United States;
2. protecting U.S. aircraft and passengers from hijacking;
3. identifying and suppressing smuggling; and
4. controlling carriers and merchandise imported/exported to or from the U.S.

The Service's Office of Investigations is charged with conducting relevant investigations when alleged violations have occurred.[4] The Fraud Investigations Division maintains an intelligence system to facilitate discovery of import/export violations. The General Investigations Division is responsible for recruiting, training, and supervising a force of Customs Security Officers. These officers fly on flights and conduct certain predeparture inspections in accordance with the Air Security Program of 1970.

The Compliance Section has an enforcement responsibility in relation to imported merchandise. The section develops proce-

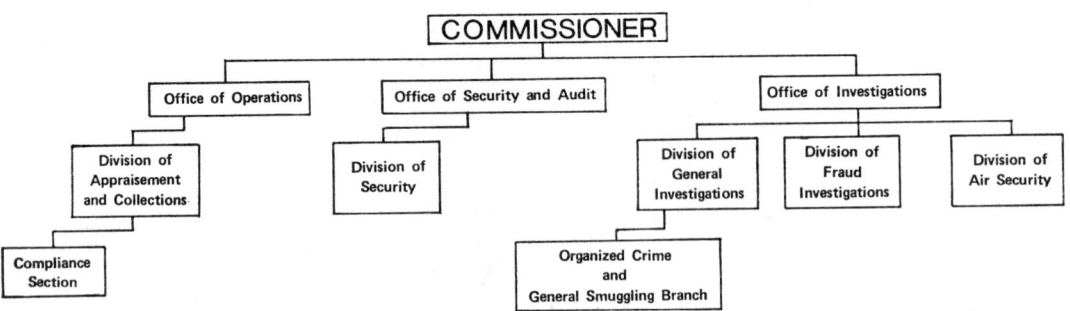

Figure 4-9. Organization of U.S. Customs Service.

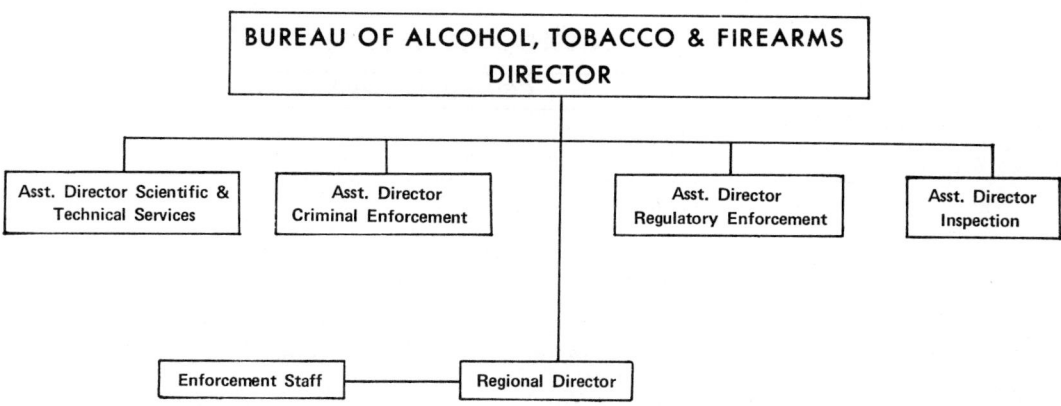

Figure 4-10. Organization of the Bureau of Alcohol, Tobacco and Firearms.

dures for control and coordination of enforcement activities in this area.

Bureau of Alcohol, Tobacco, and Firearms

This bureau was established July 1, 1972, when its functions, powers and duties of the Alcohol, Tobacco Unit (ATU) were transferred from the IRS to the bureau. ATF is responsible for the administration and enforcement of Federal laws and regulations relating to liquor, tobacco, firearms, and explosives. The criminal enforcement function of the bureau is carrying out an investigative and preventative program relating to violations of those Federal laws for which ATF has jurisdiction.

Internal Revenue Service Inspection Division

IRS Inspection Division was established in 1952 as an office reporting to the Assistant Commissioner. Inspection is authorized to (1) conduct criminal and administrative investigations; (2) execute and serve search and arrest warrants; and (3) serve Federal subpoenas and summonses. Inspection is organized into two divisions to facilitate accomplishment of its mission.

The Internal Security Division plans, develops, and controls the IRS's internal security program. This division centers its activities on the integrity of the IRS, attempted corruption of employees, the investigation of incidents involving IRS employees, and criminal and administrative sanctions imposed upon offenders. The Field Coordination Branch controls and coordinates all investigations conducted by Internal Security. The second major branch, Investigations, is responsible for conducting extremely confidential investigations of alleged employee misconduct and irregularities.

The Internal Audit Division is responsible for a systematic verification and analysis of financial transactions and a review and appraisal of the protective measures and controls established at all operating levels. The division provides support to the Intelligence Division by detecting potential civil actions and audit and tax law expertise.

Intelligence Division

The Intelligence Division was established on July 1, 1919, with the purpose of directing the supervision of important investigations demanding more exhaustive inquiry than could advantageously be made by officers assigned to work of a general nature. The division is responsible for:

1. identifying the areas and extent of willful noncompliance with tax laws;
2. investigating alleged or possible violations of such laws, and recommending, when warranted, prosecution or assertion of appropriate civil penalties; and

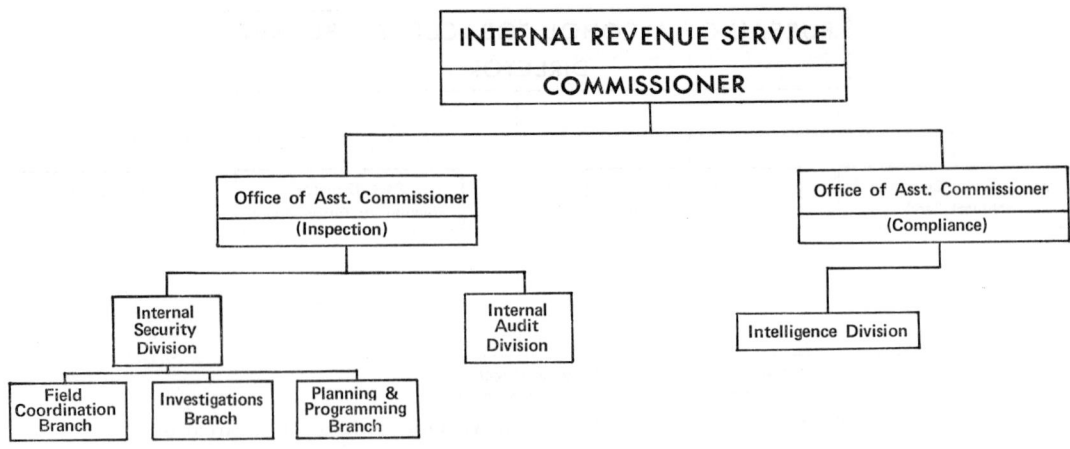

Figure 4-11. Organization of the Internal Revenue Service.

3. protecting IRS employees and property and investigating threats and assaults of employees.

The division operates two major enforcement programs to accomplish its mission. These programs are:

1. Special Enforcement Program aimed at persons engaged in illicit activities, especially organized crime figures and middle- and upper-echelon narcotics traffickers and dope financiers. This program may recommend prosecution for criminal violations or recommend invoking civil sanctions against taxpayers; and
2. General Enforcement Program aimed at cases involving legitimate businesses and professional occupations.

Consolidated Federal Law Enforcement Training Center (CFLETC)

The CFLETC was established by the Treasury Department on June 30, 1970. Its primary function is to serve as an interagency training facility for participating Federal agencies and a limited number of state and local agencies having law enforcement responsibilities. The center has two operating divisions:

1. Criminal Investigator School for agents not having previous Federal training or experience; and
2. Police School for training uniformed law enforcement officers from various participating agencies.

THE DEPARTMENT OF DEFENSE

Among the activities of the Department of Defense are activities relating to the collection of intelligence and internal security. The major components of the Department of Defense responsible for these activities are the Department of Army, Department of Navy, Department of Air Force, National Security Agency, and the Joint Chiefs of Staff.

DEPARTMENT OF DEFENSE

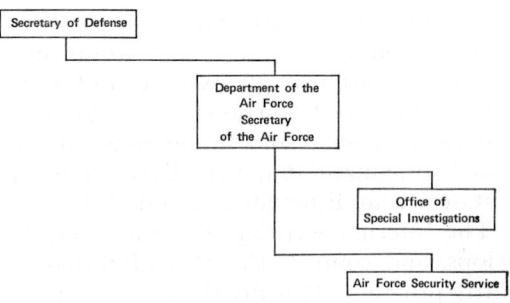

Figure 4-12. Organization of the Department of Defense.

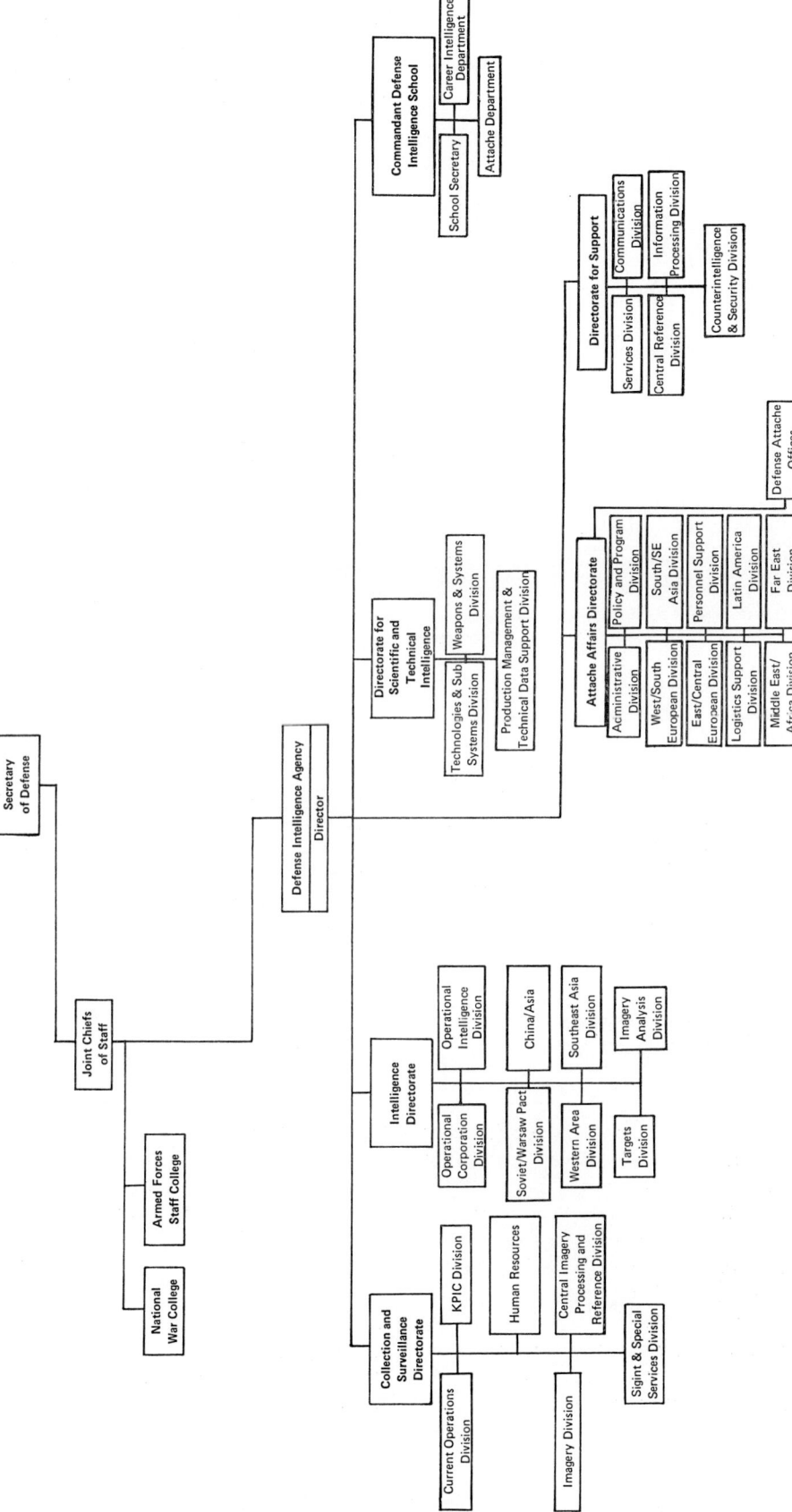

Figure 4-13. Organization of JCS functions (partial).

Assistant Secretary of Defense

International Security Affairs

ISA is responsible for Department of Defense participation in National Security Council Affairs, including:
1. development and coordination of defense positions, policies, plans, and procedures in the field of international politics—military and foreign economic affairs, including arms control and disarmament; and
2. with respect to negotiating and the monitoring of agreements with foreign governments and international organizations on military facilities, operating rights, status of forces, and other international politics-military matters.

ISA provides policy guidance, as appropriate to:
1. defense representatives on U.S. Missions and at international organizations and conferences;
2. the Security Assistance Program; and
3. other activities of interest to the Department of Defense under the Mutual Security Program.

Intelligence

Intelligence is responsible for the management of intelligence resources programs, and activities. These activities include intelligence, warning, reconnaisance, and net threat assessment.

Joint Chiefs of Staff (JCS)

The Joint Chiefs of Staff are responsible for the administration of the DIA and are the chief source of advisement to the President, the National Security Council, and the Secretary of Defense. Members of the Chiefs of Staff are the chairman, the chiefs of staff of the Army and Air Force, and the Chief of Naval Operations. The Commandant of the Marine Corps attends the meetings regularly when matters concerning the Marine Corps are discussed. Most military intelligence is channeled through the Joint Chiefs to the National Security Council.

Department of Defense Colleges

National War College

The National War College was established on July 1, 1946, under the supervision of the Joint Chiefs of Staff as the senior service school in the field of politics-military affairs. The college services highly selected, senior military officers and civilian career officials. The college is responsible for conducting a course of study for those agencies of government and the military, economic, scientific, political, psychological, and social factors of power potential, which are essential parts of national security.

Army and Navy Staff College

The college was the predecessor of the National War College having been established April 23, 1943, under operation of the Joint Chiefs of Staff. The college was redesignated as the National War College, July 1, 1946. The college's responsibility was to train specially selected Army, Navy, and Marine Corps officers for command and staff duties in joint operations.

Armed Forces Staff College

Established August 13, 1946, as a joint educational institution operating under the direction of the Joint Chiefs of Staff. The college was charged with preparing selected military officers for duty in all echelons of joint and combined commands. To achieve its objective, the college offered a course in (1) joint and combined organization, planning, and operations, and (2) related aspects of national and international security.

Industrial College of the Armed Forces

This is a joint, advanced-level, educational institution operating under the direction of the Joint Chiefs of Staff. The mission of the college is to prepare selected military officers and civilian personnel for positions of high trust in the national and international security structure. To accomplish this mission the college conducts graduate-level courses of study in national security. Emphasis is placed on management of national resources under current and predicted environments.

Defense Intelligence Agency (DIA)

The DIA was established as an agency in the Department of Defense on August 1, 1961. Under provisions of the National Security Act of 1947, as amended, the agency operates under the direction, authority, and control of the Secretary of Defense. This agency is responsible for:

1. the organization, direction, management, and control of Department of Defense intelligence resources assigned to or included within DIA;
2. the review and coordination of those Department of Defense intelligence functions retained by or assigned to the military departments;
3. the development of guidance for the conduct and management of such functions for review, approval, and promulgation by the Secretary of Defense; and
4. the supervision of the execution of all approved plans, programs, policies, and procedures for those Department of Defense general activities and functions for which DIA has management responsibility.

The Attache Affairs Directorate, within the DIA, coordinates and directs the Defense Attache System. Also, this Directorate provides all administrative and logistic support for the Attache System.

The Intelligence Civilian Career Program (Career Intelligence Department) of DIA's Defense Intelligence School provides Training Programs for Military and Civilian Intelligence Specialists.

Defense Investigative Service (DIS)

DIS was established by the Secretary of Defense on January 1, 1972, to consolidate certain investigative activities within the Department of Defense. The service operates under the authority and control of the Secretary of Defense. (See Fig. 4-14.) All investigations are limited to the United States and Puerto Rico. DIS provides Department of Defense components, and other U.S. Government activities when authorized by the Secretary of Defense, with a single centrally directed personnel security investigative service.

The specific responsibilities of DIS are as follows:

1. conducting all Personnel Security Investigations (PSI's) for Department of Defense components and, when authorized by the Secretary of Defense, other U.S. Government agencies;
2. providing Department of Defense components and other U.S. Government agencies with the results of these investigations as appropriate;
3. when feasible and within available resources, providing investigative assistance, upon request to supplement the investigative efforts of other Department of Defense components;
4. when authorized, maintaining liaison on matters of mutual interest with and, within limits of established policy, render appropriate assistance to

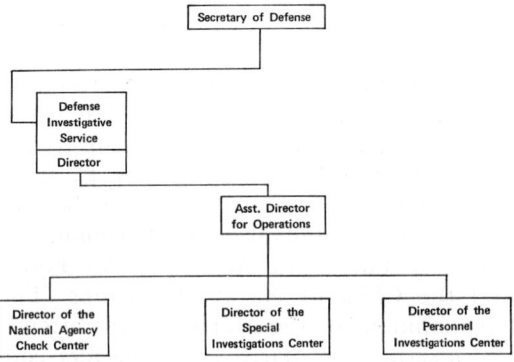

Figure 4-14. Organization of Defense Investigative Service.

investigative, law enforcement, intelligence, counterintelligence, and other U.S. and foreign government activities. Keep the Secretary of Defense informed on such activities;

5. conducting surveys and preparing analyses, special studies and estimates on investigative matters within the purview of DIS;
6. obtaining from requesting Department of Defense components and other U.S. Government activities, for record and statistical purposes, information on actions taken and final disposition of matters investigated by DIS;
7. Referring all matters developed as a result of PSI's which have a significant counterintelligence or criminal aspect to the appropriate civilian or military investigative agency;
8. Establishing standards and procedures for certification and accreditation of civilian and military personnel assigned to DIS investigative departments; and
9. Such other special investigations as the Secretary of Defense may direct.

Defense Supply Agency Contract Administration Services Regions (DCASR)

The DSA is responsible for providing (1) effective and economical support; and (2) Contract Administration Services. Contract Administration Services Regions provide the following services:

1. performance of contract administration, production, quality assurance, and data and financial management activities; and
2. administration of the industrial security, contracts compliance, and small business/labor surplus programs.

The Defense Supply Agency is responsible in its security functions for the monitoring of compliance with the Industrial Security Regulations and the development and revision of the Industrial Security Manual. The ISM/ISR establish the guidelines for the DOD program for the production of materials of use by the defense establishment.

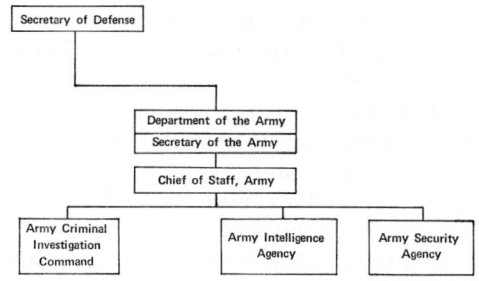

Figure 4-15. Organization of the Department of Army Intelligence, Security and Investigative functions.

THE DEPARTMENT OF ARMY

The Assistant Chief of Staff for Intelligence (ACSI) is the Army staff officer for matters pertaining to intelligence activities in the Army. He has the responsibility for the planning, coordinating, and supervising of the intelligence collection, production, and dissemination effort in the Army. He also has similar functions for counterintelligence and internal security. The total responsibility for intelligence collection rests jointly with the ACSI and the DIA. While a great portion of the intelligence collection of ACSI and DIA is formed by active military personnel the employees of the Intelligence Civilian Career Program (ICCP) perform many intelligence and counterintelligence functions for the military.

There are three major organizations within the U.S. Army which have security or intelligence functions. They are the Army Intelligence Agency, Army Security Agency, and the Army Criminal Investigation Command.

U.S. Army Security Agency (ASA)

The Commanding General, ASA, is responsible for:

1. operations, training, administration, services, and supply for all units, personnel, activities, and installations under his command throughout the world;
2. assisting Army commanders, as required, in the formulation and implementation of communications security plans, policies, and operational procedures; and
3. maintaining liaison with counterpart activities of the Departments of the Navy and Air Force and other governmental agencies for coordination of activities.

U.S. Army Intelligence Agency

Effective July 1, 1974, the U.S. Army Intelligence Command was changed to its current title. The Commanding General of the Agency is responsible for:
1. exercising central control of continental U.S. counterintelligence activities concerned with Department of the Army military and civilian personnel security programs. The industrial security program and designated Department of Defense agencies, to include conduct of counterintelligence investigations, operations, and services in support of these programs in the continental U.S., Puerto Rico, and U.S. Virgin Islands;
2. exercising central control for the initiation and conduct of personnel security investigations worldwide;
3. operating U.S. Army Investigative Records Repository, Department of Defense National Agency Check Center, and Defense Control Index of Investigations; and
4. performing other intelligence-counterintelligence support as may be assigned by headquarters, Department of the Army.

U.S. Army Criminal Investigative Command (USACIC)

The Commander, USIDC, is responsible for exercising centralized command, authority, direction, and control of Army criminal investigative activities. Also, the Command provides investigative support to all U.S. Army elements. The specific responsibilities of the command are as follows:
1. conducting, controlling, and monitoring Army criminal investigations;
2. developing investigative standards, procedures, and doctrinal policies;
3. operating a criminal intelligence element;
4. operating the U.S. Army Crime Records Repository;
5. maintaining centralized records of criminal investigative agents;
6. reviewing all investigative reports;
7. operating investigative crime laboratories;
8. planning and conducting protective service operations;
9. conducting the accreditation/certification program of investigative agents.

U.S. Naval Intelligence (USNI)

The Commander, USNI, is responsible for insuring the fulfillment of the intelligence, counterintelligence, investigative, and security requirements and responsibilities of the Department of the Navy.

The Assistant Chief of Naval Operations for Intelligence is responsible for overall supervision of security and intelligence matters. He has the Naval Investi-

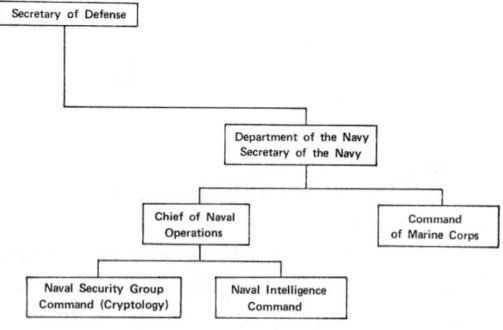

Figure 4-16. Navy Intelligence and Security units.

gative Service under his direct supervision. The Naval Investigative Service collects, processes, and disseminates intelligence of naval interest. It maintains liaison with all intelligence and counterintelligence agencies and controls the release of classified information under NIS control. Likewise, the Marine Corps has its own intelligence arm which collects counterintelligence of immediate importance to the corps.

U.S. Navy Cryptology

The Commander, Naval Security Group Command, is responsible for the:
1. direction and management of assigned shore activities which perform various support functions for the Operating Forces of the Navy, including:
   a. monitoring of naval communications to assure compliance with established procedures; and
   b. improvement of communications security;
2. provision of cryptographic devices to fleet units and shore activities;
3. support and participation in the production of certain intelligence information;
4. research into communications phenomena as a part of the continuing effort to improve naval communications.

U.S. Marine Corps (USMC)

The Marine Corps was established on November 10, 1775, by a resolution of the Continental Congress. Its present structure, missions, and functions were set forth in the National Security Act of 1947, as amended in 1952. The collateral missions of the corps are:
1. to provide security forces for the protection of naval stations or bases; and
2. to provide security guards at American embassies, legations, and consulates in various countries.

THE DEPARTMENT OF AIR FORCE

The Air Force makes a distinction between intelligence and counterintelligence operations. The Chief of Staff of intelligence matters coordinates the collection and production of their intelligence by Air Force activities, and he is responsible for operation and maintenance of the Air Force attache system. Counterintelligence activities fall under the jurisdiction of the Inspector General.

U.S. Air Force Office of Special Investigations

This office is charged with providing criminal, counterintelligence, personnel security, and special investigative services to Air Force activities. Responsible for col-

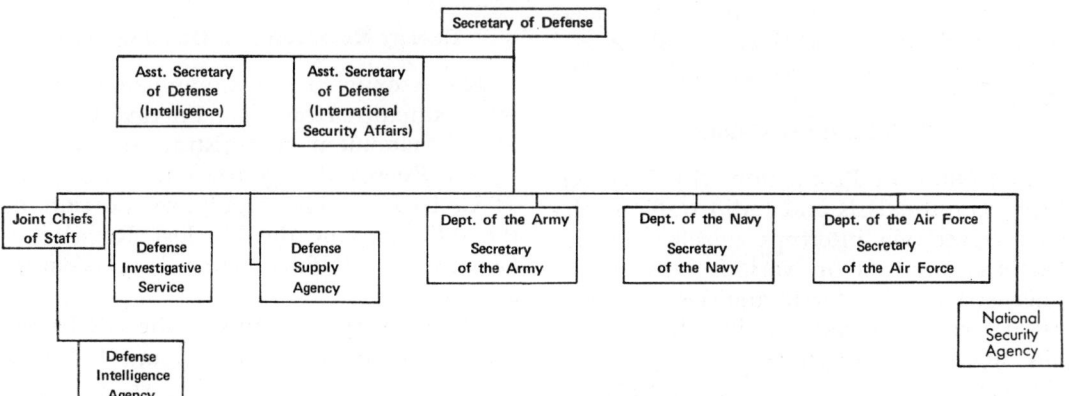

Figure 4-17. U.S. Air Force Intelligence and Security units.

lecting, analyzing, and reporting significant information about these matters.

U.S. Air Force Security Service

The service is responsible for monitoring Air Force communications in all parts of the world to insure compliance with established communication security practices and procedures. The service occasionally conducts research in communications phenomena in support of various elements of the U.S. Government.

NATIONAL SECURITY AGENCY (NSA)

The NSA was established by Presidential Directive in 1952 in the Department of Defense to provide highly specialized technical function in support of the United States intelligence activity. NSA has both a secondary and intelligence function. It specifically has the following responsibilities:

1. prescribing certain security principles, doctrines, and procedures;
2. organizing, operating, and maintaining certain activities for the production of intelligence information;
3. organizing and coordinating research and engineering activities of the government in support of the agency's assigned functions; and
4. regulating certain communications in support of agency missions.

Its major activities in the areas of intelligence are through the collection of electronic and communications emissions and security through its cryptographic operations.

NSA Major Divisions

The Office of Production (PROD) performs cryptonalysis and radio traffic analysis received via intercept stations. It also performs analysis on various types of cipher systems. Research and Development (R&D) conducts research in advanced radio and cryptographic technology. The *Communication Security* (COM SEC) office is responsible for the security and production of United States cipher systems. The Security Office (SEC) investigates applicants and maintains informational files.

Atomic Energy Commission (AEC)

The AEC was established by the Atomic Energy Act of 1954 and transferred into the Energy Research and Development Administration in 1975. The Commission has licensing and regulatory authority over persons possessing, using, and transferring certain radioactive material and nuclear facilities. The law enforcement mission of the AEC is as follows:

1. protection of classified data on atomic energy matters;
2. regulation of the atomic energy industry; and
3. application of radiation research to forensic medicine and crime analysis.

The Directorate of Regulatory Operations, formerly the Division of Compliance and the Division of Nuclear Materials Safeguards, generally carries out the enforcement functions of the AEC. This division reports possible violations of licensing and regulatory provisions of the Atomic Energy Act to the Department of Justice for possible prosecution. Possible criminal violations are reported to the FBI for investigation. The Division of Security makes referrals to the Department of Justice of violations pertaining to communication of, receipt of, tampering with, or disclosure of restricted data.

Energy Research and Development

The Atomic Energy Commission (AEC) was reorganized into the Energy Research and Development Administration (ERDA) by the Energy Reorganization Act of 1974 (P.C. 93-438). The regulatory portion of the AEC became the Nuclear Regulatory Commission (NRC) which is a separate agency.

The security programs of the AEC were transferred to the ERDA. Figure 4-18 provides an organizational chart of their functions.

Governmental Security Programs

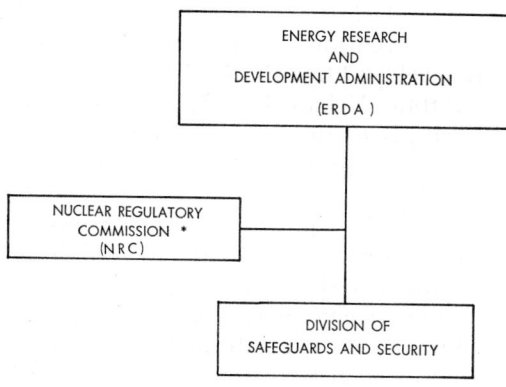

*FORMERLY AEC
(ATOMIC ENERGY COMMISSION)

Figure 4-18. Division of Safeguards and Security, Nuclear Regulator Commission, ERDA.

Energy Research and Development Administration

As a portion of the Energy Reorganization Act of 1975, the separate Security and Safeguard functions of the Atomic Energy Commission (AEC) have been consolidated into a Division of Safeguards and Security, within the Nuclear Regulatory Commission (NRC).

The Director, Division of Safeguards and Security:

1. develops and establishes policy, standards and procedures for the physical protection of unclassified Special Nuclear Materials (SNM);
2. provides staff assistance on the physical protection of SNM;
3. reports immediately to the FBI any actual or attempted theft or sabotage of SNM or any other circumstances indicating a violation of Federal law concerning such material;
4. reports to the Director, Division of Waste Management and Transportation, any such incidents involving SNM in transit, and to the Nuclear Regulatory Commission when such incidents are known to involve licensed facilities; and
5. grants exceptions to this chapter and appendix.

THE CIVIL SERVICE COMMISSION

The fundamental purpose of the Civil Service Commission is to administer the law establishing the employment merit service. The one office within the commission which formulates plans dealing with the national operation of the commission's investigative program is the Bureau of Personnel Investigations. It develops procedures and negotiates agreements with other governmental agencies concerning the delegation of the commission's authority to agencies to conduct their own personnel investigation. It also has the added responsibility of evaluating all nonsensitive cases under the commission's jurisdiction when full loyalty investigations have been conducted by the FBI. It maintains liaison both with Army intelligence agencies and the Federal Bureau of Investigation.

RELATED FEDERAL AGENCIES

In addition to the *user* agencies, there are additional federal agencies which have significant protective functions. While none of these agencies has a primary law enforcement or security function, they do have organizational programs which affect the general public.

Department of Agriculture

The Office of the Inspector General is the law enforcement arm of the Department of Agriculture. This office is respon-

DEPARTMENT OF AGRICULTURE

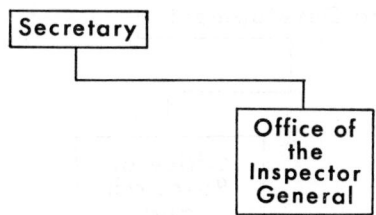

Figure 4-19. Enforcement Organization of the USDA.

sible for investigating the following crimes:
1. theft and unauthorized use and sale of food stamps;
2. mislabeling of meat and poultry destined for sale to the public;
3. commodity exchange fraud; and
4. crimes on national forest lands.

Department of Housing and Urban Development (HUD) Security Forces

HUD's public housing projects' security requirements are provided by Local Housing Authorities. Operating under HUD's Assistant Secretary for Housing Management, LHA's are responsible for assuming adequate security services through:
1. providing housing police/security forces;
2. securing local police services through a system of payments in lieu of taxes (PILOT); or
3. eliciting tenant involvement in security programs.

Crime Prevention Systems

The Assistant Secretary for Community Planning and Development is responsible for formulation of the Department's Crime Prevention Systems. Among the systems developed by this division are:
1. Public Urban Locating Services—Attempts to develop an optimum routing or control of public service vehicles.
2. Street Equipment Project—A system of street equipment to minimize crimes, reduce traffic problems and improve emergency communications.
3. Operation Breakthrough—Develops a performance criteria for home security.
4. Park Safety—Determine to what extent crime and fear of crime can be controlled in parks.
5. Youth Training—Seeks to determine if the level of vandalism in public housing projects can be lowered by teaching disadvantaged youth skills in project maintenance and management.

Residential Security

The Office of Research and Technology is responsible for studying the patterns and categories of crime in residential

DEPARTMENT OF HOUSING AND URBAN DEVELOPMENT

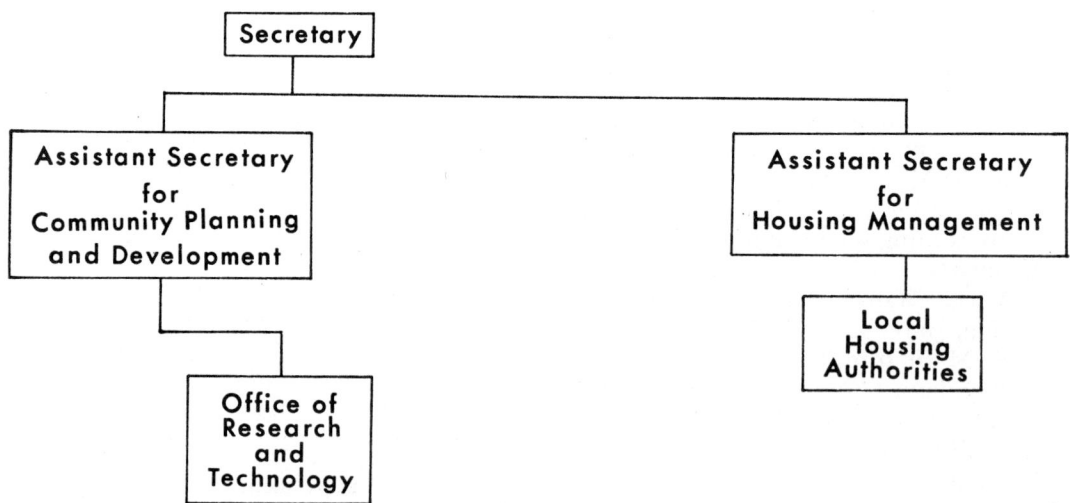

Figure 4-20. HUD Security Services.

DEPARTMENT OF THE INTERIOR

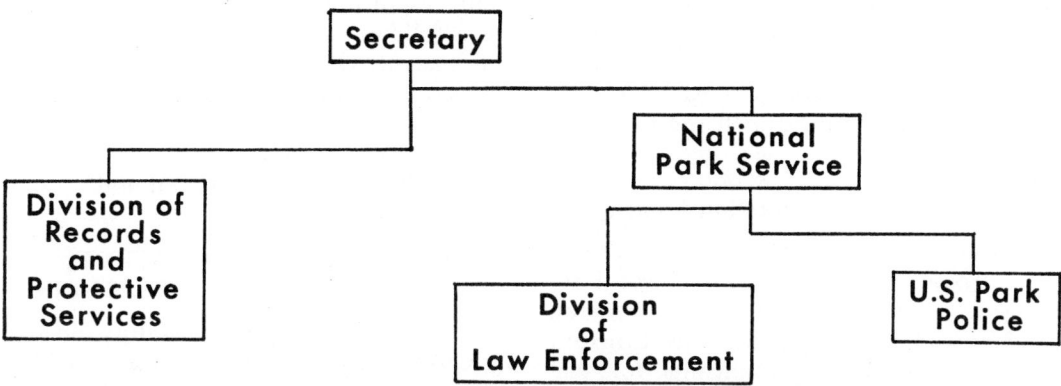

Figure 4-21. Department of Interior Protective Organization.

areas. The office attempts to develop a security system to reverse the number and severity of such crimes. Phases of the Residential Security Project include:

1. determining the nature and pattern of crimes occurring in and around residential areas;
2. developing a total security system to reduce the number and severity of crimes in different residential settings;
3. developing innovative architecture to minimize the risk to residents of crimes committed in or near dwellings; and
4. developing security systems standards for both new and existing dwellings.

Department of the Interior National Park Service (NPS)

The chief enforcement arm of the NPS is the U.S. Park Police. Park police have jurisdiction and may make arrests on Federal park lands in the District of Columbia. The primary responsibilities of the Park Police are as follows:

1. playing a major role in coping with civil disorders in the District of Columbia; and
2. providing protection for the President and visiting foreign dignitaries.

The NPS has created a Division of Law Enforcement to further its efforts to cope with law enforcement problems. This division is responsible for the following:

1. developing standards;
2. inspecting, staffing, and training; and
3. executing the services law enforcement program.

DEPARTMENT OF TRANSPORTATION

Federal Aviation Agency (FAA)

This agency was established by the Federal Aviation Act of 1958. The agency was charged with providing for the regulation and promotion of civil aviation in such manner as to best foster its development

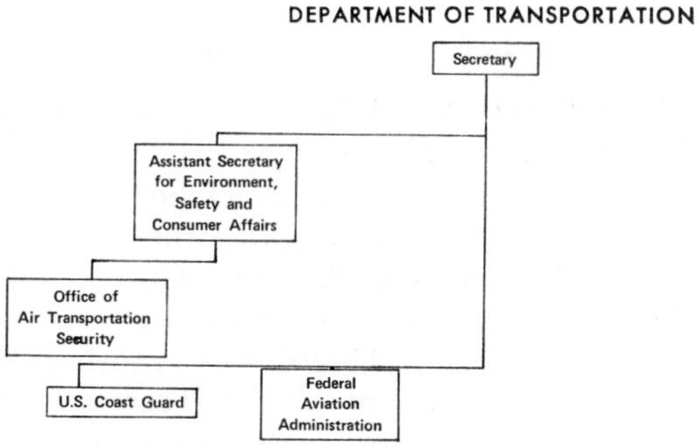

Figure 4-22. DOT Security and Protection units.

and safety. An additional responsibility was the provision of safe and efficient use of airspace by both civil and military aircraft. In 1967, the agency was made a part of the Department of Transportation.

The 1961 amended version of the Federal Aviation Act of 1958 prohibited aerial piracy. This action placed the FAA directly in the area of coping with the unlawful activity.

Office of Air Transportation Security

This office was established in 1971 for the purpose of providing a reciprocal intelligence function with other Federal, civilian, and military units. The function relates to employees, applicants for employment, contractors to the FAA, airmen, and air carriers, either currently certified or seeking certification from the FAA. Confidential background resumes, identification, and documentation resulting from investigations are furnished to those organizations requiring intelligence.

Office of Civil Aviation Security (OCAS)

This office was established in October, 1970, under the direction of the Assistant Secretary for Safety and Consumer Affairs. The office was responsible for formulating policies and providing broad overall guidance to the FAA, which assumed operational responsibility of the antihijacking program. OCAS was incorporated into a new office, Office of Air Transportation Security, in 1971.

United States Coast Guard

The Coast Guard is the primary maritime enforcement agency of the U.S. Government. The Revenue Cutter Service was established in 1790. In 1915, the Lifesaving Service and the Revenue Cutter Service were merged to form the Coast Guard. The Coast Guard was transferred from the Department of the Treasury to the Department of Transportation in 1967.

The Coast Guard is authorized to make inquiries, examinations, inspections, searches, seizures, and arrests upon the high seas and waters over which the United States has jurisdiction. The law enforcement responsibilities of the Coast Guard are as follows:

1. marine traffic control and safety;
2. ships, boats, and offshore structure safety;
3. port safety and security;
4. environmental protection; and
5. conservation.

REGULATORY COMMISSIONS, BOARDS, AND ADMINISTRATIONS

There are numerous Regulatory Boards, Commissions and Administrations within the federal government which have enforcement or protective functions affecting the general public welfare.

Civil Aeronautics Board (CAB)

The CAB is an independent agency established by the Civil Aeronautics Act of 1932 (52 STAT. 973) and continued by the Federal Aviation Act of August 23, 1958, as amended (72 STAT. 731, as amended; 49 U.S.C. 1301, et seq). The CAB is responsible for the encouragement, promotion, and regulation of a sound and vigorous air transportation system designed to serve the domestic and international needs of the traveling and shipping public, of the U.S. Postal Service, and of national defense.

The Bureau of Enforcement is the enforcement arm of the CAB. The Bureau's enforcement responsibilities are as follows:

CIVIL AERONAUTICS BOARD

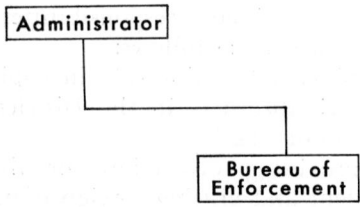

Figure 4-23. CAB enforcement.

1. enforcement of the economic provisions of the Federal Aviation Act of 1958, as amended;
2. enforcement of the relevant provisions of the Clayton Anti-trust Act and the Railway Labor Act; and
3. enforcement of all economic order, regulations, and other requirements promulgated by the board.

The bureau cooperates with other agencies in enforcement activities. This cooperation may take the form of exchange of information derived from their respective enforcement work and/or in bringing to the attention of the agency most directly concerned with violations which have been discovered.

Environmental Protection Agency (EPA)

The EPA was established as an independent agency in the executive branch pursuant to Reorganization Plan No. 3 of 1970, effective December 2, 1970. The purpose of the agency is to permit coordinated and effective governmental action on behalf of the environment.

The Office of the Assistant Administrator for Enforcement and General Counsel is the enforcement arm of the EPA. The office's responsibilities are as follows:
1. supplies legal support services for environmental control programs;
2. provides policy direction to enforcement activities in the program areas;
3. plans and coordinates enforcement conferences, public hearings, and other legal proceedings; and
4. engages in other activities related to enforcement of standards to protect the nation's environment.

ENVIRONMENTAL PROTECTION AGENCY

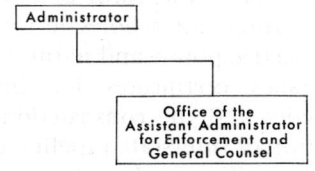

Figure 4-24. EPA enforcement.

FEDERAL COMMUNICATIONS COMMISSION

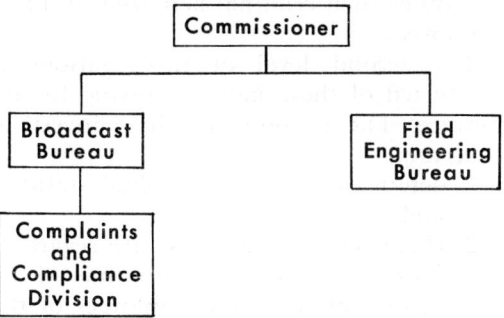

Figure 4-25. FCC enforcement.

Federal Communications Commission (FCC)

The FCC is an independent agency created by Congress to regulate non-Federal, interstate, and foreign communications by radio and wire. The Commission is responsible for enforcing the Communications Act of 1934, as amended; the Communications Satellite Act of 1962; and rules and regulations it has issued.

Information secured during investigations is used primarily in administrative proceedings. One investigative unit within the FCCA is the Field Engineering Bureau. The Bureau is responsible for investigating and attempting to eliminate all sources of interference. The second investigative unit is the Complaints and Compliance Division of the Broadcast Bureau.

Federal Deposit Insurance Corporation (FDIC)

The FDIC was established in 1933 under section 126 of the Federal Reserve Act. This authority was enacted separately as the Federal Deposit Insurance Act in 1950. The mission of the FDIC is to insure deposits of commercial and mutual savings banks.

To insure the integrity of insured banks and those applying for insurance, the FDIC conducts two levels of investigations. The first level is the investigation, in the form of bank examinations, of insured banks. The banks are examined to assure they have adequate internal controls to prevent and deter defalcation, fraud, and other such criminal activities by bank employees.

The second level of investigations is conducted of those banks applying for insurance. The purposes of these investigations are:
1. detect violations of criminal statutes; and
2. character investigations with regard to possible criminal backgrounds of directors, officers, and employees of the bank.

Federal Home Loan Bank Board (FHLBB)

The principal law enforcement activity of the FHLBB is the examination of savings and loan associations, and savings and loan holding companies and their affiliates for improper practices. The examinations are conducted as part of the board's mission to supervise the following:
1. Federal Home Loan Bank System
2. Federal Savings and Loan System
3. Federal Savings and Loan Insurance Corporation

The principal enforcement responsibilities of the board are as follows:
1. utilize administrative sanctions/remedies to force compliance;
2. refer all cases involving possible violations of Federal criminal law to the Department of Justice for appropriate disposition;
3. evaluate personal and financial backgrounds in the process of granting Federal charters and insurance of accounts;
4. enforcement of applicable Department of the Treasury regulations; and
5. as a result of the Bank Protection Act of 1968, evaluate security programs. The programs are a result of the board's authority to establish minimum security standards against robbery and burglary.

The Office of Examinations and Supervision is responsible for conducting the board's examination activities. The examinations evaluate the quality of management performance in the areas of objectives, policies, procedures, and internal controls. Also, the examinations check association compliance with Federal and state laws and regulations and with its own charter provisions and bylaws.

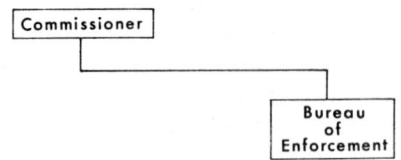

Figure 4-26. FMC enforcement.

Federal Maritime Commission (FMC)

The FMC was established by Reorganization Plan 7, effective August 12, 1961, as an independent agency. The primary enforcement responsibility of the Commission is the regulation of the United States' waterborne foreign and domestic offshore shipping and related enterprises. The enforcement arm of the FMC, the Bureau of Enforcement, makes limited inspections of containers in both the foreign and domestic trades. The primary emphasis, however, is on carrier self-inspection.

Federal Power Commission (FPC)

The FPC regulates interstate activities of the electric power and gas industries. The commission operates under the Federal Water Power Act of 1920, as amended in 1935, and the Natural Gas Act of 1938. The FPC's responsibilities are as follows:
1. issues permits and licenses for non-Federal hydroelectric power projects;
2. regulates the rates and other aspects of interstate wholesale transactions in electric power and natural gas;
3. issues certificates for interstate gas sales and the construction and operation of interstate pipeline facilities;
4. prescribes and enforces a uniform

system of accounts for regulated electric utilities and natural gas companies; and
5. regulates securities, mergers, consolidations, acquisitions, and accounts of electric utilities.

The FPC audits all companies within its jurisdiction every five to seven years. The commission is empowered to petition Federal courts to require compliance with commission hearings.

Federal Reserve System (FRS)

The FRS is comprised of (1) Board of Governors; (2) Federal Open Market Committee; (3) the Twelve Federal Reserve Banks and their twenty-four branches; (4) Federal Advisory Council; and (5) such State Banks and trust companies which have been admitted to the system. The system was established to (1) provide for the establishment of Federal Reserve Banks; (2) furnish an elastic currency; (3) afford means of rediscounting commercial paper; (4) establish a more effective supervision of banking in the United States; and (5) for other purposes.

If, during the course of bank examinations the FRS discovers possible violations of Federal criminal laws, it reports the circumstances to the Department of Justice. The system enforces provisions of the Bank Protection Act of 1968 relating to the installation, maintenance, and operation of security devices and procedures.

Federal Trade Commission (FTC)

The FTC is responsible for administering the following acts which carry criminal penalties for violation:
1. Federal Trade Commission Act (15 U.S.C. 50 and 54a)
2. Wool Product Labeling Act (15 U.S.C. 68i)
3. Fur Products Labeling Act (15 U.S.C. 69i)
4. Flammable Fabrics Act (15 U.S.C. 1196)
5. Textile Fiber Production Identification Act (15 U.S.C. 70i)
6. Consumer Credit Protection Act (15 U.S.C. 1601 *et seq*)

FTC staff members participate on Consumer Protection Coordinating Committees. These independent committees consist of representatives of Federal, state, and local governmental organizations involved with consumer protection.

General Services Administration (GSA)

The GSA was established by section 101 of the Federal Property and Administration Act of 1949 (63 Stat. 379), effective July 1, 1949. The GSA Task Force on the Federal Protective Service Program brings together government officials concerned with the protection of Federal buildings. In addition to GSA, the Task Force includes the U.S. Postal Service, U.S. Marshals Service, and the Administrative Office of the U.S. Courts.

The principal law enforcement activities of the GSA are aimed at reducing the threats of arson and bombings, bomb threats, vandalism, thefts, and damage from demonstrations and destructive acts which threaten Federal employees and buildings. The Public Buildings service of GSA is responsible for enforcement activities directed toward the enforcement objectives.

GENERAL SERVICES ADMINISTRATION

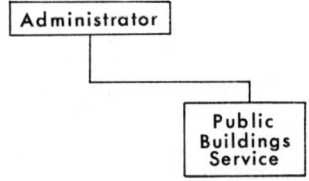

Figure 4-27. GSA Protective Services.

Interstate Commerce Commission (ICC)

The ICC was created in 1887 by the Act to Regulate Commerce, now known as the Interstate Commerce Act. The Commission derives most of its law enforcement authority from parts I, II, III, and IV of

INTERSTATE COMMERCE COMMISSION

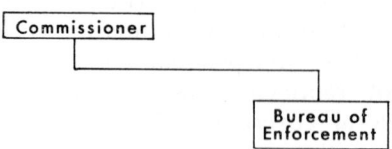

Figure 4-28. ICC enforcement.

that act and from the Elkins Act of 1903, with major amendments in 1906. The mission of the ICC is to promote the security of people and cargo in transit in the United States. The major responsibilities of the ICC are:
1. enforcement of laws and regulations relating to rail, truck, and bus transportation;
2. investigation of possible violations of Federal law in these areas and refers cases to the Department of Justice for appropriate disposition, including prosecution.

The Bureau of Enforcement is charged with being the law enforcement arm of the ICC. This bureau obtains information for its investigations from the ICC's Bureau of Operations and Bureau of Accounts. The bureau exchanges information with the Organized Crime and Racketeering Section of the Department of Justice and with other relevant Federal agencies. The bureau also cooperates under written agreement with forty-seven states, pursuant to P.L. 89-170, for:
1. exchange of information;
2. mutual assistance;
3. conduct of joint examinations;
4. investigations;
5. inspections; and
6. administrative activities.

National Labor Relations Board (NLRB)

The NLRB was created as an independent agency by the National Labor Relations Act of 1935 (Wagner Act), as amended by the Taft-Hartley Act of 1947 and the Landrum-Griffin Act of 1959. The NLRB is a quasi-judicial agency with two principal responsibilities:

1. to investigate questions concerning employee union representation and to resolve them through elections; and
2. to investigate and prosecute unfair labor practice charges brought against employers and unions.

When the NLRB uncovers actual or potention criminal activities during hearings or investigations, it contacts the appropriate Federal, state, or local law enforcement agency.

National Science Foundation (NSF)

The NSF was established by the National Science Foundation Act of 1950 (64 Stat. 149; 42 U.S.C. 1861-1875) and was given additional authority by the National Defense Education Act of 1958 (72 Stat. 1601; 42 U.S.C. 1876-1879), as amended. The foundation conducts two research programs related to crime prevention and control. These programs are outlined below:
1. Research Applied to National Needs (RANN) operated by Research Applications Directorate. Research is directed at specific environmental, societal, and technological problems, including crime; and
2. Social Science Research operated by Division of Social Sciences. Research explores social problems, including crime, from the perspective of economics, sociology, and other social sciences.

Other areas under study through research projects supported by the NSF are:
1. the causes and nature of crime and delinquency,
2. new methods for applying science and technology to law enforcement,

NATIONAL SCIENCE FOUNDATION

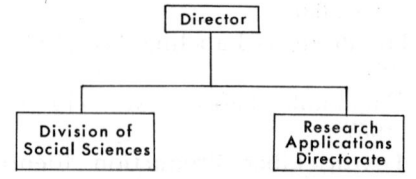

Figure 4-29. NSF Crime Control Research units.

SECURITIES AND EXCHANGE COMMISSION

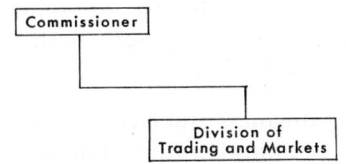

Figure 4-30. SEC Protective Services.

3. effectiveness of criminal justice methodology, and
4. the economic effects of crime on society.

Securities and Exchange Commission (SEC)

The SEC was created under authority of the Securities Exchange Act of 1924 (48 Stat. 881; 15 U.S.C. 78a to 78jj) as amended. The function of the SEC is to protect the public against wrongdoings in the securities and financial markets. The commission conducts investigations into alleged violations of the Federal securities laws. When alleged violations are discovered, the SEC may:

1. conduct administrative proceedings; or
2. seek civil injunctions; or
3. refer the alleged violation to the Department of Justice.

The commission maintains constant liaison with the Organized Crime and Racketeering Section of the Department of Justice and with Federal and State agencies concerned with organized crime. The SEC's Division of Trading and Markets conducts an enforcement training program with the purpose of alerting participants to activities which may violate Federal securities laws.

Selective Service System

The Selective Service System was established by Act of Congress of June 24, 1948 (62 Stat. 604; 50 U.S.C. app. 451-471), to supply the Armed Forces of the United States with adequate manpower to assume the security of the nation. The system administers a program of processing potential violators. When an individual violates

SMALL BUSINESS ADMINISTRATION

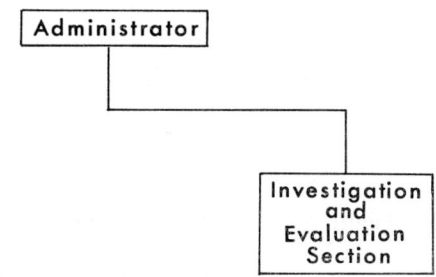

Figure 4-31. SBA Protective Services.

the Act, the system submits the case to the appropriate U.S. Attorney for action.

Small Business Administration (SBA)

The SBA is designed for the purpose of making loans to small business concerns, state and local development companies, and victims of floods and other disasters. Major enforcement responsibilities are (1) monitoring loans for possible irregularities; (2) cooperating with other Federal agencies which may be investigating the borrower; and (3) examining small business investment companies each year. If illegal activity is discovered, the evidence is reported to the FBI, the Secret Service, or the Organized Crime and Racketeering Section.

The Investigation and Evaluation Section is charged with setting out the guidelines for investigators. Also, this section describes the evaluation process of reports resulting from the investigations. The actual evaluation of reports lies with the Administrator, the General Counsel, and the appropriate associate administrators.

U.S. Postal Service

The U.S. Postal Service was created on August 12, 1970, by the Postal Reorganization Act (84 Stat. 719, 39 U.S.C. 101 et seq). The Postal Inspection Service, headed by the Assistant Postmaster General-Chief Postal Inspector, is the law enforcement arm of the Postal Service. The PIS is responsible for (1) protecting the mails;

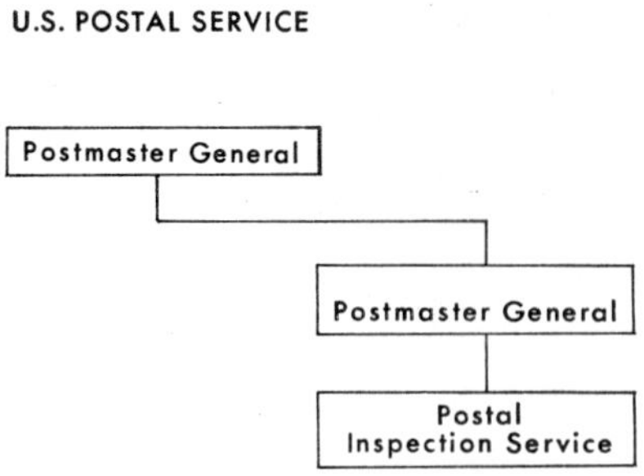

U.S. POSTAL SERVICE

Figure 4-32. Postal Service enforcement.

(2) enforcing postal laws; (3) conducting internal audits and postal inspections; and (4) maintaining plant and personnel security. The service is organized along functional lines to facilitate accomplishment of its mission. The organization is as follows:

1. criminal investigation,
2. security,
3. audit, and
4. administration.

Veterans Administration (VA)

The VA operates under Executive Order 5398 in accordance with the Act of July 3, 1930 (46 Stat. 1016). The administration's major responsibilities are:

1. to provide disability, health, and education benefits and rehabilitation to service veterans; and
2. to administer other programs of benefits provided by law for veterans and their dependents.

The VA has no direct law enforcement responsibility except for the security and safety of its own stations. This is accomplished by the maintenance of a large guard force.

COMMITTEES, COUNCILS, AND PROGRAMS

Various problems facing the Federal Government have resulted in the establishment of a variety of committees, councils, and programs to deal with them. Numerous such problems relate to protections and security. There have been many committees whose functions have been permanently transferred to permanent federal agencies while others have disbanded after the problem was solved. This section provides a brief listing of some of these groups, past and present with a summary of their activities.

Business and Defense Services Administration (Commerce)

The administration was established by the Secretary of Commerce on October 1, 1953, and operated under Department Organization Order 40-1. Department Organization Order 40-1A of September 15, 1970, abolished the administration and transferred to the Bureau of Domestic Commerce. Responsible for the promotion and development of the growth of United States industry and commerce and to prepare and execute plans for industrial mobilization readiness.

Commission on Government Security (CGS)

The commission was established by act of August 9, 1955, and terminated September 22, 1957, pursuant to terms of the act. The CGS was established to study and investigate the entire government security program.

Facility Security Program (FSP) (Office of Civilian Defense)

The FSP was established on May 19, 1942, and abolished April 18, 1944. The program was responsible for supplementing the protective programs of the Army, Navy, and Federal Power Commission. The FSP was also charged with correlating antisabotage activities of other government agencies.

Federal Fire Council

The council was established by executive order on March 13, 1972. Organized to serve as an official advisory agency in matters relating to the protection of Federal employees and property from fire.

The present council was preceded by the "old" Federal Fire Council which was established on July 1, 1939. The "Old" council was transferred with the Federal

Works Agency to the General Services Administration on June 30, 1949.

Interagency Classification Review Committee (ICRC)

The ICRC was created on March 8, 1972, to assist the National Security Council in ensuring compliance with provisions of the establishing executive order. The committee is responsible for (1) developing means to prevent overclassification of national security information; (2) ensuring prompt declassification of such material; (3) facilitating access to declassified material; and (4) eliminating unauthorized disclosure of classified information.

Loyalty Review Board (LRB) (Civil Service Commission)

The LRB was created on November 10, 1947, by the Civil Service Commission, and abolished on April 27, 1953. The board was established as a reviewing and coordinating body concerned with adjudication of cares involving loyalty of employees in the executive branch.

National Security Resources Board (NSRB)

The NSRB was established by the National Security Act of 1947 with the responsibility of advising the President concerning coordination of military, industrial, and civilian mobilization. The board has undergone several organizational changes, among which were:
1. transferred to the Executive Office of the President on August 20, 1949;
2. functions of the board transferred to the chairman and the board became advisory to him on July 10, 1950;
3. on March 13, 1953, functions transferred to the Office of Defense Mobilization; and
4. board and offices of the chairman and vice-chairman abolished and remaining functions transferred to the Office of Defense Mobilization on June 12, 1953.

Office of Censorship

The office was established by executive order on December 19, 1941, and terminated by executive order on November 15, 1945. Responsible for censoring communications by mail, cable, radio, and other means of transmission passing between the United States and any foreign country.

President's Foreign Intelligence Advisory Board

The board was established by executive order on March 20, 1969. Responsibilities include (1) advising the President concerning the various activities making up the overall national intelligence effort; (2) conducting a continuing review and assessment of foreign intelligence and related activities in which the CIA and other Government departments and agencies are engaged; and (3) reports to the President concerning the board's findings and appraisals, and makes appropriate recommendations for actions to achieve increased effectiveness of the government's foreign intelligence effort in meeting national intelligence needs.

SUBFEDERAL LAW ENFORCEMENT AGENCIES
State Agencies

The state is the primary law enforcement jurisdiction in the United States. All general police and peace officer powers at the state, county, or municipal level are delegated by the state to each substate jurisdiction. State police officers, municipal police officers, and county sheriffs' offices all receive their power to act relative to their health and safety of the community from a legislative enactment by the state. Likewise, all regulatory and supervisory units are of state origin. However, municipalities do enforce specific ordinances and regulations at the substate level. At the state level there are a number of different law enforcement or law enforcement related agencies in existence. These fall into a number of categories based upon the types of activities they perform.

Law Enforcement

There are generally two types of law enforcement agencies at the state level. The difference between state police agencies exists in the amount of police power that they possess. Some state police agencies have general police or peace officer powers to enforce all state laws. The other type of state agency is restricted in the types of activities which they may perform. Normally, the enforcement of traffic laws or operations of vehicles on public highways is their sole responsibility.

Records and Identification

Administrative and operational support is often provided within states to the state police and other law enforcement agencies through a centralized identification and records system. These units collect information on crime statistics, fingerprints, central records files, methods of operations, and disseminate this information on a need-to-know basis to law enforcement and law enforcement related agencies.

Crime Laboratories

States have laboratory facilities to process, examine, and identify physical evidence and provide laboratory testimony in criminal cases.

Miscellaneous Regulatory Agencies

Every state has some form of board or commission charged with the licensing and regulation of various types of activities. The major enforcement organizations are those dealing with alcoholic beverages. The investigative enforcement units normally are located in the units of agriculture, finance, commerce, health, education and welfare, motor vehicles, civil service, racing commissions, medical, fire, safety, fish and wildlife.

County

All states have some substate political jurisdiction; this is normally the county. The sheriff is normally the primary law enforcement official in this jurisdiction. However, the sheriff is not always the chief law enforcement officer, this normally being the prosecuting attorney, district attorney, or coroner depending upon the particular jurisdiction. The county sheriff generally has peace officer powers for the county in which he has jurisdiction. The county sheriff's department normally provides police protection for unincorporated parts of the county and provides highway patrol and traffic enforcement in the county. It normally operates and maintains a county detention facility.

Municipal

Municipal police departments normally provide a wide range of services to the community in the area of general police services and protection. There are 33,000 such agencies in operation in the United States.

SUMMARY AND CONCLUSION

There are a wide variety of intelligence, security, and law enforcement agencies within the governmental structure of the United States. In addition to the twenty units of government that have intelligence functions, there are fifty federal law enforcement agencies, 200 state law enforcement agencies, and 39,750 agencies at the county, city, state, and village level. In 1973, these agencies totalled 40,000. Also in 1973, the number of full-time personnel engaged exclusively in law enforcement activities at the federal level were 63,786; at the state level, 79,478; and local, 418,385—for a total of 561,646 individuals. In addition to this number, an additional 35,000 are engaged in internal security and intelligence functions at the federal level. This would increase the total of full-time internal security, intelligence, and law enforcement personnel in the United States to approximately 596,646. The total expenditures for law enforcement in the United States in 1973 was approximately eight billion dollars.[6]

The budget for intelligence and internal security operations are not available. It would be safe to estimate that this amount would be increased substantially if these figures could be determined. Because of the large numbers of agencies and the wide variety of areas that they deal with, formal and informal agreements exist between them which allow for information which they have collected or may need to flow freely between them.

At the federal level, such things as interdepartmental committee on internal security, interdepartmental seminar, and various types of agreements call for a free exchange of information among agencies when such information is of proper concern to the agency requesting this information. This flow of information can take place at all levels. This exchange of information also takes place between sub-federal agencies, such as municipal police departments and the federal government. The Federal Bureau of Investigation regularly provides information to law enforcement and law enforcement related agencies regarding individuals on whom the bureau has information. Likewise, at the state level, state police and state investigations units make information available to other law enforcement agencies at the state, municipal, or federal level. Many of these systems are computerized and the trend toward direct access via computer terminals to the National Crime Information Center will increase the flow and coordination of law enforcement activities at the federal, state, and municipal levels.

APPENDIX 4A

ORGANIZATION FOR SECURITY*

PRESENT CANADIAN STRUCTURE

General Considerations

Our inquiries suggest that the problem of devising an entirely satisfactory and rational security structure is one of extreme difficulty. In the first place, the government is functionally organized by departments, and there are obvious problems in superimposing and enforcing consistent standards in a given "service" area across the entire structure; the long history of issues concerned with financial control is sufficient to indicate the magnitude of this problem, which is of especial significance when organizational structures are decentralized. Also, security is an area which lies on the boundaries between general administration and professional specialty; it is easy—perhaps too easy—to regard security merely as an aspect of departmental administration and on this basis to disregard or evade its importance and complexities. Further, security structures, like police structures, must be closely related to the nature and quality of the national societies in which they operate if they are to achieve a reasonable measure of public acceptance. This relationship and this acceptance are perhaps somewhat easier to achieve in countries with long histories of threats to internal security and traditional requirements for defensive measures, or in great powers where the risks are clear and the stakes high.

The present Canadian security structure is diffuse, and consists of a number of disparate elements, including the Cabinet Committee on Security and Intelligence, the Security Panel, the Privy Council Office, the Solicitor General and his Department, the Minister of Justice and the Department of Justice, and the RCMP. In addition, all departments and agencies of the government have responsibilities for security which vary widely in scope and importance.

Policy-Making and Coordination

The Cabinet Committee on Security and Intelligence, the Security Panel, and two members of the staff of the Privy Council Office are all concerned with the formula-

---

* Report of the Royal Commission on Security, Queen's Printer, Ottawa, Canada, 1968, pp. 13-16.

tion of policy, the issuance of directives and regulations and the coordination of Canadian security policy and procedures. The Cabinet Committee was established in April 1963, and in its security role deals with important policy issues referred to it by the Security Panel. On the official level, the Security Panel is the senior body, consisting of selected deputy ministers. It was originally formed in 1946, and was reconstituted in 1963. Its terms of reference are "to advise on the coordination of the planning, organization and execution of security measures which affect government departments, and to advise on such other security questions as might be referred to it." Two officials of the Privy Council Office (an Assistant Secretary to the Cabinet and a member of the Cabinet Secretariat) together devote part of their time to providing the central point of reference and coordination for this committee structure. Theoretically, the first point of contact for departments and agencies seeking advice on security matters is this secretariat. Generally problems are dealt with by the officials of the secretariat themselves, but if necessary the issues can appear on the agenda of the Security Panel and the Cabinet Committee.

The Royal Canadian Mounted Police

The RCMP is the main federal operational and investigative body in the field of security. The Force assumed this role during World War I because, as the existing federal police force, it was at that time the natural federal instrument in this area. There is however no explicit statutory authority for the security role. Such authority as does exist is derived from certain Sections of the RCMP Act (S.C. 1959, c. 54). Section 17(3) of the Act provides *inter alia* that "every officer, and every person appointed by the Commissioner under this Act to be a peace officer, is a peace officer in every part of Canada and has all the powers, authority, protection and privileges that a peace officer has by law." Section 18 of the Act makes it "the duty of members of the force who are peace officers, subject to the orders of the Commissioner,

(a) to perform all duties that are assigned to peace officers in relation to the preservation of the peace, the prevention of crime, and of offences against the laws of Canada and the laws in force in any province in which they may be employed, and the apprehension of criminals and offenders and others who may be lawfully taken into custody . . . and

(b) to perform such other duties and functions are are prescribed by the Governor in Council or the Commissioner."

Section 21 of the Act provides that:

"(1) The Governor in Council may make regulations for the organization, training, discipline, efficiency, administration and good government of the force and generally for carrying the purposes and provisions of this Act into effect.

(2) Subject to the provisions of this Act and the regulations made under subsection (1), the Commissioner may make rules, to be known as standing orders, for the organization, training, discipline, efficiency, administration and good government of the force."

Section 44(c) of the RCMP Regulations and Orders (1960), which have been proclaimed by the Governor in Council under the Act, provides that "in addition to the duties prescribed by the Act, it is the duty of the force . . . (e) to maintain and operate such security and intelligence services as may be required by the Minister." The Commissioner's Standing Orders include orders relating to security and intelligence activities. Section 1331 of these Orders outlines the organization of the Directorate of Security and Intelligence of the RCMP and Section 1366 states the responsibilities of its Director:

> The Director, Security and Intelligence is responsible for the direction and correlation of activities in respect to coun-

ter-espionage and subversive activities against the State, for security investigations regarding personnel employed by the Government and others as required, for co-operation with Commonwealth countries and foreign nations in matters concerning the internal security of the State, co-operation with the internal intelligence organizations, both service and civilian, and for the direction of security and intelligence investigations generally.

Additional authority for the RCMP's security and intelligence operations is contained in certain instructions and directives issued by the Government, which in effect authorize the RCMP to conduct the investigations necessary for a security screening programme, and in addition make the Force responsible for various measures concerned with internal security in the event of a national emergency as proclaimed under the War Measures Act. In practice the RCMP is concerned with the following security functions:

(a) all security and security intelligence operations, and "police" operations related to security;
(b) maintenance and examination of records and field inquiries concerning personnel, but not evaluation of reports, nor decision-making in individual cases;
(c) advice concerning departmental security (this function appears to be somewhat ill-defined);
(d) record keeping; and
(e) certain staff functions (membership of interdepartmental committees, etc.) related to the management and planning of the national security effort.

These functions are performed by a headquarters Directorate of Security and Intelligence, which maintains representatives at RCMP regional headquarters and detachments and at certain locations overseas, and which operates in close liaison with the other directorates of the Force. Nearly 60 percent of the personnel of the security and intelligence directorate (including all the senior officers, all but three of the branch heads and almost all the officers responsible for operations or investigation) are Regular Members of the Force; the remaining personnel include some special constables employed on surveillance duties, Civilian Members employed as translators, technicians or researchers and public servants on clerical duties.

Until September 1966 the Minister of Justice, by virtue of the RCMP Act, was the minister responsible for the RCMP and the minister who reported to Parliament for the Force. The Government Organization Act (S.C. 1966-67, c. 25), which came into force in September 1966, transferred this responsibility to the Solicitor General, under whom a new Department of the Solicitor General was created. The duties and functions of the Solicitor General formally "extend to and include all matters over which the Parliament of Canada has jurisdiction, not by law assigned to any other department, branch or agency of the Government of Canada, relating to (a) reformatories, prisons and penitentiaries; (b) pardons and remissions; and (c) the Royal Canadian Mounted Police." The Department of Justice is now only concerned with security matters when the Minister is required to rule on the propriety of prosecutions or to provide legal advice relating to security.

Departments of Government

All departments and agencies of the Canadian Government have a responsibility to protect the confidentiality that attaches to official material. In addition, almost all have a requirement for more formal procedures to protect classified material. For example, at the very least all ministers' offices and all offices concerned with briefing ministers on Cabinet matters hold classified documents, and in addition most departments involved in any way with scientific, sociological, economic or other research require at least occasional access to classified material.

Within this broad framework, however,

departmental requirements vary widely. Some sensitive departments generate and hold large amounts of classified material, and have special security responsibilities. These include, for example, the Departments of National Defence, External Affairs and Defence Production, and the Privy Council Office and Cabinet Secretariat. Other departments are less sensitive. Such departments as the Department of Agriculture and the Department of Energy, Mines and Resources generate and hold smaller amounts of classified material, often in specific branches or sections, and their personnel generally require less frequent access to such material. Certain departments and agencies (such as the Department of National Revenue, the Dominion Bureau of Statistics and the Post Office) are responsible for the "privacy" of much material they hold; these departments are often the subject of special statutory provisions concerning confidentiality.

APPENDIX 4B

CUSTOMS SERVICE ENFORCEMENT RESPONSIBILITIES

The mission of the Customs Service is to assess and collect customs duties on imported merchandise, to prevent fraud and smuggling, and to control carriers, persons, and articles entering and departing from the United States—in accordance with the Tariff Act of 1930 and other statutes and regulations.

Laws Administered and Enforced

Tariff Act of 1930, as amended (19 U.S.C. 1202, *et seq*). This is the basic law for ascertainment, assessment, and collection of import duties, for procedures for entry of goods into the United States, and for certain restrictions on imports.

Foreign Trade Zones Act, as amended *19, U.S.C.* Customs also exercises functions establishment and operation of Foreign Trade Zones and procedures for admitting foreign and domestic merchandise into such zones and for transferring such merchandise out of them.

Anti-smuggling Act (19 U.S.C. 1701-1711). This Act provides certain special enforcement authority to cope with smuggling.

Antidumping Act, 1921, as amended (19 U.S.C. 160-173). This Act provides procedures for determining whether foreign merchandise is being dumped in the United States; i.e. being sold at less than fair value to the injury of American industry, and provides for assessment and collection of dumping duties in appropriate cases.

Miscellaneous statutes codified in Title 19, U.S.C. Customs also exercise functions under other provisions of law that are not part of the Tariff Act or the other acts already mentioned, but which are also codified in Title 19, U.S.C. such as 19 U.S.C. 1 and 2 (organization of Customs Service, arrangement and location of districts), 19 U.S.C. 68 (enforcement of laws in Guam and the Virgin Islands and along Canadian and Mexican borders; erection of buildings), 19 U.S.C. 267 (overtime compensation of Customs officers), and many others.

Controlled Substances Import and Export Act (21 U.S.C. 843, 951-966). Customs administers and enforces this act as it relates generally to the smuggling of narcotics and other controlled substances.

Quotas (usually under Presidential Proclamation). Customs administers quotas of various kinds on such products as cattle, certain cheeses, cream, milk, butter substitutes, fish, potatoes, brooms, cotton, certain cotton, wool and manmade textile fiber products, and peanuts.

Navigation Laws (Codified in Title 46, U.S.C.). Customs exercises functions provided for in the navigation laws (codified in Title 46) transferred to the U.S. Customs Service by section 102 of Reorganization Plan No. 3 of 1946 (3 CFR, 1946, Supp. C. IV) pertaining to the entry and clearance of vessels in domestic trades, including fisheries.

Internal Revenue Code (Title 26, U.S.C.). Customs collects certain import taxes imposed by the Internal Revenue Code on such products as sugar, oleomargarine, and alcohol and tobacco products.

Neutrality Act, as amended (22 U.S.C. 401, et seq). Customs assists in enforcing controls on exportation of arms and munitions of war.

Export Control (50 U.S.C. App. 2401-2419). Customs exercises important functions in assisting the Department of Commerce in the enforcement of the Export Control laws and regulations.

Criminal Code (18 U.S.C. 541-552). Customs enforces a number of miscellaneous criminal laws which are codified in 18 U.S.C.

Foreign and Cuban Assets Control Regulations (31 CFR 500.101, 515.101, and 530). Customs assists in the enforcement of regulations of Foreign and Cuban Assets Control relating to import restrictions on merchandise of North Vietnam, North Korea, Rhodesia, and Cuban origin and certain other merchandise.

Currency and Foreign Transactions Reporting Act (31 U.S.C. 1051-1122). Customs assists in enforcing controls on the importation and exportation of monetary instruments valued at over $5,000, or totaling more than $100,000 in one year.

Federal Aviation Act Provisions (49 U.S.C. 1474 and 1509). Customs enforces laws relating to the entry and clearance of civil aircraft.

Fish or Wildlife—Endangered Species—Protection (16 U.S.C. 1531 et seq).

Pre-Columbian Monumental or Architectural Sculpture or Murals (19 U.S.C. 2091).

Some of the laws and regulations enforced for other agencies are listed below.

Agriculture
Poultry Products Inspection Act
21 U.S.C., chapter 10
Foot and Mouth Disease Controls
21 U.S.C. 113a
Meat and Meat Products
9 CFR Part 327
Federal Seed Act Regulations
7 CFR Part 201
Adult Honeybees
7 U.S.C. 281

Army
Deposit of Refuse in Navigable Waters and Oil Pollution of Coastal Waters
33 U.S.C., Chapter 9

Atomic Energy Commission
10 CFR 30, 40, and 70

Commerce
Trade Mark Act
15 U.S.C. 1124
Foreign Excess Property
40 U.S.C. 512

Consumer Product Safety Commission
Hazardous Substances
15 U.S.C. 1261 et seq, 15 U.S.C. 2079
Flammable Fabrics Act
15 U.S.C. 1191-1200; 15 U.S.C. 2079
Consumer Product Safety Act
15 U.S.C. 2051 et seq

Environmental Protection Agency
Noise Control Act
42 U.S.C. 4901
Insecticides, Fungicides, & Rodenticides
7 U.S.C. 135h

Federal Trade Commission
Wool Products Labelling Act
15 U.S.C. 68 et seq
Fur Products Labelling Act
15 U.S.C. 69 et seq
Textile Fiber Products Identification
15 U.S.C. 70-70k

Hobby Protection Act
P.L. 93-167, 1973

Health, Education and Welfare
Importation of Psittacine birds, monkeys, dogs, cats, lather brushes, etiological agents and vectors, and dead bodies (Public Health Service)
42 CFR 71.151-71.157

Importation of Teas
21 CFR Part 281

Federal Import Milk Act
21 U.S.C. 141-149

Viruses, Serums, etc.
42 U.S.C. 262

Clean Air Act
42 U.S.C. 1857 *et seq*

Food, Drug and Cosmetic Act
21 U.S.C. 381

Interior
Marine Mammal Protection Act
16 U.S.C. 1361

Wild Animals, Birds, Fish, Reptiles or Eggs Thereof
18 U.S.C. 42; 50 CFR Part 13

Justice
Foreign Agents and Propaganda Registration Act
22 U.S.C. 611

Immigration and Nationality Act
8 U.S.C. (Immigration and Naturalization Service)

Library of Congress
Copyright Act
17 U.S.C. 105-109

Treasury
Importation and Exportation of Gold
31 CFR Part 54

Federal Alcohol Admn. Act (Labeling)
27 U.S.C. 201-212
(Bureau of Alcohol, Tobacco and Firearms)

Counterfeit Coins, etc.
18 U.S.C., Chapter 25 (Secret Service)

Omnibus Crime Control & Safe Streets Act
18 U.S.C. 921 *et seq* (Internal Revenue Service)

Transportation
National Traffic & Motor Vehicle Safety Act
15 U.S.C. Chapter 38

FOOTNOTES

1. This is not only true in the United States, but in other nations as well. Appendix 4A, this chapter, provides a brief description of the governmental security structure in Canada.
2. The National Intelligence Authority (NIA) was established by a Presidential directive on January 22, 1946. The authority was responsible for planning, developing, and coordinating Federal foreign intelligence activities related to national security.
    Upon creation of the CIA, the NIA was abolished and personnel, property, and records were transferred to the CIA.
3. *U.S. Government Organization Manual.* Washington, Government Printing Office, 1966, p. 60.
4. The Customs Service, for example, is charged with the enforcement of a variety of Federal statutes and regulations. A complete listing may be found in Appendix 4B, this chapter.
5. ERDAM chapter 2405-031, Volume 2400 deals with Security Regulations for ERDA.
6. Historical Statistics on Expenditures and Employment for the Criminal Justice System 1971 to 1973. U.S. Department of Justice, Washington, D.C., SD-EE No. 6. (July 1975), pp. 2-19.

BOOKS

Abshagen, Karl Heinz: *Canaris.* London, Hutchinson, 1956.
Agabekov, George: *OGPU: The Secret Russian Terror.* Bretano's Inc., 1931.
Alcon, Robert H.: *No Bugles for Spies.* New York, McKay, 1962.
———: *Allied Intelligence Bureau: Our Secret Weapon in the War Against Japan.* New York, McKay, 1958.
Almond, G.: *The Appeals of Communism.* Princeton University Press, 1954.
Alson, Stewart L., and Braden, Thomas: *Sub Rose: The OSS and American Espionage.* New York, Revnal and Hitchcock, 1946.
Aston, Sir George: *Secret Service.* London, Faber and Faber Ltd., 1930.

Babington-Smith, Constance: *Air Spy.* New York, Harper, 1957.
Baden-Powell: *My Adventures as a Spy.* London, 1915.
Bailey, Geoffrey: *The Conspirators.* New York, Harper, 1960.
Bakeless, John: *Turncoats, Traitors and Heroes.* Philadelphia, Lippencott, 1959.
Baker, Gen. Lafayette: *History of the United States Secret Service.* Philadelphia, L. C. Baker, 1867.
Barber, Willard Foster: *Internal Security & Military Power: Counter Insurgency & Civic Action in Latin America.* Columbus, Ohio State University Press, 1966.
Barron, John: *KGB The Secret Work of Soviet Secret Agents.* New York, E. P. Dutton & Co., Inc., 1974.
Bar-Zohar, Michael: *Spies in the Promised Land.* Boston, Houghton Mifflin Company, 1972.
Batchelor, James H.: *Operations Research: An Annotated Bibliography.* St. Louis University Press, 1959.
Bauermeister, Lieut. A.: *Spies Break Through.* London, Constable and Company, Ltd., 1934.
Bentley, Elizabeth: *Out of Bondage.* New York, Devin-Adair, 1951.
Best, Capt. S. Payne: *The Venlo Incident.* London, Hutchinson and Co., Ltd., 1949.
Bialoguski, Michael, M.D.: *The Case of Colonel Petrov.* New York, McGraw-Hill Book Co., Inc., 1955.
Blackstock, Paul W.: *The Strategy of Subversion.* Chicago, Quadrangle, 1964.
Blackwook, Beatrice: *Both Sides of the Buka Passage.* Oxford, 1935.
Bouscaren, Anthony T.: *The Security Aspects of Immigration Work.* Milwaukee Marquette University, 1969.
Boveri, Margaret: *Treason in the Twentieth Century.* 1963.
Brenton, Myron: *The Privacy Invaders.* New York, 1964.
Bulloch, John, and Miller, Henry: *Spy Ring.* London, Secker and Warburg, 1961.
Bulloch, John: *M.I.5.* London, Arthur Barker Ltd., 1963.
Burnham, James: *The Web of Subversion.* New York, John Day Co., 1954.
Busch, Tristan: *Secret Service Unmasked.* London, Hutchinson and Co., Ltd., 1950.
Butow, Robert J. C.: *Japan's Decision to Surrender.* Stanford, Stanford University Press, 1954.

Caldwell, John: *Communism in Our World.* John Day, 1956.
Carlson, J. R.: *The Plotters.* New York, Dutton, 1946.
———: *Undercover.* New York, Dutton, 1943.
Carr, Robert K.: *The House Committee on Un-American Activities 1945-1950.* Cornell University Press, 1952.
Carroll, John M.: *Secrets of Electronic Espionage.* New York, Dutton, 1966.
———: *The Third Listener.* New York, Dutton, 1969.
Castleman, John E.: *On Active Service.* Louisville, 1927.
Chambers, Robert: *Secret Service Operator 13.* New York, 1934.
Churchill, P.: *Duel of Wits.* New York, Putnam, 1955.
———: *Of Their Own Choice.* London, 1951.
Churchill, Winston S.: *Their Finest Hour. Vol. II: The Second World War,* 6 vols. Boston, Houghton Mifflin, 1949.
Ciano, Count: *Ciano's Hidden Diary.* New York, 1953.
Clayton, Tom: *The Protectors.* London, Oldbourne Book Co. Ltd., 1967.
Cloak and Dagger. New York, Random House, 1946.
Collins, Frederick L.: *The FBI in Peace and War.* New York, Putnam, 1943.
Colvin, I.: *Chief of Intelligence.* London, 1951.
———: *Master Spy.* New York, McGraw-Hill, 1951.
Commager, Henry S.: *Freedom, Loyalty, Dissent.* Oxford University Press, 1954.
Cookridge, E. H.: *Secrets of the British Secret Service.* London, 1948.
———: *Soviet Espionage.* New Haven, Frederick Muller, Ltd., 1955.
———: *Soviet Spy Net.* London, Miller, 1955.
———: *The Net That Covers the World.* New York, Henry Holt and Company, 1955.
———: *Sisters of Delilah.* London, Oldbourne, 1959.
———: *Gehlen, Spy of the Century.* New York, Random House, 1971.
———: *Set Europe Ablaze.* New York, Crowell, 1967.
Coulson, Thomas, Major: *Mata Hari.* New York, Harper, 1930.
———: *Queen of Spies.* London, Constable and Co., Ltd., 1935.
Crankshaw, E. H.: *Gestapo.* New York, Viking, 1956.
Craven, Wesley, and Cate, James L., eds.: *The*

Army Air Forces in World War II. University of Chicago Press, 1958.

Cronyn, George W.: *A Primer of Communism.* New York, E. P. Dutton, 1957.

D'Agapeyeff, Alexander: *Codes and Ciphers.* London, Oxford U. Press, 1939.

Dallin, David: *Soviet Espionage.* New Haven, Yale, 1955.

Daniels, Norman: *Spy Hunt.* Pyramid Books, 1960.

Dasch, George J.: *Eight Spies Against America.* New York, McBride, 1959.

Datta, Daijnath: *Weapons for War, Prescription for Peace: National Security.* Bombay, India Publishing House, 1968.

Daugherty, William E., and Janowitz, Morris: *A Psychological Warfare Casebook.* The Johns Hopkins Press, 1958.

Deakin, F. W., and Storry, G. R.: *The Case of Richard Sorge.* New York, Harper & Row, 1966.

Debray, R.: *Revolution in the Revolution.* Grove Press, 1967.

De Gramont, Sanche: *The Secret War.* New York, G. P. Putnam's Sons.

De Jong, Louis: *The German Fifth Column.* University of Chicago, 1956.

Delaney, Robert Finley: *The Literature of Communism in America.* Catholic University of America Press, 1962.

Deriabin, Peter, and Gibney, Frank: *The Secret World.* New York, Doubleday, 1959.

De Toledane, Ralph: *The Greatest Plot in History.* New York, Duell, Sloan, and Pearce, 1963.

Devons, Ely: *Planning in Practice: Essays in Aircraft Planning in Wartime.* Cambridge, Eng., at the University Press, 1950.

Dillon, Walter: *Little Brother Is Watching.* Boston, Houghton, 1962.

Dodd, William E., Jr.: *Ambassador Dodd's Diary.* London, Victor Gollancz, 1941.

Donovan, James B.: *Strangers on a Bridge: The Case of Col. Abel.* New York, Atheneum, 1964.

Downs, Donald: *The Scarlet Thread.* London, British Book Center, 1953.

Downs, Edward C.: *Four Years a Scout and Spy.* Zanesville, Ohio, 1866.

Drobutt, Richard: *I Spy for the Empire.* Low, Sampson, Marston and Co., Ltd., 1939.

Dulles, Allen W.: *The Craft of Intelligence.* New York, Harper, 1963.

————: *Germany's Underground.* New York, Macmillan, 1947.

Ebenstein, William: *Today's Isms: Communism, Facism, Socialism, Capitalism.* Englewood Cliffs, Prentice-Hall, 1964.

Edmonds, S., and Edmonds, Emma E.: *Nurse and Spy in the Union Army.* Hartford, 1865.

————: *Intelligence Establishment, The.* Cambridge, Harvard, 1970.

————: *Intelligence, political and military. International Encyclopedia of the Social Sciences,* Vol. 7, New York, Macmillan and Free Press, pp. 415-421, 1968.

Eniaudi, Mario, et al.: *Communism in the Western World.* New York, Cornell, 1951.

Essad-Bey: *OGPU.* New York, Viking, 1933.

Evans, Medford: *Secret War for the "A" Bomb.* Chicago, Regnery, 1953.

Farago, Ladislas: *Burn After Reading.* New York, Walker, 1961.

————: *War of Wits.* New York, Pap. Lib., 1962.

————: *War of Wits, The Anatomy of Espionage and Intelligence.* New York, Funk and Wagnalls, 1954.

Felix, Christopher: *A Short Course in the Secret War.* New York, Dutton, 1964.

Fineburg, S. Andhil: *The Rosenburg Case.* New York, Oceana, 1953.

Firmin, Stanley: *They Came to Spy.* London, Hutchinson and Co., Ltd., 1947.

Fleming, Peter: *Operation Sea Lion.* New York, Simon and Schuster, 1957.

Foote, Alexander: *Handbook for Spies.* Garden City, New York, Doubleday and Co., Inc., 1949.

Ford, Corey, and MacBain, Alastair: *Cloak and Dagger: The Secret Story of the OSS.* New York, Random House, 1946.

————: *Foreign Intelligence and the Social Sciences.* N.J. Research Monograph No. 17, Princeton University Press, 1964.

Frischauer, Willi: *Berlin Betrayal.* New York, Belmont Books, 1958.

————: *Himmler,* Boston, Beacon Press, 1954.

————: *Man Who Came Back.* London, Muller Ltd., 1958.

Fromm. E.: *Escape from Freedom.* New York, Holt, Rinehart & Winston, 1941.

Gaines, Helen F.: *Cryptanalysis: A Study of Ciphers and Their Solutions.* New York, Dover, 1956.

Galang, Richard C.: *Secret Mission to the Philippines.* Manila, University Publication Co., 1918.

George, Alexander L.: *Propaganda Analysis: A Study of Inferences Made from Nazi Propaganda in World War II.* Evanston,

Row, Peterson & Co., 1959.
George, W.: *Surreptitious Entry*. New York, Appleton, 1946.
Gillow, Benjamin: *The Whole of Their Lives*. New York, Scribner, 1948.
Giskes, H. J.: *London Calling North Pole*. Kimber, 1953.
Golembiewski, Robert: *Men, Management and Morality*. New York, McGraw, 1965.
Gollomb, Joseph: *Spies*. New York, Macmillan, 1942.
———: *Armies of Spies*. New York, Macmillan, 1939.
Gouzenko, Ignoe: *The Iron Curtain*. New York, Dutton, 1948.
Goudsmit, Samuel A.: *Alsos*. New York, Henry Schuman, 1947.
Griffith, Samuel: *Sun Tzu, the Art of War*. Oxford, Clarendon Press, 1968.
Gross, M. W.: *The Brain Watchers*. New York, 1962.
Gowenlock, Thomas R.: *Soldiers of Darkness*. New York, Doubleday, 1937.
Gramont, Sanche De: *The Secret War*. New York, Dell, 1963.
Granick, D.: *The Red Executive: A Study of the Organization Man in Russian Industry*. New York, Anchor Books, 1961.
Gray, Wook: *The Hidden Civil War*. New York, Viking, 1942.
Hall, Roger: *You're Stepping on My Cloak and Dagger*. New York, Bantam, 1957.
Hammond, Paul Y.: *Organizing for Defense: The American Military Establishment in the Twentieth Century*. Princeton, Princeton University Press, 1961.
Hart, H. L. A.: *Law, Liberty and Morality*. Stanford, Stanford University Press, 1963.
Hearn, Ct. V.: *Russian Assignment*. London, Hale, 1962.
Heilbrunn, Otto: *The Soviet Secret Services*. New York, Praeger, 1956.
Henderson, Nevile: *Failure of A Mission*. London, Hodder and Stoughton, 1940.
Herling, Alfred K.: *Soviet Slave Empire*. New York, Funk, 1951.
Herron, Lowell W.: *Executive Action Simulation*. Englewood Cliffs, New Jersey, Prentice-Hall, Inc., 1960.
———: *To Move A Nation*. New York, Doubleday, 1967.
———: *Strategic Intelligence and National Decisions*. Glencoe, Ill., The Free Press, 1956.
Hinchley, Vernon: *Spy Mysteries Unveiled*. New York, Dodd, 1964.

Hirsch, Richard: *The Soviet Spies*. New York, Duell, Sloan, and Pearce, 1947.
Hitch, Charles J.: *Decision-Making for Defense*. Berkeley and Los Angeles, University of California Press, 1965.
Hoettl, Wilhelm: *The Secret Front*. New York, Prager, 1954.
———: *Hitler's Paper Weapon*. London, Rupert-Davis, 1955.
Hoke, Henry: *It's A Secret*. New York, 1946.
Holbrook, James: *Ten Years Among the Mail Bags*. Philadelphia, 1855.
Hoover, J. Edgar: *A Study of Communism*. New York, Holt, 1962.
Humphreys, Andrew: *Heroes and Spies of the Civil War*. New York, 1903.
Huss, Pierre J., and Darpozi, George J.: *Red Spies in the U.N.* Coward, 1964.
Hutton, J. Bernard: *Frogman Spy*. New York, McDowell-Obolenshy, 1960.
———: *School for Spies*. New York, Coward, 1962.
Hyde, Montgomery H.: *The Quiet Canadian*. London, Hamish Hamilton, 1962.
Hyman, H. H., et al.: *Interviewing in Social Research*. Chicago, University of Chicago Press, 1954.
Hynd, Alan: *Passport to Treason*. New York, 1942.
———: *Betrayal for the East*. New York, McBride, 1943.
Icardi, Aldo: *Aldo Icardi: American Master Spy*. New York, University Books, 1956.
Ind, Allison: *A Short History of Espionage*. New York, McKay, 1963.
Inkeles, A., and Bauer, R. A.: *The Soviet Citizen. Daily Life in a Totalitarian Society*. Cambridge, Harvard University Press, 1961.
———, and Geiger, K.: *Soviet Society: A Book of Readings*. Boston, Houghton Mifflin Company, 1961.
Jaszi, O., and Lewis, J. D.: *Against the Tyrant*. Glencoe, Ill., Free Press, 1957.
Kilpatrick, Jeane (ed.): *The Strategy of Deception*. New York, Farrar Strauss.
Jensen, Carla: *I Spy*. New York, 1930.
Joesten, Joachim: *They Call It Intelligence*. New York, Abelard, 1963.
John, George S.: *Philip Henson, the Southern Union Spy*. St. Louis, 1887.
Johnson, Chalmers: *An Instant of Treason*. Stanford, Stanford Univ. Press, 1964.
Johnson, Haynes: *The Bay of Pigs*. New York, W. W. Norton, 1964.
Jowitt, Earl: *Some Were Spies*. London, Hodder and Stoughton, 1954.

———: *Strange Case of Alger Hiss, The.* New York, Doubleday, 1953.

Kahn, David: *The Code Breakers. History of Secret Communication.* New York, Macmillan, 1967.

Kaledin, Victor K.: *The Moscow-Berlin Secret Service.* London, 1940.

———: *Underground Diplomacy.* London, Hurst and Blackett, 1939.

Kane, Harnett T.: *Spies of the Blue and Grey.* New York, Country Life Press, 1954.

Kaznacheev, Aleksndr: *Inside a Soviet Embassy.* Philadelphia, Lippincott, 1962.

Kent, Sherman: *Strategic Intelligence for American World Policy.* Princeton, N.J., Princeton University Press, 1949, 1951.

———: *In the Name of Conscience.* New York, McKay, 1959.

Kibbee, J. M., et al.: *Management Games.* New York, Reinhold Publishing Corp., 1961.

Kim, Young Hum, (ed.): *The Central Intelligence Agency: Problems of Secrecy in a Democracy.* Lexington, Mass., D. C. Heath, 1968.

Kirkpatrick, Lyman B., Jr.: *The Real CIA.* New York, Macmillan, 1968.

Klein, Alexander: *The Counterfeit Traitor.* New York, Holt, 1958.

Landau, H.: *The Enemy Within—The Inside Story of German Sabotage in America.* New York, Putman, 1937.

Langelaan, George: *Masks of War.* Garden City, Doubleday, 1959.

Larus, J.: *From Collective Security to Preventive Diplomacy.* Wiley, 1965.

LeBreton, Preston P., and Henning, D. A.: *Planning Theory.* Englewood Cliffs, New Jersey, Prentice-Hall, Inc., 1961.

LeCarre, John: *The Spy Who Came in From the Cold.* New York, Coward, 1963.

Leverkuehn, Paul: *German Military Intelligence.* New York, Praeger, 1954.

Lowenthal, Max: *The Federal Bureau of Investigation.* New York, William Sloane Associates, 1950.

Mac Bain, Ford C.: *Cloak and Dagger: The Secret Story of the OSS.* New York, 1946.

Mac Donald, E. P.: *Undercover Girl.* New York, 1957.

———: *Women Spies I Have Known.* London, 1939.

Makin, William J.: *Brigade of Spies.* New York, Dutton, 1939.

Mannix, P. Daniel: *The History of Torture.* New York, Dell, 1964.

Mao Tse Tung: *Guerrilla Warfare.* New York, Praeger, 1961.

Marshall, S. L. A.: *The River and the Gauntlet: Defeat of the Eighth Army by the Chinese Community Forces, November, 1950, in the Battle of the Congchon River, Korea.* New York, William Morrow & Co., 1953.

Martelli, George: *The Man Who Saved London.* New York, Doubleday, 1961.

Massing, Hede: *The Deception.* New York, Dell, 1951.

———: *Masters of Deceit.* New York, Holt, 1958.

Matthews, Herbert Lionel: *The Cuban Story.* New York, G. Braziller, 1961.

Maugham, Somerset: *Ashenden.* New York, Doubleday.

McCloskey, H.: *Personality and Attitude Correlates of Foreign Policy Orientation in Political Inquiry.* New York, Macmillan, 1969, pp. 70-125.

McGovern, James: *Crossbow and Overcast.* London, Hutchinson, 1965.

McLaughlan, Donald: *Room 39. A Study in Naval Intelligence.* New York, Atheneum, 1968.

Meissnerx, Han Otto: *The Man With Three Faces.* New York, Reinhart, 1955.

Miksche, F. O.: *Secret Forces.* London, Faber and Faber, Ltd.

Miller, Norman C.: *The Great Salad Oil Swindle.* New York, Coward-McCann, 1965.

Milosz, C.: *The Captive Mind.* London, Secher & Warburg, 1953.

Milton, George F.: *Abraham Lincoln and the Fifth Column.* New York, Collier, 1962.

Monat, Pawel: *Spy in the U.S.* New York, Harper, 1961.

Montagu, Ewen: *The Man Who Never Was.* Philadelphia and New York, J. P. Lippincott Co., 1954.

Moyzisch, L. S.: *Operation 'Cicero.'* London, Wingate Press, 1950.

Noel-Baken, Francis: *The Spy Webb.* London, Batchworth, 1954.

Office of Strategic Services Assessment Staff: *Assessment of Men.* New York, Rinehart, 1957, 1950.

O'Neal, James and Werner: *American Communism.* New York, Dutton, 1947.

Orlov, Alexander: *Handbook of Intelligence and Guerrilla Warfare.* Ann Arbor, U. of Michigan, 1963.

Ottenberg, Miriam: *The Federal Investigators.* Englewood Cliffs, N.J., 1962.

Page, Bruce, Leitch, David, and Knightley, Phillip: *The Philby Conspiracy.* Garden City, N.Y., Doubleday, 1968.

Peers, William R., and Brelis, Dean: *Behind*

the Burma Road. New York, The Hearst Corporation, 1963.
Peis, Gunter, and Wighton, Charles: *Hitler's Spies and Saboteurs.* New York, Holt, 1958.
Penkovskiy, Oleg Vladimerovich: *The Penkovskiy Papers.* Garden City, N.Y., Doubleday and Co., Inc., 1965.
Petrov, V., and Evdokia: *Empire of Fear.* New York, Praeger, 1957.
Philbrick, Herbert: *I Led Three Lives.* Dunlap and Dunlap.
Philby, Kim (H.A.R.): *My Silent War.* London, MacGivvon and Kee, 1968.
Pilat, Oliver: *The Atom Spies.* New York, G. P. Putnam's Sons, 1952.
Pinto, Lt. Col. Oresti: *Spy-Catcher.* London, Werner Laurie, 1933.
———: *Friend or Foe?* London, Werner Laurie, 1953.
Pittenger, W.: *The Secret Service.* Philadelphia, 1882.
Platt, Washington: *Strategic Intelligence Production.* New York, Praeger, 1957.
———: *National Character in Action: Intelligence Factors in Foreign Relations.* New Brunswick, Rutgers, 1961.
Pomeroy, William: *Guerrilla Warfare.* New York, Int. Publ., 1964.
Poole, Lynne, and Poole, G.: *The Magnificent Traitor.* New York, Dodd, Mead, 1968.
Pratt, Fletcher: *Secret and Urgent: the Story of Codes and Ciphers.* Garden City, Blue Ribbon Books.
Rachles, Eugene: *They Came to Kill.* New York, Random, 1961.
Ransom, Harry Howe: *Can American Democracy Survive Cold War?* Garden City, N.Y., Doubleday, 1963.
Read, Conyers: *Mr. Secretary Walsingham and the Policy of Queen Elizabeth.* 3 vols. Cambridge, Mass., Harvard University Press, 1925.
Reilly, Sidney: *Britain's Master Spy.* New York and London, Harper and Brothers, 1933.
Re'my: *Memoirs of a Secret Agent.* New York, McGraw, 1948.
Renault-Rouliet, Gilbert: *Memoirs of a Secret Agent of Free France.* London, Whittlesey House, 1948.
Richer, Marthe: *I Spied for France.* London, John Long, Ltd., 1935.
Richings, M. G.: Espionage: *The Story of the Secret Service of the English Crown.* London, Hutchinson and Co., 1934.
Riess, Curt: *Total Espionage.* New York, Putnam, 1941.
Rivera, Joseph H. de: *The Psychological Dimension of Foreign Policy.* Columbus, Ohio, Merrill, 1968.
Root, Jonathan: *The Betrayers.* New York, Coward, 1963.
Roskill, Captain S. W.: *The War at Sea, 1939-1945.* 3 vols. London, H.M.S.O., 1956.
Rowan, Richard W.: *Modern Spies Tell Their Stories.* New York, Robert M. McBride and Co., 1934.
———: *Pinkertons, The.* Boston, 1931.
Russell, C. E.: *True Adventures of the Secret Service.* New York, Doubleday, 1923.
Sargant, William: *Battle for the Mind.* New York, Pelican Books, 1957.
Sarlat, Noah: *Spy in Black Lace.* New York, Linder Books, 1964.
Sayers, Michael and Kahn, Albert E.: *Sabotage—The Secret War Against America.* New York and London, Harper and Brothers, 1942.
Schollenberg, Walter: *Labyrinth.* New York, Harper and Brothers, 1956.
Schofield, William G.: *Treason Trail.* New York, Rand McNally, 1964.
Schumpeter, Joseph A.: *Capitalism, Socialism and Democracy.* New York, Harper and Brothers, 1947.
Scott, John: *Political Warfare, A Guide to Competitive Co-Existence.* New York, John Day, 1955.
Seabury, Paul: *The Wilhelmstrasse: A Study of German Diplomats Under the Nazi Regime.* Berkeley and Los Angeles, University of California Press, 1954.
———: *Secret Agents Against America.* New York, Doubleday, 1939.
Selznick, P.: *The Organizational Weapon. A Study of Bolshevik Strategy.* Glencoe, Ill., Free Press, 1960.
Seth, Ronald: *Anatomy of Spying.* New York, E. P. Dutton, 1963.
———: *Art of Spying, The.* New York, Philosophical Library, Inc., 1957.
———: *Secret Servants.* New York, Farrar, Straus, 1957.
———: *Spies at Work: A History of Espionage.* New York, Philosophical Library, 1954.
Shaplen, R.: *Kreuger: Genius and Swindler.* New York, Knopf, 1960.
Sherman, Kent: *Strategic Intelligence for American World Policy,* Princeton University Press, 1949.
Siguad, Louis A.: *Belle Boyd, Confederate Spy.* Richmond, Dietz, 1945.
Sillitoe, Sir Percy: *Cloak Without Dagger.* London, Cassell and Company, Ltd., 1955.

Sinevirski, N.: *Smersh.* New York, 1950.

Singer, Kurt, and Sherrod, Jane: *Spies for Democracy.* Minneapolis, Denison, 1960.

———: *Spies Who Changed History.* New York, Ace Book Company, 1960.

———: *World's Thirty Greatest Women Spies, The.* New York, Funk and Wagnall's, 1951.

———: *Men in the Trojan Horse, The.* Boston, Deacon, 1953.

———: *Spy Omnibus.* Minneapolis, Dennison, 1960.

———: *Three Thousand Years of Espionage.* New York, 1948.

———: *Spies and Traitors of World War II.* New York, Prentice-Hall, 1948.

———: *Duel for the Northland.* New York, McBride, 1943.

———: *Spies Over Asia.* London, W. H. Allen, 1956.

Sinkov, Abraham: *Elementary Cryptanalysis.* New York, Random House, 1968.

Skorzeny, Otto: *Skorzeny's Secret Missions.* New York, E. P. Dutton, 1950.

Skousen, W. Cleon: *The Naked Communist.* Salt Lake City, Utah, Ensign Publishing Company.

Smith, A.: *The Wealth of Nations.* New York, P. F. Collier, 1909.

Smith, W. C.: *Sabotage: Its History, Philosophy and Function.* Spokane, 1913.

Snowden, Nicholas: *Memoirs of a Spy.* New York and London, Charles Scribner's Sons, 1933.

Snyder, Richard M.: *Measuring Business Changes.* New York, John Wiley & Sons, Inc., 1955.

———: *Spies and the Next War.* Garden City, N.Y., Garden City Books, 1936.

———: *Spy and Counter-Spy.* New York, The Viking Press, 1928.

———: *Spy Has No Friend, A.* London, Andre Deutsch, 1952.

———: *Spy Secrets.* New York, 1946.

Spiro, Edward: *Soviet Spy Net, the Net That Covers the World,* London, Muller, 1955.

Steinhauser, G., and Felstead, S. T.: *Steinhauser: The Kaiser's Master Spy.* London, 1924.

———: *Steinhauser: The Kaiser's Master Spy.* New York, D. Appleton and Company, 1931.

Stephen, Eno: *Spies in Ireland.* Harrisburg, Stackpole, 1965.

Stern, Van Doren, Philip: *Secret Missions of the Civil War.* New York, Rand McNally, 1959.

Stewart, Alsop, and Braden, Thomas: *Subrosa: the OSS and American Espionage.* New York, Reynal and Hitchcock, 1946.

Stranger, Roland J.: *Essays on Espionage.* Ohio University Press, 1962.

Strong, Sir Kenneth: *Intelligence at the Top, the Recollections of an Intelligence Officer.* London, Cassell, 1968.

———: *Story of the Secret Service, The.* New York, Doubleday, 1937.

Sweeney, Walter C.: *Military Intelligence.* New York, Frederick Stokes, 1924.

Sweet-Escott, Bickham: *Baker Street Irregulars.* London, Methuen, 1965.

Thayer, Charles W.: *Guerrilla.* New York, Signet Books, New Am. Lib., 1963.

Thompson, B.: *Allied Secret Service in Greece, The.* London, Hutchinson and Co., 1931.

Thompson, J. W., and Padoverm, S. K.: *Secret Diplomacy: A Record of Espionage and Double Dealing.* London, 1937.

Thorwald, Jurgen: *The Century of the Detective.* New York, Harcourt, Brace & World, 1965.

Tietzen, Arthur: *Soviet Spy Ring.* New York, Coward, 1961.

Toledano, Ralph D.: *Seeds of Treason.* New York, Funk, 1950.

Tompkins, Peter: *A Spy in Rome.* New York, S. and S., 1962.

Tompkins, W. C.: *Sabotage and Its Prevention.* New York, Berkeley, 1942.

Townsend, Elias: *Risks: The Key to Combat Intelligence.* Military Service Publishing Co., 1955.

Trautman, W. E.: *Direct Action and Sabotage.* Pittsburgh, 1912.

Trevor-Roper, Hugh: *The Philby Affair.* London, Kimber, 1968.

Tuchmann, Barbara W.: *The Zimmerman Telegraph.* New York, Viking, 1958.

Tully, Andrew: *CIA—The Inside Story.* New York, William Morrow and Company, 1962.

Turrou, Leon: *Nazi Spys in America.* New York, Random, 1939.

Uranov, S.: *Espionage.* New York, Int. Pub., 1937.

Valtin, Jan: *Out of the Night.* Blue Ribbon Books, 1941.

Vesper, Amelto: *Secret Agent of Japan.* London, Victor Gollancz, Ltd., 1938.

Vogeler, Robert J.: *I Was Stalin's Prisoner.* New York, Harcourt, 1952.

Von Ritelen, Captain: *The Dark Invader.* New York, Macmillan, 1933.

———: *The Return of the Dark Invader.* London, Peter Davies, 1939.
Von Schlabrendorff, Fabian: *They Almost Killed Hitler.* New York, The Macmillan Company, 1947.
Voska, Emanuel Victor, and Irwin, Will H.: *Spy and Counterspy.* New York, Doubleday, Doran and Company, Inc., 1940.
West, Rebecca: *The New Meaning of Treason.* New York, The Viking Press, 1964.
White, John Baker: *The Soviet Spy System.* London, The Falcon Press, Ltd., 1948.
Whitehead, Donald: *The FBI Story.* New York, Random, 1956.
Whitehouse, Arch.: *Espionage and Counterespionage.* New York, Doubleday, 1964.
Wild, Max: *Secret Service on the Russian Front.* New York, G. P. Putnam's Sons, 1932.
Williams, Wythe, and Narvig, Van: *Secret Sources.* Chicago and New York, Ziff-Davis Publishing Company, 1943.
Willoughby, Maj. Gen. Charles A.: *Shanghai Conspiracy.* New York, E. P. Dutton and Co., 1952.
Wise, David, and Ross: *The Espionage Establishment.* New York, Random.
Wise, David, and Rose, Thomas: *The Invisible Government.* New York, Random, 1964.
———: *Pearl Harbor, Warning and Decision.* Stanford University Press, 1962.
Wolin, Simon and Slusser, Robert: *The Soviet Secret Police.* New York, Praeger, 1957.
Woodhall, Edwin T.: *Spies of the Great War.* London, John Long, Ltd., 1932.
Wraith, Ronald, and Simpkins, Edgar: *Corruption in Developing Countries.* London, G. Allen, 1963.
Yardley, Herbert O.: *The American Black Chamber.* Indianapolis, Ind., The Bobbs-Merrill Company, 1931.
Zacharias, Ellis M.: *Behind Closed Doors.* New York, 1950.
———: *Secret Missions.* New York, Pap. Lib., 1961.

PUBLICATIONS OF THE GOVERNMENT, LEARNED SOCIETIES, AND OTHER ORGANIZATIONS

Alexander, Sidney S.: Economics and business planning. *Economics and the Policy Maker: The Brookings Lectures,* 1959.
American Bar Association, Proceedings of the 77th Annual Meeting. 1954, p. 60.
Assessment of men, the, Office of Strategic Services Assessment Staff, 1948.
Barth: House Committee on Un-American Activities, 1952.
Bureau of National Affairs. Report of the commission on government security, Washington, Gov. Printing Office, 1957.
Combating subversion and sabotage, Studies in Business Policy #60, 1952.
Commission on Organization of the Executive Branch of the Government, Intelligence activities, June 1955.
Conduct of Espionage Within the United States by Agents of Foreign Communist Governments, 90th Cong., 1st Sess., April 6 and 7, May 10, June 15, November 15, 1967.
Confrontation. Lemberg Center for the Study of Violence, Massachusetts.
Congressional Committee on Un-American Activities Hearings in Pittsburgh, Pa. Sabotage, U.S. Congressional Record, Vol. CV, No. 114, July 8, 1959.
Counterintelligence Activities. Annual report by J. Edgar Hoover, 1968.
Davidson, Phillip B. Col.: Intelligence for commanders. Military Service Publication, 1952.
Department of State Bulletin. Coordination of foreign intelligence activities, February 3, 1946.
———: Intelligence objectives, May 12, 1946.
Evans, Allen and Gatewood, R. D.: Intelligence and research: Sentinel and scholar in foreign relations. Washington, Gov. Print. Office, 1960.
Farren, Darry D.: *Sabotage—How to Guard Against It.* New York, National Foremen's Institute, Inc., 1941.
Federal Bureau of Investigation. Expose of Soviet espionage May 1960. Committee on the Judiciary, U.S. Senate, 86th Cong., 2nd Sess., May 1960.
Fifteenth annual report—fiscal year ended June 30, 1965. Subversive Activities Control Board, Washington, D.C., U.S. Government Printing Office.
Florida, Attorney General. Report on investigation of subversive activities in Florida, 1955.
Foot, M. R. D.: SOE in France: An account of the work of the British special operations executive in France, 1940-1944. London HMSO, 1966.
Geertz, Clifford: Indonesian societies in Bali and Java. Center for Advanced Study in

the Behavioral Sciences, Stanford, California, 1959.

General Services Administration: U.S. Government Organization Manual 1966-67. Office of the Federal Register, National Archives, Washington Government Printing Office.

Great Britain, Prime Minister. B notice system. London H.M.S.O., 1967.

Great Britain, Standing Security Commission Report. London H.M.S.O., 1965.

Hans, Theordor: Soviet Terrorism in Free Germany. Washington, Government Printing Office, June 13, 1961.

Hobbs, Edward H.: Behind the President: A Study of Executive Office Agencies. Washington, Public Affairs Press, 1954.

Horowitz, Irving Louis: The life and death of Project Camelot. *Trans-action*, Vol. III, pp. 3-7 and 44-49, November-December, 1965.

House Report 157. Second report by the committee on government operations, 85th Cong., 1st Sess., 1957.

Huntington, Samuel P.: The soldier and the state: The theory of politics of civil-military relations. The Belknap Press of Harvard University Press, 1964.

Kasnakheyev, Aleksandr: Soviet Intelligence in Asia. Washington, Government Printing Office, 1959.

Kast, R., and Rosenzeig, James: Survey of intra-company impact of weapon system management. *Engineer's Management*, March, 1962, pp. 37-40.

Khokhlov, Nikolai: Activities: Soviet Secret Service. Washington Government Printing Office, 1954.

Know Your Money. United States Secret Service, Treasury Dept., Washington, D.C., Superintendent of Documents.

———: The Kremlins Espionage and Terror Organizations House Committee on Un-American Activities. Washington, Government Printing Office, March 17, 1959.

Lindsay, Franklin A.: The growth of Soviet economic power and its consequences for Canada and the United States. *National Planning Association and Private Planning Association of Canada*, October, 1959, p. 27.

Lipp, James E., and Stewart, Robert F.: Managing, planning, and operations research. Proceedings Sixteenth National ORSA Meeting, 1959.

McLeod, Robert Walter Scott: American political democracy and the problem of personnel security. WA Dept. of State, 1955.

National Industrial Conference Board: Combating Subversion and Sabotage. No. 60, New York, Nat. Indust. Conf. Board, 1952.

———: Industrial Security: Combating Subversion and Sabotage, Part I, 1953.

———: Industrial Security: Plant Guard Handbook, Part II, 1953.

New methods in mathematical programming; Texts of papers from the Second International Conference on Operational Research," *Operations Research,* July 1962, pp. 437-99.

New York State, Office of the Counsel. Combatting organized crime. Report of the 1965 Oyster Bay, New York, Conference, Albany, 1966.

Royal Commission on Espionage: *Espionage in Australia*. Sydney, A. H. Pettifer, 1955.

Senate Committee on Government Operations, Subcommittee on National Policy Machinery. Intelligence and National Security Report. 86th Cong., 2nd Sess. Washington, D.C., Government Printing Office, 1960.

Senate Committee on the Judiciary, Subcommittee to Investigate the Administration of the Internal Security Act. Hearings, Communist Forgeries. Washington, D.C., 1961.

Silverman, Corinne: *The President's Economic Advisers*. Inter-university Case Program Case Series, No. 48, Alabama, University of Alabama Press, 1959.

Snyder, Richard C., and Paige, Glenn D.: The United States decision to resist aggression in Korea. Foreign Policy Decision-Making: An Approach to the Study of International Politics, 1962, pp. 206-249.

Spevack, Jerome S. vs. Lewis S. Strauss: AEC—Release of Information on Heavy Water. United States Supreme Court. Washington, Government Printing Office, October, 1957.

State Department Security, 1963-65: The Otepka Case, II. Hearings before the Senate Subcommittee on Internal Security. Judiciary Committee, 89th Cong., 1st Sess., 1965.

Summers, J. A.: Protection Against Sabotage. General Electric Company, Nela Park Engineering Department, Cleveland, Ohio, September, 1942.

Tisler, Frantisek: Communist Espionage in the U.S. House Committee on Un-American Activities Hearings. Washington, Government Printing Office.

Tompkins, Dorothy Campbell: Sabotage and Its Prevention. University of California, Bureau of Public Administration, Berkeley, California, August, 1942.

U.S. Agency for International Development, Office of Public Safety, Police resources control operations, October 1, 1964, December 31, 1965.

U.S. Bureau of the Budget: Intelligence and security activities of the government, September 20, 1945.

U.S. Congress Senate Committee on the Judiciary: Scope of Soviet activity in the United States. Hearing 84th Cong., 2nd Sess.

U.S. Department of the Air Force: Doctrine and requirements for security system. Washington, Government Printing Office.

U.S. Department of Defense: Continuity of corporate management in event of nuclear attack. Washington, Government Printing Office, 1963.

———: Facts about fallout protection. Washington, Government Printing Office, 1963.

———: Fallout shelter program. Washington, Government Printing Office, June 1964.

———: Field manual 91-15. Washington, Government Printing Office, 1945.

U.S. Department of Interior: Federal records of World War II military report of the under secretary of war. Washington, Government Printing Office, 1951.

U.S. Department of State: Foreign affairs manual. Vol. V, Washington, Government Printing Office, November 30, 1965.

———: Foreign service security regulations. Washington, Government Printing Office, 1955.

U.S. House of Representatives, Committee on Un-American Activities: Patterns of Communist Espionage. 85th Cong., 2nd Sess., 1959.

———: Guide to Subversive Organizations and Publication. 87th Cong., 2nd Sess., Washington, Government Printing Office, December 1, 1961.

———: Chronicle of Treason. 85th Cong., 2nd Sess., March 1958.

———: Hearing Before a Subcommittee of the Committee on Government Operations. 87th Cong., 2nd Sess., 1962.

———: Departments of State and Justice, the Judiciary, and Related Agencies, Appropriations for 1956. 85th Cong., 1st Sess., 1955.

U.S. Library of Congress, Congressional Research Service: Soviet intelligence & security service, 1964-67.

———: Federal case law concerning the security of the U.S., 1954.

———: Legislation for the protection of the state & various European countries, 1956.

U.S. Office of the Provost Marshall General. World War II: A brief history. Washington, Government Printing Office, January 15, 1946.

U.S. Senate, Legislative Reference Service Library. World communism. 88th Cong., 2nd Sess., Parts I and II, 1964.

U.S. Senate, Committee on the Judiciary. The Wennerstroem spy case. 88th Cong., 2nd Sess., 1964.

U.S. Senate, Joint Committee on Atomic Energy. Soviet atomic espionage. 82nd Cong., 1st Sess., 1951.

U.S. Senate, 84th Congress. Report of the commission of government security 1957, 1957.

U.S. Senate. Interlocking subversion in government department, 1953-55.

———: Scope of Soviet activity in the U.S. Washington, Government Printing Office, 1962.

———: Internal security and subversion, Principal state laws and cases, 1965.

Walter, Francis E.: Chronicle of Treason. Washington, Government Printing Office, 1958.

———: International communism. Washington, Government Printing Office, 1957.

Wennerstroem: The Wennerstroem Spy Case. Washington, Government Printing Office.

Zlotnick, Jack: National Intelligence. Industrial College of Armed Forces, Washington, Government Printing Office, 1960.

PERIODICALS

Act to provide for the administration of the CIA . . . and for other purposes. Public Law 110, 81st Cong., 1st Sess., June 20, 1949.

Alderson, Wroe: Statistical training for marketing research. *The American Statistician*, February 1953, pp. 9-11.

Aldredge, E. S.: Monte Carlo technique of operations research. *Chemical Engineering*, June 1962, pp. 109-13.

Alert for sabotage. *Chemical Week*, December 1, 1962.

Alexander, Raphael (ed.): *Sources of Information and Unusual Services*. New York, N.Y., Informational Directory Company, 1958-59.

All the spies aren't on the air. *Television*, January, 1962, pp. 52-55.

Allis-Chalmers: Patent background for engineers. Allis-Chalmers Manufacturing Company, 1957.

Ansoff, H. Igor: Expert advice on diversification. *Missiles and Rockets,* July 1960.

Bacon, Frank R., and Lewis, Richard: Progress in the development of quantitative market requirements models for use in long-range product planning. Seventh International Meeting, Institute of Management Sciences, October 1960.

Baldwin, Hanson: Broader control set-up is held need with a "Watch-Dog" committee for Congress. *The New York Times,* July 25.

———: Competent personnel held key to success—reforms suggested. *New York Times,* July 24, 1948.

———: Intelligence—one of the weakest links in our security—shows omissions, duplications. *New York Times,* July 20, 1948.

———: Intelligence III: Errors in collecting data held exceeded by evaluation weakness. *New York Times,* July 23, 1948.

———: Older agencies resent a successor and try to restrict scope of action. *New York Times,* July 22, 1948.

Barnds, William J.: Intelligence and foreign policy: Dilemmas of a democracy. *Foreign Affairs, 47*:281-195, January 1969.

Behrman, Jack N.: The challenge of national and international programs of the U.S. government. *Industry Security, 6*:28, December 1965.

Brownell, Attorney General: The fight against communism. *New York Times,* April 9, 1954, p. 8.

Bugging is bugging the justice department. *Labor,* December 10, 1966.

Bug Thy Neighbor. *Time,* March 6, 1964, pp. 55-56.

Bunker, Paul: Twenty-four hour monitoring. *Law and Order,* January 1958, p. 14.

Bureau of Labor Standards, U.S. Dept. of Labor, Bulletin #67, 1953.

Burkhouse, Frank X.: Sabotage. *Industrial Security,* July, 1964.

———: Sabotage—The benign weapon. *Industrial Security,* July 1963.

California business: Penetration of defense industry documented. *Security World,* May 1969.

Chamberlain, John: OSS. *Life,* November 19, 1945, pp. 119-130.

Coffin: Communists in industry. *Factory Management and Maintenance,* October 1954.

———: Communists on the job. *Fortune,* September, 1950.

Collins, John R.: Internal security and subversion. *Industrial Security,* December 1964.

———: Student view of democracy: Block that speech. *Combat, 1:* No. 13, March 1, 1969.

———: Black Panthers "Free Breakfasts" and coloring book. *Combat, 1:* No. 22, July 15, 1969.

———: Communist illusion and democratic reality. *Industrial Security,* January 1960.

Conference on security draws 400. *Detroiter,* October 18, 1971, p. 5.

Cragg, Lionel C.: International and overseas security panel. *Industrial Security, 8*:60, December 1964.

Dash, Samuel, Knowlton, Robert, and Schwartz, Richard: The eavesdroppers. *Security World, 2:* No. 2, March 1965.

Davidson, W.: The Nth country problem and arms control. *NRA Planning Pamphlet,* January 1960, p. 41.

———: Do you really understand communism? *Industrial Security,* April 1962.

Donovan, Robert and Jones: Program for a democratic counter attack to communist penetration of government service. *Yale Law Journal, 58:* No. 1211, 1956.

Donovan, William J.: Intelligence: Key to defense. *Life,* September 30, 1946, pp. 108-120.

Dossier banks, the. *Wall Street Journal,* May 16, 1966.

Eavesdropping modern style. *Christian Science Monitor,* October 26, 1965.

Eight ways to prevent plant sabotage. *Steel Magazine,* August 21, 1950.

Embarrassing new events. *Newsweek,* December 12, 1966.

Evans, Allan: Intelligence and policy formation. *World Politics, XII*:84-91, October 1959.

Ewing, Ann: Lie detection at a distance. *Science News Letter,* August 14, 1965, p. 106.

Felago, Ernest E.: A sentinel against communism. *Industrial Security,* July 1961.

Fischel, Edwin C.: The mythology of Civil War intelligence. *Civil War History, 10:* No. 4, December 1964.

Former Clark Cable officer found guilty in stolen papers case. *Wall Street Journal,* November 4, 1963.

Franklin, Frank: The CP thrust into SDS. *Combat, 1:* No. 23, August 1, 1969.

———: Government watch on 200 million Americans. *U.S. News & World Report,* May 16, 1966.

———: Great Britain's secret, secret service. *Midway, 8:* No. 1, June 1967, pp. 19-35.

———: How effective is Central Intelligence? *Christian Science Monitor,* December 1, 1958, p. 13.

———: How intelligent is intelligence? *The New York Times Magazine,* May 22, 1960, pp. 26, 80-83.

Greig, Ian: Security and subversion. *Top Security,* June 1975, pp. 57-60.

Haire, Mason: Biological models and empirical histories of the growth of organizations. *Modern Organization Theory,* 1959, pp. 272-306.

Hall, Richard: What we must know about communism. *Industrial Security,* October, 1959.

Hanscom, C. B.: NARCO interrogation. *Police,* Nov.-Dec., 1957, pp. 44-50.

Hansen, Paul: Sabotage and espionage in American industry. *Industrial Security,* July 1960.

Harkness, Richard, and Harkness, Gladys: The mysterious doings of CIA. *Saturday Evening Post,* October 30, November 6 and 13, 1954.

———: How about those security cases? *Reader's Digest,* September, November 1955.

———: Treason in the United States. III: The constitution. *Harvard Law Review, 58:*806-846.

Haver, Martin: The snoopers. *Man's Magazine,* July 1960.

Heenan, F. E.: Bureau of ships work study program. *Naval Engineer,* May 1962, pp. 287-90.

Held, Virginia: PPBS comes to Washington. *The Public Interest,* No. 4, Summer 1966, pp. 102-115.

Henley, Arthur: Muggers of the mind. *Today's Health, 49:*38-41, February 1971.

Hilsman, Roger: Intelligence and policy making in foreign affairs. *World Politics,* October 1952, pp. 1-45.

Hindman, Jo: Secret cum files: A leftist wedge. *American Mercury,* October 1958, pp. 118-26.

Hobbing, Enno: CIA: Hottest role in the cold war. *Esquire,* September 1957, pp. 31-34.

Hollander, P. J.: Privacy: A bastion stormed, in: Mores and morality in communist China. *Problems of Communism,* November-December 1963, p. 1.

Horelick, Arnold L.: The Cuban missle crisis: An analysis of Soviet calculations and behavior. *World Politics, XVI:*363-389, April 1964.

House Committee on Armed Services. Amending the Central Intelligence Act of 1959. *Report,* August 11, 1966.

———: Inquiry into the U.S.S. Pueblo and EC-121 plane incidents. *Report,* July 28, 1969.

Hurst, W.: Treason in the United States. II: *The Constitution.*

Industrial Security. Combating Subversion and sabotage. No. 60, 1952, pp. 58-59.

Ingersoll, R.: *Top Secret.* New York, 1946.

Kendall, Willmoore: The function of intelligence. *World Politics, I:*542-552, July 1949.

Ketchum, H. W.: Finding new markets: How the government can help. *Iron Age,* January 1957, pp. 173-76.

Kilbridge, M. D., and Wester, L.: Review of analytical systems of line balancing. *Operations Research,* September 1962, pp. 628-38.

Kirchheimer, Otto, and Menges, Constantine: A free press in a democratic state: The Spiegel case. *Politicals in Europe,* New York, 1965.

Knorr, Klaus: Failure in national estimates: The case of the Cuban missiles. *World Politics,* April 1964, pp. 455-456.

Kukucka, Andrew J. Major: Soviet intelligence. *Industrial Security, 3:*6, January 1959.

Lamb, Edward: Trial by Battle: The Case of a Washington Witchhunt. Occasional Paper, April 1964.

Lauterpacht, H.: Allegiance, diplomatic protection and criminal jurisdiction over aliens. *Cambridge Law Journal, 9:*330-371, 1947.

MacDougall, A. Kent: Sabotage. *Security World, 4:* No. 4, April 1967.

———: The modern day Soviet spy—A profile. *Industrial Security,* August 1966.

Munford, Dilard: Competition or communism. *Industrial Security,* October, 1962.

National data center—A threat or a promise? *Occupational Hazards,* June 1969, p. 43.

Norton, John: International visit control procedures. *Industrial Security,* October 1967.

Pettee, George S.: The Future of American Secret Intelligence. Washington, Infantry.

Philipson, Moris (ed.): *Automation: Implications for the Future.* New York, Vintage Books, 1962.

Possony, Stefan T.: Organized intelligence: The problem of the French general staff.

Social Research, VIII, May 1941, pp. 213-237.
Ptacek, Bernarr M.: New communist party program. *Industrial Security*, 10:12, October 1966.
Radicals: Waiting for Armageddon. *Newsweek*, April 26, 1965.
Ranstad, Harold: Ninety miles to Moscow. *Industrial Security*, 5:10, April 1961.
———: The face of the enemy. *Industrial Security*, January 1963.
Reber, Jan R.: The essence of espionage. *Assets Protection*, Spring 1974, pp. 7-8.
Ridgeway, James: The snoops: Private lives and public service. *New Republic*, December 19, 1964, pp. 13-17.
Rising wages of fear, the. *Time*, May 24, 1971, p. 80.
Robert Kennedy defines the menace. *The New York Times Magazine*, October 13, 1963.
Robinson, Donald: They fight the cold war under cover. *Saturday Evening Post*, November 20, 1948, pp. 30ff.
Rovere, Richard: The invasion of privacy: Technology and the claims of community. *American Scholar*, 1958, p. 416.
Ruggles, Richard, and Brodie, Henry: An empirical approach to economic intelligence in World War II. *Journal of the American Statistical Association*, March 1947, pp. 72-90.
Secret files for secret purposes. *Educational News Service*, October 1958.
———: Secret mission in an open society. *The New York Times Magazine*, May 21, 1961, pp. 20, 77-79.
Security Through Snooping. Hitchcock Publishing Co., Infosystems, 20: No. 9, 84-85, September 1973.
Segal, H. A.: Initial psychiatric findings of recently repatriated prisoners of war. *American Journal of Psychiatry*, III, 1954.
Simon, W. G.: The evolution of treason. *Tulane Law Review*, 35: No. 4, 669-704, 1961.
Shannon, Donald E.: Internal security and subversion. *Industrial Security*, December 1964.
Smith, Bruce L. R.: Strategic expertise and national security policy: A case study. *Public Policy*, 1964, pp. 69-106.
Snooping electronic invasion of privacy. *Life*, May 20, 1966, pp. 38-47.
Sommers, Albert T.: The challenge of domestic and foreign economic development. *Industrial Security*, 6:8, October 1962.
Spindel, B. B., and Davidson, W.: Who else is listening? *Colliers*, 135:25-29, June 10, 1955.

———: Who else is listening? How to stop wiretapping. *Colliers*, 135:48-55, June 24, 1955.
Steinemann, Robert: If we would reduce crime. *National Civil Service League*, 76: No. 51-52, November-December 1959.
Story of the FBI, The. *Look*, New York, 1947.
Sullivan, William C.: The continued threat of espionage and sabotage in the United States. *Industrial Security*, October 1962.
Sweeney, Gerald A.: Internal security and subversion. *Industrial Security*, December 1964.
Todd, William M.: Espionage and sabotage. *Industrial Security*, October 1962.
Trail that leads to spy charges. *Business Week*, November 25, 1961.
Tulsa panel probes professional conduct. *Chemical Engineering*, December 12, 1960.
Unna, Warren: CIA: Who watches the watchman? *Harper's*, April 1958, pp. 46-53.
Vagts, Alfred: Defense and diplomacy. *Diplomacy, Military Intelligence, and Espionage*, 1956, Chap. 3, pp. 61-67.
Vetter, Charles T., Jr.: Knowing the Communist mind. *Industrial Security*, October, 1961.
Wasserman, Benno: The failure of intelligence prediction. *Political Studies, VIII*: 156-169, June 1960.
Watts, Rowland: The draftee & internal security; a study of the army military personnel security program & its effect on persons inducted under the provisions of the universal military training and service act. *NY Workers Defense League*, 1955.
Espionage in business and industry. *Wall Street Journal*, March 3, 1959.
Walmsley, Alan: Defense security: Lessons of the cold war. *Security World*, November 1970.
Wedgwood, C. H.: The nature and functions of secret societies. *Oceania*, 1940, p. 129.
Weeks, S.: Uses of world trade information service. *Foreign Commerce Weekly*, September 1957, p. 21.
Westin, A. E.: Wire-tap: The house approves. *New Republic*, April 9, 1954.
Westley, William A.: Secrecy and the police. *Social Forces, XXXIV*:254-257, March 1956.
What about safeguards against industrial sabotage? Survey by Mill and Factory, *Industrial Security*, April 1951.
When walls have ears, call a de-bugging man. *Business Week*, October 31, 1965.
Wicker, Tom and other members of the Washington, D.C. staff: CIA—Maker of policy or tool? *New York Times*.

Willis, Edwin E. Honorable: The Communist party and industrial security. *Industrial Security, 8:* No. 3, July 1964.

Wohlstetter, Roberta: Cuba and Pearl Harbor. *Foreign Affairs, 43:* No. 4, 691-707, July 1965.

Wortis, J.: Some recent developments in Soviet psychiatry. *American Journal of Psychiatry, 109,* 1953.

Yerxa, Fendall, and Ogden, R.: The threat of Red sabotage. New York, *New York Herald Tribune,* 1950.

Chapter 5

BUSINESS SECURITY

EFFECTIVE SECURITY and loss prevention programs are essential for the survival of the business organization. *Direct* losses due to theft, waste, and negligence are increased by the *indirect costs* of losses due to the time involved in processing claims, disruption in operations, or court appearances. The indirect costs of losses can often far exceed the actual direct loss itself.

Business loss prevention is complex and requires policy guidance from top management if effective programs are to be developed and implemented. Policies must integrate the interests of management, employees, customers, employee organizations and the community. These various groups must be considered since all aspects of the organization and its operations will be affected by security policy. Likewise, the nature of loss prevention problems tends to change over time and procedures which are not subject to timely review at the policy level cease to be effective.

Adequate coordination of loss prevention procedures and techniques in business is impossible except at the policy level. Personnel in most areas of an organization are required for successful implementation of security programs. The specialized loss prevention knowledge of the security manager must be used by the entire organization. As a manager, he must "work through" other people to accomplish his goal of reduced losses. He cannot be everywhere and must therefore establish acceptance for his program and seek voluntary cooperation from all other management personnel and employees. The major element in any security director's success is the degree of acceptance and support shown for the protective services function by top management. If recognition of a need is present and articulated to the management and employees, the security director will have the necessary support and cooperation for his programs.

THE PROBLEM

There are more than 5.5 million firms[1] engaged in business in the United States. These concerns are engaged in retailing and wholesaling and include all service and transportation activities.

The total amount of loss for these businesses was estimated in excess of 11.4 billion dollars. While exact figures are not available as to how many businesses were forced into bankruptcy because of either internal or external losses, it is safe to assume that many of these bankruptcies were the result of dishonesty factors.[2]

Many studies have begun to shed light on the amount of annual theft losses. One such study in 1958 indicated that dishonest employees account for more than 500 million dollars a year in losses. Estimates placed the total direct retail losses in 1975 at 5.8 billion dollars. These figures account for an average of 2 percent of the total retail sales in the United States.[3]

The figure of 2 percent of retail sales is considered the annual cost of internal employee theft. Insurance companies dealing with fidelity losses indicate that this loss figure is double the annual national United States fire losses. It is, therefore, reasonable to assume that the business community is faced with a serious problem in controlling its property and reducing its losses.

Both large and small businesses equally suffer from dishonesty losses. Many large chain stores have been forced out of business due to excessive losses. This is particularly true of high-volume, low-profit retailing establishments which cannot contain or reduce their losses to an "acceptable" degree. Retailing firms operating on a 1.5 to 2 percent net profit margin cannot survive if faced with a consistent inventory loss, which has been estimated on a national basis to be 1.1 percent of total sales by the National Retail Merchants Association. Similarly, this is true of small retailers who have a higher relative overhead than a large volume store operating on a 2 percent net profit. These businesses cannot readily absorb high losses for an extended period of time. The alternatives to high percentages of theft loss for either large or small businesses are either bankruptcy or the utilization of theft loss control techniques.

WHITE COLLAR CRIME

One of the most serious aspects of loss prevention is that of internal theft by employees. This form of loss, while currently referred to as "white collar crime," is responsible for a major portion of lost business profits. In white collar crime the following elements are present:
1. intent to commit a wrongful act or to achieve a purpose inconsistent with law or public policy;
2. disguise of purpose or intent;
3. reliance by perpetrator on ignorance or carelessness of victim;
4. acquiescence by victim in what he believes to be the true nature and content of the transaction; and
5. concealment of the crime by:
    a. preventing the victim from realizing that he has been victimized, or
    b. relying on the fact that only a small percentage of victims will react to what has happened; and making for provisions for restitution to or other handling of the disgruntled victim, or,
    c. creation of a deceptive paper, organization, or transactional facade to disguise the true nature of what has occurred.[4]

White collar crime is the most difficult to deal with since it raises questions about the basic integrity of trusted employees. It has often been said that "the only employees who can steal from you are those you trust!" This is considered true since it is only trusted employees who have access to anything worth stealing.

Trust is basic to employer-employee relations and while prevention programs and internal control systems will assist in keeping honest employees honest, they will not always prevent the determined dishonest employee from stealing.

LOSS PREVENTION IN BUSINESS

A loss prevention program is intended to reduce or eliminate shortages in inventory or cash. Losses can be identified as being either "actual" or "paper." Actual losses are those losses occurring from theft, waste of materials or time, and employee errors. Paper losses are those resulting from inadequate or careless record keeping or the lack of or nonutilization of established control systems. Losses from either of these two types of occurrence decrease net profits for the business experiencing them. Anything which can be done to eliminate or reduce their occurrence is desirable since it contributes directly to increased net profit.

Owners or managers of businesses are not really interested in security. They are in business to make a profit. Anything which contributes to their ability to either increase sales or profits is most often considered valuable while those activities which contribute to overhead cost without apparent effect on profitability are reduced or eliminated. Security and loss prevention activities have historically been viewed as falling in the latter category. They were viewed as overhead and a cost to be absorbed by management out of profits rather than a profit making or retaining function for the firm.

In reality, business loss prevention *is a profit retaining function*. Actual or paper losses are subtracted from the net profit of the firm. Measures taken which can reduce the amount to be subtracted contributes directly to an increased profit figure. The more effective the protective function can be within its budgetary constraints, the more net profit it will produce for the company.

Security and loss prevention programs most often will not make any more money for their firms. They will, however, allow them to retain more of what they make by controlling excessive costs due to actual or paper losses.

The development of an effective security or loss prevention program consists of integrating the personal, physical, and procedural aspects of the company. If a program is to be functioning smoothly, the employee motivation and executive leadership aspects of it, as well as the space layout and merchandising plans and control procedures must be considered. Those programs which rely on controls without considering the effects of those controls on employee morale; or that emphasize motivation at the expense of management controls will fail. Comprehensive asset protection applies sound system management concepts to the reduction or elimination of losses. In order to do this, all relevant functions within the operating system must be included.

Losses can occur from both internal and external sources. Since no two firms are identical in objectives, organization or philosophy, the application of a single approach to loss prevention is impossible. There are, however, a wide array of procedures and techniques which have proven successful in the reduction of various types of losses.[5] Many of these procedures are often viewed as "specifics" for solving various types of problems. "When x happens, you should do y," is an approach often used for problem solving. It does occasionally prove effective. Often, however, the effects of their application is unclear. At times they are marginally effective. What is unknown and most important, is what loss might have been reduced *if* a well-reasoned program had been instituted instead of applying general techniques.

The development of security or loss prevention programs cannot be viewed in terms of establishing *maximum amounts of protection*. Maximum security programs are the preserve of prisons, high-risk governmental facilities and the like. Business loss prevention programs should be developed, organized, and operated to provide *optimum amounts of protection*. Security programs are limited in what they can legally do. Likewise, the business wishes to maintain an environment conducive to the buying and selling of merchandise. An open, unrestricted selling area is not practically providing maximum security. *Optimal security is that amount of protection consistent with the objectives of the firm*. Risk is obviously assumed by the company in establishing such a protective program at a higher rate than under a maximum security program. The offsetting benefits of potentially increased customer traffic and sales is considered more important than restricting access and controlling the environment oppressively.

A considerable amount of loss in business is not attributable to "security problems." Much of the losses experienced by business is due to poor management! Many of the personnel, physical, and procedural safeguards established by security personnel, or identified by security surveys as being loss problem areas result from management inattention or carelessness. A company which "makes it too easy to steal" and sustains high losses does not have a basic security problem. They have a *loss* problem caused by poor management. The controls imposed are often implemented by security personnel, but the basic problem which "set up" the loss condition was not a security but a management problem.[6]

Loss prevention programs are only one dimension of total management of the business enterprise and must be viewed as such by its program managers. While tensions will exist between the goals of selling and protecting, both functions serve the same master—*profit*. Rather than conflict between loss prevention and other functions, there should be cooperation and complementing of each of the other's activities.

RISK MANAGEMENT-
INSURANCE REQUIREMENTS

The development of a comprehensive business loss prevention program requires a systematic assessment of the degree of risk involved in conducting day-to-day operations and the initiation of procedures to reduce exposure to risk to the optimal level. The amount of risk to be assumed by the business will be determined by the insurance coverage purchased as well as the requirements of the insurance company for protective measures. Specific types of devices or procedures are often required of a business for insurance coverage to be provided. Losses which are incurred by the business in excess of the amounts specified in the policy or for events of loss not covered are the financial responsibility of the insured business.

The amount of budget often spent for business protection includes the cost of insurance plus all the active measures taken to reduce loss. Most programs of active measures will not be budgeted at more than the anticipated or experienced losses. Thus the cost of protective programs are often geared to those expenditures required for insurance coverage plus those additional costs which are expected to reduce or prevent losses of a corresponding value. Risk and its reduction is the essence of preventive programs. The financial incentives inherent in the insurance concept of a group of insurees "sharing risk" collectively is sound. The decision to follow risk prevention guidelines is mandatory if coverage is expected. The installation of preventative measures exceeding those required is voluntary. There is often, however, sufficient financial incentive to have more than the minimum of protection.

The operation of a security program does not always involve high expenditures; business management for crime prevention often is accomplished at low cost once proper planning is completed. The *recognition, anticipation* and *action* to eliminate risks of loss can be accomplished once a decision is made to identify problems and work toward their solution. Insurance requirement guidelines often point the direction the program should take. Management must establish how much will be spent and how extensive management of the program will be.

CATEGORIES OF LOSS

There are two major categories of losses which are of concern to all business organizations. These are *internal* and *external* losses. Internal losses include theft losses, such as embezzlement and pilferage, mistakes caused by incompetence, carelessness, or errors, and disasters caused by fire, water, or other natural means. External losses include burglary, shoplifting, fraud, robbery, and business espionage.

Since all of the internal and external losses dealing with theft or disasters are treated in the discussion of the problem areas in Chapters 18 and 19, the organizational response to security and security controls, which are an integral part of the organizations' activities, will be treated in this Chapter.

Conditions Promoting High Losses

Many losses result from conditions existing within an organization which are conducive to high theft loss occurrences. Such things as poor control procedures, poor employee morale, overconfidence in employees, overlooking basic management and security principles for the type of business activity engaged in by the organization, and inattention to theft danger signals promote an atmosphere which results in high theft losses. In addition to these, a number of additional internal factors, while not all associated with theft losses, tend to reduce net profit earnings. These include (1) invoice errors from sources, (2) short shipments and duplicate shipments, (3) errors in pricing or marking, (4) accounts payable errors, (5) faulty handling and receiving, (6) poor stock checking and control, (7) unaccounted-for breakage, (8) unrecorded markdowns or adjustments, (9) sales check errors, (10) credit errors, (11) faulty handling in delivering, and (12) faulty handling in returns departments.

IMPLEMENTING SECURITY

Ideally, the need for security should be understood by all engaged in management activities within an organization. An organization must establish programs internally to prevent internal or external damage to the organization and its ability to attain its goals. This is done by the establishment of appropriate preventative measures.

Management's first line of defense against unwanted or undesired activities is the establishment of an internal protective program. The component parts of such a program are (1) an effective internal control mechanism, (2) fidelity bonding, (3) independent auditing or security surveying.[7]

These three activities must exist and be fully integrated for an internal prevention program to be meaningful. Such programs, once established, must be fully supported by management. There is little question that preventative measures, which reduce the organization loss potential, clearly increase the profit potential.

The implementation of control techniques vary according to the nature, cost, and type of protection desired by management. The development of internal controls must make use of physical, information, and personnel security techniques. They must be combined with originality and common sense for an organizational reduction of high risk factors. The reluctance to establish definite policies and measures have been cited as the two most common factors involved with high internal losses and unsuccessful preventative programs.

PREVENTION OF INTERNAL LOSS

Specific internal programs can be developed which reduce the possibilities of internal losses. These involve the development and implementation of definite policies and procedures for organizational activities most closely associated with high losses. The development of specific procedures should encompass all areas of organizational activity which could result in financial loss. Such areas as purchasing, receiving, warehousing, processing incoming funds, credit, payroll, petty cash, and shipping must be very closely controlled.

Activities	Require
Purchasing	
Payroll	Outgoing funds procedures
Petty cash	
Incoming funds	
Credit	Incoming funds procedures
Receiving	
Shipping	Inventory procedures
Warehousing	

Outgoing and incoming funds, as well as inventory, must be protected through procedural controls. Figures 5-2 and 5-3 indicate how a set of procedures for handling incoming funds, payments by mail or in person, might be established.

The establishment of procedures for the control of outgoing funds can be seen in Figure 5-4 which indicates such a program for controlling the processing of customer refunds.

The control of inventory is shown in Figure 5-5. The sample procedure was developed for issuance of warehouse receipts.

The control of incoming and outgoing funds requires particular attention through the establishment of programmed security measures. Internal control measures must be developed and reviewed continually to ensure that the possibility for loss is held to a minimum. Specific loss control policies should be developed which include such things as:

1. a regular schedule of cash receipt depositing;
2. all purchases made by check, and countersigned;
3. statements should be verified against accounts receivable ledgers;
4. bookkeeping personnel should not be permitted to do any receiving or shipping activities;
5. all journal entries involving sales allowances or bad debts should have managerial approval;
6. mail should be received and opened by personnel other than cash receivable cashiers or bookkeepers;
7. records should be kept of all mail

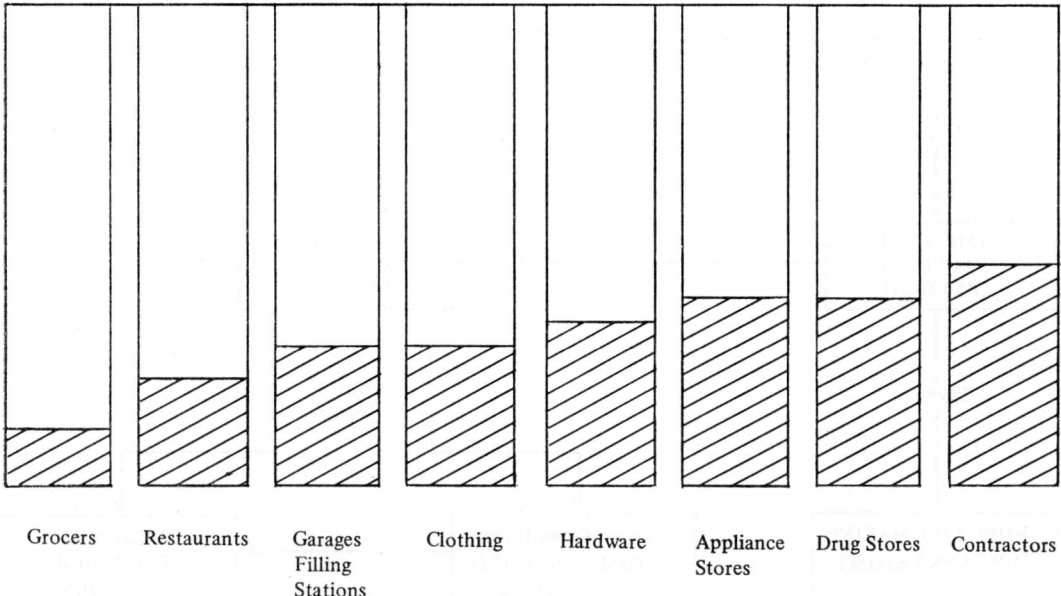

Figure 5-1. Percent of commercial concerns protected by fidelity bonds. (Courtesy of The Surety Association of America)

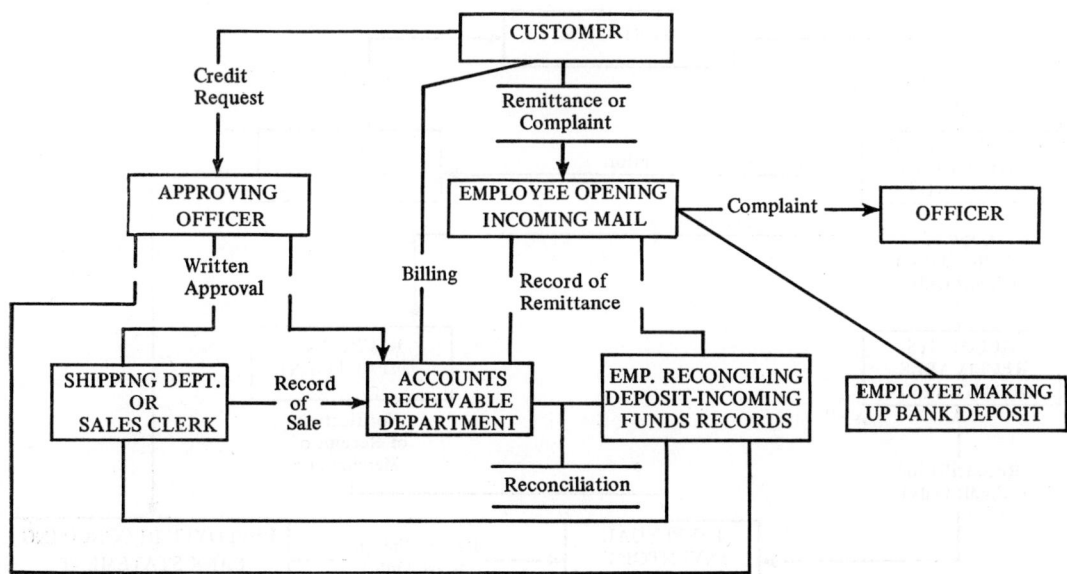

Figure 5-2. Procedures for handling incoming funds—payment by mail. (Courtesy of the Royal-Globe Insurance Companies)

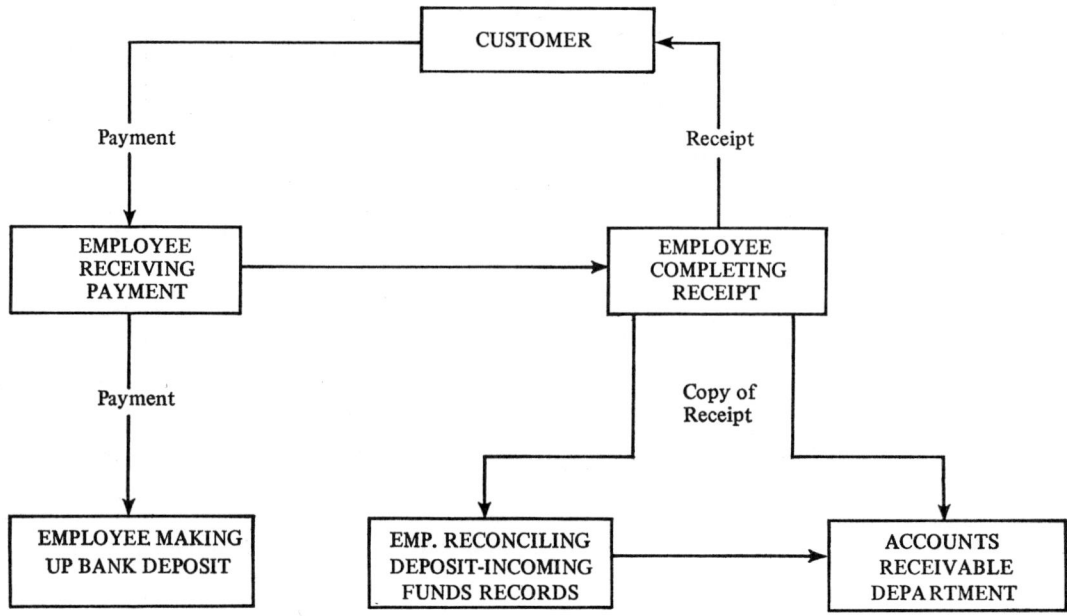

Figure 5-3. Procedure for handling incoming funds—payments in person. (Courtesy of the Royal-Globe Insurance Companies)

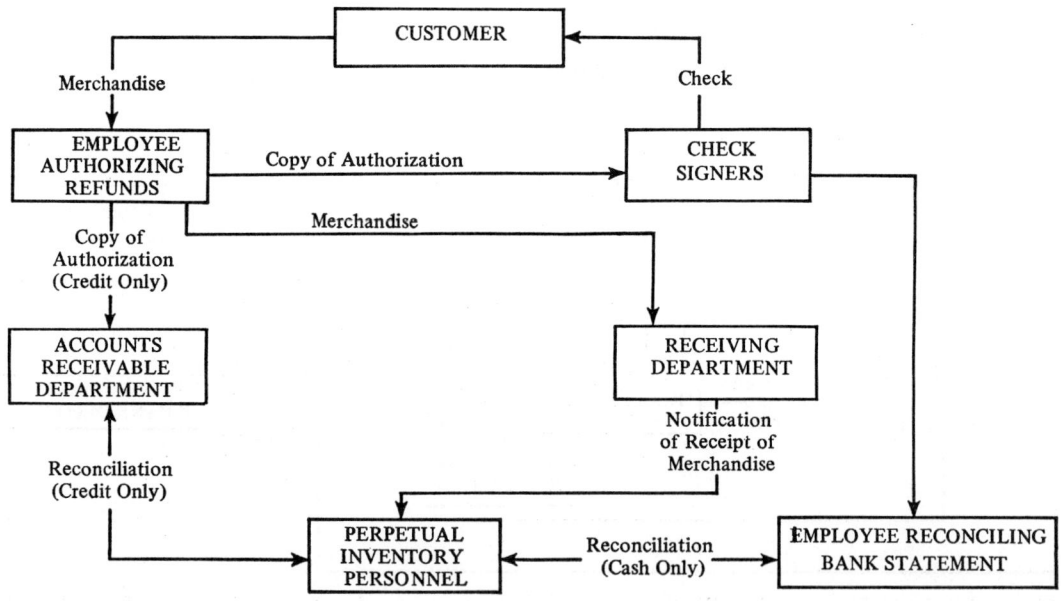

Figure 5-4. Procedures for processing customer returns. (Courtesy of the Royal-Globe Insurance Companies)

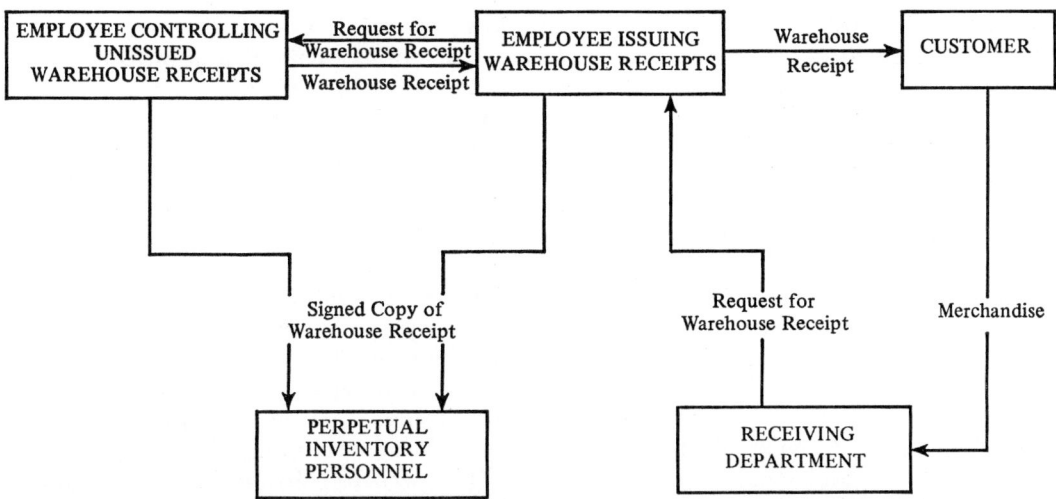

Figure 5-5. Procedure for issuance of warehouse receipts. (Courtesy of the Royal-Globe Insurance Companies)

received as well as listing of the amounts of checks, money orders, and cash;
8. lists of received funds should be compared regularly with cash receipts ledgers;
9. bookkeepers and cashiers should be covered by fidelity bonds;
10. announced and unannounced audits of all cash funds should be made periodically;
11. ensure that cash is provided with appropriate physical security; and
12. payroll distribution duties should be occasionally rotated.

EXTERNAL LOSSES

All organizations are vulnerable to external losses. Many of these may be minimized by the proper assessment of potential problems and the application of the appropriate combination of the security processes. Such external losses as might occur through burglary and robbery can be minimized through the application of appropriate security means.[8]

For example, well-trained and properly utilized guards can prevent and detect shoplifting[9] or external pilferage. However, the degree of protection that an organization deems necessary should not exceed that of the cost of potential losses to the organization.

It is difficult, if not impossible, to separate the protection program of an organization between internal and external losses. A program of loss control, to be truly effective, must be a total effort by the organization to effectively incorporate security into the organization. For security to be totally effective, all provisions must be made for the protection of all the physical property of the organization including materials, finances, equipment, facilities, and personnel. Similarly, organizational information must be protected. This includes the protection of all sensitive information relative to formulas, research, processes, techniques, plans, marketing information, sales information, programs, and so forth.

All facets of the organization must be fully covered and integrated into a total protection program. Failure to do so would present opportunities for the circumvention of developed procedures in specific areas and lead to the erosion of the entire system established for security.[10]

The ability of the business enterprise to protect itself is codified in the statutes of most states. Protection is afforded against shoplifters and other types of overt criminal acts such as robbery and burglary. The trend has been to encourage the efforts of merchants to aggressively protect their property from theft.[11] Private protective policy blends with the public interest in reducing business losses since increased theft losses are passed along to consumers in higher prices for necessary products and services. Instances of large-scale theft rings including employees have been uncovered in various parts of the country. These rings have stolen merchandise and channeled it into legitimate business at prices far below market value. In many instances they have forced competition between their employer and the buyers of this "wholesale" merchandise to the point of driving their employer out of business.

Organized criminal infiltration into legitimate business is likewise a more common occurrence today than in the past ten years. Competition between legal and illegally funded business created artificial pressures upon the business system as well as between individual firms. Aside from the unethical or illegal activities which controlled firms might engage in, the public is given reason to be additionally mistrustful of the free enterprise system. The security function in the firm has the responsibility to determine whether illegal competition is being conducted by controlled firms and work closely with public law enforcement agencies to eliminate such activities.

Design and Security Problems

Security programs for business establishments are often built around the existing physical design features of the building. There are very few buildings either designed or remodeled to make them easier to protect from losses. The security program must, therefore, often seek to optimize its effectiveness within a poorly designed operational environment. Traffic flow, display of merchandise, shipping and receiving, perimeter barriers, lighting and glazing materials, the security implications of which are not properly considered, all complicate the operations of an effective loss prevention program.

The functional separation of various departments or activities such as shipping and receiving, stock rooms, and employee rest areas reduce the opportunity to circumvent control procedures. Lack of functional separation increases the opportunity for losses to occur. Design characteristics of the building will either increase or decrease the ability and opportunity of employees and customers to steal. The integration of barrier systems, sensory devices, and control points within the building insures that opportunity and ability are reduced or eliminated.

The security director should attempt to influence the development of the building plan for any new facility as well as the remodeling plans for existing ones. The incorporation of protective features can be most economical when plans are being formulated and still "on the drawing board." Environmental conditions affecting security, neighborhood factors or sensitivities, employee-use patterns as well as availability of protective technology must all be considered prior to, rather than after the time that plans have been finalized or construction begun.

Measuring Program Effectiveness

The development of security programs in business environments involves the application of the *basic security processes*. Moreover, it requires the adherence to cost-effective criteria to insure program acceptability to corporate management. Profitable operations are the goal of business enterprises. The security function must contribute to profitability or it will be relegated to a minor role, even if unjustifiably so, using other criteria, in the corporation.

Business security programs, particularly in retail stores, find their operations being evaluated against their ability to either reduce losses or maintain them at a given level. A specific inventory shrinkage or "shrink" figure is utilized to determine the percentage of dollar volume loss being incurred. This "shrink" figure is then compared for comparable periods throughout the year. The variance up or down, is used to measure the effectiveness of the protective measures of the security program. The yearly inventory for example might indicate a 2 percent shrink since the last inventory. The previous period registered a 1.5 percent shrink. The variance for that period is thus a .5+ percentage shrink.

Increased losses measured in this way might serve to indicate that the program was not effective in reducing losses. While this increase in loss might be real, various factors other than a poor program might have contributed to it. These factors include:

1. changes in the inventory system,
2. poor supervision of employees,
3. changes in personnel hiring practices,
4. decreases in budget for security operations,
5. department managers concealing personal errors, and
6. changes in selling or merchandising procedures.

The security program must constantly respond to changes in the operations of the company. While losses can occur because changes are not made rapidly enough, prevention requires the anticipation of risks and prior action to eliminate them. If the security manager is not included in the planning for changes in personnel or operational policy within the corporation, it is impossible for the security function to do other than react to problems which develop rather than anticipating them and preventing their occurrence.

The prevention of crime and losses in the business requires the complete support of the employees. Without their support, all the physical security features, control procedures, and personnel screening techniques become things to either be endured or circumvented by the employees. Either not obtaining or losing employee support for the security program will reduce the effectiveness considerably. In the final analysis, while employees can be coerced or forced to nominal compliance with security controls, the voluntary compliance essential to long-term "good will" and "support" rather than tolerance, will be lost.

The public police depend upon the cooperation and support of the public to be effective; so too, the security program must rely upon employee acceptance and support for effective and efficient loss prevention. Security personnel like police cannot be everywhere to "enforce" regulation; voluntary compliance is what makes society operate smoothly without a policeman on every corner. It is, however, unlikely that if there was an officer on each corner that if people wanted to violate the law, their presence would deter it. Likewise, the corporation cannot post security officers or CCTV cameras to provide total surveillance of the facility. Even in those locations which have tried to do so, their losses have not been totally eliminated.

FOOTNOTES

1. United States Bureau of Census: *Statistical Abstract of the United States 1970*, 91st ed. Washington, Government Printing Office, 1970, p. 476.
2. Appendix 5A provides a detailed report of the U.S. Dept. of Commerce; the cost of crimes against business.
3. Lewis J. Camin: Some projections for business security. *Security World*, February, 1968, Vol. 5, No. 2, p. 12.
4. Appendix 5B provides several illustrations which emphasize the breakdown of the trust relationship and the measures which can be used to reduce the opportunity of this form of theft. It also explains the above elements of white collar crime in more detail.
5. Appendix 5C presents the recommendation of the Senate Select Committee on Small Business concerning *Business Management for Loss Prevention*.
6. Appendix 5D presents an example of how a business security program in a retail setting is analyzed for management problems.
7. Surety Association of America: *How Much Honesty Insurance*. New York, 1956, p. 1.
8. See Appendix 5E for *Burglary Prevention Checklist for Business Places*.
9. The security department and its employees

are not sufficient to deter or detect shoplifters. Adequate coverage requires that all employees be involved in any anti-shoplifting program. To be completely effective, however, the employees must be trained and properly motivated to participate.
10. Appendix 5F.1 presents a sample agenda for a "shrinkage control meeting" which might be used for producing the necessary training. Appendix 5F.2 contains a sample outline for an employee discussion on shoplifting.
11. Appendix 5G presents a study on the forms in which encouragement for private protective efforts has taken place. Appendix 5H summarizes current shoplifting statutes/and detailed descriptions of the laws in selected states are in Appendix 5I.

APPENDIX 5A

THE COST OF CRIMES AGAINST BUSINESS*

A REPORT BY THE U.S. DEPARTMENT OF COMMERCE, DOMESTIC AND INTERNATIONAL BUSINESS ADMINISTRATION

INTRODUCTION

In February 1972, the Bureau of Domestic Commerce published a preliminary staff report entitled *The Economic Impact of Crimes Against Business,* indicating that the 1971 national cost of crime to American business was conservatively estimated to be $15.7 billion. The estimate reached $20.6 billion for 1974.

This continued rapid rise illustrated the need to update the 1972 report, and the present analysis was undertaken to determine in more detail and for specific industries the current economic impact of crimes against business property.

This study, like its predecessor, covers only "ordinary" crimes against business. Ordinary crimes include burglary, robbery, vandalism, shoplifting, employee theft, bad checks, credit card fraud, and arson. Information on several other types of crime that continue to victimize businessmen is also presented. Organized crime and extraordinary crimes such as airplane hi-hacking and embezzlement have been excluded wherever figures permitted a breakout. These problems were considered different in character, requiring different solutions. Costs of public prosecution and law enforcement services were also excluded, since they are borne by the general public and cannot be related directly to business.

The most serious difficulty associated with analysing the impact of crimes against business continues to be the sparseness and sporadic nature of the data available. Figures are seldom based on comparable definitions or time periods, and many data gaps exist. There is no comprehensive source for information about crimes against business property.

To gather current information, a review of articles in the trade press on crime problems within particular industries was conducted, while many industry associations supplied information and estimates based on the experiences of their memberships. Various Federal Government agencies also provided statistics on crimes.

This report, therefore, presents a detailed summary of the available knowledge of both the industries themselves and the Federal Government on the extent of the dollar loss of American business to crime in the period since 1971. In almost every case the estimates are conservatively stated. The report also demonstrates that accurate data with which to quantify the economic impact of crimes against business are either scarce or, as is most likely, not available.

There are good estimates that the cost

---

* *Security Management*, March, 1975. Reprinted courtesy of *Security Management*.

of "ordinary" crimes against business reached $20.3 billion in 1974. This figure represents an increase of about 30 percent over the $15.7 billion cost estimated for 1971. Estimates by type of business are shown in Appendix Figure 5-1.

Appendix Figure 5-1

ESTIMATED COSTS OF "ORDINARY" CRIME BY SECTOR OF BUSINESS—1971, 1973 AND 1974 (BILLIONS OF DOLLARS)

Sectors of Business	1971	1973	1974
Retailing	4.8	5.2	5.8
Manufacturing	1.8	2.6	2.8
Wholesaling	1.4	1.8	2.1
Services	2.7	3.2	3.5
Transportation	1.5	1.7	1.9
Arson	0.2	0.3	0.3
Preventive measures	3.3	3.5	3.9
	15.7	18.3	20.3

Source: Bureau of Domestic Commerce, U.S. Department of Commerce.

The cost of crimes against business must be passed on to consumers in the form of higher prices, or absorbed as costs by businessmen, with resultant lower profits. The estimated total cost of $20.3 billion is equal to $89 for every man, woman, and child in the country; if one considers only the adult population, the per capita cost is $137. The crime-related losses have a depressing effect on business, as well. The ratio of losses to total capital expenditures is in excess of 16 percent or equal to about 17 percent of total corporate profits.

Crime can affect businesses regardless of location, although the incidence varies by type of area, as the following table depicts. For example, firms in central city areas, particularly in low income sections, have the highest rate of burglaries and robberies by a substantial margin. Appendix Figure 5-2 reports categories of crime.

The Small Business Administration developed an index of the impact of losses from ordinary crimes measured in relation to receipts. The table shown as Appendix Figure 5-3 demonstrates the impact by size of firm. The average for all businesses is set at 100.

Thus, small business suffers an impact that is 3.2 times the average, and 35 times that of businesses with receipts over $5 million. These small firms are less able to afford the overhead required for extensive protective measures or to absorb these losses.

Appendix Figure 5-3

INDEX OF LOSSES FROM CRIME BY RECEIPT SIZE

Business Receipt Class	Index
All businesses	100
Under $100,000	323
$100,000 to $1,000,000	205
$1,000,000 to $5,000,000	127
$5,000,000 and over	9

Source: *Crime Against Small Business*, Small Business Administration.

Appendix Figure 5-2

CATEGORY OF CRIME

Percent of Those Reporting At Least One Incident by Location:	Shoplifting	Checks	Burglary	Robbery	Vandalism
Central city	14	41	18	3	not available
Suburbs	15	31	16	2	18
Rural	15	36	9	1	not available

Source: *Crime Against Small Business*, Small Business Administration.

Selected Types of Crimes

Bad Checks accounted for about 13 percent of all crime-related losses to business in 1967-68, according to a study conducted by the Small Business Administration. As the use of checks in place of cash grows, the losses suffered may also be expected to increase.

Businessmen should be alert to the dangers of bad check artists, and adopt standard procedures for cashing checks. These should cover suitable and specified identification such as photographic or finger-printing equipment, in order to discourage potential bad check passers. Many mail-order firms already withhold shipment until the check sent in payment for merchandise has cleared. In addition, grocery stores, and other retail establishments handling large numbers of checks, have instituted check authorization cards, which are issued only after investigation of the customer's reliability.

Counterfeiting is one of the nation's oldest criminal activities, and is on the upswing. While ultraviolet scanners and other counterfeit-detecting devices are available, they add to the cost of doing business, and none has been endorsed by the Secret Service, which is responsible for combatting counterfeiting.

The Government will not reimburse a businessman who accepts a counterfeit bill. He must therefore protect himself by learning to recognize the differences between real and bogus currency. Most counterfeits are of crude craftsmanship, printed from inferior plates on poor quality paper, and the flaws are visible to the naked eye.

Most of the counterfeit currency produced is seized by Federal agents before it gets into circulation. In fiscal year 1972, $22.9 million in uncirculated fake currency was seized; however, at least $4.8 million was recovered only after it had been passed. No estimates of the amount of counterfeit currency passed but not turned over to Federal agents are available. Businessmen should learn to distinguish counterfeit currency in order to protect themselves against loss. A pamphlet available from the Secret Service describes the features of counterfeit bills.

A closely related problem is forged Government checks. Checks are looted from the mailboxes of the legitimate recipients and endorsements forged by thieves. Businessmen who cash such checks, or take them in payment for goods or services, will suffer the loss of the amount of the checks, since the Government will not honor them. It is essential when cashing Government or other checks, that the endorser identify himself adequately.

Inventory Shortages are the major factor in crime-related losses in retail stores, and are significant contributors to losses by wholesalers and manufacturers as well. Inventory shortages result from shoplifting and employee theft, as well as honest bookkeeping errors. Ticket switching by dishonest shoppers and deliberate underrings by cashiers also contribute to inventory shortages.

Most state legislatures have enacted laws designed to combat shoplifting and where such legislation exists it can prove a valuable tool for businessmen. A typical shoplifting law specifies the actions which are illegal under the terms of the act. These often include removing merchandise from the premises with intent to steal it, switching price tags, concealing merchandise with the intention of stealing it, and other actions. It is often not necessary that the merchandise be removed from the premises in order to establish guilt. The laws of most jurisdictions give the businessman the right to detain suspected shoplifters within limits, without the danger of false arrest charges. In the absence of legislation, case law, i.e. court decisions which form legal precedents with the force of legislation, gives businessmen certain specific rights in their fight against shoplifting.

It is recommended that businessmen and their employees act in strict accord with the law when apprehending shoplifters, since false arrest suits can be costly in terms of public relations as well as in the amount of damages awarded. A firm's procedures for arresting shoplifters should be reviewed in consultation with a local attorney specializing in that sector of criminal law.

Robbery consists of theft through force or threat of force, and **burglary** is the illegal entry of a premises, usually with the intention of theft. These are growing problems for banks, retail stores, service stations, and other areas where cash or valuable merchandise is present. According to the Federal Bureau of Investigation (FBI), the number of reported chain store robberies increased 167 percent from 1968 to 1973. Reported service station robberies decreased 6 percent during the same period, while holdups of other commercial establishments rose 31 percent.

The Law Enforcement Assistance Administration (LEAA) of the Department of Justice reports that robbery and burglary of commercial premises in the nation's largest five cities totalled 469,000 incidents in 1972.

Vandalism is a problem in urban as well as suburban areas. It affects all types of businesses through its senseless destruction of property. Most of the damage is directed against public buildings, although private businessmen suffer, too. The construction industry is particularly affected since construction sites are generally in poorly lighted areas, and are virtually deserted during nonworking hours and in bad weather. Public transportation facilities are favorite targets for vandals, who frequently deface surfaces with graffiti, damage seats, break windows, or strew trash on vehicles.

Crime-Related Problems

Drug abuse and alcoholism are not crimes directly against business, but their impact cannot be ignored. The most common effects on employees are absenteeism, lack of initiative, poor attention to detail, proneness to accidents, and decreased mental and physical productivity. These problems can be reflected not only in a company's profit and loss statement, but also in its reputation and good will within its community.

Drug and alcohol addiction of employees can intensify criminal problems in business. Studies indicate that a large proportion of thefts from business are committed by addicts to support their habit. Similarly, employees in debt to loan sharks and who have lost large sums of money gambling are under a great deal of pressure to steal.

Preemployment checks, either by the personnel department or by qualified outside organization can frequently turn up a history of drug abuse, alcoholism, or other adverse information that would warn the businessman of potentially dangerous applicants for employment. In addition, each manager must continually keep alert to the possibility of crime arising from drug use or sale, gambling, loan sharking or other illicit activities which may adversely affect his profits.

Office Security is a growing problem in both public and private buildings. It has been estimated that the number of incidents of office crime is doubling every three years. The actual number is impossible to determine, since many police reports do not list office thefts separately and many instances are not reported to police at all.

Among favorite targets of office thieves are cash, employees' handbags, small calculators, typewriters, personal effects, such as coats, sweaters, expensive pen and pencil sets, cameras, radios and television sets. Multitenant buildings, where there is a sense of anonymity, and where a stranger is not likely to be noticed, are susceptible to this type of crime.

Organized Crime has a "take" as high as $50 billion per year, according to the Office of the Attorney General of the United States. The honest businessman suffers because organized crime results in increased insurance costs, inequitable tax burdens resulting from tax-dodging of racketeers, higher prices to compensate for crime losses, and investment of illicitly acquired funds in licit enterprises. Today, organized crime controls a number of multimillion-dollar enterprises, according to a study of the Chamber of Commerce of the United States.

PROBLEMS IN SELECTED SECTORS OF BUSINESS

Manufacturing

In 1974, manufacturers suffered an estimated loss of more than $2.8 billion as a result of "ordinary" crime. This figure compares with an estimated $1.8 billion in 1971. The loss each year equals about 0.03 percent of industry shipments.

The President's Commission on Crime reported that 20 percent of all manufacturing companies find employee theft of tools, equipment, materials or company products a serious problem. The percentage for large companies is even higher.

Vulnerability depends in large measure on the types of materials and tools used in the plant, the nature of the company's product, and the effectiveness of security measures. High value-to-size goods are particularly desired because they are easily concealed and readily resold. Bulky items are also susceptible, however, particularly when company drivers and loading dock personnel are involved.

A major factor in the level of employee theft is the degree of plant security. This is demonstrated by the lower level of theft in defense plants, where security is strict.

Pilferage of garments from racks and loading docks as well as hijackings of entire truckloads of merchandise are major causes of loss to manufacturing firms shipping goods from New York City's garment district. The area has become the scene of petty thefts, muggings, sexual assaults and murder, with a resultant impact on the industry. The city has lost jobs, and business to surrounding areas as well as to new apparel centers as far away as Miami, Houston and Los Angeles.

Truckers carrying apparel and furs have found their loss records resulting in ever-higher cargo liability insurance premiums, which in some cases have tripled since 1965. Many have switched to hauling less theft-prone cargoes, while others have installed expensive cargo protection gear or resorted to armed guards or helicopter patrols of routes.

Street-level showrooms now often have barred doors, and buyers must identify themselves before they are admitted. Some buyers are reluctant to visit the area, and goods must be transported to hotels or trade shows for display. These problems all add to the cost of doing business, and have an adverse effect on prices and profits.

Because of their high unit value, fur garments are a favorite target of thieves. Manufacturers and retailers are finding insurance difficult to obtain because of the industry's loss record. A federally-funded insurance program, administered through Aetna Life and Casualty Company, now makes affordable insurance available to fur merchants and other businessmen in many high-risk areas regardless of their past loss record.

Wholesaling

Inventory shortages in the wholesaling sector of business which were estimated at $2.1 billion in 1974, are largely the result of employee theft, since customers and other outsiders are usually excluded from areas in which merchandise is stored. The pilferage problem includes executives, supervisors and workers. Losses in some companies range up to one million dollars.

The level of pilferage depends, at least in part, on the attitude management takes toward employees. Some firms feel that security measures interfere with production, and save less than they cost when the lower output is considered.

Retailing

The National Retail Merchants Association estimates that in 1971 retailers' losses from inventory shortages reached about 3 percent of sales in the case of department and apparel stores. Based on sales of $117 billion in 1974, losses of such stores are expected to be $3.5 billion. This is a 44 percent increase over the $2.4 loss estimated to 1971. The table in Appendix Figure 5-4 shows estimated losses to criminals by various types of retail stores.

Appendix Figure 5-4

ESTIMATED RETAIL LOSSES DUE TO ORDINARY BUSINESS CRIME 1970, 1973 AND 1974

Types of business	Sales ($ billions)			Losses ($ billions)		
	1970	1973	1974	1970	1973	1974
General merchandise and apparel	81	108	117	2.43	3.24	3.51
Drug stores	13	16	17	.45	.56	.60
Food stores	86	106	119	.86	1.06	1.19
Other[1]	21	29	31	.25	.35	.47
Total	201	259	284	3.99	5.21	5.77

* Source: *Monthly Retail Report,* Bureau of the Census, and estimates by Bureau of Domestic Commerce.

[1] Includes home furnishings, furniture, appliances, radio and TV, and hardware stores. Not included are eating and drinking places, automobile dealers, building material dealers, or gasoline service stations.

The principal types of ordinary crime affecting retail stores are shoplifting, burglary, vandalism, bad checks, employee theft and robbery.

Retail stores are the major commercial targets for burglars and robbery according to the Law Enforcement Assistance Administration's *Crime in the Nation's Five Largest Cities.* In four of the cities, the victimization rate for retail establishments was substantially higher than other types of business. While burglary or robbery were reported by 49.7 percent of all commercial establishments in these cities, the rate was 71.4 percent for retail establishments, in 1972.

The nature of the merchandise on the shelves of a retail store has a bearing on the level of inventory shortage experienced. Items which have strong buyer appeal and are easy to resell are major targets. In department stores, sporting goods sections are particularly hard hit, with losses of about 4.6 percent of sales. Juniors' dresses, sportswear, young men's clothing, small leather goods, cosmetics, costume and genuine jewelry, watches, men's casual wear, and records all suffer higher than average shortage rates.

An official of a large security firm estimates that in retail stores, shoplifting accounts for only 20 to 25 percent of total shortages, and that employees, at every job and salary level are responsible for the remainder. It is estimated that between 8 and 10 percent of the employees comprise the hard core pilferers, while many others steal on occasion.

Despite this, retailers usually concentrate their antitheft efforts on shoplifters. In discount stores, it is estimated that for every dollar lost to a shoplifter, three are lost to employees. Although apprehensions of shoplifters outnumber those of employees by 10 to 1, one company reports that dollar losses from employee pilferage are more than seven times as great as shoplifting losses.

Stealing frequently occurs in receiving, shipping, delivery and mail order departments. It is practiced by some salespersons who underring registers and pocket the difference, or who undercharge their friends or accomplices. Executives may juggle books or pad payrolls, and buyers may demand kickbacks or expensive gifts from suppliers. It is impossible to compile a complete catalog of the schemes practiced by dishonest employees.

The climate for stealing consists of three elements: temptation, opportunity, and motive. Factors which lead employees to steal are real or imagined grievances against management, alcohol or drug addiction, gambling losses, excessive debts and living beyond their means. Most petty pilferage is a result of the desire to own an item and the opportunity to obtain it without paying for it. To guard against employee theft, a retailer must carefully screen potential employees before hiring, treat employees fairly and attempt to develop loyalty to the employer, and establish strict controls over stock and cash. In addition, the security force must have the full backing of management in their efforts to minimize employee theft.

About four million shoplifters are apprehended each year. It is estimated that only one of every thirty-five shoplifters is caught. This would indicate that about 140 million instances of shoplifting occur every year. In a study conducted by a major security service firm, 500 shoppers were followed at random in a New York City department store. Forty-two, or one of every twelve, were observed stealing some item during their visit to the store.

Retailers have found that a get-tough policy is effective in reducing inventory shortages. More than two-thirds of the menswear stores prosecute all apprehended thieves. One major Washington, D.C., department store increased its security budget to over one million dollars in 1972, and expanded its security force to 150 people. The store reported a significant reduction of shrinkage as a result of these efforts. A store in Columbus, Ohio, reported that their new, tougher policy toward shoplifters had reduced losses from outside theft, but that internal theft remained high—about $2,000 per day.

New techniques to combat inventory losses are constantly under development. One of the latest is an electronic merchandise tag, which can be removed only with a special tool. If not removed by the clerk, the tag triggers an alarm at the exit from the department or from the store. Many of the alarms triggered have caught innocent customers, however, when clerks neglected to remove the tags. Advanced techniques obviously call for additional training.

The time-consuming work involved in prosecuting a shoplifting or employee theft case is a cost factor which must be considered in any security program. Many retailers, association executives and elected officials feel that the best way to attack the shoplifting problem is through a massive public relations campaign. The multimedia Shoplifters Take Everyone's Money (STEM) program in Philadelphia is considered a model that other cities might adopt for their own use. Any such program should stress the fact that shoplifting is a crime, that it is not smart, or "in," and that it results in higher prices for everyone.

Large-scale theft from retailers would be reduced if retailers would refuse to buy liquor, cigarettes, clothing, meat or other items offered for cash at greatly reduced prices by unknown suppliers. Such "peculiar bargains" almost invariably represent stolen merchandise.

Annual losses from inventory shortages in drug stores are approximately 3 percent of sales, and in 1974 are estimated to reach $500 million. This high level of loss is due in part, at least, to the nature of drug store merchandise, such as cosmetics, costume jewelry, candy, drugs, toys and records. The level of loss is about the limit which most drug stores can tolerate, since the margin of profit is not much greater than this.

Crime-related losses of food stores are estimated at $1.2 billion on sales of $119 billion in 1974. These losses result primarily from inventory shortages, bad checks, and truck hijackings.

The two primary targets for food store thieves are cigarettes and meat, both with high value/weight ratios and both easily salable. One firm in Iowa recently lost six truckloads of meat valued at from $30,000 to $65,000 each to hijackers. Many of the firm's meat trucks arrive at their destinations with part of the cargo missing. Small food markets and restaurants are the usual purchasers of stolen meat, although this is by no means a general practice of small businessmen. They generally are unaware of the origin of the meat.

Most food stores offer check cashing services for their customers. A survey of its members by the National Association of Food Chains (NAFC) brought replies from chains with 5,038 individual stores which cashed over twenty-three million checks per month. These chains experienced a total annual loss from bad checks of just over $5 million. The Department of Commerce estimates that bad checks cashed by all food stores reached over $500 million in 1974. The NAFC survey disclosed that bad check losses of the chains which responded were an average of $100,000 each, and one large chain lost over one million dollars.

Vandalism and looting are major crime-related problems in the vending machine sector of the retailing industry. While the return to the looter consists of only a small amount of money and merchandise in the machine, the cost to the businessman is increased greatly by damage to the coin meter or to the machine itself.

It is sometimes difficult for a coin-op owner to know when he has been victimized. In some cases, thieves are able to obtain keys to coin boxes, and visit coin-op locations at times when the appearance of customers or employees is unlikely. Even if they are seen, the probability is that they will be taken as owners or employees of the company. Frequently the thieves do not empty the cash box, but merely remove a portion of its contents. The owner is at a loss to explain the "decline," and is not aware that he has been robbed. Some gangs have been known to have regular routes which they "serviced" in a very business-like way.

The use of slugs, trimmed or foreign coins, and other worthless items continues to plague coin-op machine businesses. Manufacturers of coin meters have been able to cut down the use of slugs, but not to stop the practice entirely.

Summary

Business and Government have initiated programs aimed at cutting losses, and although much remains to be done, governmental programs aimed at measuring the impact of business crimes have been developed and guidance to businessmen and local law enforcement organizations has been made available. In addition, businessmen are spending large sums in crime prevention programs, and are banding together to pool their resources and to mount cooperative publicity campaigns.

One point that should be emphasized, however, is that businessmen may consider anticrime programs in the same context as other cost-cutting and profit-maximizing efforts. The Small Business Administration reported in *Crime Against Small Business*, that

> . . . the payoff to the business and the community can be big in relation to cost. For others, there may be a close margin of benefits or the benefits may not be worth the cost. In the final analysis, it is the businessman himself who must make the calculation. . . .

Because many businessmen are unaware of the magnitude of the cost of crime or of the methods by which the cost can be reduced, they tend to avoid making the necessary calculation. Insurance representatives claim, for example, that businessmen often reduce their precautions against robbery and burglary once they have obtained insurance.

Crime is often considered to be society's problem, affecting everyone, but informed, concerned, properly motivated managers can institute measures which can significantly reduce the criminals' take, and improve profits. Some of these measures are presented below:

1. Adopt professional approaches to property protection. Utilize available security equipment techniques and programs.
2. Cooperate with industry association crime reduction efforts. This is important in the case of small firms which have no resources to develop programs of their own.
3. Prosecute thieves. Fear of imprisonment has a strong deterrent effect on minor offenders in particular.
4. Support the data collecting efforts of private and government organizations. Planning for crime control can advance only if one knows what the problems are.
5. Promote local anticrime programs.
6. Do not buy merchandise offered at ridiculously low prices by unknown vendors. Such merchandise may have been stolen.

APPENDIX 5B

THE COMMON ELEMENTS OF WHITE COLLAR CRIME*

Basic to any determination of fruitful avenues of exploration with respect to the prevention, detection, and prosecution of white collar crime is an analysis of how it operates. What are its component parts? Where the spectrum of possible criminal acts is so broad and the perpetrators so different in character, status, and motivation can there be identifiable elements of universal applicability? Can they apply to crimes so diverse as antitrust violations and bank embezzlement, or so diverse as tax fraud and the ordering of merchandise with no intention to pay?

Without implying that motivations are necessarily similar, and recognizing that the *modi operandi* may be as diverse as the activities of all mankind, it may be that there are common elements which may be basic to all white collar crimes.

In any white collar crime, we will find the following elements:
1. intent to commit a wrongful act or to achieve a purpose inconsistent with law or public policy;
2. disguise of purpose or intent;
3. reliance by perpetrator on ignorance or carelessness of victim;
4. acquiescence by victim in what he believes to be the true nature and content of the transaction;
5. concealment of crime by
   a. preventing the victim from realizing that he has been victimized, or
   b. relying on the fact that only a small percentage of victims will react to what has happened; and making provisions for restitution to or other handling of the disgruntled victim, or
   c. creation of a deceptive paper, organization, or transactional facade to disguise the true nature of what has occurred.

If there are, in fact, common elements, or even elements which are present in the greater number of white collar crimes, then awareness of this structure may help us in our search for preventive, deterrent, and prosecutive measures.

INTENT TO COMMIT A WRONGFUL ACT OR TO ACHIEVE A PURPOSE INCONSISTENT WITH LAW OR PUBLIC POLICY

The presence of this element is self-evident in the case of most white collar crimes. It may be less easily seen in criminal cases in which prosecutors do not have the burden of showing that the defendant knew his act was unlawful or wrongful. Examples would be offenses of omission, such as failure to provide heat or proper repair, or failure to register securities for lack of awareness of the requirements of the Securities Act of 1933, or misinterpretation of such requirements. It is often difficult to prove such intent where the subject has been advised by counsel that his proposed course of conduct is legal, or where there has been a history of laxity with respect to such conduct by law enforcement authorities.

There is always an intent to commit a wrongful act or to achieve a purpose inconsistent with law or public policy where there is a white collar crime or offense, even one not requiring proof of intent, notwithstanding the existence of advice of counsel, misinterpretations of law, omission rather than commission, or laxity by authorities. In these instances the intent is properly inferable from the deliberate decision to go into the gray areas, and to risk crossing the line in order to achieve some advantage. It is a calculated risk, with the risk-taker seeking to achieve immunity

---

* Common Elements of White Collar Crimes, *The Nature, Impact and Prosecution of White Collar Crime*, National Institute of Law Enforcement and Criminal Justice, U.S. Department of Justice, Washington, D. C., 1971.

from the consequences of his acts. It may be that a defendant will succeed in frustrating proof of intent where that is required, or in influencing the prosecutive evaluation in his favor, but for our purposes we should not close our eyes to the presence of intent, in whatever form it may appear.[1]

Some examples of complex problems of intent would be the following: A landlord very rarely gets in trouble for failure to provide heat or maintenance, unless he skimps and tries to provide the minimum required by law, or sets out to provide less on the theory that the penalty will only be a minor fine and therefore a supportable cost of doing business. One who seeks advice of counsel in a borderline area is often seeking to establish a future defense in case his transaction is critically examined.[2] A taxpayer plays the percentages when he takes an entertainment deduction which he knows will be disallowed if closely audited, suspecting that he risks only a 6 percent interest charge on the extra tax he should have paid. One who makes purchases on credit with no intent to pay, will use the defense that he was merely improvident, which may be a good defense, but most people know whether they have enough money to pay their bills. Even in the apparently innocuous situation where one writes a check today, knowing that by the time it clears he will have deposited a salary check to cover it, he is skirting the line in issuing a check, knowing there are no funds to cover it if immediately presented.[3] In many instances in which check kites defraud banks of the principal amounts of large checks, the true intent of the defendant is to use the bank's capital without paying interest, or because his financial condition is not good enough to justify orthodox borrowing; he fully intends to ultimately cover his checks but "unforeseen" circumstances intervene. In a remarkably high proportion of white collar crimes unforeseen circumstances do intervene. For example, money is embezzled to finance a profitable investment which will enable return of the funds prior to detection, but the venture fails; or a furnace breaks down because maintenance did not anticipate a lengthy cold spell; or a shortcut in testing a drug could not be expected to result in such horrendous side effects as in the case of Thalidomide®.

DISGUISE OF PURPOSE OR INTENT

Once again, in the conventional situation this element is obviously present. It is to the unconventional situation that we must look in order to determine whether this is an element always, or almost always present in white collar crimes.

Under discussion here is a basic misrepresentation as to the nature and purpose of the transaction which is at the heart of the violation.

In an antitrust case an agreement for reciprocal business dealings between supplier and purchaser conceals (a) the absence of price, quality and service as elements inducing the transaction, and (b) the intent of the purchaser to foreclose supplier-competitors. In the SEC case the facade of a private offering to a purported limited number of offerees, or some other device, may conceal the effort to sell without a registration statement or offering circular which fully discloses the material facts which should influence investment decisions.[4] In a commercial bribery or kickback case a buyer for a corporation is given an opportunity to buy something at far below cost, the purchase being subsidized by the corrupting supplier. An investment by a union officer in a business enterprise which employs his union members is in fact consideration for breach of his fiduciary duty to his membership. What looks like a simple purchase or sale of stock by a corporate insider, may in fact be a wrongful exploitation of inside information. The submission of an order for merchandise to a supplier may mask the intent or aim not to pay. A vanity publisher or correspondence school will attest to belief in the marketability of the victims' work or potential, concealing or failing to disclose its knowledge that the odds against any success or fulfillment are astronomical.

Disguise differs from intent. That they are distinct and separate elements can be seen if one compares a white collar crime to a common crime. In a common crime the intent once formed is followed by the implementing act which is subject to no misinterpretation. The element of disguise in white collar crime serves to blur intent to the point where it often can only be derived by interpretation.

RELIANCE BY PERPETRATOR ON IGNORANCE OR CARELESSNESS OF THE VICTIM

The white collar criminal must rely on the ignorance or carelessness of the victim, and in those areas in which regulatory agencies have a statutory mandate to protect the public, the ignorance of the public must be maintained by misleading the agency or circumventing its disclosure requirements.

One example would be the looting of an automobile liability insurance carrier. In a typical situation such a carrier will be purchased by a group which will promptly sell off good assets in its portfolios and replace these with worthless or overvalued assets. The policy holders are ignorant of these manipulations. The state insurance department either accepts these new assets at their represented value, or does no more than look at over-the-counter stock quotations which may have been manipulated for this purpose.[5] Since the state insurance department is ignorant (possibly because of less than adequate audit procedures) of the hollowness of the assets in the portfolio, it permits the company to continue in business, collecting premiums, and holding off settlement of claims against its insureds. When the collapse comes, claimants cannot collect on their claims or judgments and policy holders are helplessly exposed to the very liabilities they paid premiums to avoid.

Ignorance or carelessness of the victim is crucial to the success of the white collar criminal and is the objective sought by the disguise of purpose or intent referred to above. In a home improvement scheme the victim is ignorant of the work history of the company which solicits him and does not take the precaution of requiring or checking on references. The victim is customarily unaware of the contents of the documents he signs, generally having no idea of the true price and the credit terms, and few victims have ever even suspected that they were placing mortgages on their homes as part of the deal.

Ignorance of the victim may be the direct result of a calculated effort to keep in ignorance the regulatory agency whose procedures are designed to protect him. The Securities and Exchange Commission cannot protect the public by its disclosure requirements if the white collar criminal risks prosecution by submitting false information to the Commission in purported compliance with its registration or filing requirements. Banking agencies are hardly in a position to protect a bank against loss caused by a faithless officer who inserts false loan papers in bank records and vouches for their authenticity to bank examiners. Neither physicians nor their patients can be protected properly if a pharmaceutical manufacturer submits fraudulent test results to the Food and Drug Administration (or if the manufacturer is itself a victim because it is ignorant of the fraudulent operations of its chosen testing facility).

In some instances ignorance of the victim is almost a certainty because of the context in which the wrongful actions arise. In one case (not resulting in prosecution) a department manager in a defense industry deliberately shifted labor costs from a fixed price contract to the performance of a cost reimbursable contract. He did this without the knowledge of his employer, the sole motive being to make his department more profitable and thus enhance his career and promotion prospects. This was a case where the direct financial reward from the scheme came to an innocent, albeit ignorant party (the employer), whose ignorance promoted the ignorance of the Government which was the ultimate victim of the scheme.

Ignorance of the true facts is sometimes inevitable in the face of a calculated effort to deceive, but the perpetrator's efforts to deceive and mislead are only too often matched by the carelessness, self-deception, or cupidity of victims. It does little good to require a prospectus to be issued in connection with the sale of stock if the purchaser of stock will not read it. This raises the central question of what measures can be taken to strike at the ignorance of victims or abate their cupidity.

ACQUIESCENCE BY VICTIM IN WHAT HE BELIEVES TO BE THE TRUE NATURE AND CONTENT OF THE TRANSACTION

White collar crimes are unique. They generally require the victim to acquiesce in being victimized. In the great majority of cases we are confronted with crimes which require affirmative acts of cooperation by victims before the fraud can be completed. Put another way, victims must help to "dig their own graves."

In considering this element, the term "victim" must be broadly construed. For a white collar crime to succeed, someone with an interest, either as a direct victim or as a protector of potential victims, must affirmatively cooperate or passively acquiesce in the crime.

In its role as collector of taxes the Internal Revenue Service operates on the theory that taxpayers file honest returns. It concedes the position of the taxpayer except in those rare instances where there is an audit; the Internal Revenue Service therefore acquiesces in an act the true nature and content of which is not known to it. If an issuer of securities files a false prospectus, the Securities and Exchange Commission acquiesces by permitting the filing and public sales based thereon. The victim purchaser acquiesces by the affirmative act of making a purchase. One is cheated in buying through the mails, taking the essential affirmative step of making the purchase which is the object of the criminal intent.

With respect to many *malum prohibitum* offenses, acquiescence is negative rather than affirmative. The tenement dweller acquiesces in being deprived of heat or repairs because he does not fully comprehend that this deprivation is a transaction different from that mandated by law. Collusive pricing succeeds because purchasers believe the prices quoted have been individually arrived at and do not know that the bidders have conspired. The Food and Drug Administration acquiesces in the marketing of a drug because it believes that tests are as represented, and physicians affirmatively prescribe such drugs for their patients in reliance on drug company salesmen, drug advertising, and the presumed vigilance of the Food and Drug Administration.

Since someone's acquiescence is needed for a white collar crime to be committed (in contrast to murder, robbery, assault, or rape), a central question is how to prevent acquiescence, affirmative or negative.

CONCEALMENT OF CRIME

When a murder, robbery, burglary, assault, or rape has been committed, it is clear there has been a violation, though there may be some question as to the identity of the perpetrator or his legal or mental capacity to form the requisite criminal intent. This is not the case with white collar crimes, where victims almost never know they have been victimized until well after the executing transactions or occurrences and, in fact, may never know they have been victimized.

The ideal scheme or plan, from the point of view of the perpetrator, is one in which the victim never learns the true nature of the blow struck. Charity frauds classically illustrate such a scheme. The takings are small for each individual no matter how large cumulatively, and few victims have sufficient personal interest in their contributions to attempt to follow up. As a result, charity frauds almost always are exposed through the curiosity of news media or the vigilance of public officials, rather than as the result of investigations following victims' complaints. If prepackaged goods are marked with short weights it is highly unlikely that any customer will weigh his purchase to check the labeled weight. If the grade or quality of food is mismarked, we have the victim eating the evidence. If fabrics are mislabeled, the perpetrator runs the risk of FTC surveillance, but the victims who can spot the fact that their garments are 30 percent wool rather than 50 percent are few in number. If the price of securities is manipulated in such a way as to avoid the scrutiny of the SEC, the investor victim is more likely to blame his luck, or impersonal market forces, than the chicanery of unknown persons. If sellers collaborate to fix prices the victims will rarely know about it, and only lengthy and complex Government action will uncover the facts.

Since it is not always possible to anticipate an uninterrupted series of complacent victims, standby tactics are often employed. Thus some schemes will contemplate making immediate restitution to any victim who complains, to make the victim feel that the perpetrators acted in good faith or to ensure the victim's silence.

The most usual form of concealment is the lulling tactic, followed by silence and the collapse of a corporate entity. This works best when the scheme involves a continuity of performance. In an advance fee swindle, a businessman seeking a loan will agree to pay $2,000 to a loan broker for securing a $75,000 loan. The loan broker will ask for $500 or $750 initially, graciously offering to waive the balance until he has delivered the promised financing. The loan broker has no intention of ever earning the balance. His objective is the initial retainer. A series of lulling letters is then used to keep the victim quiet while others are being victimized, and to tire the victim out. Finally, the loan brokerage firm collapses and the drifter who closed the deal (and who was working on a commission basis) drifts on to other schemes. A year may elapse while preprogrammed lulling letters continue, with the victim sinking into bankruptcy or incapacitating despondency. If the victim still is afloat after all this, the wind is usually taken out of his sails when he learns that the loan brokerage company is no longer in existence.

Concealment is achieved by design of an organizational structure to frustrate and discourage complaint or pursuit by victims. A typical example would be a home improvement fraud in which a faceless corporation is set up, hires itinerant salesmen, and promptly negotiates its paper to so-called holders in due course. The victims are so tied in legal knots that they find it hard to even consider complaining to enforcement authorities, since they do not believe this will protect them against the "holders in due course." Attorneys will rarely take these as charity cases, and if the victim does obtain legal assistance his attorney will usually concentrate on trying to settle obligations for less than the face amount of the paper. All cooperate in muffling the outcry.

Concealment is also achieved by limiting the residue of provable facts, so that there is great difficulty in organizing a case which will meet necessary legal standards for criminal sanctions or civil process. Because of the manner in which white collar crimes are organized and executed it is possible to generalize (it is not true in all cases) that in investigating and prosecuting these cases the problem is more one of what the facts spell out than what the facts are. The key question (in a prosecution, not in a study of the problem of white collar crime) is whether criminal intent is inferable beyond a reasonable doubt from the facts unearthed by investigation; that is, was there a crime? If the answer is negative, then the crime is concealed, no matter how deep the wound.

It is not uncommon for a prosecutor to face the most difficult evaluation problem where he has what amounts to a stipulated set of facts before him. One example would be the case of the "vanity publisher" who signs a contract with a would-be author to publish his book, send copies to reviewers, advertise the book in respectable publications, and provide editing services. The publisher receives many thousands of dollars for this service and, in fact, he does provide editing service, does provide a number of hard cover copies, does advertise, and does send copies to reviewers. The victim, in such cases, is led to hope that his book is being promoted and handled as it would be by a legitimate publisher, though he has nothing in writing. There is a fraud when the publisher knows that the reviewers throw all his books in the trash can, that the advertisement in a reputable newspaper's book section is almost a classified ad in format, and that of the multitude of books published by vanity presses in recent years, only a miniscule number have recovered as much as the victim's own cash outlay.[6]

CATEGORIES OF WHITE COLLAR CRIMES (EXCLUDING ORGANIZED CRIME)

A. Crimes by persons operating on an individual, *ad hoc* basis
   1. purchases on credit with no intention to pay, or purchases by mail
   2. individual income tax violations
   3. credit card frauds
   4. bankruptcy frauds
   5. Title II home improvement loan frauds
   6. frauds with respect to social security, unemployment insurance, or welfare
   7. unorganized or occasional frauds on insurance companies (theft) (casualty, health, etc.)
   8. violations of Federal Reserve regulations by pledging stock for further purchases, flouting margin requirements
   9. unorganized "lonely hearts" appeal by mail
B. Crimes in the course of their occupations by those operating inside business, Government, or other establishments, in violation of their duty of loyalty and fidelity to employer or client
   1. commercial bribery and kickbacks, i.e. by and to buyers, insurance adjusters, contracting officers, quality inspectors, government inspectors and auditors, etc.
   2. bank violations by bank officers, employees, and directors
   3. embezzlement or self-dealing by business or union officers and employees
   4. securities fraud by insiders trading to their advantage by the use of special knowledge, or causing their firms to take positions in the market to benefit themselves
   5. employee petty larceny and expense account frauds
   6. frauds by computer, causing unauthorized payouts
   7. "sweetheart contracts" entered into by union officers
   8. embezzlement or self-dealing by attorneys, trustees, and fiduciaries
   9. fraud against the government
      a. padding of payrolls
      b. conflicts of interest
      c. false travel expense, or per diem claims
C. Crimes incidental to and in furtherance of business operations, but not the central purpose of the business
   1. tax violations
   2. antitrust violations
   3. commercial bribery of another's employee, officer or fiduciary (including union officers)
   4. food and drug violations
   5. false weights and measures by retailers
   6. violations of Truth-in-Lending Act by misrepresentation of credit terms and prices
   7. submission or publication of false financial statements to obtain credit
   8. use of fictitious or overvalued collateral
   9. check-kiting to obtain operating capital on short-term financing

10. Securities Act violations, i.e. sale of nonregistered securities, to obtain operating capital, false proxy statements, manipulation of market to support corporate credit or access to capital markets, etc.
11. collusion between physicians and pharmacists to cause the writing of unnecessary prescriptions
12. dispensing by pharmacists in violation of law, excluding narcotics traffic
13. immigration fraud in support of employment agency operations to provide domestics
14. housing code violations by landlords
15. deceptive advertising
16. fraud against the government
    a. false claims
    b. false statements
        (1) to induce contracts
        (2) AID frauds
        (3) housing frauds
        (4) SIBA frauds, such as SBIC bootstrapping, self-dealing, cross-dealing, etc., or obtaining direct loans by use of false financial statements
    c. moving contracts in urban renewal
17. labor violations (Davis-Bacon Act)
18. commercial espionage

D. White collar crime as a business, or as the central activity
1. medical or health frauds
2. advance fee swindles
3. phony contests
4. bankruptcy fraud, including schemes devised as salvage operation after insolvency of otherwise legitimate businesses
5. securities fraud and commodities fraud
6. chain referral schemes
7. home improvement schemes
8. debt consolidation schemes
9. mortgage milking
10. merchandise swindles
    a. gun and coin swindles
    b. general merchandise
    c. buying or pyramid clubs
11. land frauds
12. directory advertising schemes
13. charity and religious frauds
14. personal improvement schemes
    a. diploma mills
    b. correspondence schools
    c. modeling schools
15. fraudulent application for, use and/or sale of credit, airline tickets, etc.
16. insurance frauds
    a. phony accident rings
    b. looting of companies by purchase of overvalued assets, phony management contracts, self-dealing with agents, intercompany transfers, etc.
    c. frauds by agents writing false policies to obtain advance commissions
    d. issuance of annuities or paid-up life insurance, with no consideration so that they can be used as collateral for loans
    e. sales by misrepresentation to military personnel or those otherwise uninsurable
17. vanity and song publishing schemes
18. ponzi schemes
19. false security frauds, i.e. Billy Sol Estes or De Angelis-type schemes
20. purchase of banks, or control thereof, with deliberate intention to loot them
21. fraudulent establishment and operation of banks or savings and loan associations
22. fraud against the government
    a. organized income tax refund swindles, sometimes operated by income tax "counselors"
    b. AID frauds, i.e. where totally worthless goods are shipped
    c. FHA frauds
        (1) obtaining guarantees of mortgages on multiple-family housing far in excess of value of property with foreseeable inevitable foreclosure

(2) home improvement frauds
23. executive placement and employment agency frauds
24. coupon redemption frauds
25. money order swindles

FOOTNOTES

1. This is not say that variations in degrees of intent should not be taken into account in any sane and sensible evaluative process.
2. This defense may or may not be successful. The attorney admits the advice, but claims he did not have all the facts, or that things worked out differently than he anticipated. The client says he gave, in good faith, what he thought were all the relevant and material facts.
3. The writer does not suggest that there should be prosecution in these instances, but only that there is always a questionable intent. Many state laws and the District of Columbia Code provide that a subsequent payment, within a specified time, rebuts the inference of fraudulent intent. 22 D.C. Code 1410.
4. The U.S. Court of Appeals for the Second Circuit, in *United States v. Doyle*, 384 F.2d 715, 720 (2d Cir. 1965), cert. denied, 382 U.S. 843 (1965) held that nonregistration was tantamount to fraud since it served to deny the investing public the protections of full disclosure required by the Securities Act of 1933. Our courts recognize that fraud may well be the motivation for violation of these so-called technical registration requirements. *United States v. Abrams*, 357 F.2d 539, 546 (2d Cir. 1966), cert. denied, 384 U.S. 1001 (1966); *United States v. Wolfson*, 405 F.2d 779 (2d Cir. 1968), cert. denied, 394 U.S. 946 (1969).
5. Certain stocks have been traded for the sole purpose of establishing value so that they could be part of insurance company reserves. This is sometimes their only value.
6. In talking to other than their victims, vanity publishers are quite frank about this situation and have been quoted as justifying their operations as a worthwhile ego massage for victims.

APPENDIX 5C

BUSINESS MANAGEMENT FOR CRIME PREVENTION*

INTRODUCTION

Every small businessman is confronted with the problem of protecting his business against some phase of crime. It thus becomes his responsibility to establish policies and rules and to weigh the value of different protective devices, as measures to reduce the impact of crime on his business.

Among the courses of action open to the businessman for more effective protection against crime are (1) use of physical means such as alarm systems, antitheft devices, door and window guards, and firearms; (2) training of employees to cope with shoplifting, armed robbery, burglary, employee pilferage, and embezzlement;

* Reprint from *Crime Against Small Business*. A report of the Small Business Administration transmitted to the Select Committee on Small Business, United States Senate.

(3) reorganization of business practices and procedures; (4) modification of the physical features; (5) rethinking of the attitude of management in coping with the problem of crime, leading perhaps to a stronger stand on prosecution of persons committing crimes against the firm, including shoplifters and the small minority of dishonest employees.

The Cost of Crime and of Crime Prevention

Established management practices to combat crime are notably lacking among small firms. As a result, losses through crime may not be detected early or may not be known at all. Shoplifting and employee pilferage, two costly sources of losses, are poorly controlled in small businesses. Because they do not lend themselves to precise cost analysis, the small businessman does not generally know the cost to him of these crimes. Thus, the value of losses in burglaries may also be hard to determine if control measures are inadequate.

Any investment the businessman makes in crime prevention should be considered in relation to (1) the immediate cost of crime and the savings to be realized, and (2) the effect of his investment in discouraging future crimes against him.

Immediate investments whose value management would need to appraise might include protective devices such as antitheft mirrors to cut down on shoplifting, or photographic equipment to take pictures of checkpassers, or antiburglary devices.

Investments of a more general nature, spread over a longer period, would include training employees on how to detect shoplifting and actions in the case of a robbery.

It is entirely possible that certain kinds of investments may reduce other costs. For example, the installation of a central burglar alarm system may reduce insurance costs enough to offset the alarm cost. The businessman should explore all possibilities of cost reduction in connection with his expenditures to reduce his crime losses.

The Need for a More Positive Attitude

Research into the problem of crime against small business has brought out the fact that, while small businesses may appear concerned over the problem, little action is being taken. The smaller the store and the smaller the community, the less the effort is being applied. Most small businesses are not security conscious. Yet, ironically, a single crime such as a burglary, or a series of small, continuous losses as by pilferage may so sap the firm that failure ultimately occurs.

A part of the problem is lack of knowledge by the small businessman about how to cope with the kinds of crime discussed in this study combined with sidestepping or procrastination. He finds it hard to accept the possibility that one or two of his employees might be dishonest or that his business could be burglarized. It may take a burglary or a robbery or significant employee theft losses to jar the businessman into positive action.

Ignorance of the law adds to the businessman's frustration. Many times, he is unaware of the legal protection afforded him. Many states have laws protecting the businessman from suit for false arrest for shoplifting, providing the suspect is detained in a reasonable manner for a reasonable period of time and if the businessman is reasonably sure that he has taken merchandise with no intention of paying for it. Since the businessman is often not sure of his rights, he is reluctant to take any action, thus permitting the offender to go free to practice on other, equally unwary businessmen.

The need for a positive, direct approach by the businessman is obvious. In spite of the efforts of law enforcement agencies, the primary responsibility rests with the individual to protect his firm through appropriate devices, through indoctrination and training of employees, through cooperation with other area businessmen, and through establishing appropriate liaison with his local police.

Resources Available to the Small Businessman

Most metropolitan police departments are in an excellent position to help small businessmen through training, counseling, and crime prevention surveys.

The Small Business Administration is directly involved in certain managerial aspects of crime in business. A number of regional offices, working with police departments, and local businesses, have cosponsored workshops and clinics on special problem areas, including shoplifting, employee theft, and fraudulent checks. SBA already has several films available on these subjects. Moreover, the agency now has a film company under contract to produce four more short training films of fourteen minutes each on shoplifting in retail stores, employee theft in small manufacturing firms, and burglary in small businesses.

These films are scheduled for completion within the next year.

Finally, SBA has three small marketers aids available free to the public. These are: No. 119 "Preventing Retail Theft," No. 129 "Analyzing Shoplifting Losses," and No. 134 "Preventing Burglary and Robbery Loss."

Procedures for Controlling Crime in Business

The remainder of this material deals with five separate areas of crime. In each, a brief narrative is presented, describing the problem. This is followed by a checklist of possible remedies. Thus, the businessman can identify the kind of problems with which he is confronted and then read about courses of action to meet these problems.

The five areas are:
1. burglary,
2. robbery,
3. shoplifting,
4. bad checks, and
5. employee theft.

A sixth section summarizes a survey of the results of firearms training given to businessmen by the police department of Kansas City. It appears that perhaps the principal benefit has been the awareness acquired by the participating businessmen of other measures that might be taken to reduce crime losses, and of the limitations on the value and the difficulties of using firearms.

BURGLARY

The General Problem

From the management point of view, protection against burglary is a matter of attitude and housekeeping. It is the responsibility of the businessman to discourage burglary by maintaining the highest level of protection for his establishment.

To begin with, the businessman must accept the fact that burglary is entirely possible and no business is immune to it. While certain types of crimes are more directly related to the kind of business (shoplifting in retail stores), burglary is more likely to affect all small businesses. FBI reports show that nonresidence daytime burglaries are up 83 percent between 1960 and 1967. Nonresidence nighttime burglaries are up 47 percent for the same period.

Protection of the obvious points of entry is not sufficient. The businessman must anticipate every conceivable method by which a burglar could gain entry into the building—through doors, windows, roofs, sidewalk openings, and all other possibilities.

There are certain managerial decisions that must be reached with respect to burglary. It is advisable that the businessman counsel with all sources of assistance. These might include his local police department, his insurance agent, representatives of burglar alarm companies, safe companies, cash register companies, and building architects.

The businessman should know as much as possible about the alternatives available to him in burglary protection. Basically, he needs to make decisions relating to the following:

1. The kind of alarm system is any that best suits his kind of business. The cost of maintaining an alarm system must be measured against the expected savings in insurance for the average cost of a typical burglary in his kind of business.
2. The adequacy of locks on entrances to the building is important also. This includes locks on windows, sidewalk entrances, roof openings, as well as doors. Too often, locks are not changed for long periods of time and the businessman may not actually know all who have keys to the building. He should make it a practice to change locks or tumblers on locks as often as he feels necessary to give adequate protection. Further, the types of locks should be such as to give maximum protection. Window locks should be given the same thorough inspection as door locks. Counseling with a competent locksmith

will eliminate many unforeseen problems.
3. The establishment of a routine for total protection, with assigned responsibility to others in the owner's absence is essential. There is no substitute for good housekeeping in burglary protection. The owner should establish a fixed daily routine to assure that every precaution is being taken. This includes such measures as:
   (a) leaving the cash register open at night;
   (b) turning on lights inside and outside the building before leaving;
   (c) checking to see that no one is hiding in the building at closing time;
   (d) doublechecking all doors and windows; and
   (e) checking to see that the alarm is turned on and is operating properly.

Essentially, the businessman's function in burglary prevention is to increase the time needed to gain entry. The individual businessman's effort is the most important part of prevention. By installing adequate lights, locks, alarms, and other devices, the physical security of the business will deter at best or delay at least the efforts of even the most determined burglar.

But the business is not secure unless it is totally protected. The most intricate alarm system is of no use if it fails to cover even the smallest roof opening. The strongest door will do little good if the burglar can quickly enter an unlocked window. Anything short of total protection means inadequate protection.

This Appendix contains a detailed listing of the kinds of burglary and precautions one may take to improve his level of protection. The type of business, its location, age, and architecture of the building are all factors which govern burglary protection.

Like other matters of crime, however, protection against burglary begins with a positive attitude on the part of the owner that protection of the individual business is *his* responsibility.

Lighting
The Problem

The majority of burglaries occur at night, and naturally the criminal welcomes darkness to conceal his presence and his actions. Three out of four commercial burglaries are committed against buildings that have either no lights or inadequate lights. The would-be burglar can be discouraged and perhaps thwarted by adequate lighting inside and outside a building.

Preventive Actions

1. Place a night light over the safe.
2. Alleys and rear of business should be well lighted.
3. Illuminate all entry points well.
4. Keep night lights on inside the building.
5. Night lights should be wired so that the alarm is set to go if they go out.
6. Install inside lights to the rear so that an intruder's silhouette can be seen from the street.

Locks
The Problem

The easier the method of entry, the greater the chance of burglary. Locks that can be forced, duplicated, or easily opened increase the likelihood of burglary. Experienced would-be burglers can quickly size up the ease of entry by casual observation of locks on doors, storage, windows, etc. The burglar-proof lock has not yet been invented, but adequate locks are available and will deter even the most determined.

Preventive Actions

1. Modern, cylinder-type locks are preferable.
2. Proper installation should prevent prying, cutting, twisting.
3. Lock bolts should be protected

against being pushed back with a thin instrument
4. Control of keys is important.
5. Hinge pins and hasps should be installed to prevent removal of pins and screws.
6. High-grade steel hasps will prevent prying, twisting, cutting.
7. Padlocks should be locked in place at all times to prevent key duplicating.
8. Lock bolts should be flush and point inward.

Doors

The Problem

Most burglaries occur by forcing a natural opening in the building, such as a door or a window. Inadequate doors offer the burglar easy access to the premises. Doors too fragile for adequate protection, improper fit of doors in jambs, antiquated locking mechanisms—all these add to the problem. Strength and security can be had without sacrificing looks. Protection, however, should overweigh appearance.

Preventive Actions

1. Panels and glass should be protected against being kicked or knocked out.
2. Put bars on the inside to prevent breaking the entire door.
3. Double doors should be flush-locked with long bolt.
4. If the door has glass that can be broken, install double-cylinder lock requiring key both inside and out.
5. Install sheet metal on inside and outside of basement doors.
6. Install door frames that cannot be pried off hinges or removed.
7. Cylinder ring of lock should be recessed to discourage use of lock puller.

Windows

The Problem

Windows offer easy access to the building unless adequately protected. Display windows in retail stores are susceptible to hit-and-run tactics. Other windows, poorly protected, permit the burglar to enter the building, oftentimes undetected, particularly when the windows are poorly lighted. Windows should offer light, ventilation, and visibility, but *not* easy access.

Preventive Actions

1. Properly installed grates give maximum security.
2. Glass bricks are highly effective on windows not needed for ventilation.
3. Locks must be designed and located so they cannot be reached and opened by breaking the glass.
4. Heavy merchandise and equipment piled in front of unused windows will give some protection.
5. Cleaning windowsills periodically will assure that fingerprints are more likely to be left by a burglar.
6. Avoid, wherever possible, window displays that obstruct view into the building.
7. Expensive or small items left in the windows overnight invite burglaries.

Safes

The Problem

Given the alternative, the burglar prefers cash to other property. Far too many businesses have safes that are inadequate to company needs, have not had combinations changed for years, or are easily opened or removed by a skilled burglar. Hiding the safe will serve only to give the burglar better working conditions. Money needs more protection than records.

Preventive Actions

1. The safe should be easily visible from the outside of the building.
2. Lightweight safes should be secured to the structure to prevent being carried away.
3. Cash should be kept at a minimum by frequent banking.
4. Never leave the combination written where it can be found.
5. When employees leave the firm, change the combination of the safe.

6. Keep a light burning over the safe at night.
7. Lock safe securely when leaving the premises by turning the dial several times in the same direction.

Building Exteriors
The Problem

The enterprising burglar will take every advantage to gain entry into the building, especially if entry points are poorly protected. The alert businessman needs to ask himself, "If I were determined to gain entry to this building, what are all the possible ways I could do it?" The outward appearance and security of the building will often determine whether or not it will be attacked. Every opening represents a hazard—inspect and correct wherever possible.

Preventive Actions

1. Fences should be strong, in good repair, and kept free of debris and boxes.
2. Weeds around the outside of the building or fence provide a good hiding place.
3. Ladders should be kept locked up.
4. Blind alleys afford protection for the burglar.
5. Sidewalk openings and their frames should be securely and properly locked.
6. Skylights and ventilators on the roof are easy access points unless protected.
7. Fire escapes and exits should be designed for quick exit but difficult entry.
8. Utility poles offer easy access to roofs.

Alarms
The Problem

Twenty-four hour vigilance by the businessman is not practicable; consequently, he must rely on other means of detecting any real or attempted burglary. An adequate alarm system may give constant protection, whether the businessman is on or off the premises.

Preventive Actions

1. All openings should be covered by alarms.
2. Periodic tests will insure that the alarm is in proper working order at all times.
3. Power sources should be hidden, protected, checked, and tested regularly.
4. Designate an employee who is to notify authorities if alarm goes off.
5. Properly installed alarms can result in lower insurance premiums.
6. The type of alarm should be adequate to the needs of the business.

Other Safeguards
The Problem

Poor general housekeeping or lack of controls are invitations to burglary. The businessman should establish basic policies and operational routine to reduce the risk of burglary. Responsible employees should be able to carry out the functions in the owner's absence.

Preventive Actions

1. Keep a record of serial numbers of all merchandise and equipment.
2. Policy numbers and serial numbers of large denominations of bills should be recorded.
3. Before locking up each night, check to see that no one is hiding in the building.
4. Leave the cash register open at night to prevent unnecessary damage.
5. All checks should be logged and marked "For Deposit in Account of ─────" as soon as they are received.

BURGLARY—
SUMMARY CONSIDERATIONS
How a Burglar Enters a Place of Business
(According to Police Statistics)

Break front windows: 22 percent.
Break front door glass: 16 percent.
Break rear or side windows: 14 percent.

Enter through basements, coal chutes, or other openings: 14 percent.
Break rear door glass: 10 percent.
Force front door locks: 9 percent.
Enter roof or skylight: 2 percent.

What to Do if Burglary Occurs

1. Do not disturb anything at the scene. The chances of apprehension are greatly increased if the scene is left completely intact.
2. Preserve all clues.
3. Call the police immediately.
4. Be prepared to assist the police in every way.
5. Be prepared to provide information as to items missing.

ROBBERY

In the order of magnitude of ordinary crimes, robbery represents the smallest monetary loss, falling significantly below shoplifting, employee theft, fraudulent checks, and burglary. The one outstanding factor about robbery, however, is the personal danger the businessman and his employees are likely to face from violence.

Some sort of a weapon is used in nine out of ten robberies, with firearms used in about two out of three armed robberies. There appears to be a growing tendency for robbers to shoot or otherwise injure their victims.

Because of the sudden, often violent action of a robbery, the victims are often taken by surprise and off their guard. The typical robbery occurs in a very short period of time—less than a minute. The victim generally finds it difficult to relate details of the robbery accurately and reliably to the police.

Almost universally, police departments counsel against the victim of a robbery taking any action which might antagonize the robber. Instead, he is cautioned to cooperate fully with the robber's wishes, but, at the same time, noting factors relating to the robbery that will be useful to the police—description, escape route, property taken, etc.

The businessman needs to prepare himself in advance by reaching certain basic decisions about the possibility of a robbery, in order to give himself and his employees maximum protection against the actual occurrence. These include:

1. *Admission that a robbery is possible.* Few businesses are immune from the attention of the would-be robber. Businesses that maintain cash and/or high-value items on the premises are likely targets for robbers. For this reason, all businesses are advised to maintain the least possible level of cash exposure, to make bank deposits often but not regularly, i.e. predictably, and to give merchandise subject to robbery maximum protection. Jewelry stores are preferred targets of robbery. So are pharmaceutical departments of drug stores. Any business with cash on the premises is a prospective target and the amount of cash does not necessarily have to be large. Banks represent a special case and are not dealt with specifically here.
2. *The establishment of a definite plan to be followed in the event of a robbery.* The manager and employees should be prepared to make careful mental observations and to write them down as soon as possible. Provision needs to be made for relaying information the police will need, particularly for suspicious persons noted on or near the premises.
3. *The kinds of protective devices to be maintained on the premises.* Many businesses have worked out systems with adjoining businesses in the event of a robbery. A signal or light is flashed to alert the adjoining firms that a robbery is in process so that the police can be notified.
4. *A decision on whether or not to use force to thwart an attempted robbery.* Increasingly, businessmen are arming themselves as a precautionary measure against a possible robbery. Most police departments caution against businessmen using firearms to prevent

a robbery. The typical businessman is neither adequately trained nor mentally prepared to face up to a would-be robber.

There is a growing feeling among police departments, however, that if the business owner is insistent on using guns, then he should know how to use firearms, have a gun that is safe and operable, and have an understanding of his responsibility for using a gun. The Kansas City (Mo.) Metropolitan Police Department has conducted several programs on proper and legal firearm use.

In the final analysis, the businessman's best protection is to take no action that would provoke the robber. He should cooperate to the extent demanded, making a careful note of all details of the robbery and robber and reporting to the police at the first opportunity. There follows below an action plan for the businessman in establishing a program for dealing with robbery.

General Preventive Measures

The Problem

Robberies will occur because the businessman has made it easy and convenient for the robber because of poor housekeeping, poor cash-handling methods, and a general lack of planning toward the possibility of robbery. While it is impossible to eliminate robberies completely, it is the businessman's responsibility to deter the would-be robber as much as possible through good operational practices.

Preventive Actions

1. Keep the interior and front and rear entrances well lighted.
2. Keep advertising and merchandise out of the windows as much as possible. This will permit a clear view into the building.
3. Keep the rear and/or side doors locked at all times.
4. Maintain a record of decoy currency (bait money) by serial number and series in the cash register, to be given to a robber.
5. Be sure alarms are working at all times.
6. Do not open the place of business *before* or *after* regular business hours, as far as possible.
7. Avoid routine procedures that can be observed and used to advantage by would-be robbers.
8. Call the police if a request is received to open the place of business after regular hours.
9. Keep cash exposure and cash on the premises at the lowest possible level.
10. Keep checks separate from cash, even when making the bank deposit.
11. When making bank deposits:
    a. go directly to the bank;
    b. conceal the money, if possible;
    c. do not leave deposits or withdrawals unattended in an automobile;
    d. do not go to the bank alone;
    e. vary time and routine of bank trips; and,
    f. if possible, make deposits in daylight hours.
12. Do not keep large sums of money on the premises—bank all receipts as often as possible.
13. Do not keep large sums in the cash register or where it may be exposed to the view of others.
14. Beware of bell tapping—the procedure whereby one person distracts the attention of the cashier while an accomplice steals from the cash register.

Anticipating a Possible Robbery

The Problem

The speed with which a robbery normally takes place makes it difficult for the businessman or his employees to give helpful information to the police. Be prepared for the possibility of a robbery by deciding in advance what is to be done and who is to do it.

Preventive Actions

1. Be alert for persons attempting to hide on the premise at a closing time.

2. Instruct all employees on the use of the alarm system.
3. Call the police if a suspicious person is observed on or near the premises. If he is driving a car, get the license number.
4. Make plans in advance as to who will take certain actions if a robbery occurs:
   a. Who calls the police;
   b. Who makes observations;
   c. Who protects the evidence at the scene;
   d. Who detains witnesses.
5. Some employees are gifted in the art of observation. These persons should be alerted to make observations during a holdup.
6. Practice identification with coworkers.
7. If possible, install height markers, e.g. black plastic tape, at varying heights on door frame to identify approximate height.
8. Discuss with employees what they might do if a robbery occurs.

What to Do if Robbed

The Problem

Most robberies take place in approximately one minute. During that time, the victim must do the robber's bidding, and be observant enough to give the police useful information. To the average person, however, a robbery is a frightening experience. The robber is generally armed and should be considered capable of committing bodily harm.

Preventive Action

1. Take no action which would jeopardize personal safety.
2. If the robber displays a firearm, consider it to be loaded.
3. If possible, activate the silent alarm.
4. Attempt to alert other employees by use of prearranged signals.
5. Attempt to delay the robber if at all possible, but without sacrificing personal safety.
6. Try to maintain possession of a hold-up note, if one is used.

Observation

The Problem

The chance of apprehending a robber are considerably enhanced if the victim is able to give an accurate description of the person or persons. The victim must be prepared to observe the robber, usually within a minute or less. By remaining calm during a robbery, the victim's powers of observation will increase and danger of injury will be minimized.

Preventive Actions

1. Observe physical characteristics of the robber;
   a. race, age, height;
   b. facial characteristics, complexion and hair;
   c. clothing worn, head to foot;
   d. physical carriage;
   e. speech;
   f. marks, scars, deformities;
   g. robber's method of operation.
2. Look for accomplices.
3. Note method of escape.
4. Describe escape car, model, make, year, license number.
5. Ascertain direction of travel.
6. Describe type of weapon used.
7. If more than one robber is involved, study the nearest one. Do not try to observe all in detail.
8. Comparison of the robber with someone the victim knows aids in recalling details.

After the Robbery

The Problem

The ability of the police to apprehend the robber is dependent on the speed of notification by the victim and the clarity with which he describes the circumstances of the robbery. It is essential that the victim remain calm and collected so that he can take positive and proper action in notifying the police.

Preventive Actions

1. Notify police as soon as robbers leave the premises.
2. Give the exact time the robbers left.
3. Protect the scene of the crime; stop others from disturbing the premises.
4. Hold all witnesses until police arrive.
5. Lock the doors if possible—allow no one in except the police.
6. Do not trust to memory; jot down all information immediately.
7. Do not discuss the holdup with anyone until questioned by the police.
8. Do not touch any articles that may have been touched or left by the robber.
9. Once the police are called, stay on the line so that other vital information can be obtained.
10. The following are kinds of information generally asked by the police radio dispatcher:
    a. location of the armed robbery;
    b. whether anyone was injured;
    c. when the robbery occurred;
    d. the weapon used by the robber;
    e. direction in which the robber went;
    f. description of the vehicle;
    g. description of the robber;
    h. description of clothing;
    i. description of money or article taken;
    j. how the robber carried the loot.

SHOPLIFTING

The General Problem

Although the average shoplifting loss is small (estimated at $3 average per theft in supermarkets), the cumulative effect is quite high. SBA's field survey shows shoplifting losses to all business for the year 1967-68 of $504 million.

While no store, large or small, is entirely immune to the effects of shoplifting, certain types are particularly susceptible. Stores that trade in relatively small, high-priced merchandise experience high total dollar losses and a high average loss per theft. Many convenience goods, such as those stocked in variety stores, experience individually small but numerous losses.

Modern merchandising techniques have contributed substantially to the increasing shoplifting problem. Self-service methods, while cutting costs and increasing sales, have also made shoplifting easier. The merchant should be aware that self-service displays, large areas, and limited sales personnel facilitate shoplifting.

A factor reducing the risk to the shoplifter has been retailer hesitance to prosecute those caught. Fear of suit for false arrest, time-consuming involvement in court proceedings, and concern for retaliation are among reasons for unwillingness to prosecute.

Provable figures on the cost of shoplifting can only be derived from the value of merchandise in the hands of those caught in the act. SBA's estimate of $504 million is based on reports from merchants in a scientifically representative sample of all businesses. However, the total is at best an estimate, since it may be impossible to know the separate components of total shrinkage, which includes recordkeeping errors, unrecorded markdowns, lag in markups, and employee theft.

Another situation, which adds to the difficulty of studying the nature and extent of the shoplifting problem and its cost, is the apparent tendency to ignore the problem, especially among small firms.

It is essential that the merchant take certain specific actions to combat shoplifting. There are six major areas of action:

1. *Establish a system of controls that will help to measure shoplifting.* Inventory control procedures are not well set up in many stores. Without such controls, pinpointing merchandise losses is virtually impossible.
2. *Study types of shoplifters.* By numbers alone, amateur shoplifters far exceed professional. Yet dollar losses to professionals may exceed those to amateurs. The retailer must make intelligent decisions on how to deal

with the juvenile, the vagrant, the narcotics addict, the professional and other types. The greatest increase in shoplifting is among juveniles, both in number of incidents and amount stolen. In the past, retailers have been reluctant to sign complaints against juveniles for fear of alienating their parents, who may be customers of the store. However, storekeepers are being forced to change their position, sometimes for basic survival.

3. *Become aware of how shoplifters operate.* The amazing variety of methods is limited only by the ingenuity of those stealing.
4. *Alert the sales force and give them training in a program to detect shoplifting and apprehend offenders.* The smaller business cannot ordinarily afford its own security staff. The owner, even if knowledgeable about shoplifting methods, is generally not in a position to do much observing. His primary hope lies in an alert and observant sales force, at least minimally trained in defense against shoplifting.
5. *Adopt a physical layout for the store which will discourage shoplifting.* Avoid too many entrances and exits, merchandise too near doors, crowded aisles, display counters that obstruct the view, incorrect placement of cash registers, and other mistakes.
6. *Adopt definite policies for dealing with apprehended shoplifter suspects.* Many states now have some form of "merchant protective act" which permits the retailer to detain a person in reasonable manner for a reasonable period of time if the retailer has reason to believe that the person has taken an item without the intention to pay for it.

Large retail stores generally maintain security personnel, and apprehensions can be made only by these persons. Smaller stores usually lack security personnel and must rely on regular employees to carry out their functions. Most law enforcement agencies recommend that authority to detain a suspect be limited to a select few, possibly the owner and manager.

The pages which follow give a detailed analysis of shoplifting problems the retailer may encounter, and make recommendations on reducing losses to this source.

Types of Shoplifters

The Problem

Shoplifters come from all walks of life. However, they fall into two broad categories—ordinary customers and professionals who steal for a living. Shoplifting can be impulsive, compulsive, deliberate or desperate. It will be helpful to the businessman if he recognizes and understands the individuals who may patronize his store.

The Various Types

The following is a checklist of the more important shoplifter types:

1. *The amateur*—Steals impulsively, a simple desire for an item being the most important motive. Generally is nervous and self-conscious, but exercises caution. Takes food, clothing and many other items for actual use.
2. *The kleptomaniac*—Steals compulsively whenever he or she has the urge. Repeats whenever the urge recurs. Usually nervous and shy. There are very few true kleptomaniacs.
3. *The juvenile*—The majority of the teenage thieves are girls. Usually work in groups, but not necessarily in a formally organized gang. Steal partly for thrills, or to gain status within their group. Usually take merchandise they can use, such as records, clothing, or recreational items.
4. *The professional*—He is a skillful operator and knows all the tricks of the trade in his chosen profession. Dresses, talks, and acts so as to avoid suspicion. Interested in small, high-value items for which he can find an easy resale market. Usually very cautious and does not take unnecessary

chances. Generally steals for a living.
5. *The narcotics addict*—Needs money to procure narcotics. Will take in a brazen manner when desperate. Is sometimes frantic beyond reason, and can be dangerous when an attempt is made to apprehend him.
6. *The vagrant*—Takes from need. Usually steals food, alcoholic beverages, tobacco, and articles of clothing needed for personal use. Often is under influence of alcohol. Is almost always the hit-and-run type.

Shoplifter Signs to Watch for

The Problem

The skilled shoplifter operates with speed and deftness. The typical employee, however, is not trained to observe and recognize the telltale signs that a shoplifter is at work.

A Few Typical Shoplifter Telltale Actions

Persons observed in the following actions and modes of behavior may be suspected as possible shoplifters:
1. a person who leaves the area with undue haste
2. a person who frequents washrooms
3. people who enter the store carrying bundles, bags, boxes, topcoats over arms, brief cases, newspapers, umbrellas, or have an arm in a sling. All these can provide opportunities for concealment of merchandise
4. people who come in wearing heavy outer garments out of season, baggy clothes, full or pleated skirts
5. individuals who have unusual walks, others who tug at a sleeve, adjust socks, rub the back of neck, or are noted in various other unusual actions which might assist in hiding articles
6. customers who reach into display counters or walk behind sales counters.
7. the fussy customer who does not seem to know what she wants and interchanges articles frequently
8. those who do not appear to be interested in articles about which they have inquired
9. the disinterested roamer who waits for a friend or mate to shop
10. the nervous, flush-faced, or dry-lipped person, or the perspirer in a room with normal temperature
11. the person who keeps one hand constantly in an outer coat pocket.

Common Methods Used by Shoplifters

The Problem

Professional shoplifters adopt methods of operation that fit their unique talents and the kind of merchandise they steal. They make use of sophisticated schemes and often employ special devices to aid in stealing. By contrast, the amateur uses crude and obvious procedures such as simply putting the item in his pocket. Employees need to be trained particularly to detect the more skillful operators.

Shoplifter Methods

The following is a checklist of some of the more important methods that may be employed by a shoplifter:
1. Palms small articles—Packages, newspapers, coats, gloves, and other things carried in the hand may be used as aids.
2. Uses umbrellas, knitting bags, diaper bags, large purses, briefcases, paper sacks, booster boxes, and similar devices to conceal merchandise.
3. Has a slit in pocket of his outer garment. Places hand through slit as though hand is in pocket, and carries stolen merchandise in hand which is concealed by the outer garment.
4. Wears a skirt, trousers, or other garment with elastic waistband—or wears "shoplifter bloomers."
5. Tries on a garment, places outer garment over the stolen one, and wears it out of the store.
6. Has hooks on inside of coat, pants, dress, or slip, and uses them in much the same way as a magician.
7. Enters store without jewelry or ac-

cessories and wears or carries items of this type out of the store in the conventional manner.
8. Wears a long outercoat and conceals articles between legs.
9. Walks to an unattended section, or one near a convenient exit, grabs merchandise and hastily departs from the store.
10. Two or more shoplifters may work together as a team. One or more occupy the attention of the clerks; the others, who appear to be just waiting, actually are shoplifting.

Combating the shoplifter

The Problem

Even with the best policies and practices, some shoplifting is bound to occur in any store. However, there should be an effort to hold it to the minimum. The businessman must depend on his employees and on himself to detect shoplifting. All should be trained in alertness and effective detection.

Preventive Actions to Take

The following is a checklist of policies and practices to curb shoplifting:
1. Serve all customers as promptly as possible. Customers approached immediately will appreciate the service. Shoplifters will be served notice that this is not the time or place to attempt theft.
2. When busy with a customer and another enters the store or department, the salesperson should acknowledge his presence by saying something like, "I'll be with you in a moment."
3. The sales person should never turn his back on the customer. This is an open invitation to shoplifting, if the customer is so inclined.
4. Keep an eye on people loitering or wandering in the store.
5. Never leave the store or department unattended. This offers a golden opportunity for theft.
6. If possible, give each customer a receipt for every purchase. This will help prevent shoplifters from obtaining cash refunds for stolen merchandise.
7. Develop a warning system so that all employees can be alerted when presence of shoplifters is suspected. In a small store, this might be a code word.
8. Also develop a procedure for employees to notify the office or some clerical location when they suspect thieves are present.
9. Lock up expensive merchandise that is attractive to shoplifters in a showcase displayed in a position where it can be viewed by more than one salesperson.
10. Do not stack merchandise so high on counters or in aisles that it blocks the view of salespeople.
11. Arrange merchandise so customers must pick it up. If not, a thief can push it off the counter into some type of container.
12. When merchandise is made up of pairs, display only one of a pair.
13. Whenever possible, attach merchandise in some way to make its removal difficult.
14. Keep counters and tables neat and orderly.
15. Place telephone so that sales people can view their sales area while using the telephone.
16. Return to stock any merchandise which was taken out for customer's inspection and was not sold.
17. As a deterrent to shoplifting, keep service fast and efficient, especially when waiting on juveniles.
18. Keep each area clear of discarded saleschecks. Shoplifters may use them as apparent evidence of purchase.
19. To deter till tappers, establish definite cash register procedures: (a) Keep register open while it is actually being used to ring up a sale; (b) close the drawer before wrapping the merchandise; (c) do not allow any customer to distract the cashier

while another person is being waited on; and (d) keep registers locked when not in use.

FRAUDULENT CHECKS

It has often been said that fraudulent checkwriting is the safest crime the individual can engage in—all he has to carry is a loaded pen. It is also one of the most difficult to control.

Check cashing is a service provided by a wide range of businesses, but not necessarily associated with the purchase of merchandise. Checks are cashed as a convenience to customers, whether or not the person has made a purchase and whether or not he even intends to make a purchase. It is this "service" that has made fraudulent checks the widespread problem it is today. Laxity on the part of the businessman, combined with the desire to increase sales volume has been a principal cause of the problem.

As a matter of management, the businessman must make some very basic decisions about permitting the cashing of checks at his place of business:

1. *Whether checks will be cashed for more than the amount of purchase.* In certain kinds of stores, especially food stores, there is a tendency to permit checks to be cashed for more than the amount of purchase. Knowing this, checkpassers make small purchases as a ruse to cash a check. There is no assurance, however, that exact payment will guarantee the check's genuineness. This is particularly true for items that can be easily disposed of at a price satisfactory to the "purchaser." It must be remembered that several days may elapse before discovery that a bad check has been accepted. In the meantime, the person passing the check may have departed the area.
2. *The extent to which checks will be cashed as a service to the store's customers.* It is not uncommon for certain types of stores to cash checks for amounts totaling considerably more than the gross sales of the business. These will include payroll, pension, social security, welfare, and allotment checks. It is uncertain whether the increase in sales resulting from the check cashing service is sufficient to offset the losses experienced by accepting uncollectable checks.
3. *The kind of a procedure the business institutes to insure that the check is genuine and collectable.* Insistence on certain kinds of identification, care in examining the check for accuracy in all detail, and other pertinent factors are admonished to exercise all due caution in cases where he is not personally acquainted with the person desiring to cash a check. If he is not totally satisfied that the check is authentic, he should refuse to cash the check. There is a common fear among many businesses that a stringent check cashing policy will serve to alienate the store's customers, causing a greater loss of business than that experienced through bad checks. No substantive data has as yet been advanced to prove this claim, however.
4. *Whether or not to use protective devices in cashing checks.* There are numerous devices available to the firm cashing checks, based on the principal of photographing the check and the person simultaneously. Firms that produce this equipment assert that the device has a strong deterrent effect on professional checkpassers. Thus, the cost of the equipment is more than offset by the reduction in number of fraudulent checks presented for payment. It is this deterrent quality, more than the possibility of apprehension, that give the device its sales appeal.
5. *A decision on the action to take when an uncollectable check has been accepted.* Technically, uncollectable checks can be classified as due to (a) insufficient fund checks, or (b) defraud. The former category suggests

a significantly different treatment from the latter, yet from the businessman's point of view, both represent a loss of revenue.

In the case of fraudulent checks, the businessman must decide the course of action that he will take and the extent to which he will pursue it. This includes notifying the proper law enforcement agency, signing a complaint, and prosecuting the check passer if and when he is caught.

One of the common complaints among merchants is on the complications arising from bring a check passer to trial: Testifying in court, only to have the person acquitted; the case set aside or postponed, causing undue delay and expense on the part of the merchant. In those areas, however, where a firm stand has been taken by the merchants and the courts, reduction in fraudulent check passing has been noticeable. Laws governing fraudulent checks and strict prosecution of offenders, however, will not arbitrarily reduce bad check losses unless the business that accepts checks establishes a firm policy and adheres to it.

For every careful merchant who refuses to accept a check because it is improperly written, contains abbreviated information, or lacks sufficient identification, many other merchants will cash the same check without hesitation.

Principal Causes of Losses

1. the lack of a check-cashing procedure
2. failure to examine every check
3. failure to record certain information on the check
4. indiscriminate cashing of checks
5. fear that a sale would be lost unless checks are cashed without undue complication.

Establishing a Check Cashing Procedure

The Problem

Most small businesses have no set policy for doing business by check. It is essential that the businessman establish a procedure that will give him the greatest possible protection against bad checks and then hold to that policy without deviation.

Preventive Actions

1. Establish a firm policy regarding the cashing of checks for amounts over the cost of the merchandise (or service).
2. Assign the responsibility of cashing checks for amounts higher than the purchase only to certain employees.
3. Examine *every* check carefully.
4. *Require* a suitable amount of identification.
5. *Require* an address and telephone number of the *maker* and *endorser* of every check.
6. Record identification numbers on the check.
7. Assign the cashing of checks to new or young employees only when under the supervision of experienced employees.

Identification

The Problem

Most checks are cashed in situations where the passer is not personally known to the businessmen. He must rely on some form of proof presented that the passer is the legitimate owner of the check. Usually this decision must be reached quickly under hurried conditions and often by someone not skilled in detecting fraudulent checks. It is doubtful if foolproof identification exists anywhere in the world since all types now in use can be counterfeited.

Preventive Actions

1. Be sure to ask for identification.
2. Identification should be requested if the passer is not *personally* known.
3. The best types of identification now being used include:
   (a) driver's license,
   (b) military or government identification, or
   (c) some airline and national credit cards.
4. Always require *at least one* type of

physical description identification, such as a driver's license.
5. Never accept social security cards, lodge cards, hunting and fishing licenses, employment records, or birth certificates alone.
6. Compare the physical description on the identification to the person presenting the check.
7. Compare the signature on the identification to that on the check.
8. Record the identification number somewhere on the check.
9. Require just as much identification with certified checks, cashier's checks, money orders, government checks, and state warrants as on personal checks.
10. Try not to give the impression of suspicion when asking for identification.
11. A good customer will not object to the need to ask for identification.
12. Be sure all identification used is current.
13. In the absence of a sufficient amount of good identification or none at all, do not cash the check.
14. Be cautious if the person presenting the checks becomes angry when asked for his identification.
15. If an out-of-state driver's license is used as identification, be sure to record the name of the state issuing the license.
16. Never cash a check for a stranger until positive identification is established. Insist on local references, then check them carefully.
17. Do not accept a combination of identification that is too readily offered.
18. Ask for identification that is not ordinarily carried, such as paid utility bills, a tax statement, statement from a retail store.

Examining the Check

The Problem

Most fraudulent checks are passed because the businessman does not take the time to examine the check thoroughly. Establishment of identification is not enough. Care must be taken to be sure the check is correct in *all* respects. Examine every check before it is cashed.

Preventive Actions

1. Examine the dateline.
   (a) The check must be dated.
   (b) The check must not be postdated.
   (c) Establish a policy regarding the cashing of checks over thirty days old.
2. Examine the payee line.
   (a) Be sure name of the payee and the endorsement can be read, that the endorsement is written exactly as appears on the front and includes address and telephone number.
   (b) Be sure payee/endorser identification establishes his identity.
   (c) Do not accept checks with second endorsements from strangers (two-party checks).
3. Examine the digit and written amounts.
   (a) These amounts should correspond exactly.
   (b) Do not accept the check if either shows signs of alteration.
   (c) Do not accept the check if *any* part has been altered.
4. Examine the maker.
   (a) The maker's name should be legible and should include his address and telephone number.
   (b) Beware of checks if any part of the maker's name extends past the space allotted.
   (c) If the maker's name cannot be read, ask him to write it again.
   (d) Beware of titles preceding the maker's name. These are often meant to distract attention from the check, the passer, or his identification.
5. Examine the name of the bank section.
   (a) It should be imprinted on the

check. If not, be sure the name of the bank and city of location are written out completely—not abbreviated.
  (b) Be sure it is a bank and not a savings and loan association or some other kind of business.
6. Examine the endorsement.
  (a) The endorsement should be written exactly as it appears on the payee line on the front of the check.
  (b) The endorsement should be legible and include an address and telephone number.
  (c) If already endorsed, ask that the check be endorsed again in the receiver's presence.

Additional Precautionary Measures

The Problem

Losses in merchandising to bad check artists are a serious and costly problem. Carelessness causes most of them. The best way to keep bad check losses to a minimum is to follow sound and sensible practices, and always use caution and commonsense whenever a check is accepted. The cashing of a check for a stranger should be treated in the same way an unsecured loan would be made to him. Take nothing for granted.

Preventive Actions

1. Beware of checks that have a company name stamped with a rubber stamp or typewritten.
2. Refuse to cash a check that has the word "HOLD" written anywhere on it.
3. Watch out for the "I'm an old customer" routine.
4. Do not be misled if the passer waves to someone, particularly if it is another employee.
5. Beware of the "big name" dropper.
6. It is not good business to cash a check for an intoxicated person.
7. If a check is cashed for a juvenile, be sure he or his parents are well known to the person cashing the check.
8. Never assume a check is good because it *looks* good.
9. Beware of personal checks bearing unusually high sequence numbers.
10. Beware of checks far in excess of the amount of purchase.
11. The person cashing the check should mark it with his initials so that it can later be identified in court, if necessary.
12. Report all check law violators to the proper local law enforcement agency.
13. Follow through with prosecution on all check cases after a complaint has been signed.
14. The businessman should protect his own blank checks, cancelled checks, bank statements, and check protector from theft.
15. Review own canceled checks for unauthorized signatures or altered amounts.
16. Every businessman who cashes checks should be familiar with the laws in his state governing fraudulent checks.

EMPLOYEE THEFT

The General Problem

The great majority of employees are honest and therefore have a stake in minimizing dishonesty by other employees. Furthermore, among those who may be potentially dishonest, a great part will have no opportunity for theft. Nevertheless, there will be employees who succeed in stealing, and it may never be possible to eliminate this leakage entirely. Except in very small stores, merchandise and cash may pass through the hands of several employees, affording temptations and opportunities to steal. The discussion which follows provides checklist of methods of employee theft and proven actions to meet the problem. It must be emphasized that these are not directed against honest employees, but will serve to protect them and their jobs from the debilitating effects of inside theft.

Theft of Cash

The Problem

Cash is handled by sales personnel, cashiers, bookkeepers, and credit department personnel. Employee theft can be discouraged by a management which is alert and enforces a good system of rules.

Method of Theft

The following is a checklist of some of the principal methods of employee cash theft:

1. "Underring" the cash register. The clerk does not give the customer a sales receipt and pockets the money later.
2. Failing to ring up sales. The clerk leaves the register drawer open, puts money directly into the register without ringing up certain sales and takes out the stolen money later.
3. Ringing up "no sale" on the register, voiding the sales check after the customer has left and pocketing the money.
4. Overcharging customers so that cash overages can be stolen.
5. Taking cash from a "common drawer" register.
6. Cashing bad checks for accomplices.
7. Making false entries in store's records and books to conceal theft.
8. Giving fraudulent refunds to accomplices or putting through fictitious refunds.
9. Stealing checks made payable to cash.
10. Pocketing unclaimed wages.
11. Paying creditor's invoice twice and appropriating the second check.
12. Failing to record returned purchases and stealing an equal amount of cash.
13. Padding payrolls as to rates, time worked, or number of employees.
14. Forging checks and destroying them when returned by the bank.
15. Pocketing collections made on presumably uncollectible accounts.
16. Issuing checks on "returned" purchases not actually returned.
17. Raising the amount on checks, invoices or vouchers after they have been officially approved.
18. Invoicing goods above the established prices and getting a kickback from the supplier.

Theft of Merchandise

The Problem

Thefts of merchandise may range from simple pocketing of an item to larger-scale stealing concealed by intricate accounting manipulations. The problem becomes more difficult when there are weaknesses in stock control systems.

Methods of Theft

The more frequent methods of merchandise theft are included in the following checklist:

1. Passing out merchandise over the counter to accomplices.
2. Trading stolen merchandise with friends employed in other departments.
3. Hiding merchandise on person, in a handbag or in a parcel, and taking it out of store at lunchtime, on relief breaks, or at the end of the day.
4. Hiding goods in stairways, public lockers, and corridors for later theft.
5. Taking unlisted packages from delivery truck.
6. Stealing from warehouse with cooperation of warehouse employees.
7. Stealing from stockroom by putting goods on person or in packages.
8. Stealing from returned-goods room, layaway, and similar places where goods are kept.
9. Making false entries to pad inventories so shortages will not be noticed.

10. Giving employee discounts to friends.
11. Putting on jewelry, scarves, or jackets to model; then wearing them home and keeping them.
12. Shoplifting during lunch hour or relief periods.
13. Stealing special "property passes" to get stolen articles out of store.
14. Taking sales slips from training room or supply area to put on stolen goods.
15. Stealing trading stamps.
16. Getting stolen goods through the mailroom by slapping on "customer's own" label normally used to ship out altered goods.
17. Putting "return to manufacturer" label on goods and sending them instead to the employee's own address.
18. Picking up by sales clerk of a receipt discarded by customer and putting it on stolen goods which the clerk keeps or turns in for refund.
19. Intentional soiling of garments or damaging of merchandise so employees can buy them at reduced prices.
20. Printing of own tickets for stolen goods by marking-room employees.
21. Clerks spurring sales with unauthorized markdowns, in order to get kickback from manufacturers.
22. Employees stamping own mail with store postage meter.
23. Shipping clerks sending out stolen goods to their own disguised post office boxes.
24. Smuggling out stolen goods in trash and refuse containers.

Curbing Employee Thefts Through Preventive Measures and an Action Plan

The Problem

The turnover rate in retail stores is high under normal conditions, and increases on a seasonal basis when additional personnel are required. This rapid turnover accentuates the need for well-developed policies to curb employee thefts. Honest employees will not be outraged by efforts to prevent thievery. Meanwhile, the small group of potentially dishonest employees will find it more difficult to steal if they are confronted with an effective system of control and detection.

Preventive Actions to Take

The following is a checklist of actions and policies suggested for curbing employee thefts:

1. Screen new employees carefully, insisting on references that can be checked.
2. See that supervisors set a good leadership example, alerting them to the employee theft possibility.
3. Give special attention to employees who appear to have financial or other personal problems which might increase the temptation to be dishonest.
4. Set up retraining classes for employees who make numerous sales check errors.
5. Check employees who arrive early or stay late when there is no need to do so. (When losses by theft appear very high, consider setting up after-hour "plants." Use honest shopping, for example, Wilmark—for testing salespeople.)
6. Permit no employee to makes sales to himself.
7. Require all employee purchases to be checked in the package room.
8. Restrict all employees to a single exit if possible.
9. Give each sales person his own cash drawer, but permit no one to do final tally on his own cash register.
10. Use care about allowing employees free access to storerooms.
11. If confronted with a theft problem, do not completely eliminate the possibility that relatives of management are involved. They, too, may have

personal problems and resentments which will provoke them to dishonesty.
12. Beware of "theft contamination." Dishonesty, once it gains a foothold in a business, can spread.
13. Have fixed policies about discipline for dishonesty. Failure to take decisive action, or failure to be consistent can have an adverse effect on other employees.
14. Have a good system of controls, including an effective internal audit system.
15. Have a tight control of employee packages. Also check packages found on delivery platforms, loading docks and similar locations, to see if they have correct shipping labels.
16. Use tamper-proof packaging with all price tags inside the wrapping.
17. Have a sound refund system, and be sure it is being followed.
18. Keep valuable items locked up, with the manager in possession of the keys. Also keep all storerooms locked.
19. Keep interchangeable items, such as butter and margarine, in separate cases.
20. Have employees sign for all tools and equipment issued to them.
21. Make all deliveries through the store.
22. Double check all merchandise received at docks to assure that everything paid for is there.
23. Investigate carefully all inventory shortages, remembering it is possible that thieving employees will attribute these losses to shoplifters.
24. Probe all losses, even minor ones, at once, bearing in mind that most embezzlers start with small thefts.
25. Inventory all supplies, equipment, and merchandise systematically.
26. Change all locks and combinations when you change custodial personnel.

Special Theft Problems of Small Businesses

The Problem

In small stores, one person frequently combines all the functions of bookkeeping with the collection and disbursement of funds. Moreover, in a small business, the owner's time is so often taken up with nonsupervisory activities that he is unaware of the extent to which stealing is taking place.

Preventive Actions to Take

Good internal control requires that work be divided, so that there is little opportunity for inside theft without collusion. The following are suggestions specifically for small businesses.

1. All cash receipts should be deposited intact daily.
2. All disbursements should be by check, countersigned by the manager.
3. Each month, the manager should personally reconcile the bank accounts.
4. During the first few days of each month, the manager should receive and open all the incoming mail.
5. The manager should compare all cash receipts with the deposits shown on his bank statement.
6. Someone other than the bookkeeper should do all of the receiving and shipping of merchandise.
7. The mail should be opened by someone other than the cashier or cash receivable bookkeeper.
8. Cash registers should be locked so that employees cannot read the totals.
9. All refunds and sales checks should be numbered.
10. A control should be kept of all salesbooks and all refund books.
11. Rigid control should be maintained on petty cash disbursements.

APPENDIX 5D

A. STONE & CO. SECURITY SYSTEM: AN ANALYSIS OF INVESTIGATIVE PROCESSES

(Prepared by Joseph A. Hobbick)

INTRODUCTION AND METHODOLOGY

The primary purpose of A. Stone's Security Program is to prevent exposure of customers, visitors, employees, and other persons to injury or death. Further, it also is designed to prevent destruction or loss of assets, and to minimize nonpreventable losses which may occur as a result of various emergency conditions.

The specific subject of this field inquiry is the A. Stone Retail store located in Gerard, Penn., a midsized urban center with a population of 200,000 inhabitants. The store itself is situated in a large mall-type complex and does a gross annual business of fourteen million dollars. In terms of structure A. Stone and Co. is a corporate entity, hence, many of the procedures and policy guidelines described in this discourse can be generalized and applied to the entire organization—at least within reasonable limits since regional variations do occur.

The author's field investigation of the A. Stone Security System necessitated the use of the following methodological techniques:

1. open-ended interviews with the store security manager (frequent);
2. focused interviews with the Philadelphia Regional Security Manager (2 one-hour interviews);
3. focused interviews with Regional Investigators (1-hour interview with 3 different investigators);
4. content analysis of intraorganizational memos, documents, and company operating manuals;
5. personal experience while working as a store detective; and
6. informal rap sessions with store detectives and operating personnel.

The investigative function of the A. Stone Security Department is primarily concerned with crimes against property. The following broad policy guidelines set forth the organization's posture with respect to crimes perpetrated against the company by outside agents. It is the policy of A. Stone & Co. to:

1. Report all instances of crime to the proper prosecuting authorities as dictated by its public duty.
2. Make every effort to cooperate with the law enforcement authorities in order that the normal punishment of the law for such offenses may be imposed.
3. Pursue the law enforcement policy in any given case regardless of any fear of unpleasant publicity.
4. Let the judgement in criminal matters rest with the law enforcement authorities, regardless of any unusual or exceptional circumstances.
5. Make no recommendations for probation or for suspension of sentence in any case.

In an effort to adhere to these guidelines, the company security manager submits a "States Attorney's Questionnaire" to the local prosecutors office in a given jurisdiction. Company investigative policies may or may not be altered to meet the requirements of the prosecutor's office depending on the results of the questionnaire. The jurisdiction of company investigative agents is limited firmly to any property or premises owned or leased by the company itself.

STAFFING CONSIDERATIONS
A. Stone & Co.
(Snyder-Mall) Gerard, Penn.

The investigative staff of the Gerard store consists of the following personnel:
1. Security Services Manager (1)
2. Security Services Manager Trainee (2)
3. Store Detective—full-time (3)
4. Store Detective—part-time (3)

Additional noninvestigative personnel:
1. Secretary—part-time (1)

The store *security services manager* is the top-level supervisor at the store level of investigation. His duties and responsibilities include the following:

1. He conducts investigations involving all crimes perpetrated against the company. Refers all evidence to local authorities for their disposition. Represents the company in court when necessary.
2. He supervises the activities relating to the physical security of the premises. For example, the proper functioning of the burglar alarm; fire/water-flow alarm; fire protective equipment, etc.
3. He has responsibility for the effective implementation of the Safety-Loss Prevention Program. Investigates all public liability and motor vehicle accidents and takes appropriate corrective action in an effort to prevent similar incidents.
4. He enforces various controls initiated to strengthen the inventory recovery program.
5. He recommends to management (local) those preventive measures which discourage inventory shrinkage due to shoplifting and employee pilferage.
6. He selects, trains, and supervises store protection personnel, i.e., store detectives and management trainees and regulates and plans their work assignments. In addition, he also evaluates and appraises their work performance. The *methods* and *criteria* of evaluation will be discussed in a later section.
7. He maintains a close liaison with his regional or district security manager (in Philadelphia) to receive guidance on matters pertaining to security problems.

The *security services, manager trainees* in the Gerard store assist the security manager in conducting the above mentioned duties and responsibilities. A student-teacher relationship is maintained throughout the period of understudy. Responsibility delegated is progressive in nature and commensurate with the individual abilities of the trainee. The length of the training period is six months in duration after which the trainee becomes a security services manager and is reassigned accordingly. The nature of reassignment is generally based on the abilities of the new security manager—the common denominator being the *size* of the store commensurate with the *ability* and *progression* of the individual.

The formal selection criterion for security services manager trainees (and ultimately for security services managers) includes the following:

1. The possession of a *four-year degree* from an accredited college or university. There is no policy stipulation as to the nature of the degree required, however, there is preferential consideration given to those candidates holding B.S. or B.A. degrees in the social sciences, criminal or social justice, security administration, and business administration.

   OR

   The educational requirement may be wavered if a candidate possesses experience in the security field sufficient to offset the requirement. In practice, this option is rarely utilized.
2. A candidate must have good references from former employers which demonstrate ability to learn, initiative, leadership, and maturity.
3. A candidate's background must be

free of serious or habitual criminal violations and questionable personal behavior. There are no exemptions for this requirement.
4. A candidate must possess the potential and/or ability to conduct investigations, write complex reports, and communicate with people of all levels in the social spectrum.

The *store detective* occupies the next position in the organizational hierarchy. His duties and responsibilities include the following:

1. He blends himself into the customer population in an effort to detect thefts perpetrated against the company.
2. He makes apprehensions of those suspects committing crimes against company property.
3. He assists in conducting employee investigations—and in some instances (depending on individual talent) he may be allowed to assist in or conduct employee interviews and interrogations.
4. He writes reports and documents his investigations, interviews all customer suspects, processes evidence, and gives relevant court testimony.
5. He signs warrants charging suspects with theft of company property and issues "Notices to Appear" in court.

The selection criteria utilized for store detectives includes the following:

1. He must possess a minimum of a high school education or equivalent. In actual practice, those selected possess considerably more education. For example, in the Gerard store where six store detectives are employed on a full- or part-time basis, two have an Associates Degree (2 years) in Police Science, three possess B.S. Degrees, and one is presently a candidate for a Masters Degree.
2. He must have a background which is free from serious or habitual criminal involvement.
3. He must have demonstrated maturity, good work references, initiative, and communicative abilities. Often, store detectives are recruited for security services manager trainee positions, depending on demonstrated ability and the possession of the minimal requirements described previously. Thus, all promotions are based strictly on merit.

There are no special training courses *per se*. Most training is acquired "on the job," but in a systematic progressive fashion. New investigators are required to study training manuals provided by the Corporate Security Services Department. They are also required to read and study the content of actual reports and documents from investigative files. The new investigator is also required to work the "blinds" (two-way mirrors) until he has a "feel" for working the "floor." The first five apprehensions must be made with the assistance of fellow investigators. In addition, new investigators are encouraged to devote their attentions to juvenile offenders until they gain the experience and expertise necessary to detect the more sophisticated professional "boosters."

In a progressive manner, he gradually learns to "pick-off" prospective violators by perceiving their gestures, eye movements, and body language. He learns the techniques and paraphernalia utilized by professional "boosters" and is able to pick them out. Close supervision and corrective advice bring him along to the point where he can sufficiently perform the duties required of his position. If he does not reach a level of performance necessary to perform the required functions he is quickly terminated.

INVESTIGATIVE PURPOSES

As we can see, the investigative purposes of the total investigations unit are multidimensional. In summary, the purpose of the Security Services Program is to safeguard assets by means of *preventive* security, and to fix responsibility by investigative means for crimes perpetrated against the company, and for serious infractions of policy or procedure. The program is designed to serve management by assuring minimum dissipation of profits by virtue of criminal activity. Investigations are conducted on a highly professional level. The security services manager not only supervises, but is actively engaged in field operations along with his investigators. Coordination and updated information acquisition is accomplished by means of short briefings conducted at the beginning of the investigator's work shift as well as throughout the shift depending on the nature of the information.

POLICY FOR CONDUCT AND ETHICAL CONSIDERATIONS

Security personnel at all levels previously mentioned are required to set a standard of exemplary personal and professional conduct. The rules of conduct adhered to by the investigative staff include the following.

Conduct
Personal

Off-duty activity conforms to socially acceptable standards of behavior. The company states that: "Frequenting establishments that are considered to be operating in violation of the law or which have a sordid reputation, usage of intoxicants to excess or other misconduct conflicts with security standards."

Professional

Relationships with employees are required to be harmonious but businesslike and never reach the point where friendship will interfere with, or preclude, firm enforcement of controls and procedures, or an impartial investigation. Store detectives are not allowed to date retail employees—male or female as the case may be.

Adherence to company policy, standard procedure, and local regulations (if any) is mandatory.

Ethics
General

The A. Stone & Co. Security Services Department code of ethics governs the activity of each security employee, full- or part-time. Each security manager is accountable (to region) for implementation of the code of ethics through all direct and functional areas of his responsibility. If deviation occurs—corrective action is inevitable, and a detailed report is submitted to the regional and corporate security services managers.

Investigative

a. Information gathering and investigative processes are required to be impartial and factual.
b. Policy requires that the integrity or character of a person under investigation not be degraded by innuendo, inference, or direct statement.
c. It is required that any surveillance conducted must not invade personal privacy. For example, observation of the interior of a customer fitting room or restroom is absolutely forebidden.
d. The use of techniques defined as "entrapment" (by the law of a given jurisdiction) are prohibited.
e. Apprehension of a person who is observed attempting to commit a crime, during the perpetration of an offense, or immediately after, must be accomplished without abuse (physical or verbal) and without the use of excessive force.
f. Any rumors or factual information relating to an employee's immoral behavior while *on duty* are required, by policy, to be investigated. However, any such allegations of immoral behavior *off duty* are treated as personal rather than security problems. Requests for an investigation of this nature (on duty) require the approval of the regional security services manager.
g. Any investigative action or technique which violates the law also violates company security standards.

INVESTIGATIVE OPERATIONS

1. There is no well-articulated policy regarding one- or two-man investigative

assignments. However, during high-risk periods, e.g. Friday and Saturday afternoon or evening, no less than two store investigators are on duty.

2. Investigators do engage in selective enforcement. Since they cannot cover the entire store population at any given moment, they confine their activities to selected departments where known thefts run high. There is no formal analysis technique utilized in determining the theft ratio per department—rather, this is surmized and based on the past experiences of the total unit.

   Employee investigations may originate in the same fashion as those cases involving shoplifting, e.g. the employee may be observed stealing cash and/or merchandise. However, most offenses originate from and/or proceed to a point where systematic *operational* analysis is used by investigators. These investigative "tools" may be categorized in the following manner:
   a. examination of cash register variance charts;
   b. examination of cash register tapes and department transaction files;
   c. reviewing patterns of void transactions in relation to variance charts;
   d. scaling the dollar ranges in void transactions and analyzing approval signatures;
   e. investigators also conduct honesty "test shops" of a single and double variety. These may range from the very fundamental, to a more complex type involving many hours of tedious work and patience.

3. Investigators do not use formal checklists while conducting investigations, *except* during the interview process.

4. Investigators utilize a wide range of instrumentation devices and technological equipment during the investigative process.
   a. Cameras and related paraphernalia are approved investigative aids.
   b. Closed-circuit television is approved physical security and investigative equipment.
   c. The use of tape recorders is permissible in departments designated by the corporate security services manager to record informational interviews and interviews relating to criminal activities.
   d. Electronic eavesdropping devices such as microphones, transmitters, induction coils, or phone taps are *not* used.
   e. The polygraph is utilized in some cases of *internal* crime or serious misconduct after all other investigative techniques have been exhausted.
   f. The use of observation mirrors and surveillance booths ("blinds") is authorized *except* for installations that invade personal privacy.
   g. Investigators have ready access to two-way radio communications systems, but rarely use them.
   h. Chemical powder, pastes, and vapors, magnifying lens, portable relay alarms, and similar aids are used at the discretion of the investigator.

5. The primary responsibility of collecting, tagging and processing, and storing evidence rests with the individual investigator. The security services manager, however, has the ultimate responsibility for maintaining storage facilities where evidence is kept. In the Gerard Store facilities are very clean, orderly, systematic, and constantly inspected. Only *security employees* have access to the evidence room and its contents. There is a concerted effort to maintain chain of evidence procedures which more than comply with local court standards. Evidence is held for *three* months in juvenile cases and a minimum of *six* months in adult cases before it is returned to retail stock. Transfer slips must be signed and recorded in all cases.

INTERVIEWS

The following policy, both written and unwritten, is strongly adhered to in all cases where security personnel have the occasion to interview shoplifters, employees, and persons suspected of check fraud or credit fraud. They do *not* apply to interviews conducted with accident witnesses, victims, or similar persons.

1. Persons under suspicion are interviewed in an unlocked office with the door slightly open and with the subject seated between the investigator and the door, *except* where the person has been apprehended during the commission of a crime, will be prosecuted, and his behavior indicates the probability of an escape attempt.
2. Unauthorized persons are not permitted to listen in on interviews. Within the context of shoplifting, credit fraud, and check fraud, "authorized" personnel would include: store detectives, security services trainees, the security services manager, and any accomplices to the criminal act. However, in cases involving employee theft, "authorized" persons include: the regional security services manager (in cases of staff involvement only), the regional investigator, or the store security services manager. In some rare cases a store detective may be allowed to participate in such an interview depending on his special skills in this area. The latter requires the verbal approval of the regional office in Philadelphia.
3. One particular written policy firmly stipulates that suspects are *not* to be interviewed in a hotel or motel room. Female suspects are not interviewed by male investigators on company premises after hours, in an automobile, or in their homes *except* in cases of extreme urgency and then only with another female present, which must be a company representative or police woman if circumstances apply.
4. Suspects are allowed the use of the telephone, rest room, and are allowed to terminate the discussion and leave the interview room whenever they express a desire to do so. The latter stipulation applies only to *employee* suspects who will *not* be turned over to police authorities.
5. In the case of employee interviews, suspects are not to be detained past their reasonable lunch period or past their normal or scheduled quitting time, unless the prosecutor has advised filing of charges and the police are on the way. In the event that circumstances preclude adherence to this policy, employees *are* paid for any time spent in an interview which exceeds their work schedule.
6. It is considered unethical to threaten or coerce, directly or by inference, a person being interviewed; or to make a promise which is contrary to company policy on handling criminal matters. Any promise made, which is within the scope of the investigator's authority, such as advising the prosecutor that a person was cooperative, must be fulfilled if the person *did* in fact cooperate.
7. Policy requires that statements given by suspects and witnesses must not be dictated by the investigator. They are also witnessed by at least two persons, a copy must be given to the person submitting the statement when required by law (this is not the case in the Gerard P. A. Store). *Guidance* relative to the opening and closing paragraphs of a written statement can be given.
8. Interviews are not conducted for an unreasonable length of time. A random sample of the files indicated that the *average* time for shoplifter interviews was forty-five minutes; while the average interview time for employee suspects was two hours and forty-five minutes.

INVESTIGATIVE REPORTS

Shoplifting

Shoplifter reports are the least difficult for investigators to prepare. More of these reports are written by security services personnel than any other security report.

On this report, all items are recorded; the time of the offense, time of apprehension, time police are notified, time of police custody, etc.

The narrative of the shoplifting report includes the following information: that the investigator did or did not observe the suspect enter the department; what he did with the merchandise; a description of the merchandise by article and price and color. If the suspect concealed the article in his clothing, exactly where was it concealed? Also, where the subject exited the store and the fact that he exited without payment is noted. Information as to the subject's attempt to fight with or escape from the investigator is also included in detail. Likewise, threats, verbal abuse by the suspect, acknowledgement of the theft, etc. are included in the narrative of the report.

Shoplifting Reports

1. Shoplifting reports are segregated into "juvenile" and "adult" categories. A suspense file is maintained only in the adult category. Progress checks are conducted formally by the security services manager on a weekly basis; and *informally* by investigators interested in their own cases. Progress indicators are "open" and "closed." Disposition is always noted in writing on a "closed" report.
2. All investigators complete shoplifting reports in their own handwriting. They are *never* typewritten.
3. The security services manager audits all shoplifting case reports to determine clarity, completeness, and neatness. Corrective action is taken to bring substandard reports up to par. One particular incident came to the author's attention wherein the manager insisted that a store detective rewrite a report six times in order to meet acceptable standards of reporting.
4. Case files are numbered starting with GC01, meaning *Gerard Case one,* and continue sequentially to no preset number. They are stored in a secure filing cabinet in groups of twenty in manilla folders. There is no set policy for purging files.

Employee Theft Reports

Employee theft reports are prepared in a similar manner. However, they are more detailed and always typewritten. Further, they are filed separately and suspense indicators are "open," "pending," "closed." Open cases are those which are still being documented and have not been cleared through regional channels. Pending cases are those which are awaiting restitution by the named employee. Closed cases indicate the restitution has been made or that prosecution and subsequent disposition has taken place.

Check Fraud and Credit Fraud Reports

These reports are handled in the same manner as employee theft reports with minor variances relative to the nature of the offense. They are stored in separate files.

Activity Reports

Investigators are not required to submit activity reports except at the regional level. Security services managers do submit comprehensive unit reports on a monthly basis in all categories mentioned above.

INVESTIGATOR EVALUATION

In terms of evaluation, the present system does not *require* that periodic reports be submitted. However, the security services manager does, on occasion, make use of such reports for a number of reasons. First, he may desire to secure a pay raise for one or several of his investigators. If this is the case, he will fill out a standard form used by the personnel department to justify the proposed raise. This particular form does not ask the evaluator to evaluate the investigator concerning specific job related activities. Rather, it is used for a wide range of employees and contains only broad categories of reference such as appearance, motivation to work, productivity, ability to perform required tasks, etc. A second instance that presently necessitates the use of this standard form would involve a situation where the security services manager is not satisfied with his investigator's performance and desired to justify his dismissal.

SUMMARY AND EVALUATION OF SYSTEM

As an all-around investigative and protective system, the author was thoroughly impressed with the A. Stone and Co. security services system. As with any investigative unit, there is always the need for continuous improvement in all policy areas. The following itemized list contains some tentative suggestions for areas where deficiencies may exist.

ITEM 1: MANAGEMENT TRAINING. A standardized format should be developed to provide for internal consistency in this training area. Presently, this policy is determined solely by the individual in charge at any given location and some problems do arise. For example, unit performance is determined by the statistics compiled in a number of functional areas. Hence, the local administrator is often tempted to use the talented trainee to build statistics rather than having him learn the operating procedures (retail) of the company. A systematized training program would alleviate this tendency where it exists.

ITEM 2: RESTITUTION FOR EMPLOYEE THEFT. There is no aggressive program or policy for conducting effective follow-up in this area. The threat of "prosecution" versus "restitution" is largely a hollow one and many cases are pending that should have been reviewed by the States Attorneys Office. A fresh approach is needed in this area.

ITEM 3: PURGING OF OUT-DATED REPORTS. There is no policy guideline in this area. Old investigative reports should be purged on a regular predetermined basis.

ITEM 4: INVESTIGATOR ASSIGNMENTS. As previously mentioned, there is no set policy in this area. Systematic analysis of available data could provide a basis for assignments in a store of this size, e.g. time of day, day of week, vulnerable departments, specific items of high-risk merchandise, and seasonal adjustments.

ITEM 5: CHECK LISTS. Formal investigator check lists are recommended, especially for the more complex areas of investigation, e.g. employee theft and credit and check fraud.

ITEM 6: INVESTIGATOR EVALUATION. The author recommends the following as an alternative to the present nonspecific method.

Evaluation and appraisal will be conducted on all investigators at the end of their first thirty days of employment, three months, six months, and first annual employment date. Thereafter, the investigator will receive an appraisal on an annual basis.

In order to assist the security services manager in the preparation of an accurate, objective appraisal, the following guidelines should be followed.

PERSONAL CHARACTERISTICS

Judgement
Reliability
Punctuality and Attendance
Personal Appearance
Initiative and Aggressiveness
Cooperation with Fellow Employees
Ability to Evaluate and Make Decisions
Leadership
Effective Use of Potential
Ability to Accept Direction and Instruction
Ability to Accept Criticism
Communicative Abilities

EVALUATION OF WORK RELATED ACTIVITIES

Rate of Apprehensions
Amount of Apprehensions
Consistency in Apprehension Performance
Rate of Recovery
Amount of Recovery
Interrogation Techniques
Contribution to Employee Investigations
Knowledge of Company Security Procedure

Ability to Comply with Company Security Procedure

Knowledge of Company Safety Standards

Knowledge of Store Operations and Procedures

Quality of Security Case Reports

Accuracy and Thoroughness of Case Reports

Prompt and Accurate Handling of Reports

Accuracy and Effectiveness of Court Testimony

Ability to Deal with Emergency Situations

Understanding of Job and Necessary Related Work

Knowledge of Procedures, Methods, and Techniques

Effective Application of Procedures, Methods, and Techniques

Ability to Work Without Close Supervision

Ability to Deal Effectively with Store Management

Ability to Deal Effectively with Nonsecurity Employees

Ability to Deal Effectively with Law Enforcement Agencies

Ability to Deal Professionally with Apprehended Suspects

Interest and Contribution to Total Department Performance

This criteria should be utilized and graded with the following letter symbols:

    A—Excellent
    B—Above Average
    C—Average
    D—Below Average
    F—Unsatisfactory
    U—Unable to Evaluate

APPENDIX 5E

BURGLARY PREVENTION CHECKLIST FOR BUSINESS PLACES*

Exterior

1. Are all of the points where a break-in might occur lighted by street lights, signs, or your own "burglar" lights?
2. Have you protected blind alleys where a burglar might work unobserved?
3. Are piles of stock, crates, or merchandise placed so as not to give burglars hiding places?
4. Are windows protected under loading docks or similar structures?
5. Have the weeds or trash adjoining your building been cleared away?
6. If a fence would help your protection, do you have one?
7. Is your fence high enough or protected with barbed wire?
8. Is your fence in good repair?
9. Is your fence fixed so that an intruder cannot crawl under it?
10. Are boxes, materials, etc., that might help a burglar over the fence placed a safe distance from the fence?
11. Are the gates solid and in good repair?
12. Are gates properly locked?
13. Are the gate hinges secure?
14. Have you eliminated unused gates?
15. Have you eliminated danger from poles or similar points *outside* the fence that would help a burglar over?
16. Have you protected solid brick or wood fences that a burglar could climb and then be shielded from view?
17. Do you check regularly to see that

---

* Adapted from "Burglary Prevention Checklist for Business Places." Courtesy of Professor Richard L. Holcomb, *Protection Against Burglary*, University of Iowa, 1953, pp. 49-52.

your gates are locked?

18. Do you regularly clean out **trash or** weeds on the outside of your fence where a burglar might be concealed?

Doors

19. Have you secured all unused doors?
20. Are door panels strong enough and securely fastened in place?
21. Is the glass in back doors and similar locations protected by wire or bars?
22. Are all of your doors designed so that the **lock cannot be reached by** breaking out glass or a lightweight panel?
23. Are the hinges so designed or located that the pins cannot be pulled?
24. Is the lock bolt so designed or protected that it cannot be pushed back with a thin instrument?
25. Is the lock so designed or the door frame built so that the door cannot be forced by spreading the frame?
26. Is the bolt protected or constructed so that it cannot be cut?
27. Is the lock firmly mounted so that it cannot be pried off?
28. Is the lock a cylinder type?
29. Do you remove valuable merchandise from unprotected display windows at night?
30. Have you considered the use of glass brick in place of some windows?

Other Openings

31. Do you have a lock on manholes that give direct access to your building or to a door that a burglar could open easily?
32. Have you permanently closed manholes or similar openings that are no longer used?
33. Are your sidewalk doors or grates locked securely?
34. Are your sidewalk doors or grates securely in place so that the entire frame cannot be pried up?

Walls

35. Are your walls actually as solid as they look? Have you eliminated insecure openings in otherwise solid walls?
36. In checking walls, have you paid particular attention to points where a burglar can work unobserved?
37. Is your roof either secure or protected by an alarm system?
38. Have you eliminated weak points in your walls where entrance could be gained from an adjoining building?

Safes

39. Is your safe designed for burglary protection as well as fire protection?
40. Is your safe approved by the Underwriters Laboratories?
41. If your safe weighs less than 750 pounds, is it fastened securely to the floor, the wall, or set in concrete?
42. Is your safe located so the police can see it from outside?
43. Is your safe lighted at night?
44. If you have a vault, are the walls, as well as the door, secure?

Alarms

45. Have you investigated the use of a burglar alarm system?
46. If you have a system, is it fully approved by the Underwriters Laboratories?
47. Was it properly installed by competent workmen?
48. Is your burglar alarm system tested regularly?
49. Does the system cover your hazardous points fully?
50. When your building was remodeled, was the burglar alarm system remodeled, too?

Security Officer

51. Did you investigate your security officer when you hired him?

APPENDIX 5F.1

STORE MANAGER TALK OUTLINE*

SHOPLIFTING

A. The Shoplifting Problem
1. How big is the shoplifting problem and who does it?
   a. One out of twenty shoppers is probably a shoplifter (from nationwide surveys).
   b. The novice is the most frequent offender and takes items for personal use.
   c. The professional shoplifter is the most costly offender and takes high-value items for resale. Drug addicts very often fall into this category.
   d. The kleptomaniac has a psychological compulsion to steal and is very rare.
2. What items do shoplifters go for most?
   a. Cigarettes are taken more than anything else.
   b. Meat, health and beauty aids, and candy follow in that order.
3. When do shoplifters operate?
   a. During busy times when employees are occupied up front.
   b. When employee or management apathy to the problem is apparent.
   c. The days and times to be most watchful are Fridays between 3 P.M. and 6 P.M., and Saturdays around noon. Sundays are also frequent when store personnel is at a minimum.
4. How do shoplifters operate?
   a. Concealment of items in clothing and purses.
   b. Changing containers, lids or labels with lower priced merchandise.
   c. Transferring merchandise to bags carried in from other stores.
   d. False bottom boxes (booster box), hooks inside coats, etc., are used by professionals.
5. How can a shoplifter be recognized?
   a. A shoplifter will usually pay more attention to the people in the store than the merchandise. ("People shoppers")
   b. Bulky or unseasonal clothing to conceal items is a clue.
   c. Women shoplifters often carry their billfold separate from the purse to avoid opening the purse at the checkstand.
   d. Loitering around high-value merchandise is an indicator.
   e. Open packages or large purses in the top of the bascart are used to conceal merchandise.

B. Prevention of Shoplifting
1. Shoplifting can be prevented through vigilance, interest, and involvement of all store personnel.
2. What can a checker do to prevent shoplifting?
   a. Check the bottom of the bascart.
   b. Handle magazines by binding edge. This will cause concealed items to fall out.
   c. Beware of customer who insists on holding packages in her arms. Be firm but gentle in asking her to put these packages with other purchases.
   d. Know what items are most often pilfered in your store and their prices.
   e. Ask price checks on items that you suspect of label switching.
   f. Be alert for customers that request refund without register tape. (Use 1082 form for maximum information.)
   g. Notify the manager in all cases of *suspected* or known shoplifting.
3. How do department heads, stock clerks, and baggers aid in shoplift-

---

* Courtesy Kroger Company, Security Department.

ing prevention?
a. Scan aisles as you work, be alert for the customer that is observing people instead of merchandise.
b. Offer assistance to the "people shoppers." It is extremely difficult to steal when an employee is attentive to shoppers needs.
c. Be alert for:
   (1) loiterers,
   (2) customers carrying open packages or large purses open or in the top of the bascart;
   (3) customers with heavy or bulky clothing during warm weather;
   (4) customers who keep returning to a particular aisle and apparently are not selecting items on each visit to that aisle;
   (5) notify the manager in all cases of *suspected* or known shoplifting.

C. Apprehension and Prosecution of Shoplifters
1. What should be done if a shoplifter is observed?
   a. First, notify the management. (A prearranged signal is helpful.)
   b. Keep him in sight at all times and know WHAT has been taken and WHERE it is concealed.
   c. Stop him after he has passed the last place to pay and he still has the goods in his possession. (at the "Out" door is best)
   d. When you approach the suspect, say (for example) "There is a private matter we would like to discuss. Please come to the office with me."
   e. Question the suspect in private WITH A WITNESS PRESENT. (A woman witness, if the suspect is female.)
   f. DO NOT say "steal," "shoplift," or similar words.
   g. DO NOT attempt to physically hold the subject and DO NOT search. The police will do that.
   h. Get a written confession, if you can, on the official form.
   i. Write down exactly what happened and attach it to the report.
   j. Call the police and make a complaint on all cases where the merchandise is valued over $1.00, the suspect is uncooperative, does not sign the confession, or you expect trouble.
   k. Forward a report to Division Security on all shoplift apprehensions.

THE BEST WAY TO REDUCE LOSSES BY SHOPLIFTING
IS TO PREVENT THEM FROM OCCURRING

5F.2

**AGENDA FOR
SHRINK CONTROL MEETING***
(one day)

Informal meeting, please participate

9:00 Store Manager—The problem
   Results YTD compared to last year
   Display of items picked up from store
   Slides of material returned in salvage

9:15 Security Director—Causes of shrink
   —group participation
9:45 Stock Room Manager—Receiving Merchandise
   Backroom processing
   Stocking procedures
   D.S.D. Control
10:15 Coffee Break
10:30 Asst. Store Manager—Checking Order Register
11:15 Asst. Manager—Credits on Damage or Shortages
11:30 Floor Manager—Pricing and Inventory Control

---

* Courtesy Kroger Company, Security Department.

        Proper price marking
        Price changing
        Price checking throughout the store
        Order guide—surveys, scratches, etc.
12:00 Lunch
        Cafeteria
1:00 Comptroller—Known Loss and Gain Reporting
1:30 Acct. Manager—Front End
        Management control methods
        Attitude of checkers
        Handling of cash items
        Refund and error sheets
        Employee purchase policy
            Carry home orders
            Product consumed in the store
        Handling mispriced and damaged product
2:00 Security Director—Store Security
        Shoplifting (film & discussion)
        Shifting the emphasis from apprehension to *prevention*
        Employee theft
        Protection of property
            Audit method—slides
            Management of employees
            Physical security
2:45 Director of Training—Training for improved loss control performance
        Getting the message across to employees—film

        Employee attitude and morale
        Cooperation with unions
3:30 Corporate Staff—A Store Experience
        Method of Approach
            Price checking
            Rotation
            Pulling samples
        Activities
            Teamed up to check prices
            Covered entire store to locate areas of shrink
            Separate merchandise into groups
            Total dollars involved
        Results
            Graphic depiction of weaknesses in management
            Action taken to resolve problems
3:55 Store Manager—Getting the Job Done
        Shrink control is a management responsibility
        The store manager is primarily responsible for the entire operation
        Head grocery clerk is responsible for the grocery dept.
        Be more aggressive
        Be assured of store manager backing
        Held responsible for shrink results
        Be "hard nosed" about shrink

APPENDIX 5G

STATUTORY ENCOURAGEMENT OF MERCHANT EFFORTS TO APPREHEND SHOPLIFTERS*

JOHN D. DONNELL[†]
JOHN E. D. PEACOCK[‡]

THE SHOPLIFTING PROBLEM

One shopper in fifteen entering four eastern department stores stole merchandise of an average worth of $5.26 according to a recent study.[1] In this study more than 1,600 shoppers were selected at random as they entered the store and then kept under total surveillance. Only one of the 109 observed shoplifters was apprehended by a store detective although each store had a large security force. These figures would suggest that only a minute fraction of store thefts are reported. This might account for the substantial differ-

---

* Reprinted courtesy of *American Business Law Journal* and John D. Donnell.

† Professor of Business Administration, Indiana University.

‡ Indiana University 1971, now employed by L. S. Ayers and Co., Indianapolis, Indiana. The authors wish to thank Robert W. Lauritzson for assistance.

ence between the average observed theft in the study and the average theft reported to the Federal Bureau of Investigation, which was $26 in 1970.[2]

The findings of the department store study indicate both how serious the shoplifting problem is and how difficult it is to estimate the magnitude of total loss to merchants from this source. Individual retailers are very reluctant to disclose figures on their inventory shrinkage, which includes internal thievery and record keeping errors as well as shoplifting. The National Retail Merchants' Association estimates that inventory shrinkage for American retailers is running over $8 million per day or $3.5 billion each year.[3] A Pinkerton survey in 1970 found shoplifting to account for 45.9 percent of retail shrinkage.[4] Applying that proportion to the NRMA estimate of shrinkage would put the overall loss from shoplifting in the United States at $1.6 billion per year.[5] The Pinkerton study found shrinkage losses ranging from .8 percent to 7.5 percent of sales, a level that is frequently higher than the percentage of net profit enjoyed by the retailer.

The FBI figures show that reported shoplifting cases have increased 221 percent from 1960 to 1970.[6] As the shoplifting problem has increased in magnitude, retailers have turned to their legislatures for help in dealing with it. These statutes take several different approaches, and within each type they vary considerably in wording. The most common type seeks to protect the merchant from damage suits when he apprehends a suspected shoplifter who is acquitted or never prosecuted.[7]

This article will first discuss the risk of tort liability the merchant who apprehends a suspected shoplifter runs under the common law. Next, we shall review the legislation passed to reduce this risk. Then we shall discuss an empirical study we made to determine the effect upon merchants' behavior of one of these laws. Since the owner is liable for the acts of his employees under the doctrine of *respondeat superior,* the term merchant will be used throughout the article although the acts leading to suit are usually committed by employees.[8]

TORT LIABILITY AT COMMON LAW

A damage suit is not unlikely if the merchant discovers in his confrontation of a shoplifting suspect[9] that his suspicions were unfounded or he fails to take steps to prosecute, or the police officer he calls or the prosecuting attorney recommend against his filing of the affadavit or complaint necessary to start prosecution. Suit is even more likely if the suspect goes through the embarrassment of being booked and then efforts at prosecution are aborted or he wins an acquittal. Some shoplifting suspects, apparently operating on the theory that the best defense is an offense, begins threatening suit the moment they are detained by the merchant. In fact, merchants are occasionally met with what appears to be an entrapment routine. A customer handles the merchandise in a very suspicious manner and appears to be slipping goods into a bulging pocket or the like in full view of the manager or a clerk. When he is detained, no stolen goods are found and an investigation discloses nothing is missing and the suspect forcefully threatens suit. Such actions may be diversionary tactics to protect the activities of an accomplice or an attempt at blackmail. There is no way of knowing how frequently either ploy is successful.[10]

When successful, a tort action growing out of a confrontation of the suspect by the merchant may result in substantial damages. The highest amount of compensatory damages affirmed upon appeal which our research uncovered was $25,000.[11] Since intentional torts are involved, punitive damages are often awarded also. In another recent case the total award was $40,000, of which $30,000 was granted as punitive damages.[12]

The basis for the suit may be one or more of at least four distinct torts. They are: false imprisonment or false arrest, slander, malicious prosecution, and, if the detention involves touching by the merchant of the person or purse or parcel of the suspected shoplifter, assault and battery may be alleged. The following discussion is based upon the common law. The common law has been changed by statute to some degree in all states but California, which by court decision granted merchants the protection which many other states granted by statute.[13] The effect of these statutory changes will be discussed later in this article.

FALSE IMPRISONMENT

If a merchant confronts a shoplifting suspect and requires him to wait for the arrival of a police officer or detains the suspect for interrogation or to be searched for suspected stolen articles, he risks liability for false imprisonment. The tort consists of intentional confinement, without justification, of the plaintiff within boundaries fixed by the defendant.[14]

Prosser treats false arrest and false imprisonment as identical torts.[15] However, occasionally a plaintiff will plead both on the basis of the same incident, usually when a merchant holds a shoplifting suspect until police arrive and then the confinement is continued.[16] When the confinement is by a police officer the tort may be called false arrest, otherwise it is more likely to be called false imprisonment.

When a merchant gets a police officer to stop and question a suspected shoplifter the merchant is not assured of protection from liability for false imprisonment. A distinction is made between the situation where the merchant merely gives information to an officer who utilizes his own judgment in determining whether to apprehend a suspected shoplifter[17] and where the merchant instigates the apprehension.[18] In the former case the merchant is not a participant in the arrest and, therefore, is not liable for false arrest, although he might be liable for slander.

The advantage to the merchant in relying on a police officer, even if he is employed for store security duty, lies in the broader authority to arrest without a warrant given the police officer as compared to that granted to an ordinary citizen in most states. Generally, police officers may arrest without a warrant if they have reasonable grounds to believe a felony has been committed and that they have the right person, while the right of a private person depends upon the crime in fact having been committed and his having reasonable grounds for belief that the person arrested is the criminal. Most shoplifting involves only a misdemeanor, and for those the general rule is that neither the private person nor the officer may arrest without a warrant.[19]

An element of the tort is the use of force or threat to restrain freedom of movement. An alternative is the assertion of legal authority in making an arrest and perhaps even the appearance of legal authority is sufficient. Recovery for false imprisonment was granted in a case involving teenage girls who followed the requests of a uniformed security officer to open a train case and return to the store even though he made no threat nor physically blocked the girls' freedom of movement.[20] Generally, a mere request to the suspect to return to the office for questioning[21] or to open a purse for inspection[22] and which is acceded to voluntarily does not constitute false imprisonment. However, a sufficient restraint was found when a woman remained in the store to recover her purse which had been impounded by the merchant.[23]

There is no false imprisonment if the suspect is unaware that he is being held. Therefore, if the merchant uses a pretext, such as confirming a price or verifying a tendered check, to detain the suspect while he investigates or until the arrival of a police officer, there is no false imprisonment.[24]

Although tort law grants an owner of property a privilege to defend it against theft and even to recapture his goods if they have been wrongfully removed from his premises,[25] this privilege is closely circumscribed and frequently fails to protect the merchant in his efforts to thwart shoplifters. The privilege is dependent upon the fact of wrongful taking. If the suspicion of theft turns out to be erroneous, the privilege disappears and the merchant may be liable.

In most states, prior to statutory change, detention of a suspected shoplifter was at the merchant's peril. His good faith and the reasonableness of his suspicion did not protect him. However, courts in a few states, notably Massachusetts[26] and California,[27] have held that if a businessman has probable cause for believing that a person is intending to avoid payment for a service or goods, he has a qualified privilege to detain that suspect. So long as the detention is for the purpose of ascertaining the facts and is done in a reasonable manner and for a reasonable time, the detention is privileged, and although he is mistaken, the merchant will not be liable. The leading shoplifting case recognizing this right is *Collyer v. S. H. Kress Co.*,[28] decided in 1936 by the Supreme Court of California. This rule has been adopted by the *Restatement (Second) of Torts,* which expresses it as follows:

> One who reasonably believes that another has tortiously taken a chattel upon his premises, or has failed to make due cash payment for a chattel purchased or services rendered there, is privileged, without arresting the other, to detain him on the premises for the time necessary for a reasonable investigation of the facts.[29]

SLANDER

If the merchant accuses a suspect of shoplifting he may be liable for slander if the accusation is untrue. Since shoplifting is a crime, such an accusation is defamatory *per se.* Therefore, damages are presumed, and the plaintiff is entitled to a recovery even though he cannot prove actual injury.[30]

A direct accusation is not necessary to constitute slander. In fact, almost any question beyond a cash register clerk's inquiry, "Is that everything?" may be viewed as imputing theft. For example, the Louisiana Court of Appeals has held that "Wouldn't you like to pay for the merchandise that you have in your handbag?" was defamatory. It declared:

> The entire incident suffered by the plaintiff, taken as a whole, was slanderous. The import of the statements was that plaintiff was endeavoring to steal the firm's merchandise.[31]

In another case, the same court said:

> Although Cummings did not specifically accuse plaintiff of committing theft, we think his request that she open her purse and his subsequent inquiry as to what she had done with the bottle of Loving Care constituted an accusation that she had committed that offense.[32]

The Louisiana court does not stand alone. The Supreme Court of Appeals of Virginia held the statement of a security guard who called to two girls as they left the store, "Hey, young ladies, let me see what you have in that suitcase," sufficient to support a jury's finding that they had been falsely accused of larceny.[33]

The question as to whether the words used in dealing with a suspected shoplifter are defamatory is generally determined by the court as a matter of law. However, if the words are ambiguous, or subject to a special meaning in the context in which used, then the sense in which they were understood is an issue of fact for the jury.[34]

A requirement of the tort of slander is publication; if no third person is aware of or understands the accusation, there is no action. However, if an employee of the store[35] or the spouse[36] of the accused overhears the accusation, this constitutes publication. Publication has been found even when there was no testimony by anyone who actually heard the words, only testimony that there were persons present who could have heard[37] or perhaps understood the accusation from pantomime.[38]

At common law truth is a complete defense to a suit for defamation,[39] but even if the merchant is mistaken in his accusation, he has a qualified privilege to protect his own interests.[40] The privilege only goes to statements reasonably necessary for the privileged purpose, here the prevention of theft. Therefore, the privilege may be lost if the defendant makes the accusation in a place or manner likely to attract the attention of others.[41] Actual malice also defeats the privilege but since the accused shoplifter is usually a stranger he seldom can prove this.[42]

MALICIOUS PROSECUTION

If the merchant institutes a criminal proceeding against the suspected shoplifter and the accused is acquitted or the proceeding is terminated, he may become liable for malicious prosecution. As in the case of false arrest, if the merchant merely provides information and leaves the initiation of action to the discretion of an officer, the merchant will not be liable. However, if he gives information he knows to be false he may be vulnerable.[43]

Want of probable cause is a requirement of the tort of malicious prosecution. Therefore, a finding that the defendant had probable cause to believe the plaintiff committed the crime for which he instituted the criminal proceeding relieves the defendant of liability.[44] The burden of proof is on the plaintiff to establish want of probable cause and where the facts pertaining to the issue of probable cause are clear, the question is for the court to determine.[45] For example, the court found probable cause where a reliable clerk reported to the manager that she saw a mother take a jacket from a rack and put it on her six year old daughter before they left the area, although, in fact, the daughter had carried a jacket identical to those on the rack into the store.[46] If a merchant places all the facts before his lawyer and the lawyer then advises prosecution, this in itself will constitute probable cause.[47]

As the name of the tort implies, malice on the part of the defendant must also be proved by the plaintiff.[48] Although malice is said also to be a requirement for defamation, something more—malice in fact—is required in malicious prosecution. It need not be spite or ill will, but if the primary purpose of the prosecution is to recover property or to collect a debt, that is sufficient to constitute malice.[49] A lack of probable cause permits drawing the inference of malice so that this issue can go to the jury. Unlike the determination of the probable cause issue, the issue of malice is treated as a fact issue and is for the jury.[50]

ASSAULT AND BATTERY

In attempting to detain a suspected shoplifter, the merchant may become liable for assault and/or battery. No physical harm need be done; merely offensive contact is sufficient.[51] Of course, offensiveness is not likely to be an issue in shoplifting cases. However, a mere tap on the shoulder to gain attention as a prelude to a request to accompany the merchant to the office would appear to be within the bounds of normal personal contact and not considered offensive to an ordinary person.[52] Searching a suspect's pockets[53] or grabbing a woman's handbag constitute battery.[54]

Consent, of course, is a defense. Another defense is the privilege to defend possession of property or to recover possession promptly.[55] However, as discussed under false imprisonment, the merchant who relies upon the privilege to recover acts at his peril; if he is in error as to the facts and the suspect doesn't have stolen goods, the merchant will be liable.[56] Furthermore, he cannot use unreasonable force.[57]

Our search revealed few shoplifting cases where assault or battery were the basis of suit. This is in part due to the fact that the action is usually part of the threat or force used to prevent escape[58] or the search for goods believed to be stolen[59] and therefore is part of a false imprisonment.[60]

SUMMARY

Under the common law merchants found their rights to defend their goods from shoplifters and to recover them were dangerous to exert. In most states, even if they acted with care and upon reasonable belief, detention of the suspected shoplifter could result in liability for substantial damages for false imprisonment or assault or slander, or all three, if they could not prove the suspected theft. The individual items stolen by most shoplifters are of quite low value. Therefore, the odds against the merchant were such as to favor a policy of ignoring a suspected shoplifter unless there was nearly absolute certainty of his intent. This required not only the observation of the theft but of all subsequent movements of the suspect until he was apprehended so that the goods would be found in his possession at a point where his intent would be no longer in doubt. Even then the suspect might be able to convince the court that he had a lapse of memory and did not intend to steal, and therefore, there was no crime and so no justification for detention. Although it would seem that the suspect would fail to prove the lack of probable cause necessary for malicious prosecution, recoveries were frequent enough to deter prosecutions.

STATUTORY PROTECTION OF MERCHANTS

State legislatures, responding to pleas from merchants that they were severely handicapped by the state of the law in dealing with the increasing cost of shoplifting, have passed a variety of legislation. There are two types of shoplifting laws which have attained wide adoption, each type subject to considerable variation. One might be called the "presumption of intent" type. The second essentially codifies the rule in *Collyer v. S. H. Kress and Co.*[61] and might be described as the "detention for probable cause" type. A third category of statute might be called the "request to keep in full view" type, and this has gained a few enactments.

Presumption of Intent Statutes

Twenty-one states[62] have enacted statutes which might be called "presumption of intent" statutes. The South Carolina statute is an early and typical form. It reads:

Any person wilfully concealing unpurchased goods or merchanise of any store or other mercantile establishment either on the premises or outside the premises of such store, shall be *prima facie* presumed to have so concealed such article with the intention of converting it to his own use without paying the purchase price thereof within the meaning of §16-359.1, and the finding of such unpurchased goods or merchandise concealed upon such person or among the belongings of such person shall be *prima facie* evidence of wilful concealment. If such person conceals or causes to be concealed such unpurchased goods or merchandise upon the person or among the belongings of another, the finding of such unpurchased goods or merchandise shall also be *prima facie* evidence of wilful concealment on the part of the person so concealing such goods.[63]

There are a number of variations to this type of statute. Some, such as the Maine statute,[64] omit the presumption but include the provision making the concealment prima facie evidence of intent to steal the goods. Others by their language apply only to concealment on the store premises and Delaware's statute applies only to concealment "outside the premises."[65]

The major purpose of this type of statute appears to be to make it easier to prove intent—ordinarily a necessary element in shoplifting or larceny.[66] It was the difficulty of proving intent which led merchants to adopt the practice of delaying the apprehension of a suspected shoplifter until he had left the store or, in a self-service type operation, the checkout stand. By permitting a case to go to a jury with no proof of intent beyond concealment within the store, the statute presumably would permit the merchant to apprehend earlier as well as lighten the burden of proof of the prosecution. In view of this apparent objective, the limitation of the Delaware statute to concealment outside the premises seems anomalous. No case could be found interpreting it.

The Wisconsin statute, although of the

APPENDIX TABLE 5-I
"PRESUMPTION OF INTENT" STATUTES

State	Year Passed or Amended	Statute	Presumption of Intent	Concealment Is Prima Facie Evidence	Place of Concealment			Applies to Concealment on Others
					Premises	Outside	Either	
ALAS.	1962	ALAS. STAT. §11.20.275		X	X			
ARK.	1957	ARK. STAT. ANN. §41-3942	X	X			X	X
COLO.	1967	COLO. REV. STAT. §40-5-30		X			X	X[1]
CONN.	1961	CONN. GEN. STAT. ANN. §53-63 (c)	X	X			X	X
DEL.	1965	DEL. CODE ANN. tit. 11 §645	X	X		X		X
IDAHO	1957	IDAHO CODE §18-4626			X			
IOWA	1961	IOWA CODE ANN. §709.21		X[2]				
KY.	1968	KY. REV. STAT. §433.234 (2)		X	X		X	
ME.	1955	ME. REV. STAT. ANN. tit. 17 §3501		X	X			
MASS.	1958	MASS. ANN. LAWS c. 231, §94B	X					
MISS.	1958	MISS. CODE ANN. §2374-03		X			X	X[1]
MO.	1961	MO. ANN. STAT. §537.125 (3)	X		X		X	
N.H.	1957	N.H. REV. STAT. ANN. §582:15		X			X	X
N.J.	1962	N.J. STAT. ANN. §2A:170-99	X	X			X	X
N.M.	1965	N.M. STAT. ANN. §40A-16-21	X	X				
N.C.	1957	N.C. GEN. STAT. §14-72.1		X	X			
OKLA.	1967	OKLA. STAT. ANN. tit. 22 §1344	X				X	X
PA.	1959	PA. STAT. tit. 18 §4816.1	X	X			X	X
S.C.	1956	S.C. CODE ANN. 16-359.2	X	X			X	X
S.D.	1959	S.D. COMPILED LAWS ANN. §22-37-23	X	X			X	X
TENN.	1957	TENN. CODE ANN. §39-4236	X	X			X	X
W.VA.	1957	W.V. CODE ANN. §61-3A-3	X				X	
WIS.	1969	WIS. STAT. ANN. §943.50 (2)	X[3]	X[3]				X

1. "On his own person or otherwise."
2. Is "material evidence of" intent.
3. Is "evidence of" intent.

same genre, appears not well designed to accomplish the same purpose. It declares:

> The intentional concealment of unpurchased merchandise which continues from one floor to another or beyond the last station for receiving payments in a merchant's store is evidence of intent to deprive the merchant permanently. . . .[67]

It also declares that discovery of unpurchased merchandise concealed upon the person or among his belongings is evidence of intentional concealment. Such wording would appear, if anything, to weaken the merchant's position in dealing with shoplifters. It would seem that without any statute, concealment within the store is at least evidence of intent to steal, though certainly not prima facie evidence. To make it admissible as a basis for finding probable cause for a merchant to act only if the suspect carries the concealed goods to another floor or beyond the last checkout counter would appear more restrictive than the common law. The Iowa statute[68] puts no limitation on the place of concealment but is otherwise similar because it makes concealment only "material evidence" of intent rather than establishing a presumption.

At the other extreme is the Oklahoma statute, which declares that the finding of unpurchased goods concealed upon a person or in his belongings "shall be *conclusive* evidence of reasonable grounds."[69] Since the Oklahoma statute also specifically permits the merchant to conduct a reasonable search of the detained person, the law there appears to be weighted in favor of the merchant.

Only four states[70] have this type of statute on their books without the addition of the "detention for probable cause" statute. This and the fact we could find only one appellate decision[71] involving a "presumption of intent" type statute suggest that merchants have not heavily relied upon them for protection.

Detention for Probable Cause Statutes

The second type of statute attempts to deal directly with the merchant's exposure to liability for false imprisonment when he attempts to thwart a suspected shoplifter by detaining him to recover his property, to investigate or to hold him for the police or some combination of these objectives. It aims to immunize him from liability if he has probable cause to believe the person he detains has concealed his goods with intent to steal them.[72] Usually immunity is conditional upon the detention being for a reasonable time and being conducted in a reasonable manner. These are the requirements the court imposed in the leading case establishing this privilege for merchants, *Collyer v. S. H. Kress Co.*[73]

One of the earliest if not the earliest statute of this type is that of Florida which reads:

> (2) A peace officer, or a merchant, or a merchant's employee who has probable cause for believing that goods held for sale by the merchant have been unlawfully taken by a person and that he can recover them by taking the person into custody, may, for the purpose of attempting to effect such recovery, take the person into custody and detain him in a reasonable manner for a reasonable length of time. Such taking into custody and detention by a peace officer, merchant, or merchant's employee shall not render such police officer, merchant or merchant's employee criminally or civilly liable for false arrest, false imprisonment, or unlawful detention.
>
> (3) Any peace officer may arrest without warrant any person he has probable cause for believing has committed larceny in retail or wholesale establishments.
>
> (4) A merchant or a merchant's employee who causes such arrest as provided for in subsection (2) hereof of a person for larceny of goods held for sale shall not be criminally or civilly liable for false arrest or false imprisonment where the merchant or merchant's employee has probable cause for believing that the person arrested committed larceny of goods held for sale.[74]

The statutes vary in a number of respects.

The first clause of the Colorado and Mississippi statutes seem to create an ambiguity by making the right to question conditional upon the commission of the offense. It says,

> If any person shall commit or attempt to commit the offense of shoplifting, . . . or if any person shall willfully conceal upon his person or otherwise any unpurchased goods . . . ,

and then it goes on to permit questioning "such person." The clause appears to harken back to the common law privilege to recover goods that have been stolen which was discussed above. It is available to the merchant on an "at your peril" basis and is lost if the goods, in fact, have not been stolen. The Colorado Supreme Court had to deal with a plaintiff's argument that their statute affords no protection if the suspect is not actually guilty of shoplifting. It recognized "that the statute is certainly not a model of clarity" and "that it is ambiguous," but it held that the legislature must have intended to protect persons who stop and question another even though it later develops that the latter is not guilty of shoplifting.[75] The ambiguity of the Mississippi statute was not raised as an issue in the three reported appellate cases involving it. In each the court discusses probable cause, ignoring the apparent condition precedent.[76]

Most of the statutes specify the purposes for which detention is permitted. It will be noted from Appendix Table 5-II that most of them specify investigation as a purpose and many specify recovery of stolen goods. A few statutes list a number of purposes. For example, the Indiana statute includes the following purposes:

> to require (the subject) to identify himself, to verify such identification, to determine whether such person has in his possession unpurchased merchandise taken from such mercantile establishment, to inform the appropriate peace officers, or to inform the parents or other private persons interested in the welfare of the person detained.[77]

The Minnesota statute is the most restrictive of all in defining the purpose for which detention is permitted. It declares that detention shall be "for the sole purpose of delivering (the suspect) to a police officer."[78]

Provisions listing purposes for which detention is permitted raise policy questions which may or may not have been considered by the legislators. One important issue is whether or not the detention statutes permit the merchant to search the prospect.[79] No doubt the common law privilege to recover stolen goods permits search, but the search is at the peril of the merchant under the general common law rule. In the landmark case, *Collyer v. S. H. Kress Co.*,[80] which recognized that probable cause would protect the merchant even if he were proved to be mistaken in his suspicions, there was a forceable search by the store detective. In *Burnaman v. J. C. Penney Co.*, a federal district court declared that the merchant's employees "were acting lawfully in detaining, searching, and touching Mrs. Burnaman."[81] Here the Texas statute, apparently codifying the common law, permitted violence "in preventing or interrupting an intrusion upon lawful possession of property."

Only two statutes, those of Iowa and Oklahoma, specifically include search as one of the protected purposes. Fifteen statutes list recovery of the merchant's goods as a purpose of the permitted detention, but no case could be found which raised the search issue under any of them.[82] In fact, frequently the plaintiff brings his action only on false imprisonment without a separate count for assault, even when a search has taken place.[83] It would seem reasonable to believe that, in the fifteen states which permit detention to recover merchandise, the legislature intended to protect the merchant from tort liability for a search as well as the detention if he has probable cause, but in a time when the courts are giving increasing recognition to a right of privacy the question as to whether those which authorize detention only for investi-

APPENDIX

"DETENTION FOR PROBABLE

State	Year Passed or Amended	Statutes	Purposes					How Long		
			Recover	Investigate	Search	Summon Police	Arrest	None Stated	Reasonable	Spec. Time
ALA.	1957	Ala. Code tit. 14, §334	X						X	
ALAS.	1971	Alas. Stat. §11.20.277		X					X	
ARIZ.	1958	Ariz. Rev. Stat. Ann. §13-674-5					X			
ARK.	1957	Ark. Stat. Ann. §41-3942	X						X	
COLO.	1967	Colo. Rev. Stat. Ann. §40-5-31		X						
DEL.	1965	Del. Code Ann. tit. 11 §645-7				X			X	
FLOR.	1970	Fla. Stat. Ann. §811.022	X				X		X	
GA.	1958	Ga. Code Ann. §105-1005						X	X	
HAWAII	1967	Hawaii Rev. Stat. §663-2		X					X	
ILL.	1967	Ill. Ann. Stat. ch. 38 §10-3 (c)		X					X	
IND.	1967	Ind. Ann. Stat. §10-3042-6		X	X[2]	X[3]			X	X
IOWA	1961	Iowa Code Ann. §709.22-4		X	X[4]				X	
KAN.	1970	Kan. Stat. Ann. §21-3424		X					X	
KY.	1968	Ky. Rev. Stat. §433.236	X			X			X	
LA.	1966	La. Code Crim. Proc. art. 215		X						X
MD.	1970	Md. Ann. Code art. 27 §551A						X		
MASS.	1958	Mass. Ann. Laws ch. 231 §94B		X					X	
MICH.[5]	1963	Mich. Comp. Laws Ann. §600.2917							X	X
MINN.	1957	Minn. Stat. Ann. §629.366				X[6]			X	
MISS.	1958	Miss. Code Ann. §2374-04		X						
MO.	1961	Mo. Ann. Stat. §537.125	X	X					X	
NEB.	1957	Neb. Rev. Stat. §29-402.01	X						X	
NEV.	1969	Nev. Rev. Stat. §598.040	X		X	X			X	
N.J.	1962	N.J. Stat. Ann. §2A:170-100	X						X	
N.M.	1965	N.M. Stat. Ann. §40A-16-22	X						X	
N.Y.	1960	N.Y. Gen. Bus. §218		X					X	
N.D.	1959	N.D. Cent. Code §29-06-27	X						X	
OHIO	1969	Ohio Rev. Code Ann. §2935.041	X					X	X	
OKLA.	1967	Okla. Stat. Ann. tit. 22, §1343	X	X	X	X			X	
ORE.	1959	Ore. Rev. Stat. §164.392		X					X	
PA.	1959	Pa. Stat. tit. 18 §4816.1	X						X	
S.C.	1965	S.C. Code Ann. §16-359.4		X					X	
TENN.	1957	Tenn. Code Ann. §40-824	X						X	
TEX.	1965	Tex. Penal Code art. 1436e		X					X	
UTAH	1957	Utah Code Ann. §77-13-30, §77-13-32	X	X					X	
VA.	1960	Va. Code Ann. §18.1-127						X		
WASH.	1967	Wash. Rev. Code Ann. §9.01.116, §4.24.220		X					X	
W.VA.	1967	W.Va. Code Ann. §61-3A-4		X					X	X
WIS.	1969	Wis. Stat. Ann. §943.50 (3)				X[3]			X	
WYO.	1961	Wyo. Stat. Ann. §6-146.2, §6-146.3		X					X	

1. Concealment must be outside premises.
2. Permits detention "to determine whether such person has in his possession unpurcha merchandise."
3. Notify parent or guardian.
4. Search must be conducted by or under direction of police officer.
5. Michigan only immunizes from punitive damages and those for mental anguish.
6. Sole purpose.
7. Includes parking lot attendant.

OFFENSES COVERED

A—S., F.A., F.I., M.P.
B—A., S., C., M.P., I., F.I., F.A.
C—F.A., F.I., S., A., I.
D—S., M.P., F.I., F.A.
E—F.I., F.A., A., S.
F—S., F.A., F.I., M.P.
G—F.A., F.I., S., A., I.
H—S., F.A., F.I., A.
I—S., M.P., F.I., F.A., A.

TABLE 5-II
"...AUSE" STATUTES

Persons Protected	Crim.	Civil	Offenses Covered	Area of Detention			Word Used	
				Premises	P. or Vic.	Not Spec.	Prob. Cause	Reasonable
O., M., M.E.	X	X	F.A., F.I.				X	X
O., M. MAE	X	X	None Spec.	X				X
O., M., M.E.	X	X	F.A., F.I.				X	X
O., M., M.E.	X	X	F.A., F.I.				X	
O., M., M.E.		X	A				X	X	X
Supervisor Over 25		X	B				X¹	X
O., M., M.E.	X	X	F.A., F.I., U.D.				X	X
M.E.	X	X	F.A., F.I.				X	X
O., M., M.E.A.	X	X	C		X			X
M.E.	X	X	F.I.				X	X
O., M., adult M.E., S.A.	X	X	Any action		X		X	
O., M., M.E.	X	X	F.A., F.I.				X	X
M.E.	X	X	F.I., F.A.		X		X	
O., M., M.E., S.A.,	Not Specified						X	X
P.O., authorized M.E.	X	X	F.A.	X				X
M.E.		X	D				X	X
M.A.E.	X	X	F.A., F.I.		X			X
M.E.		X	E				X	X
E., M.	X	X	F.I., F.A.				X	X
M.E., P.O.		X	F				X	X	X
M.E.	X	X	None Spec.		X		X	X
M.E., P.O.	X	X	F.I., F.A., S., L.				X	X
M.E.	X	X	F.A., F.I., U.D., S.	X			X	
., M.	X	X	None Spec.				X	X
., M.	X	X	None Spec.				X	X
., M., M.A.E.	X	X	G		X			X
., M., M.E.	X	X	F.I., F.A.				X	
., M., M.E.	Not Specified				X		X	
M.E.	X	X	None Spec.				X	X
., M., M.E.		X	H				X	X
M.E., P.O.	X	X	F.A., F.I.				X	
M.E.	X	X	Any action		X			X
M.E., P.O.	X	X	F.A., F.I.				X	X
M.E.		X	None Spec.				X	X
M.E., P.O.	X	X	None Spec.				X	X
M.E.⁷		X	9		X		X	
., M., M.E.	X	X	Any action		X			X
M.E., P.O.		X	None Spec.				X	X
M.E. (adult)	X	X	10				X	X
M.E., P.O.	X	X	11				X	X

...ame defenses as if P.O. made arrest in line of duty.
...S., F.A., F.I., A.

ABBREVIATIONS

...ons Protected
...—merchant, owner, operator
...A.E.—authorized employee
..E.—employee or agent
..O.—police officer or peace officer or law enforcement officer
...—store security employee

...ses Covered
...—assault and battery or trespass
...A.—false arrest
...—false imprisonment or unlawful detention
...—invasion of civil rights
...P.—malicious prosecution
...—slander or defamation
...ne Spec.—not specified

gation would permit a search appears to be more difficult and as yet unanswered.

The Minnesota statute, which permits detention only for the purpose of turning the suspect over to the police, would appear likely to discourage merchants from apprehending minors, who comprise a large proportion of shoplifters, since they may not want to expose a child who appears to be a first offender to a police record. Presumably the legislatures of Minnesota and the other states limiting the right to detain to the objective of summoning the police feared overaggressiveness on the part of merchants. However, statutory requirements of probable cause and that the detention be for a reasonable time and handled in a reasonable manner would appear to discourage overaggressiveness, especially in view of the apparent tendency of courts to be reluctant to find probable cause.[84]

It might well be argued that the statutes which include no statement of purposes for detention are at least as broad as the Indiana statute. All of the purposes enumerated in that statute appear to be reasonably related to the recovery of stolen property or the deterrence of crime—the general objectives of criminal law, which presumably could be read into the statute. No cases could be found which have invalidated statutes which omit specification of purposes nor were there any located which raised the issue whether the detention was for a permitted purpose.

Most of the litigation has been on the issue of probable cause. Instead of authorizing detention upon "probable cause for believing that goods . . . have been unlawfully taken" as in the Florida statute, some statutes, as Appendix Table 5-II shows, use either the term "reasonable cause," or as Massachusetts does, "reasonable grounds to believe."[85] In a recent case, a Massachusetts court reviewed decisions of a number of courts and concluded, "Historically, the words 'reasonable grounds' and 'probable cause' have been given the same meaning by the courts."[86] And it held that the proper test is "the prudent and cautious man standard" as in the common law defense to an action for false imprisonment which was recognized in *Jacques v. Childs Dining Hall Co.*,[87] the case relied upon by the California court in *Collyer v. S. H. Kress Co.*[88] The Supreme Judicial Court of Massachusetts held that the trial court had properly denied the defendants' requested instruction which would have made the test a subjective one, "viz., whether the defendant . . . had an honest and strong suspicion that the plaintiff was committing or attempting to commit larceny."[89] It reasoned:

> If we adopt the subjective test as suggested by the defendants, the individual's right to liberty and freedom of movement would become subject to the "honest . . . suspicion" of a shopkeeper based upon his own "inarticulate hunches" without regard to any discernible facts.[90]

Most of the cases arising under the "detention for proximate cause" type statutes involve either the question whether the merchant had probable cause or whether he abused the privilege to detain. The issue of probable cause is one of law for the court if there is no dispute as to the pertinent facts.[91] The courts appear not to be in agreement as to how the issue is to be handled when the facts are disputed. The Supreme Court of Oregon has stated that when the facts are disputed, the court decides the question of probable cause (reasonableness under the Oregon statute)

> by instructing the jury that if it finds the facts to be so and so, then such facts do, or do not, constitute reasonable cause; or, the court concludes that, accepting the disputed facts in the light most favorable to the plaintiff, such facts support only one conclusion—the defendant had reasonable cause to believe the plaintiff had committed shoplifting.[92]

In several other states the jury itself makes the finding on probable cause if the relevant facts are in dispute.[93] The burden of proof as to probable cause is on the defendant.[94]

It is easier to describe from the cases what does not amount to probable cause than to state what does. Several opinions have stressed that there must be more than a "mere suspicion." In *J. C. Penney Co. v. Cox*[95] there was no testimony that anyone had seen Mrs. Cox take anything. The assistant manager had stopped her and asked her to spread out the contents of her purse on the basis of a statement to him by a sales clerk that another clerk had thought Mrs. Cox "acted suspicious" when she had approached her at the counter and a third clerk had "thought she took something." In holding that there was no probable cause to justify detention, the court said,

> The investigation should be based on more than mere conjecture or suspicion. It must be grounded on some definite information from some person who saw enough to justify the manager's belief that a theft had been made, and that a person was guilty of shoplifting.[96]

The information should come to the merchant or the one acting for him from a reliable source, not some unknown shopper.[97] Likewise, the merchant must avoid jumping to a conclusion which observed facts does not require,[98] or he will be held not to have had probable cause. Seeing a shopper pick up an item of merchandise is not alone sufficient if surveillance is not complete, since he may have put it down rather than secreted it.[99]

However, courts may apparently permit the merchant to be influenced by what might be referred to as background events, such as finding goods missing after a previous visit to the store by the plaintiff, in determining whether he has probable cause.[100] If a state has a presumption of intent type statute as well as a detention for probable cause statute, an observation of a shopper picking up an item of merchandise and putting it in a purse and closing it would constitute probable cause. The Arizona Supreme Court has reached the same result in the absence of a presumption of intent statute.[101]

Most of the "detention for probable cause" type statutes immunize the merchant from liability only if the detention is for a reasonable time and is conducted in a reasonable manner. These questions are matters of fact to be determined by the jury,[102] and the burden of proof is on the plaintiff to show unreasonableness.[103] In one case an appellate court held that a four to five hour detention, including that by the police, was not so long as to entitle the trial court to direct a verdict for the plaintiff.[104]

The legal difficulties a merchant may face under the reasonable manner requirement appear in *Chretian v. F. W. Woolworth Co.*[105] In this case the store manager had been informed by another customer that the plaintiff had secreted certain merchandise, not specified, in her translucent handbag. He then observed her and saw in the handbag articles similar to those on sale in the store. When she came to the check-out counter the manager asked in a normal voice if she had paid for all the merchandise. After an affirmative response he asked if she was sure. She again said she had paid for everything. The manager then said, "Wouldn't you like to pay for the merchandise that you have in your handbag?" She replied that she had nothing there but her own personal belongings. The manager for a fourth time asked if she had paid for everything. The plaintiff then became excited and dumped the contents of her handbag out on the counter. The Court of Appeals of Louisiana affirmed judgment for the plaintiff for slander despite a statute which permits a merchant to detain for questioning for up to sixty minutes on probable cause. The court indicated that it did not think the manager had probable cause but then added:

> The store manager may have had a perfect right to question plaintiff whom he suspected of shoplifting, but the rights and qualified privilege granted by the statute do not clothe the storekeeper with immunity when its manager resorted to slander.[106]

This statement suggests that the court believed the questioning was not done in a reasonable manner and this is the interpretation of another division of this court.[107] Yet it seems difficult to imagine how the manager could have questioned plaintiff in a less offensive manner, given the reason for the questions was the statutory one of resolving his suspicion of theft, except perhaps to have stopped with the third question. Perhaps the court thought the manager should have asked the plaintiff to go to a private office for questioning. However, once the first question had been asked at the checkout counter, such a request would seem to be as much of an accusation as the additional questions.[108] As indicated in discussing slander above, any such questioning will impute a crime and hence be slander.

However, store personnel are occasionally quite unreasonable in their manner of detaining suspected shoplifters. For example, accusing a suspect in a loud voice of stealing a specific item rather than asking if it has been paid for would appear to be properly held to result in loss of the statutory privilege.[109] So also would a refusal to permit a suspected shoplifter to leave until he has signed a release of liability.[110]

However, not all courts are as unsympathetic to the merchant as in the *Chretian* case.[111] In *Delp v. Zapp's Drug & Variety Store*,[112] the Oregon Supreme Court held that detaining a suspected shoplifter for the purpose of learning her name is reasonable as a matter of law even though the detention extended to thirty minutes because of her refusal to give her name and police were called. Likewise, a New Jersey court upheld a trial court which held that a detention of twenty-seven minutes was not unreasonable as a matter of law even though the decision to call police came after the suspect refused to sign a statement acknowledging that she had taken stretch pants for her own use without paying for them.[113] (She was acquitted upon trial for the alleged offense.)

"Keep in Full View" Statutes

A third type of statute which has been enacted by three states[114] might be called the "keep in full view" statute. Montana's[115] and Nevada's[116] statutes are identical. They give the merchant "the right to request any individual on his premises to place or keep in full view any merchandise such individual may have removed, or which the merchant has reason to believe he may have removed, from its place of display," and immunize the merchant from either criminal or civil liability for "slander, false arrest or otherwise." Vermont's statute differs in that it doesn't specifically immunize the merchant from liability and it requires him to conspicuously post a "notice of this section" on the premises.[117] We were unable to find a case involving any of these statutes.

Nevada also requires the posting of a notice in the form specified in the statute if the merchant wishes to claim immunity from liability for detention under its "detention for probable cause" statute.[118] The statute also requires the department of state printing to provide such signs at cost.

EVALUATION OF THE STATUTES

The "keep in full view" type of statute, particularly when signs are posted, may deter some shoplifting because the posted notice or a personal request may suggest vigilance on the part of the merchant. It would not seem to offer the merchant nearly the protection that the "detention for probable cause" type of statute does. Although the merchant need not worry whether he has probable cause for belief that a person has shoplifted before asking him in general terms to keep all merchandise in view, if he gets specific—"that hairspray"—then the requirement that he have "reason to believe" would appear to be operative. It is unclear what rights the merchant has after such a specific request. A literal reading suggests that he has none since his right to demand visibility would appear to apply only to goods on display in his store and not to anything a shopper may claim he brought in. But, even if a court interpreted the statute to permit the merchant to require that all packages be opened and shopping bags unloaded, it gives no right to detain or to question, so the merchant is left with only his common law defenses if he tries to find out where or when certain goods were purchased, the name of the suspect and other vital information, or if he wishes to hold him for investigation by the police.

The "presumption of intent" type of statute appears to give the merchant little encouragement to be aggressive in apprehending shoplifters, although, as we stated above, it may aid him indirectly by making conviction easier to obtain. The "detention for probable cause" type seems best calculated to change the odds in his favor in dealing with suspected shoplifters and so encourage him to be more aggressive. It appears that in this type of legislation the legislatures sought to encourage the merchant to detain and question persons he reasonably believes to have stolen his goods and to immunize him from liability for that detention and questioning so long as it is conducted in a reasonable manner, including the requirement that the detention be for a reasonable time.

EMPIRICAL STUDY OF EFFECTS

The Indiana statute[119] is one of the more comprehensive of the "detention for probable cause" type. It permits detention for a broad range of purposes, it establishes a presumption that a merchant who "informs a peace officer of the circumstantial basis for detention and any additional relevant facts" is merely providing information and not charging a crime, and it appears to immunize the merchant from all types of civil and criminal action based upon that detention. Furthermore, it provides that the merchant can rely upon information provided to him by an employee who has probable cause.

We were interested in trying to determine by empirical research whether merchants had actually changed their practices in dealing with shoplifters in response to this legislation. It had been passed by the legislature in 1967 at the strong urging of the state Retail Council, which would suggest that merchants had felt inhibited by the previous state of the law.[120] The first step consisted of interviews with managers of the security personnel in three large department stores in Indianapolis and with managers of several retail establishments in Bloomington, a small city whose economy is based upon light industry and Indiana University. It appeared from these interviews that few stores had changed their procedures as a result of the statute and some were only vaguely aware of the new law. To test further the effect of the law, we conducted a survey by mail of Bloomington retailers.

The questionnaire reproduced as Appendix Table 5-III[121] was sent to all 160 Bloomington merchants listed in the classified section of the telephone directory. Usable responses were received from fifty-four (33.6%). The questionnaires were identified only as going to one of fourteen types of retail business. Because the number of responses in each classification was so small, no statistical analysis beyond the computation of percentages was made.

The findings are consistent with the interviews held earlier. First, it appears that the 1967 statute had relatively little or no impact upon the great majority of the merchants, and a significant proportion was not aware of the legislation and the change it made in the law. Second, the survey showed that the majority of the respondents were not very aggressive in dealing with suspected shoplifters. Despite the legislation about half of the merchants surveyed continue to fear tort liability, and these merchants are less active in dealing with shoplifters than those who think the law protects them. However, the level of activity also appears to be a function of the type of business and its probable level of loss from shoplifting.

The first question asked the respondent to give the primary source of his awareness of the 1967 legislation. The fact that 48.1 percent marked newspapers as the primary source was not unexpected. Considerable publicity had been given the shoplifting problem a few months earlier by the local newspaper. It was reported that

APPENDIX TABLE 5-III

RESPONSES TO SHOPLIFTING QUESTIONNAIRE

#	%	
		1.* What was the primary source of your awareness of the 1967 change in the Indiana law with respect to the way merchants may deal with suspected shoplifters? (N = 54)
26	48.1	1. Newspaper
10	18.5	2. Indiana Retail Council Newsletter
5	9.3	3. Chamber of Commerce
8	14.8	4. Law enforcement official (policeman or prosecutor)
6	11.1	5. Word of mouth from another retailer
11	20.4	6. This letter
7	13.0	7. Other
		2. Has your business changed its procedures in dealing with suspected shoplifters since the 1967 statute became effective? (N = 52)
14	26.9	Yes
38	73.1	No
		b. If you have changed, what was the major reason for the change? (N = 14)
10	71.4	Law
4	28.6	Losses
		c. Briefly describe the major changes
		3.* Which of the following statements express your feeling about the law as it applies to a merchant who apprehends a shoplifter? (You may mark more than one.) (N = 53)
10	18.9	1. I am very much afraid of liability for false arrest or imprisonment for detaining a shoplifter.
17	32.1	2. I will be very reluctant to detain shoplifters until there is a well publicized test case which clearly protects a retailer from liability.
24	45.3	3. I think the law adequately protects me in detaining suspected shoplifters.
6	11.3	4. Other
		4.* Please check all of the following statements that are descriptive of your present policy: (N = 53)
30	56.6	1. Only the store manager or specially designated persons such as security personnel may detain a suspected shoplifter.
16	30.2	2. A suspect is not detained until he leaves the store.
14	26.4	3. A suspect is not detained until he has clearly left the department area where the suspected theft occurred even though it is believed he has concealed the goods in some manner.
18	34.0	4. In most cases the suspect is required only to pay for the goods.
26	49.1	5. Routinely, police are called and the suspect detained until their arrival.
24	45.3	6. If the suspect is a juvenile, he is detained a reasonable time or until a parent is notified.
7	13.2	7. An affadavit is filed with the county prosecutor whenever it is considered that the evidence against the suspect is strong.
		8. No effort is made to prosecute.
		5. Approximately what proportion of persons who are suspected of shoplifting in your store are stopped and detained? (N = 52)
11	21.2	1. None
17	32.7	2. Less than 5%
2	3.9	3. Between 5 and 20%
5	9.6	4. Between 20 and 50%
17	32.7	5. More than 50%
		6. Of the suspected shoplifters who have been detained by your store, in approximately what percentage of the cases is prosecution carried at least as far as the signing of an affadavit for the prosecutor? (N = 49)
22	45.0	1. None
8	16.3	2. Less than 5%
2	4.1	3. Between 5 and 20%
5	10.2	4. Between 20 and 50%
12	24.5	5. More than 50%
		7.* If you seldom or never prosecute suspected shoplifters what is the major reason? (N = 28)
12	38.7	1. Fear that I or my company will be held liable if they are not convicted.
4	12.9	2. Prosecutions are bad for customer and public relations.
2	6.5	3. I just don't like to be responsible for hurting the individual suspect.
16	51.6	4. Other.
		8.* How would you evaluate the effect of the law on your store? (Check all those applicable) (N = 53)
23	43.4	1. No effect.
19	35.9	2. We are now slightly more aggressive in detaining shoplifters.
6	11.3	3. We are slightly more aggressive in prosecuting shoplifters.
11	20.8	4. We are definitely more aggressive in detaining suspected shoplifters.
9	17.0	5. We are definitely more aggressive in prosecuting suspected shoplifters.

* Respondents marked more than one answer so the sum of percentages exceeds 100.

a few merchants had become quite aggressive in dealing with suspected shoplifters and that a local judge had started taking a hard line in sentencing offenders. Furthermore, the paper had reported that the prosecuting attorney had met with a group of retailers to discuss the problem and had said, "If the shoplifter yells at you: 'I'll sue,' say, 'Go ahead. I think I'll win this one!'"[122] Surprising was the fact that 11 (20.4%) of the respondents marked alternative #6, indicating that they had not heard of the 1967 statute at all until they had received our questionnaire.

Question 2 asked whether the merchant had changed his procedures in dealing with suspected shoplifters since the passage of the statute. Only fourteen (26.9%) of those responding gave an affirmative answer. However, ten of these (71.4%) attributed the change to the legislation; the others to increasing losses from shoplifting.

Answers to Questions 5 and 6 clustered at the low end, indicating that the majority of merchants very seldom detain a suspected shoplifter and even less frequently take steps to prosecute those they do detain. Of those responding to Question 5, 21.2 percent said they detain none of the people they suspect of shoplifting in their store. An additional 32.7 percent indicated they detain some but less than 5 percent of suspects. Thus, a total of approximately 54 percent of the merchants detain less than 5 percent of those they suspect of shoplifting. According to responses to Question 6, 45 percent never took steps to prosecute and a total of 61.3 percent take steps to prosecute in less than 5 percent of the cases in which they have detained suspected shoplifters.[123]

An effort was made in Question 7 to probe the reasons for this lack of aggressiveness. Of the twenty-eight who answered this question, twelve (38.7%)[124] are afraid of tort liability if they act; and five (17.2%) are disillusioned with efforts to prosecute either because the investment of time away from the business is too great or they fault the criminal justice system for being too lenient. An additional five (17.9%) chose alternative #2, that is these merchants thought to prosecute would damage customer and public relations. Six respondents indicated either in their answer to this or another question that they believed they had no shoplifting problem.

Answers to Question 3, which was answered by fifty-three of the respondents, indicate that twenty-seven (50.9%) are fearful of legal liability and twenty-four (45.3%) believe the law adequately protects them. The others wrote in responses which did not clearly indicate their belief concerning legal risk. Of the fearful ones, ten (18.9%) chose alternative #1[125] (very much afraid of tort liability) and seventeen (32.1%) chose #2 (very reluctant to detain shoplifters until there has been a well-publicized case which clearly protects the merchant).[126] The comment on an additional questionnaire clearly indicated fear of liability and if treated as choosing alternative #1, this would raise the percentage treated as marking #1 to 20.8 percent or a total proportion indicating fear of 52.8 percent.

When the respondents who indicated fear of tort liability on either Question 3 or on Question 7 (one respondent inconsistently marked alternative #1 on Question 7 but alternative #3 on Question 3) were compared as to reported practice in dealing with suspected shoplifters a sharp contrast was observable. The average response on Question 5 for the fearful group was 2.2; that is it would appear that on the average those respondents detained more than 5 percent of those persons they suspected as shoplifters but considerably less than 20 percent. The average response for those who reported that they felt protected by the law was 4.15; that is more than 20 percent but considerably less than 50 percent. About the same relative difference appears when responses to Question 6 are averaged. Those in the fearful group averaged a response of 1.7, indicating that an affidavit was signed in less than 5 percent of the cases in which a suspect was de-

APPENDIX TABLE 5-IV

AVERAGE RESPONSE TO QUESTIONS #5 AND #6 BY RETAIL CATEGORY

Retail category	#5	#6
Grocery	4.0	3.8
Department	3.6	4.3
Booksellers	3.5	2.8
Women's clothing	3.5	3.3
Pharmacy	3.3	3.0
Men's clothing	2.4	1.8
Gift shop	2.3	1.7
Jewelry	2.3	1.0
Liquor	2.0	1.0
Shoes	1.3	1.3
All categories	3.0	2.5

tained. The comparable average for the not afraid group was 3.4; that is steps toward prosecution were undertaken in something between 20 and 50 percent of the cases.

A similar kind of comparison between different categories of merchants indicates differences in aggressiveness between different types of retailers. Appendix Table IV shows the average response by retail classification. Grocery and department stores are most aggressive. Shoe shops, liquor stores and jewelry shops appear to be less aggressive. No doubt the physical characteristics and typical methods of displaying various goods affect differentially the shoplifting rates in the several types of retail businesses, and aggressiveness tends to be directly related to the level of loss from shoplifting.

Aggressiveness in dealing with suspected shoplifters also appears to be in part a function of individual attitudes of owners or managers. Comments of respondents varied widely. One gift shop owner declared, "The law is fine for those who wish to use it. I do not." A department store manager, on the other hand, berated fellow merchants who accept payment for goods from a shoplifter instead of prosecuting.

CONCLUSION

The hesitancy to detain suspected shoplifters expressed by almost half of the merchants who participated in the survey may represent wise caution. There has been no reported decision in Indiana of a case in which the statute was pleaded as a defense. The attitudes of courts such as that represented in the *Chretian* case[127] indicate that the merchant's position is not always viewed with understanding. However, in most of the cases in which damages have been imposed upon merchants who have questioned and/or detained suspected shoplifters under the alleged authority of a statute, the merchant has been careless or unreasonable in his actions.[128]

Merchants and their employees must learn to deal discreetly with a suspect, recognizing the need to safeguard the reputation and sensibilities of shoppers. But a merchant need not conclude from this that all attempts at apprehension are too risky, any more than the safeguards of the accused established by the Warren Court make arrests by police officers useless. On the other hand, since shoplifting is so ubiquitous as a retailing phenomenon, its cost is largely passed on to consumers as a cost of doing business and, therefore, the public at large has an interest in trying to minimize it. A necessary requirement for minimization would appear to be fear on the part of a potential shoplifter of detection, detention of suspects where there is probable cause, and prosecution and conviction where the evidence warrants. The posture of the common law discouraged the merchant from initiating these steps.

A well drafted statute such as Indiana's, in our opinion, provides a fair balance between the rights and interests of honest shoppers and those of merchants. It does not give the merchant *carte blanche* to accuse people indiscriminatingly of shoplifting, the kind of practice which resulted in the *Clark v. Kroger Co.* case.[129] The requirement of probable cause should be interpreted in a way consistent with the realities of large scale retailing. The retailer is likely to be better protected legally and the suspected shoplifter handled more discreetly if someone trained and/or experienced in such matters handles any detention. Yet decisions which seem to permit detention only by the person who actually saw the concealment of merchandise are too restrictive to be practical for the large scale merchant who, for the protection of the shopping public as well as itself, permits only designated and trained personnel to detain.[103] The Indiana act recognizes this by specifically permitting the owner or employee who detains to rely upon information from an employee who has probable cause. The requirement that the informing employee have probable cause would appear to safeguard the rights of the shopper and make it incumbent upon the person who actually detains to determine what evidence of shoplifting his informer has before acting upon it. The desirability of clearly permitting a merchant to inform parents rather than police officers in the case of juveniles was commented on earlier. The standard provision of the detention statutes requiring that the detention be handled in a reasonable manner and for only a reasonable time also seems necessary for the protection of suspects. Establishing a maximum period of detention of one hour also seems appropriate since the decision whether to call an officer should be made promptly. Perhaps it is not desirable to encourage merchants to search suspects but rather to leave that to police officers.

Given such a statute as Indiana's, we think merchants can safely be more aggressive in dealing with suspected shoplifters than the majority of merchants responding to the survey indicated that they were. It is to be hoped that a case will soon arrive at either the Appellate Court or Supreme Court of Indiana so that the statute can be authoritatively interpreted and thus reduce the uncertainty and fear that appear from our survey to be widespread among Indiana merchants. As for other detention statutes, it appears that several need redrafting, but perhaps merchants are too much frightened by the occasional case imposing tort liability. Instead of acting only when there is 100 percent certainty, merchants can, through training, prepare their employees to conform to the requirements of probable cause in dealing with shoplifters. Of course, the more reliable and effective the detection devices adopted by the merchant are, the easier it will be both to establish guilt if a suspect is apprehended and to establish probable cause if conviction is not attained or if the suspect satisfactorily explains his actions.

FOOTNOTES

1. S. Astor, *Study of 1,647 Customers Show 1 in 15 Is a Shoplifter,* (1970). The author is president of Management Safeguards, Inc., Chicago, consultants on inventory shrinkage to manufacturers and retailers. Investigators followed randomly selected shoppers as they entered two downtown stores in New York City and one store each in Boston and Philadelphia. Among 500 people followed in one of the New York stores, one out of 12 stole an average of $7.15 worth of merchandise.
2. Hoover, *Crime in the United States,* 22 and 24 (1971).
3. *Wall St. J.,* Oct. 11, 1971, at 1, col. 1.
4. *N.Y. Times,* Feb. 22, 1971, at 41, col. 4.
5. Note 3 *supra.* An estimate of $2.5 billion for shoplifting alone was made by the *N.Y. Times,* June 9, 1971, at 67, col. 3.
6. Hoover, *Crime in the United States,* 24 (1971).
7. *See* discussion *infra* and Appendix Tables 5-I and 5-II.
8. *Nelson v. R. H. Macy & Co.,* 434 S.W. 2d 767, 771 (Mo. Ct. App. 1968). The arrest was by an off-duty policeman employed by the store. It was held that he was acting under the retailer's directions rather than those of public authority. *See also Peak v. W. T. Grant Co.,* 386 S.W.2d 685 (Mo. Ct. App. 1965); *Kroger Grocery & Baking Co. v. Waller,* 208 Ark. 1063, 189 S.W. 2d 361 (1945); and *Goodyear Tire & Rubber Co. v. Paddock,* 219 Ind. 672, 40 N.E. 2d 697 (1942).
9. The term "suspect" is used although courts occasionally make the point that detention and questioning are justifiable only if based upon more than a mere suspicion, *J. C. Penney Co. v. Cox,* 246 Miss. 1, 148 So. 2d 679, 684 (1963).
10. One of the respondents to our empirical

study described such an experience but said he refused to pay and no suit was filed. It has inhibited his aggressiveness in dealing with shoplifting suspects, however.

11. *Reicheneder v. Skaggs Drug Center*, 421 F. 2d 307 (5th Cir. 1970). The manager of a self-service store watched plaintiff from his office and thought he saw plaintiff pick up two spark plugs and place them in his coat pocket. He stopped plaintiff before he reached the checkout counter and asked him if he had anything in his pocket. Plaintiff repeatedly admitted having the spark plugs, which he testified he had kept in his hand and had not put in his pocket. Police were called after the manager told employees, and later the police, that he "had a shoplifter." The police took plaintiff through the sales area of the store and handcuffed him after exiting. He was charged with shoplifting and acquitted. The damages awarded were $10,000 for malicious prosecution and $15,000 for slander.

12. *Great Atl. & Pac. Tea Co. v. Paul*, 256 Md. 643, 261 A. 2d 731 (1970). An assistant manager in a self-service store watched plaintiff, a former policeman who was convalescing from a heart attack, as he shopped, reading labels slowly and then carrying each purchase to his cart, which he left at the end of the aisle. He thought he saw plaintiff take a can of flea spray from the shelf and because he did not find it in the shopping cart a few moments later, concluded that it had been secreted by the plaintiff. The assistant manager stopped plaintiff in the middle of an aisle, roughly frisked him and called him a thief while twenty-five to thirty customers were watching. Nothing belonging to the store was found on plaintiff's person. Plaintiff's heart condition was aggravated by the experience. The damages were awarded for false imprisonment. See also *Bonkowski v. Arlan's Dept. Store*, 383 Mich. 90, 174 N.W. 2d 765 (1970), in which the jury had returned a verdict of $43,750 for false arrest and slander.

13. *Collyer v. S. H. Kress Co.*, 5 Cal. 2d 175, 54 P.2d 20 (1936).

14. *Restatement (Second) of Torts*, §35 (1965).

15. Both torts derive from the old common law action of trespass and tend to be treated as identical, W. Prosser, *Law of Torts*, §11 (4th ed. 1971).

16. See *Shaw v. May Dept. Stores Co.*, 268 A. 2d 607 (D.C. App. 1970) (store detective detained for questioning and then took plaintiff to a J.P. who issued a warrant against her); *See also J. J. Newberry & Co.*, 96 N.J.S. 9; 232 A. 2d 425 (1967) and *Lopez v. Wigwam Dept. Stores*, 49 Hawaii 416, 421 P. 2d 289 (1966). *Cf. Peak v. W. T. Grant Co.*, 386 S.W.2d 685 (Mo. Ct. App. 1965) where police were not called.

17. *Hoock v. S. S. Kresge Co.*, 230 S.W. 2d 758 (Mo. 1950).

18. *McDermott v. W. T. Grant Co.*, 313 Mass. 736, 49 N.E. 2d 115 (1943); *cf. Delp v. Zapp's Drug & Variety Stores*, 238 Ore. 538, 395 P. 2d 137 (1964).

19. W. Prosser, *Law of Torts* §26 (4th ed. 1971); *Great Atl. & Pac. Tea Co. v. Paul*, 256 Md., 643, 655-6, 261 A. 2d 731, 738-39 (1970). In some states including Indiana, shoplifting of any item, regardless of value, is made a felony by statute, Ind. Ann. Stat. §10-3030(1)d, §10-3039, *Young v. State*, —— Ind. ——, ——, 260 N.E. 2d 572, 576 (1970).

20. *Zayre of Virginia, Inc. v. Gowdy*, 207 Va. 47, 147 S.E.2d 710 (1966).

21. *Eason v. J. Weingarten, Inc.*, 219 So. 2d 516 (La. App. 1969); *Grayson Variety Store, Inc. v. Shaffer*, 402 S.W. 2d 424 (Ky. App. 1966).

22. *Martinez v. Sears, Roebuck & Co.*, 81 N.M. 371, 467 P. 2d 37 (1970).

23. *Ashland Dry Goods Co. v. Wages*, 302 Ky. 577, 195 S.W. 2d 312 (1946).

24. *Harrer v. Montgomery Ward & Co.*, 124 Mont. 295, 221 P. 2d 428 (1950).

25. W. Prosser, *Law of Torts* §11, at 47 (4th ed. 1971); *Great Atl. & Pac. Tea Co. v. Paul*, 256 Md. 643, 261 A. 2d 731 (1970).

26. *Proulx v. Pinkerton's Nat'l Detective Agency, Inc.*, 343 Mass. 390, 178 N.E. 2d 575 (1961); *Jacques v. Childs Dining Hall Co.*, 244 Mass. 438, 138 N.E. 843 (1923).

27. *Collyer v. S. H. Kress Co.*, 5 Cal. 2d 175, 54 P. 2d 20 (1936).

28. *Collyer v. S. H. Kress Co.*

29. §120A.
30. W. Prosser, *Law of Torts* §112 (4th ed. 1971).
31. *Chretien v. F. W. Woolworth Co.*, 160 So. 2d 854, 856 (La. App. 1964).
32. *Eason v. J. Weingarten, Inc.*, 219 So. 2d 516, 518 (La. App. 1969). This was dictum since the court found the statements privileged.
33. *Zayre of Virginia, Inc. v. Gowdy*, 207 Va. 47, 147 S.E. 2d 710, 712 (1966). See also *Summers v. W. T. Grant Co.*, 178 F. 2d 916 (5th Cir. 1950).
34. W. Prosser, *Law of Torts* §111 at 747-48 (4th ed. 1971).
35. *Clark v. Kroger Co.*, 382 Fed. 2d 562 (7th Cir. 1967).
36. *Bonkowski v. Arlan's Dept. Store*, 383 Mich. 90, 174 N.W. 2d 765 (1970), rev'g 12 Mich. App. 88, 162 N.W. 347 (1968).
37. *Great Atl. & Pac. Tea Co. v. Paul*, 245 Md. 643, 261 A. 2d 731 (1970). In this case the plaintiff testified that an assistant store manager accosted him in the middle of the store and in a loud voice demanded to know what he had done with the article he was suspected of secreting. The plaintiff further testified that there were twenty-five to thirty customers in the immediate vicinity. There was also testimony that word of the incident spread to the plaintiff's neighborhood some two miles distant. But cf. *Martinez v. Sears, Roebuck & Co.*, 81 N.M. 371, 467 P. 2d 37 (1970) and *Burnaman v. J. C. Penney Co.*, 181 F. Supp. 633 (S.D. Tex. 1960).
38. *Bonkowski v. Arlan's Dept. Store*, 383 Mich. 90, 174 N.W. 2d 765 (1970). See also *Reicheneder v. Skaggs Drug Center*, 421 F. 2d 307 (5th Cir. 1970) where information that a shoplifter had been apprehended was given employees by the manager and this was followed by their seeing plaintiff being taken through the store by police.
39. W. Prosser, *Law of Torts* §116 (4th ed. 1971).
40. *Holliday v. Great Atl. & Pac. Tea Co.*, 256 F. 2d 297 (8th Cir. 1958); *Scott-Burr Stores Corp. v. Edgar*, 181 Miss. 486, 177 So. 766 (1938); *Kroger Groc. & Baking Co. v. Young*, 66 F. 2d 700 (8th Cir. 1933); see annotation 29 A.L.R. 3d 961, 977 (1970).
41. *Southwest Drug Stores v. Garner*, 195 So. 2d 837 (Miss. 1967); *Perry Bros. Variety Stores, Inc. v. Layton*, 119 Tex. 130, 25 S.W. 2d 310 (1930); *Kroger Grocery & Baking Co. v. Young*, 66 F. 2d 700 (1933); cf. *Scott-Burr Stores Corp. v. Edgar*, 181 Miss. 486, 177 So. 766 (1938) in which the court declares at 503, 177 So. 766 at 770: ". . . the fact that the alleged slanderous words may have been uttered in the presence and hearing of other persons who were accidentally present and to whom the remarks were not addressed, would not overthrow the qualifiedly privileged nature of the communication."
42. *Scott-Burr Stores Corp. v. Edgar*, 181 Miss. 486, 177 Co. 486 (1938). *Quaere* if the suspect is black or wears sandals and his hair long and is accused under circumstances where a closely clipped person in a business suit would not be.
43. W. Prosser, *Law of Torts* §119 at 837 (4th ed. 1971).
44. *Reicheneder v. Skaggs Drug Center*, 421 F. 2d 307 (5th Cir. 1970); *Silvia v. Zayre Corp.*, 233 So. 2d 856 (Fla. 1970); *Shaw v. May Dept. Stores Co.*, 268 A. 2d 607 (D.C. App. 1970).
45. *Lopez v. Wigwam Dept. Stores, No. 10, Inc.*, 49 Hawaii 416, 421 P. 2d 389 (1966). Recognition by courts of the difficulty of proving a negative may result in actuality in making this requirement little different from a burden of going forward with evidence.
46. *Lopez v. Wigwam Dept. Stores*.
47. *Montgomery Ward & Co. v. Phearson*, 129 Colo. 502, 272 P. 2d 643 (1954).
48. *Reicheneder v. Skaggs Drug Center*, 421 F. 2d 307 (5th Cir. 1970).
49. W. Prosser, *Law of Torts* §119 at 848 (4th ed. 1971).
50. W. Prosser, *Law of Torts*.
51. *Restatement (Second) of Torts*, §13 (1965).
52. *Shaw v. May Dept. Stores Co.*, 268 A. 2d 607 (D.C. App. 1970).
53. *Piggly-Wiggly Ala. Co. v. Rickles*, 212 Ala. 585, 103 So. 860 (1925); *Banks v. Food Town, Inc.*, 98 So. 2d 719 (La. App. 1957).
54. *Safeway Stores, Inc. v. Harrison*, 14 Ariz. App. 439, 484 P. 2d 208 (1971); *Montgomery Ward & Co. v. Fogle*, 221 Ind. 597, 50 N.E. 2d 871 (1943).

55. *See Piggly-Wiggly Ala. Co. v. Rickles,* 212 Ala. 585, 103 So. 860 (1925).
56. *Dixon v. Harrison Naval Stores, Inc.,* 143 Miss. 638, 109 So. 605 (1926).
57. *Jefferson Stores, Inc. v. Caudell,* 228 So. 2d 99 (Fla. 1969).
58. *See Montgomery Ward & Co. v. Fogle,* 221 Ind. 597, 50 N.E. 2d 871 (1943).
59. *Burnaman v. J. C. Penney Co.,* 181 F. Supp. 633 (S.D. Tex. 1960).
60. *But see Safeway Stores, Inc. v. Harrison,* 14 Ariz. App. 439, 484 P. 2d 208 (1971), a detention case in which the jury found for the defendant on false imprisonment but for the plaintiff on assault and battery.
61. 5 Cal. 2d 175, 54 P. 2d 20 (1936). For discussion *see* Comment, The protection and recapture of merchandise from shoplifters, 47 *Nw. U.L. Rev.* 82, 84 *et seq.* (1952).
62. See discussion *infra* as to Wisconsin and Iowa statutes which are also included in Appendix Table 5-I.
63. *S.C. Code Ann.* §16-359.2 (1962).
64. *Me. Rev. Stat. Ann.* tit. 17, §3501 (1964).
65. *Del. Code Ann.* tit. 11, §645 (1953).
66. The elimination of the intent requirement was upheld against an attack on constitutional grounds in *State v. Hales,* 256 N.C. 27, 122 S.E. 2d 768 (1961). There the court declared, "It is also manifest from the language of our shoplifting statute, in view of its manifest purpose and design, that the Legislature intended that a felonious intent or a criminal intent should not be a necessary element of the statutory crime of shoplifting, and so enacted the statute, for the Legislature must have realized that the remedies heretofore provided by law for the protection of the goods and wares displayed for sale by merchants have not provided them adequate protection from sustaining very substantial losses from shoplifters," *id.* at 32, 122 S.E. 2d at 772. In discussing the purpose of the statute the court said, "The sly, stealthy, crafty nature of the crime of shoplifting and the small individual thefts make detection, prosecution and conviction of the shoplifter for larceny a most difficult and perilous matter. When a merchant accosts a shoplifter, and takes out a warrant against him for larceny, and the shoplifter is acquitted when tried, the merchant risks a lawsuit for large damages for malicious prosecution, false imprisonment, false arrest, or similar tort. Faced with such a formidable array of deterrents, many a merchant stands by and watches his property disappear without a fair, legally protected, opportunity to protect it, if his sole remedy is a successful prosecution for larceny, in which offense superadded to the wrongful taking there must be a felonious intent," *id.* at 31, 122 S.E. 2d at 771. In stating the elements of the statutory offense the court declared, " 'Willfully conceals' as used in the statute means that the concealing is done under the circumstances set forth in the statute voluntarily, intentionally, purposely and deliberately, indicating a purpose to do it without authority, and in violation of law, and this is an essential element of the statutory offense of shoplifting," *id.* at 33, 122 S.E. 2d at 773. See Comment, Criminal law—shoplifting—lack of requirement of intent does not invalidate statute, 64 *W. Va. L. Rev.* 430 (1962).
67. *Wis. Stat. Ann.* §943.50(2) (1958).
68. *Iowa Code Ann.* §709.21 (1950).
69. *Okla. Stat. Ann.* tit. 22, §1344 (1969).
70. Idaho, Maine, New Hampshire and North Carolina.
71. *State v. Hales,* 256 N.C. 27, 122 S.E. 2d 768 (1961).
72. The preamble of the original Florida statute, Laws 1955, ch. 29668, declared, "Whereas, retail merchants in Florida are now suffering great financial loss at the hands of shoplifters, this loss having been estimated by the Florida State Retailers Association to exceed the staggering annual sum of Four Million Five Hundred Thousand ($4,500,000.00) Dollars and Whereas, the retailers have been handicapped in apprehending and prosecuting shoplifters, and whereas, the majority of pilfering cases involve the theft of goods valued at less than Fifty ($50.00) Dollars and fall into a class of misdemeanors, and Whereas, a merchant now has great difficulty in securing an arrest for shoplifting

items of small value which constitute the bulk of the crime, and Whereas, it is necessary to give the merchant authority to detain a suspect and the arresting officer additional authority under the law to make an arrest for a misdemeanor, now therefore."
73. 5 Cal. 2d 175, 54 P. 2d 20 (1936).
74. *Fla. Stat. Ann.* §811.022 (1965). This was the original statute enacted in 1955. It is similar to a proposal contained in Shoplifting and the Law of Arrest, 62 *Yale L. J.* 788, 802 (1953). In 1970 a first section was inserted defining the crime of shoplifting and establishing penalties.
75. *J. S. Dillon & Sons Stores Co. v. Carrington,* 169 Colo. 242, 455 P. 2d 201, 203 (1969).
76. *Butler v. E. Walker Stores, Inc.,* 222 So. 2d 128 (Miss. 1969); *Southwest Drug Stores v. Garner,* 195 So. 2d 837 (Miss. 1967); *J. C. Penney Co. v. Cox,* 246 Miss. 1, 148 So. 2d 679 (1963).
77. *Ind. Ann. Stat.* §10-3042 (1956).
78. *Minn. Stat. Ann.* §629.366 (1947).
79. A corallary issue in criminal law, which is not here considered, is whether the evidence found is admissable in a prosecution if a search is illegal. See *Weeks v. U.S.,* 232 U.S. 383 (1914) and *Mapp v. Ohio,* 367 U.S. 643 (1961).
80. 5 Cal. 2d 175, 54 P. 2d 20 (1936). It will be noted that *Restatement (Second) of Torts,* §120 does not mention search but only investigation.
81. 181 F. Supp. 633, 636 (S.D. Tex. 1960).
82. See *Doyle v. Douglas,* 390 P. 2d 871 (Okla. 1964), where the jury found for the defendant merchant and the appeal court declared that the acts of the merchant were reasonable and, therefore protected under the Oklahoma statute. There had apparently been a conflict of testimony as to whether the detention and search had been voluntary. The jury also found for the merchant where there had been a search in *Holliday v. Great Atl. & Pac. Tea Co.,* 256 F. 2d 297 (8th Cir. 1958).
83. *Zayre of Va., Inc. v. Gowdy,* 207 Va. 47, 147 S.E. 2d 710 (1966); *Isaiah v. Great Atl. & Pac. Tea Co.,* 111 Ohio App. 537, 174 N.E. 2d 128 (1959); *Gibson v. J. C. Penney Co.,* 168 Cal. App. 2d 640, 331 P. 2d 1057 (1958); *Ashland Dry Goods Co. v. Wages,* 302 Ky. 577, 195 S. W. 2d 312 (1946).
84. *Clark v. Kroger Co.,* 382 F. 2d 562 (7th Cir. 1967).
85. The statute proposed by the Yale law students used this term, Shoplifting and the Law of Arrest, 62 *Yale L. J.* 788, 802 (1953). It is also used in *Restatement (Second) of Torts,* §120.
86. *Coblyn v. Kennedy's, Inc.,* —— Mass. ——, 268 N.E. 2d 860, 862-63 (1971).
87. 244 Mass. 438, 138 N.E. 843 (1923).
88. 5 Cal. 2d 175, 54 P. 2d 20 (1936).
89. *Coblyn v. Kennedy's, Inc.,* —— Mass. ——, 268 N.E. 2d 860, 863 (1971).
90. *Coblyn v. Kennedy's, Inc.*
91. *Gibson v. J. C. Penney Co.,* 168 Cal. App. 2d 640, 331 P.2d 1057 (1958). If there is no dispute on the facts the court can grant summary judgment, *Rothstein v. Jackson's of Coral Gables, Inc.,* 133 So. 2d 331 (Fla. 1961).
92. *Delp v. Zapp's Drug & Variety Stores,* 238 Ore. 538, 542, 395 P. 2d 137, 139 (1964); *Rothstein v. Jackson's of Coral Gables, Inc.,* 133 So. 2d 331 (Fla. 1961); accord, *Gibson v. J. C. Penney Co.,* 165 Cal. App. 2d 640, 331 P. 2d 1057 (1958).
93. *Dixon v. S. S. Kresge,* 119 Ga. App. 776, 169 S.E. 2d 189 (1969); *S. S. Kresge Co. v. Carty,* 120 Ga. App. 170, 169 S.E. 735 (1969); *Southwest Drug Co. v. Garner,* 195 So. 2d 837 (Miss. 1967).
94. *Isaiah v. Great Atl. & Pac. Tea Co.,* 111 Ohio App. 537, 174 N.E. 2d 128 (1959).
95. 246 Miss. 1, 148 So. 2d 679 (1963); accord, *Banks v. Food Town, Inc.,* 98 So. 2d 719 (La. App. 1957).
96. *Id.* at 11, 148 So. 2d at 684; accord *Butler v. W. E. Walker Stores, Inc.,* 222 So. 128 (Miss. 1969); see also *Great Atl. & Pac. Tea Co. v. Paul,* 256 Md. 643, 261 A. 2d 731, 740 (1970).
97. *Chretien v. F. W. Woolworth Co.,* 160 So. 2d 854 (La. App. 1964); *Gust v. Montgomery Ward & Co.,* 229 Mo. App. 371, 80 S.W. 2d 286 (1935); cf. *J. C. Penney Co. v. O'Daniell,* 263 F. 2d 849 (10th Cir. 1959).
98. *Wisner v. S. S. Kresge Co.,* 465 S.W. 2d 666 (Mo. App. 1971); *Great Atl. & Pac. Tea Co. v. Paul,* 256 Md. 643, 261

A. 2d 731 (1970); *Browning v. Pay-Less Self-Service Shoes, Inc.,* 373 S.W. 2d 71 (Tex. Civ. App. 1963).
99. *Wilde v. Schwegmann Bros. Giant Supermarkets, Inc.,* 100 So. 2d 839 (La. App. 1964).
100. *Burnaman v. J. C. Penney Co.,* 181 F. Supp. 633 (S.D. Tex. 1960).
101. *Safeway Stores, Inc. v. Harrison,* 14 Ariz. App. 439, 484 P. 2d 208 (1971); *see also Rothstein v. Jackson's of Coral Gables, Inc.,* 133 So. 2d 331 (Fla. 1961).
102. *Jefferson Stores, Inc. v. Caudell,* 228 So. 2d 99 (Fla. 1969); *cf. Cooke v. J. J. Newberry & Co.,* 96 N.J.S. 9, 232 A. 2d 425 (1967).
103. *Jefferson Stores, Inc. v. Caudell.*
104. *Roker v. Gertz Long Island,* —— App. Div. 2d ——, 310 N.Y.S. 2d 536 (1970).
105. 160 So. 2d 854 (1964).
106. 160 So. 2d 856 (1964).
107. *Williams v. F. W. Woolworth Co.,* 242 So. 2d 16, 18 (La. App. 1970).
108. *Cf.* cases cited at note 37.
109. *Southwest Drug Stores of Miss., Inc. v. Garner,* 195 So. 2d 837 (Miss. 1967); *J. C. Penney Co. v. Cox,* 246 Miss. 1, 148 So. 2d 679 (1963).
110. *Silvia v. Zayre Corp.,* 233 So. 2d 856 (Fla. 1970); *Jefferson Stores, Inc. v. Caudell,* 228 So. 2d 99 (Fla. 1969); *Wilde v. Schwegmann Bros. Giant Supermarkets, Inc.,* 100 So. 2d 839 (La. App. 1964); *but see Cooke v. J. J. Newberry & Co.,* 96 N.J.S. 9, 232 A. 2d 425 (1967).
111. 160 So. 2d 854 (La. App. 1964).
112. 238 Ore. 538, 395 P. 2d 137 (1964).
113. *Cooke v. J. J. Newberry & Co.,* 96 N.J.S. 9, 232 A. 2d 425 (1967); *see also, J. S. Dillon & Sons Stores Co. v. Carrington,* 169 Colo. 242, 455 P. 2d 201 (1969).
114. Montana, Nevada and Vermont.
115. *Mont. Rev. Codes Ann.* §64-213 (1947).
116. *Nev. Rev. Stat.* §598.030 (2) (1967).
117. *Vt. Stat. Ann.* tit. 13, §2566 (1958).
118. *Nev. Rev. Stat.* §598.030(4) (1967).
119. *Ind. Ann. Stat.* §10-3042-7 (1956).
120. *See Montgomery Ward & Co. v. Fogle,* 221 Ind. 597, 50 N.E. 2d 871 (1943); *Efroymson v. Smith,* 29 Ind. App. 451, 63 N.E. 328 (1902); *Clark v. Kroger Co.,* 382 F. 2d 562 (7th Cir. 1957) (the incident occurred in Indianapolis but there is no reference to Indiana law in the opinion).
121. An apparently similar survey is reported fragmentarily in Note, Shoplifting—An Analysis of Legal Controls, 32 *Ind. L. J.* 20 (1956). Unfortunately the report of the survey is no longer available in the editor's office and the author is reported to be deceased.
122. *Bloomington Herald-Telephone,* July 30, 1970, p. 8.
123. Seven of the respondents to Question 5 who said they detained no suspects also answered Question 6, all choosing #1, thus indicating that they never sought to prosecute. If these answers are omitted, 54.8 percent signed affidavits in less than 5 percent of the cases where they detain.
124. Two who marked #4, but whose comments indicated fear of liability, were treated as marking #1.
125. One merchant marked response #4 but declared that the law does not protect if the suspect does not have stolen property when detained. This was treated as response #1 since in effect he says probable cause is immaterial.
126. Three of the respondents marked both alternatives #1 and #2. Actually, only four chose #2 but the comment of another seemed consistent with this choice.
127. 160 So. 2d 854 (La. App. 1964).
128. *See e.g. Wisner v. S. S. Kresge Co.,* 465 S.W. 2d 666 (Mo. App. 1971) and *J. C. Penney Co. v. Cox,* 246 Miss. 1, 148 So. 2d 679 (1963) where no one saw plaintiff secret merchandise, and also *Southwest Drug Co. v. Garner,* 145 So. 2d 837 (Miss. 1967) where the store manager failed to check with the cashier as to whether the item in question had been paid for and then accused plaintiff of stealing in a rude and loud voice rather than making simple inquiry.
129. 382 F. 2d 562 (7th Cir. 1967).
130. *See* cases cited in notes 92 & 94.

APPENDIX 5H

SUMMARY OF SHOPLIFTING LAWS IN THE UNITED STATES

Name of State	Fine and/or Imprisonment	Detention	Felony	Misdemeanor
Alabama	Code 1940, tit. 14 §334 (1)			
Alaska	subject to local ordinances			
Arizona	up to 5 yrs.	not more than one hour	over $50.	under $50
Arkansas	up to 5 yrs.	reasonable time	depends on amount stolen	
California	up to 5 yrs.	not more than 1 hr.	over $50.	under $50
Colorado	Article 5 40-5-28, 29, 30, 31			
Connecticut	up to $1,000 and 20 yrs.	reasonable time	depends on amount stolen	
Delaware	punishable under grand larceny	reasonable time	grand larceny	petty larceny
Florida	yes	reasonable time	depends on amount stolen	
Georgia	Code Annotated §105-1005			
Hawaii	Subject to local ordinances Sec. 293-1 under larceny			
Idaho	1-14 yrs.			$300 fine
	Subject to local ordinances			
Illinois	People must prove *corpus delecti* beyond a reasonable doubt			
Indiana	subject to local ordinances		over $100 in value	under $100
Iowa	up to 5 yrs.	not more than 60 min. on premises	depends on amount stolen	
Kansas	up to 5 yrs.	reasonable length of time	grand larceny ($50 or over)	petty larceny under $50
Kentucky	up to 5 yrs.	none stated	depends on amount taken	
Louisiana	discretionary	not more than 60 min. on premises	depends on amount stolen	
Maine	not more than six months			person shall be found guilty of misdemeanor and fined not more than $100.
Maryland	Fined $1000. up to 15 yrs. subject to local ordinances		Per has to be over $100	under $100
Massachusetts	up to 5 yrs.	reasonable time	depends on amount stolen	
Michigan	common law offense classed as a felony subject to local or municipal ordinances			
Minnesota	up to 5 yrs.	not more than 60 min.	over $50	under $50
Mississippi	Code 1942	§§2374.01-2374.05		
Missouri	up to 5 yrs.	reasonable time	depends on amount stolen	
Montana	subject to local or municipal ordinances			
Nebraska	up to 5 yrs.	reasonable length of time		
Nevada	subject to local ordinances			
New Hampshire	up to 7 yrs. depends upon amount taken	subject to laws of larceny	over $100	under $100
New Jersey	depends upon amount taken			
New Mexico	depends upon amount taken		3rd degree over $2,500	between $100 to $2500
New York	Mckimeys Consol. Laws, Gen. Bus. Law 217,218			
North Dakota	Cent. Code §2906-27			
Ohio	Pages. Rev. Code §2935.04.1			
Oklahoma	up to 5 yrs.	not more than an hour	over $50	under $50
Oregon	up to 7 yrs.	reasonable time	depends on amount stolen	
Pennsylvania	up to 7 yrs.	reasonable time	depends on amount stolen	

Name of State	Fine and/or Imprisonment	Detention	Felony	Misdemeanor
Rhode Island	up to 7 yrs.	reasonable time	depends on amount stolen	
South Carolina	up to 7 yrs.	reasonable time	depends on amount stolen	
South Dakota	up to 5 yrs.	reasonable time	depends on amount stolen	
Tennessee	up to 5 yrs.	reasonable time	depends on amount stolen	
Texas	Vernon's Stat. P. C. art 1322			
Utah	Code Ann. 1953 §§77-13-30 17-13-32			
Vermont	$1000 or 1 yr.	reasonable time	over $50	under $100
Virginia	Code 1950 18.1-127			
West Virginia	punishable and disposed under laws of larceny			
Washington	Rev. Code §9.78.010			
Wisconsin	up to 5 yrs.	not more than 1 hour	depends on amount stolen	
Wyoming	Comp. Stat. 1945 9-314			

APPENDIX 5I

DETAILED SHOPLIFTING LAWS IN SELECTED STATES

Arkansas

41-3939. SHOPLIFTING—PENALTY. Any person who shall wilfully take possession of any goods, wares or merchandise offered for sale by any store or other mercantile establishment with the intention of converting the same to his own use without paying the purchase price thereof, shall be guilty of the offense of shoplifting and shall be punished by a fine of not less than twenty-five ($25.00) dollars and not more than fifty ($50.00) dollars and or imprisonment of not less than five (5) days and not more than thirty (30) days, or both for the first offense. (Acts 1957, No. 50, §1, p. 182.)

41-3940. SECOND OFFENSE—PENALTY. Any person found guilty of a second offense of shoplifting as defined in Section 1 (§41-3939) of this Act, shall be punished by a fine of not less than fifty ($50.00) dollars and not more than one hundred ($100.00) dollars and imprisonment of not less than thirty (30) days and not more than ninety (90) days. (Acts 1957, §3, p. 182.)

41-3941. THIRD OFFENSE—PENALTY. Any person found guilty of a third offense of shoplifting shall be punished by imprisonment for not less than one (1) year nor more than five (5) years. (Acts 1957, No. 50, §3, p. 182.)

Connecticut

SEC. 53-63. LARCENY. SHOPLIFTING. (a) Any person who steals any money, goods or chattels, or any bills issued by any state bank or national banking association, or any deed, lease, indenture, bond, writing obligatory, bills of exchange, promissory note, warrant or order for the payment of money or delivery of goods, receipt or discharge, or any book account or other writing being evidence of debt, adjustment or settlement, if the value of the property stolen exceeds two thousand dollars, shall be imprisoned not more than twenty years or fined not more than one thousand dollars or both; if it exceeds two hundred fifty dollars but does not exceed two thousand dollars, he shall be imprisoned not more than five years or fined not more than five hundred dollars or both; if it does not exceed two hundred fifty dollars but exceeds fifteen dollars, he shall be fined not more than two hundred dollars or imprisoned not more than six months or both; if it does not exceed fifteen dollars, he shall be fined not more than twenty-five dollars or imprisoned not more than thirty days or both.

(b) Any person who wilfully takes possession of any goods, wares or merchandise offered or exposed for sale by any store or other mercantile establishment with the intention of converting the same to his own use, without paying the purchase price thereof, if the value of the goods, wares or merchandise exceeds two thousand dollars, shall be imprisoned not more than twenty years or fined not more than one thousand dollars or both; if it exceeds fifty dollars but does not exceed two thousand dollars, he shall be imprisoned not more than five years or fined not more than five hundred dollars or both; if it does not exceed fifty dollars but exceeds fifteen dollars, he shall be fined not more than two hundred dollars or imprisoned not more than six months or both; if it does not exceed fifteen dollars, he shall, for the first offense, be fined not more than one hundred dollars or imprisoned not more than sixty days or both. For a second or subsequent offense involving goods, wares or merchandise of a value not in excess of fifty dollars, he shall be fined not less than fifty dollars nor more than two hundred dollars or imprisoned not less than thirty days nor more than one year or both.

(c) Any person wilfully concealing unpurchased goods or merchandise of any store or other mercantile establishment, either on the premises or outside the premises of such store, shall be prima facie presumed to have so concealed such article with the intention of converting the same to his own use without paying the purchase price thereof within the meaning of subsection (b) of this section. The presence of such unpurchased goods or merchandise concealed upon the person or among the belongings of such person shall be prima facie evidence of wilful concealment and, if such person conceals, or causes to be concealed, such unpurchased goods or merchandise upon the person or among the belongings of another, the presence of the same shall also be prima facie evidence of wilful concealment on the part of the person so concealing such goods. (1949 Rev., S. 8401; 1955, S. 3274d; 1959, P.A. 596; 1961, P.A. 305.)

Delaware

§644. SHOPLIFTING; PROSECUTION; PENALTIES. (a) Whoever, in a mercantile establishment in which goods, wares or merchandise are displayed for sale, removes any such goods, wares or merchandise from the immediate place of display, or from any other place within the establishment, with intent to appropriate the same to the use of the person so taking, or to deprive the owner of the use, the value or the possession thereof without paying to the owner the value thereof; or conceals any such goods, wares or merchandise with a like intent; or alters, removes, or otherwise disfigures any label, price tag or marking upon any such goods, wares or merchandise with a like intent; or transfers any goods, wares or merchandise from a container in which the same shall be displayed or packaged to any other container with a like intent, is guilty or shoplifting.

(b) When the goods, wares or merchandise shoplifted, as above defined, are of the value of $100 or more, prosecution and penalty shall be as for grand larceny under §631. When the goods, wares or merchandise shoplifted are of the value of less than $100, prosecution and penalty shall be as for petty larceny under §632. Valuation of property shoplifted shall be governed by §633.

§645. CONCEALMENT OF UNPURCHASED MERCHANDISE; PRESUMPTION. Any person wilfully concealing unpurchased merchandise of any store or other mercantile establishment, outside the premises of such store or other mercantile establishment, shall be prima facie presumed to have so concealed such merchandise with the intention of converting the same to his own use without paying the purchase price thereof within the meaning of §644(a); and the finding of such merchandise concealed upon the person or among the belongings of such person, outside of such store or other mercantile establishment, shall be prima facie evidence of wilful concealment; and if such person conceals, or causes to be concealed, such merchandise upon the person or among the belongings of another, the finding of the same shall also be prima facie evidence of wilful concealment on the part of the person so concealing such merchandise.

§646. TAKING SUSPECT INTO CUSTODY; DETENTION; ARREST WITHOUT WARRANT. A merchant or store supervisor over twenty-five years of age, who has probable cause for believing that a person has wilfully concealed unpurchased merchandise or has committed shoplifting, as defined in §644 (a) hereof, may, for the purpose of summoning a law enforcement officer, take the person into custody and detain him in a reasonable manner on the premises for a reasonable time.

§647. LIABILITY FOR DETENTION OR ARREST. A merchant or store supervisor over twenty-five years of age, who detains or causes the arrest of any person under the provisions of this Act shall not be held civilly liable for assault, trespass, unlawful detention, defamation of character, malicious prosecution, invasion of civil rights, false imprisonment or false arrest of the person so detained or arrested, provided that in detaining or in causing the arrest of such person, the merchant or store supervisor, had at the time of such detention or arrest probable cause to believe that the person committed the crime of shoplifting as defined in §644 (a).

Florida

811.022. SHOPLIFTING; EXEMPTION FROM FALSE ARREST. (1) A peace officer, or a merchant, or a merchant's employee who has probable cause for believing that goods held for sale by the merchant have been unlawfully taken by a person and that he can recover them by taking the person into custody, may, for the purpose of attempting to effect such recovery, take the person into custody and detain him in a reasonable manner for a reasonable length of time. Such taking into custody and detention by a peace officer, merchant, or merchant's employee shall not render such police officer, merchant, or merchant's employee criminally or civilly liable for false arrest, false imprisonment or unlawful detention.

(2) Any peace officer may arrest without warrant any person he has probable cause for believing has committed larceny in retail or wholesale establishments.

(3) A merchant or a merchant's employee who causes such arrest as provided for in subsection (1) hereof of a person for larceny of goods held for sale shall not be criminally or civilly liable for false arrest or false imprisonment where the merchant or merchant's employee has probable cause for believing that the person arrested committed larceny of goods held for sale.

Indiana

10-3030. THEFT IN GENERAL. A person commits theft when he (1) knowingly:

(a) obtains or exerts unauthorized control over property of the owner; or

(b) obtains by deception control over property of the owner or a signature to any written instrument; or

(c) obtains by threat control over property of the owner or a signature to any written instrument; or

(d) obtains control over stolen property knowing the property to have been stolen by another, wherever the theft may have occurred; or

(e) brings into this state property over which he has obtained control by theft, wherever the theft may have occurred; and

(2) either:

(a) intends to deprive the owner permanently of the use or benefit of the property; or

(b) uses, conceals or abandons the property in such manner as knowingly to deprive the owner permanently of such use or benefit; or

(c) uses, conceals or abandons the property knowing such use, concealment or abandonment probably will deprive the owner permanently of such use or benefit. (Acts 1963 (Spec. Sess.), ch. 10, §3, p. 10.)

A conviction for shoplifting can rest entirely upon circumstantial evidence if there is substantial evidence of probative value to support an inference of guilt, even though there were no witnesses who were able to testify that they saw defendant take anything. *Schooler v. State*, —

Ind. ——, 8 Ind. Dec. 672, 218 N.E. (2d) 135.

Kansas

21-535A. SHOPLIFTING; PENALTIES. Any person who shall wilfully take possession of any goods, wares or merchandise offered for sale by any store or other mercantile establishment with the intention of converting the same to his own use without paying the purchase price thereof, shall be guilty of the offense of shoplifting and upon conviction shall be punished as provided in section 21-535 of the General Statutes Supplement of 1957 if the value of the goods, wares or merchandise possessed was under fifty dollars ($50) or was not subject of grand larceny or as provided in section 21-534 of the General Statutes of 1949, if the value of the goods possessed was fifty dollars ($50) or more, or was the subject of grand larceny. (L. 1959, ch. 162, §1; June 30.)

Kentucky

433.234 SHOPLIFTING. (1) Any person who takes merchandise offered for sale by any store or other mercantile establishment with the intention of converting the same to his own use without paying the purchase price therefor, shall be fined not more than three hundred dollars or imprisoned not more than six months for the first offense. For a second offense he shall be fined not more than five hundred dollars and imprisoned not more than six months, and for a third offense shall be imprisoned not less than one nor more than five years.

(2) Wilful concealment of unpurchased merchandise of any store or other mercantile establishment on the premises of such store shall be prima facie evidence of concealment with intention of converting same to personal use without paying the purchase price therefor. (1958, c. 11, §1; effective June 19, 1958.)

Maine

§3501. WILFUL CONCEALMENT OF MERCHANDISE. Whoever, without authority, wilfully conceals the goods or merchandise of any store, while still upon the premises of such store, shall be guilty of a misdemeanor and, upon conviction, shall be punished by a fine of not more than $100, or by imprisonment for not more than 6 months, or by both. Goods or merchandise found concealed upon the person shall be prima facie evidence of a wilful concealment, 1955, c. 66.

Montana

64-213. RIGHT OF MERCHANT TO REQUEST INDIVIDUALS TO KEEP MERCHANDISE IN FULL VIEW—FREEDOM FROM LIABILITY. Any merchant shall have the right to request any individual on his premises to place or keep in full view any merchandise such individual may have removed, or which the merchant has reason to believe he may have removed, from its place of display or elsewhere, whether for examination, purchase, or for any other purpose. No merchant shall be criminally or civilly liable for slander, false arrest, or otherwise on account of having made such a request.

Nebraska

29-402.01. SHOPLIFTERS; DETENTION, NO CRIMINAL OR CIVIL LIABILITY. A peace officer, a merchant, or a merchant's employee who has probable cause for believing that goods held for sale by the merchant have been unlawfully taken by a person and that he can recover them by taking the person into custody may, for the purpose of attempting to effect such recovery, take the person into custody and detain him in a reasonable manner for a reasonable length of time. Such taking into custody and detention by a peace officer, merchant, or merchant's employee shall not render such peace officer, merchant, or merchant's employee criminally or civilly liable for slander, libel, false arrest, false imprisonment, or unlawful detention. *Source:* Laws 1957, c. 101, 1, p. 361; Laws 1963, c. 157, §1, p. 556.

29-402.02. SHOPLIFTERS; PEACE OFFICER; ARREST WITHOUT WARRANT. Any peace officer may arrest without warrant any person he has probable cause for believing has committed larceny in retail or wholesale establishments. *Source:* Laws 1957, c. 101, §2, p. 361.

29-402.03. SHOPLIFTERS; ARREST; MERCHANT OR EMPLOYEE NOT LIABLE. A merchant or a merchant's employee who causes the arrest of a person, as provided for in section 29-402.01, for larceny of goods held for sale shall not be criminally or civilly liable for slander, libel, false arrest, or false imprisonment where the merchant or merchant's employee has probable cause for believing that the person arrested committed larceny of goods held for sale. *Source:* Laws 1957, c. 101, §3, p. 361; c. 157, §2, p. 557.

Nevada

598.030. MERCHANTS MAY REQUEST INDIVIDUALS ON PREMISES TO KEEP MERCHANDISE IN FULL VIEW; DETENTION OF INDIVIDUALS TO RECOVER MERCHANDISE, INFORM PEACE OFFICERS; IMMUNITY OF MERCHANT FROM CRIMINAL, CIVIL LIABILITY; DISPLAY OF NOTICE.

1. As used in this section:
   (a) "Merchandise" means any personal property, capable of manual delivery, displayed, held or offered for sale by a merchant.
   (b) "Merchant" means an owner or operator, and the agent, consignee, employee, lessee, or officer of an owner or operator, of any merchant's premises.
   (c) "Premises" means any establishment or part thereto wherein merchandise is displayed, held or offered for sale.
2. Any merchant shall have the right to request any individual on his premises to place or keep in full view any merchandise such individual may have removed, or which the merchant has reason to believe he may have removed, from its place of display or elsewhere, whether for examination, purchase or for any other purpose. No merchant shall be criminally or civilly liable on account of having made such a request.
3. Any merchant who has probable cause for believing that merchandise has been wrongfully taken by an individual and that he can recover such merchandise by taking such individual into custody and detaining him may, for the purpose of attempting to effect such recovery or for the purpose of informing a peace officer of the circumstances of such detention, take the individual into custody and detain him, on the premises, in a reasonable manner and for a reasonable length of time. Such taking into custody and detention by a merchant shall not render such merchant criminally or civilly liable for false arrest, false imprisonment, slander, or unlawful detention unless such taking into custody and detention are unreasonable under all the circumstances.
4. No merchant shall be entitled to the immunity from liability provided for in this section unless there is displayed in a conspicuous place on his premises a notice in boldface type clearly legible and in substantially the following form:

> Any merchant or his agent who has probable cause for believing that merchandise has been wrongfully taken by a person may detain such person on the premises of the merchant for the purpose of notifying a peace officer. Nevada Revised Statutes, section 598.030.

Such notice shall be prepared and copies thereof supplied on demand by the superintendent of state printing. The superintendent of state printing shall be entitled to charge a fee based on cost for each copy of such notice supplied to any person. (Added to NRS by 1959, 407; A 1961, 357; 1963, 504.)

New Mexico

40A-16-1. LARCENY. Larceny consists of the stealing of anything of value which belongs to another.

Whoever commits larceny when the value of the thing stolen is one hundred dollars ($100) or less is guilty of a petty misdemeanor.

Whoever commits larceny when the value of the thing stolen is over one hundred dollars ($100) but not more than twenty-five hundred dollars ($2,500) is guilty of a fourth degree felony.

Whoever commits larceny when the value of the thing stolen exceeds twenty-five hundred dollars ($2,500) is guilty of a third degree felony.

Larceny from Store. Larceny of goods of value greater than ten dollars from a store may be punished by imprisonment of from three to five years under Laws 1869-1870, ch. 26, §§1, 2. *State v. Jones,* 34 N.M. 499, 285, p. 501.

Washington

9.78.010. SHOPLIFTING. A person who wilfully takes possession of any goods, wares or merchandise of the value of less than seventy-five dollars offered for sale by any wholesale or retail store or other mercantile establishment without the consent of the seller, with the intention of converting such goods, wares or merchandise to his own use without having paid the purchase price thereof, is guilty of a gross misdemeanor of shoplifting.

Form of Allegation Charging Shoplifting
(For general form of indictment or information, see RCWA 10.37.040, 10.37.050.)

That the said Defendant, ..., in the County of ..., State of Washington, on or about the .. day of ..., 19.., wilfully and unlawfully, without the consent of the seller thereof, to wit: ....., and with the intention of converting said ..... to Defendant's own use, did take possession of certain ....., to wit: ....., of a value of less than seventy-five dollars offered for sale by ......, a ..... store, without the consent of said seller.

9.78.020. ARREST WITHOUT WARRANT AUTHORIZED, WHEN. A peace officer may, upon a charge being made and without a warrant, arrest any person whom he has reasonable cause to believe has committed or attempted to commit the crime of shoplifting.

9.78.030. REASONABLE CAUSE DEFENSE TO CIVIL OR CRIMINAL ACTION BROUGHT BY SUSPECT. Reasonable cause shall be a defense to a civil or criminal action brought for false arrest, false imprisonment, or wrongful detention against a peace officer, by a person suspected of shoplifting.

9.78.040. "PEACE OFFICER" DEFINED. For the purposes of chapter "peace officer" means a duly appointed city, county, state law enforcement officer.

Wisconsin

939.49. DEFENSE OF PROPERTY AND PROTECTION AGAINST SHOPLIFTING. (1) A person is privileged to threaten or intentionally use force against another for the purpose of preventing or terminating what he reasonably believes to be an unlawful interference with his property. Only such degree of force or threat thereof may intentionally be used as the actor reasonably believes is necessary to prevent or terminate the interference. It is not reasonable to intentionally use force intended or likely to cause death or great bodily harm for the sole purpose of defense of one's property.

(2) A person is privileged to defend a third person's property from real or apparent unlawful interference by another under the same conditions and by the same means as those under and by which he is privileged to defend his own property from real or apparent unlawful interference, provided that he reasonably believes that the facts are such as would give the third person the privilege to defend his own property, that his intervention is necessary for the protection of the third person's property, and that the third person whose property he is protecting is a member of his immediate family or household or a person whose property he has a legal duty to protect, or is a merchant and the actor is the merchant's employee or agent.

(3) In this section "unlawful" means either tortious or expressly prohibited by criminal law or both.

BOOKS

Arey, James A.: *The Sky Pirates.* New York, Scribner, 1972.

Broom, Halsey N.: *Small Business Management.* Cincinnati, Southwestern Publishing Co., 1966.

Carlsen and Lewis: *The Systems Analysis Workbook,* Prentice-Hall, 1973.

Currer-Briggs, Noel (ed.): *Security Attitudes and Techniques for Management.* London, Hutchinson, 1968.

Cumming, John: *A Contribution Towards a Bibliography Dealing With Crime and Cognate Subjects.* Montclair, N.J., Patterson Smith Publishing Corp., 1970.

Davidson, William K., and Doody, Alton D.: *Retail Management.* New York, Ronald, 1966.

Gartner, Michael (ed.): *Crime and Business.* Princeton, Dow Jones Books, 1971.

Jaspan, Norman: *Mind Your Own Business.* Prentice-Hall, Inc. Englewood Cliffs, N.J., 1974.

Hammer, Willie: *Handbook of System & Product Safety,* Prentice-Hall, 1972.

Hemphill, Charles F., Jr., and Hemphill, Thomas: *The Secure Company.* Dow Jones-Irwin, Inc., Homewood, Ill., 1975.

Hemphill, Charles F., Jr.: *Security for Business and Industry.* Dow Jones-Irwin, Inc., Homewood, Ill., 1971.

Holcomb, Richard L.: *Cutting the Loss to Bad Checks.* Bureau of Police Science, The University of Iowa, 1973.

———: *Protection Against Burglary.* Institute of Public Affairs of the State University of Iowa, 1973.

———: *Cutting the Loss of Shoplifting.* Bureau of Police Science, The University of Iowa, 1973.

Lippert, Frederick G.: *Accident Prevention Administration.* New York, McGraw-Hill, 1947.

Manager and Security, The. G. Bell & Son Ltd., 1972.

Post, Richard S. (ed.): *Combatting Crimes Against Small Business.* Springfield, Thomas, 1972.

Rolph, C. H. (ed.): *The Police and the Public.* London, Heinemann, 1962.

Ryker, Wilbur: *Reduction of Criminal Opportunity.* National Crime Deterrence Council, Inc., Pittsburgh, Pa., 1971.

Yocum, James C.: *Information Sources for Small Business.* Columbus, Ohio, Ohio State University, B-3, 1949.

PUBLICATIONS OF THE GOVERNMENT, LEARNED SOCIETIES, AND OTHER ORGANIZATIONS

Bracy, C. W.: Vital part of your expansion planning: Plant protection. *Industrial Development and Manufacturers Records,* May 1960.

Bunn, Verne A.: Small business security, Proceedings of the Seminar on Urban Design, Security and Crime. Washington, U.S. Department of Justice, January 1972, p. 57.

Burnstein, Harvey: You are management. *Industrial Security,* 5:18, January 1961.

Burroughs Clearing House. Advice on improving security programs. New York, April, 1961.

———: How to plug holes in company security. *Business Management,* July 1962.

Crime Against Small Business. U.S. Congress Senate Select Committee on Small Business, 1969.

———: A Report of the Small Business Administration, Senate Committee on Small Business, S. Doc. 91-14, esp. pp. 2-3.

Guidelines for Effective Business Action to Help Prevent and Control Crime. Washington, U.S. Chamber of Commerce, 1969.

Impact of crime against small business. Hearings, Senate Select Committee on Small Business, 92nd Cong., 1st Sess., Parts 1-4.

Impact of crime on small business—1969-70, Part 2. Hearings Before the Select Committee on Small Business United States Senate, 1970.

Industrial security times. Bombay, India Publication, July-Sept. 1964.

Karbin, S., et al.: The deterrent effectiveness of criminal justice sanction strategies. Summary Report Prepared under NILECJ Grant 71-069, September 1972.

Marshall, M. S.: Who wants to know? *School and Society,* 385-389, 1952.

Small Business Administration: The impact of crime on small business—Part II (Air Cargo Losses), Report #91-612, Washington, Government Printing Office, 1969.

Small Business Administration: Crime against small business. Report to Select Committee on Small Business, 91st Cong., 1st Sess., Report #91-14, Washington, Government Printing Office, 1969.

Security: Is yours as good as you think? Greyhound Food Management Inc., 1967.

U.S. Congress, Senate: Small Business Select Committee. The Impact of Crime on Small Business—Part I and II, Washington, Government Printing Office, 1969.

———: Report on the Impact of Crime on Small Business in the Washington, D.C. Area. No. 564, 90th Cong., Government Printing Office, 1967.

PERIODICALS

American Cyanamid: FBI is on trail of drug pirates. *Cleveland Press and News*, June 21, 1962.

Anderson, P. H.: Security "takes a giant step" in South Africa. *Security World*, March 1970.

Anderson, R. R.: Strike force against street crime. *FBI Law Enforcement Bulletin, 41:* No. 5, 3-5 & 30, May 1972.

Annais, The: Patterns of violence. *The Annals of the American Academy of Political and Social Science, 364,* March 1966.

———: Combatting crime. *The Annals of the American Academy of Political Social Science, 374,* November 1967.

Anon: Business espionage. *Business Management*, October 1965, p. 58-66.

Anthony, Robert N.: The trouble with profit maximization. *Business Review*, November/December 1960, p. 126-34.

Armstrong, Ted, Fisher, John, and Kent, M. J.: Chiefs in conference. *Canadian Security Gazette*, Summer 1971.

A.S.I.S. initiates program study of narcotics problem in U.S. industry. *Security Systems Digest*, Washington, D.C., 1971.

Banks, James L.: Security joins the war on poverty. *Security World, 4:* No. 5, May 1967.

Barefoot, J. Kirk: Narcotic and dangerous drugs security. *Security World, 3:* No. 7, 17-19, July-August 1966.

Barrell, Clive: Modern piracy. *Top Security*, July 1975, p. 112-113.

Bastoni, Jeanne T.: The small company security officer faces big problems. *Industrial Security*, April 1968.

Bess, Coleman: Bar Assn. study: Why America can't curb crime. *International Security Review*, September, 1973, p. 1, 6, 13.

Better safe than sorry. *Security and Protection*, February 1975, p. 19-20.

Bird, Frank E., Jr.: Cues in identification of the habitual drug accuser. *Security World*, May 1975, p. 36+.

Blumstein, A., and Lanson, Richard L.: Models of a total criminal justice system. *Operations Research*, No. 2, March-April, 1969, p. 17.

———: A systems approach to the study of crime and criminal justice. *Operations Research for Public Systems*, 1967.

Boise, Robert J.: Seven steps to better security. *Industrial Security*, April 1962.

Bonk, Eugene T.: Crisis crime for small business. *Security World*, November 1971.

D'Fucci, Jean: Giant step for security: The federal crime insurance program. *Security World*, September, 1971.

Diedrich, William: Protection of the company's interest in a possible criminal situation. *Industrial Security*, October 1963.

Electronic security systems market to grow to $13 billion before 1984. *Top Security*, August 1975, p. 161.

Fairfield, W. S., and Clift, C.: The private eyes. *The Reporter, 12:*14-29, February 10, 1955.

Fortifying your business security. *The Office*, August 1969, p. 39-52.

Goddard, R. H.: Company and law enforcement relationships. *Police Chief, 31,* July 1964.

———: The anatomy of security. *Industrial Security*, April 1964.

Hale, I. B.: Cooperative efforts of police and industrial security officers—panelist. *Industrial Security*, January 1964.

How security does pay off. *The Office*, September 1971.

Hruska, C. J.: Security crackdown in the Bell System. *Industrial Security*, July 1963.

Hyman, Harold: Industrial security activities at the Ford Motor Company's San Jose assembly plant. *Law and Order, 9,* September 1961.

———: Industrial security in the space age. *Law and Order, 9,* March 1961.

Industrial security conference digest, 1965. Los Angles Chamber of Commerce, Los Angeles, 1965.

Industrial security program of Creole Petroleum Corporation. *Industrial Security, 1,* 1957.

Industrial security rises in the ranks. *Business Week*, November 10, 1962.

Industrial security seminar in San Francisco. *Police Chief, 30,* August 1963.

Inman, Bert D.: Controlling security costs—Panel moderator. *Industrial Security,* October 1960.

Know about these new developments in security systems? *Occupational Hazards,* June 1970, p. 47.

Liaison between police and industrial remarks —1963 seminar. *Industrial Security,* October 1963.

Kelly, Eugene G.: An unusual opportunity. *Industrial Security,* December 1971.

———: Seven steps to better security. *Industrial Security,* April 1962.

Kennedy, Taylor: The key to safety. *Canadian Security Gazette,* Summer, 1971.

Link, Arthur C.: Administration of security in industry. *Industrial Security,* April 1958.

Living with Crime, USA. *Newsweek, 80:* No. 5, 31-36, December 18, 1972.

Lucas, Edwin J.: Crime takes but a moment to commit. NY Society for Ethical Culture, 1947.

Lynch, Thomas C.: Protection of the company's interest in a possible criminal situation —panelist. *Industrial Security,* October 1963.

MacDonald, H. J.: Gone is the day of the gumshoe. *Weekend, 22:* No. 2, January 8, 1972.

Mandell, M.: Being safe. New York, *Saturday Review Press,* 1972.

Masterson, John M.: New security concepts. *Industrial Security,* February 1967, p. 2.

Miller, Arthur S.: The ethics of business enterprise. *Amer Acad Political Social Sci, 343*:141, September 1962.

Mobil's Ferndale Refinery stresses good security. *Industrial Security,* February 1962.

Monfort, Frank E.: Security problems by the mile. *Industrial Security,* April 1959.

Moses, F. J.: Security coordination. *Police Yearbook,* 1963.

Murphy, Michael J.: *Law Enforcement and Operation Security, 12,* October 1964.

McCormick, Robert W.: Industrial security in Europe—A multinational concept. *Security Management,* July 1974, p. 8-13.

McMahan, H. L.: The professional approach. *Industrial Security,* February 1967, p. 32.

McMurray, Robert N.: What crime costs you. *Nations Business,* June 1967.

Neeson, John V.: Public relations and security. *Security World, 9,* October 1966.

Neilson, Jerry: The case for the security consultant. *Security World,* May 1966, p. 33.

Newland, Loren E.: The why's of a security program. *Security Product News,* May-June 1975, p. 20-36.

New York Times: The night the lights went out. New York, 1965.

O'Hagen, John T.: One N.Y. plaza. *Security Management,* March 1974, p. 6-12.

Post, Richard S.: Total security capitalizes on the advantages of a center, *Shopping Center World,* December 1972.

———: Block that theft, *Nation's Business,* October 1972.

Traini, Robert: Crime losses—the "X" factor in inflation. *Security Gazette,* November 1974, p. 413+.

———: Need for comprehensive security makes blind trust incomprehensible. *Security Gazette,* October 1974, p. 375+.

———: Port wine and old masters highlight security needs. *Security Gazette,* March 1975, p. 91.

UNPUBLISHED MATERIAL

Allen, Roger: Industrial security and the motivation to work. Unpublished graduate thesis paper, Michigan State, 1968.

Capune, William Garrett. A study of private police systems: Concept and development. Unpublished Master's Thesis, University of California, Berkeley, 1965.

Jones, Paul: The Mafia: Its transition from government to business. Masters Paper, Michigan State, 1971.

Kingsbury, Arthur A.: Security: A key program in the small business. Unpublished graduate B-paper, Michigan State, 1969.

Chapter 6

COMMERCIAL PROTECTIVE SERVICES

THE GROWTH IN DEMAND for commercial protective services can be traced to three major factors: increased crime, increased criminal sophistication, and increased insurance costs and coverage. The volume of crime against property has risen almost 200 percent since 1960 with a rise of 144 percent between 1973-74. The increased sophistication of criminal attacks against property and operations has reduced the desire for corporations to have "do-it-yourself" protection programs. Furthermore, insurance companies are requiring more and increasingly sophisticated protective measures and devices *before* providing insurance coverage. Consequently, many companies have turned to the commercial protective services to fill their needs for service.

Commercial Security

Commercial security is that portion of the protective services which provides a variety of protection of life and property services to individuals or firms on a profit-making basis. Commercial security organizations (see Figure 6-1) provide the following services: investigations, guard services, alarm central station, express, money transfer services, polygraph investigation, criminal investigations, insurance investigations, recovery of stolen property, security surveys and analysis, honesty shopping. The commercial security firm provides services on a long-term or short-term contractual basis. They are hired for single events such as concerts, parties, social or athletic events on a one-time or a season-long basis. They are also retained by business and industrial firms to protect property on a long-term basis. In such situations they establish guard forces to provide security and/or night watch services on a regular basis. They can also be hired for a specific activity such as protection of corporation property during a strike. Contract guard services are provided to government agencies with many large defense related industrial programs utilizing contract guard services on a yearly contract basis.

Guard and Investigative Service

The current market for uniformed guard service is nearly $3 billion. At present, about $1.3 billion (40%) is accounted for by commercial uniformed guard service companies; the balance $1.7 billion (60%) is estimated as corporate expenditures for proprietary guard services.

Providing physical security now accounts for 85 to 90 percent of the total revenues of commercial security services firms. Investigative and miscellaneous services accounts for the balance. Guard services has gradually emerged as the major revenue-producing activity for these companies. Many of the current major national companies began as detective agencies but gradually began providing guard services for a number of reasons. These reasons included:

—requests by clients for watchmen to prevent losses.
—to stabilize income which tends to fluctuate when investigative work was not available;
—to increase revenues since guard operations are most profitable; and
—to provide employment for investigators while not on assignments.

These reasons are still valid and have created a condition that encourages new firms to enter the guard services field regularly.

Types of Firms

There are three distinct types of commercial security services firms. They are the national, regional, and local. The national

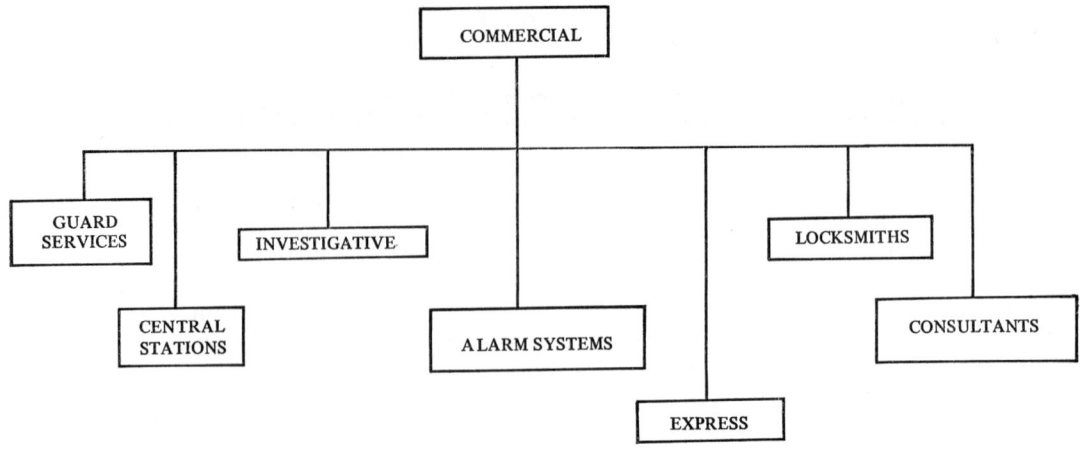

Guard Services

—maintain order
—control access
—protect people/property

Central Station

—equipment monitoring

Investigation

—information gathering

CONSULTANTS

—Provide advise on solving security problems

Express

—Transfer of cash and securities
—Check cashing

Locksmiths

—sell, install and maintain locking devices

Alarm Systems

—sell, install and maintain electronic alarm systems

Figure 6-1.

firms are the largest in total sales, manpower, and accounts handled. They are also considered the more established and have more sophistication in sales and marketing than other types of firms. The national firms are represented by:

    Allied Security
    Advanced Industrial Security
    Globe Security
    Guardsmark, Inc.
    William J. Burns International Security Services, Inc.
    Wells Fargo Security Services
    Pinkerton's, Inc.
    Wackenhut, Inc.

These companies provide a full range of protective services including guard, investigation and, in some cases, alarm services. Their clients have only one security company to deal with instead of several.

Regional Firms

There are regional security services firms throughout the United States. These companies provide protective services of various kinds in regional areas in much the same manner as national firms. Firms of regional type have fewer clients than the national companies but often provide better marketing and sales coverage in a limited geographical area. They are often located in areas not considered as major markets by the national firms. There are, however, regional firms which compete directly with nationals in the "home" areas. They are often equally if not more successful because of the owner-manager participation in sales and operations. The following firms are considered regional:

California Plant Protection
Smith Security
Per Mar Security
Kane Services

Local Firms

Local firms are those companies or sole proprietorships which provide various kinds of security services on a geographically limited basis. These firms are often begun by persons with a local reputation in law enforcement or investigations and provide protective services in their area. These companies cover the entire spectrum of quality and variety of services. The "mom and pop" security company falls in the category along with the highly professional but small local firms. Thousands of these organizations provide tens of thousands of clients with protective services.

Central Station Protection Service

It is estimated that the current revenue level for the central station protection segment of the protection industry is about $275 million with annual growth in excess of 12 percent over the last ten years. Five companies: (1) American District Telegraph Company (ADT),[1] (2) Holmes Electric Protective Company (Holmes), (3) Wells Fargo Alarm Services Group of Baker Industries, (4) AFA Protective Systems, Inc. (AFA), and (5) William J. Burns International Security Service (Burns), are believed to account for 90 percent of this volume.[2] In addition, there are about 100 other companies each providing, in most cases, service from a single central station.

The concept behind central station service is to provide both immediate notice and reaction to undesirable events through the use of electronic and electromechanical alarm systems with response by central station "runners," (guards). The alarm systems, which are generally connected over leased telephone lines, transmit indications of intrusion, fire, and other undesired events to the central station where operators notify the proper authorities, e.g. the local police, owners, or the fire department. At the same time, a runner (or guard) is dispatched from the central station to the protected facility.

Central station coverage generally provides wider protection for certain kinds of properties (for example, warehouses) at significantly lower rates than that obtainable from uniformed guard companies. The cost of a two-shift, seven-day per week uniformed guard protection is about $22,932 per guard (5096 hrs × $4.50 hr) per year. This compares very unfavorably with annual central station service charges for comparable coverage of about $1,650 to $2,100. However, just as with guard service, insurance companies allow a maximum of insurance premium discount of 70 percent to users of central station service. For similar insurance coverage, the discount is only 35 percent if local alarms are used.

The central station generally increases the response time during an intrusion and since electronic sensory devices are used it provides more complete, continuous coverage than guard personnel. It is likewise often superior to the use of a local alarm which must rely upon police or passersby to hear the alarm and call for a response.

There have been many police complaints because of the reported high incidence of "false alarms" from central stations with the national average being over 90 percent for all calls to the police. The industry in analyzing this problem indicates that the fault is not generally with the equipment or the central stations but with the users and owners of the equipment. The industry indicates that user errors in arming the system when closing a protected premises, and improper maintenance and handling of the equipment are basically the cause of the high "false alarm" rate.

Armored Car and Courier Services

The armored car business is the third and smallest of the three segments of the commercial protection industry. It is estimated that the market for armored car service was about $298 to $300 million in 1974. Four companies: (1) Brink's, (2) Wells Fargo Armored Service Group—a subsidiary of Baker Industries, (3) Loomis, Inc., and (4) American Courier Corp.—a subsidiary of Purolator, Inc. These major companies accounted for about 78 percent of the industry revenues with Brink's contributing about twice as much as the other three combined. The remaining approximately $75 million in revenues is divided amont the fifty to sixty small regional and local companies.

Armored car carriers are in the business of moving objects of value, such as cash and securities.

A number of factors have contributed to the 10 percent growth of the industry since 1960. First, inflation, among other things, has increased the need for cash. Since armored car companies base their charges to a large extent on the value of the cargo carried, inflation has led to higher revenues. Second, security procedures and robbery incidences has led businesses to make more frequent bank deposits, thereby increasing the need for armored cars to carry their deposits. Third, hijacking of and higher valuations for jewelry and other noncash valuables means a greater need for protective movement.

Electronic Protection Equipment

The electronic security system market is expected to expand from the 1974 level of $520 million to $1.3 billion by 1984. Sales and servicing of electronic burglar alarm systems are likewise anticipated to grow from $283 million to $800 million in the same period.[3] This sector of the security industry deals with the sales, installation, and servicing of electronic protective devices. The industry has direct contact with thousands of individuals and firms who have security problems and require some form of electronic protection. There have been numerous incidents of improper equipment being sold to an unsuspecting person. There have also been incidents of overprotection and underprotection occurring because of unknowledgable or unscrupulous businesses or salesmen.

There are over 1600 manufacturers of electronic protective equipment producing thousands of brands, types, and designs of hardware. Most of the equipment produced is of reasonable quality and design. the majority of problems which occur in field use result from misapplication of equipment or improper maintenance.

The increases in criminal sophistication and the rising costs of hiring security officers will undoubtedly produce continued growth in the use of electronic protective devices. The production of increasingly sophisticated hardware will continue as the ability of criminals to successfully attack current systems increases. The escalation which is taking place is comparable to the nuclear arms race between international superpowers. Countermeasures produce more sophisticated measures ad infinitum!

Consultants

Consultants provide assistance and advice in solving security problems. While consultants are not new to the security field, their number has increased significantly in the past ten years. This segment of the field has historically not been dominated by any one individual or firm. There have been numerous individuals who have been identified as "expert" in various fields of loss prevention or security. These individuals or the companies they established have responded to the needs of business, industry or government to develop solutions to security problems.

Consultants work on a fee basis with rates ranging from $200 to $750 per day, plus expenses to assist in eliminating losses. They analyze loss situations, develop alternatives, and recommend steps to reduce or eliminate losses. Their major values are the degree of objectivity they bring to problem situations plus their experience in solving problems. The average manager or security officer often has a limited amount of experience with loss problems. The consultant, however, has the experience of many problems to draw on to produce solutions or recommendations.

Consultants are normally highly experienced loss prevention professionals with extensive law enforcement or security experience in industry, government, or business. There are approximately 200 active individual or firm consultants in the United States who work full time at problem solving. Additionally, another 200 to 300 part-time consultants are active.

Locksmiths

Locksmiths are responsible for the sales, installation, and servicing of physical security devices and equipment. They primarily deal with locking devices, safes, vaults, record cabinets, and related locking equipment.

There are approximately fifteen thousand locksmith shops in the United States with 25,000 to 30,000 locksmiths actively pursuing their trade. It is estimated that the locksmith industry currently accounts for $350 million in sales with projection to $910 million estimated for 1985. Thus, locking devices account for approximately 10 percent of total security industry sales.

Growth and Development

Growth and development of the commercial security industry has been dramatic in the past ten years. The security industry was built on the provision of security services on a local basis with most commercial security firms being developed by individuals who became known in a community or a region for effective law enforcement or protective work and has built organizations to provide services in keeping with their reputation. These organizations have historically provided supplementary protective services on a basis comparable to that of public agencies. Thus, there have been a large number of commercial service firms in densely populated areas and fewer in less populous urban, rural, and semirural areas primarily because the need for protective service either in the public or private sector are more necessary in the highly populated areas. However, even in these areas police officers or private individuals have worked on a part-time, commercial basis to provide protective services on an "as needed basis" to supplement public sector protective activities.

In the last several years there has been a marked change in the operating style of the industry with large national organizations either (1) acquiring small independent operations and incorporating them into national organizations, or (2) building national organizations around the nucleus of a small, locally effective and previously independent security service organization. Since 1970 there has been a dramatic increase in the number of acquisitions of private operations in security service field. The total number of firms in the private security field has been reduced with the advent of significant acquisition programs by national commercial security service firms.

The marked increase in large security organizations on a national/international basis, is the 10 to 12 percent annual growth in dollar volume generated by the commercial security services industry, coupled with the growth projections for this field. Because of these two factors, the commercial security services industry has been singled out as an area of profitable acquisition and merger by conglomerate corporations. There have been many more conglomerates which have analyzed the security industry, acquired security service firms, built national security organizations, and have added these national organizations to their diversified holdings. While many national conglomerates which entered into the security industry in the early 1970's have divested themselves of a continued commitment in the commercial security service field, their acquisition and merger programs have nevertheless had a significant impact on the operating environment in the field.

Several organizations which were established as a result of their merger efforts continue to exist and operate as parts of national security firms. In one case, a regional firm became a national security organization by the acquisition of a firm being sold by a conglomerate,[4] e.g. Oak Security, a subsidiary of Oak Industries. Organizations which entered into the security field were firms such as Gould Industries, ITT, Oak Industries, Teleprompter Corporation, and the Westinghouse Corporation. While organizations such as these were entering or expanding into the perceived high-profit field of commercial security, the existing major security service firms were busy attempting to consolidate their respective shares of the protective services market and to increase their market position by merger and acquisition. The most noticeable examples of this effort being the significant growth of the Wackenhut Corporation to the third largest security service firm in the United States, and Guardsmark, Inc., growing to the sixth largest through significant growth and acquisition.

While this consolidation was occurring on a national basis within the industry, on the regional and local basis, acquisitions of small "mom and pop" type security operations which historically characterized the commercial security industry, were being acquired by more established, financially secure, local or regional security firms consolidating at the local level, while national security firms were attempting to accomplish the same thing on a national and international scale. Firms such as the Kane Service of Chicago acquiring other large regional service firms has become commonplace. The Kane Service, for example, acquiring in 1973 Badger Security Services and Merchants Police of Wisconsin. Regional firms such as California Plant Protection have expanded their scope of operations from merely servicing one part of California, then statewide, and recently providing their services interstate. Regional investigative and security firms such as Smith Security of San Antonio, Texas, providing services in Mexico and then opening offices to give themselves national exposure and contracts. Burns, Wackenhut and Pinkerton's also continue to expand their U.S. operations to include all principal cities in the U.S. plus acquiring significant interests in the major foreign security service firms. They now operate security offices in principal cities throughout the world.

It should be pointed out that this growth and acquisition phenomena is not only limited to the U.S. In England and in Commonwealth countries, for example, the major security service firms have likewise entered into extensive programs of growth, expansion and consolidation. The provision of security services by such firms as Securicor, Group 4, and Phillips, expanded their services not only within England but in most Commonwealth countries.

Many organizations representing high-risk businesses have retained national commercial security firms as their "official" security service organization. The National Jewelers Association, for example, has retained William A. Burns International Security Services as their official investigation force. The American Bankers Association have similarly retained Pinkertons, Inc. Expositions and trade shows similarly hire guard services on a competitive contract basis to handle security for them. In fact, many commercial guard companies have established separate operating groups for these high visibility events.[5]

Effects of Competition

The commercial security service industry is extremely competitive. Competition is based often on the rate per hour and/or on intangibles related to the quality of service offered by different firms. The national firms are experiencing the problems in marketing their services similar to those of the big three of the auto industry. The services that they offer are about equal since many of the personnel that are used to provide the services are drawn from the same labor pool and are as likely to be working for any one of the companies depending on which firm might win a contract in a competitive bidding situation.

The marketing of security services goes on at a number of different levels. The direct sales approach is used extensively with sales personnel going "door to door" to interest potential industrial or commercial clients in the use of those services of a particular company. Similarly, "national account" sales are being extensively utilized to acquire national accounts from firms having operations in several states. This would allow organizations such as Wells Fargo to provide all contracts security for Playboy Clubs, or Security 76 to provide airport screening for all American Airlines terminal facilities. This type of contract is extremely profitable to the commercial service firms and is desired by some national clients as a means of minimizing the problems of managing commercial services at a large number of facilities. This is considered to be so since one corporate security director can deal with a security service account executive to handle problems on a corporate-wide basis. In actual practice, however, the use of a national contract has little advantage over the use of the local security firms. This is primarily due to the fact that the supply of personnel to service a local contract is drawn from the local area on a competitive wage basis. Consequently, the *pool* of manpower being available to any local or national security agency is almost identical. Thus, for all practical purposes, it makes little difference which security service firm had the contract. The personnel performing the services are likely to be about the same in quality.

Example: Profit per guard hour formula

$4.50		Cost per hour (paid by client)
	$2.25	Salary paid to guard
	1.00	Cost of selling, administration
	.75	Uniforms, insurance, supervision, equipment
	$4.00	Expenses per guard hour
$ .50		Profit per hour

Figure 6-2

Training Education and Supervision

Because of the extremely competitive nature of the commercial security service firms, the rates per hour being charged to clients cannot vary significantly for similar services from agency to agency within a community. Thus, in order to maintain profitability of operations, a commercial security services firm must find ways to keep its cost of delivering services in line with the necessary or required profits to stay in business. Therefore, the areas that are normally considered as overhead to the commercial security service firm will suffer if the profit rate margin either shrinks or is too narrow to begin with. These areas are training, equipment, and supervision. Figure 6-2 shows the type of formula used by many security service firms in determining their profit per guard hour. It can be noted that the training, equipment, and supervisory costs are considered overhead and represent a significant portion of each dollaw of income generated. Thus, if the profit to rate per hour ratio shrinks these areas are the ones that will normally be reduced.

Training

The training levels of commercial security service personnel are extremely varied. There are some firms that provide a considerable degree of training or that require a great deal of previous training before being hired. Conversely, there are firms that literally provide a uniform and a gun to a man and put him on the job. The *Rand Report*[6] indicated that the security service man is poorly trained, poorly educated and considerably older than his public police counterpart. Several national organizations, most notably the National Council on Crime and Delinquency, the International Association of Chiefs of Police (IACP), and the Private Security Task Force, are attempting to establish training standards for security personnel. National Council on Crime and Delinquency (NCCD) published the first model security guard training curriculum in 1972. Similarly, the Private Security Advisory Council

(PSAC) of the Law Enforcement Administration is attempting to set meaningful training and licensing standards for private security personnel. The LEAA-IACP committees are represented by major national security service firms, law enforcement personnel and educators and are attempting to establish meaningful standards. Similarly, the American Society for Industrial Security is attempting to establish a certification program for security personnel in an effort to professionalize the security field.

Commercial Training Criteria

The criteria for training security officers for the commercial portion of the industry are more demanding than those for a proprietary guard force. The commercial guard force must compete in the marketplace profitably. To do so any training provided must produce: (1) a competitive advantage, (2) a better product, (3) quality training at low cost per man, (4) maximum training input in minimum time, (5) measureable results.

A note of caution is necessary about the speed with which any dramatic or wide-ranging changes will take place. The ability of the commercial security services to change their operational policies or procedures to accommodate the "professional security officer" will rest mainly upon economic and legislative considerations. At this stage of development of the field, it is doubtful that substantive change will occur without legislative requirement. Similarly, the ability of contract agencies to obtain personnel to fill guard service commitments is dependent upon overall economic conditions which affect employment of semi- and unskilled labor. The competition for persons in these categories will force up the wage rates required to attract and hold personnel for security services. The ability to fill positions utilizing "warm bodies" will result in it being done in the absence of either legislation prohibiting it or users disinterest in the quality of personnel supplied. The effects of a market place economy in the services field tend to produce a situation where all suppliers are at an equal disadvantage in supplying a quality service unless the user is either convinced or can be persuaded to pay a "premium" wage for contract security services. Thus, for a contract agency to make the decision to train personnel or provide a career development program for any but the management personnel, is to place itself in a poor competitive position. In light of the fiercely competitive nature of the industry, this would most likely cause short-term financial losses for it.

There are, however, contract agencies which are "selling" the concept of "premium" services quite successfully. They provide career development, training, supervision, and mobility for their employees while being compensated by the client for the services. These firms and programs are, however, the exception within the field. Several of the major national firms and some local firms are experimenting with "premium" services and have been providing it on a limited basis with clients who have demanded more than the traditional "warm body" from the agency. The personnel utilized to fill assignments are, however, far from the uniformly trained, career security officer considered a "professional." All too often, "premium" positions are considered an opportunity for the agency to make additional profits by providing the same type of relatively untrained security officer at a higher rate to the client and the same low salary to the officer.

To prevent this from occurring, clients have in a number of locations begun paying the "premium" pay directly to the officers in addition to the salary paid them by the contract agency. This insures that the continuity of service takes place and that the incentive pay is received by the officer rather than the agency. Training is also provided directly to them rather than through the contract agency thus giving the client control over the type, quantity, and quality of the training provided.

Should the economy decline and place additional semi- or unskilled workers on the employment market, the prospects for increased "professionalism" would be dimmed for a considerable period. Since the protective services technology field is viewed by the public and users as generally being a relatively low-level service, the tendency will be, as it currently is, to get the service for the lowest possible cost. It would thus be difficult if not impossible to affect any meaningful changes which would promote professionalism.

The movement toward "professionalism," at least in the commercial sector of the protective services technology field, will require optimal conditions for an extended period of time for the users to accept the concept of a "professional security officer." This is possibly not so much the case for the proprietary security organization which does not operate under the same severe financial constraints as the commercial firm. It can thus move toward the status of a profession more rapidly but will also be constrained by the lack of conception of either a profession or an emerging one by general corporate management.

Supervision

Supervision in commercial security personnel is one of the most significant problems that exists in the industry. Supervision is a costly activity. Although the cost of supervision is built into the rate per hour, since rates are competitive it is at best difficult to provide a high-quality supervisor to maintain direction control over security officers performing in the field. Consequently, the supervisor span of control and inspection of services performed is not always adequate. Consequently, poor officer performance, neglect, inappropriate actions being taken, lack of clear-cut direction to officers, are common in many segments of the commercial security field.

Equipment

Security service firms provide very little equipment to their personnel. The weapons they carry, uniforms, communications equipment, and vehicles utilized, all must be paid for by client revenues. Thus, a variety of means are utilized to keep the cost of these items as low as possible, while providing an acceptable level of service to the client. Some commercial security service firms provide uniforms and equipment such as weapons, uniforms to their employees on a no-cost basis while they are employed, others charge their employees for the cost of the uniform while providing weapons while still others require the employees to pay the full cost of equipment such as fire arms and uniforms.

Licensing and Regulations

The development of state or local licensing requirements for commercial security firms has been recommended almost universally as a means to control the industry. Protection of consumers for criminals disguised as security officers, insuring a minimum of training for officers, and registration of security personnel, are all aspects of these requirements. Very few states currently have licensing or regulatory standards which are meaningful for consumer protection.[7] The National Advisory Committee on Criminal Justice Standards and Goals—Private Police* has set advisory standards for licensing and regulations. These standards are reviewed in Appendix 6-B.

Ethics

The commercial security field like its proprietary counterpart has not developed a Code of Conduct or a Code of Ethics to govern its activities.[8] There are likewise few accepted standards for judging the effectiveness or appropriateness of recommended manpower or equipment. The ethics involved in providing commercial security are those that guide any other form of business. There are, however, a few additional dimensions which other forms of business do not have. They include ethical considerations which:

1. derive from the legal responsibilities for their actions;
2. belong to former law enforcement practitioners now in commercial protection;
3. are based on a shared sense of "morality";

4. are based on efficiency or effectiveness; and
5. are viewed as necessary for a profitable operation in the marketplace.

Most commercial protective services firms will try to provide an honest and practical protective program for their clients. Whether this is caused by marketplace competition or by the inherent sense of honesty is not known. It must be assumed, however, that there is a desire to provide conscientious protective programs.

FOOTNOTES

1. Excerpted in part from "Investment Opportunities in the Security, Protection and Investigative Services Industry." Burnham and Company, September, 1970.
2. Appendix 6A presents a historical case study of the development of the central station protective service industry.
3. Frost and Sullivan: *Analysis of Security Industry.* New York, 1975.
4. Smith Security Acquisition of certain Oak Security Operations Division, 1973.
5. Burns and Pinkerton, for example, have Special Operation units.
6. *Nature and Extent of Private Police in the U.S.,* 1971.
7. See Chapter 6, Appendix 6A for a listing of state regulatory standards.
8. See Appendix 6C for a suggested Code of Conduct for security officers.

BOOKS

Alexander, Alfred, and Moolman, Val: *Stealing.* New York, S. & S., 1969.
Crookston, Peter: *Villain.* London, Jonathon Cape, 1967.
Fletcher, J. A., and Douglas, Hugh: *Total Loss Control.* London, Associated Business Programs, Ltd.
Genet, Jean: *A Thief's Journal.* New York, Grove Press, 1964.
Hammond, Victor: *The Security Officer's Handbook of Public Relations.* Calif., Davis Publishing Co., Inc., 1975.
Haskins and Sells: *Cash Handbook.* New York, Haskins & Sells, 1940.
Mitchell, Ewan: *Coping with Crime.* London, England, Business Books Ltd., 1969.
Rogers, Keith M.: *Detection and Prevention of Business Losses,* New York, Arco, 1962.

---

* Renamed the Private Security Task Force.

Strobl, Walter M.: *Security.* New York, Industrial Press, Inc., 1973.
Woodruff, Ronald S.: *Industrial Security Techniques.* Columbus, Ohio, Charles E. Merrill Publishing Company, 1974.

PUBLICATIONS OF THE GOVERNMENT, LEARNED SOCIETIES, AND OTHER ORGANIZATIONS

Ghosh, Srikanta: The Police Want to Help You. Cutback, Orissa Olympic Association, 1950.
Gordis, Phillip: Property and Casualty Insurance. Indianapolis, 1953.
Fields, W. E.: Industrial Security Meets New Challenges. Owego, New York, IBM Corporation.
Safe Manufacturers' National Association, Inc.: Handbook for the Industry. New York, Safe Manufacturers' Nat. Assn., 1956.
Hoover, J. Edgar: What does crime mean to industry? National Association of Manufacturers, September 21, 1964.
Industrial, commercial, and residential security market. A Report by Frost and Sullivan, Inc., New York.
Journal of Insurance. The manicured hand in the till, June 1971.
Manual of Burglary Insurance. New York, National Bureau of Casualty Underwriters, 1963-64.

PERIODICALS

Ads you'll never see. *Business Week,* September 21, 1957, p. 30-31.
Arm, Walter: *Operation Security,* March 1961.
Ashton, Alfred P.: Controlling security costs—panelist. *Industrial Security,* October 1960.
Available market data: Industrial. *Advertising Age,* May 1958, p. 143-52.
Barlay, Stephen: No panic—just deadly calm. *Top Security,* August 1975, p. 140-142.
Brown, Jack: U.K. trucking executive comments on industry. *Canadian Security Gazette,* April 1971.
Boise, Robert J.: Management notebook. *Security World, 1:* No. 3, 30-31, November 1964.
Conlin, Earl E.: Internal controls at Ex-Cell-O. *Security World, 1:* No. 2, September 1964.
Crime and the convenience store. *Convenience Store Journal,* November 1971, p. 20-26.
Creating a climate for honesty. Part I and II. *Security World,* October-November 1971.

Crime Control Digest. SEI/TECH Digests, Inc., National Press Building, Washington, D.C.

Crime prevention news in Divon and Cornwall. *Crime Prevention News,* September 1975.

Crime in industry program. *Canadian Security Gazette,* April-May, 1970.

Crime in industry—Seminar III. *Canadian Security Gazette,* Summer 1971.

Darling, Don: Teleprinter, printout security. *Security World,* October 1970, p. 23.

Dauw, Betty: The high cost of vandalism. *Safety Education, 44:* No. 7, March 1965.

Davis, John R.: Plant protection pays its way. *American Business,* January 1956.

Dickson, Felice: Thieves grow bolder, plant owners seek recovery, protection. *The Miami Herald,* May 4, 1975, p. 17-K.

Doherty, Joseph F.: Seven steps to better security. *Industrial Security,* April 1962.

Donahue, Vincent J., Col. USAF: Security—a necessary evil? *Security World,* October 1966.

Brownyard, Bruce: Are you buying security—or trouble. *Security Management,* November 1974, p. 18-19.

Brown, B. N.: Security—a business asset. *Industrial Security, 9:* No. 1, 24-27, January 1965.

Buchman, Lawrence P.: The security director as well-rounded person. *Industrial Security, 4:6,* January 1962.

Buckley, John L.: Good industrial security does not cost—it pays. *Law and Order,* August, 1961.

Bunch, H. E.: Source of information, industrial security, and law enforcement. *Industrial Security, 12:27,* June 1968.

Burns, W. Sherman: Does your plant invite theft. *Management Methods,* December 1960.

Burnstein, Harvey: Security consultant—the question or answer. *Industrial Security,* August, 1971.

Buyer's guide to security services. *Business Management,* July 1962.

Byrne, John M.: Something extra—prescription for selling security to management. *Security World,* June 1975, p. 16+.

Camin, Louis J.: Some projections for business security. *Security World, V:* No. 2, February 1968.

Cantz, Marvin: Selling security to management. *Security World, VI:* No. 6, June 1967.

Carte, Gene E.: In defense of alternative policing: A reply to James Q. Wilson. *Criminal Law Bulletin, 8:207,* 1972.

Cary, Fred W.: Is the industrial security manual an adequate instrument? *Security Management,* November 1974, p. 7-9.

Celebrity watchers. *Newsweek, 36:77,* January 27, 1964.

Connolly, John: A positive police program to help small business help itself. *Industrial Security,* October 1963.

Connor, D. M.: Soviet people's guard: An experiment with civic police. *New York University Law Review, 39:579,* January 1964.

Cops patrol Grover—arrest 5 youths. *Buffalo Courier Express,* September 22, 1970.

Cost of crimes against business, the. *Security Management,* March 1975, p. 6-14.

Crime wave—what can be done about it? *U.S. News and World Report,* August 1, 1966.

Crisis Management (film). Charles S. McCrone Productions, Aptos Village.

Davis, Albert S.: Security vocation query reveals interesting data. *Industrial Security,* August 1970.

Davis, Arthur D. C.: The security as a well rounded person. *Industrial Security,* January 1961.

Davis, James A.: What is a security director? *Industrial Security,* December 1970.

Downie, John W. Col: Security background in Vietnam. *Industrial Security, 12:12,* April 1968.

Dubois, Peter C.: Specialists in security. *Barron's, XLIV,* March 16, 1964.

Dudley, W. E.: New game called industrial security. *Industrial Security, 16:* No. 2, 29-30, April 1972.

Duggan, Dennis: An all-seeing eye—security. *New York Herald Tribune,* August 18, 1968.

Elements of a complete security program. *Security Bulletin,* June 4, 1974, p. 1-4.

Elliot, Bill: Management notebook: Berkins Van and Storage. *Security World, 2:* No. 1, January 1965.

Enter a stranger. *Security and Protection,* April 1975, p. 17-18.

Eppert, Ray R.: Security and free world. *Industrial Security, 5:67,* October 1961.

Expansion all round for PPR security services. *Security and Protection,* June 1975, p. 16-17.

Facts about modern plant protection. *Mill and Factory,* May 1954.

Faltermayer, Edmund K.: Some here-and-now steps to cut crime. *Fortune,* July 1970, p. 95.

Fast, Howard: An occurrence at Republic Steel. *The Aspirin Age,* 1949.

Ferrari, Norman D.: Rescue service in industry. *Industrial Security,* July 1960.

Folz, David F.: The role of industrial security in business mergers and acquisitions. *Industrial Security, 10:* No. 2, April 1966.

Formula to maximize security awareness. *Industrial Relations News, 13:* No. 22, June 1, 1963.

Foster, Reginald: Looking in the mirror. *Security and Protection,* October 1974.

———: Food for thought. *Security and Protection,* January 1975, p. 4-6.

———: On the boards and in the field. *Security and Protection,* September 1974, p. 4-6.

———: Women in security. *Security and Protection,* March 1975, p. 4-6.

Gardner, George: The primary and secondary gains in stealing. *Nervous Child, 6:*436-446, October, 1947.

Gately, Glenn S.: Insurance looks at security. *Industrial Security,* June 1971.

Gatter, Howard: Standards for security? *Security World,* June 1970.

Gay, William O.: The man in the Quaker hat. *Police Journal, 18:*147.

Getting rid of waste. *Security and Protection,* June 1975, p. 12.

Gibson, Gwen: Over-the-counter bugging devices. *New York Herald Tribune,* October 14, 1965.

Glass firms chip in. *Security and Protection,* April 1975, p. 4-5.

Gocke, B. W.: Aspects of security protection for business and industry. *J. Criminal Law, Criminology, Police Science, 48:* No. 2, July-August, 1957.

———: Personnel aspects of industrial plant security. *Police, 2:* No. 5, May-June 1958.

Goddard, Robert J.: Wardens in industrial plants. *Industrial Security,* July 1960.

———: The anatomy of security. *Industrial Security,* April 1964.

Gorlick, Arthur: Everything's bugged as gumshoes meet. Long Island City (N.Y.) *Star-Journal,* September 7, 1965.

Gould, Leroy C.: The changing structure of property crime in an affluent society. *Social Forces, 48:*50-59, September 1969.

Griffin, G. K.: Industrial security in the automotive industry. *Canadian Security Gazette,* October 1970.

Griffin, Roger K.: Why people confess: Reflections on 1,000 interrogations. *Security World,* May 1970.

Grover, G. M.: Insurance views on security in industry and commerce. *Security World,* March 1969.

———: Insurance and the security man. *Security World,* June 1970.

Hart, R. E.: Safety and security in Aerospace Industries, panelist. *Industrial Security,* October 1963.

Hata, Shigeru: Industrial security in Japan. *Industrial Security,* December 1966, p. 8.

Hayden, Charles E.: Controlling security costs where research and development are combined with limited production panelist. *Industrial Security,* October 1960.

———: How to make your company more security conscious. *Industrial Security,* April 1962.

Hayden, Charles E., and Walsh, T.: *Industrial Security Management—A Cost-Effective Approach.* New York, American Management Association, 1971.

Hendrix, Algie A.: Industrial security—responsibilities and opportunities. *Industrial Security,* July 1964.

Hewitt, Norman: Do you really need insurance? *Security World, IV:* No. 8, September 1967.

Higgins, George D.: Is poor security costing you money? *Tooling and Production,* October 1962.

———: Safeguarding industrial security. *Industrial Security,* April 1962.

———: Seven steps to better company security. Dartnell Employee Relations Service Special Feature.

How do we stack up? *Industrial Security,* September 1963.

How security does pay off. *The Office,* September 1971, p. 22-26.

How specialists handle plant security. *Foundry,* August 1963.

Humphries, Russell J., and Beck, Sanfore E.: How to submit an insurance claim. *Security World,* December 1974, p. 12+.

Internal Control, New York, American Institute of Accountants, 1949.

Johnson, V. F.: Fire insurance coverage (Seminar Workshop Report). *Industrial Security,* October 1961.

Kaplan, Philip M.: Security: The view within. *Security World,* December 1970.

Kay, William: Security—A good investment. *Top Security,* July 1975, p. 97-99.

Kelley, Woody A.: Exercising for your own security. *Industrial Security,* April 1970.

Landau, Peter: Psst. Don't look now. . . . *Newsweek,* September 5, 1960.

Lewis, Edward, Jr.: Drug abuse effects, Part 1. *Security World,* September 1969.
———: Drug abuse effects, Part 2. *Security World,* October 1969.
Lodge logic. *Security and Protection,* January 1975, p. 8-10.
Mackey, John B.: Controls of public releases. *Industrial Security,* January 1963.
Mort, Victor C.: Discuss the value of the continuance of private police forces and the growth of commercial security organization. *Police Review,* April 4, 1975, p. 420-422, 435.
Murphey, John T.: Between patience and fortitude. *Police, 5:* No. 2, November 1960.
McCarthy, Joseph E.: Plant security guards: Outside agency or plant personnel? *Modern Manufacturing,* December 1968, p. 15.
No names, please. *Occupational Hazards,* January 1970, p. 19.
Phillips, Wm. G.: The security guard agency. *Canadian Security Gazette,* April 1972.
Peculiar position of the private guard. *National Safety News,* May 1969, p. 56.
Putting a little fun into the security program. *Business Week,* September 6, 1958, p. 64.
Reddin, Thomas: Law enforcement turned civilian: A new look at guard forces. *Security World,* June 1975, p. 18-19.
Taylor, Henry J.: Nations security measures better. *Corning Leader,* July 8, 1963.
The rent-a-cop boom. *Newsweek,* January 10, 1972.
Three cases of theft and capture. *Occupational Hazards,* June 1971, p. 39.
Walsh, Timothy J., and Healy, Richard J.: Alcohol and drug abuse. *Protection of Assets Manual,* Vol. II. California, The Merritt Company, 1974.

APPENDIX 6A

United States of America v. Grinnell Corporation. American District Telegraph Company, Holmes Electric Protective Company, and Automatic Fire Alarm Company of Delaware. Civ. A. No. 2785, United States District Court D. Rhode Island. November 27, 1964.

Action against a sprinkler company and three affiliated companies for violation of the Sherman Act. The District Court, Wyzanski, J., held, inter alia, that evidence established violations of the monopoly provisions of Section 2 of the Sherman Act by defendants.

Judgment in accordance with opinion. See also 30 F.R.D. 358.

INTRODUCTION

This is a civil suit wherein the Government complains that Sections 1 and 2 of the Sherman Act [Act of July 2, 1890, c. 647, 26 Stat. 209, 50 Stat. 693, 15 U.S.C. §§1, 2] have been violated by Grinnell Corporation, a Delaware corporation with its principal place of business in Providence, Rhode Island (hereafter called "Grinnell"), and three corporations, a majority of the capital stock of each of which is owned by Grinnell, to wit, American District Telegraph Company, a New Jersey corporation (hereafter called "ADT"), of whose capital stock Grinnell owns 76 percent, Holmes Electric Protective Company, a New York corporation (hereafter called "Holmes"), of whose capital stock Grinnell owns 100 percent, and Automatic Fire Alarm Company of Delaware, a Delaware corporation (hereafter called "AFA"), of whose capital stock Grinnell owns 89 percent. All these three corporations have their principal place of business in New York. Collectively, the three corporations are referred to sometimes as the "affiliates," and sometimes as the "alarm companies."

Complaint, seeking relief under §4 of the Sherman Act [26 Stat. 209, 15 U.S.C. §4], was filed April 13, 1961. Extensive pretrial discovery and frequent pretrial conferences with the Court, and, above all, the cooperation of informed, industrious, and experienced lawyers, who jointly took 128 depositions (totalling over 8,000 pages), to a large extent disclosed to each other proposed exhibits, entered into five stipulations (totalling 58 pages), and exchanged careful, thorough, pretrial briefs (in excess of 400 pages), enabled the parties at the outset of the trial to lay before the Court 1,181 exhibits comprising approximately 15,000 pages. The preliminary procedural and substantive steps reduced the taking of testimony in open Court to six days, from June 15, 1964 to June 24, 1964. The Court, at the conclusion of the testimony, required each party to use the summer recess to limit to forty pages the principal portion of its brief, with a right to annex appendices of unlimited length. Oral arguments upon those briefs and replies thereto took place on October 9, 1964.

Before findings, conclusions, and opinions are set forth, this Court takes this opportunity to make explicit certain aspects of its approach to this controversy.

This Court is mindful that in recent years antitrust litigation, particularly Government civil actions alleging violations of §2 of the Sherman Act, have involved an enormous, nearly cancerous, growth of exhibits, depositions, and *ore tenus* testimony. Few judges who have sat in such cases have attempted to digest the plethora of evidence, or indeed could do so and at the same time do justice to other litigation in their courts. Nor is there any sound reason to believe that such extensive presentation accomplishes any important legal or other social end.

Historically, the major explanation of prodigious records probably is the lack of certainty both at the bar and on the bench as to what was the scope of §2, and, even to some degree, of §1 of the Act. Fluctuations in Supreme Court interpretation of the statute prevented lawyers no less than laymen from having confidence as to how an antitrust controversy would be viewed in the Supreme Court, and those who were in authority were unwilling prematurely to draw sharp boundaries of relevance and materiality.

Time, however, has hardened the lines of interpretation of the Act. No doubt, courts have been aware that the text of the Sherman Act, *simpliciter*, though it withstood early attacks grounded on the charge that it was so vague as not to meet the due process standard of the Fifth Amendment to the United States Constitution, *Nash v. United States,* 229 U.S. 373, 33 S.Ct. 780, 57 L.Ed. 1232 (1913), would, if enacted in the same words today, without explicit legislative guidance from the debates, the committee reports, the exordium of the statute, and other ancillary sources, not easily hurdle the barrier of the Fifth Amendment, and comply with recent precedents requiring definiteness as a *sine qua non* of valid criminal legislation.

To satisfy both modern judicial susceptibilities, and increased awareness of the right of prospective defendants to clear warning of what constitutes criminal conduct, courts, while not abandoning the possible *in futuro* widening or deepening of the Sherman Act with the growth of experience, and while not sacrificing the possible stretch of the statute in the future to keep pace with any scheme hereafter developed to evade its deliberately prospective broad reach, have tended to lay down, especially in connection with §1 of the Sherman Act, so-called *per se* rules (as, for example, the rule invalidating price-fixing agreements made to cover a substantial market), which are relatively precise and which form the basis of virtually irrebuttable charges of violation of the Sherman Act. *United States v. Paramount Pictures,* 334 U.S. 131, 68 S.Ct. 915, 92 L.Ed. 1260; *Schwegmann Bros. v. Calvert Distillers Corp.,* 341 U.S. 384, 71 S.Ct. 745, 95 L.Ed. 1035; *United States v. New Wrinkle, Inc.,* 342 U.S. 371, 72 S.Ct. 350, 96 L.Ed. 417. Such *per se* rules go beyond presumptions, burdens of proof, or like procedural measures. They are substantive glosses constituting clear corollaries.

With regard to §2 of the Sherman Act, the trend has been less clear, and has reflected a greater degree of wariness or timidity on the part of judges. Yet in the two decades since the opinion of Judge

Learned Hand in *United States v. Aluminum Co. of America,* 2nd Cir., 148 F.2d 416 (1945), most of the cognoscenti have expected that a day would come when the Supreme Court would announce that where one or more persons acting jointly had acquired so clear a dominance in a market as to have the power to exclude competition therefrom, there was a *rebuttable* presumption that such power had been criminally acquired and was a monopolizing punishable under §2. To be sure, the putative offender would be allowed to avoid or defeat this presumption if he bore the burden of proving that this share of the market was the result of superior skill, superior products, natural advantages, technological or economic efficiency, scientific research, low margins of profit maintained permanently and without discrimination, legal licenses, or the like. *Cf. United States v. United Shoe Machinery Corp.,* D.Mass., 110 F.Supp. 295, aff'd 347 U.S. 521, 74 S.Ct. 699, 98 L.Ed. 910. Such a shifting of the burden not merely of going forward, but of proof, such a rebuttable presumption rest on the by-now dozens of court records which make it quite clear that it is the highly exceptional case, a *rara avis* more often found in academic groves than in the thickets of business, where monopoly power was thrust upon an enterprise by the economic character of the industry and by what Judge L. Hand in *Aluminum* called "superior skill, foresight and industry." More than seven decades of Sherman Act enforcement leave the informed observer with the abiding conviction that durable nonstatutory monopolies (ones created without patents or licenses or lasting beyond their term) are, to a moral certainty, due to acquisitions of competitors or restraints of trade prohibited by §1. They are the achievement of the quiet life after the enemy's capitulation or his defeat in inglorious battle.

[1, 2] To this Court it appears that the day has come for it, and more important for counsel, to proceed on the acknowledged principle that once the Government has borne the burden of proving what is the relevant market and how predominant a share of that market defendant has, it follows that there are rebuttable presumptions that defendant has monopoly power and has monopolized in violation of §2. The Government need not prove, and in a well-conducted trial ought not to be allowed to consume time in needlessly proving, defendant's predatory tactics, if any, or defendant's pricing, or production, or selling, or leasing, or marketing, or financial policies while in this predominant role. If defendant does wish to go forward, it is free to do so and to maintain the burden of showing that its eminence is traceable to such highly respectable causes as superiority in means and methods which are "honestly industrial," as Judge Hand characterized the supposititious socially desirable monopolizer.

[3] One other preliminary point does deserve note—and that is the relevance, if such there be, of defendant's intent. The issue is whether when the Government or other plaintiff charges a defendant with violation of the Sherman Act the complainant is under a duty to prove that his adversary has either *mens rea* or, more than that, the specific intent to violate the antitrust Acts. So far as concerns a charge of "monopolizing," unlike the lesser offense of "attempting to monopolize," it is not necessary, as both *Aluminum* and *United Shoe* declared, for the complainant to prove more than that defendant intended to engage in the practices which maintained its market power. A defendant who monopolizes has "first of all * * * the intent to do the act, and secondly * * * a knowledge of the circumstances that make that act a criminal offense." Patrick [Lord] Devlin, lecture on "Statutory Offenses," republished in *Samples of Lawmaking* [London, 1962], pp. 67-82, at p. 78.

FINDINGS OF FACT

1. The gist of the Government's complaint is that defendants and co-conspirators (1) have been engaged in an unlawful combination and conspiracy to re-

strain, and (2) an unlawful combination and conspiracy to monopolize, and (3) have attempted to monopolize, and (4) have monopolized, interstate trade and commerce in what the Government denominates "the accredited central protective service business."

2. The central station protective service (hereafter called "CSPS") business consists of maintenance of a central station, installation and maintenance of hazard detecting devices on the subscribers' premises, connection of these devices to the central station by wires leased from the local telephone company, and receipt and handling of alarms transmitted to the central station from the subscribers' premises.

3. CSPS, at present, involves customer concentration within specific areas. The chief reason is that alerted CSPS employees must promptly reach the scene of the alarm, and the present system does not have scattered task forces. Another reason is that high leased-line costs make it expensive to transmit signals for a great distance. Twenty-five miles from the CSPS central station has become the common radius. But this Court is not persuaded that, with the advance in technology, there are any physical obstacles, such as electrical resistance, which now stand athwart the geographical expansion of central station coverage whenever business advantage, competitive conditions, and prospect of favorable profits make the course expedient.

4. A central service station, in its normal daily service operations, is physically independent of any other central service station. As already noted, its service is summoned, responds, and is consummated usually within a twenty-five mile radius. This is *not* to imply that the ultimate supervision is not nationally directed; nor that the system of operations, nor that the equipment used, nor the checking of equipment, nor the reports about the system have a purely local, individualized character, origin, or destination. On the contrary, the enterprise in appearance and reality, in financing, selling, advertising, purchasing of equipment, processes of management, and over-all planning is national; though, of course, the impact, as is true of every enterprise, is upon local specific points.

5. By settled practice, universal in the insurance business, underwriters of fire and burglary insurance (hereinafter called "underwriters") allow a reduction of premiums to customers having approved protective signalling systems. To that end, underwriters have established testing laboratories, have adopted standards, have organized periodical inspections, and have offered premium discounts, of almost standard nature, based thereon. Discounts accorded to accredited central station service subscribers tend to be noticeably larger than for users of other systems.

6. When a prospective subscriber requests CSPS service, the service company surveys the subscriber's premises, determines the number and type of required protective devices, and estimates the cost of needed equipment and labor plus overhead expenses and often expected profit, all of which are usually used in computing an "installation charge." This charge is normally payable upon completion of the installation. Additional "annual service charges," usually payable periodically in advance, and calculated with reference to the equipment in service, telephone line rental costs, expected service including maintenance costs, and often expected profit, are also charged to the subscriber. Annual service charges are covered by contracts for periods of up to five years. Such a contract normally provides that at its termination the service company has a right to remove the equipment without refunding money or making any other payment to the subscriber.

7. In 1961 ADT provided CSPS to 121 units, located, respectively, in 115 cities, which, in turn, were located, respectively, in 35 states and the District of Columbia. In 92 of those 115 cities ADT has no CSPS competition.

8. Holmes provided CSPS to fourteen units, each located in one of three cities—eleven in New York City, two in Philadel-

phia, and one in Pittsburgh, in all of which it has CSPS competition.

9. AFA provided CSPS to three units, each located in one of three cities—one each in Boston, New York City, and Philadelphia, in all of which it has CSPS competition.

10. As of December 31, 1961 in the United States there were, in addition to the alarm company defendants, thirty-three CSPS companies approved by one or more underwriters. Of those, each of five provided CSPS from two locations; each of the remaining twenty-eight provided CSPS from only one location.

11. Of those thirty-three CSPS competing companies, each of seven operated in one of seven cities without CSPS competition; each of twenty-three operated in one of twenty-three cities where it had CSPS competition from one or more of the alarm company defendants; three operated in cities where they had CSPS competition from others.

12. Business or other enterprises located in cities where there is CSPS have the option, of course, of using watchmen, watchdogs, automatic proprietary systems confined to one site (often, but not always), alarm systems connected with some local police or fire station, often unaccredited CSPS, and often accredited CSPS. There are business or other enterprises which in the same city exercise their options differently for different sites in that city. Thus, the Government itself in Washington uses at some public buildings a proprietary system; at others, accredited CSPS.

13. Examples of a particular enterprise changing from one alarm method to another are recited in the record; but, *significantly*, the preponderant shift is from the less integrated, advanced, expensive, and safe method of proprietary systems to accredited CSPS, thus indicating a customer recognition of a difference in market. This difference, to use a popular analogy, could be compared to moving into the class of the rich by changing from a compact six-cylinder car to a chauffeur-driven sedan. That is, there is a major product difference far transcending the obvious possible minimal cross-elasticity of demand.

14. It is also true that many central station companies, including defendants, furnish other forms of protection as well as accredited central station protective service. But this entry by one company into two or more adjacent markets has no substantial significance in defining either of these markets. Here a comparison may be drawn, by analogy, with the practice of General Motors Corporation in offering customers a choice of a chauffeur-driven Cadillac or a Chevrolet Corvair. General Motors caters both to "the Privileged and the People" who, in Disraeli's phrase, form "Two Nations."

15. In 1960 the alarm company defendants together with all other CSPS companies obtained from United States manufacturers in various states over $12,000,000 worth of protection equipment for installation. According to page 6 of the Reply Brief for defendants, ADT, Holmes, and AFA, all this equipment was "manufactured * * * by the defendants" (presumably overwhelmingly by Grinnell).

16. In 1961 the alarm company defendants provided CSPS service to thousands of subscribers of whom more than 900 were each located in a state other than the one wherein the servicing station was located. For this type of service the charges in the previous year, 1960, had included $677,598.15 for interstate transmission of signals over wires.

17. Up to this point, most of what has been stated relates primarily to the three alarm company defendants which are themselves in the accredited CSPS business. Grinnell, which is not in that business, nonetheless manufactures the alarms, sprinklers, and other equipment which may be denominated the machinery of the business. It is the principal supplier of the dominant companies in the market. And its economic interest as a supplier stretches beyond the immediate purchasers, that is, beyond the alarm companies, to the subscribers for CSPS because, as will be recited in later findings, some of Grinnell's

compensation for its equipment takes the form of a share in the revenues received by the alarm companies from the fees paid to them by their subscribers.

18. Were it necessary to make such a finding, this Court would find that a primary motive of Grinnell when it acquired 89 percent of the capital stock of AFA, and 79 percent of the stock of ADT was to secure an assured outlet for part of Grinnell's manufactured products. Investment purposes were also a motive.

[4] 19. However, in courts, unlike on the stage, the drama need not concentrate upon the parties' motives. As explained in the Introduction to these Findings of Fact, the law does not require even proof of intent, much less proof of a motive, where the wrong charged is monopolization, as distinguished from an attempt to monopolize. Moreover, such intent as must be proved to sustain the charge of an attempt to monopolize, or a conspiracy to monopolize, relates to the purpose *to* monopolize, not to the purpose *of* the monopolization.

[5] 20. Hence, it is quite adequate to demonstrate that Grinnell by its mere acquisition of control of companies which in combination have 87 percent of the accredited CSPS market has shown by its actions that it is possessed by what the mediaevalists called "the lust for power." No doubt, that lust is often in modern times characterized by materialism, that is by the search for economic advantage, including financial rewards and outlets for manufactured products. Here it is indeed probable that Grinnell's and its president's acquisition of leverage over the market for accredited CSPS was in part motivated by a desire to increase Grinnell's manufacturing business and to assure a quiet market as an outlet for the sale of Grinnell's alarms, sprinklers, and other equipment. Yet, just as the Government has no obligation to prove defendants' motives, so the Court is under no duty to make findings with respect to those springs of action. Speculation as to motives may be favored by playwrights, prosecutors, and jurors, but judges do well to take in a Pickwickian sense Pascal's dictum that "the heart has reasons of which the reason knows not."

21. Grinnell's relations to ADT, AFA, and Holmes are in a formal sense so structured that each of the affiliates of Grinnell had separate officers, and separate boards of directors which have independent identity. But the form is not a true mirror of the substance of the interrelation of the four companies.

21a. Since 1919 the balance sheets of Grinnell, submitted as part of Grinnell's reports to its stockholders, have shown, though in unconsolidated form, the results of the operations of all 4 defendants.

22. Since 1948 the president of Grinnell has been James Douglas Fleming, a man who joined Grinnell in 1919, and whose competence, character, and force of leadership stand revealed throughout the whole record, and most particularly in the testimony he candidly gave in open court. Until 1964, he stood in lonely eminence; there were no vice presidents of Grinnell; nor others who had more than a closely delegated authority. A purely nominal change has been made in the current year. Four men who had been with Grinnell for thirty years but who had no other independent distinction were named vice presidents.

23. Mr. Fleming, in addition to being president of Grinnell, sits upon the board of directors of each of the four defendants; and is chairman of the board of ADT. There is only one fellow director of ADT who is not an officer of Grinnell or of ADT. The situation is not noticeably different in any of the affiliated companies. In only one instance that Mr. Fleming could recall, has a board of directors of one of the affiliated defendants ever vetoed, or rescinded, or overridden any of his or Grinnell's management proposals or plans. In all major operations of all four defendants Mr. Fleming has illustrated the Emersonian proverb that "where MacDonald [or, as misquoted in the record (Tr. 732) MacGregor] sits, is the head of the table."

24. The personnel unity achieved through the dominant character of Mr. Fleming is buttressed in many ways. Under allocation agreements, executed in 1907, Grinnell as late as 1960 received over $1,200,000 from revenue produced by contracts made by its affiliates with one or more of the subscribers located in one or more of the thirty-five states and the District of Columbia. This sum was related to the same type of revenue which in the next year, 1961, produced for ADT, AFA, and Holmes more than $57,000,000 on account of CSPS charges those companies rendered to 93,272 subscribers located in thirty-five states and the District of Columbia.

25. Moreover, there is a generic similarity, though not an identity, in the contract forms, price lists, and instruction forms used by the three alarm company defendants. Many examples of each of these documents are sent regularly from the principal or home offices of the three alarm company defendants across state lines.

26. Furthermore, across state lines move to the principal or home offices of the alarm defendants reports, proposed contracts, payments, and other communications. Travel from one state to another on company business takes a substantial part of the working time of the principal officers of each of the defendants. This travel as well as the written communications produce a high degree of integration of the major policies of the three alarm company defendants, and result in effective control by Grinnell and by Mr. Fleming of all four defendants, their company subdivisions, their district sales offices, and their personnel.

27. In considering what is the relevant market, or what are the relevant markets, of interstate commerce, many of the factors heretofore detailed require consideration.

28. From the viewpoint of the type of service or product, the relevant market is not, as defendants contend, all protective systems, ranging from individual watchmen on the premises to the elaborate accredited central station alarm systems typified by the alarm company defendants' activities.

29. Quite plainly, nonautomatic systems differ not merely in technology but in utility, efficiency, reliability, responsiveness, and continuity from all automatic arrangements. The difference between watchmen and watchdogs, at one end of the spectrum, and electrical systems at the other lies at the very heart of what is meant by such phrases as "the industrial revolution," "the machine age," "technological advance," and the "era of automation." Antitrust judges in their employment, no less than unskilled workers in their unemployment, recognize that markets for services rendered without tools and without education in the handling of them differ *toto mundo* from markets for services related to machines.

30. Nor, from the viewpoint of the type of service or product, can it be validly maintained that the relevant market embraces all automatic systems, including not merely accredited CSPS, but also local, auxiliary, and proprietary alarms and alarm systems unconnected with accredited central stations.

31. The accredited central station business is marked out by the very patterns of the alarm company defendants, who are the dominant factors in the protective industry, as being an identifiably separate market. Illustrative are the subscriber contracts with their installation and annual charges.

32. Many (though by no means all) of those who are engaged in the alarm company business, speaking in their capacity as experts (not necessarily with the special added status of authorized spokesmen making vicarious admissions binding defendants) have in correspondence and orally shown that that portion of informed opinion which this Court regards as the more credible recognizes the accredited central station business as a separate type of business.

33. Insurance companies allow their insureds a far larger reduction of premium

for accredited central station service than for other types of protective service. The annual charge paid by customers is several times as much for accredited central station service as for proprietary systems, and more than for any other types of protection service.

34. There are trade associations uniting under one roof accredited central station protective service companies, and not including the generality of all types of automatic alarm systems.

35. As already noted, customers regard a change from proprietary alarms to CSPS as an entry into what is qualitatively and monetarily a new class, as when one changes from a Volkswagen to a chauffeur-driven Rolls Royce, or from a boarding house to a hotel named in a hotel association listing.

[6] 36. Hence, for purposes of the antitrust laws there exists a specific, separate, identifiable, recognized market limited to accredited central service protective systems operated by largely automatic devices.

37. If the market, product-wise, be plainly defined as limited to accredited central station protective service connected with automatic devices, it is hardly less clear that, geographically, because of the very nature of those automatic devices and the services related thereto, and the patterns in which that industry has developed, the market has a national interstate-commerce character.

38. It is not merely a congeries of segregated local or city markets. Of course, in addition to the principal national market, there may well be local markets of limited territorial area, or city markets, which in other litigation might be found in themselves to constitute, for purposes of the antitrust laws, definable, separate markets, wherein prohibited monopolies, or prohibited monopolization, or prohibited restraints, or prohibited attempts to achieve those forbidden ends might be enjoined or punished. But regardless of such local or city markets, there exists a national interstate commerce market in accredited central station protective services.

39. The existence of a national market is underlined by defendants' own national pattern of business, with its close articulation of control from the pinnacle of Mr. Fleming down the whole range of subordinate officers, and with its constant use of interstate commerce involving equipment, officers, salesmen, correspondence, even to an appreciable extent electric current carrying alarm signals from one state to another.

40. A fire or burglary may be local. Alarms may be sent to a point only a few blocks away. Relief may be dispatched from a local fire or police station. But the system of accredited CSPS rests upon the far-flung structure (a) of national planning, (b) of lapsed agreements covering activities in many states, (c) of inspection, certification, and reduction of rates by national insurers, (d) of defendants' national schedules of prices, rates and terms, and (e) even, on occasion, of national special discount prices to a particular customer who operates multistate enterprises.

41. Grinnell and the affiliated companies included in its balance sheet do not pretend to have a purely local status when they solicit customers, or report to stockholders, or raise funds for their business purposes, or advertise their merits, or preen themselves on their conspicuous success.

42. When all the rest can see a typical American national business enterprise flourishing in an interstate market, federal courts too get the signal and can hear the alarm sounded when such a market is subject to restraint of trade or monopolization.

43. From the foregoing detailed facts there emerge as ultimate facts, found by this Court, first, that as the complaint alleges, "the business of supplying and installing protection devices on the premises of subscribers, maintaining such devices and furnishing to subscribers protection from fire, burglary, or other hazards, through central stations accredited for such service by insurance inspecting and rating organizations" is a trade or industry which constitutes part of the commerce among the several states subject to regula-

tion by Congress under Article I, Sec. 8, cl. 3 of the United States Constitution, and indeed regulated by Congress by virtue of the Sherman Act, and second, that the accredited CSPS business which is performed by each of the three alarm defendants and is controlled by defendant Grinnell is, likewise, part of interstate commerce subject to Congressional regulation and indeed regulated by the Sherman Act.

44. During the period 1957-61 defendant alarm companies' share of the total national accredited CSPS market ranged from 87 percent to 91 percent; and ADT and Holmes had 87 percent to 90 percent of so much of the central station burglar alarm business as was "certificated" by all underwriters' laboratories.

45. However, it is only fair to add that from 1957 to 1961, defendant alarm companies' shares of central station subscribers and of revenue therefrom showed a slightly declining trend. Moreover, there is no persuasive evidence that any particular prospective entrant has been unable to begin operations, or that any specific CSPS firm has failed or been driven out of business. The most that can be said is that some competitors complain of what they regard as unsatisfactory profits.

46. Defendant alarm companies do not have unfettered power to control the price of their services. Even where they have no competition from other CSPS companies, they have not always been able to receive the standard they have set for themselves, the so-called "Minimum Basic Rates," (hereafter called "MBR") or annual service charges. This is due to the fringe competition of other alarm or watchmen services.

[7] 47. Yet this ceiling imposed by fringe competition and this implied cross-elasticity of demand lose almost all their significance when set against the fact that, because of their national operations, one or more of defendant alarm companies choose in some areas to operate at a loss. ADT's own counsel, in both brief and oral argument stressed, paradoxically, that it operated 20 *"deficit"* offices in cities with no other central station. Counsel seem not to appreciate that this fact is a strong point against defendants. It conclusively proves that defendants enjoy national monopolistic power which permits one of the conspirators to occupy local markets where no independent CSPS station would long function and to extend to each of them the combination's national monopolistic control. *Hardly any indicium of monopoly power is more persuasive than the continued capacity of the asserted monopolistic combination to sustain offerings at a loss either in particular areas or in particular services or products.*

48. Defendants faced with proof of the overwhelming share of the accredited CSPS business in the hands of the alarm company defendants have offered no evidence to rebut the presumption that this share was the result of an attempt to monopolize and have not sought to maintain their burden of proving that their share is attributable primarily to their skill, efficiency, and foresight, or to like factors of obvious social utility. On the contrary, the evidence rather plainly indicates that acquisitions, both within and outside the relevant market, were important sources of the market power achieved by the alarm company defendants.

49. ADT purchased in 1946 for $282,-395.02 Reliance Alarm Co., a Detroit company. In 1955 ADT acquired for $300,000 General Alarm Corp. of Boston. It is true that before making the latter purchase ADT informed the Department of Justice; but it is incontrovertible that whatever else the Department's silence indicates, no inference can be drawn that the Department had, or attempted to exercise, authority to treat that disclosure as proof that ADT was growing larger by virtue solely of its skill, efficiency, foresight, or like factors of obvious public benefit. Nor did the Department of Justice suggest that when after acquiring General Alarm Co. of Boston, ADT dismantled it, ADT failed to reveal in the most flagrant way its intent to attempt to monopolize.

50. The Detroit and Boston concerns just mentioned operated approved central station protective systems when they were

acquired. Other companies which did not operate such systems and hence admittedly were not in the market directly at issue in this case were also acquired by ADT. In its answer ADT admitted that it acquired twenty-two protection companies. Going further, ADT's counsel at p. 9 of their Reply Brief admit that ADT purchased and dismantled in 1913 the Still Alarm Co., Cleveland, and in 1959 Federal Automatic Alarms, Albuquerque; and that ADT in 1954 purchased the subscriber equipment of Globe Electric Protection Company, Atlantic City. What is significant is *not* that those three concerns fall within or without the relevant market here in issue, for plainly they fell outside that market, but that the acquisitions indicate that ADT's growth, far from being attributable solely to superior techniques and methods of administration and like so-called honestly industrial means, owes more than a little to the special kind of appetite which has characteristically revealed a monopolistic temper and explained a monopolistic growth.

51. It is unnecessary to pause for elaborate description of the other acquisitions of ADT or Holmes or AFA, if any, or of the occasional accompanying covenants by sellers not to compete with the alarm company defendants. Enough has been noted to see why it is that defendants have not borne, and cannot bear, their burden of showing that their proportion of the market is a mere tribute to their perfect performance.

51a. Another significant factor in defendants' growth in the share of the national market was the, by-now lapsed agreements during the period 1904-1954 between the alarm company defendants before they were affiliated, or between alarm company defendants and actual or potential competitors. Those agreements provided for the orderly allocation of geographic areas and classes of CSPS. Though the agreements are no longer legally effective, their influence in forming business patterns remains to a discernible extent.

52. One of those agreements was the Burglar Alarm Agreement of September 28, 1906 between ADT and Holmes, then unaffiliated. ADT conveyed to Holmes all burglar alarm business, including subscriber contracts and installed devices, in New Jersey, Maryland, Delaware, the District of Columbia, eight counties of New York (New York, Richmond, Kings, Queens, Nassau, Suffolk, Westchester, and Rockland), six counties in Pennsylvania (Philadelphia, Delaware, Chester, Montgomery, Bucks, and Allegheny), and so much of Connecticut as lies within thirty-three miles of New York City. In those areas ADT agreed forever to refrain from engaging in central station burglar alarm service. Holmes agreed forever to refrain from engaging in any type of protection service outside of those areas; and within those areas to limit itself to burglar alarm service, and, in the case of financial institutions, to nightwatch service. Most of the pattern thus established continued, without legal buttress, at least until this suit was brought—for, as of December 31, 1961, Holmes did not render service outside of the defined area; nor within the area did it serve a single central station fire alarm subscriber.

53. Another series of agreements were the Fire Alarm Agreements of April 29, 1907 among Grinnell, ADA, AFA, and the non-defendant Automatic Fire Protection Company (hereafter called "AFP"), all then unaffiliated. (a) AFA received an exclusive right to contract with subscribers for central station sprinkler supervisory and waterflow alarm service (hereafter called "SSWF") for Greater New York, Philadelphia, Boston and Charlestown, Massachusetts. AFA agreed to refrain from engaging in nightwatch and burglar alarm service in those four cities, and to refrain from all other types of CSPS elsewhere in the United States. (b) AFP received an exclusive right to contract for SSWF service elsewhere than in the four named cities. (c) ADT received the exclusive right to render nightwatch and burglar service throughout the United States, and

agreed to permit AFA and AFP in their respective territories to connect SSWF alarm service to ADT's central stations. (d) Grinnell agreed to furnish all SSWF alarms for contracts obtained by AFA and AFP, and itself to refrain from invading the assigned territories. (e) Among the four named companies provisions were made for AFA to receive 25 percent of the SSWF alarm revenue from its territory; ADT to receive 50 percent of the SSWF alarm revenue from both AFA and AFP territories; and Grinnell to receive 25 percent of the SSWF alarm revenue from both AFA and AFP territories. January 1, 1949 ADT purchased all AFP's rights under the agreements. Much of the pattern established under the 1907 agreements continued, without legal buttress, at least until this suit was brought—for, as among defendants, ADT continues as the exclusive source of central station nightwatch and burglar alarm service, AFA as the exclusive source of central station nightwatch and burglar alarm service, AFA as the exclusive source of central station fire alarm service in New York, Philadelphia, and Boston; and Grinnell does not furnish the types of service originally precluded by the 1907 agreements.

54. Grinnell and ADT and the Rhode Island Electric Protective Company (hereafter called "RIEP") have agreed that RIEP shall have the exclusive right to render CSPS within Rhode Island, at prices not less than those established by ADT for like service, and with certain devices supplied by Grinnell. RIEP shares its revenue with ADT and Grinnell. This agreement is presently effective. Admittedly this relationship is a mere makeweight in indicating Grinnell's and ADT's role in controlling the approved CSPS market, its monopolistic power therein, and its intent to monopolize.

55. Although defendants have a "Schedule of Minimum Basic Service Charges" and "ceiling prices" which constituted their standard rates, they depart therefrom when they encounter actual or threatened effective competition. Departures have included (a) frequent quotation of prices with an annual service charge less than 100 percent but not less than 80 percent of the minimum basic rate, and with an advance service charge less than 100 percent but not less than 60 percent of the minimum basic charge, (b) occasional installation of CSPS without charge, (c) rarely, successive bids each, progressively, lower than the so-called standard minimum basic rates, (d) at least once, the reduction of rates at all locations of a chain-business account to preclude a competitor of defendants from securing in one city an account of that chain, and (e) infrequently, the deferment of prompt payment of installation charges. Sometimes these departures have merely met the prices of competitors; but, on other occasions, rare though they were, the departures undercut or beat competitors' prices.

CONCLUSIONS OF LAW

[8] 1. From the foregoing findings of fact it is transparent that during their growing period the alarm company defendants in several respects violated the restraint of trade provisions of Section 1 of the Sherman Act.

[9] 2. The alarm company defendants engaged in *per se* violations of Section 1 when they, being the most substantial and the recognized dominant enterprises in the CSPS industry, entered into written agreements, now admittedly several decades old and formally no longer effective, which allocated geographic markets and classes of customers and fixed minimum prices for types of service. Grinnell formally took a share directly or indirectly in the revenues from these agreements. When the agreements expired, defendants, including Grinnell, continued most of the arrangements once founded on those unlawful legal instruments.

[10] 3. Because of the formal expiration of the agreements, it, *conceivably*, might be too late for the Government now to seek against defendants relief from the legal instruments as such. But the folds in the industrial scene have remained even

though the technical pressure of the legal instruments has been withdrawn. Channels of activity begun under legal formalities became habitual, and, even after the formalities had expired, in most cities have continued to carry the same menace to free trade. Injunction against continued adherence to the patterns formed in violation of law is not merely appropriate but necessary to accomplish the declared objects of the antitrust laws.

[11] 4. Furthermore, the alarm company defendants resorted to restraints of trade prohibited by Section 1 of the Sherman Act when, faced with competition, they, being in a dominant position, manipulated their own prices to forestall competition. Injunction against pricing discriminatorily utilized by a monopolistic combination to forestall competition is appropriate and necessary.

5. The restraints of trade just mentioned were a chief avenue by which the alarm company defendants previously attempted to monopolize, and do monopolize and now do conspire to monopolize, the national market in the accredited CSPS industry, all in violation of Section 2 of the Sherman Act.

6. Perhaps even more effective as a means used to attempt to monopolize, and eventually to monopolize, the industry, in violation of Section 2 of the Sherman Act, were the alarm company defendants' acquisitions of competing companies, more particularly in the light of the dismantling of some of the acquired service stations.

[12] 7. Grinnell's acquisition of so high a percentage of stock of the companies engaged in the monopolization, and Grinnell's direction of the policies of the four companies as a whole toward the maintenance of the monopolization constituted a violation of Section 2.

OPINION ON LIABILITY

1. At this late date it is hardly necessary to do more than briefly recapitulate what is meant in Sherman Act parlance by the concepts "to attempt to monopolize," or a "conspiracy to monopolize," or "to monopolize."

[13] 2. To succeed in a Section 2 case plaintiff must prove that the putative monopolist, or monopolists sought to achieve or achieved the economic power, even though unexercised, to control prices or production in a relevant market, or to exclude competition therefrom. Proof may be direct or indirect.

[14] 3. One indirect method to prove the requisite power is to show defendants' occupancy of an overwhelming (but not mathematically definable) percentage of the market, unless that position—or, as it is called, "share of the market"—is shown by the supposed monopolist to be attributable exclusively to his skill, efficiency, foresight, or like affirmatively laudable business conduct. Unless he maintains the burden of proving himself within the exception, the occupant in the dominant position stands condemned.

4. What degree of occupancy, what share of the market, are indicative of monopolistic control depends on a judgment based upon all aspects of the particular market under review. But while close cases can be supposed, no informed student of prior decisions involving Section 2 of the Sherman Act could doubt that on this record defendants occupy a share of the CSPS market far beyond the line justifying a presumption of monopolistic power. The percentage is so high here as to need no explication. And, as the conclusions of law already set forth state, this share of the market rests not on the skill, efficiency, and foresight of defendants, not even on their neutrally normal business methods, but on violations of *per se* rules governing restraints of trade and on acquisitions indicating that that growth was a response to external grasp, not to internal grip.

5. Nor in this case need we consider whether there are present all the various policy considerations advanced from time to time by legislators, judges, and academicians to justify or interpret the antitrust prohibition of monopoly.

6. Of course, in the relatively small accredited CSPS industry it would be nearly ridiculous to rest the present application of the Sherman Act upon a literal interpretation of the economic and philosophical postulates offered in great antitrust cases in celebrated opinions by Mr. Justice Brandeis, Mr. Justice Black, Mr. Justice Douglas, Mr. Justice Brennan, or Judge Learned Hand. Here we are dealing with an industry which as a whole involves fewer individuals and less capital than, let us say, any one of the three leading manufacturers of automobiles, of steel, or of electronics, or any one of the three largest chain retail store systems, insurance companies, or banks. The activities of Grinnell and its affiliates, no matter what their intent or their achievement, could hardly be said to have a significant potential effect comparable to those giants' maneuvers in, at least by possibility, dwarfing men, or interfering with their creative, political, or spiritual possibilities, or determining their economic fate, or depriving them of the illusion or reality of free choice in an open society.

7. What justifies this application of Section 2 of the Sherman Act to these defendants is not avoidance of "the curse of bigness," or the fear that men will be converted into robots, or the dread that society will be stratified into trusts more burdensome than feudalism, or colonialism, or socialism, or communism, or some other despised polity, or a concern lest giants larger than the state itself shall lead us into tyranny or into the need of a socialistic regimen as an antidote to private despotism.

8. It is even doubtful whether we can say that here we have the danger of a business becoming slothful, routinized, sleepy, or wanting in alertness, initiative, and progressiveness, as a result of the quiet life sought and usually achieved by a monopolist.

9. Nor can it be truly asserted that here we are faced in microcosm with Acton's fancied or real disease—the corruption of power. Mr. Fleming and his associates are not likely to be on anyone's list of Napoleons. Nor is Grinnell's corporate power "corrupt" or "imperial."

10. In cases like this where the Sherman Act ban against monopolizing is invoked against defendants who have secured dominance of a small industry by imposing unlawful restraints of trade and by a steady stream of acquisitions of competitive enterprises, the usual rhetoric is quite out of place. All that is at stake here is the rooting out of a plant of minor importance in the rich forest of the American economy, not because it overshadows all of us, or even many of us, but because it represents an ultimate growth from seeds which have been declared unlawful. Congress in Section 1 of the Sherman Act outlawed the means and in Section 2 outlawed the end achieved by those means.

11. In most, though not all, situations relief against the monopolization forbidden by Section 2 could be achieved through a proceeding under Section 1 couched in the form of an attack on restraints of trade. But this is not always true. Proceedings under Section 1 may come too late: For example, the formal restraints may have been withdrawn, or laches may have intervened, or, as in *United States v. United Shoe Mach. Corp.*, D. Mass., 110 F.Supp. 295, hoary precedents sustaining as valid what the current law now regards as undoubted restraints of trade may preclude the Government from securing an injunction against the restraints as such. Yet though the Government may be barred from getting under Section 1 relief directly against the restraints, it may, nonetheless, under Section 2 get relief against the consequences of the restraints if, but, of course, only if, they have gone so far as to involve a continuing monopolization. Having failed to nip the bud, the Government may still pluck the flower of evil.

OPINION ON REMEDY

1. Even if this is not a "big case" in any

sense (except in the Government's unnecessary volume of exhibits disclosing chiefly the Department of Justice's unwillingness to accept the notion that in the second half of this Century, lower Courts, educated by the Supreme Court, can get the point that an antitrust case really turns on a relatively manageable set of facts, on a few by now clear legal issues, and on a presentation worthy of Mr. Justice Holmes' advice to "Strike for the jugular"), nonetheless, the findings of fact and the conclusions of law demand a decree of scope and strength.

2. This is no border-line case. A man of the capacity, sophistication, and, possibly, risk-taking temperament of Mr. Fleming cannot have been ignorant that the companies he controlled had in the most flagrant way violated the clearest aspects, the so-called *per se* rules, of the Sherman Act and were continuing to follow patterns conceived in crime. Maybe he shrewdly weighed business advantage against business disadvantage, and with keen appraising eye estimated the law's delays, the fluctuating policies of the Department of Justice, the improbability of private suitors with adequate funds and resolution, the reluctance of courts to apply surgical measures to cut deep into already established industrial patterns, and, in any event, the plausibility of the oft-cited, if strangely inept, metaphor that one cannot unscramble eggs.

3. Whatever may have been the calculation of defendants or their officers, this Court has determined that this case requires a three-pronged judgment adequate to uproot the evil of a long-effective monopolization.

[15] 4. First, defendants, including the controlling Grinnell Corporation, shall be specifically directed to cease and desist from restraints of trade in which the record shows that the alarm companies have engaged directly or indirectly. This decretal provision rests expressly on the ground that the restraints were from their first imposition a violation of Section 1, and, alternatively as well as cumulatively, on the ground that they are a continuing cause of the monopolizing prohibited by §2 of the Sherman Act. To make effective this cease and desist order which, *inter alia*, precludes the alarm company defendants from manipulating prices and terms of sale or servicing so as to bring against competitors defendants' monopolistic power, the alarm company defendants shall be required, until further order of the Court, to file with the Antitrust Division of the Department of Justice such *standard* lists of prices and terms of sale or servicing as, *at their uncontrolled pleasure,* they from time to time adopt, and also to file a record of every quotation, written or oral, departing from those standard lists, such filing to reach the Division within two weeks of the quotation. It is not the purpose of this direction to give the Government the right to fix or even to object to any alarm company's standard price list. It may select any figures it likes. And at any time it may change the standard schedule, if this is done for all customers of like kind. But if an alarm company defendant seeks to depart in individual instances from its standard schedule, then it must report the departure and run the risk of being charged with invidious, illegal discrimination. The cease and desist order shall prohibit each defendant from acquiring the stock, assets, or business of enterprises in the accredited CSPS industry. External growth is the badge of the monopolist, as internal growth is the mark of the free life.

5. Second, no later than April 1, 1966 Grinnell shall file with this Court a plan of divestiture under which it shall dispose of all of its stock in each of the other defendant companies. Such plan may, at the election of Grinnell, provide for the stock to be sold, or for the stock to be distributed to shareholders of Grinnell, or for some combination or reasonable variation of those two methods. The date is purposely set far ahead to permit an appeal to the Supreme Court, and to let defendants consider all tax and corporate problems.

6. Third, to insure that the reforms im-

posed by this decree are not thwarted by a leader of great capacity but of less than an admirable record of compliance with well-known prescriptions of antitrust law, and to guarantee that there is an entirely effective breaking-up of the channels of restraint and monopolization which the present management of Grinnell has dug so deep into the pattern of the accredited CSPS industry, and to make certain that the general public is not further prejudiced by the continued management of defendants by one who has demonstrated defiance of legal prohibitions, no defendant, after April 1, 1966, shall continue in employment as officer, director, employee, consultant, agent, or otherwise James Douglas Fleming; but nothing herein shall preclude any defendant from fulfilling any pension or like purely financial agreement it now has with Mr. Fleming. Full notice that this Court contemplated this decretal provision was given to defendants by this Court during the trial and on the very day when Mr. Fleming gave his extensive, uninterrupted account of his role in the conspiratorial combination. This provision shall not be construed as in any respect retroactive or punitive; its interpretation shall be strictly prospective and prophylactic; nor shall it be regarded as directed against Mr. Fleming, but against defendants' use of Mr. Fleming, the leader who brought them to this end and cannot be expected to regard a reversal of his policies as suitable marching orders. While this Court does not feel that it can leave Mr. Fleming in the saddle, there is intended in this decretal provision, or in any other part of this opinion and judgment any suggestion that Mr. Fleming lacks financial integrity or honesty of the usual type. He is an "honest man" after the manner of Theodore Roosevelt or Norman Hapgood. See Learned Hand, on Robert P. Patterson, in *The Spirit of Liberty,* ed. by Irving Dilliard [Vintage Books, N.Y. 1959 ed.], pp. 200-208, at page 205. He is undoubtedly a man whose virtue the Scotch would appreciate, and of a virtue the Italians would applaud. But he appears on this record to have been for well over a decade and a half the vigorous captain of the defendants' conspiracy to monopolize.

Judgment in accordance with findings of fact, conclusions of law, and opinions on liability and remedy.

APPENDIX 6B

PRIVATE SECURITY TASK FORCE TO THE NATIONAL ADVISORY COMMITTEE ON CRIMINAL JUSTICE STANDARDS AND GOALS

Position Paper on Performance Standards and Ethics

OVERVIEW

The private security industry was once described as a dubious species of law enforcement which had gone its own way for generations. Today however, the industry is being recognized as an important additional measure of protection for the community. The industry's enhanced image has come about because of a greater understanding and respect for private security services. Through improved quality and performance standards, and high levels of ethical conduct, the private security industry can continue to gain society's respect and understanding and thus become a most useful and honored part of the criminal justice system.

POSSIBLE STANDARDS

1. The areas of permissible conduct for arrest, search, interrogation, and use of force should be defined for private security personnel. Statutes outlining the powers and limitations of private security personnel should be enacted. This information should be contained in the pocket-sized manual issued to private security personnel.
2. Measures should be taken to reduce invasions of privacy and inaccurate reporting by investigative units of the private security industry.
3. Information obtained by private security personnel in an illegal search should be inadmissable in civil, criminal, or administrative proceedings.
4. Investigative agencies should be held strictly liable for injuries resulting from inaccurate or false reports.
5. The failure to control security personnel by the employers should be considered negligence and the firm should be held directly liable for such negligence.
6. Before a background investigation is commenced on an individual who has applied for some benefit (credit, insurance, employment) that individual should be fully informed of the nature and scope of the investigation. Whenever an "investigative consumer report" is given to the requesting firm, the individual being reported on should be sent a copy with the name and address of the requester.
7. Advertising of goods and/or services should always include the firm name telephone numbers, address and license number.
8. Advertising regulation should be extended to telephone directory, stationery, and company advertising.
9. The state licensing and regulatory agency should have the power to halt advertising found to be misleading or false and to impose sanctions for such acts.
10. Any badges worn by private security personnel should be cloth and there should be an indication of the wearer's private capacity.
11. The uniform of private security personnel should be designed in such a manner as to indicate the private capacity of the wearer.
12. The uniforms, equipment and badges of private security personnel should be the property of the employer. Failure to return such upon termination of employment should be classified as a misdemeanor.
13. Private security guards should be required to carry approved identification cards clearly indicating the bearer's private capacity.
14. The use of lethal weapons by private security personnel should be discouraged.
15. The private security industry should attempt to design practical alternatives to the use of lethal weapons.
16. Any personnel who must carry weapons should be required to have a special permit issued by the regulatory and licensing agency of the state.
17. Permits should be issued to private security personnel only for limited types of weapons.
18. No weapon permit should be issued until personnel have completed training in the use and safe handling of firearms and achieved a minimum qualifying score on the firing range.
19. Statutory liability should be imposed on private security organizations for the weapons abuses of their employees.

APPENDIX 6C

CODE OF ETHICS

As a security officer I have the responsibility to protect life and property from injury. In carrying out this responsibility, I will adhere to the highest standards of my profession and discharge my duties regardless of personal cost or sacrifice. I acknowledge the following hallmarks of conduct and will adhere to them without reservation;

To seek out conditions which give rise to crime or loss and insure that corrective action is taken;

To treat all persons with whom I come in contact in a professional manner;

To discharge all legal and moral duties assigned to me in a timely and effective manner;

To work with duly appointed public police officials in all reported incidents of criminal acts;

To uphold the laws of the jurisdiction in which I am assigned, consistent with my duties;

To encourage voluntary compliance with the rules and laws which govern my area of protective responsibility;

To continually seek to improve professional practices, standards and procedures;

To assist in enhancing the overall efforts of my employer in increasing productivity and eliminating losses;

To employ all available technology consistent with my responsibilities to increase my capabilities to provide the highest level of protective service;

To employ only the amount of force necessary to maintain order or take necessary enforcement actions;

To resist all efforts to violate established rules or laws by friends, associates or employers;

To strive for continuing personal growth and development to the highest level possible;

To uphold the highest levels of humanity and service in all emergency situations;

To selflessly undertake all assigned tasks without hesitation regardless of personal costs;

To prepare myself thoroughly for all assigned or assumed responsibilities through appropriate training and educational pursuits.

These guidelines for conduct mark the minimum efforts which can be expected during the discharging of my responsibilities as a professional Security Officer. I will constantly seek to exceed their main intent and will strive to raise my level of performance far above them. I will likewise insure that all Security Officers with whom I carry out my duties adhere to this Code of Conduct.

Chapter 7

INDUSTRIAL SECURITY PROGRAMS

INDUSTRIAL SECURITY is that portion of the protective services field concerned with the reduction of opportunities for losses in manufacturing, extracting, and production facilities. Current direct losses in industry are $2.8 billion which is an increase from the $1.8 billion in 1971. The duties, responsibilities, and organizational approaches to protective services in industry vary considerably from plant to plant. Various specific requirements for protection have organized the protective function to report to comptrollers, vice-presidents, corporate counsels, labor relations specialists, and physical plant managers.

The requirements for protection have also been assigned in response to governmental regulations as in the Department of Defense and NASA contracts. Likewise, they have evolved in response to crisis situations without long-range planning.[1] The merger of security and safety responsibilities has also become a commonplace occurrence in response to the requirements of the Occupational Health and Safety Act of 1968 (OSHA). It is, therefore, impossible to detail precise statements about what an industrial security program should do. Industrial programs, however, seek to employ all available protective technology, processes and procedures to develop and maintain a safe and secure work environment for management, employees, facilities, and production.

The term "industrial security" is in constant use. It is often used to refer to security programs and plant protection programs for industry as well as a special type of program in firms which have government contracts.

"Industrial security" is a broad concept which includes a totally integrated program of protection for the organization through the application of the physical, information, and personnel security processes. It is, therefore, incorrect to use synonomously the terms "industrial security" and "plant protection," since plant protection is merely the application of the physical security process in an industrial context.

The application of security in an industrial context involves property protection, protection of information, and insuring the suitability of employees. The needs of a particular organization will dictate the emphasis placed on a particular type of activity.

Similarly, the program of the federal government entitled "Industrial Security" is defined as "that portion of internal security which is concerned with the protection of classified information in the hands of United States industry."[2]

The term, *industrial security,* as used in the governmental sense, is, for example, a unitary program concerned with the protection of classified information made available to industrial firms contracting with the government. In this regard, the "Federal Industrial Security Program" has set up a standard format for the application of these security processes and has insured their uniform application in all industrial firms under contract to the federal government. They insure adherence to rules and regulations established by the government and serve to provide governmental production programs with the type of security which they require.

Since industrial security programs have a varied format, the diverse program styles will be treated individually. This will permit the discussion of each without confusion. These program formats include:

Governmental Industrial Security
Proprietary Industrial Security
—Combination Governmental-Proprietary
—Large Proprietary Programs

—Medium-sized Corporate Programs
—Small Company Industrial Programs

Each of these programs has specific features which the others do not; likewise, each has common features.

Like all security programs, industrial security places barriers between what the company does not want to happen and the things to be protected. Each program attempts to produce "time delays" and lower the risk of the organization to undesired events. The organizational techniques each program approach uses to accomplish these ends are, however, varied. Case study descriptions will be utilized to provide an in-depth analysis of the specific features of each type of program.

GOVERNMENT PROGRAMS FOR SECURITY IN INDUSTRY

The industrial security program of the government applies only to companies who have access to classified information in the production of materials for the various branches of the government. Industrial firms which seek or are awarded government contracts must comply with the regulations of the Department of Defense Industrial Security Program as a matter of contractural obligation. In order to insure that the industrial firms comply with the terms of the contractual obligation, mechanisms have been established through the Department of Defense to provide for contract administration services.

Contract administration services include a variety of functions such as preaward survey of contractor's facilities, quality assurance, production progress, and administration of the Industrial Security Program for protection of classified information.

Development

Early in 1962, the Secretary of Defense initiated a study to determine whether increased efficiency and economy in contract administration were possible. The study, known as *Project 60,* was conducted under the policy guidance of the Department of Defense military and civilian personnel with participation by representatives throughout the military service. As a result of detailed analysis, the Secretary of Defense, in the fall of 1963, directed certain changes in procedures and establishment of a pilot organization to test the feasibility of consolidating existing contract administration activities. The objective of the test was to confirm the belief that consolidated operations would be provided efficiently. Evaluation of the pilot test proved the feasibility of this consolidation, and by the end of 1965, the responsibility for all contract administration service functions within the United States was assigned to the Director of the Defense Supply Agency.

The object of the consolidation was to achieve efficiency and improve management and operation through establishment of uniform policies, procedures, and organization. Prior to this reorganization, the policies and procedures were being issued by three military services through over one hundred separate cognizant security offices. Today, there are eleven cognizant security offices operating under one agency, the Defense Supply Agency. Eleven offices administer the Industrial Security Program throughout the country.

Organization

The nationwide contract administration services field organization is composed of eleven geographic regions, each with a regional headquarters office established in a selected major city. To exercise maximum delegation of authority, regions are further subdivided into Defense Contract Administration Services Districts and Defense Contract Administration Services Offices.

The Secretary of Defense, as head of the Department of Defense, is at the top of the organizational structure of this program.

Assistant Secretary of Defense (Administration) provides overall policy guidance and advice with respect to the Defense Industrial Security Program. He is required to review the various directives, such as the Industrial Security Manual and the Industrial Security Regulations, for consistency

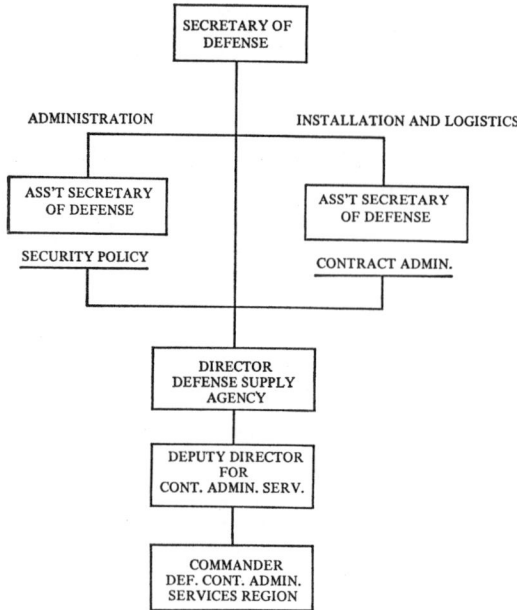

Figure 7-1. Organization of Defense Industrial Security Program (DISP).

with Department of Defense security policies. The Assistant Secretary of Defense (Installation and Logistics) reviews these various directives for consistency with Department of Defense procurement policies. The Assistant Secretary of Defense (Public Affairs) reviews various directives for consistency with Department of Defense public affairs policies.

The Secretary of Defense has assigned the responsibility of administering the Department of Defense Industrial Security Program to the Director, Defense Supply Agency, or his designee. The Director of the Defense Supply Agency is required to assume security cognizance for all United States industrial facilities under the Defense Industrial Security Program. Contract administration service functions have been delegated to the Deputy Director, Contract Administration Services, Defense Supply Agency (DDCAS/DSA). He administers the Department of Defense Industrial Security Program in behalf of all user agencies. Service Region (DCASR) assumes security cognizance of all contractor facilities within his region and with few exceptions, performs cognizant security office functions prescribed by the Department of Defense Industrial Security Regulation with respect to all contractor facilities within his region.

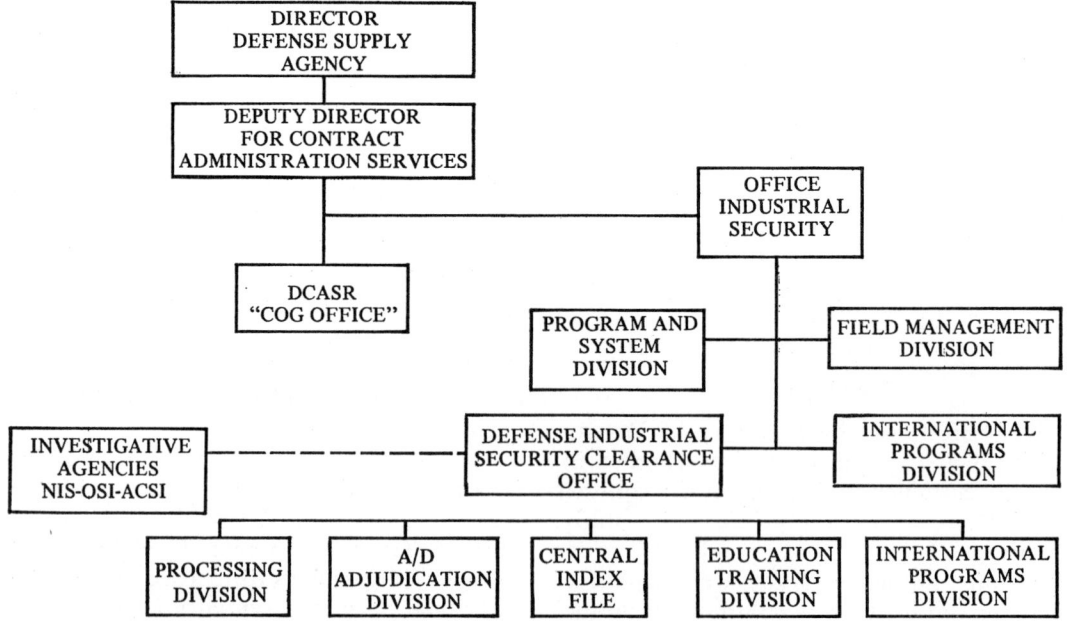

Figure 7-2. DDCAS/DSA. Office of Industrial Security Defense Supply Agency.

The Secretary of Defense is authorized to act on behalf of a number of other agencies, user agencies, in rendering industrial security services.

Two principal directives, which have been previously discussed, are utilized in the Defense Industrial Security Program. They are the *Department of Defense Industrial Security Regulation* (DODISR) and the *Industrial Security Manual for Safeguarding Classified Information* (ISM). DODISR covers the essential policies and procedures with respect to safeguarding classified information. It is for use and guidance of industrial security and procurement activities of the user agency, and is distributed through the normal channels to staff and operating activities concerned with industrial security in procurement matters. The regulation is not applicable to industrial management and is not intended for distribution to industry. The Industrial Security Manual (ISM) is an attachment to the Security Agreement (DD Form 441). It is part of the basic agreement signed between government and industry. The manual establishes the requirements for safeguarding all classified information to which contractors and their subcontractors, vendors, and suppliers have access or possession. In addition to these two basic directives, other guidance letters are issued periodically by the Deputy Director for Contract Administration Services, the *Industrial Security Bulletin* issued to the user agencies, and *The Industrial Security Letter* issued to user agencies and industrial management.

These bulletins and letters are intended to provide the recipient with information concerning developments relating to the Industrial Security Program. In the case of the bulletin, they provide technical direction to cognizant security officers. They also provide notices of current developments and pending changes within the Industrial Security Program which affect user agencies.

The Chief, Office of Industrial Security, Defense Supply Agency, at Cameron Station in Alexandria, Virginia, is the princi-

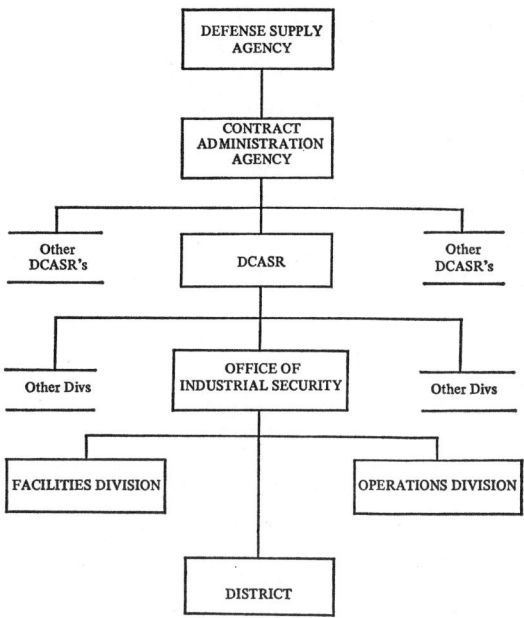

Figure 7-3. Defense Contract Administration —Services Region (DCASR).

pal staff advisor to the Deputy Director for Contract Administration Services in the field of industrial security. He establishes, develops, promulgates, and maintains within the framework of the Department of Defense directive system industrial security procedures, systems, standards, and regulations. He provides staff supervision over the industrial security activities of the various Defense Contract Administration Services Regional Offices. He insures protection of U.S. classified contractual information overseas and foreign classified contractual information in the United States. He provides staff supervision over the industrial security activities of the Defense Industrial Security Clearance Office (DISCO). These functions are performed by four divisions of the office of Industrial Security: the Program and Systems Division, the International Programs Division, the Field Management Division, and the Defense Industrial Security Clearance Office, which is a field activity of the Office of Industrial Security at headquarters level. The Programs and Systems Division is concerned with developing and publishing directives, manuals, and forms pertaining to the Industrial Security Program; new basic Industrial Security policy originates here. The Field Management Division's principal functions are to monitor, evaluate and recommend corrective action as required towards maximum effectiveness of the industrial security operations at the regional level. They exercise surveillance over security violations. They process all facility clearances which require headquarters action. The International Programs Division is concerned with coordination of industrial security activities of United States contractors abroad.

1. They develop procedures for implementation of International Industrial Security Agreements, establish channels of transmission of classified information, coordinate security aspects placed under foreign classified contracts in the United States, and United States classified contracts overseas.
2. This office is concerned with policies in the International Program.

Defense Industrial Security Clearance Office

The Defense Industrial Security Clearance Office (DISCO) was established as a field extension office of the Office of Industrial Security (DDCAS/DSA). DISCO is located in Columbus, Ohio. Its principal function is to process all requests for industrial personnel security clearance received in accordance with various Department of Defense policies and procedures. Its mission is, on a nationally centralized basis, to determine the eligibility of industrial personnel for access to classified information, foreign and domestic; maintain a record of these determinations and the legal documents related to determination of eligibility of industrial facilities for access to classified information; to process *visit requests* involving NATO and foreign classified information, or for access by foreign nationals to United States classified information. This mission is accomplished through five separate divisions.

Processing Division

1. It is responsible for processing contract requests for access authorization or transfer of such access authorization.
2. It obtains the necessary investigation, issues the letter of consent, and executes a record of access authorization actions under the Department of Defense Industrial Security Program.

Adjudication Division

1. It determines eligibility of industrial personnel for access to classified information when and if significant adverse information exists.
2. It recommends denial, revocation and/or suspension of access authorization, as appropriate.

Central Index File Division

It maintains a record file of all personnel access authorization, overseas security eligibility determinations, and a central file of legal documents for all facility security clearances issued, denied or revoked under the Department of Defense Industrial Security Program.

International Programs Division

It processes security assurance, eligibility determinations, and request for visits involving access to United States classified information by foreign nationals or access to foreign classified information by United States contractor personnel.

Administrative Division

It provides administrative, mail, classified document accountability, supply, and property support for DISCO.

DISCO's responsibilities include that of processing requests for personnel security clearances and issuing letters of consent. This function is performed in the Processing Division of DISCO; however, DISCO does not perform investigations. Required investigations of personnel, be they National Agency Checks or Background Investigations, are conducted by the investigative agencies of the three military services, namely, Naval Investigation Services, Office of Special Investigations (OSI), and United States Army Intelligence Command (USAINTC) in the United States and Army Intelligence (CI) overseas. DISCO requests investigations through these agencies. The results of the investigations are returned to DISCO, where a determination is made concerning the issuance of the letter of consent.

Defense Contract Administration Services Region (DCASR)

The authority for security cognizance is delegated to the directors of the various regional offices for all contractor facilities physically located within geographical boundaries of the regions. The Director of each of the Defense Contract Administration Services Regions is assisted by a number of policy and operational staffs, such as Contract Administration division, Quality Assurance Division, Production Division, Office of Public Affairs, Office of Command Management, and Office of Industrial Security (ROIS). Security cognizance functions are actually carried out through this office. This office provides complete coverage of industrial security functions for all facilities physically located within their region. This office does the work of a cognizant security office as defined in the Industrial Security Manual and the Industrial Security Regional.

The Office of Industrial Security, at region level, is divided into two divisions, a Facilities Division (ROIS-FD) and an Operations Division (ROIS-OD).

The activities of the Facilities Division are:
1. to insure compliance by the contractors of the Department of Defense Industrial Security agreement and contractual security requirements;
2. to furnish security guidance to contractors and process facility clearances;
3. to insure adequacy of classification guidance furnished contractors; and
4. to implement the Department of Defense Industrial Security Education Program and the Defense within the region.

The activities of the Operations Division include administering within an assigned geographical area, the Department of Defense Industrial Security Program under the guidance of the region. District Offices of Industrial Security (DOIS), are nothing more than extensions of the Operations Division in the field.

Functions and Responsibilities of Cognizant Security Offices

The cognizant security office is the one point of contact between user agencies and the contractor, and as the manual (ISM) emphasizes, all relationships between the two on industrial security matters shall be handled through, or coordinated with, the cognizant security office. Questions of interpretation of the manual or problems involving industrial security procedures as they pertain to the contractor, will be referred to the cognizant security office. The management of each facility will be notified in writing at the time the Industrial Security Program is initiated at the facility, to which of the region offices they have been assigned for security cognizance.

The cognizant security office discharges industrial security responsibilities for all user agencies at a given facility.

The Industrial Security Regulation states that the procedures set forth in the regulation prescribe the functions and responsibilities of security office. There are certain functions which may be performed under certain conditions by the head of a user agency installation for what is referred to as on-base facilities.

The mission of the Region Office of Industrial Security is to serve as the principal source of management advice to the director in the field of industrial and command security. This office is responsible for implementing and administering the Department of Defense Industrial Security Program within the region. More specifically, this office, or the extension of this office at the district level, has the following functions:

1. Processes contractor's facilities for facility security clearance to determine eligibility for access to classified information.
2. Inspects contractor facilities to insure compliance with the Department of Defense Industrial Security Agreement.
3. Assures adequacy of contractor's standard practice procedures for safeguarding classified information.
4. Terminates, inactivates, suspends, and recommends revocation of facility security clearances in appropriate situations.
5. Takes action with regard to industrial security violations and compromises.
6. Processes visit requests as appropriate.
7. Technically assists Department of Defense contractors and government activities in industrial security matters.
8. Evaluates contractor security indoctrination and education programs for the Region Security Education Division.
9. Evaluates need for and adequacy of controlled areas and associated control systems for approval. Coordinates expenditures of funds with the administrative contracting officer.
10. Conducts cryptographic security inspections.
11. Furnishes information to user agencies for Department of Defense contractors relative to the current facility security clearance and safeguarding capability of a contractor.
12. Maintains liaison with investigative and law enforcement agencies within the region.
13. Reviews Security Requirements Checklist (DD Form 254) on contracts and subcontracts being performed within the region.
14. Assures that DD Form 254, or letters in lieu thereof, are in effect for all classified contracts.
15. Insures that all periodic reviews of DD Form 254 are conducted.
16. Provides guidance in the application of the automatic time-phased downgrading and declassification system.
17. Advises and counsels on problems of classifications.
18. Monitors contractor's security education programs for effectiveness and comprehensiveness.
19. Determines requirements and assures distribution of security education material to be used by Defense Contract Administration Services Region activities and contractors.
20. Organizes and conducts Regional Industrial Security Educational Seminars and Training courses for Defense Contract Administration Services Region and contractor personnel.
21. Develops and supervises the security education programs for Defense Contract Administration Services Region personnel.

The designation of a Defense Contract Administration Services Region to exercise security cognizance at a facility does not prevent user agencies from inspecting at reasonable intervals the procedures utilized by the contractor in complying with the security requirements of its contract. The user agency is responsible for the safeguarding of its own classified material; however, in conducting inspections of a facility, the user agency shall coordinate with cognizant security office. The cognizant security office has no authority to veto the security requirements of the user agency, nor refuse to establish the security safeguards that the user agency considers necessary to adequately protect its classified information.

National Aeronautics and Space Administration (NASA)

In June, 1959, the administrator of NASA and the acting Secretary of Defense, by an exchange of letters, entered into an agreement which extended the Industrial Security Program of the DOD to include all NASA classified contracts. Subsequent to the issuance of Executive Order 10865, the administrator of NASA and the Deputy Secretary of Defense exchange letters confirming the original agreement subsequent to the modifications reflected in certain provisions of the executive order.

Under the terms of the agreement, NASA accepted the Armed Forces Industrial Security Regulations (AFISR) and the Industrial Security Manual for Safeguarding Classified Information (ISM) as satisfying NASA requirements for the security of classified information entrusted in NASA contractors. Necessary security service such as inspection and clearance of the contractor facilities and investigation and clearance of employees were performed for NASA by one of the military services, which was assigned security cognizance over the contractor facilities by DOD. NASA, as a contracting agency, retained the authority and responsibility for those security functions specified in the AFISR and the ISM performed by a contracting military department.[3]

NASA Organization for Security

The NASA Security Division and Security Office is responsible for the development of policies, procedures, and standards relating to all elements of the NASA Security Program. It is authorized to take such action as necessary to carry out the functions assigned to the office within such limitations as may be established by the Director of Administration or higher authority.

PROPRIETARY INDUSTRIAL PROGRAM OF SECURITY

Unlike governmental programs of security, proprietary programs do not follow one established pattern. While this is true, it is possible to isolate a number of activities which are performed in providing for security of industrial organizations. Protection and control of the organization through the use of security processes are basic to any security program. In organizing a particular industrial program for security, a number of specific procedures or techniques are utilized to provide security. These include the following:

1. the use of watchmen or guards,

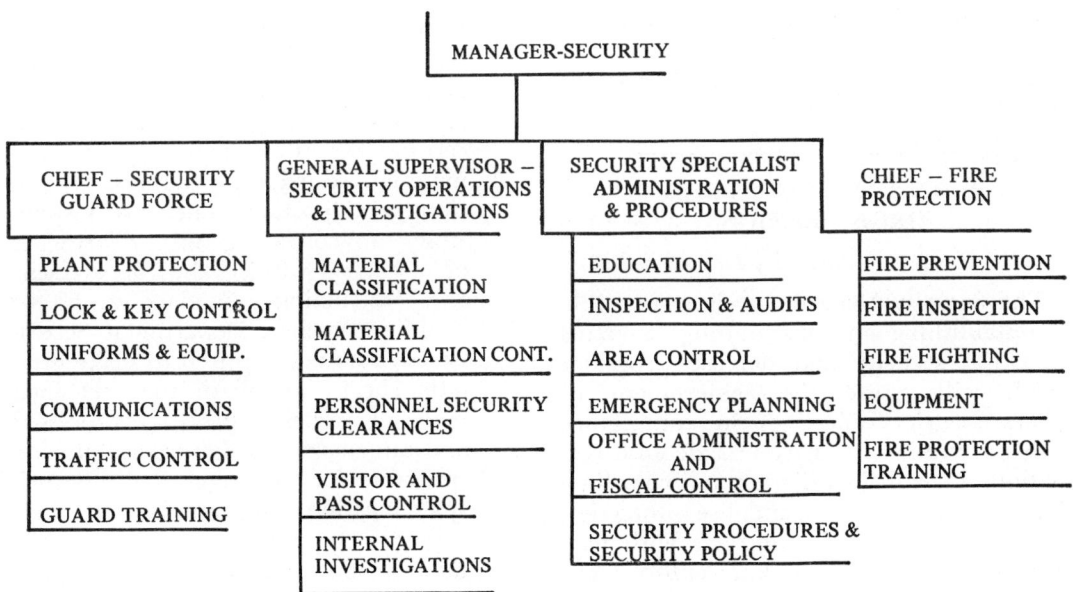

Figure 7-4. Functional organization of the security department of ABC Company.

2. plant safety,
3. disaster planning,[4]
4. riot planning and procedures,[5]
5. use of alarms and intrusion devices,
6. control of authorized area, access and egress, and
7. package, parking, and visitor controls.

The three major activities that provide security within the organizational framework are:

(1) administration,
(2) prevention, and
(3) investigation.

A wide variety of specific applications of security processes exist. Very often, an organization, in addition to providing security for its own property, is involved in production, which requires the maintenance of governmental security standards. Thus, dual responsibilities for security can exist within the same organization so that the relationship between governmental and proprietary security, which exists in a company, can be more fully appreciated. Four examples of organizational security structures will be presented. The first of these will deal with a dual security program involving proprietary and governmental standards. The second will involve only proprietary security responsibilities for a large industrial organization. The third will describe security for a medium size organization and the fourth provides a program for a small industrial concern.

COMBINATION GOVERNMENT-PROPRIETARY SECURITY ORGANIZATION

(ABC Company—Security Department)

The Security Department is responsible for establishing and maintaining effective industrial security and defense programs which will assure the achievement of ABC's security objectives in compliance with government security and defense regulations. Overall jurisdiction and coordination of security, guard and fire protection matters, including all contracts with government agencies and law enforcement agencies concerning these matters, is solely the responsibility of the Security Department.

ABC, as a prime contractor to the Department of Defense, is contractually, legally, and morally obligated to comply with those government regulations and laws which protect our nation's security and safeguard defense information. The Security Agreement is a basic and legal part of each of our contracts with the Department of Defense and with other prime contractors. The Security Department has been delegated the responsibility to insure compliance with the provisions of this agreement.

Organization

1. The Security Department is organized under a Security Manager into four major areas of responsibility: Operations and Investigations, Administration and Procedures, Security Guard Force, and Fire Protection Force.

2. In support of the above organization, each department in the company handling classified information is required to appoint a Security Advisor who has a supervisory, professional, or administrative job classification and is appropriately cleared. Department Security Advisors maintain close liaison with the Security Department and are responsible for disseminating and coordinating information concerning the safeguarding of classified material, security procedures, regulations, and policies. They are responsible for correcting minor breaches of security within the department and instructing offenders so as to avoid recurrence. Department Security Advisors are required to make periodic checks to assure that classified material is being properly processed and safeguarded. Advisors are also responsible for implementing a Security Education Program within their area of responsibility. They assist management and the Security Department in conducting preliminary inves-

tigations which may be required due to security violations.

3. Each supervisor is required to conduct continuing checks to assure that subordinate personnel fully understand and comply with security regulations and to assure that classified material and information in his area of responsibility is properly protected.

Activities

Security Manager

1. The Security Manager is charged with the overall responsibility for implementation of a security program consistent with company needs and contractual requirements and reports to the Personnel Relations Manager. This is accomplished through coordination of the efforts of four major elements of the department and liaison with other divisions of the corporation and government offices on policy matters.
2. In addition, the Security Manager is responsible for the development, maintenance, implementation of Industrial Defense, emergency and disaster plans, and for liaison with company off-site facilities in all security matters.

Operations and Investigations

1. The Security General Supervisor for Operations and Investigations is responsible for the supervision and coordination of those security matters included in Material Classification, Classified Material Control, Personnel Security Clearance, Visitor and Pass Control, and Internal Investigation.
2. A major function of this activity is material and document classification. The Security Requirements Checklist, DD Form 254, or Security Classification Guide, are prepared by government contracting activities to identify and indicate to the company the areas of classified information involved in classified contracts. In many cases, assistance is provided the government contracting officer in preparing this guide. The activity also prepares and distributes Security Requirements Checklists, as required, to subcontractors. A determination must be made that the subcontractor possesses a valid facility clearance, that the prospective recipient(s) of the information has a security clearance, that the subcontractor has approved storage facilities, and that a prospective subcontractor condition actually exists. If the prospective subcontractor is not cleared, action is initiated through the Defense Contract Administration Services Region (DCASR) office for an appropriate facility security clearance survey and issuance of clearance. Public release of contract information is a responsibility of this activity.
3. This activity is also responsible for the preparation of reports of DCASR on all changes in management, ownership, organization, or expansion that would require amendment or updating ABC's facility clearance as a prime contractor. These reports are prepared for signature of the company secretary. When a new company facility is established at a remote location, action is initiated to obtain a facility clearance from the appropriate DCASR office.
4. The Classified Material Control activity is responsible for maintaining an adequate control system within the company to assure the proper receipt, transmission, storage, and disposition of classified documents or material. This responsibility involves the formulation of policy, providing guidance to personnel handling classified documents or material, monitoring operations of Document Control stations, and designation of couriers. Guidance for marking all classified material, security procedures to be placed in effect on special projects

and programs, and requirements for safeguarding cryptographic information are also disseminated by this activity. It is also responsible for establishing company procedures concerning the removal of classified documents from company premises in connection with business visits.

5. Under the Personnel Security Clearance program, authorization for access to classified information of a specific classification is granted or continued only when it is determined that such access is clearly consistent with the national interest, and that access to classified information is essential in the performance of a classified contract. In implementing this program, close liaison is maintained with Defense Industrial Security Clearance Office which is responsible for granting and issuing clearance for access to Secret information and above. This function entails the screening and processing of all clearance requests for access to Secret, Top Secret, and NATO information to assure that the request is justified and that the required forms and questionnaires are complete and accurate. It further requires the processing and recording of all incoming letters of consent for clearance. A master file of all active and pending clearances is maintained by this activity, and change reports are forwarded to Personnel Records for inclusion in the employee's file. Authority for access authorization to Confidential information has been delegated to the company and is administered by the Employment Department as a part of the company employment program for those employees filling a position which requires Confidential clearance. If an employment application or inquiry reveals derogatory information bearing either on the suitability or loyalty of the applicant, the file is referred to the Security Department for review and recommendation as to final disposition. The Department of Defense security clearance standard and criteria is used as a guide in evaluating these cases.

6. The Visitor and Pass Control activity processes all incoming and outgoing classified visit requests and verifies whether the purpose of the visit falls within a contractual relationship and whether all personnel and facilities concerned are cleared for the degree of access requested. Classified visits outside a contractual relationship require special handling and are processed through appropriate Department of Defense agencies for authorization. The Visitor and Pass Control activity also maintains a badge and pass system to control the movement of all visitors inside company buildings and fenced areas. Lobby receptionists, who are under the direct supervision of this activity, are responsible for the identification and badging of visitors to buildings. Other persons requiring access to fenced areas, such as construction workers, are identified and badged at Security Guard Force Headquarters. Other functions performed by this activity are the approval and issuance of camera, property, and vehicle passes, and supervising assignment of reserved parking spaces.

7. The Internal Investigations activity conducts preliminary investigations of all reports of suspected espionage, sabotage, or subversive activity. It is also responsible for making complete investigations of all reported violations of company rules and regulations, accidents involving company vehicles or property, thefts within company facilities, unauthorized entry, malicious acts, and immoral conduct. This activity maintains liaison with all federal, state, county and municipal law enforcement agencies in the area and assists these agencies in investigation involving company personnel or property.

Administration and Procedures

1. The Security Specialist—Administration and Procedures is responsible for the supervision, coordination, and formalization of all security administrative and procedural matters and specifically, responsible for Security Education, Inspection and Audits, Area Control, and Emergency Planning.
2. The Department of Defense Industrial Security Manual requires that all defense facilities engaged in classified contracts maintain active security education programs. The Security Education program is planned in advance with the objective of developing a security awareness on the part of all employees. Training is designed to create a high state of security consciousness in the mind of the employee by making him aware of security objectives, principles, and measures. The program is implemented through the various departmental security advisors.
3. The Area Control function is performed by security specialists under rigid requirements established by the Department of Defense. When it is impractical to prevent visual or physical access to classified material by normal controls or proper storage, protection must be afforded through control of the area in which the material is located. These controlled areas are designated either *Closed* or *Restricted* areas. This function involves the making of detailed surveys, arranging for construction of physical barriers, or installation of electronic surveillance devices, preparation of security operating procedures, and submission of recommendations for establishment of guard posts or administrative controls. This activity is also responsible for the establishment of perimeter controls on all areas and buildings occupied by ABC personnel.
4. The Inspection and Audits activity is responsible for conducting periodic inspections within the company to ensure effectiveness of the overall security programs, to ensure adherence by all personnel to security procedures, and to ensure accountability records are maintained and accountable material is properly safeguarded. These inspections detect lax security procedures. They provide additional information for the Security Department regarding the employee knowledge and compliance with security procedures. Audits ensure accountability of classified material as required by the government.
5. The Emergency Planning activity is responsible for maintaining a working knowledge of company and Department of Defense industrial defense, civil defense, and disaster planning procedures, regulations, guides, and special instructions. This activity develops, reviews, coordinates, and integrates all emergency plans and procedures which, when implemented, cause the prompt mobilization of all available personnel and equipment to cope with any emergency situation.

Security Guard Force

1. The Security Guard Force constitutes the main operating and enforcement element of the Security Department. Its functions include the protection of all ABC owned and leased areas, buildings, and activities against espionage, sabotage, and unauthorized entry, and the protection of company and government property against theft, loss, or destruction. It is responsible for the direction and control of all vehicular and pedestrian traffic on, or adjacent to, company premises. The Force is also responsible for the enforcement of company policies and rules governing the orderly conduct of employees and visitors.
2. In performing these functions, the Security Guard Force maintains nu-

merous fixed and temporary guard posts, roving foot and motor patrols, a communications center, and fingerprinting facilities. It also maintains a master file and controls the issue of all security keys and lock combinations utilized within the company. The Force issues employee temporary badges, badges to nonofficial visitors entering company premises through guard posts. The Guard Force Communications Center monitors the electronic surveillance system, operates vehicular, emergency and civil defense radio networks, and controls the company telephone switchboard during nonoperational hours.

3. The Force is organized under a Chief with three guard shifts. Each guard shift is in the charge of a Lieutenant, who is assisted by three guard Sergeants. The lock and key control and supply sections are an integral part of the Force.

Fire Protection Force

1. Because of the widespread physical arrangement of ABC facilities containing material and equipment of considerable value, a full-time Fire Protection Force is maintained to meet emergencies resulting from fire or other causes of disaster. The primary functions of the Force are preventing and suppressing fires.

2. To perform these functions, the Fire Protection Force maintains three fire stations, a communications center, and a manual fire alarm system, and is supplemented by a volunteer fire brigade. All buildings and installations are inspected for fire safety on a monthly basis, and reports of violations and hazards are submitted to the supervisory level for correction. A continuing system of drills and classroom training is maintained to assure the Force is prepared to cope with any emergency.

3. The Force is organized under a Fire Chief with three shifts, an Administration and Supply Section, a Fire Prevention Section, and Inspection Section. Each shift is in the charge of a Captain, who is assisted by a Lieutenant. Crews are rotated periodically to provide equitable distribution of training and experience. The Fire Prevention and Inspection Section is responsible for detecting and posting hazardous areas, establishment of *Smoking* and *No Smoking* areas, and checking all fire fighting and prevention equipment on a weekly basis. It is also responsible for making detailed arson investigations when the source of fires cannot be determined readily.

THE SECURITY ORGANIZATION OF A LARGE PROPRIETARY CORPORATION

XYZ Company—Security Department

The Security Organization is responsible for coordinating all company efforts in the prevention of fraud against the company or theft of company assets. The organization is also responsible for coordinating the apprehension and conviction of individuals who attempt such fraud or theft.

This organization is the prime source of ideas concerning (1) the prevention of asset and revenue loss, (2) requirements for asset and revenue protection, (3) security inspections of buildings, and (4) other matters relating to the asset and revenue security of company operations. In this connection, sufficient statistical information should be maintained in order to keep the various operating departments informed of the areas where losses may be indicated.

Security, however, remains the responsibility of the individual departments. Each department and each employee of each department is responsible for the protection of the company's assets and revenues, and this responsibility cannot be assumed by the Security Department. The Security Organization should coordinate and integrate the efforts of all departments to ensure that security policies and practices are effective and are followed uniformly throughout the company.

The interdepartmental nature of security requires the full cooperation of each department to solve its problems. Accordingly, each company should establish an interdepartmental security committee. The purpose of the committee should be to ensure the expeditious exchange of security information among the departments in the company. The committee should include the Defense Coordinator.

Departmental Location

The Security Organization's functions are closely associated with the internal auditing functions; therefore, close liaison with the internal auditors is imperative. For this reason, the Security Organization can operate most effectively as part of the Comptroller's Department. Furthermore, having the Security Organization located in the Comptroller's Department will tend to promote the same independent, unbiased, and objective operation that is the hallmark of internal auditing activities.

The job of safeguarding the company's assets and revenues falls into three areas:

1. Verifying that the operating methods and procedures used in all departments are so designed as to prevent fraud or theft.
2. Making checks or audits to be sure, among other things, that fraud or theft is not taking place.
3. Apprehending the culprit when a fraud or theft has occurred.

Verifying that operating methods and procedures are satisfactory from a security viewpoint is the responsibility of both the Security Organization and the internal auditors. Making whatever checks and audits are necessary to ensure that fraud or theft is not occurring is the responsibility of the auditors, and to a certain extent the Security Organization. Apprehending, identifying, and prosecuting individuals engaged in fraud or theft is the responsibility of the Security Organization. Since security and internal auditing are so close-

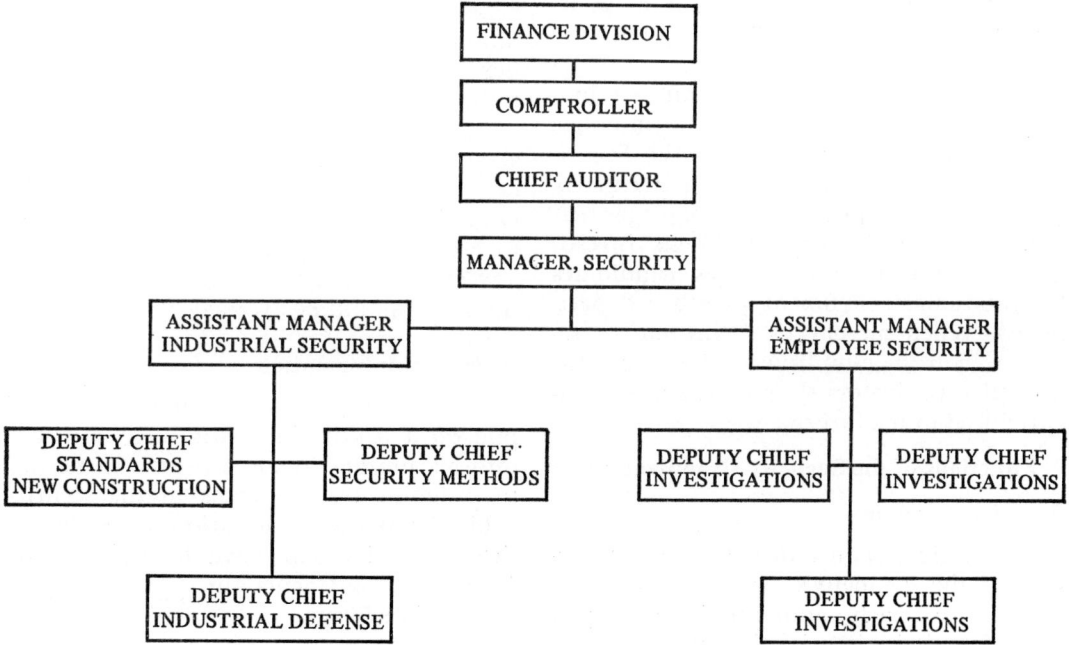

Figure 7-5. Corporate security organization of XYZ Company.

ly related, the two functions should parallel each other in organization, should be at the same level, and should report to the same department head.

Staffing of the Security Organization

Because of the interdepartmental nature of the security function, its effectiveness will be enhanced by staffing the Security Organization with representatives of the various departments. Moreover, the Security Organization should consist primarily of career employees, particularly in the upper and middle management of the organization. Most of the security losses are in areas which involve complex company procedures and equipment. A man can be taught police investigative work in a relatively short period of time, but it takes many years to develop a man with the broad knowledge of the company's business necessary for security investigations.

Most of the personnel in the Security Organization should be assigned there for two or three years and then rotated to another department. This rotation serves two purposes: First, it prevents security personnel from getting stale, second, it helps to spread "security thinking" throughout the company.

It is desirable, and in certain circumstances, essential, to have a certain number of experienced investigators in the Security Organization. They are particularly needed in the apprehension of organized gangs. Whenever possible, these investigators should be qualified as candidates for rotational assignment in other departments. When an effective "preventive security" job is established, the need for trained investigators should be substantially reduced in the future.

Functions of the Security Organization

Security Education

The basic responsibility for the protection of the company's assets and revenue rests with each individual employee. Therefore, it is essential that all employees be made aware of their responsibility in this area through the medium of a program of security education.

The Security Organization can be most helpful in assisting in this program. It should inform the various departments regarding the security problems facing the company and should recommend solutions. Departmental management, in turn, can disseminate the necessary information to all employees in order that their security obligations to the company may be completely understood.

Publicity

Publicity is an effective deterrent against illegal activity on the part of the public or our own employees. In the case of the arrest and conviction of members of the public who have committed crimes against the company, newspaper publicity should be sought in most cases. However, in the case of dishonest employees, means generally should be found to disseminate pertinent information within the company without necessarily disclosing names or locations. Such intra-company publicity will further demonstrate that the company will not tolerate dishonesty.

The Security Organization should work closely with the Personnel and Public Relations departments in each company in order to devise plans to use extracompany publicity to the best advantage.

The Public Relations Department should be informed of the progress of any case which, in the opinion of the Security Organization, is of sufficient importance to warrant newspaper publicity, and of any case where circumstances indicate that publicity should be avoided.

Statistical Information

The Security Organization should compile on a continuing basis such statistical information and reports as may be needed to pinpoint trouble areas in the security field. Often the information necessary for these statistics will have to be obtained from all departments through the use of appropriate forms.

Maintenance of Records

The Security Organization should maintain (1) sufficient records to ensure continuity of operations when personnel change, (2) supporting data for prosecution of cases, and (3) other records (such as investigative files) as are needed for the orderly and expeditious handling of the daily activities of the organization.

Physical Security Surveys

The Security Organization should conduct periodic physical security surveys of company buildings in order to ensure that they are properly protected against unauthorized access, sabotage, or other potential source of damage. These surveys should be coordinated with the Defense Director in each company. The Defense Coordinator can provide guidance for these surveys in the form of criteria established by the Department of Defense in conjunction with the Industrial Defense Program.

Investigative Principles

Because of the interdepartmental nature of the problems, the Security Organization should, in discharging its responsibilities, coordinate its activities with other interested departments.

The following specific principles will help achieve maximum effectiveness in carrying out investigative responsibilities:

1. The Security Organization should not undertake any investigation, of either an internal or external nature in which another department has a direct interest, without consultation with the interested department head.
2. Once an investigation has been initiated, the Security Organization should have the responsibility for the investigation, the authority to direct and coordinate the efforts and forces of other interested departments, and freedom of action and judgement.
3. If an investigation might adversely affect the internal operations of another department, or involve legal, public relations, personnel relations, or the Defense Coordinator, the Security Organization should consult the department head concerned before proceeding further.
4. The Security Organization should keep the department head concerned advised of the progress and final results of an investigation whenever possible.
5. Any disciplinary action taken as a result of an investigation involving company personnel should be undertaken by the department in which the employees work. The Personnel Department should be notified of such possible action and should insure that consistent disciplinary action is used for similar offenses.
6. The Security Organization has the primary responsibility for interrogating employees or nonemployees and for obtaining signed statements in connection with security investigations. Prior interrogation by company personnel, including management, outside the Security Organization should not be permitted. Management personnel will have ample opportunity to conduct any necessary interviews upon completion of the investigation and will take necessary disciplinary action.
7. The Security Organization should have a mutual understanding with the Legal Department regarding the established company policy pertaining to prosecution.

Areas of Security Activity

Within the framework of the foregoing principles, the Security Organization's scope of activity should cover the areas outlined in the following paragraphs. These include situations where the Security Organization should function on a purely investigative basis, and other areas where the Security Organization's responsibilities are more extensive.

Areas Restricted to Investigative Action

The following areas should be handled by the Security Organization on a purely investigative basis:

PERSONNEL INVESTIGATIONS. All departments in the company should refer to the Security Organization for investigation every complaint or incident which directly or indirectly may involve the theft of company assets or defrauding the company of its revenues and which appears to implicate either directly or indirectly any employee (management or nonmanagement) of the company. The following types of activity are included:

1. *Criminal activity*—Suspected criminal activity by a company employee against the company or its property, including suspected embezzlement of company funds, or by a company employee against noncompany personnel or property.
2. *Misuse of company property*—Suspected misuse of company property and use of company services or property for personal gain.
3. *Conflict of interest*—Situations where an individual, while acting on behalf of the company, engages in business transactions (such as placing orders for goods or service) with an outside firm in which he, or the members of his immediate family, have an important financial interest.
4. *Falsification of records*—Suspected falsification of, destruction of, or failure to prepare company records relating to assets or revenue.
5. *Subversive activity*—Suspected subversive activity consisting of active participation in or association with organizations cited as subversive by the Attorney General of the United States. The Security Organization should review these cases with the Defense Coordinator.
6. *Employee irregularities*—Suspected employee irregularities as defined by company personnel policy.

GRIEVANCES—ARBITRATION INVESTIGATIONS. The Security Organization should not participate in the investigation of grievances, as such, unless requested to do so by the Personnel or Legal departments, and then only in grievance arising out of alleged criminal or quasi-criminal conduct engaged in by the grieving employee or employees. The Security Organization should not be used to investigate grievances based on the supervisor-employee relationship, such as absenteeism, insubordination, or poor work performance.

PREEMPLOYMENT CHECK OF EMPLOYEES. A background investigation should be made by the Security Organization of all prospective new employees, both management and nonmanagement, males or females, who will be employed at sensitive locations. (For example, locations where employees handle money, bonds, or stock certifications.) This investigation should include, but not be limited to, inquiries and checks of previous employers, credit and criminal records, reputation in the community, etc.

Areas of Broader Responsibility

The following are additional responsibilities within the purview of the Security Organization operation. In many of these areas the Security Organization would perform not only an investigative function, but also would recommend, advise, and consult as to the appropriate action to be taken by the company:

CRIMINAL ACTIVITY BY EXTERNAL FORCES. The Security Organization should have the responsibility for directing and coordinating overall company efforts to combat any type of criminal activity directed against the company by external forces. This should include the authority to sign criminal complaints, make appearances at arraignments, testify in court and before grand juries, and to take other necessary action required by prosecuting authorities for successful culmination of the case. In this area of operation, the Security Organization should obtain the concurrence of the company Legal Department.

LAW ENFORCEMENT LIAISON. The Security Organization should have the primary responsibility for liaison with Federal, state, and local law enforcement agencies. This includes all official inquiries, attendance at bona fide law enforcement meetings, preparation of articles for law enforcement magazines, etc. Within the framework of the policy established by the Security Organization regarding such inquiries, routine police checks may be handled by other departments such as the Commercial or Personnel departments.

USE OF PRIVATE INVESTIGATIVE AGENCIES, GUARD SERVICES, AND SECURITY EQUIPMENT. It is the responsibility of the Security Organization to be knowledgeable as to the relative merits and uses of private investigative agencies, suppliers of guard services, and available security equipment. Accordingly, the Security Organization should approve arrangements for such services or equipment before their use by any department in the company.

ADMINISTRATION OF FIDELITY BOND REPORTING PROCEDURES. Since the Security Organization will be responsible for investigations concerning personnel reportable under the company fidelity bond, it should also be responsible for administering and coordinating reports to the bonding company.

FINANCIAL SETTLEMENT OF SUSPECTED FRAUD. When an executive decision is made to seek a civil remedy rather than or in addition to a criminal prosecution, the Security Organization should assist the Legal Department in conducting all necessary investigations.

MEDIUM-SIZED CORPORATION

The third Security Organization to be considered represents a medium-sized manufacturing company. The security function will be represented at the corporate level. The operating policies and procedures for company-wide security have been previously discussed. The activities in medium-sized organizations mirror those of the large company. Therefore, only broad operating policy is presented.

Organizational Security Responsibilities

1. Through the Internal Auditing Division, audits are conducted of the different line and staff areas of the corporation to insure the general compliance with the corporate financial policies and procedures, that corporate assets are safeguarded, that the reports and records of the company truly and accurately reflect the results of the company's operations and to make recommendation to management for an improved efficiency and an economical operation, and to supplement the work of the public accountants in their annual examination and certification of the corporate accounts.

2. It maintains relations with the public accountant appointed by the Board of Directors to insure complete and early knowledge of information disclosed by their examinations, and to coordinate the efforts of the Internal Auditing Staff with the work of the public accountants.

3. It reviews, appraises, and evaluates our present methods and practices for

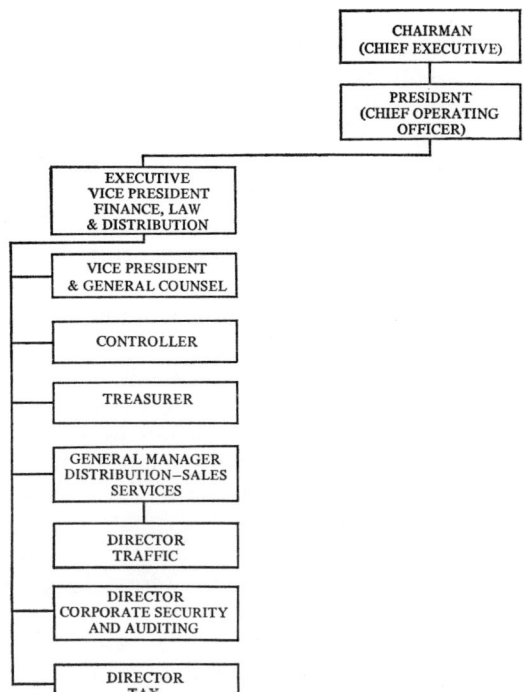

Figure 7-6. Organization for corporate security. (Courtesy of General Foods Corporation)

the protection of tangible company property including inventories, equipment, and property of any kind. It also recommends revised measures of control where indicated.
4. It reviews, appraises, and recommends methods and practices for the protection of the company's nontangible property such as patents, copyrights, trademarks, trade secrets, processes, plans, techniques, and general knowhow.
5. It maintains liaison with governmental, industrial, and other law enforcement or investigative agencies to keep abreast of the development of techniques and knowhow in the general security field.
6. It works with the different operating areas of the company—research, production, personnel, marketing, finance, and law—and aids in identifying information and data which, because of its inherent value to the company, requires special classification, treatment, security, or control, and where necessary, develops a system of document control and accountability.
7. It reviews, appraises, and recommends methods and procedures for the protection of the operating records and data of the company, as would be necessary for the continued and orderly operation of the company in the event of any loss or destruction of a major unit of the company's organization.
8. It reviews, appraises, and evaluates present procedures for the employment and investigation of prospective employees, as well as the resignation and release of departing employees with regard to the potential loss of information and data, which could result in a financial loss to the corporation.
9. It undertakes such other assignments as may from time to time be designated. This is particularly in the area of suspected or potential violations relating to antitrust matters, conflict of interest, the fair treatment of customers, and of any other matters which could involve any improper, unethical, or illegal conduct on the part of an employee, officer, or director, of the corporation on a matter related to the business of the company.[6]

Another approach to plant security is presented in the following Case Study of Builtrite Motors. The security controls developed by Builtrite are much different but equally effective of those in the previous example.

Builtrite Motors

General Description

Builtrite Motors is a division of Midamerica Corporation and is located at 1815 North Lincolnway Street in Chicago, Illinois. Builtrite is a manufacturing plant which makes parts, machines parts, receives parts, assembles the finished product and ships this product to various dealers throughout the United States and foreign countries.

Employing slightly over six hundred factory personnel and two hundred office workers, Builtrite provides twenty-four-hour production, utilizing three shifts in the plant, and two shifts of administrative workers.

Total area occupied by this facility covers approximately nine city blocks (see Fig. 7-7). This includes the manufacturing division; administrative division (located on second floor); and sufficient parking area for all employees.

Although Builtrite is located in an industrial city where hiring and retaining skilled machinists is very competitive, their turnover rate is lowest in the area. The obvious reason for this is observed when touring the plant and its various departments, working conditions are excellent. Clean, well-supervised, and apparently satisfied employees perform their duties in a conscientious manner. Builtrite's policy of an attractive environment is considered by management to be an important asset

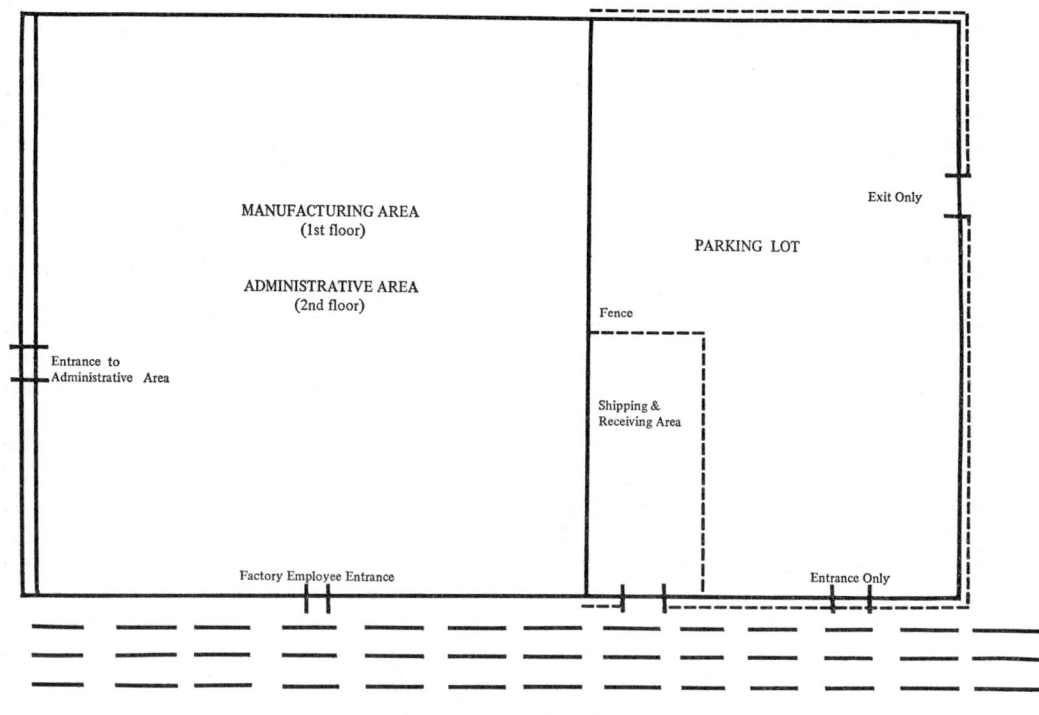

Figure 7-7.

in a successful security program. They feel that above average working conditions promotes satisfaction and helps to eliminate hostility towards the company.

Physical Layout and Security

The security guards at Builtrite are company employees, not personnel provided by commercial agencies. Nine men compose this guard force and although this number appears comparatively small, they have proven their effectiveness over the past years. Operating on a three-shift day, three men provide adequate plant security six days a week. However, it should be noted that these guards are only required to control entrances and exits, patrol the perimeter fence periodically, and observe any discrepancies in the employee's parking area.

Surrounding the employee's parking lot, which has only one entrance and one exit (four lanes to handle the flow of traffic between shifts) is a seven-foot wire mesh fence with several strands of barbed wire on the top. This parking area is well lighted with the lights strategically placed for effective lighting. The guardhouse at the entrance to this area is also elevated, allowing the guard on duty to observe the entire parking lot without leaving his post. Builtrite has not had one complaint of larceny or vandalism since installing the present system three years ago.

Another effective feature of Builtrite's "maximum security" is the handling of shipping and receiving. The entrance and exit to this area is through an administratively controlled gate. When a truck attempts to enter Builtrite's property, access must be cleared through one of the foremen in shipping and receiving who maintains communication with traffic controller and the guard at the gate. No vehicle is allowed to enter or leave without the proper credentials and a thorough shakedown.

Once material has been received and inventoried, it is disbursed to its proper storage place. Also, when material is to be shipped, it is moved to the dock area just prior to departure time. This eliminates the opportunity for pilferage provided when a dock area is overloaded with unaccountable material.

Builtrite coordinates production control with anticipated orders to eliminate providing storage for the finished product. If, however, circumstances warrant the storage of these products, the product is removed to warehouse areas separate from the plant where security is provided by foremen who assume the extra duty. This however, is very rare, as usual procedure with surplus stock is to overload the dealer, thus transferring the storage problem to many smaller areas and eliminating the cost of warehousing.

Entrance to the factory requires producing an I.D. card with photo intact. There is only one main entrance and exit and two guards are present at each shift change to observe workers and handle problems which may arise.

Entrance to the administrative section requires clearance with the receptionist. This area also has only one main entrance and exit.

The walls of the factory complex itself provides the perimeter for the manufacturing and administrative area (see Figure 7-7). Windows on the first floor are twelve feet from the base of the structure, thus eliminating the possibility of removing items in this manner.

From all observances, physical security appears to be very tight at this particular plant. However, it should be noted that Builtrite has almost ideal conditions in this particular area and consequently does not suffer from some of the problems that plague other manufacturing concerns. In addition to location, Builtrite is comparatively small, and enjoys the workability of the system being utilized.

Internal Controls

Builtrite's Internal Control program begins at the time a prospective employee submits his application for employment. A thorough testing program is administered to insure that the experience and abilities listed on the application are valid. References are checked along with records of previous employment. Immediately after being hired, employees are photographed and fingerprinted, these records are kept on a permanent basis.

Once inside the organization, the new employee is assigned to an area which requires his particular talents. This area, whether it be the turret lathe section, milling machine section, grinder section, assembly section etc., will be the only area he is authorized to be in without specific permission from his supervisor.

What is unique about this procedure is the fact that each section has a foreman who is always present, and these areas are kept small enough that the foreman is familiar with all his employees and will question anyone who does not belong in the area. This smallness of departments and restrictions combines to make the men in any given area a very close-knit group since they seldom associate with anyone outside their area. This also aids in the reduction of pilferage, since the only person an employee can steal from is the individual he associates with on a daily basis. The possibility of arranging deals with employees of other departments is also lessened because the opportunity for association with workers outside the related area is very remote.

Another feature of Builtrite's program is the unique arrangement of locker rooms. Located directly above the area to which an employee is assigned, these lockers are arranged in such a manner that each section is organized into a cubicle and only those employees familiar to everyone in the section are allowed in these cubicles. These lockers which often contain valuables, are left unattended while the individual showered or worked knowing that only those employees he knew and trusted would be in the area of his locker.

Tool cribs are located in areas closely related to the machines and men they serve. Positive control is maintained by an attendant who disburses only the specific amount listed on a request which must be signed by the foreman of the respective department. Expendable items are replaced on a one-for-one basis. Control to these areas is very limited, with only foremen allowed entrance and then only under circumstances such as preparing set-up kits when the attendant is in need of assistance.

Builtrite has a greater number of foremen than other manufacturing plants of similar size and equal number of employees. However, foremen at Builtrite perform various duties of supervision, control, are very skilled in the area of machinery to which they are assigned and also aid the employees in setting up their machines for runs on new parts. This eliminates the need for hiring highly paid set-up men. As mentioned previously, they also assume duties of security when the situation demands it, which would otherwise necessitate the employment of additional personnel.

Administrative Policy

The administration feels that their policy of hiring only the highly skilled and reliable worker, and checking his background thoroughly, has paid substantial dividends to the organization. Although retaining this "elite" work force requires that Builtrite compensate with above average earnings for the area around Chicago, the administration feels that this is insignificant when compared to annual savings on reduced pilferage, requirements for only a skeleton guard force, and increased production because of related skills and minimum loss of manpower hours.

From interviews with Builtrite officials it was established that annual inventories revealed losses attributed to employee pilferage at less than $2,500 on sales in excess of $70 million. Management firmly believes that providing an attractive environment for carefully screened personnel, in addition to their internal control program of assignment to small areas, is directly responsible for this figure. In this type of program, all personnel are involved, receiving the benefits of increased earnings, by promoting a feeling of cohesiveness and solidarity among employees.

SMALL COMPANY PROGRAMS

The final organizational security structure to be considered is applicable to many corporations which, because of size, finances, or needs, do not require extensive security coverage.

RST COMPANY

Responsibilities

The Security Organization is responsible for physical security measures. They are to prevent physical access or entry to the company's grounds or facilities by unauthorized persons. They are responsible for performing firewatch duties, making regular rounds of the company's facilities, and stopping at the clock stations to show at what time a particular station was checked. They would likewise check for employee violation of company regulations and procedures; check for safety and fire hazards in the facility; insure compliance with company policy and procedure, and insure that doors are locked and offices and equipment secured.

Many small companies cannot afford the maintenance of an internal security force and seek ways to transfer their risk to a third party. One way in which this can be accomplished is through the development of a bonding system for their facility. Through this means, protection is provided at a fixed cost with the bonding company being responsible for all covered losses. An example of how such a system

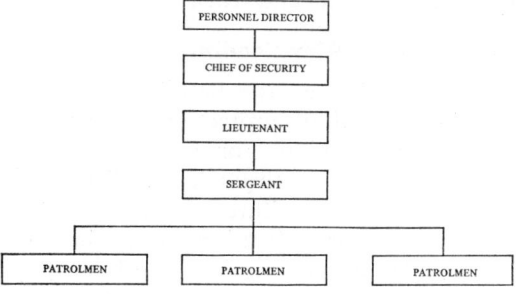

Figure 7-8. R.S.T. Company.

operates is presented in the Case Study of the Farley Manufacturing Corporation.

FARLEY MANUFACTURING CORPORATION AND AJAX BONDING CORPORATION

General

Farley Manufacturing Company is located in Farley, North Carolina, which has a population of 3,500. At the present time, its employment is approximately 450 men and women. This employment figure is lower than its normal 650 due to the fact that they have just finished two government contracts and have laid-off many employees.

The plant consists of two large, modern buildings; one for production and one warehouse. They manufacture dresses, robes, and other garments for large mail order firms. They also manufacture sleeping bags, uniforms, tents, insect net and related equipment for the United States government.

Security is not only important to management and employees but is of great economical importance to Farley and Johnston County because Farley Manufacturing is the major industry in the county. The warehouse contains $250,000 to $3 million worth of raw materials at any given time.

Farley Manufacturing has two major types of security, the local police and the Ajax System.

The local police have the responsibility of nighttime security. They make nightly patrols, checking all entrances to the building and reporting all suspicious activity both around the buildings and the area in which the boxcars and trucks are kept.

Ajax Bonding

The Ajax System is a bonding company. It is a method of protecting the government, employees, and/or Farley Manufacturing against loss through the dishonesty of others or the failure of others to fulfill contracts or obligations authorized by law. Under a bonding system of security, the company requiring protection employs a number of persons who are bonded.

These people are approved by and are responsible to the bonding company as well as to the company. The bonded employees at Farley Manufacturing are in charge of the raw materials and control their entering and leaving of the warehouse and at the other end of the line, they handle and count the finished product.

There are several types of bonding, but fidelity or surety bonds are the type that Farley utilizes. This type of bond is obtained by employers for employees who handle money or company property. The company which issues the bond will usually investigate the employee to determine credibility before bonding him. If the employee steals money or commits any other dishonest act which causes financial loss to his employer, the bonding company must pay the amount of the loss as provided for in the terms of the bond.

When an employee is selected to become a bonded employee, he must resign from employment with Farley Manufacturing to enter employment of the Ajax Warehouse Company. The employee agrees to deliver to Ajax on demand, all correspondence, instructions, stock records, releases, receipts for commodities and all other writings and supplies and equipment pertaining to Ajax operations.

Under the Ajax System, Farley Manufacturing has a bonded warehouse. This provides a guarantee that the government will be paid a tax or duty before goods are released from storage. The warehouse operator bonds himself through a bonding company to secure this payment before he releases the goods from storage. This system of the bonded warehouse is important to the company which has large inventories in storage because it allows for storage and payment of taxes and duties only when goods are withdrawn. The warehouse itself is a public building established on the premises of a business concern for the purpose of enabling the warehouse to take possession of the plant's commodities upon issuance of receipt. In most cases, these receipts are pledged by the plant as collateral.

To validate a pledge, a bonded warehouseman's possession must be *continuous, exclusive,* and *notorious*. The warehouseman accomplishes this by leasing the space necessary for storage, posts signs to make his possession notorious and hires an employee to control the movement of merchandise into and out of the warehouse. The signs posted in and around the premises give public notice of the company's occupancy to persons approaching the area and of the fact that the stored commodities are in the bonding company's possession. They have the exclusive right to enter and occupy the leased premises and the legal obligation to maintain exclusive and continuous control of all stored commodities.

Since the Ajax Company has control of all the commodities in the warehouse, they are entitled to set and execute the security measures taken. The security program starts with the warehouse manager. He is under written contract with Ajax Warehouse Bonding Company, and in so far as the control and operation of the warehouse is concerned, he must take instructions only from Ajax. Should anyone request or order him to act contrary to his instructions or to evade doing anything required by his instructions, he is to notify the Ajax Company at once.

Since the closest control of the warehouse security lies with the warehouse manager, there must be strict regulations controlling his actions in management. The following are the responsibilities and duties of the warehouse manager:

(a) Supervise all warehouse operations and direct all other bonded employees in their duties.

(b) Exercise the same care of stored commodities as a reasonably careful owner of similar goods would. Keep the warehouse clean, orderly, and in safe condition to avoid damage to commodities from a leaking roof, frost, flood, and similar causes. If the warehouse or stored commodities are damaged by fire, flood, or other casualty, telephone the operating office immediately advising them of the extent of the damage and full details of the cause.

(c) Keep personal possession of all keys to the warehouse, except such as are delivered by them to other bonded employees. Make sure all entrances to their premises are locked or fastened at all times except when necessary to receive or deliver commodities. When the warehouse is open, either you or one of the other Lawrence employees must be present to supervise the movement of commodities.

(d) The employees at the warehouse have no authority to permit withdrawals of any merchandise from the premises unless first authorized in writing by the warehouse manager.

(e) Be certain all signs posted by Lawrence are kept in place and if damaged, replaced.

(f) Keep all persons out of the warehouse unless you are satisfied their call is for legitimate purpose.

(g) Receive commodities for storage and deliver them only in accordance with these instructions.

(h) If the warehouse manager should become aware of conditions at the warehouse which are detrimental to our interests or those of the warehouse receipt holder, or if he acquires knowledge of the filing of a suit, legal proceeding, or petition in bankruptcy against the depositor, notify your operating office at once, furnishing all details available.

(i) Do not accept any legal papers unless addressed to Ajax Warehouse Company and when accepted forward immediately. Under such circumstances do not permit the removal of any commodities from the warehouse unless instructed otherwise.

(j) Their warehouse examiners will call on the warehouse manager from time to time to examine the condition and operation of the warehouse. Managers will not permit an examiner to review records or enter the warehouse unless requested, and has first exhibited his identification card bearing his photograph and signature and the signature of an officer of the company verified by an imprint of their corporate seal.

As a bonded employee, the manager must see that all requirements of the bonding contract are carried out properly. To insure security of goods, there are several regulations to be followed with the passing in and out of goods from the warehouse. When receiving commodities into the warehouse, the bonded employee must:

(1) Check all commodities being received for storage in the warehouse by listing each item as it is received and giving a proper description of it along with the date of its arrival.

(2) All the lists should then be initialed by the checker and kept by the bonded employee in charge of the warehouse. These initialed receiving lists indicate that the commodities were safely received and deposited in the Ajax warehouse.

(3) The Ajax Company will issue special blank receipts which contain twenty-five copies. Each location is issued a complete copy with all items numbered consecutively and in numerical order. Any errors made will be marked "void" but must still be a part of the completed report.

(4) At the time of preparation of a warehouse receipt, great care must be taken to give complete and accurate descriptions, unit values and extended values. These receipts are the only check the bonding company has on the stored commodities.

Since the bonding company must have exact knowledge of all commodities, there must also be regulations governing the release of said commodities from the warehouse. Since all the goods have been issued *warehouse receipts* which are pledged to the Ajax Company, they can be delivered only with express authorization from the bonded employee who holds the receipts. This authority can be granted in one of two ways: (1) before delivery, in which case a release form is filed, or (2) a letter made directly to Ajax Company. Deliveries are then confirmed by the filling out of delivery forms. When commodities, which are to be withdrawn from the warehouse, are represented by nonnegotiated warehouse receipts, the delivery of these commodities is unauthorized by any previous authority. The persons requiring delivery will then be required to prepare a special order for warehouse release including all numbers and proper descriptions. These orders must be numbered separately and consecutively for each warehouse receipt holder and are to be prepared by them in the same manner as a sample. They must double-check this information and the extended values when the documents are completed, as it will avoid many errors and assist in providing accurate records and reports. To avoid overreleasing against any particular item, double checks of the quantities being requested for release must be made. The balance remaining available for delivery must be accurate.

When the release has been prepared, they deliver all copies to the depositor, or to the holder of nonnegotiable warehouse receipts for execution by the warehouse receipt holder. No delivery of the listed commodities may be made until the order for warehouse release has been executed by an authorized party, returned to him and a signed copy is in his possession. They must post the delivery of the commodities listed on the release to the appropriate warehouse receipt on the date the document is executed.

These receiving forms and delivery forms are a very good form of security and can be used in inventory. With this system of checks and records, nothing can be misplaced or taken without someone finding out about it and investigating the loss.

Another method of checks in the bonding system is the *stack card*. Before surrendering any completed warehouse commodities, the manager will prepare stack cards. He must prepare a separate stack card for each item shown on each warehouse receipt issued and post them on or adjacent to the commodities described. Such stack cards must be posted at all times until the particular item described has been delivered and released in full. If two stacks of merchandise are represented by a single item on a warehouse receipt, a duplicate stack card must be made and posted on the second stack.

Stack cards are used so that items may be readily found and so that there is no confusion as to what is what and so that the wrong item is not sent out by mistake. They are used in case of an employment change, a new employee can step in and know where to find things.

The posting of signs is also an important part of the bonding company security. As previously pointed out, the posting of signs in and around the warehouse area, giving public notice of their occupancy, is a legal requirement in establishing a valid bailment. All signs posted by them either inside or outside the warehouse, are to be kept in place, not obscured or covered up and replaced if damaged.

To assist them in this regard, they have been furnished a drawing showing the boundaries of the warehouse premises placed in the manager's charge and the locations of all the signs to be posted. In the event it should be necessary to increase or decrease the size of the warehouse area, he must first contact a warehouse examiner or his operating office who will arrange to secure a new lease on the additional space or suspend the unnecessary area. Subsequently, the operating office will furnish blueprints reflecting the areas and location of the signs correctly. The most recent drawing will govern.

The Ajax System specifies that certain types of locks are to be used and that they are to be checked every night. Under this system, there is no alarm system provided.

The Ajax Warehouse Company provides a good system of checks upon receiving and delivering of the product. This decreases the chance of employees walking off with material and if someone does walk off with something, the management knows that it is missing. The most important advantage of the Ajax System is that the company has to be security minded. Most of the time, small industrial plants tend to be very lax, although Farley Manufacturing is at times lax, they are security minded to the extent that they have employed the Ajax System of protection.

SUMMARY

Security for the small manufacturing plant is often of low priority for company management. Unless there are motivating factors internally such as high losses or dangerous or high-value products; or external pressures from insurance underwriters or the location of the plant, little will be done about protective services. Historically, the small manufacturing plant has tried to "make do" with the minimums in security and safety. The enactment of OSHA requirements as well as increased emphasis on protection by insurance carriers has developed an atmosphere of acceptance or necessity to provide at least minimums of service. The directives of organizations such as Factory Mutual (FM) Insurance often become the guidelines for plant protection.[7]

Guard services are often considered important only to meet insurance requirements and are used as a "firewatch" rather than a "security" force. In such situations the amount of protection over the minimums required to comply with insurance company requests are considered excessive overhead cost and not implemented.

The development and management of

protective services programs is challenging in the industrial environment but especially so in the small company. It requires creativity and the ability to accomplish many things with limited resources. Security in manufacturing has never been a glamorous occupation; it has not changed. It requires dedication, hard work, and an intimate and detailed understanding of the industrial manufacturing process. Without these traits, the security manager will develop unrealistic expectations for himself and his officers. The policy statements he develops, the procedures established, and the allocation of protective resources must reflect his best assessment of the needs of the organization and his appraisal of how they might best be accomplished. If his estimate of the risk is too high, he will overprotect or overreact to problems. If it is too low, excessive losses will occur. Neither of these conditions is desirable.

Industrial protective services must insure that the day-to-day activities of the plant are conducted smoothly and kept risk-free. It must also insure that future plans of the company are not jeopardized by inadequate planning for future needs and problems. In a very real sense, the industrial security program must seek to develop a risk-free work environment in the same manner that any other managerial function seeks to avoid risk and maximize all available resources. Asset protection is essential in a programmed and responsive manner. All aspects of the company's operations must be reviewed for areas of possible loss.

FOOTNOTES

1. The organization, duties, and function of industrial security organizations have changed very little in the past forty years. Appendix 7A presents the corporate organization of Republic Steel Corporation Police Department for 1936. The striking similarity to contemporary functions is readily apparent.
2. Department of Defense. Industrial Security Regulations DOD 5220.22-R, July 1966, p. 14.
3. Committee on Judiciary. *United States Personnel Security Practices: Department and Agency Rules and Regulations.* Appendix, Vol. II, 88th Cong., 1st Sess., 1963, p. 1001.
4. Appendix 7B provides a planning guide for Industrial Defense and Community of Plant operations in emergency or disaster situations.
5. Appendix 7C provides a sample industrial civil disorder-emergency planning outline.
6. Courtesy of General Foods Corporation.
7. Appendix 7D provides an example of an approved list of Central Station Systems.

BOOKS

Allen, A. L.: *Personal Descriptions.* London, Butterworth, 1950.

Brandt, Allen D.: *Industrial Health Engineering.* New York, Wiley, 1947.

PUBLICATIONS OF THE GOVERNMENT, LEARNED SOCIETIES, AND OTHER ORGANIZATIONS

Barlow, Myron F., Col.: The Civil Defense Program, M59-8. Washington, Industrial College of the Armed Forces, November 12, 1958.

California Office of Civil Defense. Manual for Civil Defense in Governmental Buildings and Institutions, Practical plant protection and policing, 1952.

Committee on Government Operations. National Fallout Shelter Program, Washington, Government Printing Office, 1962.

Executive Office of the President. The national plan for emergency preparedness, Washington, Government Printing Office, December 1964.

Gustafson, Harry: Organization for Administration and Enforcement of Security Regulations in Defense Industries, June 1969.

Handbook on the industrial security program of the Dept. of Defense. Industrial Union Dept., AFL-CIO, Washington, D.C., p. 5-6.

Industrial defense. American Ordinance Association, Washington, D.C., 1956.

Industrial Defense Against Civil Disturbances, Bombings, Sabotage. New York, Burns International Security Services, Inc.

Industrial Defense Newsletter No. 1 dated 25 April 1969. Department of the Army, Office of The Provost Marshall General, Washington, D.C., Attn: PMGS-D.

Industrial defense planning manual—iron and steel. American Iron and Steel Institute, New York, September 1954.

Industrial Defense Survey Standards, 15 April 1969. Department of the Army, Office of The Provost Marshall General, Washington, D.C., 20315.

———: Industrial defense program. Address before National Seminar of the National Institute of Disaster Mobilization, 1964.

Industrial Union Dept., AFL-CIO. Handbook on the industrial security program of the Dept. of Defense. *Industrial Security*, 1955.

Letter from John Daley, Major, Office of the Provost Marshall, Continental Army Command, May 2, 1965.

Metropolitan Washington, D.C. Crime Conference. Office of Criminal Justice Plans and Analysis, 129 pages, 1971.

Munitions Board, Department of Defense, Industrial Security Manual for Safeguarding Classified Security Information. Washington, D.C., U.S. Government Printing Office, 1951.

Shive, Donald W., Col. USA: Military support of civil defense. Address before the Governor's Industrial Defense Seminar, June 12, 1964.

U.S. Army Military Police School, Industrial defense, Special Text, 19-172. Fort Gordon, Georgia, 1965.

U.S. Army Regulation 580-20, Armed forces industrial regulation, Revised #2. Washington, Government Printing Office, 1960.

U.S. Army Regulation 580-21, Armed forces industrial defense activities. Washington, Government Printing Office, 1963.

U.S. Department of Defense, Armed forces industrial defense regulations. Washington, Government Printing Office, 1955.

U.S. Department of Defense Munitions Board, Industrial security manual for safeguarding classified security information. Washington, Government Printing Office, 1952.

U.S. Department of Defense, Office of Civil Defense, Continuity of corporate management in event of nuclear attack. Washington, Government Printing Office, 1963.

U.S. Department of Defense, Office of Civil Defense, Information Bulletin No. 58. Washington, Government Printing Office, January 30, 1963.

———: Information Bulletin No. 59. Washington, Government Printing Office, February 1963.

———: Information Bulletin No. 65. Washington, Government Printing Office, April 19, 1963.

———: Information Bulletin No. 67. Washington, Government Printing Office, May 8, 1963.

U.S. Department of Labor, Suggested standards for industrial safeguards, Special Bulletin No. 7. Washington, Government Printing Office, 1942.

U.S. Department of Munitions Board, Standards for plant protection. Washington, Government Printing Office, 1952.

United States Government, Internal Security Manual, rev. ed., 86th Cong., 2nd Sess. Washington, Government Printing Office, 1961.

U.S. Law, Statutes, etc., U.S. Statutes of general interest to security officers in the atomic energy program. Prepared by the Division of Security, U.S. Atomic Energy Commission. Washington, Government Printing Office, 1964-1965.

U.S. Office of Emergency Planning, Executive orders prescribing preparedness responsibility of the federal government. Washington, Government Printing Office, 1963.

Weaver, Leon: *The Civil Defense Debate*. East Lansing, Michigan, Social Science Research Bureau, 75 pages, 1967.

Webster, A.: A general study of the department of defense industrial security program. Los Angeles, School of Public Administration, USC, August 1960.

PERIODICALS

Adkins, E. H., Jr.: Industrial security in Creole. *Industrial Security Magazine*, July 1957.

A factory besieged. *National Safety News*, May 1969, p. 50-55.

Bartimo, Frank A.: Current developments in the Dept. of Defense, panelist. *Industrial Security*, December 1964.

Barr, Eric L., Jr.: Industrial defense and disaster preparedness at General Dynamics electric boat division. *Industrial Security*, January 1961.

———: The role of the professional security director in industrial disaster preparedness. *Industrial Security*, July 1960.

Bixton, Richard W., and Dupell, Alfred E.: Classification and management—a joint effort of DOD and industrial. *Industrial Security*, 10:8, 1966.

Boeing emergency preparedness program, the. NIDM Conference Report 64-5, The National Institute for Disaster Mobilization, Inc. *Industrial Security*, N.Y., N.Y.

Boyatt, James W.: Plan for survival—industrial mutual aid. *Industrial Security*, July 1964.

Brandstatter, Arthur F.: Mobilization. *Industrial Security*, July 1960.

Business survival in nuclear war. Administrative Management. *Industrial Security*, January 1962.

Butchers, R. J., Maj. Gen.: Industrial security. *Industrial Security, 8:* No. 2, April 1964.

———: Industrial defense. *Industrial Security*, August 1964.

Chatham, Elbert: These 3—security, safety, civil defense. *Industrial Security, 8:* No. 3, July 1964.

Coddington, Dean C., and Gilmore, John S.: Diversification guides for defense firms. *Harvard Business Review*, May-June, 1966, p. 144-150.

Couch, Virgil: Industry and non-military defense. *Industrial Security*, April 1959.

———: National program for survival and continuity of industry. *Industrial Security*, July 1960.

———: How to overcome objections to civil defense. *Industrial Security*, October 1965.

———: How to prepare for civil defense in industry. *Industrial Security*, July 1962.

Cragg, Lionel C.: Industrial security in Canada, panelist. *Industrial Security*, 49 Por Bio, October 1962.

Cross, Richard F.: ASIS survey—department of defense industrial security clearance program. *Industrial Security*, February 1968, p. 37.

———: Report on safeguarding classified information committee. *Industrial Security*, August 1967, p. 22.

Darling, Don D.: DOD defense of the new PSQ: Fact or fiction. *Security World*, 5, July-August, 1968.

Davis, Gilbert H.: Fair employment practices under the DOD industrial security program (Seminar Workshop Report). *Industrial Security*, October 1961.

———: Security at meeting—defense security briefings, panelist. *Industrial Security*, October, 1963.

Davis, James A.: The industrial security program of the Dept. of Defense, panel moderator. *Industrial Security*, 61 Por Bio, October 1962.

———: Plant security. *Industrial Security*, July 1961, p. 10.

———: Report—emergency planning session GAMA semi-annual meeting. *Industrial Security, 7:* No. 1, January 1963.

Donovan, Robert: New charges in DOD's industrial security manual. *Security World, 4:* No. 4, October 1967.

Dupell, Alfred E.: Classification management —a joint effort of DOD and industry. *Industrial Security*, June 1966.

Gate, Charles V.: Establishing industrial security in the canal zone, 1949-1952. *Industrial Security*, April 1967, p. 4.

Greer, Jack E.: A program of self-inspection. *Industrial Security*, December 1971.

Halfmann, Robert: Self-testing for government. *Industrial Security, V:* No. 4, April 1968.

Hass, Charles F.: Emergency and civil defense planning—panelist. *Industrial Security*, October 1962.

High points of an address, NIDM Conference Report 64-3, The National Institute for Disaster Mobilization, Inc. *Industrial Security*, New York, N.Y.

Horowitz, Harold: Industrial personnel security review program 12. *Stanford Law Review*, 1956.

Industrial disaster control (a special report). *American Machinist*, February 27, 1956.

Industrial Security. (A series of articles pertaining to industrial defense and survival), 6: (No. 3, July 1962.

Kavanaugh, Gerald P.: Observations on the report of the commission on government security. *Industrial Security, 2:*14, January 1958.

Klass, P. J.: Navy revises security plan to cut costs. *Aviation Week, Industrial Security*, March 2, 1959.

Liebling, Joseph J.: Security and technology. *Industrial Security*, April 1958.

Metz, Louis F.: Civil defense and emergency planning—panel moderator. *Industrial Security*, December 1964.

National plan for emergency preparedness, the. Edited and published by the Office of Emergency Planning, Executive Office of the President. *Industrial Security*, December 1964.

Nieman, R.: Industrial security as seen through eyes of the cognizant security officer. *Law and Order, 12,* October 1964.

North, Robin: Industrial civil defense service in Great Britain. *Industrial Security, 5:*16, April 1961.

Norton, John Joseph: The industrial security officer abroad—panelist. *Industrial Security*, 50 Por Bio, October 1962.

Nugent, James E.: International and overseas security—panel moderator. *Industrial Security*, 58 Por Bio, December 1964.

O'Brien, Thomas: Today's industrial security (Seminar Workshop Report). *Industrial Security,* October 1961.

O'Brien, Thomas J.: Is the industrial security manual an adequate instrument? *Security Management,* November 1974, p. 10-15.

———: Judicial decisions affecting the defense industrial security program. *Industrial Security,* December 1971.

Pittman, Stuart L.: The national civil defense program. *Industrial Security,* July 1962.

Proposed: Defense facilities and industrial security act of 1969, the. Part 1. *Security World,* January 1970.

Proposed: Defense facilities and industrial security act of 1969, the. Part 2. *Security World,* February 1970.

Redmond, John H.: Industry defense preparedness pays off in peacetime emergencies. *Industrial Security,* July 1960.

Roberts, W. W., Lt. Col.: DISCO's third anniversary. *Industrial Security,* June 1968.

Role of decontamination in industrial recov. NIDM Conference Report 64-2, The National Institute for Disaster Mobilization, Inc., *Industrial Security,* New York.

Rubenstein, Sidney S.: The U.S. government and industrial security abroad. *Industrial Security,* October 1962.

Sanford, Hugo C.: Industrial defense—what it means to you. *Industrial Security,* October 1957.

Seebode, O.: Industrial security in Germany—panelist. *Industrial Security,* 59 Por Bio, December 1964.

Seeger, Raymond J.: Security and United States technological progress. *Industrial Security,* October 1958.

Sen, K.: Indian industrial security faces sabotage. *Industrial Security,* August 1966, p. 36.

Tagnon, Robert J.: Industrial security in Belgium—panelist. *Industrial Security,* 59 Por Bio, December 1964.

Tracy, Stanley J.: Industrial defense in other countries. *Industrial Security,* 4:125, July 1960.

———: How necessary is industrial civil defense. *Industrial Security,* January 1964.

Turner, Carl C.: The industrial defense program. *Industrial Security,* April 1965.

Walsh, Timothy J.: Public information and defense contracts. *Industrial Security,* 5 Por Bio, July 1958.

Waters, J. A., Jr., Rear Adm. USN (Ret): Remarks on security program of AEC—panelist. *Industrial Security,* 101 Por Bio, October 1962.

Wright, Loyd: The industrial security program of the DOD. *Industrial Security,* October 1963.

UNPUBLISHED MATERIAL

Grazioli, Albert J.: Industrial defense program, unpublished thesis. East Lansing, Michigan State Univ., 1965.

Moss, Forrest Mayo: The House Committee on Internal Security: An American dilemma (B Paper). Michigan State Univ., 1970.

Nicholson, Thomas Gerald: A study of the use of closed circuit television and intrusion detection equipment in industrial plant security (thesis). Michigan State Univ., 1963.

Small, George M.: A model employee identification system for National Defense Industries (thesis). Michigan State Univ., 1964.

Ueltschi, Donald R.: Comparative study of nonmilitary defense measures and their relationship to the facilities protection program of industry, unpublished B-paper. Michigan State Univ., 1965.

APPENDIX 7A

MISCELLANEOUS DOCUMENTS OBTAINED UNDER SUBPOENA FROM THE FILES OF THE POLICE DEPARTMENT OF THE REPUBLIC STEEL CORPORATION, PERTAINING GENERALLY TO THE ORGANIZATION OF THE POLICE DEPARTMENT*

Exhibit 7349

[Copied from the files of the Republic Steel Corporation 6-6-38]

Organization of Police Department File No. HQ 1.19

The Corporation Police Department is set-up with the following personnel:

---

* *Violations of Free Speech and Rights of Labor,* U.S. Senate, La Follette Committee Report, 1939, pp. 13525-13540.

Superintendent of Police—In charge of the department.
Captain of Police—One assigned to each district.
Lieutenants of Police—In the larger district we have one for each eight hour turn.
Sergeants of Police—One for each eight hour turn.
Patrolmen—As are required.

DUTIES

Superintendent: In complete charge of all phases of the work.

Captain: Handles all department business connected with his district and is directly responsible to the District Manager and the Supt. of Police.

Lieutenant: Is the equivalent of turn foreman.

Sergeant: Assistant to the Lieutenant and in sections where the number of men employed and the size of the plant would not warrant a Lieutenant he is utilized in this capacity.

Patrolman: These men are utilized, in addition to regular patrol work, as Gate-man.

It is almost impossible to definitely define the duties of any particular individual due to the fact that conditions where they work naturally have a distinct bearing on their duties. In general, however, the Lieutenant is assigned to his turn, together with his Sergeant, and they in turn place the patrolmen on their turn where needed.

Quite naturally the main duties are the preservation of order, elimination of thieving, and the checking of individuals entering and leaving the plant.

Under present conditions it is also well to have patrolmen available to escort visitors when in the plant.

Exhibit 7350

[Copied from the files of the Republic Steel Corporation 6-7-38]

Republic Steel Corporation, Buffalo District, General Orders, Police Department

Revised January 1, 1936

Personal Appearance

A neat appearance is necessary and compulsory on the part of all officers, whether on either of the gates or patrolling. During daylight hours especially, men on duty at the gates should be on their feet all the time and refrain from making a lazy appearance. If your hands are cold, get a pair of gloves; don't use your pockets in your breeches to keep your hands in all the time.

Main & Clock House Gate

Be courteous and attentive at all times while on duty. Do your utmost to accommodate but remember this is a police department and reserve your authority as same without offending.

Entrance around Main and Clock House Gates will be kept clear of loiterers at all times.

Traffic Light

Think and act safely in operating traffic light, keeping in mind the safety of the men above everything else.

It is proper that the light be used to protect employees while alighting or boarding street cars. Avoid shouting at the men when changing the light from red to green.

Telephone

When answering the telephone, use the term "Police Headquarters" and give your name.

Time Cards

Any employee failing to ring their card IN or OUT will have one half hour deducted. The Police Department is not exempt from this rule.

No one is permitted to ring the card of another employee.

Confidential Reports

Confidential reports will be made on a separate report blank and given personally to the officer in charge of the turn. Neither complaint sheets or the telephone will be used for anything of a confidential nature.

Complaint Sheet Reports

Be observing at all times, because you may be questioned upon matters of importance at various times. The best procedure of putting yourself on record of reporting anything of general importance is to make out a complaint sheet. We insist upon you using these complaint sheets in making a report of general nature, because this is our record and protection.

Orders & Investigations

Orders received and investigations made by the officers will at no time be discussed with anyone, anytime, or any place excepting those in charge of the department.

Trespassers

Be cautious and alert while talking to trespassers, be able to defend yourself in case of attack, treat them as courteously as possible, and escort them to the nearest and safest point of exit. See that your instructions are obeyed.

Lights

All lights burning in the Plant and not being used will be turned out. Exercise extreme caution as to whether the lights are being used, so as to avoid injury or interruption to employees at their work. Keep count of the number of lights turned out, location and time, making note of same on your daily report.

Be sure that all Safety lights are in working order and turned on.

Guns

Your gun has but one purpose and that is the protection of your life. While on duty it must be kept in its holster at all times. Under no condition attempt to frighten or intimidate a person either by the handling or firing of your gun. An infraction of these orders will mean your instant dismissal.

Passes

Representatives of competitive organizations will not be permitted to enter the Plant unless escorted by an officer or the department superintendent requesting their admittance. A visitor's pass will be required.

Visitors to the Die Roll Sales will not require a pass. The officer will, however, secure their name and the company or organization they represent and note same on extended pass sheet with time IN and time OUT.

Visitor's passes shall never be issued to anyone seeking to enter the Plant for the purpose of securing employment.

Employees will not be permitted to enter the Plant at irregular hours unless general pass is requested and approved by their superintendent. Observe the same procedure as that used in issuing a visitor's pass, and advise the employee that it will be necessary for him to have his superintendent sign pass before leaving Plant. Irregular hours will mean other than regular working hours.

Particular care should be taken upon issuing these visitors and general passes, and before issuing same the officer on duty at the Main Gate will contact the party whom the visitor wishes to see.

Employees attending bonus meetings will not require a pass.

Employee representatives will be permitted to enter the Plant to attend meetings without a pass. If an employee representative desires to make a visit in Plant during the night hours, permission from Mr. Stearns must be secured.

Intoxicating Beverages

Any employee of any company or corporation attempting to enter this Plant while under the influence of intoxicating beverages will be refused admission.

Any person or persons found on corporation property with intoxicating beverages in their possession shall be brought immediately to Headquarters.

Approved:

APPENDIX 7B

SAMPLE: INDUSTRIAL DEFENSE STATUS REPORT*

I. Security Officer (or Defense Director)

   Name ....
   Title ....

II. On Duty Population

   Approximate number of employees (including other employees for whom the Defense Director is responsible).

   First shift ....
   Second shift ....
   Third shift ....
   Total ....

III. Control Center and Equipment

   A. Is your control center fully equipped and operational? (See checklist, Part I, following this report)   Yes ....   No ....
   B. Protection factor in control center   ....

IV. Shelters

   A. Has your building been surveyed by the Department of Defense as part of the National Fallout Shelter Survey?   Yes ....   No ....
   B. Protection factor in shelter area (s)   ....
   C. Has a Shelter License Agreement been offered?   Yes ....   No ....

---

* The replies, where appropriate, should cover all work shifts. Circle the number *(shift)* in the "Yes and/or No" blank when applicable.

D. Has a Shelter License Agreement been signed?   Yes .... No ....
E. Has the facility been fully stocked with Federal supplies?   Yes .... No ....
F. Total number of persons to be sheltered under the Shelter License Agreement.   ....
G. Have fallout shelter spaces been assigned to personnel?   Yes .... No ....
    1. Total number of spaces provided for employees?   ....
H. Does your location have a shelter management team?   Yes 1 2 3   No 1 2 3
I. Date of last training session?   ....

V. Warden Organization

A. Is your Warden Organization fully manned and trained?   Yes .... No ....
B. Date of last training session?   ....
C. Date of last drill?   ....
    1. Fire   ....
    2. Shelter   ....

VI. Rescue Organization

A. Is your Rescue Organization fully manned and trained?   Yes 1 2 3   No 1 2 3
B. Date of last drill?   ....

VII. Disaster Medical Organization

A. Number of employees holding valid first aid certificates?   ....
B. Have first aid teams been organized for Disaster Medical Organization?   Yes 1 2 3   No 1 2 3
C. Number of Disaster Medical Aides
    1 ....
    2 ....
    3 ....
D. Do you have emergency medical supplies (first aid supplies, etc.)?   Yes .... No ....
    If yes, how often are they checked?   ....
E. Are they Federal supplies?   ....
F. Date of last drill?   ....

VIII. Emergency Shutdown Drills and Procedures

A. How frequently are actual or simulated shutdown drills conducted?   First shift ....
    Second shift ....
    Third shift ....
B. Date of last drill?   ....

IX. Fire Fighting Organization

A. Is your Fire Fighting Organization manned and trained?   Yes 1 2 3   No 1 2 3

B. Number of trained personnel? ....
C. How frequently are actual or simulated fire fighting drills conducted? First shift ....
Second shift ....
Third shift ....

D. Date of last drill? ....

X. Radiological Monitoring Organization

A. Are your Radiological Monitors completely trained and organized? Yes 1 2 3 No 1 2 3
B. Number of trained Radiological Monitors? ....
C. Are review training sessions of the Radiological Monitoring Organization held every six months? Yes 1 2 3 No 1 2 3
D. Date of last training session? ....
E. Number of monitoring instrument sets on hand consisting of: (See checklist, Part II) ....
F. How frequently are the instruments operationally checked? ....

XI. Radiological Decontamination

A. Have Radiological Decontamination Squad Leaders been trained? Yes 1 2 3 No 1 2 3
B. Number of trained R.D. Squad Leaders ....
C. Has a survey of plant and equipment available for decontamination been completed? Yes .... No ....

XII. Emergency Power, Water, and Food

A. Has emergency power been provided? Yes .... No ....
 1. How is emergency power provided?
  a. Generator Yes .... No ....
  b. Battery Yes .... No ....
  c. Other ....
 2. For what length of time is emergency power sufficient? ....
 3. Is power sufficient for emergency lighting and ventilation in sheltered areas? Yes .... No ....
B. Are seven gallons of emergency potable water available for the maximum number of people to be sheltered? Yes .... No ....
If no, how many gallons per person are available? ....
C. Is survival food available for a two

week emergency period for the maximum number of people to be sheltered? Yes .... No ....
   1. What type
     a. Federal ....
     b. Private ....
   If no, how many days of food are available? ....

XIII. Continuity of Management and Operations
   A. Has an emergency relocation been established? Yes .... No ....
     If yes, where?
       Address ....
       Telephone Number ....
   B. Alternate emergency relocation? Yes .... No ....
     If yes, where?
       Address ....
       Telephone Number ....
   C. Have employee reporting centers been established? Yes .... No ....
     If yes, where?
       Address ....
       Telephone Number ....
       Address ....
       Telephone Number ....
   E. Have cards been issued to employees listing these centers? Yes .... No ....
   F. Has a record site been established for the preservation of essential records? Yes .... No ....
     Name ....
     If yes, where?
       Address ....
       Telephone Number ....

XIV. Emergency Communications
   A. Are alternate means for external communications available? Yes .... No ....
     If yes, please indicate method:
     1. Via alternate central office (Foreign Exch. Line) ....
     2. Mobile radio ....
     3. Amateur radio (Call Letters) ....
     4. Other ....
     5. Telephone numbers for Emergency Control Center ....
   B. Has essential telephone line service priority on line load control been established? Yes .... No ....
   C. Has an alternate means of internal

communications been established? Yes .... No ....
If yes, please indicate method
....
....

Other ....

XV. Other Emergencies
- A. Has a Civil Disturbance Plan been developed and submitted to headquarters? Yes .... No ....
- B. Has a Bomb Threat Plan been developed and submitted to headquarters? Yes .... No ....
  1. Have any bomb threats been received? Yes .... No ....
  2. If yes, how many? ....
  3. Were the premises evacuated? Yes .... No ....
- C. Has an Internal Disturbance or Sit-In Plan been developed? Yes .... No ....
- D. What provisions have been made for natural or man-made emergencies?
  1. Fire
  2. Explosion
  3. Flood
  4. Blackout
  5. Windstorm
  6. Riot
  7. Earthquake
  8. Strike
  9. Nuclear attack

Please attach a current copy of the Industrial Defense Organizational Chart for the location.

Has Industrial Defense been integrated with the location's Security Plan and other emergency programs? Yes .... No ....

Additional Remarks:

Signed ....
                                                      Company Organization

CHECKLIST
Part I
Control Center and Equipment

To be operational, a control center must
(1) be in a sheltered area and provide a reasonable degree of privacy;
(2) have an available telephone or telephone hook-up; and
(3) have a file containing the facility's emergency plans; up-to-date defense team rosters; floor plans of the facility and shelter areas; maps of the geographical area; and any other papers, plans, and publications useful in an emergency.

It would also be desirable to have alternate means of communications capability either in the control center itself or in the immediate area.

Part II

Monitoring instruments sets on hand consisting of
   1—Geiger Counter, CDV-700 ....
   1—Medium Range Gamma Survey Meter, CDV-710, or ....
   1—High Range Beta-Gamma Survey Meter, CDV-720, CDV-715* ....
   4—Dosimeters, CDV-740 or CDV-742 ....
   1—Dosimeter Charger, CDV-756 or CDV-750 ....

If there is an odd number of instruments to a set, insert the word surplus or need on the right.

SAMPLES: EMERGENCY PREPAREDNESS QUESTIONNAIRE

I. Civil Disorder and Strike Planning

  A. Emergency plan organization and communications
    1. Who is delegated decision making authority?
      Name ....
      Title ....
      Home Address ....
      Home Telephone Number ....
    2. List order of delegation as indicated in plan.

      Name ....
      Title ....
      Home Address ....
      Home Telephone Number ....
      a. Who is delegated authority after normal working hours and how is he contacted?
      b. Is one organization functional for decison making with regard to to all emergencies?
    3. What are criteria for determining at the time of the emergency whether the location will be
      a. operated on a normal basis (possibly with some modification to guard and protection tours)?
      b. operated on a limited basis (designate functions that will operate)?
      c. closed down and manned solely by supervisory and plant protection personnel?
      d. closed down and unmanned for the duration of the emergency?
    4. Is there a procedure for advising employees of the decision to activate the plan?
      a. What method is used?
      b. Describe.
    5. What is the possibility of a civil disturbance?
      a. What is the geographic proximity to troubled area?
      b. What is the proximity to parks or other locations where a civil rights rally might be held?

---

* The CDV-715 has the capability of performing the functions of the CDV-710 and CDV-720.

c. What is the probability of an internal conflict?
d. Can plant be reached by public transportation?
6. Have all sources of advance civil disturbance information been explored?
   a. Local Government agencies.
      Name ....
      Title ....
      Agency ....
      Telephone Number ....
   b. Civil Defense
      Name ....
      Title ....
      Agency ....
      Telephone Number ....
   c. Local law enforcement and state police
      Name ....
      Title ....
      Agency ....
      Telephone Number ....
   d. Local fire department
      Name ....
      Title ....
      Agency ....
      Telephone Number ....
   e. National Guard
      Name ....
      Title ....
      Agency ....
      Telephone Number ....
7. Is location working with neighboring industry in the formation of a united front? Is information gathering associated with the civil disorder threat part of the organization's goal and how it is disseminated?
   a. List neighboring industry involved.
   b. Who is the company's representative?
   c. What significant items have developed from these discussions?
8. Have employees who might participate in an internal or external demonstration directed against the location been identified?
   a. Have employees who might incite backlash violence against demonstrations been identified?
B. Security personnel
   1. Are all security personnel fully instructed on their responsibilities prior to a disorder threat both externally and internally?
      a. Verbally?
      b. Written instructions?
   2. Are training sessions scheduled so that personnel will be thoroughly familiar with all safeguards and protective devices provided in the plant?
   3. Are security personnel sufficiently equipped for self protection?
      a. Emergency lights?
      b. Battery operated megaphones? Public address system?
      c. Transceivers (walkie-talkies)?

    d. First-aid kits?
    e. Battery powered AM/FM receiver tunable to police band?
    f. Metal helmets?
    g. Bulletproof vests?
    h. Other?
 4. Have emergency procedures been established for both operating and nonoperating hours? Do security personnel fully understand these procedures?
 5. Have provisions been made for full utilization of uniformed guard service?
 6. Can guard force be supplemented rapidly?
    a. Guard contractors?
    b. Other company locations?
    c. Company management employees?
 7. Are living arrangements and supplies available on the premises?
 8. Are guards who normally carry sidearms cautioned as to the use of discretion during periods of disturbance? Guards normally wearing firearms should continue to do so during the disturbance period.
 9. Has line load service been provided for key personnel? Who?
C. Regular personnel
 1. Are supervisory personnel alerted to report suspicious persons or activities to security?
 2. Have lines of communication to security personnel been well established and supervisory personnel so informed?
 3. Is there a plant fire brigade?
    a. How many on brigade?
    b. Is the brigade trained in combating fires of the types frequent to civil disorders?
    c. Do all personnel understand how to report a fire?
    d. Is the location prepared to provide its own fire control if local fire protection is unavailable?
 4. Is there an intraplant emergency signaling system?
    a. Do all personnel understand its significance and know what is required of them?
 5. Have selected supervisory personnel been instructed about the possibility of an emergency and the details of the plan?
 6. Is photographic equipment available to obtain documentation?
    a. Still pictures?
    b. Motion pictures?
    c. Sound recording facilities?
    d. Who has been designated this function?
 7. Have provisions been made for employees to reside at location?
    a. Sleeping accommodations?
    b. Food and sanitary facilities?
    c. Medical facilities?
 8. Will employees' families be allowed to reside at the location if their homes are uninhabitable?
D. Yard areas
 1. Are yard areas enclosed by fencing?
    a. Are areas properly secured during nonoperating hours?
    b. Are areas included in security patrols?

c. Are they surveyed by sight, CCTV, or by monitored intrusion devices?
  2. Is the area adequately lighted?
  3. Are materials stored at the fence capable of creating a ladder-like condition to bypass fence security?
  4. Are combustible materials or structures sufficiently removed from the fenced perimeter to preclude easy ignition from outside?
  5. Are all external control valves checked daily?
     a. Has the emergency plan included the removal of hand wrenches from the valves to preclude their shutoff?
     b. Are valves equipped with central station or proprietary supervision?
E. Building and perimeter considerations
  1. Has a means been provided for locking all normal entries, such as doors, windows, and penthouses from inside the building?
     a. Are the normal entrances alarmed?
     b. What action is planned for door security during a disorder situation?
  2. Are exposed windows and glass along unfenced perimeters protected by:
     a. Shatterproof glass? Impact resistant glass?
     b. Shutters, heavy wire screening?
     c. Wooden covers for emergency window protection?
  3. Is fencing secure and are gates in good condition?
     a. Can gates be locked?
     b. Are there additional locks and chains for gates?
  4. Are parking lots fenced?
     a. Can the parking lots be locked?
     b. Is monitoring of automobiles into the lots possible?
     c. What is the procedure planned for parking lot control and operation during a disorder condition?
  5. Have all employees been issued proper credentials?
     a. Is there a control or employee entrance through use of the company pass?
     b. Is there effective inspection of incoming packages and lockers?
  6. Are door hinges so located or designed that the pins cannot be pulled or broken?
     a. Are lock bolts so protected that these cannot be disengaged by a thin instrument, or pried off, or readily cut?
     b. Are keys in the hands of a few designated trustworthy personnel only?
        Who?
     c. Are padlock hasps of the type that the holding screws cannot be removed when doors are locked?
     d. Are all unused doors permanently secured?
  7. Are easily accessible windows protected by suitable bars or grating?
     a. Are plans in progress to dispatch fire fighting brigades quickly to these areas?
     b. Is fire fighting equipment readily available opposite the window area?
  8. Are security patrols maintained during nonoperating hours?

9. Are roof hatches firmly secured?
    a. Are there any skylights and how are they protected?
    b. Are ventilator openings protected and how?
10. If the building is low with a flat roof, is there access to the roof from inside and a means of containing roof fires?
    a. What equipment is available?
    b. Are there fire escapes designed to permit ready egress from the building but make entry difficult?

F. Interior protection
  1. Are all areas protected by automatic sprinkler systems, including outside receiving and shipping platforms?
     a. What protection is afforded flammables?
     b. Are flammables inaccessible from outside the building?
     c. Are extra sprinkler heads on hand to meet any foreseeable replacement requirement?
  2. Are control valves equipped with central station or proprietary supervision?
     a. Are these checked daily?
  3. Is there adequate distribution of fire extinguishers and hand hoses?
     a. Are locations accessible and distinctly marked?
  4. How are important services secured?

     Normal   Emergency
     a. Electricity
     b. Gas
     c. Water
     d. Steam
     e. Chilled water
     f. Oil
     g. Boiler equipment
     h. Air conditioning equipment
     i. Switchgear and transformers
     j. Telephone equipment
     k. Vital business records
     l. Data processing and tape storage
     m. Internal communications
     n. Fire protection devices
     o. Fire pumps
  5. Has an emergency generator been provided?
     a. Does the generator have capacity to handle essential services?
        elevators?
        lighting (emergency)?
        security systems? CCTV?
        fallout shelter area?
        public address systems?
        fire alarms?
  6. Are supplies of materials maintained on hand—plywood, lumber, etc.—sufficient to repair broken doors, windows, roofs, etc.?
  7. Does location have sensitive manufacturing operations?
     a. How would these be protected during a disorder?
     b. Can areas be isolated?

c. Have plans been made to close off any area where an internal demonstration is staged and isolate demonstrators?
G. Emergency operations
1. Has an alternate operating location been established away from the disturbance area for use by key management personnel?
   a. Where?
2. Has the emergency plan included diversion of incoming shipments to locations outside the disturbance areas?
3. Has alternate means of carrying on critical functions been established?
4. Has a means been established for emergency evacuations?
   a. Have emergency routes been established for personnel access and egress?
   b. Have designated employee entrances been established?
5. When would force be used to disburse an internal demonstration?
   a. What techniques would be used?
   b. How would sit-ins be handled?
6. Has the means been provided for an orderly termination of the emergency measures after cessation of the disturbance?

II. Bomb Threat

A. Has a plan been developed?
   1. Has it been disseminated to all supervisors and key personnel?
   2. Has it been distributed to telephone operators, receptionists, mailroom personnel, secretaries, etc.?
B. Have responsibilities and authority for making decisions on the conduct of search and on the plant's evacuation been clearly defined?

III. Industrial Defense Planning

Refer to latest Industrial Defense Planning Program Status Report
A. General
   1. Defense director   Alternate
      Name   ....
      Title   ....
      Home Address   ....
      Home Telephone Number ....
   2. What are geographic coordinates of location?
      a. Longitude
      b. Latitude
   3. Is control center fully equipped and operational?
      a. Where is control center located?
      b. Is it in shelter area?
      c. How is area secured?
      d. What additional equipment, if any, is available in control center?
      e. When were instruments last calibrated?
   4. Has facility been fully stocked with Federal supplies?
      a. Where are they stored?
   5. What is building rating? at the shelter area?

APPENDIX 7C

SAMPLE: INDUSTRIAL EMERGENCY PLAN OUTLINE AGAINST CIVIL DISORDERS*

I. Purpose

The following items should be included in the purpose of the plan:
A. Orderly and efficient transition from normal to emergency operations.
B. Delegate emergency authority.
C. Assign emergency responsibilities.
D. Assure continuity of operations.
E. Indicate authority by company executives for actions contained in the plan.

II. Execution Instructions

This should include the elements of who, what, when, where, and how for executing the plan:
A. Individual(s) having authority to execute the plan.
B. Conditions under which the plan may be partially executed.
C. Conditions under which the plan may be fully executed.
D. Coordination between all responsible individuals to assure an efficient sequence of execution.

III. Command Control Center

The command control center is the plan command post—the focal point for directing all emergency actions. If more than one control center is established, for decentralized operations, all emergency actions should be coordinated through the central control center.
A. Location. The primary location should be in a well-protected area of the plant where access can be controlled with a minimum of manpower. An alternate location, also well protected, should be selected in the event of damage or inaccessibility to the primary location.
   1. Primary.
   2. Alternate.
B. Chain of command. Assure the legal continuity of leadership and direction. Prepare a management succession list to assure leadership and supervision in the event executive and administrative personnel and key employees are incapacitated or unable to report to work. Assure that management continuity and other emergency modifications of the organization are in accord with state corporate laws and the charter or bylaws of the company.
   1. Emergency organization.
   2. Continuity of management and key employees.
   3. Designation of successors.
   4. Prepublished company orders constituting emergency authority.
   5. Establish in accordance with state corporate laws and charter or bylaws of the company.
C. Planning coordination and liaison. This element of the plan is designed to assure mutual planning approaches and objectives. It also provides a means of keeping you abreast of the social climate and receiving advance warning of the imminence and possible magnitude of a disturbance. Coordination and liaison should be maintained with:
   1. Local and state officials.
   2. Fire departments.
   3. Adjacent plants and business firms.
   4. Local utilities.
   5. Employee union officials.
   6. Local news media for news release policy.
D. Communications. Internal and external for command control units.
   1. Internal.

---

* Reprinted courtesy of the National Association of Manufacturers.

a. Adequate coverage of plant area.
   b. Complement primary system with two-way radios, walkie-talkies, field telephones, or megaphones (bull-horns).
   c. Controlled usage.
  2. External. Local and state law enforcement agencies (consider police radio monitor).
   a. Fire departments.
   b. Hospitals.
   c. Adjacent plants and business firms.
   d. Management and key employees.
   e. Train switchboard operators in emergency procedures.
 E. Maintain a log of all emergency actions taken.

IV. Personnel

A. An inventory should be made of employee secondary skills.
  1. Apply secondary skills to possible emergency requirements based on emergency organization.
  2. Determine degree of competence.
  3. Develop accelerated training where necessary.
B. Availability. Keep switchboards open and operators available. Designate male operators as alternates for female operators who may not report. The cascade system of notification is very effective. This is accomplished by having the switchboard operator, or whatever means are available, notify two or more key persons. They in turn will notify a designated number of employees, who in turn will notify others until all employees have been notified.
  1. System of notification (cascade system).
   a. Recall to work.
   b. Reporting instructions.
  2. Rendexvous or reporting points. Central assembly points for employees should be preselected, if possible, in areas of relative safety. Employees should be informed of the location of these areas and instructed to report to them if routes to the plan are inaccessible. Transportation, i.e., busses, trucks, should be provided from assembly points to the plant.
   a. Primary—out of emergency area.
   b. Secondary—alternate for primary.
   c. Inform employees of location(s).
  3. Transportation.
   a. Company-owned.
   b. Contract needs.
   c. Mutual needs with other plants.
   d. Escort by law enforcement agencies.
   e. Preselected routes from reporting points to and from the plant. (Plan for escort of female personnel—car pools should be considered.)
  4. Training
   a. Emergency functions.
    1) Primary and/or secondary skills (related to inventory of secondary skills).
    2) Immediate emergency repairs.
     a) Internal.
     b) External.
   b. Situation briefings. Employees should be briefed daily as to the impact of the riot on the plant and the overall status of the community. These briefings must be factual in order to dispel rumors and speculation.
    1) Pre-emergency.
    2) During the emergency.
     a) Reacting to crowd pressure. Employees should be prepared psychologically to remain on the job. They should be advised that management needs their loyalty. Self-

restraint must be emphasized: don't irritate the mob; don't associate with rioters; in essence, "ignore the overtures of the mob." Act only upon and as directed by management or local law enforcement officials. The pressures of the mob may be overwhelming; thus this type of training is essential.
1) Psychological preparation.
2) Self-restraint.
3) Act only upon direction of management of law enforcement officials.
4) Report all rumors to supervisor.
5) Loyalty to the organization.
   b) Postemergency.
   1) Recognition of exemplary performance.
   2) Impact of emergency on plant.
   3) Employment continuity.

V. Evacuation Routes

Predesignated routes to evacuate buildings and/or the plant should be included. All employees should be informed of these routes and procedures for evacuation.
A. Buildings.
   1. Evacuate by departments if practicable.
   2. Exits.
      a. Primaries.
      b. Alternates.
B. Plant.
   1. Primary—away from emergency area.
   2. Alternates—away from emergency area.

VI. Electric Power

Coordinate this portion of the plan with local electric power companies.
A. Transmission lines.
   1. Location of transformer banks.
   2. Availability of alternate distribution lines.
B. Emergency power. An auxiliary source for providing sufficient emergency power for lighting and other essentials. This should not be construed to mean a stand-by capability to continue full production operations. The following items are suggested:
   1. Generators.
      a. Show size and location.
      b. Fuel supply.
      c. Operators.
   2. Battery-powered.
      a. Flashlights.
      b. Lanterns.
      c. Other battery-powered sources of illumination.

VII. Plant Security

The essential elements of this portion of the plan are:
A. Organizational plans.
   1. Develop plant security organization.
   2. Put security plans and procedures in writing.
   3. Provide for reporting promptly to the FBI any actual or suspected acts of espionage or sabotage.
   4. Liaison with the local and state law enforcement agencies.
   5. Have supervisory personnel attend plant protection training courses.
B. Guard force.
   1. Organize the guard force.
   2. Prescribe qualification standards.
   3. Insure that guards are:
      a. Trained.
      b. Uniformed.
      c. Armed. (Examine the authority and legal liability during civil disturbances. Check with local officials.)

d. Deputized (if necessary).
4. Assure that the guard force is on duty at all times.
5. Issue written orders to the guard force.
6. Have an internal communication system for the exclusive use of the guard force.
7. Plan for an auxiliary guard force for use in an emergency. (This may be accomplished by designating and training company employees. If contract guards are to be used, advance arrangement should be made.)

C. Perimeter barriers.
1. Check the facility security fence (or other perimeter barriers) to insure that it is:
   a. Properly maintained.
   b. Inspected regularly.
2. Vehicle parking should be located outside of the security fence or wall. (This reduces the fire potential from gasoline in vehicle tanks and minimizes the hazard of explosives and incendiary devices which are easily concealed in a vehicle.)
3. There should be adequate protective lighting to illuminate critical areas.
4. Intrusion detection devices may also be used.

D. Control of entry.
1. Develop procedures for positive identification and control of employees. (Samples of identification media should be given to local law enforcement officials. This is essential for getting through police lines and during times of curfew.)
   a. Identification cards (sample to police).
   b. Badges (sample to police).
   c. Personal recognition (may be used for routine admission of employees to plants with less than thirty employees per shift).
2. Develop procedures for control of visitors.
3. Admittance to the facility should be controlled by the guard force.
4. Exercise control over movement and parking of vehicles.

E. Protection of critical areas. Identify and list critical areas within the plant.
1. Enclose critical areas with physical barriers.
2. Designate specific personnel who are to have access to critical areas.
3. Admittance to critical areas should be controlled by:
   a. The guard force or
   b. Supervisory personnel.
4. Protect unattended critical areas by:
   a. Locks. (Locks should be rotated upon notification of impending civil disorder or other emergency.)
   b. Intrusion detection devices.
5. Develop a key control system.
6. Develop package and material control procedures.
7. Institute procedures to protect gasoline pumps and other dispensers of flammable material. (Disconnect power source to electrically operated pumps.)

F. Arms rooms.
1. Keep arms rooms:
   a. Locked.
   b. Under 24-hour surveillance.
2. Keep ammunition:
   a. Stored in locked separate location.
   b. Under 24-hour surveillance.

G. Personnel security.
1. Conduct pre-employment investigations of applicants.
2. Make personnel checks of persons who are authorized access to critical areas.

3. Brief employees regarding the importance of plant security and the need for exercising vigilance.

VIII. Fire Prevention

These measures are of utmost importance in preventing or minimizing fire damage resulting from civil disorders.
  A. Post and enforce fire prevention regulations.
  B. Extend fire alarm system to all areas of the facility.
  C. Determine whether the municipal fire department can arrive at the facility within:
     1. Five minutes after the report of an alarm.
     2. Ten minutes after the report of an alarm.
  D. Have a secondary water supply system for fire protection.
  E. Have facility fire protection equipment on-site and insure that it is properly maintained.
  F. Determine from local fire department the feasibility of using mesh wire or other screening material to protect roofs from fire bombs, Molotov cocktails, or other incendiary devices.
  G. Organize employees into fire-fighting brigades and rescue squads.
  H. Store combustible material in a well-protected area.
  I. Instruct employees in the use of fire extinguishers.
  J. Conduct fire drills periodically.
  K. Maintain good housekeeping standards.
  L. Implement recommendations in the latest fire insurance inspection report.

IX. Vital Records Protection

Develop procedures for classification and protection of vital corporate records and protection of cash and other valuable items.

X. Property and Liability Insurance

Review property and liability insurance against potential loss or obligation resulting from riots and other destructive acts.

XI. Emergency Requirements

These requirements should be based on estimated needs for the duration of the emergency. These items should be prestocked because conditions may preclude their procurement during the emergency. Unused portions can be carried over for postemergency use.
  A. Food.
  B. Water.
  C. Medical supplies.
  D. Quarters.
     1. Sleeping.
     2. Separate male and female employees.
  E. Sanitation.
  F. Administrative supplies (office equipment).
  G. Emergency repair tools and equipment).
  H. Develop procedures for employees to purchase gasoline from plant supply in case local stations are closed.

XII. Testing the Plan

Frequent testing and correcting the plan will improve its effectiveness upon implementation under actual conditions. An emergency plan, like a chain, is no stronger than its weakest link.
  A. Types of tests:
     1. Partial—testing individual segments of the plan.
     2. Complete—testing entire plan.
  B. Tests should be unannounced.
  C. Weaknesses should be noted and the plan revised to include corrective actions.

APPENDIX 7D

CENTRAL-STATION SYSTEMS*

A central-station system consists of electrically operated circuits, instruments and devices, together with necessary electrical energy supply, designed to transmit alarms, supervisory and trouble signals to a constantly attended central station where signals are recorded, and experienced operators will take proper action in accordance with prescribed regulations. Such systems are independently controlled and operated by a person, firm or corporation, whose principal business is the furnishing and maintenance of supervised protective signaling service. These systems are adaptable to plants of any size and may consist of a simple manually operated fire alarm system, an extensive system including manual fire alarm boxes and automatic fire and smoke detecting devices with coded signals, and supervision of the sprinkler system, waterflow alarms and watchman's tours.

Ace Alarm Co., Inc., 57 Blundell St., Providence, RI 02905

Acme Central Station Alarm Co., Inc., 25 Jerusalem Ave., Hempstead, NY 11550

Action Alarm & Signal Co., 835 S.E. 17th Ave., Portland, OR 97214

AFA Protective Systems, Inc., 519 8th Ave., New York, NY 10018

AFA Protective System, Inc., Boston Automatic Fire Alarm Div., 61 Batterymarch St., Boston, MA 02110

AFA Protective Systems, Inc., Consolidated Fire Alarm Co. Div., 127 N. 4th St., Philadelphia, PA 19106

Alarmco, Inc., 6569 S. Vermont Ave., Los Angeles, CA 90044

Alarmco, Inc., 708 6th St., Las Vegas, NV 89101

Alarmtec International Corp., 2032 Scott St., Hollywood, FL 33032

Albany Protective Service, Inc., 547 2nd St., Albany, NY 12206

Altronics, Inc., 747 Main St., Bethlehem, PA 18018

American Alarm Co., Inc., 1121 Race St., Cincinnati, OH 45210

American Alarm Co., Inc., 912 N. Delaware St., Indianapolis, IN 46202

American Automatic Alarm Co., 3780 5th Ave., San Diego, CA 92103

American District Telegraph Co. (ADT Security Systems), 155 6th Ave., New York, NY 10013

Central offices located at:
  335 S. Main St., Akron, OH 44308
  11 N. Pearl St., Albany, NY 12207
  735 Gordon St., Allentown, PA 18102
  326 S. Lemon St., Anaheim, CA 92805
  89 Ellis St., N.E., Atlanta, GA 30303
  108 E. Baltimore St., Baltimore, MD 21202
  5616 Belair Rd., Baltimore, MD 21206
  2029 First Ave., North, Birmingham, AL 35203
  100 Charles River Plaza, Boston, MA 02114
  1016 Broad St., Bridgeport, CT 06603
  173 Green St., Brockton, MA 02403
  295 Main St., Buffalo, NY 14203
  130 N. Broadway, Camden, NJ 08102
  803 2nd St., N.W., Canton, OH 44703
  306 Meeting St., Charleston, SC 29401
  325 E. 9th St., Charlotte, NC 28202
  1307 Carter St., Chattanooga, TN 37402
  29 S. LaSalle St., Chicago, IL 60603
  157 W. 75th St., Chicago, IL 60620
  3235 W. Montrose Ave., Chicago, IL 60618
  2439 Warren Ave., Chicago, IL 60612
  6119 W. 26th St., Cicero, IL 60650
  18 W. 7th St., Cincinnati, OH 45202
  812 Huron Rd., Cleveland, OH 44115
  2067 E. 102 St., Cleveland, OH 44106
  14835 Emery Ave., Cleveland, OH 44135
  333 E. Livingston Ave., Columbus, OH 43215
  1500 Jackson St., Dallas, TX 75201
  333 W. 1st St., Dayton, OH 45402
  5486 Schaefer Rd., Dearborn, MI 48126
  817 17th St., Denver, CO 80202

---

* Factory Mutual Approval Guide, 1975. pp. 228-30. Courtesy Factory Mutual System.

406 6th Ave., Des Moines, IA 50309
611 W. Philadelphia Ave., Detroit, MI 48202
301 W. 1st St., Duluth, MN 55802
1587 Lincoln Highway, Edison, NJ 08817
900 Newman St., El Paso, TX 79902
810 Sassafras St., Erie, PA 16501
419 N.W. 6th St., Evansville, IN 47708
432 N. Saginaw St., Flint, MI 48502
621 E. Wayne St., Fort Wayne, IN 46802
106 W. 5th St., Ft. Worth, TX 76102
2201 Market St., Galveston, TX 77550
60 Division Ave., N. Grand Rapids, MI 49502
242 Trumbull St., Hartford, CT 06103
Calle Marginal (Ave. F. D. Roosevelt), Hato Rey, Puerto Rico 00919
1717 Rusk A., Houston, TX 77003
2561 Saturn Ave., Huntington Park, CA 90255
702 N. Capitol Ave., Indianapolis, IN 46204
46 W. Duval St., Jacksonville, FL 32202
591 Summit Ave., Jersey City, NJ 07306
124 Lake St., Kalamazoo, MI 49001
823 Walnut St., Kansas City, MO 64106
1212 Pierce Parkway, Knoxville, TN 37921
503 N. Larch St., Lansing, MI 48912
29-27 41st Ave., Long Island City, NY 11101
745 S. Flower St., Los Angeles, CA 90017
600 S. 7th St., Louisville, KY 40203
32 S. 2nd St., Memphis, TN 38103
135 N.E. 8th St., Miami, FL 33132
623 N. 2nd St., Milwaukee, WI 53203
155 Mineola Blvd., Mineola, NY 11501
9 N. 4th St., Minneapolis, MN 55401
203 E. Adams St., Muncie, IN 47305
315 Union St., Nashville, TN 37201
9 Clinton St., Newark, NJ 07102
109 Church St., New Haven, CT 06510
228 St. Charles Ave., New Orleans, LA 70130
44 E. 23rd St., New York, NY 10010
1624 Franklin St., Oakland, CA 94612
129 N.W. 6th St., Oklahoma City, OK 73102
1319 Farnam St., Omaha, NB 68102
738 High Ave., Oshkosh, WI 54901
454 Getty Ave., Paterson, NJ 07503
331 Fulton St., Peoria, IL 61601
125-27 S. 13th St., Philadelphia, PA 19107 (Philadelphia Local Telegraph Co.)
355 5th Ave., Pittsburgh, PA 15222
244 Middle St., Portland, ME 04111
333 S.W. 5th Ave., Portland, OR 97204
447 Penn St., Reading, PA 19601
201 E. Cary St., Richmond, VA 23219
16-18 Capron St., Rochester, NY 14607
510 Lafayette Ave., Rockford, IL 61105
1516 6th Ave., Rock Island, IL 61202
128 Eye St., Sacramento, CA 95814
420 N. Washington Ave., Saginaw, MI 48607
338 Missouri Ave., East St. Louis, IL 62201
910 Chestnut St., St. Louis, MO 63101
446 University Ave., St. Paul, MN 55103
35 New Derby St., Salem, MA 01970
175 S. Main St., Salt Lake City, UT 84111
305 N. Presa St., San Antonio, TX 78205
1416 Market St., San Diego, CA 92101
717 Market St., San Francisco, CA 94103
505 Asbury St., San Jose, CA 95110
24 Drayton St., Savannah, GA 31401
538 Spruce St., Scranton, PA 18503
1326 5th Ave., Seattle, WA 98101
224 West Jefferson Blvd., South Bend, IN 46601
25 W. Pacific Ave., Spokane, WA 99204
280 Chestnut St., Springfield, MA 01104
25 S. Yellow Springs St., Springfield, OH 45506
351 S. Warren St., Syracuse, NY 13202
19 S. 6th St., Terre Haute, IN 47801
140 N. Huron St., Toledo, OH 43604
143 E. State St., Trenton, NJ 08608

262 Genesee St., Utica, NY 13502
14536 Archwood St., Van Nuys, CA 91405
1012 14th St. N.W., Washington, DC 20005
73 Field St., Waterbury, CT 06702
35 Lake St., White Plains, NY 10603
431 N. Washington St., Wichita, KS 67202
913 King St., Wilmington, DE 19801
401 N. Main St., Winston-Salem, NC 27101
778 Main St., Worcester, MA 01608
412 Belmont Ave., Youngstown, OH 44502

American Protection Industries
Kings Alarm Systems, 2025 E. Curry St., Long Beach, CA 90805
Monarch Burglar Alarm Co., 2909 Beverly Blvd., Los Angeles, CA 90057

Apex Signal Service Div., Apex-Genie, Inc., 530 Bloy St., Hillside, NJ 07205
Ascot, Inc., Box 6247, Pasadena, TX 77502
A-Sonic-Guard, Inc., 745 South St., Louisville, KY 40203
Automatic Burglar Alarm Systems, Inc., 205 Wallace St., New Haven, CN 06511
Bay Alarm Co., 325 7th St., Oakland, CA 94607
Burns Electronic Security Services, Inc., 320 Briarcliff Rd., Braircliff Manor, NY 10510
Central Offices located at:
  42 S. Frank Blvd., Akron, OH 44313
  1300 Soldiers Field Rd., Brighton, MA 02135
  30 Church St., Buffalo, NY 14202
  221 N. LaSalle St., Chicago, IL 60601
  530 W. 2nd St., Dayton, OH 45405
  9500 W. Belmont, Franklin Park, IL 60131
  80 State St., Hartford, CT 06103
  20974 Corsair Blvd., Hayward, CA 94545
  3911 28th St., Long Island City, NY 11101
  4530 N.W. 7th Ave., Miami, FL 33127
  6 Orange St., New Haven, CT 06510
  1518 Walnut St., Philadelphia, PA 19102
  7th & Main Sts., Richmond, VA 23219
  1531 3rd Ave., Seattle, WA 98101
  627 Crockett St., Shreveport, LA 71101
  111 Northfield Ave., West Orange, NJ 07052
  600 N. Broadway, White Plains, NY 10603

Cal Crim Central Alarm, Inc., 3625 Hauck Rd., Cincinnati, OH 45241
Capitol Alarm Co., Box 1888, Sacramento, CA 95809
Central Alarm Co., 916 W. Adams St., Phoenix, AZ 85005
Central District Alarm, Inc., 6450 Clayton Ave., St. Louis, MO 63139
Central Ohio Alarm Systems Div., Moling & Assoc., Inc., 3379 E. Main St., Columbus, OH 43213
Central Protective Alarm Systems, Inc., 308 Burnet Ave., Syracuse, NY 13203
Certified Alarm & Signal Co., 1810 Jefferson Ave., Toledo, OH 43620
Chubb-Mosier & Taylor Alarms Div. of Chubb Industries, Ltd., 950 Yonge, Toronto 530 Ont.
Chubb-Mosier & Taylor Alarms, Ltd., 351 De Louvain St., W & 1450 City Councillors St., Montreal, PQ
Circle Alarm Corp., 1500 Madison Ave., Indianapolis, IN 46225
Columbus Electronic Protection Co., 108 Cleveland Ave., Columbus, OII 43215
Damon Alarm Corp., 307 Admiral Blvd., Kansas City, MO 64106
Denver Fire Reporter & Protective Co., Inc., Denver Burglar Alarm Co., Inc., 1955 Shermer, Denver, CO 80203
Electric Protection Services (Div. American District Telegraph Co., Inc.)
  Hillgate House 26 Old Bailey, London E C 4, Eng.
  Several European Central Stations
Electro-Protective Corp., 25 Eastmans Rd., Parsippany, NJ 07054
Central Offices located at:
  12 W. 22nd St., Baltimore, MD 21218
  60 Lafayette St., Newark, NJ 07102
  30 E. Livingston St., Orlando, FL 32801

25 Eastmans Rd., Parsippany, NJ 07054
1102 N. B St., Tampa, FL 33606
14 E. 6th Ave., Trenton, NJ 08619
Engineered Protection Systems, Inc., 950 Ionia N.W., Grand Rapids, MI 49503
Engineered Systems, 311-313 Smyres Pl., Greensboro, NC 27402
Flint Electronic Security Service, Inc., G-3163 Flushing Rd., Flint, MI 48504
General Protection Service Corp., 6727 Odessa Ave., Van Nuys, CA 91406
Guardian Alarm Div., Colbert's Security Services, Inc.
   15 John Williams St., Attleboro, MA 02703
   333 Smith St., Providence, RI 02908
Holmes Protection, Inc., 229 W. 36th St., New York, NY 10013
   Central Offices located at:
      229 W. 36th St., New York, NY 10013
      11th & Sansom Sts., Philadelphia, PA 19107
      521 6th Ave., Pittsburgh, PA 15219
      27 State Highway No. 17, Rutherford, NJ 07070
Honeywell Protection Services (Honeywell, Inc.), 2701 4th Ave., S., Minneapolis, MN 55408
   Central Offices located at:
      135 Memorial Drive, S.W., Atlanta, GA 30310
      76 Shirley St., Boston, MA 02119
      85 R Hoffman Lane, S. Central, Islip, NY 11722
      301 N. Kedzie, Chicago, IL 60612
      6707 Carnegie Ave., Cleveland, OH 44103
      1107 Peters St., Dallas, TX 75215
      35 Gaylord St., Elk Grove Village, IL 60007
      2032 Scott St., Hollywood, FL 33020
      37-08 Greenpoint Ave., Long Island City, NY 11101 (Honeywell Mutual Protection Services)
      50 W. North St., Manchester, NH 03104
      525 N. 6th St., Milwaukee, WI 53203
      415 E. 27th St., Minneapolis, MN 55408
      461 Park Ave., S., New York, NY 10016 (Honeywell Mutual Protection Services)
      1725 Linwood Blvd., Oklahoma City, OK 73106
      120 N. Camac St., Philadelphia, PA 19107 (Honeywell Owl Protective Services)
      5993 Penn Circle S., Pittsburgh, PA 15206
      4585 Allstate Drive, Riverside, CA 92502
      3055 47th Ave., N., St. Petersburg, FL 33714
      514 S. Lyon St., Santa Ana, CA 92702
      7723 24th Ave., N.W., Seattle, WA 98117
      527 S. Kenosha Ave., Tulsa, OK 74120
      1720 Upland Rd., West Palm Beach, FL 33401
      311 Amboy Ave., Woodbridge, NJ 07095
Instant Alarms, 303 Highland Ave., Salem, MA 01970
Johnson Controls, Central Station of Milwaukee, 507 E. Michigan Ave., Milwaukee, WI 53201
Laclede Gas Security Systems, Inc., 6108 Madison Ave., St. Louis, MO 63134
Master Alarm Co., 5719 York Rd., Baltimore, MD 21212
McCane-Sondock Protection Systems, 1612 Austin St., Houston, TX 77052
Michigan Still Alarm Co., 10410 W. Chicago St., Detroit, MI 48204
Midnight Burglar Alarm Systems, Inc., 1026 Ann Arbor St., Flint, MI 48503
Miley Protective Alarms, Inc., West Point Pike, West Point, PA 19486
Minuteman Alarm Co., Inc., 1901 Beverly Blvd., Los Angeles, CA 90057
Morse Signal Devices, 6601 Santa Monica Blvd., Los Angeles, CA 90038
Morse Signal Devices of California, 121 Broadway, San Diego, CA 92101
Mossie Alarm Co., Inc., 1049 Central, Kansas City, MO 64105
Multra-Guard, Inc., Central Station Div., 2714 W. Mercury Blvd., Hampton, VA 23369

National Protection Div., National Kinney Security, Inc., 3500 W. 1st St., Los Angeles, CA 90054

Northern Burglar Alarm, Inc., 7820 N. 27 Ave., Phoenix, AZ 85021

Pacific Fire Extinguisher Co., American Burglar Alarm Div., 165 Jessie St., San Francisco, CA 94105

Per Mar Security & Research Corp., 425 W. 2nd St., Davenport, IA 52808

Potter Electric Signal Co., 1211 Pine St., St. Louis, MO 63103

Protection Div., ISC Industries, Inc., 601 Westport Rd., Kansas City, MO 64111

Protection Alarms, Inc., 725 W. Main St., Peoria, IL 61606

Protelec Ltd., 138 Portage Ave., E., Winnipeg, Manitoba

Rhode Island Electric Protective Co., 111 Mathewson St., Providence, RI 02901

Robinson Protective Alarm Co., 15th & Chestnut Sts., Philadelphia, PA 19102

Rochester Central Alarms, Inc., 180 Clinton Ave., Rochester, NY 14604

Rodgers Police Patrol, 3780 5th Ave., San Diego, CA 92103

San Diego Alarm Co., Inc., 2054 State St., San Diego, CA 92101

Schirmer-National Alarm Div., The Schirmer National Co., 100 Portland Ave., Bergenfield, NJ 07621

Security Alarms, Inc., 470 N. Cleveland Ave., St. Paul, MN 55104

Security Alarms & Services, 300 Church St., Nashville, TN 37219

Security Controls, Inc., 16143 Wyoming Ave., Detroit, MI 48221

Security Instrument Corp. of Delaware, 309 W. Newport Pike, Wilmington, DE 19804

Security International, Inc., 4622 S. 88th, Omaha, NB 68127

Security Systems, Inc., 2512 S. Carleton Ave., Appleton, WI 54911

SIS Alarms Ltd., 164 Ave. Rd., Toronto 5, Ont.

Smith Alarm Systems, 2627 Flora St., Dallas, TX 75221

Southern Burglar Alarm Co. of Georgia, Inc., 178 Broad St., S.W., Atlanta, GA 30303

Southern Burglar Alarm of Virginia, Inc., 2400 Granby St., Norfolk, VA 23517

Standard Alarm Co., Inc., 5765-67 W. Pico Blvd., Los Angeles, CA 90019

Systems for Security, Inc., 3010 N.W. 17th Ave., Miami, FL 33142

3M Alarm Services, 3M Center, St. Paul, MN 55101
Central Offices located at:
63-69 Washington Ave., Chelsea, MA 02150
969 River Dr., Methuen, MA 01844
2404 Lyndale Ave., S., Minneapolis, MN 55405

United Protective Corp., 299 Main St. (Rear), Hackensack, NJ 07601

United States Burglar Alarm Co., Inc., 1322 W. 12th Pl., Los Angeles, CA 90015

Valley Alarm Co., 362 E. 4th St., Pomona, CA 91766

Valley Burglar & Fire Alarm Co., 1543 O St., Fresno, CA 93721

Wackenhut Protective Systems, Inc., 3280 Ponce de Leon Blvd., Coral Gables, FL 33134
Central Offices located at:
6900 Hoover St., Los Angeles, CA 90044
2550 N.W. 39th St., Miami, FL 33142
12117 Nebel St., Rockville, MD 20852

Wells Fargo Alarm Services Div., Baker Protective Services, Inc., 104 Witmer Rd., Horsham, PA 19044
Central Offices located at:
180 Memorial Drive, S.W., Atlanta, GA 30303
214 W. Ohio St., Chicago, IL 60610
341 Broad St., Clifton, NJ 07013
171 Addison St., Elmhurst, IL 60126
6317 S. Figueroa St., Los Angeles, CA 90003
337 Court St., Memphis, TN 38103
2931 N.E. 2nd Ave., Miami, FL 33137
372 University Ave., Newark, NJ 07102
1023 Baronne St., New Orleans, LA 70113
53 W. 23rd St., New York, NY 10010

825 S. 20th St., Omaha, NB 68108
10th & Vine Sts., Philadelphia, PA 19106
27 Sheer Plaza, Plainview, NY 11803
29 Emmons Dr., Princeton, NJ 08540
50 N. 5th St., Reading, PA 19601
9313 Manchester Rd., St. Louis, MO 63119
466 Vendome St. at Coleman, San Jose, CA 95106
1004 6th St., Washington, DC 20001
Wise Security, Inc., 524 Franklin St., Buffalo, NY 14202

Chapter 8

INSTITUTIONAL PROTECTIVE SERVICES

INSTITUTIONAL PROTECTIVE services has grown dramatically in the past ten years. Severe crime and direct loss stresses have been placed on those activities which serve the needs of the public for a wide variety of services. Institutions include:

- Banks
- Colleges, Universities, and Schools
- Courts
- Hospitals, Nursing Homes, and Health Care Facilities
- Libraries
- Museums
- Parks
- Public Buildings
- Public Utilities
- Sports Facilities
- Transportation Terminals

The indirect costs of crime and disruption of operations are difficult to assess. What are the indirect costs to students being deprived of educational opportunities by closed facilities; transportation terminals closed because of high crime rates or uncontrolled internal losses; the extra costs of airport security screening of passengers to prevent hijacking of flights; the losses of books and materials from public libraries; and increased health care costs because of required security measures to protect patients and employees from criminal attack or losses of hospital property? It is the indirect cost of losses which place the greatest stress on the system of institutional services provided the public.

The protective services program for each institutional setting responds to various unique situations which often are not a part of other institutional requirements. The goals and services rendered by each institution are different as are the environmental conditions they seek to maintain. The open, academic atmosphere of the school or university is difficult to maintain if criminal losses or attacks are commonplace. Conversely, the development of a "police state" atmosphere is equally destructive of an open learning environment.

The wide variations in types of services performed and the clientele served presents difficulties in categorizing procedures and techniques for "institutional security." It is, however, possible to establish general principles of institutional protective services operations.

PRINCIPLES OF INSTITUTIONAL PROTECTIVE SERVICES

Security, in order to be effective in an institutional environment, must provide its protective and preventative programs in an acceptable and nonrepressive manner. It must provide its functional services in a manner that does not interfere with institutional activities, while providing control activities in an unintrusive manner. The problems of security, protection, and prevention transcend the operating responsibilities of the institutions protective services department. The effective security department evaluates and assesses its needs and develops a total organizational response to protection that will insure the greatest amount of support for programs developed while providing the highest degree of protection and control at the best possible cost-benefit ratio.

The prevention of losses is a total institutional effort. The institution is a community and the many members of the community are responsible for, and share responsibility in, the protection of property and prevention of losses. While the security office bears the formal responsibility; all members equally share this responsibility for the provision of an environment which will materially decrease the probability of acts which disturb the institution's operations.

Security services, much like the activities

of a formal law enforcement unit, must direct their activities and attention towards the conditions, places or the persons that contribute to, or are involved in, acts which are considered disruptive, illegal, or unsafe. Security officers are much more than policemen. They can arrest individuals who, by their activities, either contribute to the committing of undesirable acts or engage in these activities themselves. This, however, does not limit or circumscribe the total responsibility of the members of the protective service function. Like other professionals, they must go beyond the symptomatic act and deal with the things, conditions, or thinking that prompted the individual to commit unacceptable acts. The security officer, therefore, deals with *threats* and must understand the causes of these activities and develop programs to reduce their happening, both by actions within the physical confines of this area of responsibility and also in the broader community in which the institution is located.

The control and prevention of undesired, illegal, or unsafe acts, as with other types of problems of social deviancy, is not and should not be the sole responsibility of the security department, nor of the police department, nor for that matter of any one agency. These are the responsibilities incumbent upon the entire institution. The security office should work closely with the various institutional departments and with the local police department in an effort to bring together all of social, moral, and community resources possible to provide protection and prevention. It should take an active role to coordinate, stimulate, and direct their activities to the specific types of problems which must be solved.

Departmental managers and their employees who experience crime losses often inadvertently contributed or increased the likelihood of their losses because of their lack of knowledge in property protection or their failure to assume departmental and individual responsibility for the protection of their own property. "Community" complacency and apathy in the protection of individual property has become common in most institutions of the United States. Individuals have, over a period of years, transferred the primary responsibility of property protection to public and/or private agencies and specialized departments. As a result, protective agencies are expected to assume both the individual's primary and their own secondary responsibility of property protection. Security departments were never intended to assume the entire *burden* of property protection. Consequently, after a theft occurs, the department managers, employees, or employer blames the local security or public law enforcement agency for their own failure to protect and secure the property under their care.

Institutional security departments are in a position to initiate active prevention programs. The involvement and development of prevention programs can generate a considerable degree of favorable sentiment and support for programs which seek to help the institution develop resistance to incidents of criminality, destruction of property and restrictions on individual activities and liberties.

The development of an active prevention program in the institution or the area surrounding it would provide a relatively secure working and service environment.

The security department is responsible for the establishment of liaison and coordination of activities with local police agencies and other civic and community groups. The active participation in community action programs often fosters rapport between the institution and local community agencies of social control, as well as grass roots community leaders and citizens in the area immediately adjacent to the facility. This would tend to insure a greater understanding of and sympathy for the problems of social control, crime prevention, and protection of property which are unique to the institution.

The institutional security department occupies a critical role in prevention activities. Since the responsibility for protection is placed in it, every means available must be used to provide protection for the institution. If the security department is able to develop programs of prevention through the involvement in community and institutional activities and is successful in gaining the support and respect of local civic groups and community leaders, the responsibility of protecting life and property will be met and the possibility of increased losses will be significantly reduced.

THE ROLE OF SECURITY

Security is a functional activity which should be organic to institutional management.

The responsibility of prevention and enforcement activities should rest with the security department. All policies dealing with law enforcement, crime and violence prevention, and general police or security services should be the responsibility of the security department. All departments of the institution should provide input for the maintenance of records and development of reports for the security department with respect to prevention and protection problems.

A comprehensive institutional program requires the integration, in the right proportion, of the security processes so that the resulting program is acceptable to the institution and the surrounding community. Security personnel, technical equipment, and architecture design must be integrated into the environment so that it becomes a part of the institution and thus illicits the support of the community. Without support, the security program of any installation will not be able to operate effectively and/or efficiently. The key to success of a comprehensive protective service program is the interweaving of a community relations program into the various components of the overall security program.

"Good" community relations do not just happen, rather they are the result of a sound plan and the implementation of that plan by a well-trained and fully informed security staff.

The caliber of security personnel utilized in the implementation of an institutional security program must be above that of the usual security guard who too often is ill-equipped by training and experience to cope with its special communications and human relations problems. These positions are extremely important and only sensitive, people-oriented, well-trained personnel attracted by a higher salary can do an effective job.

TYPES OF INSTITUTIONAL PROTECTIVE SERVICES
Bank Security

The Bank Protection Act of 1968 (Public Law 90-389, July 7, 1968, 90th Congress) was adopted on January 14, 1969 by the Comptroller of the Currency and the Directors of the Federal Deposit Insurance Corporation.[1]

The regulations cover the approximately 13,000 banking offices of insured state banks not members of the Federal Reserve System and the 15,000 offices of national banks. The Board of Governors of the Federal Reserve System and the Federal Home Loan Bank Board have also issued similar regulations for their respective institutions.

The Bank Protection Act was passed by the Congress in an effort to stem the growing incidence of crimes against financial institutions. The Federal Bureau of Investigation has estimated that there were approximately 2,658 violations of the Federal bank robbery statute, including armed robbery, burglary, and larceny during 1968.

Under the law, the four Federal supervisory agencies for financial institutions were directed to promulgate regulations establishing standards with which their member institutions must comply with respect to the installation, maintenance, and operation of security devices and procedures, reasonable in cost, to discourage robberies, burglaries, and larcenies and to assist in the identification and apprehension of persons who commit such acts. The Act also expressly directed the four agencies to require the submission of periodic reports from financial institutions relating to the installation, maintenance, and operation of security devices and procedures.

The regulations required each bank to designate a security officer by February 15, 1969, to implement a security program by July 15, 1969, and to develop a plan for installation of appropriate security devices by January 1, 1970.

The regulations require bank security officers to seek the advice of law enforcement officers in determining the specific protective needs of each banking office. The regulations allow varying degrees of protection for individual banking offices in accordance with the incidence of crimes in the area and other bank-related factors.

In addition to specifying certain minimum requirements, the regulations contain an appendix which lists some standards for various other security devices that will be appropriate for banks in areas with high incidences of crimes against financial institutions.

The Act also indicated that specific training should be provided to employees to insure proper action during robberies.[2] It likewise specified that a security audit be conducted of the facility and a copy submitted to FDIC, FRS, or FHLB.[3] After every criminal attack against an insured banking institution a Report of Crime[4] is to be filed.

Bank security which was a matter of concern only at the individual bank is now prescribed by federal regulation. Compliance is not uniform throughout the banking community, but the entire level of protection has been raised greatly by the efforts to date.

Campus Security

The incidents of campus unrest, bombings, disruptions, riots, increased theft, and vandalism losses greatly change the scale of the campus protective function. Campus protection had historically been a watchman function to detect and prevent fires and insure buildings were secured. Today, campus protection has changed its character with its officers ranking among the best trained, educated, and technologically equipped protective personnel in the United States.

University protective services are a function basic to the university administrative and management necessary to maintain twenty-four hours a day, 365 days a year. Campus security provides the following operational services for the university:

1. inspectional services (fire equipment, maintenance equipment security and hazards);
2. emergency services (fire, first aid, and ambulance, etc.);
3. traffic regulations and supervision of the university property;
4. protection of lives as well as state property and records;
5. enforcement of Board of Regents and university regulations;
6. collecting and disseminating information regarding student activities;
7. investigate infractions of rules and regulations;
8. coordination between university and local community law enforcement agencies; and
9. the development of overall security policy for the university.

The objective of protective services is to provide an environment which allows for activities of the university to continue in an uninterrupted and orderly fashion without interference with university academic and administrative activities. To accomplish this, security must be fully integrated into university administration. It can thus best provide its protective and preventative activities within the university community.

The major problem that University Protective Programs currently deal with are:

1. access control for buildings and faculties;
2. identification of staff, faculty and students;
3. parking/traffic;
4. key control;
5. inventory control;
6. theft control;
7. fire prevention; and
8. safety.

Civil disorder, violence, and bombings are

still possible areas of concern but not to the degree they were during the 1960's and early 1970's.

The problem of integration of electronic sensory devices to perform protective functions has likewise become a major administrative and operation problem. Due to rising costs of personnel services and the increased effectiveness and reliability of electronic devices, their use has become widespread. Devices often supplement personnel functions for protection. Thus, the balancing of university protective needs with administrative realities places a need for greater accountability and operational planning of security personnel.

The increasingly widespread use of "designed-in" protective features and devices as well as better site planning has likewise streamlined protective practices on the campus. The development of protective priorities is likewise necessary to allocate resources most effectively.[5]

Since the university is a community, efforts are made to educate faculty, staff and students of their responsibilities for self-protection. Likewise, services provided by the university are made known to all concerned to reduce misunderstanding and increase mutual respect.[6]

Courts

The ability of the justice system to operate is basic to the U.S. system of government. The protection of judges, juries, witnesses while the responsibility of the public police agencies is likewise a function of court security.

The design of buildings, the procedures for transferring prisoners, the configuration of the court room, the presence of surveillance equipment and the use of protective equipment all add to the protective capability of the facility. There have been many instances of judges being assaulted in their courtrooms, prisoners escaping because of inadequate security measures for bringing them to court; and witnesses being intimidated by suspects during breaks in court proceedings. Events such as these can be prevented by design and/or redesign to insure adequate security control over the court and its operations.

Many court facilities were designed prior to the recent increase of attacks on the criminal justice system and its operations. Most programs which have been developed in the past several years have been remedial in nature. Their main focus has been to improve procedures and make those structural and/or hardware additions economically feasible.[7] Many court security problems can be reduced significantly by procedural modifications and judicial support. As in other applications, the opportunity for improvements in protective services is limited by the recognition of risk and the willingness to support meaningful changes.

Hospital, Health Care and Nursing Home Protection

Hospital and health care facilities are basic institutions which provide services to broad segments of society. They tend to bring together diverse segments of the community under conditions of stress and emotional upheaval. Many of the health care facilities are now located in areas of communities which have deteriorated but, because of the high cost associated with relocation and community pressures, have remained. In these areas it becomes a high-visibility, high-risk facility which requires proportionate protective measures.

Health care facilities require protective programs which treat the unique problems associated with delivering highly technical services to patients while maintaining a controlled environment. Historically, hospitals have been slow to develop effective protective programs. Few hospitals currently utilize sophisticated protective technology except in new construction or rennovation areas. Protective services programs tend to be personnel-based rather than hardware-based. In a recent study, 75 percent of the hospitals were not utilizing antiintrusion equipment and 86 percent were not using CCTV surveillance technology.[8]

Protection and security are relative concepts. The perceptions of users and developers of programs often set the level of acceptable services from the function. Health care facilities in the above-cited study further indicated that one in ten institutions characterized their protective programs as being "strict" while only one in three characterized their program as being "adequate."

Existing Protective Programs

There are four major areas of protective concern in the health care facility. They are:
1. security,
2. fire control and prevention,
3. safety, and
4. disaster planning and control.

Historically, hospital protective services were of the nightwatch variety with fire and safety hazards of primary concern. The introduction of rigid safety requirements and the increased need for protection from internal and external losses has greatly modified the traditional view of the function.

Security

Contemporary security programs deal with the full range of loss prevention concerns common to all organizations. Access control is, however, a major problem because of the special nature of the services performed. Procedural controls are required to insure identification of employees, staff, visitors, and patients. Escort services are required for employees to parking lots or for visitors to various locations within the facility. Likewise, controls for access to equipment or supply storage areas must be maintained. Losses of linen, food, drugs and equipment are the major areas of loss prevention concern.

The monitoring of intrusion detection and surveillance equipment, where it is used, required a considerable investment in manpower. It is used to protect against unauthorized entry or use of facility premises, and to protect specific high-risk areas such as narcotics storage, food lockers, and administrative areas. Investigative services as well as preemployment screening assistance are also portions of the security responsibility. Likewise, the patrol of facility grounds and buildings is performed.

Safety

Safety responsibilities include insuring compliance with OHSA standards as well as related state and local safety codes and standards such as the National Fire Protection Association (NFPA) Standards for Health Care Facilities. The safety function involves inspections for hazards in equipment, procedures or practices which could be hazardous to life or property. The large amount of electrical and gas systems make the prevention of conditions conducive to sparking or shocking particularly important.

Safety education of employees and the establishment of safe working practices are often done by the protective services function.

Fire Control and Prevention

Fire is a major area of concern for the health care facility. Patients are under the complete care of the facility and must have every precaution taken to prevent or minimize their exposure to fire risks. The regular inspections of all conditions and aspects of the facility, as well as fire control equipment, is a primary function. Likewise, the proper training of all facility employees of their duties and necessary skills in fighting fires is essential.

Disaster Planning and Control

The development of evacuation plans and their testing is the role of protective services. Likewise, the preparation of plans for reacting to emergency or disaster conditions is integral to the protective services function. Bomb threats or bombings, fires, natural disasters, civil disturbances can all cause disruptions of facility operations. Adequate plans must be developed for their occurrence. Likewise, mutual assistance programs should be developed with other health care facilities in the area as well as with appropriate state, federal and local public protective services agencies such as police, fire, civil defense, and military agencies. Likewise, testing of plans and programs must be done to insure that the plans actually are effective.

Facility Design

The contemporary health care facility of new construction is being designed from a prevention perspective. Adherence to new life safety codes and standards are making the development of a safe and secure environment more possible now and in the future. The impact of "designed-in" features is yet to be experienced on a wide scale throughout the health care field. Where it has been introduced however, the lowering in losses and costs has justified the higher initial expense for construction and equipment.

Library Protection

The library is a valuable community resource that requires protection of its holdings of books and materials while making them readily available to the public. Protecting books and materials in the facility as well as while circulating is essential to insure that they are available in useable condition for extended periods of time. Libraries list the following as the major protection problems that they face:[9]
1. book theft,
2. vandalism,
3. book and materials mutilation,
4. nonreturn of books and materials,
5. theft of library and personal property of employees,
6. burglary,
7. disorderly conduct, assault, sex offenses,
8. key control,
9. poor facility design for security, and
10. fire protection systems and equipment.

The development of library security programs thus requires a blending of safety and security procedures and techniques to provide the desired environment. The use of exit book detection equipment might stop the loss of entire books, but might also increase the mutilation rate with pages being removed rather than the entire book. Positive identification systems for library uses likewise increases protection against losses but does little to reduce disorderly conduct. The integration of all protective requirements into a comprehensive program will balance the approach and allocate resources to accomplish more than one specific security function.

Museums

Museums contain the physical history of the world and try to make its wonders available to the general public. Irreplaceable objects must be placed before anyone who wishes to see them; damage from fire, theft, vandalism, carelessness, and accidental destruction must all be provided by the protective services function.

The development of an open yet fully protected environment for public use requires the integration of highly technological equipment coupled with well-trained personnel. Unfortunately, this has not been the case historically with museums. Protection has often been limited to firewatch or minimal protection except for obvious high-value items or exhibits. This is no longer totally true. Museum protection has developed into a sophisticated and highly technological branch of protective services.

Fire protection, safety and protection are all a portion of the function. Since many artifacts housed in a museum are not replaceable, every precaution must be taken to prevent damage as well as plans made for removal in emergency situations. Evacuation procedures, fire control are, however, two after-the-fact situations which must be treated in any museum protection programs. The identification of museum property as well as accountability must be undertaken to insure against loss or to aid in recovery of lost or stolen items.

Parks

Parks and recreational areas constitute a difficult security challenge for program development. Crime, disorder, and losses both to users of these public facilities and the communities or agencies who support them has grown to tremendous proportions. Campers raped and killed, property stolen, vehicles stripped, wildlife killed, historical sites vandalized or looted, drugs distributed and used are common occurrences. State, local, and Federal agencies have attempted to develop protective programs which would insure safety and security for the general public yet not make access so restrictive that the parks seem like prisons.

Park rangers, police officers, and security officers as well as high-technology sensory devices are now utilized to implement crime control programs to reduce or eliminate various types of criminal conduct and losses in these areas.

The area of greatest optimism is in the development of planning and space utilization which reduces criminal opportunity. Improved lighting, shrubbery and floral design, walkway construction coupled with surveillance and sensory equipment reduce the problem areas to be patrolled by security personnel. The cost, both in social as well as monetary terms for even minimal preventative programs of other than an educational or random patrol basis are prohibitive. Consequently, as criminal attacks get more difficult in urban, highly protected areas, the rural park which has a high usage will become an attractive target.

Public Buildings

The protection of public buildings and the functions of government which are conducted inside of them are critical to the continuity of governmental operations. The attacks on public buildings increased dramatically during the 1960's with bombings, bomb threats, civil disturbances and disorders being commonplace activities. Government at all levels determined that protective programs were required and a variety of programs were initiated to lower the risk of loss.

The General Services Administration of the Federal government began a program of upgrading or installing security systems in major Federal facilities across the country. This program conducted by the newly redesignated *Federal Protective Service* was also charged with retraining and upgrading the security officers assigned to protect these facilities. Given sixteen weeks of training, the Federal Protective Service police officer was assigned to supplement the newly installed and improved protective measures, screen for bombs, control access to buildings, and implement the intent of the program to reduce losses in Federal facilities.

Similar programs were developed in many major cities to protect city and county facilities from damage or disruption. Several of these programs were established by police departments to provide additional preventative measures above those offered on a regular basis by the police department. The Dallas Security Force (Dallas, Texas) is an example of a protective services program being developed specifically to protect public facilities and supplement police services for the continuity of governmental operations.

Public Utilities

Utilities which serve the basic requirements for electrical power, telecommunications, gas, water, and entertainment are most often public corporations. These utilities are carefully regulated by governmental agencies and seek to provide services to the public at reasonable costs. Utility rates are regulated by the government with increases justified only by increases in operating costs. These operating costs and associated rates have increased tremendously in the past several years because of the increases in fuel and labor costs. They have additionally been increased by the direct and indirect cost of losses due to theft, waste, and damage. The theft of tools and equipment, transmission lines for their copper as well as theft of gas, water, and electrical current have substantially increased operating costs.

The advent of nuclear power has likewise increased the security requirements of the public utility. The requirements of the Atomic Energy Commission (AEC) regarding the hiring of guards, their training, selection, and operations has increased protective costs.

Security measures for communications facilities include:
1. safeguarding Microwave sites,
2. computer terminals,
3. telemetering equipment,
4. transmission facilities,
5. switching facilities,
6. long-line transmission facilities,
7. client-owned or leased equipment, and
8. fraudulent or abusive use of equipment.

Security measures of other utilities include:
1. power generating facilities and equipment,
2. power transmission facilities and equipment,
3. relay equipment,
4. switching or regulating equipment, and
5. monitoring and metering equipment.

Protective services programs for all utilities include the application of the three security processes of physical, information, and personnel measures to prevent or reduce losses. These processes are focused on the reduction of losses in the following areas:
1. employee theft of equipment, revenue or time,
2. external theft of equipment, products or services,
3. abuse of equipment or services by employees or public,
4. insurance of compliance with all regulatory body requirements for operational security,
5. losses through employee lack of knowledge about security procedures or policies,
6. damage or destruction of facilities or operation by sabotage or vandalism.
7. transmission of confidential information of the company or transmitted by its facilities,
8. protection of key corporate personnel,
9. developing plans for continuity of operations in emergency or disaster situations,
10. conducting investigations into improper conduct of employees,
11. limiting access to company facilities to those persons essential for operations, and
12. insuring liaison with appropriate law enforcement agencies in areas of mutual concern.

The individual utility will organize their protective program to provide the appropriate amount of emphasis on their highest priority problems. There will be, however, some measure of attention given to each of the above areas in a total protective program.

Sports Facilities

Protective services at sports facilities involve crowd control, parking and traffic control, first aid, protection of ticket holders and buyers from dishonest actions, prevention of illegal gambling, protection of facility revenues, and equipment and protection of team personnel and equipment.

The protective function maintains order and insures the operations of sporting events within a safe and secure environment. The problems of protection are often a joint public-private protective services responsibility. Professional sporting events normally have significant numbers of public law enforcement personnel available for police duties while private security officer performs crowd control and revenue protective duties. Large sporting events normally have commercial security agencies under contract to provide overall planning and direction for the sports season. Liaison with the public agencies is coordinated with the agency to insure complete coverage and understanding between all protective personnel assigned to the facility.

Transportation Terminals

Transportation facilities require the entire spectrum of protective services. Protection is often jointly provided by public and private personnel since these facilities are considered public in nature. Bus and rail as well as port and airport facilities all have unique protective requirements which are provided by some combination of public and private services. Public agencies provide police services and supplement the preventative activities of the private officers.

Protection of the traveling public, transportation equipment, suppression of illegal activities, insuring the safety and well being of passengers, protection cargo and baggage, crowd control, and providing passenger assistance are but a few of the many responsibilities of transportation facility protective services personnel.

Airport Security

Airport security programs involve the protection of property of persons in transit between U.S. domestic and foreign air terminals. Duties of airport security personnel involve the screening of passengers, the protection of facilities, the provision of police services in the airport areas, and the securing of cargo moving through the airport terminal.

Management programs include development of airport security programs that comply with Federal Aviation Agency (FAA) (Part 107.B), requirements for protection, development, and maintenance of mutual aid programs with local law enforcement agencies, supervision of contract security personnel, liaison with local law enforcement agencies, development and maintenance of meaningful airport security programs, and standards for protection of life and property in the facility.

Bus Terminal Security

Bus terminals are normally protected in urban centers by both public police and private protective personnel. Since large numbers of travelers pass through these facilities there must be protection both for them and for the terminal. Travelers are particularly vulnerable for criminal attacks of all types ranging from crimes against persons to fraud, theft, and kidnapping. The protection of baggage and personal effects is likewise essential. The provision of information, general and emergency assistance is likewise necessary.

The protection of terminal, of bus company equipment, or cargo in transit is likewise the responsibility of the police or security personnel. Personnel screening of employees is often conducted by bus company security offices in cooperation with the personnel office. Likewise, the investigation of losses is conducted by corporate protective personnel. Contract guard services are most often hired for terminal protection.

Rail Terminal Security

Rail terminal facilities in major urban centers is provided by railroad police. These are special-purpose private personnel who most often have "peace officer" powers to arrest and take official police actions on railroad property. Protection needs are similar to those for bus facilities.

FOOTNOTES

1. See Appendix 8A for a copy of Regulation P(12 CFR 216) **Minimum Security Devices and Procedures for Federal Reserve Banks and State Member Banks.** This Regulation is a detailed procedural guide for the implementation of the intent of PL 90-389.
2. See the Bank Protection Act in Appendix 8A this chapter.
3. See Appendix 8B for a copy of Report of Security Devices.
4. See Appendix 8C for a copy of **Report of Crime Form.**
5. Appendix 8D presents a methodology for obtaining the information required for developing protective priorities on the campus.
6. See Appendix 8E for a sample guide to university safety and security services.

7. An example of procedural changes possible for increased court protection is presented in Appendix 8F. The changes described are possible at a minimum of cost and greatly increase court protection in high-risk trial situations.
8. National Survey on Hospital Security, Burns Security Institute, October 1972. This study is based on a sample of 196 hospitals in thirty states which employ a total of over 2,300 guards.
9. Adapted from Burns Security Institute, National Survey on Library Security, Briar Cliff Manor, New York, September 1973, p. 10.

BOOKS

Bank Administration Institute. *Bank Security Manual.* Park Ridge, Ill., 1973.

———: *Your Bank and Armed Robbery.* Park Ridge, Ill., 1973.

———: *A Study of Internal Frauds in Banks.* Park Ridge, Ill., 1972.

Burns Security Institute. *National Survey on Hospital Security.* Briarcliff Manor, N.Y., October 1972.

Colling, Russell L. (ed.): *Hospital Security and Safety Journal Articles.* New York, Medical Examination Publishing Company, Inc., 1970.

Conference Institute, The: *Procedures for Security Control.* Division of James O. Rice Associates, Inc., New York, 1970.

Dewhurst, H. S.: *The Railroad Police.* Springfield, Thomas, 1955.

Fish, Donald E.: *Airline Detective: The Fight Against Internation Air Crime.* London, Collins, 1962.

———: *The Lawless Skies: The Fight Against Internation Air Crime.* New York, Putnam, 1962.

McNew, Bennie B.: *Solid Controls for Commercial Banks.* Homewood, Ill., R. D. Irwin, 1962.

McNew, Bennie B., and Prathe, Charles L.: *Fraud Control for Commercial Banks.* Homewood, Ill., R. D. Irwin, 1962.

Morse, George, and Morse, Robert F., II: *Protecting the Health Care Facility.* Baltimore, The Williams & Wilkens Co., 1974, 318 p.

Neilson, Swen C.: *General Organizational and Administrative Concepts for University Police.* Springfield, Thomas, 1971.

Pratt, Lester A.: *Bank Frauds Their Detection and Prevention.* New York, Ronald Press.

San Luis, Ed.: *Office and Office Building Security.* Los Angeles, Security World Publishing Co.

PUBLICATIONS OF THE GOVERNMENT, LEARNED SOCIETIES, AND OTHER ORGANIZATIONS

Adams, Gary and Rogers, Percy G.: *Campus Policing—State of the Art 1971.* Center for Justice Administration, University of Southern California, June 1971.

Advice on improving security programs. *Burroughs Clearing House,* XLV:38-39, April 1961.

American Library Association. Protecting the library and its resources. Chicago, American Library Association, 1963.

American Security Council, Washington Report, April 7, 1969, Wr 69-14. SDS and the campus revolt. Published by Washington Report, 123 North Wacker Drive, Chicago, Ill. 60606.

Anarchy in the academy. Report of the Temporary Commission to Study the Cause of Campus Unrest, March 1971.

Banking. Bank crimes and their prevention. Report by House Committee on Government Operation, April 1964.

Brown, William P.: *Order and Justice on Campus.* School of Criminal Justice, State University of New York at Albany, 1971.

Crimes against banking institutions; eighteenth report, 41 p. (88:2, H.Rep. no. 1147). Washington, D.C., February 20, 1964.

Dissent and disruption in the schools: A handbook for school administrators. I.D.E.A., 1971.

Gelber, S.: Role of campus security in the college setting. NILECJ, p. 209. 1-0877-J-LEAA (Grant), 1972.

Griffin, R. K.: Shoplifting in supermarkets. TPC, *31,* July 1964.

Harris, O. L., Jr.: Methodology for developing security design criteria for subways. Carnegie-Mellon Univ., Pittsburgh, 1971, p. 114.

How Much Blanket Bond Insurance? New York, American Bankers Assoc., 1939.

International Association of Chiefs of Police. Campus unrest: Dialogue of destruction. Proceedings of the IACP workshop for state police officers and campus security directors. Washington, D.C., 1970.

Johnson, Edward M. (ed.): *American Library Association, Protecting the Library and Its Resources: A Guide to Physical Protection and Insurance.* Chicago, 1963.

Kassinger, Edward T.: Law enforcement assistance administration supports new directors in law enforcement on campus. *LEAA Newsletter,* 2: No. 10, 8, November 1972.

Miami Beach Police Department. Manual on hotel and apartment security. Miami Beach, Fla., July 19, 1956.

Minami, Manshū. Tetsudō Kabushiki Kaisha. (Shina Tetsudō Keisatsu Seido) (Rational Police—China), 1935.

Miura, Keiichi: (Chigai hōkei Teppaigo Ni Okeru Mantetsu Shasenio No Keisatsu Ni Kansuru iken).

National School Relations Association. *Vandalism and Violence.* Arlington, Va., 1971, 56 p.

National Survey on Exhibition Hall Security. New York, Burns Security Institute, 1975.

National Survey on Hospital Security. Burns Security Institute, 1972, p. 43.

National Survey on Library Security. Briarcliff Manor, N.Y., Burns Security Institute, 1973.

Post, Richard S.: *Campus Security Data Collection Handbook.* Sterling Heights, Mich., Security Management Services, Ltd., 1971, 20 p.

Public Safety Institute. Plant Patrol Problems; Industrial Traffic Control; Safety Inspections; Plant Police Tactics; Plant Geography; Rules and Regulations for Plant Police; Safety Rules and Regulations; Plant Police Training Guide; Public Relations; Legal Problems of Plant Police; Small Arms Training; Panic Control; Personnel Investigation. Lafayette, Ind., Purdue University, 1942.

Recommended Practice for the Protection of Library Collections from Fire. Boston, NFPA, 1970.

Report: A Panel Discussion on Campus Crimes and Security. New York, Burns Security Institute, 1973.

Report: A Panel Discussion on Security for Big Crowd Sports Events. New York, Burns Security Institute, 1974.

Role of campus security in the college setting. National Institute of Law Enforcement and Criminal Justice, LEAA., December 1972.

Sowler, Martin K.: The cause & prevention of bank desalcations. New York, The Bankers Publishing Co., 1924.

Special Libraries Association, Financial Division. *Handbook of Commercial, Financial and Information Services* (comp. by Walter Hausdorfer). New York, Special Library Association, 1956.

Tested devices to prevent holdup. *Protective Bulletin.* New York, American Bankers Association, January-February 1966.

U.S. Congress, House of Representatives. Crime Against Banking Institutions, 18th Report. Washington, Government Printing Office, February 20, 1964.

U.S. Congress, House Committee on Banking and Currency. Security Measures for Financial Institutions, Hearing, 90th Cong., 2d Sess. Washington, Government Printing Office, April 1968.

U.S. Congress, House, Government Operations Committee. Crimes Against Banking Institutions, 18th Report. Washington, Government Printing Office, 1964.

U.S. House of Representatives. Eighteenth Report by the Committee on Government Operations. Crimes Against Banking Institutions. Report No. 1147, 88th Cong., 1st Sess., February 20, 1964.

U.S. House of Representatives. Subcommittee on Government Operations. Hearings on Crimes Against Banking Institutions. 88th Cong., 1st Sess., October 15, 1963.

Yamashita, Masami. (Tetsudō Keisatsu), Japan, 1924.

PERIODICALS

Adams, A. B.: What to do till the policeman arrives. *Banking, LII,* October 1959.

Advice on holdups of banks. *Newsweek, XL:* 62ff, August 18, 1952.

Allen, Tom: Snoopers in our schools. *New York Daily News,* June 11 and 18, 1961.

Allison, Thomas H.: The emotionally troubled employee. *Industrial Security, 11*:8, October 1967.

Allsbrook, D. N., Jr.: Elements of a good security program. *Burroughs Clearing House, XLVI*:44-45, September 1961.

Amateurs; bank robberies in the U.S. *Time, XXCII*:19, August 23, 1963.

American Bankers Association testifies on bank crime before House Subcommittee. *Banking, LVI*:125, December 1963.

Arkansas banker's four hold-up adventures. *Literary Digest, XCII*:56-60, March 26, 1927.

As violence spreads in high schools. *U.S. News and World Report,* 69, November 30, 1970.

Bagley, Gerald L.: Collimation of the campus cop. *Industrial Security,* April 1971.

Bank crimes and their prevention. *Banking, LVI*:54ff, April 1964.

Bank fraud and embezzlement. *F.B.I. Law Enforcement Bulletin,* February 1975.

Bank guards; private gunmen. *Economist, CCI*:1273, December 30, 1961.

Bank holdup losses show decrease. *Banking LII*:184, June 1960.

Bank personnel—citizen teamwork foils holdups. *Banking, LII*:128, February 1960.

Bank robber sees his picture and gives up, United California Bank. *Life, LVIII*:39, March 12, 1965.

Bank robbers—why? *New Era, XV:* No. 2, 1-28.

Bank security consultant advises on how to better reduce exposure to loss. *Security Letter,* July 16, 1975, p. 3-4.

Bankers advise on office security precautions. *The Office,* September 1971, p. 28.

Bartram, John L., and Smith, Larry E.: A survey of campus police forces. *Journal of College and University Personnel Association,* 2:34, November 1969.

Big rise in bank robberies and the F.B.I. response. *U.S. News and World Report, LIV*:14, April 15, 1963.

Black, J.: Two crimes of 1928. *New Republic, LVII*:293-295, January 30, 1929.

Bloom, M. T.: Crime we can wipe out tomorrow. *Reader's Digest, LXXII*:96-98, April 1958.

Boom in bank robbery, the. *Fortune, LXI:* 115ff, January 1960.

Booth, E.: We rob a bank. *American Mercury, XII*:1-11, September 1927.

Brabury, M.: Can we bring back the old-fashioned bank robber? *Harper, CCXXII*:37-39, April 1961.

Bratter, H.: Know the man's habits: Average bank robber. *Banking, LVI*:78, September 1963.

Breaking the banks. *Economist, CCXIV*:1137-1138, March 13, 1965.

Brundidge, H. T.: Take the profit out of bank robbery. *American Mercury, XXCI*:129-133, August 1955.

Bungled bank robbery: Roseville, Ohio. *Life, XXXVII*:24-25, November 1, 1954.

Buono, John L.: School security, how basic? *Industrial Security,* October 1971.

Business at the Bank: Woodsid Branch, Chase Manhattan Bank. *Newsweek, XLV*:25-26, April 18, 1955.

Buxby, Walter J.: Management notebook: Hotel security. *Security World, 2:* No. 2, March 1965.

Callahan, George A.: Railroad police. *Industrial Security, 1*:6, July 1957.

Carle, Stanley M.: Security goes collegiate. *Industrial Security,* June 1971.

Carlson, Paul: A bank protects its memory. *Banking,* April 1971, p. 38-39.

Carroll, Leslie H.: Bomb scare: A medical center's program. *Security World,* December 1970.

Cauper, David C.: The need for excellence in campus policing, some specific organizational and behavioral recommendations. *Police Chief,* January 1971, p. 58.

Clamen, Michael: Museums and the theft of works of art. *International Criminal Police Review,* February 1975, p. 51-58.

Coleman, C. W.: A look at the International Association of Museums security. *Assets Protection,* Spring 1975, p. 63-66.

Colling, Russell L.: Hospital security problems. *Police, 6:* No. 5, May-June 1962.

———: Hospital security. *Security World, 3:* No. 6, June 1966.

Cooke, Walter C.: Airport security searches: A rationale. *American Journal of Criminal Law,* 2:128-145.

Cooley, J. L.: Look, listen, remember. *Banking, LI*:46-47ff, May 1959.

Coon, Thomas F.: The railroad police—the world's largest privately supported police system. *Police, 9:* No. 2, November-December 1964.

Crime prevention on the railways. *Security Gazette, 11:* No. 7, 286, 1969.

Dallas, Robert F.: Hotel thefts. *The Miami Herald,* 3-J, May 4, 1975.

Danzansky, Joseph: Supermarket chain needs security. *Industrial Security,* June 1966.

Davies, Helen: Security devices in the library. *Assistant Librarian,* January 1975, p. 6-8.

Davis, Albert S.: The bank protection act after one year. *Industrial Security,* April 1970.

Dews, Edgar B., Jr.: School Security: A humanistic approach. *Security World,* December 1974, p. 39-40.

Dorian, George: Airport security at Los Angeles International. *Security World, 2:* No. 3, May 1965.

Duke, William: Security in harness racing. *Industrial Security,* October 1970.

Einreinhof, Emrey L.: Anti-robbery program for bank employees. *Law and Order, 11:* No. 3, March 1963.

Electronic security features in new L.A. county museum. *Security World, 2:* No. 5, July 1965.

Epstein, David G.: Combating campus terrorism. *Police Chief,* January 1971, p. 47.

Erven, C. C.: The argument for a national school identification code. *Security World,* October 1974, p. 64.

Evans, C. A.: Bank robbers take to the air in 1958-1959, robberies reach all-time high of 764. *National Underwriter, LXIV*:9ff, January 22, 1960.

———: FBI and the banks. *Banking, LVI,* October 1963.
Exhibit hall security. *Top Security,* May 1975, p. 36.
Fay, Bill: I am a supermarket detective. *Collier's,* March 29, 1952.
Federal bank protection rules. *Security World,* January 1969.
FBI: A look at bank robbery statistics. *F.B.I. Law Enforcement Bulletin,* April 1967.
FBI. Profile of a bank robber. *F.B.I. Law Enforcement Bulletin,* November 1965.
FBI. Increasing problems on bank robberies matter for study. *F.B.I. Law Enforcement Bulletin,* August 1961.
Final federal bank security rules. *Security World,* March 1969.
Footlick, Jerrold K.: Campus stealing rises rapidly. *National Observer,* November 7, 1966.
———: Campus stealing rises rapidly. *College Store Journal, 34:* No. 2, 78, 1967.
Foster, Reginald: The "nuts and bolts" of railway security. *Security and Protection,* May 1975, p. 4-6.
Garvin, John D.: Windows: Factors to consider before buying. *American School and University,* March 1975, p. 18-20.
Gay, William O.: Railway luggage thefts. *Police Journal, 16:*280.
Glassman, Stanley A., and Fitzgerald, William J.: Contemporary changes in hospital security. *Security Management,* September 1974, p. 18-19.
———: Contemporary changes that improve your hospital security. *Security World,* October 1974, p. 30-35.
Grant, Robert J.: How good hotel security can increase profits. *Security Gazette,* August 1974, p. 291.
Grealy, Joseph I.: Nature and extent of school violence and vandalism. *Security World,* May 1975, p. 51-54.
Grutzner, Charles: Private security flies high under new F.A.A. rules. *Security News,* January 30, 1973, p. 3, 17.
Hames, Lee N.: Industrial medical services in disaster. *Industrial Security,* July 1960.
Hamilton, Margaret: Health and safety at work. *Security Gazette,* February 1975, p. 45-46.
Handley, J. L.: F.B.I.'s concern about bank losses. *National Underwriter, LXVI:*36, March 16, 1962.
Hane, W. H.: Striking back at bank robbers. *Banking, LVII:*62, January 1965.
Hartnack, Carl E.: Internal and external crimes against banks. *Security World,* September 1969.
Hayes, P. J.: On the initial interpreting of the Federal Bank Protection Act of 1969. *Security World,* July-August, 1969, p. 41.
High schools, too, have a crime problem. *U.S. News and World Report,* 71, No. 21, November 22, 1971.
Holloman, Frank C.: The new breed: College and university police. *Police Chief, 39:*41, 1972.
Hoover, J. Edgar: Banks can be protected. *Banking, LI:*42-43ff, June 1959.
How Southern Baptist Hospital in New Orleans meets pilferage problems with patrols and good inventory control. *Industrial Security,* July 1963.
How to keep vandals off guard. *School Management, 9:* No. 8, August 1965.
How schools combat vandalism. *Nation's Schools, 81:* No. 4, April 1968.
Howell, James A.: Hotel security. *Industrial Security,* October 1963.
Hynd, A.: How to rob a bank. *Cosmopolitan, CXLVIII:*42-47, March 1960.
Increasing problems of bank robberies matter for study. *F.B.I. Law Enforcement Bulletin,* August 1961.
International Association for Hospital Security. Security standards for large institutions. *Security World,* December 1971.
Jackson, R. G.: Dealing with bomb threats in hospitals. *Security Gazette,* May 1975, p. 168.
James, Howard: The police—friends or enemies? *PTA Magazine, 64:* No. 10, June 1970.
Janssen, J. J.: *Hotel Security, VI:* No. 1, January 1967.
Jones, J. Elroy: New image for the campus security officer. *Security World,* May 1970.
Kaufman, Louis: College security: Its problems and solutions. *Industrial Security,* October 1971.
Keller, E. J.: School security—the role of the police. *Law and Order, 20:* No. 12, 50-52, December 1972.
Kiefer, Norvin C.: Industrial medical services in disaster. *Industrial Security,* July 1960.
Kingsbury, Arthur: Guidelines for security education, the investigation and prevention of vandalism. *Security Management, 17:* No. 4, 50, September 1973.
Kneen, O. H.: Robberies on a gigantic scale. *Popular Science, CXXI,* October 1932.
Lawrence, N. L.: Bank security—what is

enough? *F.B.I. Law Enforcement Bulletin, 42:* No. 11, 2-7, November 1973.

———: Lessen robbery violence hazards. *Banking, LV:*54ff, July 1962.

Lee, Robert J.: Pharmacy crime: Recent research. *Security World,* May 1975, p. 38+.

Lowering the toll of vandalism. *American School and University,* 38, No. 12, July 1967.

Lunden, W. A.: Bank robbery boom. *Security World, 10:* No. 1, 26-30, January 1973.

Malcolmson, Robert E.: Law and security for public recreation. *Canadian Police Chief,* January 1975, p. 21, 22, 38.

Malvesta, Daniel and Ronayne, E.: Cops in the classroom. *NEA Journal, 56:* No. 9, December 1967.

Mann, Floyd H., Col.: A new image for campus police. *F.B.I. Law Enforcement Bulletin, 42:* No. 2, 13, February 1973.

Mannen, H. W.: Preparedness prevents holdups. *Bankers Monthly, LXXVIII,* March 1961.

McCoy, Kathleen: A quandary: Should campus police be armed? *College and University Business, 42:*60, May 1967.

Mendenhall, W. K.: Bank losses. *National Audiogram,* February 1947.

Metzger, G. M.: No bank holdups in Philadelphia. *Banking, LII:*118ff, October 1959.

Meyer, Edward J.: Design for a hospital fire safety program. *Security World, 1:* No. 1, July 1964.

———: Pre-riot retail fire training. *Security World, 3:* No. 8, September 1966.

Middlemas, Keith: Double market: Art theft and art thieves. Farnborough, D. C. Heath, 1975.

Milborrow, E. A.: Security on the campus. *Security Gazette,* August 1974, p. 296-298.

Miller, William L.: Hotel and motel security. *Industrial Security,* April 1971.

Modern decor blamed for increase in bank robbers. *Science Digest, XXXIX:*30, March 1956.

More muscle in the fight to stop violence in schools. *U.S. News and World Report,* 1974, p. 4.

Morris, Thomas, A.: A case in point: Hotel masterkeying. *Security World, 2:* No. 6, September 1965.

New Jersey's biggest bank robbery, and how the cover paid off. *National Underwriter, LXIX:*14, March 26, 1965.

New threats and new defenses. *Banking,* August 1970, p. 69-70.

New ways to keep our city schools secure. *American School and University, 39:* No. 11, July 1967.

Nolan, F.: New boom in banks: Today's holdup men. *N.Y. Times Magazine,* May 8, 1955, p. 19ff.

———: On thwarting bank crimes. *Banking, LVII:*44ff, April 1965.

Nwamefor, Ralph: Security problems of university libraries in Nigeria. *Library Association Record,* December 1974, p. 244-245.

O'Grince, Syl and Hodgins, H.: Public school vandalism: How Baltimore fights it. *American School and University, 40:* No. 7, July 1963.

Olmstead, John A.: The legal aspects of hospital protection. *Security Management,* September 1974, p. 25-26.

O'Neil, James W.: Strengthen your security posture. *Library Security,* March 1975, p. 1, 7, 8.

Pakalik, Michael J.: Security and protection in the museum. *Police, 3:* No. 1, 26-29, September-October, 1958.

Peterson, Ronald A.: Authority to arrest on campus. *Canadian Security Gazette,* October 1971.

Piez, Gladys T.: Insurance and the protection of library resources. *American Library Association Bulletin,* May 1962, p. 421-423.

Pizer, Harry: Transporting art objects. *Assets Protection,* Summer 1975, p. 33-34.

Pointer, Homer F.: Bomb threats to public schools. *Industrial Security,* June 1966.

Post, Richard S.: "Security's Critical Factor: Design," *Building Operating Management,* January 1973.

———: "Role and Philosophy of School District Security," *Wisconsin School News,* October 1971.

———: "Campus Security: Security Takes more than Locks and Cops," *College and University Business,* August 1971.

———: "Designing Security In," *American School and University,* July, 1971.

Powell, John W.: History and the power role of campus security. Part I and II. *Security World,* March-April, 1971.

———: The future of campus security. *Security World, 3:* No. 2, February 1966.

Probst, Tom: Electronic security protects art museum. *Industrial Security,* December 1965.

Richmond, Michael L.: Attitudes of law librarians to theft and mutilation control methods. *Law Library Journal,* February 1975, p. 60-70.

Role of campus disorders in civil disturbance losses, the. *Security World,* February 1970.

Roth, Harold: A case study in library-police relations. *Library Security,* March 1975, p. 2, 3, 6.

Ryan, Richard A.: Travelers will pay for skyjack protection. *The Detroit News,* December 6, 1972, p. 2A.

Safe schools act. *School Management, 15:* No. 4, April 1971.

Sager, Donald: Vandalism in libraries: How senseless is it? *Library Security,* January 1975, p. 5.

Sander, Frederick: The new style bank robber. *Reader's Digest,* December 1963.

Sanders, M., and Welton, J. H.: *Vandalism.* Naval Ammunition Depot, Crane, Ind., 1972, p. 199.

UNPUBLISHED MATERIAL

Carso, Martin B.: A study of the theft control implications of the bank credit card plan & the future electronic monetary exchange system. Unpublished graduate thesis, Michigan State Univ., 1970.

Fandt, Edward L.: A study of the practices of New Jersey boards of education in protecting school property against losses due to vandalism and malicious mischief. Ph.D. dissertation, Rutgers Univ., 1961.

Nowakowski, Rodney E.: Vandals and vandalism in the schools: An analysis of vandalism in large school systems and a description of ninety-three vandals in Dade County schools. Ph.D. dissertation, University of Miami, 1966.

Pan, Cheng-Dah: A study of personnel identification systems and proposals for Taipei International Airport. (B Paper), Michigan State Univ., 1967.

Pugliese, Nicholas R.: A study of planning requirements for security services at special events. Unpublished thesis, Michigan State Univ., Fall 1964.

Quinney, Earl R.: Retail pharmacy as marginal occupation: A study of prescription violation. Ph.D. dissertation, University of Wisconsin, 1962.

Smith, Donald D.: Vandalism in selected Southern California school districts: Nature, extent and preventive measures. Ph.D. dissertation, University of Southern California, 1966.

Steen, William Joseph: A model security services department for the board of education of the City of Buffalo. Michigan, University Microfilms, Inc., 1972, 165 p.

Stewart, Phillip C.: Campus security planning. (B Paper), Michigan State Univ., 1972.

Zioner, Adolf A.: Airline hijackings. Unpublished thesis, Michigan State Univ., Fall 1974.

APPENDIX 8A

MINIMUM SECURITY DEVICES AND PROCEDURES FOR FEDERAL RESERVE BANKS AND STATE MEMBER BANKS*

SECTION 216.0—SCOPE OF PART

Pursuant to the authority conferred upon the Board of Governors of the Federal Reserve System by section 3 of the Bank Protection Act of 1968 (82 Stat. 295) with respect to State banks which are members of the Federal Reserve System and to Federal Reserve Banks[1] the rules contained in this Part—

(a) establish minimum standards for the installation, maintenance, and operation of security devices and procedures to discourage robberies, burglaries, and larcenies and to assist in the identification and apprehension of persons who commit such acts;

(b) establish time limits for compliance; and

(c) require the submission of reports.

SECTION 216.1—DEFINITIONS

For the purposes of this Part—

(a) The term **"State member bank"** means any bank that is a member

---

* This text corresponds to the Code of Federal Regulations, Title 12, Chapter II, Part 216, cited as 12 CFR 216. The words "this Part," as used herein, mean Regulation P.

of the Federal Reserve System (other than a national bank or a District of Columbia bank).

(b) The term **"banking hours"** means the time during which a banking office is open for the normal transaction of business with the banking public.

(c) The term **"banking office"** includes the main office of any State member bank and any branch thereof.

(d) The term **"branch"** includes any branch bank, branch office, branch agency, additional office, or any branch place of business located in any State of the United States or in any Territory of the United States, Puerto Rico, Guam, or the Virgin Islands at which deposits are received or checks paid or money lent.

(e) The term **"Board"** means the Board of Governors of the Federal Reserve System.

(f) The term **"teller's station or window"** means a location in a banking office at which bank customers routinely conduct transactions with the bank which involve the exchange of funds, including a walk-up or drive-in teller's station or window.

SECTION 216.2—DESIGNATION OF SECURITY OFFICER

On or before February 15, 1969 (or within thirty days after a State bank becomes a member of the Federal Reserve System, whichever is later), the board of directors of each State member bank shall designate an officer or other employee of the bank who shall be charged, subject to supervision by the bank's board of directors, with responsibility for the installation, maintenance, and operation of security devices and for the development and administration of a security program which equal or exceed the standards prescribed by this Part.

SECTION 216.3—SECURITY DEVICES

(a) INSTALLATION, MAINTENANCE, AND OPERATION OF APPROPRIATE SECURITY DEVICES. Before January 1, 1970 (or within thirty days after a State bank becomes a member of the Federal Reserve System, whichever is later), the security officer of each State member bank, under such directions as shall be given him by the bank's board of directors, shall survey the need for security devices in each of the bank's banking offices and shall provide for the installation, maintenance, and operation, in each such office, of—

(1) a lighting system for illuminating, during the hours of darkness the area around the vault if the vault, is visible from outside the banking office;

(2) tamper-resistant locks on exterior doors and exterior windows designed to be opened;

(3) an alarm system or other appropriate device for promptly notifying the nearest responsible law enforcement officers of an attempted or perpetrated robbery or burglary; and

(4) such other devices as the security officer, after seeking the advice of law enforcement officers, shall determine to be appropriate for discouraging robberies, burglaries, and larcenies and for assisting in the identification and apprehension of persons who commit such acts.

(b) CONSIDERATIONS RELEVANT TO DETERMINING APPROPRIATENESS. For the purposes of subparagraph (4) of paragraph (a) of this section, considerations relevant to determining appropriateness include, but are not limited to—

(1) the incidence of crimes against the particular banking office and/or against financial institutions in the area in which the banking office is or will be located;

(2) the amount of currency or other valuables exposed to robbery, burglary, or larceny;

(3) the distance of the banking office from the nearest responsible law enforcement officers and the time required for

such law enforcement officers ordinarily to arrive at the banking office;

(4) the cost of the security devices;

(5) other security measures in effect at the banking office; and

(6) the physical characteristics of the banking office structure and its surroundings.

(c) IMPLEMENTATION. It is appropriate for banking offices in areas with a high incidence of crime to install many devices which would not be particable because of costs for small banking offices in areas substantially free of crimes against financial institutions. Each bank shall consider the appropriateness of installing, maintaining, and operating security devices which are expected to give a general level of bank protection at least equivalent to the standards described in Appendix A of this Part. In any case in which (on the basis of the factors listed in paragraph (b) or similar ones, the use of other measures, or the decision that technological change allows the use of other measures judged to give equivalent protection) it is decided not to install, maintain, and operate devices at least equivalent to these standards, the bank shall preserve in its records a statement of the reasons for such decision and forward a copy of that statement to the Federal Reserve Bank for the District in which its main office is located.

SECTION 216.4—SECURITY PROCEDURES

(a) DEVELOPMENT AND ADMINISTRATION. On or before July 15, 1969 (or within thirty days after a State bank becomes a member of the Federal Reserve System, whichever is later), each State member bank shall develop and provide for the administration of a security program to protect each of its banking offices from robberies, burglaries, and larcenies and to assist in the identification and apprehension of persons who commit such acts. This security program shall be reduced to writing, approved by the bank's board of directors, and retained by the bank in such form as will readily permit determination of its adequacy and effectiveness, and a copy shall be filed with the Federal Reserve Bank for the District in which the main office of the bank is located.

(b) CONTENTS OF SECURITY PROGRAMS. Such security programs shall—

(1) provide for establishing a schedule for the inspection, testing, and servicing of all security devices installed in each banking office; provide for designating the officer or other employee who shall be responsible for seeing that such devices are inspected, tested, serviced, and kept in good working order; and require such officer or other employee to keep a record of such inspections, testings, and servicings;

(2) require that each banking office's currency be kept at a reasonable minimum and provide procedures for safely removing excess currency;

(3) require that the currency at each teller's station or window be kept at a reasonable minimum and provide procedures for safely removing excess currency and other valuables to a locked safe, vault, or other protected place;

(4) require that the currency at each teller's station or window include "bait" money, i.e. used Federal Reserve notes, the denominations, banks of issue, serial numbers, and series years of which are recorded, verified by a second officer or employee, and kept in a safe place;

(5) require that all currency, negotiable securities, and similar valuables be kept in a locked vault or safe during nonbusiness hours, that the vault or safe be opened at the latest time practicable before banking hours, and that the vault or safe be locked at the earliest time practicable after banking hours;

(6) provide, where practicable, for designation of a person or persons to open each banking office and require him or them to inspect the premises, to ascertain that no unauthorized persons are present, and to signal other employees that the premises are safe before permitting them to enter:

(7) provide for designation of a person or persons who will assure that all security devices are turned on and are operating during the periods in which such devices are intended to be used;

(8) provide for designation of a person or persons to inspect, after the closing hour, all areas of each banking office where currency, negotiable securities, or similar valuables are normally handled or stored in order to assure that such currency, securities, and valuables have been put away, that no unauthorized persons are present in such areas, and that the vault or safe and all doors and windows are securely locked; and

(9) provide for training, and periodic retraining, of employees in their responsibilities under the security program, including the proper use of security devices and proper employee conduct during and after a robbery, in accordance with the procedures listed under the heading, *Proper Employee Conduct During and After a Robbery* in this Part.

SECTION 216.5—FILING OF REPORTS

(a) COMPLIANCE REPORTS. As of the last business day in June of 1970, and as of the last business day in June of each calendar year thereafter, each State member bank shall file with the Federal Reserve Bank for the District in which its main office is located a statement certifying to its compliance with the requirements of this Part. The statement shall be dated and signed by the president, or cashier, or other managing officer of the bank and may be in a form substantially as follows:

> I hereby certify, to the best of my knowledge and belief, that this bank has developed and administers a security program that equals or exceeds the standards prescribed by §216.4 of Regulation P; that such security program has been reduced to writing, approved by the bank's board of directors, and retained by the bank in such form as will readily permit determination of its adequacy and effectiveness; and that the bank security officer, after seeking the advice of law enforcement officers, has provided for the installation, maintenance, and operation of appropriate security devices, as prescribed by §216.3 of Regulation P, in each of the bank's banking offices.

(b) REPORTS ON SECURITY DEVICES. On or before March 15, 1969, and upon such other occasions as the Board may specify, each State member bank shall file with the Federal Reserve Bank for the District in which it is located a report on Form P-1 (in duplicate) for each of its offices that is subject to this Part.

(c) EXTERNAL CRIME REPORTS. Each time a robbery, burglary, or nonbank-employee larceny is perpetrated or attempted at a banking office operated by a State member bank, the bank shall, within a reasonable time, file a report in conformity with the requirements of Form P-2. One copy of such report shall be filed with the appropriate State supervisory authority and three copies of such report shall be filed with the Federal Reserve Bank for the District in which the head office of the reporting bank is located.

(d) SPECIAL REPORTS. Each State member bank shall file such other reports as the Board may require.

SECTION 216.6—CORRECTIVE ACTION

Whenever the Board determines that the security devices or procedures used by a State member bank are deficient in meeting the requirements of this Part, or that the requirements of this Part should be varied in the circumstances of a particular banking office, it may take or require the bank to take necessary corrective action. If the Board determines that such corrective action is appropriate or necessary, the bank will be so notified and will be furnished a statement of what the bank must do to comply with the requirements of this Part.

SECTION 216.7—APPLICABILITY TO FEDERAL RESERVE BANKS

The provisions of this Part apply to each Federal Reserve Bank and its branches, except that reports and other writings required or permitted to be filed by a State member bank with the Federal Reserve Bank for the District in which it is located must, in the case of a Federal Reserve Bank, be filed with the Board; provided, however, that the applicability of the Bank Protection Act of 1968 and of this Part to Federal Reserve Banks and their branches does not preclude the Board from requiring, by virtue of its authority under other provisions of law, that Federal Reserve Banks and their branches comply with higher standards respecting the installation, maintenance, and operation of security devices and procedures than those that are prescribed by this Part.

SECTION 216.8—PENALTY PROVISION

Pursuant to Section 5 of the Bank Protection Act of 1968, a State member bank or Federal Reserve Bank that violates any provision of this Part shall be subject to a civil penalty not to exceed $100 for each day of the violation.

MINIMUM STANDARDS FOR SECURITY DEVICES

(1) *Surveillance systems.* (i) *General.* Surveillance systems should be:

(A) equipped with one or more photographic, recording, monitoring, or like devices capable of reproducing images of persons in the banking office with sufficient clarity to facilitate (through photographs capable of being enlarged to produce a one-inch vertical head-size of persons whose images have been reproduced) the identification and apprehension of robbers or other suspicious persons;

(B) reasonably silent in operation;

(C) so designed and constructed that necessary services, repairs, or inspections can readily be made. Any camera used in such a system should be capable of taking at least one picture every two seconds and, if it uses film, should contain enough unexposed film at all times to be capable of operating for not less than three minutes, and the film should be at least 16 mm.

(ii) *Installation, maintenance, and operation of surveillance systems providing surveillance of other than walk-up or drive-in teller's stations or windows.* Surveillance devices for other than walk-up or drive-in windows should be:

(A) located so as to reproduce identifiable images of persons either leaving the banking office or in a position to transact business at each such station or window; and

(B) capable of activation by initiating devices located at each teller's station.

(iii) *Installation, maintenance, and operation of surveillance systems providing surveillance of walk-up or drive-in teller's stations or windows.* Surveillance devices for walk-up and drive-in teller's stations or windows should be located in such a manner as to reproduce identifiable images of persons in a position to transact business at each such station or window and areas of such station or window that are vulnerable to robbery or larceny. Such devices should be capable of activation by one or more initiating devices located within or in close proximity to such station or window. Such devices could be omitted in the case of walk-up or drive-in teller's station or window in which the teller is effectively protected by a bullet-resistant barrier from persons outside the station or window, but if the teller is vulnerable to larceny or robbery by members of the public who enter the banking office, the teller should have access to a device to activate a surveillance system that covers the area of vulnerability or the exits to the banking office.

(2) *Robbery alarm systems.* A robbery alarm should be provided for each banking office at which the police ordinarily can arrive within five minutes after an alarm is activated. Robbery alarm systems should be:

(i) designed to transmit to the police, either directly or through an intermediary, a signal (not detectable by unauthorized persons) indicating that a crime against the banking office has occurred or is in progress;

(ii) capable of activation by initiating devices located at each teller's station (except walk-up or drive-in teller's stations or windows in which the teller is effectively protected by a bullet-resistant barrier and effectively isolated from persons, other than fellow employees, inside a banking office of which such station or window may be a part);

(iii) safeguarded against accidental transmission of an alarm;

(iv) equipped with a visual and audible signal capable of indicating improper functioning of or tampering with the system; and

(v) equipped with an independent source of power (such as a battery) sufficient to assure continuously reliable operation of the system for at least twenty-four hours in the event of failure of the usual source of power.

(3) *Burglar alarm systems.* Burglar alarm systems should be:

(i) capable of detecting promptly an attack on the outer door, walls, floor or ceiling of each fault, and each safe not stored in a vault, in which currency, negotiable securities, or similar valuables are stored when the office is closed, and any attempt to move any such safe;

(ii) designed to transmit, to the police, either directly or through an intermediary, a signal (not detectable by unauthorized persons) indicating that any such attempt is in progress; and in the case of a banking office at which the police ordinarily cannot arrive within five minutes after an alarm is activated, designed to activate a loud sounding bell or other device that is audible inside the banking office and for a distance of approximately 500 feet outside the banking office;

(iii) safeguarded against accidental transmission of an alarm;

(iv) equipped with a visual and audible signal, capable of indicating improper functioning of or tampering with the system; and

(v) equipped with an independent source of power (such as a battery) sufficient to assure continuously reliable operation of the system for at least eighty hours in the event of failure of the usual source of power.

(4) *Walk-up and drive-in teller's stations or windows.* Walk-up and drive-in teller's stations or windows contracted for after February 15, 1969, should be constructed in such a manner that tellers are effectively protected by bullet-resistant barriers from robbery or larceny by persons outside such stations or windows. Such barriers should be of glass at least one and three-sixteenths inches thick,[3] or of material of at least equivalent bullet-resistance. Pass-through devices should be so designed and constructed as not to afford a person outside the station a direct line of fire at a person inside the station.

(5) *Vaults, safes, and night depositories.* Vaults and safes (if not to be stored in a vault) in which currency, negotiable securities, or similar valuables are to be stored when the office is closed, and night depositories, contracted for after February 15, 1969, should meet or exceed the following standards:

(A) *Vaults.* Vault walls, roof and floor contracted for after February 15, 1969, should be made of steel-reinforced concrete, at least 18 inches thick; vault doors should be made of steel or other drill and torch-resistant material, at least three and one-half inches thick, and be equipped with a dial combination lock and a time lock and a substantial, lockable day-gate; or vaults and vault doors should be constructed of materials that afford at least equivalent burglary-resistance.

(B) *Safes.* Safes contracted for after February 15, 1969, should weigh at least 750 pounds empty, or be securely anchored to the premises where located. The door should be equipped with a combination lock, and with a relocking device that will effectively lock the door if the combination lock is punched. The body should consist of steel, at least one inch in thickness, with an ultimate tensile strength of 50,000 pounds per square inch, either cast or fabricated, and be fastened in a manner equal to a continuous one-fourth inch penetration weld having an ultimate tensile strength of 50,000 pounds per square inch. One hole not exceeding $\frac{3}{16}$-inch diameter may be provided in the body to permit insertion of electrical conductors, but should be located so as not to permit a direct view of the door or locking mechanism. The door should be made of steel that is at least one and one-half inches thick, and at least equivalent in strength to that specified for the body; or safes should be constructed of materials that afford at least equivalent burglary-resistance.

(C) *Night depositories.* Night depositories (excluding envelope drops not used to receive substantial amounts of currency) contracted for after February 15, 1969, should consist of a receptacle chest having cast, or welded, steel walls, top and bottom, at least one inch thick; a combination locked steel door at least one and one-half inches thick; and a chute, made of steel that is at least one inch thick, securely bolted or welded to the receptacle and to a depository entrance of strength similar to the chute; or night depositories should be constructed of materials that afford at least equivalent burglary-resistance. The depository entrance should be equipped with a lock. Night depositories should be equipped with a burglary alarm and be designed to protect against the "fishing" of a deposit from the deposit receptacle, and to protect against the "trapping" of a deposit for extraction.

Each device mentioned in this Appendix should be installed and regularly inspected, tested, and serviced by competent persons, so as to assure realization of its maximum performance capabilities. Activating devices for surveillance systems and robbery alarms should be operable with the least risk of detection by unauthorized persons that can be practicably achieved.

PROPER EMPLOYEE CONDUCT DURING AND AFTER A ROBBERY

With respect to proper employee conduct during and after a robbery, employees should be instructed:

(1) to avoid actions that might increase danger to themselves or others;

(2) to activate the robbery alarm system and the surveillance system during the robbery, if it appears that such activation can be accomplished safely;

(3) to observe the robber's physical features, voice, accent, mannerisms, dress, the kind of weapon he has, and any other characteristics that would be useful for identification purposes;

(4) that if the robber leaves evidence (such as a note) try to put it aside and out of sight, if it appears that this can be done safely; retain the evidence, do not handle it unnecessarily, and give it to the police when they arrive; and refrain from touching, and assist in preventing others from touching, articles or places the robber may have touched or evidence he may have left, in order that fingerprints of the robber may be obtained;

(5) to give the robber no more money than the amount he demands, and include "bait" money in the amount given;

(6) that if it can be done safely, observe the direction of the robber's escape and the description and license plate number of the vehicle used, if any;

(7) to telephone the local police, if they have not arrived, and the nearest office of the Federal Bureau of Investigation, or inform a designated officer or other employee who has this responsibility, that a robbery has been committed;

(8) that if the robber leaves before the police arrive, assure that a designated officer or other employee waits outside the office, if it is safe to do so, to inform the police when they arrive that the robber has left;

(9) to attempt to determine the names and addresses of other persons who witnessed the robbery or the escape, and request them to record their observations or to assist a designated officer or other employee in so doing;

(10) to refrain from discussing the details of the robbery with others before recording the observations respecting the robber's physical features and other characteristics as hereinabove described and the direction of escape and description of vehicle used, if any.

BANK PROTECTION ACT OF 1968
Act of July 7, 1968 (82 Stat. 294)

To provide security measures for banks and other financial institutions, and to provide for the appointment of the Federal Savings and Loan Insurance Corporation as receiver.

Be it enacted by the Senate and House of Representatives of the United States of America in Congress assembled, That this Act may be cited as the "Bank Protection Act of 1968."

SEC. 2. As used in this Act the term "Federal supervisory agency" means—

(1) The Comptroller of the Currency with respect to national banks and district banks.

(2) The Board of Governors of the Federal Reserve System with respect to Federal Reserve banks and State banks which are members of the Federal Reserve System,

(3) The Federal Deposit Insurance Corporation with respect to State banks which are not members of the Federal Reserve System but the deposits of which are insured by the Federal Deposit Insurance Corporation, and

(4) The Federal Home Loan Bank Board with respect to Federal savings and loan associations, and institutions the accounts of which are insured by the Federal Savings and Loan Insurance Corporation.

SEC. 3. (a) Within six months from the date of this Act, each Federal supervisory agency shall promulgate rules establishing minimum standards with which each bank or savings and loan association must comply with respect to the installation, maintenance, and operation of security devices and procedures, reasonable in cost, to discourage robberies, burglaries, and larcenies and to assist in the identification and apprehension of persons who commit such acts.

(b) The rules shall establish the time limits within which banks and savings and loan associations shall comply with the standards and shall require the submission of periodic reports with respect to the installation, maintenance, and operation of security devices and procedures.

SEC. 4. The Federal supervisory agencies shall consult with

(1) insures furnishing insurance protection against losses resulting from robberies, burglaries, and larcenies committed against financial institutions referred to in section 2, and

(2) State agencies having supervisory or regulatory responsibilities with respect to such insurers

to determine the feasibility and desirability of premium rate differentials based on the installation, maintenance, and operation of security devices and procedures. The Federal supervisory agencies shall report to the Congress the results of their consultations pursuant to this section not later than two years after the date of enactment of this Act.

SEC. 5. A bank or savings and loan association which violates a rule promulgated pursuant to this Act shall be subject to a civil penalty which shall not exceed $100 for each day of the violation.

***

[U.S.C., title 12, sec. 1881-1884.]

MINIMUM SECURITY DEVICES AND PROCEDURES FOR FEDERAL RESERVE BANKS AND STATE MEMBER BANKS

BOARD OF GOVERNORS OF THE FEDERAL RESERVE SYSTEM

Amendment to Regulation P

1. Effective November 1, 1973, the section entitled *Minimum Standards for Security Devices* of Section P is as follows:

In order to assure realization of maximum performance capabilities, all security devices utilized by a bank should be regularly inspected, tested, and serviced by competent persons. Actuating devices for surveillance systems and robbery alarms should be operable with the least risk of detection by unauthorized persons that can be practicably achieved.

(1) *Surveillance systems.* (i) *General.* Surveillance systems should be:

(A) equipped with one or more photographic, recording, monitoring, or like devices capable of reproducing images of persons in the banking office with sufficient clarity to facilitate (through photographs capable of being enlarged to produce a one-inch vertical head-size of persons whose images have been reproduced) the identification and apprehension of robbers or other suspicious persons;

(B) reasonably silent in operation; and

(C) so designed and constructed that necessary services, repairs or inspections can readily be made.

Any camera used in such a system should be capable of taking at least one picture every 2 seconds and, if it uses film, should contain enough unexposed film at all times to be capable of operating for not less than 3 minutes, and the film should be at least 16 mm.

(ii) *Installation and operation of surveillance systems providing surveillance of other than walk-up or drive-in teller's stations or windows.* Surveillance devices for other than walk-up or drive-in teller's stations or windows should be:

(A) located so as to reproduce identifiable images of persons either leaving the banking office or in a position to transact business at each such station or window; and

(B) capable of actuation by initiating devices located at each teller's station or window.

(iii) *Installation and operation of surveillance systems providing surveillance of walk-up or drive-in teller's stations or windows.* Surveillance devices for walk-up or drive-in teller's stations or windows should be located in such a manner as to reproduce identifiable images of persons in a position to transact business at each such station or window and areas of such station or window that are vulnerable to robbery or larceny. Such devices should be capable of actuation by one or more initiating devices located within or in close proximity to such station or window. Such devices may be omitted in the case of a walk-up or drive-in teller's station or window in which the teller is effectively protected by a bullet-resistant barrier from persons outside the station or window. However, if the teller is vulnerable to larceny or robbery by members of the public who enter the banking office, the teller should have access to a device to actuate a surveillance system that covers the area of vulnerability or the exits to the banking office.

(2) *Robbery and burglary alarm systems.* (i) *Robbery alarm systems.* A robbery alarm system should be provided for each banking office at which the police ordinarily can arrive within 5 minutes after an alarm is actuated; all other banking offices should be provided with appropriate devices for promptly notifying the police that a robbery has occurred or is in progress. Robbery alarm systems should be:

(A) designed to transmit to the police, either directly or through an intermediary, a signal (not detectable by unauthorized persons) indicating that a crime against the banking office has occurred or is in progress;

(B) capable of actuation by initiating devices located at each teller's station or window (except walk-up or drive-in teller's stations or windows in which the teller is effectively protected by a bullet-resistant barrier and effectively isolated from persons, other than fellow employees, inside a banking office of which such station or window may be a part);

(C) safeguarded against accidental transmission of an alarm;

(D) equipped with a visual and audible signal capable of indicating improper functioning of or tampering with the system; and

(E) equipped with an independent source of power (such as a battery) sufficient to assure continuously reliable operation of the system for at least twenty-four hours in the event of failure of the usual source of power.

(ii) *Burglary alarm systems.* A burglary alarm system should be provided for each banking office. Burglary alarm systems should be:

(A) capable of detecting promptly an attack on the outer door, walls, floor, or ceiling of each vault, and each safe not stored in a vault, in which currency, negotiable securities, or similar valuables are stored when the office is closed, and any attempt to move any such safe;

(B) designed to transmit to the police, either directly or through an intermediary, a signal indicating that any such attempt is in progress; and for banking offices at which the police ordinarily cannot arrive within 5 minutes after an alarm is actuated, designed to actuate a loud sounding bell or other device that is audible inside the banking office and for a distance of approximately 500 feet outside the banking office;

(C) safeguarded against accidental transmission of an alarm;

(D) equipped with a visual and audible signal capable of indicating improper functioning of or tampering with the system; and

(E) equipped with an independent source of power (such as a battery) sufficient to assure continuously reliable operation of the system for at least eighty hours in the event of failure of the usual source of power.

(3) *Walk-up and drive-in teller's stations or windows.* Walk-up and drive-in teller's stations or windows contracted for after February 15, 1969, should be constructed in such a manner that tellers are effectively protected by bullet-resistant barriers from robbery or larceny by persons outside such stations or windows. Such barriers should be of glass at least 1 3/16 inches in thickness,[4] or of material of at least equivalent bullet-resistance. Pass-through devices should be so designed and constructed as not to afford a person outside the station or window a direct line of fire at a person inside the station.

(4) *Vaults, safes, safe deposit boxes, night depositories, and automated paying or receiving machines.* Vaults, safes (if not to be stored in a vault), safe deposit boxes, night depositories, and automated paying or receiving machines, in any of which currency, negotiable securities, or similar valuables are to be stored when banking offices are closed, should meet or exceed the standards expressed in this section.

(i) *Vaults.* A vault is defined as a room or compartment that is designed for the storage and safekeeping of valuables and which has a size and shape which permits entrance and movement within by one or more persons. Other asset storage units which do not meet this definition of a vault will be considered as safes. Vaults contracted for after November 1, 1973,[5] should have walls, floor, and ceiling of reinforced concrete at least twelve inches in thickness.[6] The vault door should be made of steel at least $3\frac{1}{2}$ inches in thickness, or other drill and torch resistant material, and be equipped with a dial combination lock, a time lock, and a substantial lockable daygate. Electrical conduits into the vault should not exceed $1\frac{1}{2}$ inches in diameter and should be offset within the walls, floor, or ceiling at least once so as not to form a direct path of entry. A vault ventilator, if provided, should be designed with consideration of safety to life without significant reduction of the strength of the vault wall to burglary attack. Alternatively, vaults should be so designed and constructed as to afford at least equivalent burglary resistance.[7]

(ii) *Safes.* Safes contracted for after February 15, 1969, should weigh at least 750 pounds empty, or be securely anchored to the premises where located. The body should consist of steel, at least 1 inch in thickness, either cast or fabricated, with an ultimate tensile strength of 50,000 pounds per square inch and be fastened in a manner equal to a continuous $\frac{1}{4}$-inch penetration weld having an ultimate tensile strength of 50,000 pounds per square inch. The door should be made of steel that is at least $1\frac{1}{2}$ inches in thickness, and at least equivalent in strength to that specified for the body; and the door should be equipped with a combination lock, or time lock, and with a relocking device that will effectively lock the door if the combination lock or time lock is punched. One hole not exceeding ½-inch diameter may be provided in the body to permit insertion of electrical conductors, but should be located so as not to permit a direct view of the door or locking mechanism. Alternatively, safes should be constructed of materials that will afford at least equivalent burglary resistance.

(iii) *Safe deposit boxes.* Safe deposit boxes used to safeguard customer valuables should be enclosed in a vault or safe meeting at least the above-specified minimum protection standards.

(iv) *Night depositories.* Night depositories (excluding envelope drops not used to receive substantial amounts of currency) contracted for after February 15, 1969, should consist of a receptacle chest having cast or welded steel walls, top, and bottom, at least 1 inch in thickness; a steel door at least 1½ inches in thickness, with a combination lock; and a chute, made of steel that is at least one inch in thickness, securely bolted or welded to the receptacle and to a depository entrance of strength similar to the chute. Alternatively, night depositories should be so designed and constructed as to afford at least equivalent burglary resistance.[8] Each depository entrance (other than an envelope drop slot) should be equipped with a lock. Night depositories should be equipped with a burglary alarm and be designed to protect against the "fishing" of a deposit from the deposit receptacle, and to protect against the "trapping" of a deposit for extraction.

(v) *Automated paying or receiving machines.* Except as hereinafter provided, cash dispensing machines (automated paying machines), including those machines which also accept deposits (automated receiving machines) contracted for after November 1, 1973, should weigh at least 750 pounds empty, or be securely anchored to the premises where located. Cash dispensing machines should contain, among other features, a storage chest having cast or welded steel walls, top, and bottom, at least one inch in thickness, with a tensile strength of at least 50,000 pounds per square inch. Any doors should be constructed of steel at least equivalent in strength to the storage chest and be equipped with a combination lock and with a relocking device that will effectively lock the door if the combination lock is punched. The housing covering the cash dispensing opening in the storage chest and the housing covering the mechanism for removing the cash from the storage chest, should be so designed as to provide burglary resistance at least equivalent to the storage chest and should also be designed to protect against the "fishing" of cash from the storage chest. The cash dispensing control and delivering mechanism (and, when applicable, cash deposit receipt mechanism) should be protected by steel, at least ½ inch in thickness, securely attached to the storage chest. A cash dispensing machine which also receives deposits should have a receptable chest having the same burglary resistant characteristics as that of a cash dispensing storage chest and should be designed to protect against the fishing and trapping of deposits. Necessary ventilation for the automated machines should be designed so as to avoid significantly reducing the burglary resistance of the machines. The cash dispensing machine should also be designed so as to be protected against actuation by unauthorized persons, should be protected by a burglary alarm, and should be located in a well-lighted area. Alternatively, cash dispensing machines should be so designed and constructed as to afford at least equivalent burglary resistance.[9] A cash dispensing machine which is used inside a bank's premises only during bank business hours, and which is empty of currency and coin at all other times, should at least provide safeguards against "jimmying," unauthorized opening of the storage chest door, and against actuation by unauthorized persons.

FOOTNOTES

1. See section 216.7 regarding the applicability of this Part to Federal Reserve Banks.
2. A branch of a Federal Reserve Bank means an office established pursuant to section 3 of the Federal Reserve Act (12 U.S.C. §521).
3. It should be emphasized that this thickness is merely bullet-resistant and not bullet-proof.
4. It should be emphasized that this thickness is merely bullet-resistant and not bullet-proof.
5. Vaults contracted for previous to this date should be constructed in conformance with all applicable specifications then in effect.
6. The reinforced concrete should have: two grids of #5 (⅝" diameter) deformed steel bars located in horizontal and vertical rows in each direction to form grids not more than 4 inches on center; or two grids of expanded steel bank vault mesh placed parallel to the face of the walls, weighing at least 6 pounds per square foot to each grid, having a diamond pattern not more than 3" x 8"; or two grids of any other fabricated steel placed parallel to the face of the walls, weighing at least 6 pounds per square foot to each grid and having an open area not exceeding 4 inches on center. Grids are to be located not less than 6 inches apart and staggered in each direction. The concrete should develop an ultimate compression strength of at least 3,000 pounds per square inch.
7. Equivalent burglary-resistant materials for vaults do *not* include the use of a steel lining, either inside or outside a vault wall, in lieu of the specified reinforcement and thickness of concrete. Nonetheless, there may be instances, particularly where the construction of a vault of the specified reinforcement and thickness of concrete would require substantial structural modification of an existing building, where compliance with the specified standards would be unreasonable in cost. In those instances, the bank should comply with the procedure set forth in section 216.3 (c) of Regulation P.
8. Equivalent burglary-resistant materials for night depositories include the use of one-fourth inch steel plate encased in 6 inches or more of concrete or masonry building wall.
9. Equivalent burglary-resistant materials for cash dispensing machines include the use of ⅜-inch thick nickel stainless steel meeting American Society of Testing Materials (ASTM) Designation A 167-70, Type 304, in place of 1 inch thick steel, if other criteria are satisfied.

APPENDIX 8B

FORM CC-9030-01 (1/69)

COMPTROLLER OF THE CURRENCY REPORT ON SECURITY DEVICES

Pursuant to the Bank Protection Act of 1968 and 12 C.F.R. 21.5(c)

Each national and district bank must file this report in duplicate with the Regional Administrator of National Banks for the region in which it is located at the times provided for in 12 C.F.R. 21.5 (c).

I. PHYSICAL SPECIFICATIONS OF BANKING OFFICE

Name of bank ....
Name and address of reporting office ....
....

1. (12) ...... Dimensions of lobby (length and width in whole feet of banking floor)
2. (18) ...... Number of entrances or exits to lobby
3. (20) ...... Number of teller stations serving lobby
4. (22) ...... Number of teller stations serving walk-up or drive-in stations or windows
5. Design of inside teller stations (check one)
    a. (24) .... Open style, low partition
    b. (25) .... Closed, with open grill barriers
    c. (26) .... Protected with bullet-resistant materials

d. (27) .... Other, specify ....
6. Design of walk-up or drive-in teller windows (check all applicable items)
   a. (28) .... Protected by bullet-resistant materials
   b. (29) .... Protected by camera
   c. (30) .... Protected by alarms
   d. (31) .... Not protected by any of these
7. Location (check one only)
   a. (32) .... Center city
   b. (33) .... Suburban
   c. (34) .... Town
   d. (35) .... Rural
   e. (36) .... Other, specify ....

II. CRIMES AGAINST REPORTING OFFICE

8. (37) ...... Number of robberies or attempted robberies last 5 years if initial report, otherwise since last report. Check here if initial report. ......
9. (39) ...... Number of burglaries or attempted burglaries last 5 years if initial report, otherwise since last report. Check here if initial report. ......
10. (41) ...... Number of nonemployee larcenies or attempted larcenies last 5 years if initial report, otherwise since last report. Check here if initial report. ......

III. ARMED GUARD PROTECTION

11. (43) ...... Number of guards in lobby during banking hours
12. (45) ...... Number of guards in office during nonbanking hours

IV. SURVEILLANCE SYSTEM

13. Type of equipment
    a. (47) .... Number of photographic cameras used
    b. (49) .... Number of television cameras and recorders used
    c. (51) .... Number others used, specify type
    ....
    ....
14. Specifications, size of film
    a. (53) .... 16 mm or larger
    b. (54) .... Other, specify ....
15. Specifications, photographing capabilities
    a. (55) .... Number of frames per minute, rapid speed photographing
    b. (58) .... Number of frames per hour, slow speed continuous surveillance
16. Coverage (check all applicable categories)
    a. (61) .... Number of devices used at exit
    b. (63) .... Number of devices used at teller positions
17. Method of activation (check all applicable items)
    a. (65) .... Automatic and continuous
    b. (66) .... Activating device at each teller position
    c. (67) .... Other, specify ....
18. Audibility of system when in operation
    a. (68) .... Relatively silent so as not to attract attention
    b. (69) .... Clearly audible
19. System visible to public view?
    a. (70) .... Yes
    b. (71) .... No
20. Public informed through decals or other means of use of surveillance system?
    a. (72) .... Yes
    b. (73) .... No
21. Installation by
    a. (74) Equipment supplier
    b. (75) .... Central station alarm service
    c. (76) .... Other, specify ....
22. Maintenance by
    a. (12) .... Bank employee
    b. (13) .... Installer
    c. (14) .... Other, specify ....

Leave nonapplicable items blank.

V. ACCESSIBILITY OF LAW ENFORCEMENT OFFICERS

(In developing this information, it may be necessary to consult with local law enforcement officials.)

23. (15) ...... Distance from banking office to nearest local law enforcement station having jurisdiction. (in miles)
24. (17) ...... Estimate of shortest time within which enforcement officers could be expected to arrive at banking office after being summoned. (in minutes)

VI. ALARM SYSTEMS

25. Installation
    a. (19) .... By equipment supplier
    b. (20) .... By central station alarm company
    c. (21) .... By other, specify ....
26. Signal transmission method
    a. (22) .... Wires or cables
    b. (23) .... Wireless equipment (for some or all signals)
    c. Means to instantly indicate circuit failure, malfunction or tampering attempts in system?
       1. (24) .... Yes
       2. (25) .... No
    d. Emergency power supply for use in case of failure of regular power supply?
       1. (26) .... Yes
       2. (27) .... No
27. Reporting location for alarms
    a. (28) .... At central station alarm company that is in service 24 hours per day
    b. (29) .... At local law enforcement office that is in service 24 hours per day
    c. (30) .... Other, specify ....
28. Activation of robbery alarms
    a. (31) .... At teller stations
    b. (32) .... Elsewhere, specify ....
29. Does burglary alarm system have a loud bell outside the banking office?
    a. (33) .... Yes
    b. (34) .... No
30. Can activating devices be inconspicuously operated?
    a. (35) .... Yes
    b. (36) .... No
31. Door-type, window-type, or other intrusion detection alarms
    a. (37) .... Yes, specify type ....
    b. (38) .... No
32. Noise-generating device audible outside banking office?
    a. (39) .... Yes
    b. (40) .... No

VII. VAULTS AND SAFES

33. Vault construction
    a. (41) ...... Thickness (in inches) if concrete and steel
    b. (43) ...... Thickness (in inches) other construction, specify ....
    ....
    c. (45) ...... Thickness of vault door (in inches)
34. Vault equipment
    a. Combination, dial locks
       1. (47) .... Yes
       2. (48) .... No
    b. "Time" lock
       1. (49) .... Yes
       2. (50) .... No
    c. Lockable day-gate
       1. (51) .... Yes
       2. (52) .... No
    d. Alarm
       1. (53) .... Yes
       2. (54) .... No
35. Vault is visible from outside office
    a. (55) .... Yes
    b. (56) .... No
36. Vault is in illuminated area
    a. (57) .... Yes
    b. (58) .... No
37. Safes
    a. Construction in conformance with standards in Appendix A?
       1. (59) .... Yes
       2. (60) .... No
    b. Alarm
       1. (61) .... Yes
       2. (62) .... No

VIII. OTHER SECURITY DEVICES

38. Night depository
    a. Alarm
       1. (63) .... Yes
       2. (64) .... No
    b. Construction
       1. (65) .... In conformance with standards in Appendix A
       2. (66) .... Other (specify) ....
       ....
39. Safe deposit boxes
    a. (67) .... Yes
    b. (68) .... No
40. Are all exterior doors and windows that can be opened equipped with tamper-resistant locks?
    a. (69) .... Yes
    b. (70) .... No

APPENDIX 8C

FORM CC-9030-02 (1/69)

COMPTROLLER OF THE CURRENCY
REPORT OF CRIME

Pursuant to the Bank Protection Act of 1969 and 12 C.F.R. 21.5(d)

FOR OFFICE USE ONLY

State .... Bank ....
Branch (7) ... Card (11) ...

This report must be filed within a reasonable time after a robbery, burglary or non-employee larceny is perpetrated or attempted at an office of national or district bank that is subject to 12 C.F.R. Part 21. Copies of the report must be filed with the Regional Administrator of National Banks for the region in which the reporting bank is located.

(Mark or enter the appropriate information. Leave blank non-applicable items.)

1. Name and address of bank head office:
   ....
   ....

2. If crime being reported occurred at a branch office, give name and address:
   ....
   ....

3. Type of crime:
   a. (12) .... Robbery
   b. (13) .... Burglary
   c. (14) .... Non-employee larceny

4. (15) (...........) .... 19.. Date of crime
   (For office use only)

5. (21) (...........) .... Day of week
   (For office use only)

6. (22) (...........) .... Time of day
   (For office use only) (If actual not known, estimate)

7. Amount of loss (to the nearest dollar):
   a. (26) $.......... Currency loss
   b. (35) $.......... Securities loss
   c. (44) $.......... Damage to bank property. (May be estimated)
   d. (53) $.......... Other, specify

(IF CRIME OF ROBBERY HAS BEEN PERPETRATED OR ATTEMPTED ANSWER THIS SECTION)

8. (62) .... Number of robbers participating in crime

9. Weapons:
   a. Did robber(s) have weapon(s) or did it appear they may have had weapons?
      (64) .... No
      (65) .... Yes, specify kind ....
   b. Was other intimidation used?
      (66) .... No
      (67) .... Yes, specify ....

10. Were robber(s) wearing masks or otherwise disguised?
    a. (68) .... No
    b. (69) .... Yes, indicate how ....

11. Was a description of the robber(s) obtained and recorded?
    a. (70) .... Yes
    b. (71) .... No, why ....

12. Was a description and/or license number of vehicle(s) obtained?
    a. (72) .... Yes
    b. (73) .... No, why ....

13. (74) .... Estimated minutes between beginning and end of robbery

14. Modus operandi:
    a. Did robber(s) pass a note to teller demanding money?
       (77) .... Yes
       (78) .... No
    b. Did robber(s) vocally demand money?
       (79) .... Yes
       (80) .... No
       Card (11) ...... (For official use only)
    c. Did robber(s) subdue employee(s) and take money from containers?
       (12) .... Yes
       (13) .... No
    d. (14) .... Other, specify ....

15. Harm to persons:
    a. Were either employees or customers physically harmed?
       (15) .... Yes
       (16) .... No
    b. Were other persons harmed?
       (17) .... No
       (18) .... Yes, give details ....
       ....

16. Was a hostage or threat of holding a hostage used?
    a. (19) .... No

b. (20) .... Yes, give details ....
....
17. Was cash or valuables taken from other than teller drawers?
    a. (21) .... No
    b. (22) .... Yes, specify ....
18. Was "bait" money given out or taken during the robbery?
    a. (23) .... No
    b. (24) .... Yes
    If yes, was the identification of this money furnished to the law enforcement officers?
    c. (25) .... Yes
    d. (26) .... No, why ....
    ....
19. Was the cash contained in the teller drawer(s) within the maximum permitted by the bank's security program?
    a. (27) .... Yes
    b. (28) .... No, why ....
20. Cameras (or other surveillance device) (check one):
    a. (29) .... Camera(s) recorded useful pictures during this robbery
    b. (30) .... Camera(s) did not record useful pictures during this robbery, why ....
21. Robbery alarm (check one):
    a. (31) .... Alarm was effective during this robbery, how ....
    b. (32) .... Alarm was not effective during this robbery, why? ....
22. Did robber(s) leave note or other item which was retained and preserved for use of enforcement officers?
    a. (33) .... Yes, what
    b. (34) .... No, explain if necessary ....
    ....
23. Was conduct and performance of employees in conformance with 12 C.F.R. 21.5 and the bank's security procedure?
    a. (35) .... Yes
    b. (36) .... No, explain ....
    ....

(IF CRIME OF BURGLARY HAS BEEN PERPETRATED OR ATTEMPTED ANSWER THIS SECTION)

24. How did burglars gain entrance to the premises?
    a. (37) .... Break-in, where and how ....
    ....
    b. (38) .... Other, specify ....

25. Vault (check one);
    a. (39) .... No apparent attempt was made to gain access to vault
    b. (40) .... Penetration of vault wall, floor or ceiling was made or attempted, how ....
    c. (41) .... Vault door was opened or penetrated, how ....
    ....
    d. (42) .... Other, specify ....
    ....
26. Were the lights required by 12 C.F.R. 21.3 in good working order and turned on?
    a (43) .... Yes
    b. (44) .... No, explain ....
27. Were safe deposit boxes broken into or opened?
    a. (45) .... No
    b. (46) .... Yes, indicate extent and how
    ....
28. Money safe (check one)
    a. (47) .... No apparent attempt made to gain access to contents
    b. (48) .... A penetration or an attempted penetration of safe was made, how ....
    c. (49) .... Safe door opened or an attempt made to open. How? ....
    d. (50) .... Other, specify ....
    ....
29. Night depository (check one)
    a. (51) .... No attempt was made to gain access to contents
    b. (52) .... Contents taken or attempted by "Fishing" or "Trapping" methods, how, if known ......
    ....
    c. (53) .... Night depository penetrated or access door opened, explain ..
    ....
    d. (54) .... Other, specify ....
    ....
30. Burglary alarms (check one)
    a. (55) .... Alarms were of value in connection with this crime, how
    ....
    b. (56) .... Alarms were not of value in connection with this crime, why
    ....
31. (57) ...... Estimated length of time during which burglary was being committed (in minutes)

(IF CRIME OF NONEMPLOYEE LARCENY HAS BEEN PERPETRATED OR ATTEMPTED ANSWER THIS SECTION)

32. Modus operandi or larceny (check one):
    a. (60) .... Money or valuables where thief had access, explain ....
    ....
    b. (61) .... Theft by trick or pretext; explain ....
    ....
    c. (62) .... Other, specify ....
    ....

(ALL BANKS ANSWER THIS SECTION)

33. (63) ...... Length of time after beginning of crime when call for help was transmitted to appropriate law enforcement agency. (in minutes)
34. (66) ...... Length of time after beginning of crime before first law enforcement personnel arrived at the bank office. (in minutes)
35. Did law enforcement personnel arrive at bank office before violators had departed?
    a. (69) .... Yes
    b. (70) .... No

36. Arrests of violators (check all applicable categories):
    a. (71) .... None have been arrested as of the date of this report
    b. (72) .... Some or all arrested before they escaped from the bank office
    c. (73) .... Some or all arrested subsequent to leaving the bank office
37. Would improvements in protection facilities or employee performance be helpful in preventing or handling any future similar occurrences?
    a. (74) .... No
    b. (75) .... Yes, indicate what plans the bank has to take corrective action
38. Use additional pages to set forth any information about the crime or the protection measures that is not adequately covered previously. Furnish photographs or sketches if necessary to completely describe the crime being reported.

Signature ....
                    (Security Officer)
Name (typed) ....
Title ....
Date ....

APPENDIX 8D

CAMPUS SECURITY DATA COLLECTION HANDBOOK*

Prepared by
RICHARD S. POST
(Then) Chairman, Dept. of Police Science & Administration
Wisconsin State University-Platteville

I. INTRODUCTION

In an effort to assist university administrators collect data relative to security needs on their campus, this handbook has been prepared to assist in initial data collection. The data collected by individual university faculty and staff members ideally should be supplemented by information supplied from external sources as well as the findings of a security survey team.

II.

Four areas are included for preliminary data collection. These are:

1. *Building Priority List.* The Building Priority Worksheet is provided to assist in the establishment of building priorities relative to protective needs.

© 1971 Security Management Services, Ltd., P.O. Box 215, Sterling Heights, Michigan. All rights reserved. (Reprinted courtesy of Post and Associates.)

2. *Alarm and Equipment Monitoring Inventory.* The collection of information relative to existing alarms and equipment monitoring devices to determine the need for additional or supplementary equipment.
3. *Crime Incidence Inventory.* This provides for the collection of information from the Campus Security Office, Dean of Students and police agencies relative to various categories of criminal activity and unauthorized building and equipment use.
4. *The Attitude Survey Form.* This questionnaire is provided for administration to faculty, staff and students to determine general attitudes relative to the University security program.

A sample form is provided in each area. A detailed explanation of the form and its administration is also provided.

III.

If an independent security survey is to be conducted, prior to the arrival of a security survey team, the following information should be collected. It will reduce costs and the amount of time required for data collection.

Security Office

1. Number of security officers on campus;
2. The Director of Security should insure that a detailed list of duties performed by each security officer on each duty shift is prepared;
3. An inventory of security equipment including radios, weaponry, and vehicles;
4. A list and location of any detex and time clock stations or routes on campus;
5. A detailed list by type and location of all fire fighting equipment on campus.

Vice President of Business Affairs

1. A complete description of university identification card programs;
2. A list of all the university parking regulations;
3. A detailed description of the university's lock and key system;
4. A copy of the university emergency operating plan;
5. A copy of the university master plan (or development plan);
6. A list of students enrolled on campus and enrollment projections through 1980.

President

1. If a survey team is to be utilized, notification should be sent to all university staff to extend cooperation to the survey team and inform them of their presence on campus.

Superintendent of Buildings and Grounds and Campus Planner

I. Blueprints, Maps, and Line Drawings*
   A. Overall Campus map showing property lines
   B. All major highways and access roads leading to and from campus
   C. Blueprints showing all electrical routing, underground and overhead wiring, master control panels, telephone systems, and location of the source of electrical power supply for the campus

---

* If possible, insure that maps, drawings, etc. are of uniform scale.

D. Obtain maps or blueprints showing all sanitary and storm sewer systems
   1. Show all underground tunnels, manholes, and controlling tunnels
E. Obtain maps showing steam, water, and gas systems on campus
   1. Indicate all master cut offs, meters, underground tunnels showing access points, above ground systems, monitoring systems, and sources of power
F. Obtain information on how, who, and when each of the above systems are monitored
G. Obtain floor plans of each building on campus
   1. Plot location of each of the following items:
      a. Doors—General condition
         (1) Indicate fire doors
         (2) Indicate exit doors
         (3) Indicate if locked
         (4) Indicate if monitored
      b. Windows
         (1) Indicate if locked
         (2) Monitored—General condition
         (3) Indicate if basement and first floor windows are screened
      c. Fire Alarms
         (1) Indicate condition
      d. Lights (exit)
         (1) Indicate condition
      e. Lights (security)
         (1) Indicate condition
      f. Fire fighting equipment
         (1) Check tags for inspection validation initials of fire inspector
         (2) Note if equipment is operative
      g. Hazards
         (1) List on separate paper all building hazards
         (2) Indicate if hazards are for security, fire, or safety
      h. Time Clock Stations
         (1) Indicate the use of time clock stations by campus security personnel
      i. Secured areas within each building
         (1) Administrative Offices
         (2) Equipment storage areas
         (3) Boiler rooms
         (4) Laboratories
         (5) Vaults and Safes
         (6) Master cut off switches
            (a) Electric
            (b) Fire Alarms
            (c) Gas
            (d) Water
            (e) Heat or Steam
            Indicate if these systems are monitored
H. Make provisions to have copies made of blueprints, maps and line drawings from:

1. Administrative offices
    a. It may be necessary to have line drawings made for floor plans*

II. Physical building inspections will be necessary in order to plot the following information:
   A. Fire Alarms
   B. Exit Lights
   C. Security Lights
   D. Doors (Locked)
   E. Windows (Locked)
   F. Window Screens
   G. Fire fighting equipment
   H. Time Clock Stations
   I. Security areas (within each building)
   J. Hazards (very important)
   K. Master control panels or access to such areas
   L. Cut off switches for:
      1. Gas
      2. Water
      3. Fire Alarms
      4. Heat
      5. Electricity
      6. Telephones
      7. Steam
      8. Sewerage

III. Blue prints for each campus will be found with the following information listed on them:
   A. ——— University (Name of School)
   B. Board of Regents of University
      1. Sequence
         a. Sanitary and storm sewer
         b. Steam–water–gas
         c. Electrical
         d. Property map
         e. Property map (showing major highways)
         f. Aerial map

IV. Inventory and building inspection forms may be obtained from the following departments:
   A. Maintenance
   B. Security

IV.

Building priorities were determined through the use of a Building Priority Worksheet, which was developed by the survey staff. This instrument was used to provide an objective means of collecting and evaluating empirical dates relating to buildings on the campus and their relative importance to the overall University mission.

The problem of a subjective evaluation of building importance was pointed

---

* Suggestion: Enlist the aid of students under workstudy, if these drawings are not available.

out through the use of a forced choice ordering survey conducted of key university officials. This required the top six administrators at a given university to rank order a list of buildings according to their perception of the building's importance to the overall university mission. The results of these studies indicated that very little consensus could be accurately determined as to building priorities.

The established priorities are specifically in relation to the day-to-day activities of the university and their importance to the overall university mission. In emergency, conflict or disaster situations, the only changes in priorities that could be envisioned would be the provision of additional protective measures to the Central Heating Plant and Emergency Control Center due to their service value to the entire university mission.

It should be pointed out that the priorities established are on the basis of existing conditions and the existing status of the building, either temporary or permanent. Should these conditions change, the building priority should likewise change. Similarly, should the crime incidence or the evaluation of the building change, the priorities should likewise be changed. (See Appendix A for forms.)

The collection of materials for the establishment of building priorities should include, at a minimum, the following individuals: (1) University President, (2) Vice-President for Development & Services, (3) Vice-President for Business Affairs, (4) Campus Planner, (5) Chief Security Officer and (6) Head of Buildings and Grounds.

The usefulness of the building priority form is based on the following assumptions:

1. establishment is necessary because some buildings are more important to the university than others;
2. hazards, functional use of buildings, insurance value (building plus contents) and whether or not a building is permanent or temporary will indicate the importance of a building to the university;
3. to collect data needed so that statistical formulas can be used;
4. need for objective and empirically verifiable method for determining building priorities;
5. *weighted factors:* key offices contained in a building and "vital" services performed within a building are two special factors which indicate the importance of a building;
6. it is possible to obtain a fair and impartial rating for each building;
7. buildings can be given a priority rating based upon the total number of points they are given; and
8. buildings were grouped into the following categories:
   A. Permanent
   B. Temporary
9. Federal space allocation standards can be applied in utilizing the forms.

Definition of Terms

The following definition of terms are provided in an effort to explain the Building Priority Worksheet and so that the correct data can be collected and applied to categories listed.
1. Hazards
   A. Fire (12 points) —Buildings are evaluated as good (3 points), fair (4 points), or poor (5 points). This evaluation is based upon the following data:

1. Current fire ratings
2. Inventory of fire equipment and its condition
3. Highly combustible materials stored in building
4. Building equipment inventory
5. Fire drill information (time required to evaluate building)
6. Condition of electrical wiring
7. Condition of storage areas in buildings (in lieu of data, use on-sight inspection)
8. Building location within the university complex

B. Safety (10 points)—Buildings are to be evaluated as good (2 points), fair (3 points), or poor (5 points). This evaluation is based upon the following data:
1. Engineering structural study
2. Building accident records
3. Building equipment inventory
4. First aid stations and personnel in building
5. Condition of storage areas in building (in lieu of data, use on-sight inspections)

C. Thefts (9 points)—Buildings are to be evaluated as the type of theft that would occur. Currency (4 points), equipment—TV, Recorders, etc. (3 points), and records (2 points). This evaluation is based upon the following data:
1. List of safes, vaults, etc. and their general content
2. Inventory contents of building
3. Tabulations of thefts reported to the:
   a. Security department,
   b. Local police, and
   c. Other university offices such as the Dean of Students
   d. This tabulation should include the sum and total dollar costs of all thefts by building. In lieu of accurate records, on-sight inspections and interviews with personnel responsible for the investigations of thefts should be conducted.

D. Vandalism and Criminal Damage (7 points)—Buildings are to be evaluated by tabulations of incidents from such sources as local police, security department, and other university offices such as Dean of Students, Business Office and from the publicity value of the building. (Example—Main or Administration Building would have more publicity value than the Maintenance Building.) In lieu of accurate records, this hazard can be assigned points using the following data:
1. Publicity value of building
2. Number of large windows on its perimeter
3. Contents in either building
4. Information gathered from interviews with maintenance and security personnel. Each building should be rated: Good (1 point), Fair (2 points), and Poor (3 points). (Example, a building with a high publicity value and/or a large number of good sized windows would be rated poor.)

E. Communication Interruption (5 points)—Buildings are allotted points on the following formula: maximum number of points for this hazard

multiplied by the building percentage of the total square footage on campus.
F. Trespass (Unauthorized Building Use) (4 points) —Buildings are allotted points using this formula (building percentage of total number of open doors—found by the security department of the university and the local police department—multiplied by the total maximum number of points allocated for the trespass hazard).
G. Environmental (3 points) —Buildings are allotted points using this formula: Building percentage of the total damage (in dollars) which occurred as a result of storms, severe weather, or natural disasters multiplied by maximum points for hazard. In lieu of adequate data, leave blank.

2. Building Function
   A. *Building Function Category.* Federal space allocation program has established a function code which applies to all buildings on campus with federal government programs. The United States Office of Education system of space allocation which unified the definitions of type and function classification of building space for the purpose of giving a comparative data base for statistical studies at the national and state level. It is recommended that this standardized space allocation or functional coding be utilized in the collection of data relative to campus security needs.

   It is the purpose of this form to offer a working guide for personnel charged with the responsibility for making physical security surveys of a university. It will insure that all data necessary for objectively evaluating the "state of the arts" of security on a given campus can be quickly, efficiently and effectively collected.

   The use of a standardized collection format allows the use of this data in conjunction with other materials, e.g. student services, registrar, etc. relating to campus needs as might be collected. It will also insure that data can be integrated into a computerized storage for analysis or used in manual analysis (which might be desirable for smaller institutions). The basic methods of gathering original data, classifying and utilizing the information will, however, remain the same regardless of the size of the institution.

   The use of this code will allow access to other information compiled by the university. Many of the buildings have several functions and points are allocated accordingly. Several functional categories were grouped together for simplification thus reducing the number of categories as are found in Attachment A. In order to establish building priority in functions 050, 010, 020, and 040, the following information was collected and tabulated:

   1. Total square footage of function 010 and 020 per building
      a. Square footage of function 010 and 020 on campus
      b. % of total square footage of functions 010 and 020 on campus
   2. Total square footage of function 050 on campus
      a. points were allotted for function 050 by the following math formula: Building percentage of total square footage of function 050 on campus multiplied by the maximum points allotted for function 050 (25 points). Points were allotted for functions 010,

020, by this math formula: Building % of the combined total square footage of functions 010, 020 (35 points). Points were allocated for function 040 by this math formula: Building percentage of total square footage of function 040 on campus multiplied by the maximum number of points allotted for function 040 (15 points). Points were allocated for functions 060, 030 and 070 by this math formula—percentage of total square footage of functions 060, 030 and 070 in a building multiplied by the maximum number of points allotted for functions 060, 030 and 070 (10 points). Points were allotted for function 080 by this math formula in building percentage of total square footage of function 080 on camp is multiplied by maximum number of points allotted for function 080 (5 points).

3. Insurance Value—Key Offices—Service Buildings
   A. Insurance Value (10 points)—Buildings are allotted one point per million dollars. This figure includes the building and the value of its content.
   B. Key Offices (Varied points)—Each administrative office is allotted one point with the total maximum points determined by the number in the building. The key administrative offices include the following offices:
      1. President, Academic Affairs Vice-President, Admissions, Registrar, Dean of Students, Financial Aids, Mail Room, Public Relations, Television and Radio Stations, Security Office, Central Heating Plant Office, Student Affairs Office.
   C. Service (10 points)—Some buildings have few, if any classrooms, laboratories and offices yet they provide "vital" services to other buildings. These vital services would include shipping, receiving, computers, purchasing, business records, heating, etc. This is a weighted factor which is used in taking into consideration alternate sources of supply.

The means of allocating points to the building are for the purpose of establishing a building priority. It is felt that this will also indicate what building will have priority in any phase of development of additional security measures.

V. ALARM AND MONITORING EQUIPMENT INVENTORY

The purpose of the alarm and monitoring equipment inventory is to accurately determine technical data relating to the presence of various types of alarm and equipment monitoring devices on the campus. This inventory should be completed under the direction of the Superintendent of Buildings and Grounds. The chart includes the majority of equipment monitoring devices commonly used in physical plant maintenance.

All categories of equipment should be checked as either not present, or specific detail data given if a particular item of monitoring equipment is present.

If additional monitoring equipment is present which is not included on the inventory list, insure that this information is provided by attaching supplementary sheets to the inventory.

One inventory sheet should be used for each building on campus; use coding designated by the university in addition to building names. This will aid in converting the data into EDP format.

VI. CRIME INCIDENCE INVENTORY

In gathering crime information, it is necessary to contact the Director of the Security Department of the campus, the Dean of Student Affairs, and the Chief of Police of local jurisdictions.

EXAMPLE OF TOTAL INCIDENT CHART

Time	Sunday	Monday	Tuesday	Wednesday	Thursday	Friday	Saturday	Total	% of Total
12 am to 1 pm	TCM	V	M		IC	TA		9	5.77
1 - 12	C	V		C		M		3	1.92
2 - 3									
3 - 4	MC								
4 - 5									
etc.									
Total	12							156	
% of Total	7.69							100%	

KEY: C—Crimes; M—Maintenance; V—Vandalism; T—Thefts; I—Student Incidents; H—Accidents.

Appendix 8D—Figure 1.

The purpose of contacting these people is to gather all of the pertinent information relating to the services performed by the security officers, as well as the frequency, location, and types of incidents occurring on campus. The need for all pertinent data relating to the security function is to determine what are the best hours for shifts, how many men are needed for each shift and where trouble areas are located.

The data to be collected should be categorized as follows:
1. Incident, i.e., thefts, vandalisms, accidents, calls for services, student incidents, etc.
2. Location of incident
3. Hour of day
4. Day of week
5. Vandalism and theft amounts

Data for one calendar year should be gathered so as to show a complete view of security problems. Once the data has been collected, it should be presented in chart form. This chart would consist of nine spaces on the vertical axis and 27 on the horizontal axis. The vertical axis is for the days of the week, total for that day and percentage of total for each day of all incidents for the year. The horizontal axis contains the 20 hours of the day, the total for each hour and the percentage per hour of the number of incidents for the year. To find the percentage for each day of the year and each hour of the day, each column is added separately to find the total for each day and each hour. This total is then divided by the total number of incidents for the year. The percentage found is then placed in its proper column for each day and each hour of the day. Once these percentages are found, the two percentage columns should add up to 100 percent. Once this chart is completed, it will present which are the most active days of the week and which are the most active hours per day with this information, the Security Department can prepare a realistic man power distribution chart.

Proportionate need is used to determine the shift hours and the number of men per shift. The chart previously referred to is used to determine proportionate need. In determining proportionate need, the researchers should try a variety of possible shift arrangements, i.e. 8 to 4, 4 to 12, and 12 to 8 or 7 to 3, 3 to 11, and 11 to 7, or any other combination as long as there is always eight continuous hours to a shift. For each combination, the percentage of the total for each hour of the eight hour shift should be added. The shift to be utilized is the one with the smallest deviation between the three shifts, i.e.

8 to 4 —32% of total	7 to 3 —20% of total
4 to 12—33% of total	3 to 11—30% of total
12 to 8—35% of total	11 to 7—50% of total

The deviation per shift is found by subtracting the percentage of the total from 33.33 which is the average percentage of need for each shift. These three figures are then added together. Once this is accomplished, the shift with the smallest percentage of deviation is selected as the desirable shift.

If the problem should arise that two or more shift possibilities are close together in their percentage of total deviation, the highest percentage of total from one of the shift possibilities should be choosen and subtract each of the other shifts in that shift possibility from it. Once these two figures are found, they must be added to find a total. This is to be done for all the shift possibilities.

Once the shift hours have been determined, the number of men that will be needed for each shift should be determined. This is accomplished by finding the total number of men on the force and subtracting from this figure the number of men used for auxiliary purposes, i.e. radio dispatch, etc. This figure is the total men assignments to security activities. Next this figure is multiplied by 7, which represents a complete week. The product, called man day per week, is found and put into a percentage basis. Then the percentage of need per day, which must be carried out only to the tenths place, is multiplied by the man day per week. This product is the total men for a shift. Finally, the total men figure is divided by 1.5 to find the number of men available for duty that day. 1.5 is a relief factor. A relief factor is the number of men on vacation, sick leave, off-duty, etc. This process must be repeated for each day of the week.

To determine how many men are needed for each shift, first multiply the number of men available for that day by three (3), which represents the number of shifts per day. This will find the man hours per day and place this product into a percentage. Then multiply the percentage of need per shift by the man hours per day. This will find the total manpower per shift. Then divide the manpower per shift by three (3) to find how many men are needed for a shift. This must be repeated for each shift of the day.

Dean of Students and Police Department Information

The security office must be informed of *all* incidents that occur on the campus. To insure compliance with restricted student data, the names of the students involved in activities would not be essential (although it would be very beneficial for the security office to know who the students are who are causing problems), but the incident should be reported to the office with the type of incident, date, time of day, and the location. The head of the security office should periodically confer with the police and the Dean of Student Affairs to collect all of the information needed for each incident that he does not have on record.

VII. CAMPUS SECURITY ATTITUDE SURVEY

Background

In order to provide the survey team with information relating the perceptions of the university security function by the clientele group which it services, it is necessary that a cross-section of opinion be sampled from the university community. It is of great interest and need to the study team to have the information relative to perceptions and expectations relating to the university security force. The enclosed "short" form will give us a method of obtaining some of this necessary data.

The responsibility for obtaining a composite picture of attitudes and perceptions of campus security will be borne by both the university and the survey staff. The survey staff will be responsible for (1) interviewing administrative officials on campus, (2) interviewing city officials, and (3) developing a composite inventory.

The university will be responsible for distributing questionnaires to faculty, staff, and students. The questionnaires should be distributed proportionately among faculty, staff, and students (maximum number needed is 50 per 1000 on campus). The date for return should be set to allow the complete return of all forms prior to the completion of on-sight work by a study team. It is recommended that self-addressed stamped envelopes be utilized for the return of these letters. It is also recommended that they be addressed to the security study team liaison officer appointed by the university.

BUILDING PRIORITY WORKSHEET

1. INSTRUCTION, RESEARCH 35 Points Maximum ....
   (010) + (020)
   Total Square Footage of Function in Building ....
   Total Square Footage of Function on Campus ....
   Building percentage of square footage ....
   35 × Building percentage equals points allotted for Functions

2. PUBLIC SERVICES, AUXILIARY ENTERPRISES, NONINSTITUTIONAL AGENCIES 10 Points Maximum ....
   (030), (060) + (070)
   Total Square Footage of Function in Building ....
   Total Square Footage of Function on Campus ....
   Building percentage of square footage ....
   10 × Building percentage equals points allotted for Functions

3. LIBRARY 15 Points Maximum ....
   (040)
   Total Square Footage of Function in Building ....
   Total Square Footage of Function on Campus ....
   Building percentage of square footage ....
   15 × Building percentage equals points allotted for Function (030)

4. ADMINISTRATIVE AND GENERAL SERVICES
   25 Points Maximum   ....
   (050)
   Total Square Footage of Function in Building   ....
   Total Square Footage of Function on Campus   ....
   Building percentage of square footage   ....
   25 × Building percentage of square footage equals points allotted for Function
5. NON-ASSIGNED 5 Points Maximum   ....
   (080)
   Total Square Footage of Function in Building   ....
   Total Square Footage of Function on Campus   ....
   Building percentage of square footage   ....
   5 × Building percentage of square footage equals points allotted for Function

   TOTAL POINTS FOR SECTION I   ....

II. HAZARDS
   1. FIRE—12 Points: Good (3 pt.) Fair (4pt.) Poor (5 pt.)   ....
   2. SAFETY—10 Points: Good (2 pt.) Fair (3 pt.) Poor (5 pt.)   ....
   3. THEFT—9 Points: records (2 pt.) equipment (3 pt.) currency (4 pt.)   ....
   4. VANDALISM AND CRIMINAL DAMAGE—7 Points
      Total cost of vandalism and damage on campus   ....
      Percentage of total cost for building   ....
      Seven times percentage of total cost for building   ....
      Good (1 pt.) Fair (2 pt.) Poor (3 pt.)   ....
   5. COMMUNICATION INTERRUPTION—5 points
      Summation of following products
      a. Five times fraction or decimal of Total building square footage   ....
   6. TRESPASS (UNAUTHORIZED BUILDING USE) —4 Points
      Total number of open doors on campus   ....
      Building percentage of total number of open doors on campus   ....
      Four times building percentage   ....
   7. ENVIRONMENT—3 Points
      Total Cost of damage to buildings as a result of storms, severe weather, or natural disasters   ....
      Three times Building % of cost   ....

      TOTAL POINTS FOR SECTION II   ....

III. INSURANCE VALUE—10 Points
   (One Point per million dollars)
   Value $....................

   TOTAL POINTS FOR SECTION III   ....

IV. KEY ADMINISTRATIVE OFFICES
(One point for every administrative office in building)

V. VITAL SERVICES ....
   TOTAL POINTS FOR SECTION V ....
   TOTAL POINTS FOR BUILDING ....

AVAILABLE SERVICES

1. 050 ADMINISTRATION AND GENERAL SERVICES (15 points)

   All areas used by personnel having responsibility for general administration and supporting services.

   Included in this category are offices, work areas, and related service areas used by the following:

President	Director of public relations
Vice president	Dean of student affairs
Business manager	Maintenance
Planning officer	Security personnel

   Also included in this category are areas used for the provision of such student services as:

Health services	Student activities
Placement	

2.   ACADEMIC (15 points)

   010 INSTRUCTION

   All areas used for the transmission or dissemination of knowledge to college students on a group or incidental basis.

   Included in this category are:

Classrooms	Offices used by academic deans,
Class laboratories	department chairmen, and
Faculty offices	related service facilities

   Also included in this category are the following areas, sometimes referred to as *organized activities related to instruction* operated for the primary purpose of providing professional training for students:

Laboratory schools	Physical education facilities
Demonstration facilities	

   020 RESEARCH

   All areas used by faculty, staff or students for *activities whose primary objective is the discovery or application of knowledge.*

   Areas used for activities which are primarily instructional and only secondarily involve research should be classified as Instruction. Conversely, areas used primarily for research and only secondarily involving instruction should be classified as Research. Prorate between the function of research and instruction where necessary.

   Institutions desiring to identify areas used for the following will need to subdivide this category:
   Organized research
   Contact research   **Sponsored research**

Building by Name _____
Building by Number _____

CODE:
- RR — Rate of Rise
- SD — Smoke Detectors
- RC — Remote Control
- I — Interior Devices
- A — Audio Identification
- BP — By-Pass Switch
- VP — Variable Parameters
- SC — Supervisory Circuit (failure of circuit line)

CATEGORIES	SENSORS			TRANSMISSION		ANNUNCIATORS			DATE OF INST.
1) Fire Safety	No.	Type	Brand	Wire	Other	No.	Type	Brand	
2) Exhaust Ventilators									
3) Gas Meter On-Off									
4) Exit Doors									
5) Transformer Hi-Temp Alarm									
6) Power Dist. Air Flow									
7) Exhaust Fans Air Flow									
8) Compressed Air Low Alarm									
9) Steam Main Pressure Drop Low Side									

CATEGORIES	SENSORS			TRANSMISSION		ANNUNCIATORS			DATE OF INST.
	No.	Type	Brand	Wire	Other	No.	Type	Brand	
10) Air Supply On - Off Control									
11) Area Temp.									
12) Water Pumps									
13) Environment Chambers Temp.									
14) Environment Chambers Humidity									
15) Storage Vault Vent Fans Air Flow									
16) Water Distillers (Low Level)									
17) Intercomm Stations Mech. Equip. "to Where"									
18) Sump Pumps Hi - Level									
19) Elevators Alarm (Where sound?)									
20) Environment Chambers Personal Alarms									
21) Emergency Lighting									
22) Emergency Generators									
23) Air Conditioners									
24) Intrusion Devices									

Appendix 8D—Figure 2.

040 LIBRARY
All areas used for the orderly location, storage and retrieval of knowledge.
This category includes all rooms under the general supervision or control of a central or departmental librarian, such as:

Reading rooms   Library offices
Study rooms   Work areas
Stack areas

This category does NOT include areas used primarily for library science instruction.

3. 060 AUXILIARY ENTERPRISES (6 points)
All areas housing activities operated by the institution (or provided by contact with the institution), usually on a self-supporting basis, primarily for the purpose of providing auxiliary services to students, faculty and staff.
Included in this category are the following:

Food service facilities   Recreation facilities
Residence facilities   Student unions
Merchandising facilities   Faculty clubs

Also included n this category are areas used primarily for intercollegiate or varsity athletics. This category does NOT include areas used primarily for academic instruction in *physical education*.

4. 080 NONASSIGNABLE AND UNASSIGNED (3 points)
All areas which are not available for assignment either because of the nature of the space or because of its present condition.

5. 030 PUBLIC SERVICE (1 point)
All areas designed to house activities serving the general public.
Included in this category are the following:
Assembly facilities used for concert and lecture series and dramatic presentations
Exhibition facilities, including
  museums
  art galleries
areas used for adult and continuing education

070 NONINSTITUTIONAL AGENCIES
All areas used by public or private agencies not under the supervision or control of the institutional administration.
Included in this category are the following:

ROTC   Civil Defense

STUDENT QUESTIONNAIRE

A Security Survey Committee now working at (your university) is seeking your assistance in learning about attitudes concerning the campus security force. All answers are strictly confidential, so we ask that you not sign your name to the questionnaire.

This Committee is functioning in conjunction with an attempt by the university to evaluate opinions relating to its Campus Security Program.

We wish to thank you for your cooperation in filling out this opinion questionnaire. Your immediate attention is requested to enable us to complete this survey by ...., 19.....

1. Are you aware of the functions of campus security? Yes .... No .... If yes, what in your opinion does security do? ....
   ....
2. Do you feel it is necessary to maintain a campus security force? Yes .... No .... Comments: ....
   ....
3. Do you feel the University would be better served by local police or by campus security personnel? Comments: ....
   ....
4. Do you feel that training courses should be conducted for campus security personnel? Yes .... No .... Comments: ....
   ....
5. What powers should be granted to campus security personnel?
   Power of Arrest ....
   Traffic and Investigation Only ....
   Other .... Comment: ....
   ....
6. Do you feel that there should be a student review board in considering decisions made by campus security? Yes .. No .... Comments: ....
   ....
7. In general, what do you think of the treatment of students by campus security?
   ....
8. In the event of riots or disturbances, what role should the campus police play in their control? Comments: ....
   ....
9. Have you had any dealings with campus security? Yes .... No ....
   Area of Satisfaction: ....
   ....
   Area of Dissatisfaction: ....
   ....
   ....

APPENDIX 8E

SYRACUSE UNIVERSITY SECURITY AND SAFETY GUIDE*

A MESSAGE FROM THE CHIEF OF POLICE CITY OF SYRACUSE

College students, especially those who have matriculated at large universities, and who are living away from home for the first time in an atmosphere of reduced restraints, are extremely vulnerable to the criminal element who stalk this type of environment in search of easy prey. If you, the student, ignore the common sense rules of safety that have been promulgated for your benefit, you leave yourself available to these merchants of crime and violence. If you refuse to do your part by disregarding normal safety precautions, you are inviting these predators to use you as their next target. Do not be so foolish as to believe, "It can't happen to me." I urge you, with all the sense of emergency I command, to use the good common sense that God gave you and take the normal precau-

* From SECURITY AND SAFETY GUIDE. Courtesy of Syracuse University, Syracuse, New York.

tions that are required so that the odds of your becoming a "crime statistic" will be materially reduced.

Thomas J. Sardino
Chief of Police

A MESSAGE FROM THE DIRECTOR OF SAFETY AND SECURITY

Today, students across the land on college campuses are faced with an increase in the crime rate. They are the ones who are exposed and are the potential prey of lawbreakers who may be armed and dangerous.

What Can Be Done About It?

The prime function of the Syracuse University Security Department is to provide a safe environment for students, staff, and faculty. Our department works closely with the well equipped and highly efficient Syracuse Police Department. They provide special units for high crime rate areas known as Crime Control Teams. These officers are charged with selective enforcement duties and are dedicated and well trained. We share with the Police Department a grave concern for the welfare and life safety of the students in our community.

What Can You Students Do to Reduce Your Changes of Becoming a Statistic?

Following the simple precautions outlined for you in this pamphlet and using good common sense should greatly increase your personal safety and the safety of your possessions.

R. D. Flaherty
Director

SELF-PROTECTION

The following suggestions are offered to you in order that you may help protect yourself.

1. Report all strangers loitering in the building to Security, ext. 2224.
2. When alone in your office or room after hours or in the evening, keep your door locked.
3. If you leave your office or room for only a few minutes, close and lock the door. This way you will know that no one is inside waiting for you when you return.
4. When taking a midafternoon siesta in your room, lock the door.
5. Never prop open a door for someone who will be joining you later who does not have a key to the building. A propped open door destroys the best security plans and is an open invitation to undesirables.
6. When walking at night:
    a. Avoid shortcuts—walk where there is plenty of light and traffic. Avoid the dark vacant areas on campus.
    b. Never walk alone.
    c. Don't hitchhike—it's illegal and dangerous.
    d. Best of all—**TAKE THE SHUTTLE BUS.**

PROTECT PERSONAL AND UNIVERSITY PROPERTY

1. Record serial numbers, model and brand name, and description of valuable items in your room or office. Keep a duplicate copy in a separate location.
2. Never leave a wallet or purse lying on top of a desk or dresser. Keep it in a drawer or somewhere out of sight. Keep a record of your credit card numbers.
3. Require identification and authorization from service men wanting to do work in your room or office or wanting to remove items for servicing.
4. Never lend equipment or keys to strangers.
5. When leaving your room or office make sure:
    a. All windows are closed and locked.
    b. All valuable items are removed from the tops of desks and stored out of sight.
    c. All desks and files are locked, if so equipped.
    d. All doors are closed and locked—

EVEN IF YOU WILL ONLY BE GONE FOR A MINUTE.

6. Keep your car locked and **TAKE THE KEYS**. Don't leave packages and personal items in full view inside—put them in the trunk.

With staff in police departments and other law enforcement agencies at an all time high, crime continues to increase across the nation. Therefore, any agency charged with the protection of life and property and the prevention and detection of crime welcomes assistance from every and any source. This is particularly true on college campuses and it is certainly true at Syracuse University. Therefore, the following suggestions are being offered so **YOU CAN HELP US HELP YOU.**

HOW YOU CAN HELP

1. Be alert and observant. Make note of persons committing suspicious acts, and notify your supervisor, R.A., or call Security direct, ext. 2224 or 2225.
2. Report all locks, windows, and doors in need of repair, and lights which are out to Physical Plant.
3. Be security-conscious at all times.

Remember, effective safety and protection of life and property is the responsibility of all members of the University community. So, if you witness a violation of the rights of yourself or others, call Security and provide as much identifying information as possible:

> Car license, make, model, color, individual marks, i.e. rust spots or dents; description of property stolen; persons involved—names, sex, age, height, weight, clothing, distinguishing features; facial scars, large nose, hair style, etc.; method and direction of travel.

Once a crime has been committed, the best aid in its solution is the rapid and accurate reporting of all pertinent information. Go to or call Security in Sims IV, ext. 2224, immediately after the commission of a crime.

Too often young people fail to report an offender because they are too kind hearted, and dislike the idea of getting the offender into trouble. Just remember that if you don't report him, he'll probably get into worse trouble later on—to say nothing of the harm he may cause.

OPERATION IDENTIFICATION

Operation Identification is an anti-theft program that is spreading rapidly across the country. Its aim is to deter burglars, and—failing that—to help police recover and identify stolen property.

The program operates this way:

1. Electric engravers can be checked out from the Security Office for a 12-hour period. An ID card or a deposit of $5.00 must be left.
2. Engrave your social security number on all easily movable or portable items of value.
3. List your valuables room by room so that you don't overlook anything. Keep the following record of articles with brand, model, and serial number, where it is safe and easily found.
4. When you return the engraver to Security you may purchase stickers that identify your valuables as being marked and readily identifiable by law enforcement agencies.

To help identify stolen property recovered by the police the Security Office will keep a list of students by social security number. Thus, recovered property can be matched with the student master list.

BICYCLE REGISTRATION

To help deter bicycle thefts and to aid in identifying and returning lost or stolen bicycles, all members of the University community are encouraged to register their bikes with the Security Office. Bikes may be registered Monday through Friday from 9:00 AM to 4:30 PM. There is a $1.00 registration fee to cover the costs of administering the program. Each bicycle will be issued a numbered sticker. When registering bicycles you must have the following information: make, model, serial number, color, number of gears, special equipment, and wheel size.

It should be noted that bicycles must be licensed with the Syracuse Police Department. If the bicycle is not licensed it is a violation of the City Ordinance, subject to a $10.00 fine under Article 14, Section 10: Article 19, Section 2.

Bicycle Safety Regulations

1. Have adequate brakes in good working order.
2. Have reflectors, front and rear.
3. Have your bicycle equipped with a horn or bell, in good working order.
4. If you ride one-half hour after sunset or one-half hour before sunrise your bicycle must be equipped with a white or yellow light on front which will be visible from a point 500 feet ahead of you.
5. Always use proper arm signals when turning.
6. OBEY all Traffic Regulations, such as red and green lights, one-way streets, and stop signs.
7. Don't ride out of alleys or driveways until you have slowed up and looked both ways.
8. Don't carry extra riders on your bicycle.
9. Don't try to beat an automobile across an intersection.
10. Don't race or speed.
11. Don't hitch onto buses or other motor vehicles.
12. Don't attempt to ride your bicycle without at least one hand on the handle bars.
13. Don't cut in and out of traffic.
14. Don't forget to lock your bicycle.
15. Always ride carefully.

TRANSPORTATION OF STUDENTS

Security patrol cars will transport sick or injured students within the following guidelines:

1. Emergency requests for students who are sick or injured to be transported to the infirmary must come from an R.A. or director if one is available. If a student is unable to meet the car at the sidewalk an ambulance will be called. The individual student will be responsible for the cost of the ambulance service.
2. Upon written authorization from Dr. Day, Director of Student Health Service, injured or ailing students will be transported from the living center to campus and back once a day. More than one round trip per day per individual would pose an undue strain on Security manpower.
3. It must be realized that transportation requests cannot be Security's first priority. If an individual has to meet a time schedule he should call a few minutes in advance. Requests will be accommodated as quickly as possible.
4. When you call be sure to specify if this is a transportation request to class, a sick or injured student, or a medical emergency requiring an ambulance. Give as much information as possible. Security does not transport students to doctor or dentist appointments.
5. Remember you may not be the only person in need of help and in all but emergency cases you will have to wait your turn.

FIRE SAFETY

Recommended Good Practices for Fire Safety and Loss Prevention

Fire Reporting Procedures

If you smell smoke or suspect a fire, sound the fire alarm if there is one in your building. If there is no internal warning system, warn occupants by shouting or in some other manner if possible.

If a campus phone is available, call University Fire Control—ext. 2323 or Security ext. 2224. If only an outside line is available, call Syracuse Fire Control—471-1161.

If it is possible to get out of your room into the hallway, leave immediately by the nearest exit. To determine if you can get out into the hall from your room, feel the door or doorknob. If they are too hot to the touch, do not go out of your room. If you can, upon leaving your room, close the door.

When outside of the building stand clear of it so that fire and police and emergency crews may operate without hindrance.

Unless the fire is of a minor nature or in a beginning stage, do not attempt to fight the fire yourself.

Fire Drills

Fire drills are mandated by the State Education Law to orient occupants of buildings, residence halls, and classrooms to conditions of fire, smoke, and other emergencies. Records must be kept and sent to the Fire Marshall's Office for inspection by the state authorities.

Other Suggestions for Life Safety

1. Keep all fire corridors and stairwell doors closed at all times. This will prevent smoke, fire, and toxic gases from entering and spreading to other areas of the building. Smoke and toxic gases kill more people than the actual flames themselves. Rugs, drapes, and furniture are now made of material that gives off acrid smoke and very poisonous gases.
2. Fire alarms and other warning systems are in buildings for the safety of the people living or working therein. Tampering with the fire bells, pulling boxes, wiring or power supply panels is prohibited by law. It is a misdemeanor according to the city fire codes. Pulling fire alarm boxes to cause a false alarm may eventually cause a false sense of security, cause injury or loss of life to firemen, or result in the failure to evacuate the building in case of fire.
3. Tampering with, discharging, or stealing fire extinguishers may cause loss of life or injury to the occupants of the building.
4. Never use an elevator in case of fire or during a fire drill. In case of fire the elevator shaft acts as a chimney flue.
5. Keep all corridors and stairwells free of storage and other obstructions.
6. Do not use candles, hot plates, or unapproved electrical items. Use approved electrical appliances safely and properly.
7. Do not store flammable liquids, turpentine, paint, gasoline, or ether in residential buildings.
8. Do not use the "octopus" type of electrical cords. This overloads the circuits and may cause a serious fire.

Please be careful. Fire not only hurts, but kills. Anything that has happened can happen again.

FIRE ALARM PROCEDURE

Whenever a fire alarm rings in any University building it must be assumed to be real. Any other assumption could lead to a tragic loss of life and/or property. Therefore, the following procedure is to be followed in all cases of fire alarms.

1. If you hear a fire alarm sound, call University Fire Control—ext. 2323 or University Security—ext. 2224. If you are at an outside phone, call City Fire Control—471-1161.
2. Evacuate the building **IMMEDIATELY,** and await the arrival of the Fire Department.
3. No one is to re-enter a building or shut off the alarm system until approval is received from the University Fire Marshall or City Fire Department. These are the only people qualified to determine the cause of the alarm.
4. After approval is received to re-enter the building, the building staff may silence the bells. At this time, Security will notify the University electrician who will reset the system and check for any possible malfunction.

OPERATION IDENTIFICATION

Use this form to record serial and model numbers of items of value in your household.

ITEM	BRAND NAME MODEL	SERIAL NO.

KEEP IN A SAFE PLACE WHERE IT CAN BE FOUND IN CASE OF NEED

Appendix 8E—Figure 1.

If you find a fire in a building, the following procedure is to be followed:
1. Sound the building's fire alarm system and notify Security.
2. If the fire is small and an extinguisher is present, an attempt to control the fire may be made. However, Security must be notified first even if the fire is minor.

EMERGENCY PHONES

Outside emergency telephone call boxes are located, as shown below, throughout the Hill area. These may be used to call either the Syracuse Police or Fire Department. Give the exact location and nature of the incident to the dispatcher. If the trouble is not located at the call box, wait there to direct the police or firemen to the trouble area.

Call Box #	Location
324	Madison and Walnut
328	Harrison and University
332	Adams and Comstock
339	Veterans Hospital—800 Irving
340	Crouse-Irving Memorial Hospital, 820 S. Crouse
341	Huntington Hall, University and Marshall
342	Waverly and Walnut
343	University Place and University Ave.
344	University Place and Comstock
345	College Place, 100 block—Slocum Hall
346	Euclid and Ostrom
347	Euclid and Lancaster
362	Colvin and Comstock
363	Colvin and Lancaster

Note: The fire alarm systems in the residence halls and campus buildings are not connected to the City Fire Alarm System. The Syracuse Fire Department must be called by a phone or by using a fire alarm box.

Campus Phones:

Security	ext. 2224
Fire	ext. 2323

City Departments:

Police 473-3555
Fire 471-1161

BAIL PROCEDURE

Any SU student who is in need of borrowing bail monies in the Syracuse area should contact the Student Association—476-5541 ext. 2650. The Student Association has a limited bail fund available to students.

APPENDIX 8F

SECURE CORRIDOR PROCEDURES FOR HIGH-RISK CRIMINAL COURT PROCEEDINGS*

When it has been determined that a secure corridor and courtroom are necessary the following procedures will be implemented:

1. The 4th floor west corridor and 4th floor west courtroom will be used when it is necessary to set up a secure corridor and courtroom.
2. One hour before court time, the corridor and courtroom will be cleared of all persons by two uniformed deputies.
3. All restrooms and closets will be checked and cleared.
4. The doors leading from the corridor to the stairway will be secured to entry from the stairway.
5. A barricade will be placed across the corridor at the Sheriff's elevator entrance.
6. A metal sensing device will be placed at the barricade.
7. The deputy at the barricade will check each person entering the secure corridor with the metal sensing device.
8. The deputy at the barricade will check all packages, handbags and briefcases that are being brought into the secure corridor for weapons.
9. The deputy at the barricade will issue colored badges to all persons entering the secure corridor. He will instruct the persons that the badge must be worn at all times while in the secure corridor.
10. One color badge will be issued to those persons working and/or having business with the offices which open off the secure corridor.
11. A second color badge will be issued to those persons having business in the courtroom.
12. All colored badges will be returned to the deputy at the barricade upon exiting the secure corridor.
13. If it is necessary to set up a secure corridor for more than one day, different colored badges will be issued for each new day.
14. Any person found on the secure corridor without a badge will be immediately removed from the secure corridor.
15. The offices which open off of the secure corridor will furnish each day the deputy at the secure corridor entrance with a list of persons having an appointment with that office.
16. If the offices cannot furnish an appointment list, a Bailiff or Bailiffs will escort persons having an appointment to that office.
17. One deputy will be stationed at the entrance to the courtroom door to check and insure that all persons entering the courtroom are wearing the proper badge.
18. The deputy stationed at the courtroom door will provide any assistance to the prisoner escort in the courtroom.
19. The deputy stationed at the entrance to the courtroom will assist the prisoner escort from the Sheriff's elevator to the courtroom and from the courtroom to the Sheriff's elevator.
20. Ten (10) minutes before court time,

---

* Courtesy Post and Associates.

the prisoner escort will escort the prisoner to the courtroom by using the Sheriff's elevator to the secure corridor and then following the secure corridor to the courtroom.
21. The deputy at the barricade will close the entrance to the secure corridor when the prisoner is being escorted from the Sheriff's elevator to the courtroom and again when the prisoner is escorted from the courtroom back to the Sheriff's elevator.
22. At the end of each day's court session, after the prisoner has been escorted back to the jail and all the badges have been collected, the metal sensing device and barricade will be removed and the corridor will be opened.

These procedures were established after a physical security survey of the court complex and a determination made of the courtroom most readily adaptable to procedural protective changes. Cost considerations were considered most important and modifications to the facility were prohibited. Likewise, the availability of additional personnel was not possible due to cost limitations. These procedures will significantly reduce risk for the high-risk trial situation and increase security control over the areas involved.

Chapter 9

INDIVIDUAL PROTECTION PROGRAMS

THE BASIS FOR collective protective programs of all types and descriptions has been the perceived needs felt by individuals who arranged for or developed the means for its provision. Collective programs evolved from the benefits observed when more than one person acted to protect himself or employed whatever technology level there was available at a given point in time for protection. The responsibility for protection has been and still remains a matter for individual concern.

The role that the individual has played in devising or implementing protective programs has been governed by the society in which he lived and the governmental system present. Western civilization, and particularily those counties which assumed Roman law for a heritage and possess a Common Law basis for these legal institutions and social fabric, has placed protective responsibility upon the individual. As late as 1735 in England, the individual citizen living in a community had the legal responsibility to provide his service as a community protective agent on a regular basis. The assumption being that individual and collective protection flow from the same basic need to be safe and secure in person and property and that to insure this protection collectively was to provide it individually.

The involvement of the individual in active protection programs for collective purposes was changing between 1735 and 1829 which was the period before the development of the "new era in policing" developed by Sir Robert Peel, 1829. During this period, the individual could be called on to act as constable for a year or watchman for his street for the night.[1] However, money could be used to pay a substitute to perform this service. The responsibility for collective policing gradually was eroded by the hiring of paid substitutes to carry out the responsibilities of the individual who did not wish to be actively involved in collective protective measures. He paid a fee directly to the persons taking his place. This system of paid substitutes gradually became so institutionalized that various "parishes" or "districts" within cities, especially large urban areas such as London, levied a general fee tax upon all living in the affected area which paid for collective protection agents on a regular basis. The choice to become actively involved in protection had passed as an option to be exercised as a regularized private but collective protection effort to be paid for on a fee basis.

The development of the reform measures of Peel further removed the option of personal and individual involvement in protection programs even further from the individual to the collective body of a community. The option for protection programs were now virtually eliminated and replaced by the new police force for the city, with a tax levied for their payment. Individual "districts" could now "supplement" the regular police with additional, privately paid agents, but they must use and pay for the new "public" police for general protective needs. The responsibility of the individual and his involvement was thus weakened and subsequently eliminated from direct participation in protective programming on a collective basis.

The development of protection in the U.S. in urban areas, particularly Boston, New York, and other major cities was based on the British model which stressed collective and paid public systems. Individual protective programs were, however, employed prior to the development of formalized agencies on a basis similar to that of England. The fact that many of the early settlers were from England and brought the systems of social control and organization to the U.S. provided a similar background for development.

The rapidly expanding and industrializing United States, however, provided a stimulus for a more rapid testing of the British system. The ineffectualness of public paid policing programs and the growing demands for more protection paved the way for individuals to once again provide their own protection. Responsibility became an individual matter from the urban areas to the frontier. Protection became the province of those best able to provide it, whether they were public or private persons. Protection from Indian attacks, cattle rustling, bank robbers, and all other forms of violence and lawlessness was the province of all affected. The responsibility for prevention of crime and the protection of life and property was once again with the individual. The watchman, constable, and sheriff systems brought from England were narrow, unreliable, and ineffectual in providing the protection necessary for a growing nation.

The development of an effective policing system for the cities and rural areas gradually shifted the responsibility for protection from the individual to collective, public bodies, e.g., the police. Taxes were levied for their support and services were provided for all parties. The levels of protection were not always up to the

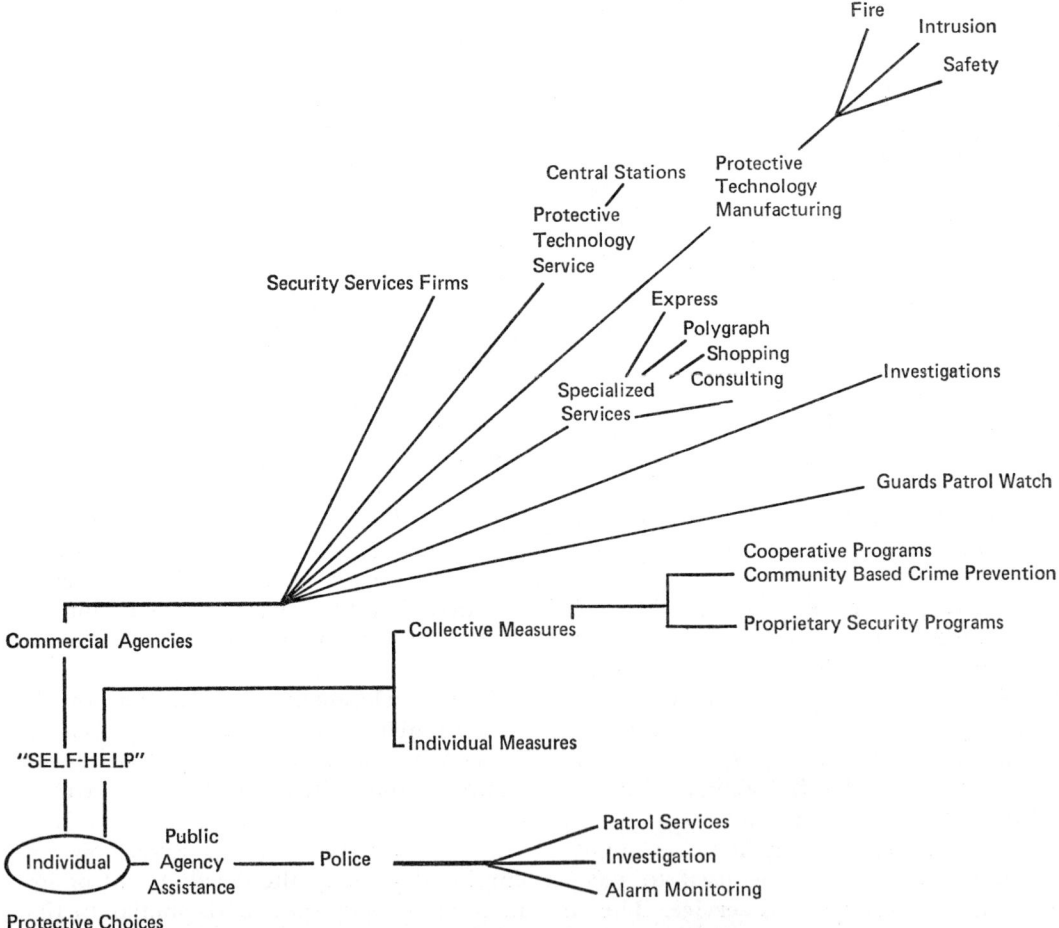

Figure 9-1. Individual protective services choices.

level required by many in the community who wanted more services or specialized protection. Consequently, the development of the private protection industry paralleled that of public policing.

The individual in contemporary society has a vast array of resources to choose from in protecting himself and property. (Figure 9-1 describes individual choices.) He may act as an individual or within a group to develop protective programs for his own use or for a community. The selection of alternatives available for the individual is made more difficult by the number of competing and conflicting methods and programs currently available. These programs include:[2]

1. Community-based crime prevention programs,
2. defensible space programs,
3. police crime prevention programs,
4. private security programs, and
5. individual measures.

COMMUNITY CRIME PREVENTION

Crime reduction is a difficult and complex task. It is not solely the responsibility of criminal justice agencies. In fact, one of the most important ingredients in controlling crime is the community itself. Each citizen can take simple measures to protect himself and his property, can be willing to report crimes to police and to serve as a witness or juror.

As part of the new Citizen's Initiative Program launched recently by LEAA, the National Institute is working to devise sound programs to stimulate citizen action in crime reduction. To provide more integrated and comprehensive research planning, the Institute has established a Community Crime Prevention Division to study ways to:

1. reduce opportunities for crime to occur;
2. develop a comprehensive community approach to crime prevention;
3. promote more effective citizen participation in the criminal justice process; and
4. develop a more meaningful response to the citizen as a victim of crime and client of the criminal justice system.

The National Advisory Commission on Standards and Goals defines community crime prevention as "any public or private activity outside the conventional criminal justice system that is directed toward reducing crime." The field encompasses a wide range of activities—including:

1. volunteer citizen participation in criminal justice programs;
2. improving locks, burglar alarms, and other security hardware;
3. programs to help crime victims to architectural design standards for improving building safety; and
4. neighborhood patrols to national media campaigns for educating the public on criminal justice issues.

Institute-sponsored research has addressed many of these issues. Various efforts in crime prevention include reducing criminal opportunities, increased use of security equipment, environmental design, lighting, and citizen and community action.

Reducing Criminal Opportunity

Research shows that the greatest portion of street crime and burglary—is the result of opportunity rather than of careful and professional planning. For example:

1. Someone sees an "opportunity"—in an open window, an empty house, a person alone in a dark alley—and acts on it.
2. Muggers look for likely victims, not specific individuals.
3. Burglars look for a house they can enter, not a particular address.
4. Preselected targets frequently are chosen precisely because they are seen as "easy marks."

Eliminating or reducing these opportunities is one way to cut crime. In particular crimes in which opportunity is a major factor, prevention and control strategies tailored to those crimes or lighting physical design for example, can be used to reduce a range of criminal opportunities.

Crimes of Opportunity—Burglary

Much of the Institute's research into specific crimes of opportunity has focused on burglary, which now accounts for 40 to 50 percent of all major crime. Institute projects have investigated what makes a site vulnerable to burglary, who commits burglaries, and how to protect or eliminate targets.

Patterns of Burglary, a study conducted in metropolitan Washington, D.C., found that very simple precautions—locking doors and windows, keeping lights on during an absence—could greatly decrease the risk of burglary. Interviews with burglary victims and nonvictims disclosed that victims were less likely to have taken extra precautionary measures. Researchers also found that burglary victims tend to be victims of other crimes as well.

The study identified *fencing*—the sale or disposition of stolen property—as the most critical element in the burglary cycle. It suggested that thwarting the fencing process could simultaneously increase the risks in burglary and reduce the monetary rewards.

The first Institute study of fencing (70-065-PG-10) explored the feasibility of using conventional marketing theory to understand the behavior of thieves and fences. The distribution of stolen goods was addressed, not as a criminal or societal problem, but as a straightforward marketing issue. This study indicated that thieves and fences face many of the same problems a legitimate businessman faces in matching supply and demand. Thus, their behavior is economically motivated.

The report included suggestions for methods to impede the fencing process. Lengthening distribution channels, for example, would raise the price of stolen goods. As prices approached the legitimate retail level, according to the report, the fence would lose his competitive advantage. Any action increasing the number of middlemen, the study concluded, not only would lower the incentive to trade in stolen goods but also would increase the probability of detection. Another deterrent recommended by the study was permanent marking of personal and household goods with individual identification numbers.

Merely reducing the profit margin in burglary, however, is not enough. Institute research has studied ways to prevent burglary by defeating the initial entry attempt. In conjunction with the Department of Housing and Urban Development, the Institute sponsored a two-phase project to develop architectural security guidelines for government-supported housing.

Because burglary is the most common residential offense, the project's first phase focused on its characteristics in *Crimes In and Around Residences*. Fear was found to be the most significant consequence of burglary. While burglaries are less dangerous than most stranger-to-stranger crimes, their lack of visibility and wide geographic range make them difficult to control. An effective attack on residential crime, the study concluded, must address the offender's motivation as well as methods to reduce opportunities by improving doors and locks and installing burglar alarms.

In the second phase (71-026-C2), the project surveyed and assessed different ways to prevent residential burglaries. It summarized the advantages and disadvantages of private group action such as tenant patrols and emphasized the importance of environmental design for safety—topics already under investigation in other Institute crime prevention grants discussed below.

The study concluded that *the individual was responsible for the security of his own residence.*

The government's role, according to the study, was to provide information for builders, landlords and residents, and to sponsor research. To help the construction industry build safety into homes, the National Bureau of Standards, under the Institute's Advanced Technology Program, is developing detailed performance and building standards for burglar-deterrent doors, windows and alarm systems. Research has shown that if entry can be delayed for only four minutes, a burglar generally will give up on that entry and often can be caught.

The Bureau of Standards also tests the effectiveness of burglar alarms in signaling the entrance of an intruder. Because false alarms are a major problem, special attention is being given to the sensitivity of alarms: can they distinguish a burglar from the neighbor's cat?

Burglar Alarms

Burglar alarms have long been available as a deterrent, but not until 1970 was their effectiveness systematically measured. In that year, an Institute-funded program in Cedar Rapids, Iowa placed alarms in 300 businesses and evaluated their performance over a one-year period. The inexpensive, silent alarm systems were connected directly to the police station through telephone lines.

Results were impressive. The probability of an arrest at the scene was five times greater at alarm-equipped sites. Total clearance rates were 46 percent for locations with alarms compared to 27 percent for those without. In only 11 percent of the cases where an alarm was received was there loss through theft, and those losses were minimal. Equipment failure never occurred during a burglary, although it did cause some false alarms. Human error was the major cause of false alarms, but researchers concluded that education could reduce it to an acceptable level.

The study reported the cost of the alarm system could be justified in a city's budget by the decrease in investigation and prosecution time alone. It pointed out that a system need not be complex or expensive to be effective: Very simple devices were used in the test. At the conclusion of the study, the system was given to the city of Cedar Rapids where it has been expanded and is in use today.

DEFENSIBLE SPACE PROGRAMS

Environmental Design

Environmental design is the development of coordinated design standards—for architecture, land use, street layout and street lighting—which improve security. Its goal is to create environments which reduce the opportunities for crime while encouraging people to use public space in ways that contribute to their safety and enhance their sense of community.

The first Institute project in this area examined the possibility of developing physical planning principles to reduce urban crime. Though limited in scope, its results are interesting in light of the findings of later, more extensive research. The study found:

1. Visibility in commercial store fronts had a direct relationship to the incidence of crime.
2. Physical design could encourage residents of housing projects to monitor outside activity.
3. Street lighting affected crime rates.
4. Building condition and maintenance appeared related to the incidence of crime.

In 1969, an Institute grant to Oscar Newman of New York University began the landmark research into the relationship between architectural design and crime prevention which culminated in the concept of *defensible space*. Analyzing over 100 housing projects across the country, the research identified four critical design elements that inhibit criminal activity:

1. the subdivision of public space into strongly defined zones of territorial influence, because the fewer people who use a hall or door the more protective their attitude toward it and one another;
2. the creation of opportunities for natural surveillance through the placement of doors, lobbies and windows because the knowledge—by both resident and intruder—that they can constantly be seen can allay fear and deter crime;

3. the removal of the stigma of institutional appearance; and
4. the recognition that the character of the surrounding neighborhood affects the safety of the project's public areas.

Environmental design, however, involves more than simply redesigning space. It includes the process through which redesign changes the residents' use of and attitude toward their "territory."

The most publicized evidence from Newman's work is the comparison of two New York City housing projects where the anonymous high-rises had 66 percent more total crime than the small, walk-up buildings. These findings were corroborated in other redesigned and renovated housing projects. Attractive exterior designs incorporated both visual and physical territorial barriers, such as curbs, lawns, porches, and fences. Each building was given its own clearly defined common area, and the grounds were redesigned to restrict access and discourage commuter traffic. Finding the grounds both comfortable and safer, residents began using them.[3] That use itself increased their security. As a result, both actual crime and the fear of crime declined by more than one-third.

Street Lighting

While improved street lighting has long been promoted as one way to reduce night street crime, large-scale, comparative evaluations have been lacking. To meet this need, in 1971, the Institute began to study the effect of improved street lighting in Kansas City, Missouri.

Using test and control sites, comparisons were made of the extent of night street crime before and after improved lighting was installed. Results showed that improved lighting can significantly reduce night street crime. Night robberies, for example, decreased 54 percent in urban test sites, while the decrease was 23 percent in control sites. In two years preceding the study, night robberies at the test sites had been increasing by over 50 percent while at the control sites they had been decreasing by over 30 percent. The study also found indications of crime displacement as a result of improved lighting and some suggestion that street crime may move not only to new locations, but also move off the street into residences and commercial establishments.

Citizen and Community Action

Meaningful crime reduction requires citizen cooperation and involvement. The formation of a Community Crime Prevention Division within the Institute is one recognition of the critical nature of the citizen and community role in effective crime prevention and control. In its early years, the Institute focused on research designed to reduce the physical vulnerability of various targets of crime and sought to improve elements of the criminal justice system. However, it soon became apparent that the most sophisticated locks and alarms were effective only if the public accepted and used them. Speedy trials were of little benefit if the crucial witness did not testify.

Recent surveys show that witness problems are one of the major reasons for the dismissal of criminal cases. To explore the factors involved in witness cooperation, a survey of witnesses in criminal cases in Washington, D.C. was conducted by the Institute of Law and Social Research. Preliminary results indicate that faulty communication is a major issue. Most witnesses interviewed told researchers that they wanted to cooperate, but some were unsure if their cooperation was needed. Many individuals recorded as witnesses by the police were unaware that they needed to appear in court, or did not know the time and place. Clearer communication from criminal justice personnel and greater public understanding of the criminal justice process are needed, the project concluded.

Encouraging public understanding of, and involvement in, the criminal justice system is one goal of the Community Crime Prevention Division's efforts to promote effective citizen and community action to deal with crime.

Volunteers

A direct form of citizen involvement is the use of volunteers in criminal justice programs. Studies have shown that volunteers can enhance the operation of the system if properly trained and if their programs enjoy the support of the host agency or institution.

The University of Nebraska evaluated the use of citizen volunteers as probation counselors for high-risk misdemeanant youth. Young, high-risk offenders were assigned on a one-to-one basis to carefully-selected volunteers who met with them weekly and were available for emergencies.

At the end of the one-year probation period, the volunteer-counseled youths had committed 45 percent fewer offenses than a control group of high-risk misdemeanants on regular probation. The offenses committed by the experimental group were less serious than those of the control group. The experimental group also had more successful ratings in personality and social competence measures than the high-risk control group. However, both high-risk groups were less successful than a control group of low-risk misdemeanants assigned to traditional probation.

The study found successful volunteers met regularly with their probationers, submitted monthly progress reports, were liked by their probationers and participated in planned activities with them.

Community Patrols

Community patrol is a somewhat controversial form of citizen action to deal with crime. To identify both the strengths and problems of such patrols, the Institute sponsored an examination of citizen patrols in the Boston, Massachusetts area. Researchers found these self-defense groups had organized in the belief that law enforcement was inadequate. This appeared to reflect community feelings because over half the citizens sampled in an opinion survey reacted favorably to the idea of private patrols.

Analysis of twenty-eight such groups showed that the longest-lived citizen patrols were those that worked cooperatively with the police, were subject to some police regulation, and were able to draw financial aid and legitimacy from the local government. Most groups, however, had appeared in response to a crisis and disappeared once that crisis was past.

New Research Programs

Citizen and community involvement in criminal justice and the responsiveness of the criminal justice system to citizen needs are receiving important new emphasis in the community crime prevention research now being planned and funded.

POLICE CRIME PREVENTION

In this era of limited resources and increased demands for service, all communities face a difficult task in determining their community protective services priorities. Community public security is one of many services offered by suburban communities. However, security is generally given the highest priority, and rightfully so, because it is a basic community need.

The improvement of any community security system must be designed to strengthen existing facilities and procedures which are useful, establish new procedures where none exist, and introduce new equipment and additional manpower when needed to supplement ongoing security activities.

Police Sidetracked

Traditionally, community security has been supplied by a law enforcement agency. While it is traditionally and legally required to fulfill a crime control function, studies indicate that an increasing amount of their time is spent providing a variety of general services. In many cities, the police department spends 75 to 85 percent of their active patrol time in noncrime control endeavors.

Crime control, as practiced by the police, embraces both crime prevention and crime suppression. Police Departments attempt to prevent crime through a wide range of activities. Educational programs for adults and juveniles, recreational activities for the young, and the extensive use of general visible patrol units are just three of the approaches that have been employed by the police in an effort to prevent crime. In the area of crime suppression, early detection of offenders and their rapid apprehension is the goal of police work.

It should be noted that even if police adopt these crime control procedures, there is no evidence that their efforts will be effective. Many suggest that the basic causes of crime are so deeply rooted that the police have only a minimal effect in its elimination. Research indicates that the ability of police to deter crime is limited. They are clearing major crimes against property at a rate of only one in five.

Recent analysis of the police function strongly suggests that other police tasks are also extremely important. Certainly, order maintenance (that is, the ability to keep the peace) is recognized as a significant part of police work, and the fact that service activities require such a large part of police department time makes it mandatory that police officials give them careful attention and high priority.

This multiplicity of competing and sometimes conflicting priorities has created a serious dilemma for police department administrators. Given a choice, most chiefs of police would probably place crime control as their primary goal, relegating the others to positions of low priority or even eliminating them. Yet, this is in some ways out of touch with reality.

Shortage of Resources

Most police departments are given too much responsibility without adequate resources. Public and private security chiefs have for years said, "Give us more men, equipment and facilities so we can do the job." While there is a necessity for security departments to have sufficient manpower, equipment and facilities, this is not the total answer to community security.

Security in the community is everyone's responsibility, not just that of the police department. It is unfortunate but true that citizens who suffer crime losses often inadvertently contribute or increase the likelihood of these losses because of their lack of knowledge in property protection or their failure to assume responsibility for the protection of their property. Citizen complacency and apathy in the protection of individual property is common in most communities.

Citizens have, over a period of years, transferred the main responsibility of property protection to public agencies. As a result, public law enforcement agencies are expected to assume both the citizen's primary and their own secondary responsibility of property protection.

Public law enforcement agencies were never intended to assume the entire burden of protection. Consequently, after a theft occurs, the businessman or citizen blames the local public law enforcement agency for their failure to protect and secure their property.

Complex Age

In today's complex society, the traditional approaches to community security no longer assure that a citizen or a group of citizens will function free of anxieties, disruptions, and loss of life and property. The answer lies in collective community security programs.

In a residential environment, the opportunities for assuring cost-effective security programs are not only possible, but imperative. The piecemeal approach to protection by individual residents in a community is not enough.

The use of security by one resident provides a greater risk for other residents since crime has been shown to move from areas of great security to areas employing a lesser degree of security. Thus, those not utilizing security services in the village are more likely to be victimized. It is an established fact that displacement of crime takes place when security is tightened. Crime is not prevented; it is moved to a more vulnerable location.

Team Effort

The development of Collective Security Programs on a community basis reflects the "economies of scale" concept. The value of a security program must be sold to residential owners who feel a need to participate in the program. All too often, participation on the part of reluctant home owners is the result of a tragedy or the discovery of a sizeable loss. Failure of all residents to participate in the security program can only result in increased losses for those who are reluctant to join.

The "ripple" effect is well documented in police work. In metropolitan public housing areas, for example, it is graphically illustrated. As "site hardening" took place through improved locks, lights and screens, crimes were shown to move to those locations without the added protection. The same effect is shown through saturation patrolling by police. When large numbers of police operate in an area, the crime rate drops. It is not reduced for the entire city, however, but merely shifted to another area. In sites where high-value items are protected, they are seldom lost; where security is lax or vigilance is low, high losses occur.

Three categories of crime and delinquency prevention have been identified. They include: *punitive prevention, corrective prevention,* and *mechanical prevention.*[4] Punitive prevention attempts to reduce crime by making the threat of punishment more evident. The enactment of new "tough" laws, reduced time for prosecution, etc. are examples of this concept being operationalized. Corrective prevention attempts to deal with the causal factors of criminal conduct as they apply to the individual criminal. Mechanical prevention attempts to place barriers between the criminal and the things most likely to be attacked. Target hardening, security devices, and increasing the probability of arrest are examples of this concept.

These three concepts are all utilized by public protective agencies and the criminal justice system to reduce criminal incidence. Mechanical prevention has, however, become the focus of police crime prevention programs and of most private protective services efforts.

The individual in society who wishes to protect himself has the option of selection of all or any of these three concepts for protection. As a practical matter, mechanical preventative methods are the easiest to implement for the private citizen and require the least amount of effort while demonstrating the most results. These concepts are, therefore, most often employed.

Public agencies of protection and justice, however, attempt to employ all three concepts since they have control over the processing of offenders as well as control over the system of apprehension. Even in this area the process of bringing about broad acceptance of alternatives to criminal activity is difficult at best with results not readily evident. Consequently, the principal thrust of prevention programs at the individual and community level has been mechanical through the mechanism of police crime prevention.

Police crime prevention programs in the U.S. were originally designed in 1971[5] to provide technical assistance to individuals to reduce the opportunity for criminal attack by mechanical means. Site hardening, security devices and procedures were stressed to displace criminal attack to other less well protected locations. Efforts have gradually shifted from solely mechanical protection to the development of community organizations to promote punitive and corrective prevention through collective efforts of the community. It was assumed, and correctly so, that collective community action would not be viable or sustainable without demonstrating the limited usefulness of individual measures which only stress mechanical means.

The involvement of police crime prevention programs with community development and civic organizations has increased considerably. The provision of police resources for community based programming allow a sustainable level of participation by citizens to develop. Programs such as Neighborhood Watch, Safe City, Nosey Neighbor, Crime Stop, Crime Check, and Community Action, all attempt to involve citizens in crime prevention and reduction in their own neighborhoods. These programs offered in conjunction with the mechanical prevention services of the department tend to promote a total community effort at loss prevention.

PRIVATE SECURITY PROGRAMS

The entire range of private protective services are available to the individual for his protection. The decision as to the type of protective services or devices is an individual matter. Guards, alarms, express services, investigators, locksmiths, and consultants are all available and willing to provide those services requested. The use of a guard for a private party, to patrol a neighborhood, or check a home while a person is on vacation are all regular uses of additional protection.

The installation of alarm systems, sophisticated locking devices, use of glass substitutes or increased lighting, all are attempts to lower the risk of criminal attack for the individual.

Likewise, the purchase of a guard dog or individual weapons provides the perception of protection for many people.

The relative degree of protection afforded by any or all of these private means for protection relates to the threat present and the potentiality of criminal attack against the protected property. A property or person who is well protected and not attacked might assume that his protection prevented loss. It might well be, however, that the person was never considered as a potential target for attack and thus the question of his protection being a deterrent was never tested. Conversely, a person with minimal protection might be subject to repeated attacks and assumes that more protection might have eliminated the losses. While this is possible, the determination of the attacker might have been increased by additional protective means and encouraged increased physical harm to the victim. Instead of just robbing a person, an assault might occur to overpower or disarm the victim.

The selection of private protective means must be done with care and consideration of the consequences of their use. Calling attention to property by having guard patrol might as well attract potential criminals as it provides a slightly higher degree of immediate protection. It is, therefore, important that the measures selected be appropriate and viewed not only for their short-range effectiveness, but their overall effect on the level of protection.

Individual Programs

In the final analysis, it is the choice of the individual in society that determines the type of protection he will utilize. Choices are made either with or without sufficient data to accurately assess the implication of the choices made. The degree of knowledge available at a given point in time directly affects the type of decision as does the quality of the available data. Consumer education in protective service choices is essential and critical in competent decision making.

The person who utilizes the public police as a means of protection because he does not know about any other is potentially not as well protected as the informed person who voluntarily choses the police to serve his protective needs. The ability to assess risk, anticipate and recognize potential loss situations, and select the appropriate alternative means for protection is essential.

The decision to prosecute an employee who steals or to terminate him is individual. The choices are either to use the public remedies available through the criminal justice system, or to just "get rid of him." In the first instance, the public system might be employed; in the latter, private means of resolving the problem was employed. The decision was individual. The choice to use public or private police or the decision not to prosecute but terminate the involved employees. If, however, public police are brought in to conduct the initial investigation, individual control over the outcome is lost or impaired. Consequently, the choice made directly affects the outcomes for the individual making the choice.

FOOTNOTES

1. See Appendix 9A for a comprehensive discussion of the history of crime prevention.
2. Adapted from a report by the National Institute of Law Enforcement and Criminal Justice, U.S. Department of Justice, Washington, D.C., 1975.
3. Defensible Space: Crime prevention through environmental design (CPTED).
4. Peter Lejins, "Recent Changes in the Concept of Prevention," 95th Annual Congress of Corrections, Boston, 1965.
5. The National Crime Prevention Institute, University of Louisville (Ky.) under an initial LEAA Grant of $200,000 to train U.S. Police Officers on the principles of crime prevention developed in England.

APPENDIX 9A

THE STORY OF CRIME PREVENTION*

THE EARLY DAYS

Before the seventeenth century, the idea of a disciplined preventive police had not seriously been considered. Despite the growth of serious crime in London and our larger towns, where it was unsafe to walk the streets in daylight as well as at night, action against crime was taken only after the crime had actually been committed and the rusty, creaking machinery of the criminal law was set into motion. All citizens were expected to play their part in the maintenance of law and order and in the pursuit and capture of criminals, and this underlines the first principle of crime prevention—public cooperation.

At that time, our only weapon against crime was terror, or the severity of punishment, and this can perhaps best be illustrated by the number of capital offences that were still on the Statute Book in the early part of the nineteenth century. In 1819, no fewer than 223 offences were still punishable by death, but, because of the continuing absence of an organised police system, the prospect of a criminal being detected was very slim and the likelihood of his conviction even more remote.

The idea of an organised body of professional police had always met with strong resistance.

Even Cromwell, the Lord Protector, was rebuffed in his efforts to set up a police system when, in 1655, he divided England and Wales into twelve police districts each under the command of a Major General. But such was the strength of the opposition to his idea that even the all-powerful Oliver Cromwell had to abandon his plans. When his system failed, the old practices were taken up again.

In the seventeenth century an attempt was made to improve London's security against crime by the establishment of a professional nightly watch. This consisted of about 1000 watchmen who were paid 1/- (5p) or less per night, and this rate of pay naturally limited recruitment to those who were least able to carry out their simple duties.

The eighteenth-century writer, Henry Fielding, described these watchmen, or "Charlies" as they were called after King Charles II who introduced the system, in these terms:

> They were chosen out of those poor, old, decrepit people who, from their want of bodily strength, were rendered incapable of getting a living by work. These men, armed only with a pole, which some of them are scarce able to lift, are to secure the persons and houses of His Majesty's subjects from the attacks of young, bold, stout, desperate and well-armed villains.

---

* F. W. Hudson, MBE (Courtesy Group 4 Security, London, England)

Contemporary records indicate that this is probably a very charitable description.

Henry Fielding was the originator of the change of attitude towards policing generally and towards the prevention of crime in particular. He was born in 1707 and in the first forty years of his life he made his name as a novelist, historian, and playwright.

In December 1748, Henry Fielding was appointed a Magistrate at Bow Street, London, and immediately made his presence felt in his efforts to deal with a worsening crime situation. He set himself two tasks: to stamp out existing crime; and the revolutionary one—not seriously thought of before—to prevent outbreaks of crime in the future.

To achieve these twin objectives, Fielding made it clear that three things were necessary: a strong police force; the active cooperation of the public—what he called a body of citizen house-holders, and the removal of the causes of crime and the conditions in which it flourishes.

These remain to this day the basic principles for the prevention of crime. Fielding spent the remaining years of his life in the pursuit of these objectives. Through advertisements in the press, he appealed to the public to report robberies and burglaries "to Henry Fielding Esquire at his house in Bow Street" and this began what might be called our first crime prevention publicity campaign. It was also the first time that a formal crime reporting agency came to exist in Britain, giving the authorities a clearer picture of the true crime situation.

In 1750 he took his first positive steps towards the formation of the police in London by appointing six carefully chosen ex-Parish Constables, later to be called the Bow Street Runners. But crime continued to increase and when, in 1750, Parliament set up a committee of enquiry to look into the causes of crime, Fielding prepared "an enquiry into the causes of the late increase of Robberies etc. with some proposals for remedying this growing evil."

In his recommendations, however, he deliberately ignored the main reason for the prevalence of crime—the lack of a preventive police. England was not yet ready for its police.

Before his plans for a police force could be completed, he died in 1754 and his blind half-brother, John Fielding succeeded him at Bow Street and furthered his brother's plans for a preventive police. He emphasised their preventive nature and wrote in a pamphlet on the subject: "It is much better to prevent even one man from being a rogue than apprehending and bringing forty to justice." While we might today quarrel with the proportions here, it is difficult to resist the good sense of the idea.

When John Fielding died in 1780, crime was still prevalent and it seemed that Henry and John Fielding's efforts had been in vain. But they must be credited with the idea of the prevention of crime as an alternative to the severity of punishment.

A Police Bill was introduced to the House of Commons in 1785 but was not approved, again because of the fear of a repressive police. Government Committees thirty years later rejected the idea for the same reason and it was not until 1822, when Sir Robert Peel was appointed Home Secretary, that we took a further step forward.

After six years, in 1828, Peel felt he had gained sufficient support for his plans and he induced a committee of the House of Commons to recommend the formation of the police. In 1829 the Metropolitan Police Bill passed through Parliament and became the Metropolitan Police Act—the first police marched out on to the streets of London and Fielding's hopes for the creation of a preventive police were realised. For the basic concept was preventive and the very first orders of the Metropolitan Police contained these words:

> It should be understood, at the outset, that the principal object to be attained is the prevention of crime. To this great end every effort of the police is to be directed. The security of **person** and property, the preservation of public tran-

quillity, and all the other objects of a Police Establishment will thus be better effected than by the detection and punishment of the offender, after he has succeeded in committing the crime.

In Britain, the extension of police forces to the whole of the country was completed by 1856 and, in pursuit of a policy of first things first, it can be no surprise that the increasing crime rate of the early part of the nineteenth century, and the anxieties of the law-abiding public, were responsible for the diversion of the major part of police attention to the detection of crime, to the discovery of the authors of offences against the law, to their apprehension, and to their appearance before a Court to be dealt with in accordance with the laws which created the offences which they had committed.

The police had to be, and had to be seen to be, in control of a worsening situation and sought to obtain the public expression of disapproval of crime by the detection and prosecution of criminals.

For the criminal element was an active, visible and threatening presence in the community. Moreover, the effects of criminal activity were—and are—measurable, not only in terms of loss to the individual or to the nation, but also perhaps in terms of human misery. In the same way, detective activity is measurable in terms of the numbers of crimes which are investigated, of the number that the police can clear up, of the number of criminals whom the police are able to arrest, the amount and value of the stolen property which can be recovered, and the sentences which can be imposed by courts of law, whether by way of withdrawal of liberty or by financial or other penalty. Furthermore, people everywhere prefer to direct their activities towards results which can be measured in some way or other, and the British police have been no exception.

THE EMERGENCE OF MODERN TECHNIQUES

For more than a century—from the mid-nineteenth century to about 1950—the trend in police work was away from organised preventive activity, whose effects, at least in the short term, are intangible and not capable of precise measurement. It is not easy to assess or record accurately how much crime our police services prevent; no reliable yardstock has yet been found.

The result has been that the end-product of detective activity, expressed as a ratio of crime cleared up to crime which is reported to the police, has for many people become a gauge of police success or failure, and public opinion about the police tends to rise or fall with the result of this calculation.

Throughout and beyond the nineteenth century, therefore, the strength and the efficiency of the detective agencies of the police service increased steadily and have continued to increase to this day, to such effect that last year saw the police of Great Britain clearing up more than 40 percent of the total crimes reported to them, and four times the amount of crime that was being cleared up thirty years ago.

Advanced technical aids of all kinds; the development of forensic science and medicine; the establishment and development of a crime squad network which now covers the whole country; vastly improved training methods; better communications; the more effective deployment of police strengths, which are now greater than they have ever been; and a higher degree of mobility, have combined to produce in Britain, and in other countries, the most active and the most efficient crime-fighting machines that the world has ever seen.

In Britain today, more crime is being solved, more criminals are being arrested, more property is being recovered, and more criminals are being dealt with by our courts than ever before in our entire history.

But is this enough? For, despite all the evidence of growing police efficiency and effectiveness, we have never had more crime. In 1971 the record and sinister total of one and one-half million serious crimes was reported to the police representing in terms of property values more than £74 million.

If there is one lesson to be learned from the outline which I have drawn, it is that the detection of crime by the police will not itself provide a total solution to the problem of crime. The whole history of law enforcement, if it teaches us anything, teaches us that detective activity cannot be relied on to put the majority of our criminals out of circulation.

Its value as a first step in the protection of society and its possessions is, to a large extent, related to the measures that society is prepared and able to take in dealing with criminals whom the police arrest and place before the courts of law.

Nevertheless, I believe it to be beyond argument that the detection and arrest of criminals is one of the most powerful deterrents against crime. Its real value lies not so much in the fact of criminals who appear before our courts as in the transformation of criminal success into failure.

We have a saying: "Nothing succeeds like success," and this has a special application to crime, for successful crime breeds further crime. It is a fundamental duty of the police to ensure that crime is an activity which is doomed to failure, that it cannot and will not succeed. Detection is one of the most effective, visible, and measurable branches of police activity which can decisively achieve this. The certainty of criminal failure by means of an equal certainty of detection must remain an objective of police forces everywhere.

But it is not their only objective, nor must it ever be.

Despite the preventive concept of the police organisation in Britain, and in contrast to the continuing development of detective skills and techniques, the prevention of crime did not even begin to emerge as a specialised and organised police activity until the end of the second world war.

At this time, in the face of a crime rate which was increasing more swiftly than our population, and which was fully engaging our detective staffs, there should be no surprise that police thinking turned towards prevention as at least a partial solution of an intractable crime problem.

But it was not until 1943 that the police in Britain began seriously to consider this. In that year, a small crime prevention exhibition was arranged by one of our police forces. Six years later, 1949 saw what was probably the first police publicity campaign to be held in Britain, when the chief constable of Brighton, recognising the value of good relations with the public, organised a public relations and crime prevention exhibition.

In 1950, the Home Secretary launched the first National Crime Prevention Campaign with the object of enlisting public support against crime. This continued into 1951 but was not to be repeated until 1965, and during the period 1950 to 1965, the use of central publicity was on a very limited scale.

However, in 1951, the police of the city of London and a small number of chief constables of county forces created small departments which were exclusively preventive in purpose, with the task of advising the public about protection against crime. Encouraged by initial successes and by the public response, the idea of a specialised preventive activity within the police service began to gain ground.

The attention of central government was attracted to these early activities and, in 1954, a working group was called together under the auspices of the Home Office, to "enquire into the methods adopted by police forces in the field of crime prevention and to report on the need for the appointment and training of police officers in this aspect of police work."

In 1956, this group presented the first government report on crime prevention methods and, for the first time in Britain, the outlines of a preventive policy were to appear: the police service was provided with a framework within which preventive thinking could develop. The report gave impetus to the growth of small but fully committed crime prevention units in police forces and the following years have seen a quickening pace in police preventive activity. Today there is not a police force without its own preventive organisation, staffed by trained officers, and able to meet a growing public demand for sound advice about the methods that can be used to give protection to people and their property against criminals.

This process was accelerated in 1960 by the appointment of a Home Office Departmental Committee to study the prevention and detection of crime under the chairmanship of Major Cornish. The study carried out by this committee dealt first with the question of the formal training of the police in the prevention of crime. There was unanimous agreement that such training should be provided without delay and, with the cooperation of the Chief Constable of Staffordshire, the Home Office Crime Prevention Centre was established in 1963 at Police Headquarters, Stafford. The Centre *(for which Mr. Hudson was responsible, as director, from 1964 until 1970. Ed.)* has gone from strength to strength.

Twenty-one officers attend each four-week course, and no substantial responsibility of the police in the prevention of crime is omitted from the syllabus. The course is constantly reviewed and updated to ensure that the most up-to-date preventive techniques are quickly brought into use, in order to meet the challenge of the changing pattern of risks. In this relatively new training arm, the police have the ready support of outside speakers from commercial and professional organisations.

All the technical aspects of prevention are dealt with in lectures on locks; safes; strongrooms; security glass; vehicle and premises protection; and intruder alarm systems, and each course visits a lock and safe factory and an intruder alarm central station.

Representatives of the security industry talk on the equipment and services which are commercially available and there are speakers from the fields of architecture, banking, and insurance.

Students are taught the skills of communication and the basic training at Stafford is extended throughout the police by the nucelus of trained officers who can impart their colleagues. By means of discussions and seminars for senior officers in the service, police "management" has been closely involved in the spread of the preventive idea.

In addition to training police officers, the facilities at Stafford were made available in 1966 for the first time to the insurance profession. As a result of liaison between the Home Office and the British Insurance Association, a short course in Crime Prevention was prepared for twenty-one Burglary Insurance Surveyors. This proved to be a highly successful venture and has helped to establish much closer contact between the police and insurance companies. A second course, with an extended syllabus, was arranged in 1967, and has been repeated each year since that time.

All aspects of publicity, using all available media, are discussed at Stafford and are brought into the service of crime prevention. Since 1965, the Government has financed regular national crime prevention campaigns which have been enthusiastically supported by intense local activity. The last of these was launched in December 1972 at a cost of around £300,000.

The immediate aim of the police crime prevention activity in Britain has nevertheless been a modest one: to raise security levels in property by persuading occupiers of vulnerable premises to take simple, sensible—and often inexpensive—measures to reduce the risk of crime. In a word, to secure the cooperation of occupiers to remove the opportunities which are available to criminals in this time of affluence, when such opportunities and incentives for crime are more numerous and greater than ever before.

By far the greater volume of crime, as statistically recorded, is small crime which is so often committed because owners of property by, their carelessness, ignorance or negligence, present the criminal with the opportunity he is forever seeking. In the long term, the objective is to initiate a changing attitude in the public towards the security of both people and property.

Now, because of specialised training in the principles and application of preventive techniques, the police in Great Britain are better equipped and organised than at any time in our history to assume and discharge their preventive responsibilities but, so far at least as crimes against property are concerned—and these form by far the majority of crimes which are reported to the police—it must surely be relevant to ask: Can the police anywhere, acting alone, bear the total responsibility for the removal of criminal opportunity and the prevention of crime?

THE POLICE AND THE PUBLIC

Every owner of property has a fundamental duty, not only to himself and his family or his employees, but also to the society in which he lives and of which he is a part, to take sensible, basic precautions to prevent the commission of a crime. To fail to do so is not only a rejection of a clear social duty, but also places him in the category of a passive, if involuntary, participant in crime.

Police crime prevention activity in Britain, therefore, seeks firstly to train the police in prevention and, by means of their efforts, to enlist the cooperation of the public. This preventive partnership is developed by example, by persuasion, by teaching, and by making available to the public sound advice about security at all levels of life—advice which is related to the risk of crime, and which is given by police officers who have received specialised training in all aspects of preventive work.

Perhaps the most convincing evidence of such a partnership has been the establishment of the Home Office Standing Committee on Crime Prevention, which was formed early in 1967. The organisations represented on this Committee include the Confederation of British Industry, the British Insurance Association, the Committee of Lloyds, the Trades Union Congress, the Association of British Chambers of Commerce, the National Chamber of Trade, the British Security Industry Association, the Committee of Clearing Banks, the Road Haulage Association, the Tobacco Industry, the Police Service and the Home Office.

All sections of the nation's commercial and industrial life are able to meet formally to discuss the problems of crime prevention and to make their recommendations both to the Government and to the organisations which are represented on the Committee.

Two subcommittees, one to consider "mobile" property and the other to consider "static" property have been established to report and make recommendations to the Standing Committee, and working parties have been created to consider and report on particular aspects of crime prevention in commerce and industry.

Reports from this committee structure are circulated to all interested organisations and a great deal of valuable work is in progress. Perhaps the most significant step so far has been the cooperation with the motor industry to install security devices on motor vehicles. From the beginning of 1970, antitheft devices were fitted to all the new models of motorcars and light vans and, from the beginning of 1971, all motorcars and light vans have been so fitted with such equipment as a standard accessory.

The Standing Committee has done much of the planning of our National Crime Prevention publicity campaigns and in 1968 strengthened the police/public partnership by the creation of crime prevention panels in all towns with a population of more than 150,000. These discussion panels, under the guidance of the police, are another example of public involvement in the prevention of crime.

But a partnership between the police and the public, however necessary I believe this to be to ensure the success of our policies of prevention, is unlikely to be fully realised or fully effective without a true sense of partnership within the police service itself.

Some people believe that there is, in police activity, a clear choice to be made between detection and prevention, and from it we are invited to draw the inference that a solution of our crime problem is to be found in the department of our selection. Such a choice cannot be seriously entertained.

Each of these activities makes its contribution to the success of the other. Who will deny the preventive implications of the detective activities of the police service? Who will ignore the effects of police preventive activity which, by making its contribution to the control of "opportunist" crime, aims not only at reducing its volume, but by doing so also permits a more concentrated and powerful attack to be made by the detective agencies of police everywhere against those infinitely more serious crimes in which the criminal, because of his skill, his determination, his resources, his ruthlessness, or his brutality, creates the opportunity for himself?

No such choice exists. The partnership of detection and prevention is already with us and must become even stronger, each extending and expanding its influence and effectiveness in support of the other. New methods of detecting crime are, almost daily, becoming available to the police as the result of developments being made in science and technology. Everywhere, developing preventive techniques are seeking to take the initiative from criminals who have held it far too long. This work must go on. New and more effective measures to influence and change public attitudes about security must be found.

In order to be fully effective, Crime Prevention Departments of the police must be sensitive to changes in patterns of crime and risk and must at all times strive to work ahead of the criminals. This is not the place to attempt an analysis of the causes of crime but there are many factors which can influence the crime rate and of which the police, in the discharge of their preventive responsibilities, must be aware.

It has been said that poverty is a basic cause of crime, but today's crime statistics have established beyond argument that crime is also the companion of prosperity. Today there is more cash and property capable of being stolen and quickly removed—all of which is a constant attraction not merely a skillful and well-organised criminals, but to persons of all ages in a widening range of society. Unemployment will affect crime figures; the economic situation will affect crime figures. A few years ago, a shift in world prices turned copper into a semiprecious metal overnight and accounted for the thefts of millions of pounds worth of this metal.

Population shifts will affect crime. The migration of populations from urban industrial areas into surrounding rural communities will change the patterns of crime and must be anticipated by the police.

Among all these variables, there is one constant: the desire of the criminal to get something for nothing.

The fight against crime in the future will be on a much wider front than many people outside the police service now realise. Today, police everywhere have to do their work without an entrenched moral demand from the public for high standards of behaviour and public order, and one of the greatest dangers which we face today is public acceptance of a high and rising crime rate as a normal adjunct to modern living. An element of violence has crept into crime today which was not there ten years ago and there are now unmistakable signs of criminal organisation which, a few years ago, were not even contemplated.

We live in a society which puts a high value on success, and consequently there are inevitable problems with society's failures. Many of these become police problems.

Our society is daily becoming more mobile, and the more mobile a society becomes, the less personal it is; the less it cares about what the neighbours think. Years ago, this was a brake on human conduct: it set standards of human behaviour and with its disappearance we have increasing problems for the police.

In Britain the prevention of crime has been defined as the anticipation, the recognition and the appraisal of a crime risk, and the initiation of action to remove or reduce it. This clearly states the functions of the police in crime prevention so far as the protection of property is concerned.

In practical terms, this imposes heavy burdens upon the preventive staffs of the police. Every police force today has its force crime prevention officer, usually based at Police Headquarters. It is his duty to organise the preventive effort of his colleagues within the framework of the policy which has been decided by the Chief Constable. He is, moreover, expected to advise on the formation of such policy and to communicate these decisions to all levels.

He must maintain a study of preventive and protective techniques. This will lead to much more than mere technical competence. It will lead to a real understanding of security problems and a thorough recognition and appreciation of the risks of crime. This in its turn will ensure the selection of the correct action and realistic recommendations which will always be related to the risks.

He will be supported by divisional crime prevention officers who are responsible for the implementation of decided policy, the survey of vulnerable properties and advice to members of the public. They will maintain contact with architects and others about the security of new buildings and advise on the inclusion of security measures as the building progresses.

The police are expected to organise a sustained programme of local crime prevention publicity, using all media, directed at all levels of the public, and will organise exhibitions with a similar purpose. They will present talks to the public on general crime prevention themes or specialised talks to selected audiences.

Contact is maintained with banks, the Post Office, architects, builders, insurance companies, industrial police, and all sections of the security industry, as well as with fire prevention officers to ensure that there is no conflict between the security of property and the safety of people.

But perhaps the time is coming when the social role of the police should be reexamined.

Some years ago, I visited several countries in Europe to examine police crime prevention policies. One of the most impressive features of my tour was the striking difference which I found in the continental police approach to young people. Besides a policy of the protection of property, the continental aim has been to prevent people from becoming criminals and not merely to prevent people from committing crime.

Already much valuable work has been done in Britain along these lines and I hope that the future will see a much closer cooperation between our countries.

I have devoted a large part of my professional life to this work both in the police and now in the security industry with Group 4 and I hope that others will share my view that the prevention of crime is not an end in itself. Nor is it some abstract or academic activity. It is concerned with the realities of crime and with people. Above all, its ultimate purpose is the improvement of the quality of life, and this is a matter of immediate and serious concern to everyone. We all want to be able to walk alone or with our families in the streets of our cities without fear of molestation.

We want to live in homes, not in fortresses, and to work in premises without risk of intrusion.

For these reasons the preventive policies of the police must be encouraged and allowed to develop. We must create the necessary partnerships between detection and prevention within our police organizations, between the police and the public and between the police of different countries. It is my belief that the powerful combination of these elements will bring us all a little nearer to the goal which we seek—the goal which Henry Fielding began to seek more than 200 years ago—freedom from crime.

BOOKS

Arnold, Peter: *Burglar-Proof Your Home and Car*. Nash Publishing, Los Angeles, 1971.

Barnes, R. E.: *Are You Safe from Burglars?* Garden City, N.Y., Doubleday and Co., 1971.

Brown, R. M.: *The American Vigilante Tradition: The History of Violence in America*. New York, Bantam, 1969.

Currer-Briggs, Noel, Hamilton, Peter, and Norman, Adrian: *Handbook of Security (Basic Work: Part II)*. London, Kluwer-Harraps Handbooks, August 1975.

Ellison, Bob, and Shipstap, Jill: *This Book Can Save Your Life!* New York, New American Library, 1968.

Feagin, Joe R.: *Home-Defense and the Police: Police in Urban Society*. Beverly Hills, Calif., Sage Publications, 1971.

Finesilver, Sherman G.: *Protect Your Life*. New York, Grosset & Dunlap, 1968.

Hair, Robert A., and Baker, Sam Sinclair: *How to Protect Yourself Today*. New York, Stein and Day, 1970.

Hall, Edward T.: *The Hidden Dimension*. Garden City, N.Y., Doubleday, 1966.

Kaufman, Ulrich: *How to Avoid Burglary, Housebreaking, and Other Crimes*. New York, Crown, 1967.

Levinson, Harry: *Organizational Diagnosis*. Cambridge, Mass., Harvard University Press, 1972.

Moolman, V.: *Practical Ways to Prevent Burglary and Illegal Entry*. New York, Cornerstone Library, 1970.

Newman, O.: *Defensible Space—Crime Prevention Through Urban Design*. New York, Macmillan, 1972.

Nonte, George C., Jr.: *To Stop a Thief: The Complete Guide to House, Apartment, and Property Protection*. New Jersey, Stoeger Industries, 1974.

Pawley, Martin: *Architecture versus Housing*. New York, Praeger, 1971.

Scheuer, James H.: *To Walk the Streets Safely*. New York, Doubleday and Co., 1969.

Ursic, Henry S., and Pagano, Leroy E.: *Security Management Systems*. Springfield, Thomas, 1974.

Wallard, Willis, T.: *How to Defend Yourself, Your Family and Your Home; A Complete Guide to Self Protection*. New York, Demakay Co., 1967.

PUBLICATIONS OF THE GOVERNMENT, LEARNED SOCIETIES, AND OTHER ORGANIZATIONS

Angel, Schlomo: *Discouraging Crime Through City Planning*. Berkeley, University of California, 1968.

Atlanta Commission on Crime and Juvenile Delinquency. Opportunity for urban excellence. Atlanta, February 1966.

Atlanta Urban Observatory. Impact City's Report presented by Sam Massell, Mayor of Atlanta, 1972.

Brill, William H.: Security in public housing: A synergistic approach. *Deterrence of Crime in and Around Residences*. Criminal Justice Monograph Series. Washington, Government Printing Office, June 1973, p. 26.

Burglary Prevention—Oakland City Ordinance. Police Department, Oakland, Calif.

Carman, John, LoPinto, Stephen, Morrissey, Ann, Rivera, Jose, and Fisher, Andrew: Housing Authority Security Patrol Evaluation Report. Hartford, Conn., Evaluation Unit, October 20, 1972.

Carter, Thomas Stanley: Crime prevention; notes for the guidance of police officers and security officers. London, Police Review Publishing Co., 1965.

Chamber of Commerce, United States of America. Marshalling citizen power against crime. Washington, Government Printing Office, 1970.

Crime Prevention. A paper evaluating crime prevention measures at local level by Ch. Supt. C. V. Hewett, Metropolitan Police.

Coates, Joseph: Overview: Urban community and its relation to security and crime. Proceedings of the Seminar on Urban Design, Security and Crime. Washington, U.S. Department of Justice, January 1973, p. 2.

Community Crime Prevention—Report of the National Advisory Commission on Criminal Justice Standards and Goals. National Advisory Commission on Criminal Justice Standards and Goals. 72-DF-99-0008 (Grant), 1973, p. 382.

De Vines, Hollis: Building security codes and ordinances. Proceedings of the Seminar on Urban Design, Security and Crime. Washington, U.S. Department of Justice, January 1973, p. 48.

———: Building security. Proceedings of the

Seminar on Urban Design, Security and Crime. Washington, U.S. Department of Justice, January 1973, p. 44.

———: Identification of personal property. Proceedings of the Seminar on Urban Design, Security and Crime. Washington, U.S. Department of Justice, January 1973, p. 68.

Eagle, Alyn Bush: Family and home protection against crime. College Park, Md., Executive House-Services, 1968.

Erie County N.Y. Technical Institute: Some ideas on crime prevention and public safety. A research project under the locational education act of 1963. Directed by George A. Lankes, Buffalo, N.Y., 1967.

Fairley, W.: *Improving Public Safety in Urban Apartment Dwellings: Security Concepts and Experimental Design for New York Housing Authority Buildings.* M. I. Liechenstein, A. F. Westia, 155 pp., ref., June 1971.

Home Office: Home office crime prevention newsletter. London, 1971.

Liechenstein, M. I.: *Reducing Crime in Apartment Dwellings: A Methodology for Comparing Security Alternatives.* 31 pp., June 1971.

———: *Designing for Security.* 17 pp., ref., April 1971.

Luedtke, Gerald and Associates: *Crime and the Physical City: A Pilot Study Prepared for the National Institute of Law Enforcement and Criminal Justice.* Detroit, 1970.

Malty, Michael D.: Evaluation of crime control programs. Washington, Government Printing Office, 1972.

Michigan State Housing Development Authority: *Security Guidelines.* Lansing, Mich., 63 pp., 1975.

Misner, Gordon E.: Community involvement in crime prevention. *Deterrence of Crime in and Around Residences.* Criminal Justice Monograph Series. Washington, Government Printing Office, June 1973, p. 44.

Na Strazhe, Poriadka: Leningrad—Auxiliary Police—Crime Prevention, USSR, 1969.

Nash, G.: The Community Patrol Corps: A descriptive evaluation of the one-week experiment. Columbia University Bureau of Applied Social Research. Mimeographed. May 1968.

National Institute of Law Enforcement and Criminal Justice: Crime and the physical city: Neighborhood design techniques for crime reduction. Springfield, Va., National Technical Information Services, 1971.

Newman, Oscar: Defensible space: Architectural design for crime prevention. *Deterrence of Crime in and Around Residences.* Criminal Justice Monograph Series. Washington, Government Printing Office, June 1973, p. 52.

———: Environment design. Proceedings of the Seminar on Urban Design, Security and Crime. Washington, U.S. Department of Justice, January 1973, p. 9.

———: Security personnel. Proceedings of the Seminar on Urban Design, Security and Crime. Washington, U.S. Department of Justice, January 1973, p. 21.

———: Architectural design for crime prevention. Inst. of Planning and Housing, 1973, p. 214.

O'Dell/Hewlett and Luckenbach, Inc.: Security Handbook for Residential Developments financed by the Michigan State Housing Development Authority. Birmingham, Mich., May 1974, 65 p.

Oregon Department of Justice: A plan, statewide, community-wide, country-wide crime prevention program for the state of Oregon. Salem, 1958.

O'Rourke, J. K.: Need for and projected contents of a suggested property security code. National League of Cities, 1967, 39 p.

Repetto, Thomas: Building security: Crime in and around residence study. Proceedings of the Seminar on Urban Design, Security and Crime. Washington, U.S. Department of Justice, January 1973, p. 32.

———: Future research directions. Proceedings of the Seminar on Urban Design, Security and Crime. Washington, U.S. Department of Justice, January 1973, p. 71.

Residential Security. Security Planning Corp. (J-LEAA-007-72) (Grant), 1973, p. 218.

"Rip Off" (film). Aptos Film Enterprises, Aptos, Calif.

Rothblatt, J.: New York City Housing Police Dept.—Vertical Patrol Procedures. New York Police Dept., 1969, p. 36.

Rykert, Wilbur: *Reduction of Criminal Opportunity.* Pittsburgh, National Crime Deterrence Council, Inc., 1971.

Rykert, Wilbur: Crime is a thief's business: Prevention is yours. *Deterrence of Crime in and Around Residences.* Criminal Justice Monograph Series. Washington, Government Printing Office, June 1973, p. 66.

Scarr, Harry A.: The nature and patterning of residential and non-residential burglaries. *Deterrence of Crime in and Around Resi-*

dences. Criminal Justice Monograph Series. Washington, Government Printing Office, June 1973, p. 78.

Sears, P. M., and Wilson, S.: *Crime Reduction in Albuquerque—Evaluation of Three Police Projects.* New Mexico Univ., 1973, 136 p.

Selected crime prevention programs in California. California Council on Criminal Justice, 1973, p. 106.

Stevens, Richard C.: Building security: Burglary prevention study. Proceedings of the Seminar on Urban Design, Security and Crime. Washington, U.S. Department of Justice, January 1973, p. 25.

Urban design, security and crime. Proceedings of a National Institute of Law Enforcement and Criminal Justice Law Enforcement Assistance Administration Seminar on April 12 and 13, 1972. Washington, D.C. 20530, NILECJ, 1973.

U.S. Army Material Command—Military Police; Crime Prevention Activities. Washington, 1968.

U.S. Congress—House Committee on Education and Labor, General Subcommittee on Education. The Safe School Act. Washington, Government Printing Office, 1972.

Wood, Elizabeth: Housing design: A social theory. New York, Citizen's Housing and Planning Council of New York, Inc., 1961.

Zane, Thomas Leeds: A crime prevention program for an Army post. September 1966.

PERIODICALS

Blanchard, Janette: Proposal for a model residential building security code. *Deterrence of Crime in and Around Residences.* Criminal Justice Monograph Series. Washington, Government Printing Office, June 1973, p. 1.

Blanchard, Janette for Oscar Newman: Building security codes and ordinances. Proceedings of the Seminar on Urban Design, Security and Crime. Washington, U.S. Department of Justice, January 1973, p. 52.

Barnard, Charles: The fortification of suburbia against the burglar in the bushes. *Saturday Review of the Society, 1:* No. 4, April 21, 1973.

Barnes, Robert Earl: How safe is your home from burglars? *The Washington Star.* (Reprinted in booklet by the *Washington Star,* n.d.) August 19-22, 1968.

Burglary boom, the. *Newsweek, 71:*73-74, January 29, 1968.

Calame, Byron E.: Community patrol: Los Angeles' planned police-Slumdweller "Buffer" in disputes. *Wall Street Journal,* August 2, 1967.

Carpenter, C. R.: Territoriality: A review of concepts and problems. In Roe, A., and Simpson, G. G. (eds.), *Behavior and Evolution.* New Haven, Conn., 1958.

Carter, Thomas: How I'd rob your house. *Life,* May 31, 1966, p. 29.

Church, Orin: Crime prevention: A stitch in time. *Police Chief,* March 1970, p. 52-54.

Cole, Richard B.: Crime prevention loss control. *Security Management,* May 1975, p. 46-47.

Conklin, J. E., and Bittner, E.: Burglary in a suburb. *American Society of Criminology,* 1973, p. 27.

Conner, Lawrence S.: When mourners go to services, robbers break in some homes. *National Observer,* January 4, 1971, p. 12.

Crime and crime prevention in England and Wales. *Security Gazette,* July 1971.

Crime Prevention—Part 2. *The Police Chief,* June 1967, p. 10-24.

Cummings, Harry: Protect your home from theft. *News and Views,* November 1963.

Darnton, John: Outwitting burglars is now universal pastime: Strategems in suburbs. *New York Times,* April 23, 1970, p. 39.

Davis, John E.: Preparedness and the vertical city. *Security Management,* March 1974, p. 14-15.

Fabbri, J.: Crime prevention—before or after the fact. *FBI Law Enforcement Bulletin, 42:* No. 1, 20-24, January 1973.

Furlong, William Barry: How to keep thieves out of your home. *Good Housekeeping, 167:*68-69+, July 1968.

Galub, Jack: Burglars will get you, if you don't watch out. *American Home, 73:*108+, September 1970.

Gulinello, Leo J.: What is the responsibility of local housing authorities for the safety and security of their residents? *The Journal of Housing,* Feb.-March 1973, p. 72-77.

Henley, Arthur: Making your home safe against intruders. *Ladies Home Journal, 85:* 66+, July 1968.

Home security program important. *Sentinel Star,* Sunday, September 29, 1974, p. D-38.

"Hot Cards" (film). Muyshens Madison, Inc., New York, N.Y.

How safe and secure is your home? *Better Homes and Gardens,* August 1969.

Hudson, F. W.: Crime prevention—the new

challenge. *Security Gazette,* September 1974, p. 332-333.

———: Crime prevention—past and present. *Security Gazette,* August 1974, p. 292-295.

Hudson, Fred: Story of crime prevention. *Canadian Police Chief,* April 1975, p. 34-35, 37-40.

Hughes, Donald R.: California answer to building security legislation. *Security Management,* July 1974, p. 33-41.

Jack, Robert C.: Burglary and security ordinances. *Industrial Security,* December 1965, p. 24.

Jewish vigilantes, the. *Newsweek,* January 12, 1970.

Kearns, Jack: Inviting burglars is illegal. *Security World,* January 1967, p. 29-30.

———: Oakland: Inviting burglars is illegal (two parts). *Security World,* December 1966, p. 23; January 1967, p. 29.

Kelley, Bennie: A pound of prevention! *Security World,* April 1975, p. 85-87.

Laurin, C. J.: The St. Johns volunteers. *Canadian Security Gazette,* May 1971.

Legal controls on neighbourhood defense organizations. *University of Pennsylvania Law Review, 120:*952, May 1972.

Marx, G. T., and Archer, D.: Citizen involvement in the law enforcement process: The case of community police patrols. *American Behavioural Scientist, 15:*52, September 1971.

Matorin, R.: Jewish defense league. *Phoenix,* July 11, 1970.

McArdle, E. C., and Betjemann, W. N.: Return to neighborhood police. *F.B.I. Law Enforcement Bulletin, 41:* No. 7, 8-11, 28, July 1972.

Miraval, A. J.: Skyscraper protection. *Security Management, 18:* No. 2, 8, 9, 13, 14, May 1973.

Neighborhood patrols and the law: Citizen's response to urban crime. *Fordham Law Review, 41:*973, May 1973.

Neighbourhood security program, the. *Signal One, 3:* No. 3, 2, 1971.

"Neighborhood Watch" (film). Charles S. MacCrone Productions. Aptos Village, Calif.

"Ounce of Prevention" (film). Aptos Film Enterprises, Aptos, Calif.

Penland, Jack: A community service challenge to ASIS members. *Industrial Security,* April 1971.

Powell, John G.: Denver meets high rise dilemma. *Security Management,* May 1975, p. 56.

Reading parental awareness campaign. *Crime Prevention News,* September 1975, p. 15-16.

Repetto, T. A.: *Residential Crime.* Cambridge, Mass., Ballinger Publishing Co., 1974.

Simons, J.: Security at the world's tallest buildings. *Security Management, 17:* no. 5, 34-36, October-November 1972.

Skyscraper security pitted against bomb terrorists. *Occupational Hazards,* March 1971, p. 50.

Stea, David: Space, territory and human movements. *Landscape, 15:*13-16, Autumn 1965.

Stevens, Richard E.: The high-rise building dilemma. *Fire Journal,* July 1971, p. 5-7.

Strobel Walter M.: The architect and security. *Industrial Security,* December 1970.

———: The architect and industrial security. *Security Gazette,* September 1970.

———: Planned security for high-rise buildings. *Industrial Security,* June 1971.

Stuart, Peter C.: Are the police helped? Security groups gain support but controversy still smolders. *The Christian Science Monitor,* January 16, 1970.

Teaching householders how to "buy time." *Security and Protection,* June 1975, p. 10-11.

Tronrud, J. J.: Some problems and solutions in policing high-rise buildings. *Canadian Police Chief,* July 1972, p. 19.

Weeks, J. K.: A concept of law enforcement: Vertical policing. *Police, 11:* No. 1, 1966.

UNPUBLISHED MATERIAL

Avery, James R.: Policing the vertical city, Master's paper. Michigan State Univ., 1970.

Boles, Jerome C.: Crime prevention: Premise security surveys, unpublished thesis paper. Michigan State Univ., 1973.

Corbett, Wm. R.: Crime prevention police investigations. Paper presented to C. C. Mahoney, Southwest Texas State Univ., August 3, 1971.

Gutknecht, Raymond: The community services unit—a new approach to crime prevention, Master's paper. Michigan State Univ., 1970.

Kingsbury, Arthur A.: A comparative study of educational programs for crime prevention in England and the United States, unpublished PhD. dissertation, Wayne State University, 1976.

Thompson, John T.: Preventive policing a modern concept of crime prevention, unpublished graduate B paper. Michigan State Univ., 1966.

Chapter 10

GENERIC SECURITY FUNCTIONS*

SECURITY, IN ITS BROADEST SENSE, fulfills man's basic biosocially expressed need for protection against the elements and frees him from the uncertainty of facing danger. It has been identified as a basic need of individuals, essential for their growth and development. Unless this basic human requirement is fulfilled, alienation, communication failures, and social disintegration will occur. The development of legal structures, contractual relationships, and family groups are attempts to provide a secure environment for the individual.

Security may be applied at a personal, national, or international level. The concept is essentially the same: to insure that the individual or group is free to conduct its activities in an atmosphere which maintains freedom from fear.

When viewed in the framework of an organization, security displays many of the same characteristics as it does in the case of an individual, state, or nation. It provides the mechanism for promoting the ability of the organization to conduct its activities in a state of freedom from internal or external shocks to its operating systems. Security should be considered as:

1. an institutional system for insuring freedom from fear for the organization and the individuals who comprise it, and
2. a management function.

PREVENTING SHOCKS

As a "shock preventing" mechanism, security must display those characteristics which insure the desired operational environment for the organization, as well as for the individuals working within it. Organizations have been established to optimize the performance of specified tasks. The environment can contribute to task performance to an extent, but if the environment reduces the ability of the employees to be productive, creative, and self-fulfilled, it is detrimental. It is essential that the most appropriate methods and techniques are utilized to create an appropriate productive environment. Figure 10-1 presents these functions within the content of an operating system.

Systems theory interrelates seemingly unrelated management functions. The application of systems theory to security functions presents the relationships between the various functions performed to *prevent shocks* and *minimize fear* within the organization. As a management function, security must be provided on a controllable, cost-effective basis utilizing appropriate security principles in the analysis, planning, implementation, and evaluation of programs of environmental design.

"Preventing shocks" and "minimizing

---

* Much of the material in this chapter originally appeared in the *Security Program Development Notebook of the International Association of Chiefs of Police, Public Security Center*. It was prepared by Richard S. Post and this revision is presented courtesy of IACP.

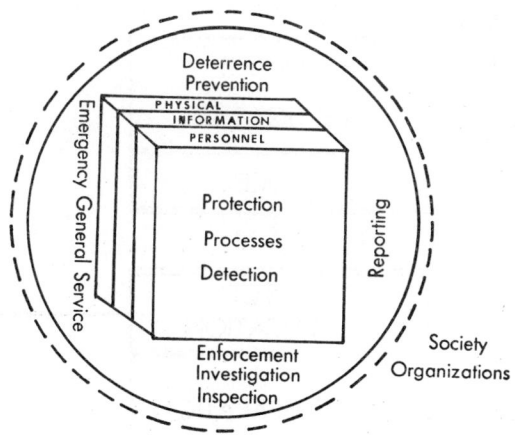

Figure 10-1. Basic protective functions.

fear" seem to be rather sterile terms to treat the contemporary problems of crime and safety which plague communities and corporations. The use of such terms, however, has a purpose. It is through their use that a systems perspective can be developed to assist in the reevaluation, review, and development of comprehensive security plans and programs. The narrow view of security as being a theft control, safety promotion, or employee protection program does not allow for a panoramic view of the opportunities to deal with *all* security needs comprehensively.

SIMILARITY BETWEEN SECURITY AND POLICE FUNCTIONS

The security function within an organization is similar to that of the internal security or police functions provided within the governmental framework of a nation or state. This function is to insure that the organization will continue its operations over a period of time without disruption or interference, and maintain a relative degree of freedom from fear. The laws which authorize role, function, and scope of action may differ according to time and place. The basic functions and the techniques utilized by the corporate security program generally follow standard procedure. The overall general function is to insure the continuation of the organization and its ability to operate.

When the actual tasks performed by either the public or private protective ser-

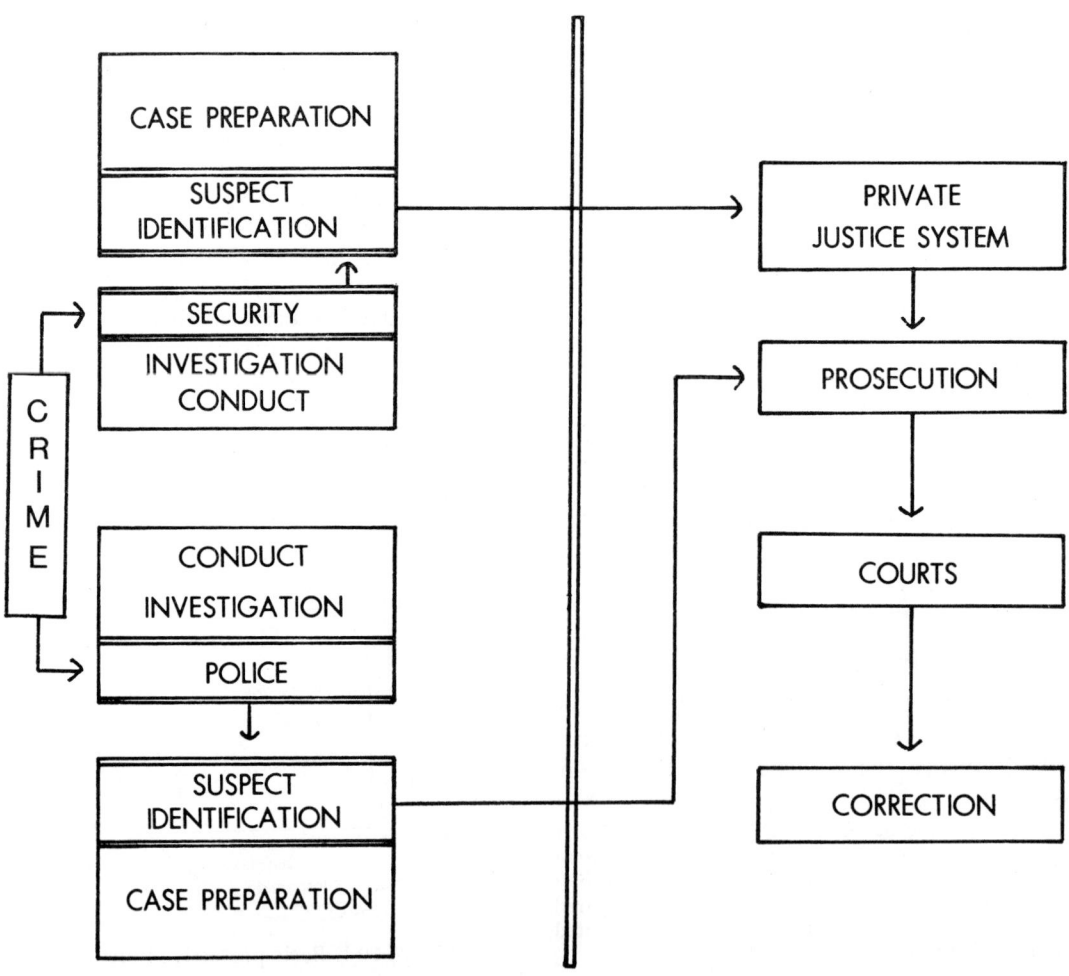

Figure 10-2. Comparison of current public and private systems.

vice practitioners are looked at, the current artificial distinctions based on the origin of "authority to act" becomes apparent. Figure 10-2 indicates the similarity between functions performed in either the public or private sector in relationship to an act designated as a crime. It can be noted that the steps followed by either of these groups of practitioners are quite similar. Their authority, however, is different. One derives its power from an individual, the other derives its power from a formal justice system operating in the United States. Clearly, the latter is in a superior position vis-à-vis the general public and the formal system of justice. It is thus considered to be a formally superior position and has in fact developed into a much more visible (although it is questionable that it is more efficient) and definitely more vocal grouping of protective agencies.

When, however, the tasks performed by these agencies in performing protective services are analyzed in either of the sectors, a significant number of the tasks performed are identical. With the exception of those tasks which are peculiar to the public agencies (because of their legal or statutory operating requirements), all tasks performed are identical in areas where the private protective service practitioner does perform, e.g. collection of evidence, following constitutional safeguards, interrogation, following accepted procedures, etc. Where there is no private counterpart, the tasks are most likely dissimilar.

The protective tasks performed are only different because of operating policies and procedures which form the context in which they are performed. In either the public or the private sectors these policies would form the major variables in distinguishing one set of tasks performed in the public from the same set of tasks performed by a private agency. Admittedly, the manner in which the tasks are combined in either the public or the private sector is quite different. Because of the varied operating responsibilities of the private and public agencies, the grouping of tasks for public police officers are likely to be much different than a grouping of tasks for a private protective services officer. This, however, does not weaken the position that the tasks performed are similar. It merely indicates that the way in which the individual tasks, the *work units,* are combined to form a total job is different.

PROTECTIVE SERVICES

Protective Services Technology

There exists what might be labeled a *protective services technology* in which the tasks, resources, processes, functions, applications, and organization of activities currently labeled "security" can be placed. Protective service technology is an inclusive term which provides a framework for analyzing the interrelationships between the components within the field currently labeled can be placed. Protective services technology is an inclusive term which provides a framework for analyzing the interrelationships between the components within the field currently termed "security," "loss prevention," etc. It refers to all those tasks described in the *Dictionary of Occupational Titles* as "protective services," thus including both the private and public forms of protection including law enforcement and policing. Thus, protective services technology becomes an "umbrella" term or the framework under which a systematic analysis of all dimensions of the function can take place.

Protective Service Functions

Protective service functions are those activities performed in furtherance of some specified protective services goal. The major categories are:

    Prevention
    Protection
    Enforcement
    Detection
    Investigation
    Deterrence
    Emergency Services
    Reporting
    Inspections
    General Service

Within these major categories all tasks performed by those engaged in the furtherance of a protective service goal take place.

Protective Services Processes

There are three basic processes within which all protective services take place. They are *physical security, information security, and personnel* (or personal) *security.* These processes constitute the framework for the series of actions or operation taken to produce protection for a specific thing, person, piece of information, corporation, state, or nation. These processes are applied in many different ways and with differing resources being utilized. Similarly, they are organized in a variety of ways, often reflecting ideology or policy guidance.

Protective Service Resources

Protective services resources are those persons, devices, techniques, procedures, design features, materials, and educational programs utilized to construct a protection program. They can be viewed in a systems theory analogy as the components of the program. As the program's "building blocks" or subsystems they can be integrated into a program in a variety of ways. Similarly, if the system's analogy is followed, it follows implicitly that as components, more than one of the resources is required to construct a viable program, thus making it imperative that the subsystem interface be correct and the system application be appropriate to the need.

Protective Services Organization

The organization of protective services resources into effective responsive protective services requires appropriate program design, administration, and management. The coordination, direction, supervision, inspection, and development of resources to provide programs is critical to program effectiveness. These management functions while closely related to ideology and policy are an integral portion of the protective services technology. Program management is thus a major point of interface between the political or ideological control structure for the technology and the technology itself. It must be considered as a facet of the technology in those areas which directly relate to the operational skills, resource decisions, process structuring and functional application of programs.

The three dimensions of processes, resources, and organization are inextricably related in a protective service technology. Clearly, the protective services functions are carried out through the processes of physical, information, and personal security. They are carried out utilizing the re-

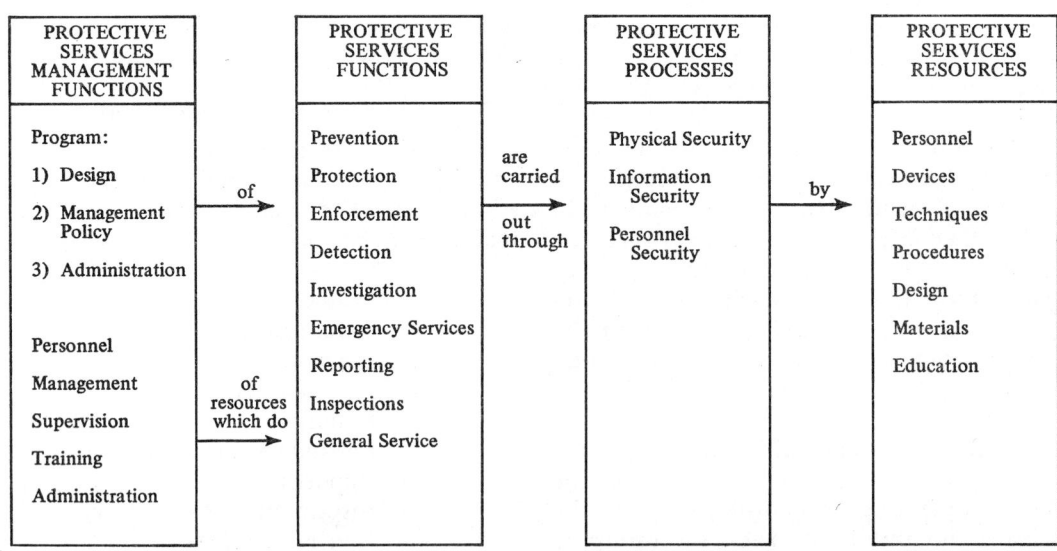

Figure 10-3. Protective services technology.

Prevention — establish risk-free environment

Protection — physical, information, personnel processes

Enforcement — environmental maintenance, (compliance with laws, rules)

Inspection — systematic reevaluation of environment (redundancy)

Detection — problem identification/location

Investigation — collection, evaluation, dissemination of information to users

Emergency Service — immediate response to problems
- Fire
- Safety

Reporting — systematic/logical transmission of information to users

Deterrence — barriers, time delay, features (procedures, equipment, personnel)

General Services — providing general services.

Figure 10-4. Protective services functions.

sources of personnel, devices, techniques, procedures, design features, materials, and educational programs. Figure 10-3 presents the relationships between the major components of the technology with the addition of the management functions associated with the technology. The emphasis of the management functions in this figure is on the personnel resources in the technology, and therefore is depicted by dotted lines rather than as a portion of the major conceptual framework.

Generic Functions of Protective Services Technology

The framework of the protective services technology provides a way of looking at the operations of the function. It is also necessary to delineate the generic functions performed by protective services for the individuals who are using it. The generic functions are presented in Figure 10-4. Each function has a specific role in maintaining the level of security desired.

Prevention

The prevention function attempts to avoid crimes, accidents, and other undesirable events which affect persons and property adversely. Prevention planning in anticipation of such hazards will remove or reduce losses due to these risks, creating blocks against such threats and occurrences, and maintaining a secure environment. A secure situation insures low-risk, low-loss, and effective production. Simply stated, the purpose of a prevention program is to reduce exposure and minimize loss.

Target

A potential target is an item, asset, procedure, policy, or facility that has real or relative value. The degree of a target's criticality varies greatly depending on the level of its value. This value is relative depending on the degree of the impact that the loss will have on other activities. Production loss due to theft is an example.

Vulnerability is measured in terms of ease of penetration. This is exposure and it must be measured to predict the loss potential. Criticality is weighed against vulnerability to determine risk. Items which have a high rating are to be treated with an equivalent level of preventive maintenance.

Target Environment

The environment surrounding a target determines its degree of risk. The target environment includes: (1) *physical factors* —buildings, building configuration, lighting, fencing, locking devices, alarm systems, and landscaping; (2) *procedural factors*—the guard system, the I.D. system, package inspection techniques, shipping and receiving procedures, and accounting methods; and (3) *actions of personnel*— careless acts and unnecessary exposure to personal risk. The latter is illustrated by the actions of victims of kidnappings which were not appropriate for their environment.

Desires

The desire for the commission of a criminal act against a target is necessary for it to successfully occur. The motivation and conditioning factors which lead to a person's desire to commit a criminal act are many and varied. In an organization it might range the entire socio-psychological gamut of not being promoted to being fired, not feeling properly compensated for work being done down to and including a feeling that a more equitable distribution of corporate assets is desirable.

Volumes have been written on criminal desire, causative factors, etiology, and desire reduction programs. Each prevention program must take into consideration those factors within the organization which might give rise to employee or other types of criminal desires and insure that appropriate safeguards are erected to minimize risk exposure from such desire.

Ability

The presence of a high-risk target and the desire to commit a criminal act against it are not sufficient for a successful criminal action. The added necessity is the ability of the criminal or perpetrator to successfully complete his act. The degree to which the ability to commit the act exists has a considerable bearing on the probability of a successful criminal attack.

The opportunist in committing an attack or the unskilled amateurs' ability normally does not pose as great a risk as the professional criminal or the determined perpetrator who sets out to attack the target. The amount of protection provided within an organization and the type of preventative programs established must take into consideration the costs involved in deterring those with different degrees of ability. The costs for reducing an attack by the opportunist and the amateur are normally quite low, but the costs accelerate in direct proportion to the increased professionalism of the attacker.

Figure 10-5 presents the elements neces-

> DESIRE + ABILITY + OPPORTUNITY = CRIMINAL ATTACK

Figure 10-5.

sary for a criminal attack. It can be seen that desire, ability, and opportunity are all essential elements. If one of these is removed, as in Figure 10-6, an attack will ei-

> DESIRE + ABILITY = NO ATTACK OR UNSUCCESSFUL ATTACK

Figure 10-6.

ther not occur or is less likely to be successful.

Opportunity

Figure 10-7 presents the relationship between opportunity and target. The combination of these two presents the degree of

Generic Security Functions

> OPPORTUNITY + TARGET = RISK

Figure 10-7.

risk that is present within the organization. Opportunities are those shortcomings, exploitable problems, deficiencies, weaknesses, inappropriate or inoperative policies, procedures or systems that are being used to conduct operations within the organization. Inasmuch as in general organization theory each asset can be a liability so each liability can be an asset. In the case of criminal attack, the organization's liability becomes an asset to the person with criminal desire. Each shortcoming becomes an opportunity for criminal attack.

It is, therefore, critical that these liabilities be minimized in a protection program through the preventative programming of the organization and the risk reduced or eliminated consistent with budgetary and policy considerations.

PROGRAM COMPONENTS

Time delay and protection-in-depth are the two major dimensions of a crime prevention program.

Time Delay

Program Components

The concept of time delay is essentially that of prolonging the amount of time that it would take for a successful criminal attack to take place. Each component of the prevention program should prolong the amount of time required for a criminal to successfully penetrate the organization, the target environment, and be successful in his criminal attack.

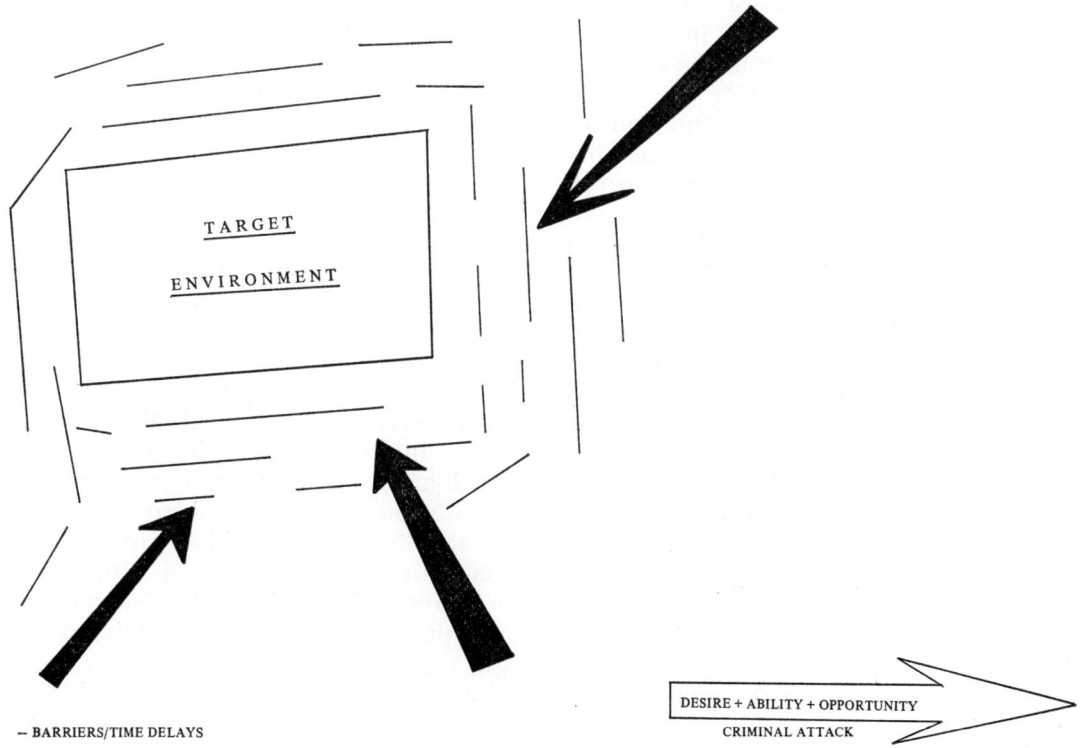

-- BARRIERS/TIME DELAYS
EFFECT OF BARRIERS/TIME DELAYS ON CRIMINAL ATTACK

DESIRE + ABILITY + OPPORTUNITY
CRIMINAL ATTACK

Figure 10-8.

Protection-in-depth

Protection-in-depth provides the system of barriers between the organization or the target environment and a person seeking to commit a criminal attack against it. These barriers may consist of the architecture of the building, fences, gates, locks, window materials, screening, alarm devices, guards, or any other physical or personal factors which would prevent or deter attack. Similarly, they include policy procedures, rules, and regulations, which insure the reduced ability of an unauthorized person to penetrate high-risk targets within a protected environment. Figure 10-8 presents the effects of time delay and protection-in-depth. It describes the value of these principles in deterring a criminal attack against a target environment.

TYPES OF SECURITY PROTECTION

The protection function involves the safeguarding of all assets of an organization. Assets are the physical, informational, and personnel elements of the organization. The traditional kinds of security equated with the protection function are physical security, information security, and personnel (or personal) security. These three types of protective subsystems provide the framework of barriers between the organization and internal and external shocks. They provide the subsystem organizational focal points for resource organization and allocation to fulfill security program requirements.

Each of the three types of security processes have specific characteristics which are examined in depth in subsequent chapters. A brief overview of their operations is, however, necessary in their section.

Physical Security

Physical security is often considered the "hardware" aspect of security programs. It most often relates to the establishment of barriers such as fences, locks, gates, vaults, alarm systems, sensory devices, as well as the use of guards, watchmen, dogs, and lighting. This security process (physical security) is to be considered as a key portion "time delay" factor in the establishment of a security program. Physical barriers make entry or penetration of a facility more difficult and thus more time consuming. The *criticality* of the assets to be protected, the *vulnerability* of the premises to be safeguarded will determine the amount of physical security to be provided and, thereby, the total cost of the physical security aspects of the program.

The application of physical security principles, techniques, and hardware to a specific asset protection (loss control or security) problem or situation requires that the balance be one of cost versus effectiveness. A $50,000 alarm system is not required to provide physical security for a facility that has a low replacement cost and contains little high-value material. On the other hand, the same system might be very inexpensive in a high-value, high-risk situation. The question is one, then, of determining the degree of physical security or amount of time delay required by the organization and insuring that the physical security process is tailored to that requirement.

Information Security

The information security process provides the organization with an ability to protect such things as product design, patent information, trade secrets, mailing lists, client and customer lists, and any other type of information which insures the growth and/or continued safe or profitable operation of the organization.

The information security process has traditionally made use of a combination of both the process of physical security and the fixing of individual personal responsibility for protection of proprietary (or secret) information. The Federal government has, for example, established the rigid regulations (industrial security regulations) for handling the safeguarding of classified information. These regulations prescribe the conditions under which information will be handled, stored, and safeguarded while not in use. While the kind and type of alarm systems, guard tours to be conducted, and specification of safes, vaults and cabinets are highly prescribed, the primary responsibility for safeguarding rests with the individual.

In the governmental programs for information protection, the legal sanction Title 18 of the U.S. Criminal Code with its severe penalties for violation provides and has provided a positive deterrent, in equal measure the patriotism and desire for protection of national secrets have worked together to minimize the risk to the protected information.

Both factors must have been present for the program to be as successful as it has in the past. The hardware and procedural aspects of the federal program have served as assistance to those charged with safeguarding the materials.

Industrial experience has shown that, unless both dimensions are present (hardware and procedures as well as employee motivation), a program will not be effective. The acceptance of the organization's needs for protection by the employee and his consideration that it is in fact important that protection be afforded the materials make the real difference between an effective program and one which is merely adequate. In many states, it is required that for a case of industrial espionage or theft of trade secrets to be proven, it is necessary for the company to show that the materials were in fact treated as secrets. This treatment includes the provision of security storage facilities, secrecy agreements, and knowledge by the employees that the materials were proprietary.

The information security process involves (1) the development of an accepting attitude by the employee to protect the information, (2) providing the environment in which the information can readily be protected, (3) insuring compliance with established policies and procedures for information handling, (4) providing sufficient barriers against attack or compromise by outsiders, and (5) investigation of all compromises and suspected compromises so that corrective actions may be taken to prevent future loss.

Personnel Security

Personnel security seeks to insure that the organization hires only those employees who can assist it in achieving its goals. It further assists employees to protect themselves and their property while participating in the life of the organization. The personnel security process is concerned with the welfare of the organization and that of the employee. It serves as a screening device to assist the organization in screening employees to ensure their desirability in meeting the goals of the organization. Typical activities include personnel and background investigation of prospective employees and that of current employees suspected of violation of company regulations. It further assists employees in protecting themselves through security awareness and educational programs. The activities of those engaged in personnel security are regulated by a myriad of federal and state regulations about the nature and extent to which background investigation can be conducted. Similarly, the rationale for nonhiring based on derogatory information developed during an investigation are subject to review.

The "wholeman" concept enunciated by the AEC and the DOD provide stepping-off places for the test of whether or not a particular person should be hired or deemed unqualified for reasons developed during an investigation. The organization's philosophy on personnel security should reflect its desire to get the most qualified person available to do the job desired. Compliance with the Equal Opportunities Act and local Non-Discriminatory Hiring Practices Legislation require that flexibility and trade-offs be made between the kinds of barriers previously afforded throughout the personnel security process.

The Three Processes Together

Physical, information, and personnel security form the organizational barrier system to shield itself against internal and external shocks. These three processes are always present in some combination or mix in every organization. The balance between them or the emphasis on one as opposed to another will vary considerably between organizations. They are, however, all present to some degree.

The decision regarding program emphasis on any one of the processes must be based on the need for specific types of safeguards, the criticality and vulnerability of the organization, its activities and assets. The relative costs and effects of each possible combination of the processes must be examined prior to application of them.

ENFORCEMENT

Introduction

Enforcement is a systems maintenance effort directed at obtaining compliance to loss control prevention regulations after all other means have failed. An enforcement program consists of highly complex activities, far more extensive than the application of rules and regulations to a given situation. Enforcement deals with human factors. The necessity to take any kind of enforcement action in the corporation results from the action of individuals or groups which deviate from established norms. These norms might be either criminal or civil law. Likewise, it might be a failure to comply with corporate policy. It is this failure to comply which necessitates the enforcement action.

The need for enforcement action in system terms is that the security program is not automatically self-correcting. That is, it is necessary since, as expected, people do not automatically correct their mistakes. In the nature of a police-man interface, the intervention of protective services officers to deal with and correct the situation is essential. It is through this function that the maintenance of the system is assured.

Enforcement Process

The enforcement process involves various levels of the organization, management, employees, unions, and protective services personnel. The efforts initiated at the protective services level deal directly with intervention in employee activities. Typical of these intervention activities are enforcement of loss or theft control procedures, badge control, parking, vehicle search, and the like. Similarly, protective services officers seek to reduce the possibility of violations by following well-conceived patrol plans and the strict adherence to established policy and procedural control systems.

Protective services administrators must, in turn, plan, organize, and support the activities of the officers. It is the responsibility of protective services management to (1) maintain, establish, and update record systems, (2) establish clear and meaningful policy, and (3) to provide other necessary tools to accomplish the job. This includes trained personnel, equipment, and related security hardware to insure "protection in depth."

Corporate management must further display its leadership by actively seeking out the cooperation of employees and insuring them of active support to assist in their individual loss control responsibilities. It also should establish a meaningful prosecution program and insist that all violators of the criminal law be prosecuted and that all violators of company policy be censured in some way.

Philosophy of Enforcement

Enforcement efforts are designed to create an awareness of the consequences of violating rules, regulations, and criminal laws within the organization. Enforcement measures are generally repressive. They aim to deter the potential violator by making the outcome of violating an established norm an unpleasant experience.

An enforcement program is based on the psychological principle that acknowledges the reluctance of man to willfully expose himself to unpleasant or painful experiences. For instance, if an employee is apprehended in stealing something from the corporation, some form of punitive measure by a court, a union hearing, or other administrative vehicle would normally provide unpleasant memories which would cause him to avoid a similar situation in the future.

Seeing other employees arrested or detained for committing an offense might deter other employees from committing similar acts. Similarly, the reputation enjoyed by protective services department for its energetic enforcement activities may have a bearing on the behavior of other employees in general.

Objectives of Enforcement

The immediate objective of an enforcement program is to deter the potential violator. A long-range goal is to increase the level of voluntary compliance with established regulations. These goals, though obtainable, cannot be reached until all employees become aware that enforcement efforts by the protective services department are continuous and are followed by prompt and decisive legal or administrative action.

The protective services responsibility is to effect action against violators and to apprehend as many offenders as possible. The organization, in turn, must be known for imposing penalties and insuring prosecutions in all cases.

Enforcement Deters

The high risk of being detected and apprehended, coupled with the certainty of an imposed penalty upon those found guilty, influences general compliance to established regulations. This belief is instilled in the following manner.

1. *Observation*

   When an employee sees an officer engaged in patrol he is reminded that he must obey regulations or suffer the consequences. Few people violate laws or policy in the presence of a security officer. The risk of apprehension is just too great. The reminder of the presence of security personnel will linger after the officer is no longer in sight. However, the length of time the deterrent affect will last depends upon:

   a. *The strength of the symbol of enforcement*—this is directly proportional to the enforcement practice of the department. The more vigorous are these activities and the more conscious of violations the officers are, the longer lasting will be the deterrent influences of the officer. If, on the other hand, enforcement practices are minimized or haphazardly conducted, the lingering effect will be short-lived.

   b. *The frequency with which the employee sees such symbols*—the frequency of observations relates to the availability of manpower to carry out the activity. Consideration of this fact contributes to the administrative decision of making enforcement the responsibility of the total protective services department rather than the task of a few individuals assigned to gates or other fixed posts.

Reputation of the Department

Long-range, comprehensive enforcement programs seek to generate voluntary compliance to existing regulations. When enforcement practices are uniformally and extensively conducted over a long period of time, employees become aware of the risks of apprehension. When these circumstances exist, an employee no longer depends upon seeing a protective services officer to be aware of the enforcement practices within the organization. The "reputation of the security of department will then become the most significant control factor."

The impact of an enforcement program depends in part upon what the employee thinks the security department will do and upon the reputation of the department. When the employee knows that security officers are actively engaged in enforcement, the observance of a security officer performing his normal duties will prevent some employees from committing an offense.

The Need for an Enforcement Policy

The enforcement program of progressive protective services departments should follow policy guidelines that stress uniformity of policy and procedure. This is especially important in large departments or in those organizations having a wide number of posts covering a broad geographical area.

Policy must be set by the security director. It must be explicit and inclusive. Ambiguous or equivocal policy statements detract from a total effort. Policy must deal in particular with all situations and be well communicated to all officers.

Officers in the field must know what is expected of them in "borderline" decisions and in the execution of their responsibilities.

Selective enforcement policies for a department can be used to concentrate enforcement activities at a particular time and place where losses have been determined to occur. The policy of conducting selective enforcement activities is a key to effective loss prevention.

Records and Enforcement Programs

Loss records must be available to provide necessary information to determine (1) the scope of loss control problems, (2) the identification of hazards and (3) the type of losses occurring.

Loss records assist enforcement effort by providing:
1. locations of hazardous or high-risk areas;
2. identifying the times at which losses are known to occur; or
3. identifying possible means by which losses are taking place.

In order to provide information for these records and for the use by officers in an enforcement situation, the observation of alertness of patrol officers is essential. Alertness, sensitivity to the unusual, the suspicious and obvious as well as initiative are needed qualities for effective performance.

Effects of Enforcement

The achievement of the Protective Services Department objectives require the support and interest of the entire organization. Many people, despite their normal patterns of conduct, violate established rules and regulations. The reasons for these violations are as varied as the individuals themselves. They range from a deliberate attempt to get away with excuses or shown disregard for policies or procedures, to those who inadvertently commit a violation. Nevertheless, the average violator is still a member of the organization and as such must be respected for the support and interest which he generally provides to the organization and its activities.

The act of violating a policy, rule, or law does not mean the individual has a total disregard for the organization. It is important that he be treated as a valued member of the organization. Thus, the responsibility for taking the enforcement action rests upon the officer and his businesslike and effective treatment of the violator.

Policies, rules, and laws are enacted to discourage types of behavior or conditions which give rise to losses, problems, or incidents. Disregard for these regulations represents an immediate threat to either the health, safety, or security of the organization. Enforcement action is taken so that (1) an evaluation of the improper action may be conducted, (2) appropriate measures to prevent future conduct can be initiated.

One of the desirable side effects of enforcement activities by the security department is their impact upon persons other than the violator. The sight or the story of an employee being stopped and apprehended for improper behavior serves to remind others in the organization that compliance with the law with regulations are fully expected. Awareness of the enforcement action taken by the protective services department serves as a reminder to pay strict attention to established rules and regulations. Many employees who are apprehended tell their friends and relatives about it. If there is no basis for criticizing the officer, all those who learn of the incident from the violator have greater reason to respect the efforts of the organization.

INVESTIGATIONS

Introduction

The process of investigation involves the collection of information regarding specific subjects, incidents, or occurrences. The investigative process deals with (1) activities and incidents involving personnel; (2) the interrelationships of personnel in a specific policy, legal, or procedural framework; (3) the conduct of inquiry regarding the ability of personnel to interact according to established rules, policies, or procedures.

Investigative process involves the collection, evaluation, and dissemination of information for a specific purpose. In a security or loss prevention program, there are many purposes for conducting an investigation. They, however, are used primarily to determine the reasons for a system failure, documenting the problems involved, the causative factors, and current conditions.

It should be noted that the investigative process is intended to be an objective subsystem for information collection. Normally, in the investigative process, evaluation of collected data takes place. The specific needs of the organization, the type of the incident or event being investigated will determine the type and extent of evaluation to be conducted. It may however, be conducted in the following three ways:
1. placing of raw data into some logical form for decision-making or legal action;
2. making an assessment of data regarding a situation, event, or activity, or
3. presenting the data along with an evaluation for correction of a specific situation.

Legal Considerations

The security department is subject to both policy and procedural as well as legal constraints in the conduct of its investigation. While not a duly constituted law enforcement agency with police powers, the security department nevertheless, must abide by court decisions and legal considerations in the collecting of its information. This is particularly critical in areas where possible prosecution, terminations, or law suits might be involved.

The use of illegal electronic surveillance, searches and seizures, or entrapment must be done in a way which can only be construed to be strictly according to established legal precedent. Similarly, policies and procedures for the administrative conduct of an internal security system must be such that they can be investigated using acceptable and legal means.

The security office is often required to conduct investigations and become involved in case building activities in acts that are both overt and covert. Overt acts such as a theft or an accident would be examples of the first category. Covert acts would be things such as industrial espionage or sabotage of corporate activities or property. In either of these two types of incidents, the activities of data collection using surveillance, undercover operations, intelligence collection, counterintelligence activities, and file building must be done using legal acceptable means.

Ethical

The ethical considerations in the conduct of an investigation closely *parallel* those of the department's legal responsibilities. While many areas of investigation differ as to their legality between the public and private sector, the ethics of utilizing various borderline or marginal techniques must be considered from an ethical viewpoint. The concept of "the ends justifying the means" while a cornerstone of many intelligence, undercover, and counterintelligence operations must be considered in the light of the responsibilities to individuals as well as the organization being served. This is a particularly difficult problem since statements by labor leaders have been made to the effect that the "greatest crime against labor is corporation of not making a profit." Thus, the *profitability* of the organization, its continued ability to exist and provide work opportunities for a large number of employees is of a very high priority. Thus, it becomes a question of how far an organization is willing to go to protect itself and its property against threats against its internal operations.

Similarly, the question of ethical conduct becomes a very direct issue for the security administrator in authorizing use of techniques during an investigation that is less than meeting the ethical standards which have been established for the organization, or that he personally establishes for the conduct of security investigations.

Administrative Aspects

The administrative aspects of the investigative process directly effect the organization's ability to deal with problems to be solved. The control and supervision over investigative case loads, assignment of personnel to handle cases, and the way in which organizations are divided between a generalist and specialist investigators all serve to affect positively or negatively the organization's ability to carry out its assigned tasks. It is directly related to the effects of legal and ethical considerations on types of methods used in investigations. At one time, the use of "third degree" tactics were considered very effective for obtaining information. Similarly, the use of truth verification devices such as polygraph, PSE and drugs are today considered to be investigative aids to reduce investigative time and to produce effective results. Their application, however, is governed by both legal and ethical as well as administrative considerations in the conduct of an investigation.

Cost-effectiveness

The cost-effectiveness of any policy, procedure, or aid used in the investigative process must be closely considered. The election of investigative resource, either internally or externally, has a direct effect on cost and efficiency of the conducting of the investigation. The proper selection is, therefore, critical. Similarly, the use of any investigative aid or, for that matter, the decision to conduct the investigation initially must be determined prior to its initiation. This, therefore, requires that planning for the investigative process takes place to determine, (1) if investigation should be done, (2) under what conditions they should be done, (3) what resources should be employed, (4) how they should be employed, (5) to whom should the investigator report, (6) what methods will be used for verification of investigative information, (7) how will the information from the investigative report be used in the decision-making process, and (8) what legal and ethical guidelines will be used during the conduct of an investigation.

Each one of these questions must be answered in advance of the use of this function within the organization. Not to do so would result in serious damage to the organization as a whole and cause irreparable damage to the security function within the organization.

INSPECTIONS

The inspection capability generates a cross-checking program within the system. It is a form of system redundancy which insures that the two related functions of enforcement and detection are operational and performing properly.

Inspections deal with the human-technology interface within an organization as well as with either of them separately. Inspections create the mechanism for an in-depth review of actual operations by evaluating their compliance with established policy and procedural norms. They, likewise, insure that all technology systems are operating according to established

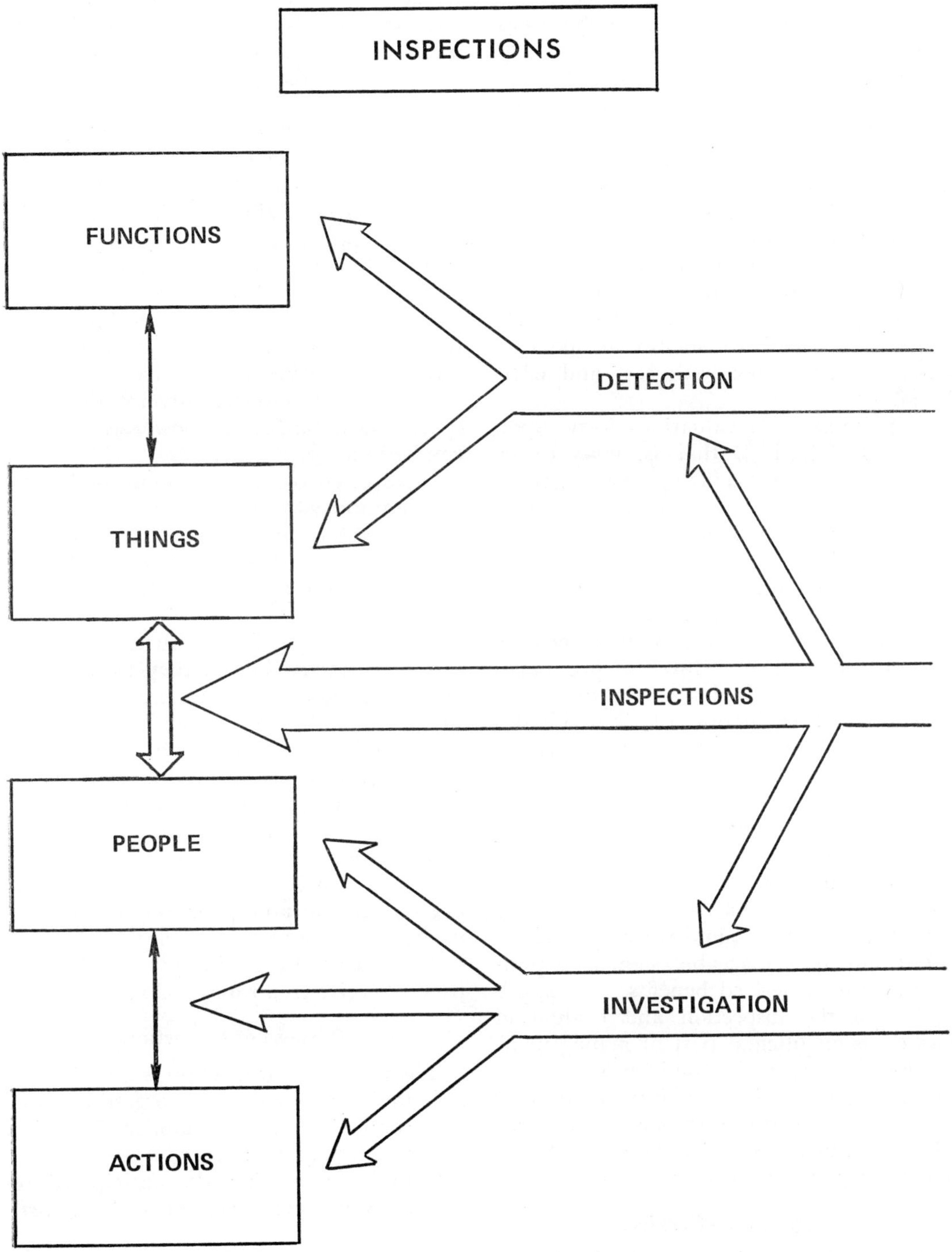

Generic Security Functions 485

Relationship of Inspections to Investigation and Detection

Figure 10-9.

specifications and tolerances. Moreover, it relates to the interface between these subsystems and attempts to determine the effectiveness and efficiency of the system design in light of operational problems (see Figure 10-9).

Inspections insure that a nonemotional, objective evaluation is conducted on a regular and systematic basis for all operations in the security (or loss prevention) program. All aspects must be evaluated on a regular basis. Such things as policy, procedures, equipment and its application, techniques, and methods should all be constantly reviewed for both cost and effectiveness.

Inspections and evaluations focus upon a set of objectives; that is, ends to be achieved to solve specific operational problems. These objectives are derived through a detailed analysis of operational commitments, and how those commitments are implemented within an organization.

Traditionally, inspections and evaluations have used the "system" to determine who should be fired or hired or promoted. This concept of thinking results in the inspection and evaluation of "individuals" as separate entities and fails to recognize the basic reasons they do or do not function as expected. To be effective, the inspection and evaluation process must encompass the total organization and its mission, and in some instances, reach beyond into other support activities. It must be realistic, meaningful, and workable. Above all, it must be manageable, showing a positive return in derived benefits. The application of the inspection and evaluation process is an integral part of management in the planning and organizing stages. Beyond that point, it is an integral part of directing and controlling operations at all levels within the security department and for the organization as well.

Inspectional Services

Just as the exercise of authority is not restricted to the top-level management, so inspection and control are not performed exclusively by any one person. These activities are constantly in process throughout the protective services organization. At each level of the organization, there must be planning, direction, and control over the activities.

The extent of each individual officer's participation cannot be defined exactly because it is influenced by the organizational structure, size of the organization, and its goals. In practice, the manager's participation is influenced by the amount of time he has available, his concept of the importance of a particular task, his interest in it, his confidence in the ability of his subordinates, and his own ability to delegate authority. His responsibility for their successful accomplishment, however, must be emphasized.

Misdirected or capricious use of inspection authority may do great harm. Controls, therefore, must be provided so that all who exercise this authority will be held responsible for the consequences of their actions.

Through inspections, failures and errors can be corrected before they assume serious proportions. Weaknesses can be ferreted out and often overcome by modifying procedures or instructions. Even more importantly, a proper inspection has a motivating effect on employees who know that their work will be scrutinized critically. They become more painstaking in their efforts and are less likely to be guilty of neglect, errors, and poor judgment. Similarly, inspections raise morale because members of the organization can then expect to receive credit for good work.

Purpose of Inspection

The purpose of inspection is fivefold:
1. to learn whether an organization is complying with established policy, regulations, and procedures;
2. to learn whether the anticipated results of these directives are being realized;
3. to discover whether the resources of the organization are being utilized to the best advantage;
4. to reveal a need to alert management

to problems which cannot be solved at the line level; and
5. to discover the solution to those problems by improving and/or updating operational requirements.

The inspection process then covers the spectrum of an organization as it applies to the:
1. *Situation*—the working environment, the capability and limitations of the organizations, and the constraining factors affecting goal accomplishment.
2. *Goal(s)*—the operational commitment imposed upon the organization; its charter.
3. *Implementation*—the concept of operation and coordinating activities.
4. *Administration and Logistics*—the administration of equipment, facilities, manpower, money, and administrative support functions.
5. *Management Supervision*—the operational commitments; how they are implemented under normal and emergency conditions; and the visual management techniques which are present, and how they are employed.

Although an inspection may uncover an unsatisfactory condition, the condition itself may not reveal the exact nature of the cause. For example, noncompliance with regulations and procedures, unsatisfactory results and conditions, and failure to use available resources to their best advantage may indicate a need for additional or modified:

    Organizations
    Regulations
    Procedures
    Equipment
    Manpower
    Training
    Direction
    Leadership

Further study will be necessary to determine in which of these areas the primary weakness lies and the causes that contribute to the weakness. These actions will be discussed in the evaluation process.

The inspection process will give management an added benefit when used in another light. While inspections directly by management insure compliance with security procedures at a specific work center, work centers should also have the opportunity to approach management and request a "special inspection" of their activity when:
1. the work center identifies a specific problem and does have the solution to the problem; or
2. when the problem stems from outside the center and it is beyond their capability to solve.

Scope of Inspection

Everything relating to the organization must be subject to control. Consequently, inspection must be universally applied. Conditions, situations, functions, and actions that contribute to the success or failure of operations are exposed by the inspection of persons, things, procedures, and resultant actions. Obviously, the inspection process cannot be divided sharply among these four classifications because of the interrelationships involved. The following examples of inspection make this fact apparent.

Things

The physical inspection of things indicates whether repair and replacement are needed. It may also reveal the extent of compliance with regulations. This inspection cannot be allowed to become casual or superficial. Every item and aspect of a building, its offices, corridors, storage spaces, firefighting, detection devices, locks, windows, doors, fences, lighting, loading docks, etc., must be inspected regularly to determine whether or not they are in good condition and suitably maintained in accordance with applicable policies governing their maintenance and use.

Procedures and Actions

Conformity with procedures designed for use during the tour of duty such as the inspection of hazards, the challenge of unknown individuals, response actions and reporting procedures may be determined more accurately by observation of acts than by analysis of the results or study of reports covering the particular incident. Obviously, not every action can be observed. The observation of some, such as the stopping of suspicious persons and the detention of suspects, will be difficult and infrequent during the inspection process. Nevertheless, it is important that inspections be made as often as possible to determine whether or not operational procedures are being followed.

Results

An examination of results indicates the success of the operation. It also indicates whether each task was performed in the manner outlined and whether or not the resources of the organization were used to best advantage. In addition, it may reveal needs not previously discovered. The inspection of results includes the analysis of statistics, the examination of reports and follow-through actions. In research and analysis, inspection goes hand in hand with planning.

Line and Staff Inspections

Inspection can be divided into two segments: *line* and *staff*. An authoritative or line inspection is conducted by those in direct control of the persons and things being inspected. A staff inspection is conducted by those who lack direct control, but who have the responsibility to determine whether the job has been satisfactorily completed and whether subordinate line inspections are effective.

In the latter case, the results and findings of the inspection must ordinarily be reported by the security officer to a supervisor for having action taken. In the former, the one making the inspection should exercise his control. It is his duty to take direct action to see that the job is performed properly.

Line Inspection

Line inspection, made by supervisors and management personnel charged with responsibility for the completion of an operation, is basically the continuous inspection of the process of work performance and is essential if the supervisor is to ensure satisfactory accomplishment by holding his subordinates accountable. As long as the manager gives his order directly to the person who is to execute it, as in a small organization, inspection is fairly simple, since it is a part of the task of supervision. As the chain of command is lengthened, however, the inspection process becomes more complicated because those responsible for seeing that the task is accomplished are further removed from the one who actually performs it.

For example, in a small organization, the manager might order an individual to enforce a certain regulation, and he would ascertain by personal inspection that his order was being carried out. In a large organization, such an order, emanating from the manager, would not be given directly to the individual but would descend through several levels of authority—an assistant director, supervisor and officer—and each level is accountable to a superior. This delegation down the chain of command involves no complications; but knowledge of the results, necessary if those at each level of authority are to be held accountable, becomes difficult to obtain.

Staff Inspection

Staff inspection itself is divided into two categories. The first is informal and/or functional and is most commonly encountered when line personnel lack the necessary technical skill (sometimes accompanied by a lack of interest) to inspect all aspects of their assigned post. The second is the formal responsibility to evaluate operations assigned to a staff unit. Technical and specialized inspections are conducted without line authority by such a unit.

Inspections by the Security Department

Inspections should not duplicate efforts accomplished by other parts of the organization. The security officer is interested in discovering and pinpointing specific areas where irregularities or weaknesses occur and in keeping supervisory officers informed about them. Corrective action can then be taken. Because of the nature of his duties, the security officer has flexible working areas. Each day he should select the areas that will prove most productive. He should submit in advance a weekly schedule showing the places to be worked within his assigned part, but deviations from this should be expected when conditions require it. Officers should report matters that deserve present or future attention as they are discovered.

A Direct Inspectional Process

Inspection conducted by the line supervisor is usually informal, without a sharp distinction between the casual observation involved in supervision and the somewhat more orderly observation which is characteristic of officers' inspections. When inspection is formalized, however, it should be conducted in an open, straightforward way regardless of whether it is carried out by line supervisors or staff inspectors. Inspection carried out by stealth is certain to have undesirable effects on employee morale, and the conditions which the inspector tries to observe will go underground. Also, the staff inspector should not operate in such a manner as to lull personnel into thinking he is not actually inspecting.

The Relationships and Responsibilities of Security to the Organization

The inspector's relationship with management is best established by regularly scheduled conferences. It is influenced by the extent to which management wishes to participate in the corrective action process. The inspector should reveal all information which he believes management might require. To establish an optimum relationship, the inspector must learn to recognize the type of information needed and should err on the side of providing too much rather than too little.

The security officer must also establish a relationship with operating personnel that will enable him to deal frankly and directly with them. The use of persuasion, suggestion, and personal request rather than citing command authority is effective in correcting irregularities that establish a workable relationship. To establish this relationship, it is necessary that inspectional duties be performed by individuals whose maturity, experience, integrity, judgment, and personality win for them the respect of all members of the organization. A desirable relationship is promoted further by an attitude of friendliness reflected in demeanor, voice, and facial expression and is evidenced by a willingness to be helpful and an interest in the welfare and personal problems of the members of the organization.

The security officer must be circumspect in his relations with both management and employees in order to avoid friction. Findings and any resultant suggestions or requests to employees should be reported punctiliously to the superior officer with further recommendations. Care must be employed in making suggestions to employees in order to avoid creating the impression that they are commands.

A good relationship between nonsecurity employees and those engaged in security inspections is vital to the success of this kind of supervision. The relationship is a delicate one, and desirable results are achieved only when it is worked out to the satisfaction of all concerned.

Since the individual who makes a security inspection often lacks direct control over the subject of his inspection, what is he to do when he discovers an unsatisfactory condition? In order to effect correction, he must first call the matter to the attention of the individual at fault. The inspector cannot be limited to communication only with his supervisor. He must discuss the problem with the employees to determine if his observation is valid and can be corrected on the spot. In most organizations, limiting the exchange between the officer and employee will result in management's exposure to trivial information and impede the problem-solving process.

The evils of unnecessarily delaying corrective action and the desirability of lessening the routine duties of operating personnel justify cutting across the lines of direct authority when this may be done without jeopardizing harmonious working relationships. For example, the inspector cuts across the lines of authority by reporting matters directly to the immediate superior of the individual involved. The lines can be successfully cut only by a friendly discussion—not by command. The level at which this can be done without friction is influenced by the personalities involved, their willingness to give and take, and the *esprit de corps* of the organization.

Operating personnel should understand the nature of staff inspection and should recognize that it is a service rather than a device for catching them in a dereliction of duty.

DETECTION

Detection involves the primary responsibility for insuring that all the physical dimensions of the protective (or security) program are operational. The monitoring of all physical security subsystems, surveillance of all barrier subsystems, the review of lighting and locking subsystems and the inspection of all other electrical, mechanical, electromechanical and physical plant systems which affect the organizational environment and its ability to reduce the possibility of internal or external shocks.

Detection deals primarily with the hardware or technological aspects of the organization's security program. The checking of doors, locks, closing open windows, turning lights on or off, unplugging coffee pots, checking steam or water pressure, monitoring heating or cooling systems are all a part of this function. Similarly, the monitoring of an alarm annunciation panel, or conducting a watch tour are detection type activities.

REPORTING

Reporting systems within a security department provide the medium to transfer data between various functions in the security department and to other parts of the organization. The security department must provide for the transfer of records regarding each phase of work that they perform. The nature of a security operation requires information about:

- accidents
- fires
- arrests
- investigations
- safety and security hazards
- alarm calls
- visitor logs
- equipment checklists
- time clock tapes
- daily activity reports
- inspectional reports
- evaluation reports

The transfer of data in a timely manner between functions and within the entire organization is critical. Without this ability, the organization cannot properly plan for preventive measures and response. Management elements of the organization, likewise, require this kind of information in their decision-making processes.

Importance of Records to the Security Department

Reports serve as the department's memory system by documenting the services provided, the tasks performed, and the measures taken to satisfy party responsibilities. They are the administrative tools that insure that information regarding all incidents, activities, or observations are reported to individuals responsible for taking corrective action or making relevant decisions.

There are several administrative uses for reporting. These include the coordination of interdepartmental activities and the coordination with other departments in the organization. Similarly, they provide factual data to answer management's requests for information related to planning and budgeting and also to provide assistance to persons outside the organization who have a need for such information. These outside organizations include the newspapers, law enforcement agencies, insurance companies, and prosecutors.

Operational Requirements

Reports are necessary to assist in the apprehension of criminals, the recovery of lost or stolen property and the preparation of accident data. Similarly, they are essential for distribution and deployment of manpower and to assist in the establishment of equal work loads. They likewise permit the preparation and justification of budgets based on actual data about the job performance. Reporting also serves to pinpoint problem areas within the corporation. It points out trends and the need for greater attention to specific areas or activities within the organization. Thus, they permit the study and solution of specialized problems in both the security department and by outside departments.

Uses of Reports

The report provides an opportunity for the exchange and interchange of information between members in the security departments. This permits transferring of cases from officer to officer, thus allowing preliminary and specialized investigative reports to be accomplished by more than one person. It similarly provides an opportunity for investigative work to be reviewed by supervisory and management personnel.

Reports, likewise, provide a factual record of work done and thus show the capabilities and abilities of officer's performance. The reports provide an important evaluation tool.

The development of well-written, clear, concise, and accurate reports permits the evaluation of policies and procedures. Thus, they permit the study and solution of specialized problems. Reports also can be used for accident reduction and public relations uses.

Submission of Reports

Reporting in some form should occur whenever the department is called upon to take some action. Reports are normally generated in one of two ways:

1. when required by policy, procedure, or practice; or
2. when deemed necessary by the individual in preparing the report.

Examples of reports required by policy procedure are:

1. activity reports,
2. investigative reports,
3. reports on specific assigned activities,
4. personnel evaluations,
5. patrol reports,
6. time clock tapes, and
7. equipment inspections.

Examples of those reports initiated by an individual are:

1. self-directed staff study,
2. requests for additional equipment,
3. information of general interest, and
4. training documents.

EMERGENCY SERVICES

Emergency Service

This function provides the security program (or organization) with the ability to respond to problem situations in a timely way. Crisis management and immediate response to emergency or serious incidents within the organization must be provided by the security program. Much of the organization, training, and operations of the security program relate to this type of function. It represents the delivery of "security and protection" to individuals and the organization on a very personal and meaningful basis. Consequently, it must be provided with efficiency, effectiveness and sensitivity.

Emergency services include such things as fire protection, ambulance services, holdup response, bomb threats, disturbances, panic alarms, first aid, or other calls for service of an immediate nature. An example of emergency service for a fire might involve the procedure detailed below.

Emergency Reaction (Fire)

Once a fire has been discovered, the first act to be accomplished is sound the alarm.

Fire fighting procedures must start immediately to preclude the spread of the fire. Small fires must be checked as soon as they start. The first five minutes will be crucial if the fire is to be contained. Procedures should be established to:

1. Evacuate all nonessential personnel.
2. Notify the fire department, giving the exact location. If possible, send someone to guide the firemen to the scene.
3. Fight the fire (if properly trained).

Preplanning and inspection can provide extra protection by:

1. insuring all escape routes are identified, well lighted and kept free of obstruction;
2. insure certain personnel are designed to man fire equipment and are trained in their uses;
3. conduct fire drills which will familiarize all personnel with exits, thus reducing possible panic in the event of an actual fire;
4. insure all exit doors are equipped with panic hardware.

Fire

Activities of the security program in fire protection and prevention assist the organization in maintaining and promoting a risk-free working environment. The reduction of fire risk lowers costs, promotes good working conditions, and insures the continued operations of the organization with low risk of interruption.

Inspection of equipment for fire fighting, fire notification and reporting, physical plant conditions, emergency plans, testing and evaluation of evacuation plans, records dispersal, training of fire brigades and wardens, and the conducting of fire drills and exercise are all a part of the function of this subsystem.

Safety

The purpose of subsystem activities in any safety program is to assist in the development and maintenance of a risk-free environment for the organization. Activities include the inspection of equipment, conditions, activities, procedures, and policies to insure their functioning to promote appropriate work conditions. They also must insure compliance with state, Federal and corporate policies concerned with occupational safety. Notably, the requirement of the Occupational Health and Safety Act of 1968 (OSHA) compliance is included in this function.

Total Loss Control

Total Loss Control (TLC) is the concept for insuring that all aspects of protection for an organization are considered as a part of comprehensive planning. The Total Loss Control procedure involves the following:

1. identification of possible loss situations,
2. measurement of such losses or potential losses,
3. the selection of methods to minimize losses, and
4. implementation of such methods necessary to eliminate or significantly reduce loss potential.

The major significance of using terms such as TLC is the emphasis on integration and inclusiveness in the development of the plan. The planning for only one element of an organization, while not treating other areas which are subject to the same kinds of risks and problems, is not an acceptable format for planning under a system approach.

Although normally independent of each other, the safety and fire functions have similar objectives. There is a cost-benefit from an exchange of information and services. One common objective is the development of a risk-free environment through the application of rules and policies.

Role of Protective Services

The Protective Service Department is responsible for all loss control in an organization. Its functions include the following:

1. accident prevention programs and promotional activities;
2. consultation with management and employees in matters of safety;
3. recommendation of codes and standards, and the interpretation of standards in corporate operations;
4. review of plans for new construction and building alterations for points of safety and fire prevention;
5. problem-solving services;
6. investigation of injuries, losses, and accidents with a view toward improvement of methods, equipment and removal of hazards;
7. cooperation with committees concerned with matters of safety and security;
8. inspectional services related to safety and fire prevention;
9. liaison with state and local safety and fire prevention authorities; and
10. maintenance of a complete library of technical literature.

To insure maximum effectiveness for fire protection and safety efforts, mandatory compliance with all codes and standards published by the authorities listed below should be required.

1. The National Building Code (The American Insurance Association)
2. The National Fire Protection Association (Fire Codes)
3. The American Standards Association (USA Standards)
4. The Associated General Contractors of America, Inc. (Manual of Accident Prevention in Construction)
5. The National Safety Council (Accident Prevention Manual)
6. The Underwriters Laboratories, Inc.
7. The American Association of Mechanical Engineers
8. The Compressed Gas Association, Inc.
9. All applicable state laws, acts or codes pertaining to fire prevention, construction, exits, fire escapes, fire extinguishers, excavations, scaffolding, hoists, elevators.

Fire protection and safety responsibilities are often divided between specific departments or organizations. This might be accomplished with a division between a safety or maintenance department and the Protective Services Department.

Maintenance

The maintenance organization is responsible for designating and marking all fire lanes and fire hydrants as prohibited parking areas. It is also responsible for the selection, installation, inspection, and maintenance of all portable fire extinguishers and related fire fighting hand tools. The procurement, posting, and maintenance of signs pertaining to "SMOKING" or "NO SMOKING" are other related fire responsibilities. As an operational organization it is responsible for the testing and operational maintenance of all fire protection systems and related equipment as: fire alarms and fire detection systems, fire sprinkler systems, fire standpipe systems, fire water pumps, fire hydrants, fire escape chutes and ladders, fire exit signs and emergency lighting, fire- and smoke-stop doors, emergency generators, emergency showers and emergency eye baths, air circulating and exhaust systems, laboratory exhaust hoods, elevators, escalators, heating and air conditioning equipment, etc. (*Note:* These functions are to be divided to allow inspections by the Protective Services Department and maintenance by the Maintenance Department.)

As a housekeeping function, maintenance is responsible for maintaining all utility and equipment rooms in a safe and orderly manner. All such rooms should be kept locked and should not be used for storage.

Protective Services Department

As a traffic control, enforcement and emergency services agency, the Protective Services Department is responsible for all vehicular and pedestrian traffic safety on the grounds. In this respect, it is responsible for designating the type and location of all traffic control signs and pedestrian crosswalk markings. It is also responsible for establishing visibility clearance at intersections involving all sidewalks, roads, or both. With reference to the maintenance of traffic control devices, it is responsible for initiating necessary work orders to insure that all signs and markings are readable at all times. With reference to fire lanes and hydrants, security is responsible for keeping the prohibited areas clear and accessible for fire emergency at all times.

As a communications activity, the Protective Services control center is responsible for prompt notification of appropriate personnel in case of fire or other emergencies in any corporate facility.

The Protective Services Department is responsible for providing twenty-four-hour emergency medical service. This service includes an emergency ambulance-type vehicle available for local medical transportation. Emergency first aid medical equipment will be properly maintained in this vehicle at all times.

It is additionally responsible for conducting sufficient fire drills throughout the year to insure that all employees are properly oriented for any emergency evacuation. Reports and results of all evacuation drills will be recorded and filed as a permanent record. Protective Services is also responsible for selecting, distributing, inspecting, and maintaining their own portable fire extinguishers located within all buildings.

Types of Plans Required

The development of plans for fire prevention and protection can be divided into eleven elements. These areas cover the most critical activities within an organization. Each of these areas should have an action plan developed. These areas include:

1. during the design stage, expansion or modification of facilities to insure that fire prevention and protection aspects are included;
2. insure that plans and schedules exist for the maintenance of all fire extinguishment in all organizational facilities;
3. insure that fire prevention and protection policies and procedures are developed and disseminated to all responsible personnel;
4. plans and procedures for detailed inspections of fire control equipment and facilities;
5. a plan for fighting fire should exist;
6. a plan for use of community fire-fighting equipment should be developed;
7. planning should be done to insure that all employees are trained in use of fire-fighting equipment;
8. plans to provide for emergency evacuation of employees;
9. plans for the dispersal of records;
10. fire prevention regulations posted, enforced and emphasized in training; and
11. the development of disaster and emergency plans.

Prevention Concepts

It is of primary importance to inspect buildings to determine if they are fire resistant or not, and if not, what actions can be taken to improve the structure to recent building codes and suggested agency standards. Regarding the protection and safety of human lives, it is important to see that adequate exits are provided. It is not only important to inspect the building, but it is also important to inspect the contents of the building and determine whether or not content materials are a fire hazard. Measures should be taken to isolate and confine those materials which may be highly flammable and present a fire hazard.

A good sprinkler system is an excellent asset for fire prevention and protection. Fire extinguishers should be placed throughout the building in accordance with local code requirements. Do not purchase and place fire extinguishers until first determining the type of fire that must be provided in specific areas. After this has been done, the proper type of extinguisher may be purchased and placed in that specific area.

Warning devices provide additional safety for human life and result in an earlier response to an actual fire. This warning device could range from a local alarm on the premises to a sophisticated alarm system wired directly to a central station or local fire department.

The Three I's of Fire Prevention

The three I's of fire prevention consist of *instruction, inspection* and *investigation*. When a fire occurs, it is very important that an investigation be conducted to find out the cause of the fire. The resulting information obtained from the investigation can possibly help to determine what steps need to be taken to provide future fire protection and prevention.

Follow-Up

Prevention and protection devices, equipment or alarms must be tested on a routine basis to see that they are functioning properly. It is also essential to conduct training and fire drills for employees. The short amount of time spent in training employees how to handle fire extinguishers properly could result in a saving of both lives and property. New employees, during their indoctrination period, should be provided with thorough explanations of fire safety, rules and information plus instruction on how to use fire and/or safety equipment. It is also advisable to inform the local fire department of any potential fire hazards that may be present in the building. This information would enable the fire department to efficiently fight a fire on the premises, should one occur.

The primary goal in fire prevention and safety is to save human lives. Saving property is a secondary consideration. Every effort should, however, be made during planning and inspection to reduce the risk of damage to property.

General Services

The protective services department is often called upon to provide a wide variety of services of a general nature. These services include delivery of mail, chauffeur, supplying information, cleaning up after a fire or spill, delivering messages, etc. These services often do not seem related to the protective function but must be performed. The typical protective service department often spends a considerable portion of the time providing such services. These services can be performed if they do not interfere with the performance of required protective duties.

General services often can create support and good will for the department if they are not demeaning to the role of the protective services officer required to perform them.

Deterrence

The deterrence function provided by protective services is accomplished through the systems of barriers and time-delay features incorporated into the protective program. The procedures, equipment, and personnel utilized in developing the physical, information, and personnel processes provide the deterrent effect. The deterrence of an individual from creating loss situations or committing crimes is the desired result of the deterrent program.

The actions of the security officer on patrol often deters the commission of a criminal or unsafe act. Surveillance or access control equipment likewise reduce the opportunity for undesired acts to take place in protected areas.

BOOKS

Walsh, Timothy J., and Healy, Richard J.: *Protecting Your Business Against Espionage.* New York, Amacom, 1973.

Healy, Richard J., and Walsh, Timothy J.: *Industrial Security Management: A Cost-Effective Approach.* Pub. American Management Assn., Inc., 1971.

PUBLICATIONS OF THE GOVERNMENT, LEARNED SOCIETIES, AND OTHER ORGANIZATIONS

Aspley, John C.: The handbook of employee relations. Section 68. Company Security Programs. Chicago, The Dartnell Corporation, 1955.

Institute for Local Self Government. *Private Security and the Public Interest*. Institute for Local Self Government, Berkeley, Calif.

PERIODICALS

Coster, Clarence M.: Security objectives of the LEAA. *Security World,* September 1971.

Davis, Albert S.: Toward professionalism. *Industrial Security, 12*:18, June 1968.

Fleming, George: Security police at Picatinny. *Law and Order, 12,* September 1964.

PART III

COMPONENTS OF SECURITY

INTRODUCTION

The first two parts of this text set forth the broad guidelines for the development and operation of protective services programs. Part III begins the detailed analysis of the processes of security and loss prevention. The components of security operate within the framework of systems, subsystems, and programs and provide the methods, techniques, and means utilized to maintain a security operating environment. The discussion will begin with a treatment of the three security processes of physical, information and personnel security. The planning, organization, training, and operations of protection programs will follow in subsequent chapters.

Chapter 11

PHYSICAL SECURITY

Physical security is the most common form of protection known to man. It is the basis for most protective programs and is almost "second nature" to human beings in their search for stability and protection. The locking of a door, the very decision to have a lock on the door is a manifestation of the desire for physical protection. Likewise, the use of fencing, lighting, guards, sentry dogs, alarm systems, moats, and the various structures and devices conceived of by man have been attempts at various times in history to insure physical security.

In contemporary protective services terminology, however, physical security has a very specific meaning.[1] While it provides the same function of establishing some type of barrier or barrier system between what was to be protection and the things to be protected against; there is something more.

It should be pointed out that the three processes of physical, personnel, and information security are interrelated parts of a security environment. All three parts are used simultaneously in any security program. The emphasis placed on each specific one is determined by the type of program requirements for security. Therefore, physical security must be viewed as a part of a total protection program; the portion that provides the physical means for security of a particular area. No one single process is perfect. Each system backs up the other or provides additional support to the other processes.

The physical security process is a functional subactivity within the total framework of the concept of security. The physical security process is utilized *to deny access to organizational or governmental facilities for unauthorized purposes by the establishment of barrier systems.* The physical security process is, then, a protective device against security hazards. Security hazards are conditions which may result in the loss of life, damage or destruction of property, the compromise of information, or the disruption of the activities of the organization's facilities. Recognition of all risks is mandatory so that appropriate security measures can be instituted to control or eliminate them.

A physical security program consists of many components. Figure 11-1 outlines these components and the functions that each performs. It can be noted that these components relate directly to the ability of the entire process to operate effectively. Much like the interrelatedness of the three processes (physical, information, and personnel) the components of the physical

---

PHYSICAL SECURITY SYSTEM COMPONENTS

1. *Procedures*
   collective information systems involving what, who, when, where
2. *Personnel*
   collective people to administer/implement system
3. *Barriers*
   access control or prevention devices/structures
4. *Sensors*
   any detector of unusual activity
   (access I.D.)
5. *Annunciators*
   the recognition of unusual activity that has been detected
   (access verify)
6. *Communication Subsystem*
   any transmission methods to convey the detection of unusual activity to its recognition
7. *Bookkeeping*
   data of time records of detection/recognition and use procedures

---

Figure 11-1

security must be present and operate efficiently for adequate protection to take place.²

PHYSICAL SECURITY SURVEY

Before conducting a survey, several preliminary steps are taken to provide an estimate of the security situation:
(1) Preliminary contacts are made with appropriate personnel to arrange time and other details.
(2) Previous surveys, if any reports are checked for background information and action taken on noted deficiencies.
(3) The reasons for the survey are determined.
(4) Survey personnel are familiar with the mission and history of the installation or intended use of the area, or any changes in the mission or use since previous surveys.
(5) Installation floor and ground plans are evaluated.
(6) Installation regulations and operating procedures are reviewed.
(7) A checklist³ is prepared and used as a guide in making the survey.

The development of a comprehensive protective program for an installation requires that all aspects of its needs and operations be considered (see Figure 11-2).

The determination of the degree of risk for a particular organization requires the analysis of two factors: (1) relative criticality and (2) relative vulnerability. In other words, how critical is the need for a degree of security measures due to the degree of risk involved? and secondly, how vulnerable is the facility or installation in terms of security hazards which might exist? The hazards which may exist may be divided into three general categories: (1) theft of assets or property, (2) sabotage, or man-caused emergencies, and (3) espionage of proprietary or governmental documentary information.⁴

FACTORS INFLUENCING THE PHYSICAL SECURITY PROCESS

The physical security process includes all the measures employed to protect personnel, products, materials, services, and premises from hazards inherent in the organization's operations. The scope and character of physical security measures can be determined by evaluating the following criteria:
1. Is there evidence of a clear and present security hazard?
2. Is there a definite security risk to the organization's security?
3. What is the effect of physical security measures on organizational efficiency?
4. Do the physical characteristics of the facility impose limitations on the security program?
5. How much do budgetary limitations affect physical security measures?
6. Evaluation of potential damage.
7. Alternate measures or techniques which will provide a minimal level of physical security.
8. The capabilities of providing a minimal level of physical security with the currently available resources.

Governmental or proprietary security must continually evaluate its effectiveness in terms of the foregoing factors and utilize appropriate physical security measures which would be consistent with these criteria.

SITUATIONAL SECURITY

It must be pointed out that in determining a trade-off between security and budgetary limitations, the *criticality* and the *vulnerability* of facilities may vary from one point in time to another depending upon many internal and external factors. Should the organization's situation change, the type of security measures must likewise change in order to continue to be appropriate for the organization's activities.

In determining what type of physical security measures to be employed, surveys or analysis should be conducted to determine the physical security hazards. A physical security survey would include a study and analysis of the organization, its facilities, and its operations with the purpose of testing the organizational security, noting deficiencies, and attempting to provide appropriate revised security measures.

BARRIER SYSTEMS

The concept of barrier systems, a subarea of physical security, provides a series of defensive barriers surrounding facilities or a portion of facilities which produce an entry time-delay. The purpose of the barriers is to insure that access to areas within the facility cannot be made readily. When the term "facility" is used in reference to the provision of physical security, two different terminologies are used depending upon the system in which the definition is applied.

Barrier systems provide the first line of protection for the organization. They "buy time" by delaying an intruder or criminal long enough for the related protective features of the security program to capture or deter him. Barrier systems in a very real sense reflect the basic concept of security programs. . . . *For every dollar spent, you buy time in which you are not criminally attacked or that losses do not occur.* Barriers, and the entire security program for that matter, provide a mechanism for organizations or individuals to trade "delay time" for dollars.

The government definition for *facilities* in which physical security is provided includes government and privately owned plants, mines, buildings, occupied in whole or in part by any federal department or agency which houses their essential emergency functions, materials, products, processes, and those government and privately provided services which are of importance to defense mobilization, defense production, or of the essential civilian economy and are located and provided in the continental United States, its territories or possession.[5] The proprietary definition of facilities would include building structures, mines, and plants which the organization determines to need protection. There are, obviously then, needs for the establishment of security areas within either governmental or proprietary organizations.

A *security area* is a physically defined area, access to which is subject to special restrictions and controls. Physical security barriers are imposed in an effort to prohibit or control the access into these areas and/or the access to sensitive matters contained within these areas. Various areas and various portions of one general area often have varying degrees of security requirements depending upon their purpose and the nature of the operations performed in the area. It is necessary, therefore, to compartmentalize or segregate physically those areas having different levels of security needs. Segregation or compartmentalization provides an efficient and effective basis for applying various degrees of restrictions of access. This arrangement often facilitates operational control of movement and results in reduced cost of construction and guard protection and still provides each area with the security required for the particular need.[6]

Security Areas

Security areas can be categorized into three subclassifications:

1. EXCLUSION AREA. An exclusion area is an area containing an activity which is of such a nature that access to the area constitutes, for all practical purposes, access to the operation itself or an operation or activity of such vital importance that the proximity resulting from access to the area will be treated as the equivalent of gaining access to the operation or activity.

2. LIMITED AREA. A limited area is one containing an operation or activity in which uncontrolled movement would permit access to the operation or activity, but within which area such access may be prevented by escort or other internal restrictions or controls.

3. CONTROL AREA. A control area is an area adjacent to, or encompassing limited or exclusion areas and within which area, uncontrolled movement would permit access to the restricted operations or activities. It is designed for the principal purpose of providing administrative control, safety or a buffer area of security for limited or exclusion areas. The controlled area, as well as the limited or exclusion area, must be completely within the same

area under the control of organizational security.[7]

These types of restricted areas depend on the nature, sensitivity or importance of the activities, operations, or information involved. Restricted areas are established to provide the following:

1. Effective application of necessary security measures and exclusion of unauthorized personnel.
2. Intensified controls over those areas requiring special protection.
3. Conditions for compartmentalization of operational activities or information with a minimum of interference to other organizational activities.[8]

PHYSICAL SECURITY BARRIERS

The physical security process utilizes a number of barrier systems, all of which serve specific needs. These systems include natural, structural, human, animal, and energy barriers. In providing these barrier protections, a number of specific activities are performed by physical security personnel. While the barrier systems and guard systems prevent unauthorized persons from gaining access to the facilities of the organization, they likewise perform activities to control authorized entry. Such things as employee identification programs, control of visitors, package or material control systems, vehicle control systems, and lock and key control systems all provide for control of authorized entry. However, the maintenance of protective alarm and communication systems and providing fire-fighting facilities also insure the physical security of the organization and its operations.

Natural Barriers

Natural barriers are natural topographic features that lend themselves to denying or hindering access to a facility. These features may include such things as rivers, cliffs, canyons, oceans, or other terrain which is difficult to traverse.

Natural topographic features seldom provide ideal security barriers without the addition of other features. Where rough terrain features slow down a potential intruder, or where they tend to screen the facility from observation, they will usually provide a great deal of cover and concealment for a person wishing to gain access. Natural barriers therefore must be evaluated from the standpoint of possible aid to the intruder as well as the time-delay advantages.

Structural Barriers

A structural barrier is a permanent or semipermanent feature which has been constructed to deny or hinder access to an installation. In most cases, the structural barrier will be an actual construction; however, the term could include certain reductions (such as removal of vegetation, disposal of inflamable waste, etc.) that create an actual or theoretical barrier to access to the facility.[9]

The structural barrier differs from the natural barrier basically in the way it is created. Structural barriers include a system of barriers which create a physical or psychological deterrent to authorized or unauthorized entry. These barriers can be divided into four major categories:

1. perimeter barriers (fences, etc.),
2. outer walls and openings of buildings,
3. the inside walls and doors of buildings, and
4. safes and vaults.[10]

These barriers delay intrusion, either authorized or unauthorized, into an area making detection and apprehension by guards more probable. They also facilitate the effective utilization of guards by economizing the number of guards and guard locations required. They also direct the flow of persons and vehicles through designated entrances and facilitate identification and control procedures within the facility.

The architectural arrangement and design of buildings are an integral part of the total physical security process. Likewise, in the provision of barriers, locking devices play a major role. Locking devices are mechanical features designed to secure objects in the same relative position in such a way that one or more of the objects can be moved from its position only by a person authorized to do so. The term "locking devices" refers to all devices designed to fulfill this function regardless of their simplicity or complexity.[11]

Human Barriers

The human barrier system is the systematic employment of humans as barriers between potential intruders and the area to be protected in order to give an alarm in the event a situation arises which would threaten security. They also apprehend unauthorized persons and serve to aid the identification of personnel entering or leaving the facility. Guard personnel can generally be divided into two general types according to the manner in which duties are performed: First, persons who are assigned guarding duties as a primary function and who have been selected and trained to take appropriate actions in situations creating hazards to security; secondly, those persons who, by the nature of their location or as an additional duty, fulfill a security function.[12]

Animal Barriers

This category applies to the use of any type of animal to provide a security barrier. Many types of animals are used for this process; however, the most commonly used is the sentry dog. A sentry dog can perform two of the three basic functions of a human guard. He can detect the presence of persons and things, and he can apprehend. The third (the ability to check and identify) is a relatively weak point although dogs can identify those persons whom he personally knows or substances such as drugs. For this reason, sentry dogs are handled in all respects by one man and are taught to trust no other person. This limits the use of a dog to working alone in an area at a time in which there is to be no entrance. When working with a handler, the handler makes the necessary identification or verification of animal observation.[13]

Energy Barriers

Energy systems can be divided into two general subclassifications: (1) protective lighting, and (2) protective alarm systems.

Protective Lighting

The function of protective lighting is to provide uninterrupted light during periods of darkness and low visibility. Light generally has some value as a deterrent but its primary purpose is to provide illumination. Normally, protective lighting utilizes less candle power than a working area light except at identification and inspection points.

Protective lighting is utilized to increase the effectiveness of guard systems by increasing the visual range of the guards during periods of darkness or in an area where natural light does not reach or is in-

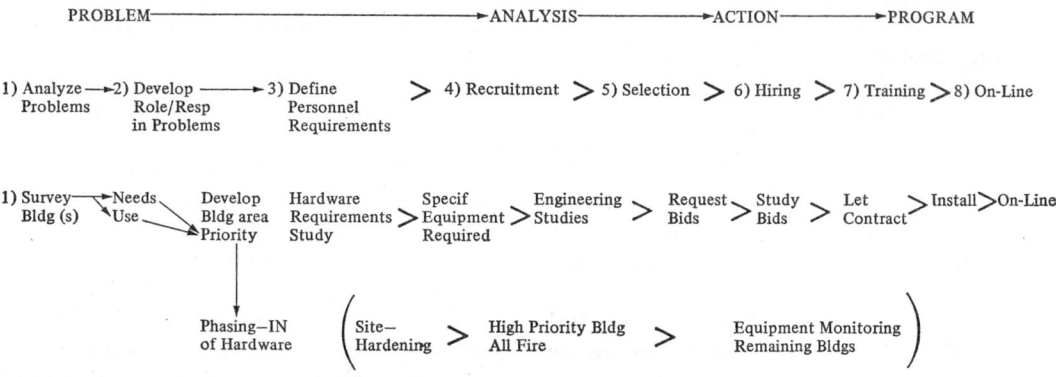

Figure 11-2. Security Survey Flow Chart.

sufficient to provide illumination necessary for visibility. Outdoor protective lighting has considerable value as a deterrent to thieves and vandals and makes the job of a person desiring unauthorized access much more difficult. Lighting is necessary for a minimal level of visibility so that guards may perform inspectional duties properly. They must be able to see badges, people at gates, inspect vehicles, stop attempts at illegal entry, detection of intruders inside and outside of buildings and structures, and inspection of unusual or suspicious circumstances.

Each facility presents its particular problems based on physical layout, terrain, atmospheric conditions, and security requirements. There are four types of protective lighting: (1) continuous, (2) standby, (3) movable, and (4) emergency.

Various types of lighting sources have been developed and are used for illumination. The type of use, location, budget, and space layout will determine the type of lighting to be used. Figure 11-3 provides a comparison of various lighting sources from a cost-effective viewpoint.

Alarm Procedures

Protective alarm systems provide an electrical or mechanical means of detecting and announcing the proximity of or intrusion by unauthorized persons to the area of an installation. Protective alarm systems are utilized to accomplish one or more of the following purposes:
1. to permit more economical and efficient use of manpower by substituting mobile responding guard units for large numbers of fixed post guards or patrols;
2. to take the place of other necessary elements of security which cannot be used because of building layouts, safety regulations, appearance, operating requirements, costs, or other reasons; and
3. to provide additional controls at vital areas as insurance against human or mechanical failure.[14]

Alarms can be classified into four major systems:[15]
1. LOCAL ALARM SYSTEM. A system in which the alarm sounding device is located in the immediate vicinity of the protected area. Applications: shops, stores, or small retail establishments and homes.
2. CENTRAL STATION ALARM SYSTEM. A system in which the alarm is relayed to a remote panel located at some centralized facility owned by an outside agency; usually a commercial agency. Applications: same as local alarm systems.
3. PROPRIETARY ALARM SYSTEM. A system in which the alarm signal is relayed to a headquarters location owned, manned, and operated by an internal security organization. Applications: school systems, industrial facilities, municipalities, office buildings, military bases, large institutional buildings and others.
4. DIRECT CONNECT ALARM SYSTEM. A system where the alarm signal is carried to an alarm annunciator at a remote location such as a police station. (This system is often combined with proprietary alarm systems to provide additional protection.) Applications: same as proprietary alarm systems.

Alarm Components

All electronic and mechanical alarms are comprised basically of three component parts. They consist of a triggering mechanism, transmission line, and a monitor or annunciator. All such systems require that the annunciator provide a signal which will alert an individual or group. The triggering mechanism for activating the alarm system may be activated by one or a combination of the five following methods:
1. breaking of an electric circuit,
2. interruption of motion,
3. detection of sound and vibration,
4. detection of motion, and
5. penetration of an electronic field.[16]

Intrusion/detection systems differ slightly from alarm systems. The intrusion system is the triggering mechanism for the total

Physical Security

COST-EFFECTIVE LIGHT SOURCE COMPARISON

MERCURY VAPOR VS SODIUM VAPOR

	100 watt lumens output	150 watt lumens output	175 watt lumens output	250 watt lumens output	400 watt lumens output	1000 watt lumens output
MERCURY LAMP	3,850	----	7,450	11,200	21,000	57,000
SODIUM LAMP	9,500	16,000	----	25,500	50,000	130,000
MERCURY LAMP LIFE	24,000 hours	----	24,000 hours	24,000 hours	24,000 hours	24,000 hours
SODIUM LAMP LIFE	12,000 hours	12,000 hours	----	15,000 hours	20,000 hours	10,000 hours

Figure 11-3. Prepared by Larry Vardell for The National Crime Prevention Institute.

SYSTEMS	AREA	Rooms	Storage Areas	Court Yards	Shipping Areas	Ventilators (vents)	Elevator Shafts	Stairwell	Halls	Spaces (Interior)	Garage	PERIMETER	Property Lines	Accessways	Fence	Space (Exterior)	OBJECT	Metal Cabinets	Safes	Doors	Windows	Walls	Ceilings	Floors	Sky Lights	Gates	Traps	Vaults	Desk	Cabinets	Cashier's Cages	Critical Eq.	Display Cases	Storage Bins
ACTIVATED BY SOUND:																																		
Audio	●										●																		●					
ACTIVATED BY MOTION:																																		
Ultrasonic	●	●					●	●																										
Pressure/Release															●	●								●		●								
Electro-magnetic (Exterior)																●			●	●					●									
Radiomatic															●																			
Seismic	●	●	●	●			●	●							●	●																		
Capacitance																		●	●							●		●	●	●				
ACTIVATED BY VIBRATION:																																		
Vibration Detector																																		
Contacts																				●	●								●					
ACTIV. BY BREAKING ALTERING ELEC. FIELD																																		
Electro-mechanical					●	●												●	●	●						●	●			●		●	●	
Spring Contacts				●														●	●	●	●					●	●			●	●	●	●	
Wired Areas			●							●					●					●	●					●	●			●		●		
Metallic Foil					●															●	●													●
ACTIV. BY INTERRUPTION OF LIGHT. BEAM																																		
Photo-electric	●		●	●		●	●	●	●			●	●	●	●	●				●														
Laser	●		●	●								●	●	●	●	●																		
VISUAL DETECTION																																		
Infrared	●					●																								●				
Closed Circuit T.V.	●					●																								●				
Photographic (movie-still)	●					●																												

Figure 11-4. Sensor intrusion detection guide.

alarm system. The entire intrusion/detection system provides the triggering for the total alarm system. Based on the operation of the detector or the intrusion/detection systems, the devices may be grouped into five families:
1. electrical-mechanical,
2. photo-electric,
3. proximity,
4. motion detection, and
5. acoustic-seismic.

Approximately 1,600 companies currently manufacture security sensory devices and related equipment. This equipment can be identified according to the various methods of operation used for sensing an undesired condition. The uses of such equipment are presented in Figures 11-4 and 11-5.

Detection Devices[17]

Detection devices are usually designed to detect a single phenomenon. The choice of what type of detection device is to be employed is based upon what will be most readily detectable in the given situation. It may be desirable, in some cases, to employ more than one type of detection device to protect against all possible methods of entry. Usually, similar equipment is manufactured by several companies. Such equipment will operate on the same basic principles, but may very well differ in refinements. These differences may, under certain circumstances, alter the degree of security provided.

Electro-mechanical or current continuity devices are designed to effectively place a current-carrying conductor in a position which stands between the intruder and the enclosed area to be protected. In each case, the conductor carries current that retains a holding relay in the open position. A cessation of the current flow in the device releases the relay and allows its contacts to close, which closes the alarm circuit and provides the alarm. Current continuity devices may be of the following types:

1. METAL TAPE OR FOIL. This type of device is a thin strip of metal foil which is applied to the glass surface of windows or glass doors in a continuous pattern beginning and ending at contact points providing voltage of opposite polarities to cause a continuous flow or current through the foil. The object is to arrange the pattern of the foil in such a way that breaking the window will break the foil and will as a result, cause a cessation of current flow.

2. SCREEN. This device is simply a lacing of a conductor back and forth across an area of structural weakness in such a way that cutting or breaking through this area would also break through the conductor and cause a cessation of current flow.

3. CONTACT SWITCHES. Contact switches are electrical contacts placed in such a position on doors, windows, or other openings capable of being closed, that when the opening is closed the switch is closed and there is a continuity of current flow in the circuit to the holding relay. However, when the door or window is opened, the contacts separate and the resultant cessation of current flow activates the alarm. These devices are sometimes manufactured with two separate circuits operated by two switches. The first circuit is a current continuity circuit as described above. The second circuit is a portion of the alarm circuit which provides a normally open switch connected parallel to the open holding relay and physically located at the door or window in such a way that opening the door or window will cause the switch to close. Thus, opening the door or window will open the current continuity circuit and close the alarm circuit in two ways: first, by means of the holding relay; secondly, by physical closure of the switch parallel to the relay contacts.

4. VIBRATION DETECTOR. The vibration detector is a self-contained switch in a small housing unit. One contact of the switch is a small pendulum-like weight which is held under slight spring tension in contact with the other switch contact and maintains a closed electrical circuit.

Photoelectric detection devices are designed to transmit a beam of light from a light source to a light-sensitive receiver, which will, in turn, react to a cessation or substantial decrease of received light. This reaction results in the initiation of the alarm signal. The components are arranged in such a way that the beam of light crosses the approach to the area to be protected.

1. THE ELECTROMAGNETIC DETECTION DEVICE. This is basically a low-power, low-frequency radio transmitter with two receivers tuned to mesh with the transmitter. The detection field is established by three parallel wires, the center wire being an antenna to the transmitter and the two outer wires antennae to the two receivers. In operation, the transmitter transmits a radio signal causing electromagnetic waves to radiate concentrically from its antenna. These waves are inductively received by the outer wires and create voltages which are received and detected by the receivers. In tuning the system the output voltage from the receivers are made equal and opposite and fed into a comparison circuit where they cancel, causing an absence of output from the comparison circuit. If a conducting body moves into the radiating field of the transmitter so that some of the energy destined for one of the receiving antennae is absorbed, the output from that receiver is less than the output from the other receiver. This reflects itself at the comparison circuit as a difference in voltage and an output from the comparison circuit results. This output is the initial alarm signal. Since the center wire is a transmitting antenna, it must be properly turned to effectively radiate its generated energy. Ideally, this is accomplished by achieving a precise length of wire which will cause resonance for the frequency at which the transmitter is being operated. Since this is not always practical, a loading coil is used to create a theoretical length to provide the resonance. Loading the antenna in this manner greatly attenuates the transmitted signal at shorter wire lengths to a point at which the device is no longer effective. The minimum length of antenna will depend, of course, upon the transmitting frequency and the turning arrangement, but it can be safely said that an antenna shorter than fifty feet would render the device ineffective.

2. THE CAPACITIVE (ELECTROSTATIC) DETECTION DEVICE. This device operates superficially in the same manner as the electromagnetic, including in some applications, the use of the three parallel wires. The capacitive alarm is generally more versatile than the electromagnetic since it does not require the lengths for tuning that the electromagnetic device does. The capacitive detection device is, in effect, a large electric condenser which radiates, as does its smaller counterpart, electrostatic lines of force created by a buildup of electrons on one plate which is separated from its second plate by air as a dielectric. There are several possible applications, but each involves a conduction body moving into the electrostatic field after a balance has been achieved and absorbing some of the energy to disrupt this balance. The situation resulting here is similar to that brought about when one moves close to an old radio and hears a squeal. The energy absorbed by the body causes an internal circuit to break into oscillation creating an initial alarm signal.

Acoustic detection devices are those devices that are actuated by the sound or vibrations made by the intruder during his approach or as a result of his attempt to gain entry.

Movement detection devices are designed to create an alarm when there is movement of any sort within the established limits of the device. There are two types of movement detection devices; ultrasonic and radar.

Fire Protection and Safety Equipment

The physical security of a facility or organization is not provided only by deterrence or detection equipment.[18] It is provided by fire protection and safety equipment. Fire protection equipment includes: sensory devices, sprinkler systems, and extinguishing equipment in conjunction with notification and annunciation systems for monitoring the other protective functions in the facility. Safety equipment includes:

Company	Ultrasonic	Microwave	Infrared-Passive	Infrared-Active	Photoelectric	Perimeter	Audio	Proximity	Radio Signaling	Fire Sensors	Proprietary	CENTRAL STATION EQUIPMENT	McCulloh Trans.	McCulloh Recv.	Tape Registers	Direct Wire Equip.
Mini Computer Corp.													x			
Damon Alarm Corp.													x	x	x	x
Remote Control Devices									x							
Renco Corp.									x							
Kerux Systems									x							
Surcom-Div. of Yorklite									x							
Technology Systems, Inc.									x							
Three B Electronics										x						
Pyrotector Inc.										x						
Elan Industries										x						
Fire Alarm Thermostat Corp.										x						
Fire-Lite Alarms										x			x	x	x	x
Infinetics							x									
Delphi Corporation							x									
Transcience							x									
Multi-Elmac Co.							x									
Linear Corporation							x									
Record-o-Fone						x										
Executone Inc.						x										
Multra Guard						x										
ADT, Inc.						x										
Design Controls Inc.					x							x				
Electronic Locator Corp.					x											
Alarm Device Mfg. Co.				.	x											
Allied Radio				x												
Worner Electronic Devices				x												
Knight Watch			x													
Optronix			x													
Nebetco Engineering			x													
Optical Controls Inc.			x													
Arrowhead Enterprises			x													
P.M.C. Company		x														
Microtech Associates		x														
Metropole Electronics		x														
Radar Devices		x														
Larson Industries		x														
Pinkerton		x														
Johnson Service Co.		x														
Northern Electric	x															
Artolier Lighting	x															
Acron Corp.	x															
Anaren Security Inc.	x															
Bourns	x															
Sontrix	x															
Emergency Products Corp.	x															
Aerospace Research, Inc.	x															
Benedict Electronics									x	x						
Notifier Company									x	x		x				
Mosler					.	x	x									
Decatur Electronics	x	x														
Detectron Security Systems	x	x														
Detection Systems, Inc.	x			x												
James Electronics	x								x							
Alarmtronics Engineering, Inc.		x		x												
Morse Products Mfg. Co.					x	x				x			x	x		x
Diebold, Inc.						x	x									
Advanced Devices Laboratories		x	x													
Motorola			x					x								
ATA Control Systems						x		x		x						
Honeywell						x	x			x	x				x	
Systron-Donner	x	x				x	x						x	x	x	
Walter Kidde & Company	x			x	x		x		x							

Figure 11-5. Representative companies and types of equipment.

Physical Security

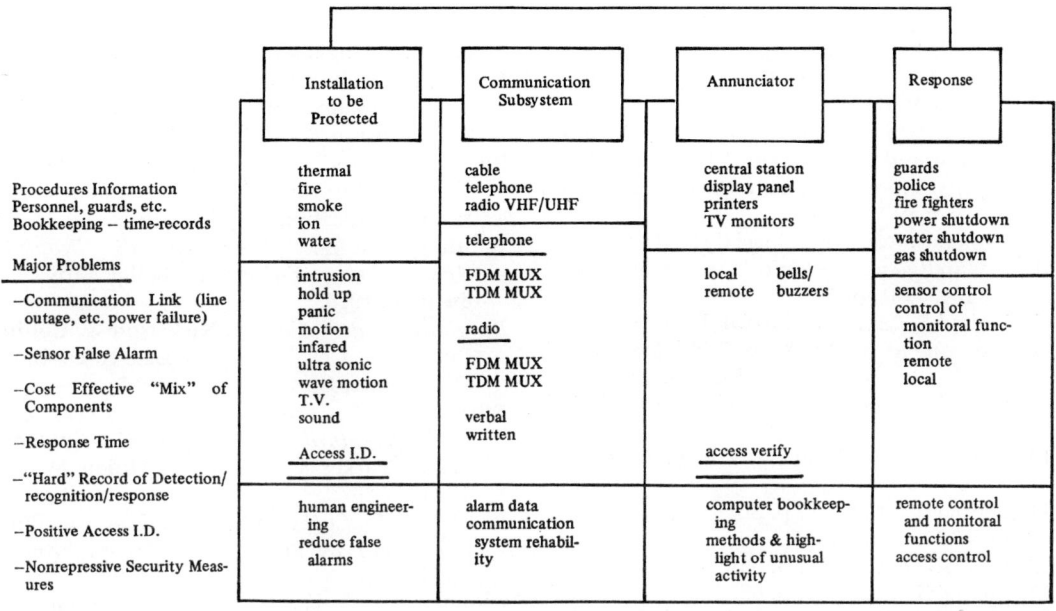

Figure 11-6.

sensors for monitoring production processes, life-safety systems, temperatures, pressures, stresses, noise, and related functions.

Fire losses for an organization are often more serious than all other physical hazards. Consequently, if a total loss control program is to be developed for physical protection, fire and safety devices must be an integral portion of program development.

Protection against physical hazards must include risks from all potential causes. Fires and industrial accidents are a major cause of direct and indirect losses in organizations. The protective services program must seek methods of employing the basic protective services concept and the physical security process to minimize risk and increase the potential for profitable or secure operations (see Figure 11-6).

FOOTNOTES

1. The problems of security have been treated by many authors. Included among them are Harry Soderman and John J. O'Connel, *Modern Criminal Investigation* (Funk & Wagnalls, New York, 1962), pp. 506-520. Chapter 29, "Plant Protection," represents the problems of industrial security from a principles viewpoint. The emphasis is on the organization and operations of a protective nature. While the approaches in some areas are drawn from military sources, the historical value of the presentation is good. It should be noted that the Security Plan Check List (p. 510-511) corresponds closely in concept and detail to those developed for the protection of defense production facilities. These plans, in updated version, are currently in use in both the civilian and military security program.

2. Various Federal programs have specific requirements to insure high-risk operations with high security. An example is the Energy Research and Development Administration (ERDA) for physical protection. Appendix 11A provides the ERDA guidelines for its protection.

The similarity in principle to the basic concepts of physical security is evident. The methodology for implementation is, however, unique to the items being protected.
3. Appendix 11B presents selected physical security survey formats.
4. Richard J. Healy: *Design for Security*. New York, Wiley, 1968, p. 9.
5. *Standards for Physical Security of Industrial and Governmental Facilities*. Washington, Government Printing Office, 1958, p. 1.
6. United States Army Special Text: *Counterintelligence Physical Security*, ST30-154, 1961, p. 5.
7. United States Army Special Text, p. 6.
8. Department of the Army Field Manual 19-30: *Physical Security*, 1965, p. 30.
9. United States Army Special Text 30-154, p. 14.
10. Suggested by Richard J. Healy: *Design for Security*, p. 74.
11. Special Text 30-154, p. 30.
12. Special Text 30-154, p. 46.
13. Special Text 30-154, p. 56.
14. Special Text 30-154, p. 68.
15. Minneapolis Honeywell Corporation: *Honeywell Planning Guide—Building Security*, 1968.
16. Department of the Army Field Manual 19-30, p. 49.
17. Special Text 30-154, pp. 69-77.
18. Appendix 11C presents a Glossary of Protective Services Terminology currently used within the industry. Section I presents alarm terminology and Section II presents general physical security terms.

BOOKS
PHYSICAL SECURITY

Abbat, John D., et al.: *Protection Radiation: A Practical Handbook*. Springfield, Thomas, 1961.
Adams, Thomas: *Police Patrol Tactics and Techniques*. New Jersey, Prentice-Hall, 1971.
Alth, Max: *All About Locks and Locksmithing*. New York, Hawthorn Books, Inc., 1972.
Bassiouni, M. Cherif: *The Law of Dissent and Riots*. Springfield, Thomas.
Bebie, J.: *Manual of Explosives, Military Pyrotechnics and Chemical Warfare Agents—Composition-Properties*. New York, Macmillan, 1943.
Best's Loss Control and Underwriting. New Jersey, A. M. Best Co., 1968.
Brann, D. R.: *How to Install Protective Alarm Devices*. Directions Simplified. New York, 1972, p. 130.
Brodie, Thomas G.: *Bombs and Bombings: A Handbook to Detection, Disposal, and Investigation for Police and Fire Departments*. Springfield, Thomas, 1971.
Brownell, Adon Hamilton: *Architectural Hardware Specifications Handbook*. Philadelphia, Chilton Book Company, 1971.
Bugg, D. E., and Bridges, C.: *Burglary Protection and Insurance Surveys*. London, Stone & Cox Publications Limited, 1974.
Bunting, James: *Protection of Property Against Crime*. Folkestone, Bailey Bros. and Swinfen, 1975.
Burglar-Intruder Alarms. Nickerson & Collins Publishing Co., Locksmith Ledger Division, Des Plaines, Illinois, 1972.
Butters, F. J.: *Encyclopedia of Locks and Builders Hardware*. Wilennall, England, Josiah Parkes & Sons, Ltd., Union Works, 1958.
Butters, F. J.: *Locks and Lockmaking . . . With a Forward by Sir George Hayter*. Pitman, 1926.
Chubb, George Hayter: *Protection from Fire and Thieves, Including the Construction of Locks, Safes and Strongrooms. . . .* Longmans Green and Co., 1875.
Cole, Richard B.: *Protect Your Property: The Application of Burglar Alarm Hardware*. Springfield, Thomas, 1971.
———: *The Application of Security Systems and Hardware*. Springfield, Thomas, 1970.
———: *Security Systems and Hardware*. Springfield, Thomas, 1970.
———: *Protection Management & Crime Prevention*. Cincinnati, W. H. Anderson Company, 1974.
Conway, James V. P.: *Evidential Documents*. Springfield, Thomas, 1959.
Crichton, Whitcomb: *Locksmithing*. Chicago, Nelson-Hall, 1943.
Crichton, Whitcomb: *Practical Course in Modern Locksmithing*. Chicago, Nelson-Hall Company, 1962.
Criminal Investigation, Basic Perspectives. Englewood Cliffs, Prentice-Hall, 1970.
Cruikshank, George: *Stop Thief: or Hints to Housekeepers to Prevent Housebreaking*. London, Bradbury and Evans for the Author, 1851.
Curran, Charles: *Handbook of Radio and TV Technique*. New York, Peligin and Cudahy, 1953.

Davis, John R.: *Industrial Plant Protection.* Springfield, Thomas, 1957.

Detex Watchclock Corporation: *Plant Protection Manual.* New York, 1965.

Dienstein, W.: *Technics for the Crime Investigator.* Springfield, Thomas, 1952.

Elseman, J. S.: *Elements of Investigative Techniques.* Bloomington, Illinois, McKnight & McKnight, 1949.

Eras, Vincent, J. M.: *Locks and Keys Throughout the Ages.* Dordrecht (Holland), Lips' Safe and Lock Manufacturing Company, 1957.

———: *Espionage and Subversion in an Industrial Society.* Hutchingson, 1967.

Eton, William: *A Survey of the Turkish Empire.* London, Cadell and Davies, 1798.

Fitzgerald, Maurice J.: *Handbook of Criminal Investigation.* New York, Greenberg, 1951.

Franzheim, Horst Werkshutzrecht: *Handbuch fur die Betriebspaaxis Koln,* Missener, 1966.

Gavzer, Bernard: *On Guard: Protect Yourself Against the Criminal.* New York, The Associated Press, 1970.

General Services Administration: *Handbooks for Guards.* Washington, Government Printing Office, 1952.

Giedion, Siegfried: *Mechanization Takes Command: A Contribution to Anonymous History.* New York, Oxford W.P., 1948.

Gocke, B. W.: *Practical Plant Protection and Policing.* Springfield, Thomas.

Greenwood, H. W.: *Infra-Red for Everyone.* New York, Chemical Publishing Co., 1940.

Gross, H.: *Criminal Investigation.* London, Sweet and Maxwell, 1949.

Hair, Robert, and Baker, Sam Sinclair: *How to Protect Yourself Today.* New York, Stein and Day Publishers, 1970.

Handbook of Security. Middlesex, England, Kluwer-Harrop Handbooks, 1975.

Hasler, Gordon: *Integrated Alarm Systems.* London, Chubb, 1971.

Healy, Richard J.: *Design for Security.* New York, John Wiley & Sons, Inc., 1968.

———: *Emergency and Disaster Planning.* New York, John Wiley & Sons, Inc., 1969.

Japp, Alexander Hay: *Industrial Curiosities: Glances Here and There in the World of Labour.* T. Fisher Unwin, 1882.

Julian, Pick, and Critchley: *Collective Security.* Praeger, 1973.

Katschke, Walter (ed.): *Der Bewachungsveitriag.* Berlin, 1929.

Kidd, W. R.: *Police Interrogation.* Brooklyn, R. V. Basuino, 1940.

Kingsbury, Arthur A.: *Introduction to Security and Crime Prevention Surveys.* Springfield, Thomas, 1974.

Kirk, L.: *Crime Investigation.* New York, Interscience, 1953.

Kraske, Robert: *Silent Sentinels: the Story of Locks, Vaults, and Burglar Alarms.* New York, Doubleday and Company, 1969.

Lanzi, Lawrence H.: *Radiation Accidents and Emergencies in Medicine, Research and Industry.* Springfield, Thomas, 1965.

Lee, Albert: *Crime-Free.* Maryland, Penguin Books, Inc., 1974.

Lenz, Robert: *Explosives and Bomb Disposal Guide.* Springfield, Thomas, 1965.

Lueder, Donald R.: *Aerial Photographic Interpretation Principles and Applications.* New York, McGraw-Hill Book Co., Inc., 1960.

Luis, Ed San: *Office & Office Building Security.* Los Angeles, Security World Publications, 1973.

Mandell, Mel: *Being Safe.* New York, Saturday Review Press, 1972.

Manual Describing Explosives and Practical Methods for Using Them. Wilmington, Delaware, E. I. DuPont de Nemours, 1952.

McLean, Stanley Allen: *The Manual of Locksmithing.* Denver, National Locksmith Building, 1953.

Mercer, G. Lawrence: *Security Systems and Equipment.* 3rd ed. Macdonald and Evans, 1969.

Momboisse, Raymond M.: *Industrial Security for Strikes, Riots, and Disasters,* Springfield, Thomas, 1968.

Mulbar, Harold: *Interrogation.* Springfield, Thomas, 1951.

National Fire Protection Association: *Guard Operations in Fire Loss Prevention.* Boston, NFPA, 1968.

Nelken, Sigmund: *Das Bewachungsgewerbe.* Koln, 1927.

New York State Civil Defense Committee: *Industrial Plant Protection Operations,* Plant Protection Operations Manual, Albany, 1961.

O'Hara, Charles E.: *Fundamentals of Criminal Investigation.* Springfield, Thomas, 1975.

Oliver, Eric, and Wilson, John: *Security Manual.* London, Gower, 1969.

———: *Practical Security in Commerce and Industry,* 2nd ed. London, Halstead Press, 1973.

Ottenburg, Miriam: *The Federal Investigators.* New York, Pocket Books, 1963.

Own, Wilfred: *The Metropolitan Transporta-*

tion Problem. Garden City, NY, Doubleday and Co., Inc., 1966.

Paine, David: *Industrial Security.* Oak Security Publications Division, May 1972.

Parkes, Josiah and Sons Limited: *The History of Locks.* Willenhall, J. Parkes, 1970.

Pike, E.: *Protection Against Bomb and Incendiaries for Business, Industrial and Educational Institutions.* Springfield, Thomas, 1972.

Pitt-Rivers, A. H. L. Fox: *On the Development and Distribution of Primitive Locks and Keys.* Chatto and Windus, 1883.

Plant Protection Manual. New York, Detex Watchclock Corporation Publishers, 1965.

Pollock, David A.: *Methods of Electronic Audio Surveillance.* Springfield, Thomas, 1973.

Post, Richard S.: *Determining Security Needs,* Nickerson-Collins, Des Plaines, Ill. 1973.

Price, George: *A Treatise on Fire and Thiefproof Depositories and Locks and Keys.* London, Simpkin, Marshall, 1856.

Reppetto, Thomas A.: *Residential Crime.* Cambridge, Mass., Ballinger Publishing Company, 1974.

Robinson, Clark: *Explosions, Their Anatomy and Destructiveness.* New York, Doubleday, 1939.

Safety in Handling of Explosives. Institute of Makers of Explosives, Technical Manual, 1940.

Schurman, E. A.: *Plant Protection.* Cornell Maritime Press, 1942.

Scott, W. R.: *Fingerprint Mechanics.* Springfield, Thomas, 1951.

Simms, Robert: *Bibliographical Account of Books and Other Printed Matter . . . Relating to the County of Stafford.* Lichfield, A. C. Lomas, 1894.

Skehan, J. J.: *Modern Police Work Including Detective Duty.* Brooklyn, R. V. Basuino, 1939.

Stoffel, Joseph F.: *Explosives and Homemade Bombs.* Springfield, Thomas, 1962.

The Thief You Pay. Los Angeles, Security World Publishing Co., Inc., 1969.

Tobias, Marc Weber: *Locks, Safes and Security: A Handbook for Law Envorcement Personnel.* Springfield, Thomas, 1971.

Towne, Henry R.: *Locks and Builders Hardware: a Handbook for Architects.* New York, John Wiley and Sons. London, Chapman and Hall, 1904 .

Treves, Ralph: *Do-It-Yourself Home Protection.* New York and London, Popular Science Publishing Company, Harper & Row.

Tsurumaki, Ya Suo: *Gadoman-no sekai.* 1971.

U.S. General Services Administration: *Handbook for Guards.* Washington, Government Printing Office, 1952.

———: Vulnerability, *Protection of Assets Manual,* Vol. 1, California, The Merritt Company, 1974.

Wade, G. A.: *Factory Defense.* London, Gale and Polden, Ltd., 1942.

Weber, Thad L.: *Alarm Systems & Theft Prevention.* Security World Publishing Company, Inc., 1973.

White, Michael (ed.): *New Trends in Office Management.* Business Publications Limited, 1967.

Wolff, Curt: *Die Zirilrechtleche stillung der wach- und schliessgesillschaften unter Besonderer Berucksichtigung der Parteiabreden.* Suttin Druck, General-Anyergir, 1926.

Woodruff, Ronald S.: *Industrial Security Techniques.* Columbus, Ohio, Charles E. Merrill Publishing Company, 1974.

Yale, Lines, Jr.: *A Dissertation on Locks and the Principles of Burglar Proofing, Showing the Advantages Attending the Use of the Magic Infallible Bank Lock, the Infallible Safe Lock and the Patent Door Lock Invented by Linue Yale.* Philadelphia, Collins, 1856.

PUBLICATIONS OF THE GOVERNMENT, LEARNED SOCIETIES, AND OTHER ORGANIZATIONS

African Insurance Record. Extract from a paper on "lock security," read by the Hon. G. C. H. Chubb to the Royal Society of Arts, *38:*31-32, March 1962.

American District Telegraph. The watchman's handbook. 3rd ed. New York, A.D.T., 1949.

American Society for Industrial Security. A guide to security investigations. Washington, D.C., Publications Department, ASIS, 1970.

American standard practice for lighting outdoor locations of central station properties. *Illuminating Engineering Society,* New York, April 1957.

American standard practice for protective lighting. *Illuminating Engineering Society,* New York, 1957.

American standard practice for lighting for

the Vidicon camera. *Illuminating Engineering Society*, New York, July 1963.

American Standards Association. American standard practice for protective lighting. New York, Am. Standards Assn., December 1956.

Amos, A. G.: History of safe making. *Association of Burglary Insurance Surveyors: Journal. 5:*29-35, 1958.

Annoying telephone calls, Training Key, No. 192, Maryland. Professional Standards Division, International Association of Chiefs of Police, 1973.

Ancient keys. *Journal of Design and Manufactures, 3:*160, 1850.

Association of Burglary Insurance Surveyors: Journal, 8, 1962-63. (Ceased Publication.)

ATF issues guidelines for dealing with bomb threat; calls and search procedures. *National Sheriff, 24:*1, 12, 20, 21, and 30, February 1972.

Atomic Energy Commission, U.S. AEC security manual.

Baker, J. Stannard: *Traffic Accident Investigator's Manual for Police*. Evanston, The Traffic Institute of Northwestern University, 1963.

Barnard, LeRoy: Techniques of detection—panelist. *Industrial Security,* October 1960.

Biles, David: *Car Stealing in Australia: Facts and Figures.* Canberra, Australian Institute of Criminology, 1975.

Bomb Conveying Vehicle. Milwaukee (Wisconsin), Police Department.

Bramah, Joseph: The petition and case of Jospeh Bramah of Piccadilly in the county of Middlesex, engineer, inventor of the patent locks for the security of life and property. London, 5 p. folio, 1798.

Briggs and Stratton Corporation. Automotive lock servicing manual. New York, Briggs and Stratton Corp., 1964.

British Museum. A guide to the antiquities of Roman Britain in the Department of British and Medieval Antiquities. Trustees of the British Museum, 1922, p. 42, 44-45, 67.

British Standards Institution. British Standard Code of Practice: Building. Precautions against fire. Part I, Chapter 4, 1962.

Building security, Honeywell planning guide. Research in education. H.E.W., Government Printing Office, January/December, 1970.

Check list for plant security, A. New York, National Association of Manufacturers.

Chleboun, T. P., and Chleboun, K. M.: Evaluation of small business and residential alarm systems, *1:* NTIS PB 214 975, Springfield, VA, 1972, p. 295.

Combination locks. Underwriters Laboratories, Inc., Chicago, National Board of Fire Underwriters, May 1961.

Courtney, J.: *Con Artist, Crime Rates and Confusion—Alarm Systems, Frequencies and the FCC.* (Communications), November 1972, p. 28-42.

Criminal investigation. War Department Field Manual FM 19-20, Washington, Government Printing Office, 1945.

Crockett, T. S., and Goering, G. G.: *Bomb Security Guidelines—The Protective Response.* IACP, 63 pages.

Dahlin, Roy E.: Seeing. Western Institute of Light and Vision, Official Bulletin #1, 63d Printing, Southern California Edison Company, 1935.

Department of the Army. Physical security of military and industrial installations. FM 19-30, Washington, Government Printing Office, 1952.

Domestic Disturbances. War Department Field Manual FM 19-15, Washington, Government Printing Office, 1945.

Eastman Kodak Company. Law enforcement series. Rochester, N.Y., 1961.

Electronic Engineering. Electronic watch dog sees all, hears all, gives industrial plants 100 percent protection. January 1960, p. 110.

Ennis, P.: Field Surveys II, criminal investigation in the United States: A report of a national survey. Washington, Government Printing Office, 1967.

Feasibility report and recommendations for a New York identification and intelligence system. Systems Development Corporation Report, TM-LO-1000/600/00, November 1, 1963.

Freimuth, Kenneth C. (Capt.): Lock security. Davis Publishing Co., Inc., Santa Cruz, Calif.

General investigative principles for security officers, vandalism prevention and investigation. Security Education Briefs, OAK Security, Inc., *1:* no. 7.

Hall, Earl, and Chappell, Jack E.: Man-machine systems to aid in the apprehension of career criminals. In Proceedings of the 2nd National Symposium of Police Science and Technology. Chicago, Illinois Institute of Technology, April 18, 1968.

Hamilton, Peter James Sidney: *Security Is an*

Attitude. London, Chubb and Son's Lock and Safe Company, 1965.

———: *The Philosophy of Security and the Technology of Burglary Prevention*, 1966.

Handbook for guards. U.S. Public Buildings Service, General Services Administration. Washington, Government Printing Office, April 1962.

Handbook of correctional institution design and construction. Federal Bureau of Prisons. Washington, D.C., Department of Justice, 1949.

Helicopter patrol—aerial surveillance in law enforcement. Los Angeles County Sheriff's Dept.

Higgen, Edward: Sketch of the history of the ancient modes of fastening doors. Lancashire and Cheshire Historic Society Proceedings, *3*:57-68, 1850.

Hogg, Garry: *Safe Bind, Safe Find; The Story of Locks, Bolts and Bars*. Phoenix House, 1961.

Hold-up alarm systems. Underwriter's Laboratories, Inc. 4th ed., Chicago.

Hopkins, Albert A.: *The Lure of the Lock*. New York, The General Society of Mechanics and Tradesmen, 1928.

Howington, Jon: Prevention and enforcement concepts of the patrol function for security officers. *Security Education Briefs*, OAK Security, Inc., *2*: no. 1.

———: Patrol concepts for security officers. *Security Education Briefs*, OAK Security Inc., *2*: no. 2.

———: Patrol techniques for security officers. *Security Education Briefs*, OAK Security, Inc., *2*: no. 8.

———: Introduction to technical investigative aids. *Security Education Briefs*, OAK Security, Inc., *2*: no. 3.

Illuminating Engineering Society. American standard practice for lighting outdoor locations of central station properties. New York, Illuminating Engineering Society, April 1957.

———. American standard practice for lighting for the Vidicon camera. New York, Illuminating Engineering Society, July 1963.

———. American standard practice for protection lighting. New York, Illuminating Engineering Society, 1957.

———. American standard practice for roadway lighting. New York, Illuminating Engineering Society, 1964.

Industrial defense against civil disturbances, bombings, sabotage. Washington, D.C., Department of the Army, Office of the Provost Marshall General, 1971.

Industrial security, part II, plant guard handbook. New York, National Industrial Conference Board, 1953.

Janowitz, Morris: *Social Control of Escalated Riots*. University of Chicago, Chicago Press.

Joyce, R. P.: Developing and implementing a burglary prevention program summary report. American Technical Assistance Corporation Grant, 1973, p. 41.

Kelling, George L., Pate, Tony, and Dieckman, Duane, and Brown, Charles: The Kansas City Preventive Patrol Experiment. A Technical Report. Police Foundation, 1974.

Krendel, E. S.: Protective device systems, 1970, p. 41.

Kubie, J. F.: Instrumental chemical, and psychological aids in the interrogation of witnesses. *Journal of Social Issues*, *13*:44-46, 1957.

Kubis, J. W.: Studies in lie detection. Report No. RADC-TR, United States Air Force, 1962, pp. 62-205.

Lomax, Joe B.: Parking lot security. *Security Education Briefs*, OAK Security, Inc., *2*: no. 4.

———: Parking security, part II. *Security Education Briefs*, OAK Security, Inc., *2*: no. 5.

Manual of Traffic Engineering Studies. New York, The Accident Prevention Department of the Association of Casualty and Companies, 1953.

Martinek, F.: *Plant Protection Guide*. Chicago, Chicago Association of Commerce and Industry.

McNamara, Joseph M.: *Burglary Protection Devices*. New York, New York Metropolitan Loss Prevention and Control Department, 1971.

Military explosives. Technical Manual TM 9-2900, Washington, United States War Department.

Moore, R. T.: Penetration tests on J-Siids Barriers. Measurements Automation Section, Information Processing Technology Division, Institute for Computer Sciences and Technology, Washington, D.C., June 4, 1973.

———: Penetration tests of reinforced concrete barriers. Measurement Automation Section, Information Processing Technology Division, Institute for Computer Sciences and Technology, Washington, D.C.

Morris, H. B.: *Elements of Successful Plant*

Protection. Louisville, Industrial Training Associated, 1953.

National Industrial Conference Board, Inc. Industrial Security I, *Combatting Subversion and Sabotage.* New York, 1952.

———. Industrial II, *Theft Control Procedures.* New York, 1954.

———. Industrial III, *Plant Guard Handbook.* New York, 1953.

National Technical Information Service. Proceedings of the 1973 Carnahan Conference on Electronic Crime Counter-measures, April 25-27, 1973, U.S. Dept. of Commerce.

Northwestern National Life Insurance Company. Amateur thieves outstealing "pros." Minneapolis, Family Economics Bureau, December, 1966.

PA Military & Civil Defense Commission. Principles of plant protection. Harrisburg, 1950.

Personal patrol car program, evaluation report. Prince George's County Police Dept., 1973.

Physical security for security officers. *Security Education Briefs,* OAK Security, Inc., *1:* no. 10.

Physical Security Committee. Security in a can? *Industrial Security,* October 1970.

Planning for the bomb threat. Guardsmark, Inc.

Planning guidelines and programs to reduce crime. 1972.

Principles of electricity and magnetism applies to telephone and telegraph work. New York, American Telephone and Telegraph Co., 1953.

Principles of plant protection. Munitions Board, Washington, Government Printing Office, 1950.

Principles of plant protection. U.S. Department of Defense Munitions Board, Washington, D.C., Government Printing Office, 1950.

Project Sky Knight—A demonstration in aerial surveillance and crime control—final report and evaluation report. Los Angeles County Sheriff's Dept., 1968.

Provost Marshal General's School. Industrial Defense Symposium, U.S. Army, Fort Gordon, Georgia, May 1960.

Sabotage and plant protection, panel discussion. 39th Annual Meeting, Chamber of Commerce of the United States, Washington, May 1951.

Sander, Wm.: What to do in handling bombs. Washington, National Bureau of Standards, U.S. Dept. of Commerce, 1940.

Scott, C.: *Photographic Evidence, Preparation and Presentation.* Kansas City, Vernon Law Book Co., 1942.

Secured Cable Alarm Systems. Mosler Safe Co. Bulletin, 1970, p. 16.

Security Control Conference. Procedures for Security Control. New York, The Conference Institute, 1969.

Security file containers. Underwriters' Laboratories, Chicago, National Board of Fire Underwriters.

Space management and courthouse security. *Architectural Design,* 1972.

Staley, Karl A.: Fundamentals of light and lighting. Large Lamp Department, General Electric Bulletin LD-2, Cleveland, Ohio, August 1960.

Standard for tamper-resistant doors, T-20. Underwriters Laboratories.

Standards for physical security of industrial and governmental facilities. Office of Civil Defense Mobilization. Supt. of Documents, Washington, D.C., Government Printing Office, 1958.

Standards for plant protection. U.S. Department of Munitions Board, Washington, D.C., Government Printing Office, 1952.

Traffic safety. Washington, National Safety Council, August 1967.

Underwriters' Laboratories, Inc. Accident automotive burglary protection equipment lists. Chicago, Illinois, Underwriters' Laboratories, Inc., September 1971.

———. Building materials list. Northbrook, Ill., Underwriters' Laboratories.

———. Burglary resistant safes. Chicago, National Board of Fire Underwriters.

———. Central station burglar alarm systems. Chicago, National Board of Fire Underwriters, December 1962.

———. Combination locks. Chicago, National Board of Fire Underwriters, May 1961.

———. Connectors and switches for use with burglar alarm systems.

———. 1970 field service record (Certified Burglar Alarms), Bulletin No. 31. Chicago, Underwriter's Laboratories, Inc.

———. Holdup alarm systems. Chicago, Underwriters' Laboratories, Inc.

———. Installation classification and certification of burglar alarms. Chicago, National Board Fire Underwriters, June 1965.

———. Intrusion detection units. Chicago, National Board Fire Underwriters, April 1964.

———. Key locks. Chicago, National Board Fire Underwriters.

———. Local burglar alarm systems. Chicago, National Board Fire Underwriters, November 1950.

———. Standard for tamper-resistant doors. Chicago, National Board Fire Underwriters.

U.S. Atomic Energy Commission. AEC security manual. Washington, Government Printing Office.

U.S. Congress Senate Committee on the Judiciary. Safeguard communication facilities hearing. 87th Cong., 1st Sess., June 7, 1961.

U.S. Department of the Army. Criminal investigation. Washington, Government Printing Office, 1951.

———. Counterintelligence survey, part II. 1964.

U.S. Department of Army Field Manual FM 19-30. Physical security of military and industrial installation.

U.S. Department of Army. Physical security surveys.

U.S. Department of Commerce. Emergency and disaster planning for the water and sewage utilities. Washington, D.C., Government Printing Office.

———. Iron and steel—industrial defense planning manual. Washington, D.C., Government Printing Office, October 1965.

———. Barrier penetration tests. Washington, Government Printing Office, 1974.

———. National Technical Information Service. Penetration tests on J-SIIDS barriers. NTIS, Washington, D.C., June 4, 1974.

U.S. Department of Commerce Area Development Division. Industrial dispersion guidebook for communities; a technique for more secure location of new defense-supporting plants. 1952.

U.S. Department of Defense. Fallout protection. *Industrial Security*, 1961.

———. Industry guide to planning for restoration of production. Washington, Government Printing Office, 1954.

———. Military support of civil defense. Washington, Government Printing Office, August, 1964.

U.S. Department of Defense Munitions Board. Principles of plant protection. Washington, Government Printing Office, 1952.

U.S. Department of Defense, Office of Civil Defense. Standards for physical security of industrial and government facilities. Washington, Government Printing Office.

U.S. Department of Health, Education, and Welfare. An outline guide concerning sanitation aspects of mass evacuations. Washington, Government Printing Office, 1956.

U.S. Department of Interior. Civil defense in the minerals and solids fuels industries. Washington, Government Printing Office, June 1964.

———. Federal records of World War II civilian report to the under secretary of war. *1:* Washington, Government Printing Office, July 1, 1945 to June 30, 1947.

———. Civil defense preparedness in the electric power industry. Washington, Government Printing Office, March 1966.

U.S. Department of Justice. Law Enforcement Assistance Administration Emergency Energy Committee. Street lighting, energy conservation and crime. No. 2, Washington, Government Printing Office, March 1974.

U.S. Federal Bureau of Investigation. Suggestions for protection of industrial facilities, 1941.

U.S. General Services Administration. Building organization for self protection. Washington, Government Printing Office, September 1959.

U.S. House of Representatives. Security practices in national security agency. Washington, Government Printing Office, 1962.

U.S. Munitions Board Office of Industrial Security. Standards for plant protection, 1962.

U.S. Office of Civil and Defense Mobilization. 10 steps to industrial survival. Washington, Government Printing Office, April 1960.

———. Emergency sanitation at home—a family handbook. Washington, Government Printing Office.

———. Plant protection fact sheet. 1943.

———. Protection of industrial plants and public buildings. 1941.

———. Sabotage and preventive measures. July 1943.

U.S. Office of Emergency Planning. The national plan for emergency preparedness. Washington, Government Printing Office, December 1964.

Use of electronic data processing equipment in the federal government. House Report No. 858, Committee on Post Office and Civil Service, 88th Cong., 1st Sess., 1963.

Wallace, B. B.: *Co-Working for Public Safety.* 1967.

Westinghouse Electric Corp., Lighting Division. Light for plant safety and security. Cleveland, Westinghouse, 1951.

Wright, Roger, et al.: The impact of street lighting on street crime: Technical appendices. May 1974.

———: The impact of street lighting on street crime: Results. May 1974.

PERIODICALS

Alarm manufacturers in national meeting. *Security World*, December 1969.

Altenburg, Arthur J.: Notes on building security. *The Police Chief*, November 1970.

Amatt, Leslie Keith: Padlocks; a brief history. *Ingersoll News*, June 1973, p. 3.

Amoroso, Louis J.: Parking lot security. *Security World*, September 1970.

Anonymous and Staff Written: CCTV has as yet only been lightly touched upon. *Security World*, V: no. 3, March 1968.

———: 114,000 credentials: Access control at the democratic convention. *Security World*, V: no. 9, October 1968.

Anrider, S. A.: Cameras shoot first: Answer questions later. *Banking*, June 1964.

Anti-intruder fencing. *Security Gazette*, March 1975, p. 88-89.

Antonucci, Joseph T.: Effective communication means good security. *Industrial Security*, February 1968, p. 16.

Attack on plant protection. *Factory*, November 1961.

August, Kendall: New complexities in plant security. *Dun's Review and Modern Industry*, March 1965.

Automate plant protection. *Factory*, December 1961.

Avondale shipyards utilizes pros for varied plant security program. *Maritime Reporter/ Engineering News*.

Baking Industry, Plant security at Nabisco. December 12, 1959, p. 47.

Bannon, J. D.: Foot patrol—the litany of law enforcement. *Police Chief*, 39: no. 4, 44-45, April 1972.

———: Stress, zero visibility policing. *Police Chief*, 39: no. 6, 32-34, 36, June 1972.

Barnham, E. S.: The development of modern locks. *Security Gazette*, February 1961, p. 47-48.

Barrell, Clive: Tear gas: Offensive or Defensive. *Top Security*, June 1975, p. 82-83.

Bastoni, Jeannie: The small company faces visitor control. *Industrial Security*, February 1970.

Beck, E. W.: The keys of S. Peter at Liege and Maestritcht. *Archaeological Journal*, 47, 1890.

Benham, Edward: The economics of electronic surveillance. *Security World*, 5: no. 9, October, 1968.

Bennett, Charles C.: A case for closed circuit television. *Industrial Security*, January 1965.

Benton, J. R.: How the safemaker meets changing methods of criminal attack. *Security Gazette*, November 1974, p. 406-407.

Berkovitch, Israil: The danger within—some guidelines for vehicle safety on factory premises. *Security Gazette*, May 1974, p. 186.

———: Fingerprint identification of gems. *Security Gazette*, November 1974, p. 409-410.

Berry, Alfred B., and Buckley, John L.: Security officer's notebook: Industrial security and law enforcement. *Law and Order*, 10: no. 12, December 1962.

Betz, G. M., and Edner, E. A.: Locks and keys; sentinels of plant security. *Plant Engineering*, June 1964.

Big brother is listening: Ultrasonic alarm guards classified papers. *Mill and Factory*, February 1958.

Bilhorn, Robert: Card key access control at Chicago convention. *Industrial Security*, December 1968.

———: Access control. *Security Register*, November 1973, p. 30+.

Birminghan, Michael G.: Field operations for security. *Police Yearbook*, 1963.

Bishop, Vernon R.: Choose the right protective lighting arrangement. *Plant Engineering*, October, 1964.

Bisset, George: Sabotage threat in the electric utility industry. *Industrial Security*, October 1958.

Blackwell, Robert R.: The art of detection, by Jack Fisher. *Industrial Security*, April 1964.

Bloom, E. Kenneth: Security barriers. *Security and Protection*, December 1974, p. 9-10.

Bomb threats: A practical procedure. *Industrial Security*, June 1970, p. 5-6.

Bomb threats and personnel safety, part II. *Industrial Security*, February 1971.

Borell, Clive: The search for the ultimate in a non-lethal weaponry. *Top Security*, August 1975, p. 152-154.

Boyle, William M.: The application of infrared technology of intrusion detection system. *Industrial Security*, February 1968.

Brace, E. M.: The T.V. camera as witness: Video tape recording in internal theft. *Security World*, 4: no. 19, 14, 1967.

Brooks, D. C. T.: Perimeter lighting. *Security and Protection*, September 1974, p. 26-27.

Brookshire, Leo A.: Five "E's" for industrial security. *Industrial Security*, April 1961.

Brown, J. W. M.: Preserving the scene. *Security and Protection*, January 1975, p. 11-12.

Brown, Carlton E.: Shelters in industrial plants. *Industrial Security,* July 1960.

Brownell, Adon H.: Builders' hardware handbook. *Hardware Age,* 1956/1961.

———: Taking the mystery out of builders hardware. New York and Philadelphia, *Hardware Age,* 1940.

Browster, K.: Ultrasonic intruder detection. *Security Gazette,* November 1974, p. 404+.

Buckley, F. S.: New methods make an exact science of anti-trusion protection. *Law and Order,* March 1964.

Buckley, John L.: Security officers' notebook. *Law and Order,* March 1962.

———: How to select the proper security and equipment surveillance systems to protect your facilities. *Law and Order,* May 1964.

Bugg, Donald Ernest, and Price, R. S.: Protection against burglary and housebreaking: The importance of action at the planning stage of building. *The Builder, 186:* June 1954.

Builder. Door locks. 207:946, October 3, 1964.

Bunt, Cyril G. E.: Old wooden locks. *Architectural Review,* March 1915, p. 53-55.

Burglary resistant safes. *Standard for Safety,* July 20, 1972.

Burglary resisting glazing material. *Standard for Safety,* June 2, 1972.

Burkett, Frank E.: A plant engineer's guide for protecting plant personnel from tornadoes. *Plant Engineering,* February 1963.

Burton, Lucius W.: Model school security system cuts crime. *Security World,* June 1975, p. 12+.

CCTV and safety. *National Safety News,* January 1971.

Caffey, J. J.: Protection by electronic watchmen. *Law and Order, 2:* no. 12, December 1963.

Cahill, Carl: Planning security in new buildings. *Security Management,* November 1974, p. 20-21.

Cameras shoot first; answer questions later. *Banking,* June 1964.

Capshaw, Roy E., Jr.: Capabilities and limitations of electrical devices used in plant protection. *Industrial Security,* April 1964.

———: Electrical devices used in plant protection. *Industrial Security,* April 1964, p. 50.

Carley, William M.: Electronic advances aid snoopers using eavesdropping devices. *Wall Street Journal,* April 9, 1963.

Carmody, Deirdre: Outwitting burglars is now universal pastime—variety of locks tried by apartment dwellers. *New York Times,* April 23, 1970, p. 39.

Carry, Edward J.: Technical development in security equipment panel moderator. *Industrial Security,* December 1964.

Cassell's Illustrated Exhibitor and Magazine of Art. Locks and keys, 2:59-64, 1852.

Central station: A case in point. *The Security World,* October 1970.

Central Station burglar-alarm units and systems. *Standard for Safety,* June 19, 1972.

CCTV has as yet only been lightly touched upon. *Security World, 5:* no. 3, March 1968.

Chamberlain, Gary M.: Improve your city's street lighting. *The American City,* November 1974.

Chamber's Journal. Locks and keys, 48:107-110, 1871.

Chatwood, Samuel: Locks and safes. *Journal of the Society of Arts, 36:*704-718, May 11, 1888.

Cherico, Phillip J.: The physical protection of nuclear power plants and materials. *Security Management,* March 1975, p. 48-51.

Christiansen, Douglas G.: A critical analysis of interrogation for investigators, by Richard O. Arthur. *Industrial Security,* April 1964.

Chubb, George Charles Hayter: Good security is beating the thief. *Security Gazette, 8:* no. 1, 8, January 1966.

Chubb, Harry Withers: The construction of locks and safes. *Journal of the Society of Arts,* April 14, 1893, p. 510-528.

———: On locks and keys: report of a paper read at Glasgow to members of the architectural section of Glasgow Philosophical Society. *Scotsman,* February 1888, p. 4.

Chubb, John: On the construction of locks and keys. *Institution of Civil Engineers: Proceedings,* April 9, 1850, p. 310-343.

———: On the history and construction of latches and locks. *Society of Arts Transactions,* January 1851, p. 37-41.

Chubb-Mosler and Taylor Safes Ltd.: The burning bar reviewed. *Security World,* March 1969.

Closed circuit TV system plays role of guard. *Factory,* September 1964.

Cochran, Murray O.: Police science traps lone wolf burglar. *Law and Order, 12:*12-13, 1964.

Cohen, Kenneth: Nuclear power plants—new challenge to security. *Security Management,* January 1975, p. 30-31.

Coile, R. C.: Parabolic sound concentrators. *J. Soc. Motion Picture Eng., 51:*298-311, September 1948.

Coin telephone larceny. *Training Key,* No. 194, Maryland: Professional Standards Division,

International Association of Chiefs of Police, 1973.

Combination locks. *Standard for Safety*, July 20, 1972.

Communications explosion: Extending security electronically, The. *Occupational Hazards*, April 1971, p. 53.

Connectors and switches for use with burglar alarm systems. *Standard for Safety*, June 29, 1973.

Construction of safes, The. *Security Surveyor*, *1:* no. 2, 21-30, July 1970.

Cooper, Clare C.: Fenced back yard—unfenced front yard—enclosed front porch. *The Journal of Housing*, no. 5, 1967, p. 268-274.

Copeland, Joseph H.: Technics for the crime investigation by William Dienstein. *Industrial Security*, April 1964.

Cousins, Margaret: How to foil the burglar. *House Beautiful*, no. 111, February 1969, p. 72-73+.

Cowie, Donald, and Henshaw, Keith: Antique collectors' dictionary. *Arco*, 1962, p. 117-118.

Crompton, R.: Crime prevention squad experiment makes impact on premises security. *Security Gazette*, December 1970.

Cummings, Brian W.: After the lights go out. *Security and Protection*, September 1974, p. 23-24.

Curtis, Bob: Barefoot in broken glass. *Top Security*, May 1975, p. 26-30.

Daskam, Samuel W.: Electronic surveillance. *Industrial Security*, August 1971.

Davis, Derek: Handwriting analysis. *Security and Protection*, February 1975, p. 8-10.

Delayed-action timelocks. *Standard for Safety*, July 20, 1972.

Delmage, Sherman: Electronic countermeasures, part II. *Security World*, *3:* no. 2, February 1966.

———: Design for security. *Progressive Architecture*, *51:*84-89, November 1970.

Dessin, G. H., et al.: Drug induced revelation and criminal investigation. *Yale Law Journal*, February 1953, p. 315-47.

Dial plate updates combination lock. *Machine Design*, *36:*212-213, 1964.

Dickie, Robert: Locks: A basic element of security. *Security Register*, January-February 1974, p. 15+.

Dierst, Glenn V.: Alert warning and communication system for industry. *Industrial Security*, July 1960.

Donald, I. A.: Architects journal technical study: Aspects of security-safes. *Architects' Journal*, no. 139, 611, March 1964.

Donovan, Robert: A manual for bomb attacks. *Security World*, *5:* no. 5, May 1968.

Dudley, W. E.: Electronics guards plants. *Electronics*, November 13, 1959.

———: Interrogation takes practice. *Industrial Security*, August 1970.

———: Electronic watchdog sees all, hears all, gives industrial plants 100 percent protection. *Electronic Engineering*, January 1960.

———: How to guard a plant. *Duns Review and Modern Industry*, March 15, 1964.

———: This sound system tests itself. *Duns Review and Modern Industry*, July 1956.

Dussia, Joseph, Lt. Col.: Safe burglary investigation, part I. *Security World*, *5:* no. 7, July-August 1968.

———: Safe burglary investigation, part II. *Security World*, *5:* no. 8, September 1968.

Edings, C. A.: Collecting locks and keys. *Connoisseur*, no. 105, September 1940, p. 106-110.

Eicholz, Jack, and Allen, Louise: Crackdown on the telephone crank. *Family Weekly, Elmira Star Gazette*, Elmira, N.Y., August 18, 1963.

Electronic watchdogs guard Minnesota campus. *Security Management*, January 1974, p. 45-46.

Electronic guard. *Business Week*, April 13, 1963, p. 104.

Elements of effective gate control. *Security Gazette*, April 1963.

Engineer. Concerning locks and keys. December 1, 1865, p. 347.

———: New locking attachment for bicycles. August 2, 1890, p. 66.

———: A new form of lock. September 1906, p. 279.

———: Locks for sliding doors of railway carriages. January 2, 1914, p. 27.

———: Zippman and Furthmann's railway carriage lock. February 13, 1891, p. 199.

———: Burglar-proof locks and safes. March 9, 1866, p. 178.

Equipment advances, losses retreat. *Occupational Hazards*, October 1963.

Eskin, David H.: Lock and key control and master keying. *Industrial Security*, December 1964.

Everist, J. A.: Infra-red burglar alarm using OCP71 phototransister. *Mullard Technical Communications*, *5:* August 1961.

Eversull, Kennen: Automatic plant protection. *Factory*, December 1961.

———: Consideration in radio communications. *Security World*, *5:* no. 9, October 1968.

———: Planning the communications center, part I. *Security World,* 5: no. 8, September 1968.

———: Planning the communications center, part II. *Security World,* 5: no. 11, December 1968.

Farrell, R. J.: Exterminating the electronic "bugs." *New York Herald Tribune,* March 29, 1964.

Field, Franklin: New techniques in processing latent fingerprints—panelist. *Industrial Security,* December 1964.

Fleming, George: Effective plant security via electronics. *Food Engineering,* February 1962.

Fogle, Robert C.: Detection of transmitting devices, part 1. *Security World,* October 1970.

———: Detection of transmitting devices, part 2. *Security World,* November 1970.

Food Engineering. Effective plant security via electronics. February 1962.

Forbin, John C.: Alarm system that protects 16 ways. *Factory Management,* October 1954.

Foster, Reginald: In the lab. *Security and Protection,* February 1975, p. 4-6.

Foster, H.: Street lighting in crime prevention. *Crime Prevention News,* June 1975, p. 16-19.

Freimuth, K. C.: *Military Police Journal,* 21: no. 10, 21-24, May 1972.

Fry, Phillip L.: Redstone: Its eyes and ears. *Military Police Journal,* August 1964.

Fulton, W. K.: Physical security measures. *Research Management, VII:* no. 5, 1964.

Fuss, Eugene: Security equipment and technology: Protection systems' contact devices, part III. *Security World,* January 1975, p. 41+.

———: Security equipment and technology; protection systems' contact devices, Part IV. *Security World,* February 1975, p. 21+.

———: Security equipment and technology; protection systems' contact devices, part IV (cont'd). *Security World,* March 1975, p. 42-44.

———: Security equipment and technology, protection systems, part VI. Audio detection systems. *Security World,* April 1975, p. 44-45.

———: Security equipment and technology; protection systems, part VII. Vibration detection. *Security World,* May 1975, p. 40+.

———: Security equipment and technology; capacitance direction systems. *Security World,* June 1975, p. 42+.

———: Ultrasonics: How the Doppler effect works. *Security World,* July-August 1975, p. 20, 134, 135.

———: Handbook of modern alarm systems, part II. *Security World,* September 1974, p. 40-45.

———: Handbook of modern alarm systems, part III. *Security World,* October 1974, p. 44-48.

———: Security equipment and technology. *Security World,* October 1974, p. 26.

———: Security equipment and technology, protection systems, contact devices. *Security World,* November 1974, p. 69.

———: Security equipment and technology, protection systems, contact devices, part II. *Security World,* December 1974, p. 24+.

Gadgets with big ears; an end to privacy? *U.S. News and World Report,* April 22, 1955.

Galvin, Aaron A.: Devices for premises protection. *Security News,* January 30, 1974, p. 1, 19.

Gambling, W. A.: The laser: A new power in communications and security. *Security Gazette,* April 1966.

Gardner, John: Shooting your bolt in home protection. *Security Gazette,* August 1974, p. 300.

Gate guarding is a science at Cyanamid. *Occupational Hazards,* June 1965.

GE guards are tuned into TV. *Occupational Hazards,* March 1970, p. 50.

General Foods Corporation Guide: Protection of company information and property. *Industrial Security,* December 1967, p. 30.

George, Harold S.: Practical limitations to master keying. *Security World,* April 1971.

Germann, A. C.: Guard force. *Factory,* October 1962.

Gilbert, James B.: Plant security overseas—panelist. *Industrial Security,* December 1954.

Glass in the service of security. *Security Gazette,* May 1962.

Glenn, Todd R.: Voice print identification—panelist. *Industrial Security,* December 1964.

Goering, George B.: Basics for bomb emergency planning. *Security World,* June 1970.

Gorman, R. G.: Electronic devices—the pros and cons. *Top Security,* August 1975, p. 169.

Grazeler, F. J.: Growing guards. *Life,* March 10, 1952.

Greany, William: Parking lot security. *Security Management,* Washington, D.C., 1973.

Green, Victor: Surveillance equipment—its

functions and uses. *Security World*, February 1975, p. 15+.

Gregg, James R.: Visual needs of policemen. *Law and Order*, 7: October 1959.

Grumbach, A. T.: Locks and locking. *Security World*, June 1975, p. 32+.

Guard service handles plant security at low cost. *Iron Age*, August 27, 1959.

Guarded Lorry parks and in-transit cover. *Security Gazette*, December 1971.

Hale, D. A.: The comparative security of locks. *Brokers Monthly*, no. 15, October 1965, p. 5935-5939.

Halm, Eugene J.: A vehicle theft ring—how it operates—ditto. *Industrial Security*, October 1964.

Hamilton, Peter: The courage not to react violently. *Security World*, November 1969.

Handling strikes in industry. *Security Register*, November 1973, p. 59-60.

Handwriting comparison for the security officer. *Security World*, 1: no. 1, July 1964.

Hampton, R. A.: Alarming dangers. *Security and Protection*, December 1974, p. 14-19.

Hart, John W.: Guarding the city within the mountain. *Security World*, July/August, 1970.

Haswell, William S.: Lock and key control and master keying—a neglected area of security. *Industrial Security*, December 1964.

Hata, Bill: Demonstration halts employee in plant stealing. *Law and Order*, 12: October 1965.

Hatschek, R. L.: Fast communication can lift your plant efficiency. *Mill and Factory*, November 1962.

Hazardous materials board amends rules for fire extinguisher shipping. *Industrial Fire and Security Report*, January 10, 1975, p. 5.

Healy, R.: Disaster planning. *Security World*, November 1965, p. 24.

Healy, Richard: Industrial security and physical controls. *Industrial Security*, April 1968, p. 3.

———: Systems engineering and industrial security. *Industrial Security*, June 1965, p. 10.

———: Riot and premise protection. *Security World*, 2: no. 6, September 1965.

Heaton, H. A.: Old locks and keys. *Leisure Hour*, no. 47, March 1899, p. 323-329.

Hedin, Robert A.: Electronic identification of the human population for controlled access and automatic crediting. *Industrial Security*, April 1967.

———: Space age security. *Industrial Security*, June 1971.

Heims, Peter: Control of the private investigator. *Top Security*, July 1975, p. 128-129.

Hendrickson, D. G.: An experiment in low-cost time lapse photography. *Industrial Security*, January 1962.

Higgins, George D., Jr.: Background investigations—panel moderator. *Industrial Security*, October 1960, p. 14.

High cost of civil disorders, The. *Security World*, July/August 1970.

Hinman, David B.: Combating copper theft. *Security Management*, January 1975, p. 14-17.

Hobbs, Ian J.: Emergency lighting. *Security and Protection*, December, 1974, p. 20.

Hoefler, Don: The hardware of eavesdropping. *Security World*, May 1966, p. 18A.

Holcomb, Richard L.: *Police Patrol*. Springfield, Thomas, 1952.

Hold-up camera "does its thing" . . . and more. *Security World*, April 1970.

Hold-up alarm units and systems. *Standard for Safety*, July 27, 1973.

Hollinger, J. R., and Mulligan, J. E.: Build the shotgun sound snooper. *Popular Electronics*, June 1964, p. 51, 84.

Holmes, Warren: Interrogation—the guilty reveal themselves. *Security World*, 1: no. 2, September 1964.

———: New hope for the innocent: APA's case review committee. *Security World*, 5: no. 8, September 1968.

Hopkins, Pascal B.: Applicant investigations. *Industrial Security*, January 1963.

Horner, William J.: A secret lock for a cabinet door. *Scientific American*, 105:497, December 1911.

Hotel master keying. *Security World*, September 1965.

Household burglar-alarm system units. *Standard for Safety*, September 12, 1972.

How closed-circuit TV cuts pilferage losses for chain store. *Chain Store Age*, September 1959.

How good is your building security? *Buildings*, June 1962.

How to guard a plant. *Duns Review and Modern Industry*, March 1962.

How your firm can guard against riots. *Business Management*, September 1968, p. 36-38.

How to make security fundamentals pay off. *Occupational Hazards*, April 1970, p. 76.

How to protect your plant during a strike. *Occupational Hazards*, August 1970, p. 45.

How to select the proper security and equipment surveillance systems to protect your facilities. *Honeywell Planning Guide*, Honeywell.

Howard, Seymour: Hardware—19: Lock functions. *Architectural Record,* January 1951, p. 143, 157.

Hubbard, David: Extortion threats: The possibility of analysis. *Assets Protection,* Summer, 1975, p. 17-19.

Hughes, Mary Margaret: High-rise safety and security in action. *Security World,* March 1975, p. 12-16.

———: High-rise safety and security in action, part II. *Security World,* April 1975, p. 20+.

———: High-rise safety and security in action, part III. *Security World,* May 1974, p. 16-19.

———: High-rise safety and security in action, part IV. *Security World,* June 1975, p. 20+.

———: Loss prevention in riot-affected area, part I. *Security World, 2:* no. 4, April 1968.

———: Loss prevention in riot-affected areas, part II. *Security World, 5:* no. 5, May 1968.

———: Loss prevention in riot-affected areas, part III. *Abstracts on Criminology and Penology.*

Hutchinson, William James: A policeman puts you wise. *Police Journal,* 1951.

Ideas you can use. *Occupational Hazards,* April 1960.

In the 70's—the posture of plant security. *Occupational Hazards,* October 1969, p. 113.

In-office ID systems aid company security. *Administrative Management,* April 1969, p. 24.

Industrial communications, part VI. *Plant Security,* December 1962, p. 139.

Industrial cops shooting it out. *Printer's Ink,* May 22, 1964.

Industrial security II, plant guard handbook. *Studies in Business Policy,* no. 64, New York, 1953.

Industrial security and plant engineering—a joint responsibility. *Plant Engineering,* February 1962.

Industry cracks down on plant pilferage. *Management Review,* November 1957.

In office ID systems aid company security. *Administrative Management,* April 1969, p. 24-28.

Inside the Ford design center. *Occupational Hazards,* February 1967.

Installations and classification of mercantile and bank burglar-alarm systems. *Standard for Safety,* June 27, 1972.

In-house investigative resources. *Protection of Assets Manual,* September 1975, p. 23-46.

Intrusion detection units. *Standard for Safety,* December 29, 1971.

Investigative casebook. *Security World,* January 1969.

Jensen, C. F.: Techniques for exposing the industrial saboteur. *Illuminating Engineer,* no. 47, July 1952. p. 367-368.

Jerome, Frank A.: Inventory and maintenance management of audiovisual equipment. *Audiovisual Instruction,* March 1975, p. 98-99.

Jesse, John P.: Photo instrumentation—Investigation aid. *Industrial Security,* April 1965.

Jeweler's circular—Keystone: A century of jeweler's security. *Security World,* November 1969.

Jewitt, Llewellynn: Gravemounds and their contents: A manual of archaeology, as exemplified in the burials of the celtic, the Romano British, and the Anglo-Saxon periods. London. *Groombridge,* 1870, p. 189, 201.

Johnson, P. J.: Streetlighting brings results. *Nation's Cities,* August 1973.

Johnstone, K. T.: The accident-repeater. *Security World,* February 1969.

Johnstone, T. H.: Locks and locking mechanisms. *Industrial Security,* July 1963, p. 21.

———: Lock and key control and master keying. *Industrial Security,* December 1964.

Jones, C. N.: New retail electronics system aids industry environment. *Security World, 10:* no. 9, 25, 72, October 1973.

Judd, R. E.: Latching on. *Security and Protection,* October 1974, p. 8-10.

———: The role of locks in security. *Security and Protection, 3:*35-37, May 1971.

Jupiter, Robert M.: Security and guards and firearms. *Industrial Security,* October 1970.

Kadlec, J.: The construction of safes and bank strongrooms. *International Criminal Police Review, 37:*101, 1950.

Kasky, Frank: Explosives and bomb investigation. *Security Register,* November 1973, p. 40+.

Kearns, John G.: Legislation in the field of crime prevention. *Security World, 6:* June 1969.

Keeler, Frank J.: Determining vulnerability of your plant to attack. *Industrial Security,* July 1960.

Kempf, Vern: Industrial communications, part VI, plant security. *Plant Engineering,* December 1962, p. 132+.

Key locked safes, Class KL. *Standard for Safety,* May 31, 1973.

Key locks. *Standard for Safety,* May 15, 1973.

King, Tracy: Recovery boiler efficiency, control

and safety through TV monitoring. *Paper Trade Journal,* October 19, 1964.

King, Warren, Lt.: Technical casebook: Rapport is the key to communication. *Security World, 1:* no. 1, July 1964.

———: Technical course: An introduction to interrogation. *Security World, 1:* no. 1, July 1964.

Kingsbury, Arthur: Guidelines for security education, physical security. *Industrial Security, 16:* no. 2, 12, April 1972.

———: Guidelines for security education, basic investigation. *Industrial Security, 16:* no. 4, 12, August/September 1972.

———: Guidelines for security education, report writing. *Security Management, 17:* no. 5, 31, October/November 1972.

———: Guidelines for security education, introduction to security and crime prevention surveys, part I. *Security Management, 18:* no. 1, 30, April, 1973.

———: Guidelines for security education, introduction to security and crime prevention surveys, part II. *Security Management, 18:* no. 2, 16, May 1973.

Komaroff, Samuel: Fidelity claims require further proof. *Weekly Underwriter,* December 10, 1960.

Krill, Jackson N.: The secret service and plant protection. *Industrial Security,* April 1970.

Lauback, Alexander C.: The future of CCTV in security. *Security Management,* September 1974, p. 16-17.

———: Industrial survival in nuclear war. *Industrial Security,* December 1968.

Laughlin, R. W.: On the beach—low-key law enforcement. *Police Chief, 39:* no. 3, 22-23, March 1972.

Lawless, T., and Peterman, J.: Auto theft and the uniformed officer. *Police Chief, 40:* no. 6, 58-62, June 1973.

Lederer, F. J.: Exclusive inside story of Philadelphia vice investigation. *Official Detective Stories,* February 1957, p. 10.

Lee, E. L.: Back to bikes for Baltimore. *Police Chief, 39:* no. 5, 22, May 1972.

Leedham, Charles: The "Chip" revolutionizes electronics. *The New York Times,* September 19, 1965, p. 56.

LeNay, Tom W.: Alarm sensors and systems, parts I and II. *Security World,* March-April 1971.

Letter and package bombs. *Security Register,* November 1973, p. 50.

Lewe, William T.: Warehouse security. *Security World, 1:* no. 3, November 1964.

Lewin, Thomas M.: How serious are false alarms? *Security World,* October 1969. 2:62-63, January 1972.

Liardet, V. H.: Key security. *Security Surveyor,*

Lighting decreases crime rate. *Municipal South,* reprinted by Street and Highway Safety Lighting Bureau, New York.

Lindrose, C. W., Jr.: Emergency power for vault alarm systems. *Industrial Security,* February 1966.

Linn, William C.: Guarding celebrities. *Security World, 4:* no. 2, February 1967.

Local burglar-alarm units and systems. *Standard for Safety,* January 12, 1972.

Lock it. *Crime Prevention Newsletter,* no. 22, June 1974.

Lock opened by electronic combination. *Machine Design, 44:*114, January 1972.

Lorence, Earl F.: Undercover investigation: Management's periscope. *Security World, 2:* no. 4, June 1965.

Lorenz, Robert: A chain-wide store alarm program. *Security World, 5:* no. 9, October 1968.

Launges, Ernest: Paper fraud deterred by photographic system. *Security World,* June 1974, p. 15+.

Love, Robert C.: The security officer's notebook. *Law and Order, 9:* January/April 1961.

———: Employment of audio-visual techniques—panelist. *Industrial Security,* October 1963.

Lyons, Stanley: Security lighting economics. *Security Gazette,* September 1974, p. 331.

MacHayes, John: Faulty burglar alarms. *Top Security,* September 1975, p. 183-185.

Mackey, John B.: Implementing security requirements check list. *Industrial Security,* December 1961.

Mahoney, Bill: Electronic eyes. *Law and Order, 9:* November 1961.

———: Inside the Ford design center. *Occupational Hazards,* February 1967.

———: Protecting life and property with TV. *Law and Order, 12:* May 1964.

———: Security checklist. *Occupational Hazards,* November 1966.

———: Trespassers, beware: Electronic watchdogs on duty! *Occupational Hazards,* June 1963.

———: TV and radio: Security standbys. *Occupational Hazards,* September 1966.

Major security challenge solved through use of I.D. system. *Security Industry and Product News,* July/August 1975, p. 28, 30.

Malone, Lee F.: Control of closed areas by

closed-circuit television. *Industrial Security,* April 1958.

———: Modern electronics assisted by the age-old mirror. *Industrial Security,* January 1963.

Manthey, R. F.: Electronic watchdog: Closed circuit TV tightens plant security. *Plant Engineer,* December 1965.

Marcher, Sterling R.: Evidential photography. *Security World, 2:* no. 7, October 1965.

Markel, N. M., Meisels, J., and Houck, J. E.: Judging personality from voice quality. *Journal Abnormal Social Psychology, 69:* no. 4, 458-462, 1964.

Mason, Andrew: Security lighting, basic requirements. *Security and Protection,* September 1974, p. 19-20.

Matias, G. J.: Better criminal mousetrap. *FBI Law Enforcement Bulletin, 40:* no. 5, 11-14, 29, September 1971.

Mattson, John H.: Security aspects of the dewline. *Industrial Security,* January 1960.

McCollum, David, Jr.: Some points on bombs and bomb threats. *FBI Law Enforcement Bulletin,* April 1971, p. 13.

McGrady, Mike: You, too, can bug people ... and their phones. *Miami Herald,* September 23, 1965.

McGuire, Patrick E.: When bombing threatens. *The Conference Board Record,* September 1971.

McKeon, Joseph M., Jr.: Mechanical and electronic security measures. *Industrial Security,* August 1965.

McMurray, Robert N.: Who riots and why. *Nation's Business,* October 1967, p. 72-75.

McPhail, W. A.: Security in factory. *Metal Industry, CIII:* September 12, 1963.

Meisler, Stanley: Trial by gadget. *Nation, 199:* September 1964.

Metarelis, George S.: Lawmen for the reservation. *FBI Law and Enforcement Bulletin, 40:* 1971.

Mertz, Louis F.: We're playing a losing game. *Industrial Security, 9:* no. 1, January 1965.

Miles, W. P.: Locks and fastenings. *Society of Engineers,* March 5, 1860, p. 26-29.

Miller, A. Ross: Physical security: The in depth concept. *Industrial Security,* October 1967.

Miller, Floyd G.: Practical solutions to entrance security for large area, multiple buildings. *Security World,* October 1974, p. 16-20.

———: Practical solutions to entrance security for large area, multiple buildings, part II. *Security World,* November 1974, p. 36.

Miller, John: Lorry thefts and the insurer. *Security Gazette,* October 1974, p. 37.

Miller, Joseph G.: Infrared burglar alarm. *Police Chief, 29:* no. 3, March 1962.

Miller, Stewart E.: Communication by laser—the unique properties of the light produced by lasers open the way to the eventual exploitation of light waves for the long-distance transmission of electrical signals. *Scientific American, 214:*19-27, January 1966.

Milsom, H. W.: Old locks. *The Ironmonger,* no. 249, November 8, 1958, p. 647.

Mingus, R.: Thief you pay. *Security World Publications,* California, 1969, p. 64.

Miniature microphones. *Advertisement in The New York Times International Edition,* June 30, 1964.

Modern myths in thief resistance: A look at some popular fallacies by "Accident Surveyor." *The Police,* February 1970, p. 143.

Moloney, D. J.: Electronics aid in defeating crime. *Security Gazette,* June 1963.

Momboisse, Raymond M.: Crowd assembly and riot prevention. *Security World, 2:* no. 1, January 1965.

———: Management and the "ad hoc" committee. *Security World,* June 1967, p. 12-16.

More security at less cost. *Mill and Factory,* October 1960.

More uses for closed-circuit TV. *Management Review,* September 1963.

Morris, Thomas A.: Lock and key control and master keying—panelist. *Industrial Security,* December, 1964.

Muir, C. B.: Setting up a plant lock and key system. *Factory Management,* no. 166, November 1948, p. 114-116.

Murray, W. E.: A word from the front line. *Security World,* October 1969.

Mustick, J. A.: Build for the future. *FBI Law Enforcement Bulletin, 38:* no. 4, 6-8, 23, April 1965.

Nash, Captain J. D., and Pate, Sergeant B. R.: The history of a community provides related reasons for selection of an alarm system. *Security World,* February 1975, p. 50-51.

———: The history of a community provides related reasons for selection of an alarm system, part II. *Security World,* March 1975, p. 63-64.

NBS to issue user guideline for photographic surveillance systems. *Industrial Fire and Security Report,* January 10, 1975, p. 3-4.

Neereamer, L. H.: Lock and key control and

master keying—panelist. *Industrial Security*, December 1964.
Neville, Henry C.: Foiling the electronic eavesdroppers: A surveyor of available countermeasures. *Security World*, April 1975, p. 20.
New aries whodunnit. *Chemical Week*, June 1962, p. 20-21.
New device to cut hijacking. *Top Security*, September 1975, p. 187.
New high in plant protection. *Factory Management*, May 1956.
New weapons to protect you against crime. *Nation's Business*, April 1969.
New York Times Magazine, The. October 28, 1962, 74 (voice prints).
Network TV riot policy. *Security World*, October 1970.
Nielson, George E.: Computerized access control. *Industrial Security*, October 1968.
Night depositories. *Standard for Safety*, March 29, 1972.
Night Watchman. *ILR Research, 10:* no. 2, 3, 1964.
Norpell, Bradley F.: Gate guarding is a science at Cynamid. *Occupational Hazards*, June 1965.
———: Equipment advances, losses retreat. *Occupational Hazards*, October 1963.
———: Identification cards and badges. *Security World, 3:* no. 2, February 1966.
Norton, Jerry E.: Electronic surveillance and the fourth amendment. *Security World*, January 1970.
Norton, R. E.: The uses of video. *Top Security*, July 1975, p. 129.
NY nuclear attack policy: Stay put! *Industrial Security*, May 12, 1963.
O'Brian, Thomas: Lockheed points the way to plant security. *Occupational Hazards*, January 1962.
———: Protection and security issue. *Occupational Hazards*, June 1966.
———: Today's industrial security. *Industrial Security*, October 1961.
O'Connor, Thomas: Security responsibility during work stoppage—panel moderator. *Industrial Security*, October 1962.
O'Farrell, Ralph W.: Lost, strayed or stolen . . . Mr. Inventory. *Police Chief*, The Kane Service, Chicago, Illinois.
Office, The, A push-button lock for computer room security, 74:161-163, March 1971.
On all the shelves at once. *Business Week*, October 5, 1963, p. 127.
Once too often. *Security World, 2:* no. 2, March 1965.
Operations manual for crowd control. *Police Division*, Cincinnati, Ohio, 1963.
Orne, M. T.: The potential use of hypnosis in interrogation. *The Manipulation of Human Behavior*, New York, 1961, p. 169-215.
Otterbourg, Robert K.: Efficient and accurate tape monitor aids public safety. *Law and Order, 7:* October 1959.
Oursler, W.: Plant security assured two ways, guard force plus automatic equipment. *Plant Engineering*, March 1965.
———: Primer for plant protection. *Nation Business Magazine*, June 1951.
———: TV camera helps guard remote gate. *Plant Engineering*, July 1961.
Palmer, Sidney E.: Historical development of locks and the lockmaking industry in England. *Association of Burglary Insurance Surveyors: Journal, 3:*15-21, 1956.
Pannell, W. N.: Telecommunications for security. *Security Gazette*, July 1970.
Parker, W. H.: Play ball—security at Dodger stadium. *Security World, 2:* no. 2, March/April 1965.
———: Port security at Matson Terminals, Inc. *Law and Order, 9:* April 1961.
Parloto, L. P.: The recording eye. *Law and Order*, April 1964, p. 30.
Payne, J. C.: Protective screens for cashiers. *Security World*, September 1970.
Pearsy, George: Security organization in a large vehicle plant. *Security Gazette*, October 1961.
Pedelty, Supt. J.: Protecting metals against theft. *Security Gazette*, September 1970.
Peek, Edward: Care in the use of electrical equipment. *Security Gazette*, September 1974, p. 335.
———: Danger—men at work. *Security Gazette*, February 1975, p. 57.
Peel, John D.: You can't win: The plight of security personnel. *Security World*, March 1970.
Penland, Jack D.: Investigation committee report—responsibilities and purposes. *Industrial Security*, April 1965.
———: Operation LPAP pays out. *Industrial Security*, August 1971.
Pennsylvania State Council of Civil Defense: Principles of plant protection. Harrisburg, Pennsylvania.
Penny, W. E. W.: The medieval keys in Salisbury museum. *Connosseur*, 29:11-16, January 1911.
Perlmutter, E.: Patrols cut subway crime: Attacks aboard buses rise. *The New York Times*, June 6, 1975.

Perry, Sue: Risk management and the security habit. *Security Gazette,* June 1975, p. 206-207.

Petrowski, Robert: Status—an alternative to extra channels or how fast is your RF system. *Canadian Police Chief,* January 1975, 17-19.

Pettigrew, R. C., and Shields, Allan F.: Security guard administration. *Canadian Security Gazette,* April/May 1970.

Petzinger, Clarence F.: Industrial shelters for employees. *Industrial Security,* July 1960.

Phantom gateman operates via TV. *Factory,* June 1961.

Pharmaceutical's fabulous security system. *Occupational Hazards,* October 1966.

Phillips, Hal: Investigating auto accidents. *Industrial Security,* April 1961.

Pilgrim, R.: Burglar alarms, locks, and hardware. *The Manager and Security, Bell,* 1972, p. 34-38.

Plainfield lights up to catch a thief. *New Jersey Municipalities,* June 1973.

Plant security by telephone. *Factory,* December 1962.

Plants are pushovers for thieves. *Occupational Hazards,* July 1963.

Plant guard: New image for a new era. *Occupational Hazards,* October 1963.

Plant protection. Fire Department of the City of Los Angeles with Office of Civil Defense, City of Los Angeles, January 1958.

Plant protection. *Industrial Security,* October 1963, p. 25.

Plant protection—a bomb is the problem. *Business Week,* March 21, 1951.

Plant protection. Civil Defense Office, Los Angeles, California, January 1958.

Plant protection demands split-second communications. *Occupational Hazards,* March 1966.

Plant protection guide for Chicago. Civil Defense Committee, Chicago Association of Commerce and Industry.

Plant security assured two ways: Guard force plus automatic equipment. *Plant Engineering,* March 1965.

Plant security at Nabisco. *Banking Industry,* December 12, 1959, p. 47.

Portable "eye" spots intruders. *Business Week,* April 13, 1963.

Potter, Anthony N., Jr.: A battalion of security men? *Industrial Security,* February 1971.

Poulson, Norval: False alarms: An industry view. *Security World,* October 1969.

Powell, E. A.: The evolution of the lock. *Craftsman,* 2:80-90, 1902.

Powell, Gregg E.: Identification cards: How to select them, part II. *Security World,* April 1975, p .38-41.

———: Identification cards: How to select them. *Security World,* February 1975, p. 22-23.

Powell, John: Plant operations. *College and University Business,* 48:94, April 1970.

Predicasts, Inc. Security Systems. Special Study 56, March 5, 1970.

Prial, F. S.: Holdups increasing sharply at subway change booths. *The New York Times,* May 12, 1971.

Price, R. S.: New lock will defy burglars. *Security Gazette,* June 1963.

———: Protection against burglary and housebreaking. *Security Gazette,* July 1960.

———: Protecting the company's marketing secrets, part II. *Security World,* 5: no. 7, July/August 1968.

———: Reference charts for handwriting, check and check protector comparison. *Security World, 1:* no. 2, September 1964.

Infrared light devices. *Princeton Engineer,* 26: no. 4, January 1966, p. 26.

Printing works where every day is another "D" Day. *Security Gazette,* October 1974, p. 368-369.

Problems of business survival. *Industrial Security,* January 1962.

Progressive Pritchard. *Security and Protection,* March 1975, p. 14-15.

Protecting personnel in wartime. National Industrial Conference Board. *Studies in Business Policy,* no. 55, New York, 1952.

Protection and security issue. *Occupational Hazards,* June 1966.

Protection you can bank on. *Security World,* May 1970.

Push-button lock for computer room security, A. *The Office,* March 1971, p. 161-163.

Pushbuttons open a lock. *Business Week, 103:* April 13, 1963.

Quinn, Dennis E.: The clandestine listening device threat. *Industrial Security,* October 1968, p. 40.

Rail clamp point locks. *Railway Magazine, 118:*293, June 1972.

Ratcliff, W. R. C.: Some useful notes on safes and their construction. *Association of Burglary Insurance Surveyors: Journal,* 4:43-49, 1957.

Ravey, J.: Strike-breaking—a Canadian problem. *Canadian Labour, 17:* no. 10, 9, 1972.

Rawnsley, John E.: Lock of long ago and the men who made them. *Copper*, 19:24-26, Winter 1963.

Relocking devices. *Standard for Safety*, October 17, 1972.

Report of a paper on ancient keys read to the association by Mr. Wills. *Journal of the British Archaeological Association*, 13:335-339, 1857.

Report of the remarks made by Mr. Syer Cuming on a lock and key of the fifteenth century. *Journal of the British Archaeological Association*, 14:288-289, 1858.

Richards, Guy: Is your telephone tapped? *New York Star*, September 27-30, 1948.

Ridout, J. E.: Standley power: generators. *Security and Protection*, September 1974, p. 29-31.

Riot and premise protection. *Security World*, September 1965, p. 10.

Riot loss prevention: One corporate program. *Security World*, 5: no. 3, March 1968.

Riot precautions for retail merchants, Detroit's Retail Merchant Association. *Security World*, 5: no. 7, July/August 1968.

Riots: Tension mounts in our cities. *Occupational Hazards*, May 1967.

Roberts, Matt: Electronic security in dollars and cents. *Library Security*, January 1975, p. 1-3.

Robinson, Robert L.: Masterkeying: The decision to masterkey. *Security World*, October, 1974, p. 50-54.

Robinson, Worrick G., III: Radio relay security. *Industrial Security*, June 1971.

Rockwell, Robert R.: Private guards: A viewpoint. *Security Management*, September 1975, p. 24-25.

Roetter, Charles: Self protection: Commonsense security for yourself, your home, your possessions and your business. *Newness*, 1967, p. 16-28.

Romaine, Lawrence B.: The wooden box lock. *Antiques*, 34:32-34, July 1938.

Ronayne, John A.: Bomb threats in industrial plants. *Industrial Security*, October 1969, p. 11.

Rose, Richard P.: The dollars and sense of precaution, part II. *Security World*, 2: no. 1, January 1965.

———: The dollars and sense of precaution, part III. *Security World*, 2: no. 2, March 1965.

———: The dollars and sense of precaution, part IV. *Security World*, 2: no. 3, May 1965.

Rosenzweig, M. L.: The law of wiretapping. *Cornell Law Quarterly*, 73:33, 1947.

Ross, Gorden E.: Dogs: The patrolling officer's sixth sense. *Security World*, 2: no. 3, May 1965.

Roterus, Victor: Dispersal of industrial plants. *Industrial Security*, July 1960.

Rouner, Thomas J.: The electric utility industry in peacetime and wartime emergencies. *Industrial Security*, October 1957.

Rubenstein, Col. Sidney S.: Your safe can save your company. *Security World*, 2: no. 1, January 1965.

Rudolphs, Willem (ed.): *Industrial Wastes—Disposal and Treatment*. New York, Ronald, 1953.

Russell, J. L., Jr.: New roles for industrial closed circuit TV. *Supervision*, May 1964.

Rutz, Donald E.: Guard force report forms. *Industrial Security*, December 1965.

Ryan, James H.: Construction site security. *Security Management*, September 1975, p. 18, 21, 26.

Rydz, John S.: Recent advances in electronics for security and protection. *Industrial Security*, February 1968.

Rytten, J. E.: How to tap a telephone and Recording and the police profession. *Law and Order*, January, March and May 1955.

Safe as a bank vault. *Factory Management*, July 1956.

Safe delivery. *Security and Protection*, March 1975, p. 19.

Sams, Crawford, F.: Bio-medical effects of radiation. *Industrial Security*, October 1963.

Saunders, George W.: Classifying physical security equipment. *Police Yearbook*, 1965.

Santa Barbara News-Press: Fatal bombing on California campus. *Security World*, May 1969.

Saving big with robot guards. *Factory Management*, May 1957.

Schiedermayer, Philip L.: The security program of the atomic energy commission—panel moderator. *Industrial Security*, October 1962.

An improved alarm lock. *Scientific American*, 62:212, April 5, 1890.

Schlachtmeyer, A.: Security system features closed circuit TV. *Actual Specifying Engineer*, January 1965.

———: Glass in the service of security. *Security Gazette*, May 1962.

———: Elements of effective gate control. *Security Gazette*, April 1963.

Schnabolk, Charles: Protection against a guard force. *Security World,* May 1971.

———: Open and shut case for sound security. *Buildings,* March 1972, p. 48-53.

Schoech, W. A., Rear Adm., USN: Naval airpower and security. *Industrial Security,* January 1958.

Schumann, Walter: The prototype of the Yale lock. *Scientific American, 109:*125, August 16, 1913.

Scientific American: Versatile recording lock, *147:*376, December 1932.

Security and the architect. *Security Gazette,* November 1974, p. 412.

Surveyor's casebook: Key security. *Security and Protection, 5:* no. 5, 19, May 1973.

Locks, safes and strongrooms: A current appraisal, *2:* no. 4, 12-19, April 1970.

Security Report, The. Minneapolis, Foresight Security, Inc., 1972.

Security at a western atomic plant. *Western Manufacturing,* May 1962.

Security checklist. *Occupational Hazards,* November 1966.

Security—companies besieged. *Time,* April 13, 1970, p. 92.

Security Gazette, 1: November 1958.

Lessons from America—on systems, locks and standards. *Security Gazette, 5:* no. 1, 15, January 1963.

Security Gazette. Lessons from America—on systems, locks and standards. *5:* no. 1, 15, January 1963.

———. Crime thrives on carelessness: Security and architects. *5:* no. 5, 195, May 1963.

———. Trends in lock and key design: Safeguard for the purchaser. *3:* no. 10, 320, 325, October 1961.

———. London locksmiths hold their first convention: Exhibition and lectures. *3:* no. 12, 427, December 1961.

———. Card-key systems are designed to control access: Roneo-Neopost introduce new security technique. *7:* no. 9, 333, September 1965.

———. Security panic bolts and locks: Some devices for securing fire exit doors against intruders. *13:* no. 2, 78, February 1971.

———. Security essentials made plain: Review of cost-effective security. *14:* no. 12, 499, December 1972.

Security Surveyor. Journal of the Association of Burglary Insurance Surveyors, *3:* no. 1, May 1972.

———. *1:* May 1970.

———: Are the keys secure? *2:* no. 1, 78-80, May 1971.

Security systems electronics to the rescue. Quantum Science Corp., 1973.

Sen, K.: Indian industrial security faces espionage. *Industrial Security, 9:* no. 24, August 1965.

———: Indian industrial security faces sabotage. *Industrial Security, 10:*36, August 1966.

SDS target for 1970: The industrial plant. *Occupational Hazards,* December 1969, p. 37.

Shanley, Donald J.: A systems approach to nuclear power plant security. *Security Management,* January 1975, p. 26-27.

Shanley, John F.: Objectives of the police intelligence unit. *The Police Chief, XXXI:* 10-11, 50, May 1964.

Shaw, W.: An introduction to law enforcement electronics and communications, part III (TV surveillance). *Law and Order,* May 1965, p. 36.

———: Applied electronics . . . radio beacon trails. *Law and Order,* November 1962, 34 ff.

Shiken, John: The card that's a guard. *Factory Management and Maintenance, 114:*106-108, November 1956.

Shrader, R. L.: *Electronic Communication,* 2d ed., New York, McGraw-Hill, 1967.

Shredding services fill the gap. *Top Security,* June 1975, p. 77.

Signaling victory. *Security and Protection,* February 1975, p. 14-16.

Silvaman, Allan B. I.: Building and use of a mobile surveillance unit. *Industrial Security,* June 1965.

Skolnick, J. M.: Scientific theory and scientific evidence: An analysis of lie-detection. *Yale Law Journal,* April 1961, p. 715.

Smith, Charles L.: Riding shotgun. *Security Register, 1:* no. 4, 17-19, November 1974.

Sloane, Charles F.: Dogs in war, police work and on patrol. *The Journal of Criminal Law, Criminology and Police Science, 46:* no. 3, September-October 1955.

Smith, J. L.: False alarms—a major problem to law enforcement. *FBI,* 1973, p. 4.

Sodium vapor attack on street crime, A. *Business Week,* May 12, 1973.

Sodium vapor lights enjoy rave reviews. *Kahala Sun Press,* June 6, 1973.

Southworth, Robert: The master of evidence. *Industrial Security,* October 1963.

Spot the gaps in plant security. *Factory,* August, 1961.

Stacy, Marvin J.: Controlled areas. *Industrial Security,* August 1967.
Stanier, Harold: Security in a large factory. *Security Gazette,* January 1961.
Straus, Lyle B.: New alarm standards for Britain. *Security Gazette,* June 1961.
Streeter, Donald: Early American stock locks. *Antiques, 98:*251-255, August 1970.
Strom, Maggie: Selecting suitable hardware: Three school systems, rationale. *Security World,* April 1975, p. 87-88.
Stuart, Charles: Locks. *Ramsey, 4,* 1959.
Study of state requirements for investigating, A. *The Police Chief, 36:* no. 7, 31, 1969.
Sullivan, William T.: Citation enforcement of parking and traffic regulations. *Industrial Security,* December 1967.
Surveillance cameras. *Standard for Safety,* January 18, 1973.
Surveillance equipment—its functions and uses, part II. *Security World,* May 1975, p. 28-30.
Survival—a plan. *Industrial Security,* January 1962.
Survival plans your company can use. *Nation's Business,* December 1961.
Taylor, J. W.: Assessing the security value of locks. *Assn. of Burglary Insurance Surveyors, Journal, 3:*42-51, 1956.
———: The security value of locks. *Security Surveyor, 3:* no. 1, 36-40, May 1972.
Taylor, L. B., Jr.: Guardians of the space frontier. *Law and Order, 9:* December 1961.
Taylor, Lawrence M.: Security cost control in the manufacture of classified hardware—panelist. *Industrial Security,* October 1960.
Tenex: Man Bites Dog. *Top Security,* June 1975, p. 63-64.
———: Spotting and swatting the squatter. *Top Security,* September 1975, p. 190-193.
———: Vive la difference (Whatever that means!). *Top Security,* July 1975, p. 105-107.
Ten code and radio communications, The. *Security World,* 1970.
Testing time at Shorrocks. *Security and Protection,* October 1974, p. 12-13.
There is a bomb planted in your office! *Administrative Management,* August 1970, p. 18-25.
They're on candid camera. *Occupational Hazards,* January 1971, p. 24.
Thorsen, J. E.: Dictionary of anti-intrusion devices for architects and builders basic information for the security layman in easy-to-use reference and question-answer format. *Security World, 10:* no. 10, p. 30-33, 35, November 1973.
———: Security devices, systems and electronic technology. *Security World, 10:* no. 9, 20, 21, 50, 52, 53, 57, October 1973.
———: Security applications, needs and trends. *Security World, 10:* no. 9, 18-19, October 1973.
———: Sub-miniaturization, new electronics that could revolutionize security! *Security World,* September 1974, p. 22-29.
Thieves versus locks and safes. *The Strand Magazine, 8:*497-506, 1894.
Tibbles, H. V.: They had a false alarm problem in those days too. *Security Gazette,* October 1974, p. 376-377.
Tightening up industrial security. *Business Week,* October 15, 1960.
Tildesley, Norman V.: Locks and keys. *Greenslade,* 1967.
Todd, William M.: Industrial investigation. *Industrial Security,* January 1964.
Toepfer, Edwin F.: The doors that locks must go on. *Security World,* November 1974, p. 22.
———: Lock security: Cylinders, keys and keying, part I. *Security World, 2:* no. 5, July/August 1965.
———: Lock security: Cylinders, keys and keying, part II. *Security World, 2:* no. 6, September 1965.
———: Insuring entrance door security. *Architectural Record, 135:*221-222, April 1964.
———: A new building look at locks and keys. *Security World,* November 1974, p. 20-21.
———: Safe cracking . . . how is it taught? *Security World,* July/August 1971.
Tracy, William J.: Revlon gives more than lip service to security. *Industrial Security,* August 1970, p. 10.
Traini, Robert: Bomb hoaxers are dangerous criminals. *Security Gazette,* December 1971.
———: Robbery trials reveal criminal tactics. *Security Gazette,* September 1974, p. 337+.
———: So primitive these jemmy and sledgehammer attacks. *Security Gazette,* August 1974, p. 303-304.
———: When will they ever learn? *Security Gazette,* February 1975, p. 55.
Trends in lock and key design. *Security Gazette,* October 1961.
Trespassers, beware: Electronic watchdogs on duty! *Occupational Hazards,* June 1963.

TV camera helps guard remote gate. *Plant Engineering,* July 1961.

TV for industry: Uses unlimited. *Steel Magazine,* October 12, 1959.

TV and radio: Security standbys. *Occupational Hazards,* September 1966.

TV stands watch for busy utility garage. *Fleet Owner,* April 1962.

Twelve gains in out-contracting plant security. *Rubber Age,* October 1959.

Ungar, Paul E.: Eyesight testing for security. *Security Gazette,* November 1970.

Unique liquor store installation, A:. *Security World,* April 1970.

Unique premise protection squad surveys and studies security techniques, A. *Security Product News,* November 1971.

University directors' first meeting. *Canadian Security Gazette,* Summer 1971.

Ward, Ralph V.: Alarm line security. *Security World, 4:* no. 3, March 1967.

———: Technical surveillance and detection equipment—panelist. *Industrial Security,* December 1964.

———: Why alarm the burglar? *Industrial Security,* June 1966.

Ward, William: Technical surveillance: Some basic considerations. *Assets Protection,* Spring 1975, p. 35-40.

Wattenberg, William W., and Balistrieri, James: Automobile theft: A favored group delinquency. *Amer J Sociol, 57:*575-579, May 1952.

Wearstler, Earl F.: Vault, safe and depository security. *Security World,* October 1969.

Wechter, Harry L.: Security at a Western atomic plant. *Western Manufacturing,* May 1962.

———: What a guard should know. *Safety Maintenance,* January 1962.

Wecksler, A. N.: This is not the time to laugh off plant security. *Mill and Factory,* July 1951.

Weiland, Paul E.: Long-range planning for public utilities. *Public Utilities Fortnightly,* December 1959, p. 969-76.

We never sleep. *Newsweek,* July 31, 1961.

Wessel, Milton R.: Legal protection of computer programs. *Howard Business Rev, 43:* March 1965.

Wessler, Myers and Gardner: Physical security —facts and fancies. *Datamation,* July 1, 1971.

West, J.: Better alarmed than sorry: Alarm systems. *Yachting,* February 1973.

West, P. Casner: Breaking the security barriers. *Security Management,* March 1975, p. 44-47.

Western Union links teletypewriter system and computer center. *Wall Street Journal,* March 17, 1966.

What a guard should know. *Safety Maintenance,* January 1962.

What a tangled web. *Top Security,* September 1975, p. 177-178.

Wheatley, George E.: Lock and key control and master keying panelist. *Industrial Security,* December 1964.

When failure to do the impossible becomes an offense. *Security Gazette,* August 1974, p. 301.

Where is the army explosive ordnance disposal team nearest you? *Occupational Hazards,* October 1970.

Whipperman, Robert F.: ABC's of private patrol. *Security World, 2:* no. 4, June 1965.

White house answer to bombings. *Occupational Hazards,* June 1970.

White, Russell E.: Seven steps to better security. *Industrial Security,* August 1962.

———: Background investigations—panelist. *Industrial Security,* October 1960, p. 14.

Who fired it? *Top Security,* July 1975, p. 123.

Wile ones, The. *Security and Protection,* September 1974, p. 12-14.

Williams, C. N.: Telephone reporting devices: An overview. *Security World,* June 1970.

Williams, H. E.: The art of surveillance. *Industrial Security,* April 1970.

Williams, Jack, the Hon.: Return to decency. *Industrial Security, 12:*4, August 1968.

Wilson, Thomas J.: The fingertip alarm system. *Law and Order, 8:* no. 8, 1960.

Windeler, Peter: Two hundred years of safemaking. *Security and Protection, 4:* no. 6, 7-11, June 1972.

Window grille lock device. *Security Gazette,* May 1974, p. 195.

Wiren, S.: Intrusion detection by electromagnetic fields. *Siemens Review,* October 1961.

Wlamsley, Alan: Defense security: Lessons of the cold war. *Security World,* November 1970.

Wolz, Phillip C.: Photography in emergencies. *Industrial Security,* July 1959.

Wombach, Albrecht G.: Planning for emergency repair and restoration. *Industrial Security,* July 1960.

Works Defense Manual. Cleveland, American Steel and Wire Co., 1951.

Wright, Edwin: Disaster control (seminar

workshop report). *Industrial Security,* October 1961, p. 20.

Van Emden, Bernard: A checklist for photo identification card security. *Security Management,* January 1974, p. 42-44.

Varrelman, David A.: Thoughts on police buildings design. *The Police Chief,* November 1970.

Vehicle alarm systems and units. *Standard for Safety,* August 30, 1973.

Visit to an underground storage vault. *Banking,* April 1964.

Wackenhut, George R.: Business is the target of bombings and bomb hoaxes. *The Office,* September 1971, p. 14.

Waisanin, W. V.: Plant security guards: Outside agency or plant personnel? *Modern Manufacturing,* December 1968, p. 14.

Walker, T. B.: Light is a security aid. *Security Gazette,* September 1974, p. 329-330.

Walsh, Timothy J.: A machine record method for maintaining lock and key accountability. *Industrial Security,* April 1965, p. 10.

———: The year of the cheat. *Top Security,* August 1975, p. 144-146.

Walsh, Timothy J., and Healy, Richard J.: Alarm sensors. *Protection of Assets Manual, 1:* California, The Merritt Company, 1974.

———: Barriers. *Protection of Assets Manual, 1:* California, The Merritt Company, 1974.

———: Bombs and bomb threats. *Protection of Assets Manual, 2:* California, The Merritt Company, 1974.

———: Communications. *Protection of Assets Manual, 1:* California, The Merritt Company, 1974.

———: Guard operation. *Protection of Assets Manual, 1:* California, The Merritt Company, 1974.

———: Identification. *Protection of Assets Manual, 1:* California, The Merritt Company, 1974.

———: Investigations. *Protection of Assets Manual, 2:* California, The Merritt Company, 1974.

———: Locking. *Protections of Assets Manual, 1:* California, The Merritt Company, 1974.

———: Parking and traffic control. *Protection of Assets Manual, 1:* California, The Merritt Company, 1974.

———: Physical security planning. *Protection of Assets Manual, 2:* California, The Merritt Company, 1974.

———: Sensor integration. *Protection of Assets Manual, 1:* California, The Merritt Company, 1974.

UNPUBLISHED MATERIAL

Amatt, Leslie Keith: Locks and lockmaking an annotated bibliography. Thesis submitted for Fellowship of its Library Ass'n. University Microfilms, Ltd.

Heckaman, F. C.: A survey of security alarm systems and their operational characteristics. Unpublished graduate thesis paper, Michigan State Univ., 1973.

Jorgensen, Ronald Eugene: Theft control on shipping and receiving docks (thesis), Michigan State Univ., 1965.

McQuinn: Thesis on the lock industry. Willenhall, the author, 1957.

Pollock, David Allen: Aspects of electronic audio surveillance. University of California, Berkeley, D. Criminology, 1968.

MOTOR THEFT
BOOKS

Chilimidos, R. S.: *Auto Theft Investigation.* Los Angeles, Legal Book Corp., 1971.

Kingsbury, A.: *Motor Vehicular Theft, A Bibliography.* M.V.M.A. 1974.

Powis, David: *Thieves in Wheels.* London, Police Publishing, 1971.

Tobias, M. W.: *Locks, Safes, and Security—A Handbook for Law Enforcement Personnel.* Springfield, Thomas, 1971.

PUBLICATIONS OF THE GOVERNMENT, LEARNED SOCIETIES, AND OTHER ORGANIZATIONS

Collins, D. E.: Auto theft investigations—an outline. Oregon Board on Police Standards and Training, 1968.

PERIODICALS

American Automobile Association Traffic Engineering and Safety Department. *Parking Programs.* Washington, D.C., 1954.

Auto industry works to cope with auto theft problem, part II. Work on theft prevention. Sherman, W. F. Automobile Manufacturing Association. Texas Police Association. *Texas Police Journal,* February 1956.

Auto Theft Prevention Act of 1968. Congressional House Committee. The Judiciary, Washington, D. C., Serial 15, 1968.

Auto theft rings. *FBI Law Enforcement Bulletin, 40*:6, 9, 28, August 1971.

Bicycle circulation and safety study. City of Davis. University of California. DeLeuw, Cather and Company, D06000, 1972.

Braien, Richard L.: Stolen cars—highway hazard. *Traffic Safety,* September 1967.

Cars and Trucks. November 1973.

Cars and Trucks. April 1974.

Chilimidos, R. S.: *Auto Theft Investigation.* Los Angeles, Legal Book Corp., 1971.

Collins, D. E.: Auto theft investigations—an outline. Oregon Board on Police Standards and Training, 1968.

Davis, W. J.: An investigative aid to the law enforcement profession. National Automobile Theft Bureau.

Frese, R., and Heller, N. B.: Measuring auto theft and the effectiveness of auto theft control programs, 1970.

Greater Cleveland Auto Theft Prevention Program Phase 2—1972 Progress Report. (July 1, 1971-June 30, 1972). Administration of Justice Committee. 79 pages. 1973.

Hill, Comp., Y. J.: Transportation security: Literature survey, preliminary draft. Department of Transportation, D17400, 1972.

Juvenile delinquency, auto theft and juvenile delinquency. U.S. Congress Senate Committee on the Judiciary, 1968.

Keenan, B., and Kerin, K.: Patterns of auto theft and criminal careers—a supplemental study for the ALPS Socio-Economic Evaluation Report, final report. IIT Es. Inst. 120 pages.

———: Socio-economic valuation study for automatic license plate scanning system. NYSIS. 278 pages.

The key to stopping car thieves is in your hand. *California Highway Patrolman,* May **1970.**

Lawless, T., and Peterman, J.: Auto theft and the uniformed officer. *Police Chief, 40:* no. 6, 58-62, June 1973.

Lock it or lose it. Greater Cleveland Auto Theft Prevention Program, eighteen month report. July 1970-December 1971. Administration of Justice Committee. 39 pages. 1972.

McGrath, J. H.: Comparative study of adolescent drug users, assaulters and auto thieves. University Microfilms, Ann Arbor, Mich., 1967.

Moolman, V.: Practical ways to prevent burglary and illegal entry. Cornerstone Library, New York City, 192 pages, 1970.

Murphy, Harry J.: Security of airport parking lots. *Security Management,* May 1975, p. 22-25.

Oliver, Clyde W., Jr.: The stolen car—increasing highway menace. *Traffic Safety,* May 1970.

Price, R. S.: Car thefts and how to prevent them. *Auto Car,* December 1959, p. 726-727.

Stage, A. G.: Repairing windows, winders and door locks. *Practical Motorist, 10:*1325-1327, August 1964.

Theft prevention devices. *Society of Motor Manufacturers and Traders,* March 1969.

Theft protection. Presented by R. P. Trowbridge to AAMVA Committee on Engineering and Vehicle Inspection. Trowbridge, R. P., General Motors Corporation. American Association of Motor Vehicle Administration, August 29, 1967.

Thorpe, John: Door locks—a look at modern lock mechanism. *Practical Motorist, 7:*386-387, December 1961.

Wattenberg, W. W., and Balistrieri, J.: Automatic theft—a favored group delinquency. *Crime and Delinquency, A Reader,* 1970.

West, A. S.: Collection and analysis of auto theft in Denver, July 1971. Denver Research Institute, 65 pages, 1971.

Williams, Thomas, Sr.: Auto theft: The problems and the challenge. *FBI Law Enforcement Bulletin,* December 1968.

UNPUBLISHED MATERIAL

Bash, Abdullatif A.: Saudi Arabian traffic law system. Thesis for Indiana University, May 1974.

Daily, Bruce W.: The breathalyzer: It's use in industry. Unpublished graduate thesis, Michigan State Univ., 1969.

Dosick, Martin L.: Behavior patterns of young Dyer Act offenders. Ann Arbor, University of Michigan, Ph.D. dissertation.

McGrath, J. H.: Comparative study of adolescent drug users, assaulters and auto thieves. University Microfilms, Ann Arbor, Mich., 1967.

APPENDIX 11A

PHYSICAL PROTECTION*

GENERAL PROVISIONS

Sec.
73.1 Purpose and scope.
73.2 Definitions.
73.3 Interpretations.
73.4 Communications.
73.5 Specific exemptions.
73.6 Exemptions for certain quantities and kinds of special nuclear material.

PHYSICAL PROTECTION OF SPECIAL NUCLEAR MATERIAL IN TRANSIT

73.30 General requirements.
73.31 Shipment by road.
73.32 Shipment by air.
73.33 Shipment by rail.
73.34 Shipment by sea.
73.35 Transfer of special nuclear material.
73.36 Miscellaneous requirements.

PHYSICAL PROTECTION REQUIREMENTS AT FIXED SITES

73.40 Physical protection: General requirements at fixed sites.
73.50 Requirements for physical protection of licensed activities.
73.60 Additional requirements for the physical protection of special nuclear material at fixed sites.

RECORDS AND REPORTS

73.70 Records.
73.71 Reports of unaccounted for shipments, suspected theft, unlawful diversion, or industrial sabotage.

ENFORCEMENT

73.80 Violations.
Appendix A—United States Atomic Energy Commission Regulatory Operations Regional Offices.

GENERAL PROVISIONS

§73.1 Purpose and scope.

(a) *Purpose.* This part prescribes requirements for physical protection of special nuclear material at fixed sites and in transit and of plants in which special nuclear material is used, for the purpose of protection against acts of industrial sabotage and protection of special nuclear material against theft by establishment and maintenance of a physical protection system of: (1) Protective barriers and intrusion detection devices at fixed sites to provide early detection of an attack, (2) deterrence to attack by means of armed guards and escorts, and (3) liaison and communication with law enforcement authorities capable of rendering assistance to counter such attacks.

(b) *Scope.* (1) This part prescribes requirements for (i) the physical protection of production and utilization facilities licensed pursuant to Part 50 of this chapter; (ii) the physical protection of plants in which activities licensed pursuant to Part 70 of this chapter are conducted, and the physical protection of special nuclear material, by any person who pursuant to the regulations in Part 70 of this chapter possesses or uses at any site or contiguous sites subject to control by the licensee, uranium-235 (contained in uranium enriched to 20 percent or more in the U-235 isotope), uranium-233, or plutonium alone or in any combination in a quantity of 5,000 grams or more computed by the formula, grams = (grams contained U-235) + 2.5 (grams U-233 + grams plutonium).†

(2) This part prescribes requirements for the physical protection of special nuclear material in transportation by any per-

---

* United States Atomic Energy Commission. Rules and Regulations. Title 10—Atomic Energy, Part 73. *Physical Protection of Plants and Materials.* Republished December 28, 1973.

† Corrected.

son who is licensed pursuant to the regulations in Part 70 of this chapter who imports, exports, transports, delivers to a carrier for transport in a single shipment, or takes delivery of a single shipment free on board at the point where it is delivered to a carrier, either uranium-235 (contained in uranium enriched to 20 percent or more in the U-235 isotope), uranium-233, or plutonium, or any combination of these materials, which is 5,000 grams or more computed by the formula, grams = (grams contained U-235) + 2.5 (grams U-233 + grams plutonium).

(3) This part also applies to shipments by air of special nuclear material in quantities exceeding (i) 20 grams or 20 curies, whichever is less, of plutonium or uranium-233, or (ii) 350 grams of uranium-235 (contained in uranium enriched to 20 percent or more in the U-235 isotope).

(4) Special nuclear material subject to this part may also be protected pursuant to security procedures prescribed by the Commission or another Government agency for the protection of classified materials. The provisions and requirements of this part are in addition to, and not in substitution for, any such security procedures. Compliance with the requirements of this part does not relieve any licensee from any requirement or obligation to protect special nuclear material pursuant to security procedures prescribed by the Commission or other Government agency for the protection of classified materials.

§73.2 **Definitions.**

As used in this part.

(a) Terms defined in Parts 50 and 70 of this chapter have the same meaning when used in this part.

(b) "Authorized individual" means any individual, including an employee, a consultant, or an agent of a licensee, who has been designated in writing by a licensee to have responsibility for surveillance of special nuclear material.

(c) "Guard" means a uniformed individual armed with a firearm whose primary duty is the protection of special nuclear material against theft and/or the protection of a plant against industrial sabotage.

(d) "Watchman" means an individual, not necessarily uniformed or armed with a firearm, who provides protection for a plant and the special nuclear material therein in the course of performing other duties.

(e) "Continuous visual surveillance" means unobstructed view at all times of a shipment of special nuclear material, and of all access to a temporary storage area or cargo compartment containing the shipment.

(f) "Physical barrier" means

(1) Fences constructed of No. 11 American wire gauge, or heavier wire fabric topped by, three strands or more of barbed wire or similar material on brackets angled outward between 30° and 45° from the vertical, with an overall height of not less than eight feet, including the barbed topping.

(2) Building walls constructed of stone, brick, cinder block, concrete, steel or comparable materials (openings in which are secured by grates, doors, or covers of construction and fastening of sufficient strength such that the integrity of the wall is not lessened by any opening), or walls of similar construction, not part of a building, provided with a barbed topping described in paragraph (f) (1) of this section of a height of not less than 8 feet.

(3) Ceilings and floors constructed to offer resistance to penetration equivalent to that of building walls described in paragraph (f) (2) of this section.

(g) "Protected area" means an area encompassed by physical barriers and to which access is controlled.

(h) "Vital area" means any area which contains vital equipment within a structure, the walls, roof, and floor of which constitute physical barriers of construction at least as substantial as walls as described in paragraph (f) (2) of this section.

(i) "Vital equipment" means any equipment, system, device, or material, the failure, destruction, or release of which could directly or indirectly endanger the public health and safety by exposure to radiation. Equipment or systems which would be required to function to protect public health and safety following such failure, destruction or release, are also considered to be vital.

(j) "Material access area" means any location which contains special nuclear material, within a vault or a building, the roof, walls, and floor of which each constitute a physical barrier.

(k) "Isolation zone" means any area, clear of all objects which could conceal or shield an individual, adjacent to a physical barrier, which is monitored to detect the presence of individuals or vehicles within that area.

(l) "Intrusion alarm" means a tamper indicating electrical, electromechanical, electrooptical, electronic or similar device which will detect intrusion by an individual into a building, protected area, vital area, or material access area, and alert guards or watchmen by means of actuated visible and audible signals.

(m) "Lock" in the case of vaults or vault type rooms means a three-position, manipulation resistant, dial type, built-in combination lock or combination padlock and in the case of fences, walls, and buildings means an integral door lock or padlock which provides protection equivalent to a six-tumbler cylinder lock. "Lock" in the case of a vault or vault type room also means any manipulation resistant, electromechanical device which provides the same function as a built-in combination lock or combination padlock, which can be operated remotely or by the "reading" or insertion of information, which can be uniquely characterized, and which allows operation of the device. "Locked" means protected by an operable lock.

(n) "Vault" means a burglar-resistant windowless enclosure with walls, floor and roof of: (1) Steel at least one-half inch thick, (2) reinforced concrete or stone at least 8 inches thick, (3) nonreinforced concrete or stone at least 12 inches thick, or (4) monolithic floor or roof construction of equivalent resistance to entry, with a built-in lock in a steel door at least 1 inch thick, exclusive of the locking mechanism.

(o) "Vault-type room" means a room with one or more doors, all capable of being locked, protected by an intrusion alarm which creates an alarm upon the entry of a person anywhere into the room and upon exit from the room or upon movement of an individual within the room.

(p) "Industrial sabotage" means any deliberate act directed against a plant in which an activity licensed pursuant to the regulations in this chapter is conducted, or to any component of such a plant, which could directly or indirectly endanger the public health and safety by exposure to radiation, other than such acts by an enemy of the United States, whether foreign government or other person.

§73.3 **Interpretations.**

Except as specifically authorized by the Commission in writing, no interpretation of the meaning of the regulations in this part by any officer or employee of the Commission other than a written interpretation by the General Counsel will be recognized as binding upon the Commission.

§73.4 **Communications.**

Except where otherwise specified, all communications and reports concerning the regulations in this part should be addressed to the Director of Licensing, U.S. Atomic Energy Commission, Washington, D.C. 20545, or may be delivered in person at the Commission's offices at 1717 H Street, NW, Washington, D.C.; at 7920 Norfolk Avenue, Bethesda, Maryland; or at Germantown, Maryland.

§73.5 **Specific Exemptions.**

The Commission may, upon application of any interested person or upon its own initiative, grant such exemptions from the requirements of the regulations in this part as it determines are authorized by law and will not endanger life or property or the common defense and security, and are otherwise in the public interest.

§73.6 Exemptions for Certain Quantities and Kinds of Special Nuclear Material.

A licensee is exempt from the requirements of §§73.30 through 73.36 and of §§73.60 and 73.70 of this part, with respect to the following special nuclear material:

(a) Uranium-235 contained in uranium enriched to less than 20 percent in the U-235 isotope:

(b) Special nuclear material which is not readily separable from other radioactive material and which has a total external radiation dose rate in excess of 100 rems per hour at a distance of 3 feet from any accessible surface without intervening shielding; and

(c) Special nuclear material in a quantity not exceeding 350 grams of uranium-235, uranium-233, plutonium, or a combination thereof, possessed in any analytical, research, quality control, metallurgical or electronic laboratory.

PHYSICAL PROTECTION OF SPECIAL NUCLEAR MATERIAL IN TRANSIT

§73.30 General Requirements.

(a) Except as specified in §73.36 (a) or as otherwise authorized pursuant to §73.30 (f), each licensee who transports or who delivers to a carrier for transport either uranium-235 (contained in uranium enriched to 20 percent or more in the U-235 isotope), uranium-233, or plutonium, or any combination of these materials, which is 5,000 grams or more computed by the formula, grams = (grams contained U-235) + 2.5 (grams U-233 + grams plutonium), shall make arrangements to assure that such special nuclear material will, if a common or contract carrier is used, be transported under the established procedures of a carrier which provides a system for the physical protection of valuable material in transit and requires an exchange of hand-to-hand receipts at origin and destination and at all points enroute where there is a transfer of custody.

(b) Transit times of shipments other than those specified in §73.1 (b) (3) shall be minimized and routes shall be selected to avoid areas of natural disaster or civil disorders. Such shipments shall be preplanned to assure that deliveries occur at a time when the receiver at the final delivery point is present to accept receipt of shipment.

(c) Special nuclear material shall be shipped in containers which are sealed by tamper indicating type seals. The container shall also be locked if it is not in another container or vehicle which is locked. If inspection of the container or vehicle is not required by State or local authorities before final destination, the outermost container or vehicle shall also be sealed by tamper indicating type seals. No container weighing 500 pounds or less shall be shipped in open trucks, railroad flat cars or box cars and ships. This paragraph does not apply to shipments of quantities specified in §73.1 (b) (3).

(d) When guards are used pursuant to §§73.31 (c) (1), 73.31 (c) (2), 73.33 and 73.35, the licensee shall not permit an individual to act as a guard unless there is documentation that the individual has been qualified by demonstrating an understanding of his duties and responsibilities. The licensee or his agent shall have documentation that guards have been requalified annually.

(e) By January 7, 1974, each licensee shall submit a plan outlining the procedures that will be used to meet the requirements of §§73.30 through 73.36 and 73.70 (g) including a plan for the selection, qualification, and training of armed escorts, or the specification and design of a specially designed truck or trailer as appropriate. This plan shall be followed by the licensee after March 6, 1974.

(f) A licensee or applicant for a license may apply to the Commission for approval of proposed procedures for transport of special nuclear material in a manner not otherwise authorized by the regulations of this part. Such application shall include a description and quantity of the special nuclear material involved, the origin and destination, the carriers to be used, the expected time in transit, the number of transfer points, the communications to be used, the vehicle visual identification, and the cargo security and surveillance measures to be used.

(g) Paragraphs (b), (c), (d), and (f) of this section are effective March 6, 1974.

§73.31 Shipment by Road.

(a) All shipments by road shall be made without any scheduled intermediate stops to transfer special nuclear material or other cargo between the facility from which it is shipped and the facility of the receiver.

39 FR 2352

(b) All motor vehicles used to transport special nuclear material shall be equipped with a radiotelephone which can communicate with a licensee or his agent. The licensee or agent with whom communications shall be maintained for different segments of the shipment shall be predesignated before a shipment is made. Calls to such licensee or agent shall be made at least every 2 hours when radiotelephone or conventional telephone coverage along the route is available to relay position and projected route. Call frequency may extend up to 5 hours when radiotelephone or conventional telephone coverage is not available along the preplanned route, at which time a conventional telephone call shall be made. In the event no call is received in accordance with these requirements, the licensee or his agent shall immediately notify an appropriate law enforcement authority and the appropriate Atomic Energy Commission Regulatory Operations Regional Office listed in Appendix A of this part.

(c) A shipment shall be accompanied by at least two people in the vehicle containing the shipment, which may be two drivers or one driver and an authorized individual. The vehicle containing the shipment shall be under continuous visual surveillance, or one of the drivers or authorized individuals shall be in the cab of the vehicle, awake, and not in a sleeper berth. The shipment shall be further protected by one of the following methods:

(1) An armed escort consisting of a least two guards shall accompany the shipment in a separate escort vehicle. Escorts shall maintain continuous vigilance for the presence of conditions or situations which might threaten the security of the shipment, take such action as circumstances might require to avoid interference with continuous safe passage of the cargo vehicle, provide assistance to, or summon aid for crew of cargo vehicles in case of emergency, check seals and locks at each stop where time permits, and observe the cargo vehicle and adjacent areas during stops or layovers. Continuous radio communication capability shall be provided between the cargo vehicle and the escort vehicle. Escort vehicles shall also be equipped with a radiotelephone. The licensee may use his own employees as armed escorts or he may use an agent. Only the driver is required in the vehicle containing special nuclear material for shipments involving an average of less than an hour in transportation, if communication is maintained during the course of the shipment with the licensee or agent monitoring the shipment.

(2) The shipment shall be made in a specially designed truck or trailer which reduces the vulnerability to diversion. Design features of the truck or trailer shall permit immobilization of the van and provide barriers or deterrents to physical penetration of the cargo compartment unless armed guards are also used in which case immobilization of the vehicle is not required.

(d) Transfers to and from other modes of transportation shall be in accordance with §73.35.

(e) Vehicles shall be marked on top with identifying letters or numbers which will permit identification of the vehicle under daylight conditions from the air in clear weather at 1,000 feet above ground level. The same code of letters and numbers as those used on the top shall also be marked on the sides and rear of the vehicle to permit identification from the ground.

(f) This section is effective March 6, 1974.

§73.32 Shipment by Air.

(a) Except as specifically approved by the Atomic Energy Commission, no shipment of special nuclear material shall be made in passenger aircraft in excess of (1) 20 grams or 20 curies, whichever is less, of plutonium or uranium-233, or (2) 350 grams of uranium-235 (contained in uranium enriched to 20 percent or more in the U-235 isotope).

(b) In shipments on cargo aircraft of either uranium-235 (contained in uranium enriched to 20 percent or more in the U-235 isotope), uranium-233 or plutonium, or any combination of these materials which is 5,000 grams or more computed by the formula, grams = (grams contained U-235) + 2.5 (grams U-233 + grams plutonium), transfers shall be in accordance with §73.35. Transfers shall be minimized.

(c) Export shipments shall be escorted by an unarmed authorized individual, who may be a crew member, from the last terminal in the United States until the shipment is unloaded at a foreign terminal. He shall perform monitoring duties at foreign terminals as described in §73.35.

(d) Paragraph (c) of this section is effective March 6, 1974.

§73.33 Shipment by Rail.

(a) A shipment by rail shall be escorted by two guards, in the shipment car or an escort car of the train, who shall keep the shipment cars under observation and who shall detrain at stops when practicable and time permits to guard the shipment cars under observation, and check car or container locks and seals. Radiotelephone communication shall be maintained with a licensee or his agent to relay position every 2 hours or less, and at scheduled stops in the event that radiotelephone coverage was not available in the last 5 hours before the stop. The licensee or agent with whom communications shall be maintained for different segments of the shipment shall be predesignated before a shipment is made. In the event no call is received in accordance with these requirements, the licensee or his agent shall immediately notify an appropriate law enforcement authority and the appropriate Atomic Energy Commission Regulatory Operations Regional Office listed in Appendix A of this part.

(b) Transfers shall be in accordance with §73.35.

(c) This section is effective March 6, 1974.

§73.34 Shipment by Sea.

(a) Shipments shall be made on vessels making the minimum ports of call. Transfers to and from other modes of transportation shall be in accordance with §73.35. There shall be no scheduled transfers to other ships. At domestic ports of call where other cargo is transferred, the shipments shall be protected in accordance with §73.35 (a).

(b) The shipment shall be placed in a secure compartment which is locked and sealed. Locks and seals shall be periodically inspected in transit, if accessible, by an escort or crew member.

(c) Export shipments shall be escorted by an unarmed authorized individual, who may be a crew member, from the last port in the United States until the shipment is unloaded at a foreign port. He shall perform monitoring duties at foreign ports as described in §73.35.

(d) Ship-to-shore communications shall be available, and a ship-to-shore contact shall be made every twenty-four hours to relay position information, and the status of the shipment, which shall be determined by a daily inspection where possible. This information shall be sent, as often as it is available, to the licensee or his agent who makes the arrangements for the protection of the shipment.

(e) This section is effective March 6, 1974.

§73.35 Transfer of Special Nuclear Material.

All transfers shall be monitored by a guard. An alternate guard shall be designated at all transfer points to substitute, if necessary. Monitoring of special nuclear material transfers shall be conducted as follows:

(a) At scheduled intermediate stops where special nuclear material is not scheduled for transfer, the guard shall observe the opening of the cargo compartment and assure that the shipment is not removed. The guard shall maintain continuous visual surveillance of the cargo compartment. Continuous visual surveillance of the cargo compartment shall be maintained up to the time the vehicle is ready to depart. The guard shall observe the vehicle until it has departed, and shall notify the licensee or his agent of the latest status immediately thereafter.

(b) At points where special nuclear material is transferred from a vehicle to storage, from one vehicle to another, or from storage to a vehicle, the guard shall keep the shipment under continuous visual surveillance by observing the opening of the cargo compartment of the incoming vehicle and assuring that the shipment is complete by checking locks and/or seals. Continuous visual surveillance of a shipment shall be maintained at all times it is in the terminal or in storage. Shipments shall be preplanned in order to avoid storage times in excess of 24 hours. Continuous visual surveillance of the cargo compartment shall be maintained up to the time the vehicle is ready to depart from the terminal. The guard shall observe the vehicle until it has departed, and shall notify the licensee or his agent of the latest status immediately thereafter.

(c) The guard shall be required to immediately notify the carrier and the licensee who made the arrangements for protection of special nuclear material of any deviation from or attempted interference with schedule or routing.

(d) This section is effective March 6, 1974.

§73.36 Miscellaneous Requirements.

(a) Each licensee who takes delivery of special nuclear material free on board (f.o.b.) the point at which it is delivered to a carrier for transport shall make the arrangements to assure that such special nuclear material will be protected in transit as prescribed in §§73.30 through 73.35, rather than the person who delivers such shipment to the carrier for transport.

(b) Each licensee who imports special nuclear material shall make arrangements to assure that such material will be protected in transit as follows:

(1) An individual designated by the licensee or his agent, or as specified by a contract of carriage, shall confirm the container count and examine locks and/or seals for evidence of tampering, at the first place in the United States at which the shipment is discharged from the arriving carrier.

(2) The shipment shall be protected at the first terminal at which it arrives in the United States and all subsequent terminals as provided in §§73.30 through 73.35 and paragraphs (c) and (f) of this section.

(c) (1) Each licensee who delivers special nuclear material to a carrier for transport shall immediately notify the consignee by telephone, telegraph, or teletype, of the time of departure of the shipment, and shall notify or confirm with the consignee the method of transportation, including the names of carriers, and the estimated time of arrival of the shipment at its destination. (2) In the case of a shipment free on board (f.o.b.) the point where it is delivered to a carrier for transport, each licensee shall, before the shipment is delivered to the carrier, obtain written certification from the licensee who is to take delivery of the shipment at the f.o.b. point that the physical protection arrangements required by §§73.30 through 73.35 for licensed shipment have been made. When an AEC license-exempt contractor is the consignee of a shipment, the licensee shall, before the shipment is delivered to the carrier, obtain written certification from the contractor who is to take delivery of the shipment at the f.o.b. point that the physical protection arrangements required by AEC Manual Chapters 2401 or 2405 have been made.

(3) Each licensee who delivers special

nuclear material to a carrier for transport shall also make arrangements with the consignee to be notified immediately by telephone, telegraph, or teletype, of the arrival of the shipment at its destination.

(d) In addition to complying with the requirements specified in paragraphs (c) and (f) of this section, each licensee who exports special nuclear material shall comply with the requirements specified in §§73.30 through 73.35, as applicable, up to the first point where the shipment is taken off the vehicle outside the United States. The licensee shall also make arrangements with the consignee to be notified immediately by telephone and telegraph, teletype, or cable, of the arrival of the shipment at its destination, or of any such shipment that is lost or unaccounted for after the estimated time of arrival at its destination.

(e) Each licensee who receives a shipment of special nuclear material shall immediately notify the person who delivered the material to a carrier for transport of the arrival of the shipment at its destination. In the event such a shipment fails to arrive at its destination at the estimated time, the consignee, if a licensee, or in the case of an export shipment, the licensee who exported the shipment, shall immediately notify by telephone and telegraph, or teletype, the Director of the appropriate Atomic Energy Commission Regulatory Operations Regional Office listed in Appendix A of this part, and the licensee or other person who delivered the material to a carrier for transport. The licensee who made the physical protection arrangements shall also immediately notify by telephone and telegraph, or teletype the Director of the appropriate Atomic Energy Commission Regulatory Operations Regional Office listed in Appendix A of the action being taken to trace the shipment.

(f) Each licensee who makes arrangements for physical protection of a shipment of special nuclear material as required by §§73.30 through 73.36 shall immediately conduct a trace investigation of any shipment that is lost or unaccounted for after the estimated arrival time and file a report with the Commission as specified in §73.71. If the licensee who conducts the trace investigation is not the consignee, he shall also immediately report the results of his investigation by telephone and telegraph, or teletype to the consignee.

(g) Paragraphs (a), (b), (c) and (d) of this section are effective March 6, 1974.

PHYSICAL PROTECTION REQUIREMENTS AT FIXED SITES

§73.40 Physical Protection: General Requirements at Fixed Sites.

Each licensee shall provide physical protection against industrial sabotage and against theft of special nuclear material at the fixed sites where licensed activities are conducted. Security plans submitted to the Commission for approval shall be followed by the licensee after March 6, 1974.

§73.50 Requirements for Physical Protection of Licensed Activities.

In addition to any other requirements of this part, each licensee who is authorized to operate a fuel reprocessing plant pursuant to Part 50 of this chapter or who possesses or uses uranium-235 (contained in uranium enriched to 20 percent or more in the U-235 isotope), uranium-233, or plutonium alone or in any combination in a quality of 5000 grams or more computed by the formula, grams = (grams contained U-235) + 2.5 (grams U-233 + grams plutonium), other than in the operation of a nuclear reactor licensed pursuant to Part 50 of this chapter, shall comply with the following.

(a) *Physical security organization.* (1) The licensee shall establish a security organization, including guards, to protect his facility against industrial sabotage and the special nuclear material in his possession against theft.

(2) At least one supervisor of the security organization shall be on site at all times.

(3) The licensee shall establish, maintain and follow written security procedures which document the structure of the security organization and which detail the

duties of guards, watchmen, and other individuals responsible for security.

(4) The licensee shall not permit an individual to act as a guard or watchman unless such individual has been properly trained and equipped and has qualified by demonstrating: (i) An understanding of the licensee's security procedures, and (ii) the ability to execute all duties required of him by such procedures. Each guard and watchman shall be requalified at least annually. Such requalification shall be documented.

(b) *Physical barriers.* (1) The licensee shall locate vital equipment only within a vital area, which, in turn, shall be located within a protected area such that access to vital equipment requires passage through at least two physical barriers. More than one vital area may be within a single protected area.

(2) The licensee shall locate material access areas only within protected areas such that access to the material access area requires passage through at least two physical barriers. More than one material access area may be within a single protected area.

(3) The physical barrier at the perimeter of the protected area shall be separated from any other barrier designated as a physical barrier within the protected area, and the intervening space monitored or periodically checked to detect the presence of persons or vehicles so that the facility security organization can respond to suspicious activity or to the breaching of any physical barrier.

(4) An isolation zone shall be maintained around the physical barrier at the perimeter of the protected area and any part of a building used as part of that physical barrier. The isolation zone shall be monitored to detect the presence of individuals or vehicles within the zone so as to allow response by armed members of the licensee security organization to be initiated at the time of penetration of the protected area. Parking facilities, both for employees and visitors, shall be located outside the isolation zone.

(5) Isolation zones and clear areas between barriers shall be provided with illumination sufficient for the monitoring required by paragraph (b) (3) and (4) of this section, but not less than 0.2 foot candles.

(c) *Access requirements.* The licensee shall control all points of personnel and vehicle access into a protected area, including shipping or receiving areas, and into each vital area. Identification of personnel and vehicles shall be made and authorization shall be checked at such points.

(1) At the point of personnel and vehicle access into a protected area, all individuals, except employees who possess an AEC personnel security clearance, and all hand-carried packages shall be searched for devices such as firearms, explosives, and incendiary devices, or other items which could be used for industrial sabotage. The search shall be conducted either by a physical search or by the use of equipment capable of detecting such devices. Employees who possess an AEC personnel security clearance shall be searched at random intervals. Subsequent to search, drivers of delivery and service vehicles shall be escorted at all times while within the protected area.

(2) All packages being delivered into the protected area shall be checked for proper identification and authorization. Packages other than hand-carried packages shall be searched at random intervals.

(3) A picture badge identification system shall be used for all individuals who are authorized access to protected areas without escort.

(4) Access to vital areas and material access areas shall be limited to individuals who are authorized access to vital equipment or special nuclear material and who require such access to perform their duties. Authorization for such individuals shall be provided by the issuance of specially coded numbered badges indicating vital areas and material access areas to which access is authorized. Unoccupied vital areas and material access areas shall be protected by an active intrusion alarm system.

(5) Individuals not employed by the licensee shall be escorted by a watchman, or other individual designated by the licensee, while in a protected area and shall be badged to indicate that an escort is required. In addition, each individual not employed by the licensee shall be required to register his name, date, time, purpose of visit, employment affiliation, citizenship, name and badge number of the escort, and name of the individual to be visited. Except for a driver of a delivery or service vehicle, an individual not employed by the licensee who requires frequent and extended access to a protected area or a vital area need not be escorted provided such individual is provided with a picture badge, which he must receive upon entrance into the protected area and which he must return each time he leaves the protected area, which indicates (i) nonemployee—no escort required, (ii) areas to which access is authorized, and (iii) the period for which access has been authorized.

(6) No vehicles used primarily for the conveyance of individuals shall be permitted within a protected area except under emergency conditions.

(7) Keys, locks, combinations, and related equipment shall be controlled to minimize the possibility of compromise and promptly changed whenever there is evidence that they have been compromised. Upon termination of employment of any employee, keys, locks, combinations, and related equipment to which that employee had access shall be changed.

(d) *Detection aids.* (1) All alarms required pursuant to this part shall annunciate in a continuously manned central alarm station located within the protected area and in at least one other continuously manned station, not necessarily within the protected area, such that a single act cannot remove the capability of calling for assistance or otherwise responding to an alarm. All alarms shall be self-checking and tamper indicating. The annunciation of an alarm at the onsite central alarm station shall indicate the type of alarm (e.g., intrusion alarm, emergency exit alarm, etc.) and location. All intrusion alarms, emergency exit alarms, alarm systems, and line supervisory systems shall at minimum meet the performance and reliability levels indicated by GSA Interim Federal Specification W-A-00450 B (GSA-FSS).

(2) All emergency exits in each protected area and each vital area shall be alarmed.

(e) *Communication requirements.* (1) Each guard or watchman on duty shall be capable of maintaining continuous communication with an individual in a continuously manned central alarm station within the protected area, who shall be capable of calling for assistance from other guards and watchmen and from local law enforcement authorities.

(2) The alarm stations required by paragraph (d)(1) of this section shall have conventional telephone service for communication with the law enforcement authorities as described in paragraph (e)(1) of this section.

(3) To provide the capability of continuous communication, two-way radio voice communication shall be established in addition to conventional telephone service between local law enforcement authorities and the facility and shall terminate at the facility in a continuously manned central alarm station within the protected area.

(4) All communications equipment, including offsite equipment, shall remain operable from independent power sources in the event of loss of primary power.

(f) *Testing and maintenance.* Each licensee shall test and maintain intrusion alarms, emergency alarms, communications equipment, physical barriers, and other security related devices or equipment utilized pursuant to this section as follows:

(1) All alarms, communications equipment, physical barriers, and other security related devices or equipment shall be maintained in operable and effective condition.

(2) Each intrusion alarm shall be functionally tested for operability and required performance at the beginning and end of each interval during which it is

used for security, but not less frequently than once every seven (7) days.

(3) Communications equipment shall be tested for operability and performance not less frequently than once at the beginning of each security personnel work shift.

(g) *Response requirement.* (1) The licensee shall establish liaison with local law enforcement authorities. In developing his physical security plan, the licensee shall take account of the probable size and response time of the local law enforcement authority assistance.

(2) Upon detection of abnormal presence or activity of persons or vehicles within an isolation zone, a protected area, a material access area or a vital area, or upon evidence of intrusion into a protected area, a material access area or a vital area, the facility security organization shall (i) determine whether or not a threat exists, (ii) assess the extent of the threat, if any, and (iii) take immediate measures to neutralize the threat, either by appropriate action by facility guards or by calling for assistance from local law enforcement authorities, or both.

(h) This section is effective March 6, 1974.

§73.60 Additional Requirements for the Physical Protection of Special Nuclear Material at Fixed Sites.

In addition to the applicable requirements of §73.50, each licensee who pursuant to the regulations in Part 70 of this chapter possesses at any site or contiguous sites subject to control by the licensee uranium-235 (contained in uranium enriched to 20 percent or more in the U-235 isotope), uranium-233, or plutonium alone or in any combination in a quantity of 5,000 grams or more computed by the formula, grams = (grams contained U-235) + 2.5 (grams U-233 + grams plutonium) shall protect the special nuclear material from theft or diversion as follows:

(a) *Access requirements.* (1) Special nuclear material shall be stored or processed only in a material access area. No activities other than those which require access to special nuclear material or equipment employed in the process, use, or storage of special nuclear material, shall be permitted within a material access area.

(2) Material access areas shall be located only within a protected area to which access is controlled.

(3) Special nuclear material not in process shall be stored in a vault equipped with an intrusion alarm or in a vault-type room, and each such vault or vault-type room shall be controlled as a separate material access area.

(4) Enriched uranium scrap in the form of small pieces, cuttings, chips, solutions or in other forms which result from a manufacturing process, contained in 30-gallon or larger containers, with a uranium-235 content of less than 0.25 grams per liter, may be stored within a locked and separately fenced area which is within a larger protected area provided that the storage area is no closer than 25 feet to the perimeter of the protected area. The storage area when unoccupied shall be protected by a guard or watchman who shall patrol at intervals not exceeding 4 hours, or by intrusion alarms.

(5) Admittance to a material access area shall be under the control of authorized individuals and limited to individuals who require such access to perform their duties.

(6) Prior to entry into a material access area, packages shall be searched for devices such as firearms, explosives, incendiary devices, or counterfeit substitute items which could be used for theft or diversion of special nuclear material.

(7) Methods to observe individuals within material access areas to assure that special nuclear material is not diverted shall be provided and used on a continuing basis.

(b) *Exit requirement.* Each individual, package, and vehicle shall be searched for concealed special nuclear material before exiting from a material access area unless exit is into a contiguous material access area. The search may be carried out by a physical search or by use of equipment capable of detecting the presence of concealed special nuclear material.

(c) *Detection aid requirement.* Each unoccupied material access area shall be locked and protected by an intrusion alarm on active status. All emergency exits shall be continuously alarmed.

(d) *Testing and maintenance.* Each licensee shall test and maintain intrusion alarms, physical barriers, and other devices utilized pursuant to the requirements of this section as follows:

(1) Intrusion alarms, physical barriers, and other devices used for material protection shall be maintained in operable condition.

(2) Each intrusion alarm shall be inspected and tested for operability and required functional performance at the beginning and end of each interval during which it is used for material protection, but not less frequently than once every seven (7) days.

(e) This section is effective March 6, 1974.

RECORDS AND REPORTS

§73.70 **Records.**

Each licensee subject to the provisions of §§73.30 through 73.36 and/or §73.50 and/or §73.60 shall keep the following records:

(a) Names and addresses of all individuals who have been designated as authorized individuals.

(b) Names, addresses, and badge numbers of all individuals authorized to have access to vital equipment or special nuclear material, and the vital areas and material access areas to which authorization is granted.

(c) A register of visitors, vendors, and other individuals not employed by the licensee recorded pursuant to §73.50 (c) (5).

(d) A log indicating name, badge number, time of entry, reason for entry, and time of exit of all individuals granted access to a normally unoccupied vital area.

(e) Documentation of all routine security tours and inspections, and of all tests, inspections, and maintenance performed on physical barriers, intrusion alarms, communications equipment, and other security related equipment used pursuant to the requirements of this part.

(f) A record at each onsite alarm annunciation location of each alarm, false alarm, alarm check, and tamper indication that identifies the type of alarm, location, alarm circuit, date, and time. In addition, details of response by facility guards and watchmen to each alarm, intrusion, or other security incident shall be recorded.

(g) Shipments of special nuclear material subject to the requirements of this part, including names of carriers, major roads to be used, flight numbers in the case of air shipments, dates and expected times of departure and arrival of shipments, names and addresses of the monitor and one alternate monitor at each transfer point, verification of communication equipment on board the transfer vehicle, names of individuals who are to communicate with the transport vehicle, container seal descriptions and identification, and any other information to confirm the means utilized to comply with §§73.30 through 73.36. Such information shall be recorded prior to shipment. Information obtained during the course of the shipment such as reports of all communications, change of shipping plan including monitor changes, trace investigations and others shall also be recorded.

(h) Procedures for controlling access to protected areas and for controlling access to keys for locks used to protect special nuclear material.

§73.71 **Reports of Unaccounted for Shipments, Suspected Theft, Unlawful Diversion, or Industrial Sabotage.**

(a) Each licensee who conducts a trace investigation of a lost or unaccounted for shipment pursuant to §73.36 (f) shall immediately report to the Director of the appropriate Atomic Energy Commission Regulatory Operations Regional Office listed in Appendix A, by telephone, telegram, or teletype, the details and results of his trace investigation and shall file within a period of fifteen (15) days a written report to the Director of the appropriate Regulatory Operations Regional Office with a copy to the Director of Regulatory Operations, U.S. Atomic Energy Commission, Washington, D.C. 20545, setting forth the details and results of the trace investigation.

(b) Each licensee shall report immediately to the Director of the appropriate Atomic Energy Commission Regulatory Operations Regional Office listed in Appendix A, by telephone, telegram, or teletype, any incident in which an attempt has been made, or is believed to have been made, to commit a theft or unlawful diversion of special nuclear material which he is licensed to possess, or to commit an act of industrial sabotage against his plant. The initial report shall be followed within a period of fifteen (15) days by a written report submitted to the Director of the appropriate Regulatory Operations Regional Office, with a copy to the Director of Regulatory Operations, U.S. Atomic Energy Commission, Washington, D.C. 20545, setting forth the details of the incident. Subsequent to the submission of the written report required by this paragraph, a licensee shall immediately inform the Director of the appropriate Regulatory Operations Regional Office by means of a written report of any substantive additional information, which becomes available to the licensee, concerning the incident.

ENFORCEMENT

§73.80 **Violations.**

An injunction or other court order may be obtained prohibiting any violation of any provision of the Act or any regulation or order issued thereunder. A court order may be obtained for the payment of a

U.S. ATOMIC ENERGY COMMISSION REGULATORY OPERATIONS
REGIONAL OFFICES*

Region and Address
Region I, Directorate of Regulatory Operations, USAEC, 631 Park Ave., King of Prussia, Pa. 19406
Region II, Directorate of Regulatory Operations, USAEC, Suite 818, 230 Peachtree St. N.W., Atlanta, Ga. 30303
Region III, Directorate of Regulatory Operations, USAEC, 799 Roosevelt Rd., Glen Ellyn, Ill. 60137
Region V, Directorate of Regulatory Operations, USAEC, P.O. Box 1515, Berkeley, Calif. 94701

* For the purposes of this regulation, the geographical areas assigned to the regional offices are as follows:

REGION I

Connecticut, Delaware, District of Columbia, Maine, Maryland, Massachusetts, New Hampshire, New Jersey, New York, Pennsylvania, Rhode Island, and Vermont.

REGION II

Alabama, Arkansas, Florida, Georgia, Kentucky, Louisiana, Mississippi, North Carolina, Puerto Rico, South Carolina, Tennessee, Virginia, and West Virginia.

REGION III

Illinois, Indiana, Iowa, Kansas, Michigan, Minnesota, Missouri, Nebraska, North Dakota, Ohio, Oklahoma, South Dakota, and Wisconsin.

REGION V

Alaska, Arizona, California, Colorado, Hawaii, Idaho, Montana, Nevada, New Mexico, Oregon, Texas, Utah, Washington, and Wyoming.

civil penalty imposed pursuant to section 234 of the Act for violation of sections 53, 57, 62, 63, 81, 82, 101, 103, 104, 107, or 109 of the Act or any rule, regulation, or order issued thereunder, or any term, condition, or limitation of any license issued thereunder, or for any violation for which a license may be revoked under section 186 of the Act. Any person who willfully violates any provision of the Act or any regulation or order issued thereunder may be guilty of a crime and upon conviction, may be punished by fine or imprisonment or both, as provided by law.

APPENDIX 11B

PHYSICAL SECURITY SURVEYS: GOVERNMENTAL*

PERIMETER BARRIERS

1. Is the perimeter of the activity defined by a fence or other type physical barrier?
2. If a fence is utilized as the perimeter barrier, does it meet the minimum specifications for security fencing?
    a. Is it of chain link design?
    b. Is it constructed of No. 11 gauge or heavier wire?
    c. Is mesh opening no longer than two inches square?
    d. Is selvage twisted and barbed at top and bottom?
    e. Is bottom of the fence within two inches of solid ground?
    f. Is the top guard strung with barbed wire and angled outward and upward at a 45-degree angle?
    g. Is it free from damage and deterioration?
3. If masonry wall is used, does it meet minimum specifications for security fencing? Is it at least seven feet high with a top guard similar to that required on a chain link fence or at least eight feet high with broken glass set on edge and cemented to top surfaces?
4. If building walls, floors, and roofs form a part of the perimeter barrier, do they provide security equivalent at least to that provided by chain link fence? Are all openings with an area of ninety-six square inches or greater and located less than eighteen feet above the level of the ground outside the perimeter barrier or less than fourteen feet from uncontrolled structures outside the perimeter barrier, properly secured?
5. If a building forms a part of the perimeter barrier, does it present a hazard at the point of juncture with the perimeter fence? If so, is the fence height increased 100 percent at the point of juncture?
6. If a river, lake, or other body of water forms any part of the perimeter barrier, are additional security measures provided?
7. Are openings such as culverts, tunnels, manholes for sewers and utility access, and sidewalk elevators which permit access to the installation properly secured?
8. Are there sufficient entrances through the barrier?
9. Do the gates and/or other entrances in perimeters exceed the number required for safe and efficient operation?
10. Are all perimeter entrances equipped with secure locking devices and are they always locked when not in active use?
11. Are all perimeter gates of such material and installation as to provide protection equivalent to the perimeter barriers of which they are a part?
12. Are gates and/or other perimeter entrances which are not in active use frequently inspected by guards or other personnel?

---

* *Army Crime Prevention Program:* Department of the Army Technical Bulletin, May, 1968.

13. Is the provost marshall (or security officer) responsible for security of keys to perimeter entrances?
14. Are keys to perimeter entrances issued to other than installation personnel?
15. Has any perimeter gate or entrance been newly established or permanently closed since the last survey?
16. Are all normally used pedestrian and vehicle gates and other perimeter entrances effectively and adequately lighted so as to assure (1) that there is proper identification of individuals and examination of credentials, (2) that interiors of vehicles are clearly lighted, and (3) that glare from luminaries is not in guard's eyes?
17. Are appropriate signs setting forth the provisions of entry conspicuously posted at all principal entrances? Are "No Trespassing" signs posted on or adjacent to perimeter barriers at such intervals that at least one sign is visible at any approach to the barrier for a minimum distance of fifty yards?
18. Are clear zones maintained on both sides of the perimeter barrier? Or if clear zone requirements cannot be met, are there additional security measures in use?
19. Are automobiles permitted to park against or too close to perimeter barrier?
20. Are lumber, boxes, or other material allowed to be stacked against, or in close proximity to, perimeter barrier?
21. Are frequent checks made by maintenance crews of condition of perimeter barriers and do guards patrol perimeter areas and report insecure factors related to perimeter barriers?
22. Are reports of inadequate perimeter security by maintenance crews or guards immediately acted upon and the necessary repairs made?
23. Is an interior all-weather road provided for the use of guard patrol cars? If so, what is its condition?
24. Are perimeters protected by intrusion alarm devices?
25. Does any relocated function, newly designated restricted area, physical expansion, or other like reason indicate necessity for installation of additional perimeter barriers or additional perimeter lighting?

Appendix 11B—Part II
SAMPLE SURVEY: PROPRIETARY
(To Be Used for Survey Purposes Only)

A. Physical plant
  1. Critical service areas
     a. How are key plant facilities protected from unauthorized entry and describe physical means of protection, including exterior lighting? What services are outside fenceline?
        switchgear or power plant
        transformer bank
        fire protection equipment
        fire pump
        water valves and lines
        gas meter house and gas lines
        oil storage tanks
        computer facilities
        boiler room
        incoming electric power
        water towers
        post indicator valves
        emergency generator
        Tel switching equipment
        computer tapes
     b. Comments
  2. Buildings
     a. Type of construction? Door construction?
     b. Leased or owned? Details if leased.
     c. Guardhouse construction and location.
     d. Comments
  3. Perimeter security
     Describe:
     a. Terrain?
     b. Type of fence?
        Height of fence?
        Distance from buildings?
        Cleared area both sides of fence?
        Accessible for viewing during patrols?
        Other

c. Gates and their use? Control of gates?
d. Patrols and their frequency?
e. Barbed wire on top of fence? How installed?
f. Are there any overpasses or subterranean passageways near the fence?
g. Height of windows from ground?
   a. Are windows located at perimeters?
   b. Critical operations opposite windows?
h. Are windows locked? Screened?
i. How are entrances controlled? Day? Night?

4. Parking areas
   a. What type of fences are used? Are cars parked abutting interior fences? Describe
   b. Are gates closed during work periods?
   c. Are unauthorized visits made to parking lot?
   d. Do employees have access to cars during their standard shift?
   e. Vehicle passes or decals?
   f. Are guards involved in traffic control?
   g. Are employees parking areas separated from company buildings?
   h. How are visitors' cars handled?
   i. What controls are in effect for vehicles parked within interior fence?
   j. Comments

5. Perimeter lighting
   a. Is the lighting provided adequate for the area?
      Foot candles on horizontal at ground level?
      (Estimate if possible)
   b. Is there a system of emergency lights that will automatically take over?
   c. Are doorways given sufficient light?
   d. Is lighting in use during all night hours?
   e. Is lighting directed toward perimeter? Can lighting hamper guard surveillance?
   f. Comments

6. Interior lighting
   a. Is the lighting provided during the day adequate?
   b. Is the light at night adequate?
   c. Is there a system of emergency lights?
   d. Is the night lighting sufficient for good surveillance by the night guard (or by municipal law enforcement agents)?
   e. Are guard shelter areas properly illuminated?
   f. Comments

B. Interior operations
   1. Key control
      a. Is a master key system in use?
      b. Does local security organization control or keep informed on lock installation and key distribution?
      c. Is the key locker and record files in order and current? Are issued keys cross-referenced?
      d. How long since the last visual inventory?
      e. Who is responsible for ascertaining the possession of keys?
      f. Have locks been changed when keys were lost?
      g. Are the holders of master keys authorized?
      h. Is locksmith's cage on master key system? If so, why?
   2. Safeguarding cash
      a. How much cash is maintained on the premises?
      b. What is the location and type of the repository? What protection? Is location of teller's area acceptable?
      c. What protective measures are taken when a money delivery is made?
      d. Burglar alarm devices? Describe. Tone controls?
      e. Employee background checks? Are tellers instructed in security measures?

f. How are blank employee checks handled? Blank bonds? Blank airline tickets? Travelers checks?
g. Comments
3. Shipping and receiving
   a. Supervision in attendance at all times? Guard checks?
   b. Are materials promptly removed from dock area?
   c. Are truck drivers allowed to wander about the area? Is there a waiting area outside a company office? Is there toilet facilities nearby? Water cooler? Pay telephone?
   d. Are truck court doors used by persons arriving and leaving their job?
   e. Are attractive items removed from the dock immediately? Storage area at platforms? Are there paperwork controls?
   f. Can each area be separately secured? Are they physically separated? Are they secured during lunch and break periods?
   g. Are shipping and receiving office fully enclosed? How are they secured?
   h. What type seals are in use? Are they tamper proof? How are seals controlled?
   i. Are truck interiors checked?
   j. Comments
4. Scrap operations
   a. What physical controls are available for area? Can area be secured?
   b. What type checks are made?
   c. How is station equipment scrap controlled?
   d. How is scrap contractor controlled?
5. Rubbish operations
   a. What are the physical controls?
   b. What type checks are being made?
   c. How is contractor controlled?
   d. What are hours of pickup?
6. Proprietary information
   a. Is management aware of need for protecting private information?
   b. How is this information handled?
   c. What are the safeguards for paper waste and how is it collected and destroyed?
C. Guards and company employees
   1. Guards
      a. How many are on the force?
      b. Are shifts arranged to provide adequate coverage at shift change?
      c. Is guard force own or contract?
      d. Review guard patrols and frequency. Is there complete penetration of all areas? Are yard areas reviewed?
      e. Are all clock stations being recorded? How?
      f. What are other duties?
      g. What reports do they make out? Follow up?
      h. Do they make adequate checks of parcels and lunch boxes being carried out of the building? Carried into the building?
      i. Are the guards alert and capable of carrying out their duties?
      j. Weapons carried? Are guards trained when to use weapons?
      k. Are instructions issued on weapons use?
      l. Does location have written instructions for guards?
      m. Do guards receive training? What type of training?
      n. How much training?
      o. Comments
   2. Employee credentials and passes
      a. Do employees have passes for identification purposes?
      b. Is it serially numbered and recorded? How are blanks stored?
      c. What are the routines for lost passes? What approval required to replace?
      d. How frequently are passes checked?
      e. What is the frequency of lost

passes at termination? What measures can be taken to insure receipt of pass?
    f. Visitor credentials?
    g. Personal and company property passes?
    h. Comments
D. Alarm system
  1. Specific type
    a. Is it owned or leased?
    b. Is it tied to outside agencies or does it report to an in-plant local board?
    c. What type of alarms are in the system? Electrical? Mechanical? Photoelectric? Capacitance? Other?
    d. Is this a full time system or part time? Describe.
    e. What is the response time?—(Minute) (Seconds) Which device tested?
    f. Do vulnerable areas have special alarm devices? Type?
    g. Are electrical or mechanical alarms given frequent checks or tests?
    h. Comments
E. General
  1. No. of employees? On each shift?
  2. Male? Female?
  3. Security Administration
  4. Responsibility for location's security?
     Name ....
          Department Chief
     Name ....
          Section Chief

APPENDIX 11C

GLOSSARY

SECTION I: ALARM SYSTEMS

Access control—The control of pedestrian and vehicular traffic through entrances and exits of a *protected area* or premises.

Access mode—The operation of an *alarm system* such that no *alarm signal* is given when the *protected area* is entered; however, a signal may be given if the *sensor, annunciator,* or *control unit* is tampered with or opened.

Access/secure control unit—(See *control unit.*)

Access switch—(See *authorized access switch.*)

Accumulator—A circuit which accumulates a sum. For example, in an audio alarm control unit, the accumulator sums the amplitudes of a series of pulses, which are larger than some threshold level, subtracts from the sum at a predetermined rate to account for random background pulses, and initiates an alarm signal when the sum exceeds some predetermined level. This circuit is also called an integrator; in digital circuits it may be called a counter.

Active intrusion sensor—An active sensor which detects the presence of an intruder within the range of the sensor. Examples are an *ultrasonic motion detector,* a *radio frequency motion detector,* and a *photoelectric alarm system.* (See also *passive intrusion sensor.*)

Active sensor—A sensor which detects the disturbance of a radiation field which is generated by the sensor. (See also *passive sensor.*)

Actuating device—(See *actuator.*)

Actuator—A manual or automatic switch or sensor such as *holdup button, magnetic switch,* or thermostat which causes a system to transmit an *alarm signal* when manually activated or when the device automatically senses an intruder or other unwanted condition.

Air gap—The distance between two magnetic elements in a magnetic or electromagnetic circuit, such as between the core and the armature of a relay.

Alarm—An *alarm device* or an *alarm signal*.

Alarm circuit—An electrical circuit of an alarm system which produces or transmits an *alarm signal*.

Alarm condition—A threatening condition, such as an intrusion, fire, or holdup, sensed by a *detector*.

Alarm device—A device which signals a warning in response to a *alarm condition*, such as a bell, siren, or *annunciator*.

Alarm discrimination—The ability of an alarm system to distinguish between those stimuli caused by an *intrusion* and those which are a part of the environment.

Alarm line—A wired electrical circuit used for the transmission of *alarm signals* from the protected premises to a *monitoring station*.

Alarm receiver— (See *annunciator*.)

Alarm sensor— (See *sensor*.)

Alarm signal—A signal produced by a *control unit* indicating the existence of an *alarm condition*.

Alarm state—The condition of a *detector* which causes a *control unit* in the *secure mode* to transmit an *alarm signal*.

Alarm station— (1) A manually actuated device installed at a fixed location to transmit an *alarm signal* in response to an *alarm condition*, such as a concealed *holdup button* in a bank teller's case. (2) A well-marked emergency control unit, installed in fixed locations usually accessible to the public, used to summon help in response to an *alarm condition*. The *control unit* contains either a manually actuated switch or telephone connected to fire or police headquarters, or a telephone answering service. (See also *remote station alarm system*.)

Alarm system—An assembly of equipment and devices designated and arranged to signal the presence of an *alarm condition* requiring urgent attention such as unauthorized entry, fire, temperature rise, etc. The system may be *local, police connection, central station* or *proprietary*. (For individual alarm systems see alphabetical listing by type, e.g. *intrusion alarm system*.)

Annunciator—An alarm monitoring device which consists of a number of visible signals such as "flags" or lamps indicating the status of the *detectors* in an alarm system or systems. Each circuit in the device is usually labelled to identify the location and condition being monitored. In addition to the visible signal, an audible signal is usually associated with the device. When an alarm condition is reported, a signal is indicated visibly, audibly, or both. The visible signal is generally maintained until reset either manually or automatically.

Answering service—A business which contracts with subscribers to answer incoming telephone calls after a specified delay or when scheduled to do so. It may also provide other services such as relaying fire or intrusion alarm signals to proper authorities.

Area protection—Protection of the inner space or volume of a secured area by means of a *volumetric sensor*.

Area sensor—A sensor with a detection zone which approximates an area, such as a wall surface or the exterior of a safe.

Audible alarm device— (1) A noisemaking device such as a siren, bell, or horn used as part of a local alarm system to indicate an *alarm condition*. (2) A bell, buzzer, horn or other noisemaking device used as a part of an *annunciator* to indicate a change in the status or operating mode of an alarm system.

Audio detection system— (See *sound sensing detection system*.)

Audio frequency (sonic)—Sound frequencies within the range of human hearing, approximately 15 to 20,000 Hz.

Audio monitor—An arrangement of amplifiers and speakers designed to monitor the sounds transmitted by microphones located in the *protected area*. Similar to an *annunciator*, except that supervisory personnel can monitor the protected area to interpret the sounds.

Authorized access switch—A device used

to make an alarm system or some portion or zone of a system inoperative in order to permit authorized access through a *protected port*. A *shunt* is an example of such a device.

BA—Burglar alarm.

Beam divergence—In a *photoelectric alarm system*, the angular spread of the light beam.

Break alarm—(1) An *alarm condition* signaled by the opening or breaking of an electrical circuit. (2) The signal produced by a break alarm condition (sometimes referred to as an open circuit alarm or trouble signal, designed to indicate possible system failure).

Bug—(1) To plant a microphone or other *sound sensor* or to tap a communication line for the purpose of *surreptitious* listening or *audio monitoring;* loosely, to install a sensor in a specified location. (2) The microphone or other sensor used for the purpose of surreptitious listening.

Building security alarm system—The system of *protective signaling* devices installed at a premise.

Burglar alarm (BA) pad—A supporting frame laced with fine wire or a fragile panel located with *foil* or fine wire and installed so as to cover an exterior opening in a building, such as a door, or skylight. Entrance through the opening breaks the wire or foil and initiates an *alarm signal*. (See also *grid*.)

Burglar alarm system—(See *intrusion alarm system*.)

Burglary—The unlawful entering of a structure with the intent to commit a felony or theft therein.

Cabinet-for-safe—A wooden enclosure having closely spaced electrical *grids* on all inner surfaces and *contacts* on the doors. It surrounds a safe and initiates an alarm signal if an attempt is made to open or penetrate the cabinet.

Capacitance—The property of two or more objects which enables them to store electrical energy in an electric field between them. The basic measurement unit is the farad. Capacitance varies inversely with the distance between the objects, hence the change of capacitance with relative motion is greater the nearer one object is to the other.

Capacitance alarm system—An alarm system in which a protected object is electrically connected as a *capacitance sensor*. The approach of an intruder causes sufficient change in *capacitance* to upset the balance of the system and initiate an *alarm signal*. Also called proximity alarm system.

Capacitance detector—(See *capacitance sensor*.

Capacitance sensor—A sensor which responds to a change in *capacitance* in a field containing a protected object or in a field within a protected area.

Carrier current transmitter—A device which transmits *alarm signals* from a sensor to a *control unit* via the standard AC power lines.

Central station—A control center to which alarm systems in a subscriber's premises are connected, where circuits are supervised, and where personnel are maintained continuously to record and investigate alarm or trouble signals. Facilities are provided for the reporting of alarms to police and fire departments or to other outside agencies.

Central station alarm system—An alarm system, or group of systems, the activities of which are transmitted to, recorded in, maintained by, and supervised from a *central station*. This differs from *proprietary alarm systems* in that the central station is owned and operated independently of the subscriber.

Circumvention—The defeat of an alarm system by the avoidance of its detection devices, such as by jumping over a pressure sensitive mat, by entering through a hole cut in an unprotected wall rather than through a protected door, or by keeping outside the range of an *ultrasonic motion detector*. Circumvention contrasts with *spoofing*.

Closed circuit alarm—(See *cross alarm*.)

Closed circuit system—A system in which the sensors of each zone are connected in series so that the same current exists in each sensor. When an activated sensor breaks the circuit or the connecting wire is cut, an alarm is transmitted for that zone.

Clutch head screw—A mounting screw with a uniquely designed head for which the installation and removal tool is not commonly available. They are used to install alarm system components so that removal is inhibited.

Coded-alarm system—An alarm system in which the source of each signal is identifiable. This is usually accomplished by means of a series of current pulses which operate audible or visible *annunciators* or recorders or both, to yield a recognizable signal. This is usually used to allow the transmission of multiple signals on a common circuit.

Coded cable—A multiconductor cable in which the insulation on each conductor is distinguishable from all others by color or design. This assists in identification of the point of origin or final destination of a wire.

Coded transmitter—A device for transmitting a coded signal when manually or automatically operated by an *actuator*. The actuator may be housed with the transmitter or a number of actuators may operate a common transmitter.

Coding siren—A siren which has an auxiliary mechanism to interrupt the flow of air through its principal mechanism, enabling it to produce a controllable series of sharp blasts.

Combination sensor alarm system—An alarm system which requires the simultaneous activation of two or more sensors to initiate an *alarm signal*.

Compromise— (See *defeat*.)

Constant ringing drop (CRD)—A relay which when activated even momentarily will remain in an *alarm condition* until *reset*. A key is often required to reset the relay and turn off the alarm.

Constant ringing relay (CRR)— (See *constant ringing drop*.)

Contact— (1) Each of the pair of metallic parts of a switch or relay which by touching or separating make or break the electrical current path. (2) A switch-type sensor.

Contact device—A device which when actuated opens or closes a set of electrical contacts; a switch or relay.

Contact microphone—A microphone designed for attachment directly to a surface of a *protected area* or object; usually used to detect surface vibrations.

Contact vibration sensor— (See *vibration sensor*.)

Contactless vibrating bell—A *vibrating bell* whose continuous operation depends upon application of an alternating current, without circuit-interrupting contacts such as those used in vibrating bells operated by direct current.

Control cabinet— (See *control unit*.)

Control unit—A device, usually *electronic*, which provides the interface between the alarm system and the human operator and produces an *alarm signal* when its programmed response indicates an *alarm condition*. Some or all of the following may be provided for: power for sensors, sensitivity adjustments, means to select and indicate *access mode* or *secure mode*, monitoring for *line supervision* and *tamper devices*, timing circuits, for *entrance* and *exit delays*, transmission of an alarm signal, etc.

Covert—Hidden and protected.

CRD— (See *constant ringing drop*.)

Cross alarm— (1) An *alarm condition* signaled by crossing or shorting an electrical circuit (2) The signal produced due to a cross alarm condition.

Crossover—An insulated electrical path used to connect foil across window dividers, such as those found on multiple pane windows, to prevent grounding and to make a more durable connection.

CRR—Constant ringing relay. (See *constant ringing drop*.)

Dark current—The current output of a *photoelectric sensor* when no light is entering the sensor.

Day setting— (See *access mode*.)

Defeat—The frustration, counteraction, or thwarting of an *alarm device* so that it fails to signal an alarm when a protected area is entered. Defeat includes both *circumvention* and *spoofing*.

Detection range—The greatest distance at which a sensor will consistently detect an intruder under a standard set of conditions.

Detector—(1) A sensor such as those used to detect *intrusion,* equipment malfunctions or failure, rate of temperature rise, smoke or fire. (2) A demodulator, a device for recovering the modulating function or signal from a modulated wave, such as that used in a modulated photoelectric alarm system. (See also *photoelectric alarm system, modulated.*)

Dialer—(See *telephone dialer, automatic.*)

Differential pressure sensor—A sensor used for *perimeter protection* which responds to the difference between the hydraulic pressures in two liquid-filled tubes buried just below the surface of the earth around the exterior perimeter of the *protected area.* The pressure difference can indicate an intruder walking or driving over the buried tubes.

Digital telephone dialer—(See *telephone dialer, digital.*)

Direct connect—(See *police connection.*)

Direct wire burglar alarm circuit (DWBA)—(See *alarm line.*)

Direct wire circuit—(See *alarm line.*)

Door cord—A short, insulated cable with an attaching block and terminals at each end used to conduct current to a device, such as *foil,* mounted on the movable portion of a door or window.

Door trip switch—A *mechanical switch* mounted so that movement of the door will operate the switch.

Doppler effect (shift)—The apparent change in frequency of sound or radio waves when reflected from or originating from a moving object. Utilized in some types of *motion sensors.*

Double-circuit system—An *alarm circuit* in which two wires enter and two wires leave each sensor.

Double drop—An alarm signaling method often used in *central station alarm systems* in which the line is first opened to produce a *break alarm* and then shorted to produce a *cross alarm.*

Drop—(1) (See *annunciator.*) (2) A light indicator on an annunciator.

Duress alarm device—A device which produces either a *silent alarm* or *local alarm* under a condition of personnel stress such as holdup, fire, illness, or other panic or emergency. The device is normally manually operated and may be fixed or portable.

Duress alarm system—An alarm system which employs a *duress alarm device.*

DWBA—Direct wire burglar alarm. (See *alarm line.*)

E-field sensor—A *passive sensor* which detects changes in the earth's ambient electric field caused by the movement of an intruder. (See also *H-field sensor.*)

Electrical—Related to, pertaining to, or associated with electricity.

Electromagnetic—Pertaining to the relationship between current flow and magnetic field.

Electromagnetic interference (EMI)—Impairment of the reception of a wanted electromagnetic signal by an electromagnetic disturbance. This can be caused by lightning, radio transmitters, power line noise and other electrical devices.

Electromechanical bell—A bell with a prewound spring-driven striking mechanism, the operation of which is initiated by the activation of an electric tripping mechanism.

Electronic—Related to, or pertaining to, devices which utilize electrons moving through a vacuum, gas, or semiconductor, and to circuits or systems containing such devices.

EMI—(See *electromagnetic interference.*)

End of line resistor—(See *terminal resistor.*)

Entrance delay—The time between actuating a sensor on an entrance door or gate and the sounding of a *local alarm* or transmission of an *alarm signal* by the *control unit.* This delay is used if the *authorized access switch* is located within the *protected area* and permits a person with the control key to enter without causing an alarm. The delay is provided by a timer within the *control unit*.

E.O.L.—End of line.

Exit delay—The time between turning on a control unit and the sounding of a *local alarm* or transmission of an *alarm signal* upon actuation of a sensor on an exit door. This delay is used if the *authorized access switch* is located within the *protected area* and permits a person with the control key to turn on the alarm system and to leave through a protected door or gate without causing an alarm. The delay is provided by a timer within the *control unit*.

Fail safe—A feature of a system or device which initiates an alarm or trouble signal when the system or device either malfunctions or loses power.

False alarm—An alarm signal transmitted in the absence of an *alarm condition*. These may be classified according to causes: environmental, e.g. rain, fog, wind, hail, lightning, temperature, etc.; animals, e.g. rats, dogs, cats, insects, etc.; man-made disturbances, e.g. sonic booms, EMI, vehicles, etc.; equipment malfunction, e.g. transmission errors, component failure, etc.; operator error; and unknown.

False alarm rate, monthly—The number of false alarms per installation per month.

False alarm ratio—The ratio of *false alarms* to total alarms; may be expressed as a percentage or as a simple ratio.

Fence alarm—Any of several types of sensors used to detect the presence of an intruder near a fence or any attempt by him to climb over, go under, or cut through the fence.

Field—The space or area in which there exists a force such as that produced by an electrically charged object, a current, or a magnet.

Fire detector (sensor)—(See *heat sensor* and *smoke detector*.)

Floor mat—(See *mat switch*.)

Floor trap—A *trap* installed so as to detect the movement of a person across a floor space, such as a *trip wire switch* or *mat switch*.

Foil—Thin metallic strips which are cemented to a protected surface (usually glass in a window or door), and connected to a closed electrical circuit. If the protected material is broken so as to break the foil, the circuit opens, initiating an alarm signal. Also called tape. A window, door, or other surface to which foil has been applied is said to be taped or foiled.

Foil connector—An electrical terminal block used on the edge of a window to join interconnecting wire to window *foil*.

Foot rail—A *holdup alarm device*, often used at cashiers' windows, in which a foot is placed under the rail, lifting it, to initiate an *alarm signal*.

Frequency division multiplexing (FDM)—(See *multiplexing, frequency division*.)

Glassbreak vibration detector—A *vibration detection system* which employs a *contact microphone* attached to a glass window to detect cutting or breakage of the glass.

Grid—(1) An arrangement of electrically conducting wire, screen, or tubing placed in front of doors or windows or both which is used as a part of a *capacitance sensor*. (2) A lattice of wooden dowels or slats concealing fine wires in a closed circuit which initiates an *alarm signal* when forcing or cutting the lattice breaks the wires. Used over accessible openings. Sometimes called a protective screen. (See also *burglar alarm pad*.) (3) A screen or metal plate, connected to earth ground, sometimes used to provide a stable ground reference for objects protected by a *capacitance sensor*. If placed against the walls near the protected object, it prevents the sensor sensitivity from extending through the walls into areas of activity.

Heat detector—(See *heat sensor*.)

Heat sensor—(1) A sensor which responds

to either a local temperature above a selected value, a local temperature increase which is at a rate of increase greator than a preselected rate (rate of rise), or both. (2) A sensor which responds to infrared radiation from a remote source such as a person.

H-field sensor—A *passive sensor* which detects changes in the earth's ambient magnetic field caused by the movement of an intruder. (See also *E-field sensor*.)

Holdup—A *robbery* involving the threat to use a weapon.

Holdup alarm device—A device which signals a holdup. The device is usually *surreptitious* and may be manually or automatically actuated, fixed or portable. (See *duress alarm device*.)

Holdup alarm system, automatic—An alarm system which employs a holdup alarm device, in which the signal transmission is initiated solely by the action of the intruder, such as a money clip in a cash drawer.

Holdup alarm system, manual—A holdup alarm system in which the signal transmission is initiated by the direct action of the person attacked or of an observer of the attack.

Holdup button—A manually actuated *mechanical switch* used to initiate a duress alarm signal; usually constructed to minimize accidental activation.

Hood contact—A switch which is used for the supervision of a closed safe or vault door. Usually installed on the outside surface of the protected door.

Impedance—The opposition to the flow of alternating current in a circuit. May be determined by the ratio of an input voltage to the resultant current.

Impedance matching—Making the *impedance* of a *terminating device* equal to the impedance of the circuit to which it is connected in order to achieve optimum signal transfer.

Infrared (IR) motion detector—A sensor which detects changes in the infrared light radiation from parts of the *protected area*. Presence of an intruder in the area changes the infrared light intensity from his direction.

Infrared (IR) motion sensor—(See *infrared motion detector*.)

Infrared sensor—(See *heat sensor, infrared motion detector*, and *photoelectric sensor*.)

Inking register—(See *register, inking*.)

Interior perimeter protection—A line of protection along the interior boundary of a *protected area* including all points through which entry can be effected.

Intrusion—Unauthorized entry into the property of another.

Intrusion alarm system—An alarm system for signaling the entry or attempted entry of a person or an object into the area or volume protected by the system.

Ionization smoke detector—A *smoke detector* in which a small amount of radioactive material ionizes the air in the sensing chamber, thus rendering it conductive and permitting a current to flow through the air between two charged electrodes. This effectively gives the sensing chamber an electrical conductance. When smoke particles enter the ionization area, they decrease the conductance of the air by attaching themselves to the ions causing a reduction in mobility. When the conductance is less than a predetermined level, the detector circuit responds.

IR—Infrared.

Jack—An electrical connector which is used for frequent connect and disconnect operations; for example, to connect an alarm circuit at an overhang door.

Lacing—A network of fine wire surrounding or covering an area to be protected, such as a safe, vault, or glass panel, and connected into a *closed circuit system*. The network of wire is concealed by a shield such as concrete or paneling in such a manner that an attempt to break through the shield breaks the wire and initiates an alarm.

Light intensity cutoff—In a *photoelectric alarm system*, the percent reduction of light which initiates an *alarm signal* at the photoelectric receiver unit.

Line amplifier—An audio amplifier which is used to provide preamplification of an audio *alarm signal* before transmission of the signal over an *alarm line*. Use of an amplifier extends the range of signal transmission.

Line sensor (detector)—A sensor with a detection zone which approximates a line or series of lines, such as a *photoelectric sensor* which senses a direct or reflected light beam.

Line supervision—Electronic protection of an *alarm line* accomplished by sending a continuous or coded signal through the circuit. A change in the circuit characteristics, such as a change in *impedance* due to the circuit's having been tampered with, will be detected by a monitor. The monitor initiates an alarm if the change exceeds a predetermined amount.

Local alarm—An alarm which when activated makes a loud noise (see *audible alarm device*) at or near the *protected area* or floods the site with light or both.

Local alarm system—An alarm system which when activated produces an audible or visible signal in the immediate vicinity of the protected premises or object. This term usually applies to systems designed to provide only a local warning of *intrusion* and not to transmit to a remote *monitoring station*. However, local alarm systems are sometimes used in conjunction with a *remote alarm*.

Loop—An electric circuit consisting of several elements, usually switches, connected in series.

Magnetic Alarm System—An alarm system which will initiate an alarm when it detects changes in the local magnetic field. The changes could be caused by motion of ferrous objects such as guns or tools near the *magnetic sensor*.

Magnetic contact— (See *magnetic switch*.)

Magnetic sensor—A sensor which responds to changes in magnetic field. (See also *magnetic alarm system*.)

Magnetic switch—A switch which consists of two separate units: a magnetically-actuated switch, and a magnet. The switch is usually mounted in a fixed position (door jamb or window frame) opposing the magnet, which is fastened to a hinged or sliding door, window, etc. When the movable section is opened, the magnet moves with it, actuating the switch.

Magnetic switch, balanced—A *magnetic switch* which operates using a balanced magnetic field in such a manner as to resist *defeat* with an external magnet. It signals an alarm when it detects either an increase or decrease in magnetic field strength.

Matching network—A circuit used to achieve *impedance matching*. It may also allow audio signals to be transmitted to an *alarm line* while blocking direct current used locally for *line supervision*.

Mat switch—A flat area switch used on open floors or under carpeting. It may be sensitive over an area of a few square feet or several square yards.

McCulloh Circuit (Loop)—A supervised single wire *loop* connecting a number of *coded transmitters* located in different *protected areas* to a *central station* receiver.

Mechanical switch—A switch in which the *contacts* are opened and closed by means of a depressible plunger or button.

Mercury fence alarm—A type of *mercury switch* which is sensitive to the vibration caused by an intruder climbing on a fence.

Mercury switch—A switch operated by tilting or vibrating which causes an enclosed pool of mercury to move, making or breaking physical and electrical contact with conductors. These are used on tilting doors and windows, and on fences.

Microwave alarm system—An alarm system which employs *radio frequency motion detectors* operating in the *microwave frequency* region of the electromagnetic spectrum.

Microwave frequency—Radio frequencies in the range of approximately 1.0 to 300 GHz.

Microwave motion detector— (See *radio frequency motion detector*.)

Modulated photoelectric alarm system—(See *photoelectric alarm system, modulated.*)

Monitor cabinet—An enclosure which houses the *annunciator* and associated equipment.

Monitor panel—(See *Annunciator.*)

Monitoring station—The *central station* or other area at which guards, police, or commercial service personnel observe *annunciators* and *registers* reporting on the condition of alarm systems.

Motion detection system—(See *motion sensor.*)

Motion detector—(See *Motion sensor.*)

Motion sensor—A sensor which responds to the motion of an intruder. (See also *radio frequency motion detector, sonic motion detector, ultrasonic motion detector,* and *infrared motion detector.*)

Multiplexing—A technique for the concurrent transmission of two or more signals in either or both directions, over the same wire, carrier, or other communication channel. The two basic multiplexing techniques are time division multiplexing and frequency division multiplexing.

Multiplexing, frequency division (FDM)—The multiplexing technique which assigns to each signal a specific set of frequencies (called a channel) within the larger block of frequencies available on the main transmission path in much the same way that many radio stations broadcast at the same time but can be separately received.

Multiplexing, time division (TDM)—The multiplexing technique which provides for the independent transmission of several pieces of information on a time-sharing basis by sampling, at frequent intervals, the data to be transmitted.

Neutralization—(See *defeat.*)

NiCad—(Contraction of "nickel cadmium.") A high performance, long-lasting rechargeable battery, with electrodes made of nickel and cadmium, which may be used as an emergency power supply for an alarm system.

Night setting—(See *secure mode.*)

Nonretractable (one-way) screw—A screw with a head designed to permit installation with an ordinary flat bit screwdriver but which resists removal. They are used to install alarm system components so that removal is inhibited.

Normally closed (NC) switch—A switch in which the *contacts* are closed when no external forces act upon the switch.

Normally open (NO) switch—A switch in which the *contacts* are open (separated) when no external forces act upon the switch.

Nuisance alarm—(See *false alarm.*)

Object protection—(See *spot protection.*)

Open-circuit alarm—(See *break alarm.*)

Open-circuit system—A system in which the sensors are connected in parallel. When a sensor is activated, the circuit is closed, permitting a current which activates an *alarm signal.*

Panic alarm—(See *duress alarm device.*)

Panic button—(See *duress alarm device.*)

Passive intrusion sensor—A passive sensor in an *intrusion alarm system* which detects an intruder within the range of the sensor. Examples are a *sound sensing detection system,* a *vibration detection system,* an *infrared motion detector,* and an *E-field sensor.*

Passive sensor—A sensor which detects natural radiation or radiation disturbances, but does not itself emit the radiation on which its operation depends.

Passive ultrasonic alarm system—An alarm system which detects the sounds in the *ultrasonic frequency* range caused by an attempted forcible entry into a protected structure. The system consists of microphones, a *control unit* containing an amplifier, filters, an *accumulator,* and a power supply. The unit's sensitivity is adjustable so that ambient noises or normal sounds will not initiate an *alarm signal;* however, noise above the preset level or a sufficient accumulation of impulses will initiate an alarm.

Percentage supervision—A method of *line supervision* in which the current in or resistance of a supervised line is monitored for changes. When the change exceeds a selected percentage of the normal operating current or resistance in the line, an *alarm signal* is produced.

Perimeter alarm system—An alarm system which provides perimeter protection.

Perimeter protection—Protection of access to the outer limits of a *protected area,* by means of physical barriers, sensors on physical barriers, or exterior sensors not associated with a physical barrier.

Permanent circuit—An *alarm circuit* which is capable of transmitting an *alarm signal* whether the alarm control is in *access mode* or *secure mode.* Used, for example, on foiled fixed windows, *tamper switches,* and supervisory lines. (See also *supervisory alarm system, supervisory circuit,* and *permanent protection.*)

Permanent protection—A system of alarm devices such as *foil, burglar alarm pads,* or *lacings* connected in a permanent circuit to provide protection whether the *control unit* is in the *access mode* or *secure mode.*

Photoelectric alarm system—An alarm system which employs a light beam and *photoelectric sensor* to provide a line of protection. Any interruption of the beam by an intruder is sensed by the sensor. Mirrors may be used to change the direction of the beam. The maximum beam length is limited by many factors, some of which are the light source intensity, number of mirror reflections, detector sensitivity, *beam divergence,* fog, and haze.

Photoelectric alarm system, modulated—A photoelectric alarm system in which the transmitted light beam is modulated in a predetermined manner and in which the receiving equipment will signal an alarm unless it receives the properly modulated light.

Photoelectric beam type smoke detector—A *smoke detector* which has a light source which projects a light beam across the area to be protected onto a photoelectric cell. Smoke between the light source and the receiving cell reduces the light reaching the cell, causing actuation.

Photoelectric detector— (See *photoelectric sensor.*)

Photoelectric sensor—A device which detects a visible or invisible beam of light and responds to its complete or nearly complete interruption. (See also *photoelectric alarm system* and *photoelectric alarm system, modulated.*)

Photoelectric spot type smoke detector—A *smoke detector* which contains a chamber with covers which prevent the entrance of light but allow the entrance of smoke. The chamber contains a light source and a photosensitive cell so placed that light is blocked from it. When smoke enters, the smoke particles scatter and reflect the light into the photosensitive cell, causing an alarm.

Point protection— (See *spot protection.*)

Police connection—The direct link by which an alarm system is connected to an *annunciator* installed in a police station. Examples of a police connection are an *alarm line,* or a radio communications channel.

Police panel— (See *police station unit.*)

Police station unit—An *annunciator* which can be placed in operation in a police station.

Portable duress sensor—A device carried on a person which may be activated in an emergency to send an *alarm signal* to a *monitoring station.*

Portable intrusion sensor—A sensor which can be installed quickly and which does not require the installation of dedicated wiring for the transmission of its *alarm signal.*

Positive noninterfering (PNI) and successive alarm system—An alarm system which employs multiple alarm transmitters on each *alarm line* (like *McCulloh loop*) such that in the event of simultaneous operation of several transmitters, one of them takes control of the alarm line, transmits its full signal, then release the alarm line for successive transmission by other transmitters which are held inoperative until they gain control.

Pressure alarm system—An alarm system which protects a vault or other enclosed space by maintaining and monitoring a predetermined air pressure differential between the inside and outside of the space. Equalization of pressure resulting from opening the vault or cutting through the enclosure will be sensed and will initiate an *alarm signal*.

Printing recorder—An electromechanical device used at a *monitoring station* which accepts coded signals from alarm lines and converts them to an alphanumeric printed record of the signal received.

Proprietary alarm system—An alarm system which is similar to a *central station alarm system* except that the *annunciator* is located in a constantly manned guard room maintained by the owner for his own internal security operations. The guards monitor the system and respond to all *alarm signals* or alert local law enforcement agencies or both.

Protected area—An area monitored by an alarm system or guards, or enclosed by a suitable barrier.

Protected port—A point of entry such as a door, window, or corridor which is monitored by sensors connected to an alarm system.

Protection device—(1) A sensor such as a *grid, foil, contact,* or *photoelectric sensor* connected into an *intrusion alarm system*. (2) A barrier which inhibits *intrusion*, such as a grille, lock, fence or wall.

Protection, exterior perimeter—A line of protection surrounding but somewhat removed from a facility. Examples are fences, barrier walls, or patrolled points of a perimeter.

Protection off— (See *access mode*.)

Protection on— (See *secure mode*.)

Protective screen— (See *grid*.)

Protective signaling—The initiation, transmission, and reception of signals involved in the detection and prevention of property loss due to fire, burglary, or other destructive conditions. Also, the electronic supervision of persons and equipment concerned with this detection and prevention. (See also *line supervision* and *supervisory alarm system*.)

Proximity alarm system— (See *capacitance alarm system*.)

Punching register— (See *register, punch*.)

Radar alarm system—An alarm system which employs *radio frequency motion detectors*.

Radar (radio detecting and ranging)— (See *radio frequency motion detector*.)

Radio frequency interference (RFI)—*electromagnetic interference* in the radio frequency range.

Radio frequency motion detector—A sensor which detects the motion of an intruder through the use of a radiated radio frequency electromagnetic field. The device operates by sensing a disturbance in the generated RF field caused by intruder motion, typically a modulation of the field referred to as a *doppler effect*, which is used to initiate an *alarm signal*. Most radio frequency motion detectors are certified by the FCC for operation as "field disturbance sensors" at one of the following frequencies: 0.915 GHz (L-Band), 2.45 GHz (S-Band), 5.8 GHz (X-Band), 10.525 GHz (X-Band), and 22.125 GHz (K-Band). Units operating in the *Microwave frequency* range are usually called *microwave motion detectors*.

Reed switch—A type of *magnetic switch* consisting of contacts formed by two thin moveable magnetically actuated metal vanes or reeds, held in a normally open position within a sealed glass envelope.

Register—An electromechanical device which marks a paper tape in response to signal impulses received from transmitting circuits. A register may be driven by a prewound spring mechanism, an electric motor, or a combination of these.

Register, inking—A register which marks the tape with ink.

Register, punch—A register which marks the tape by cutting holes in it.

Register, slashing—A register which marks the tape by cutting V-shaped slashes in it.

Remote alarm—An *alarm signal* which is transmitted to a remote *monitoring station*. (See also *local alarm*.)

Remote station alarm system—An alarm system which employs remote *alarm stations* usually located in building hallways or on city streets.

Reporting line— (See *alarm line*.)

Reset—To restore a device to its original (normal) condition after an alarm or trouble signal.

Resistance bridge smoke detector—A *smoke detector* which responds to the particles and moisture present in smoke. These substances reduce the resistance of an electrical bridge grid and cause the detector to respond.

Retard transmitter—A *coded transmitter* in which a delay period is introduced between the time of actuation and the time of signal transmission.

RFI—*Radio frequency interference.*

Rf motion detector— (See *radio frequency motion detector*.)

Robbery—The felonious or forcible taking of property by violence, threat, or other overt felonious act in the presence of the victim.

Secure mode—The condition of an alarm system in which all sensors and *control units* are ready to respond to an intrusion.

Security monitor— (See *annunciator*.)

Seismic sensor—A sensor, generally buried under the surface of the ground for *perimeter protection*, which responds to minute vibrations of the earth generated as an intruder walks or drives within its *detection range*.

Sensor—A device which is designed to produce a signal or offer indication in response to an event or stimulus within its detection zone.

Sensor, combustion— (See *ionization smoke detector, photoelectric beam type smoke detector, photoelectric spot type smoke detector* and *resistance bridge smoke detector*.

Sensor, smoke— (See *ionization smoke detector, photoelectric beam type smoke detector, photoelectric spot type smoke detector* and *resistance bridge smoke detector*.)

Shunt— (1) A deliberate shorting-out of a portion of an electric circuit. (2) A key-operated switch which removes some portion of an alarm system for operation, allowing entry into a *protected area* without initiating an *alarm signal*. A type of *authorized access switch*.

Shunt switch— (See *shunt*.)

Signal recorder— (See *register*.)

Silent alarm—A *remote alarm* without an obvious local indication that an alarm has been transmitted.

Silent alarm system—An alarm system which signals a remote station by means of a silent alarm.

Single circuit system—An *alarm circuit* which routes only one side of the circuit through each sensor. The return may be through either ground or a separate wire.

Single-stroke bell—A bell which is struck once each time its mechanism is activated.

Slashing register— (See *register, slashing*.)

Smoke detector—A device which detects visible or invisible products of combustion. (See also *ionization smoke detector, photoelectric beam type smoke detector, photoelectric spot type smoke detector,* and *resistance bridge smoke detector*.)

Solid state— (1) An adjective used to describe a device such as a semiconductor transistor or diode. (2) A circuit or system which does not rely on vacuum or gas-filled tubes to control or modify voltages and currents.

Sonic motion detector—A sensor which detects the motion of an intruder by his disturbance of an audible sound pattern generated within the protected area.

Sound sensing detection system—An alarm system which detects the audible sound caused by an attempted forcible entry into a protected structure. The system consists of microphones and a *control unit* containing an amplifier, *accumulator*, and a power supply. The unit's sensitivity is adjustable so that ambient noises or normal sounds will not initiate an *alarm signal*. However, noises above this preset level or a sufficient accumulation of impulses will initiate an alarm.

Sound sensor—A sensor which responds to sound; a microphone.

Space protection— (See *area protection*.)

Spoofing—The defeat or compromise of an alarm system by "tricking" or "fooling" its detection devices such as by short circuiting part or all of a series circuit, cutting wires in a parallel circuit, reducing the sensitivity of a sensor, or entering false signals into the system. Spoofing contrasts with *circumvention*.

Spot protection—Protection of objects such as safes, art objects, or anything of value which could be damaged or removed from the premises.

Spring contact—A device employing a current-carrying cantilever spring which monitors the position of a door or window.

Standby power supply—Equipment which supplies power to a system in the event the primary power is lost. It may consist of batteries, charging circuits, auxiliary motor generators or a combination of these devices.

Strain gauge alarm system—An alarm system which detects the stress caused by the weight of an intruder as he moves about a building. Typical uses include placement of the strain gauge sensor under a floor joist or under a stairway tread.

Strain gauge sensor—A sensor which, when attached to an object, will provide an electrical response to an applied stress upon the object, such as a bending, stretching or compressive force.

Strain sensitive cable—An electrical cable which is designed to produce a signal whenever the cable is strained by a change in applied force. Typical uses including mounting it in a wall to detect an attempted forced entry through the wall, or fastening it to a fence to detect climbing on the fence, or burying it around a perimeter to detect walking or driving across the perimeter.

Subscriber's equipment—That portion of a *central station alarm system* installed in the protected premises.

Subscriber's unit—A *control unit* of a *central station alarm system*.

Supervised lines—Interconnecting lines in an alarm system which are electrically supervised against tampering. (See also *line supervision*.)

Supervisory alarm system—An alarm system which monitors conditions or persons or both and signals any deviation from an established norm or schedule. Examples are the monitoring of signals from guard patrol stations for irregularities in the progression along a prescribed patrol route, and the monitoring of production or safety conditions such as sprinkler water pressure, temperature, or liquid level.

Supervisory circuit—An electrical circuit or radio path which sends information on the status of a sensor or guard patrol to an *annunciator*. For *intrusion alarm systems*, this circuit provides *line supervision* and monitors *tamper devices*. (See also *supervisory alarm system*.)

Surreptitious—*covert*, hidden, concealed, or disguised.

Surveillance— (1) Control of premises for security purposes through alarm systems, closed circuit television (CCTV), or other monitoring methods. (2) Supervision or inspection of industrial processes by monitoring those conditions which could cause damage if not corrected. (See also *supervisory alarm system*.)

Tamper device— (1) Any device, usually a switch, which is used to detect an attempt to gain access to intrusion alarm circuitry, such as by removing a switch cover. (2) A monitor circuit to detect any attempt to modify the alarm circuitry, such as by cutting a wire.

Tamper switch—A switch which is installed in such a way as to detect attempts to remove the enclosure of some alarm system components such as control box doors, switch covers, junction box covers, or bell housings. The alarm component is then often described as being "tampered."

Tape— (See *foil.*)

Tamper bell—A *single-stroke bell* designed to produce a sound of low intensity and relatively high pitch.

Telephone dialer, automatic—A device which, when activated, automatically dials one or more pre-programmed telephone numbers (e.g., police, fire department) and relays a recorded voice or coded message giving the location and nature of the alarm.

Telephone dialer, digital—An automatic telephone dialer which sends its message as a digital code.

Terminal resistor—A resistor used as a *terminating device.*

Terminating capacitor—A capacitor sometimes used as a terminating device for a *capacitance sensor* antenna. The capacitor allows the supervision of the sensor antenna, especially if a long wire is used as the sensor.

Terminating device—A device which is used to terminate an electrically supervised circuit. It makes the electrical circuit continuous and provides a fixed *impedance* reference (end of line resistor) against which changes are measured to detect an *alarm condition.* The impedance changes may be caused by a sensor, tampering, or circuit trouble.

Time delay— (See *entrance delay* and *exit delay.*)

Time division multiplexing (TDM)— (See *multiplexing, time division.*)

Timing table—That portion of *central station* equipment which provides a means for checking incoming signals from *McCulloh circuits.*

Touch sensitivity—The sensitivity of a *capacitance sensor* at which the *alarm device* will be activated only if an intruder touches or comes in very close proximity (about 1 cm or ½ in.) to the protected object.

Trap— (1) A device, usually a switch, installed within a protected area, which serves as secondary protection in the event a *perimeter alarm system* is successfully penetrated. Examples are a *trip wire switch* placed across a likely path for an intruder, a *mat switch* hidden under a rug, or a *magnetic switch* mounted on an inner door. (2) A *volumetric sensor* installed so as to detect an intruder in a likely traveled corridor or pathway within a security area.

Trickle charge—A continuous direct current, usually very low, which is applied to a battery to maintain it at peak charge or to recharge it after it has been partially or completely discharged. Usually applied to nickel cadmium (NiCad) or wet cell batteries.

Trip wire switch—A switch which is actuated by breaking or moving a wire or cord installed across a floor space.

Trouble signal— (See *break alarm.*)

UL— (See *Underwriters Laboratories, Inc.*)

UL certificated—For certain types of products which have met UL requirements, for which it is impractical to apply the UL Listing Mark or Classification Marking to the individual product, a certificate is provided which the manufacturer may use to identify quantities of material for specific job sites or to identify field installed systems.

UL listed—Signifies that production samples of the product have been found to comply with established Underwriters Laboratories requirements and that the manufacturer is authorized to use the Laboratories' Listing Marks on the listed products which comply with the requirements, contingent upon the follow-up services as a check of compliance.

Ultrasonic—Pertaining to a sound wave having a frequency above that of audible sound (approximately 20,000 Hz). Ultrasonic sound is used in ultrasonic detection systems.

Ultrasonic detection system— (See *ultrasonic motion detector* and *passive ultrasonic alarm system.*)

Ultrasonic frequency—Sound frequencies which are above the range of human hearing; approximately 20,000 Hz and higher.

Ultrasonic motion detector—A sensor which detects the motion of an intruder through the use of *ultrasonic* generating and receiving equipment. The device operates by filling a space with a pattern of ultrasonic waves; the modulation of these waves by a moving object is detected and initiates an *alarm signal*.

Underdome bell—A bell most of whose mechanism is concealed by its gong.

Underwriters Laboratories, Inc. (UL)—A private independent research and testing laboratory which tests and lists various items meeting good practice and safety standards.

Vibrating bell—A bell whose mechanism is designed to strike repeatedly and for as long as it is activated.

Vibrating contact—(See *vibration sensor*.)

Vibration detection system—An alarm system which employs one or more *contact microphones* or *vibration sensors* which are fastened to the surfaces of the area or object being protected to detect excessive levels of vibration. The contact microphone system consists of microphones, a *control unit* containing an amplifier and an *accumulator,* and a power supply. The unit's sensitivity is adjustable so that ambient noises or normal vibrations will not initiate an *alarm signal*. In the vibration sensor system, the sensor responds to excessive vibration by opening a switch in a *closed circuit system*.

Vibration detector—(See *vibration sensor*.)

Vibration sensor—A sensor which responds to vibrations of the surface on which it is mounted. It has a *normally closed switch* which will momentarily open when it is subjected to a vibration with sufficiently large amplitude. Its sensitivity is adjustable to allow for the different levels of normal vibration, to which the sensor should not respond, at different locations. (See also *vibration detection system*.)

Visual signal device—A pilot light, *annunciator* or other device which provides a visual indication of the condition of the circuit or system being supervised.

Volumetric detector—(See *volumetric sensor*.)

Volumetric sensor—A sensor with a detection zone which extends over a volume such as an entire room, part of a room, or a passageway. *Ultrasonic motion detectors* and *sonic motion detectors* are examples of volumetric sensors.

Walk test light—A light on motion detectors which comes on when the detector senses motion in the area. It is used while setting the sensitivity of the detector and during routine checking and maintenance.

Watchman's reporting system—A *supervisory alarm system* arranged for the transmission of a patrolling watchman's regularly recurrent report signals from stations along his patrol route to a central supervisory agency.

Zoned circuit—A circuit which provides continual protection for parts or zones of the *protected area* while normally used doors and windows or zones may be released for access.

Zones—Smaller subdivisions into which large areas are divided to permit selective access to some zones while maintaining other zones secure and to permit pinpointing the specific location from which an *alarm signal* is transmitted.

SECTION II: PHYSICAL SECURITY

Access, accessibility—The ability and opportunity to obtain knowledge of classified information. An individual may have access to classified information by being in a place where such information is kept, if the security measures which are in force do not prevent him from gaining knowledge of the classified information.

Accountable document control system—A system of formal records and receipts to control the following types of material: Top Secret, Secret, Restricted Data, NATO, SEATO, CENTO, classified proposal information, crytographic material regardless of classification, registered publications of a military department, classified loan documents, classified foreign documents, selected company documents marked "Proprietary," and classified material marked "Special Access Required."

Alarm, break—Alarm signal produced by opening an electrical circuit.

Alarm cross—Alarm signal generated when the wires of an alarm system are shorted together.

Alarm, holdup—Device which generates an alarm when a concealed switch is opened or closed.

Alarm, local—System which causes a local bell or horn to sound when an alarm condition exists.

Alarm, nuisance—Any alarm signal caused by a factor other than an intrusion or circuit malfunction.

Alien—Any person not a citizen or national of the United States.

Annunciator—Device which provides visual and/or audible indications of the existence of an alarm condition.

Authorized persons—Those persons who (1) have a "need-to-know" for the classified information involved and (2) have been cleared for the receipt of such information. Responsibility for determining whether a person's duties require that he possess or have access to any classified information, and whether he is authorized to receive it rests upon the individual who has possession, knowledge, or control of the information involved, and not upon the prospective recipient.

Bolt—Projectable member of lock or latch mechanism which engages door frame and strike. *Deadbolt:* bolt which has no automatic spring action and which is operated (both projected and retracted) by key cylinder or lever handle. *Latchbolt:* A bevelled spring-loaded bolt that automatically seats in strike on contact, retracted by key cylinder or lever handle.

Bound documents—Books or pamphlets the pages of which cannot be removed without damage or mutilation; they must be sewed and have the glued binding common to the art of bookbinding. A bound document does not include those documents fastened only with staples, brads, or other commercial paper fasteners.

Button, holdup—Pushbutton switch for activating a holdup alarm.

Central office of record—The Department of Defense or user agency activity to which an accounting and reports for accountable COMSEC material is required for a particular contract.

Change-key—A mechanical device for operating a specific mechanical lock.

Classification—The determination that official information requires, in the interests of national defense, a specific degree of protection against unauthorized disclosure coupled with designation signifying that such a determination has been made.

Classified contract—Any contract or purchase order that requires or will require access to classified information by the contractor or his employees in the performance of the contract. A contract may be classified even though the contractual document is unclassified.

Classified document—Any recorded information regardless of its physical form or characteristics, exclusive of machinery, apparatus, equipment, or other items of material which incorporates classified information and which requires protection in the interest of national defense. The term includes, but is not limited to, the following: all written material, whether handwritten, printed, or typed; magnetic recordings; all photographs, negatives, exposed or printed films and still or motion pictures; all punched cards or tapes; and all reproduction of the foregoing by whatever process reproduced.

Classified information—Official information, including foreign classified information, which requires protection in the interest of national defense and which has been so designated.

Classified material—Any document, product, or substance on or in which classified information may be recorded or embodied and which requires protection in the interest of national defense. Material shall include everything, regardless of its physical character or makeup. Machinery, documents, apparatus, devices, models, photographs, recordings, reproductions, notes, sketches, maps, and letters as well as all other products, substances or materials shall fall within the general term of material.

Clearance—An administrative determination by competent authority that an individual has been adjudged eligible for access to defense information of a specified category should his duties so require.

Closed areas—Controlled areas established to safeguard classified material which because of its size or nature cannot be adequately protected during working and nonworking hours in accordance with the provisions of Section 5, Safeguarding Classified Information.

Cognizant security office—The Defense Contract Administration Services Region (DCASR) having contract administration services jurisdiction over the geographical area in which a facility is located.

Communications analysis—The analysis of communications signals and the results of that analysis.

Communications intelligence—Technical and intelligence information derived from foreign communications by other than the intended recipients.

Communications security—Protection resulting from all measures designed to deny to unauthorized persons information of value which might be derived from communications. Communications security includes transmission security, cryptographic security, and physical security.

Communist-action organization—Any organization in the United States other than a diplomatic representative or mission of a foreign government accredited as such by the Department of State which is substantially directed, dominated, or controlled by the foreign government or foreign organization controlling the world Communist movement.

Communist-front organization—Any organization in the United States (other than a Communist-action organization as defined) which (a) is substantially directed, dominated, or controlled by a Communist-action organization, and (b) is primarily operated for the purpose of giving aid and support to a Communist-action organization, a Communist foreign government, or the world Communist movement.

Compromise—A loss of security resulting from an unauthorized person obtaining knowledge of classified information.

Confidential—Defense information and material, the unauthorized disclosure of which could be prejudicial to the defense interest of the nation.

Connection, police—Pair of wires connected to a monitor in a police station in addition to the connection to a proprietary monitor or central station monitor.

Consultant—A person, not a specific employee, who may be retained on a consultant contract to perform special tasks for the company premises or at his own facility.

Contact—Door or window switch which opens an electrical circuit when the door or window to which it is attached is opened.

Contract security classification specification (DD form 254)—A classification guide furnished by the contracting officer, or in the case of subcontracts, by the prime contractor, which stipulates the degree of classification that shall be assigned to information developed or produced under a contract or purchase order.

Contracting officer—Any person who, in accordance with departmental or agency procedures, is currently designated a contracting officer with the authority to enter into and administer contracts and make determinations and findings with respect thereto, or any part of such authority. The term also includes the authorized representative of the contracting officer acting within the limits of his authority.

Contractor—Any industrial, educational, commercial, or other entity which has executed a contract with a user agency or a department of Defense Security Agreement (DD Form 441) with a Department of Defense agency or activity.

Control, active standby—Control used with constantly supervised security systems to allow normal traffic through a protected area during the hours of occupancy. This control renders the system inoperative, but allows tamper and line supervision to be maintained to prevent the system from being compromised during the hours of normal occupancy.

Control, day-night—Control which turns a security system on or off to allow normal occupancy and to provide protection during the hours of closure.

Cosmic top secret—Marking used on a NATO Top Secret document to signify that it is the property of NATO and that it is subject to special security control.

Courier—An authorized person designated to hand-carry classified material.

Crypto—A bold marking used to identify correspondence, documents, and material which contain classified cryptographic information. The designation Crypto replaces the designations Crypto Clearance Required and Crypto Clearance Not Required.

Cryptographic information—Information pertaining to the various means and methods for rendering plain text unintelligible and reconverting cipher text into intelligible form.

Custodian—An individual to whom classified material is assigned and who is responsible for its protection in accordance with company and Department of Defense regulations.

Cylinder—Housing containing a tumbler mechanism and a keyway plug which can be turned on only by the correct key. Includes a cam or spindle to transmit rotary action to a lock or latch mechanism. For security and keying versatility, authorities generally specify a pin tumbler cylinder of no less than five pins. These are available in the mortise cylinder (round, threaded housing) or the bored lock cylinder (sometimes called a cylinder "insert"). Both types offer the same functional value of security and convenience and are often included in the same keying system. (See *keying*.)

Declassification—The determination that classified information no longer requires, in the interests of national defense, any degree of protection against unauthorized disclosure, couples with a removal or cancellation of the classification designation.

Declassify—To cancel the security classification of an item of classified material.

Defense information—Official information which requires protection in the interests of national defense, which is not common knowledge, and which would be of intelligence value to an enemy or potential enemy in the planning or waging of war against the United States or its allies. There are three categories of defense information which, in descending order of importance, shall carry one of the following designations: Top Secret, Secret, or Confidential.

Derivative classification—That requirement to classify material created as a result of, in connection with, or in response to existing material, dealing with the same subject which already bears a classification.

Detector—Any device which senses the

presence of an intruder and causes an alarm indication to be given.

Detector, audio—System for detecting an intruder by the noise he makes; usually a microphone and amplifier are used to operate a relay. When the sound level rises by a preset amount above the normal ambient level, alarm indication is given. Most systems of this type permit a guard at a remote location to listen to the sounds in the protected area, and some have an intercom feature.

Detector, capacitance—Device which protects a metal fixture such as a safe or filing cabinet by using the fixture as one plate of a capacitor and the earth or a metal mat as the other place. Approaching or touching the protected fixture changes the capacitance in the circuit. This change in capacitance activates the alarm.

Detector, heat—Thermostatic type switch designed for installation on metal doors. The heat from a cutting torch will open the switch, causing an alarm.

Detector, infrared—Photoelectric intrusion detection device which uses infra red light rather than visible light. (See *photoelectric detector.*)

Detector, microwave—Detector which operates at microwave frequencies (above 2000 megacycles) rather than lower radio frequencies. The unit detects apparent frequency shifts caused by movement of an intruder in the protected area and generates an alarm indication.

Detector, photoelectric—Detection device which utilizes a beam of light projected into a photocell to detect an intruder. A person walking through the light beam blocks the light from the photocell which causes an alarm to be generated.

Detector, proximity—Device such as a capacitance detector which initiates an alarm if a person comes near or touches a protected object.

Detector, radar—Device which detects an intruder by his movement in a field of electromagnetic energy. Such a device may operate at microwave frequencies or lower radio frequencies. (See *microwave detector.*)

Detector, sonic—Device which sets up a field of sound waves in the air inside a protected area. A moving intruder causes an apparent frequency shift in the sound waves returning to the detector from the protected area. The detector senses this apparent frequency shift and generates an alarm indication.

Detector, vibration—Device which contains a pair of contacts that open (or close) if the unit is jarred. Vibration detectors are used to protect soft walls, floors, ceilings, etc., against forcible entry. They are usually used in perimeter systems.

Detector, ultrasonic—Device similar in operation to the sonic detector, except that the frequency at which it operates is in the ultrasonic rather than sonic range.

DISCO—Defense Industrial Security Clearance Office. An element of the Defense Supply Agency located at Columbus, Ohio, to which requests are made for personnel security clearance, and where the Central Index File of all existing industrial personnel security clearance granted by Department of Defense is located.

Disaffection—The alienation or estrangement from those in authority or lack of loyalty to the Government and Constitution of the United States.

Disclosure—An officially authorized release or dissemination by competent authority whereby the information is furnished to a specific individual, group, or activity.

Disseminate—To furnish classified material under continued control of the United States Government to persons having a proper clearance and a "need-to-know," e.g. to another United States governmental agency or department or to defense contractor.

Document—Any recorded information regardless of its physical form or characteristics, exclusive of machinery, apparatus, equipment or other items of material. The term includes, but is not limited to the following: all written material, whether handwritten, printed or typed; magnetic recordings; all photographic negatives, exposed or printed films, and still or motion pictures; all punched cards or tapes and all reproductions of the foregoing by whatever process reproduced.

Document control station—An activity designated to record the receipt and dispatch, and to maintain accountability, of classified documents.

Downgrade—To determine that classified information requires, in the interests of national defense, a lower degree of protection against unauthorized disclosure than currently provided, coupled with a changing of the classification designation to reflect such lower degree.

Downgrading—The assigning of a lower defense classification than that previously assigned.

Espionage—The obtaining, transmitting, communicating or receiving in respect to the national defense with intent or reason to believe that the information may be used to the injury of the United States or to the advantage of any foreign nation.

Espionage notation—Notation which must be applied to all classified material, except Restricted Data, which is furnished to authorized persons other than those in the executive branch of the Government, to inform recipients of Federal prohibition against unauthorized disclosure.

Executive personnel—Those individuals in managerial positions other than owners, officers, or directors who administer the operations of the facility. (This category includes such designations as general manager, plant manager, plant superintendent, or similar designations and facility security supervisors.)

Facility—A plant, laboratory, office, college, university or commercial structure with associated warehouses, storage areas, utilities and components, which, when related by functions and locations, form an operating entity. (A business or educational organization may consist of one or more facilities as defined above.) For purpose of industrial security, the term does not include user agency installations.

Facility security clearance—An administrative determination by the Department of Defense that a facility is eligible for access to classified information of a certain category (and all lower categories). Facility clearance may be revoked by the Cognizant Security Officer for security deficiencies and appeal of such revocation is not authorized.

Fence, capacitance—A fence which uses insulated wires as plates of a capacitor. A person's body close to or touching the wires, changes the capacitance of the fence, activating an alarm circuit.

Fence, electromagnetic—A fence composed of wires which are electrically insulated from the fence posts. Electronic equipment is employed to set up an electromagnetic field between the wires. A person disturbing the field by going close to or through the fence causes an alarm to be initiated.

Field of interest—One or more categories of classified material which relate directly to the work assignment of an individual.

Foil—Very thin metal strips which are cemented to a glass window or door. The foil is connected into a closed electrical circuit. If the glass is broken and breaks the foil, the circuit will be opened, causing an alarm.

Foreign classified information—Official information of a foreign government which (1) it has classified; (2) the United States Government has determined requires protection in the interest of national defense; (3) has been furnished to a United States contractor in connection with a contract, subcontract, precontract negotiations or other arrangement approved by the United States Government, (4) the United States Government is obligated to protect pursuant to an agreement with that government.

Foreign classified material—Official material of a foreign government which is classified or the United States Government has determined requires protection in the interest of national defense.

Foreign national—All persons not citizens of, not nationals of, nor immigrant aliens to the United States. (Except for citizens of Canada and the United Kingdom, such persons cannot be granted a security clearance.

Formerly Restricted Data—Information which has been removed from the Restricted Data category by joint action of the Atomic Energy Commission and Department of Defense under section 142D, Atomic Energy Act of 1954, as amended. This action is based upon a determination by these agencies that the information relates primarily to the military utilization of atomic weapons and that the information can be adequately safeguarded as classified defense information. Formerly Restricted Data may not be transmitted or otherwise made available to any regional defense organization or foreign nation while it remains classified defense information except under the provisions of the Act.

For official use only—A restrictive marking used by the military to indicate information which is of a sensitive nature and which is restricted in the public interest because widespread dissemination could unduly interfere with the efficient and effective functions of the Government, violate a legal or moral obligation, or result in injury to an innocent person.

Graphic arts—Facilities and individuals engaged in performing any consultation service or the production of any component or end-product which contributes to, or results in the reproduction of classified information. Regardless of trade names of specialized processes, it includes writing, illustrating, advertising services, copy preparation, all methods of printing, finishing services, duplicating, photocopying, and film processing activities.

Immigrant alien—A person lawfully admitted into the United States under an immigration visa for permanent residence. (Such persons may be granted a DOD clearance but are not eligible for company Confidential clearance.)

Industrial defense—Refers to all nonmilitary measures to assure the uninterrupted productive capability of vital facilities and attendant resources essential to mobilization. These measures are designed to prevent or minimize loss of disruption of productive capability from any cause or hazard and to provide for the rapid restoration of production after any damage.

Industrial security government—That portion of internal security which is concerned with the protection of classified information in the hands of United States industry.

Information—Knowledge which can be communicated, either orally, visually, or by means of material.

Information security—Safeguarding all information, ideas, correspondence, etc., which have been printed, written, or verbalized by a person, organization, or government.

Intelligence—The product resulting from the collection, evaluation, analysis, integration and interpretation of all available information which concerns one or more aspects of foreign nations or of areas of foreign operations and which is immediately or potentially significant to military planning and operations.

Internal security—The prevention of action against United States resources, industries, and institutions; and the protection of life and property in the event of a domestic emergency by the employment of all measures in peace or war, other than military defense.

Inventory—A procedure employed to verify accountability of Classified material by comparing entries on the register against the document or entry on the record of destruction or a signed receipt.

Keying—Pin-tumbler cylinders offer the possibility of very complex keying arrangements. These are the basic terms: *Individual key:* Key for an individual cylinder. *Keyed alike:* All cylinders may be operated by same key. (Not to be confused with *master-keyed*.) *Keyed different:* Different individual key operated each cylinder (or group of cylinders). *Master key:* Key to operate a group of cylinders, each of which may be set to a different individual. *Master keyed:* All cylinders in a group can be operated by one master key, although all cylinders may be keyed different. (Not to be confused with *keyed-alike*.)

Line, alarm—An electrically supervised pair of wires connected between the intrusion detection equipment in a protected area and the alarm indicating equipment for the purpose of transmitting alarm indications.

Line, reporting—Same as *alarm line*.

Long title—The full title of name assigned to a publication, an item of equipment or device.

Loss—Classified material which is out of the control of its custodian or which cannot be located.

Marking—Placement of security classifications and required security notations on or to classified material.

Mat, contact—Rubber mat which has switches built into it. Contact mats are often used as floor traps. Stepping on the mat operates a switch which causes an alarm to be given.

Material—Any document, product, or substance on or in which information may be recorded or embodied.

Material control station—An activity designated to record the receipt and dispatch, and to maintain accountability of classified material, other than documents, at a location.

Microphone, contact—Microphone which can be mounted directly on a protected wall, safe, etc. The microphone is usually insensitive to ambient room noises and detects the sound of the wall or safe being breached to activate an alarm circuit.

Monitor—Remote indicating device which provides audible and/or visual indications of alarm conditions.

Need-to-know—A determination made by the possessor of classified information that the prospective recipient, in the interest of national defense, has a requirement for access to, knowledge of, or possession of the classified information in order to perform tasks or services essential to the fulfillment of a classified contract or program approved by a user agency.

Official information—Information which is owned by, produced by, or is subject to the control of the United States Government.

Original classification authority—That authority required to classify independently any type of material.

Originator—The commander by whose authority an item of information is created and disseminated.

Personnel security government—Measures taken to ensure thorough and careful selection of personnel for employment who may have access to or use of classified information or devices.

Personnel security access authorization—This term has the same meaning as personnel security clearance.

Personnel security clearance—An administrative determination by DOD or a qualified contractor that from a security viewpoint, an individual is eligible for access to classified material of a certain category and all lower categories. (Also known as *personnel security access authorization*.)

Physical security—Protective device against security hazards. Physical measures designed to safeguard personnel and to prevent unauthorized access to facilities, material, and documents.

Picking, lock—The process of operating a lock into a locked or unlocked condition by using means other than the specifically designed key to operate said lock, with said means being operated or manipulated within the lock in lieu of said operating key.

Private—Marking assigned to sensitive information, the premature disclosure of which might jeopardize financial planning, manufacturing methods, employee relations, investigations or would be detrimental to the best interest of the company or employees concerned.

Project—A project is an undertaking to (1) develop or procure an item, piece of equipment, system, device, material, or component, together with any and all required test facilities including technical buildings or (2) explore a field in search of knowledge.

Protection, area—Protection of the inner space or volume of a security area, rather than the perimeter of the area.

Protection, object—System for protecting a specific object such as a safe or file cabinet, by the use of a capacitance detector or similar device.

Pro-Tex—The procedures and techniques used in the actual implementation of security processes within a particular program system. A derivative of procedures and techniques. It designates the means which can be utilized in the actual establishment of a security program.

Receipt—A written acknowledgement of the change of custody of classified documents or material.

Regrade—To determine that certain classified information requires, in the interests of national defense, a higher or a lower degree of protection against unauthorized disclosure than currently provided, coupled with a changing of the classification designation to reflect such higher or lower degree.

Relay alarm—A high resistance, sensitive relay used in security equipment. On most burglar alarms, this relay is held in by current through the alarm circuit. Opening the circuit deenergizes the relay which usually pilots another (drop) relay. The second relay is equipped with heavier contact than the alarm relay and operates alarm horns, bells, etc., for local alarm systems. On more sophisticated systems, the alarm relay causes a line current change upon opening which is detected by the monitor, then provides alarm indications.

Relay, drop—An electrically latching relay which activates alarm indicating devices in a burglar alarm system. The operation of this relay is usually controlled by a low current (alarm) relay which has a coil connected in series with the foil, door, and window contacts used in a burglar alarm system.

Release—Passage of information to another individual or agency by any means.

Reproduction—Any duplicating process including photography and typed or manual copying; also the product of a duplicating process.

Reproduction center—An activity designated to reproduce and maintain a record of the reproduction of accountable classified and proprietary material.

Research—All effort directed toward increased knowledge of natural phenomena and environment and toward the solution of problems in all fields of science. This includes basic and applied research. (1) Basic research—the type of research directed toward the increase of knowledge, the primary aim being a greater knowledge or understanding of the subject under study. (2) Applied research—The type of research concerned with the practical application of knowledge, material and/or techniques directed toward a solution to an existent or anticipated military or technological requirement.

Resistor, end-of-line—Resistance connected in an alarm line circuit to provide a required value of alarm line current. The more sophisticated security systems use a change in alarm line current rather than a simple break or cross indication to activate the alarm indicating device.

Resistor, termination—Same as *end-of-line resistor*.

Reference material—Documentary material over which a user agency does not have classification jurisdiction, and does not have jurisdiction at the time the material was originated.

Restricted data—All data (information) concerning (1) design, manufacture, or utilization of atomic weapons, (2) the production of special nuclear material, (3) the use of special nuclear material in the production of energy, but not to include data declassified or removed from the Restricted Data category pursuant to Section 142 of the Atomic Energy Act. (See Section 11W, Atomic Energy Act of 1954, as amended, and *formerly restricted data*.)

Sabotage—An act with an intent to injure, interfere with, or obstruct the national defense of the United States by willfully injuring or destroying, or attempting to injure or destroy any national defense material, premises, or utilities.

Screen—Covering for a window or similar opening usually consisting of light wooden strips or dowels with fine wire cemented to or inside them. The wire of the screen is a continuous circuit and is connected in series with the circuit of a burglar alarm. Cutting or breaking through the screen opens the circuit and activates the alarm.

Secret—Information or material, the unauthorized disclosure of which could result in serious damage to the nation, such as, by jeopardizing the international relations of the United States, endangering the effectiveness of a program or policy of vital importance to the national defense, a comprising important military or defense plans, scientific or technological developments important to national defense, or information revealing important intelligence operations.

Secure room—A room that offers the same or greater security than a security container authorized for the storage of classified material through the use of guards, alarms, or locking devices.

Sedition—Any act to overthrow, put down, or destory by force, the government of the United States, to oppose by force the authority of the United States to prevent, hinder, or delay by force, the execution of any law of the United States, or by force, to seize, take, or possess any property of the United States contrary to its authority.

Short title—A designation applied to a classified document, project, material, or device for purposes of security and brevity. It consists of figures, letters, words, or combinations thereof, without giving any information relative to classification or content of the document, project, material, or device. It may include, for example, the first letter of each word of the subject of a document.

Signal trouble—Signal which indicates some defined abnormal condition or conditions such as a circuit malfunction, loss of power, or tampering with alarm circuitry.

Special access required—A notation to be placed at least once on material which contains or reveals military space information designated by the Secretary of the Air Force.

Storage center—Any department or division maintaining a collection of reference material for issue on a returnable basis.

Strategic Intelligence—Refers to information regarding the capabilities, vulnerabilities, and intentions of foreign nations required by planners in establishing the basis for an adequate national security policy in time of peace.

Strongroom—An interior space enclosed by or separated from other similar spaces by four walls, a ceiling and a floor, constructed of solid building materials, and used for storage of classified material.

Subversion—All willful acts which do not fit the categories of treason, espionage, sabotage, or sedition, but which are intended to lend aid, comfort, or moral support to individuals, groups, or organizations advocating the overthrow of the United States government by force and violence, or are otherwise intended to be detrimental to the national security of the United States.

Supervision, electronic—Pertains to the supervision of the security equipment itself, rather than the alarm line. An electronically supervised security device incorporates fail-safe electronic circuits to warn of equipment malfunction.

Supervision, line—Electrical protection of an alarm line. This is accomplished by having a continuous flow of current through the circuit. A change of current will be detected by the monitor. The monitor gives an alarm if the change exceeds the allowable amount for a given percentage of line supervision. (See *percentage of supervision*.)

Supervision, mechanical—Protection of security equipment against tampering by use of tamper switches connected in series with an electrically supervised alarm line.

Supervision, percentage of—Percentage by which the supervisory current in an alarm line can be varied without causing an alarm. The lower the percentage of supervision, the more difficult the alarm line is to compromise.

Surveillance—To closely watch, supervise, or guard.

Switch, balanced magnetic—Magnetic door or gate switch which operates in a balanced magnetic field. This switch is built in such a manner as to make it difficult to compromise by the application of an external magnet. These switches usually consist of one or more reed switches held closed by a magnet on the protected door. Application of an external magnet causes a second set of contacts to close, causing an alarm to be given. Opening the door, of course, also opens the switch.

Switch, day-night—Switch used to deactivate a security system to allow access to the protected area during hours of normal occupancy.

Switch, door—A switch, usually magnetically operated, which opens its contacts when the door which it is protecting opens. The switch is usually mounted on the door frame and the magnet which operates it is usually mounted on the door. The switch is connected in series with a closed alarm circuit. Opening the circuit causes an alarm to be given.

Switch, gate—This switch operates in the same manner as a door switch. It is usually enclosed in a weatherproof housing to permit outdoor use. (See *door switch*.)

Switch, tamper—A switch in security equipment enclosures which opens the alarm line circuit if the enclosure is opened, causing an alarm to be given.

System, active security—Security system employing a detector which generates the energy used to detect the presence of an intruder and receives the same energy when it is reflected back to the detector from the protected area.

System, building security—Protective apparatus of a building. This can include electrical and electronic security equipment as well as a guard force. Equipment surveillance devices can also be included in the security system, along with the intrusion detection equipment. Security systems are generally more sophisticated than the simple burglar alarm and they provide more comprehensive protection.

System, central station—Alarm system connected to a central guard station. These are usually owned and operated by the installing company which also furnishes the guard personnel at the central station.

System, circuit alarm—Same as *code transmission system*.

System, code transmission—A type of alarm system which has several customers' premises on a single alarm loop connected to a central station. The circuit on each property sends a different coded signal to the central station upon alarm. Such systems are intended to minimize the cost of leased wires.

System, combination central station and local alarm—Alarm system which sounds a local alarm (e.g. horn or bell) and also transmits an alarm to a central station.

System, direct connected—Alarm system having an individual supervised connection to the central station or police headquarters, guard shack, etc.

System, direct wire—Alarm system connected directly to police headquarters.

System, electro-mechanical—Alarm system consisting of a closed electrical loop which runs around a protected area. In the loop are protective switching devices such as door switches, window foil, screens, etc. All of these components are connected in series. Opening any one of them to enter the protected area breaks the circuit and de-energizes an alarm relay. This activates the alarm. The alarm can be either local or remote.

System, passive security—Security system such as an audio system which employs a detector that depends on energy (audio or vibration) produced by an intruder to detect his presence.

System, proprietary—Alarm system owned by the customer rather than by the installing company.

Technical information—Information, including scientific information, which relates to research, development, engineering, test, evaluation, production, operation, use and maintenance of munitions and other military supplies and equipment.

Technical intelligence—The product resulting from the collection, evaluation, analysis, and interpretation of foreign scientific and technical information which covers (1) foreign developments in basic and applied research and in applied engineering techniques, (2) scientific and technical characteristics, capabilities, and limitations of all foreign military systems, weapons, weapon systems, and material, the research and development related thereto, and the production methods used in their manufacture.

Top Secret—Information or material the defense aspect of which is paramount and the unauthorized disclosure of which could result in exceptionally grave damage to the nation, such as (1) lead to a definite break in diplomatic relations affecting the defense of the United States or its allies, an armed attack against the United States or its allies, a war, or (2) the compromise of military or defense plans, intelligence operations, or scientific or technological developments vital to the national defense.

Trap—This can consist of fine wire lacing in the opening of a skylight or similar opening. The operation is the same as that of a screen. (See *screen*.)

Treason—The levying of war against the United States or the adherence to their enemies, giving them aid and comfort, by a person owing allegiance to the United States.

Unauthorized person—Any person not authorized to have access to specific classified information or to a security area in accordance with the provisions of company procedures.

Unbound documents—Material such as letters, memoranda, reports, telegrams, and similar documents, the pages of which are not permanently and securely fastened together.

Upgrading—The assigning of a higher classification than previously assigned. Notification to recipients of the information is a part of this process.

User agencies—The Office of the Secretary of Defense (including all boards, councils, staffs, and commands); Department of Defense agencies and departments of the Army, Navy, and Air Force (including all of their activities); National Aeronautics and Space Administration; Federal Aviation Agency; General Services Administration; departments of State and Commerce; the Small Business Administration; the National Science Foundation; and the Department of Treasury.

Vault—A room or area which has a masonry wall extending from the floor to the floor above. Ceiling and floors must be constructed in such a way as to offer the same protection as do the walls. Doors to the vault must be of drill resistant steel construction with a built-in, three position combination lock. If classified material is placed in open storage within the vault a sprinkler system should be installed. Additional security may be obtained by the use of two separate alarm systems.

Weapon system—A general term used to describe a weapon and those components required for its operation.

Work-in process—Classified material being made ready for typing and/or final reproduction. These documents consist of preliminary rough drafts, notes, work papers, etc., which precede release of finally approved copy.

Zone—Divisions within a security area. A zone can contain more than one intrusion detection device, but the protective devices are connected to a single monitor or monitor module, providing an indication of the area of the building in which an intrusion occurs.

Zone, clear—Cleared area around an electromagnetic or capacitance fence. Its purpose is to minimize nuisance alarms caused by falling limbs, blowing rubbish, small animals, etc.

Chapter 12

INFORMATION SECURITY

THE PROTECTION of the information produced by an organization or individual is often vital to the success or failure of an activity. The ability to prevent the unauthorized disclosure of sensitive information requires the application of a wide variety of protective measures and procedures which are a portion of the information security process.

The functional subarea of security which is specifically the information security process involves the safeguarding of all information, ideas, correspondence, and materials of an organization which have been written or verbalized. It involves all the administrative and control techniques utilized by the organization to protect its interest. The process of information security is also a specific process in the total security function that works in conjunction with physical and personnel security to provide a totally safe environment for the organization. It has applications in both governmental and proprietary programs.

The need for information security is similar in governmental and proprietary organizations. The specific requirements for processing, control, and handling of information which must be safeguarded do however differ sufficiently to treat them individually. It should be stressed that many proprietary systems are based on standards established in governmental programs. Therefore, the procedures and controls utilized by governmental programs find counterpart applications in proprietary systems.

As we have seen, governmental programs are necessary to protect information of national security and importance, such things as research and development for new military systems, supply and logistic information, transportation routes, new weapons systems, deployments, troop concentration. These types of information require governmental measures to insure protection of the government and its operations.

Likewise, private individuals, business, and industrial corporations require measures to provide for security of various operations. For example, in an industrial manufacturing or business context, the following types of information generally require protection:

1. basic manufacturing data,
2. design manuals,
3. plant operating instructions,
4. sample manuals,
5. plant test results,
6. raw material specifications,
7. analysis,
8. technical report of experiments,
9. reports of production,
10. process evaluation,
11. engineering drawings,
12. flow chart,
13. production and process research,
14. research notebooks,
15. plant training manuals,
16. quality control charts,
17. market analysis data,
18. cost data,
19. distribution techniques,
20. market strategies,
21. customer lists,
22. accounting data,
23. forecasting budgets,
24. mailing lists,
25. company records of any type.

In other words, any information relating to the organization, operation, or maintenance of the business or industry requires security.

GOVERNMENT APPLICATIONS

Procedures to handle and protect classified documents are required for the national defense.[1] In the interest of national defense, the United States is required to preserve its ability to protect itself against all hostile or destructive actions including espionage as well as military. Thus, it is essential that certain official information which affects the national defense be protected uniformly against unauthorized disclosures. Obviously, security regulations do not guarantee protection and they cannot be written to cover all conceivable situations. However, basic security principles in a logical interpretation of the existing regulations would be normally applicable to the safeguarding of defense information. It is possible to achieve an acceptable degree of security with a minimum of sacrifice of operating efficiency. This is obtained by indoctrination of individuals dealing with information to be safeguarded that they discharge their security responsibilities automatically with logic and discretion without thinking of the specific rules as a burden imposed upon them.

Criteria for Safeguarding Information

It is possible to select a number of criteria which are useful in determining when information is to be safeguarded and administrative protection or control exerted over it. These should include when:

1. The information provides the United States with a scientific, engineering, technical, operational, intelligence, tactical, strategic, advantage compared to other nations in relation to national defense.
2. Disclosure of the information would weaken the international posture of the United States, could create or increase international tensions involving United States' interests and result in a break in diplomatic relations or lead to hostile, political, economic, or military action against the United States or its allies thereby adversely affecting the national defense.
3. The disclosure of the information would weaken the ability of the United States to defend itself or to wage war successfully, limit the effectiveness of the armed forces and make the United States vulnerable to attack.
4. There is a sound reason to believe that other nations do not know that the United States has or is capable of obtaining certain information or materials of importance to the international posture and the national defense of the United States related to those nations.

Country	Top Secret	Secret	Confidential
Argentina	Estrictamente Secreto	Secreto	Confidencial
Austria	Streng Geheim	Geheim	Verschluss
Cambodia	Tres Secret	Secret	Secret/Confidential
Cuba	Muy Secreto	Secreto	Confidencial
Denmark	Yderst Hemmeligt	Hemmeligt	Fortroligt
Ethiopia	Yemiaz Birtuo Mistir	Mistir	Kilkil
France	Tres Secret	Secret	Secret/Confidential
Germany	Streng Geheim	Geheim	Vertraulich
Hungary	Szigoruan Titkos	Titkos	Bizalmas
Israel	Sorde Beyoter	Sorde	Shimor
Norway	Strengt Hemmlig	Hemmelig	Fortrolig
Portugal	Muito Secreto	Secreto	Confidencial
Sweden	Helmig	Helmig	Helmig
United Kingdom	Top Secret	Secret	Confidential
Uruguay	Secreto	Secreto	Confidencial

Figure 12-1. Security classification of foreign countries.

5. The information represents a significant breakthrough, basic research which has an inherent military application potential in a new field, or make radical changes in existing fields.
6. There is sound reason to believe that knowledge of the information would provide a foreign nation with an insight into the war potential or defense plans of the United States, allowing a foreign nation to develop or improve or refine a similar item of war potential, provide a foreign nation with a base upon which to develop effective countermeasures, weaken or nullify the effectiveness of a defense or military plan, project, operation or activity which is essential to the national defense.[2]

This official information or material referred to as classified information or material is expressly exempted from public disclosure by Section 552(b)(1) of Title 5, United States Code. Wrongful disclosure of such information or material is recognized in the Federal Criminal Code as providing a basis for prosecution.

To ensure that such information and material is protected, but only to the extent and for such period as is necessary; an executive order[3] identifies the information to be protected, prescribe classification, downgrading, declassification, and safeguarding procedures to be followed, and establishes a monitoring system to ensure its effectiveness.

SECTION 1. SECURITY CLASSIFICATION CATEGORIES. Official information or material which requires protection against unauthorized disclosure in the interest of the national defense or foreign relations of the United States (hereinafter collectively termed "national security") is classified in one of three categories, namely "Top Secret," "Secret," or "Confidential," depending upon the degree of its significance to national security. No other category is used to identify official information or material as requiring protection in the interest of national security, except as otherwise expressly provided by statute. These classification categories are defined as follows:

(A) *"Top Secret."* "Top Secret" refers to that national security information or material which requires the highest degree of protection. The test for assigning "Top Secret" classification shall be whether its unauthorized disclosure could reasonably be expected to cause *exceptionally grave damage* to the national security. Examples of "exceptionally grave damage" include armed hostilities against the United States or its allies; disruption of foreign relations vitally affecting the national security; the compromise of vital national defense plans or complex cryptologic and communications intelligence systems; *the revelation of "sensitive" intelligence operations;* and the disclosure of scientific or technological developments vital to national security. This classification shall be used with the utmost restraint.

(B) *"Secret."* "Secret" refers to that national security information or material which requires a substantial degree of protection. The test for assigning "Secret" classification is whether its unauthorized disclosure could reasonably be expected to cause *serious damage* to the national security. Examples of "serious damage" include disruption of foreign relations significantly affecting the national security; significant impairment of a program or policy directly related to the national security; *revelation of "significant" military plans or intelligence operations;* and compromise of significant scientific or technological developments relating to national security. The classification "Secret" is to be sparingly used.

(C) *"Confidential."* "Confidential" refers to that national security information or material which requires protection. The test for assigning "Confidential" classification is whether its unauthorized disclosure could reasonably be expected *to cause damage* to the national security.

SECTION 2. AUTHORITY TO CLASSIFY. The authority to originally classify information or material under this order shall be restricted solely to those offices *within the executive branch* which are concerned with matters of national security, and shall be limited to the minimum number absolutely required for efficient administration. Except as the context may otherwise indicate, the term "Department" as used in this order shall include agency or other governmental unit.

(A) The authority to originally classify information or material under this order as "Top Secret" is exercised only by such officials as the President may designate in writing and by:
(1) The heads of the Departments listed below:
(2) Such of their senior principal deputies and assistants as the heads of such Departments may designate in writing; and
(3) Such heads and senior principal deputies and assistants of major elements of such Departments, as the heads of such Departments may *designate in writing.*
> Such offices in the Executive office of the President as the President may designate in writing
> *Central Intelligence Agency*
> Nuclear Regulatory Commission
> Department of State
> Department of the Treasury
> Department of Defense
> Department of the Army
> Department of the Navy
> Department of the Air Force
> United States Arms Control and Disarmament Agency
> Department of Justice
> National Aeronautics and Space Administration
> Agency for International Development

(B) The authority to originally classify information or material under this order as "Secret" is exercised only by:
(1) Officials who have "Top Secret" classification authority;
(2) Such subordinates as officials with "Top Secret" classification authority under (A) (1) and (2) above may designate in writing; and
(3) The heads of the following named Departments and such senior principal deputies or assistants *as they may designate in writing.*
> Department of Transportation
> Federal Communications Commission
> Export-Import Bank of the United States
> Department of Commerce
> United States Civil Service Commission
> United States Information Agency
> General Services Administration
> Department of Health, Education, and Welfare
> Civil Aeronautics Board
> Federal Maritime Commission
> Federal Power Commission
> National Science Foundation
> Overseas Private Investment Corporation

(C) The authority to originally classify information or material under this order as "Confidential" may be exercised by officials who have "Top Secret" or "Secret" classification authority and such officials *as they may designate in writing.*

(D) Any Department not referred to herein and any Department or unit established hereafter shall not have authority to originally classify information or material under this order, unless specifically authorized hereafter by an Executive order.

SECTION 3. AUTHORITY TO DOWNGRADE AND DECLASSIFY. The authority to downgrade and declassify national security information or material is exercised as follows:

(A) Information or material may be downgraded or declassified by the official authorizing the original classification, by a successor in capacity or by a supervisory official of either.

(B) Downgrading and declassification authority may also be exercised by an official specifically authorized under regulations issued by the head of the Department listed in Sections 2(A) or (B) hereof.

(C) In the case of classified information or material officially transferred by or pursuant to statute or Executive order in conjunction with a transfer of function and not merely for storage purposes, the receiving Department shall be deemed to be the originating Department for all purposes under this order including downgrading and declassification.

(D) In the case of classified information or material not officially transferred within (C) above, but originated in a Department which has since ceased to exist, each Department in possession shall be deemed to be the originating Department for all purposes under this order. Such information or material may be downgraded and declassified by the Department in possession after consulting with any other Departments having an interest in the subject matter.

(E) Classified information or material transferred to the General Services Administration for accession into the Archives of the United

States shall be downgraded and declassified by the Archivist of the United States in accordance with this order, directives of the President issued through the National Security Council and pertinent regulations of the Departments.

(F) Classified information or material with special markings, as described in Section 8, shall be downgraded and declassified as required by law and governing regulations.

SECTION 4. CLASSIFICATION. Each person possessing classifying authority is *held accountable* for the propriety of the classifications attributed to him. *Both unnecessary classification and overclassification shall be avoided. Classification shall be solely on the basis of national security considerations.* In no case shall information be classified in order to conceal inefficiency or administrative error, to prevent embarrassment to a person or Department, to restrain competition or independent initiative, or to prevent for any other reason the release of information which does not require protection in the interest of national security. The following rules shall apply to classification of information under this order:

(A) *Documents in General.* Each classified document shall show on its face its classification and whether it is subject to or exempt from the General Declassification Schedule. It shall also show the office of origin, the date of preparation and classification and, to the extent practicable, be so marked as to indicate which portions are classified, at what level, and which portions are not classified in order to facilitate excerpting and other use. Material containing references to classified materials, which references do not reveal classified information, shall not be classified.

(B) *Identification of Classifying Authority.* Unless the Department involved shall have provided some other method of identifying the individual at the highest level that authorized classification in each case, material classified under this order shall indicate on its face the identity of the highest authority authorizing the classification. Where the individual who signs or otherwise authenticates a document or item has also authorized the classification, no further annotation as to his identity is required.

(C) *Information or Material Furnished by a Foreign Government or International Organization.* Classified information or material furnished to the United States by a foreign government or international organization shall either retain its original classification or be assigned a United States classification. In either case, the classification shall assure a degree of protection equivalent to that required by the government or international organization which furnished the information or material.

(D) *Classification Responsibilities.* A holder of classified information or material shall observe and respect the classification assigned by the originator. If a holder believes that there is unnecessary classification, that the assigned classification is improper, or that the document is subject to declassification under this order, he shall so inform the originator who shall thereupon reexamine the classification.

SECTION 5. DECLASSIFICATION AND DOWNGRADING. Classified information and material, unless declassified earlier by the original classifying authority, shall be declassified and downgraded in accordance with the following rules:

(A) *General Declassification Schedule.*
(1) *"Top Secret."* Information or material originally classified "Top Secret" shall become automatically downgraded to "Secret" at the end of the second full calendar year following the year in which it was originated, downgraded to "Confidential" at the end of the fourth full calendar year following the year in which it was originated, and declassified at the end of the *tenth* full calendar year following the year in which it was originated.
(2) *"Secret."* Information and material originally classified "Secret" shall become automatically downgraded to "Confidential" at the end of the second full calendar year following the year in which it was originated, and declassified at the end of the *eighth* full calendar year following the year in which it was originated.
(3) *"Confidential."* Information and material originally classified "Confidential" shall become automatically declassified at the end of the *sixth* full calendar year following the year in which it was originated.
(B) *"Exemptions from" General Declassification Schedule.* Certain classified information or material may warrant some degree of protection for a period exceeding that provided

in the General Declassification Schedule. An official authorized to originally classify information or material "Top Secret" may exempt from the General Declassification Schedule any level of classified information or material originated by him or under his supervision if it falls within one of the categories described below. In each case such official shall specify in writing on the material the exemption category being claimed and, unless impossible, a date or event for automatic declassification. The use of the exemption authority shall be kept to the absolute minimum consistent with national security requirements and shall be restricted to the following categories:

(1) Classified information or material furnished by foreign governments or international organizations and held by the United States on the understanding that it be kept in confidence.

(2) Classified information or material specifically covered by statute, or pertaining to cryptography, *or disclosing intelligence sources or methods.*

(3) Classified information, or material disclosing a system, plan, installation, project or specific foreign relations matter the continuing protection of which is essential to the national security.

(4) Classified information or material the disclosure of which would *place a person in immediate jeopardy.*

(C) *Mandatory Review of Exempted Material.* All classified information and material originated after the effective date of this order which is exempted under (B) above from the General Declassification Schedule shall be subject to a classification review by the originating Department at any time after the expiration of ten years from the date of origin provided:

(1) A Department or member of the public requests a review:

(2) The request describes the record with sufficient particularity to enable the Department to identify it; and

(3) The record can be obtained *with only a reasonable amount of effort.* Information or material which no longer qualifies for exemption under (B) above shall be declassified. Information or material continuing to qualify under (B) shall be so marked and, unless impossible, a date for automatic declassification shall be set.

(D) *Applicability of the General Declassification Schedule to Previously Classified Material.* Information or material classified before the effective date of this order and which is assigned to Group 4 under Executive Order No. 10501, as amended by Executive Order No. 10964, shall be subject to the General Declassification Schedule. All other information or material classified before the effective date of this order, whether or not assigned to Groups 1, 2, or 3 of Executive Order No. 10501, as amended, shall be excluded from the General Declassification Schedule. However, at any time after the expiration of ten years from the date of origin it shall be subject to a mandatory classification review and disposition under the same conditions and criteria that apply to classified information and material created after the effective date of this order as set forth in (B) and (C) above.

(E) *Declassification of Classified Information or Material After Thirty Years.* All classified information, or material which is *thirty years old or more,* whether originating before or after the effective date of this order, shall be declassified under the following conditions:

(1) All information and material classified after the effective date of this order shall, whether or not declassification has been requested, become *automatically declassified* at the end of thirty full calendar years after the date of its original classification *except for* such specifically identified information or material which the head of the originating Department personally determines in writing at that time to require continued protection because such continued protection is essential to the national security or disclosure would place a person in immediate jeopardy. In such case, the head of the Department shall also specify the period of continued classification.

(2) All information and material classified before the effective date of this order and more than thirty years old shall be systematically reviewed for declassification by the Archivist of the United States by the end of the thirtieth full calendar year following the year in which it was originated. In his review, the Archivist will separate and keep protected only such information or material as is specifically identified by the head of the Department in accordance with (E)(1) above. In such case, the head of the Department shall also specify the period of continued classification.

(F) *Departments Which Do Not Have Authority for Original Classification.* The provisions of this section relating to the declassification of national security information or material shall apply to Departments which, under the terms of this order, do not have current authority to originally classify information or material, but which formerly had such authority under previous Executive orders.

SECTION 6. POLICY DIRECTIVES ON ACCESS, MARKING, SAFEKEEPING, ACCOUNTABILITY, TRANSMISSION, DISPOSITION AND DESTRUCTION OF CLASSIFIED INFORMATION AND MATERIAL. The President acting through the National Security Council shall issue directives which shall be binding on all Departments to protect classified information from loss or compromise. Such directives shall conform to the following policies:

(A) No person shall be given access to classified information or material unless such person has been determined to be trustworthy and unless access to such information is *necessary for the performance of his duties.*

(B) All classified information and material shall be appropriately and conspicuously marked to put all persons on clear notice of its classified contents.

(C) Classified information and material shall be used, possessed, and stored only under conditions which will prevent access by unauthorized persons or dissemination to unauthorized persons.

(D) All classified information and material disseminated outside the executive branch under Executive Order No. 10865 or otherwise shall be properly protected.

(E) Appropriate accountability records for classified information shall be established and maintained and such information and material shall be protected adequately during all transmissions.

(F) Classified information and material no longer needed in current working files or for reference or record purposes shall be destroyed or disposed of in accordance with the records disposal provisions contained in Chapter 33 of Title 44 of the United States Code and other applicable statutes.

(G) Classified information or material shall be reviewed on a systematic basis for the purpose of accomplishing downgrading, declassification, transfer, retirement and destruction at the earliest practicable date.

SECTION 7. IMPLEMENTATION AND REVIEW RESPONSIBILITIES.

(A) The National Security Council shall monitor the implementation of this order. To assist the National Security Council, an Interagency Classification Review Committee shall be established, composed of representatives of the Departments of State, Defense and Justice, the Atomic Energy Commission, the Central Intelligence Agency and the National Security Council Staff and a Chairman designated by the President. Representatives of other Departments in the executive branch may be invited to meet with the Committee on matters of particular interest to those Departments. This Committee shall meet regularly and on a continuing basis shall review and take action to ensure compliance with this order, and in particular:

(1) The Committee shall oversee Department actions to ensure compliance with the provisions of this order and implementing directives issued by the President through the National Security Council.

(2) The Committee shall, subject to procedures to be established by it, receive, consider and take action on suggestions and complaints from persons within or without the government with respect to the administration of this order, and in consultation with the affected Department or Departments assure that appropriate action is taken on such suggestions and complaints.

(3) Upon request of the Committee Chairman, any Department shall furnish to the Committee any particular information or material needed by the Committee in carrying out its functions.

(B) To promote the basic purposes of this order, the head of each Department originating or handling classified information or material shall:

(1) Prior to the effective date of this order submit to the Interagency Classification Review Committee for approval a copy of the regulations it proposes to adopt pursuant to this order.

(2) Designate a senior member of his staff who shall ensure effective compliance with and implementation of this order and shall also chair a Departmental committee which shall have authority to act on all suggestions

and complaints with respect to the Department's administration of this order.

(3) Undertake an initial program to familiarize the employees of his Department with the provisions of this order. He shall also establish and maintain active training and orientation programs for employees concerned with classified information or material. Such programs include, as a minimum, the *Briefing of new employees and periodic reorientation during employment to impress* upon each individual his responsibility for exercising vigilance and care in complying with the provisions of this order. Additionally, upon *termination of employment or contemplated temporary separation for a sixty-day period* or more, employees shall be debriefed and each reminded of the provisions of the Criminal Code and other applicable provisions of law relating to penalties for unauthorized disclosure.

(C) The Attorney General, upon request of the head of a Department, his duty designated representative, or the Chairman of the above described Committee, shall personally or through authorized representatives of the Department of Justice render an interpretation of this order with respect to any question arising in the course of its administration.

SECTION 8. MATERIAL COVERED BY THE ATOMIC ENERGY ACT. Nothing in this order supersedes any requirements made by or under the Atomic Energy Act of August 30, 1954, as amended. "Restricted Data," and material designated as "Formerly Restricted Data," shall be handled, protected, classified, downgraded and declassified in conformity with the provisions of the Atomic Energy Act of 1954, as amended, and the regulations of the Atomic Energy Commission [Nuclear Regulatory Commission (1975)].

SECTION 9. SPECIAL DEPARTMENTAL ARRANGEMENTS. The originating Department or other appropriate authority may impose, in conformity with the provisions of this order, special requirements with respect to access, distribution and protection of classified information and material, including those which presently relate to communications, intelligence, *intelligence sources and methods* and cryptography.

SECTION 10. EXCEPTIONAL CASES. In an exceptional case when a person or Department not authorized to classify information originates information which is believed to require classification, such person or Department shall protect that information in the manner prescribed by this order. Such persons or Department shall transmit the information forthwith, under appropriate safeguards, to the Department having primary interest in the subject matter with a request that a determination be made as to classification.

SECTION 11. DECLASSIFICATION OF PRESIDENTIAL PAPERS. The Archivist of the United States has the authority to review and declassify information and material which has been classified by a President, his White House Staff or special committee or commission appointed by him and which the Archivist has in his custody at any archival depository, including a Presidential Library. Such declassification shall only be undertaken in accord with: (i) the terms of the donor's deed of gift, (ii) consultations with the Departments having a primary subject-matter interest, and (iii) the provisions of Section 5.

SECTION 12. HISTORICAL RESEARCH AND ACCESS BY FORMER GOVERNMENT OFFICIALS. The requirement in Section 6 (A) that access to classified information or material be granted only as is necessary for the performance of one's duties shall not apply to persons outside the executive branch who are engaged in historical research projects or who have previously occupied policy-making positions to which they were appointed by the President; *Provided,* however, that in each case the head of the originating Department shall:

(i) determine that access is clearly consistent with the interests of national security; and
(ii) take appropriate steps to assure that classified information or material is not published or otherwise compromised.

Access granted a person by reason of his having previously occupied a policymaking position shall be limited to those papers which the former official originated, reviewed, signed or received while in public office.

SECTION 13. ADMINISTRATIVE AND JUDICIAL ACTION.

(A) Any officer or employee of the United States who unnecessarily classifies or over-classifies information or material shall be notified that his actions are in violation of the terms of this order or of a directive of the President issued through the National Security Council. Repeated abuse of the classification process shall be grounds for an administrative reprimand. In any case where the Departmental committee or the Interagency Classification Review Committee finds that unnecessary classification or over-classification has occurred, it shall make a report to the head of the Department concerned in order that corrective steps may be taken.

(B) The head of each Department is directed to take prompt and stringent administrative action against any officer or employee of the United States, at any level of employment, determined to have been responsible for any release or disclosure of national security information or material in a manner not authorized by or under this order or a directive of the President issued through the National Security Council. Where a violation of criminal statutes may be involved, Departments will refer any such case promptly to the Department of Justice.

Initial Marking

After a determination is made of the classification to be applied, classified materials are marked with the date of origin, the name and address of the facility responsible for its preparation, and it is plainly and conspicuously marked or stamped with the appropriate classification. The document is classified in accordance with its contents. Documents will normally be classified according to their own content and not necessarily according to their relationship to other documents.[4]

Documents separated from a file or a group, however, will be handled in accordance with their individual defense classification. Similarly, a document, product, or substance will bear a classification at least as high as that of its highest classified component and will bear only one overall classification. Pages, paragraphs, sections, or components thereof, may bear different classifications. Compilations of defense information are normally protected and classified in accordance with the protection which is required by the sensitivity of the combined information even though individual items may be unclassified separately or bear a lower classification.

Transmittance of Classified Information

When classified documents are transmitted, all paragraphs or contents which are of different classifications must be so marked and a statement must be included on the front of the document or in the text identifying the parts of the document that are classified and as to their degree. A letter, endorsement, or other correspondence which transmits the classified material must also be classified at least as high as that of the highest classified attachment. If the transmittal document does not contain defense information or if the information in it is classified lower than in a proceeding element or enclosure, a downgrading notation must be included.

Overclassification of Information

The practice of overclassifying defense information prejudices the integrity of the classification system. It depreciates the importance of correctly classified materials and creates unnecessary expense, delay, and administrative burden. To avoid overclassification, the original classification authority is required to review the contents of the information to be classified specifically to insure that the assigned classification is appropriate for the defense information contained therein.

Handling of Classified Information

The application of security classification information developed by an organization engaged in defense production is based upon the classification guidance by the contracting officer of the user agency. User agencies include the Office of the Secretary of the Defense, including all boards, councils, staffs, and commands; the Department of Defense agencies and the Department of the Army, Navy, and Air Force, including all of their activities; the National Aeronautics and Space Administration; the Federal Aviation Agency; the General Service Administration; the Department of State, the Department of Commerce; the Small Business Administration; the Department of Treasury, and the National Science Foundation. Security classification to be applied to information involved in user agency contracts and programs will be supplied by the contracting officer or his designated representative within the user agency concerned.

Responsibility for Safeguarding Classified Information

Those organizations and persons delegated the responsibility for safeguarding classified information must provide the means for safeguarding it. These include the physical security, containers, the accounting system, the security education program, and the inspection and checks necessary to insure that all the requirements are being fulfilled. A company or firm must have a facility security clearance before it can negotiate a bid, prepare a quotation, or perform work on a contract, if for these purposes the company will require access to classified information. The clearance that they receive must be at least as high as the classification of the materials which must be utilized. Likewise, an individual is only permitted to have access to classified information when cleared by the government or by the contractor, or when a contractor determines that access is necessary in performance of tasks or services essentially for the fulfillment of a contract or program.

To be eligible for a security clearance, a person must have reached the age of sixteen for Confidential information and the age of eighteen for Secret or Top Secret material. Merely because a person or a facility is cleared for a certain level of information is not a guarantee that any information classified at that level will be made available to him. A major criteria is one of "need to know." This requirement is that the individual who is to receive the information actually needs it in order to perform his assigned functions properly.

Restrictions on Classified Information

Classified information normally may not be transmitted by telephone, teletype (TWX), television, intercommunications, or other electrical means except on approved cryptographic systems. It may never be discussed in public areas or in the presence of unauthorized individuals or at locations other than United States government cleared facilities or facilities specifically approved for classified conferences, meetings, or seminars by the United States government. Additionally, it may never be left unattended by users, messengers, or couriers unless stored in approved classified storage areas or containers. Likewise, it may not be disclosed by the incorporation of any special features of design or construction in any project other than that for which the features were furnished by, designed for, or developed for the government unless prior written authorization of the contracting officer has been obtained. Lastly, it may never be removed to an individual's residence for after-hours work for the convenience in connection with travel.

Accountability of Classified Material

Governmental programs require that an accountability record be maintained of all Top Secret and Secret materials and all cryptographic material regardless of classification. The record normally includes all such classified materials received or produced by or in the possession or custody of the contractor or the organization and shall reflect as a minimum the date of receipt or origin; the activity for which received or by which it originated; the classification of the material; a brief unclassified description of the materials; and the disposition of the materials and the date thereof (for example, destroyed, downgraded to confidential; declassified, or dispensed outside the facility). Contractors are required to maintain these records for a minimum of four years from the date of the last item recorded thereon.

Safeguarding Classified Information in Use

A great deal of classification is utilized in government programs. It is therefore necessary that control and procedural mechanisms be established which insure the protection of these materials. When classified information is being used it must generally be protected. It must be protected according to specific procedures.

Classified material must be conspicuously identified as classified material, for example, by the use of classified material cover-sheets, markings, and so forth. It must be kept under constant surveillance by an authorized person in a physical position to exercise direct security controls over the material. Likewise, it should be covered, turned face down, or placed in approved classified storage containers or otherwise protected when unauthorized persons are in the immediate area. It can only be released to persons who are cleared to receive the information and have a proper "need to know." The responsibility for determining whether a person's duties require that he possess or have access to any classified information rests upon the individual who has possession, knowledge, or control of the information involved. Regarded materials, until the regarding marks are actually made by the proper authority, shall be safeguarded in the prescribed manner for the level of classification that appears thereon. Drawings containing classified information while being prepared may be left on the drawing board during brief absences if they are covered and left under the personal control and constant surveillance of an authorized person. Materials must be returned to approved storage containers as soon as practicable when not in use.

Storage[5]

Organizations providing services to the government are not eligible to receive or have possession of classified material unless adequate storage for safeguarding is available at the cleared facility. All classified material while not in actual use must be safeguarded and stored as follows:

Top Secret. When not in use, Top Secret material shall be stored in a security filing cabinet as listed in the Federal Supply Schedule of the General Services Administration or in a Class-A vault. In addition to the cabinets and vaults specified by the government during nonworking hours, the following types of control are required:

(1) Entry to the room, building or structure in which the container is located shall be controlled by a properly cleared authorized employee or guard stationed so as to control admittance to room, building, or structure, or by a lock which provides reasonable protection against surreptitious entry.

(2) For the purpose of detecting unauthorized personnel, or persons attempting illegal entry to the container, the interior of the room, building, or structure shall be controlled and each container inspected at least once during each two-hour period by a guard, one of whose principal duties is to safeguard classified information and who is supervised by a system which provides a written record of the coverage of key points within the area.

(3) The room, building, or structure in which the container is located or the container itself shall be equipped with an alarm system as prescribed by government regulations which has a response time to an activated alarm not exceeding fifteen minutes.

Special Requirements for Top Secret. It is mandatory that an up-to-date record be maintained of all persons who are afforded access to Top Secret information. For each item of Top Secret material a record shall be maintained which identifies the item of material and which shows the names of all the individuals given access to the items and the first date on which their access occurred. Subsequent access to the same material by the same individual may not be reflected on the access record. Such access shall be maintained in the appropriate control station for a period of four years from the date that the material was destroyed, dispatched to outside facilities, declassified or downgraded. This record requirement also applies to those employees to whom visual or aural access to Top Secret information is made available.

Secret Information. When not in use Secret materials shall be stored in a cabinet or vault authorized for the storage of Top Secret information or in security vaults or cabinets. These vaults or cabinets include filing cabinets listed in the Federal Supply Schedule, General Services Administration. A Class-B vault constructed in accordance with these specifications, a steel file cabinet, or safe-type steel file container having an automatic unit locking mechanism and a built-in three position dial-type changeable combination lock, or a steel file cabinet sealed by a steel bar and a three position dial-type changeable combination padlock, or a Class-C vault constructed in accordance with established specifications may be used, other vaults and storage rooms provided that the vault or storage room is under surveillance by a regularly scheduled hourly guard patrol or is equipped with an alarm system as prescribed in government regulations. The response time to an activated alarm shall in no case exceed fifteen minutes. Likewise, a steel container in a pedestal desk which encloses a drawer of five sides and is riveted or bolted to the desk provided that the drawer is secured by a steel bar and a three position dial-type changeable combination padlock may also be used.

Supplemental Controls. In addition to the cabinets and vaults indicated, during nonworking hours the following area controls may be required:

(1) Entry into the room, building, or structure in which the container is located shall be controlled by a properly cleared authorized employee or guard stationed so as to control admittance to the room, building, or structure or by a lock which provides reasonable protection against surreptitious entry or by a properly cleared guard stationed at each unsecured perimeter entrance to a complex which is enclosed by a physical barrier and provided further that the area is patrolled adequately to provide reasonable opportunity to detect unauthorized personnel.

(2) For the purpose of detecting unauthorized personnel or attempted illegal entry to the container, the interior of the room, building, or structure in which the container is located shall be patrolled at least once during each four-hour period by the properly cleared authorized employee or guard one of whose principal duties is safeguarding of classified information and who is supervised by a system which provides a written record of the coverage of key points within the area, or,

(3) The room, building, or structure in which the container is located or the container itself shall be equipped with an alarm system that is prescribed by government regulations and the response time to an activated alarm shall in no case exceed fifteen minutes.

Confidential. When not in use, Confidential materials must be stored in the same manner as Top Secret and Secret material. However, supplementary controls are only required in the case of confidential cryptographic materials.

Confidential—Modified Handling Authorized. When not in use, Confidential—Modified Handling Authorized materials shall be stored in the same manner as Top Secret, Secret or Confidential materials.

Security Checks. Security Checks are normally performed within the facility to insure that at all times security precautions are taken to protect classified materials in the possession of the facility. Individuals are designated to make such checks on a room or area basis to insure that all material has been properly stored and secured.

Compromise or Suspected Compromise of Classified Information

Defense or classified material access to which may have been given to unauthorized persons creates serious problems to government and to government contracting agencies or organizations. Contractors are required to establish a procedure to insure that each loss, compromise or suspected compromise of classified information and each failure to comply with a security requirement is immediately reported to the facility security supervisor. Classified material which is out of the control of its custodian or which cannot be located shall be presumed to be lost until an investigation can determine otherwise.[6]

Security Violations

If classified information is compromised or regulations concerning the handling and maintenance of classified information are not complied with, security violations are noted by the security officer to the cognizant security office. Serious breaches in security can result in termination of contracts, payment of fines, or imprisonment.

Locking Devices

Locking devices of the three position dial-type changeable combination variety are used on storage containers for classified information. Only a minimum number of authorized persons should have knowledge of the combination to a classified storage container or have access to the materials stored therein.

Security of Combinations. To insure that information is kept secure, agencies shall insure that the combinations to the safes, containers, and three position dial-type changeable combination padlocks used to lock containers holding classified information are classified in accordance of the highest classified materials stored in the containers. This further requires that combinations be changed at intervals of at least every year and at the earliest practicable time following (1) the relief, transfer, or discharge of any person having knowledge of the combination; (2) the compromise or suspected compromise of the safes and containers or their combinations; (3) the initial receipt of safes, containers, or three position dial-type changeable combination padlocks.

Destruction of Classified Materials

With few exceptions agencies or contractors to the government are required to destroy classified information in its possession as soon as practicable after it has served the purpose for which it was released by the government, developed or prepared by the contractor, or retained after completion by the contractor.

Record of Destruction. An accurate record of destruction of classified materials is an important as is their destruction. Proper accounting procedures together with accurate records of the destruction provide official information as to the status of classified materials.

Security in Destruction. The destruction of all categories of classified documents and classified materials must be accomplished by a responsible official in the presence of a disinterested witnessing officer, civilian official, governmental official in the grade of GS-4 or above, or a noncommissioned officer or a specialist in grade E-5 or above designated by a responsible official. Such designation will be in writing when Top Secret or Secret information is destroyed. Witnessing personnel must have a security clearance at least as high as the category of the material being destroyed. The destruction official must insure that the destruction is complete and thorough and that legible scraps or recognizable parts of documents do not remain in the incinerator, pulping machine, or at the site of destruction.[7]

Methods of Destruction. The responsible agency shall destroy classified material beyond recognition so as to preclude reconstruction of the classified information in whole or in part. He shall insure the effectiveness of the destruction process and shall specify in the standard practice procedure the requirement to examine the residue for remnants of understroyed classified material. Destruction shall be by burning or by one of the following alternative methods: (1) burning, pulping, or pulverizing; (2) other classified materials may be destroyed by melting, chemical decomposition, pulverizing or mutilation.

Security Termination Statement and Debriefing Procedure

A security termination statement is normally executed by personnel who have had access to defense information prior to their termination of active military service, or civilian employment by a User agency or contractor, or their reassignment from significantly sensitive duties to other duties not requiring the same access. The statement normally includes and is limited to the following:

1. An acknowledgment that the individual executing the statement has read the appropriate provisions of the Espionage Act and other criminal statutes and regulations applicable to the level of classified information to which he has access and understands their implications.
2. The statement that the individual does not have classified material or documents in his possession.
3. A statement that the individual will not divulge orally or in writing defense information to unauthorized persons or agencies.
4. A statement that the individual will report to the Federal Bureau of Investigation immediately any attempt by an unauthorized person to obtain defense information.
5. An acknowledgment that the individual received an oral debriefing.[8]

General Requirements for Facility Information Security

Contractors to the government are responsible for safeguarding all classified information under their control. In furtherance of this requirement, the contractor is required to appoint an individual to supervise and direct security measures necessary for the safeguarding of classified information. He is also required to insure that classified information is furnished or disclosed only to authorized persons. To this end, the contractor shall determine to what extent an employee, subcontractor, vendors, and suppliers require access to classified information, in the performance of tasks or services essential to fulfill the contract. He shall take all reasonable measures to adjust plant layout and organized work so that to limit such access to the least number of individuals or firms consistent with the efficient performance of the classified contract.

Contractors are also required to provide suitable protective measures within their facilities for the safeguarding of classified information. A contractor performing work within the confines of a user agency installation must safeguard classified information in accordance with regulations unless his responsibilities for security are modified by his contract. He is also required to exclude from those parts of his plant, facilities, or sites where classified work is being performed any person or persons whom the head of the User agency concerned may, in the interest of security, designate in writing. Exclusion does not mean that such an employee must be dismissed or denied employment in another part of the plant or facility.[9]

Individual Responsibility for Safeguarding

The contractor shall bring to the attention of his employees engaged in the preparation of bids, quotations, or in the performance of work on contracts or programs which involve access to classified information their continuing *individual responsibilities* for safeguarding classified information. The employee who has knowledge or possession of an element or item of classified information shall be informed that he is responsible for determining whether a prospective recipient is an authorized person. The employee shall be informed that he is required to advise the recipient of the classification of the information which he discloses. The contractor shall also inform his employees that unauthorized disclosures of classified information violates the Department of Defense regulations and contractual obligations and is punishable under the provision of the Federal Criminal Statutes, Title 18.

Security Debriefing and Termination

Contractors prior to permitting an employee to have access to classified information shall inform him of his obligation to safeguard classified information so advising him of its importance and informing him of the required security procedures and have him read or have read to him the portion of the Espionage Laws, Conspiracy Laws, and Federal Criminal Statutes applicable to the safeguarding of classified information. In addition, the employee must be advised that he must report to the contractor if he becomes a representative of foreign interests and must likewise fulfill the security termination and debriefing procedure as previously outlined upon the termination of having access to classified information.[10]

Privacy Act of 1974

The information security process not only protects governmental operation against those with criminal intent but is also used by government to protect citizens against unauthorized disclosure of harmful or possibly derogatory information. The "Privacy Act of 1974" (Public Law 93-579, 93rd Congress, December 31, 1974) was, for example, enacted to protect personal information in collection by Federal Agencies.[11]

SEC. 2. (a) The Congress finds that—

(1) the privacy of an individual is directly affected by the collection, maintenance, use, and dissemination of personal information by Federal agencies;

(2) the increasing use of computers and sophisticated information technology, while essential to the efficient operations of the Government, has greatly magnified the harm to individual privacy that can occur from any collection, maintenance, use, or dissemination of personal information;

(3) the opportunities for an individual to secure employment, insurance, and credit, and his right to due process, and other legal protections are endangered by the misuse of certain information systems;

(4) the right to privacy is a personal and fundamental right protected by the Constitution of the United States; and

(5) in order to protect the privacy of individuals identified in information systems maintained by Federal agencies, it is necessary and proper for the Congress to regulate the collection, maintenance, use, and dissemination of information by such agencies.

(b) The purpose of this Act is to provide certain safeguards for an individual against an invasion of personal privacy by requiring Federal agencies, except as otherwise provided by law to—

(1) permit an individual to determine what records pertaining to him are collected, maintained, used, or disseminated by such agencies;

(2) permit an individual to prevent records pertaining to him obtained by such agencies for a particular purpose from being used or made available for another purpose without his consent;

(3) permit an individual to gain access to information pertaining to him in Federal agency records, to have a copy made of all or any portion thereof, and to correct or amend such records;

(4) collect, maintain, use, or disseminate any record of identifiable personal information in a manner that assures that such action is for a necessary and lawful purpose, that the information is current and accurate for its intended use, and that adequate safeguards are provided to prevent misuse of such information;

(5) permit exemptions from the requirements with respect to records provided in this Act only in those cases where there is an important public policy need for such exemption as has been determined by specific statutory authority; and

(6) be subject to civil suit for any damages which occur as a result of willful or intentional action which violates any individual's rights under this Act.

SECURITY TERMINATION STATEMENT AND DEBRIEFING CERTIFICATE (AR 380-5)		DATE
PART I — BASIC INFORMATION		
FROM *(Originating Headquarters)*		DOSSIER NO.
LAST NAME - FIRST NAME - MIDDLE INITIAL	GRADE *(Mil or Civ)*	SVC NO. *(Mil)* - SOCIAL SCTY NO. *(Civ)*
DATE OF BIRTH *(Day, Mo. Yr)*	PLACE OF BIRTH *(City, State, Country)*	

PART II — REFERENCES

a. APPLICABLE TO ALL PERSONNEL WHO HAVE HAD ACCESS TO DEFENSE INFORMATION:
   (1) ESPIONAGE LAWS: TITLE 18, U.S. CODE, SECTIONS 793, 794 AND 798 *("temporary extension of Section 794")*.
   (2) INTERNAL SECURITY LAWS: TITLE 50, U.S. CODE, SECTION 783.
   (3) AR 380-5
b. ADDITIONALLY APPLICABLE TO PERSONNEL WHO HAVE HAD ACCESS TO RESTRICTED DATA:
   (1) ATOMIC ENERGY ACT OF 1954: TITLE 42, U.S. CODE, SECTIONS 2014, 2162, 2274, 2275, 2276 AND 2277.
   (2) AR 380-150
   (3) AR 380-157
c. ADDITIONALLY APPLICABLE TO PERSONNEL WHO HAVE HAD ACCESS TO *CRYPTOGRAPHIC* MATERIAL OR INFORMATION:
   (1) ESPIONAGE LAWS: TITLE 18, U.S. CODE, SECTION 798.
   (2) AR 380-40
d. ADDITIONALLY APPLICABLE TO PERSONNEL WHO HAVE HAD ACCESS TO INFORMATION SPECIALLY COMPARTMENTED BY DOD OR DA DIRECTIVE:
   (1) LETTER, AGAM-P *(M) (16 Mar 64)* ACSFOR, HQ DA, 8 APR 64, SUBJECT: "SPECIAL SECURITY POLICY FOR THE NIKE X SYSTEM".
   (2) LETTER, AGAM-P *(M) (23 Dec 63)* ACSI-AS, HQ DA, 2 Jan 64, SUBJECT: "SECURITY POLICY FOR MILITARY SPACE PROGRAMS".
   (3) AR 380-34.
e. OTHER: *(Specify)* *

PART III — SECURITY TERMINATION AND DEBRIEFING STATEMENT

1. I acknowledge that I have read the applicable material for the level of classified information to which I have had access, and I understand that the revelation of classified information to an unauthorized person or agency is prohibited and punishable by law. My initials below attest to the level of access which I have had and to the applicable material, as identified in References, which I have read.

INITIALS	EXTENT OF ACCESS
_____	a. TOP SECRET - SECRET - CONFIDENTIAL defense information *(Reference a)*.
_____	b. RESTRICTED DATA *(Reference b)*.
_____	c. CRYPTOGRAPHIC material or information *(Reference c)*.
_____	d. Information specially compartmented by Department of Defense or Department of the Army directives. *(SPECIAL ACCESS Information)*
_____	e. Other: *(Specify)* *

2. I do not have classified material or documents in my possession.
3. I will not divulge classified information orally, in writing, or by any other means, to an unauthorized person or agency.
4. I will immediately report to the Federal Bureau of Investigation, my supervisor/commander, or other military authority, as appropriate, any attempt by an unauthorized person or agency to obtain classified information.
5. I received an oral debriefing, immediately prior to the execution *(i.e., signature)* of this Security Termination Statement.

*Can include access to information covered by treaties involving the U.S. (i.e., AR 380-15, AR 380-16, or AR 380-17) and access to critical stockpile and production information (i.e., AR 380-157), plus any travel restrictions.

DISTRIBUTION:	SIGNATURE
FIELD 201 FILE *(Military)* CO. USACRF *(Civilian)* LOCAL SECURITY FILE	

DA FORM 2962, 1 APR 65

OUTLINE OF GENERAL CONTENT ORAL SECURITY DEBRIEFING
(Part of Security Termination Statement)

1. PURPOSE OF DEBRIEFING. a. To establish that the individual does in fact understand the implications, to national security, and to himself, of the statutes and regulations which he has read.

    b. To emphasize to the individual that he was afforded access to classified information solely because of his "need-to-know" in the performance of official duties; that this information was entrusted, as well as officially charged to him; and that his impending retirement, separation, transfer, or reassignment (*as applicable*) in no way lessens his responsibilities – and liabilities – for ensuring that the classified knowledge acquired in his most recent position is not divulged in any manner to an unauthorized person or agency.

2. SERIOUS NATURE OF THE SUBJECT MATTER WHICH REQUIRES PROTECTION. Emphasize to the individual that classified information is defined and described in the pertinent statutes and regulations which he has read. As an illustration, cite the fact that *SECRET* defense information is "information or material the unauthorized disclosure of which *COULD RESULT IN SERIOUS DAMAGE TO THE NATION" (AR 380-5)*. Where the individual has had access to *TOP SECRET, RESTRICTED DATA, CRYPTOGRAPHIC* information, compartmented information, etc., cite the specific definition(s) and description(s) and emphasize that such material is even more serious in nature.

3. NEED FOR CAUTION AND DISCRETION. a. Emphasize to the individual that the responsibility is *HIS* to specifically establish that a person or agency requesting any classified information is officially authorized (*NEED-TO-KNOW*) that information; that if he is leaving the service (*includes civilian employees*), absolutely no other person or agency is authorized the classified information; that if he is being transferred or reassigned, any classified information known by the individual being debriefed cannot be divulged by that individual unless the new organization and specific person requesting the information has established the need-to-know.

    b. Emphasize to the individual that the mere fact that he reads a news article which appears to contain classified information in no way authorizes him to confirm the item. Explain that good "guesses" frequently are reflected by the news media, but bad "guesses" and incorrect information also are included.

    c. Caution the individual that history records a number of cases involving unauthorized disclosures in clubs and at social gatherings which have been reported and which resulted in punitive action.

4. ANY TRAVEL RESTRICTIONS IMPOSED. Cite the specific travel restrictions (*e.g., as set forth in AR 380-15, AR 380-16 and AR 380-17, for NATO, CENTO, and SEATO information*) AR 380-157, for restricted data.

5. SUMMARY. Specifically ask the individual if he understands what he has read and what he is about to sign. Based on his response (*and questions he may raise*) re-emphasize the content of the Security Termination Statement.

Figure 12-2.

Criminal Justice Information Systems[12]

The U.S. Department of Justice requires that specific procedural and hardware guidelines be followed in the processing and handling of Criminal History Record Information (CHRI). Its purpose is to afford greater protection of the privacy of individuals who may be included in the records of the Federal Bureau of Investigation, criminal justice agencies receiving funds directly or indirectly from the Law Enforcement Assistance Administration, and interstate, state or local criminal justice agencies exchanging records with the FBI or these federally funded systems.

The major objective of these guidelines is to insure that information collected about individuals are not disclosed to unauthorized parties. It further insures that the information included in the system is complete and accurate and that it is not disseminated without adequate records being maintained. It essentially establishes accountability for collected information and requires that protection be afforded to those included in it.

Educational Requirement for Information Protection

A wide variety of organizations and activities require the protection of the information they use. Schools, for example, must maintain protection over student files and related information such as counselors records. The Family Educational Rights and Privacy Act of 1974 for example, requires that schools follow specific procedures for protection of student data.

The Act assures students (or their parents, if the student has not attained the age of 18 years) "the right to inspect any and all official records, files, and data directly related . . ." to themselves, and assures the student an opportunity for a hearing to challenge the content of the record, and assures the student an opportunity for debate or correction of inaccurate, misleading or otherwise inappropriate data in the student's file.

The Act provides that no party may review a student's record with the exception of the following, who do not need written consent of the student to view a record:

(a) other school officials
(b) officials of other schools or systems in which the student intends to enroll. The student must be notified of the transfer of records, must be given a copy of records transferred if requested and must have the opportunity for a hearing to challenge the content of the record.
(c) authorized representatives of (1) the Comptroller General of the U.S., (2) the Secretary of H.E.W., (3) administrative head of an educational agency and (4) State educational authorities.
(d) persons working in connection with a student's application for, or receipt of, financial aid.

Any others than those listed above can obtain access to a student's file only upon written release from the student. A record of requests for a file will be kept with each student's file. This record will be available for inspection only by the student.

The following procedures has been established at many schools for students (or parents of students who have not attained the age of 18 years) to inspect records relating to themselves:

1. A written request, signed by the student (or parent) shall be presented to the Registrar.
2. The Registrar will arrange a time and date, within 45 days after receipt of the request for the student (or parent) to review the record.

The Act also provides that all students must be notified of the rights accorded to them by the Act.

Proprietary Applications

The safeguarding of information developed by private organizations for their own use also involves the information security process. The requirements of the Federal government for contracting agencies or organizations are in many cases utilized with very little modification to protect proprietary information; the principles involved in both types of programs being very similar. Both programs promote individual employee responsibility for information security, develop internal protection, control and administrative devices which control the access to classified information by unauthorized employees. They also develop similar techniques which protect this information when it is not in use; they control dissemination through a system of accountability; and provide for its destruction or dissemination when it is no longer of value.

The protection of information in both government and industry requires the development and use of information management programs. The amount of information which requires protection will determine the scope of the program but *all* programs will have the following characteristics:[13]

(1) a limited number of people can initiate classification of information;
(2) limit the duration during which classification will be in effect;
(3) hold the original classifier accountable for the correctness of his designation and require him to initial or sign all documents he classifies;
(4) require a management committee to monitor the effectiveness of the information protection program and to hear and take action on suggestions and complaints;
(5) write and use a checklist[14] to periodically assess and evaluate the management of the information security program; and
(6) have an employee education program to promote better decision-making by those who utilize the classification system.

The fewer documents which require protection, the better the protection. The sophistication and extent of programs are directly related to the sensitivity of the information as well as the volume of data or access required to it. Thus, the fewer personnel who can classify, require access to it, and the smaller the volume of information requiring protection, the higher the protection level provided and the lower the cost.

The rapid growth of technology has increased the desirability of stealing information about processes and improvements. The incidence of industrial espionage by employees as well as competitors has grown proportionally. The protection of information for the private business or industrial concern has become an extremely sophisticated function. The information security process is employed in a manner very similar to that in government. While the requirements for protection are not specified in the detail described for their programs, nevertheless, extensive protection is provided.

The principles of protection employed are identical in design and intent to those of the government. The protection of proprietary information requires attention to both internal and external threats. Internal threats can come from personal contacts within an organization with access to sensitive information who wish to sell it more and make it available outside the company. It likewise includes published data about processes or developments; and design or composition characteristics obtained through "reverse engineering" or chemical analysis. Protection of information, like other security processes, attempts to "buy time" before secrets are released to the public. In the case of a new product, it must be kept secure until it is introduced to the market. Once it is released, it becomes "fair game" for anyone who wishes to analyze and reproduce it.

The major concern of an information protection program is involved with the early release or disclosure of sensitive information through illegal or questionable means. The three most common methods of information loss to be guarded against are:

- Employee hirings
- Patent infringements
- Business or industrial espionage

Protection against employee hiring is most

often provided through the use of *Employee Patent and Secrecy Agreements.* These agreements are contracts between a company and an employee which restricts the employee's ability to disclose information provided or generated by the employee for purposes not designated by the company.

The elements of an information security program include:
(1) designation of certain information as sensitive;
(2) informing employees that the information is to be protected;
(3) establishing the use of patent or secrecy agreements;
(4) providing the means for employees to protect the information; and
(5) treating the designated sensitive information as proprietary.

Unless these criteria are met, the company not only does not have an adequate program of protection but does not have justification to prosecute an employee for disclosing sensitive information. Most courts in ruling on theft of trade secrets have indicated that "for something to be secret, it must not only be treated as secret, but the employees must be given the means to protect the information."

The information security program must include:
(1) preemployment screening of prospective employees;
(2) instituting a "need to know" program for information distribution;
(3) instituting procedures for control of information in use;
(4) development of procedures for classification of information;
(5) introduction of adequate physical security measures for access control and protection;
(6) use of secrecy agreements; and
(7) security education programs for employees.

To insure protection from professional industrial espionage agents, it is additionally necessary to:
(1) provide for the supervised destruction of waste or discarded documents;
(2) institute periodic electronic countermeasure "sweeps" to prevent or eliminate electronic surveillance;
(3) investigate high-risk or high-vulnerability employees; and
(4) minimize the number of employees given access to sensitive information.

The protection afforded against losses by either internal or external sources is at best limited. If an employee or "spy" is determined to obtain the information the best program will at best slow down the effort, it will not eliminate it. The information security process is an attempt to "slow it down," make it difficult, and prevent the otherwise honest employee from succumbing to temptation. Since employees must have sensitive information to conduct their activities, a trust relationship must be established and maintained. The good will and compliance of employees with the intent as well as the form of the program is, however, crucial for success.

Records Protection

Information security programs in proprietary settings are also concerned with the protection of vital organizational records. Vital records are those necessary to insure the survival of a business. They normally constitute about 2 percent of the company's total records.

The purpose of a vital records program is to protect the essential information contained in the records but not the protection of the records themselves. There are five methods of protecting vital records: (1) natural dispersal of copy of vital records either within or outside the organization; (2) planned dispersal of copy of the vital records; (3) on-sight storage in vaults or safes; (4) protection of original vital records, and (5) duplication of vital records.[15]

Planning to protect vital business records is normally part of a larger plan designed to protect all of the company and property in the event of man-made or natural disasters.

FOOTNOTES

1. It should be noted that protection of defense information is not limited to the activities of the U.S. Government. The governments of all U.K. countries and Canada have similar provisions for governmental protection. The basic legislation for this protection is often referred to as the "Official Secrets Act." The Canadian "Act Respecting Official Secrets" is presented in Appendix 12A. It is similar in intent and purpose to those of other countries.
2. Department of the Army: Safeguarding Defense Information AR 380-5. May 1965, p. 61.
3. This executive order (11652) is the *Revocation of Executive Order No. 10501*. Executive Order No. 10501 of November 5, 1953, as amended by Executive Orders No. 10816 of May 8, 1959, No. 10901 of January 11, 1961, No. 10964 of September 20, 1961, No. 10985 of January 15, 1962, No. 11097 of March 6, 1963 and by Section 1(a) of No. 11382 of November 28, 1967, are superseded as of the effective date of this order (June 1, 1972).
4. *Industrial Security Manual for Safeguarding Classified Information.* DOD 5220.22-M, July 1966, p. 26.
5. *Industrial Security Manual for Safeguarding Classified Information*, pp. 36-38.
6. *Industrial Security Manual for Safeguarding Classified Information*, p. 19.
7. *Industrial Security Manual for Safeguarding Classified Information*, p. 43.
8. Department of the Army: AR 380-5, p. 22.
9. ISM 5220.22-M, pp. 7-8.
10. ISM 5220.22-M, p. 8.
11. See Appendix 12B for the complete text of PL 93-579.
12. Pursuant to the authority vested in the Attorney General by 28 U.S.C. 509, 510, 534, and Pub. L. 92-544, 86 Stat. 1115, and 5 U.S.C. 301 and the authority vested in the Law Enforcement Assistance Administration by sections 501 and 524 of the Omnibus Crime Control and Safe Streets Act of 1966, as amended by the Crime Control Act of 1973, Pub. L. 93-83, 37 Stat. 197 (42 U.S.C. §3701 *et seq* (Aug. 6, 1973), this addition to Chapter I of Title 28 of the Code of Federal Regulations is issued as Part 20 by the Department of Justice, effective June 19, 1975.
13. This list is adapted from "DOD shows you how to manage company secrets." *Occupational Hazards*, September 1973, p. 57.
14. The Department of Defense Information Security Checklist is presented in Appendix 12C. Courtesy of Joseph J. Liebling, Assistant Secretary of Defense—Security Policy.
15. Office of Civil Defense, Protection of Vital Records. Washington, D.C., 1966, pp. 11-13.

BOOKS

Carroll, John M.: *Confidential Information Sources: Public & Private*. Security World Publishing Co., Inc., 1975.

Chermayeff, S., and Alexander, C.: *Community and Privacy*. Garden City, Anchor Books, 1965.

Lecrute, Jacques: *Secret Militaire et Liberte de la Presse*. France, 1957.

Lysenko, Georgit: *Isidorovich*. (Khrani Voennuiu Tainu) Russian, 1972.

PUBLICATIONS OF THE GOVERNMENT, LEARNED SOCIETIES, AND OTHER ORGANIZATIONS

A link system for assuring confidentiality of research data in longitudial studies. ACE Research Reports, 4: no. 3, Research in Education. H.E.W., Washington, Government Printing Office, Jan./Dec. 1970.

Applying records schedules. General Services Administration. Washington, Government Printing Office, 1956.

A survey of the present and probable future state of technology affecting privacy. Report No. 1008, prepared for the Special Committee on Science and Law by Bolt, Beranek, and Newman, Inc., Cambridge, Mass., June 20, 1963.

Archer, James E.: Guarding confidential information. American Society of Mechanical Engineers, New York, N.Y.

Clark, Lyle R.: Course of study on records management. USC, Los Angeles, Calif., May 1961.

Classified documents—RCA. Paperwork Simplification. The Standard Register Company, Dayton, Ohio.

Comprehensive security program for vital corporate records. Recordak Corp., New York, August 1956.

Department of the Army. Information Security Program. Washington, Government Printing Office, February 1973.

Department of the Defense. Carrier Supplement to Industrial Security Manual for Safeguarding Classified Information. Washington, Government Printing Office, April 1970.

———. Comsec Supplement to Industrial Security Manual for Safeguarding Classified Information. Washington, Government Printing Office, January 1973.

———. Information Security Program Regulation. Washington, Government Printing Office, November 1973.

———. Industrial Security Manual. par. 19, as revised.

Development of criteria and procedures for management of classified documents collecting. *Research in Education,* Washington, Government Printing Office, Jan.-Dec. 1970.

Federal paperwork jungle, The. Hearings before the House Subcommittee on Census and Government Statistics, Committee on Post Office and Civil Service, 86th Cong., 1st Sess., 1964.

Freedman, Samuel B.: *Vital Records Protection Through the Use of Microfilms.* Cleveland, Ohio, Bell and Howell Company, December 1963.

Fujii, Hacuo: *Ninon no Koklia Kimitsu.* Japan, 1972.

Garrick, B. John, and Williamson: *Protection of Vital Records Against Nuclear Weapons Effects.* Los Angeles, National Conference of the American Records Management Association, October 9-10.

General Services Administration. Applying Records Schedules. Washington, Government Printing Office, 1956.

———. Guide to Records Retention Requirements—Part 1. Washington, Government Printing Office, January 1, 1964.

———. Protecting Vital Operating Records Service. Washington, Government Printing Office, 1958.

Graham, Robert A.: *Developing Record Retention and Disposal Programs.* Cleveland, Ohio, Deibold, Inc., December 10, 1963.

Guarding Confidential Information. New York, American Society of Mechanical Engineers, 1950.

Henderson, Robert P.: Record keeping in the space age. Speech before National Symposium of the National Archives and Record Services (GSA), Washington, June 8, 1970.

Industrial Security Manual for Safeguarding Classified Information. May 1, 1968.

McGuire, John E.: The dreaded shredder. *National Association of Accountants Bulletin,* June 1964.

McGuire, Martin C.: *Secrecy and the Arms Race.* Cambridge, Mass., Harvard Univ. Press, 1965.

National Industrial Conference Board. Protecting records in wartime. New York, April 1951.

National Records Management Council, Inc. Guide to selected readings in records management. New York, 1954.

N.A.T.O. Agreement for the mutual safeguarding of secrecy of invention relating to defense and for which applications for patents have been made. London, 1962.

New approach to classified document control. General Electric Co., Advanced Electronics Center, Light Military Electronics Dept., Ithaca, N.Y.

O'Connor, Eugene T.: *Value of Records and Protection of Records.* New York, The Mosler Safe Company, December 1962.

Office of Civil Defense Mobilization. Protection of vital records and documents. Technical Bulletin 16-2, Battle Creek, Mich., January 1959.

Protecting vital operating records service. General Services Administration. Washington, Government Printing Office, 1958.

Protection of proprietary information—panel moderator, Robert J. Boise. *Industrial Security,* October 1960, p. 23-24.

Protection of vital corporate records. *Bulletin of National Assn. of Cost Accountants,* New York, January 1954.

Protection of vital industrial records and documents. *Civil Defense Technical Bulletin 12-2,* May 1956.

Rates, R. D.: CICRIS: Cooperatives, Industrial and Commercial Reference and Information Service. ASLIB Proceedings, March 1957, p. 83-84.

Raymone, Morton, M.: Improving your record disposition—a mental hygiene approach to records management. Cleveland, Ohio, Diebold, Inc., December 11, 1962.

Records Retention Program. Moore Business Forms, Inc., Emeryville, Calif.

Richman, Leo: Practical records management—a case history study. Hawthorne, Calif., Northrop Corporation, October 28, 1959.

———: Retention of records. Hawthorne, Calif., Northrop Corporation, October 10, 1961.

Safeguarding confidential information. *Management Record,* December 1960.

Sampson, Rod: Records maintenance: Coding. Visual Communications Congress, December 1962.

Task Force on Paperwork Management. Report on paperwork management, Part 1. Washington, Government Printing Office, January 1955.

U.S. Army Corps of Engineers. Military security. Washington, Government Printing Office, 1962.

U.S. Congress, House Committee on Un-American Activities. Protection of classified information, released to U.S. industry and defense contractors. Report No. 11363. Washington, Government Printing Office, 1962.

U.S. Department of Air Force. Armed forces and censorship. Washington, 1957.

U.S. Department of the Army. Commander's handbook of security. Washington, 1962.

U.S. Department of the Defense. Safeguarding defense information—why we need security. Audio-Visual Support Center, Ft. Sheridan, Ill.

U.S. Department of Defense. Office of Civil Defense. Protection of vital industrial records and documents. Washington, Government Printing Office, July 1966.

U.S. Department of Defense Industrial Security Division. Questions and answers on safeguarding classified information. Washington, Government Printing Office, 1954.

U.S. Department of Defense. Safeguarding military information. Audio-Visual Support Center, Ft. Sheridan, Ill.

———. D.O.D. implementation of recommendations of Coolidge committee on classified information. Washington, 1957.

———. Industrial security manual for safeguarding classified information. Washington, Government Printing Office, March 1, 1965.

———. Industrial personnel access authorization review regulation. Washington, Government Printing Office, April 7, 1954.

———. The Hollow Coin. Audio-Visual Support Center, Ft. Sheridan, Ill.

———. The Daily Enemy. Audio-Visual Support Center, Ft. Sheridan, Ill.

U.S. Department of Justice. Regulations relating to defense information, under E.O. 10501. Atlanta, 1954.

U.S. General Accounting Office. Unnecessary cost received for commercial protective security used for shipments of classified material. Washington, 1965.

U.S. House of Representatives, Committee on Un-American Activities. Protection of classified information released to U.S. industry and defense contractors. 87th Cong., 2nd Sess., Washington, Government Printing Office, June 28, 1962.

U.S. Navy Department. Security manual for classified information. Office of Naval Intelligence, Washington, 1958.

Utt, Charles T.: *A Program for Records Survival.* Scranton, Pa., International Textbook Company, August 1957.

Wheelan, Robert B.: *Corporate Records Retention—A Guide to United States Federal Requirements,* New York, Controllers Institute Research Foundation, 1958, vol. 1.

———: *Corporate Records Retention—A Guide to Requirements of State Governments of the United States.* New York, Controllers Institute Research Foundation, 1960, vol. III.

Wise, John W.: The company classification management program—Panelist. *Industrial Security,* October 1963, p. 55.

PERIODICALS

Adkinson, B. W.: United States scientific and technical information services. *Special Libraries,* November 1958, p. 407-414.

Aho, K. Alferd, Capt. USN.: Evolution in declassification procedures. *Industrial Security,* April 1959, p. 10.

Baaden, Walter J., Jr.: Management role in document reduction. *Industrial Security,* July 1963, p. 8.

Caffey, J. J.: Case for security classification. *Law and Order, 9:* December 1961.

Calvert, R.: When you use secret information. *Chemical Industry, 66:*518, April 20, 1950.

Carroll, John Millar: *Confidential Information Sources: Public and Private.* Los Angeles, Security World Publishing Co., 1975.

Carson, Hilbert M.: Transmission of classified information and material. *Industrial Security, 3:5,* July 1959.

Casey, J. E.: The automatic time-phased downgrading and declassification system (seminar workshop report). *Industrial Security,* October 1961, p. 61.

Chafee: Thirty-five years with freedom of speech. *Industrial Personnel Security,* 1952.

Classification management—a plan for industry. *Industrial Security,* October 1970.

Comprehensive system provides reliable, economical protection for Scott paper office facility. *Security Industry and Product News,* July/August 1975, p. 26-27.

Coulton, John D.: Records protection in the cold war. *Records Review, 4:* no. 5, December 1963.

Curtis, Bob: I never promised you a secret room. *Assets Protection,* Spring 1975, p. 49-60.

Defense information bulletin. U.S. Office of Education. Washington, Government Printing Office.

DeLorenzo, Anthony: Whose lip are you trying to zip? *Industrial Security,* October 1961.

Donovan, Robert: Automated document control: Boon or bane? *Security World, 3:* no. 4, April 1966.

―――: Classification management. *Security World, 3:* no. 6, June 1966.

―――: Management techniques for loss prevention. *Industrial Security,* August 1965, p. 35.

―――: New DOD manual revisions. *Security World, 3:* no. 5, May 1966.

―――: New DOD manual preview. *Security World, 3:* no. 5, March 1966.

―――: Security and "individual privacy," Part I. *Security World, 4:*24, no. 6, June 1967.

―――: Security and "individual privacy," Part II. *Security World, 4:* no. 7, July/August 1967.

―――: Security and "individual privacy," Part III. *Security World, 4:* no. 29, September 1967.

―――: The security classification dilemma: Part I. *Security World, 3:* no. 10, November 1966.

―――: The security classification dilemma: Part II. *Security World, 3:* no. 11, December 1966.

Draper, T.: Classifiers of classified documents are breaking their own classification rules. *New York Times Magazine,* February 1973.

Dudley, W. E.: Need to know: The commonsense essence of security. *Industrial Security,* October 1967.

Emergency Planning Committee. Safeguarding classified information in an emergency. *Industrial Security,* August 1968.

Everybody's listening. *Nation,* January 5, 1957.

Gardner, John: A matter of disclosure of material fact. *Security Gazette,* October 1974, p. 378.

General Electric Company. New approach to classified document control. *Industrial Security,* 7:60, 1960.

Grand, David A.: The 1968 storage requirements. *Industrial Security,* June 1967.

Greenstein, Raymond: Can we lessen vandalism? *Instructor, 79:* no. 5, January 1970.

Grimes, R. C.: Protecting vital industrial records. *Industrial Security,* July 1960.

Guise, Robert F.: File security. *Data Systems News,* November 1969, p. 6.

Harwood, Richard: Is your name on a secret dossier? *Washington Post,* May 29, 1966.

Hayes, Donald, Lt. USNR: Evolution in classification procedures. *Industrial Security, 16:* July 1958.

Herald, Virgil H.: Is classification confusion really necessary? *Industrial Security,* June 1968, p. 39.

Hinkle, Charles W.: Security of information proposed for public release—panelist. *Industrial Security,* October 1962, p. 69.

Hughes, Charles E.: Better records management. *Factory Magazine,* December 1960, p. 16.

Kelly, Eugene: An innovation in document control. *Industrial Security,* 7:10, July 1963.

―――: An innovation in securing files. *Industrial Security, 7:* no. 3, July 1963.

La Rue, Roger L.: Protecting your business information. *Security Management,* September 1975, p. 14-16.

Lett, S. H.: Exhibits for document examination. *Security World,* April 1969.

Lewis, James M.: Security violations and compromises of classified information. *Industrial Security,* October 1968, p. 47.

Loomis, Roland L.: Safeguarding proprietary information (industrial espionage)—panel moderator. *Industrial Security,* December 1964, p. 28.

Moran, James D.: Let's cooperate on security classification. *Security Management,* March 1974, p. 30-32.

Musolino, Anthony B.: Document accountability via the punched card. *Industrial Security,* December 1970.

National industrial conference board: Safeguarding confidential information. *Management Record,* December 1960.

New foreign technical information center offers service to business. *Commerce Week,* August 1958, p. 2.

O'Leary, C. J.: Classified document control—panel moderator. *Industrial Security,* October 1963, p. 60.

One hundred fourteen thousand credentials: Access control at the Democratic Convention. *Security World, 5:* no. 9, October 1968.

Parker, W. H.: Problems confronting small business in classified defense work. *Law and Order, 9:* October 1961.

Patton, A. Gordon: The significance of the confidential classification (seminar workshop report). *Industrial Security,* October 1961, p. 49.

Powell, Marlyn R.: A machine-based document control system—panelist. *Industrial Security,* October 1962, p. 34.

Rasmussen, R. F.: Functions and operations of central index file Europe—panelist. *Industrial Security,* October 1963, p. 57.

Richardson, Dean C.: God, motherhood and classification. *Industrial Security,* October 1970.

Rubenstein, Sidney S.: What price publicity? *Industrial Security,* January 1962, p. 18.

Rubenstein, Sidney S.: Classification management—panelist. *Industrial Security,* October 1962, p. 63.

Ruitschle, William H.: Movement of U.S. classified material in the international environment. *Security Management,* July 1974, p. 17-20.

Rushing, Robert J.: Classification management—panelist. *Industrial Security,* December 1964, p. 40.

Safeguard confidential information. *Management Record,* October 1960.

Schisgall, Oscar: Our defense secrets are for sale cheap. *Look,* August 27, 1963.

Scruton, Robert A.: Security of classified material shipments. *TPC, 30:* May 1963.

Security of classified information (seminar workshop report). *Industrial Security,* October 1961, p. 47.

Shiff, Robert A.: Protect your records against disaster. *Harvard Business Review,* July/August 1956.

Someone knows all about you. *Esquire,* May 1966, p. 101.

Stubbs, James: Filing and finding computer tapes. *Administrative Management,* May 1969, p. 56.

Sue, Newton: Inspecting document control. *Industrial Security, 7:*46, January 1964.

Tanner, Shelby G.: Controlling classified working papers. *Industrial Security,* February 1970.

Thompson, George D.: Address before National Classification Management Society. *Industrial Security,* October 1966, p. 28.

Wilkie, Francis E.: Classification management. *Industrial Security, 9:*20, June 1965.

Wilson, Kenneth E.: Classification management's role in cost savings. *Security World, 5:* no. 9, October 1968.

Yaspan, Arthur J., and Halbert, Michael: Information-collection procedures. *Operations Research,* August 1957, p. 582.

UNPUBLISHED MATERIAL

Bailey, Charles: National survey of United States Air Force directors/chiefs of security police attitudes pertaining to United States Air Force security, police-community relations. (Thesis), Michigan State Univ., 1968.

Yang, Show-Shan: The fundamentals of personnel security programs. (B Paper), Michigan State Univ., 1966.

APPENDIX 12A

AN ACT RESPECTING OFFICIAL SECRETS

SHORT TITLE.

1. This Act may be cited as the *Official Secrets Act.* 1939, c. 49, s. 1.   Short title.

INTERPRETATION.

2. In this Act,   Definitions.

(a) "Attorney General" means the Attorney General of Canada;   "Attorney General."

(b) "document" includes part of a document;   "Document."

(c) "model" includes design, pattern and specimen;

(d) "munitions of war" means arms, ammunition, implements or munitions of war, army, naval or air stores, or any articles deemed capable of being converted thereinto, or made useful in the production thereof;

(e) "offence under this Act" includes any act, ommission, or other thing that is punishable hereunder;

(f) "office under Her Majesty" includes any office or employment in or under any department or branch of the Government of Canada or of any province, and any office or employment in, on or under any board, commission, corporation or other body that is an agent of Her Majesty in right of Canada or any province;

(g) "prohibited place" means
   (i) any work of defence belonging to or occupied or used by or on behalf of Her Majesty including arsenals, naval, army or air force establishments or stations, factories, dockyards, mines, minefields, camps, ships, aircraft, telegraph, telephone, wireless or signal stations or offices, and places used for the purpose of building, repairing, making or storing any munitions of war or any sketches, plans, models, or documents relating thereto, or for the purpose of getting any metals, oil or minerals of use in time of war,
   (ii) any place not belonging to Her Majesty where any munitions of war or any sketches, models, plans or documents relating thereto, are being made, repaired, gotten or stored under contract with, or with any person on behalf of, Her Majesty, or otherwise on behalf of Her Majesty, and
   (iii) any place that is for the time being declared by order of the Governor in Council to be a prohibited place on the ground that information with respect thereto or damage thereto would be useful to a foreign power;

(h) "sketch" includes any mode of representing any place or thing;

(i) "senior police officer" means any officer of the Royal Canadian Mounted Police not below the rank of Inspector; any officer of any provincial police force of a like or superior rank; the chief constable of any city or town with a population of not less than ten thousand; or any person upon whom the powers of a senior police officer are for the purposes of this Act conferred by the Governor in Council;

(j) any reference to Her Majesty means Her Majesty in right of Canada or of any province; and

(k) expressions referring to communicating or receiving include any communicating or receiving, whether in whole or in part, and whether the sketch, plan, model, article, note, document or information itself or the substance, effect, or description thereof only is communicated or received; expressions referring to obtaining or retaining any sketch, plan, model, article, note, or document, include the copying or causing to be copied the whole or any part of any sketch, plan, model, article, note, or document; and expressions referring to the communication of any sketch, plan, model, article, note or document include the transfer or transmission of the sketch, plan, model, article, note or document. 1939, c. 49, s. 2; 1950, c. 46, s. 1.

3. (1) Every person who, for any purpose prejudicial to the safety or interests of the State,
   (a) approaches, inspects, passes over, or is in the neighbourhood of, or enters any prohibited place;
   (b) makes any sketch, plan, model or note that is calculated to be or might

be or is intended to be directly or indirectly useful to a foreign power; or
- (c) obtains, collects, records, or publishes, or communicates to any other person any secret official code word, or pass word, or any sketch, plan, model, article, or note, or other document or information that is calculated to be or might be or is intended to be directly or indirectly useful to a foreign power;

is guilty of an offence under this Act.

(2) On a prosecution under this section, it is not necessary to show that the accused person was guilty of any particular act tending to show a purpose prejudicial to the safety or interests of the State, and, notwithstanding that no such act is proved against him, he may be convicted if, from the circumstances of the case, or his conduct, or his known character as proved, it appears that his purpose was a purpose prejudicial to the safety or interests of the State; and if any sketch, plan, model, article, note, document or information relating to or used in any prohibited place, or anything in such a place, or any secret official code word or pass word is made, obtained, collected, recorded, published or communicated by any person other than a person acting under lawful authority, it shall be deemed to have been made, obtained, collected, recorded, published or communicated for a purpose prejudicial to the safety or interests of the State unless the contrary is proved. *[Accused person may be convicted if purpose prejudicial to the safety of the State.]*

(3) In any proceedings against a person for an offence under this section, the fact that he has been in communication with, or attempted to communicate with, an agent of a foreign power, whether within or without Canada, is evidence that he has, for a purpose prejudicial to the safety or interests of the State, obtained or attempted to obtain information that is calculated to be or might be or is intended to be directly or indirectly useful to a foreign power. *[Communication with agent of foreign power, etc., sufficient evidence.]*

(4) For the purpose of this section, but without prejudice to the generality of the foregoing provision *[When person deemed to have been in communication with agent of a foreign power.]*
- (a) a person shall, unless he proves the contrary, be deemed to have been in communication with an agent of a foreign power if
  - (i) he has, either within or without Canada, visited the address of an agent of a foreign power or consorted or associated with such agent, or
  - (ii) either within or without Canada, the name or address of, or any other information regarding such an agent has been found in his possession, or has been supplied by him to any other person, or has been obtained by him from any other person;
- (b) "an agent of a foreign power" includes any person who is or has been or is reasonably suspected of being or having been employed by a foreign power either directly or indirectly for the purpose of committing an act, either within or without Canada, prejudicial to the safety or interests of the State, or who has or is reasonably suspected of having, either within or without Canada, committed, or attempted to commit, such an act in the interests of a foreign power; and *["An agent of a foreign power" defined.]*
- (c) any address, whether within or without Canada, reasonably suspected of being an address used for the receipt of communications intended for an agent of a foreign power, or any address at which such an agent resides, or to which he resorts for the purpose of giving or receiving communications, or at which he carries on any business, shall be deemed to be the address of an agent of a foreign power, and communications addressed to such an address to be communications with such an agent. 1939, c. 49, s. 3. *[When address deemed that of an agent of a foreign power.]*

4. (1) Every person who, having in his possession or control any secret official code word, or pass word, or any sketch, plan, model, article, note, document or

Wrongful communication, etc., of information.

information that relates to or is used in a prohibited place or anything in such a place, or that has been made or obtained in contravention of this Act, or that has been entrusted in confidence to him by any person holding office under Her Majesty, or that he has obtained or to which he has had access while subject to the Code of Service Discipline within the meaning of the *National Defence Act* or owing to his position as a person who holds or has held office under Her Majesty, or as a person who holds or has held a contract made on behalf of Her Majesty, or a contract the performance of which in whole or in part is carried out in a prohibited place, or as a person who is or has been employed under a person who holds or has held such an office or contract,

    (*a*) communicates the code word, pass word, sketch, plan, model, article, note, document or information to any person, other than a person to whom he is authorized to communicate with, or a person to whom it is in the interest of the State his duty to communicate it;

    (*b*) uses the information in his possession for the benefit of any foreign power or in any other manner prejudicial to the safety or interests of the State;

    (*c*) retains the sketch, plan, model, article, note, or document in his possession or control when he has no right to retain it or when it is contrary to his duty to retain it or fails to comply with all directions issued by lawful authority with regard to the return or disposal thereof; or

    (*d*) fails to take reasonable care of, or so conducts himself as to endanger the safety of the sketch, plan, model, article, note, document, secret official code word or pass word or information;

is guilty of an offence under this Act.

Communication of sketch, plan, model, etc.

(2) Every person who, having in his possession or control any sketch, plan, model, article, note, document or information that relates to munitions of war, communicates it directly or indirectly to any foreign power, or in any other manner prejudicial to the safety or interests of the State, is guilty of an offence under this Act.

Receiving code word, sketch, etc.

(3) Every person who receives any secret official code word, or pass word, or sketch, plan, model, article, note, document or information, knowing, or having reasonable ground to believe, at the time when he receives it, that the code word, pass word, sketch, plan, model, article, note, document or information is communicated to him in contravention of this Act, is guilty of an offence under this Act, unless he proves that the communication to him of the code word, pass word, sketch, plan, model, article, note, document or information was contrary to his desire.

(4) Every person who

Retaining official document, etc.

    (*a*) retains for any purpose prejudicial to the safety or interest of the State any official document, whether or not completed or issued for use, when he has no right to retain it, or when it is contrary to his duty to retain it, or fails to comply with any directions issued by any Government department or any person authorized by such department with regard to the return or disposal thereof; or

Allowing other to have possession.

    (*b*) allows any other person to have possession of any official document issued for his use alone, or communicates any secret official code word or pass word so issued, or, without lawful authority or excuse, has in his possession any official document or secret official code word or pass word issued for the use of some person other than himself, or on obtaining possession of any official document by finding or otherwise, neglects or fails to restore it to the person or authority by whom or for whose use it was issued, or to a police constable;

is guilty of an offence under this Act. 1939, c. 49, s. 4; 1951 (2nd Sess.), c. 7, s. 28.

5. (1) Every person who, for the purpose of gaining admission, or of assisting

any other person to gain admission, to a prohibited place, or for any other purpose prejudicial to the safety or interests of the State,

(a) uses or wears, without lawful authority, any naval, army, air force, police or other official uniform or any uniform so nearly resembling the same as to be calculated to deceive, or falsely represents himself to be a person who is or has been entitled to use or wear any such uniform;

(b) orally, or in writing in any declaration or application, or in any document signed by him or on his behalf, knowingly makes or connives at the making of any false statement or any omission;

(c) forges, alters, or tampers with any passport or any naval, army, air force, police or official pass, permit, certificate, licence or other document of a similar character, (hereinafter in this section referred to as an official document), or uses or has in his possession any such forged, altered, or irregular official document;

(d) personates, or falsely represents himself to be a person holding, or in the employment of a person holding, office under Her Majesty, or to be or not to be a person to whom an official document or secret official code word or pass word has been duly issued or communicated, or with intent to obtain an official document, secret official code word or pass word, whether for himself or any other person, knowingly makes any false statement; or

(e) uses, or has in his possession or under his control, without the authority of the Government department or the authority concerned, any die, seal, or stamp of or belonging to, or used, made, or provided by any Government department, or by any diplomatic, naval, army, or air force authority appointed by or acting under the authority of Her Majesty, or any die, seal or stamp, so nearly resembling any such die, seal or stamp as to be calculated to deceive, or counterfeits any such die, seal or stamp, or uses, or has in his possession, or under his control, any such counterfeited die, seal or stamp;

is guilty of an offence under this Act.

Unauthorized use of uniforms; falsification of reports, forgery, personation and false documents.

(2) Every person who, without lawful authority or excuse, manufactures or sells, or has in his possession for sale any such die, seal or stamp as aforesaid, is guilty of an offence under this Act. 1939, c. 49, s. 5.

Unlawful dealing with dies, seals, etc.

6. No person in the vicinity of any prohibited place shall obstruct, knowingly mislead or otherwise interfere with or impede any constable or police officer, or any member of Her Majesty's forces engaged on guard, sentry, patrol, or other similar duty in relation to the prohibited place, and every person who acts in contravention of, or fails to comply with, this provision, is guilty of an offence under this Act. 1939, c. 49, s. 6.

Interfering with officers of the police or members of Her Majesty's forces.

7. (1) Where it appears to the Minister of Justice that such a course is expedient in the public interest, he may, by warrant under his hand, require any person who owns or controls any telegraphic cable or wire, or any apparatus for wireless telegraphy, used for the sending or receipt of telegrams to or from any place out of Canada, to produce to him, or to any person named in the warrant, the originals and transcripts, either of all telegrams, or of telegrams of any specified class or description, or of telegrams sent from or addressed to any specified person or place, sent to or received from any place out of Canada by means of any such cable, wire, or apparatus and all other papers relating to any such telegram as aforesaid.

Power to require the production of telegrams.

(2) Every person who, on being required to produce any such original or transcript or paper as aforesaid, refuses or neglects to do so is guilty of an offence under this Act, and is for each offence, liable on summary conviction to imprisonment, with or without hard labour, for a term not exceeding three months, or to a fine not exceeding two hundred dollars, or to both such imprisonment and fine. 1939, c. 49, s. 7.

Refusing or neglecting to produce original, etc.

Penalty.

8. Every person who knowingly harbours any person whom he knows, or has

Harbouring spies.

reasonable grounds for supposing, to be a person who is about to commit or who has committed an offence under this Act, or knowingly permits to meet or assemble in any premises in his occupation or under his control any such persons, and every person who, having harboured any such person, or permitted to meet or assemble in any premises in his occupation or under his control any such persons, wilfully omits or refuses to disclose to a senior police officer any information that it is in his power to give in relation to any such person, is guilty of an offence under this Act. 1939, c. 49, s. 8.

Attempts, incitements, etc.

9. Every person who attempts to commit any offence under this Act, or solicits or incites or endeavours to persuade another person to commit an offence, or aids or abets and does any act preparatory to the commission of an offence under this Act, is guilty of an offence under this Act and is liable to the same punishment, and to be proceeded against in the same manner, as if he had committed the offence. 1939, c. 49, s. 9.

Power to arrest without warrant.

10. Every person who is found committing an offence under this Act, or who is reasonably suspected of having committed, or having attempted to commit, or being about to commit, such an offence, may be arrested without a warrant and detained by any constable or police officer. 1939, c. 49, s. 10.

Search warrants.

11. (1) If a justice of the peace is satisfied by information on oath that there is reasonable ground for suspecting that an offence under this Act has been or is about to be committed, he may grant a search warrant authorizing any constable named therein, to enter at any time any premises or place named in the warrant, if necessary by force, and to search the premises or place and every person found therein, and to seize any sketch, plan, model, article, note or document, or anything that is evidence of an offence under this Act having been or being about to be committed, that he may find on the premises or place or on any such person, and with regard to or in connection with which he has reasonable ground for suspecting that an offence under this Act has been or is about to be committed.

In case of great emergency.

(2) Where it appears to an officer of the Royal Canadian Mounted Police not below the rank of Superintendent that the case is one of great emergency and that in the interest of the State immediate action is necessary, he may by a written order under his hand give to any constable the like authority as may be given by the warrant of a justice under this section. 1939, c. 49, s. 11.

Prosecution only with consent of Attorney General.

12. A prosecution for an offence under this Act shall not be instituted except by or with the consent of the Attorney General; except that a person charged with such an offence may be arrested, or a warrant for his arrest may be issued and executed, and any such person may be remanded in custody or on bail, notwithstanding that the consent of the Attorney General to the institution of a prosecution for the offence has not been obtained, but no further or other proceedings shall be taken until that consent has been obtained. 1939, c. 49, s. 12.

Offences committed outside Canada triable in Canada.

13. An Act, omission or thing that would, by reason of this Act, be punishable as an offence if committed in Canada, is, if committed outside Canada, an offence against this Act, triable and punishable in Canada, in the following cases:

   (a) where the offender at the time of the commission was a Canadian citizen within the meaning of the *Canadian Citizenship Act;* or

   (b) where any code word, pass word, sketch, plan, model, article, note, document, information or other thing whatsoever in respect of which an offender is charged was obtained by him, or depends upon information that he obtained, while owing allegiance to Her Majesty. 1950, c. 46, s. 2.

14. (1) For the purposes of the trial of a person for an offence under this Act, the offence shall be deemed to have been committed either at the place in

which the same actually was committed, or at any place in Canada in which the offender may be found.

(2) In addition and without prejudice to any powers that a court may possess to order the exclusion of the public from any proceedings if, in the course of proceedings before a court against any person for an offence under this Act or the proceedings on appeal, application is made by the prosecution, on the ground that the publication of any evidence to be given or of any statement to be made in the course of the proceedings would be prejudicial to the interest of the State, that all or any portion of the public shall be excluded during any part of the hearing, the court may make an order to that effect, but the passing of sentence shall in any case take place in public.

(3) Where the person guilty of an offence under this Act is a company or corporation, every director and officer of the company or corporation is guilty of the like offence unless he proves that the act or omission constituting the offence took place without his knowledge or consent. 1939, c. 49, s. 13.

15. (1) Where no specific penalty is provided in this Act, any person who is guilty of an offence under this Act shall be deemed to be guilty of an indictable offence and is, on conviction, punishable by imprisonment for a term not exceeding fourteen years; but such person may, at the election of the Attorney General, be prosecuted summarily in the manner provided by the provisions of the *Criminal Code* relating to summary convictions, and, if so prosecuted, is punishable by fine not exceeding five hundred dollars, or by imprisonment not exceeding twelve months, or by both fine and imprisonment.

(2) Any person charged with or convicted for an offence under this Act shall, for the purposes of the *Identification of Criminals Act,* be deemed to be charged with or convicted of an indictable offence notwithstanding that such person is prosecuted summarily in the manner provided by the provisions of the *Criminal Code* relating to summary convictions. 1950, c. 46, s. 3.

Where offence deemed to have been committed. Public may be excluded from trial.

Where guilty person a company or corporation. Penalties.

General. Indictable offence.

Summary conviction. Application of the Identification of Criminals Act.

APPENDIX 12B

Public Law 93-579
93rd Congress, S. 3418
December 31, 1974

AN ACT

To amend title 5, United States Code, by adding a section 552a to safeguard individual privacy from the misuse of Federal records, to provide that individuals be granted access to records concerning them which are maintained by Federal agencies, to establish a Privacy Protection Study Commission, and for other purposes.

Be it enacted by the Senate and House of Representatives of the United States of America in Congress assembled, That this Act may be cited as the "Privacy Act of 1974."

SEC. 2. (a) The Congress finds that—

(1) the privacy of an individual is directly affected by the collection, maintenance, use, and dissemination of personal information by Federal agencies;

(2) the increasing use of computers and sophisticated information technology, while essential to the efficient operations of the Government, has

Privacy Act of 1974.
5 USC 552a note.
Congressional findings.
5 USC 552a note.

greatly magnified the harm to individual privacy that can occur from any collection, maintenance, use, or dissemination of personal information;

(3) the opportunities for an individual to secure employment, insurance, and credit, and his right to due process, and other legal protections are endangered by the misuse of certain information systems;

(4) the right to privacy is a personal and fundamental right protected by the Constitution of the United States; and

(5) in order to protect the privacy of individuals identified in information systems maintained by Federal agencies, it is necessary and proper for the Congress to regulate the collection, maintenance, use, and dissemination of information by such agencies.

Statement of purpose.

(b) The purpose of this Act is to provide certain safeguards for an individual against an invasion of personal privacy by requiring Federal agencies, except as otherwise provided by law, to—

(1) permit an individual to determine what records pertaining to him are collected, maintained, used, or disseminated by such agencies;

(2) permit an individual to prevent records pertaining to him obtained by such agencies for a particular purpose from being used or made available for another purpose without his consent;

(3) permit an individual to gain access to information pertaining to him in Federal agency records, to have a copy made of all or any portion thereof, and to correct or amend such records;

(4) collect, maintain, use, or disseminate any record of identifiable personal information in a manner that assures that such action is for a necessary and lawful purpose, that the information is current and accurate for its intended use, and that adequate safeguards are provided to prevent misuse of such information;

(5) permit exemptions from the requirements with respect to records provided in this Act only in those cases where there is an important public policy need for such exemption as has been determined by specific statutory authority; and

(6) be subject to civil suit for any damages which occur as a result of willful or intentional action which violates any individual's rights under this Act.

SEC. 3. Title 5, United States Code, is amended by adding after section 552 the following new section:

"§552a. Records maintained on individuals

"(a) DEFINITIONS.—For purposes of this section—

"(1) the term 'agency' means agency as defined in section 552(e) of this title;

"(2) the term 'individual' means a citizen of the United States or an alien lawfully admitted for permanent residence;

"(3) the term 'maintain' includes maintain, collect, use, or disseminate;

"(4) the term 'record' means any item, collection, or grouping of information about an individual that is maintained by an agency, including, but not limited to, his education, financial transactions, medical history, and criminal or employment history and that contains his name, or the identifying number, symbol, or other identifying particular assigned to the individual, such as a finger or voice print or a photograph;

"(5) the term 'system of records' means a group of any records under the control of any agency from which information is retrieved by the name of the individual or by some identifying number, symbol, or other identifying particular assigned to the individual;

"(6) the term 'statistical record' means a record in a system of records maintained for statistical research or reporting purposes only and not used in whole or in part in making any determination about an identifiable individual, except as provided by section 8 of title 13; and

8 USC 8.

"(7) the term 'routine use' means, with respect to the disclosure of a record, the use of such record for a purpose which is compatible with the purpose for which it was collected.

"(b) CONDITIONS OF DISCLOSURE.—No agency shall disclose any record which is contained in a system of records by any means of communication to any person, or to another agency, except pursuant to a written request by, or with the prior written consent of, the individual to whom the record pertains, unless disclosure of the record would be—

"(1) to those officers and employees of the agency which maintains the record who have a need for the record in the performance of their duties;

"(2) required under section 552 of this title;

"(3) for a routine use as defined in subsection (a) (7) of this section and described under subsection (e) (4) (D) of this section;

"(4) to the Bureau of the Census for purposes of planning or carrying out a census or survey or related activity pursuant to the provisions of title 13;

"(5) to a recipient who has provided the agency with advance adequate written assurance that the record will be used solely as a statistical research or reporting record, and the record is to be transferred in a form that is not individually identifiable;

"(6) to the National Archives of the United States as a record which has sufficient historical or other value to warrant its continued preservation by the United States Government, or for evaluation by the Administrator of General Services or his designee to determine whether the record has such value;

"(7) to another agency or to an instrumentality of any governmental jurisdiction within or under the control of the United States for a civil or criminal law enforcement activity if the activity is authorized by law, and if the head of the agency or instrumentality has made a written request to the agency which maintains the record specifying the particular portion desired and the law enforcement activity for which the record is sought;

"(8) to a person pursuant to a showing of compelling circumstances affecting the health or safety of an individual if upon such disclosure notification is transmitted to the last known address of such individual;

88 STAT. 1898

"(9) to either House of Congress, or, to the extent of matter within its jurisdiction, any committee or subcommittee thereof, any joint committee of Congress or subcommittee of any such joint committee;

"(10) to the Comptroller General, or any of his authorized representatives, in the course of the performance of the duties of the General Accounting Office; or

"(11) pursuant to the order of a court of competent jurisdiction.

"(c) ACCOUNTING OF CERTAIN DISCLOSURES.—Each agency, with respect to each system of records under its control, shall—

"(1) except for disclosures made under subsections (b) (1) or (b) (2) of this section, keep an accurate accounting of—

"(A) the date, nature, and purpose of each disclosure of a record to any person or to another agency made under subsection (b) of this section; and

"(B) the name and address of the person or agency to whom the disclosure is made;

"(2) retain the accounting made under paragraph (1) of this subsection for at least five years or the life of the record, whichever is longer, after the disclosure for which the accounting is made;

"(3) except for disclosures made under subsection (b) (7) of this section, make the accounting made under paragraph (1) of this subsection available to the individual named in the record at his request; and

"(4) inform any person or other agency about any correction or notation of dispute made by the agency in accordance with subsection (d) of this section of any record that has been disclosed to the person or agency if an accounting of the disclosure was made.

"(d) ACCESS TO RECORDS.—Each agency that maintains a system of records shall—

Personal review.

"(1) upon request by any individual to gain access to his record or to any information pertaining to him which is contained in the system, permit him and upon his request, a person of his own choosing to accompany him, to review the record and have a copy made of all or any portion thereof in a form comprehensible to him, except that the agency may require the individual to furnish a written statement authorizing discussion of that individual's record in the accompanying person's presence;

Amendment request.

"(2) permit the individual to request amendment of a record pertaining to him and—

"(A) not later than 10 days (excluding Saturdays, Sundays, and legal public holidays) after the date of receipt of such request, acknowledge in writing such receipt; and

"(B) promptly, either—

"(i) make any correction of any portion thereof which the individual believes is not accurate, relevant, timely, or complete; or

"(ii) inform the individual of its refusal to amend the record in accordance with his request, the reason for the refusal, the procedures established by the agency for the individual to request a review of that refusal by the head of the agency or an officer designated by the head of the agency, and the name and business address of that official;

Review.

"(3) permit the individual who disagrees with the refusal of the agency to amend his record to request a review of such refusal, and not later than 30 days (excluding Saturdays, Sundays, and legal public holidays) from the date on which the individual requests such review, complete such review and make a final determination unless, for good cause shown, the head of the agency extends such 30-day period; and if, after his review, the reviewing official also refuses to amend the record in accordance with the request, permit the individual to file with the agency a concise statement setting forth the reasons for his disagreement with the refusal of the agency, and notify the individual of the provisions for judicial review of the reviewing official's determination under subsection (g) (1) (A) of this section;

Notation of dispute.

"(4) in any disclosure, containing information about which the individual has filed a statement of disagreement, occurring after the filing of the statement under paragraph (3) of this subsection, clearly note any portion of the record which is disputed and provide copies of the statement and, if the agency deems it appropriate, copies of a concise statement of the reasons of the agency for not making the amendments requested, to persons or other agencies to whom the disputed record has been disclosed; and

"(5) nothing in this section shall allow an individual access to any information compiled in reasonable anticipation of a civil action or proceeding.

"(e) AGENCY REQUIREMENTS.—Each agency that maintains a system of records shall—

"(1) maintain in its records only such information about an individual as is relevant and necessary to accomplish a purpose of the agency required to be accomplished by statute or by executive order of the President;

"(2) collect information to the greatest extent practicable directly from the subject individual when the information may result in adverse determinations about an individual's rights, benefits, and privileges under Federal programs;

"(3) inform each individual whom it asks to supply information, on the form which it uses to collect the information or on a separate form that can be retained by the individual—

"(A) the authority (whether granted by statute, or by executive order of the President) which authorizes the solicitation of the information and whether disclosure of such information is mandatory or voluntary;

"(B) the principal purpose or purposes for which the information is intended to be used;

"(C) the routine uses which may be made of the information, as published pursuant to paragraph (4) (D) of this subsection; and

"(D) the effects on him, if any, of not providing all or any part of the requested information;

"(4) subject to the provisions of paragraph (11) of this subsection, publish in the Federal Register at least annually a notice of the existence and character of the system of records, which notice shall include—

"(A) the name and location of the system;

"(B) the categories of individuals on whom records are maintained in the system;

"(C) the categories of records maintained in the system;

"(D) each routine use of the records contained in the system, including the categories of users and the purpose of such use;

"(E) the policies and practices of the agency regarding storage, retrievability, access controls, retention, and disposal of the records;

"(F) the title and business address of the agency official who is responsible for the system of records;

"(G) the agency procedures whereby an individual can be notified at his request if the system of records contains a record pertaining to him;

"(H) the agency procedures whereby an individual can be notified at his request how he can gain access to any record pertaining to him contained in the system of records, and how he can contest its content; and

"(I) the categories of sources of records in the system;

"(5) maintain all records which are used by the agency in making any determination about any individual with such accuracy, relevance, timeliness, and completeness as is reasonably necessary to assure fairness to the individual in the determination;

"(6) prior to disseminating any record about an individual to any person other than an agency, unless the dissemination is made pursuant to subsection (b) (2) of this section, make reasonable efforts to assure that such records are accurate, complete, timely, and relevant for agency purposes;

"(7) maintain no record describing how any individual exercises rights guaranteed by the First Amendment unless expressly authorized by statute or by the individual about whom the record is maintained or unless pertinent to and within the scope of an authorized law enforcement activity;

"(8) make reasonable efforts to serve notice on an individual when any record on such individual is made available to any person under compulsory legal process when such process becomes a matter of public record;

"(9) establish rules of conduct for persons involved in the design, development, operation, or maintenance of any system of records, or in maintaining any record, and instruct each such person with respect to such rules and the requirements of this section, including any other rules and procedures adopted pursuant to this section and the penalties for noncompliance;

Confidentiality of records.

"(10) establish appropriate administrative, technical, and physical safeguards to insure the security and confidentiality of records and to protect against any anticipated threats or hazards to their security or integrity which could result in substantial harm, embarrassment, inconvenience, or unfairness to any individual on whom information is maintained; and

Publication in Federal Register.

"(11) at least 30 days prior to publication of information under paragraph (4) (D) of this subsection, publish in the Federal Register notice of any new use or intended use of the information in the system, and provide an opportunity for interested persons to submit written data, views, or arguments to the agency.

"(f) AGENCY RULES.—In order to carry out the provisions of this section, each agency that maintains a system of records shall promulgate rules, in accordance with the requirements (including general notice) of section 553 of this title, which shall—

5 USC 553.

"(1) establish procedures whereby an individual can be notified in response to his request if any system of records named by the individual contains a record pertaining to him;

88 STAT. 1901

"(2) define reasonable times, places, and requirements for identifying an individual who requests his record or information pertaining to him before the agency shall make the record or information available to the individual;

"(3) establish procedures for the disclosure to an individual upon his request of his record or information pertaining to him, including special procedure, if deemed necessary, for the disclosure to an individual of medical records, including psychological records, pertaining to him;

"(4) establish procedures for reviewing a request from an individual concerning the amendment of any record or information pertaining to the individual, for making a determination on the request, for an appeal within the agency of an initial adverse agency determination, and for whatever additional means may be necessary for each individual to be able to exercise fully his rights under this section; and

Fees.

"(5) establish fees to be charged, if any, to any individual for making copies of his record, excluding the cost of any search for and review of the record.

Publication in Federal Register.

The Office of the Federal Register shall annually compile and publish the rules promulgated under this subsection and agency notices published under subsection (e) (4) of this section in a form available to the public at low cost.

"(g) (1) CIVIL REMEDIES.—Whenever any agency

"(A) makes a determination under subsection (d) (3) of this section not to amend an individual's record in accordance with his request, or fails to make such review in conformity with that subsection;

"(B) refuses to comply with an individual request under subsection (d) (1) of this section;

"(C) fails to maintain any record concerning any individual with such accuracy, relevance, timeliness, and completeness as is necessary to assure fairness in any determination relating to the qualifications, character, rights, or opportunities of, or benefits to the individual that may be made on the basis of such record, and consequently a determination is made which is adverse to the individual; or

"(D) fails to comply with any other provision of this section, or any rule promulgated thereunder, in such a way as to have an adverse effect on an individual, the individual may bring a civil action against the agency, and the district courts of the United States shall have jurisdiction in the matters under the provisions of this subsection.

Jurisdiction.

"(2) (A) In any suit brought under the provisions of subsection (g) (1) (A) of this section, the court may order the agency to amend the individual's record in accordance with his request or in such other way as the court may direct. In such a case the court shall determine the matter de novo.

Amendment of record.

"(B) The court may assess against the United States reasonable attorney fees and other litigation costs reasonably incurred in any case under this paragraph in which the complainant has substantially prevailed.

"(3) (A) In any suit brought under the provisions of subsection (g) (1) (B) of this section, the court may enjoin the agency from withholding the records and order the production to the complainant of any agency records improperly withheld from him. In such a case the court shall determine the matter de novo, and may examine the contents of any agency records in camera to determine whether the records or any portion thereof may be withheld under any of the exemptions set forth in subsection (k) of this section, and the burden is on the agency to sustain its action.

Injunction.

"(B) The court may assess against the United States reasonable attorney fees and other litigation costs reasonably incurred in any case under this paragraph in which the complainant has substantially prevailed.

"(4) In any suit brought under the provisions of subsection (g) (1) (C) or (D) of this section in which the court determines that the agency acted in a manner which was intentional or willful, the United States shall be liable to the individual in an amount equal to the sum of—

Damages.

"(A) actual damages sustained by the individual as a result of the refusal or failure, but in no case shall a person entitled to recovery receive less than the sum of $1,000; and

"(B) the cost of the action together with reasonable attorney fees as determined by the court.

"(5) An action to enforce any liability created under this section may be brought in the district court of the United States in the district in which the complainant resides, or has his principal place of business, or in which the agency records are situated, or in the District of Columbia, without regard to the amount in controversy, within two years from the date on which the cause of action arises, except that where an agency has materially and willfully misrepresented any information required under this section to be disclosed to an individual and the information so misrepresented is material to establishment of the liability of the agency to the individual under this section, the action may be brought at any time within two years after discovery by the individual of the misrepresentation. Nothing in this section shall be construed to authorize any civil action by reason of any injury sustained as the result of a disclosure of a record prior to the effective date of this section.

"(h) RIGHTS OF LEGAL GUARDIANS.—For the purposes of this section, the parent of any minor, or the legal guardian of any individual who has been declared to be incompetent due to physical or mental incapacity or age by a court of competent jurisdiction, may act on behalf of the individual.

"(i) (1) CRIMINAL PENALTIES.—Any officer or employee of an agency, who by virtue of his employment or official position, has possession of, or access to, agency records which contain individually identifiable information the disclosure of which is prohibited by this section or by rules or regulations established thereunder, and who knowing that disclosure of the specific material is so prohibited, willfully discloses the material in any manner to any person or agency not entitled to receive it, shall be guilty of a misdemeanor and fined not more than $5,000.

"(2) Any officer or employee of any agency who willfully maintains a system of records without meeting the notice requirements of subsection (e) (4) of this section shall be guilty of a misdemeanor and fined not more than $5,000.

"(3) Any person who knowingly and willfully requests or obtains any record concerning an individual from an agency under false pretenses shall be guilty of a misdemeanor and fined not more than $5,000.

"(j) GENERAL EXEMPTIONS.—The head of any agency may promulgate rules, in accordance with the requirements (including general notice) of sections 553 (b) (1), (2), and (3), (c), and (e) of this title, to exempt any system of records within the agency from any part of this section except subsections (b), (c) (1) and (2), (e) (4) (A) through (F), (e) (6), (7), (9), (10), and (11), and (i) if the system of records is—

"(1) maintained by the Central Intelligence Agency; or

"(2) maintained by an agency or component thereof which performs as its principal function any activity pertaining to the enforcement of criminal laws, including police efforts to prevent, control, or reduce crime or to apprehend criminals, and the activities of prosecutors, courts, correctional, probation, pardon, or parole authorities, and which consists of (A) information compiled for the purpose of identifying individual criminal offenders and alleged offenders and consisting only of identifying data and notations of arrests, the nature and disposition of criminal charges, sentencing, confinement, release, and parole and probation status; (B) information compiled for the purpose of a criminal investigation, including reports of informants and investigators, and associated with an identifiable individual; or (C) reports identifiable to an individual compiled at any stage of the process of enforcement of the criminal laws from arrest or indictment through release from supervision.

At the time rules are adopted under this subsection, the agency shall include in the statement required under section 553 (c) of this title, the reasons why the system of records is to be exempted from a provision of this section.

"(k) SPECIFIC EXEMPTIONS.—The head of any agency may promulgate rules, in accordance with the requirements (including general notice) of sections 553 (b) (1), (2), and (3), (c), and (e) of this title, to exempt any system of records within the agency from subsections (c) (3), (d), (e) (1), (e) (4) (G), (H), and (I) and (f) of this section if the system of records is—

"(1) subject to the provisions of section 552 (b) (1) of this title;

"(2) investigatory material compiled for law enforcement purposes, other than material within the scope of subsection (j) (2) of this section: *Provided, however,* That if any individual is denied any right, privilege, or benefit that he would otherwise be entitled by Federal law, or for which he would otherwise be eligible, as a result of the maintenance of such material, such material shall be provided to such individual, except to the extent that the disclosure of such material would reveal the identity of a source who furnished information to the Government under an express promise that the identity of the source would be held in confidence, or, prior to the effective date of this section, under an implied promise that the identity of the source would be held in confidence;

"(3) maintained in connection with providing protective services to the President of the United States or other individuals pursuant to section 3056 of title 18;

"(4) required by statute to be maintained and used solely as statistical records;

"(5) investigatory material compiled solely for the purpose of determining suitability, eligibility, or qualifications for Federal civilian employment, military service, Federal contracts, or access to classified information, but only to the extent that the disclosure of such material would reveal the identity of a source who furnished information to the Government under an express promise that the identity of the source would be held in confidence, or, prior to the effective date of this section, under an implied promise that the identity of the source would be held in confidence;

"(6) testing or examination material used solely to determine individual qualifications for appointment or promotion in the Federal service the disclosure of which would compromise the objectivity or fairness of the testing or examination process; or

"(7) evaluation material used to determine potential for promotion in the armed services, but only to the extent that the disclosure of such material would reveal the identity of a source who furnished information to the Government under an express promise that the identity of the source would be held in confidence, or, prior to the effective date of this section, under an implied promise that the identity of the source would be held in confidence.

At the time rules are adopted under this subsection, the agency shall include in the statement required under section 553(c) of this title, the reasons why the system of records is to be exempted from a provision of this section.

"(l) (1) ARCHIVAL RECORDS.—Each agency record which is accepted by the Administrator of General Services for storage, processing, and servicing in accordance with section 3103 of title 44 shall, for the purposes of this section, be considered to be maintained by the agency which deposited the record and shall be subject to the provisions of this section. The Administrator of General Services shall not disclose the record except to the agency which maintains the record, or under rules established by that agency which are not inconsistent with the provisions of this section.

"(2) Each agency record pertaining to an identifiable individual which was transferred to the National Archives of the United States as a record which has sufficient historical or other value to warrant its continued preservation by the United State Government, prior to the effective date of this section, shall, for the purposes of this section, be considered to be maintained by the National Archives and shall not be subject to the provisions of this section, except that a statement generally describing such records (modeled after the requirements relating to records subject to subsections (e) (4) (A) through (G) of this section) shall be published in the Federal Register.

"(3) Each agency record pertaining to an identifiable individual which is transferred to the National Archives of the United States as a record which has sufficient historical or other value to warrant its continued preservation by the United States Government, on or after the effective date of this section, shall, for the purposes of this section, be considered to be maintained by the National Archives and shall be exempt from the requirements of this section except subsections (e) (4) (A) through (G) and (e) (9) of this section.

"(m) GOVERNMENT CONTRACTORS.—When an agency provides by a contract for the operation by or on behalf of the agency of a system of records to accomplish an agency function, the agency shall, consistent with its authority, cause the requirements of this section to be applied to such system. For purposes of subsection (i) of this section any such contractor and any employee of such contractor, if such contract is agreed to on or after the effective date of this section, shall be considered to be an employee of an agency.

"(n) MAILING LISTS.—An individual's name and address may not be sold or rented by an agency unless such action is specifically authorized by law. This provision shall not be construed to require the withholding of names and addresses otherwise permitted to be made public.

"(o) REPORT ON NEW SYSTEMS.—Each agency shall provide adequate advance notice to Congress and the Office of Management and Budget of any proposal to establish or alter any system of records in order to permit an evaluation of the probable or potential effect of such proposal on the privacy and other personal or property rights of individuals or the disclosure of information relating to such individuals, and its effect on the preservation of the constitutional principles of federalism and separation of powers.

"(p) ANNUAL REPORT.—The President shall submit to the Speaker of the House and the President of the Senate, by June 30 of each calendar year, a consolidated report, separately listing for each Federal agency the number of records contained in any system of records which were exempted from the application of this section under the provisions of subsections (j) and (k) of this section during the preceding calendar year, and the reasons for the exemptions, and such other information as indicates efforts to administer fully this section.

"(q) EFFECT OF OTHER LAWS.—No agency shall rely on any exemption contained in section 552 of this title to withhold from an individual any record which is otherwise accessible to such individual under the provisions of this section."

SEC. 4. The chapter analysis of chapter 5 of title 5, United States Code, is amended by inserting:
"552a. Records about individuals."
immediately below:
"552. Public information; agency rules, opinions, orders, and proceedings."

SEC. 5. (a) (1) There is established a Privacy Protection Study Commission (hereinafter referred to as the "Commission") which shall be composed of seven members as follows:

(A) three appointed by the President of the United States,
(B) two appointed by the President of the Senate, and
(C) two appointed by the Speaker of the House of Representatives.

Members of the Commision shall be chosen from among persons who, by reason of their knowledge and expertise in any of the following areas—civil rights and liberties, law, social sciences, computer technology, business, records management, and State and local government—are well qualified for service on the Commission.

(2) The members of the Commission shall elect a Chairman from among themselves.

(3) Any vacancy in the membership of the Commission, as long as there are four members in office, shall not impair the power of the Commission but shall be filled in the same manner in which the original appointment was made.

(4) A quorum of the Commission shall consist of a majority of the members, except that the Commission may establish a lower number as a quorum for the purpose of taking testimony. The Commission is authorized to establish such committees and delegate such authority to them as may be necessary to carry out its functions. Each member of the Commission, including the Chairman, shall have equal responsibility and authority in all decisions and actions of the Commission, shall have full access to all information necessary to the performance of their functions, and shall have one vote. Action of the Commission shall be determined by a majority vote of the members present. The Chairman (or a member designated by the Chairman to be acting Chairman) shall be the official spokesman of the Commission in its relations with the Congress, Government agencies, other persons, and the public, and, on behalf of the Commission, shall see to the faithful execution of the administrative policies and decisions of the Commission, and shall report thereon to the Commission from time to time or as the Commission may direct.

(5) (A) Whenever the Commission submits any budget estimate or request

to the President or the Office of Management and Budget, it shall concurrently transmit a copy of that request to Congress.

Budget requests.

(B) Whenever the Commission submits any legislative recommendations, or testimony, or comments on legislation to the President or Office of Management and Budget, it shall concurrently transmit a copy thereof to the Congress. No officer or agency of the United States shall have any authority to require the Commission to submit its legislative recommendations, or testimony, or comments on legislation, to any officer or agency of the United States for approval, comments, or review, prior to the submission of such recommendations, testimony, or comments to the Congress.

Legislative recommendations.

(b) The Commission shall—

Study.

(1) make a study of the data banks, automated data processing programs, and information systems of governmental, regional, and private organizations, in order to determine the standards and procedures in force for the protection of personal information; and

(2) recommend to the President and the Congress the extent, if any, to which the requirements and principles of section 552a of title 5, United States Code, should be applied to the information practices of those organizations by legislation, administrative action, or voluntary adoption of such requirements and principles, and report on such other legislative recommendations as it may determine to be necessary to protect the privacy of individuals while meeting the legitimate needs of government and society for information.

Ante, p. 1897.

(c) (1) In the course of conducting the study required under subsection (b) (1) of this section, and in its reports thereon, the Commission may research, examine, and analyze—

(A) interstate transfer of information about individuals that is undertaken through manual files or by computer or other electronic or telecommunications means;

(B) data banks and information programs and systems the operation of which significantly or substantially affect the enjoyment of the privacy and other personal and property rights of individuals;

(C) the use of social security numbers, license plate numbers, universal identifiers, and other symbols to identify individuals in data banks and to gain access to, integrate, or centralize information systems and files; and

(D) the matching and analysis of statistical data, such as Federal census data, with other sources of personal data, such as automobile registries and telephone directories, in order to reconstruct individual responses to statistical questionnaires for commercial or other purposes, in a way which results in a violation of the implied or explicitly recognized confidentiality of such information.

(2) (A) The Commission may include in its examination personal information activities in the following areas: medical; insurance; education; employment and personnel; credit, banking and financial institutions; credit bureaus; the commercial reporting industry; cable television and other telecommunications media; travel, hotel and entertainment reservations; and electronic clerk processing.

(B) The Commission shall include in its examination a study of—

(i) whether a person engaged in interstate commerce who maintains a mailing list should be required to remove an individual's name and address from such list upon request of that individual;

(ii) whether the Internal Revenue Service should be prohibited from transfering individually identifiable data to other agencies and to agencies of State governments;

(iii) whether the Federal Government should be liable for general damages incurred by an individual as the result of a willful or intentional violation of the provisions of sections 552a (g) (1) (C) or (D) of title 5, United States Code; and

Ante, p. 1897.

(iv) whether and how the standards for security and confidentiality of records required under sections 552a (e) (10) of such title should be applied when a record is disclosed to a person other than an agency.

Religious organizations, exception.

(C) The Commission may study such other personal information activities necessary to carry out the congressional policy embodied in this Act, except that the Commission shall not investigate information systems maintained by religious organizations.

Guidelines for study.

(3) In conducting such study, the Commission shall—

(A) determine what laws, Executive orders, regulations, directives, and judicial decisions govern the activities under study and the extent to which they are consistent with the rights of privacy, due process of law, and other guarantees in the Constitution;

(B) determine to what extent government and private information systems affect Federal-State relations or the principle of separation of powers;

(C) examine the standards and criteria governing programs, policies, and practices relating to the collection, soliciting, processing, use, access, integration, dissemination, and transmission of personal information; and

(D) to the maximum extent practicable, collect and utilize findings, reports, studies, hearing transcripts, and recommendations of governmental, legislative and private bodies, institutions, organizations, and individuals which pertain to the problems under study by the Commission.

(d) In addition to its other functions the Commission may—

(1) request assistance of the heads of appropriate departments, agencies, and instrumentalities of the Federal Government, of State and local governments, and other persons in carrying out its functions under this Act;

(2) upon request, assist Federal agencies in complying with the requirements of section 552a of title 5, United States Code;

(3) determine what specific categories of information, the collection of which would violate an individual's right of privacy, should be prohibited by statute from collection by Federal agencies; and

(4) upon request, prepare model legislation for use by State and local governments in establishing procedures for handling, maintaining, and disseminating personal information at the State and local level and provide such technical assistance to State and local governments as they may require in the preparation and implementation of such legislation.

(e) (1) The Commission may, in carrying out its functions under this section, conduct such inspections, sit and act at such times and places, hold such hearings, take such testimony, require by subpena the attendance of such witnesses and the production of such books, records, papers, correspondence, and documents, administer such oaths, have such printing and binding done, and make such expenditures as the Commission deems advisable. A subpena shall be issued only upon an affirmative vote of a majority of all members of the Commission. Subpenas shall be issued under the signature of the Chairman or any member of the Commission designated by the Chairman and shall be served by any person designated by the Chairman or any such member. Any member of the Commission may administer oaths or affirmations to witnesses appearing before the Commission.

(2) (A) Each department, agency, and instrumentality of the executive branch of the Government is authorized to furnish to the Commission, upon request made by the Chairman, such information, data, reports and such other assistance as the Commission deems necessary to carry out its functions under

this section. Whenever the head of any such department, agency, or instrumentality submits a report pursuant to section 552a (o) of title 5, United States Code, a copy of such report shall be transmitted to the Commission.

(B) In carrying out its functions and exercising its powers under this section, the Commission may accept from any such department, agency, independent instrumentality, or other person any individually identifiable data if such data is necessary to carry out such powers and functions. In any case, in which the Commission accepts any such information, it shall assure that the information is used only for the purpose for which it is provided, and upon completion of that purpose such information shall be destroyed or returned to such department, agency, independent instrumentality, or person from which it is obtained, as appropriate.

(3) The Commission shall have the power to—

(A) appoint and fix the compensation of an executive director, and such additional staff personnel as may be necessary, without regard to the provisions of title 5, United States Code, governing appointments in the competitive service, and without regard to chapter 51 and subchapter III of chapter 53 of such title relating to classification and General Schedule pay rates, but at rates not in excess of the maximum rate for GS-18 of the General Schedule under section 5332 of such title; and

(B) procure temporary and intermittent services to the same extent as is authorized by section 3109 of title 5, United States Code.

The Commission may delegate any of its functions to such personnel of the Commission as the Commission may designate and may authorize such successive redelegations of such functions as it may deem desirable.

(4) The Commission is authorized—

(A) to adopt, amend, and repeal rules and regulations governing the manner of its operations, organization, and personnel;

(B) to enter into contracts or other arrangements or modifications thereof, with any government, any department, agency, or independent instrumentality of the United States, or with any person, firm, association, or corporation, and such contracts or other arrangements, or modifications thereof, may be entered into without legal consideration, without performance or other bonds, and without regard to section 3709 of the Revised Statutes, as amended (41 U.S.C.5);

(C) to make advance, progress, and other payments which the Commission deems necessary under this Act without regard to the provisions of section 3648 of the Revised Statutes, as amended (31 U.S.C. 529); and

(D) to take such other action as may be necessary to carry out its functions under this section.

(f) (1) Each [the] member of the Commission who is an officer or employee of the United States shall serve without additional compensation, but shall continue to receive the salary of his regular position when engaged in the performance of the duties vested in the Commission.

(2) A member of the Commission other than one to whom paragraph (1) applies shall receive per diem at the maximum daily rate for GS-18 of the General Schedule when engaged in the actual performance of the duties vested in the Commission.

(3) All members of the Commission shall be reimbursed for travel, subsistence, and other necessary expenses incurred by them in the performance of the duties vested in the Commission.

(g) The Commission shall, from time to time, and in an annual report, report to the President and the Congress on its activities in carrying out the provisions of this section. The Commission shall make a final report to the President and to the Congress on its findings pursuant to the study required to be made

Security Administration

Report to President and Congress.

under subsection (b) (1) of this section not later than two years from the date on which all of the members of the Commission are appointed. The Commission shall cease to exist thirty days after the date on which its final report is submitted to the President and the Congress.

Penalties.

(h) (1) Any member, officer, or employee of the Commission, who by virtue of his employment or official position, has possession of, or access to, agency records which contain individually identifiable information the disclosure of which is prohibited by this section, and who knowing that disclosure of the specific material is so prohibited, willfully discloses the material in any manner to any person or agency not entitled to receive it, shall be guilty of a misdemeanor and fined not more than $5,000.

(2) Any person who knowingly and willfully requests or obtains any record concerning an individual from the Commission under false pretenses shall be guilty of a misdemeanor and fined not more than $5,000.

USC 552a note.

SEC. 6. The Office of Management and Budget shall—

Ante, p. 1897.

(1) develop guidelines and regulations for the use of agencies in implementing the provisions of section 552a of title 5, United States Code, as added by section 3 of this Act; and

(2) provide continuing assistance to and oversight of the implementation of the provisions of such section by agencies.

USC 552a note.

SEC. 7. (a) (1) It shall be unlawful for any Federal, State or local government agency to deny to any individual any right, benefit, or privilege provided by law because of such individual's refusal to disclose his social security account number.

(2) the provisions of paragraph (1) of this subsection shall not apply with respect to—

(A) any disclosure which is required by Federal statute, or

(B) the disclosure of a social security number to any Federal, State, or local agency maintaining a system of records in existence and operating before January 1, 1975, if such disclosure was required under statute or regulation adopted prior to such date to verify the identity of an individual.

(b) Any Federal, State, or local government agency which requests an individual to disclose his social security account number shall inform that individual whether that disclosure is mandatory or voluntary, by what statutory or other authority such number is solicited, and what uses will be made of it.

88 STAT. 1910
Effective date 5 USC 552a note.

SEC. 8. The provisions of this Act shall be effective on and after the date of enactment, except that the amendments made by sections 3 and 4 shall become effective 270 days following the day on which this Act is enacted.

Appropriation. 5 USC 552a note.

SEC. 9. There is authorized to be appropriated to carry out the provisions of section 5 of this Act for fiscal years 1975, 1976, and 1977 the sum of $1,500,000, except that not more than $750,000 may be expended during any such fiscal year.

Approved December 31, 1974.

* * *

LEGISLATIVE HISTORY:

HOUSE REPORT No. 93-1416 accompanying H.R. 16373 (Comm. on Government Operations).
SENATE REPORT No. 93-1183 (Comm. on Government Operations).
CONGRESSIONAL RECORD, Vol. 120 (1974):
    Nov. 21, considered and passed Senate.
    Dec. 11, considered and passed House, amended, in lieu of H.R. 16373.

Dec. 17, Senate concurred in House amendment with amendments.
Dec. 18, House concurred in Senate amendments.
WEEKLY COMPILATION OF PRESIDENTIAL DOCUMENTS, Vol. 11, No. 1:
Jan. 1, Presidential statement.

APPENDIX 12C

DEPARTMENT OF DEFENSE INFORMATION SECURITY PROGRAM

INSPECTION CHECK LIST*

The following questions are designed to assist inspectors in determining the degree of compliance with the provisions of DoD Information Security Regulation 5200.1-R. They are also designed to assist DoD activities in educating and training personnel involved in day-to-day operations which relate to the Defense Information Security Program.

References shown in parentheses following questions are to pertinent paragraphs of the DoD Information Security Program Regulation 5200.1-R.†

	N/A	Yes	No

General Provisions

1. Does the Activity have on hand or on requisition DoD Regulation 5200.1-R and all changes thereto?
2. Are procedures established to assure that unnecessary classification and higher than necessary classification is avoided? (1-400)
3. Is official information and material afforded protection commensurate with the level of classification assigned? (1-402)
4. Have procedures been established for detecting, reporting and correcting classification abuses? (13-201d)
5. Is a listing of TOP SECRET, SECRET and CONFIDENTIAL original classification authorities maintained in a current status by written action of authorized officials? (1-601)
6. Are quarterly reports on changes to original classification authorities submitted as required? (1-601)
7. Have officials been designated at headquarters levels who are authorized to downgrade and declassify information under the classification jurisdiction of that organization or to resolve doubts or conflicts as to appropriate classifications? (1-603)

---

* Used with permission of the Office of the Assistant Secretary of Defense, Office of Information Security, Washington, D.C.
† November 1973 edition.

Classification

		N/A	Yes	No
8.	Are adequate records maintained to support classification actions of classifiers? (2-100)	....	....	....
9.	Are classification guides prepared for each classified plan, program or project? (2-102) (2-405)	....	....	....
	a. Do the guides isolate and identify classifiable items of information?		....	....
	b. Are levels of classification to be applied to identified items in information or material specified?		....	....
	c. Is a phased downgrading and declassification schedule included?		....	....
	d. Are provisions made for reviews to be conducted at least annually?		....	....
	e. Have guides been updated to conform to the provisions of DoD Regulation 5200.1-R?		....	....
10.	Is the classifier of each document or piece of material identified? (2-200)	....	....	....
11.	Are copies of each classification guide and changes thereto submitted to OASD (PA) and ODASD (SP) as required? (2-407)	....	....	....
12.	Are disagreements on classification, declassification or regrading problems referred to the next higher echelon if not resolved within 30 days? (2-503)	....	....	....
13.	Is classification in industrial operations based strictly on security classification guidance furnished by the Government? (2-900)	....	....	....
14.	Is a reevaluation of classification conducted on information subjected to compromise or possible compromise? (2-313)	....	....	....

Downgrading & Declassification

		N/A	Yes	No
15.	Has the original classification authority to the maximum extent possible, predetermined at the time of origination dates or events on which downgrading and declassification shall occur? (3-101)	....	....	....
16.	Are downgrading and declassification dates or events carried forward whenever classified information is incorporated in later documents or other material? (3-102)	....	....	....
17.	Where appropriate, is downgrading and declassification accomplished in accordance with the GDS? (3-200)	....	....	....
18.	Is former Group 3 Information or material automatically downgraded at 12 year intervals to CONFIDENTIAL but not automatically declassified? (3-201)	....	....	....
19.	Are all recipients of information notified when the information or material is redesignated as exempt? (3-202)	....	....	....
20.	Are exemptions made *only* by an original TOP SECRET classification authority? (3-300)	....	....	....
21.	On exempted material, has the original TOP SECRET classification authority specified in writing on the ma-			

	N/A	Yes	No

terial or by means of advance written policy the exemption category claimed and the date or event, unless impossible, for automatic declassification? (3-300)

22. Are procedures established for the mandatory classification review, upon request, of all exempted or excluded classified information or material which is 10 years old to include acknowledgement to the requester and notification of his right to appeal? (3-400) (3-403)

23. Have provisions been made to review information and material approaching 30 years of age to determine if there is a continued requirement for protection against unauthorized disclosure? (3-501)

24. Are periodic reviews conducted on all information and material classified after 1 June 1972 for the purpose of making such information or material publicly available after it is declassified? (3-702)

Marking

25. Are markings properly applied to all information and material? (4-200 thru 4-604)

26. Does each document show on its face its overall classification and whether it is subject to or exempt from the GDS? (4-200)

27. Is paragraph marking used when there are differences in the classifications of section, parts, paragraphs, etc.? (4-202)

28. Are files, folders or groups of documents conspicuously marked to assure their protection in accordance with the highest level of classification of material contained therein? (4-205)

29. Does the last paragraph of electrically transmitted messages show the appropriate downgrading and declassification instructions? (4-207)

30. Has former Group 4 material been subjected to the GDS and remarked as appropriate? (4-500)

31. Has action been taken to remark a document "Excluded from General Declassification Schedule," that was classified prior to June 1, 1972, and marked Group 1 or Group 2 or not marked at all when it is removed from file or storage for any purpose? (4-501)

32. Have additional warning notices been applied as appropriate? (4-600)

Safe Keeping & Storage

33. Is classified information properly guarded or stored in approved containers? (5-102)

34. Has a current physical survey been made of on-hand security equipment and classified records prior to the procurement of new security storage equipment? (5-103a)

	N/A	Yes	No

35. Have security containers other than those listed on the Federal Schedule, GSA, been procured without the granting of an exemption by Component Heads? (5-103b)
36. Have vaults or containers to be used for the storage of classified information or material been designated and a number or symbol affixed to each container? (5-104a and b3)
37. Are combinations to security containers changed at least annually? (5-104b1)
38. Are records of combinations assigned a security classification equal to the highest category of classified material authorized to be stored therein? (5-104b2)
39. Are combinations to security containers disseminated on a need-to-know basis? (5-104b4)
40. Do procedures provide for only cleared and trained persons to change and correct lockouts? (5-104b5)
41. Are procedures established controlling the removal of classified material from the physical confines of the facility? (5-200)
42. Are preliminary drafts, carbon sheets, work sheets, stencils, etc. protected according to their content and destroyed after they have served their purpose? (5-201b)
43. Are carbon, plastic typewriter ribbons and carbon papers used in the production of classified information destroyed after initial usage? (5-201c)
44. Is an adequate system of security checks at the close of the working day established? (5-202)
45. Have emergency destruction and evacuation plans been developed and tested? (5-204)
46. Is security practiced in telephone conversations on non-secure communications circuits? (5-205)
47. Are appropriate security precautions taken prior to convening a conference or other meetings wherein classified information is to be discussed or disclosed? (5-206)

Compromise of Classified Information

48. Are personnel aware of their responsibilities in the event of a compromise or suspected compromise? (6-102)
49. Have cases of espionage or deliberate compromise been reported in accordance with DoD Instruction 5200.22, DoD Directive 5210.50 and implementing issuance? (6-107)
50. Has a "knowledgeable" AWOL program been established? (6-108)

Access, Dissemination and Accountability

51. Is classified information disseminated on a need-to-know basis? (7-100)

Information Security 625

	N/A	Yes	No

52. Has a program been established for the periodic evaluation of clearances? (7-102)
53. Are provisions made and used to administratively withdrawn security clearances of persons for whom there is no foreseeable need for access to classified information or material in connection with the performance of their official duties? (7-104)
54. Have appropriate officials been designated to determine, prior to the release of classified information to persons outside the Executive Branch, the propriety of such action in the interests of national security, the assurance of the individual's trustworthiness and need-to-know? (7-106)
55. Have procedures been established for the dissemination of classified information or material originated or received by the organization? (7-200)
56. Is TOP SECRET information reproduced without the consent of the originating activity or higher authority? (7-209)
57. Have officials been designated to approve the reproduction of TOP SECRET and SECRET information and material? (7-209b)
58. Has specific equipment been designated for the reproduction of classified material and rules and warning notices posted thereon? (7-209c)
59. Have TOP SECRET control officers and alternates been designated? (7-300a)
60. Are TOP SECRET accountability registers and access rosters maintained? (7-300b)
61. Are TOP SECRET documents physically sighted or accounted for at least annually? (7-300c)
62. Is a continuous chain of receipts established for all TOP SECRET documents? (7-300e)
63. Are administrative procedures established for controlling SECRET and CONFIDENTIAL material to include origination, receipt, distribution, transfer and destruction? (7-301, 302)
64. Have screening points been established to insure incoming classified material is properly controlled and access is limited to cleared personnel? (7-303)
65. Are working papers marked, protected and destroyed in accordance with the classification level of the material? (7-304)
66. Are working papers accounted for or controlled in the same manner as a finished document or comparable classification when:
    a. Released by the organization to an agency or activity outside the headquarters or transmitted through mes-

	N/A	Yes	No

sage center channels within the headquarters? (7-304e1)
67. b. Placed permanently in the file system? (7-304e2)
   c. Retained for more than 180 days? (7-304e3)
68. Have downgrading or exemption instructions been placed on working papers permanently filed? (7-304f)

Method of Transmission or Transportation

69. Is classified information or material transmitted or transported in accordance with the provisions of DoD Regulation 5200.1-R and Component supplements? (8-100 thru 8-107)
70. Are personnel permitted to carry classified material while in a travel status except in exceptional and approved circumstances? (8-200)
71. Are individuals authorized to carry classified information while in a travel status fully informed of appropriate security regulations and requirements? (8-301c)
72. Is classified information transmitted in two opaque wrappings or equivalent? (8-400b)
73. Is a documentary record for classified information or material made or received by a DoD Component disposed of or destroyed in accordance with DoD Record Management regulations? (9-100)
74. Is classified material destroyed in the presence of approved officials and by an approved method? (9-102)
75. Are destruction records maintained for TOP SECRET and SECRET material? (9-102)

Security Education

76. Are effective security education programs established to familiarize personnel with the purposes and provisions of DoD Regulation 5200.1-R and all changes thereto? (Chapter X)
77. Has an effective and viable security education program been establishment? (10-100)
78. Are personnel who have had access to classified information given foreign travel briefings prior to such travel? (10-104)
79. Are debriefings conducted for personnel upon termination of employment or contemplated temporary separation for a 60-day period or more? (10-105)

Foreign Origin Material

80. Has classified information furnished by a foreign government or international organization retained its original assigned classification or been assigned a U. S. classification that assures equivalent protection? (11-100)

	N/A	Yes	No

81. Has NATO, CENTO or SEATO classified information been marked with appropriate classification in English? (11-100, 101 and Annex B)
82. Is NATO, CENTO or SEATO classified information safeguarded in accordance with DoD Instructions C5210.-21, C5210.35 and 5210.54? (11-200)

Special Access Program

83. Are Special Access Programs held to a minimum? (12-100)
84. Are Military Department Special Access Programs submitted to the Secretary of the respective Department for approval and subsequently reported to the ASD (C)? (12-103)
85. Are other DoD Component Special Access Programs submitted to the ASD (C) for approval? (12-103)

Program Management

86. Has an official been assigned as Security Manager for the Activity? (13-304)

Administrative & Judicial Action

87. Has administrative action been taken against repeated abusers of the Information Security Program? (14-101)

Chapter 13

PERSONNEL SECURITY

"It should be remembered that the best system in the world falls apart, if the individuals involved in its operation become biased, careless or corrupt."

C. W. Anderson
The Portland (Oregon) Clinic
in *Administrative Management*, Dec., 1967.

THE DISCUSSION thus far concerning the functional subareas of security have focused on security of information and the security of physical things. Every type of employer, governmental or proprietary, has a minimal degree of security required in regard to personnel to be employed. The needs of the employer and the type of employment dictate the exact policies to be established and procedures to be used in providing a program of personnel security.

The policies surrounding personnel security vary considerably. The government's program for personnel security establishes one set of policy and criteria for hiring and retention, whereas a myriad of different policies exist in the proprietary sector.

Although policies vary widely, procedures utilized in performing personnel security functions are alike in many respects. Identical investigative and decision-making techniques are used in determining the acceptability of a person for employment or the retention of a person in an organization. These techniques will be viewed in the governmental and proprietary context.

Personnel security is the most critical of the three security processes. An organization cannot run without personnel. The activities that comprise the day-to-day operations are conducted by people. People must be hired, trained, and given trust to carry out their responsibilities. The personnel security process attempts to protect the organization against undesirable persons through the application of appropriate recruitment and hiring safeguards. It likewise attempts to protect the organization against the dishonest, illegal, as well as unethical and nonperformance of assigned duties by employees through the use of appropriate investigative techniques and control safeguards. Likewise, the process attempts to insure the protection of employees from discriminatory hiring or terminating procedures, ill-founded allegations of illegal or unethical conduct and insure that employees are provided with appropriate safeguards for meeting their protective safeguards commensurate with their organizational responsibility.

Federal agencies as well as proprietary organizations have developed various programs to insure that only those applicants who are of value to the organization are hired and retained and that those who are not suitable for working in a particular organization are not hired or if hired, terminated before they can cause significant damage to the overall operations of the company.

PROPRIETARY APPLICATIONS OF THE PROCESS

It was noted that the information security process places the responsibility for protecting information directly upon the individual. In Federal programs as well as proprietary programs, the employee is told that the information being given to him must be protected. He is given adequate means to protect it and is given the responsibility for insuring that such information remains protected while in his care. The responsibility is thus very *directly fixed with the individual.* The personnel security process, on the other hand, is the reverse of that process.

In the case of the personnel security process, the responsibility for protecting either information or the organization from disloyal or unsuitable employees is placed on the organization. The management of a company or the government of a nation has the responsibility to insure the continuity of its operations. It does this through instituting appropriate safeguards to protect its activities, information, and operations from internal and external threats to its continuity.

A comprehensive personnel security program includes the following:
(1) adequate job specifications and performance standards;
(2) appropriate recruitment and selection criteria;
(3) background applicant screening procedures and standards;
(4) background investigation standards;
(5) truth verification standards;
(6) criteria for employee conduct;
(7) investigation of questionable employee conduct;
(8) disciplinary procedures; and
(9) termination procedures.

The personnel security process is to insure that the corporate or organizational work environment is kept secure both for employee and from employees. The development of a climate of honesty, integrity, and security is the goal of this process. Employee problems or misconduct and use of violence are problems of concern. Likewise, direct losses to the company through theft, misuse of company property, disclosure of trade secrets, absenteeism as well as problems of moonlighting and conflict of interest are of equal concern.

Personnel provide the organization with its primary resource for goal attainment. The protection of the organization from any harmful act by the employees and the protection of employees from danger or risk is basic to sound management. The development of comprehensive protective programs must include provisions for monitoring employee performance, attitudes, and activities in a positive but not repressive manner. The program which utilizes proper hiring and screening measures, adequate job performance standards, appropriate internal safeguards and reasonable disciplinary standards coupled with acceptable salary and benefits has done all possible to establish a protective working environment.

The personnel security process provides the administrative ability for the organization to properly screen potential employees to determine their suitability for employment. Within legally prescribed standards, the process attempts to insure that employees who can contribute to the goals of the organization are hired, while those not considered acceptable risks are not hired. Since personnel contribute to organizational success directly, the process must insure that valuable potential employees are not rejected because of minor problems or concerns. The tension which exists between security programs and organizational goals is present in this process. A balancing of risk versus value within the organization is inherent in the process. A strict interpretation of security requirements can exclude "good" employees while a too liberal approach can introduce "risky" employees.

Employee Screening

The screening of individuals during the hiring process is required to insure suitability and desirability within the guidelines established by the organization. These guidelines should comply with appropriate state, local, and Federal hiring laws and regulations including: Federal Civil Rights Act (1964) (Title VII); Consumer Credit

Protection Act (1971);[1] Fair Employment Practices legislation;[1] as well as Federal Information Privacy[2] and the Security Regulation for specific agencies.

Each law or Executive Order establishes specific procedural guidelines for the collection, dissemination, processing, and decision-making using information collected about an applicant or employee. Existing legislation was enacted to reduce the possibility of discrimination in hiring and to insure that decisions were made based on accurate and reliable information. This legislation provides a review opportunity for applicants to have access to collected information about their backgrounds and activities which is being used for hiring decisions. They further require organizations requesting information as well as those collecting it to keep records of their actions and sources.

The intent of consumer protection legislation was to provide an opportunity for a balanced approach to personnel investigations, insure reliability of data, and provide redress for applicants not hired or denied credit based on erroneous information. While insuring that the applicant is protected, it has placed an additional burden upon the security department and the personnel office. The cost, time required, and the records system have all been increased as has the surveillance of the hiring process by state and Federal Employment Practices Agencies. A preponderance of information is required to preclude employment based solely on derogatory information collected during the preemployment screening or investigative process.

Information protection legislation precludes the disclosure of a considerable portion of information required for adequate background investigation and screening without the consent of the applicant. The use of authorizations which grant permission from the applicant to the prospective employer to obtain information to verify statements made on an employment application are being regularly used. These authorizations also release the employer from any liability concerning the results of using the information collected.

The restrictiveness of legislation coupled with the cost of conducting an investigation makes it imperative that complete information be obtained from the applicant and reviewed prior to beginning the investigative process. A brief review of the application form with the applicant is often enough to locate omissions or false statements which would have taken more time and cost had they been discovered during the field investigation. The initial application form and interview is therefore critical to the entire process. It must be well designed and properly administered.

TECHNIQUES

Personnel security programs are involved with the development and use of predictive technology to assist organizations to properly screen potential employees prior to hiring. The use of polygraph or Psychological Stress Evaluators (PSE) as well as Reid Reports, Personnel Security Previews, and other types of related psychological evaluation techniques is common. The primary purpose of these devices and techniques is to determine the truthfulness of statements about applicant background and qualifications. They are also used to determine the degree of risk the potential employee will be to the organization by analyzing reaction patterns and providing predictions about future actions.

The screening of potential employees is accomplished through the use of structured application forms, interviews, testing, and investigations. The type of position, the sensitivity of the employment, and the legal limitations or requirements for security will determine the measures employed. The decisions about the level of security required or the degree of risk that the organization is willing to assume will determine the amount of time, effort and money to be expended in performing the personnel security function.

PROTECTIVE MEASURES

The protection of personnel both within the organization and the protection of the organization from the employees does not stop when the hiring process is completed. The supervisory and management practices of the organization directly affect the ability of the organization to protect itself against internal losses of equipment, materials, time, and work quality. Security problems are often the result of poor management and organizational practices. The improvement of personnel attitudes through well-developed and implemented personnel procedures and policies is basic to effective security programs.

Personnel security measures often place responsibilities on employees beyond their capacity. When such a situation occurs, the organization exposes itself to extraordinary risk which is not anticipated or planned into the protective program. The amount of protective responsibility given to employees must be subject to the same type of review as all other aspects of the program. In fidelity bonding programs for example, company employees are insured against loss to the organization through dishonesty or negligence. The amount of loss to be covered is clearly specified so that the company knows what risk it is incurring when the employee is allowed to exceed his coverage limits. Unfortunately, the same principle is not often applied to other loss areas in which bonding is not used.

The use of investigative techniques to evaluate the adherence of employees to organizational policies is commonplace. Undercover operatives are hired to investigate suspected losses or wrongdoing by employees. The posting of rewards for information about wrongful employee activities is also used. Regular audits of operations likewise seeks to eliminate losses and expose risks.

The use of truth verification devices such as the polygraph or PSE as a regular employee rescreening method is also popular. These devices must, however, be used within the regulations covering their employment. If, for example, the polygraph is to be used it must be made known to the employee at the time of employment and made a condition of employment. It further must be used on a regular basis and for all covered employees. It may not be used for "fishing expeditions" indiscriminately, but rather as either a regular process or after an investigation has "focused" on particular individuals.

The reduction of losses basic to any protective program is implicit in personnel security measures. Anything which lowers productivity, profitability, efficiency, or effectiveness of the organization or affects its ability to survive is not acceptable from a protective viewpoint. As in any security decision-making process, there is always a trade-off between protection and utility to the organization. A prudent manager must exercise his responsibility to provide all necessary protective measures while not inhibiting the attainment of organizational goals.

GOVERNMENTAL APPLICATIONS OF THE PROCESS

Executive Order 10450[3] establishes the standards for employment of military and government personnel. The essence of the governmental policy in personnel security is to keep those considered to be "risky" from obtaining jobs which are considered sensitive. It is the long-range and immediate aim of governmental personnel security practices to screen out those who could be potentially dangerous before they have an opportunity to harm the government.[4] These practices are considered a primary defense against breaches in security. The employees of governmental agencies and those proprietary firms dealing directly with governmental contracts are required to obtain personnel clearances. Individuals are not permitted to have access to classified governmental information beyond the extent of the particular individual's security clearance. He must have a *need to know* the information in order to perform his assigned task. Authorization for access to classified information of a specific classification and all lower classifications shall be granted or continued only when it is determined that such access is consistent with the national interest.

A personnel security clearance is an administrative determination that an individual is eligible to have access to classified information. These security clearances are issued in four categories of classification, the most familiar of which, as previously discussed are Top Secret, Secret, and Confidential. There are however, cryptographic clearances and *Q Clearance* which is identified with the atomic energy program.

In order to accomplish the requirements established by the governmental security program, the personnel selection process normally involves three specific parts. The first of these is an investigation, which is essentially a fact finding activity. The second step is an evaluation of the information obtained, and the third is an adjudication based on these materials to determine whether or not a security clearance will be issued and the scope of the clearance. Obviously then, the personnel security investigation is the most critical part of these three steps.

A personnel security investigation is essentially an inquiry into the loyalty, character, integrity, morals, and descretion of an individual to determine whether or not the individual should be employed in this specific capacity or whether his clearance for access to classified defense information is clearly consistent with the national security. Similarly, no person is granted a security clearance unless it is clearly determined that such clearance is consistent with the national security interest.

If no derogatory information is developed during the clearance process which is of sufficient gravity to justify an overall common sense determination that a clearance is not consistent with the national security, the clearance will be issued. The National Agency Check (N.A.C.) provides a general outline for the background check of the individual to determine gross inconsistencies or uncovered bits of derogatory information about the individual available with one or more of the various federal agencies involved in the National Security Check. The background investigation (B.I.) provides a more detailed appraisal of the person's loyalty and suitability for a particular type of federal employment.

Derogatory Information

Derogatory loyalty information reflects adversely upon the loyalty of an individual to the United States. It indicates that the person's conduct or attitude toward the United States is deficient, possibly even extending into one of the following areas: treason, espionage, sabotage, sedition, subversion, or disaffection. Derogatory suitability information reflects adversely on a person's character, integrity, reliability, and trustworthiness. Obviously, the sensitivity of the position will determine the amount of risk the government is willing to take in hiring a person for a particular job. When the nature of the job is such that a person holding it could substantially injure the national security, the employment is sensitive and therefore requires a thorough investigation and a well balanced adjudication of the person in relation to the job that needs to be done.

Loyalty and Suitability Criteria

The terms loyalty and suitability have been used throughout this chapter as being the goals of a governmental personnel security program. What specifically are loyalty and suitability criteria? Why exactly are these governmental personnel security criteria?

Loyalty

Unquestionably loyalty means many things, depending upon the particular context in which it is used. Loyalty is something more than formal duty to obey, or obligation, and usually limited in its scope to political authority. It is viewed as that type of attitude based upon personal beliefs and values formed throughout life. It must be viewed in terms of the needs of a particular organization. The beliefs, actions, and motivations of a person must be compatible with those of the organization. Loyalty criteria are therefore designed to provide investigators and those charged with adjudicating the security clearance with information to prove or disprove loyalty. This evidence can be divided into three categories:

1. Evidence of association with known subversives.
2. Evidence of membership in subversive organizations.
3. Evidence of unusual individual activities.

Based on the particular job to be done, the person's loyalty may be viewed and determination made.

Suitability

Suitability, likewise can be viewed from a wide variety of viewpoints. This concerns the needs of the organization and the job to be done by the person being considered. The integrity, discretion, morals, and character of the person all interact and are all considered when determining whether or not this is the *right* person for the job.

The "Clearance" Process

The Department of Defense is responsible for the processing of requests for security clearances. Under the following circumstances, requests for confidential clearance must be processed by the Department of Defense:

1. The employee is not a United States citizen.
2. The employee is a representative of a foreign interest.
3. The employee indicates that a prior clearance has been suspended, revoked or denied.
4. The employee indicates affiliation with organization designated under Executive Order 10450.
5. The employee indicates that he has resided in Sino-Soviet countries in the last fifteen years.
6. The employee lists relatives residing in Sino-Soviet countries.

Secret and Top Secret Clearances

All Secret and Top Secret clearances are granted by the Department of Defense. When responsible supervision determines that a United States citizen employee, not previously cleared above Confidential level, will have need for access to Secret and Top Secret information, the following forms must be completed and submitted:

1. Personnel Security Questionnaire (Industrial) (DD Form 48), for Secret or Personnel Security Questionnaire (Industrial) (Multiple Purpose) (DD Form 49) for Top Secret.
2. Fingerprint card (DD Form 258).

Interim Clearance

Interim Top Secret and Secret clearance are granted by the Department of Defense in cases of extreme emergency:

1. The employee, because of unique qualifications, is the only person available for assignment to the work. The nature of the work, identification of the classified contract involved and the qualifications shall be included.
2. Because of the above factors, a crucial delay will be incurred in the performance of the contract (or other classified contracts) unless the interim clearance is granted.

Requests for Interim Secret clearance must be submitted through the Government contracting officer responsible for the classified contract to which access is required. Request for Interim Top Secret clearances must be forwarded through the Government contracting officer as outlined above, but will usually not be authorized without the approval of the Secretary of Defense. Such approval is very seldom granted.

Interim clearance is not valid for access to *Restricted Data* unless specifically authorized by the Department of Defense. In specific instances, other restrictions may be imposed. Interim clearances above Confidential level of immigrant aliens are not authorized.

Clearance of (Nonmilitary) Service Representative

Clearance of Immigrant Aliens

All clearances for immigrant aliens are granted by the Department of Defense. Inasmuch as the clearance will take a minimum of nine to twelve months, the employment of aliens in positions requiring clearance is not ordinarily practical. Prior to

DEPARTMENT OF DEFENSE PERSONNEL SECURITY QUESTIONNAIRE	DATE	*Form Approved* *Budget Bureau No. 22-R046*

INSTRUCTIONS: Five (5) copies of accomplished form will be submitted for U.S. Citizens by the Contractor when investigation for clearance by a military department is required. TYPE OR PRINT ALL ANSWERS. If more space is required, attach additional sheets, identifying by corresponding block number. FORM WILL NOT BE ACCEPTED UNLESS COMPLETELY AND PROPERLY EXECUTED. Questions which do not apply will be marked "None".

NOTE. PENALTY FOR MISREPRESENTATION — Failure to answer all questions, or any misrepresentation (by omission or concealment, or by misleading, false, or partial answers) may serve as a basis for denial of clearance for access to classified Department of Defense information. In addition, Title 18 United States Code 1001 makes it a criminal offense, punishable by a maximum of 5 years' imprisonment, $10,000 fine, or both, knowingly and willfully to make a false statement or representation to any Department or Agency of the United States as to any matter within the jurisdiction of any Department or Agency of the United States. This includes any statement knowingly and willfully made by employer or employee herein which is knowingly incorrect, incomplete or misleading in any important particular — Title 18 United States Code 911 states "whoever falsely and willfully represents himself to be a citizen of the United States shall be fined not more than $1,000 or imprisoned not more than three years, or both".

TO BE COMPLETED BY EMPLOYER

TO: *(Cognizant Security Office)*	NAME AND ADDRESS OF EMPLOYER *(If a subsidiary, include name of parent company)*
JOB TITLE AND DESCRIPTION OF EMPLOYEE'S DUTIES WHICH REQUIRE ACCESS TO CLASSIFIED INFORMATION	CONTRACT NUMBER, WHEN APPLICABLE
	SECURITY CLASSIFICATION OF MATERIALS OR INFORMATION EMPLOYEE WILL HAVE ACCESS TO
I CERTIFY THAT THE ENTRIES MADE BY ME ABOVE ARE TRUE, COMPLETE, AND CORRECT TO THE BEST OF MY KNOWLEDGE AND BELIEF AND ARE MADE IN GOOD FAITH.	SIGNATURE OF EMPLOYER OR DESIGNATED REPRESENTATIVE

TO BE COMPLETED BY EMPLOYEE

1. LAST NAME - FIRST NAME - MIDDLE NAME	2. ANY OTHER NAME BY WHICH KNOWN *(Alias, maiden or former legal name; designate which)*				
3. DATE OF BIRTH	4. PLACE OF BIRTH			5. SOCIAL SECURITY NO.	
6. MARITAL STATUS	7. SEX	8. HEIGHT	9. WEIGHT	10. COLOR EYES	11. COLOR HAIR

12. EDUCATION *(Account for all civilian schools)*

YEARS *(Include month if known)*		NAME AND LOCATION SCHOOL	GRADUATE		DEGREE
FROM	TO		YES	NO	

CITIZENSHIP

13. ARE YOU A CITIZEN OF THE UNITED STATES? ☐ YES ☐ NO *(If answer is "Yes", complete following: If answer is "No", return this form to your employer.)*

☐ I AM A CITIZEN OF THE UNITED STATES BY REASON OF MY BIRTH IN THE UNITED STATES
☐ MY NATURALIZED CITIZENSHIP*
☐ MY BIRTH IN A FOREIGN COUNTRY OF UNITED STATES PARENTS ☐ MY DERIVATIVE CITIZENSHIP**

If checked complete either "Citizenship by Naturalization" or "Citizenship by Derivation" Section below.

CITIZENSHIP BY NATURALIZATION*

WHERE NATURALIZED *(City, County, State)*	DATE NATURALIZED
COURT	CERTIFICATE NO.

CITIZENSHIP BY DERIVATION**

PARENT'S NAME	PARENT'S CERTIFICATE NO.

DD FORM 48
1 MAR 64

REPLACES EDITION OF 1 JUL 63 WHICH MAY BE USED.

14a.		MILITARY SERVICE			
COUNTRY	BRANCH OF SERVICE		SERVICE NUMBER	FROM *(Date)*	TO *(Date)*

b. ARE YOU A MEMBER OF A RESERVE COMPONENT ☐ YES ☐ NO *(If answer is yes, furnish service, component and current status under Item 27, Remarks).*

c. TYPE OF DISCHARGE	d. LOCAL DRAFT BOARD *(United States)* AND ADDRESS	e. ORDER NO.	f. CLASSIFI-CATION

15. RESIDENCE *(List all places of residence from 1 Jan 1937, starting with present)*

STREET, CITY, STATE, OR OTHER POLITICAL SUBDIVISION AND COUNTRY	FROM *(Date)*	TO *(Date)*

16. ORGANIZATIONAL MEMBERSHIP

LIST ALL ORGANIZATIONS EXCEPT LABOR UNIONS AND EXCEPT ORGANIZATIONS LISTED ON DD FORM 48-1 IN WHICH YOU HOLD OR HAVE HELD MEMBERSHIP

NAME AND ADDRESS	TYPE	OFFICE HELD	FROM *(Date)*	TO *(Date)*

17.	FOREIGN COUNTRIES VISITED OR RESIDED IN			
	CITY AND COUNTRY	DATE LEFT U.S.	DATE RETURNED U.S.	PURPOSE

18.		RELATIVES		
a. LIST PARENTS, SPOUSE, *(including maiden name)* CHILDREN, BROTHERS, AND SISTERS *(16 years and older)* LIST EVEN THOUGH DECEASED.				
RELATION	NAME	ADDRESS *(Enter "deceased" if no longer living)*	PLACE & DATE OF BIRTH	PRESENT CITIZENSHIP
b. LIST OTHER LIVING RELATIVES AND RELATIVES OF SPOUSE KNOWN TO BE RESIDING OUTSIDE THE U.S. REGARDLESS OF AGE.				

19. EMPLOYMENT *(Show every employment you have had since 1 Jan 1937 and account for all periods of unemployment)*				
POSITION HELD	EMPLOYER AND IMMEDIATE SUPERVISOR	ADDRESS	FROM (Date)	TO (Date)

20.	HAVE YOU EVER BEEN ARRESTED, CHARGED, OR HELD BY FEDERAL, STATE, OR OTHER LAW ENFORCEMENT AUTHORITIES, FOR ANY VIOLATION OF ANY FEDERAL LAW, STATE LAW, COUNTY OR MUNICIPAL LAW, REGULATION OR ORDINANCE? INCLUDE ALL COURT MARTIALS, WHILE IN MILITARY SERVICE. DO NOT INCLUDE ANYTHING THAT HAPPENED BEFORE YOUR 16TH BIRTHDAY. DO NOT INCLUDE TRAFFIC VIOLATIONS FOR WHICH THE ONLY PENALTY IMPOSED WAS A FINE OF $25.00 OR LESS. ALL OTHER CHARGES MUST BE INCLUDED EVEN IF THEY WERE DISMISSED. ☐ YES ☐ NO IF "YES", GIVE DATE AND PLACE, CHARGE, AND DISPOSITION:
21.	HAVE YOU EVER BEEN PREVIOUSLY GRANTED AN ACCESS AUTHORIZATION *(Security clearance)*? *(If answer is "Yes", indicate level of clearance, when granted, by whom and where employed at that time under Item 27, "Remarks".)* ☐ YES ☐ NO
22.	HAVE YOU EVER HAD AN ACCESS AUTHORIZATION *(Security clearance)* SUSPENDED, DENIED, OR REVOKED? *(If answer is "Yes" indicate level of clearance, when suspended, denied or revoked, by whom and where employed under Item 27, "Remarks".)* *(If you since have been granted a clearance by the government, indicate under under Item 27, "Remarks" the date, level of clearance and activity which restored the clearance.)* ☐ YES ☐ NO
23.	HAVE YOU EVER TERMINATED EMPLOYMENT WHILE A REQUEST OR APPLICATION FOR AN ACCESS AUTHORIZATION *(Security clearance)* WAS PENDING? *(If answer is "Yes", furnish name and address of employer under "Remarks". If you since have been granted a clearance by the Government indicate under Item 27, "Remarks" the date, level of clearance and where employed.)* ☐ YES ☐ NO
24.	REFERENCES *(Give five personal references, stating business address of all references, if known. Do not include relatives, former employers, or persons living outside the United States.)*

NAME	YEARS KNOWN	STREET AND NUMBER	CITY	STATE

25. LIST EACH FOREIGN GOVERNMENT, FIRM, CORPORATION OR PERSON FOR WHOM YOU ACT AS A REPRESENTATIVE, OFFICIAL OR EMPLOYEE. *(If none, so indicate)*

26. REMARKS *(Use the space provided below and attach additional sheets, if necessary, for a full statement)*

CERTIFICATION

WARNING: Read every sentence of Certification before signing.

I certify that the entries made by me above are true, complete, and correct to the best of my knowledge and belief, and are made in good faith.

I certify that I am a citizen of the United States.

I certify that I know that any misrepresentation or false statement made by me here in may subject me to prosecution under Title 18, United States Criminal Code, Sections 911 and 1001, with penalties up to five (5) years imprisonment and $10,000 fine.

I certify that I have read and understand each sentence of this Certification.

SIGNATURE OF WITNESS	SIGNATURE OF EMPLOYEE	DATE OF SIGNATURE
ADDRESS OF WITNESS *(City, County, State)*		

Figure 13-1.

processing for a personnel security clearance, the immigrant alien shall be required to produce the Alien Registration Card (Form Number I-151) which has been issued to him upon lawful admission to the United States under an immigration visa for permanent residence. Immigrant aliens are not eligible for Interim Secret or Top Secret personnel security clearance, not for access to Secret or Top Secret Cryptographic information, or Communications Analysis information.

Clearance of Foreign Nationals

Persons who are not United States citizens or immigrant aliens are not eligible for a personnel security clearance except under Canadian and United Kingdom Reciprocal Clearance.

Canadian and United Kingdom Clearance

Under the terms of agreements between the United States Government and the Governments of Canada and the United Kingdom, citizens of Canada and the United Kingdom (other than those who have immigrant alien status) may be processed for Canadian Reciprocal or United Kingdom Reciprocal clearance through the Department of Defense when authorization for access to classified information is required in connection with the performance of classified duties. Application for Canadian Reciprocal or United Kingdom Reciprocal clearances shall be made for each individual concerned by submitting the following executed forms:
1. Personnel Security Questionnaire (Industrial) (Multiple Purpose) (DD Form 49) and
2. Fingerprint card (DD Form 258), 3 copies.

The provisions of these agreements are not valid for access authorization to:
1. Restricted Data, as defined in the United States Atomic Energy Act of 1954, as amended;
2. Formerly Restricted Data removed from the Restricted Data category pursuant to Section 142 (d) of the United States Atomic Energy Act of 1954, as amended;
3. Classified Atomic Energy Data as defined in the Atomic Energy Control Regulations, Order-in Council PC 1959-1643;
4. cryptographic information;
5. any classified information for which a special access authorization is needed; and
6. other information that would not be released under applicable disclosure policies.

Position Sensitivity

The clearances indicated and the types of jobs for which these clearances are required are normally considered to be sensitive positions. The duties and jobs for individuals engaged in various types of sensitive positions can vary. An example of these can be found in the sensitive duty listings of the United States Army. These duties consist of individuals involved in the following types of activities:
1. those with Top Secret, Secret, or Confidential clearances;
2. those in troop information assignments;
3. those in crypto systems and equipments;
4. those in research and development;
5. members of boards which pass upon security cases;
6. those involved in the Atomic Energy program;
7. warrant officers or officers;
8. those in GS-14 and above government service ratings; and
9. all other positions so designated by the Secretary of the Army or authorized commanders.

As defined by the Army, a sensitive position is:
1. any position the duties or responsibilities of which requires access to defense information, classified Top Secret, Secret, or Confidential;
2. positions filled by commissioned or warrant officers;

3. personnel engaged in troop information or troop education activities; and
4. any other position so designated by the Secretary of the Army or by any of the officials so authorized.

Sensitive Position (Critical)

This is a position that in addition to meeting the criteria above involves the responsibility for the following:
1. the development of war plans;
2. the development or approval of critical and extremely important items of war;
3. the development or approval of plans or particulars of future major special operations; and
4. the development or approval of policies or programs which effect the overall operations of the Department of the Army, Department of Defense, or the other military department, or is a member of security screening, hearing, or review board, regardless of the degree of clearance required.

Ultrasensitive Positions (USP)

These are positions which afford the individual continuing access to Top Secret information of the highest level of sensitivity requiring special handling and protection. Such positions will be designated in writing by the official so authorized. Care will be exercised to restrict the USP designation to positions which are genuinely the most sensitive. Included in this category are high level critical sensitive positions. This includes the individuals who have nuclear stock-pile weapons data.

Personnel Security Investigation

There are two types of personnel security investigation. These are the National Agency Check and the background investigation. These two types of investigation are used either singularly on in conjunction with each other for the purpose of providing an adequate picture of the person through the use of various investigatory techniques. These techniques provide a set of composite information about a person under investigation which are transmitted to an adjudicating officer.

National Agency Check (NAC)

A National Agency Check consists of checking records of appropriate federal agencies for information bearing on the loyalty and suitability of a person under investigation. The National Agency Check generally referred to as the NAC can be completed in one of four standard ways:
1. the standard NAC;
2. entrance NAC (ENTNAC);
3. expanded NAC (ENAC); and
4. NAC plus written inquiry (NACI).

The purpose of the established format of the NAC is to provide for a series of crosschecks on information provided by the individual as to his background and activities. The NAC specifically makes inquiries to the following:
1. FBI records;
2. Department of Defense Central Index;
3. Department of Army, Navy, Coast Guard, USMC, and Air Force records;
4. Civil Service Commission records;
5. Immigration and Naturalization Service records;
6. Central Intelligence Agency records;
7. Department of State; and
8. Treasury Department.

Background Investigation (BI)[5]

In conducting a background investigation, the following areas are checked to provide information about an individual's activities:

NATIONAL AGENCY CHECK (NAC). A wide vareity of sources are utilized in the National Agency Check to obtain information which will provide cross-checking abilities. The National Agency Check provides for the collection of many types of initial information.

BIRTH RECORDS. An individual's data and place of birth are verified through the examination of school, employment, and other records. If a discrepancy appears, or

NACC Use Only	DEPARTMENT OF DEFENSE NATIONAL AGENCY CHECK REQUEST		REQUEST DATE	
	1. LAST NAME - FIRST NAME - MIDDLE NAME		2. SEX	
	3. ALIAS(ES) AND ALL FORMER NAME(S)		4. SOCIAL SECURITY NUMBER	
5. MONTH, DAY, YEAR OF BIRTH	6. PLACE OF BIRTH		7. SERVICE NUMBER	
RETURN RESULTS TO: *(Include ZIP Code)*			8.a. SECURITY PROGRAM ☐ MILITARY ☐ CIVILIAN ☐ INDUSTRIAL	
			b. ☐ LOCAL FILES CHECKED WITH FAVORABLE RESULTS	
			c. INITIATOR OF REQUEST	
9. RELATIVES	10. DATE AND PLACE OF BIRTH	11. PRESENT ADDRESS		12. CITIZENSHIP
a. FATHER				
b. MOTHER *(Full Maiden Name)*				
c. SPOUSE *(Full Maiden Name)*				

13. RESIDENCES *(List all from 18th birthday or during past 15 years, whichever is shorter. If under 18, list present and most recent addresses.)*

a. FROM	b. TO	c. NUMBER AND STREET	d. CITY	e. STATE

14. EMPLOYMENT *(List all from 18th birthday or during past 15 years, whichever is shorter. If under 18, list present and most recent employment)*

a. FROM	b. TO	c. EMPLOYER	d. PLACE

15. LAST CIVILIAN SCHOOL

a. FROM	b. TO	c. NAME	d. PLACE

YES	NO	16. *("Yes" answers must be explained in Item 18, below.)*	17. REQUEST DATA	
			a. REQUESTER DESIGNATOR	
		a. Is the subject an alien or naturalized citizen?	ARMY	DASA
		b. Has the subject any foreign connections, employment or military service?	NAVY	DCA
		c. Has the subject travelled or resided abroad other than for the U.S. Government?	AIR FORCE	DCAA
		d. Has the subject had employment requiring a security clearance or investigation?	OSD	DIA
		e. Is the subject now or has he ever been in the Federal Civil Service or Armed Forces?	JCS	DSA
		f. Has the subject qualified DD Form 398, 98, 48-1, or similar security form?	NSA	DISCO
		g. Has the subject ever been addicted to drugs?		
18. REMARKS *(If additional space is needed, continue on plain paper.)*			b. REASON	
			BASIC TRAINEE	
			PRE-COMMISSION	
			NUCLEAR	
			BI.	
			SECRET CLEARANCE	

DD FORM 1584, 1 DEC 66   REPLACES DA FORM 3027, 1 AUG 65, WHICH IS OBSOLETE   ☆GPO; 1967 O 279-094

Figure 13-2.

pertinent records are unavailable, it is necessary to examine vital statistics records to establish the individual's correct data and place of birth.

EDUCATION. College attendance, or if no college, then last secondary school. The last secondary, business, or trade school whose course of instruction exceeding ninety days, if within the last fifteen years, should be verified. In addition to examining school records, persons in a position to know the individual's activities while in attendance at the school are interviewed if available. The results of attendance service schools will as a rule appear on an individual's service record and are not reconfirmed. Checking the educational institution the following information is obtained.

 a. The subject's full name and any alias or maiden name
 b. The date of birth
 c. The place of birth
 d. Address
 e. Previous address
 f. Name of parents
 g. Dates of attendance
 h. Courses studied
 i. Grades
 j. Explanation of the grade structure
 k. Status (graduate or nongraduate)
 l. Honors or the degree granted
 m. Character traits
 n. Student activities
 o. Attendance records
 p. Transcript of grades
 q. Previous education
 r. Employment, if any
 s. Additional pertinent information
 t. Cross-references

EMPLOYMENT. Records of present and former full-time employment during last fifteen years, or subsequent to the person's eighteenth birthday, whichever involves the shorter period should be checked. One former employer, supervisor, or coworker for periods of employment in excess of ninety days will be interviewed to determine loyalty, character, behavior traits, and reputation of the individual. Part-time employment and summer employment only requires a record check, except to substantiate or disprove derogatory information.

REFERENCES. Three listed character references will be interviewed in addition to three developed character references. Developed references are persons who are determined during the course of the investigation to be people who know the individual quite well but are not listed as references on the National Agency Check work sheet. Developed references are used to substitute for listed references so long as a minimum of six references are interviewed. Each reference must be familiar with the background of the individual and his activities.

NEIGHBORHOOD OR RESIDENCE INVESTIGATION. Normally these investigations will be conducted only when necessary to substantiate or disprove derogatory information. These involve making inquiries in former places or present places of residence of individuals requesting a security clearance.

CRIMINAL RECORDS. The records of a police department or other law enforcement agencies in the vicinities of residence, or places of employment involving periods of ninety days will be checked whenever information develops in the National Agency Check which is not considered adequate. The records of local FBI offices are not checked unless special circumstances warrant doing so.

MILITARY SERVICE. The service of an individual in the Armed Forces of the United States will be verified. The type of service and discharge, service schools, foreign travel, associates, etc., will be checked if considered necessary.

FOREIGN ACTION. Any close or continuous connection with foreign persons, organizations or in countries with interest believed inimical to those of the United States will be investigated. Extent and purpose of any such connections will be ascertained as well as the relationship of the person to such individual or organization.

CITIZENSHIP STATUS. In all cases the citi-

WORKSHEET FOR NATIONAL AGENCY CHECK REQUEST
(AR 381 - 130)

1. LAST NAME – FIRST NAME – MIDDLE NAME	2. SEX
3. ALIAS(ES) – FORMER NAME(S)	4. SOCIAL SECURITY NUMBER
5. MONTH, DAY, YEAR OF BIRTH / 6. PLACE OF BIRTH	7. SERVICE NUMBER

8. RELATIVES	9. DATE AND PLACE OF BIRTH	10. PRESENT ADDRESS *(Include ZIP Code)*	11. CITIZENSHIP
a. FATHER			
b. MOTHER *(Full Maiden Name)*			
c. SPOUSE *(Full Maiden Name)*			

12. RESIDENCES *(List all from 18th birthday or during past 15 years, whichever is shorter. If under 18, list present and most recent addresses.)*

a. FROM	b. TO	c. NUMBER AND STREET	d. CITY	e. STATE

13. EMPLOYMENT *(List all from 18th birthday or during past 15 years, whichever is shorter. If under 18, list present and most recent employment.)*

a. FROM	b. TO	c. EMPLOYER	d. PLACE

14. LAST CIVILIAN SCHOOL

a. FROM	b. TO	c. NAME	d. PLACE

YES	NO	15. *("Yes" answers must be explained in Item 17 below. See reverse for instructions.)*
		a. ARE YOU AN ALIEN OR A NATURALIZED CITIZEN OF THE UNITED STATES?
		b. DO YOU HAVE ANY FOREIGN CONNECTIONS OR RELATIVES, OR HAVE YOU HAD FOREIGN EMPLOYMENT OR MILITARY SERVICE?
		c. HAVE YOU TRAVELED OR RESIDED ABROAD OTHER THAN FOR THE UNITED STATES GOVERNMENT?
		d. HAVE YOU HAD ANY EMPLOYMENT REQUIRING A SECURITY CLEARANCE OR INVESTIGATION?
		e. ARE YOU NOW OR HAVE YOU EVER BEEN IN THE FEDERAL CIVIL SERVICE OR THE ARMED FORCES?
		f. DID YOU QUALIFY DO FORMS 398, 98, 48-1, OR A SIMILAR SECUIRTY FORM?
		g. HAVE YOU EVER BEEN ADDICTED TO DRUGS?

16. HAVE YOU EVER BEEN DETAINED, HELD, ARRESTED, INDICTED OR SUMMONED INTO COURT AS A DEFENDANT IN A CRIMINAL PROCEEDING, OR CONVICTED, FINED, OR IMPRISONED OR PLACED ON PROBATION, OR HAVE YOU EVER BEEN ORDERED TO DEPOSIT BAIL OR COLLATERAL FOR THE VIOLATION OF ANY LAW, POLICE REGULATION OR ORDINANCE *(excluding minor traffic violations for which a fine or forfeiture of $25 or less was imposed)?* INCLUDE ALL COURTS MARTIAL WHILE IN MILITARY SERVICE. *(If "Yes", enter in Item 17 below the date, the nature of the offense or violation, the name and location of the court or place of hearing, and the penalty imposed or other disposition of each case.)*

☐ YES ☐ NO

17. REMARKS *(If additional space is needed, continue on plain paper.)*

THE ENTRIES ON THIS FORM ARE TRUE, COMPLETE, AND CORRECT TO THE BEST OF MY KNOWLEDGE AND BELIEF AND ARE MADE IN GOOD FAITH. I UNDERSTAND THAT A KNOWING AND WILLFUL FALSE STATEMENT ON THIS FORM CAN BE PUNISHED BY FINE OR IMPRISONMENT OR BOTH *(see U.S. Code, Title 18, Section 1001).*

DATE	SIGNATURE OF PERSON COMPLETING FORM

DA FORM 3208, 1 DEC 66   REPLACES DA FORM 3027 WS, 1 AUG 65, WHICH IS OBSOLETE.

INSTRUCTIONS FOR EXPANDING ITEMS a THRU g, BLOCK 15

If a "Yes" answer is given to any item in Block 15, a full explanation will be made in the "Remarks" section of the form and on additional sheets of plain paper if more space is needed.

a. Are you an alien or a naturalized citizen of the United States?

If an alien, provide alien registration number, the date and port of entry, the last Immigration and Naturalization Service Office with which registered. If a minor at the time of immigration, provide parents registration number, the date and port of entry. If a naturalized citizen, provide naturalization number, the date, place and court where certificate was issued. If citizenship was acquired through the naturalization of parent(s) provide the name of the parent(s) and the date, place and court.

b. Do you have any foreign connections or relations, or have you had foreign employment or military service?

Foreign Connections: Identify all foreigners or foreign organizations in the United States or abroad with whom you have been connected.

Foreign Relatives: List name, address and relationship of each relative residing either in a foreign country or in the United States as an alien. (*Relatives to be included are brothers, sisters, children, guardians, step-parents and former spouses.*)

Foreign Employment: Indicate name and location of the foreign firm or governmental agency which employed you and list the inclusive dates of employment (*if this information has not been provided in response to Item 12*).

Foreign Military Service: State the country, branch of military service, grade, service number, inclusive dates of service and type of discharge.

c. Have you traveled or resided abroad other than for the United States Government?

List each foreign country traveled in or resided in other than as a direct result of United States Government duties. Provide the inclusive dates and the purpose of all foreign travel.

d. Have you had any employment requiring a security clearance or investigation?

Name the employer and indicate the type of security clearance issued and/or investigation conducted. If known, provide the name of the agency that completed the investigation and the date.

e. Are you now or have you ever been in the Federal Civil Service or the Armed Forces?

If currently or previously employed by the United States Government, state the employing Federal agency or department, and the location and inclusive dates if not listed in Item 12.

If formerly a member of the United States Armed Forces, provide branch of service, rank/grade, service number, inclusive dates of service and nature of discharge or separation. Include as military service Coast Guard and Merchant Marine duty.

If currently on active duty in the United States Armed Forces, state "ACTIVE DUTY" and list branch of service, rank/grade and date active service started.

If currently a member of a Reserve Component of the Armed Forces, indicate whether Reserve or National Guard, the branch of service, the Unit designation and location, individual status, rank/grade, service number and date of initial entry.

f. Did you qualify DD Forms 398, 98, 48-1 or a similar security form?

Qualification of security form refers to answers or remarks of security significance entered on the form, or to a refusal to complete the security form in its entirety. Specifically, qualification of DD Form 398 means a "Yes" answer was given to any question in Item 17 of the form. Qualification of DD Form 98 means a "Yes" answer was given to any question in paragraph 2, Section IV, except 2r. Qualification of DD Form 48-1 means a "Yes" answer was given to any question in the "Provisions" section of that form.

g. Have you ever been addicted to drugs?

If you are now or have ever been addicted to the use of habit forming drugs, such as narcotics or barbituates, explain in detail, providing dates and residences.

Figure 13-3.

zenship status of individuals shall be established. These status classifications are as follows:
  a. United States citizen and
  b. Immigrant aliens. The records of the Immigration Naturalization Service of the United States Department of Justice are searched to verify the date, place of birth, date of entry into the United States and to determine if the individual has indicated any intention to become a citizen of the United States.
  c. Naturalized citizens. The naturalization and date and place of birth shall be verified through the records of the appropriate United States District Court. If the place of naturalization cannot be determined, the Immigration Naturalization Records in Washington, D.C. are examined.

FOREIGN TRAVEL. If during the past fifteen years the individual has traveled outside the United States in which a passport or visa was required, the Department of State Passport Records will be checked. Exceptions to this are military personnel or government employees performing travel in an official capacity. Other foreign travel may be checked if there is reason to believe that the travel was not in the best interest of the security of the United States.

CREDIT RECORDS. A credit bureau check will be made when possible in areas where the individual has most recently resided for a period of nine months or more, subsequent to his twenty-first birthday. Interviews in connection with other phases of the investigation will stress the subject's financial responsibility. The credit record phase is often expanded if investigation reveals a lack of financial responsibility on the part of the subject.

ORGANIZATIONS. During the investigation, efforts will be made to determine if the individual has

> ... membership, affiliation, or sympathetic associations with any movement, group, or combinations of persons that is totalitarian, Fascist, communist, or subversive, or which has adopted or shown a policy of advocating or approving the commission of acts of force or violence to deny other persons their rights under the Constitution of the United States, or which seeks to alter the form of government of the United States through unconstitutional means.[6]

DIVORCE RECORDS. When it has been determined that an individual has been divorced, the appropriate court records shall be reviewed for verification of pertinent information and the former spouse interviewed.

Once these particular investigations have been completed, the results are sent, to the responsible security officer who may take one of the following courses of action:

1. He may grant security clearance.
2. He may deny security clearance.
3. He may revoke a security clearance already issued.
4. He may reinstate a suspended clearance.

Investigations Required for Access

In order for a person to be declared eligible for access to classified defense information and granted a personnel security clearance, the minimum investigatory requirements must be met for each specific category of defense information. These categories are as follows:

Top Secret

I. Final clearance.
  A. Civilian personnel—United States citizens.
    1. Background investigation.
  B. Military personnel—United States citizens.
    1. Background investigation, or
    2. National Agency Check, plus
      a. Continuous honorable active duty as a member of the Armed Forces, or a combination of such active duty and civilian employment in the Federal Government service on a continuous basis, with no break greater than six months, for a minimum of fifteen

consecutive years immediately preceding the date of the current investigation, or current need for clearance.

Secret

I. Final clearance.
   A. Civilian personnel—United States citizens.
      1. National Agency Check, plus written inquiries to appropriate local law enforcement agencies, former employers and supervisors, references and schools attended. The written inquiries portion of the investigation may be dispensed with for employees who have been continuously employed for a period of five years (with no break greater than six months) immediately preceding the date of the current investigation, or current need for clearance.
      2. A background investigation is required for employees occupying sensitive positions (critical).
   B. Military personnel—United States citizen.
      1. National Agency Check or an Entrance National Agency Check. An ENTNAC is not valid for the purpose of granting an interim TOP SECRET clearance.
      2. Check of the military field 201 file, local military intelligence files, provost marshal files, and medical records.
   C. Immigrant aliens—Civilian or military.
      1. Background investigation.
   D. Nonappropriated fund employees and employees of the Army National Guard—United States citizens.
      1. National Agency Check.
II. Interim clearance.
   A. Civilian personnel—United States citizens.
      1. National Agency Check.
      2. In case of emergency, interim clearance for access to SECRET may be granted for a limited period provided a National Agency Check has been initiated and that such action is necessary in the interests of national security and a record of such findings is made. In every case, this action will be based upon a check of available records.
   B. Military personnel—United States citizens.
      1. Continuous honorable active duty as a member of the Armed Forces for a minimum of two consecutive years immediately preceding the date of the current investigation, plus a check of the military field 201 files, local intelligence files, provost marshal files, and medical records; or, in the case of personnel with less than two years service.
      2. A check of files required above, plus
      3. A check of the Federal Bureau of Investigation, Investigative and Identification files.
   C. Immigrant aliens—Civilian or military.
      1. No interim clearance authorized.

Confidential

I. Final clearance.
   A. Civilian personnel—United States citizens.
      1. National Agency Check, plus written inquiries to appropriate local law enforcement agencies, former employers and supervisors references, and schools attended, except that the written inquiries portion of the investigation may be dispensed with for employees who have been continuously employed for a period of five years (with no break greater than six months) immediately preceding the date of the current investigation.

B. Military personnel—United States citizens.
   1. A check of the military field 201 file, local intelligence files, provost marshal files, and medical records.
C. Immigrant aliens—Civilian or military.
   1. Background investigation.

Summary

Obviously, the standards and policies for governmental needs are not transferable across the board to proprietary programs. The common factors, however, are the needs on the part of both government and proprietary interests to protect their organizations from potential employees, or employees, who do not have the best interests of the organization at heart. In order to do this, guidelines for employment and retention are established. These guidelines are both concerned with a composite picture of the *whole man*. Both systems attempt or should attempt to obtain as much information as possible about the person and make decisions relative to employment or retention based on this information. If a job is more important or more sensitive, a great deal of expense and many sources should be employed to get a total picture of the person and his particular personal makeup. If, however, a job is relatively minor, fewer resources and less expense is in order.

A wide variety of personnel screening devices are available to insure that organization only hire those individuals which it considers filling its needs. Various personnel tests, both intelligence and psychological, can screen out certain types of undesirable individuals. Likewise, the use of the background investigations or polygraph examinations are useful in making these loyalty and suitability types of investigations.

It must be remembered that one of the major purposes of the personnel security process is to insure that the organization does not hire those which it considers undesirable. But moreover, it is charged with the responsibility of including those which might be useful to the organization.

FOOTNOTES

1. See Appendix 13A for the Title VI Amendment, "Consumer Credit Reporting" for the guidelines relating to personnel security.
2. Equal Employment Opportunity Act (1972) PL 92-261, House Report No. 92-238.
3. As amended by Executive Order 11785 (June 4, 1974). This order eliminated the "Attorney General's List" and revoked Executive Order 9835 (1947) and Executive Order 11605 (1971).
4. The Attorney General's List is presented in Appendix 13B.
5. Department of Army Regulations 381-130, 1965, p. 34-35.
6. Executive Order 10450, Section 8A5.

BOOKS

Association of the Bar of the City of New York. *The Federal Loyalty—Security Program*. New York, Dodd, 1956.

Barth, Alan: *Government by Investigation*. New York, Viking Press, 1955.

———: *The Loyalty of Free Men*. New York, Viking Press, 1951.

———: *When Congress Investigates*. Public Affairs Pamphlet No. 227, 1955.

Blum, R. H.: *Police Selection*. Springfield, Thomas, 1964.

Bontecou, Eleanor: *The Federal Loyalty-Security Program*. Ithaca, N.Y., Cornell University Press, 1952.

Chambers, Whittaker: *Witness*. New York, Brandt and Brandt, 1952.

Chase, Harold W.: *Security and Liberty: The Problem of Native Communists 1947-1955*. New York, Doubleday, 1955.

Cronbach, L. J.: *Essentials of Psychological Testing*. 2nd ed., New York, 1961.

Davis, R. C.: *Physiological Responses as a Means of Evaluation Information*. New York, Wiley, 1961.

Gelhorn, Walter: *Security, Loyalty, and Science*. Ithaca, N.Y., Cornell University Press, 1950.

Gorrill, B. E.: *How to Prevent Losses and Improve Profits with Effective Personnel Security Procedures*. Illinois, Dow Jones-Irwin, Inc., 1974.

Grodzins, M.: *The Loyal and the Disloyal*. Chicago, University of Chicago Press, 1956.

Hayman, Harold M.: *To Try Men's Souls: Loyalty Tests in American History*. Berkeley, University of California, 1966.

Inbau, Fred E., and Reid, John E.: *Lie Detector and Criminal Interrogation. Baltimore, The Williams and Wilkins Co., 1953.

Larson, J. A.: *Lying and Its Detection—A Study of Deception and Deception Tests.* Haney and Keeler. Chicago, University of Chicago Press, 1932.

Lasswell, H.D.: The threat to privacy. In: *The Conflict of Loyalties.* New York, Harper & Row, 1952.

Lee, C. L.: *The Instrumental Detection of Deception. The Lie Test.* Springfield, Thomas, 1953.

Leonard, V. A.: *Academy on Lie Detection.* Springfield, Thomas, 1957, Vol. 1.

———: *Academy Lectures on Lie Detection.* Springfield, Thomas, 1957, Vol. 2.

Long, Edward V., Senator: *The Intruders: The Invasion of Privacy by Government and Industry.* New York, Praeger, 1966.

Rourke, Francis E.: *Secrecy and Publicity: Dilemmas of Democracy.* Baltimore, Johns Hopkins Press, 1961.

Weaver, Leon H.: *Industrial Personnel Security.* Springfield, Thomas, 1964.

Westin, Alan F.: *Privacy and Freedom.* New York, The Association of the Bar of the City of New York, 1967.

PUBLICATIONS OF THE GOVERNMENT, LEARNED SOCIETIES, AND OTHER ORGANIZATIONS

Ayllon, Teodoro, and Michael, Jack: The psychiatric nurse as a behavioral engineer. *The Control of Human Behavior,* New York, Scott, Foresman and Co., 1966.

Backster, Cleve: Is the polygraph profession's greatest danger from within? Paper presented at Fifth Annual Polygraph Examiners Clinic, University of Oklahoma, March 9-11, 1964.

Brown, Ralph S., Jr., and Fassett, John D.: Security tests for maritime workers: Due process under the port security program. *Yale Law Journal,* no. 8, 1163-1165, 1953.

Bureau of National Affairs. *Government Security and Loyalty: A Manual of Laws, Regulations, and Procedures.* 2: Washington, Government Printing Office, 1957.

———. *Personnel Security Program in U.S. Industry.* Washington, Government Printing Office, 1966.

Commission on Government Security. Federal Civilian Loyalty Program. Report of the Commission on Government Security, Washington, Government Printing Office, 1957.

Federal Loyalty-Security Program. Report of The Special Committee of the Association of the Bar of the City of New York. Dodd, Mead, 1956.

Government Security and Loyalty, A Manual of Laws, Regulations, and Procedures. Bureau of National Affairs. Washington, 2: 21, 29-31, 26 and no. 29, 1-29, 6.

Guide to Subversive Organizations and Publications. 87th Cong., 2nd Sess., House Document No. 398. Washington, Government Printing Office.

Hollow Coin, The. Documentary concerning the events leading up to, during and following the Abel Trail. Cognizant Offices.

House Report 198. Use of polygraphs as "Lie Detectors" by the federal government. 85th Cong., 1st Sess., March 22, 1965.

Morse, W.: Remarks on rights of privacy as a constitutional guarantee. *Congressional Record,* February 2, 1967.

Personnel security programs in U.S. industry. Proceedings of a conference held in Washington, D.C. Bureau of National Affairs, Inc., June 3, 1955.

Report of the Special Committee on the Federal Loyalty-Security Program of the Association of the Bar of the City of New York. New York, Dodd, 1956.

Rights of government employees, psychiatric exams, and psychological tests. Hearings before the Subcommittee on Constitutional Rights of the Senate Judiciary Committee, 89th Cong., 1st Sess., 1965.

Security Man, The. (film) Cognizant Offices. Free loan.

Spy Next Door, The. (film) Armstrong Cork Co.

Statement regarding psychological testing in the schools. Committee on Testing of the California Association of School Psychologists and Psychometricists, 1963.

U.S. Congress, Senate. Internal security manual. Washington, Government Printing Office, August 31, 1960.

———. U.S. Personnel Security Practices. Hearings before the subcommittee to investigate the administration of the Internal Security Act and other internal security laws of the Committee on the Judiciary. Washington, Government Printing Office, 1963.

———. Invasion of Privacy. Hearing before the Subcommittee on Administrative Practice and Proceedure of the Committee on the Judiciary. 89th Cong., Washington, Government Printing Office, 1966.

———. Procedures and the Rights of Federal

Employees. Hearing before the Subcommittee on Constitutional Rights of the Committee on the Judiciary. 89th Cong., Washington, Government Printing Office, June 1965.

United States Congress, Subcommittee of the Senate Committee on Post Office and Civil Service. Administration of the Federal Employee's Security Program. Report No. 2750, 84th Cong., 2nd Sess., Washington, Government Printing Office, 1956.

U.S. Department of Defense. The Secret Underworld. Audio-Visual Support Center, Ft. Sheridan, Ill.

———. Unauthorized Disclosure. Audio-Visual Support Center, Ft. Sheridan, Ill.

———. Army Secretary Aides Addresses State CD Directors. DOD Information Bulletin 113. Washington, Government Printing Office, May 13, 1964.

———. Civil Defense and Emergency Planning for the Petroleum and Gas Industries. Washington, Government Printing Office, March 19, 1964.

———. Civil Defense Critical to Security. DOD Information Bulletin 115. Washington, Government Printing Office, May 18, 1964.

———. Civil Defense for National Security. Washington, Government Printing Office, 1948.

———. Civil Military Relationships. Washington, Government Printing Office, February 1963.

———. Congressman Herbert Discusses Civil Defense. DOD Information Bulletin 114. Washington, Government Printing Office, May 15, 1964.

———. Memorandum on Security. (film) Audio-Visual Support Center, Ft. Sheridan, Ill.

———. The Enemy Agent and You. (film) Audio-Visual Support Center, Ft. Sheridan, Ill.

———. The Security Man. (film) Audio Visual Support Center, Ft. Sheridan.

———. Bureau of National Affairs. *Government Loyalty and Security.* Vol. 2. Washington, Government Printing Office, April 7, 1954.

———. Industrial Personnel Access Authorization Review Regulation. Washington, Government Printing Office, April 7, 1954.

———. Rules for the Avoidance of Organization Conflicts of Interest. Washington, Government Printing Office, 1963.

———. The Case of Comrade T. (film) Audio-Visual Support Center, Ft. Sheridan, Ill.

———. Office of Personnel Security Policy. Industrial Personnel Security Review Program. Washington, Government Printing Office, September 4, 1956.

United States Government. U.S. Personnel Security Practices, 88th Cong., 1st Sess., Washington, Government Printing Office, 1963.

U.S. Library of Congress, Legislative Reference Service. Internal security and subversion: Principal state laws and cases. Library of Congress. Washington, Government Printing Office, 1965.

U.S. Munitions Board, Office of Industrial Security. How to be cleared to handle classified military information within industry. Washington, Government Printing Office, 1951.

U.S. Senate, Eighty-Eighth Congress. U.S. personnel security practices. Washington, Government Printing Office, February 20, 26, and March 12, 1963.

Yarmolinsky, Adam: *Case Studies in Personnel Security.* Bureau of National Affairs. Washington, Government Printing Office, August 1955.

PERIODICALS

Arthur, R. O.: Polygraph picks potential policemen. *Law and Order,* September 1964.

The attack on invasion of employee's privacy. *Industrial Relations News,* April 1965.

Auerback, Alexander: Big brother and you. *Sky,* June 1963, p. 8-9.

Avison, David: How to tell a bad apple. *International Security Review,* September 18, 1972, p. 16.

Allen, Chester R. Major: Industrial defense—what it means to you. *Industrial Security,* October, 1957.

Baaden, Walter J., Jr.: Improve your procedure for obtaining security clearances. *Industrial Security,* January 1964, p. 26.

Baker, H. Lee: Is visitor control out of control. *Industrial Security,* February 1966.

Barclay, Dorothy: Supervision without snooping. *The New York Times,* November 1, 1959.

Barlay, Stephen: Everyone has a price. *Top Security,* June 1975, p. 101-103.

———: Total information security is a myth. *Top Security,* May 1975, p. 12-15.

Barrett, R. S.: Guide to using psychological tests. *Harvard Business Review,* September 1963, p. 138-146.

Bates, Alan: Privacy—a useful concept? *Social Forces, 52:*429, 1964.

Bates, Edward Bryant: The polygraph. *Top Security,* May 1975, p. 3-9.

Beech, Paul Cole: The vertification problem. *Industrial Security,* December 1966, p. 38.

Backerley, J. G.: The impact of government information and security controls on competitive industry. *Industrial Personnel Security, 11:* 1955.

Benedon, William: What makes up an adequate records program? *Nat Assn of Costs Accountants Bull,* August 1956.

Benge, Eugene J.: The expanded sales manager and long-range planning. *Sales Management,* November 1960, p. 54-56.

Berlie, R. F.: The ad hoc committee on social impact of psychological assessment. *American Psychologist, 20:*143-146, 1965.

Berle, A. A., Jr.: The protection of privacy. *Political Science Quarterly, 79:*162-168, 1964.

Bier, W. C.: Psychological testing of candidates and the theology of vocation. *Review for Religious, 12:*296, 1953.

Blank-Leahy and Company. Paperwork: It's smothering us. *Nation's Business,* August 1954.

Blum, R. H., and Osterloh, W. J.: Keeping policemen on the job: Some recommendations arising from a study of men and morale. *Police,* May/June 1966.

Bontecou, Eleanor: The federal employee loyalty program 132. *Industrial Personnel Security,* 1953.

Boyd, J. Edwin: Assessing a policeman's performance. *Canadian Police Chief,* January 1975, p. 8-10.

Boyle, Joseph H.: Security test on industrial security manual. *Industrial Security,* January 1959, p. 18.

Bracy, Clarence: Our experience in using the polygraph for background investigations—panelist. *Industrial Security,* October 1960, p. 16.

Brenner, Leonard M.: Classification management—panelist. *Industrial Security,* December 1964, p. 39.

Brim, O. G., Jr.: American attitudes toward intelligence tests. *American Psychologist,* February 1965.

Brown, Ralph S., Jr., and Fassett, John D.: Security tests for maritime workers: Due process under the port security program. *Yale Law Journal, 62:*1165-1953, 1953.

———: The operation of personnel security programs. *Industrial Personnel Security,* 1955.

———: Loyalty and security. *Industrial Personnel Security,* 1958, p. 179-180.

Brunswick, R.: The accepted lie. *Psychoanalytic Quarterly, 12:*458-464, 1943.

Burkey, L. M.: Lie detectors in industrial relations. *Continuing Legal Education, 1:* no. 2, 107-118, April 1963.

Business use of the lie detector. *Business Week,* June 18, 1960.

Buxton, Richard W.: Classification management—a joint effort of DOD and industry. *Industrial Security,* June 1966, p. 8.

Carley, William M.: To tell the truth: Lie detectors used increasingly by business. *Wall Street Journal,* October 1961, p. 1.

Carlson, Gus R.: The uses of agency investigation. *Security World,* May 1969.

Carratu, Vincent: The right man in the right job. *Top Security,* June 1975, p. 70-73.

Chandler, Lee: Do you know the polygraph? *Security World, 1:* no. 1, July 1964.

Coffin: Industry goal: Simple employee loyalty check. *Nation's Business,* December 1955.

Cogley, John: *Report on Blacklisting.* New York, Fund for the Republic, Vol. 1, 1956.

Connell, John G., Jr.: Loyalty-security parallels in history. *Industrial Security, 7:*140, April 1958.

Council of Polygraph Examiners Newsletter, Nos. 1-2, April 21, and May 3, 1964.

Critics of psychological tests—basic assumptions—how good. *Psychology in the School,* January 1964.

Curtis, Robert: Interview techniques that work. *Industrial Security, 5:*26, October 1961.

Davis, Albert S.: Adjustication of security clearance cases. *Industrial Security,* August 1968.

Davis, James A.: Greetings to national classification management society. *Industrial Security,* August 1965, p. 3.

Do you violate your teen-ager's privacy? *Parent's Magazine, 48:* 1960.

Doherty, Joseph F.: Security risk discharge. *Industrial Security,* July 1964.

Dudley, William E.: Reviewing the personal history statement. *Security Management,* May 1974, p. 31-33.

Emerson and Helfeld: Loyalty among government employees. *Yale Law Journal, 58:* 1948.

Ferguson, Col. J. P.: The polygraph knockers. *The Police Chief,* May 1962, p. 26-27.

Goldwater, Barry: Tax supported testing. *Los Angeles Times,* December 1963.

Goulden, Joseph C.: *Truth Is the First Casualty.* Chicago, Rand McNally, 1969.

Halfmann, Robert H.: Temporary help firms—a new source for cleared personnel. *Industrial Security, 8:*38, July 1964.

Halman, Robert: Self-testing for government-industrial security. *Security World, 5:* no. 4, April 1968.

Harkness, Richard: Hiring key people. *Nation's Business,* September 1963.

Harman, George W.: Anti-polygraph smear. *Security World, 1:* no. 2, September 1964.

———: How the polygraph can aid industry—panelist. *Industrial Security,* October 1963, p. 41.

Heckscher, August: The invasion of privacy: The reshaping of privacy. *American Scholar, 28:*13, 1959.

Heims, Peter A.: The right of privacy: England 1970. *Security World,* July/August 1970.

Hire a thief. *Top Security,* May 1975, p. 47.

Hiring key people. *Nation's Business,* September 1963.

Horowitz, Harold: It's hard to get rid of a Communist. *Business Week,* November 11, 1950.

———: How you can find out about a man's subversive affiliations before hiring him. *Public Relations News,* August 9, 1958.

———: Loyalty tests for employment in the motion picture industry. *Stanford Law Review, 6:*466-472, 1954.

Job application form with a built-in integrity test. *Auditgram, 38:* no. 4, 32-35, April 1962.

Johnson, Michael: How many criminals do you employ? *Industry Week,* September 22, 1974, p. 23-30.

Kaufman, Victor: Drugs can't defeat the lie detector. *Security News,* January 30, 1975, p. 18.

Kelin, Stanley: The polygraph: Boom in testing is on. *International Security Review,* September 18, 1972, p. 1, 14, 16.

Klump, Carl S.: Initiating polygraph examination of employees. *Security World, 2:* no. 4, March 1965.

———: So you want to beat the polygraph. *Security World, 2:* no. 6, June 1965.

———: Understanding polygraph use in retail security. *Security World,* December 1974, p. 29+.

Knapp, Harold D.: The case I can't forget. *Industrial Security,* October 1961, p. 91.

Know the man you hire. *Occupational Hazards,* September 1966.

Kruetzer, June: To tell the truth. *Buffalo Business,* February 1963.

La Forge, C. A.: Personnel security pays off. *Industrial Security, 3:*10, January 1959.

Lawrence, C. H., M.D.: Our readers inform us: The psychiatrist, the polygraph, and police selection. *Security World, 3:* no. 3, March 1966.

Lear, John: Whither personal privacy. *Saturday Review,* July 23, 1966, p. 36.

Lewis, Herbert: The industrial personnel access authorization review program—panelist. *Industrial Security,* October 1963, p. 67.

Lindberg, George W.: A test for truth. *Security World, 5:* no. 1, January 1968.

Long, Edward V., Senator: Loyalty and private employment: The right of employees to discharge suspected subversives. *Yale Law Journal, 62:*854, 1953.

———: You ought to be left alone. *Security World, 4:* no. 1, January 1967.

Loomis, Roland L.: Industrial security and professional personnel. *Personnel Journal,* April 1962, p. 193-197.

Loyalty and Security in a Democracy, A Roundtable Report. Public Affairs Pamphlet No. 170, Public Affairs Committee, Inc., 1957.

MacClain, George: Security clearance—protection of the rights of individuals—panelist. *Industrial Security,* October 1962, p. 71.

———: DOD classification management program—panelist. *Industrial Security,* October 1963, p. 53.

———: Classification management—panelist. *Industrial Security,* December 1964, p. 39.

Maines, Howard G.: Classification management—panelist. *Industrial Security,* December 1964, p. 39.

Mental exams said firing tool at Kelly (Air Force Base). *San Antonio News,* June 11, 1965.

Mitchell, Maurice B.: This right of freedom to seek the truth. *Security World, 5:* no. 10, November 1968.

Moley: The Hays office. *Industrial Personnel Security,* 1945.

O'Brian: Know the man you hire. *Occupational Hazards,* September 1966.

———: New encroachments on individual freedom. *Harvard Law Review, 66:* 1952.

O'Keefe, Jack A.: Background check on polygraph interview a useful tool. *Police Chief,* August 1960.

Pell, Arthur R.: Reference checks: The best guide to employee selection. *Industrial Security, 4:* April 1960.

Personal check on employees paying off for big business. *Elmira Star Gazette,* Elmira, N.Y., August 1962.

Personality testing: An invasion of privacy? *Industrial Relations News, 14:* no. 30, July 25, 1964.

Powdermaker: Hollywood the dream factory. *Industrial Security,* 1950.

Price, Carroll S.: An instrumental approach to applicant evaluation. *Industrial Security, 5:* January 1961.

———: Post-polygraph interrogation techniques. *Security World, 3:* no. 4, April 1966.

Psychological testing for workers: Is industry buying a fad? *Business Week,* July 19, 1952, p. 32-86.

Psychology and the crooked employee. *Management Rev,* April 1950.

Robbins, Edgar L.: Loss prevention by selection of personnel. *Industrial Security, 9:* no. 2, April 1965.

Sheehan, Robert: Running a company personnel investigation. *Industrial Security, 4:* no. 16, January 1960.

———: The use of testing. *Industrial Security, 3:* no. 2, 4, April 1959.

Slotnick, Michael C.: The anathema of security risk: Arbitrary dismissals of federal government civilian employees and civilian employees of private contractors doing business with the federal government. *Industrial Security,* 1962.

Soloman, Morris J.: A statistician looks at the employee security problem. *Personnel Administration,* 1955.

Sternbeck, Richard A.: Don't trust the lie detector. *Harvard Business Review,* November-December 1962.

Stessin, Lawrence, and Wit, Ira: The disloyal employee. New York, *Man & Manager,* 1967.

Student unrest in Michigan: One state's protest profile. *College and University Business, 48:* March 1970.

Sutor, David: The gray flannel couch—the lie detector in business. *U.S. Catholic,* July 1964, p. 11.

Testing and discrimination. *Wall Street Journal,* April 1964.

Thompson, George D.: Security evaluations—panel moderator. *Industrial Security,* December 1964, p. 68.

Thompson, G. W., et al.: An empirical evaluation of the watchers of brainwatchers: A survey of testers, tests, and testing programs. *Journal of School Psychology, 3:* no. 4, 49-57, Summer 1965.

Walsh, Thomas J.: The case for applicant investigation. *Industrial Security, 10:* no. 5, 3-6, 8, 10, 43-48, October 1966.

Walsh, Timothy J., and Healy, Richard J.: *Proprietary Data. Protection of Assets Manual, 1:* California, The Merritt Company, 1974.

Warren, S., and Brandeis, L. D.: The right to privacy. *Harvard Law Review,* 1890, p. 193.

Weaver, Leon: Character assessment in personnel administration. *Industrial Security, 9:* December 1965.

Wetzig, Mina: Have personality testers gone too far? *Information,* November 1961, p. 2-9.

White, Charles M.: Security clearance requirements. *Industrial Security,* 2:14, July 1958.

White, Joseph: An argument for pre-employment tests. *Auditgram. 40:* no. 8, 7-9, August, 1964.

Wise, Robert W.: New procedure for industrial security hearings. *Industrial Security,* January 1961, p. 4.

Wright, Loyd: Securing America's productive strength. *Industrial Security, 2:* no. 9, January 1958.

———: Security America's productive strength. *Industrial Security,* 5:4, July 1961.

———: Toward a working program for industrial security. *Industrial Security,* October 1957, p. 5.

UNPUBLISHED MATERIAL

Gafreed, Robert B.: The Use of the Polygraph as a Method of Screening Applicants. Unpublished graduate thesis paper, Michigan State Univ., 1966.

Moris, Lawrence: A Critical View on Oaths of Loyalty in America. Unpublished graduate thesis paper, Michigan State Univ., 1968.

Yang, Shaw-Shan: The Fundamentals of Personnel Security Programs. Unpublished graduate thesis paper, Michigan State Univ., 1966.

APPENDIX 13A

TITLE VI—PROVISIONS RELATING TO CREDIT REPORTING AGENCIES

AMENDMENT OF CONSUMER CREDIT PROTECTION ACT

Sec. 601. The Consumer Credit Protection Act* is amended by adding at the end thereof the following new title:

"TITLE VI—CONSUMER CREDIT REPORTING

"Sec.
"601. Short title.
"602. Findings and purpose.
"603. Definitions and rules of construction.
"604. Permissible purposes of reports.
"605. Obsolete information.
"606. Disclosure of investigative consumer reports.
"607. Compliance procedures.
"608. Disclosures of governmental agencies.
"609. Disclosure to consumers.
"610. Conditions of disclosure to consumers.
"611. Procedure in case of disputed accuracy.
"612. Charges for certain disclosures.
"613. Public record information for employment purposes.
"614. Restrictions on investigative consumer reports.
"615. Requirements on users of consumer reports.
"616. Civil liability for willful noncompliance.
"617. Civil liability for negligent noncompliance.
"618. Jurisdiction of courts; limitation of actions.
"619. Obtaining information under false pretenses.
"620. Unauthorized disclosures by officers or employees.
"621. Administrative enforcement.
"622. Relation to State laws.

---

* 15 U.S.C.A. §1601 *et seq.*

"§601. Short title

"This title may be cited as the Fair Credit Reporting Act.

"§602. Findings and purpose

"(a) The Congress makes the following findings:

"(1) The banking system is dependent upon fair and accurate credit reporting. Inaccurate credit reports directly impair the efficiency of the banking system, and unfair credit reporting methods undermine the public confidence which is essential to the continued functioning of the banking system.

"(2) An elaborate mechanism has been developed for investigating and evaluating the credit worthiness, credit standing, credit capacity, character, and general reputation of consumers.

"(3) Consumer reporting agencies have assumed a vital role in assembling and evaluating consumer credit and other information on consumers.

"(4) There is a need to insure that consumer reporting agencies exercise their grave responsibilities with fairness, impartiality, and a respect for the consumer's right to privacy.

"(b) It is the purpose of this title to require that consumer reporting agencies adopt reasonable procedures for meeting the needs of commerce for consumer credit, personnel, insurance, and other information in a manner which is fair and equitable to the consumer, with regard to the confidentiality, accuracy, relevancy, and proper utilization of such information in accordance with the requirements of this title.

"§603. Definitions and rules of construction

"(a) Definitions and rules of construction set forth in this section are applicable for the purposes of this title.

"(b) The term 'person' means any individual, partnership, corporation, trust, estate, cooperative, association, government or governmental subdivision or agency, or other entity.

"(c) The term 'consumer' means an individual.

"(d) The term 'consumer report' means any written, oral, or other communication of any information by a consumer reporting agency bearing on a consumer's credit worthiness, credit standing, credit capacity, character, general reputation, personal characteristics, or mode of living which is used or expected to be used or collected in whole or in part for the purpose of serving as a factor in establishing the consumer's eligibility for (1) credit or insurance to be used primarily for personal, family, or household purposes, or (2) employment purposes, or (3) other purposes authorized under section 604. The term does not include (A) any report containing information solely as to transactions or experiences between the consumer and the person making the report; (B) any authorization or approval of a specific extension of credit directly or indirectly by the issuer of a credit card or similar device; or (C) any report in which a person who has been requested by a third party to make a specific extension of credit directly or indirectly to a consumer conveys his decision with respect to such request, if the third party advises the consumer of the name and address of the person to whom the request was made and such person makes the disclosures to the consumer required under section 615.

"(e) The term 'investigative consumer report' means a consumer report or portion thereof in which information on a consumer's character, general reputation, personal characteristics, or mode of living is obtained through personal interviews with neighbors, friends, or associates of the consumer reported on or with others with whom he is acquainted or who may have knowledge concerning any such items of information. However, such information shall not include specific factual information on a consumer's credit record obtained directly from a creditor of the consumer or from a consumer reporting agency when such information was obtained directly from a creditor of the consumer or from the consumer.

"(f) The term 'consumer reporting agency' means any person which, for monetary fees, dues, or on a cooperative nonprofit basis, regularly engages in whole or in part in the practice of assembling or evaluating consumer credit information or other information on consumers for the purpose of furnishing consumer reports to third parties, and which uses any means or facility of interstate commerce for the purpose of preparing or furnishing consumer reports.

"(g) The term 'file,' when used in connection with information on any consumer, means all of the information on that consumer recorded and retained by a consumer reporting agency regardless of how the information is stored.

"(h) The term 'employment purposes' when used in connection with a consumer report means a report used for the purpose of evaluating a consumer for employment, promotion, reassignment or retention as an employee.

"(i) The term 'medical information' means information or records obtained, with the consent of the individual to whom it relates, from licensed physicians or medical practitioners, hospitals, clinics, or other medical or medically related facilities.

"**§604. Permissible purposes of reports**

"A consumer reporting agency may furnish a consumer report under the following circumstances and no other:

"(1) In response to the order of a court having jurisdiction to issue such an order.

"(2) In accordance with the written instructions of the consumer to whom it relates.

"(3) To a person which it has reason to believe—

"(A) intends to use the information in connection with a credit transaction involving the consumer on whom the information is to be furnished and involving the extension of credit to, or review or collection of an account of, the consumer; or

"(B) intends to use the information for employment purposes; or

"(C) intends to use the information in connection with the underwriting of insurance involving the consumer; or

"(D) intends to use the information in connection with a determination of the consumer's eligibility for a license or other benefit granted by a governmental instrumentality required by law to consider an applicant's financial responsibility or status; or

"(E) otherwise has a legitimate business need for the information in connection with a business transaction involving the consumer.

"§605. Obsolete information

"(a) Except as authorized under subsection (b), no consumer reporting agency may make any consumer report containing any of the following items of information:

"(1) Bankruptcies which, from date of adjudication of the most recent bankruptcy, antedate the report by more than fourteen years.

"(2) Suits and judgments which, from date of entry, antedate the report by more than seven years or until the governing statute of limitations has expired, whichever is the longer period.

"(3) Paid tax liens which, from date of payment, antedate the report by more than seven years.

"(4) Accounts placed for collection or charged to profit and loss which antedate the report by more than seven years.

"(5) Records of arrest, indictment, or conviction of crime which, from date of disposition, release, or parole, antedate the report by more than seven years.

"(6) Any other adverse item of information which antedates the report by more than seven years.

"(b) The provisions of subsection (a) are not applicable in the case of any consumer credit report to be used in connection with—

"(1) a credit transaction involving, or which may reasonably be expected to involve, a principal amount of $50,000 or more;

"(2) the underwriting of life insurance involving, or which may reasonably be expected to involve, a face amount of $50,000 or more; or

"(3) the employment of any individual at an annual salary which equals, or which may reasonably be expected to equal $20,000, or more.

"§606. Disclosure of investigative consumer reports

"(a) A person may not procure or cause to be prepared an investigative consumer report on any consumer unless—

"(1) it is clearly and accurately disclosed to the consumer that an investigative consumer report including information as to his character, general reputation, personal characteristics, and mode of living, whichever are applicable, may be made, and such disclosure (A) is made in a writing mailed, or otherwise delivered, to the consumer, not later than three days after the date on which the report was first requested, and (B) includes a statement informing the consumer of his right to request the additional disclosures provided for under subsection (b) of this section; or

"(2) the report is to be used for employment purposes for which the consumer has not specifically applied.

"(b) Any person who procures or causes to be prepared an investigative consumer report on any consumer shall, upon written request made by the consumer within a reasonable period of time after the receipt by him of the disclosure required by subsection (a)(1), shall make a complete and accurate disclosure of the nature and scope of the investigation requested. This disclosure shall be made in a writing mailed, or otherwise delivered, to the consumer not later than five days after the date on which the request for such disclosure was received from the consumer or such report was first requested, whichever is the later.

"(c) No person may be held liable for any violation or subsection (a) or (b) of this section if he shows by a preponderance of the evidence that at the time of the violation he maintained reasonable procedures to assure compliance with subsection (a) or (b).

"§607. Compliance procedures

"(a) Every consumer reporting agency shall maintain reasonable procedures designed to avoid violations of section 605 and to limit the furnishing of consumer reports to the purposes listed under section 604. These procedures shall require that prospective users of the information identify themselves, certify the purposes for which the information is sought, and certify that the information will be used for no other purpose. Every consumer reporting agency shall make a reasonable effort to verify the identity of a new prospective user and the uses certified by such prospective user prior to furnishing such user a consumer report.

Individuals and firms preparing credit and informational reports shall follow reasonable procedures to assure maximum possible accuracy of the information concerning the individual about whom the report relates.

"§608. Disclosures to governmental agencies

"Notwithstanding the provisions of section 604, a consumer reporting agency may furnish identifying information respecting any consumer, limited to his name, address, former addresses, places of employment, or former places of employment, to a governmental agency.

"§609. Disclosures to consumers

"(a) Every consumer reporting agency shall, upon request and proper identification of any consumer, clearly and accurately disclose to the consumer:

"(1) The nature and substance of all information (except medical information) in its files on the consumer at the time of the request.

"(2) The sources of the information; except that the sources of information acquired solely for use in preparing an investigative consumer report and actually used for no other purpose need not be disclosed: *Provided,* That in the event an action is brought under this title, such sources shall be available to the plaintiff under appropriate discovery procedures in the court in which the action is brought.

"(3) The recipients of any consumer report on the consumer which it has furnished—

"(A) for employment purposes within the two-year period preceding the request, and

"(B) for any other purpose within the six-month period preceding the request.

"(b) The requirements of subsection (a) respecting the disclosure of sources of information and the recipients of consumer reports do not apply to information received or consumer reports furnished prior to the effective date of this title except to the extent that the matter involved is contained in the files of the consumer reporting agency on that date.

"§610. Conditions of disclosure to consumers

"(a) A consumer reporting agency shall make the disclosures required under section 609 during normal business hours and on reasonable notice.

"(b) The disclosures required under section 609 shall be made to the consumer—

"(1) in person if he appears in person and furnishes proper identification; or

"(2) by telephone if he has made a written request, with proper identification, for telephone disclosure and the toll charge, if any, for the telephone call is prepaid by or charged directly to the consumer.

"(c) Any consumer reporting agency

shall provide trained personnel to explain to the consumer any information furnished to him pursuant to section 609.

"(d) The consumer shall be permitted to be accompanied by one other person of his choosing, who shall furnish reasonable identification. A consumer reporting agency may require the consumer to furnish a written statement granting permission to the consumer reporting agency to discuss the consumer's file in such person's presence.

"(e) Except as provided in sections 616 and 617, no consumer may bring any action or proceeding in the nature of defamation, invasion of privacy, or negligence with respect to the reporting of information against any consumer reporting agency, any user of information, or any person who furnishes information to a consumer reporting agency, based on information disclosed pursuant to section 609, 610, or 615, except as to false information furnished with malice or willful intent to injure such consumer.

"§6.11. Procedure in case of disputed accuracy

"(a) If the completeness or accuracy of any item of information contained in his file is disputed by a consumer, and such dispute is directly conveyed to the consumer reporting agency by the consumer, the consumer reporting agency shall within a reasonable period of time reinvestigate and record the current status of that information unless it has reasonable grounds to believe that the dispute by the consumer is frivolous or irrelevant. If after such reinvestigation such information is found to be inaccurate or can no longer be verified, the consumer reporting agency shall promptly delete such information. The presence of contradictory information in the consumer's file does not in and of itself constitute reasonable grounds for believing the dispute is frivolous or irrelevant.

"(b) If the reinvestigation does not resolve the dispute, the consumer may file a brief statement setting forth the nature of the dispute. The consumer reporting agency may limit such statements to not more than one hundred words if it provides the consumer with assistance in writing a clear summary of the dispute.

"(c) Whenever a statement of a dispute is filed, unless there is reasonable grounds to believe that it is frivolous or irrelevant, the consumer reporting agency shall, in any subsequent consumer report containing the information in question, clearly note that it is disputed by the consumer and provide either the consumer's statement or a clear and accurate codification or summary thereof.

"(d) Following any deletion of information which is found to be inaccurate or whose accuracy can no longer be verified or any notation as to disputed information, the consumer reporting agency shall, at the request of the consumer, furnish notification that the item has been deleted or the statement, codification or summary pursuant to subsection (b) or (c) to any person specifically designated by the consumer who has within two years prior thereto received a consumer report for employment purposes, or within six months prior thereto received a consumer report for any other purpose, which contained the deleted or disputed information: The consumer reporting agency shall clearly and conspicuously disclose to the consumer his rights to make such a request. Such disclosure shall be made at or prior to the time the information is deleted or the consumer's statement regarding the disputed information is received.

"§612. Charges for certain disclosure

"A consumer reporting agency shall make all disclosures pursuant to section 609 and furnish all consumer reports pursuant to section 611 (d) without charge to the consumer if, within thirty days after receipt by such consumer of a notification pursuant to section 615 or notification from a debt collection agency affiliated with such consumer reporting agency stating that the consumer's credit rating may be or has been adversely affected, the consumer makes a request under section 609 or 611 (d). Otherwise, the consumer reporting agency may impose a reasonable charge on the consumer for making disclosure to such consumer pursuant to section 609, the charge for which shall be indicated to the consumer prior to making disclosure; and for furnishing notifications, statements, summaries, or codifications to person designated by the consumer pursuant to section 611(d), the charge for which shall be indicated to the consumer prior to furnishing such information and shall not exceed the charge that the consumer reporting agency would impose on each designated recipient for a consumer report except that no charge may be made for notifying such persons of the deletion of information which is found to be inaccurate or which can no longer be verified.

"§613. Public record information for employment purposes

"A consumer reporting agency which furnishes a consumer report for employment purposes and which for that purpose compiles and reports items of information on consumers which are matters of public record and are likely to have an adverse effect upon a consumer's ability to obtain employment shall—

"(1) at the time such public record information is reported to the user of such consumer report, notify the consumer of the fact that public record information is being reported by the consumer reporting agency, together with the name and address of the person to whom such information is being reported; or

"(2) maintain strict procedures designed to insure that whenever public record information which is likely to have an adverse effect on a consumer's ability to obtain employment is reported it is complete and up to date. For purposes of this paragraph, items of public record relating to arrests, indictments, convictions, suits, tax liens, and outstanding judgments shall be considered up to date if the current public record status of the item at the time of the report is reported.

"§614. Restrictions on investigative consumer reports

"Whenever a consumer reporting agency prepares an investigative consumer report, no adverse information in the consumer report (other than information which is a matter of public record) may be included in a subsequent consumer report unless such adverse information has been verified in the process of making such subsequent consumer report, or the adverse information was received within the three-month period preceding the date the subsequent report is furnished.

"§615. Requirements on users of consumer reports

"(a) Whenever credit or insurance for personal, family, or household purposes, or employment involving a consumer is denied or the charge for such credit or insurance is increased either wholly or partly because of information contained in a consumer report from a consumer reporting agency, the user of the consumer report shall so advise the consumer against whom such adverse action has been taken and supply the name and address of the consumer reporting agency making the report.

"(b) Whenever credit for personal, family, or household purposes involving a consumer is denied or the charge for such credit is increased either wholly or partly because of information obtained from a person other than a consumer reporting agency bearing upon the consumer's credit worthiness, credit standing, credit capacity, character, general reputation, personal characteristics, or mode of living, the user of such information shall, within a reasonable period of time, upon the consumer's written request for the reasons for such adverse action received within sixty days after learning of such adverse action, disclose the nature of the information to the consumer. The user of such information shall clearly and accurately disclose to the consumer his right to make such written request at the time such adverse action is communicated to the consumer.

"(c) No person shall be held liable for any violation of this section if he shows by a preponderance of the evidence that at the time of the alleged violation he maintained reasonable procedures to assure compliance with the provisions of subsections (a) and (b).

"**§616. Civil liability for willful noncompliance**

"Any consumer reporting agency or user of information which willfully fails to comply with any requirement imposed under this title with respect to any consumer is liable to that consumer in an amount equal to the sum of—

"(1) any actual damages sustained by the consumer as a result of the failure;

"(2) such amount of punitive damages as the court may allow; and

"(3) in the case of any successful action to enforce any liability under this section, the costs of the action together with reasonable attorney's fees as determined by the court.

"**§617. Civil liability for negligent noncompliance**

"Any consumer reporting agency or user of information which is negligent in failing to comply with any requirement imposed under this title with respect to any consumer is liable to that consumer in an amount equal to the sum of—

"(1) any actual damages sustained by the consumer as a result of the failure;

"(2) in the case of any successful action to enforce any liability under this section, the costs of the action together with reasonable attorney's fees is determined by the court.

"**§618. Jurisdiction of courts; limitation of actions**

"An action to enforce any liability created under this title may be brought in any appropriate United States district court without regard to the amount in controversy, or in any other court of competent jurisdiction, within two years from the date on which the liability arises, except that where a defendant has materially and willfully misrepresented any information required under this title to be disclosed to an individual and the information so misrepresented is material to the establishment of the defendant's liability to that individual under this title, the action may be brought at any time within two years after discovery by the individual of the misrepresentation.

"**§619. Obtaining information under false pretenses**

"Any person who knowingly and willfully obtains information on a consumer from a consumer reporting agency under false pretenses shall be fined not more than $5,000 or imprisoned not more than one year, or both.

"**§620. Unauthorized disclosures by officers or employees**

"Any officer or employee of a consumer reporting agency who knowingly and willfully provides information concerning an individual from the agency's files to a person not authorized to receive that information shall be fined not more than $5,000 or imprisoned not more than one year, or both.

"**§621. Administrative enforcement**

"(a) Compliance with the requirements imposed under this title shall be enforced under the Federal Trade Commission Act by the Federal Trade Commission with respect to consumer reporting agencies and all other persons subject thereto, except to the extent that enforcement of the requirements imposed under this title is specifically committed to some other government agency under subsection (b) hereof. For the purpose of the exercise by the Federal Trade Commission of its functions and powers under the Federal Trade Commission Act, a violation of any requirement or prohibition imposed under this title shall constitute an unfair or deceptive act or practice in commerce in violation of section 5(a) of the Federal

Trade Commission Act and shall be subject to enforcement by the Federal Trade Commission under section 5 (b) thereof with respect to any consumer reporting agency or person subject to enforcement by the Federal Trade Commission pursuant to this subsection, irrespective of whether that person is engaged in commerce or meets any other jurisdictional tests in the Federal Trade Commission Act. The Federal Trade Commission shall have such procedural, investigative, and enforcement powers, including the power to issue procedural rules in enforcing compliance with the requirements imposed under this title and to require the filing of reports, the production of documents, and the appearance of witnesses as though the applicable terms and conditions of the Federal Trade Commission Act were part of this title. Any person violating any of the provisions of this title shall be subject to the penalties and entitled to the privileges and immunities provided in the Federal Trade Commission Act as though the applicable terms and provisions thereof were part of this title.

" (b) Compliance with the requirements imposed under this title with respect to consumer reporting agencies and persons who use consumer reports from such agencies shall be enforced under—

"(1) section 8 of the Federal Deposit Insurance Act, in the case of:

" (A) national banks, by the Comptroller of the Currency;

" (B) member banks of the Federal Reserve System (other than national banks), by the Federal Reserve Board; and

" (C) banks insured by the Federal Deposit Insurance Corporation (other than members of the Federal Reserve System), by the Board of Directors of the Federal Deposit Insurance Corporation.

" (2) section 5 (d) of the Home Owners Loan Act of 1933, section 407 of the National Housing Act, and sections 6 (i) and 17 of the Federal Home Loan Bank Act, by the Federal Home Loan Bank Board (acting directly or through the Federal Savings and Loan Insurance Corporation), in the case of any institution subject to any of those provisions;

" (3) the Federal Credit Union Act, by the Administrator of the National Credit Union Administration with respect to any Federal credit union;

" (4) the Acts to regulate commerce, by the Interstate Commerce Commission with respect to any common carrier subject to those Acts;

" (5) the Federal Aviation Act of 1958, by the Civil Aeronautics Board with respect to any air carrier or foreign air carrier subject to that Act; and

" (6) the Packers and Stockyards Act, 1921 (except as provided in section 406 of that Act), by the Secretary of Agriculture with respect to any activities subject to that Act.

" (c) For the purpose of the exercise by any agency referred to in subsection (b) of its powers under any Act referred to in that subsection, a violation of any requirement imposed under this title shall be deemed to be a violation of a requirement imposed under that Act. In addition to its powers under any provision of law specifically referred to in subsection (b), each of the agencies referred to in that subsection may exercise, for the purpose of enforcing compliance with any requirement imposed under this title any other authority conferred on it by law.

"§622. Relation to State laws

"This title does not annul, alter, affect, or exempt any person subject to the provisions of this title from complying with the laws of any State with respect to the collection, distribution, or use of any information on consumers, except to the extent that those laws are inconsistent with any provision of this title, and then only to the extent of the inconsistency."

EFFECTIVE DATE

Sec. 602. Section 504 of the Consumer Credit Protection Act* is amended by add-

---

* 15 U.S.C.A. §1601 *et seq.*

APPENDIX 13B

Organizations designated by the Attorney General, pursuant to Executive Order 10450, are listed below:

Communist Party, U.S.A., its subdivisions, subsidiaries and affiliates.
Communist Political Association, its subdivisions, subsidiaries and affiliates, including—
  Alabama People's Educational Association.
  Florida Press and Educational League.
  Oklahoma League for Political Education.
  People's Educational and Press Association of Texas.
  Virginia League for People's Education.
Young Communist League.
Abraham Lincoln Brigade.
Abraham Lincoln School, Chicago, Illinois.
Action Committee to Free Spain Now.
American Association for Reconstruction in Yugoslavia, Inc.
American Branch of the Federation of Greek Maritime Unions.
American Christian Nationalist Party.
American Committee for European Workers' Relief.
American Committee for Protection of Foreign Born.
American Committee for the Settlement of Jews in Birobidjan, Inc.
American Committee for Spanish Freedom.
American Committee for Yugoslav Relief, Inc.
American Committee to Survey Labor Conditions in Europe.
American Council for a Democratic Greece, formerly known as the Greek American Council; Greek American Committee for National Unity.
American Council on Soviet Relations.
American Croatian Congress.
American Jewish Labor Council.
American League Against War and Fascism.
American League for Peace and Democracy.
American National Labor Party.
American National Socialist League.
American National Socialist Party.
American Nationalist Party.
American Patriots, Inc.
American Peace Crusade.
American Peace Mobilization.
American Poles for Peace.
American Polish Labor Council.
American Polish League.
American Rescue Ship Mission (*a project of the United American Spanish Aid Committee*).
American-Russian Fraternal Society.
American-Russian Institute, New York (*also known as the American Russian Institute for Cultural Relations with the Soviet Union*).
American Russian Institute, Philadelphia.
American Russian Institute of San Francisco.
American Russian Institute of Southern California, Los Angeles.
American Slav Congress.
American Women for Peace.
American Youth Congress.
American Youth for Democracy.
Armenian Progressive League of America.
Associated Klans of America.
Association of Georgia Klans.
Association of German Nationals (*Reichsdeutsche Vereinigung*).
Ausland-Organization der NSDAP, Overseas Branch of Nazi Party.

Baltimore Forum.
Benjamin Davis Freedom Committee.
Black Dragon Society.
Boston School for Marxist Studies, Boston, Massachusetts.

Bridges-Robertson-Schmidt Defense Committee.
Bulgarian American People's League of the United States of America.
California Emergency Defense Committee.
California Labor School, Inc., 321 Davisadero Street, San Francisco, California.
Carpatho-Russian People's Society.
Central Council of American Women of Croatian Descent *(also known as Central Council of American Croatian Women, National Council of Croatian Women)*
Central Japanese Association *(Beikoku Chuo Nipponjin Kai).*
Central Japanese Association of Southern California.
Central Organization of the German-American National Alliance *(Deutsche-Amerikanische Einbeitsfront).*
Cervantes Fraternal Society.
China Welfare Appeal, Inc.
Chopin Cultural Center.
Citizens Committee to Free Earl Browder.
Citizens Committee for Harry Bridges.
Citizens Committee of the Upper West Side *(New York City).*
Citizens Emergency Defense Conference.
Citizens Protective League.
Civil Liberties Sponsoring Committee of Pittsburgh.
Civil Rights Congress and its affiliated organizations, including Civil Rights Congress for Texas.
  Veterans Against Discrimination of Civil Rights Congress of New York.
Columbians.
Comite Coordinador Pro Republica Espanola.
Comite Pro Derechos Civiles.
Committee to Abolish Discrimination in Maryland.
Committee to Aid the Fighting South.
Committee to Defend the Rights and Freedom of Pittsburgh's Political Prisoners.
Committee for a Democratic Far Eastern Policy.
Committee for Constitutional and Political Freedom.
Committee for the Defense of the Pittsburgh Six.
Committee for Nationalist Action.
Committee for the Negro in the Arts.
Committee for Peace and Brotherhood Festival in Philadelphia.
Committee for the Protection of the Bill of Rights.
Committee for World Youth Friendship and Cultural Exchange.
Committee to Defend Marie Richardson.
Committee to Uphold the Bill of Rights.
Commonwealth College, Mena, Arkansas.
Congress Against Discrimination.
Congress of the Unemployed.
Connecticut Committee to Aid Victims of the Smith Act.
Connecticut State Youth Conference.
Congress of American Revolutionary Writers.
Congress of American Women.
Council on African Affairs.
Council of Greek Americans.
Council for Jobs, Relief, and Housing.
Council for Pan-American Democracy.
Croatian Benevolent Fraternity.

Dai Nippon Butoku Kai *(Military Virtue Society of Japan or Military Art Society of Japan).*
Daily Worker Press Club.
Daniels Defense Committee.
Dante Alighieri Society *(Between 1935 and 1940).*
Dennis Defense Committee.
Detroit Youth Assembly.

East Bay Peace Committee.
Elsinore Progressive League.
Emergency Conference to Save Spanish Refugees *(founding body of the North American Spanish Aid Committee).*
Everybody's Committee to Outlaw War.

Families of the Baltimore Smith Act Victims.
Families of the Smith Act Victims.
Federation of Italian War Veterans in the U.S.A., Inc. *(Associazione Nazionale Combattenti Italiani, Federazione degli Stati Uniti d' America).*
Finnish-American Mutual Aid Society.

Florida Press and Educational League.
Frederick Douglass Educational Center.
Freedom Stage, Inc.
Friends of the New Germany *(Freunde des Neuen Deutschlands)*.
Friends of the Soviet Union.

Garibaldi American Fraternal Society.
George Washington Carver School, New York City.
German-American Bund *(Amerikadeutscher Volksbund)*.
German-American Republican League.
German-American Vocational League *(Deutsche-Amerikanische Berufsgemeinschaft)*.
Guardian Club.

Harlem Trade Union Council.
Hawaii Civil Liberties Committee.
Heimuska Kai, also known as Nokubei Heieki, Gimusha Kai, Zaibel Nihonjin, Heiyaku Gimusha Kai and Zaibei Heimusha Kai *(Japanese Residing in America Military Conscripts Association)*.
Hellenic-American Brotherhood.
Hinode Kai *(Imperial Japanese Reservists)*.
Hinomaru Kai *(Rising Sun Flag Society—a group of Japanese War Veterans)*.
Hokubei Zaigo Shoke Dan *(North American Reserve Officers Association)*.
Hollywood Writers Mobilization for Defense.
Hungarian-American Council for Democracy.
Hungarian Brotherhood.

Idaho Pension Union.
Independent Party *(Seattle, Washington)*.
Independent People's Party.
Industrial Workers of the World.
International Labor Defense.
International Workers Order, its subdivisions, subsidiaries and affiliates.

Japanese Association of America.
Japanese Overseas Central Society *(Kaigai Doho Chuo Kai)*.
Japanese Overseas Convention, Tokyo, Japan, 1940.
Japanese Protective Association *(Recruiting Organization)*.
Jefferson School of Social Science, New York City.
Jewish Culture Society.
Jewish People's Committee.
Jewish People's Fraternal Order.
Jikyoku Lin Kai *(The Committee for the Crisis)*.
Johnson-Forest Group.
Johnsonites.
Joint Anti-Fascist Refugee Committee.
Joint Council of Progressive Italian-Americans, Inc.
Joseph Weydemeyer School of Social Science, St. Louis, Missouri.

Kibei Seinen Kai *(Association of U.S. citizens of Japanese ancestry who have returned to America after studying in Japan)*.
Knights of the White Camelia.
Ku Klux Klan.
Kyffhaeuser, also known as Kyffhaeuser League *(Kyffhaeuser Bund)*. Kyffhaeuser Fellowship *(Kyffhaeuser Kameradschaft)*.
Kyffhaeuser War Relief *(Kyffhaeuser Kriegshilfswerk)*.

Labor Council for Negro Rights.
Labor Research Association, Inc.
Labor Youth League.
League for Common Sense.
League of American Writers.
Lictor Society *(Italian Black Shirts)*.

Macedonian-American People's League.
Mario Morgantini Circle.
Maritime Labor Committee to Defend Al Lannon.
Maryland Congress Against Discrimination.
Massachusetts Committee for the Bill of Rights.
Massachusetts Minute Women for Peace (not connected with the Minute Women of the U.S.A. Inc.).
Maurice Braverman Defense Committee.
Michigan Civil Rights Federation.

Michigan Council for Peace.
Michigan School of Social Science.
Nanka Teikoku Gunyudan *(Imperial Military Friends Group or Southern California War Veterans)*.
National Association of Mexican Americans *(also known as Asociacion Nacional Mexico-Americana)*.
National Blue Star Mothers of America (not to be confused with the Blue Star Mothers of America organized in February 1942).
National Committee for the Defense of Political Prisoners.
National Committee for Freedom of the Press.
National Committee to Win Amnesty for Smith Act Victims.
National Committee to Win the Peace.
National Conference on American Policy in China and the Far East *(a Conference called by the Committee for a Democratic Far Eastern Policy)*.
National Council of Americans of Croatian Descent.
National Council of American-Soviet Friendship.
National Federation for Constitutional Liberties.
National Labor Conference for Peace.
National Negro Congress.
National Negro Labor Council.
Nationalist Action League.
Nationalist Party of Puerto Rico.
Nature Friends of America *(Since 1935)*.
Negro Labor Victory Committee.
New Committee for Publications.
Nichibei Kogyo Kaisha *(The Great Fujii Theatre)*.
North American Committee to Aid Spanish Democracy.
North American Spanish Aid Committee.
North Philadelphia Forum.
Northwest Japanese Association.

Ohio School of Social Sciences.
Oklahoma Committee to Defend Political Prisoners.
Oklahoma League for Political Education.
Original Southern Klans, Incorporated.

Pacific Northwest Labor School, Seattle, Washington.
Palo Alto Peace Club.
Partido del Pueblo of Panama *(operating in the Canal Zone)*.
Peace Information Center.
Peace Movement of Ethiopia.
People's Drama, Inc.
People's Educational and Press Association of Texas.
People's Educational Association *(Incorporated under name Los Angeles Educational Association, Inc.)*, also known as People's Educational Center, People's University, People's School.
People's Institute of Applied Religion.
People's Programs *(Seattle, Washington)*
People's Radio Foundation, Inc.
People's Rights Party.
Philadelphia Labor Committee for Negro Rights.
Philadelphia School of Social Science and Art.
Photo League *(New York City)*.
Pittsburgh Arts Club.
Political Prisoners' Welfare Committee.
Polonia Society of the IWO.
Progressive German-Americans, also known as Progressive German-Americans of Chicago.
Proletarian Party of America.
Protestant War Veterans of the United States, Inc.
Provisional Committee of Citizens for Peace, Southwest Area.
Provisional Committee on Latin American Affairs.
Provisional Committee to Abolish Discrimination in the State of Maryland.
Puerto Rican Comite Pro Libertades Civiles (CLC).
Puertorriquenos Unidos *(Puerto Ricans United)*.

Quad City Committee for Peace.
Queensbridge Tenants League.

Revolutionary Workers League.
Romanian-American Fraternal Society.
Russian American Society, Inc.

Sakura Kai *(Patriotic Society, or Cherry Association, composed of veterans of Russo-Japanese War)*.
Samuel Adams School, Boston, Massachussetts.
Santa Barbara Peace Forum.
Schappes Defense Committee.
Schneiderman-Darcy Defense Committee.
School of Jewish Studies, New York City.
Seattle Labor School, Seattle, Washington.
Serbian-American Fraternal Society.
Serbian Vidovdan Council.
Shinto Temples (Limited to State Shinto abolished in 1945).
Silver Shirt Legion of America.
Slavic Council of Southern California.
Slovak Workers Society.
Slovenian-American National Council.
Socialist Workers Party, including American Committee for European Workers Relief.
Sokoku Kai *(Fatherland Society)*.
Southern Negro Youth Congress.
Suiko Sha *(Reserve Officers Association, Los Angeles)*.
Syracuse Women for Peace.

Tom Paine School of Social Science, Philadelphia, Pennsylvania.
Tom Paine School of Westchester, New York.
Trade Union Committee for Peace.
Trade Unionists for Peace.
Tri-State Negro Trade Union Council.

Ukranian-American Fraternal Union.
Union of American Croatians.
Union of New York Veterans.

United American Spanish Aid Committee.
United Committee of Jewish Societies and Landsmanschaft Federations, also known as Coordination Committee of Jewish Landsmanschaften and Fraternal Organizations.
United Committee of South Slavic Americans.
United Defense Council of Southern California.
United Harlem Tenants and Consumers Organization.
United May Day Committee.
United Negro and Allied Veterans of America.

Veterans Against Discrimination of Civil Rights Congress of New York.
Veterans of the Abraham Lincoln Brigade.
Virginia League for People's Education.
Voice of Freedom Committee.

Walt Whitman School of Social Science, Newark, New Jersey.
Washington Bookshop Association.
Washington Committee to Defend the Bill of Rights.
Washington Committee for Democratic Action.
Washington Commonwealth Federation.
Washington Pension Union.
Wisconsin Conference on Social Legislation.
Workers Alliance *(since April 1936)*.

Yiddisher Kultur Farband.
Yugoslav-American Cooperative Home, Inc.
Yugoslav Seamen's Club, Inc.

Chapter 14

THE PLANNING PROCESS

"Often, a wide gap exists between ... actions and company goals. But, there is a practical and tested solution to this universal goal-action gap. It starts with three vital and searching questions ...
1. *What are the factors vital to your company's successful operation?*
2. *How have your activities related to these vital success factors?*
3. *What contributions have you made to the company's profitability in the past year in terms of these factors?"*

(*Nation's Business,* May 1969, pages 75-77)

THERE IS NOTHING more vital to the long-term success of a security program than planning. Planning is essential to develop and implement policy and the resulting programs. The development of either the policy to govern operations or the plan for its execution is the result of a process involving the entire organization and its operating environment. Plans and policies are developed as a result of the interaction of:
- People
- Needs
- Data

in order to:
- Evaluate data
- Formulate alternatives
- Develop appropriate responses

The decision-making involved in the planning or policy development process is described in Figure 14.1.

Security policy must be sufficiently broad and flexible as well as dynamic to insure its effectiveness over a period of time. The policy established must be a guide for action and the framework within which program decisions are made. Policy should reflect consistency with organizational objectives and adequate planning for or evaluation of their consequences. In this regard, policy should not be confused with rules and regulations. Policy is the framework which directs the development of appropriate rules and procedures for attaining stated policy or organizational goals. These policies should be written and distributed to all concerned with compliance or supervision of them. They should also be subject to regular review and revision to insure relevance.

POLICY DEVELOPMENT

Security in an organization must provide policy direction for total organizational activities; it functions properly, and the organization is cognizant of and responds to the policies formulated when standard operating procedures (SOP's) are developed which include all aspects of organizational activities which require security pol-

---

DECISION MAKING

individual	*democratic*	*authoritative*
(individual makes his decision based on his own authority and power over himself)	(decision is made by a vote of the people involved)	(person has both the authority and power to make the decision)

Figure 14-1

icy and guidance. The development of policies for security should include a number of basic considerations. These can be outlined as follows:

Guidelines for the Development of Security Policy

1. Essentials of security policy formation
   a. Definite, positive, and clear
   b. Translatable into practice
   c. Flexible, yet highly permanent
   d. Cover all foreseeable situations
   e. Founded on facts and sound judgment
   f. Conform to laws and organizational interest
   g. General statement rather than detailed procedure
2. Why policy should be reduced to writing (SOP)?
   a. Lessens misinterpretations and error
   b. Provides a check list
   c. Constitutes useful instructional device
   d. Failure to write is admission of weakness
3. Who is responsible for formulating policy?
   a. Control must lie at top management level
   b. Policymaker seeks staff aid and guidance
   c. Security not only a concern of security officer
4. Steps in the development of security policies
   a. Determine objectives
   b. Outline problems
   c. Consider practical aspects
   d. Test and analyze
5. Prescribing procedures and rules
   a. Consider objectives, problems and policies
   b. Make it a job analysis
   c. Make it extensive enough to maintain uniformity
   d. Be brief as possible with clarity
   e. Follow standard pattern
6. Examples of operational areas needing policy formulation
   a. Visitor control
   b. Loading dock area
   c. Document control
   d. Political activities
   e. Disbursement of funds
   f. Check cashing

Examples of such policy development can be seen in Appendix 14B which contains standard operating procedures for governmental and proprietary security operations.

Security organizations should make extensive use of the SOP in all levels of operation. The objectives of these procedures are to avoid uncertainty and to narrow the range of alternatives to be considered in a given set of conditions. They are used to make and implement choices. The SOP is extensively utilized in four major operational areas: (1) task performance rules, (2) continuing records and reports, (3) information handling rules, and (4) plans.

One of the major policy decisions by the security organizational management is the means by which actual security activities will be provided for the organization. Security may be provided within the organization by two categories of activity: internal means or external means. Internal means would entail the use of regular organizational personnel for the provision of security, whereas external means would employ the use of commercial guard or protective services to provide organizational security.

Security policy development does not take place in a vacuum but rather as a desire to rationally establish appropriate protective responses for the organization. Policy development is subject to both internal and external pressures which affect the decisions made and the methods selected for implementation. Figure 14-2 presents the framework for security policy development. It can be noted that various pressures affect both the decision-making and the implementation processes; both of

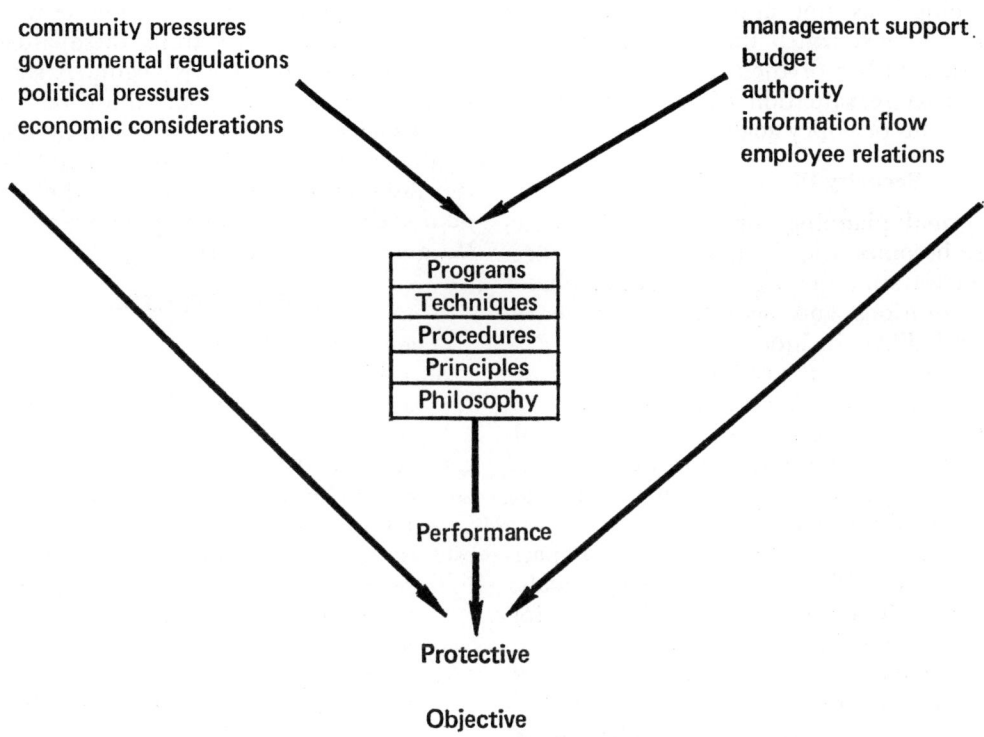

Figure 14-2. Framework of Security Policy Development.

which directly affects departmental or protective objectives.

Security policy directly affects the establishment of security program measures. These measures, in turn affect morale, risk levels, productivity, and cost as well as any legal requirements for protection. Policy must relate to the main purposes and functions of the organization or facility being protected. Moreover, it must take into consideration the desired outcome of the program and the impact of program operations on the organization and its environment. Consequently, for the development of meaningful and realistic policy, planning is a key factor.

Well-conceived and implemented plans are a cornerstone of modern protective services. The planning function is basic to good administration and management to control execution and results of activities. Plans developed provide the program of action to attain a particular goal.

Planning Assumptions

Planning is concerned with the future goals, alternatives, and methods as well as with programs of action. Plans are a prediction about the future based on information and probabilities of events occurring as anticipated. It is not an exact "science" or totally reliable but is rather a management tool for reducing conflict and directing activities into those areas considered most desirable based on all available information. While every plan limits activity, it maximizes effort in a particular direction. If the direction chosen is appropriate, the plan is useful and should be followed. If, however, the stated goals are not being attained, either the planning was not appropriate or conditions changed which the plan did not anticipate. Either of these conditions should initiate a reevaluation of the plan, its assumptions or policy goals.

Security Planning Steps[1]

Formal planning for the security function becomes the foundation for protective efforts for the organization. All phases of operations and activities must be reviewed. These include:
 a. a review of pertinent historical security data from similar systems;
 b. a continuing review of the gross hardware and personnel requirements and concepts, to maintain an understanding of the evolving system;
 c. a review of the proposed departmental objectives and corporate goals;
 d. completion of the planning for follow-on security activities; and
 e. the completion of preliminary hazard analyses to identify potentially hazardous systems and to develop initial security requirements and criteria.

GOALS AND OBJECTIVES

Any planning exercise requires that certain basic decisions be made. Security goals and objectives should be established, and the type of security input that is to be furnished to the overall program should be determined prior to initiating the planning effort. Goals should be measurable in every case and should state what security would intend to accomplish as a result of having performed the various protective tasks.

Once these security goals and objectives have been established and agreed upon by the appropriate program management level, the planner can begin to become familiar with both the evolving hardware and personality systems within which the security program is to operate. Having equated the security goals to the needs of the plan being developed, the planner considers all the alternative methods and analyses that can be used to meet these protective goals and objectives. The optimum methods are selected from these alternatives and the planning is begun. It should be noted that these goals must be structured such that security tasks can be selected that will accomplish the goals and when the tasks have been completed, the result of the effort will clearly demonstrate that the goals have been met.

PLAN CONTENTS

The Security Plan[2] should include:
 a. a description of the security tasks that are to be continued throughout the foreseeable future of the program;
 b. additional tasks that are to be initiated during ensuing program growth or modification;
 c. an estimate of numbers and types of personnel and equipment required for normal as well as emergency situations;
 d. a description of the methods that will be used to perform these security tasks, control the effort, and accomplish the objectives;
 e. scheduling the security effort including milestone identification, program activities, phasing, and integration with other organizational functions; and
 f. identification of the security output that will result from the effort, the expected application of the output, with provisions for the documentation of specific results of the effort.

PLAN REVISIONS

It should be noted that if the planned security program is to have sufficient flexibility, the individual plan must be revised as required to satisfy the changing needs of the program. The plan, however, should not be revised for the sake of change alone. Further, it can readily be seen in the appendices of this chapter that a plan may vary in size from one page in length to a detailed multipage document, depending on the size, complexity, and needs of the program.

Planning tends to be a repetitive process as the plan is refined and expanded as milestones of the system's development are accomplished. Emphases are realigned, nonproductive tasks are abandoned, and new tasks developed as required to accomplish the protective goals. Analysis methods are evaluated against program needs and specific techniques are selected and implemented, as required to provide a sufficient depth of security/risk evaluation.

ORGANIZATIONAL FACTORS AFFECTING PROTECTIVE SERVICES PLANNING

Security plans are developed to control and prevent losses. Performance of this function requires policies, organization, and programs which direct and guide this operation. Planning for protective services operations is effective only if plans are carried out with a broad understanding of the organization, its needs, and the many factors which influence its operations. These factors include the following:

Product or Service

The product or service provided by the organization directly affects the type of protective program required. High-risk or high-value products demand significantly greater protection than low-risk items. Similarly, the provision of highly vulnerable or sensitive service operations such as data processing requires a greater degree of protection than a rug or furniture cleaning service.

Facility Configuration

The physical design or layout of the facility to be protected will determine the type of site hardening or hardware and personnel requirements as well as procedural controls. Some facilities, because of their design, are relatively simple to protect while others are almost impossible to provide with more than a minimal amount of protection.[3]

Neighborhood or Site Plan

The area in which the facility is located can vary in both demographic and geographic characteristics which can aid or reduce the protective capabilities of an organization. The location of a high-risk plant in an inner-city, high-crime area would certainly compound the protective problems for the security department. The location of the same facility in a semirural environment would, on the other hand, most likely reduce the risks of certain types of losses occurring significantly.

Employee Attitudes

The attitudes of workers toward their jobs and the organization directly affects the type and quantity of losses which will occur. Highly motivated employees with good morale are low-risk, high-value assets. If they are present in significant numbers, the security precautions taken are more likely to be minimal and service-oriented rather than performing enforcement and deterrence functions.

Employee Relations

The type of relationship that exists between the organization and its employees will often determine protective measures required. The presence of labor-management disputes, strikes, slowdowns, etc. can lead to violence and losses. Consequently, protective measures are required at a higher level to maintain order and protect employees as well as company property from damage or disruption.

Budget and Sunk Costs

The development or maintenance of security programs is directly related to the amount of funding available for protection. The budget determines the levels of protection that can be provided as well as what services can be offered. The existence of security devices or equipment often limit the type of protection which can be given because of program decision made and the high cost of revising existing operational plans. If, for example, CCTV was installed in an area to eliminate the need for personnel, but the equipment does not provide the type of coverage desired, it might be impossible to add manpower to supplement the equipment. The cost of removing or replacing the CCTV would be more than the budget would allow, particularly since the purchase of the equipment was based on reduced operational costs.

Policy

Planning can be limited by rigid organizational policies. For example, organizational policy might preclude the development of certain types of security measures which would be desirable or useful but are not acceptable in a particular organization. Consequently, the policy of the organization is the guiding factor for the development of operational plans.[4]

Procedures

The existence of procedures can give direction to the planning process. They likewise can indicate the manner in which things are done and provide a standard against which new procedural plans can be evaluated.

Politics

Politics has been characterized as the "art of the possible." Likewise, the security program must deal with what is possible. The development of adequate plans for protection requires that those political forces in the organization or the community be consulted or taken into consideration when plans are developed.

Police Relations

The Private Security Task Force has suggested close working relations between the public police and private security organizations. The security planner must take into consideration the availability of local public police to support his activities or assist in dealing with crime problems in the community. Similarly, the security plan should consider how the security program can assist the local police in carrying out their responsibilities.

Facility Design

The physical plan for a facility as well as the design of security systems are portions of the planning process. The development of facilities which are "protectable" greatly facilitate the overall ability of the security department to be cost-effective in operations. The development of adequate "defensive perimeters" for the facility or organization to be protected can be accomplished most economically during the design phase. The security director or crime prevention specialist should provide input for planning all new facilities or major remodeling of existing ones.

The introduction of protective measures and devices are most often installed after buildings have been completed and occupied. In new construction, the cost and ease of installation of protective features are relatively low when done during the installation of other electrical and mechanical systems. However, their installation at this point in construction requires that the security program for the facility be completed or at least have protective goals established. Unless these goals and the resulting program are preplanned, the placement and installation of protective devices is impossible.

PLANNING CYCLE

Security programs are based on the premise that definable conditions cause loss problems to develop. The security planning process seeks to identify those conditions during a hazard analysis (security survey) and introduce control systems which will eliminate or reduce the possibility of their occurrence. The planning for protection is cyclical in that information collected about problems becomes the input for developing new approaches for eliminating future events. There is a constant flow of information for decision-making about the actions to be taken in the future. The planning process, in practice, allows the organization to "learn by its mistakes" while preventing their reoccurrence. The planning cycle includes the activities shown in Figure 14-3.

The security plan itself becomes a goal for the function which it hopes to accomplish. It is the "predetermined course of action" which directs the efforts of the function in relation to overall organizational goals. Since plans must be controlled, there must be criteria to determine if the plan is being carried out as anticipated. To compare performance with expectations, standards are used. For example, if there is a security department goal of reducing theft losses by 20 percent for the current year and there are losses within the first quarter which are 75 percent of the previous year, a potential problem is readily apparent. Since only 5 percent more loss will equal the maximum loss for the year, more activity is required to prevent higher loss. It might be that the goal established was unrealistic, but the standard of a reduced loss rate of approximately 2 percent per month was exceeded within the first quarter indicating that the plan was not working properly. Then a reevaluation of the plan or goal is indicated.

Types of Security Plans

There are two basic categories or protective services plans: long-range and short-range. Within these two categories are a variety of plans which are essential to effective operations.

Operational Plans

Operational plans include both "Single Use" and Standard Operating Procedures (SOP's) which outline day-to-day usage of personnel and equipment. These plans include the instructions provided to security officers for conducting their patrol rounds, access control procedures, alarm system response, etc.[5]

Procedural plans assure continuity of operations and action and facilitates the interaction of personnel and equipment. They also lessen the possibility of errors and confusion by preplanned coordination.

Management Plans

These plans relate to the operations of the security department and involve the paperwork, reporting, filing, and all administrative control.[6] Management planning insures that all administrative activities are analyzed regularly and improved to increase efficiency and effectiveness of the security function.

Emergency Plans

Plans to direct action in the event of natural or man-caused disasters is essential to effective protective services operations. Mutual aid agreements, evacuation and disaster operations programs, records and personnel dispersal,[7] as well as fire control, bomb search, civil disturbances and sabotage,[8] weather watch, and related life safety needs all require adequate preplanning and testing.[9]

These three types of plans have common dimensions in that they increase the control over the security response to attaining its goals. Performance standards can be determined for each type of action for which plans are in effect and their degree of compliance measures. Variance will result in either a change in the plan or insure more managerial control.

Planning Steps	Functions	Activities
• Risk Analysis	Security Survey	Problem Identification
• Set Goals	Policy Developed	Select Alternatives
• Develop Plan	Program Development	Design Response
• Implementation		
• Evaluate Plan		
• Modify if Necessary		

Figure 14-3. Planning activities.

DISASTER PLANNING

A sound disaster plan should be based on the needs of the facility. All possible disasters which could involve the facility should be incorporated into the development of a disaster plan. These include floods, fires, tornadoes, enemy attack, sabotage, or other types of natural or manmade disasters. While it is impossible to develop a perfect plan, a plan must be developed which insures survival of operation of the facility and the protection of vital records and documents.

A realistic disaster plan is one which includes maximum use of existing personnel and facilities. Except in extreme disaster conditions, most facilities would not sustain more than medium to moderate damage; therefore, the plan should include provisions for management personnel to provide leadership and direction, skilled technicians to handle special tasks, and the marshalling of equipment and materials to provide the necessary hardware.

Content of a Disaster Plan

The basic disaster plan should contain a number of basic items of consideration. These include a statement as to the purpose and authority of the plan, assumptions as to the types of consideration when the plan was developed, data and an outline of the basic plan, and a set of criteria which would measure the effectiveness of training in disaster control procedures.

Organization for Disaster Control

The organization's plan for coping with disaster should be coordinated with the local Civil Defense organization. The company disaster control plan is the counterpart of the government's Civil Defense program at the facility level. The plan should parallel the day-to-day operations and procedures of the organization. The ultimate goal of plan development and procedure training should be the orderly transition from normal to emergency conditions. It should also provide for a return to the normal operations when the emergency has passed. In order to accomplish the development of transition from disaster control, a number of specific criteria should be considered in reference to overall plan development.

Procedural Checklist[10]

Administrative

1. Assess vulnerabilities of physical facilities.
2. Obtain details of area civil defense plans.
3. Appoint organizational disaster control coordinators.
4. Select organizational disaster control advisors.
5. Issue organization policy directives establishing a disaster control program.
6. Train disaster control coordinators.
7. Establish a control center communications system.
8. Develop emergency shutdown procedures.
9. Plan for mass movements of employees in disaster situations.
10. Organize employees into self-help groups.
11. Establish emergency organizational headquarters in an alternative location.
12. Protect vital company records and documents.
13. Plan for continuity of each organizational major function.
14. Assign emergency duties to department heads.
15. Hire appropriate employees.
16. Develop emergency financial procedures.
17. Develop plans to quickly assess and report damage following a disaster.
18. Develop plans for emergency repair and resumption of operations.
19. Establish executive succession lists to insure continuity of management.
20. Amend corporation bylaws and regulations as necessary for post-disaster operations.
21. Develop mutual aid programs with related organizations.

Operations

1. Prepare manuals for organizational disaster control plans.
2. Inform employees about organizational disaster plans.
3. Test disaster control plans with drills and exercises.
4. Train employees for first aid, self-help, rescue, fire fighting, ideological monitoring, decontamination, and shelter management.
5. Designate postdisaster assembly points for employees.
6. Develop an employee education program relating to the benefits and provisions of the disaster control plan. Appendix 14G presents additional disaster planning materials.

FOOTNOTES

1. Suggested by N.A.S.A. Systems Safety Report, Washington, D.C., 1974.
2. A typical security plan outline is shown as Appendix 14A. The specific planning mechanics are outlined in Appendix 14B.
3. Appendix 14C provides a Sample Plan and Bid on specific facility configuration requirements.
4. Appendix 14D presents typical work rules which explain corporate policy toward employees and their work related activities.
5. Appendix 14E presents a typical plan for security operations.
6. Appendix 14F presents an Emergency Records Management Plan.
7. Appendix 14G presents an Emergency Management Plan for Corporations.
8. Appendix 14H provides an Industrial Defense Plan against civil disturbances; 14I deals with sabotage.
9. Appendix 14J provides a Guide to Industrial Civil Defense Planning.
10. *Industrial Civil Defense Workbook.* Department of Defense, Office of Civil Defense, May 1966, p. 28.

BOOKS

Currer-Briggs, Noel (ed.): *Security, Attitudes and Techniques for Management.* London, Hutchinson & Co., Ltd., 1968.

DeCelle, Jack: *The Safety Strategy.* Los Angeles, Frank Publishers, 1971.

Dimock, Marshall S.: *Administrative Vitality.* New York, Harper, 1959.

Dowling, William, and Sayles, Leonard. *How Managers Motivate—The Imperatives of Supervision.* New York, McGraw-Hill, 1971.

Healy, Richard J., and Walsh, Dr. Timothy J.: *Industrial Security Management: A Cost-Effective Approach.* AMA, Inc., 1971.

Levinson, Harry: *Organizational Diagnosis.* Cambridge, Harvard University Press, 1972.

Leonard, V. A.: *Police Organization and Management.* Brooklyn, The Foundation Press, 1951.

Momboisse, Raymond M.: *Community Relations and Riot Prevention.* Springfield, Thomas, 1968.

Now's The Time to Prepare for Hurricanes. New York, McGraw-Hill, August, 1956.

Nuclear Attack and Industrial Survival. New York, McGraw-Hill, 1962.

Pfiffner, John M., and Sherwood, Frank P.: *Administrative Organization,* Englewood, Prentice-Hall, 1960.

Rawls, Robert: *Tested Techniques for Developing and Selling Your Ideas.* Littleton, New Hampshire, Executive Development Press, 1960.

Robinson and Bresso: *Store Organization and Operations.* New York, Prentice-Hall, 1949.

Schlaifer, Robert: *Probability and Statistics for Business Decisions.* New York, McGraw-Hill, 1959.

Spencer, M. H., and Siegelman, Louis: *Managerial Economics: Decision Making and Forward Planning.* Homewood, Ill., Richard D. Irwin, Inc., 1959.

Stahl, O. Glenn: *Public Personnel Administration.* New York, Harper, 1962.

Steiner, George A.: *Managerial Long-Range Planning.* New York, McGraw-Hill, 1963.

Ursic, Henry S., and Pagano, Leroy E., *Security Management Systems.* Springfield, Thomas, 1974.

Vance, Stanley: *Management Decision Simulation.* New York, McGraw-Hill, 1960.

Wessel, Robert H., and Willett, Edward R.: *Statistics as Applied to Economics and Business.* New York, Henry Holt and Co., 1959.

Whyte, William H., Jr.: *The Organization Man.* New York, Simon and Schuster, 1956, p. 12.

Wiltson, H. Hubert: *The Problem of Internal Security in Great Britain,* Garden City, N.Y., Doubleday, 1954.

Wood, Sterling A., Colonel: *Riot Control.* 2nd ed. New York, Military Service Publishing Co., 1942.

Zwelling, Marc: *The Strike Breakers.* Toronto, New Press, 1972.

PUBLICATIONS OF THE GOVERNMENT, LEARNED SOCIETIES, AND OTHER ORGANIZATIONS

Civil Disturbances and Riot Control. 14th Annual Seminar, Industrial Security, 1968.

Company continuity in case of disaster. *Business Record.* New York, National Industrial Conference Borad, November 1961.

Cramer, E. H.: Integrity, are you protecting this risk asset? April 10, 1951.

———: Mr. Jones? A bomb will go off in your building in just seven minutes. But you don't know where it is, do you, Mr. Jones? *Professional Management Bulletins,* Administrative Management Society, August 1971.

Dartnell Corporation. The handbook of employee relations, section 68: "Company Security Programs." Chicago, 1955, p. 1243.

Disaster control and civil defense in federal buildings. General Services Administration, GSA Handbook. Washington, D.C., October 4, 1965.

Disaster planning for the oil and gas industries. National Petroleum Council, Washington, D.C., May 5, 1955.

Doherty, Joseph F.: Community relations and riot prevention (book review). *Industrial Security,* June 1968, p. 46.

Ekey, David C.: *The Use of Consultants by Manufacturers.* Washington, University of Richmond, Bureau of Business Res., 1964.

Emergency Preparedness Progress in the Electric Utility Industry. Department of Defense, Interior, 1973.

Emergency Resources Management and Continuity of Government. New York, National Institute for Disaster Mobilization.

Factory-Protection. Mass., Committee on Public Safety, Emergency Organization for Industrial Plants; Boston; Protective Division, 1941.

Federal Civil Defense Administration. *Civil Defense Industry and Institutions.* Washington, Government Printing Office, 1951.

Fendrock, John J.: *Managing in Times of Radical Change.* New York, American Management Association, 1971.

Gallagher, Frank S.: Fundamentals of emergency and disaster planning—panelist. *Industrial Security,* October 1960, p. 30.

Governor's conference—industrial survival. The Bureau of Labor and Management, College of Business Administration, State Univ. of Iowa, January 1963.

Guide to Developing a Company Industrial Civil Defense Manual, A. Office of Civil Defense, Washington, D.C.

Hauser, N., and Gordon, G. R.: *Computer Simulation of a Police Emergency Response System.* Polytechnic Inst. of Brooklyn, N.Y., 1969.

Hill, H. L.: *Information Systems for Decision-Making and Program Evaluation in the Prevention and Control of Crime International Preview of Criminal Policy.* No. 28, UN Sales Section, New York, 1970.

Hoggatt, Austin: Business games as tools for research. Proceedings ORSA National Meeting, 1960.

In an emergency. *Old Bullion,* Chemical Bank New York Trust Co.

International City Manager's Association. *Municipal Police Administration.* Chicago, Brock and Ranking, 1961.

Johnson and Johnson, and Affiliated Companies. *Minimum Standards for Disaster Controlled.* November 1961.

Lucas, Joseph W.: More profit-less paper through work simplification. Standard Oil Co. of California, 1953.

Meehl, P. E.: *Clinical vs. Statistical Prediction.* Minneapolis, Univ. of Minnesota Press, 1954.

Meerloo, J. A. M., M.D.: *Patterns of Panic.* New York, International Universities Press, 1950.

Miller, A. Ross: Fundamentals of emergency and disaster planning—panelist. *Industrial Security,* October 1960, p. 27.

Minimum Standards for Disaster Control. Johnson & Johnson and Affiliated Companies, January 10, 1957. Revised, November 13, 1961.

Model of a Problem Encountered in a Metropolitan Police Department—Final Report. Wayne State Univ. Press, 1971.

Municipal Police Administration, 5th ed. Chicago, The International City Manager's Association, 1961.

National Advisory Commission on Civil Disorders, March 1, 1968, p. 305-313.

National Commission on the Causes and Prevention of Violence. Washington, D.C., 1969.

Nichols, W., and Heitman, R. E.: Decision making in engineering. ASME Paper Number 59, 1959.

Norton, O. Perry: Industrial security planning—a total protection concept—panelist. *Industrial Security,* October 1963, p. 26.

Osteen, Carl: *Forms Analysis: A Management Tool for Design and Control.* Stanford, Conn., Office Publications Inc., 1969.

Policy Approach to Planning and Social Defense. New York, United Nations, 1972.

Plant Protection Handbook. Studies in business policy, no. 64. National Industrial Conference Board, Inc., New York, 1953.

Richards, Paul M. Human reliability in industry—panelist. *Industrial Security,* October 1963.

Rothenbuer, Jerome. *Cost-Benefit Analysis of Urban Renewal.* Washington, D.C., The Brookings Institution, 1964.

Shiff, Robert, and Steere, Ralph E.: *How One Small Agency Cut Costs, Speeded Paperwork Without Mechanics from Office Management.* New York, Naremco Services, April 1959.

Simulation and gaming: A symposium. American Management Association, 1961.

Stout, Harold: Fundamentals of emergency and disaster planning—panelist. *Industrial Security,* October 1960, p. 30.

Theft Control Procedures. Studies in business policy, no. 70. National Industrial Conference Board, Inc., New York, 1954.

Thirty-Third Annual Report of the National Labor Relations Board. Washington, Government Printing Office.

U.S. Army Military Police School, Fort Gordon, Georgia. *Civil Defense and Disaster Relief.* Washington, Government Printing Office, February 1, 1963, p. 19-173.

U.S. Business and Defense Services Administration. Emergency and Disaster Planning for Chemical and Allied Industries. Washington, U.S. Department of Commerce, 1954.

U.S. Department of Agriculture. Guide to Civil Defense Management in the Food Industry. Washington, Government Printing Office, November 1963.

U.S. Department of the Army. Civil Disturbances and Disasters, Field Manual FM 19-15. Washington, Government Printing Office, December 1964.

U.S. Department of Defense. Responsibility for Civil Defense and Other Domestic Emergencies. Washington, Government Printing Office, July 14, 1956.

———. The Civil Defense System. Washington, Government Printing Office, July 1963.

U.S. Federal Civil Defense Administration. Civil Defense in Industry & Institution. Washington, Government Printing Office, 1951.

U.S. General Services Administration, Industrial College of the Armed Forces. Civil Defense Planning for Survival and Recovery. Washington, Government Printing Office, April 1962.

U.S. Office of Civilian Defense. The Defense Coordinator in an Industrial Plant. Washington, U.S. Office of Civilian Defense, 1943.

———. Passive Protection for Industrial Plants. Washington, U.S. Office of Civil Defense, 1942.

Wasserman, Paul: *Information for Administrators: A Guide to Publications and Services for Management in Business and Government.* Ithaca, N.Y., Cornell University Press, 1956.

What Is Management Security Control? W. J. Burns International Detective Agency, Inc., New York—a brochure on the subject.

Workshop for Supervisory Boards (Southeastern Region LEAA). North Carolina Dept. of Local Affairs, 1971.

Your Guide to Safety in Mercantile Establishments. National Conservation Bureau, Accident Prevention Division, Association of Casualty and Surety Companies, New York City.

PERIODICALS

Carter, Carl L.: Guard force S.O.P. *Security Management,* January 1975, p. 22-23.

Caskey, C. C.: How to get more benefit from training. *Supervision,* October 1964.

Cate, Charles V.: Establishing industrial security in the Canal Zone, 1949-1952. *Industrial Security,* April 1967, p. 4.

Chaden, Sydney: Security in a condition. *Security World,* 5: no. 3, March 1968.

Chambers, Paul: Security as a management function. *Security Gazette,* 13: no. 1, 21, 1971.

Chisholm, Wesley: Emergency planning for life safety. *Security Management,* March 1974, p. 23-24.

Colline, William P.: Trends in planning plant security. *The Plant Corp.,* October 1961, p. 14.

Communications revolution: Its impact on safety/security management. *Occupational Hazards,* September 1970, p. 37.

Company organization for expansion planning. *Industrial Development,* May 1959.

Coon, Thomas F.: Basic minimums of plant security blended with proper leadership. *Police, 9:* no. 4, March-April 1965.

Coping with the increased perils of the modern world. *Top Security,* August 1975, p. 171.

Corsbie, Robert L.: Effects of attack by nuclear weapons. *Industrial Security,* July 1962.

Couch, Virgil L.: Emergency and civil defense planning—panelist. *Industrial Security,* October 1962, p. 29.

———: Is your company ready for civil disturbances? *Canadian Police Chief,* January 1975, p. 23, 24, 34, 35, 37.

———: Status of civil defense—panelist. *Industrial Security,* December 1964, p. 49.

Crowd dispersal without injury. *Canadian Security Gazette,* Summer, 1971.

Curtis, Sargent J.: How to make your company more security conscious. *Industrial Security,* April 1962.

Daly, James W.: Security supervision: What does it require? *Security Industry and Product News,* July/August 1975, p. 8.

Darling, Don D.: Action security for "That Crisis Situation". . . . *Security World,* November 1974, p. 14+.

Davis, James A., and Lockhart, John E.: Preplanning for industrial emergencies and disaster. *Plant Engineering Magazine,* May/June 1960.

Davis, R. H.: Business simulator-management game. *Westinghouse Engineer,* November 1963, p. 55-60.

Directory of market research organizations. *Advertising Agency Magazine,* May 1957, p. 57-58.

Disaster and the role of the loss assessor. *Security and Protection,* June 1975, p. 18-20.

Doberstein, Robert R.: Why salary administration. *Auditgram, 41:* no. 11, November 1965.

Doherty, Joseph F.: Inventory of employees' skills for disaster control. *Industrial Security,* July 1960.

Dolan, Henry P.: Audits of human resources. *Industrial Security,* June 1965.

Draper, Jean, and Strother, George: Testing a model for organizational growth. *Human Organization, XXII:* Summer, 1963.

Drinker, The. *Security Management Plant and Property Protection,* April 1973, p. 3-4.

Drucker, Peter F.: The way to industrial peace: Citizenship in the plant. *Harpers,* December 1946.

Emergency, riot and disaster control, plan for retailers. *Industrial Security,* August 1966, p. 12.

Faneuf, Leston: Industrial security—A management responsibility. *Industrial Security,* January 1958, p. 8.

Flagle, Charles D.: Probability: Based tolerances in forecasting and planning. *Journal of Industrial Engineering,* March/April 1961.

Flanigan, J. C., and King, R. E.: Testing in management selection, state of the art. *Personnel Administration,* March/April 1964, p. 3-39.

Fleishman, Aerom: A survey of problems, techniques, schools of thought in market research. *Industrial Design,* 1958, p. 26-43.

Fox, Robert W.: Anti-panic planning against sudden disaster. *Security World,* December 1971.

Galbraith, D. L.: Emergency control procedures: Are we really prepared? *Industrial Security,* October 1968.

Getty, W. P.: Emergency reporting centers. *Industrial Security,* July 1960.

Glasstone, Samuel: The effects of nuclear weapons. *Industrial Security,* April 1962.

Goddard, R. J.: Good industrial security does not cost—it pays industrial security. *Law and Order, 9:* July 1961.

Griffin, Roger K.: Failure to record. *Security World,* April 1968, p. 38.

Griffin, Robert W.: Buying plant protection. *New England Purchaser,* August 1962.

Haas, Charles F.: How federal experts appraise plant security. *Occupational Hazards,* June 1965.

Hansen, Paul: The security director as a well-rounded person. *Industrial Security,* January 1962, p. 4.

Healy, Richard J.: Industrial security and the scientist. *Industrial Security,* January 1958.

———: Industrial security and physical controls. *Industrial Security,* April 1968.

———: Putting security on the management team. *Security World, 2:* no. 5, July/August 1965.

———: Systems engineering and industrial security. *Industrial Security,* June 1965.

———: Coordinated system in industrial security. *Security World, 1:* no. 2, September 1964.

———: Enlightened industrial security. *Ad-*

vanced *Management Office Executive,* August 1962.

———: Facility planning. *Security World,* 1: no. 1, July 1964.

———: Industrial security—An important element in business. *Southern Calif. Industrial News,* February 7, 1966.

Heap, Dale A.: Modern industrial security challenge to management. *Industrial Security,* January 1963, p. 20.

Heinlein, Karl W.: Why we think professional guards cut security costs. *Pulp and Paper,* April 1964.

Here security keeps pace with growth. *Industrial Security,* January 1964.

Hibbs, Irvin E.: The ASIS program for emergency and disaster control. *Industrial Security,* July 1960.

Highlights of a security plan devised by experts. *Occupational Hazards,* March 1969, p. 35.

Holstein, D.: Decision tables; A technique for minimization routine, repetitive design. *Machine Design,* August 1962, p. 76-79.

Horton, H. Burke: Assessment and reporting of damage following an attack. *Industrial Security,* July 1960, p. 116.

How to make your management security conscious. *Industrial Security,* 6: no. 2.

Humphries, Russell J.: Quiz. *Security Management,* May 1975, p. 32-33.

———: Quiz, management employee relations. *Security Management,* May 1974, p. 21-23.

Humphries, Russell J., and Beck, Sanford E.: What every security director should know. *Security Product News,* January/February 1975, p. 16, 17, 27.

Imberman, A. A.: These executive traits cause personnel problems. *The Office,* December 1971, p. 12-16.

Industrial security and plant engineering—A joint responsibility. *Plant Engineering,* February 1962.

Jackson, Stephen S.: Manpower and security. *Industrial Security,* January 1958, p. 9.

Knapp, Harold D.: Industrial radiological defense services. *Industrial Security,* July 1960.

———: Selling security. *Industrial Security,* April 1958.

Koch, Edward G.: A practical approach to management planning and control. *Advanced Management,* July 1959.

Lauback, Alexander C.: Gaining acceptance for a security program. *Industrial Security,* 11:20, April 1967.

Levitt, Theodore: Marketing myopia. *Harvard Business Review,* July/August, 1960, p. 45-46.

Lighten your security burden. *Security Gazette,* December 1969.

Llewellyn, R. W.: Game information theoretic decision model. *Industrial Engineer,* May/June 1961, p. 150-154.

Long range planning for new plant payoff. *Dun's Review and Modern Industry,* March 1959, p. 61-62.

Loomis, Roland L.: Selling security. *Industrial Security,* January 1959.

McDonald, W. J.: Security pays in many ways. *Industrial Security,* 5:16, January 1961.

McPherson, L. W.: Assistance from industry to local government in pre-attack planning. *Industrial Security,* July 1960.

Malone, Lee F.: Planning continuity of management. *Industrial Security,* July 1960, p. 100.

Marketing planned five years ahead. *Sales Management,* May 1961, p. 75-76.

Marcus, Stanley: Management looks at security. *Industrial Security,* October 1960.

Marshall, Robert R.: The role of market research in the security industry. *Top Security,* May 1975, p. 42-43.

Meisler, Stanley: Charge of civil defense. *The Nation,* June 11, 1960.

Moral power of shareholders, The. *Business Week,* May 1, 1971, p. 6.

Morse, George P.: Protection must be a plan. *Industrial Security,* February 1971.

Mumford, J. K.: Planning for the retail security seminar. *Security World,* 3: no. 1, December/January 1956.

Munnich, Michael: Testing the plant emergency plan and organization. *Industrial Security,* July 1960.

National Institute for Disaster Mobilization, Inc. The Boeing emergency preparedness program. *Industrial Security.*

———. The role of decontamination in industrial recovery. *Industrial Security.*

New products: The push is on marketing. *Business Week,* March 4, 1972, p. 72.

Neilson, P. Raymond: Telling employees about the company emergency plan. *Industrial Security,* July 1960.

Norton, John Joseph: The responsibility of management in the industrial security program. *Industrial Security,* February 1966, p. 28.

———: The security executive must be a business man too. *Industrial Security*, 5:18, April 1961.

Overseas market research: A do-it-yourself guide: with data by subject, by publication source. *Dun's Review and Modern Industry*, May 1957, p. 105-113.

Paradiso, L. J.: How can business analyze its markets? *Survey of Current Business*, March 1945, p. 6-13.

Payne, Bruce: Long-range planning: Special report. *Chemical Week*, January 1960, p. 78-84.

Pearson, Andrall E.: An approach to successful marketing planning. *Business Horizons*, Winter, 1959.

Planning in today's management; panel discussion. *Aerospace Engineering*, April 1962, p. 56-57.

Pre-riot retail planning. *Security World*, June 1966, p. 34-36.

Price, Ralph A.: Why emergency or disaster planning—panelist. *Industrial Security*, October 1960, p. 26.

Role of the professional security director in industrial disaster preparedness, The. *Industrial Security*, July 1960, p. 9.

Role of the security department in human and community relations. *Security Education Briefs*, OAK Security, Inc., 2: no. 12.

Rompilla, Michael, Capt. USA: Security aids for management. *Industrial Security*, July 1959.

Ronald, Daniel D.: Reorganization for results. *Harvard Business Rev.*, 44: November-December 1966.

Rose, A. M.: Motivation research and subliminal advertising. *Social Research*, 25:271-284, 1958.

Russell, A. Lewis: Leadership responsibility for disaster control—where should it be placed in the company organization. *Industrial Security*, July 1960.

Rutherford, Ivan H.: Expanding the role of the industrial security department in modern industry. *Industrial Security*, 8:22, December 1964.

Schurr, Robert C.: Management of the corporate security program. *Security Management*, May 1975, p. 14-21.

Schwartz, Dr. Howard: Business security in a "Lay Off" economy. *Security World*, January 1975, p. 24+.

Schweyer, H. E., and May, F. P.: Criteria for capital expenditure. *Industrial and Engineering Chemical*, August 1963, p. 46-50.

Scott, Don H.: A brief look at management in 1970. *Sales Management*, November 1960.

———: Failure to make long-range plans. *Sales Management*, 1960, p. 38-40.

Security is your business, too. *Administrative Management*, December 1962.

Seney, Wilson: Financing for the long-range. *Dun's Review and Modern Industry*, July 1960, p. 43-45.

Shelley, Tully, Jr., and Pearson, Andrall W.: A blueprint for long-range planning. *Business Horizons*, Winter, 1958, p. 77-84.

Shoemaker, Donald R.: Disaster control—survival of the fittest. *Industrial Security*, June 1971.

Simon, Herbert A.: Management by machine: how much and how soon. *Management Review*, November 1960.

Smith, E. M.: Planning emergency shutdown procedures. *Industrial Security*, July 1960, p. 33.

Spencer, Ivor: Security consultants. *Top Security*, June 1975, p. 74.

Spalding, W. F.: Don't suffocate the security forces. *Security World*, 9: no. 6, 13, 15-16, June 1973.

Starratt, Charles A.: Industrial security management programs. *Industrial Security*, June 1966, p. 26.

Steiner, George A.: How to forecast defense expenditures. *California Management Review*, Summer, 1960.

Stessin, Lawrence: Right or wrong in labor relations. *Mill and Factory*, November 1962.

Stocker, William M., Jr.: Industrial disaster control. *Special Report 416. The American Machinist, C:* no. 5, February 27, 1956.

Stolze, W. J.: Decision making as it affects long-range planning. *Engineering Management*, March 1962, p. 33-36.

———: Measuring effectiveness of technical proposals and marketing effort in military electronics. *IRE Transactions on Engineering Management*, June 1960, p. 62-66.

Thompson, Kenneth L.: Long-range planning for a growing firm. *Journal of Accounting*, March 1960, p. 39-43.

Turner, Carl C.: Industrial preparedness against civil disorder. *Security World*, March 1969.

Veiana, Anthony F.: Discharge for dishonesty and theft. *Labor Law J.*, May 1959.

Vogel, T. E.: Management in security. *Security World*, May 1969.

Vorenberg, J.: Experiments with innovative procedure. *Police Chief, 23:* no. 12, 44-47, December 1966.

Walsh, Timothy J.: Functions of the security director in industry. *Industrial Security,* October 1957.

———: Law enforcement controls in labor disturbances. *Police, 8:* no. 6, 25-31, July-August 1964.

Walsh, Timothy J., and Healy, Richard J.: Disaster controls, *Protection of Assets Manual. 1.* California, The Merritt Company, 1974.

———: Human relationship. *Protection of Assets Manual. 3.* California, The Merritt Company, 1974.

———: Organization and management. *Protection of Assets Manual. 2.* California, The Merritt Company, 1974.

Wanbach, Albrecht G.: Planning for emergency repair and restoration. *Industrial Security,* July 1960, p. 111.

Whitney, Mary: Greater freedom for women has increased the security problem. *College and University Business, 48:*97, April 1970.

Wihstron, Walter S.: Factors in manpower planning. *Management Record,* September 1960, p. 2-5.

Wilkins, J. W.: Blueprint for corporate security. *Canadian Security Gazette,* Summer, 1971.

Wilson, John: Security during works shutdown with contractors on site. *Security Gazette,* July 1975, p. 242-243.

Williams, Joseph D.: A business executive's view of the drug problem. *Security Management,* May 1975, p. 7-10.

Wortham, Dr. A. W.: Planning is essential to corporate success. *News Front,* April 1961.

Wris, Anren: Are you tomorrow minded? *Management Review,* February 1961.

Yale, Jorden P.: Elements in long-range planning. *Advanced Management,* May 1961.

Yont, L. W.: Developing a vital records protection program. *Records Review, 5:* no. 1, January 1964.

Young, Frank E.: Civil disturbance in business and industry a case in point. *Industrial Security,* June 1966, p. 29.

Ziegler, Edward F.: Emergency and civil defense planning—panelist. *Industrial Security,* October 1962, p. 28.

UNPUBLISHED MATERIAL

Harringan, A. W.: The role of the supervisor in an asset protection program. Presentation to Western Electric Security Conference, Princeton, November 10, 1965.

Kettler, Gordon W.: A comparative study of the effectiveness and efficiency of integrated and non-integrated industrial security organizations. Unpublished Michigan State Univ. thesis, Fall 1964.

Walsh, Timothy J.: Security standards and the national labor policy in private industry. Thesis, New York University School of Law, 1962.

APPENDIX 14A

CHECKLIST OF REQUIREMENTS FOR DEVELOPING A SECURITY PLAN

1.0 PURPOSE AND SCOPE OF PLAN
2.0 SECURITY ORGANIZATION
    2.1 Relationship to total organization
    2.2 Organizational structure
    2.3 Responsibilities
    2.4 Interfaces
3.0 SECURITY TASKS TO BE COMPLETED
    3.1 Criteria development
    3.2 Analyses
    3.3 Design/program review participation
    3.4 Reporting

3.5 Documentation
3.6 Planning
3.7 Evaluations
4.0 METHODS FOR ACCOMPLISHING SECURITY TASKS
    4.1 Criteria—development, documentation, and monitoring
    4.2 Analysis technique
    4.3 Other program activities
5.0 SCHEDULE FOR TASK COMPLETION
    5.1 Key to major planning milestones
6.0 Appendices

APPENDIX 14B

MECHANICS OF PROTECTIVE SERVICES PLANNING*

The eight steps discussed below provide an orderly means for the development of plans, be they large or small, long-range or short-range. These provide a framework for the consideration of all aspects of planning.

1. *"Frame of reference."* This is based on a careful review of the literature relating to the situation for which plans are being developed, and opinions or ideas of persons who may speak with authority on the subject of concern. Definitive views of the security director, officers, and other organizational officials are important. In effect, data gathered provide an outline of the best available information on the situation.

2. *Clarifying the problem.* This requires identifying the problem, and understanding both its record and its possible solution. A situation must exist for which something must and can be done. For example, a plant is victimized by a series of burglaries. There is need for reaching the preliminary decision that burglaries may be reduced at the plant and that the security department can reduce them.

3. *Collecting all pertinent facts.* No attempt should be made to develop a plan until all facts relating to it have been gathered. In the series of burglaries, all information should be carefully reviewed to determine *modus operandi,* suspects, and such other information as the public police will provide. Facts relating to such matters as availability, deployment, and use of present personnel should be gathered, as well as such other data as may be needed—a review of general and special orders, and pertinent literature on burglary coverage.

4. *Analyzing the facts.* After all data has been gathered, a careful analysis and evaluation must be made. This provides the basis from which a plan or plans are evolved. Only such facts as may have relevance should be considered.

5. *Developing alternative plans.* In the initial phases of plans development, several alternative measures will appear to be logically comparable to the needs of a situation. Persons responsible for plans development may overlook or fail to consider important factors. As the alternative solu-

---

* Adapted from *Mechanics of Planning,* U.S. Dept. of Justice Planning Project, #122, 1967.

tions are evaluated, one of the proposed plans will usually prove more logical than the others. For example, one plan may call for the use of off-duty patrolmen for "stake-outs" in the burglary case, and the other plans for use of increase site-hardening and electronic protection with better communication to response force personnel.

6. *Selecting the most appropriate alternative.* A careful consideration of all facts usually leads to the selection of a "best" alternative proposal. In the burglary case, the "best" plan may call for use of hardware with maximum use of police for "stake-outs." Sometimes a synthesis or a compromise between plans may provide a solution.

7. *"Selling" the plan.* A plan, to be effectively carried out, must be accepted by all personnel concerned. This requires involving all persons concerned at the appropriate level of the plan's development. For example, in the burglary case, the security director may be preparing the plan. At the outset, the police department is concerned and should be consulted. As the planning develops, there may be need to involve the patrol personnel, and communication units, as well as company officials. Lastly, all personnel to be involved in the "stake-out" should be thoroughly briefed and involved, as necessary, in the making of minor (and, in some instances, major) changes in the plan. In other words, all involved personnel should be "sold" on the merits of the plan before its implementation in order to reduce possible chances of failure or disruption.

8. *Arranging for execution of the plan.* The execution of a plan requires the issuance of orders and directives to involved units and personnel, the establishment of a schedule, and the provision of manpower and equipment for carrying out the plan. Briefings must be held and assurance must be received that all involved personnel understand when, how, and what is to be done.

9. *Evaluating the effectiveness of the plan.* The results of plans should be determined. This is necessary in order to know whether a correct alternative was chosen, whether the plan was correct, which phase (if any) was poorly implemented, and whether additional planning may be necessary. Also, the effects of the executed plan on other operations and on total departmental operations must be determined. Follow-up is the control factor essential for effective departmental management.

APPENDIX 14C

SECURITY SURVEY AND PROPOSAL TO ABC INDUSTRIES

ALARMS AND PROTECTIVE EQUIPMENT INSTALLATIONS

Security manpower requirements in Plant 1 can be reduced and a higher level of security achieved without inconvenience to ABC Personnel by the following modificatons and installations.

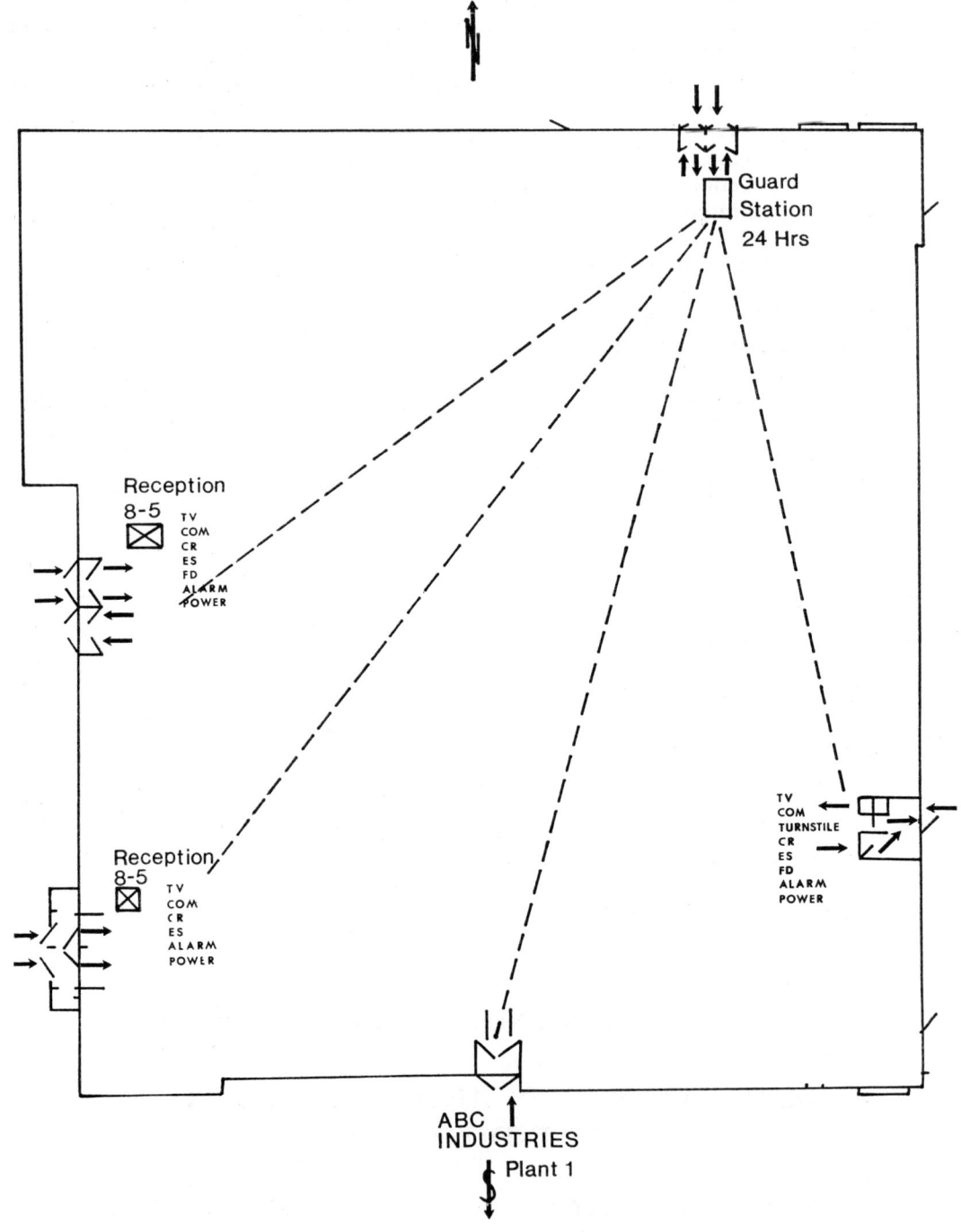

I. IDENTIFICATION CARD SYSTEM

Modify present ABC I.D. system by use of a color photo, electronically encoded I.D. card system. The new card better serves the I.D. purpose. It is now in common use in industry; a duplicate photo can be retained for file and/or replacement of lost cards; it is not alterable; it is tamper-proof; it is readily recognized; it can be modified for use as an outer-wear badge in some applications; it can be encoded for use as a control for reader devices used to control access to or to record other data such as time and attendance and material control; and its cost is nominal as set forth in the following section. Each card can be serial numbered with its own code.

A. *Suppliers:*
   1. General Binding Corporation
      a. Produces a camera utilizing Polaroid® film which renders four separate photos per sheet and which can be cut to varying size specifications.
      b. Manufactures and supplies key cards, laminating sheets and pouches for key cards.
      c. Assembles a portable kit camera, laminator, cutter and supplies so it may be quickly set up for I.D.'s issued as rapidly as subjects can sit down for a photo.
      d. Prices:

Complete photo I.D. system with carrying case	$1,045.00
Supplies—Uncoded electronic key cards per C	80.00
Laminating pouches per C	10.00
Company would supply the printed I.D. portion	
Cost per card in quantities of supplies for 1,000	1.02

      e. Availability—4 to 6 weeks. Service anywhere in the U. S.
   2. Polaroid I.D.-3 System
      a. Produces a kit in a single unit to produce and issue I.D. cards and badges utilizing fixed Polaroid camera and film. The card key codes and laminating material are supplied through General Binding or Card Key Co.

b. Cost of the I.D.-3 System (2 to 4 weeks delivery)	$1,890.00
c. Utilizing source material from other suppliers than Polaroid the cost per I.D. card with card key capability is	1.13
Simple photo I.D. card is	.34

B. *Application*

ABC Security can:

(1) Acquire the equipment from either supplier through lease purchase and outright purchase plans. No special savings accrue through purchase of film through Polaroid although some savings are realized by purchasing supplies through General Binding. Both companies offer different colored photo backdrops which produce a better photo and give color code aspects to the picture.

(2) We can then supply a service to a client by making up I.D. badges and cards on a regular and special basis and for other clients as needed for an appropriate prorated fee.

(3) We can provide the I.D. system to clients as a part of our equip-

ment package. I.D.s can be made up in minutes by a personnel clerk while the subject is being otherwise processed and lost cards can be quickly replaced.

(4) The I.D. card should have the company (client) data on the front with the photo and the rear would have the ABC Security serial number and return address for lost cards. The serial number would help in re-coding cards. This will be covered more thoroughly.

C. *Recommendation*

Both suppliers use systems commonly accepted in industry. They are both portable and to some extent use interchangeable material and equipment. The General Binding system uses a $\frac{1}{4}$ film sheet photo which is affixed to the client's choice of I.D. card styles. Four separate photos can be taken on one sheet of film.

Polaroid utilizes one half of a film sheet to produce a complete photo I.D. card, reduced in size which can have an overlay of the client's logo or authorized signature. Two cards may be made to the film sheet. The resultant card will still be laminated and/or affixed to the type of carrier required by the client.

Both systems can be readily color-coded for recognizable access control, both are tamper-proof. Polaroid is completely self-contained in one kit and is more attractive. General Binding is equally portable, can be used better by two operators for speed and is more adaptable to a variety of uses.

General Binding equipment is recommended for purchase and use by ABC Security on basis of cost and adaptability.

Polaroid is recommended for lease or issuance to client on the basis of its simplicity of operation for training client's personnel to use it. It is self-contained and easily and safely stored and is more attractive.

II. CARD KEY ACCESS SYSTEM

A. *System Capability*

1. Essentially the photo *I.D. card* is electronically encoded. When it is inserted into a *card reader,* an *electric strike* is activated allowing the key holder to open a door for entry or departure from the plant or to other controlled areas. A separate card reader is required for each entry or departure application as it is permanently installed at the access point. A key or safety latch can be used to override the system.
2. The key card can be coded to a large variety of access possibilities, and systems can be installed which will cancel out lost, stolen or otherwise invalidated cards. Master key controls are more stringent than conventional lock systems and allow closer restriction possibilities.
3. A single system can be further modified to include time clock punching when interfaced with a computer. Remote control operation by valid card control holders can be used. Access to restricted equipment such as computers, Xerox®, alarm controls, etc., is also a standard application of the system.

B. *System Components*

1. *Card Reader:* a switch device. It can be a simple magnet control or more sophisticated electronic circuit board which can be recoded by

installation of a new cartridge, replacement of a coded matrix card, change of circuitry, or application of a coding machine. The card reader can be mounted in a variety of housings for esthetic appeal, unit security and environmental protection.
2. *Electric Strike:* a remote control electromagnet. Replaces the face plate in a door frame into which the latch bolt of any standard mechanism slides. It is activated by electric current supplied via the card reader circuit. Electric strikes are manufactured in a variety of qualities dependent upon the level of security desired.
3. *Transformer:* a simple item in this case to reduce standard AC line voltage to the electric strike. It can be purchased independently.
4. *Emergency Power Supply:* used for conversion to battery power for standby operation of the electric strike in event of a power loss. This can be supplanted by an automatic device on the electric strike causing it to lock in an open or closed position in event of power failure.

C. *Suppliers*
1. *Cardkey Systems (TM):* Security-Lock Access Control systems and machine readable credentials.
    a. Utilizes the identification card described previously with their patented magnetically coded key card which is inserted into a slot in the face plate of the "Securiti-Lock." This activates the electric strike. Variations are available according to customer needs for mounting, security levels, environment and type of application, e.g. door, machine control, time clock recorder, etc. More sophisticated systems are available with "Memori-Lock" (pushbutton controlled system) adaptation, Securiti Card Data System which provides perimeter security control in areas with multiple access points by means of high-speed data collection and transmittal to a central control unit which records the data, allows closeout and false card alarm indication, etc.
    b. At the level of security required for general application at ABC Industries, this supplier's "Programmable Security Cartridge" with changeable matrix code cards and utilizing the "Programmable Identification Card" (P.S.L. and K-3) is appropriate. The supplier has proposed contractual price agreements to Oak Security and as authorized distributor of their equipment in quantity lots.
    c. Increased sophistication of the system requires increasing the sophistication of the Card Key, Card Reader, housing, etc. A complete file of all adaptations is attached. Cost factors will be outlined here for a simple door system without central data control or recording equipment.
    d. Cost for each Programmable Security-Lock (PSL) $ 115.00
        (1) Less 30% for quantities over five
            Assembly including Housings—e.g.   81.00
            Flush Type   91.00*
            Aluminum box type for doors, etc.   101.50
            All Weather type for outside use   126.00

---

* Estimated cost indicated they might vary depending on cost at the time of installation.

(2) Electric Strike (Not recommended by Company)   (13.00)
    Suggested outside purchase   45.00*
(3) Transformer also better if bought separate   7.00*
(4) Emergency Power Supply with battery   28.00*
(5) Security Card Data System less 30%   3357.00

e. Cost of one door installation . . . . $ 181.00

2. *General Binding Corporation & Eaton Corporation, Electronic Security Division.*

   a. "Identi-Logic (TM)": Single Door Entry Control. Researching this equipment revealed that General Binding Corporation, a previously referenced supplier of Key Cards and I.D. Photo Systems is including a system of electronic access controls in their capability. We were referred to G.B.C. by the "Card Key" representative who recommended their camera and supplies over Polaroid. G.B.C. in turn demonstrated the following items and submitted the referenced price schedules. Further research revealed that the "Identi-Logic" system is actually manufactured and distributed by E.T.N. or Eaton Corporation as above. We have requested pricing and distribution information from Eaton on an Original Equipment Manufacturer's basis.

   b. This system utilizes the General Binding Corp. Identi-Logic coded key card which is inserted into a key way mounted on or at the exterior of the door or system to be controlled. The key way is connected to the logic control panel located on the interior of the doorway in a secure place (usually not more than 6' distance), by a color coded cable. The power is supplied to the control panel and subsequently, to the electric strike from a standard A.C. outlet via a step down transformer. Several levels of security are available. Sophisticated systems with additional features such as central control station with voiding and lock-out systems and printers are also available. In one year these systems will be equipped to interface with computers to record time and attendance or other desired features.

   c. At the current level of security required for ABC Industries, the Series No. 1001 Identi-Lock or the Series No. 1155 (with printer control interface) is most applicable. The second unit (1155) is capable of being adapted to a more sophisticated system at a later date without additional cost. No change in basic or ancillary equipment is necessary.

   d. Cost for each Identi-Lock® (Series No. 1155) $ 295.00
      (1) Less 30 percent estimated discount for O.E.M. rate   206.50*
      (2) Includes key way, transformer, alarm, etc.
      (3) Emergency Power Supply (est.)   30.00*
      (4) Electric Strike (est.)   45.00*
      (5) Identi-Coder (Key Machine) O.E.M. est.   822.50
      (6) Central Control & Printer O.E.M. est.   4760.00
      (7) Multiplex for five remote stations est.   1645.00

   e. Estimated cost for a single Identi-Lock door installation, with printer interface $ 281.50
      Estimated cost for Series No. 1001   210.50

3. *Detex Entry Control Systems (DENTCO TM).*
   a. This company makes a wide range of exit and entry controls and alarms. The "DENTCO" system utilizes the same components referred to in previous suppliers. The key card is coded and supplied by DENTCO with a registered serial number. Operation is simple: Coded I.D. card is inserted into card reader in one of several available mountings, electric circuit activates electric strike and door is opened, AC line voltage is reduced to 24 volts to the components. Single and Multiple code applications are available. Master coding can be accomplished according to specifications submitted to DENTCO. This equipment is the simplest studied, has the least degree of further available sophistication or adaptability and appears to be the easiest to install.
   b. As noted, the DENTCO System is not adaptable to further sophistication, however, it is certain that knowledgeable installers could interface the card reader with other functions. Catalog price of available equipment are reduced by 50%.
   c. Cost of Standard Card Reader (Multi-Code) $150.00
      Cost of Electric Strike, list from $36 to $240
      Suggested outside purchase estimate   45.00
      Transformer varies—approx.   8.50
      Standby power supply (requires DC adapter)   60.00
      Total cost for each door installation $263.50

4. *Other Suppliers.*
   a. Inquiries to other suppliers of card key type access control systems are still being received. Cost data has been requested and specific answers about the ability to further sophisticate and initial installation are being sought. Variations in systems make very few functions interchangeable. For example, the Polaroid Co. and General Binding Co. both make I.D. cards which may be used or included with any card key, but one card key is not interchangeable with another manufacturers reader.
   b. This report will be amended with additional data as it is received. No dramatic changes in price or capability is expected.
   c. Other suppliers whose prices and/or equipment is to be considered are Eaton (E.T.N.), Unit Systems, Inc., Alarm-Lock Corp., and Rusco Access Control Systems, Inc.

D. *Recommendation*
   1. Pending receipt of additional data as above, no specific supplier is recommended. At present, Identi-Logic and Card Key Systems appear to offer the most adaptable available systems.

   It is recommended that card reader access control be installed at the following locations:
   a. East Detex Emergency Door. Two (2); one for exit and one at entrance.
   b. South Office or Accounting Door. One on outside wall by safety latch controlled door.
   c. Main Lobby, Corporate Division. One on outside glass lobby door.
   d. Personnel Entrance. One on outside, safety-latch, glass door.
   2. Further adaptions of the Card Reader System would be effective for time clock readings when interfaced with the data center for payroll

purposes. A higher level of security could be accorded to the data center with one card reader for access (possibly in lieu of the exit reader for the East Detex door). Several systems have a central control capability which produces a printout record of card usage at each location.

Card Keys can be master keyed to allow access to all points, several points only at certain hours, only one access point, etc. For example, the Security Officer would have Total Access. Assembly workers would have access only to the East Detex Door and the time clocks. Office workers would have access to the locale of their time clocks. The north employees entrance would be controlled by a guard-monitor on post in any case, so a card reader would afford measurable degree of additional security with a printout record.

In short, in addition to being an effective, efficient and nominally inexpensive method of achieving a high level of security and personnel control, the additional features offer a wide range of desirable accessory functions.

III. CLOSED CIRCUIT TELEVISION MONITORING SYSTEM

A. The ability of a single security officer to visually monitor a number of different locations at a central location affords a recognizable advantage over hourly foot patrols. Not only can he effectively supervise access and egress at controlled locations, but other adaptations include general interior and exterior scanning which affords security for the company and safety for personnel. Localized CCTV can afford an executive the opportunity to scan his reception area and interior operations and check out persons requesting access to restricted areas.

It is recommended that Closed Circuit TV be installed at the following locations to be monitored at the ABC Security Central Station on a 24-hour a day basis:
1. East Detex Door Entrance for plant employees.
2. South Accounting Office Entrance.
3. Main Lobby (Corporate Reception Area).
4. Personnel Entrance.

B. *Equipment and Costs:*
New equipment and new suppliers of industrial closed circuit TV become available daily. A study of proposed equipment suggests the following as relatively inexpensive, highly reliable and adaptable to the requirements of this installation:

ASACA, A.S.A. 500 CCTV Camera w/fixed lens	$180.00
ASACA, A.V.M. 090 9" Monitor	123.00
Remote Switcher	59.40
Switch Control Head	62.65
Mounting Hardware	12.00

C. *Supplier:*
This equipment is available through RST Services. They advised the best method of transmittal from Plant #1 to the Central Station is via Modulated Laser Communication. This is of limited availability. By means of the remote switcher and control head, all four cameras would

be connected to a convenient central switch point and run via one coaxial cable to the switch control. Installation costs are reduced and one monitor can easily handle up to six cameras.

IV. TURNSTILE ACCESS & EGRESS CONTROL
   A. In conjunction with a card reader, this device offers a high level of impersonal control of employees' entry and departure via a specific exit in the plant.

   In this instance, it is recommended that a floor-to-ceiling (7') turnstile be installed at the East Detex door utilized by employees moving to and from the Main-East Parking Lot. Presently the door is used twice daily for shift changes. A security officer monitors this door from 6:00 A.M. to 8:10 A.M. and again from 4:15 P.M. to 5:15 P.M. Employees needing access at other times must depart via the main guard desk at the North Entrance or via the Personnel exit, both of which require that they walk around the plant to their cars.

   The exit is identified as "Fire Exit" by illuminated signs and has a crash bar alarmed locking device which is deactivated by the guard when the door is used.

   B. A physical count of employees using this door was taken on April 6, 7, 8, 1976. On Tuesday, April 6, eighty-seven (87) employees used the employees exit. On Wednesday, April 7, a security officer was posted at the Shipping walk-in door (not an employee exit). He did not challenge anyone, but one hundred fifty-seven (157) employees used the Detex door exit. On Thursday, April 8, a security officer was posted at the Receiving exit, which had previously been locked and two hundred (200) employees used the East Detex door.

   Utilizing a two-way turnstile, monitored by card key access and CCTV and a remote electric strike, authorized employees can use this exit at all appropriate times, still maintaining a high control level of security.

   C. Other types of door controls, such as glass revolving doors, were found to be impracticable and not adaptable to security utilizations. A full-stride single entry turnstile will require employees to use their identification for access under TV observation. The door cannot be held open or left ajar for everyone else behind them.

   Two types of turnstiles are appropriate.
   1. Perey type AA "Full-Stride Roto Gate" is esthetically appealing. It can be painted to match the decor to allow a less ominous appearance. The cost is $940 F.O.B. New York, plus $50 for additional safety switches.
   2. A less expensive model is available, the type B, at $765. It is principally used for outside application.

   The current vestibule at the East Detex door would have to be remodeled to accept this installation. The exit itself is deficient in that it does not meet current OSHA requirements, nor does it meet Life Safety Code Standards, pertaining to "Means of Egress, general," Section 1910-37, Code of Federal Regulations, Title 29.

   D. Research on this matter, with the OSHA Training Institute, disclosed that a turnstile does not invalidate a fire exit as long as an alternate "crash bar or gate" is immediately available for emergency use.

APPENDIX 14D

COMPANY RULES AND POLICIES

XYZ Industries tries to operate with a minimum of rules and regulations, and attempts to create as pleasant a working environment as possible for all its people. History has shown XYZ employees have a high sense of responsibility and understanding for each other's rights.

However, when a group as large as ours must be in close contact every working day, we must have certain rules to be sure our employees are protected from occasional trouble-makers. (Infractions of these rules may bring disciplinary action up to and including discharge.)

1. Reporting for or working while under influence of alcohol or drugs, or possession of any alcoholic beverage or narcotics on company property.
2. Theft of company or personal property.
3. Gambling or playing games of chance on company premises.
4. Destroying company property deliberately.
5. Fighting, horseplay, and disorderly or immoral conduct.
6. Providing false or misleading information intentionally to obtain employment, or the making of false, vicious or malicious statements concerning any employee, the company or its products.
7. Being late or absent from work without notifying your supervisor or the Personnel Department, continued tardiness or absenteeism.
8. Loafing on the job or idling in washrooms or other areas during working hours.
9. Stopping work before quitting time, inefficient or unsatisfactory performance of duties.
10. Punching a fellow employee's time card, or permitting another to punch your card.
11. Submitting false time records or work tickets.
12. Entering unauthorized areas or walking in plant without proper authorization.
13. Smoking in unauthorized areas or otherwise violating fire regulations.
14. Violating safety or health regulations.
15. Failing to carry out instructions of your supervisor.
16. Collecting money for gifts, flowers, parties, memorials or selling of merchandise or tickets without written approval of the Personnel Department.
17. Unauthorized carrying or possession of firearms on company property.
18. Bringing cameras on company property without the approval of the Personnel Department.
19. Playing radios or T.V. except as approved by the Personnel Department.
20. Oral solicitation or distributing literature on company premises by nonemployees.
21. Oral solicitation by employees during working time.
22. Distributing literature by employees during working time.
23. Distributing literature by employees in working areas at all times, posting or the removal of notices, signs or writing in any form on bulletin boards or company property without specific company approval.
24. Receipt by the company of three or more wage assignments or deductions in a twelve-month period.
25. Using company telephones for personal calls without proper authorization.

APPENDIX 14E

SECURITY PROCEDURES

A. This manual is for the use of Security Officers assigned to tours of duty at XYZ Industries, Plants 1 and 2. It is the property of ABC Security, Inc. The information contained herein is proprietory both to ABC Security and the client, XYZ Industries. The interest of both could be seriously compromised should this information be divulged to unauthorized persons.

The security officer is responsible for all duties set forth applicable to his tour. He will understand the procedures outlined herein for carrying out those duties, and he will find reference material and further explanatory data to assist him. The manual is subject to revision and amendment and should be continually reviewed.

B. *Description of Responsibilities*
   1. At least one guard is required to be present, 24 hours a day, 365 days a year in each plant.
   2. He will implement the policies of XYZ Industries relative to persons entering or departing the premises. In this connection, an officer will ordinarily be assigned at any exit allowing regular access and egress such as the main employee entrances not controlled by a receptionist.
   3. Since there will always be at least two officers on duty between both plants, safety and good sense dictate they must coordinate their activities.
   4. Breaches of security, hazards, safety violations, injuries, fires and other emergencies will always involve the security officer in one manner or another. He must render assistance when asked, and initiate such requests at the time of an incident. His sole presence at times other than normal plant operations is the only means of detection, deterrence, and disclosure of an occurrence. Thus, security officers will patrol the interior and exterior of both plants at appropriate times.
   5. In order to effectively communicate his observations and occurrences he will have knowledge of appropriate personnel and agencies to contact in any circumstance. In addition, in order to maintain an intelligent and legal record of the security officers activities while on an assignment, he will make a daily report on the prescribed form and will utilize a "Detex Guard Clock" and "Station Keys" while making his patrol.
   6. Certain ancillary duties will be performed at various times, seemingly unrelated to his primary responsibilities; such as turning on and off lights and equipment, handling telephone messages, shipping and receiving after hours, facilitation of interplant mail, and assistance with time and work cards. These duties and their explanatory instructions will be incorporated in this manual as they are received from proper authority.
   7. Since security personnel are only assigned when actually required to perform the duties and to provide the scheduled coverage as needed and without overlap, each tour is critical. The assigned person is responsible for meeting his obligation to report when scheduled and remain until relieved or the tour terminates.

   Changes, substitutions, adjustments etc. must be coordinated and approved by the supervisor.
   8. A security officer's purpose and re-

sponsibilities are unique as applied to other personnel at the installation and it is important that they be recognizable to others as the person on whom certain responsibilities fall. A uniform is, therefore, required. It must be worn while on duty and may be worn while directly en route to or from assignment. The uniform will be supplied by ABC Security, Inc., and will be approved by the client. Deviations are not warranted.

9. The uniform, I.D. card, and other issued equipment remain the property of ABC Security, Inc. They will be maintained with care and be returned at the time of termination or transfer as requested.
10. The security officer will abide by the code of rules and regulations for employees of XYZ Industries.

APPENDIX 14F

SUBJECT: Emergency Planning Program

REFERENCE: Records Management

INTRODUCTION:

This document outlines the emergency planning requirements for the protection of employees and company property during and after local or national emergencies.

All operating units (groups, divisions, subsidiaries and corporate staffs) will maintain active emergency planning programs at all of their locations for the protection of employees and company property during and after local or national emergencies.

The widespread dispersal of company facilities and personnel increases the potential exposure to numerous types of emergencies, e.g. natural, accidental, civil disorder, hostile attack.

Experience has proved that casualties and damage can be minimized through careful advance planning. Effective planning and implementation of an emergency program begin with an evaluation and utilization of existing resources. The major areas requiring consideration and written plans are:

1. Administration
2. Personnel
3. Emergency Services
4. Vital Records
5. Continuity of Operations
6. Restoration of Facilities
7. Reporting

I. *Administration*

Emergency plans should provide for:
A. Appointment of coordinators having responsibility for the overall program on a divisional/subsidiary as well as location level.
B. Appointment of an advisory staff (where appropriate) to represent various departments of the plant, laboratory, office or other locations involved.

C. Designation of management succession to provide for continuity of management.
D. Training of personnel in fire, first aid, rescue, damage control, and evacuation.
E. Designation of both on-site and off-site emergency control headquarters.
F. Procedures for emergency shutdown of the facility.

II. *Personnel*

Plans should provide for locating, accounting for, communicating with and rendering appropriate assistance to employees both at home and at work. The following should be considered:
A. Personnel reporting centers or procedures as required by local conditions.
B. Assignment of personnel to operate such centers.
C. Advice to employees concerning the plan as it affects their welfare. Employee-manager meetings, individual letters to employees, location and divisional newspapers, bulletin boards, etc., may be used.
D. Evacuation procedures and the designation of available shelter areas.
E. Provision for emergency funds to meet payroll and operating expenses.
F. Immediate availability of current personnel listings. Listings should include home addresses, position classification, and other personnel data which would be helpful in an emergency.
G. Ability of a manager to reach employees at home during nonworking hours.

III. *Emergency Services*

It is necessary to provide adequate personnel and proper equipment to respond immediately to an emergency condition. This is the first line of defense in limiting the extent and seriousness of injuries and damage. Advance training, specific assignments and planning are necessary for:
A. Facility protection,
B. fire protection,
C. vital services and supplies,
D. emergency alert system,
E. rescue teams,
F. damage control units,
G. first aid teams,
H. emergency equipment,
I. emergency transportation,
J. communications facilities,
K. medical personnel and supplies, and
L. emergency numbers for police, fire, hospital, ambulance, utility companies, military or police explosive unit.

IV. *Vital Records*

To ensure continued operations, each location should provide a program for the protection of vital records, based on requirements to reconstruct and resume activities after a disaster.

V. *Continuity of Operations*

Planning should include the resumption of operations (production, development, marketing, etc.) at an alternate site. While immediate standby facilities cannot be provided, contingency plans should be maintained in respect to leased space or other company facilities. This, together with a planned course of action and off-site vital records, will assist in the prompt resumption of operations.

Particular attention should be given to back-up facilities for data processing, reproduction, and microfilming operations, as well as supporting parts procurement, manufacturing processes, and engineering records which support manufacturing. Current documentation to support vital data processing operations is critical to disaster recovery.

VI. *Restoration of Facilities*

Necessary plans or blueprints for each facility should be protected as part of the vital records program. Provision should be made locally for restoring a damaged facility as quickly as possible with consideration given to availability of extra contractors and materials. Personnel should be assigned responsibility for liaison with supply and service organizations, e.g. power, light, water, telephone.

VII. *Reporting*

In addition to offering every reasonable assistance to company employees, we should consider our customers and the general population in the event of an emergency. When a disaster occurs it should be promptly reported up the line organization. When appropriate, it will be the line organization's responsibility to advise the corporate office. Preliminary reports must be made at the earliest possible time and should be followed by detailed information as soon as it becomes available. Reports should include the nature and extent of the emergency, injuries to employees, damage to employees' property, damage to property, damage to customers' offices and installed equipment, and action taken or recommended.

Conditions affecting the welfare and/or safety of employees should be reported by the division director of personnel to the Vice President, Personnel Plans and Programs and the Vice President, Personnel Relations.

Other company locations in the immediate area, and those that normally have day to day business with the affected location, should be advised any time a location closes due to an emergency, or experiences a serious situation that could cause injuries or otherwise affect nearby company locations. (Examples: fire, explosion, threat, inclement weather, power failure, etc.)

GENERAL

A. *Testing*

Periodic tests of the vital records program will be conducted at locations by Company Records Management and/or divisions. These tests will necessarily include those parts of emergency planning that bear on resumption/continuity of operations. Operating units should encourage locations to test their own vital records/emergency planning programs. These tests are especially effective for emergency services, continuity of operations, facility restoration, etc.

B. *Review of Planning*

Corporate Legal is responsible for reviewing division and subsidiary planning. Company Records Management is responsible for reviewing and testing the vital records program. Division and subsidiaries are responsible for reviewing plans of their locations.

C. *Fallout Shelters in Company Buildings*

Local managers will cooperate with the federal government program of surveying, licensing and marketing of existing areas of company space as public shelters. There will not be fallout shelters as such included in the design of new buildings. Only those supplies furnished by the government will be stocked.

Licensing and stocking of company space as a public shelter should not interfere with the continuous use of that space for business purposes.

D. *Emergency Work Rotation Plan*

The purpose of this plan is to permit dispersal of employees during a grave national emergency, i.e. imminent threat of nuclear attack. The emergency plans at each location must contain procedures which give assurance that this plan can be activated within 24 hours.

Upon receipt of instructions from corporate headquarters to activate the plan, locations will assign employees to one of five teams on either a departmental or functional basis. Each team, representing 20 percent of the normal work force, will work its customary hours two days a week, so that 40 percent of the work force will be at work and 60 percent will be dispersed. During the period that the plan is in operation, no employee is to receive less than his regular salary.

This will be maintained as an "on the shelf" plan with no ongoing implementation.

E. *Disaster Support*

Financial or other assistance to company employees affected by a disaster should be based on individual consideration. Immediate action should be taken if the situation is urgent. If time permits, recommendations should be made by the location manager and approved by the division.

In cases of disaster, organizations such as the American Red Cross and Salvation Army usually mobilize their forces in the area to alleviate hardships to the best of their ability. We should cooperate with such organizations, and contribute to relief funds occasioned by the disaster. (Follow corporate guidelines on making contributions.)

F. *Emergencies Affecting the Community*

Areas containing a number of company facilities require additional planning for emergencies (e.g., civil disorders) affecting part or all of the community.

An area coordinator should be appointed to act as a central point for communications and to serve in an advisory capacity to the other locations.

Liaison with police and other authorities should be centralized to avoid unnecessary inquiries from company locations.

On an individual location basis:

—Designate those individuals authorized to declare an emergency closing.
—Assure that employees understand they are not required to enter a

troubled area and may leave such an area if a disturbance develops.
—Make sure that employees are not placed in a position of violating curfews imposed by the authorities.

APPLICABILITY
Applicable to all operating groups, divisions, subsidiaries and corporate staffs.

EFFECTIVE DATE
Immediately

DISTRIBUTION
Distribution List "A"

APPENDIX 14G

EMERGENCY MANAGEMENT PLAN*
FOR A UTILITY COMPANY

	Paragraph
Objective	1
General	2
Succession of Company Officers	
Chief Executive Officer	3
Secretary	4
Treasurer	5
General	6
Directors	
The XYZ Company	7
Major Subsidiaries	8
Retired Personnel and Management Development Replacement Charts of Major Subsidiaries	9
Procedures with Depositaries and Trustees	10
Alternate Headquarters	11
Records Protection Committee	12
Maintenance of Plan	13

Attachments
No. 1 Sample Resolutions re Contingent Officers
No. 2 Instructions to be put on Contingent Officer Envelopes
No. 3 Instructions to be put on Management Development Chart Envelopes

---

* Continuity of Corporate Management in Event of Major Disaster Office of Civil Defense, Washington, D.C., 1970, pp. 29-36.

THE XYZ COMPANY EMERGENCY MANAGEMENT PLAN

1. OBJECTIVE

To be able to continue the business of the XYZ Co. and its subsidiaries following the destruction of company officials, company headquarters, or both, by a severe catastrophe resulting from explosions, accidents, fires, floods, windstorms, riots, and the like, under a plan which provides for continuity of management and for alternate headquarters.

2. GENERAL

The XYZ Co. Emergency Management Plan sets forth procedures whereby the XYZ Co. and each of its three major subsidiaries, X Power Co., Y Power Co., and Z Power Co., shall have at all times officers with the powers and duties of the chief executive officer, the secretary, and the treasurer if a catastrophe should result in the death or incapacity of any such officer. This plan also sets forth the procedures for reconstituting the boards of directors of such companies in case a catastrophe results in the death or incapacity of some or all of their members. Alternate headquarters are designated for each of these companies and provision is made for the establishment of XYZ Co. Records Protection Committee to prepare a program for the protection of all vital records. Provision is also made for the maintenance of the plan.

3. SUCCESSION OF COMPANY OFFICERS—CHIEF EXECUTIVE OFFICER

THE XYZ CO. If the chairman does not survive or is incapacitated, the president will have the powers and duties of the chief executive officer. If neither the chairman nor the president survives or both are incapacitated, the surviving and not incapacitated regular vice president with the longest tenure in office will have such powers and duties. If neither the chairman nor the president and no regular vice president survives or all are incapacitated, a vice president previously elected by the board for this contingency will have such powers and duties.

MAJOR SUBSIDIARIES. If the president does not survive or is incapacitated, the surviving and not incapacitated regular company-paid vice president with the longest tenure in office will have the powers and duties of the president. If neither the president nor any regular company-paid vice president survives or all are incapacitated, the remaining surviving and not incapacitated regular vice president with the longest tenure in office, if there be any, will have such powers and duties, or if there be none, a vice president previously elected by the board for this contingency will have such powers and duties.

4. SUCCESSION OF COMPANY OFFICERS—SECRETARY OF XYZ OR A MAJOR SUBSIDIARY

If the secretary does not survive or is incapacitated, the surviving and not incapacitated company-paid assistant secretary with the longest tenure in office will have the powers and duties of the secretary. If neither the secretary nor any regular company-paid assistant secretary survives or all are incapacitated, the remaining surviving and not incapacitated regular assistant secretary with the longest tenure in office, if there be any, will have such powers and duties, or if there be none, an assistant secretary previously elected by the board for this contingency will have such powers and duties.

5. SUCCESSION OF COMPANY OFFICERS—TREASURER OF XYZ OR A MAJOR SUBSIDIARY

If the treasurer does not survive or is incapacitated, the surviving and not incapacitated company-paid assistant treasurer with the longest tenure in office will have the powers and duties of the treasurer. If neither the treasurer nor any regular company-paid assistant treasurer survives or all are incapacitated, the remaining surviving and not incapacitated regular assistant treasurer with the longest tenure in office, if there be any, will have such powers and duties, or if there be none, an assistant treasurer previously elected by the board for this contingency will have such powers and duties.

6. SUCCESSION OF COMPANY OFFICERS—GENERAL

Officers who serve in an emergency will have powers and duties which they would not otherwise have only in the absence of the officer who regularly has such powers and duties and only until that officer is replaced or his office filled by the board.

The officers of a company elected to serve in the contingencies referred to in paragraphs 3, 4, and 5 will be elected at the annual meeting of the board of that company, the person elected to each of the offices being the first person on a list of not less than seven who is not incapacitated and can serve without prior regulatory approval. Samples of the resolutions used are attached as attachment No. 1. No person will be named in more than one resolution of any one company, but the same person may be named in one of the resolutions of two or more companies. Generally, not more than four persons who customarily work in the same place or are in the New York office at the time of the monthly meetings will be named in any one resolution. To assist the board of a company, tentative lists will be prepared by the president of that company before each annual meeting. These tentative lists will show the present position and location of each individual named and may include retired persons.

Promptly after each annual meeting of a company board, four certified copies of each board resolution electing an officer for an emergency will be prepared by the secretary. Each copy will be placed by the secretary in a separate envelope, and all the envelopes will be delivered by him to the president of his company. The president will seal and identify with his signature each such envelope and he will then cause one of the four envelopes for each such office to be placed in a safe or vault in his company's headquarters office and one to be forwarded to the president of each of the other three companies to be placed in a safe or vault in his company's headquarters office. Each envelope will bear the information shown on attachment No. 2.

Each officer and director of XYZ and the major subsidiaries will be given a copy of this plan, and will thereby know of the existence, location, purpose, and general content of these envelopes, but not the names on the lists in each. Each envelope shall be opened only in the event of the contingency noted on it and then only upon authorization from any director, president, or vice president of XYZ or the major subsidiaries.

Each envelope shall be returned to the president of the company to which it relates upon receipt of a corresponding envelope from the president of that company after the succeeding annual board meeting.

7. DIRECTORS—THE XYZ CO.

Vacancies in the board will be filled by the remaining directors or director. If no director survives or if all who survive are incapacitated and thereby unavailable (which would mean the incapacity of the chairman and the president since each is required to be a director), the vice president with the powers and duties of the chief executive officer will call a special meeting of stockholders to elect directors. During any period in which the board is unable to act because of a lack of members, the officers will manage the company.

8. DIRECTORS—MAJOR SUBSIDIARIES

Vacancies in the board will be filled by the remaining directors, after consultation with the officers of XYZ, if there is a sufficient number of them to do so. The number of directors required to fill vacancies will have to be determined at the time by an examination of the charter and bylaws of the company in question. If the number of directors remaining is not sufficient to fill vacancies, the company's parent, as stockholder, will do it. During any period during which the board is unable to act because of a lack of members, the officers will manage the company.

9. RETIRED PERSONNEL AND MANAGEMENT DEVELOPMENT REPLACEMENT CHARTS OF MAJOR SUBSIDIARIES

In planning for contingent officers and in filling other vacancies that might arise in an emergency, the board and officers of XYZ and the major subsidiaries will consider persons employed in or retired from any such company. The president of each major subsidiary will have prepared in

January of each year a list of his company's retired personnel who are physically and mentally able and who, at retirement, held positions as superintendents or higher. Each list will show the addresses of the persons retired, the position each occupied prior to retirement, and the position which he might fill in an emergency.

A management development replacement chart for each major subsidiary will be prepared in quadruplicate under the direction of the president of that company. One copy of the retired list mentioned above and one copy of the management development replacement chart will be placed in each of four separate envelopes to be sealed and identified with the signature of the president of the subsidiary. The president will then cause one of these envelopes to be placed in a safe or vault in his company's headquarters office and will forward one to the president of XYZ and the president of each other major subsidiary to be placed in a safe or vault in each company's headquarters office. These envelopes will bear the information shown on attachment No. 3. Each envelope will be replaced annually by an envelope containing an up-to-date retired personnel list and management development replacement chart. The envelope being replaced will be returned to the president of the company to which it relates.

10. PROCEDURES WITH DEPOSITARIES AND TRUSTEES

If a person who because of a catastrophe becomes an officer under the contingent officer arrangements of this plan is required to satisfy a depositary or a trustee of his incumbency, he will furnish the depositary or trustee a certificate of his election to his office and, if considered necessary or desirable, a copy of this plan and any other pertinent papers. After contingent officers have taken office, specimens of their signatures will be furnished to the depositaries and trustees.

Depositary resolutions should in all cases cover positions.

11. ALTERNATE HEADQUARTERS—XYZ AND MAJOR SUBSIDIARIES

In the event a catastrophe makes the New York office of XYZ or the headquarters office of a major subsidiary unusable, alternate headquarters are designated as follows:

The XYZ Co.—Headquarters office of the Y Power Co.
The X Power Co.—Alpha Division office
The Y Power Co.—Beta district office
The Z Power Co.—Gamma Division office

It is recommended that no attempt be made to designate particular persons to report to alternate headquarters if a regular headquarters becomes unusable. It is to be understood that if a headquarters becomes unusable, business will, to the extent feasible, be carried on at the alternate headquarters. It is also recommended that the alternate headquarters not be supplied with food, water, beds, and so forth.

The Records Protection Committee (see paragraph 12) shall make recommendations to the company presidents concerning the vital records, if any, to be maintained at the alternate headquarters of each of these companies.

The alternate headquarters of each company is in existing office facilities of the XYZ Co. and is provided with minimum office equipment, supplies and communications facilities. It is assumed that reasonable quantities of needed items can be made available in an emergency from other locations of the company facilities or can be procured. Each secretary will see that the alternate headquarters of his company is provided with the minimum essentials for emergency operations.

12. RECORDS PROTECTION COMMITTEE

A Records Protection Committee responsible for the preparation of a program for the protection of vital records has been created by the president of XYZ and the presidents of the three major subsidiaries. Each president has designated a representative of his company to serve on this committee. The committee will formulate the criteria to determine the records which are vital, and will coordinate the preparation by the head of each department in each company of an inventory of his department's existing records, the type of records which it retains, and a list of the records which each department head deems vital in the light of the committee's criteria.

The Records Protection Committee shall recommend the steps necessary to prepare the program for the protection of vital records, including the items referred to in the preceding paragraph. When the program has been prepared, approved by the presidents, and placed in operation in each of the companies for a period of not more than three years, the company presidents shall determine if the program should be extended to provide for special protection or duplication of any records other than those considered to be vital.

13. MAINTENANCE OF PLAN

The secretary of XYZ and the secretary of each of the major subsidiaries will assist his company president in maintaining the continuity and effectiveness of the emergency management plan. In order to maintain the plan on a current basis, the following will be performed annually:

a. The president of each major subsidiary will have prepared a list of retired personnel as described in paragraph 9.

b. The president of each major subsidiary will have prepared four copies of the management development chart as described in **paragraph 9.**

c. Company presidents will prepare tentative contingent officer lists as described in paragraph 6.

d. Each company board at its regular annual meeting will elect officers to serve in the contingencies referred to in paragraphs 3, 4, and 5 as set forth in paragraph 6.

e. After the annual meeting of the board of XYZ and each of the major subsidiaries, certified resolutions electing contingent officers will be replaced as described in paragraph 6.

f. The secretary of XYZ and each of the three major subsidiaries will periodically inspect the alternate headquarters of his company to ensure that minimum essentials, described in paragraph 11, and vital records, described in paragraph 12, are maintained and available.

g. Company presidents will review the emergency management plan and records protection program prior to each annual board meeting and will recommend any changes deemed desirable.

SAMPLE OF EMERGENCY MANAGEMENT PLAN CONTINGENT OFFICER RESOLUTIONS

ATTACHMENT NO. 1

THE XYZ CO.

RESOLVED, that this board hereby elects to be a vice president of this company during such time, if any, as this company shall have no chairman of the board, president or other vice president living and not incapacitated, the first person named on the list marked "Election of vice presidents, ..., 19..," submitted to this meeting (which shall be identified by the secretary and kept with the records of the company) who at such time is not incapacitated and can serve as such without prior regulatory approval and that during such time such person shall have all the powers and duties of the chief executive officer of this company.

THE X POWER CO., THE Y POWER CO., AND THE Z POWER CO.

RESOLVED, that this board hereby elects to be a vice president of this company during such time, if any, as this company shall have no president or other vice

```
                    (Face of Envelope)
┌─────────────────────────────────────────────────────────────────┐
│                                                                 │
│                   EMERGENCY MANAGEMENT PLAN                     │
│                                                                 │
│                                    Envelope No. ____ of 4       │
│   This envelope contains a resolution electing a _____ of ___ │
│                                                      (Company)  │
│   NOTICE: To be opened only in the event of the death or        │
│           incapacity of _____ and then only upon           │
│           authorization from a member of the Board, the         │
│           President, or any Vice President of the XYZ Co.       │
│           or any of the three principal operating companies     │
│           of The XYZ Co. This envelope is to be returned        │
│           unopened to the President of _____ upon          │
│                                        (Company)                │
│           receipt of a corresponding envelope after the         │
│           19__ annual Board meeting of that Company.            │
│                                                                 │
│                              _____            │
│                                (Signature and Title)            │
│   Date: _____, 19__                                        │
│                                                                 │
└─────────────────────────────────────────────────────────────────┘
                                                    ATTACHMENT NO. 2
```

Appendix 14G—Figure 1.

president living and not incapacitated, the first person named on the list marked "Election of vice president, ..., 19..," submitted to this meeting (which shall be identified by the secretary and kept with the records of the company) who at such time is not incapacitated and can serve as such without prior regulatory approval and that during such time such person shall have all the powers and duties of the chief executive officer of this company.

THE XYZ CO., THE X POWER CO., THE Y POWER CO., AND THE Z POWER CO.

RESOLVED, that this board hereby elects to be an assistant secretary of this company during such time, if any, as this company shall have no secretary or other assistant secretary living and not incapacitated, the first person named on the list marked "Election of assistant secretary, ..., 19..," submitted to this

```
                    (Face of Envelope)
┌─────────────────────────────────────────────────────────────────┐
│                   EMERGENCY MANAGEMENT PLAN                     │
│                                                                 │
│                                    Envelope No. ____ of 4       │
│   This envelope contains the management development             │
│   replacement chart of _____                               │
│                          (Company)                              │
│   and a list of retired personnel available for emergency       │
│   management positions.                                         │
│   NOTICE: To be opened only upon authorization from a           │
│           member of the Board, the President, or any Vice       │
│           President of The XYZ Co. or any of the three          │
│           principal operating companies of The XYZ Co.          │
│           This envelope is to be returned unopened to the       │
│           President of _____ upon receipt of a             │
│                         (Company)                               │
│           corresponding envelope for 19__.                      │
│                                                                 │
│                              _____            │
│                                (Signature and Title)            │
│   Date: _____, 19__                                        │
│                                                                 │
└─────────────────────────────────────────────────────────────────┘
                                                    ATTACHMENT NO. 3
```

Appendix 14G—Figure 2.

meeting (which shall be identified by the secretary and kept with the records of the company) who at such time is not incapacitated and can serve as such without prior regulatory approval and that during such time such person shall have all the powers and duties of the secretary of this company.

THE XYZ CO., THE X POWER CO., THE Y POWER CO., AND THE Z POWER CO.

RESOLVED, that this board hereby elects to be an assistant treasurer of this company during such time if any as this company shall have no treasurer or other assistant treasurer living and not incapacitated, the first person named on the list marked "Election of assistant treasurer, ..., 19..," submitted to this meeting (which shall be identified by the secretary and kept with the records of the company) who at such time is not incapacitated and can serve as such without prior regulatory approval and that during such time such person shall have all the powers and duties of the treasurer of this company.

APPENDIX 14H

INDUSTRIAL DEFENSE PLAN AGAINST CIVIL DISTURBANCES AND SABOTAGE

Office of the Provost Marshal General,
U.S. Dept. of Army, Washington, D.C., 1969.

INTRODUCTION TO THE PLAN
This presents the foundation on which the plan is based.

1. PURPOSE. (This paragraph should include a statement or statements comparable in scope to the following: "To establish a continuing program of preparation for protection against civil disturbances and sabotage, and to insure the continuation or restoration of essential operations in the event of other hostile or destructive acts.")

2. ASSUMPTIONS. (Assumptions stating in substance the premises shown below should appear in this paragraph.)
    a. National.
        (1) Potential civil disturbances in the United States could, with little or no warning, seriously endanger selected areas within the U.S. industrial base.
        (2) Widespread sabotage against U.S. industry is not inconceivable.
    b. Local. Each facility is vulnerable and subject to sabotage, civil disturbances, and other hostile or destructive acts.

3. BASIC PLANNING DATA. (This paragraph should include information as listed below.)
    a. Maps. (Attach as appendix a topographical map showing the facility and surrounding areas, including the road and rail nets, the locations of neighboring industrial facilities, power plants pumping stations, etc. Indicate on the map the location of residence of key employees residing in each area. Indicate the distance most of the employees live from the plant, i.e. 11-25 miles or whether there is no general pattern.)

b. Vulnerability. (The degree of vulnerability to civil disturbances is contingent primarily upon sociological, environmental, and geographic factors. Vulnerability to sabotage may in addition to these factors include criticality of the plant, criticality of the product and accessibility to the plant. Answers to the following questions should provide indicators to the relative degree of vulnerability):

(1) Is the facility located in an urban area?

(2) Is the facility located in close proximity, 5 to 10 miles, to an urban area?

(3) Is the facility located near other industries or near military installations?

(4) Is the facility in a remote location?

(5) Have there been previous incidents of civil disturbances, fire bombing or similar acts by dissident groups? At what frequency? To what degree of destruction?

(6) Are environmental and sociological conditions conducive to incidents which might erupt into a riot situations?

(7) Are there good plant/police/community relations?

(8) Is there good plant management-employee relations?

(9) Has a determination been made whether hostile factors exist among plant employees?

(10) Is the plant producing "war materials" under defense contract and has there been employee opposition to this endeavor?

(11) Have there been incidents of employee disfavor to the U.S. involvement in Vietnam, or other areas?

(12) If producing war materials, are they "critical" to the defense effort?

(13) Have there been unexplained incidents of production stoppage? Slowdown? Defective end items?

(14) Have there been incidents of unexplained small fires in the plant?

(15) Have there been internal labor disputes which have not been completely reconciled?

c. Physical layout. (Maps, blueprints, and schematic drawings of production and/or assembly lines.)

d. Operational data.

(1) Personnel. (Indicate the total number of employees and specify the number of contractual or vendor personnel present daily.)

(2) Shift operation. (Indicate the total number of employees and contractual personnel, male and female, assigned to each shift.)

4. EMPLOYEE TRANSPORTATION. (Indicate the mode of transportation used by employees for getting to and from work, i.e. 60 percent bus, 30 percent private auto, 10 percent subway.)

5. TRAINING AND TESTS. (This paragraph should contain instructions for training and rehearsing personnel and testing the plan.)

6. IMPLEMENTING INSTRUCTIONS. (Include a statement to the effect—this plan is effective immediately for training purposes. It will be effective for emergency actions when ordered by (specify the job title(s) of the person(s) with authority to partially or completely implement the plan under emergency conditions.))

SIGNATURE
(Senior Executive)

ANNEXES

I Emergency Organization
II Personnel Protection
III Fire Prevention
IV Plant Security
V Utilities and Services
VI Planning Coordination and Liaison
VII Records Protection
VIII Damage Reduction
IX Restoration
X Emergency Requirements
XI Testing

APPENDIX I Industrial Sabotage

ANNEX I
Emergency Control Organization

1. CHAIN OF COMMAND
   a. Legal aspects.
      (1) State and local laws should be examined to determine the legality of the management succession list.
      (2) Company by laws should be adopted or revised to provide adequate authority for successors during a civil disturbance.
   b. Succession list.
      (1) A management succession list should be developed to provide alternates or successors for key positions. The plan should provide for at least two or three successors for each position.
      (2) Provision should be made for succession or emergency utilization of key operational personnel.
      (3) Geographic employment location and residence data should be carefully considered in preparing succession lists for both management and operational personnel.
      (4) Consider effect of military Reserve and National Guard membership of key personnel on operations if they are called to duty.

2. PERSONNEL UTILIZATION
   a. Employee registration.
      (1) Prepare registration card on each employee for file at control center. Registration cards should contain information regarding secondary skills, pay data, other personnel identifying data and emergency assignment. Emergency assignments should consider training and degrees of competence in secondary skill.
   b. Recall of former employees. (Provide for recall of retired personnel.)

3. MEDICAL REQUIREMENTS. Based upon the existing medical organization, the following should be taken into consideration in preparing medical requirements:
   a. Is there a physician on duty at all times?
   b. Alternately or additionally, is there a nurse on duty?
   c. Has a plant emergency first aid station (s) been established?
   d. Have first aid teams been organized?
   e. Have litter-bearer teams been organized?
   f. Have ambulance services been organized?
   g. Has the plant health service plan been coordinated with the local health programs?
   h. Has the American National Red Cross first aid course been offered to plant employees?

i. Is the plant health service organization a part of a coordinated mutual aid organization of several plants?

j. Have emergency first aid supplies been stocked in sufficient quantities?

k. Have employees been blood-typed?

l. Does the plant have a blood program?

4. WELFARE SERVICES

a. Provisions should be made for the following services to be available during and immediately after a civil disturbance or other emergency:

(1) Emergency feeding of employees.
(2) Emergency sleeping quarters.
(3) Emergency transportation.
(4) Registration and information service for employees.
(5) Emergency financial assistance for employees.
(6) Individual counselling services for employees.

b. Situation Briefings. (Designate a management official to brief employees daily on the impact of the civil disturbance on plant operations and the impact on the community. These briefings must be factual in order to dispel fear, rumors and speculation.)

5. CONTROL CENTERS.
(The control center is the plant command post. The focal point for directing all emergency actions.)

a. Location. (Indicate the location of an adequately protected site within the facility to be designated as the primary control center for the facility. To augment or replace the primary control center, an alternate control center should be selected. Include schematic drawing of internal layout of control center, to include location of equipment, communications, supplies and personnel.)

b. Equipment. (Indicate equipment to be habitually maintained in the control center, e.g., communication equipment, public address system, emergency power, maps, plant layout, food blank forms, office supplies.) **Note:** Necessary supplies and equipment can best be determined by testing the operation of the control center.

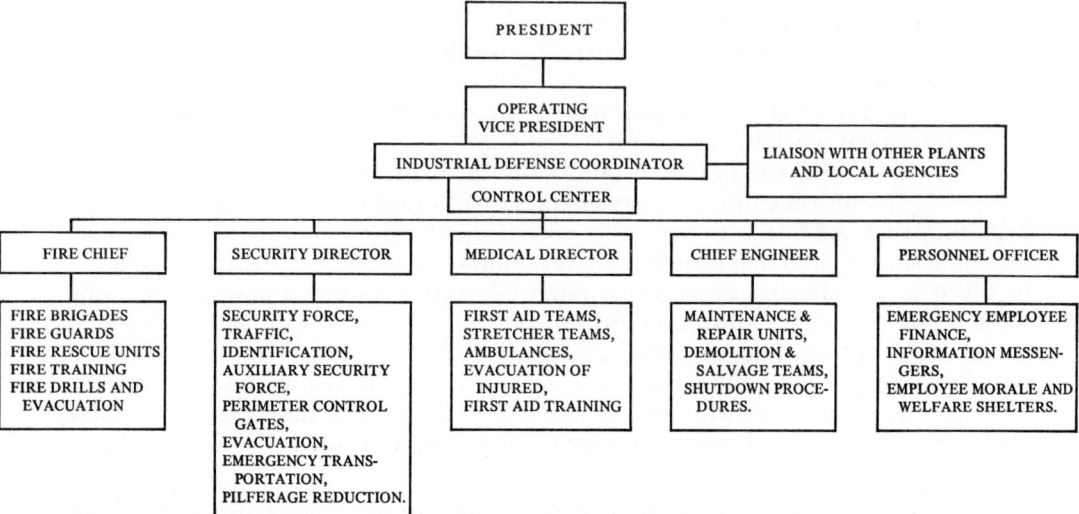

Appendix 14H—Figure 1.

c. **Operation.** (List duties, responsibility, authority and hours of operation of personnel. Indicate restrictions or limitations on use of equipment. Require that a log be maintained of all emergency actions taken. All actions and damages should be photographed.)

6. **EMERGENCY NOTIFICATION.** (Indicate personnel who will be notified in case of various types of disaster or incidents and method of notification. Consideration should be given to the use of chain or progressive (cascade) system of notification, i.e., two or three key personnel receive initial notification, they each in turn notify three or four other key personnel—this progression should continue until all key personnel have received notification.)

7. **ORGANIZATION.** (Appendix 14H—Figure 1 is a type of emergency organization which may be modified to meet the needs of your facility.)

ANNEX II
Personnel Protection

1. **EVACUATION.** (The question to be resolved on this subject is whether to evacuate the plant during a civil disturbance. The decision must be made by management based on such things as the magnitude and severity of the disturbance, danger to employees, and availability of evacation routes away from the danger area. This decision should be coordinated with local law enforcement, fire and civil defense officials. If the decision is to evacuate, consideration should be given to leaving a skeleton force at the plant. There are numerous incidents of buildings not being burned or looted because of the appearance that personnel were in the building. A skeleton force would provide a continuing capability of spotting, reporting and fighting fires, emergency shutdown and liaison.) The following factors should be considered:

   a. **Buildings.**

   (1) Evacuate by departments if practicable.

   (2) Exits.

   (a) Primaries.

   (b) Alternates.

   b. **Plant.**

   (1) Away from the emergency area.

   (2) Toward evacuation routes if possible.

   c. **Routes.**

   (1) Preselect evacuation routes in coordination with local law enforcement officials.

   (2) Emphasize the importance of following these routes.

   (3) Inform employees, preemergency, of evacuation procedures.

2. **ASSEMBLY AREAS.** (Conceivably a civil disturbance could continue in considerable severity and magnitude for several days. Therefore, consideration should be given to preselecting areas where essential employees could assemble for safe transport to the plant. This also requires coordination with local law enforcement and civil defense officials. Obviously, if the disturbance changes course and denies use of preselected areas, other areas would have to be selected, perhaps during the disturbance. Again, coordination is essential. Employees designated for these areas must be notified of primary areas and any changes made.)

3. **SHELTERS**

   a. **Requirements.** (List the total shelter requirements based upon the maximum number of personnel at the facility at any one time. It is recommended that an allowance be made of fifteen square feet per person. A comprehensive survey should be conducted of the facility to determine **those areas which would afford the best shelter for employees against any type of an emergency.** Every facility has a shelter capability of some kind. Any structure regardless of its construction will provide shelter **better than** being out in the open. These shelter areas will possibly protect some personnel, if, due to time and type of emergency, personnel cannot be evacuated. The assistance of the local civil defense authorities and plant engineers who have completed the Shelter Analysis Courses, should be used in making a survey for **best** shelter areas. Upon the identification of these areas, they should be stocked with emergency supplies, communications, and other equipment essential to rapid activation and operation. Shelter areas should be properly marked, and directional signs posted throughout the plant directing employees to these areas. Management should also be aware of the National Fallout Shelter Program sponsored by the Office of Civil Defense.)

(1) Have shelter managers been appointed and trained? Civil defense conducts courses in shelter management. The proper operation of a shelter requires such instruction by at least the designated manager and an alternate.

(2) Have buildings been licensed by civil defense as public shelters? The benefits to be gained by the facility having a licensed shelter should be considered.

(3) Are shelters marked and stocked? Unmarked or unstocked shelters may be detrimental, rather than advantageous, in the facility's emergency program. Shelter areas within the facility must be available for immediate use and not simply unused or undeveloped areas of basements or such places as storage areas which would need extensive clearing before occupancy. Marking and stocking may be had at no cost to the facility by participating in the national shelter program.

(4) Are instructions for the movement of personnel to shelters posted? Such instructions may be very brief, as for example, in a small facility with few personnel and single shelter, or may be quite detailed in a larger facility or one with more than one shelter area. In the latter case, it may be necessary to designate personnel by section or other groups to specific shelters, and to post the routes with directive signs to be followed.

(5) Have communications been established from shelter areas to the facility control center and to local government? The need for internal and external communication is obvious. The shelter cannot exist as an entity completely isolated from others inside and outside of the facility. Communication with local government and with civil defense and similar agencies is extremely important since developments in the community will dictate actions to be taken by shelter management and occupants. Proper planning of cafeterias, lounges, and similar areas which can be used for dual purposes will provide shelter areas with no sacrifice of space and at no additional cost.

(6) Has management assured that all employees know the location of those public fallout shelters closest to the facility and to the homes of employees? (Lists of public shelters are available from the local civil defense organization, and should be provided to employees for their use when they are away from the facility.)

(7) Are all employees encouraged to participate in the facility's emer-

gency preparedness program? (Instruction and training programs should include explanations of the advantages to the employee of participating in and supporting the emergency program. His livelihood and his life, along with those of his family and friends, may depend on the success of the program.)

b. Operations. (List personnel responsible for leadership in each shelter. Indicate health, welfare items and communications for each shelter.)

ANNEX III
Fire Prevention

These measures are of utmost importance in preventing or minimizing fire damage resulting from civil disorders. While the possibility of arson, stemming from riot, commands your attention, don't casually over-look those little fires of unknown origin. You may have an arsonist inside the gate—in fact, he might even be a member of the brigade. Investigate every fire, no matter how small. If you eliminate all possible accidental causes, start looking for an arsonist. Some ways to do it:

    a. Find out who turned in the report. Has he reported several fires?

    b. Ask supervisors about workers who have had bad relations with the company, are easily upset, or are around the plant at odd hours?

    c. Has anyone turned up consistently at scenes of in-plant fires?

    d. Has anyone been seen hurrying from the fire, or acting suspiciously?

    e. Look for these physical clues:

        (1) Piles of wood shavings, debris, paint, or turpentine.

        (2) Strands of gasoline-soaked cotton leading to flammables.

        (3) Heating system which has been tampered with.

        (4) Doors or windows forced open to provide a draft.

If you suspect arson, call your insurance company or the local fire department. They can provide you with a trained investigator. Don't allow mop-up operations to start until the investigator arrives, and post a guard at the scene so evidence won't be tampered with.

f. Ready Defenses. If your plant already has solid fire defenses, you may simply have to set up a plan of action to use them effectively in case of civil disturbance or other emergency. If you're not up to standards on the basics, don't loiter. When riots hit Detroit, the fire department battled as many fires in five (5) days as it normally does in a month. Could you have waited your turn for help?

    (1) An adequate, properly maintained sprinkler system is a "must" in the fire prevention program. (The location of the shutoff valve for the system should be known by all key personnel and the security force.)

    (2) Post and enforce fire prevention regulations.

    (3) Place buckets of sand throughout the plant.

    (4) Extend fire alarm systems to all areas of the facility.

    (5) Provide a secondary water supply system for fire protection.

    (6) Have facility fire protection equipment on-site and insure that it is inspected regularly and properly maintained.

    (7) Determine from local fire department, the feasibility of using mesh wire or other screening material to protect roofs from fire bombs, molotov cocktails, or other incendiary devices.

    (9) Organize employees into fire fighting brigades (for building if possible and rescue squads).

(10) Store combustible material in a well-protected area.

(11) Instruct employees in the use of fire extinguishers.

(12) Place fire extinguishers near exposed windows. The use of Class B—(foam, dry chemical or inert gas) or multipurpose Class ABC (dry chemical extinguishers) should be used for extinguishing gasoline.

(13) Conduct fire drills periodically.

(14) Put people on roofs as spotters for fires. Provide them with special clothing for identification; radio or other means of communication—advise police of this action.

(15) Maintain good housekeeping standards.

(16) Assure that the following areas are adequately protected against fire bombs and other incendiaries. Some protective measures to consider are wood or metal shutters, shatterproof (unbreakable) glass, wired glass, protective screening or mesh.)

    (a) Package and trash chutes.
    (b) Skylights.
    (c) Roof hatches.
    (d) Ventilator shafts.
    (e) Windows and other glass areas.
    (f) Entrances to sewers and service tunnels.
    (g) Computer rooms.

(17) Consider flooding flat roofs (depending on roof decking and building structure to carry the extra weight) with at least 2 inches of water. If flooding is impractical consider installing extinguishers on or near roofs.

(18) Implement recommendations in the latest fire insurance inspection report.

ANNEX IV
Plant Security

1. SECURITY PLAN. (Outline the emergency organization and responsibilities of the plant security force. The normal organization and responsibilities should be adapted to meet the requirements imposed by a civil disturbance, sabotage, bomb threat, unexploded ordnance or other hostile or destructive acts. The security plan should include all actions and techniques to be employed to protect personnel, materials, products or services, premises and process from hazards inherent in operations and other acts mentioned above. The security organization of a facility will depend almost entirely on the size, criticality and vulnerability of the facility.

2. LEGAL RIGHTS AND RESTRICTIONS. (This is a most important element and must be understood by management and members of the security force. The facility legal counsel must coordinate with the city attorney, district attorney or other legal offices to determine the authority of the property owner, and his employees, in protecting property and life.) Some factors to be considered are:

    a. What are the geographic limits within the authority of management?

    b. What are local laws and statutes concerning security force being armed? Their use of weapons?

    c. With what type weapons can they be armed?

    d. What actions can the security force take during a civil disturbance? What is the limit of their jurisdiction (authority) in protecting property? Life?

    e. How and under what conditions might they exercise "citizens arrest"?

    f. Under what conditions can force be used? How much force can be used?

g. The advisability of **deputizing** the security force?

h. Are the legal limits of authority the same for "normal" (day to day) conditions and "emergency" conditions?

3. LIAISON AND COORDINATION. (List the names (positions), telephone numbers, law enforcement agencies (local, State and Federal), with whom the plan has been coordinated and liaison should be maintained.

4. SECURITY FORCE. (The organization of the security force should be tailored to meet the requirements of a specific facility. The security force is the most effective and important element of security planning. It is the only in-house element capable of physically responding, utilizing judgment in an incident. The following factors should be considered relative to the security force.

    a. Qualification Standards.
        (1) Age.
        (2) Loyalty.
        (3) Intelligence.
        (4) Physical Qualifications.
        (5) Dependability.
        (6) Cooperativeness.
        (7) Ability to exercise good judgment, possess courage; alertness, self-reliance, tact and even temper.
        (8) Security clearances may be required in some instances.

    b. Training. These are basic essentials:
        (1) Discipline.
        (2) Familiarization firing of weapons.
        (3) Use and safe practices and maintenance of weapons.
        (4) Legal limits of authority.
        (5) Procedures for apprehension and restraint to include citizens arrest.
        (6) Self defense.
        (7) Actions during civil disturbances.
        (8) Actions in event of bomb threats.
        (9) Actions upon discovery of unexploded ordnance.
      (10) Elementary first aid and fire protection.
      (11) Communications procedures.
      (12) Report writing.
      (13) Employee and public relations.
      (14) Basic rescue techniques.

    c. Uniforms. (It is recommended that security personnel wear uniforms or clothing with distinctive markings. This facilitates identification and minimizes problems which could arise from lack of immediate recognition of the individual as a member of the security force.)

    d. Weapons. (The matter of arming the security force is quite controversial. The decision must be made by management. Consideration should be given to the mission of the security force. If the mission is to protect life and property, can this be accomplished without firearms? Will the presence of an armed security force deter the omission or commission of destructive acts? Will the presence of firearms incite trouble? Management may decide not to arm the security force during normal operations, but rather to have weapons available to arm the force during an emergency.) The following factors should be considered and included in the plan:

(1) Type of weapon.
(2) Registration of firearms (check with local Internal Revenue Service, Alcohol, Tobacco, and Firearms Division).
(3) Procedures for issue and turn-in of weapons and ammunition.
(4) Maintenance.
(5) Inspections.
(6) Frequency of familiarization firing (at least annually). If your security force is armed, the question is: What should their orders be? First of all, armed security people must be thoroughly trained in the use of and when they are legally authorized to use the weapon with which they are armed. Even when they're proficient with weapons, be sure they know the consequences of firing. In all cases, the byword is discretion. Minimum force should be standard, for instance, a member of the security force shoots an escaping felon who turns out to be 15 years old. He was still an escaping felon, but once it happens, he'll never again be known as anything but a defenseless, young boy. You'll have a case on your hands. If you even restrain a trespasser, you can be in trouble.

e. Organization. (Security forces may be organized in any one or any combination of the following.)
(1) Regular force.
  (a) Fixed post deployment.
  (b) Patrol deployment.
(2) Auxiliary force. (An auxiliary security force should be established to supplement the regular force during an emergency. Personnel should be selected from the employee population and trained in their emergency security function.)

f. Shift Changes. (Show the times of shift changes and tours of duty. Shift change times should not be the same as the time for employee shift changes. It is well to consider establishing shift changes of the security force at least one hour in advance of or one hour after employee changes.

g. Communications. Adequate communications are essential to the effective operation of the security force during normal times and especially in the event of a civil disturbance or other emergency. Consideration should be given to a communication system for the exclusive use of the security force. The type and comprehensiveness of the system will vary with the size of the facility and the size of the force.

h. Limitations of Security Force Functions. (Members of the regular force should have no "fire fighting" or other duties. Cross training to provide an in depth-dual capability is acceptable. Such emergencies offer an excellent diversion to cover the entrance of a saboteur, or dissident groups. During such incidents the security force should be more than normally alert in the performance of its primary mission.)

5. PERIMETER BARRIERS. (Fences and other antipersonnel barriers are the physical media by which the boundaries of a facility, or restricted area with a facility, are physically defined for protection and control. The fundamental purposes of such barriers are to define the area, impede access or intrusion, aid security personnel, channel the flow of personnel and vehicles, and provide a psychological deterrent.)

a. Type of barriers:
(1) Natural (body of water, cliffs, canyons or other terrain difficult to traverse.)

(2) Structural (buildings, chain link fence, barbed wire). Natural barriers should be reinforced by a structural system of barriers.
  b. Construction:
   (1) Chain link fence
      (a) Minimum height of chain link portion—7 feet
      (b) Mesh openings not larger than 2" square
      (c) Number 11 gauge or heavier wire
      (d) Twisted barbed selvage—top and bottom
      (e) Extend to within 2" of firm ground or below the surface if soil is sandy and easily wind blown or shifted.
      (f) Fence mesh should be drawn taut and securely fastened to rigid metal posts set in concrete. Additional bracing, as necessary, should be placed at corners and gate openings.
      (g) Topped with a 45° outward and upward extending arm bearing 3 strands of barbed wire stretched taut and spaced to increase the vertical height of the fence by approximately 1 foot.
      (h) Provided with culverts, troughs, or other openings, where necessary, to prevent washouts in the barrier. If such openings are larger than 96 square inches in area they should be provided additional protection.
      (i) Checked (inspected) periodically for undergrowth, damage or deterioration.
   (2) Masonry walls when used as perimeter barriers should have a minimum height of 7 feet and topped by a barbed wire guard as indicated, or have a minimum height of 8 feet and be topped by a layer of broken glass set on edge and cemented to the top surface.
   (3) Building walls, floors, roofs, and dikes, when serving as perimeter barriers, should in general be of such construction and so arranged as to provide uniform protection equivalent to that provided by chain link fencing as specified. **If buildings form a part of the perimeter barrier—protective grill work or laminated shatter proof glass should be installed to increase the protection for windows, doors or other openings.**
   (4) Bodies of water. If a lake or stream forms one side or any part of the perimeter, it in itself should not be considered an adequate perimeter barrier. Additional security measures must be provided for that portion of the perimeter, such as a fence or frequent guard patrol and flood lighting.
  c. Posting. Post with "no trespass" signs in accordance with criminal laws of of the state.
  d. Protective Lighting.
   (1) Inspect the perimeter barrier to insure that it is properly maintained and properly lighted.
   (2) Without doubt, lighting is the best security bargain available. Most riot and firebomb damage occurs after dark, and nothing discourages hit and run types like full coverage, glare lighting. They don't know whether a camera—or an armed guard—may be waiting beyond the glare. You may be able to reposition existing lighting for this purpose, but light the villain, not the target. And don't make the mistake of relying entirely on mercury protective lighting, because even a momentary power dip can mean several minutes of darkness.
   (3) One solution is direct substitution of instant-starting combination

incandescent mercury lamps for some of the straight mercuries. Light output and wattage remain essentially the same. Another is to use low-cost quartz iodine fixtures on weighted pedestals on the plant roof.

(4) Insure continuous lighting in parking lots, and on ground floors.

(5) Use screening to protect lighting fixtures against rocks and other objects.

e. Vehicle Parking. Vehicle parking should be located outside of the security fence or wall. (This reduces the fire potential from gasoline in vehicle tanks and minimizes the hazard of explosives and incendiary devices which are easily concealed in a vehicle.)

f. Intrusion Detection Devices.

Antiintrusion alarm devices are employed for the purpose of detecting an announcing proximity or intrusion which endangers or may endanger the security of a facility. These systems are utilized to accomplish one or more of the following purposes:

(1) To permit more economical and efficient use of manpower by substituting mobile responding security units for larger numbers of fixed security posts and/or patrols.

(2) To take the place of other necessary elements of plant security which cannot be used because of building layout, safety regulations, operating requirements, appearance, cost, or other reasons.

(3) To provide additional controls at vital areas as insurance against human or mechanical failure.

The advantage of a protective alarm is measurable reliability. While there is a wide range of complexity between the various alarm systems, each can be tested and evaluated to determine what degree of security can be expected from the device.

Detection devices are usually designed to detect a single phenomenon. The choice of the type detection device to be employed is based upon what will be most readily detectable in the given situation. It may be desirable in some cases to employ more than one type of detection device to protect against all possible methods of entry. Usually, similar equipment is manufactured by several companies. Such equipment will operate on the same basic principles, but may well differ in refinements. These differences may, under certain circumstances, alter the degree of security provided.

The most common detection devices are:

a. Electro-Mechanical Devices are designed to effectively place a current carrying conductor between the intruder and the area to be protected. The most common in this category are foils, screens, contact switches and vibration detections which are damaged or disturbed by penetration (usually used for protection of doors, windows, ducts, and nonsubstantial walls or partitions.)

b. Photoelectric Device, whereby interruption of virtually invisible beam of light is detected. This device is highly effective in detecting vehicular movement since it is impractical to move vehicles over or under the beam.

c. Proximity Detection Device operates by surrounding an object with an electrical field in such a balance that, any disturbance of the field creates imbalance in the system that results in the initiation of an alarm. There are two different types of proximity alarms—electromagnetic and the capacitance. Both of these systems lend themselves to use as fence alarms and the capacitance device is also effective for interior use.

d. Acoustic Detection Device actuates by the sound or vibration made by the intruder during his approach or as a result of his attempt to gain entry. Environmental conditions must be carefully evaluated before applying these devices since peripheral noise will cause false alarms.

e. Movement Detection Devices are designed to create an alarm when there is movement of any sort within the established limits of the device. There are two types of movement detection devices: Ultrasonic and Radar.

(1) Ultrasonic detects movement by the reflection of sound waves which causes electronic control units to trigger an alarm signal.

(2) Radar detection is designed to serve any doppler shift in the frequency of transmitted signals. The movement of a human being within the sensitive detection field will generate an alarm signal.

The use of alarms in the protective program of a restricted area or facility may be required in certain instances because of the critical importance of the area or the facility and, in other instances, because of situations and conditions pertaining to the location or the layout of the area or facility. In some instances, their use may be justified as a more economical and efficient substitute for other necessary protective elements. In determining whether the use of alarms in a restricted area is essential or advisable, the various conditions and situations peculiar to the restricted area or facility will, of course, affect the ultimate decision. However, in general, the following criteria should form the basis for a determination of the use of alarms:

a. The critical importance and vulnerability of certain restricted areas or facilities require the additional control and insurance against human or mechanical failure which is provided by alarms systems. In this group are:

(1) Restricted areas or facilities which, because of a concentration of vital components, materials, or data, are attractive, high-priority targets for sabotage, theft, espionage, or other criminal acts.

(2) Critical processes and process controls.

(3) Very important restricted areas or facilities where it is desirable to have admission controlled by both guards and operational employees, or where it is desirable for operators to deny access to guards.

b. In certain cases due to restrictions imposed by location, layout, or construction, alarms are necessary to take the place of the more usual protective elements such as fences, lighting, patrols, etc. Included in this group are:

(1) Restricted areas or facilities which, because of proximity to adjacent structures, activities, or property lines, require the use of alarms in lieu of physical barriers to limited or exclusion areas.

(2) Restricted areas or facilities which are difficult or impossible to protect effectively due to terrain conditions, personnel hazards, or atmospheric conditions, and where other types of protection are not effective or practicable.

(3) Restricted areas or facilities, or components which are small, or remote areas requiring more than safe and lock protection but not justifying a full time guard.

c. Alarm systems, because of their cost, are justified only where their use results in a commensurate reduction or when need dictates a higher level of protection to include a more positive or fail safe method of detecting unauthorized entry. In determining the advisability of substituting alarms for other protective elements, a careful comparison of relative costs is essential. This should include service and maintenance charges. In this connection, it should be borne in mind that many alarm systems have little salvage value and, consequently, the longevity of the activity being protected is an important consideration. The advice of a competent engineer from a reputable firm dealing in protective devices and signal alarms should be obtained when considering protective alarm systems.

To afford the required degree of protection and be acceptable as protective units, alarm installations should meet the following requirements:

  a. The system should be so designed that the interval of time between the detection of activity and the achievement of the objective of such activity is sufficient to permit the application of necessary countermeasures.

  b. Central station systems should be specified for all locations where security personnel are not continually in the immediate vicinity to pick up a local alarm signal and make adequate response.

  c. All systems, materials, and equipment should meet the Underwriter's Laboratories, Inc. standards where applicable, for the purpose for which they are used.

Generally, it may be stated that there are two types of intrusion detection systems:

  a. A central station system is one in which the operation of electrical protective circuits and devices is automatically signaled to a central station which has a trained response force and operators in attendance at all times. The central station monitors the signal end of the system, provides the response to a signal, and supervises the functioning of the system.

  b. A local alarm system is one in which the protective circuits and devices are connected to a visual and/or audible signal element which is located at or in the immediate vicinity of the protected facility or component, and which is responded to by security personnel in the immediate vicinity.

6. CONTROL OF ENTRY. (Develop procedures for positive identification and control of employees, visitors and vehicles. A positive means of identifying employees is the use of a photograph identification card. Samples of the identification media should be given to law enforcement officials. (This is essential for getting through police lines and during times of curfew.) Coordinate with the police the category of personnel essential to plant operations, i.e. engineer, maintenance, etc.)

7. PROTECTION OF CRITICAL AREAS. (Identify and list critical areas within the plant. (Refer to ANNEX VIII.)

  a. Enclose critical areas with physical barriers.
  b. Designate specific personnel who are to have access to critical areas.
  c. Admittance to critical areas should be controlled by:
    (2) Supervisory personnel.
    (3) Where locks are used, they should be rotated upon notification of impending civil disorder or other emergency.
  d. Develop a key control system.
  e. Develop package and material control procedures. (All packages and materials going into or out of critical areas should be checked.)
  f. Institute procedures to protect gasoline pumps and other dispensers of flammable material. Disconnect power source to electrically operated pumps.

8. ARMS ROOMS.
  a. Keep arms rooms.
    (1) Locked.

(2) Under 24-hour surveillance.
  b. Ammunition.
    (1) Stored in locked separate location.
    (2) Under 24-hour surveillance.

9. PERSONNEL SECURITY.
  a. Conduct pre-employment check of applicants.
    (1) State and local police.
    (2) Former employers.
    (3) References (not limited to those provided by applicant).
    (4) High schools (be watchful for falsification of education and background).
    (5) Colleges and universities.
  b. Check Selective Service Classifications.
    (1) Registration certificate.
    (2) Notice of classification.
    (3) Selective service number.
    (4) Local selective service board number.
  c. Military Service and type discharge (have applicant show discharge papers).
  d. Make personnel checks, of persons who are authorized access to critical areas.
  e. Brief employees regarding the importance of plant security and the need for exercising vigilance.

10. REPORTING OF INCIDENTS. (Show procedures as to how, when, where and to whom incidents will be reported.)

11. BOMB THREATS. (List address and telephone number of):
  a. Nearest military explosive ordnance disposal team.
  b. Bomb disposal unit of local police force.
  c. Local fire department.
(Show procedures to be followed upon receipt of bomb threat. This should be coordinated with local law enforcement officials, local fire department, and the nearest military explosive ordnance disposal (EOD team.))

12. EMERGENCY NOTIFICATION. (Prepare an emergency notification list or chart of personnel to be notified in the event of civil disturbance, or other emergency.) This list must be kept current.

13. EMERGENCY SHUTDOWN. (Indicate procedures to be followed by security personnel during and after shutdown.)

14. SAFEGUARDING CLASSIFIED MATERIAL. (Specify procedures for safeguarding or removal of classified material. Security personnel should know how to contact custodians of classified material. They should also be advised of actions to be taken with regard to the Department of Defense Industrial Security Cognizant Office, if applicable.)

ANNEX V
Utilities and Services

The importance of utilities and services during an emergency cannot be overemphasized. The disruption of communications, electric power, water, transportation or fuel sources could seriously impair or stop production. It is essential that these utilities and services be considered critical to the continuity of operations; that they be properly protected and adequate emergency back-ups be developed. Essential utilities and services to be considered are listed below. The details for each should be coordinated with the respective utility or service company.

a. Communications.
  (1) Coordinate with local telephone companies.
  (2) Adequately cover plant area.
  (3) Back-up primary system with two-way radios, walkie-talkies, field telephones, or megaphones (bull horns).
  (4) Monitor local and state police radios.
  (5) Monitor fire department radios.
  (6) Monitor hospital and ambulance radios.
  (7) Establish communications with adjacent plants and businesses.
  (8) Establish communications with management and key employees.
  (9) Train switchboard operators in emergency procedures.
  (10) Inquire as to availability of telephone—radio mobile equipment—license and frequency are assigned to the common carrier.
  (11) Designate male operators as alternates for females who may not report.
  (12) Unlisted telephone numbers, at control center, for use by management and key executives. Don't have all telephone numbers plainly listed—a few determined harassing callers can keep your lines occupied.

b. Electric Power.
  (1) Coordinate this portion of the plan with local electric power companies.
  (2) Emergency power.
  (An auxiliary source for providing sufficient emergency power for lighting and other essentials. This should not be construed to mean a stand-by capability to continue full production operations. The following items are suggested):
      (a) Generators.
          1. Show size and location.
          2. Fuel supply.
          3. Operators.
      (b) Battery-powered equipment.
          1. Flashlights.
          2. Lanterns.
          3. Other battery powered sources of illumination.

c. Water.
  (1) Secondary source for fire fighting, essential operational needs, drinking, and sanitation.
  (2) Location of primary water main.

d. Transportation.
  (1) Primary routes of ingress and egress.
  (2) Alternate (emergency) routes.
  (3) Accessibility of alternate routes to suppliers.

e. Fuel Sources, i.e. pipelines, coal, and diesel fuel. (Stockpiling for emergency use should be considered.)

ANNEX VI
Planning Coordination and Liaison

This is a most important element of the plan and is designed to assure mutual planning approaches and objectives. It also provides a means of keeping you abreast of the social climate and receiving advance warning of the imminence and possible magnitude of a disturbance. Coordination and liaison should be maintained with:

a. Facility Members and Locations. (List the name and location of each industrial facility or organization of the mutual aid pact, or with which coordination has been affected. Indicate who in each facility or organization can approve the implementation of the pact during a civil disturbance. Also include any other mutual aid pacts with which you made unilateral agreements. Show restrictions, if any, on mutual aid assistance during riots or civil disturbances.)

b. Local, State and Federal Officials. (List the name, location and telephone number of each agency with which coordination has been accomplished):
   (1) Law enforcement.
   (2) Fire departments.
   (3) Adjacent plants and business firms.
   (4) Employee union officials.
   (5) Local utilities.
   (6) Local news media for news release policy.

c. Communications and Control. (List the primary and alternate methods of communications that will be used to alert the mutual-aid pact members and local state and federal agencies and your facility. Include methods of alerting during normal working hours and nonworking hours. Include the methods that will be used in controlling personnel at the scene of the emergency, including direction of police, fire and emergency vehicles and crews. Coordination must be made in advance for use of facility security personnel, state, and county police, as applicable.)

d. Facility Responsibilities.

   (1) Personnel. (List by job title the various skills that you have agreed to furnish the mutual aid organization. Maintain a current roster of these personnel by name, with alternates. Include supervisory responsibilities when aid is required.)

   (2) Equipment. (List the material and equipment that your facility will have available for mutual aid. Establish a method of having the material and equipment delivered as needed.)

e. Other Participants Responsibilities.

   (1) Personnel. (List by job title or skill, the personnel to be furnished by other mutual aid participants. Indicate procedure for their reporting, utilization and control. Indicate responsibility for control and supervision for each group.)

   (2) Equipment. (List the material and equipment that may be obtained from other mutual aid members. All items should be listed by location and include procedure for obtaining them.)

f. Operational Procedures. (List special limitations, legal aspects, feeding and transportation of personnel, prorating cost and use of any special items not covered above.)

Note.—The Mutual Aid Pact or Coordination Agreement may be substituted for part of this annex.

ANNEX VII
Records Protection

1. CLASSES OF RECORDS.

    a. Administrative. (Indicate those records needed by the administrative functions of the facility, to include as a minimum payroll, accounting, personnel and sales records.)

    b. Operational. (Indicate those records needed by the operations, engineering, or maintenance sections, and production records.)

2. REPRODUCTION METHODS AND PRIORITY. (Indicate the methods that will be used to reproduce administrative and operational records. Protection considerations should be given to microfilming, use of film sort cards, carbon copies, photocopying, and duplicate records. Specify the records in order of priority for reproduction. Reproduction of classified material must be coordinated with the issuing agency. Defense contractors are governed by the provisions of DOD 5220.22-M, 1 July 1966, subject: Industrial Security Manual for Safeguarding Classified Information.

3. PROTECTION OF RECORDS. (Indicate the location of reproduced or duplicated records. Consideration for the location of duplicate records should be given to the use of alternate headquarters, small town banks, commercial depositories, the homes of key employees living out of the probable damage area, and vaulting in special circumstances. If classified material is stored, suitable clearance from the issuing agency must be obtained for the location where the documents are to be stored. Special instructions should be included for protecting records in the hands of employees at the time of the emergency.

4. PROTECTION OF COMPUTERS. (Do not rely solely on the "machine" to protect its content. The wide use of computers today has resulted in a "false" sense of security. Computer has no immunity from the vagaries of mankind. Tapes, cards and discs require the same degree of protection provided for ledgers, journals and other hard copy records in the days preceding the computer era. Computer protection comprises physical means of protecting against tampering and misuse and protection against fire.)

5. OPERATIONS. (List special instructions for handling and storage of records. Indicate the person or persons charged with record protection responsibilities and establish his definite authority.)

6. CASH, NEGOTIABLES AND OTHER VALUABLES. (Procedures should be developed for the immediate removal and protection of these items. Items of "attraction" and value should be removed from show windows and display areas.)

ANNEX VIII
Damage Reduction

1. FUNCTIONAL AREAS.

    a. Criticality. (List functional areas, in order of priority, most critical to overall facility operations. This should include consideration for all types of emergencies.)

    b. Protection. (Functional areas most critical to the overall operation and/or production should be given priority of protection, prior to, during and after the emergency.) Refer to plant security annex.

    (1) Buildings. (Include measures for reinforcing walls, roofs, floors and protection of wall openings such as windows and doors of existing buildings. These protection factors should be considered in new constructions.)

    (2) Machinery. (Factors to be considered are dispersal, protection of one piece of equipment by use of another, and parts removal.)

    (3) Hand tools. (Indicate individual action and responsibilities for protection of hand tools. Include tool crib dispersal.)

    (4) Special equipment. (Indicate methods used or to be used to disperse on- or off-site parts, subassemblies, completed items, jigs, dies, patterns, moulds and other critical items.)

    (5) Transportation. (Indicate dispersal location of transportation equipment to protect machine tools.)

    (6) Utilities. (Indicate protection afforded utilities and include location and protection of electrical transformers at load centers and communications centers.) Refer to Plant Security Annex and Utilities Annex.

2. SHUTDOWN PROCEDURES. (Specify shutdown procedures to include methods and sequence for individual sections within the facility and the facility as a whole. Designate title (positions) of individuals responsible for implementing shutdown procedures.) Refer to Item 12, Plant Security annex.

3. FIRE CONTROL. (See Fire Prevention annex.)

4. DISPERSION. (Consider the dispersion of machinery, material and personnel.)

5. OTHER MEASURES. (List other measures peculiar to your facility that may be necessary to minimize damage.)

ANNEX IX
Restoration

1. COMMAND RESPONSIBILITIES AND CONTROL. (Show plans and control of reconstruction and restoring damaged areas.)

2. DAMAGE ASSESSMENT.

    a. Internal Reporting. (Indicate procedure for reporting damage within the facility to the control center. The damage reported should be assessed for overall effect on the facility and as a guide for restoration.)

    b. External Reporting.

        (1) (Indicate procedure for reporting damage from facility to corporate/company/system).

        (2) (Indicate procedure for reporting damage to local agencies and news media if applicable.)

3. RESTORATION MEASURES.

    a. Alternate Sources of Supply. (List the names and addresses of those firms which can be used as a source of alternate supply. List agreements that have been made with them.)

    b. Stockpile. (Cover information concerning inventory of essential raw material, component parts, parts for machine tools and maintenance, and critical machinery.)

    c. Alternate Production Method. (Indicate those processes that lend themselves to alternate methods even though they may be slower and more costly. Outline the alternate methods and indicate conditions under which they will be put into effect.)

d. Subcontracting. (Indicate those facilities or installations with which subcontract agreements have been made.)

e. Utilities. (List the requirements of each subsection for continued operation. Include agreements with local utilities and others having facilities for furnishing the following utilities: electricity, water, gas, sewage, fuel.) Refer to utilities annex.

f. Salvage Procedures. (List procedures for salvaging and rebuilding machinery, equipment and buildings.)

g. Transportation. (Based upon anticipated loss of transportation and remaining capability determine additional requirement, if any.)

ANNEX X
Emergency Requirements

These requirements should be based on estimated needs for the duration of the emergency. These items should be prestocked because conditions may preclude their procurement during the emergency. Unused portions can be carried over for postemergency use.

    a. Food, water and medical supplies.

    b. Emergency repair tools and equipment.

    c. Administrative supplies office equipment.

    d. Provide emergency sanitation facilities.

    e. Designate separate sleeping quarters for male and female employees.

    f. Maintain an inventory of 55-gallon drums to be filled with water or sand to reinforce barricades at entrances.

    g. Have on hand barbed wire to form a barrier directly in front of each row of 55-gallon drums. Concertina type wire is very effective.

    h. Maintain supply of panels or screen mesh to protect windows on ground floors.

    i. Develop procedures for employees to purchase gasoline from plant supply in case local stations are closed.

ANNEX XI
Testing the Plan

Frequent testing and correcting the plan will improve its effectiveness upon implementation under actual conditions. An emergency plan, like a chain, is no stronger than its weakest link.

a. Types of tests. (Specify type of tests, whether partial or complete and when umpires or observers are to be present. Indicate frequency of partial or complete tests.)

    (1) Partial—testing individual segments of the plan.

    (2) Complete—testing entire plan.

b. Tests should be unannounced.

c. Weaknesses should be noted and the plan revised to include corrective actions. (Include reports of test results by observers or umpires and action to be taken by designated company official to improve techniques and take corrective action on deficiencies:)

Date and Type of Test	Deficiencies	Corrective Action and Plan Change

Note.—In order not to interfere with production or operations to a great degree, it is suggested that, initially, tests to determine adequacy of the plan be conducted on an individual annex basis, i.e. control center operations, plant security, coordination, etc. When these individual annexes prove effective, an overall test should be conducted.

APPENDIX 14I

INDUSTRIAL SABOTAGE*

The scope of sabotage in which American industry has an interest is much broader than the enemy agent or foreign trained saboteur. It goes beyond the limits of the legal definition.

The saboteur is not necessarily a foreign national or of foreign parentage. He may be a highly trained professional or a rank amateur. He may be a laborer, a machinist, a foreman, a top-flight engineer, or even a member of management. He may be anyone. But one thing is certain—he is likely to be one of the least suspected members of the organization. His motives may be as varied as his personality. He may work for love of his native land; for pay; for hatred; for sincere if misguided, devotion to a cause; for revenge; to settle a real or imaginary grievance; or under threat of blackmail or fear of reprisal against relatives in a foreign country.

Generally, there are two basic types of saboteurs. The first is the enemy agent. He is usually directed, trained, supported and supplied by a sabotage organization. He coordinates his activities in an overall effort to impede or disrupt our industrial potential. He may attack his targets directly from the outside; however, if he cannot penetrate the facility's outer defenses, he may infiltrate the facility as an employee and lie dormant for a considerable period of time, withholding any act of sabotage until directed by his superiors. This "dormant" enemy agent, while an employee, will probably be industrious and outwardly apparently harmless. He will do his utmost to avoid suspicion. He will become familiar with all phases of operations by showing interest in the work of others. By thoroughly examining the entire facility for security, and for structural and functional vulnerabilities, he can fit together his complete scheme of sabotage. He will never reveal any anti-American or anti-production sympathies; he will probably be well-liked and respected, and may be

---
* U.S. Government Printing Office, 0-367-877. 1969.

considered by some to be a model employee.

The second type, the "individualist" or "independent" saboteur, commits acts of sabotage for personal reasons and has no affiliation with a foreign power or military group. He might be the disgruntled employee who commits sabotage for revenge; he might be mentally ill; he could be a person who has been duped by enemy propaganda. Since sabotage essentially is an inside job, and requires the assistance, knowingly or unknowingly, of someone inside, he may be the sabotage contact or key within a plant or facility. The efforts of this type of saboteur are exceedingly difficult to detect. In many instances, his actions cannot be predicted or anticipated. Although he will have no particular training for sabotage, his presence and familiarity with the facility pose a serious problem, for he may strike at anytime, anywhere.

No facility is immune to attack, for some form of sabotage can be committed despite efforts to prevent it. The enemy will attack production anywhere between the raw material stage and the delivery of the finished product; he will attack any facility where loss of production, even though temporary, would hinder or retard the war effort of a nation. The large industrial complexes established for the production of newly-developed weapons of war, dispersed or not, have expanded the field of operations for the saboteur. However, there is no reason for guesswork in determining the probability of attack by the saboteur. The type of targets in a given area, as well as where and how they may be attacked, can usually be predicted with reasonable accuracy. The saboteur will look for a target which is critical, vulnerable, accessible, and at least partially conducive to self-destruction.

Criticality and vulnerability are discussed in the plan outline. Target accessibility will be closely related to target vulnerability; it refers to the ease with which the saboteur may approach the target. Accessibility depends primarily on two factors; the amount and type of plant security maintained, and the geographic location. In most cases, the security of the facility is the only factor which can be controlled. It should be noted that, from a saboteur's point of view, any target which is accessible is vulnerable to attack by at least one of many methods.

The capability of self-destruction is one of the more important elements of target susceptibility to sabotage. A target is said to be capable of self-destruction when its nature is such that it will continue its own destruction as a result of even a comparatively minor act of sabotage. For example, when an explosive charge is placed on a rail line at midpoint on a timber trestle bridge, the wheels of the locomotive will detonate the explosive, thus destroying the tracks. The continuing motion and weight of the train will not only destroy the bridge, but the train as well. Still another example would be a high-speed, revolving electric motor or generator, where a comparatively small disturbance in the alignment of the shaft would cause considerable damage to the target.

The tools and methods of sabotage are limited only by the skill and ingenuity of the saboteur. A major sabotage effort may be undertaken after thorough study of the physical layout of the facility and its production processes by technical personnel fully qualified to select the most effective means to strike one or more of the most critical and/or vulnerable parts of the facility. Sabotage may, on the other hand, be improvised by the saboteur relying solely upon his own knowledge of the facility and the materials available to him. Industrial engineers are well aware of the potential acts of sabotage which may be consummated through the use of materials readily available in the normal course of operations. Examples are the periodic availability of explosives intended for industrial purposes; product and process contamination by the use of additives and spoilers; incorrect cycle time-phasing; tampering with control devices, operating equipment, and so forth. The saboteur, in such a case, may or may not possess or need high degree of technical knowledge. Hence, the selected vehicle may range from the crude or elementary to the ingenious or scientific.

The methods of sabotage may be generally classified as follows:

a. **Mechanical**—breakage or omission of parts, substitution of improper or inferior parts, failure to lubricate or properly maintain.
b. **Chemical**—the insertion or addition of destructive or polluting chemicals in supplies, raw materials, equipment, or utility systems.
c. **Explosive**—damage or destruction by explosive devices or the detonation of explosive raw materials or supplies.
d. **Fire**—ordinary means of arson, including the use of incendiary devices ignited by mechanical, chemical, electric, or electronic means.
e. **Electric or electronic**—interfering with or interrupting power, jamming communications, interfering with electric and electronic processes.
f. **Psychological**—the inciting of strikes, jurisdictional disputes, boycotts, unrest, personal animosities; inducing excessive spoilage and inferior work, causing "slowdown" of operations or work stoppage by false alarms; character assassination; on a larger scale, the instigation of false political and economic public issues and the dissemination of inflammatory propaganda so as to break morale.

The prevention of sabotage involves the reduction of target accessibility and vulnerability. This may be accomplished by:

a. Institution of security measures to prevent unauthorized access to target areas.
b. Development of an employee security education program.
c. Security screening of employees, and the removal or relocation of known or suspected security risks.
d. Development of appropriate emergency plans and organizations.
e. Protective construction and/or modification of equipment or material design where appropriate.

The program for the prevention of sabotage must be dynamic and continuous; it must receive the full support of all echelons; and it must be so designed that it will complement operational requirements and situations.

APPENDIX 14J

GUIDE TO INDUSTRIAL CIVIL DEFENSE PLANNING*

Industry must be able to withstand a nuclear attack upon our country if we are to restore a viable civilization in the aftermath of an enemy attack. The role of industry in civil defense is intimately connected with the survival of the people of the Nation, as well as the survival of our free enterprise system, both of which must be accomplished if we are to maintain our way of life.

* Continuity of Corporate Management in Event of Major Disaster, Office of Civil Defenses, Washington, D.C., 1970, pp. 45-54.

Making ready to withstand nuclear attack will involve solving many complex problems, but none so complex as those which an unprepared American industry would face following an attack.

Planning for Continuity of Corporate Management is an important element of industrial civil defense. Such planning is basic to preservation of the corporate structure in event of nuclear attack. But, industry cannot survive if it loses its most important resources—its people. All the saved buildings, equipment, supplies, records, systems, processes and blueprints in the country, important as they are, would serve little purpose if the people who use or operate them were lost.

Industry, therefore, has a special stake and exceptional responsibility in the success of the National Civil Defense Program, which is based on the concept that the foundation stones of any civil defense program are measures to protect lives, and that these stones must be firmly laid as the first action in preparing measures to protect the American population and property.

On the basis of numerous studies by the Department of Defense, it has been concluded that the saving of lives that could be expected to result from a nationwide shelter system is enough to assure survival of this country as a Nation under the worst hypothetical attack.

As a result of this finding, the emphasis of civil defense has been directed to saving lives through a nationwide system of public fallout shelters. In rural or other areas where community shelters would not be readily accessible or where other special circumstances exist, home shelters may be necessary or preferable.

Readying fallout shelters is by no means the only preparedness action which should be taken. However, it is the keystone upon which all other measures should be based; and fallout shelters must be created and made ready before other civil defense actions will have much meaning.

Industry must take the initiative and carry the load in providing for its own survival. Government can, and will, give technical and other assistance, and will cooperate in every way possible; but industry itself, in cooperation with government, must make and put into effect its own civil defense plans.

It is clear that there can be no one plan which will meet the requirements of all companies and plants. However, with information gathered from Hiroshima and Nagasaki, from the A-bomb and H-bomb tests, and from the thinking of many industrial executives who have already made certain disaster preparations within their companies and plants, government can suggest certain elements of an industrial civil defense program which can be tailored to fit each company and plant situation.

Typical actions which business and industrial managements should take to minimize the effects of attack include:

1. PROVIDE FALLOUT SHELTER FOR EMPLOYEES AND THE PUBLIC

Intensive studies of nuclear weapons and their effects have made it clear that fallout shelter is the best method for saving life at least cost in event of nuclear attack. The key effort in the National Civil Defense Program, therefore, is the establishment of a nationwide system of fallout shelters. As the first step the Federal Government has made a national fallout shelter survey to determine the amount of shelter space which already exists in major buildings all over the country. This national fallout shelter survey, as of July 1970, has found fallout shelter in existing buildings for over 193 million people, much of which is located in business and industrial establishments. In addition, fallout shelter has been found in smaller buildings and in homes for over 4½ million people. The survey of existing buildings and new construction is continuing. Industrial and business leaders should take the initiative in protecting their most important asset, their employees.

a. Cooperate with government in the National Shelter Survey, Marking and Stocking Program, by allowing all buildings to be surveyed and their shielding capacity protection factors assessed by architects and engineers who are working under contract with the Army Corps of Engineers or the Naval Facilities Engineering Command.

The minimum requirements for a *public* fallout shelter are: a protection factor of 40 (this means that the radiation measurement inside the shelter would be 1/40th of the radiation measurement outside the shelter) a capacity of 50 people, 10 square feet of space per person, storage capacity for supplies and equipment of approximately 1½ cubic feet per person, adequate ventilation (500 cubic feet per person in unventilated areas and 3 cubic feet of fresh air per minute per person in ventilated areas).

Some companies have surveyed and evaluated the fallout protection potential of their existing buildings independently. Although you may have done so, this information should be made available to the Federal Government in order that it may know how much adequate fallout protection is in being throughout the country and determine how much more is needed.

b. To the extent possible, enter into agreement to allow use as public shelter of space which meets Federal protection criteria and is needed, by signing the "Fallout Shelter License or Privilege" form.

If your buildings meet these minimum requirements and are licensed as public shelter, the Federal Government will mark and stock such shelters at no cost to the owner.

Supplies provided by the Federal Government include a special austere food ration of 10,000 calories per person, metal containers for water, medical and sanitation items, and radiation-detection and measuring instruments.

c. If it is discovered, either through the Government survey or your independent one, that your buildings do not offer adequate fallout protection, arrange to improve them by modification or by additional construction to bring them up to the minimum standards. The OCD has established courses of instruction on protective structural design which may be attended by your plant engineer. Consult your local or State civil defense office for applications (DD Form 1353), and for additional information regarding this program.

d. Include dual-use fallout shelter in the design and specifications for all your new plants and structures. Experience has shown this can be done at little or no additional cost.

e. Assist local government authorities in marking and stocking public fallout shelters. The locating, marking and stocking of public fallout shelters is a tremendous nationwide logistical undertaking, placing a severe burden on local government authorities. Business and industrial firms can be of great assistance by obtaining shelter signs from local government and installing such signs in their plants in accordance with specifications provided by the Army Corps of Engineers or the Naval Facilities Engineering Command.

Local governments are responsible for moving shelter supplies from Federal warehouses to public shelters. Business and industrial firms can assist local government in accelerating the movement of these supplies by picking up shelter supplies at Federal warehouses and transporting and placing them in shelters. In some cities, this has been accomplished by volunteer services of the trucking industry or by use of transportation provided by a variety of business and industrial firms.

f. Urge employees to arrange shelter for themselves and their families by preparing home shelters or by assuring that they and their families know where public shelter is located. Provide them with guidance and assistance on how to establish group shelters in their residential areas and in their homes. Be sure to coordinate your efforts with those of local government.

The Office of Civil Defense conducts continuing shelter research and can provide advice and assistance to industry in developing plant shelters as safeguards against radioactive fallout.

Making effective use of fallout shelter at the workplace will require industry to take other actions, especially regarding warning, shelter management, and radiological monitoring training.

2. ESTABLISH A PLANT WARNING SYSTEM

Arrangements should be made to receive attack warning information and to alert employees quickly throughout the plant. Where buildings are spread over a wide area or located beyond hearing distance of community warning signals, a separate warning system may be necessary. In many instances, the existing public address system can be used. However, the warning system must be adequate to reach all office buildings, plants, laboratories, and other places where employees are located.

Inform employees of how they will be warned of impending attack, what the signals are, what they mean, and what action should be taken upon receipt of the warning signal.

3. ESTABLISH A CONTROL CENTER AND COMMUNICATIONS SYSTEM

A control center should be set up with communications to the control center of the local government and to other plants in the community which are in mutual aid associations. It is especially important for the plant civil defense coordinator to be acquainted with the situation in order to use effectively all resources at his command during an emergency and to coordinate the activities of the different self-help services. A control center is useful in peacetime disaster such as fire or explosion when damage occurs simultaneously at several points. It is vital in event of enemy attack. Even a small plant should have an emergency control center. Some companies have located their control centers underground which provides some blast protection and excellent fallout protection.

4. ORGANIZE AND TRAIN EMPLOYEES FOR SELF-HELP

Self-help or self-protection is the concept of training each employee, and organizing and training small groups of employees, within and among industrial plants, large buildings and other facilities for specialized emergency services, such as firefighting, rescue, police, first-aid, and radiological monitoring to safeguard the building and its occupants in time of attack or other major disaster. The framework of an effective disaster control or self-help organization is already in existence in most large buildings and industrial plants. For example, most plants already have organized and trained fire brigades, guard services, rescue teams, first-aid and welfare groups. In order to be prepared in the event of nuclear attack the problem is simply one of enlarging and extending already organized groups, with the addition of perhaps a few teams concerned with radiological monitoring and shelter management.

A civil defense self-help organization does not replace the normal plant protection organization or emergency forces. Instead, it is designed to expand existing emergency and protective groups to meet large scale disasters more effectively.

Any plant is capable of organizing for its own self-protection provided that management supplies leadership and a chain of command for emergency planning and emergency action. In many ways the average industrial plant is like a small community. The plant protection or emergency organization resembles the organization structure of the local government community services.

These specialized groups can serve as auxiliaries to the various departments of local government, that is, police, fire, health, radiological defense, etc. Therefore, by enlargement and extension of normal industrial emergency protective groups, and their enrollment as auxiliaries to the various protective forces of local government, there will be a "built-in" community capability for quick action in an emergency.

  a. Shelter Management Training—Business and industrial leaders should immediately organize and train for the management and use of shelters, including radiological monitoring.

  A Corporate Shelter Manager

should be designated and trained for the overall supervision and management of shelters throughout the corporation. In addition, shelter managers should be designated for each shelter area in the plants and office buildings.

Each plant shelter should be staffed with fire prevention and control personnel, plant police, radiological monitors, personnel trained in first-aid and medical self-help, and others who can handle the rationing of food and water, supervise sanitation procedures and perform other duties which may be required during shelter occupancy.

OCD has put into effect throughout the country, concentrated short training courses in shelter management to train large numbers of people in short periods of time. Information regarding these courses is available from local civil defense authorities.

Be sure to train for the rapid movement of employees to shelters and inform employees fully of the company emergency plan.

b. Training in Radiological Monitoring and Reporting—Each industrial plant or office should organize and train radiological monitoring teams responsible for measuring radioactive fallout intensity and reporting the information to the local government emergency operating center, thereby not only serving occupants but also forming an important part of the community radiological defense system. Selected personnel should be sent to the Office of Civil Defense "Radiological Monitoring for Instructors" training course. They may then train others, both within the plant and in the community.

Be sure to inform and educate employees regarding the hazards of fallout, the procedures to be followed in detecting, measuring and reporting fallout intensity, and in decontamination procedures.

c. Tell Employees About the Company Disaster Plan—Handbooks should be prepared containing basic information regarding the company plan and information necessary for self-protection at the workplace, including description of the attack warning signal, floor plan drawings showing shelter areas and hazardous points within within the plant, maps showing mass movement or evacuation routes within and near the plant. Such information should be given to all employees.

d. Test the Plan—Conduct periodic exercises and drills. All plant employees should be tested in evacuation procedures and movement to shelter areas. Management personnel should go to the emergency headquarters and practice working under simulated attack conditions. Shutdown procedures should be simulated. When community civil defense exercises are held, management should cooperate with local government in testing survival plans in plants and office buildings.

e. Urge Employees to Prepare Their Communities and Homes for Wartime Emergency and Natural Disaster—One of the most effective ways to reach people with civil defense information is where they work. Employees often look to their employer for guidance on major issues and problems, for if the individual is to be prepared, he must be prepared in his home as well as at his workplace.

Civil defense publications are available from local government offices. These publications should be given to employees with an appropriate letter from management urging them to make survival plans at home.

This is not only important to the family, but is a vital service to the community as well. Employees should be familiar with the community shelter plan.

f. Utilize Employee Publications and Organizations—Information regarding emergency preparedness techniques should be included as a regular department in employee publications and as a regular feature of employee meetings.

5. DEVELOP EMERGENCY SHUT-DOWN PROCEDURES

Orderly and speedy shutdown in industrial plants is vital in time of emergency. Whether simply pulling a switch, closing a valve, or cooling a large furnace, all must be planned for in advance. This is true also in office buildings and institutions. In large buildings explosive gas fumes, high voltage lines, fire, and similar hazards can be almost as deadly as natural forces or military weapons. In some instances, due to manufacturing processes, disorderly shutdown could result in self-destruction of the plant—destruction almost as great as that resulting from an attack. Procedures must be planned and tested. Mock shutdown training exercises are conducted regularly in many plants.

6. PLAN FOR MASS MOVEMENT OF EMPLOYEES

Movement of civilians from dangerous and potentially dangerous areas to shelter in time of civil defense emergency requires thorough organization, timing and supervision. Great confusion can result from the spontaneous and rapid movement of employees when not properly directed. The plant civil defense coordinator should designate at least one person to be responsible for evacuation planning including movement to shelters.

A general traffic plan, both within and outside the plant must be made, directional signs and instructions posted, and if necessary, adequate transportation provided. The plans should include a priority departure for operating and maintenance personnel so that plant shutdown can be orderly, and production resumed as quickly as possible.

Arrange for employees to move to shelters in the plant or to shelter areas in nearby buildings. Cooperate with local government in planning movement to shelter facilities in safe locations outside the city or to the employees' own home shelters.

It should be kept in mind that there never will be definite assurance of a specific amount of warning time. Therefore, plant personnel must be ready to move fast if local authorities decide there is enough time for evacuation prior to attack—or after an attack when it is safe to move people out of severely damaged areas.

7. ESTABLISH A PLANT SECURITY SYSTEM FOR PREVENTION OF SABOTAGE AND ESPIONAGE

Industry should take appropriate measures to prevent the commission by misguided persons or enemy agents of any destructive act to endanger employees or impair the productive capacity of the plant. Measures also are needed to prevent the collection of information which might contribute to the enemy's knowledge of the Nation's war potential. Such information might be used to advantage by an enemy in attacking this country either through direct or covert means.

Action against sabotage and espionage is primarily the responsibility of the Federal Government, but there are many steps that industrial plants can take to reduce or eliminate this danger.

Protective measures include adequate guard or watchman services, fencing, protective lighting, investigation of applicants and employees, pass and identification systems, safeguarding classified information and reports to the FBI of suspicion of sabotage, espionage or subversion.

8. PROVIDE EMERGENCY PROTECTION FROM DELAYED OR UNCONVENTIONAL WEAPONS EFFECT

Clandestine and unexploded ordnance are hazards which must be dealt with promptly. Organization and training must ensure prompt reporting to state and local police forces which will conduct reconnaissance for unexploded ordnance and report the existence of such ordnance to the closest Department of Defense Explosive Ordnance Disposal Unit or ZI-Army Commander.

State and local authorities will provide for restriction of areas and protection of persons from such ordnance, including execution of plans for evacuation to safer areas, until arrival of the responsible explosive ordnance personnel. The Federal Bureau of Investigation will investigate reported incidents of clandestinely introduced weapons. The Department of Defense, through its Explosive Ordnance Disposal Units, will disarm atomic weapons and dispose of other unexploded weapons. The Atomic Energy Commission will take custody and dispose of fissionable materials of unexploded ordnance.

9. PARTICIPATE IN INDUSTRIAL MUTUAL-AID ASSOCIATIONS FOR EMERGENCY

This type of association is an organization of industry officials representing facilities in a particular area, united by voluntary agreement to assist each other with facilities, equipment and manpower as needed in time of disaster.

Few plants can provide all the services and equipment needed in emergency. By joining other large facilities in the neighborhood and through proper coordination with departments of local government, assistance can be provided to one another in the form of equipment, materials or personnel in time of disaster. Many mutual-aid groups have cataloged their supplies of fire hose and other firefighting materials and equipment, medical supplies, rescue items, and other emergency-use materials so that each member of the group knows what is on hand.

Although industrial mutual-aid associations are not new, the idea is especially applicable to civil defense for dealing with wartime and natural disaster problems. Experience gained in peacetime disaster is very valuable in preparing for wartime disasters.

The emergency operations plans of the industrial mutual-aid associations are part of the total emergency operations plan of local government.

10. PLAN FOR EMERGENCY REPAIR AND RESTORATION

In a natural disaster, emergency repair crews usually are available from many outside areas. In event of attack, however, outside assistance may not be available. Therefore, each plant must have its own emergency repair crews.

A good disaster plan will provide for the organization and training of selected employees to assess damage and repair damaged electrical, communication, gas, water and other vital facilities. Local utility companies can assist in this plant training program.

Plans are required for the replacement or repair of damaged production machinery and supplies, and preparations to use alternate production methods, substitute production machinery, and standby power and communication equipment.

11. PREPARE TO REPORT DAMAGE QUICKLY

Early and comprehensive assessment of damage caused by attack is vital to quick recovery. Each level of government will design systems capable of providing rapid and reasonably accurate estimates of (1) locations and degree of attack effects, especially radiological contamination and (2) what has survived the attack and is useful for recovery. Industrial managements will be expected to cooperate fully in reporting as required to local authorities the nature and extent of damage to their plants. Decentralized and dispersed multiplant corporations will need to receive such information also at their emergency company headquarters.

ESTABLISHMENT OF AN ORGANIZATIONAL AND ADMINISTRATIVE PLAN FOR CORPORATE CIVIL DEFENSE MANAGEMENT

1. GET IN TOUCH WITH THE LOCAL CIVIL DEFENSE DIRECTOR

It is the job of local governments to provide guidance and assistance, and to coordinate the emergency planning activities among the various departments of local government. This includes also the development of plans to utilize fully and coordinate the nongovernment leadership and resources into community emergency preparedness. Continuing liaison with local civil defense officials and other departments of local government must be maintained to ensure proper coordination. Up-to-date emergency planning information, guidance and assistance should be available to you from your local civil defense director and emergency planners in the various other departments of local government.

2. ESTABLISH CORPORATE AND PLANT LEADERSHIP RESPONSIBILITY FOR CIVIL DEFENSE PREPAREDNESS

One of the first steps in the actual preparation for disaster is the appointment of a single individual as civil defense coordinator at both the company level and in each plant to provide coordination and direction of the overall corporate disaster plan and the disaster plan in each plant.

3. APPOINT ADVISORY COMMITTEES

At both the company and plant levels a civil defense advisory committee may be designated representing various departments of the company, to assist in the development of the various phases of the disaster plan or civil defense program.

4. SEND THE CD COORDINATORS AND MEMBERS OF ADVISORY COMMITTEES TO OCD STAFF COLLEGE

It is imperative that coordinators and committee members maintain close liaison with civil defense officials. They should become thoroughly familiar with the authority, organization, and emergency procedures which are established by law and which become effective upon declaration of a civil defense emergency by the President or the Congress.

Coordinators, committee members, and service chiefs may obtain valuable information and training by attending the Office of Civil Defense training courses. A schedule of courses is available from local CD officials or the Office of Civil Defense in Washington, D.C. A variety of publications also are available.

5. ISSUE APPROPRIATE POLICY STATEMENTS AND ADMINISTRATIVE DIRECTIVES

Executive action is necessary in order to establish the disaster control program and thoroughly inform all executive and supervisory personnel of corporate attitudes and procedures regarding civil defense preparedness.

6. PLAN FOR CONTINUITY OF EACH MAJOR CORPORATE FUNCTION

The enlarged aspects of industrial civil defense planning must involve practically every department head in a company: the security officer, the purchasing agent, the treasurer and controller, the personnel director, the general counsel, the secretary, the production manager, the chief engineer, the head of research and, of course, the board of directors. Each department head must examine the functions for which he is responsible in peacetime and work out answers to problems involving continuity during and following attack. Unquestionably, many peacetime functions would become unnecessary while some would become extremely complicated and vitally necessary.

7. ASSIGN EMERGENCY FUNCTIONS TO EACH EMPLOYEE

A severe emergency may in a short time convert an existing peacetime organization into quite a different type of organization structure with altered or expanded functions. Such drastic changes inevitably mean that some groups of employees will become surplus at the same time that other units urgently need more workers. Plans should be made for quick emergency employee training, revised salary and wage administration, hours of work, protection of the retirement system and union relations.

8. PLAN FOR CONTINUITY OF MANAGEMENT

For each key position, replacement should be designated in order of succession. Thus the surviving person highest on the list could assume temporary position responsibilities following attack. This is an important part of disaster planning. No company is better than the people who make it work. If a company or plant is to continue production during and following attack, its key jobs must be filled.

Leadership needed for continuation of production after an attack could become virtually nonexistent, unless plans are made prior to an attack for lines of succession. Preservation of managerial skills must be a major part of the disaster control and civil defense planning.

9. AMEND BYLAWS AND ADMINISTRATIVE REGULATIONS

Appropriate legal authorization is often necessary to act in event of emergency. Each function of the various offices and departments of the company should be reviewed to determine whether the function should be continued in wartime, and alternate solutions listed. Bylaws should be amended to provide authorization for establishment of succession lists and for reestablishing the company and continuing production under conditions caused by enemy attack.

10. ESTABLISH ALTERNATE CORPORATE AND PLANT HEADQUARTERS

If continuity of management is to be preserved, managers must have a place to assemble and work, equipped and furnished for carrying on corporate operations during and after attack.

The emergency company headquarters should be located at a point outside a probable target area. Often it is possible to establish this at a point where the company already has an installation. The facility should have fallout protection and should be equipped with communications and living quarters. The objective should be development of a capacity to manage the corporation from any one of several different plant locations in less vulnerable areas by transfer or delegation of management and command authority during the survival and early recovery period.

11. DESIGNATE EMPLOYEE REPORTING CENTERS

Secondary emergency reporting points should be established for all personnel. Some companies have set up reporting centers at the homes of executive personnel who reside in suburban areas. Certain records are maintained at these locations and arrangements have been made for priority telephone service in an emergency. The cost of this type of planning is insignificant when compared with the advantages of recovery-planning which can be conducted at these locations during or after the emergency. A few companies have stored "disaster checks" which may be given to employees who report to these centers following an attack, as a source of funds until the plant can resume operations.

12. PROTECT VITAL INDUSTRIAL RECORDS AND DOCUMENTS

Protection of vital records includes the duplication and safe storage of records important to the continuation of the company and plant production. Written descriptions of activities, manufacturing processes, engineering designs, and essential legal documents and accounting records must be safeguarded to aid in continuation or restoration of production.

Even in peacetime companies are sometimes forced to go out of business due to loss of their vital records. Corporate records are in effect the corporate memory, and when a company loses its records it loses its corporate memory. This loss results in what is sometimes called "corporate amnesia." A company cannot function if its designs, formulas, accounting records, and other vital documents are destroyed.

Protection can be accomplished by duplication and storage in safe locations. Many companies have already provided their own safe record storage area. Some such record storage areas are deep underground, others are at dispersed areas.

Companies not having their own record storage facilities should consider renting suitable space from record storage companies.

13. DEVELOP EMERGENCY FINANCIAL PROCEDURES

Money may be needed promptly following attack for wage payment, cash advances to employees, payment of bills, and purchase of survival items such as medicine, food and equipment. Some companies maintain bank accounts of an unrestricted nature at scattered locations and have established lines of credit at a variety of places.

Emergency procedures for drawing company funds should be developed in advance. Banking arrangements should be adequate and corporate bylaws should have ample provisions for withdrawal of funds and should be kept current and on file in the company record storage vaults. In some instances it may be desirable to deposit a small amount of cash at the alternate company headquarters and in each of the security storage vaults.

14. DECONCENTRATE CRITICAL PRODUCTION

Deconcentration refers to the geographic decentralization of critical production so that all manufacture of a critical item is not in one location.

This type of decentralization refers not only to the production lines, but also to management and technical offices, and other departments concerned with the production of the item.

Deconcentration key departments or smaller segments of an industry will help assure continuity of production of essential items.

15. DISPERSE NEW INDUSTRIAL PLANTS

This is the employment of the simple military measure of using space and topography for defense of industrial plants against attack. By multiplying the number of targets an enemy must hit to inflict the same total damage, industrial dispersal tends to reduce the total effects of attack on our capability to produce. Consideration may be given also to placing the entire plant underground.

In addition to dispersal of production, it is evident that certain finished items, especially materials necessary for survival, be dispersed and available for immediate use following attack.

16. PREPARE CORPORATE AND PLANT DISASTER PLANS

Put the corporate and plant civil defense plans in writing. The civil defense coordinator and the employee-management civil defense advisory committee should develop an emergency operations plan including but not limited to: (1) The company policy statement regarding civil defense planning. (2) Statement describing the purpose of the plan. (3) A description of warning and emergency communications procedures. (4) The nature and responsibility of each plant protective service, a list of team leaders and members, and an outline of who does what in a disaster. (5) Drawings showing plant floor plans, shelter locations, and evacuation routes. (6) A list of available emergency equipment and a list of needed equipment.

The disaster plan should be furnished to executives and members of the protective and self-help groups. The corporate manual could serve as a guide for periodically reviewing and up-dating the plan.

17. TELL STOCKHOLDERS ABOUT THE COMPANY DISASTER PROGRAM

Some companies have included information regarding the company civil defense plan in the regular quarterly and annual reports to stockholders. This gives assurance to the investors that the company is planning to continue in business following an emergency. Some have urged stockholders to cooperate with local government in serving as auxiliaries and in preparing their homes and families for disaster.

18. TELL THE PUBLIC THAT YOU HAVE MADE PLANS FOR WARTIME EMERGENCY AND DISASTER CONTROL IN YOUR COMPANY AND IN EACH PLANT

Industrial and business employers occupy a position of prestige and influence in the community. If it is known that private industry and business are making plans for disaster, then individuals at home will be motivated to make like plans. Some companies have included information regarding their civil defense plans in advertisements, radio announcements, and TV skits.

19. SUPPORT AND ASSIST THE COMMUNITY FALLOUT SHELTER PLANNING EFFORTS

The excellent organization which industry already has and the leadership capabilities of its management can be a powerful force in assuring public awareness and support of the community shelter program. Industry has the managerial skills, facilities, communications channels and people to stimulate the development of an understanding by all citizens in a community of the need for fallout shelter and the organization and training necessary to make the best use of it.

In communities in which there is little or no civil defense, alert industrial management can motivate other community leaders and assist them in creating and developing effective civil defense.

Many industrial and business executives are serving as volunteer civil defense directors, community industrial defense coordinators, members of industrial civil defense advisory committees, volunteer leaders of fire, rescue, and police auxiliaries and reserve units. Some firms have sponsored special movies and publications as a means of furthering the objectives of emergency preparedness. Industrial executives can be of great assistance in developing community shelter plans and encouraging employees to prepare their families and homes for civil defense.

CONCLUSION

These are, in the main, the important steps which must be taken by industry in order to survive an attack or disaster.

It is again emphasized that no one plan will fit all conditions. Too much emphasis cannot be given to the importance of basing the company civil defense plans on the conditions and factors which make up the local situation. Corporate Secretaries and other corporate executives must recognize, therefore, that the steps and guides suggested here can only provide general assistance in planning and that specific plans for each company and each plant must be built upon careful analysis of the local situation and upon the creativeness, imagination, and common sense of the local planners. What happened afterward will depend largely on the kind of prior preparations or disaster.

Industry has always assumed immense tasks and taken great initiative in national emergencies. We do not anticipate war. But if an attack should come—by miscalculation or otherwise—there will probably not be sufficient warning to take even the first precautionary step—preparing places of refuge; it already must have been taken. Therefore, industry should direct its imagination and energy now to ways and means of providing adequate fallout shelter and related facilities for its employees.

Preparing to protect its own employees would be a great public contribution, but industry can widen that contribution by assistance in preparing any community in which a plant is located. Survival of all the people cannot be guaranteed. In fact, it is impossible to save enough people to assure National survival without the know-how, energy, enthusiasm, and leadership of American industry. Its great leadership should be directed to the foremost current civil defense problem—providing fallout protection for every American.

Chapter 15

ORGANIZATION FOR SECURITY

THE ORGANIZATION of a security program requires more than the combining of men, material, and budget. The protective response of the organization to its environment is most often through the security organization. While budgetary limitations, productivity requirements, and administrative responsibilities all interact to set the parameters for program operations, the creativity of the manager responsible for protection is often a deciding factor. The managers skill in analysis of problems, the adeptness with which alternatives are examined and selected; as well as the ability to "sell" a particular advantageous plan are critical to program success.

The organizational system used to "deliver" protective services is not uniform throughout the field. Organizational patterns vary considerably with centralized and decentralized management control being used to solve similar loss problems. Protection will be as effective as management planning and organization allows. The organization of the protective resources in a responsive manner is essential for good service. The direction, authority, responsibility, and performance standards established are critical to effective operations.

Organization requires that specific tasks are made the responsibility of individuals and that they be designated for completion within definite time limits following specific instructions. The specific format that the organization uses for obtaining performance is secondary to the establishment of adequate operational guidelines. Particularly in organizations which stress delegation of authority for completion of assignments, it is not enough to merely "tell" someone to "get the job done." It is also vital that the person being asked to complete the task have the necessary training and education to be successful. Thus, the organization for security services requires that organization be geared to the training and educational levels of the employees. An organizational system which anticipates great discretion and delegation should be staffed with well-trained and qualified personnel. If this is not done, high-quality results should not be expected. Conversely, if low-quality personnel are present, centralized control and administration as well as supervision are essential to reduce losses since protective services activities are often difficult to fully define for workers in advance; general rules and regulations are often provided for guidelines, but field supervision is essential to make the program operate effectively.

Organization for security can take the traditional forms of line and staff, straight line, functional, or variations based on individual organizational needs. The critical factor in establishing the pattern is the degree of compatability between organizational approach and the problems which are present.

In the preceding chapters the various component parts of security have been discussed and analyzed. The systems in which security is utilized have been presented as well as the specific processes which interact or act singly in providing security. It is now essential that the security function be viewed in its relationship to the structure and activities of an organization.

It should be readily apparent that there are a wide variety of applications for security within an organization. All organizations have needs which differ slightly and which require a different organizational security response. This being so, there are many different ways in which the component parts of security can be integrated into a workable protective device for an organization.[1]

FUNCTIONAL ACTIVITIES

Since the general goals of security can be established as being *protection and preservation of an environment,* or *environmental maintenance,* it is possible to isolate common functional activities performed by all security programs. Governmental as well as proprietary security programs perform these activities. These activities may be categorized as being administrative, investigative, and preventative.

ADMINISTRATIVE. This category includes activities such as conducting surveys of security needs, establishing policies and procedures for security, the advisement of other organizational activities about security problems, establishing security training and education programs, and providing management direction for interdepartmental cooperation for security.

INVESTIGATIVE. This category includes preemployment screening of employees of the organization, and conducting investigations of offenses committed against the organization or its policies.

PREVENTATIVE. This includes the use of guards, watchmen, and safety personnel. This also includes the use of electronic security devices, such as closed circuit television equipment monitoring devices, anti-intrusion devices, and alarms.

These activities all appear to some degree in organizations which organically include security. The relative degree of emphasis or need for these activities however does vary considerably in different organizations. This is in evidence in Figure 15-1 which illustrates the relative degree of emphasis placed upon security activities in the major categories of organizations. It should be obvious that small proprietary organizations, if security is emphasized at all, will normally emphasize some type of preventive activity, e.g. watchmen or alarms. Similarly, large proprietary organizations have all activities present within the organization; however, the emphasis being on prevention and administration with investigation being present, but not to the same relative degree of utilization as the other two activities. Governmental programs of security emphasize all three types of activities almost equally and proprietary organizations which have governmental contracts are required to maintain the same relative degree of emphasis on activities as governmental programs. While it is possible to observe the relative degree of activity emphasis based upon an organization's need for security, the form which security will take in a particular organization is dependent upon a wide variety of factors. The critical issues involved with the incorporation of security into an organization is the organization's response to it.

ORGANIZATION RESPONSE TO SECURITY

Security being introduced into an organization is the result of the realization of existing problems. A major goal of all organizations, regardless of their nature (either governmental or proprietary), is organizational survival. Unquestionably, many sub-goals such as maximizing profits and provision of public service also exist. These must be viewed in context of the organizational desire to continue its existence and to expand its operations. This being so, security and its utilization must be com-

	Government	Proprietary-Government (Contract)	Proprietary		
			Large	Medium	Small
Preventative	XXX	XXX	XXX	XXX	XX
Investigative	XXX	XXX	XX	X	
Administrative	XXX	XXX	XXX	XX	

Figure 15-1. Relative degree of emphasis on security activities in organizations of various sizes.

patible with the goals of the organization that it is serving. The means that security would use to reach its particular goal of protecting the company must be compatible with and acceptable in the goal attainment patterns of the organization. Based on the organization's perception of their needs and an evaluation and analysis of problems and vulnerabilities of the organization; security must try to integrate a program of protection into the organizational structure in such a way that it will provide *environmental maintenance* without causing any undue organizational stresses. It must *protect and preserve* the organization's environment without disrupting the company's activities unduly. Only in this way could it be considered to be fully integrated into the organization.

The goals, both broad and specific, of the organization will determine the scope of the security program and the activities which it will emphasize. The emphasis on functional activities within an organization depend on a number of factors. These include the following:

1. Historical developments with the organization.
2. Individual idiosyncrasies within an organization.
3. Growth of organizational preventative activities.
4. A natural growth of staff support activities.
5. A function of organizational growth in general.

Similarly, the organization's response to problems through the use of security is based upon an evaluation of the specific activities necessary to accomplish its goals. Specific questions to be answered would include:

1. To what extent will a particular security organizational structure contribute to the accomplishment of total organizational goals?
2. How expensive is it?
3. How many personnel will be involved?
4. How much time does it take?
5. Is it possible to integrate security into the total organization? If it cannot be integrated, what acceptable alternatives are there to accomplish the organization's goals?

INTERNAL ORGANIZATION FOR SECURITY

The establishment of security within an organization should be done in an environment which is conducive to organizational participation in the activity. The development of broad based support for, and participation in, the security program for the organization will result in many desirable effects, these include (1) reduced losses, (2) increased production, and (3) job enlargement for employees.

The promotion of actual participation is a desirable method of involving employees in the problems of security for the organization. Since problems would thus become personalized, it would be much more difficult not to respond to and deal with organizational security problems at the lowest levels in the organization. This promotion of security, rather than a coerced cooperation, provides a stable positive environment in which to develop or introduce a security program.

Once the decision is made to place security functions in an organization, a number of questions must be answered. These include the following:

1. Where should the function be located organizationally?
2. How should the function be administered by the organization?
3. What activities should this function perform in the organization?

The manner in which these questions are answered will determine the scope and shape of the organization's security function.

Location

Security functions may be placed in a variety of organizational locations. The responsibility for security can be placed in many organizational administrative functions. For example, many proprietary organizations had recognized that a security problem existed and attempted to deal with it within existing organizational control mechanisms. Since financial control was probably the most effective avenue, many finance or comptroller functions developed an administrative or investigatory security function to serve organizational needs. As additional problems developed, more emphasis was placed on preventive activities, but administration of the total security function continued through the finance activity. In other organizations, the legal department initiated the security function within the organization and continued this responsibility in much the same vein as did the finance function. It should be pointed out that while security activities are administered through a wide variety of organizational mechanisms, and the activities and the functions performed in security are not all the same, all security functions are responsible for three principal activities: (1) planning for security needs and operations, (2) implementing plans and policies, and (3) controlling results.

Regardless of the organizational location and administration of the security function, security can be internally organized in a wide variety of ways depending upon the size of the organization.

All security activities can be divided into either line or staff functions. Similarly, security may have either line, staff, or functional authority within an organization. Normally, security performs line activities in the preventive functions. It performs staff and functional activities in administrative and/or investigative matters. The distinctions between the various types of authority can be clearly understood if each activity is viewed separately.

Preventative

The preventative activities of the security organization, through the use of guards for example, represents the line authority within the organization. The guard represents the final link in an uninterrupted series of directly delegated authority from the organizational policymaker. Through this line authority, the guard exercises direct command and control over employees in all matters under his jurisdiction.

Investigative

Investigative activities in an organization can represent either line, staff, or functional authority. It is line authority if organizational policy allows for investigative activities to have direct command or control over employee activities. Staff authority would be present if it is only authority to advise, recommend, counsel, or assist others in the organization relative to security investigations, but not to directly have control or command in employee activities relative to security. Finally, functional authority would be present if no command and control exist and the authority is restricted to the right to command action only in specified procedures, policies, or practices. This functional authority is most widely used in *specialty* organizational areas, such as security.

Administrative

Administrative security activities are normally either of the staff or functional variety. The distinction would be based upon a particular organization's policy and functional security activities.

INTERNAL SECURITY

When an organization determines that an internal guard force may be developed, a set of problems is developed which includes the selection, hiring, and training of appropriate personnel and the organization of their activities into the total security function within the company. The guard force may be organized either functionally, geographically, or chronologically. Similarly, under any of these labor division categories, they may be further organized by (1) fixed-post deployment, (2) roving patrol deployment, (3) reserves, and (4) combinations of the former.

The use of guards and their overall effectiveness will be increased by the effective location of barriers, adequate protective lighting, alarm devices, adequate communication, appropriate SOP's etc. A guard force is normally used to provide some means of physical security (preventative activities) for the organization.

Since guards that are provided internally are an integral part of the organization, their total maintenance and supervision is a responsibility of the organization. The maintenance of an internal force is very often more expensive than the provision of the same types of service through commercial agencies. The needs of the organization should determine which approach to the organization's means of security should be utilized. Such factors as cost, effectiveness, neutrality, control, and supervision should be used to determine which approach is most advantageous for the organization.

EXTERNAL SECURITY

A wide variety of commercial agencies are utilized to provide the means of organizational security. A large number of companies exist to provide security for both government and proprietary needs. These organizations can be divided into two major categories. These are companies that provide security services which are (1) limited in scope, or (2) total.

TOTAL SECURITY SERVICES. Commercial agencies which provide total security are those that would offer a complete spectrum of security services within their organization. This would include guards, watchmen, patrolmen, investigative activities, alarm services, etc.

LIMITED SECURITY SERVICES. Commercial firms which are limited in scope are those which offer specialized types of services short of the wide scope activities offered by total security agencies. This would include agencies which offer investigative services only, or guard-watchmen services only, or alarm services, or combinations of these which together do not equal total security services.

CHOICE RATIONALE

External guard forces are normally less expensive for an organization to utilize, since the responsibility for maintenance of preventative guard personnel is the responsibility of the contracting agency. The supervision and administration of external forces can either be the responsibility of the contracting agency or the responsibility of the organization requiring security.

Three factors should be considered when the final determination is made as to whether internal or external guard means should be adopted.

1. Which will maximize security protection?
2. Which will maximize fire and safety protection?
3. Which will minimize security, safety, and fire protection costs while *protecting and preserving the organizational environment?*

The administrative, investigative, and preventative activities, which are the organization's means for security, are affected by the organization's perceived needs. They are also affected by the organization's ability to support these activities in terms of cost and employee support. Thus, the organization will determine how much security and of what type it is willing to accept and utilize for protection within the organization.

INSPECTION AND CONTROL[2]

No matter how well a security department is organized and coordinated, its efficiency will be nil unless it is properly controlled. The system or form of control exerted over it is of utmost importance. The whole character of the department reflects the type of control exerted over it. The control of the department must be efficient and all encompassing.

Although control of an organization may seem to be of lesser importance than planning and directing phases, the organization's very existence is dependent upon its proper application. Security service is a critical area from the standpoint of administration since being a service organization it is constantly open to criticism. Since security officers are operating in ever-increasingly delicate matters and more frequently involved in the regulation of noncriminal conduct, the manner in which duties are performed must be brought under close scrutiny. This being the case, every action either good or bad, which is performed by a member of the department in carrying out his official duties, reflects directly upon the entire organization. People tend to judge an entire organization on the basis of a possibly isolated experience which they might have had with a member. First impressions are unfortunately the most lasting and the hardest to change.

POLICY AND CONTROL

Since control is a basic organizational process there are certain elements which it must contain wherever it is found. Its objective must be made clear. Standards of performance must be established to attain the desired end results. To this end policies are established. Since policies are established for the basis of governing future actions, they must be made known to all those who are in positions which they affect. Policies are thus said to be the framework for action and as such are a method to delegate authority within the organization. Established policies give authority for action to be taken at the level of execution in repetitive situations.[3] These are commonly called Standard Operating Procedures (SOP's). Every care should be taken that these policies are fully understood, including the conditions under which they apply.

TYPES OF POLICY

There are three classifications for policies: basic, general, and departmental. Basic policies are statements of goals with guidelines established for the long-range attainment of those goals. These usually have their origin with the Security Director. General policies are those which deal with everyday operating procedure. These usually evolve at the managerial level (Shift Command, Director), but as with basic policies, they may both be formulated at any level in the organization with top level final approval. Departmental policies result from interpretations of both general and basic policies by departmental heads to provide supplemental rules and regulations for operations. It is understood, however, that these should not conflict with higher level policy. Further, all policies should be put into writing so that all personnel will have the same interpretation of them and that they will not become clouded with time. The procedure manual is probably the best reference for established policies, and Standard Operating Procedures (SOP's). Once policies have been put clearly into writing, they may be checked for compliance with relative ease.

Control of an organization may be accomplished in many different ways. The goals that an organization has set for it control it as much as they narrow its field of interest. These controls are established in the organizational framework by the administrators to insure that duties and functions are carried out in the most efficient way.[4] During the formulation of policies and procedures by the security administrator as many contingencies as possible must be considered before a particular rule became part of official departmental policy. Since much care and research went into its institution, and since a particular goal is set to be attained by using these policies, it is the responsibility of the departmental administrator to insure compliance with them. Controls are implemented in various ways. The most often used technique is some form of inspection.

INSPECTIONS

There are five basic questions that should be answered by any inspection.[5]

1. Are established policies, procedures and regulations being carried out to

the letter and spirit in which they were made?
2. Are those policies, procedures, and regulations adequate to attain those desired results?
3. Are the resources at your disposal, both personnel and equipment, being utilized to the fullest?
4. Are the physical resources adequate to carry out the job of the department?
5. Is there any deficiency in the integrity, training, morale, or supervision of the personnel which should be corrected or improved?

It should be apparent that all phases of activity are included in such inspections. This is absolutely necessary; if they were not checked, the planning and organizing done by the administrator would be to no avail.

Everything relating to the security department must be subject to control by inspections. Although they cannot be sharply defined because of inter-relationships, there are four identifiable areas which require inspections.

The personnel who perform the duties of the department must be made aware of all departmental regulations and procedures. The health and physical appearance of the officers must be maintained and compliance with the regulations can best be accomplished by a system of inspections. A constant check must be conducted at all levels to insure that the authorized number of personnel are available for duty. Work schedules must be reviewed to insure that an adequate number of personnel will be available for duty during peak work seasons. The systems of personnel relief should be examined to insure proper coverage, while affording personnel adequate time for meals and breaks. And relief policies for shift changes should also be reviewed for compliance with company sick-leave policies.

The state of departmental morale should be ascertained, since low morale, for any reason, has an adverse effect upon duty performance. The supervision that the personnel are receiving should merit close inspection as to adequacy and appropriateness. One of the most important items which should be examined is personnel integrity. It goes without saying that instances of misconduct must be thoroughly investigated. The state of personnel training must be constantly reviewed to insure that personnel will be able to perform their duties efficiently. Finally, the reports and records submitted by personnel should be reviewed for accuracy and compliance with proper procedures.

Things

Repair and replacement of items is accomplished by physical inspections. This type of inspection must of necessity be exacting in its standards; nothing must be overlooked. All items and aspects of equipment and material should be examined. This is to include all vehicles, communication equipment, safety and security related equipment, the security facility itself, as to general housekeeping of the offices; corridors and motor pool areas, the uniforms and related equipment; even the officer's personal equipment which he uses on duty must be checked, and supplies of virtually every kind must be inspected to insure compliance with departmental procedures and policies covering their maintenance and use.

Procedure and Actions

This area of inspections is probably one of the most important to the individual security officers themselves, as it is interested in improving their professional skills. Procedures used in stopping, detaining or interrogating suspects, techniques of dealing with witnesses, the handling of violators, inspection of safety or security hazards, all should be checked to see if established policies are being followed.

Established policies have been set up on the basis of surveys and studies which give the officers adequate guidelines for their own safety and for the preservation of departmental respect. Therefore, deviations must be checked as soon as possible to prevent possible injury to either officers or the department.

Results

As might be expected, the easiest method to find out if procedures are being followed is to view the success of the operation. Results will usually show if procedures are being followed. While results might indicate present compliance, long-term results might indicate just the opposite. Too often, token or official compliance with procedures is reflected in reports, while in actuality, procedures used may bear little resemblance to established policies. The inspection may reveal needs not previously discovered. These inspections should include the analysis of statistics, examination of incident reports and public reaction to the department as a whole.

TYPES OF INSPECTION

Persons to whom authority is given cannot be held accountable until the results of their actions have been determined. The results are appraised by inspections. Therefore, control may be said to be implemented by a system of inspections.

There are two types of inspections:[6] the authoritative and the staff inspection. It must be understood that neither of these two types is mutually exclusive. Rather, they are supplemental to each other.

Staff inspections are conducted by those who do not have direct control over security operations while authoritative inspections are conducted by those with a direct or line control. The line inspection is usually concerned with people while the staff is usually interested in functions.

Line Inspections

The authoritative or line inspection is conducted with direct authority. In conducting inspections there are four approaches or techniques which can be effectively used.[7] The first of these is usually employed by the lower eschelon supervisors such as the sergeants, the other three being used by higher eschelons of command.

1. PHYSICAL OBSERVATIONS. Actually examining the personnel as to physical appearance is conducted, to check their equipment and materials to ascertain proper maintenance and proper use.

2. IMMEDIATE EXAMINATION. On-the-scene examination as to how specific problems are handled.

3. CAREFUL REVIEW. Written observations and examinations in report form by subordinates to their superiors, who review overall results.

4. DAILY INQUIRY. Is work or action achieving the desired results? Do records reflect an adequate picture?

These techniques may be used successfully to determine deficiencies. Since they are employed by supervisory personnel, corrective action may be immediately initiated. This type of inspection is relatively simple since it is a function of supervision.

STAFF INSPECTIONS

The staff inspection provides administrative assistance to the superior officers. Since it is conducted out of the normal lines of authority and responsibility the detailed checks provided are of functions or tasks rather than of personnel. The normal work load of the line supervisor makes it impossible for him to adequately inspect or have knowledge of all details of a specialized operation. Very often he simply does not have the time to conduct inspections himself.

It is not uncommon for a supervisor to neglect the supervision of a particular policy because he personally disapproves of it or thinks it to be a waste of time. It is therefore necessary for specialists in a particular field to analyze operations and reveal areas of weakness or strengths. In theory, any discrepancy found in such inspections should be submitted to the security director for disposition. This, however, is very seldom done. Usually a good relationship exists between the operations personnel and the staff inspectors which allows corrective action to be taken without unnecessary delay.

Staff inspections are solely dependent upon the security director for their authority and responsibility. Normally, staff inspections are a detailed analysis of specific things with findings, opinions, and suggestions for improvement presented to the director for disposition.

FOOTNOTES

1. Appendix 15A presents a proprietary security manual which details the organization of the protective program. Appendix 15B presents a proprietary governmental security manual.
2. This material originally appeared as "Inspection and Control in the Security Department," *Security Education Briefs,* Vol. 2, No. 6, written by Richard S. Post. Reproduced courtesy of OAK Security Inc., Madison, Wisconsin.
3. Southwestern Law Enforcement Institute. *Police Management for Supervisory and Administrative Personnel.* Springfield, Thomas, 1963, p. 45.
4. Wilson, O. W.: *Police Administrator,* 2nd ed. New York, McGraw-Hill, 1963, p. 21.
5. Southwestern Law Enforcement Institute. *Police Management for Supervisory and Administrative Personnel,* p. 45.
6. *Police Management for Supervisory and Administrative Personnel,* p. 46.
7. *Police Management for Supervisory and Administrative Personnel,* p. 49.

BOOKS

Mandelbaum, R. J.: *Fundamentals of Protective Systems—Planning Evaluation, Selection.* Springfield, Thomas, 1973.
Morgan, Patrick: *Successful Handling of Casualty Claims.* New York, Prentice-Hall.
Snigel, E. O., and Ross, H. L.: *Crime Against Bureaucracy.* New York, Van Nostrand Reinhold Company, 1970.
Wright, K. G.: *Cost-Effective Security.* New York, McGraw-Hill, 1972.

PUBLICATIONS OF THE GOVERNMENT, LEARNED SOCIETIES, AND OTHER ORGANIZATIONS

Bartel, Ann, and Landes, William M.: The demand for private police in the United States. National Bureau of Economical Research, Inc., 52nd Annual Report. September 1972, p. 60.
Glertz, J. F.: *Economic Analysis of the Distribution of Police Patrol Forces.* Miami Univ., Oxford, Ohio, 1970.
National Bureau of Casualty Underwriters. Burglary insurance manual. New York.
National crime information center, A. FBI Law Enforcement Bulletin, May 1966, p. 1.
Offenders as a correctional manpower resource. Washington, D.C., Joint Commission on Correctional Manpower and Training, 1968.
Port, A. Tyler: Security policy in the department of defense. *Industrial Security,* July 1957, p. 9.
Standard Oil Company. Blueprint for industrial security. New Jersey, Standard Oil Company.
United States Senate Committee on Education and Labor. *Private Police Systems.* Police in America Series. New York, Arno Press and the *New York Times,* 1971.

PERIODICALS

Anrider, S. S.: Profits in protection: Security systems, electronic and human, have become big business. *Barron's,* February 20, 1961.
Astor, Saul D.: Contract guards: The facts as I see them. *Security World,* July-August 1975, p. 18, 19, 123, 124, 127.
Bain, Walter G.: Management and security. *Industrial Security,* January 1962, p. 10.
Buckley, John L.: Industrial security: Use of the guard force. *Best Insurance News,* January 1962.
Burstein, Harvey: The M.I.T. security guards. *Law and Order,* 7: June 1959.
Business Week. The guard you hire may be dangerous. December, 1971.
Camden, John O.: Modern plant guards are more than just gate catchment. *Mill and Factory,* October 1961.
Case for contract guards. *Occupational Hazards,* June 1969, p. 49.

Coultier, Richard L.: Goodby guard, hello Joe. *Industrial Security*, October 1971.

Davis, Albert S.: Company guards vs. subcontractor guards. *Industrial Security*, December 1967.

Deere, Albert T.: The plant security system. *Industrial Security*, July 1962.

———: The plant security system for prevention of sabotage and espionage. *Industrial Security*, July 1960.

Diamond, Harry: Operation security. *Police*, 7: no. 2, November-December 1962.

Displaced guards get other and better jobs. *Factory*, August 1959.

Donovan, Robert: Anatomy of a demonstration, Part I. *Security World*, 3: no. 7, July-August 1966.

———: Anatomy of a demonstration, Part II. *Security World*, 3: no. 8, September 1966.

Ex-convicts take aim at jobs as guards. *Top Security*, August 1975, p. 147.

Hanson, Douglass A.: The sale and acquisition of a contract guard service. *Security Register*, 1: no. 3, 11-13, 1974.

Healy, Richard J.: Enlightened industrial security. *Advanced Management Office Executive*, August 1962.

———: Industrial security—an important element in business. *Southern California Industrial News*, February 7, 1966.

———: Industrial security and the scientist. *Industrial Security*, January 1958, p. 14.

———: Our new society president speaks. *Industrial Security*, 2:13, October 1958.

Heaton, J.: Rent a guard? *Industrial Security*, February 1969.

Monahue, Vincent J., Col. USAF: Security—A necessary evil? *Industrial Security*, October 1966.

How Brink's guards its profits, too. *Business Week*, February 1965, p. 54.

How to cut theft. *Occupational Hazards*, January 1969, p. 15.

Japanese concerns beef up security; woes plague London's transit system. *Wall Street Journal*, Monday, October 28, 1974, p. 4.

Johnson, Clarence L.: The scientist, the engineer and security. *Industrial Security*, October 1958.

Knightly, John: The Conington train disaster. *Police Journal*, 42:219.

Lansdale, Edward G., Maj. Gen. USAF: Plant security overseas—panelist. *Industrial Security*, December 1964, p. 51.

Manes, A.: Insurance crimes. *J Criminal Law*, 1945.

Marson, Michael: The insurance broker. *Top Security*, September 1975, p. 200.

Nelson, Harold A.: The defenders: A case study of an informal police organization. *Social Problems*, 15: no. 2, 127, 1967.

Pati, Gopal C.: Business and ex-offenders: A case of holy alliance. *Security Management*, May 1975, p. 26-29.

———: Business and ex-offenders: a case of holy alliance. *Security Management*, March 1975, p. 20-24.

Potter, Anthony N., Jr.: Uniforms and security. *Industrial Security*, October 1970.

Royal, Robert F.: Hiring the "unemployable": It works. *Security World*, March 1970.

Salerno, Anthony J.: Can industrial security be established in a centrally organized agency? *Industrial Security*, April 1960.

Schramm, Robert W.: Where is your company headed? *Petroleum Refiner*, November 1959, p. 361-365.

Security: The "Official" and the "unofficial." *Security and Protection*, April 1975, p. 7-8.

Security is everybody's business. *Industrial Security*, October 1958, p. 8.

Security men thrive on the wages of fear. *Business Week*, June 20, 1970, p. 112-114.

Security's role in engineering total peace. *Industrial Security*, July 1957, p. 13.

Shaw, Harry L., Jr.: You think you can't fight city hall. *Industrial Security*, August 1971.

Shaw, Paul: A message from the New Year's Gang. *Security Register*, November 1973, p. 45-49.

Sheehan, John B.: National symposium on science and criminal justice. *Industrial Security*, December 1966.

Simon, Raymond F.: Strategy of confrontation, Part 2. *Security World*, January 1969.

———: Strategy of confrontation, Part 3. *Security World*, February 1969.

Skyjackings: Should security be relaxed? *Protect Newsletter*, September 1973, p. 2-3.

Smith, Jack: Ex-army man tracks down industrial spies. *Los Angeles Times*, February 11, 1962, p. 1-2.

Staffing for security. *Canadian Security Gazette*, October 1971.

Stealing is a symptom. *Parents*, January 1952.

Stephens, G. W.: Are we really security conscious? *Police Chief*, 40: no. 12, 55, December 1973.

Stevens, Arthur: Consider this. *Security and Protection*, November 1974, p. 22-23.

Storm, J. G.: Public relations. *Industrial Security*, 5: October 1961.

Supplying the world security market. *Security and Protection*, May 1975, p. 16-18.

Van Evera, Dr. B. D.: University research and security. *Industrial Security*, January 1963.

Walsh, Timothy J., and Healy, Richard J.: Insurance. *Protection and Assets Manual*, 2. California, The Merritt Company, 1974.

———: Propaganda and security—Some comparison observations. *Industrial Security*, April 1959.

Waterman, Kenneth F.: Tape recording expedites industrial security operations. *Industrial Security*, February 1966.

Weiss, Harold: The dangers of total corporate amnesia. *Financial Executive*, June 1969, p. 63-66.

UNPUBLISHED MATERIAL

Flavel, W.: Research into security organizations. Paper delivered at the Second Bristol Seminar on the Sociology of the Police, April 1973.

APPENDIX 15A

PROPRIETARY SECURITY MANUAL*

PREFACE

Inventions, technology, techniques and know-how developed or acquired by Monsanto and not freely available for the use of others are company assets.

A responsibility of management and of each employee is the protection of Monsanto assets—both tangible and intangible—against theft or against loss resulting from the unauthorized or inadvertent disclosure of trade secrets or other company confidential information to outside parties.

For protection of its assets the company must rely on the good faith of Monsanto people. To support this good faith, certain criteria and administrative practices are prescribed in this Security Manual.

I. **Security Classification Nomenclature.** Where it is desirable to denote trade secrets or other sensitive Monsanto information, the following classification identification is to be employed: *Company Confidential.*

II. **Company Confidential Information**
   A. *Definition.* All trade secrets and other information belonging to Monsanto which the company chooses to retain as a secret are classified as *Company Confidential.*
   B. *Determination.* Constant management judgment and discretion at all levels is essential
      1. to judge and determine if Monsanto has or does choose to hold secret, particular information, a document or a project; and
      2. to effectively communicate Monsanto's choice as to secrecy to all employees who are to have access to the information, document or project.
   These are the two key essentials to security of information. These need not be allowed to become a complex process. To reasonably err in favor of denoting information, a document or a project as Company Confidential can do no harm to security and in most cases causes little or no legal or administrative inconvenience. The company's choice to hold as secret and thereby classify as Company Confidential is obvious for the large bulk of Monsanto exclusive information.

---

* Courtesy of Monsanto Corporation.

Examples of information and documents which are obviously Company Confidential in most circumstances are as follows:

Technical	*Other*
Basic manufacturing data	Cost data
Design manuals	Market potential developed by market research
Plant operating instructions	
Plant tests	List of customers
Sample manuals	Distribution techniques and market strategies
Raw material specifications	
Technical reports of experimental investigations	Forecasts and budgets
	Accounting data
Analyses	Mailing lists
Reports of product and process evaluation	Stockholder records[8]
Engineering drawings and flow sheets	
Product and process research	
Research notebooks	
Plant training manuals	
Quality control charts	

Therefore, for simplicity and the least amount of administration activity, most questions as to the propriety of denoting as *Company Confidential* particular information, documents or projects should be resolved in favor of security.

In some cases where (a) the answer is not obvious or (b) to err in favor of security would cause an unnecessary administrative burden, a decision should be obtained from appropriate supervisory authority. Such authority is the general manager or his delegate of the operating division having prime interest and responsibility for the area of interest involved, or the staff department director or his delegate having prime interest and responsibility in their area of interest. Should there be an overlap of interest, the division general manager or his delegate will confer with the staff department director to resolve the problem.

III. **Company Confidential Information—How to Limit Distribution.** Where it is desirable to designate a Company Confidential document as one requiring special security care and restricted to certain specific Monsanto employees, at the author's discretion an additional legend may be added. The legend will specify the restrictions applicable to the document. Some suggestions follow:

A. This information is confidential in its nature, for use in Monsanto's business. It shall be retained by the recipient, not copied or loaned to others in the company. All requests for duplicate copies or initial copies for individuals or departments not included in the distribution tables must be forwarded to the originating department.

B. This document contains confidential information which is the property of Monsanto Company. Only those portions of the document relevant to a duly authorized individual's need to know may be excerpted for him. The report must not be sent outside the company without proper written approval and the authorized recipient is ac-

countable for its safekeeping, excerpting or otherwise disclosing its contents and for its proper disposal. The document remains the property of Monsanto Company and shall be returned within two months to its author, except upon specific authorization to hold for a longer period.
C. This document is the property of Monsanto Company and the recipient is responsible for its safekeeping and disposition. It contains confidential information which must not be reproduced, revealed to unauthorized persons or sent outside the company without proper authorization. *Destroy this document within two months.*

IV. Criteria for Access, Distribution and Circulation.
A. *Outside Parties.* For persons outside the Monsanto organization, the "need-to-know" rule is interpreted as absolute; that is, "Monsanto's need for the outside party to know" must be the criterion for disclosing Company Confidential information. Any disclosure made to satisfy a "Monsanto need for the outside party to know" must be made under conditions which impose an absolute obligation on the party to maintain the confidential nature of the information. (See Section XIII.)
B. *Monsanto employees.*
  1. Criteria. For Monsanto employees to be made privy to Company Confidential information, "need-to-know" interpreted with liberal common sense should apply. The objective is to contain Company Confidential information within the bounds of the Monsanto organization. It is not intended to restrict employees who have a necessary interest for full job performance. Thus, security is aided and the chances of disclosure outside the bounds of the Monsanto organization are lessened when employees do not concern themselves with information about which they do not have any real working interest.
  2. Who determines. Any employee who has been designated by the author to receive Company Confidential information has the responsibility and the authority to determine access by his associates, subject to any limitations imposed by his line supervision or the originating authority and subject to his own good judgment as to the propriety of expanding access at his management level. Concurrently he has the responsibility and the authority to seek advice and instructions from his supervision whenever the nature and the importance of Company Confidential subject matter to be disclosed does, in his opinion, warrant higher level attention.

V. Caption Notice on Company Classified Documents.
A. *Company Confidential Documents.* For the purpose of conveying notice with the document and to provide protective care, Company Confidential documents should be marked with the legend **Company Confidential** in bold letters on the face of the document. Any document not marked should be treated as confidential if one thinks it should be even though it is not marked.
B. *Blueprints, drawings, etc.* All company blueprints and drawings relating to manufacturing and research operations shall be prepared on sheets bearing the following legend as part of the Monsanto title block:

COMPANY CONFIDENTIAL

This drawing is the property of Monsanto Company and must be accounted for. Information hereon is confidential and must not be reproduced, revealed to unauthorized persons or sent outside the company without proper authorization.

Where it is necessary to send blueprints or drawings to outside parties such as contractors, fabricators or suppliers, and after authorization is obtained, the documents shall be stamped with the following legend before being released:

NOTICE

This drawing is the property of Monsanto Company and must be returned, without reproduction or duplication, at any time upon request, but in any event at completion of the work or job. While in the possession of the recipient, it must be properly safeguarded against revelation or disclosure to any one except those employees who require it for the work or job. The recipient must keep confidential, and require his (its) employees to keep confidential, the information contained hereon.

VI. Storage of Documents.
   A. Within the confines of a building or facility area with positive perimeter control of egress and ingress Company Confidential documents should be kept out of sight of the casual observer or curiosity seeker while not in the possession of an employee.
   B. Within the confines of a building or facility area which affords less than positive perimeter control of egress and ingress Company Confidential documents when not in the possession of an employee must be under lock and key or equivalent security safekeeping.

VII. Transmittal of Documents. Company Confidential documents and information should be transmitted in a manner selected by the sender with due regard for the contents of the document. During transmittance in the mail, outside envelopes should not display a security marking. Certified or Registered Mail should be used where appropriate.

VIII. Duplicate Copy Storage for Security Against Disaster. Copies of documents, engineering drawings, flow sheets, etc. which should be on hand after any location disaster should be stored at a separate location. Copies may be on microfilm.

IX. Unwritten Company Confidential Information.
   A. *Monsanto employees.* All discussion and verbal communication of Company Confidenntial information should be verbally identified by its appropriate security designation; that is Monsanto people involved in the discussion should be reminded that they are talking about Company Confidential information. Only those having a need-to-know interest should be allowed in the discussion.
   B. *Outside parties.* With persons outside the Monsanto organization, Company Confidential information should not be discussed except as necessary to transact the business of the company applying the criterion of "Monsanto's need for the outside party to know." (See Section XIII.)

X. Employee Administration—Security Aspects.

A. *Employee skills and experience.* There is no intention in the application of Monsanto's security program to restrict its employees in the use of their professional or other skills and experience, but only a desire to safeguard the company's property of accumulated knowledge. Accordingly, in the application of this program care must be exercised so that the individual rights of the employee are not infringed. In some instances, no precise line of demarcation can be drawn between Company Confidential information and an employee's skill and experience. In those cases where the line is narrow, the principal guide may be the individual's own sense of propriety and professional ethics. The company recognizes the obligations employees may have not to disclose any inventions, trade secrets or other confidential information which they have acquired as a result of previous employment elsewhere. Under such circumstances, employees should notify their supervisors of any such prior obligations to avoid embarrassment to the company.

B. *Preemployment investigations.* Investigation of all new employees before being placed on the payroll, or by agreement with the applicant soon thereafter, should include, as a minimum, some inquiry into his or her reliability consistent with the position being filled. This investigation may be conducted by an outside investigative organization.

C. *Employee agreements—company policy.* Monsanto, like most chemical companies, long has had a policy of entering into agreements with technical employees and other personnel in security-sensitive positions. These agreements largely represent a formal recognition of commonly accepted, time-honored ethical and legal obligations not to disclose information obtained in confidence in the course of employment. Such ethical and legal obligations have for centuries been embodied in the Anglo-American system of common law and they are carried forward and embodied in Monsanto's employee agreements. The canons of ethics for engineers, chemical engineers, lawyers and other professional groups also have incorporated these obligations as standards for such professions.

In recognition of the fact that virtually all salaried employees have contact with Company Confidential information, henceforth all salaried employees shall enter into an employee agreement with the company.

It is important that the company's purposes in requiring these agreements be fully explained to new employees and present employees executing agreements for the first time. Not only will this serve to eliminate any misunderstandings as to the nature of the agreements but it also will impress upon the employee the importance of his own activities in this field.

Two copies of the agreement shall be signed by the employee and the general manager of the division or staff department director, who may delegate this authority. A copy of the agreement shall be given to the employee and the other signed copy forwarded to the payroll and pensions section of the Accounting Department. It is the payroll section's responsibility to determine that a signed employee agreement is on file for every salaried employee. Special agreements shall be used for production, exploration and geological personnel of the Hydrocarbons Division.

D. *Transferred or reassigned employees.* Any employee who has secured Company Confidential documents at the location or in the division or department where he is employed shall, prior to being transferred or reassigned to some other plant, location or department, account for and turn in all classified documents which have been issued to him for his work at the particular location. If it is necessary that such transferred or reassigned employee take such documents with him for his succeeding position, then special arrangements should be made to account for the custody and the location of such documents.

E. *Terminating employees.* All employees who have had access to Company Confidential information, including not only design data, process information and the like, but also accounting, sales production, cost and similar data, shall upon termination account for and return all manuals, blueprints, specifications, memoranda, diaries, notebooks and other documents pertaining to the company's business which they received or prepared in the course of their employment. Also, they shall sign the following statement:

> In terminating my employment with Monsanto Company, I have returned and accounted for all material, of whatever kind, containing company information, received or prepared by me in connection with my employment, and I have retained no copies, reproductions or excerpts of such material.
>
> ....
> Date   Signature of Employee

Location managers or their designees are responsible for obtaining such signed statements except at the company's General Offices, where staff department directors, general managers or their designees have such responsibility. In addition, as a matter of notice, each terminating employee shall be given a copy of his employee agreement, even though he received one when employed.

F. *Employee education and security awareness.* Employees will be made aware of security practices through orientation at the time of employment and thereafter by retraining programs. For salaried employees, this should include a discussion of the following:

1. The law and nature of trade secret information—definition and bounds.
2. The value of Monsanto's intangible assets to the company, to employees, and to the health of the economy of the nation.
3. The need for and process of clearing technical papers before publication, speeches before delivery, and other releases of Monsanto information.
4. Some aspects of the process for obtaining patent coverage and how the mishandling of the information could prejudice Monsanto's ability to obtain a patent.

5. The value to Monsanto of commercial lead time over competitors in the marketing of a new or improved product.
6. How premature and fragmented disclosure of releasable information in advance of the Public Relations Department's handling could negate an opportunity to exploit favorable news or minimize the impact of bad news.
7. The importance of each employee's being aware of the extent and bounds of his authority and responsibility to speak for or commit Monsanto to a position while in attendance at outside meetings of associations, industry committees, civic groups or governmental committees and other meetings, as well as being alert to the extent and bounds of his authority to communicate Company Confidential information.

G. *Specific security advice.* Each employee will be advised by his supervisor that he is in a position of trust and confidence and of those particular aspects of his job which are security sensitive.

H. *Notice to subsequent employers.* When an employee leaves to go to a new employer it is necessary, in many cases, that Monsanto notify the new employer, whether or not a competitor, of the existence of his employee agreement with Monsanto and its more important items. The letter or notice also shall state that Monsanto expects the terminating employee to comply with his Monsanto employee agreement. The employee shall receive a copy of this letter.

Monsanto realizes that there are a number of situations in which the company may feel confident that (1) the new employer cannot or will not have use for Company Confidential information, or (2) the employee has no Company Confidential information, or (3) Monsanto's relationship with the new employer assures against use of any Company Confidential information which the individual may take from Monsanto. Careful judgment must be exercised by general managers and staff department directors in preparation of the letter and in making a decision whether it should be mailed at all. In cases where there is a question whether a letter should be sent, the Law Department should be consulted. Divisional personnel directors and staff department directors shall be responsible for sending such letters.

Following is a form of letter which can be used in writing to employers, competitive or potentially competitive with Monsanto, concerning former Monsanto employees who have joined their organizations:

Mr. ....
        (title)
....
        (company name)
....
        (address)
....
Dear Mr. ....:

We understand that Mr. (employee), a former employee of ours, has accepted employment with your company as a ....

(chemist) (engineer) working in the field(s) of ....
(chemistry) (engineering).

Our best wishes for success go with Mr. (employee) in his new endeavor. However, we think you should be aware that, in connection with his employment by Monsanto, Mr. (employee) executed an employee agreement, certain provisions of which remain in force after his separation from Monsanto. A copy of this agreement is attached for your review.

We wish to call to your attention the provisions which refer to information and materials considered confidential by our company. During the course of Mr. (employee)'s employment with us he was given access to confidential information and documents which are Monsanto's trade secrets.

While it is not our desire in any way to impair Mr. (employee)'s employment opportunities or performance, we do expect and feel confident that he will honor his employee agreement with Monsanto by keeping confidential and not using any of our company's classified information or trade secrets during the course of his new employment with you.

Sincerely,
Monsanto Company
By ....
(Title)

Enclosure
cc: Mr. (employee)

XI. Facility Security—Plants and Laboratories.
A. *Defense against invasion by unauthorized personnel.* At most facility locations there are three lines of defense against intrusion by information seekers, unfriendly persons, or those with criminal intent:
1. At the real property line—The minimum protection here should be the posting of "No Trespassing" signs at intervals, and reasonable enforcement should instances of trespassing occur.
2. At the perimeter encircling the immediate environment of operating structures and areas—Perimeter control of ingress and egress should be practiced at all times by fences, guards, or other means.
3. At internal areas of special security importance—Any variation from positive control of ingress and egress at the perimeter should be compensated for by internal control of access to any internal area where a process may be observed, where a piece of equipment may be observed, control rooms where Company Confidential information might be read from instrument panels or flow diagrams or operating instructions, and plant supervision offices where Company Confidential documents, including cost data, might be perused or appropriated. Internal areas of special security importance must have area access control regardless of the adequacy of the perimeter control.

Optimum protection at a facility location is obtained by the most feasible combination of ingress and egress control (1) at the real property line, (2) at the facility environs perimeter, and (3) at internal areas. The optimum combination at one facility might

favor strong perimeter control, while at another facility optimum protection might favor area control with only casual perimeter surveillance.
B. *Employee identification passes.* Each employee should be issued at his base location an identification pass or badge which he should have on his person at all times when on company property. It shall be optional with the location as to whether or not the employee's picture appears on his pass or badge or whether a badge is fastened to outside clothing. He should not be issued a second permanent identification pass or badge from a location where he is not permanently based.
C. *Visiting Monsanto employees.* All visiting employees should be personally identified by a receptionist or a guard, or the person being visited should come to the gate or reception room and receive the employee. The presentation of an off-location identification pass is not sufficient of itself to admit a visiting employee. Visiting employees should register in and out or be logged in and out and be given a "visiting" Monsanto employee's badge; it should be collected when the visit is concluded.
D. *Nonemployee visitors.*
  1. To plant or laboratory office areas only—Nonemployee visitors who will be confined to office areas must be registered in and out.
  2. To plant or laboratory areas. Nonemployee visitors who are to be admitted to plant facilities or laboratory areas must be
     a. reasonably identified, and
     b. required to register in and out and sign a nondisclosure agreement similar to the following before being issued a visitor's badge:

     > In consideration of Monsanto Company's granting me permission to enter its (insert plant or laboratory), I hereby covenant and agree that all the knowledge or information which I may acquire thereby, and particularly any knowledge or information which I may acquire with reference to the manufacture of Monsanto's products, shall and will be held by me in confidence, and that, without the prior written consent of Monsanto, I will not disclose, divulge or reveal the same or any part thereof, directly or indirectly, to any person or persons, or make any use of the same for myself or others.
     >
     > ....
     > Signature of Visitor

     c. required to wear the badge and be accompanied by a selected Monsanto employee at all times while in the plant or laboratory.
E. *Cargo vehicles and material.* Plants are responsible for security control of plant transportation traffic. Vehicles and drivers entering the environs should be regulated by means appropriate to the location.

All plants shall establish and maintain adequate security and accounting control of a plant material pass system.
F. *Contractors' employees.*
  1. Contractors and subcontractors and their employees normally will be governed by the terms of the particular contract. The terms of the contract shall be brought to the attention of the location's

supervisory personnel so that they are aware of the duties and liabilities of such employees on the site. Notices setting forth security rules applicable to contractors' personnel must be posted at the job site in prominent places.
2. In all cases, contractors' personnel must be confined to the exact area in which they are working under contract. All other areas of the plant are closed to them. The areas to which they shall confine themselves shall be designated by the plant manager, and the contractor shall be notified.
3. Wherever possible, construction employees should use a separate gate.
4. Every employee of a contractor shall be required to wear an identifying badge (including, where practical, the name of his employer) at all times when on company property.
5. Service representatives and service company employees who are not under a formal contractual arrangement containing a security agreement clause will be registered in and out or be logged in and out and will sign the visitor's nondisclosure agreement. In addition, they will be required to wear identifying badges when on company property. (See Paragraph D of this Section XI.)

XII. **Facility Security—Other Monsanto Locations.** With appropriate modifications to accomplish the intended security purposes, the practices set forth in Section XI should be followed at the General Offices and at all other facility locations (including sales and other offices) of Monsanto. Because of special problems at such locations, perimeter control is usually lacking; therefore, personnel must be particularly alert and careful with respect to the security of Company Confidential information. The supervisor at each facility should consult with the Security Manager, Central Engineering Department, St. Louis, as to security measures appropriate for his location.

XIII. **Secrecy Agreements With Outside Persons and Companies.** It sometimes is necessary in the course of the company's activities to reveal Company Confidential information to third parties or to permit them access to areas or materials from which such information can be obtained.

Of course, every effort must be made to avoid this where possible; but when it is unavoidable the information revealed should be limited to that which is absolutely necessary. In such cases, the third party shall be put on notice of his obligations to respect the confidential nature of the information and, where appropriate, a written undertaking should be obtained. Contracts shall be entered into with outside parties in the following categories to protect against unauthorized use or disclosure of Company Confidential information. Responsibility for securing signed contracts with the following shall rest with Monsanto management representatives arranging for or retaining such services.

A. Consultants.
1. Technical. Since the work of outside consultants varies considerably, contract language has to be tailored to fit the particular situation. If Monsanto is buying information from a consultant, it must be determined whether the company can restrict the consultant in his revelation of the same information to anyone else.

If his services involve exposure to Company Confidential information to be merged with technical additions by the consultant, it is necessary that he sign an agreement not to reveal or use the Monsanto information or the results of his work.

The consultant agreement should include a provision that the consultant, at the termination of his work or at any other time Monsanto requests, shall return any written, printed or other material given to him in connection with his work or prepared by him and embodying Company Confidential information. The agreement should specifically provide that he shall not duplicate any such material. Those negotiating such contracts should endeavor to obtain agreement that additions, modifications, information, design and the like which the consultant furnished to Monsanto shall become the property of Monsanto, and that the consultant shall not reveal or use such except with the prior written consent of Monsanto. As indicated, of course, any of the foregoing provisions depend upon the nature of the services to be rendered but all of them must be considered in the drafting of any agreement. The services of the Patent Department should be obtained in drafting these contracts and, where indicated, those of the Law Department.

2. Business or nontechnical. The business or nontechnical consultant shall agree that he will not, directly or indirectly, use for himself or others, or disclose to any third party, any Company Confidential information regarding any business or accounting methods, manufacturing methods, costs or any other information revealed in confidence to him. The agreement shall provide also for the return of all written or printed material, either supplied to him or which he prepares during the course of his services with the stipulation that he retain no copies. The services of the Law Department should be secured in drafting agreements of this kind.

B. *Construction contractors and subcontractors.* Secrecy provision shall be incorporated in construction contracts.[11] These provisions should be sufficiently extensive to protect Monsanto against any revelation of Company Confidential information by contractors or subcontractors or their employees. The contractor should agree to keep confidential, and require his employees and, where appropriate, subcontractors to keep confidential, all information deemed confidential by Monsanto which may come within his or their knowledge in the course of their work. The contract should also provide that, as appropriate, all drawings, specifications, manuals, notes and other documents, and copies thereof, which come into the possession of, or are compiled by, the contractor or his subcontractors during the course of their work will be returned to Monsanto upon completion of their work.

All Monsanto drawings must be prepared on sheets bearing the confidential legend described in section V as part of the Monsanto title block and, as appropriate, such documents, together with specifications, sketches, etc., issued the confidential stamp described in Section V. In certain instances a noncompetition provision, under which the contractor agrees not to design or construct for any other party a facility similar to that which he will construct for Monsanto, should be incorporated in the contract.

The services of the Law Department should be secured in drafting construction agreements. (Standard A.I.A. forms are not sufficient to protect Monsanto.)

C. *Outside designers.* In all cases where outside engineering firms are employed to execute any design work, the agreement for their services should contain a secrecy provision binding the firm and its employees. Where appropriate, employees of the design firm shall execute secrecy agreements directly with Monsanto. Design work shall be accomplished on drawing paper showing the Monsanto title block with the confidential legend described in Section V imprinted thereon. The agreement shall provide for ownership of the work produced, whether the outside designer arrives at it independently or whether he merges some Monsanto design information with that of his own. The particular type of contract, of course, must be tailored to fit the situation. As with construction contracts, it should provide for an accounting of all documents received from Monsanto and their return. In certain instances a noncompetition provision such as described in Paragraph B above should be incorporated into the agreement. Consideration should also be given to restricting an outside firm from disclosing (in any way, including advertising and other sales promotion, with or without photographs) the purchase of a plant or process by Monsanto. The services of the Law Department should be obtained in drafting these agreements.

D. *Fabricators.* Monsanto drawings sometimes are submitted to structural steel, equipment and other fabricators. The drawings and related correspondence should carry notice to the fabricator that he is under a duty to keep Monsanto's design information confidential and to return the drawings to Monsanto. (See Section V.) Where appropriate, the fabricator shall be required to impose a secrecy requirement on his employees.

E. *Salesmen and/or suppliers.* Agreements should be executed by suppliers and their salesmen to keep confidential design and other information supplied by Monsanto. Moreover, if salesmen or other suppliers' representatives are called into a plant for inspection or to aid in operating equipment, only information necessary for the accomplishment of their mission shall be given them. They should be accompanied by a plant employee and should not be permitted to go into any area of the plant other than that which their activity requires. Routinely, they shall sign an agreement not to reveal Company Confidential information. (See Section XI.)

F. *Insurance men, investigators, underwriters.* Insurance investigators, underwriters, insurance photographers and the like shall agree in writing to keep confidential any information supplied to them or observed by them in any plant or laboratory. If in the course of litigation it becomes necessary to reveal information of this nature, the Law Department should be consulted to determine the limitations and protective controls which can be established.

G. *Outside printing, art design and contract mailing.* Where this type of work involving Company Confidential information is done outside the Monsanto organization, particular selectivity should be exercised as to (1) the work that is sent outside as against that done within the Monsanto organization and (2) the selection of the outside organization doing the work. Nondisclosure agreements should be in force where any Company Confidential material, information, or mailing lists are being made available to the outside party performing the work.

XIV. **Prospective or Actual Revelation of Company Confidential Information.**
  A. Whenever a plant, division, or department has reason to believe that a terminated employee is revealing, or is in a position to reveal, Company Confidential information in violation of his obligations to the company, the Law Department should be notified to determine what action should be taken with respect to the ex-employee, his new employer or both. This also applies to current employees.
  B. Any actual or suspected security violation is to be reported to the Director of Safety and Property Protection, Central Engineering Department, St. Louis. He will consult with the particular division or staff department as the case may be and advise the Law Department if necessary.
  C. The Office of the Vice President and Corporate Secretary has authority and responsibility for investigating and following through on security violations or problems connected with any high risk that may develop.

XV. **Internal Publication and Printed Communication.** Divisions and plants are responsible for the security of information published and distributed to their employees. However, when overall company policy is involved, clearance shall be obtained from the Public Relations Department and from officers or staff department directors involved. Divisions, plants, and staff departments also have the responsibility of determining the extent of the distribution of internal material.

XVI. **Review and Clearance of Speeches and Material for External Publication.** Division general managers and staff department directors are responsible for effecting a review and clearance of speeches and material for external publication originating in their division or department. A clearance and review committee or other delegate to the general manager or staff department director, as the case may be, should possess both technical and professional competence to review and pass judgment on the material presented, referring such material or speech to the Law, Patent, and Public Relations departments for further appropriate review and clearance where technical information of substance is involved and/or any of the subject matter is of public relations importance.

XVII. **Release of General Information About Monsanto.** Such material is to be cleared by the Public Relations Department and, where applicable, by the particular officer or staff department director responsible for the function involved. Inquiries from the press are handled by the Public Relations Department.

XVIII. **Photography.**
  A. The Public Relations Department has responsibility for securing

proper divisional or plant approval before any photos are made. Additionally, the Public Relations Department has responsibility for obtaining proper divisional and/or plant clearance concerning the ultimate use of such photographic material. All photographic material is distributed to external sources by the Public Relations Department.

B. No photographs of equipment or processes shall be taken for any reason whatsoever without the approval of the location manager and, even with such approval, the actual photographs shall be examined by him or his designee before release. If any question exists, such photographs shall be submitted to the clearance and review committees referred to in Section XVI. No cameras shall be permitted within any plant or laboratory without specific approval of the location manager.

XIX. **Security Staff Assistance.** A security manager reporting to the Director of Safety and Property Protection of the Central Engineering Department will coordinate and provide staff supervision for security matters. Instances of information leaks, suspected espionage, aggressive commercial intelligence activities against the company, theft of material, unusual inventory shrinkage, and situations of suspected high-risk exposure will be reported to the director of safety and property protection by location management. The Office of the Vice President and Corporate Secretary will have final authority and responsibility for investigation and follow through on major security violations. The internal auditing group in the Accounting Department will assist in detecting violations. Management at each location, including the Creve Coeur site, will administer location security measures.

The personnel and Administrative Services Department will develop and coordinate throughout the company the conduct of indoctrination and refresher courses for all employees on the importance of company security.

XX. **Submission of Company Documents to Regulatory Boards or Authorities.** Regulatory boards or authorities for pollution control, safety, health, fire protection, zoning, etc., exist or are being established by legislation in various communities. Usually the statute or ordinance gives such authorities the right to inspect plants and to require permits to operate or to construct new units. To secure a permit, a company may have to submit plans, specifications, and process information to the authority. A submission of this nature usually supplies to such authorities or boards and their engineers Company Confidential information which such engineers could use for the benefit of others, including a subsequent employer, perhaps a competitor. On occasion, representatives of these authorities or boards demand more plans, specifications, or data than absolutely necessary.

Often the representative of such regulatory authority or board is designated a peace officer and a penalty is provided if such inspecting officer is denied entrance. If he can be shown a part of the plant involving the source of his concern without being shown Company Confidential information, no security problem exists. If, however, he demands to see Company Confidential information, he should be asked to execute a secrecy agreement along the lines set forth below. Should he refuse, his authority or board should be contacted to resolve the issue. If an impasse is reached, the Law Department should be notified promptly.

In view of the foregoing, therefore, no Company Confidential information should be made available to any such authority or board or to any of its representatives unless the following steps shall have been accomplished:

(1) A Monsanto engineer (or other representative) shall confer with a proper officer of the authority or board to determine specifically what plans, specifications or data are required. If the required documents, plans, specifications, etc., contain no Company Confidential information, they should, of course, be submitted.

(2) If the authority or board demands plans, specifications, or documents which contain Company Confidential information, such materials may be supplied only after written agreement by the authority or board and its employees that they will be kept secret and confidential. If the authority or board or its employees refuse to bind themselves, the Law Department should be informed so that proceedings to protect the secrecy of the documents can be initiated.

The following is a suggested type of agreement:

AGREEMENT

The (name of regulatory authority or board) and those of its employees who are signatory to this agreement agree to accept a relationship of trust and confidence with Monsanto Company and to use Monsanto's secret and confidential information and documents (all marked and identified by Monsanto as Company Confidential) for the specific purpose of determining compliance with the .... (in this space, put in the name of the particular ordinance or statute which controls) and for none other. The foregoing obligation shall also apply to any knowledge, information or experience secured in person or received from any employee or agent of the .... (name of regulatory board or authority) in connection with inspection of Monsanto's plants. The .... (names of regulatory authority or board) and its employees signatory hereto further agree (1) that it and they will not reveal or disclose any information supplied to it or them and covered by this agreement to any other person, firm, corporation or partnership, directly or indirectly, at any time, (2) that access to the information and to the aforementioned documents will be limited to those signatory hereto, and (3) that they will not copy or reproduce in whole or in part any of the confidential or secret information contained in the documents hereinabove referred to, and will return all such documents and information to Monsanto Company promptly when their use is no longer required for determining compliance with the above-mentioned ordiance or statute.

Dated this .... day of ...., 19......

....
On behalf of .... (name of
regulatory board or authority)
.... (name of regulatory board of authority employees)
....
....
....

XXI. **U.S. Government Security.** This Security Manual relates only to security safeguards for Monsanto trade secrets and confidential information. It does not apply to, or in any way supersede, governmental security regulations relating to U.S. Government defense or other work being carried on by the company. Violations of governmental security regulations may result in fine, imprisonment or both. Information about U.S. Government security may be obtained from the Secretary of the Executive Committee.

XXII. **Requests From Suppliers and Contractors to Use the Monsanto Name.** The following is a general policy covering requests from suppliers and contractors to use the Monsanto name in advertising, sales promotion, or publicity. As a basic policy, no supplier or contractor should be allowed to imply that Monsanto uses his product exclusively or specifically recommends it over other competitive products. However, in the interest of good trade relations, the basic policy should be flexible enough to permit judgments on a case-by-case basis since Monsanto, as a supplier, solicits cooperation from its customers in its advertising, promotion, and publicity programs. Handling of requests is by the following:

A. Requests for permission to mention Monsanto in advertising, sales promotion, or sales brochures should be referred to the Marketing Services Department with a copy to the Patent Department. The Marketing Services Department will direct the request to the Purchasing Department or Central Engineering Department, as appropriate, at St. Louis for basic clearance and then to the interested Monsanto operating division, department, or other company unit for additional clearance if necessary. Marketing Services will then assume responsibility for notifying the supplier or contractor that his proposal is approved or denied.

B. Requests for permission to use Monsanto's name in product publicity, external or internal publications, annual reports or other communications will be referred to the Public Relations Department with a copy to the Patent Department. The Public Relations Department will direct the request to the Purchasing Department or Central Engineering Department, as appropriate, at St. Louis for basic clearance and then to the interested Monsanto operating division, department, or other company unit for additional clearance if necessary. Public Relations will then assume responsibility for notifying the supplier or contractor that his proposal is approved or denied.

XXIII. **Disclosures from Outsiders.** The company receives, from time to time, letters disclosing or asking permission to disclose inventions, advertising themes, slogans and the like which the writer hopes may be of value to the company in its business.

Also, customers, contractors, inventors, designers, and other individuals and companies may desire, as a condition of some business arrangement, to impose upon Monsanto obligations with respect to secrecy or nonuse of certain information they wish to disclose to Monsanto. The situation may arise where (1) the outside individual requests Monsanto to agree to keep information confidential and upon Monsanto's agreement to do so, the information is revealed; or (2) the information may be revealed first and then a request made that Monsanto keep such information secret; or (3) the person may disclose information stating it is confidential without any agreement on Monsanto's part to keep it secret and then later claim he had secured such agreement from Monsanto.

There is danger in accepting or discussing such information or suggestions from outsiders inasmuch as similar ideas or suggestions may already be under development in the laboratories, plants, or the marketing services or public relations departments and when adopted later, may give rise to the belief of an outsider that such development was based on his suggestion.

To avoid these controversies, each such matter should be referred immediately (and prior to the revealing of any information) to the director or associate director of the Patent Department who shall handle it after consulting with appropriate company personnel. If the matter is received or arises at a location (other than St. Louis) where there is a resident patent attorney, it should be referred to such attorney.

APPENDIX 15B

PROPRIETARY—GOVERNMENTAL: SECURITY ORIENTATION AND GUIDANCE FOR ALL CLEARED EMPLOYEES

I. Purpose and Application. This guide presents a general orientation on the concepts and objects of the Department of Defense Industrial Security Program. It contains the basic information needed by anyone who has knowledge of, or access to classified information and is furnished for orientation and training purposes only.

II. The Security System

A. *Need for security.* Although the term security has many meanings, as used in this guide it refers to industrial security. Security encompasses all the various methods and operations designed to prevent vital defense information from falling into the hands of those individuals who would use this information to the detriment of the United States. Security attempts to keep our enemies, or potential enemies, from knowing what we are doing in order that we can stay ahead of them in military strength and capabilities.

B. *Responsibility for security.* The basic security responsibilities of the XYZ Corporation are set forth in the Department of Defense Industrial Security Manual for Safeguarding Classified Information (ISM). Under the terms of a security agreement between the company and the Department of Defense, the company is obligated to provide and maintain a security program consistent with the provisions of the ISM. The Security Department is responsible for developing and maintaining a security program. Administration of the security program depends upon supervision and security advisors. In the final analysis, however, security is the responsibility of each individual employee. The company Security Department is established to serve the individual employee, to enable him to fulfill his security obligations. As an employee at Corporate Headquarters, you are responsible for:

1. Complying with all company security procedures as stated in the XYZ Security Handbook.
2. Reporting immediately to the Security Department through Supervision
   a. Any knowledge of loss, compromise, or suspected compromise of classified information.
   b. Any knowledge of acts or threatened acts of espionage, sabotage, or subversive activities.
   c. Any knowledge of a security violation.
   d. Any knowledge of instances of delay in the movement of classified material by commercial carriers.
   e. Any knowledge of evidence of tampering with a shipment of classified material.
   f. You contemplated travel to or through a Communist country, or attendance at an international meeting outside the United States, where it is anticipated that representatives of Communist countries will also attend.
   g. When a member of your immediately family or your spouse takes up residence in a Communist country.
   h. When, through marriage, you acquire relatives who are citizens or residents of a Communist country.
   i. Any affiliation with a foreign interest—acting as a representative, official agent or employee of a foreign government, firm or corporation.
   j. Any change in your name.

III. **Classification Categories.** In order to safeguard information according to its degree of importance to our national defense, the information is placed in one of the following categories:
   A. *Top Secret information.* Defense information and material, the defense aspects of which are paramount, and the unauthorized disclosure of which could result in exceptionally grave damage of the nation.
   B. *Secret information.* Defense information and material, the unauthorized disclosure of which could result in serious damage to the nation.
   C. *Confidential information.* Defense information and material, the unauthorized disclosure of which could be prejudicial to the defense interest of the Nation.

IV. **Determining and Assigning a Classification**
   A. The Department of Defense determines to which of the above categories the various types of defense information will be assigned. For each classified contract awarded the company, the Department of Defense provides a Contract Security Classification Specification (DD Form 254). The Security Department distributes copies of the DD Form 254 to supervision and security advisors for dissemination to concerned employees.
   B. When any XYZ employee originates classified material, he should refer to the latest DD Form 254 to determine the correct classification of the material. An employee's responsibility

is marking, not classified. He marks the material he originates with a classification which has been determined by the Department of Defense. Information extracted from other classified documents will be classified on the basis of the source documents. If you are in doubt as to what classification applies to originated or extracted material, you should contact your security advisor or the Security Department classification management representative.

V. **Downgrading and Declassifying**
  A. Providing safeguards for material at a level of classification higher or for a longer period of time than is warranted is a waste of time, effort, and money. It is, therefore, very important not only to initially mark material with its correct classification but also to downgrade classified material when there is authority to do so. Downgrading instructions are usually received through revisions to the DD Form 254.
  B. A continuing system has been established by the Department of Defense which authorizes automatic declassification of information based upon the passage of time. The system is known as the Automatic, Time-Phased Downgrading and Declassification System (ATP). Under this system, classified material is assigned to one of four groups: Group 1, Group 2, Group 3, and Group 4, to assure that the information is automatically downgraded and declassified on a time-phased basis.

VI. **Marking Classified Material**
  A. *General.* Classified information may be contained in or revealed by various media such as reports, letters, drawings, films, models, equipment, etc.; these media would then be referred to as classified material. Classified material regardless of its form must be identified by correct classification markings to ensure that it will be handled and safeguarded according to its degree of importance to our national defense. The originator of classified material is responsible for having the material marked with its correct security classification and other required markings. Classified material received from outside the company should already be marked.
  B. *Classification marking.* Each document or item of material which contains or reveals classified information must be conspicuously marked or stamped (not typed) to show the degree of classification. The following are additional instructions for marking various types of material:
  1. Correspondence and other unbound documents (letters, memoranda, rough notes, working papers, reports, brochures, etc.). The top and bottom of each page containing classified information will be marked with the classification of that particular page. In addition, the cover page, or when applicable, the outside front and back covers and the title page, will be marked with a classification as high as the highest classified page contained therein. The first page of a letter of transmittal, which itself is not classified, must be marked with the highest classification of any of the enclosures or attachments with a notation added that the letter is unclassified when the classified enclosures or attachments are removed.
  2. *Artwork.* Original artwork

will have the classification marked on the top and bottom margins of the mounting board and on all overlays and cover sheets.
3. Charts, maps, drawings, and tracings. The classification will be marked under the legend, title block or scale, and at the top and bottom, in such a manner as to be reproduced on all copies.
4. Equipment. The classification will be marked on the equipment itself or on tags or stickers attached to the equipment.
5. Photographs, negatives, and vu-graphs. Appropriate markings shall be conspicuously placed. If the items do not lend themselves to marking, they shall be kept in containers properly secured, which shall bear the classification marking.

C. *Other required markings.* In addition to classification markings, classified material will show each of the following markings at least once:
1. Date of origin.
2. Name and address of the company.
3. Espionage law notation, except for Restricted Data. (The "AEC Act of 1954" notation shall appear on all restricted data.)
4. Automatic downgrading and declassification group assignment.
5. Subject or title classification. A "U," "C," "S," or "TS," will be placed in parentheses following the subject or title.
6. Paragraph marking. In the case of paragraphs or documents which are transmitted outside the company and which are of different classification, each paragraph must be marked to show the category of classification of information it contains. The appropriate letter, "U," "C," "S," or "TS," will be placed in parentheses immediately preceding and to the left of the paragraph involved.

D. *Downgrading.* When downgrading, the following procedures should be followed:
1. Draw lines through the old classification markings.
2. If the material is still classified, mark or stamp the material with the new classification.
3. Add the following information at least once on the downgraded material: change of classification, authority for change, date of change, and person making change.

Until the above remarking action has been taken, the material will be safeguarded according to the original classification marked on it.

VII. Dissemination. Knowledge or possession of classified information is permitted only to persons who need it to do their job (need-to-know), and who have been cleared to the level of the information. No one has a right to classified information solely by virtue of his position. The responsibility for determining whether or not a person's duties require access to classified information rests with the individual who has custody of the classified material.

VIII. Safeguarding Classified Information. When an employee has custody of classified material, it is his personal responsibility to ensure that the material is appropriately stored during all nonworking periods or any time it is not under the direct control and surveillance of an authorized person. Detailed security standards are prescribed for each category of classified material and apply to various types of facilities including file cabinets, safes, vaults, controlled areas (Closed Areas, Strongrooms, Restricted Areas).

IX. **Transmitting Classified Material.** Classified material may be transmitted only to a properly cleared person or company having a need for the material in connection with a classified contract or program. The Security Department should be contracted to verify all facility clearances.
  A. Transmittal within the plant
    1. Confidential material may be hand-carried by a properly cleared person or sent through the interoffice mail system in a sealed envelope using a special sticker to identify it as classified material.
    2. Secret material must always be hand-carried by a properly cleared person or a cleared secret messenger.
  B. Transmittal outside the plant
    1. Preparation. Classified material must be enclosed in opaque inner and outer covers. The inner cover must be sealed and plainly marked with the assigned classification and additional markings as required, and address of both the sending and receiving offices. The outer cover must be sealed and addressed, but it must not give any indication of classification.
    2. Methods. Top Secret material may be transmitted by an authorized messenger. Secret material may be transmitted by similar means, by registered mail, or by some types of protective commercial transportation services. Confidential material may be transmitted by the means authorized for Top Secret or Secret, by certified mail (in the U.S.), or by a variety of approved commercial transportation services.

X. **Control and Accounting.** In order to assure the protection of classified information and to limit its dissemination, documents and other material containing or revealing classified information must be subjected to controls such as:
  A. Restrictions on preparation, production and reproduction.
  B. Continuous accountability receipts.
  C. Maintenance of appropriate accountability records.

The company is required by the provisions of the ISM to maintain a record of all Top Secret and Secret material, and, when directed, submit an accounting to the DOD. Each custodian of accountable material must conduct a quarterly inventory of his material and certify as to its accountability. A record is maintained of the receipt into, and dispatch out of, the company of all Confidential material; however, continuous accountability while within the company is not required. It is referred to as nonaccountable material.

XI. **Destruction.** Classified material should be destroyed as soon as practical after it has served its purpose. The approved method of destruction is by burning, which is performed by members of the Security Guard Force. To destroy nonaccountable paperwork (including Confidential waste), the employee places the material in a container designated for classified waste, Secret paperwork, or other accountable classified documents are destroyed by delivering to a Document Control Station. Destruction of accountable documents must be certified in writing by those performing the destruction. Certified destruction is required to remove the material from the accountable inventory.

XII. Preparation and Reproduction. Generally, when classified information is included in a document, incorporated into any article of material, or physically recorded in any way, the security hazards increase. Further, excessive production of material containing classified information not only encourages unnecessary dissemination of the information, but also increases the volume of material which must be safeguarded. In addition, the conditions and processes involved in preparing and reproducing material are especially susceptible to a loss of security through unauthorized access or intelligence collection efforts. Therefore, the preparation or reproduction of classified material should be avoided whenever possible, and when it must be done, it must be carefully supervised and controlled. Accountable documents must be processed through a Document Control Station for reproduction.

XIII. Control of Visitors. There is a constant need for persons from other companies, military establishments, and government agencies to visit the XYZ Corporation. These visitors are received in the lobbies where they are logged in, receive badges and provided an escort to take them to the concerned employees in the plant. Employees in the area being visited are responsible for controlling access to classified information consistent with the purpose of the visit. For visitors, just as for employees, access to classified information is granted only when the requirements of clearance and need-to-know have been fulfilled.

XIV. Subjection to Compromise. Official information is assigned a classification because its unauthorized acquisition or disclosure could be damaging to the national defense. Therefore, when classified material is lost or when classified information is subjected to compromise, action must be taken to minimize the damaging effect on the national defense, and to prevent repetition of the security failure. When classified information is lost, disclosed to unauthorized persons, or otherwise subjected to compromise, the person who discovers the fact must report it promptly to his supervisor.

Security Precautions

Don't discuss classified information in the presence of or within the hearing of unauthorized persons.

Don't take it for granted that your visitor is cleared. Check the clearance and "need-to-know" before revealing any classified information to a visitor. Challenge all strangers.

Don't forget to lock up classified information or to leave it under the surveillance of an authorized person whenever you leave your work area. Lunch hours should be given special attention.

Don't place combinations to locks on calendar pads, in improper desk storage, in purses, or in billfolds. Memorize your combination.

Don't discuss classified information over the telephone.

Don't leave classified material at someone else's unattended work station. Make certain that the classified material gets into the hands of the person for whom it is intended.

Don't put classified material in waste baskets.

Don't give access to classified material until you have verified the recipient's clearance and "need-to-know."

Don't fail to mark classified notes, drafts, working papers, stencils, shorthand notes, carbons, etc. These items should be stamped when originated so that they will receive proper storage and be properly destroyed.

Don't let distractions at closing time disturb your routine of properly locking and checking all classified storage containers.

Don't remove classified material from the plant for home study or other personal reasons.

Chapter 16

SECURITY TRAINING AND EDUCATION

TRAINING AND EDUCATIONAL opportunities for the protective services field have historically been limited. The commercial and proprietary organizations providing protection have traditionally relied upon hiring experienced personnel who received training from public police or governmental agencies prior to engaging in private protective services. Limited "apprentice" or "on-the-job" training was provided in a systematic manner. Skills were obtained by practice over a lengthy period of time. Similarly, educational opportunities had been virtually nonexistent until the late 1960s. In common contemporary usage, security training relates to the imparting of skills and procedures primarily to those engaged in security work as an occupation. These skills include such things as patrol procedures, use of firearms, first aid, fire fighting, investigation, etc. Security education is, however, viewed as providing four different functions:

1. prepare individuals in an educational institution, e.g. college or community college, for entry into the security profession;
2. provide nonsecurity professionals with corporate policy regarding security, e.g. helpful hints to make them "security conscious";
3. make employees aware of their responsibility for protecting company property; and
4. prepare employees to take emergency procedures in loss situations.

In 1971 the Rand Report characterized the state of security training and education as being of general low-quality and quantity. It further stated that training and educational opportunities should be made available and mandatory for the industry. Since the publication of this report, several other studies and reports were published which set standards or made recommendations for improved training of security personnel. These organizations included:

- National Council on Crime and Delinquency (NCCD) (1972)
- Private Security Advisory Council (PSAC) (1973)
- Private Security Task Force to the National Advisory Committee on Criminal Justice Goals and Standards (PSTF) (1975)
- International Association of Chiefs of Police—Private Security Committee (1974)
- American Society for Industrial Security-Certification Program (ASIS) (1974)
- Committee of National Security Companies (CONSCO) (1972)

The basic position of these groups was that some form of training was necessary, the differences between them being the duration, type, and content of the curriculum.

The Private Security Task Force, having the responsibility for drafting National Standards was successful in obtaining agreement from the industry for a minimum standards program for security officer training which consisted of forty hours for all security officers with an additional twenty-four hours for armed personnel.[1]

Each group identified the need for providing training and educational opportunities for those engaged in the private protective service field. They also identified a wide variety of problems which have inhibited the development of high-quality training programs. These conditions include:

1. lack of formal training programs in existence,
2. low levels for training for those currently in the field,
3. lack of clearly defined job performance requirements,
4. lack of identified skill requirements,
5. low educational attainments of current practitioners,

6. low motivation for self-improvement within the field,
7. high turn-over of employees within individual organizations,
8. weak or nonexistent training standards
   a. within the industry,
   b. within individual licensing jurisdictions,
   c. and within organizations hiring protective personnel,
9. lack of clearly identifiable career patterns within the field,
10. low status, pay, and benefits traditionally associated with employment.

On a national scale, a few of the larger commercial security service firms offer limited training. The Rand Study concluded that "65 percent of the guards surveyed indicated that they received no training prior to beginning work." "Less than 7 percent received more than eight hours of prework training, while 50 percent reported receiving no retraining of any type." Furthermore, there are few states which currently require mandatory training for security officers such as Illinois with a thirty-hour requirement for armed guards, with a maximum program of 120 hours being operated in Ohio for special police commissioning.

Programs of training for private protective personnel are currently provided by a variety of organizations, agencies, and institutions. These include police departments, colleges and universities, commercial security companies, proprietary security departments, proprietary schools, and other various government and association-sponsored programs. The type, kind, duration, curriculum, content and goals of the various programs vary widely, although they all proport to be providing security officer training.

TRAINING PROGRAMS

Commercial Security Company

Effects of Competition on Training

The security field has attracted and continues to attract a large number of persons seeking to establish new guard service companies. This is possible because of the low capital requirements for beginning a business coupled with low or nonexistent licensing or regulation of the industry by state or local government. Consequently, many new companies enter the already competitive market and can underbid or underprice the more established companies who are providing a quality service.

Very few of these companies remain in business for a long period of time. However, as one goes out of business another comes along to take its place, thus not alleviating the problems for the older companies who might be trying to provide a quality service.

Very often it is difficult to justify a higher rate of pay for trained personnel even for the older companies if they must draw their officers from the same labor pool as the new company. This is the case, however, when the cost of the services is the sole criteria when hiring a security company. Training and supervision cannot be given or, if initially given, cannot be maintained. Thus, men are hired, given a uniform, and placed on the job and instructed to (1) use "common sense," (2) "don't smell of alcohol," and (3) "don't go to sleep!"

The users of security, while not always completely knowledgeable about guard operations, know that when they change firms, the guards previously employed are hired by the new firm and remain at their old posts and that a significant change in quality of service has not taken place. This occurrence within the industry is a major contributing factor in users unwillingness to pay more than the "low bid" for security services. The old adage of "you get only what you pay for" has become a self-fulfilling prophesy. "Don't pay more because you won't get more" is not only the assumption but the reality of the situation in so many of the cases that users *expect* it to happen. This expectation, in turn, cause them to accept the lowest bid, thereby keeping the chain intact.

Since this situation is the result of the interactions between both the client and the supplier, it is a difficult one for them to resolve. Certainly, creative sales and marketing techniques, backed by a quality service could alleviate some of the negativism by users. Self-regulation by the industry or imposed stringent regulation by licensing authorities could also set uniform performance standards for the users to expect. But this would have a negative effect on the industry by reducing the possibility of new firms from entering the service field. The training standards recommended by the Task Force for legislation of commercial security firms will begin reducing some of this situation.

Manpower Needs

The need for manpower within the private protective services fields will continue to increase at its current rate of 15 percent each year for the next five years. The report of the Private Security Task Force about the qualifications and motivations of those engaged in security operations will undoubtedly result in additional regulation of the industry and the individuals providing services. It likewise will increase the sophistication of the user of security products and services.

The costs of services provided have not risen as rapidly as those in other service fields. This is due to the extreme competitiveness of the industry, high employment of unskilled workers, and low wages being paid. It is not uncommon for contracts to be awarded on the basis of a 5 cent per hour difference with neither the bidders nor the user of the services concerned with the quality of the service being provided.

Effects of the Marketplace on Training

The extremes of competition have caused two major areas of reduction in the provision of private protective services. These areas are the supervision of personnel while on-the-job and preassignment training. Neither training nor supervision contribute to profit when a short-term gain in business is the goal. The very narrow profit margins which are common in the commercial security field make the addition of training an unnecessary cost which a company is not willing to assume voluntarily.

When the net profit per guard hour is 11 percent, 3 percent for training is felt to be too great a price for the company to pay. An additional dimension of this problem is that, if the training is provided because the company feels it is necessary, the guards who received it might well be working for a competitor the following week on the same job site!

Many companies have lost in net profit, trained manpower, and the contract because they were not the "low bidder" on a job which they previously held. This has, in fact, been caused by providing a quality service with well-trained and supervised personnel when all the client wanted was a "warm body" to satisfy insurance requirements for a security guard.

The "marketplace philosophy" and the use of training as a point of competition among commercial service firms are indicators of an immature industry. Security is, in fact, a relatively new "industry" considered to be of the "high-growth" variety. There have been commercial firms providing services since 1856 but never on the current or projected scale. However, the ease with which a new company can enter the industry makes "cutting corners" and eliminating "unnecessary expenditures" such as training a "necessity" from their perspective. However, from the viewpoint of the user, the public interest and the officer, this is the least desirable area to eliminate.

Proper job performance of security tasks must be learned from sources with the ability to provide the information. This must come from within the industry. This can be achieved directly in an internal training program or through programs offered by academic or proprietary institutions which obtain source material and the type of skills which must be taught from the industry. Similarly, the industry must reward and utilize the training obtained by the employee. To date, all outward appearances within the industry point to an assumption that the security officer does "common sense" things on the job and should, therefore, be paid accordingly, e.g. common wages.

Training Standards

The Private Security Task Force and the Private Security Advisory Council both developed training standards which require forty hours of mandatory training for security officers who are unarmed. Additional firearm training and qualification is required before they can be armed. The PSTF program requires that eight hours of the program be "Preassignment" with the remaining thirty-two hours be provided within the first ninety days of employment with a maximum of sixteen of these hours being provided as supervised on-the-job training. However, if the officer is to be armed, the firearms training and qualification must be completed prior to assignment. The company should choose the portions of the curriculum that most directly apply to the specific assignment of the officer. This permits the officer to have specialized information that reflects his work requirements rather than a general curriculum such as provided to public police officers.

Educational Program Development

The preparation of individuals to become security officers has been characterized by a lack of formalized training programs which are uniformly provided by commercial firms before placing an officer on a job assignment. The education and socialization of security officers to their "world of work" is done through an informal on-the-job orientation to the industry and its activities. In addition to individual agency efforts, the major sources for obtaining security protective service training and education can be categorized as follows:

1. Colleges and Universities
2. Public Police Departments
3. Proprietary Schools
4. Trade Association, Conference
5. Literature

COLLEGE AND UNIVERSITY PROGRAMS

Development

The development of security education programs at academic institutions—two-year, baccalaureate and postgraduate has increased dramatically in the last several years. Prior to 1968 and the advent of Law Enforcement Assistance Administration funds for program development, the number of police programs was exceedingly small. It now numbers 1,245 programs in 664 colleges offering police science and criminal justice programs in the United States. As late as 1970, the American Association of Junior Colleges indicated a total of two associate degree programs in the United States offering security and loss prevention programs. There was also at that time one four-year program (Michigan State University) offering specialization at the baccalaureate and master's degree level.

In universities and public community and junior colleges, on a national scale, in 1972 there were courses for the private protective services field in twenty-four institutions either being developed or offering courses. This number grew to 113 in 1975. Guidelines for the development of these programs have been completed and are presented in Appendix 16B.

Public Police Departments

Training for security officers has been conducted by public police agencies where municipal or state ordinances required it: The St. Louis (Mo.) Watchman Training Program, the Ohio Peace Officer Standards and Training Program being two major efforts in training specific types of security personnel. In addition to required training, a large number of police departments provide firearms or legal training for officer given Special Police or Special Deputy Commissions to act with Public authority in special situations or locations.

There have been, however, other police

agencies who have not provided training to security officers because of the possibility of incurring liability for the actions of the officers as a result of the training provided. New Orleans (La.), for example, refused such training on the advice of the City Attorney in 1973.

The PSTF Report has paved the way for public police agencies to begin working closely with private security agencies in many areas including training. The development of close and effective working relations between both sectors of protective services are enhanced by shared training.

Proprietary Schools

Proprietary Schools offer training and educational opportunities in a variety of vocational and occupational areas including private protective services. Private schools offer courses in either a residential or home study environment to meet the needs of students or an industry. Proprietary schools are successfully offering alternatives to police departments and community college programs to provide skill development for security officers, supervisors, private investigators, safety inspectors, firefighters, and locksmiths, and alarm technicians.

Proprietary schools serve a useful purpose in delivering needed educational services which are not available from other sources or where the clientele to be served are not attracted to more traditional institutions. Proprietary schools in the private protective services field have developed because the academic institutions were not responsive to the needs of the industry in providing programs of instruction. Proprietary schools have been able to develop and offer programs quickly, without the time required for obtaining the variety of approval necessary within colleges before a new program is offered. Consequently, the proprietary school has often been the first available school offering a necessary program, and aggressively selling it to the practitioners in the field.

Trade Associations and Conferences

Various professional and commercial associations and conferences are regularly scheduled to provide training and educational experiences for the private protective services field. The American Society for Industrial Security conducts national programs through its *Institute of Learning* while individual chapters offer seminars to upgrade local security personnel. The International Security Conference conducted by Security World Publishing Company likewise conducts three national conferences and trade exhibits yearly. There are likewise a variety of organizations which have portions of conferences or entire programs devoted to security and loss prevention.

Related professional associations such as the International Association of Chiefs of Police (IACP) has offered a variety of security programs through its Public Security Center.

Literature and Materials

The rapid development of the industry has precipitated a flood of literature and materials unprecedented in history. The volume of books being published has grown to the current point of 90 percent of all books on security ever published being published within the past two years. There are currently dozens of either regional or national periodicals, newsletters, or reports published on a regular basis. Training films, audio-visual materials, games, training aids are also present in ever growing quantities. This is in contrast to the early 1970s when little literature and fewer periodicals were available.

Security Education in Industry and Government

Security education in industry and government for nonsecurity personnel is involved with:

- Making employees "Security Conscious" or raising their "Security Awareness"
- Indoctrinating employees with "Security Policy"

- Instructing employees about "Emergency Procedures"

Governmental and industrial programs of protection must rely upon the integrity of employees to insure adequate security. In order to insure that employees are aware of their responsibilities to protect information or property as well as what they can do to reduce or lower losses; the security education process is used.

Depending upon the criticality and vulnerability of the facility, a process of educating or indoctrinating employees about their role in protecting assets is initiated when the new employee is hired. It is continued when assignment is made to a work station and continued at regular intervals throughout the period of employment. The purpose is to make the employee aware of the consequences of failure to protect information or property charged to his care and of the value of protecting company assets in general.

Security education programs often attempt to illicit support for company policy by explaining why certain security measures are employed and the value of cooperation or support.[2] Protective measures, both for personal and company property, are sometimes used to help both employees and the company in high-risk areas or situations.

In high-risk activities security education is provided employees for dealing with specific problems such as robbery, bomb threats, fire, kidnapping or terrorist activities, assaults, etc. Protective policies are provided to the employees and training sessions conducted to insure a thorough understanding.

Security education programs are presented in a variety of ways. Interviews with security officials, lectures, films, slide programs, reading manuals, problem-solving exercise, simulations and demonstrations are all effectively used, depending upon the topic, audience and physical environment and financial resources of the company. The major object of all programs and the format in which they are presented is the complete presentation of a specific message to the employee. That message, and the intent of the organization must be clear, concise, and accurate. The training medium utilized must not interfere with the message.

Many security education programs have failed because the method of presentation was not appropriate to the employees. Employees are adults and must be treated as such during the presentation of materials. If they are to assimilate and use the information being provided, such materials must encourage them to accept what is said and develop an internal motivation to comply. External pressures might bring about a short-term compliance but never internalized acceptance. Merely providing the employee with an "exposure" to the materials does not insure acceptance. The technical competence with which a program is presented is not an indication that employee learning, understanding or compliance has occurred.

Employee motivation to accept company protective goals is the desired educational outcome of the security education program. The appropriate "educational technology" can be identified and employed utilizing the appropriate principles of adult education, curriculum development and instruction. The primary task of security education is to facilitate the learning and acceptance of company policy and philosophy for protecting its assets.

If the company policy or philosophy is unacceptable to the employee, no amount of education will insure compliance. If personal pressures or convictions develop which are more motivating than the previously accepted company policy, employees can and will violate previously closely held company values. In either of these cases, the security program must be such that other safeguards are present to minimize the damage to the organization.

PROGRAM DEVELOPMENT

Any security education program can be considered effective only when individuals consciously accept security as a personal responsibility. Security consciousness is then a state of mind which implies an understanding of security objectives, principles, and procedures. It also implies a willingness to abide by these principles in order to achieve the goals of the organization.

A complete security education program can be divided into five phases. These are initial orientation, training, refresher training, promotion, and debriefing.

Initial Orientation Phase

The material for an orientation should be general in nature. This phase is not aimed at giving specific instructions in security procedures, but rather to provide a framework for the total security education program. This phase has specific objectives which can be stated as (1) to impress upon an individual the importance of guarding company information or materials, (2) to impress the individual with the importance of security to the company, and to himself, and (3) to explain the penalties for violation of established security procedures.

Training Phase

The training phase should begin as soon as possible after the general orientation. If possible, it should follow immediately before employees are allowed to actually begin work. The objective of the training phase is to provide detailed instruction concerning security procedures. In this phase the material is explained in detail and the security program can also be presented and explained.

Refresher Phase

Refresher training should be given to all employees on a regular basis, preferably on six- to twelve-month cycles. Groups receiving this training should be kept small and presentations individualized. Objectives of this phase are to review materials presented in previous phases. It is also to present new changes in security relations and procedures, and thirdly, its objective is to answer questions or give solutions to problems concerning security of the company and its facilities.

These three phases of initial orientation, training, and refresher, would complete what is considered a total education program. There are, however, a number of supplementary procedures which comprise or go into the composition of a total security education program. These are the *development of promotional reminders*, such as posters, cartoons, award programs, or related materials.

The final phase of the security education program is the debriefing phase. The debriefing should be part of the exit or employment termination interview. It is an opportunity for the organization to remind the individual of his continuing responsibility to the company. He must be impressed with the requirements of the organization that he not disclose information obtained while an employee of the company. If formal written agreements are part of this phase, they should be signed and witnessed during this debriefing. If verbal, they should be made clear to the individual prior to his termination.

A successful security education program must be planned, designed, and coordinated to fit the needs of the particular organization in which it is to be used. Similarly, a training program for regular security personnel should likewise fit the needs of a particular organization.

In determining the need for security training, a determination as to the type, duration, and scope of training must be made dependent upon the activities engaged in by the security officer. This individual officer's skills, attitudes, and knowledge will probably require certain modifications. There are, however, basic criteria which must be considered in presenting any type of training to a security officer.

Officers are required to perform activities in a wide variety of areas and situations. They do, therefore, require certain basic universal skills. These include the following:

1. a complete and thorough understanding of the purpose and scope of their position;
2. an ability to recognize and a proficiency in the handling of emergency situations;
3. an ability to evaluate and control unexpected activities;
4. an ability to express ideas and present information;
5. a knowledge of self-protection; and
6. a basic knowledge of the principles of prevention detection of crimes and violations of company policy.

FOOTNOTES

1. See Appendix 16A for the PSTF Standards for Training.
2. Appendix 16C provides an Employee Security Education Handbook which sets forth policy and procedures for an industrial plant.

BOOKS

A.S.I.S. *Academic Guidelines for Security and Loss Prevention Programs in Community and Junior Colleges.* Washington, D.C., American Society for Industrial Security, 1972. [Appendix 16B]
Center for Criminal Justice, School of Law, Case Western Reserve University. *Private Police Training Manual.* 4th ed., Cleveland, November 1974.
DePhillips, Frank: *Management of Training Program.* Homewood, Ill., Irwin, 1960.
DeSanto, John F.: *The Key Man, Training Directors Manual,* July 19, 1965.
Howington, Jon R., and Kingsbury, Arthur A.: *A Bibliographical Manual for Criminal Justice, Crime Prevention and Security for the Community College.* Detroit, M & L Associates, 1972.
Public Safety Institute. *Industrial Protection Training Series.* Lafayette, Ind., Purdue University, 1941.
McGehee, William, and Thoyer, Paul: *Training in Business and Industry.* New York, Wiley, 1961.
Parker, W. H.: *Daily Training Bulletin of the Los Angeles Police Department.* Springfield, Thomas, 1954.
Peel, John Donald: *The Training, Licensing and Guidance of Private Security Officers.* Springfield, Thomas, 1973.
———: *Fundamentals of Training for Security Officers.* Springfield, Thomas, 1970.
Wathen, Thomas W.: *Security Subjects: An Officer's Guide to Plant Protection.* Springfield, Thomas, 1972.

PUBLICATIONS OF THE GOVERNMENT, LEARNED SOCIETIES, AND OTHER ORGANIZATIONS

Adkins, Elmer H.: The police and resources control encounter insurgency: A training manual for police. Public Safety Division, U.S. Operations Mission for Viet Nam, 1964.
Audio-Visual (AV-4) Company Security Program. Narration and slides prepared to conform with any company's security program. Modern Marketing Inc., Los Angeles, Calif.
Basic Security Indoctrination. Thirty colored slides, cartoon type, with 14 minutes prekeyed tape. Northrop Nortronics, Marine Equipment Dept., Security Office, Needham Heights, Mass.
Bombs and explosives. *Industrial Protection Training Series,* Lafayette, Ind., Purdue University.
Lester, J. T.: Behavioral research during the 1963 Mt. Everest expedition. *ONR,* NR: 171-257, 1964.
PA State Capitol of Defense, Proceedings of Plant Protection School. Transcripts of lectures presented at 3 day school at Philadelphia, November 1941. Harrisburg, Pa. State Council of Defense, 1941.
Police Reserve Training, Ps. Bulletin No. 9, Salem, Ore., Division of Vocational Education, September 1941.
Public Safety Institute. Industrial protection training series. Lafayette, Purdue, 1941.
Security Education. Color slides and script available on loan from Electro-Optical Systems, Inc., Pasadena, Calif.
Security Indoctrination. McDonnel Aircraft Corp., Lambert-St. Louis Municipal Airport, St. Louis, Mo.
Security Indoctrination and Education Material. Standard Security Systems, North Hollywood, Calif.
Security Orientation. System Development Corp., Santa Monica, Calif.
U.S. Congress Senate Committee on the Judiciary. Education for survival and the strug-

gle against world Communism. Symposium, 87th Cong., 2nd Sess., Washington, Government Printing Office, 1962.

U.S. Department of the Air Force. Guides for security indoctrination. Washington, Government Printing Office, 1955.

PERIODICALS

Ayres, Prof. Loren: Educator's challenge (seminar workshop comments). *Industrial Security*, October 1961, p. 14.

Aitken, M. D.: Motivation—key to security awareness. *Industrial Security*, February 1967.

Badin, Fred P.: The security education program. *Industrial Security*, April 1968.

Basting, Alvin: Identification pre-training for potential victim. *Security World, 3:* no. 8, September 1966.

Blauvelt, Peter D.: Innovative "maximum security," program involves student security advisory council in huge responsibility for maintaining a healthy and secure learning atmosphere. *Security World*, December 1974, p. 37-39.

Burroughs Clearing House. Novel personnel techniques include programmed instruction training (for bank guards at New York Chase Manhattan Bank). *XLVII:* Issue 32, Spring, 1963.

Contract Guards West: Training and supervision make the difference. *Security World*, February 1972.

Crews, Paul H.: How to make your company more security conscious/jt auth. *Industrial Security*, April 1962, p. 14.

Crowe, J. M.: Effective training reduces fire losses. *Industrial Security, 4:* no. 2, 10-12, April 1960.

Curtis, S. J.: The psychology of security training. *Police, 4:* no. 5, May-June 1960.

Dangy, Gerald L.: Security education goals and principles. *Industrial Security*, December 1965.

Dennis, Robert L.: Report of the ASIS Education Committee. *Industrial Security*, October 1963.

———: The security poster's ten vital seconds. *Industrial Security*, April 1962.

Dudley, W. E.: Must security lectures be dull? *Industrial Security*, August 1971.

Elkins, E. V., and Reeder, J. A.: The answer to security training. *Industrial Security, 3:* no. 6, October 1959.

Flint, Dr. Calvin C.: The process of education —panelist. *Industrial Security*, October 1963, p. 69.

Fort, Dr. William E., Jr.: Education and industrial security. *Industrial Security*, July 1958, p. 8.

Germann, A. C.: Scientific training for industrial security. *Industrial Security*, January 1961.

Goddard, Robert J.: Security education and enforcement. *Industrial Security*, July 1961.

Government conducts security school. *Canadian Security Gazette*, Summer, 1971.

Government safety training. *National Safety News*, January 1971.

Hall, Wayne L.: Educator's challenge. *Industrial Security*, October 1961.

Harrigan, Arthur W., Jr.: Training and plant civil defense corps. *Industrial Security*, July 1960, p. 66.

Hayden, Charles E.: A professional approach to security education—panel moderator. *Industrial Security*, October 1963, p. 69.

Hazelton, Tom: Educator's challenge. *Industrial Security*, October 1961.

Industrial security recruitment and training. *Security Gazette, 13:* no. 2, 71, 100; no. 4, 161.

Industrial security recruitment and training. *Security Gazette*, March 1971.

Jail operations—A training course for jail officers. *Bureau of Prisons*, Wisconsin University, 1971.

Jordan, Franklin E., Lt. Col., USAR: DOD industrial security courses. *Industrial Security*, April 1958, p. 22.

Journal of the American Society of Training Directors. *Industrial Security, 15:* no. 9, 1961.

Kendall, Charles F., and Surles, Lynn: Operation training. *J. Amer. Society for Training*, June 1960.

Kingsbury, Arthur A.: Macomb College looks at security education. *Industrial Security*, August 1970.

———: Two-year effort bears fruit—guidelines for a two-year degree in security and loss prevention. *Industrial Security*, February 1972.

———: Wanted: Academically trained security personnel. *Environmental Control and Safety Management*, August 1970.

———: Security education. *Security Management*, September 1974, p. 42-45.

———: Guidelines for security education. *Industrial Security, 16:* no. 4, 7, February 1972.

———: Guidelines for security education, security library. *Security Management,* 17: no. 3, 40, July 1973.

Landers, Robert A.: Use and value of security posters. *Industrial Security,* January 1959.

Larkins, Mayes G.: The evolution of a curriculum in industrial security administration. *Industrial Security,* 11:10, February 1967.

Malone, Lee F.: Psychological and educational approaches to create favorable employee attitudes—panelist. *Industrial Security,* October 1963, p. 70.

Mansfield, J. C.: Is your plant getting full- or part time protection: How Westinghouse trains plant guards. *Factory Management,* February 1951.

Moran, James D., Maj., USA: Security education and training—panel moderator. *Industrial Security,* December 1964, p. 30.

Morse, George P.: Security education and training—panel moderator. *Industrial Security,* October 1962, p. 80.

Myren, Richard A.: Indiana University's Department of Police Administration. *Industrial Security,* 2:12, July 1958.

National Fire Protection Association. Organization, training and equipment of private fire brigades. Boston, N.F.P.A., 1967.

National Training Center of Lie Detection. Advertisement in *Law and Order,* September 1965, p. 52.

Owens, R. G.: Evaluating and training program. *Police Chief,* February 1966.

Pollock, Rose: A philosophy of training. *Journal of the American Society of Training Directors,* November 1959, p. 13-20.

Pope, Donald J.: Security education—goals and principles. *Industrial Security,* December 1965.

———: Evaluating the effectiveness of your security education program. *Industrial Security,* April 1966.

Potter, Anthony N., Jr.: Effective in-service training. *Security World,* October 1971.

Professional security and loss prevention training programs. (Pamphlet) Washington, D.C. American Society for Industrial Security.

Put realism in disaster drills. *Industrial Security,* May 1964.

Reeder, J. A.: The answer to security training. *Industrial Security,* October 1959.

Reinke, Roger W.: A pilot study of security education and training. *Industrial Security,* April 1960.

Rooney, Joseph E.: Security education with a sugar coat. *Industrial Security,* June 1971.

Rose, Homer C.: The development and supervision of training program. Washington, *Amer. Technical Society,* 1964.

Quillen, Eugene K.: A typical security education talk. *Industrial Security,* June 1967.

Savord, George H.: Teaching clerks to look at checks. *Police Chief,* January 1961.

Schiedermayer, Phillip: A suggested security curriculum, one gangs opinion. *Industrial Security,* December 1968.

Security Education Committee. ASIS survey—Department of Defense Industrial Security Clearance Program. *Industrial Security,* February 1968.

Security is your business too. *Administrative Management,* December 1962, p. 38+.

Security recruitment and training. *Security Gazette,* February 1971.

Sheehan, Robert J.: Advancement in security education—panelist. *Industrial Security,* December 1964, p. 31.

———: A pilot study for security education and training. *Industrial Security,* April 1960.

———: Teaching means helping people to learn. *Supervisory Management,* September 1964.

———: The industrial security administration curriculum at Michigan State University. *Industrial Security,* January 1959.

———: The use of testing. *Industrial Security,* 3: no. 2, April 1959.

———: Thirteen keys to better training. *Supervisory Management,* January 1966.

Smith, Robert J.: Firearms training for security officers. *Security Register,* 1: no. 4, 11+, 1974.

Streamline your training in industrial security. *Industrial Security,* March 26, 1963.

Strobl, Walter M.: The employee security education program. *Security Management,* September 1974, p. 22-24.

Survey of security instruction time. *Security World,* February 1972.

Tomberlin, John R., Jr., Major: Training: key to effective security. *Industrial Security,* August 1963.

Training in human relations essential for effective modern security staff. *Security Letter,* July 16, 1975, p. 1-2.

The training of a security guard. *National Safety News,* January 1971.

Training in industrial security. *Security Gazette,* April 1971.

Training for security responsibility. *Security Gazette,* 13: no. 5, 256.

Troy, Walter W.: A practical training program for security personnel. *Industrial Security,* April 1963.

Use slides to promote security. *Industrial Security,* May 1965.

Video aids in industrial security. *Police Chief, 30:* May 1963.

Visual aids: An important part of your safety program. *Security World,* February 1970.

Watchman training for fire protection. *National Safety News,* July 1969, p. 53.

Wathen, Thomas W.: Dial-a-number training tapes for guard force. *Industrial Security,* August 1968.

Watson, Nelson A.: Thoughts on police training. *Police Chief,* January 1965.

Weaver, Leon: Educator's challenge. *Industrial Security,* October 1961.

Weber, Rudy: Selling security to employees. *Security World, 4:* no. 4, April 1967.

Wechter, Harry L.: Economy and efficiency in police training. *Law and Order,* October 1965.

Wheland, R. L.: Informing and educating employees in home protection. *Industrial Security,* July 1960, p. 80.

Workshop for management. *Business Management,* May 1965, p. 32.

UNPUBLISHED MATERIAL

Post, Richard Stanley: Application of functional job analysis to the development of curriculum guidelines for the protective services field. Unpublished Doctoral dissertation. University of Wisconsin, 1974.

Tomberlin, John R.: Training the key to effective security. Unpublished graduate B paper, Michigan State Univ., 1967.

APPENDIX 16A

SECURITY OFFICER 40-HOUR BASIC COURSE

PRESERVICE   8 Hours

ORIENTATION   2 Hours
—What is Security
—Public Relations
—Deportment
—Appearance
—Maintenance and Safeguarding of Uniform and Equipment
—Notetaking/Reporting

LEGAL POWERS AND LIMITATIONS   2 Hours
—Prevention vs. Apprehension
—Use of Force
—Search and Seizure
—Arrest Powers

HANDLING EMERGENCIES   2 Hours
—Procedures During Fires, Explosions, Floods, Riots, Etc.
—Procedures for Bomb Threats
—Responding to Alarms

GENERAL DUTIES   2 Hours
—Patrol
—Inspections
—Fire Prevention and Control
—Safety

IN-SERVICE
32 HOURS
MINIMUM 4 HOURS FROM EACH SECTION

SECTION I
PREVENTION/PROTECTION
—Patrolling
—Checking Conditions
—Personnel Control
—I.D. Systems
—Access Control
—Types of Alarms
—Police/Security Relations

SECTION II ENFORCEMENT
—Techniques of Searching

—Crime Scene Searching
—Handling Juveniles
—Handling Mentally Disturbed Persons
—Parking and Traffic
—Enforcing Employee Work Rules/ Regulations
—Observation/Description
—Preservation of Evidence
—Criminal/Civil Law

SECTION III
GENERAL/EMERGENCY SERVICES
—First Aid
—Self Defense
—Fire Fighting
—Communications
—Crowd Control

SECTION IV   SPECIAL PROBLEMS
—Escort
—Vandalism
—Arson
—Burglary
—Robbery
—Theft
—Drugs/Alcohol
—Shoplifting
—Sabotage
—Espionage
—Terrorism
—Fire Control Systems

APPENDIX 16B

SUGGESTED CURRICULUM*

FIRST YEAR

First Semester	Credits	Second Semester	Credits
English I	3	English II	3
General Psychology	3	Introduction to Sociology	3
Criminal and Civil Law I	3	Criminal and Civil Law II	3
Introduction to Security	3	Security Administration	3
Elective	3	Elective	3
	15		15

Electives:
Accounting I	3	Accounting II	3
Economics I	3	Economics II	3
Science I	3	Science II	3
Administration of Justice	3	Civil Rights & Civil Liberties	3
Principles of Interviewing	3	Report Writing	3
Industrial Relations	3		

SECOND YEAR

First Semester	Credits	Electives:	
Fundamentals of Speech	3	Document & Personnel Security	3
Social Problems	3	Business Mathematics	3
Human Relations	3	Emergency Preparedness	3
Principles of Loss Prevention	3	Environmental Security	3
Elective	3	Physical Security	3
	15	Safety & Fire Prevention	3

* American Society for Industrial Security—American Association of Junior Colleges, Academic Guidelines for Security and Loss Prevention Programs in Community and Junior Colleges, Washington, D.C., 1972, p.19.

Second Semester	Credits	Electives:	
Criminal Investigation	3	Commercial/Retail Security	3
Criminology	3	Field Practicum	3
Labor & Management Relations	3	Industrial Fire Protection	3
Current Security Problems	3	Security Education	3
Elective	3	Special Security Problems	3
	15		

APPENDIX 16C

ALPHA CORPORATION*

Employee Security-Safety Handbook

Table of Contents

Letter From President
Introduction . . . . p. 1
Security . . . . p. 2
    Security Areas . . . . p. 2
    Need to Know . . . . p. 3
    Information Classification . . . . p. 3
    Alpha Facilities . . . . p. 4
    Reporting Incidents . . . . p. 4
    Identification Cards . . . . p. 5
    Protective Services Director . . . . p. 5
    Departmental Responsibility . . . . p. 5
    Supervisor's Responsibilities . . . . p. 6
    Employee Responsibilities . . . . p. 6
    General Regulations . . . . p. 7
    Personnel in Restricted, Limited and Controlled Areas . . . . p. 9
    Motor Vehicles . . . . p. 10
Safety Regulations . . . . p.

INTRODUCTION

Every employer has a moral responsibility to his employees to remove as much of the temptation to steal as is humanly possible. Likewise, every employee has the same moral responsibility toward his fellow employees.

Mr. S. J. Curtis, a professional security expert, feels that: "In many instances—perhaps the majority—the individual would not have turned dishonest had reasonable, precautionary methods been exercised. To take steps that will successfully prevent this dishonesty is to save many lives from ruin."

The Alpha Corporation, recognizing that its employees are its greatest asset, has developed the Employee Security-Safety Handbook. This handbook is designed to acquaint you with our security practices.

Many of the losses suffered by both the employees and the company are caused by the uninformed or neglectful employees.

Careful inspection and constant supervision are necessary on the part of supervisors and foremen; continuous observance of security practices is required of each employee to reduce the temptations that contribute to the losses of employee property and the ruination of lives.

---

* Courtesy Post & Associates Security Consultants.

Study and know your Security-Safety Handbook and remember that everyone is subject to temptation and it is the responsibility of each one of us to help each other by observing security-safety regulations.

Report all unsafe or insecurity conditions and security violations to your Supervisor; report all losses at once.

Security and Safety suggestions are welcome; in fact, requested. If you know how to improve the security system for yourself and others, inform your supervisor. It will *pay you* to do so! Additional Security Policies will be issued from time to time by the Protective Services Department or by your department.

Violation of any security regulation may subject you to disciplinary action.

Security

Security provides those means which serve to protect and preserve our environment. It allows for the conduct of our activities without disruption.

Security Areas

The security system of the Alpha Corporation has designated certain areas that only authorized personnel are permitted to enter. The areas are classified accordingly to their criticality and vulnerability to the Alpha Corporation.

The three classifications are:

Restricted—This is a tightly controlled area where only selected personnel are allowed. An example would be the R & D area.

Limited—This is an area of sensitive nature. Employees other than authorized personnel may walk past but may not enter. An example of this would be a records area.

Controlled—This is an area where all employees are permitted. It may be defined as the widest possible area with least risk of damage or loss of knowledge. Only authorized personnel are allowed.

The restricted areas are conspicuously marked with signs designating their classification.

Need-to-Know

Need-to-know—A determination made by the possessor classified information that a prospective recipient, in the interest of state defense, has a requirement for access to, knowledge of, or possession of the classified information in order to perform tasks or services essential to the fulfillment of a classified contract or program approved by a specified agency within the Alpha Corporation.

Information Classification

Certain documents and information, pertaining to the company are to be considered "Company Confidential." They are for the use of only authorized persons.

All classified documents are clearly marked on their cover sheet with the notation "Company Confidential." They are not to be shown to persons without a need for the information.

Alpha Facilities

Facilities of Alpha Corporation are defined as: Something that is built, constructed, installed, or established to perform some particular function or to serve or facilitate some particular ends. Every employee works at one or more facilities. Drawing from the above definition, a facility includes the buildings, parking lots and the ground around the buildings. Facility security begins when an employee enters the facility and ends when he leaves. Employees should only enter and leave by prescribed exits.

Upon entry into a work area, employees should first take a minute to see that it is as it was left the night before. If things are not in their proper order, call the security department and wait for their arrival. Do not touch anything or try to straighten things up.

Reporting Incidents

1. All thefts, attempted bribes, kickbacks, vandalism, mislaid property and other security regulation violations, no matter how small, shall be

reported directly to the department supervisor.

2. The supervisor will contact the Protective Services Department and complete the official form and route as directed on the form. If the supervisor is unavailable, the employee shall contact the Protective Services Department immediately and the supervisor as soon as possible.

Identification Cards

All Alpha employees must have an Alpha identification card issued under the direction of the Protective Services Department. These cards are to be used only for employment identification. When an employee transfers to a new department he may be issued a new identification card. When an individual's employment ceases, he must surrender his identification card.

Security Coordinator

The Director of Protective Services of the Alpha Corporation has the overall responsibility for formulating, directing and coordinating security-safety activities throughout the company. This is done through liaison with all department heads, personnel officers, and all management, and through the initiation of an on-going security awareness program.

Departmental Responsibility

Each department must assume responsibility for an effective employee security program which shall include the following:
a. leadership and direction,
b. periodic inspections,
c. insure that all security violations are investigated,
d. post and enforce security regulations,
e. review and sign reports of security violations and performance,
f. initiate and evaluate departmental security programs and regulations which includes the checking in and out of all equipment and classified documents,
g. cooperate with the Director of Protective Services on all programs,
h. plan and make known emergency exit routes,
i. organize a departmental security committee, and
j. compile an accurate account of all equipment within the department.

Supervisor's Responsibilities

a. Train all employees in security regulations within their department and point out security hazards.
b. Make sure that the necessary security equipment and protective devices for the department are provided, in proper working condition, and are used.
c. Take prompt corrective action whenever insecure conditions and actions are observed.
d. Investigate and report all violations of security regulations.
e. Conduct frequent, unannounced security inspections of all work areas and operations within the department.

Employee Responsibilities

a. Report any security violation to your supervisor.
b. Report any missing money, equipment, documents, or personal belongings to your supervisor.
c. Use all security equipment for your job without fail.
d. Follow all security regulations.
e. Report any vandalism to company or employee property.
f. Lock your vehicle in the prescribed parking lot.
g. Take your purse or belongings with you whenever you leave your work area.
h. Report anyone who is unauthorized to be in your work area.
i. Do not give any information to *anyone* who does not have a need to know.
j. Lock your desk and file cabinets when you leave your work area.
k. Report any bribe attempts even if the person(s) offering states they were not serious.

General Regulations

1.01 You are required to be familiar with, and to observe all security-safety regulations. Violation of any security-safety regulation may be cause for disciplinary action.

1.02 Drinking of alcoholic beverages during working hours is prohibited. Any employee reporting for work while under the influence of alcohol shall be subject to disciplinary action.

1.03 Use of illicit drugs is strictly prohibited. Any employee who is convicted of a drug offense shall be subject to disciplinary action.

1.04 Any employee who is convicted of a serious misdemeanor or felony is subject to disciplinary action.

1.05 Any employee who is charged with a serious misdemeanor, other than traffic, or a felony must report this to his supervisor. Failure to do so within seven days counting the day of arrest will subject the employee to disciplinary action.

1.06 Employees are required to park motor vehicles in assigned places or lots. Continued failure may result in disciplinary action and most assuredly involve a parking ticket.

1.07 Employees are permitted to enter and leave their places of employment at assigned exits.

1.08 Employees must produce their identification cards upon request by a security officer and wear "badges" while in the facility. Failure to do so may lead to disciplinary action.

1.09 Packages brought into and leaving buildings are subject to inspection. Passes are required to remove anything from the facility.

1.10 Entry into classified areas without proper authorization is prohibited.

1.11 All supplies must be requisitioned by supervisors.

1.12 All supply rooms must be kept locked.

1.13 Employees must report all incidents, all security violations and all hazardous actions by other employees in order to make all employees safe.

1.14 All Alpha equipment which may be lent out to employees must be checked out and back in.

Personnel in Restricted, Limited, and Controlled Areas

2.01 Transmission of information to people who do not have a need to know is prohibited.

2.02 All documents must be signed in and out.

2.03 Documents when not in use must be returned to their proper file.

2.04 All files must be locked when not in use.

2.05 All desks must be locked whenever you leave your work area.

2.06 Documents may not be removed from the building unless proper approval is given by the head of the department and documents are transported with care.

2.07 All R & D and sensitive material workers must use the paper shredder for destruction of notes.

2.08 Destruction of all "Company Confidential" documents is to be supervised by a supervisor and two witnesses.

2.09 All dials to safes or vaults must be spun three times around when the safes are locked.

2.10 All secretaries must use the paper shredder for destruction of typewriter ribbons, scrap paper and old mail envelopes.

2.11 Communication of "Company Confidential" information by telephone is prohibited.

Motor Vehicles

3.01 Unauthorized use of company vehicles is prohibited.

3.02 Employees who have vehicles assigned to them should inspect them for operating condition. No one should operate a defective vehicle.

3.03 Vehicles must be checked in and out properly.

PART IV

PROBLEM AREAS

INTRODUCTION

Losses can occur from both internal and external causes. It has often been said that "the only people who can steal are those who are trusted." It is only those persons who occupy positions of trust in an organization who have access to things worth stealing. Losses from fire, accidents, and other internal problems are also caused or assisted by employees. The protective program for the organization must, therefore, develop a system of internal control to "keep honest people honest" and insure that dishonest or careless persons do not get an opportunity to cause losses.

The prevention of both internal and external problems is accomplished by the application of basic security principles described in the chapters on physical information and personnel security. Specific problems and threats of either an internal or external nature represent applications for these processes. All problems present opportunities for the creative application of available principles, techniques, and technology. The security policymaker or manager must select the appropriate response to deal with both general protective problems and those posed by specific or unique conditions.

Chapter 17

INTERNAL PROBLEMS

There are a number of problems and concerns common to the internal operations of a variety of organizations both governmental and proprietary. These areas require the application of specific procedures, techniques, and hardware for their reduction or elimination. This Chapter treats the internal problems, while Chapter 18 deals with external ones. Each of these two chapters will present those areas and preventative and protective measures utilized in government and industry and commercial applications.

"WHITE COLLAR" CRIME

A considerable portion of the energies and activities of security officers is related to activities which have been designated as *"white collar"* crime. Considerable research and study has been done relating to the causation of *"white collar"* crime and related deviant behavior. These theories include the following:

Donald R. Cressey—"Trusted persons become trust violators when they conceive of themselves as having a financial problem which is non-sharable, are aware that this problem can be secretly resolved by violation of the position of financial trust, and are able to apply to their own conduct in that situation verbalizations which enable them to adjust their conceptions of themselves as users of the entrusted funds or property."[1]

S. J. Curtis—He cites four major factors which he considers as strongly contributory to the upswing in the breach of financial trust:

1. *The big shots are getting theirs, why shouldn't I get mine?* attitude.
2. *Santa Claus Philosophy*—The idea of getting something for nothing ever since we've been told the government owes us a living.
3. Inflation and the constantly rising cost of living.
4. Elimination of the personal relationship between and within organizational levels of the organization.[2]

M. B. Clinard—He feels more attention should be paid to certain personality traits of individual violators.[3]

Edwin H. Sutherland—The hypothesis of differential association is that criminal behavior is learned in association with those who define such behavior favorably, and that a person in an appropriate situation engages in such criminal behavior if, and only if, the weight of the favorable definitions exceed the weight of the unfavorable definition.[4]

W. Reckless—With the problem of developing a single theory to account for all types of crime, he looks into individual differences in personalities to explain the white collar violator.[5]

R. Lane—The question of corporate violation supports, in general, the differential-association hypothesis by pointing to the consistency of law violations in certain firms, even when management has changed several times.[6]

F. E. Hartung—White collar crime is a result of "social differentiation" rather than disorganization.[7]

D. Taft—Stresses the "exploitive" nature of our society and sees white collar crime as a mere social class variation of common motives and practices.[8]

G. Vold—The impossibility of gaining legal conformity unless laws are accepted and respected by most of the important power groups or elements in the organized political state.[9]

Similarly, many control procedures have been hypothesized by the same authors, which can be summarized by the following:

M. B. Clinard—(1) Effective control rests on the voluntary compliance with the regulations of society by the vast majority of the citizens. (2) Punishment, either administrative or criminal, does little to control white collar offenders, except to increase caution and cleverness in the methods of their evasions.[10]

D. Newman—Public opinion of punishment as a means of control.[11]

R. C. Fuller—The necessity of convincing the public that white collar crimes are more serious than conventional offenses and strict enforcement of the laws.[12]

R. Lane—Educational and experimental program involving the interaction of government and business management personnel. Ambiguity of many regulatory laws, therefore, propose a clarification of provisions, improved communications between business and government, a study of social pressures and community attitudes, with an eye to building respect for the law, and a "dry-run" experimental period whenever a new regulation is introduced.[13]

Professionalization—Self-policing policies based, in some cases at least, upon rather elaborate codes of ethics. Internal methods of control may make external enforcement of regulatory laws less necessary.[14]

Douglas Committee—The plea for a permanent commission on ethics in government.[15]

D. Taft—What liberties are we prepared to sacrifice in the interest of crime prevention? . . . Can criminogenic political corruption be eliminated and yet democracy be retained?[16]

Obviously, there is a wide variety of legalistic and social scientific viewpoints and hypotheses which attempt to explain the causative factors for white collar crimes and also hope to provide control of this form of deviant behavior.

Ever since the term "white collar crime" was first used by Edwin H. Sutherland in his address to the American Sociological Society in 1939, the problem has become more acute. "In 1967, the National Crime Commission estimated that the economic loss from white collar crime dwarfed that of all crimes of violence."[17]

With the proliferation of theoretical concepts of causation and prevention, the one factor which appears to permeate the discussions is that there is no one causative factor for white collar crimes as committed by Sutherland's "respectable person."[18]

Within the context of there not being one "best" approach to prevention and protection in specific theft situations, the following major headings will be treated in a principles format: (1) loss control, (2) area control, (3) fire prevention, (4) disaster planning, and (5) training and education.

Pilferage

Pilferage is a form of theft of merchandise and/or money normally committed by an employee of an organization. Billions of dollars a year in cash and merchandise are stolen by employees. Supervisors and executive personnel have consistently been attributed with pilfering amounts proportionately greater than the estimated four million dollars a day taken by employees. Thefts of merchandise have been estimated to be several times greater than that of actual cash.[19]

Because of the considerable amount of employee dishonesty many businesses have been forced into bankruptcy.

The loss of money and merchandise which is often termed *shrinkage* represents a loss in net profit to an organization. Most businesses operate at relatively low net profit margin. Department stores, for example, operate on a 2.3 percent profit on sales and inventory depreciation of 1.4 percent of sales according to the National Retails Merchants Association. Therefore, a business operating on a 2.5 percent net profit on sales must sell approximately twenty thousand dollars worth of merchandise to offset five hundred dollars in shrinkage. Thus, a store with a 75 million dollar annual sales volume, which is able to reduce its shrinkage losses by .5 percent, would see the same results in profit as if it would increase sales by 19.5 million dollars annually.

Shrinkage is an unknown quantity until the final physical inventory has been taken, tabulated and consolidated with the accounting records. It is the difference between the sum of the closing inventory, net sales, discounts, price changes, markdowns, and the starting inventory, plus purchases.[20]

How Employees Steal

The following presents a sampling of techniques utilized by employees engaged in pilferage:

1. Issuance of checks in payment of bills to fictitious suppliers and cashing them through a dummy, or by faked endorsements.
2. Invoicing goods below established prices and getting cash "kickback" from the purchasers.
3. Raising the amounts of checks, invoices, and vouchers after they have been officially approved.
4. Issuing and cashing checks for returned purchases not actually returned.
5. Pocketing the proceeds of cash sales and not recording the transactions.
6. Pocketing collections made on presumably uncollectable accounts.
7. *Lapping*, i.e. pocketing small amounts from incoming payments and applying subsequent remittance on other items to cover the shortage.
8. Forging checks and destroying them when returned by the bank, then concealing the transactions by forcing footings in the cash books or by raising the amounts of legitimate checks.
9. Charging customers more than the duplicate sales slips show and pocketing the difference.
10. Padding payrolls as to rates, time, production or number of employees.
11. Failing to record returned purchases, allowances and discounts and appropriating equivalent amounts of cash.

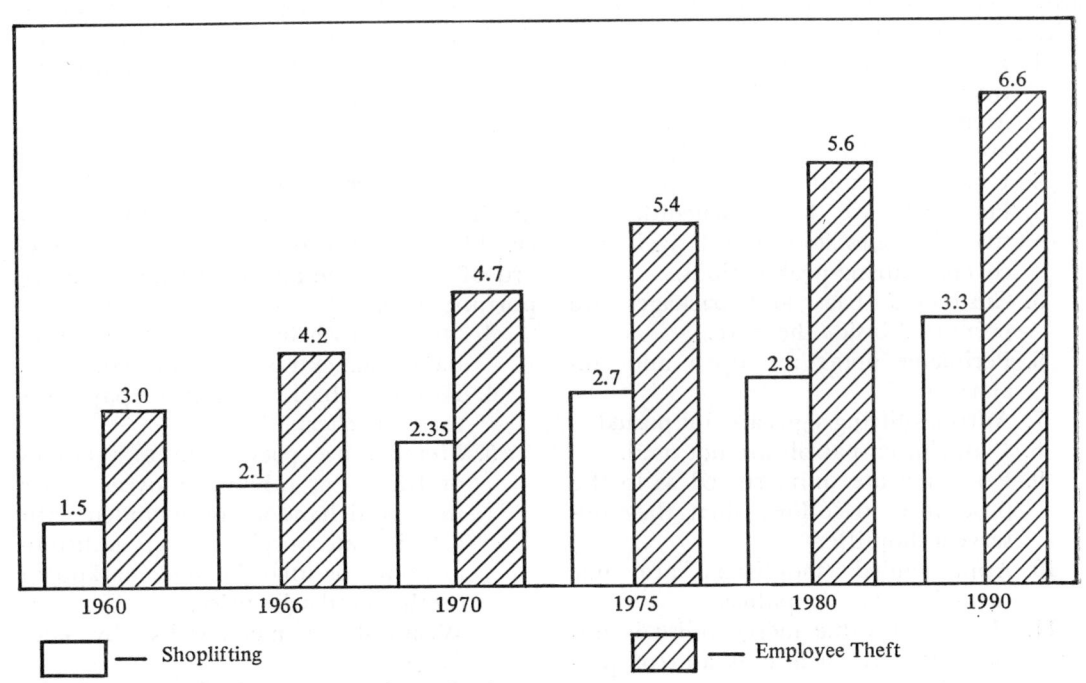

Figure 17-1. Estimated retail losses 1960-1990 (in billions). (From Louis J. Camin: Some projections for business security. *Security World,* Vol. 5, No. 2, February 1968.) **Courtesy of** *Security World.*

12. Paying creditor's invoices twice and appropriating the second check.
13. Appropriating checks made payable to "cash" or bank, supposedly for creditor's accounts, payment of notes, or other expense.
14. Stealing from the cash register and tampering with the tape.
15. Misappropriating cash and charging the amounts taken to fictitious customer's accounts.
16. Increasing the amounts of creditor's invoices and pocketing the excess or splitting with the creditors.
17. Pocketing unclaimed wages or dividends.
18. Pocketing portions of collections made from customers and offsetting them on the books by improper credits for allowances or discounts.[21]

Protection Failures

Shortages occur and shrinkage increases since many protective devices established by businesses do not function properly or are allowed to deteriorate in effectiveness. These failures include the following:

1. Fitting rooms are often uncontrolled and unchecked.
2. Men's and ladies' rooms in the stores are unlocked and uncontrolled.
3. Unused checkout aisles are open.
4. Personnel are badly scheduled so that the sales floor is relatively uncovered during peak periods.
5. Customers' bags and packages are permitted inside the store.
6. Perimeter doors are open and unalarmed.
7. Antishoplifting signs are not posted.
8. Plainclothes patrols are not used.
9. Personnel are uninstructed as to the procedure to follow when they observe a shoplifter.
10. Apprehended shoplifters are not turned over to the police.
11. High unit value merchandise is unprotected by enclosures and/or personnel.
12. Price-ticketing methods are not adequate to minimize ticket switching.
13. Cashiers are free to accept handwritten prices.[22]

These are shown as failures of a physical system. The causes of internal theft in the total context of environment, establishment, and maintenance can be seen in the failure to insure that prescribed methods are followed, coupled with lax or indifferent supervisory practices. Inadequate control of packages and shipping, and particularly negligence in checking applicants' previous employment, character, and credit background results in a substantial amount of employee pilferage.

Additional Methods of Pilfering

Since protective measures often fail, the way in which employees steal can also include the following:

1. Phony refunds.
2. Cash over or shorts. Theft from cash drawers takes several forms which basically are variations of the following methods:

THEFT OF CASH. The actual amount of sales may be rung but the employee "palms" cash from the drawer. If the shortage isn't too large and does not occur too frequently, the employee figures the shortage will be attributed to ordinary error.

UNDERRINGING CASH SALES. By this, we mean a cash reserve is built up in the drawer. The employee may keep a pocket record of the overage and take it at some opportune time.

3. Stealing cash by regulars on "heavy" days when extras are employed.
4. Holding back customer payments made by check.
5. Merchandise passed out to relatives or friends or other employees either in a bold handout or included with a legitimate purchase as an additional bonus or unauthorized markdown on the retail sales price.
6. Wear-outs of merchandise by sales force.
7. Conspiracy of "back-washing" by groups of employees, "I'll give you a break when you buy in my depart-

ment if you'll give a break to me when I buy in yours."
8. Lunch hour carry-outs by employees.
9. Dock thefts through the medium of customer pickups and by other means.
10. Phony mail-outs or deliveries made with forged slips.
11. Demonstration merchandise.
12. Employees' own goods loaded with unpaid for merchandise when employee packages are not properly checked and examined.[23]

Reduction of Pilferage

A number of weaknesses exist in the internal organizations of business and industry which tend to increase shortages. The operating policies and procedures of these organizations lend themselves to potentially great losses. If these policies and procedures were critically examined, a number of problem areas would immediately present themselves. By studying these problem areas, appropriate countermeasures for shortage reduction could be readily determined. These problem areas can be found in the general management and policy areas, control, merchandising and operations. The major weaknesses in policy or procedure leading to increased shortages are as follows:

General Management and Policy

1. Exposing merchandise and making shoplifting easier through self-selection and fixturing.
2. Failure to prosecute shoplifters and/or employees apprehended for theft.
3. Lack of enforcement of present controls and systems and lack of insistence upon high standards of adherence to prescribed procedures.
4. Failure to give concentrated attention (including frequent unannounced checks) to areas with high shortage situations.
5. Substitution of consolidated records only for individual store book inventories in multistore operations.
6. Insufficient control over transfer of merchandise to and from store, warehouse, branches, and stock areas.
7. Unnecessary exposures to fraud through too liberal or too loosely controlled refund, adjustment and allowance policies and procedures.
8. Excessive pressure on buyers through penalty factors for shortages incorporated in buyer's bonus arrangements leading to manipulation by buyers to minimize shortage figures.
9. Undue risk exposure through discontinuance of essential controls and the adoption of shortcuts as a means of reducing expense.
10. Lack of written procedures which will maintain carefully developed safeguards in the training of new employees and prevent inevitable modifications which result when no procedure and policy manuals are available.
11. Failure to educate all employees about and maintain their interest in both stock shortages and theft prevention.
12. Insufficient management emphasis on a continuous program of shortage consciousness and prevention so that this attitude will permeate the entire organization.

Control

1. Calculated risks taken to streamline systems and reduce expenses without adequate subsequent checks to determine whether such risks are warranted.
2. Weaknesses in inventory-taking procedures.
3. Lack of control over refund books and failure to limit the number of persons authorized to issue refunds, particularly in big ticket departments, such as furniture, appliances, and television.
4. Insufficient effort to achieve a maximum degree of accuracy and error prevention within the control division.

5. Lack of control and follow-up over the merchandise in lay-away.
6. Failure to establish and enforce a follow-up system to see that merchandise loan procedures are adhered to by display, advertising, workrooms, and on merchandise sent out for repair and alterations.
7. Lack of control and accountability for all sales checks, markdown books, interdepartment and interstore transfers.
8. Inadequate system of checks and balances for fraud prevention and detection, with particular emphasis on uncovering collusion of two or more employees.
9. Lack of follow-up on cash return verifications on a regular organized approach.
10. Weakness in open C.O.D. control leading to shortages.
11. Failure of cashiers and others to check carefully for authorized signatures.

Merchandising

1. Failure to recognize the importance of shortage prevention and to exert maximum effort to minimize shortages.
2. Insufficient knowledge of merchandising arithmetic leading to inability to recognize shortage consequences of deviations from regular procedures.
3. Manipulation of shortage figures to achieve a better departmental showing.
4. Permitting direct delivery of merchandise to departments.
5. Inaccuracies in retailing of invoices and ticketing.
6. Failure to adhere to store procedures in handling of repairs and alterations by vendors and other outside repair services.
7. Improper handling of returns to vendors, resulting in incorrect counts, retails, etc., with attendant errors and fraud possibilities.
8. Inaccuracies in price changes—incorrect counts, inaccurate recordings and effecting of price change, mistakes in reticketing returned goods, and in handling the problem of temporary price reductions for special sales events.
9. Overselling and failure to follow up on back-ordered merchandise, resulting in frequent duplication of customers' orders.
10. Failure to recognize shortage possibilities in connection with prepacking, nonmarking and bulk marking and to take action to minimize shortages from these causes.
11. Lack of enforcement of procedures for reporting broken, damaged, and soiled merchandise.

Operations

1. Inadequacy of personnel and procedures for proper control of merchandise transfers.
2. Lack of proper control over damaged and salvaged merchandise and its disposition.
3. Insufficient degree of insistence upon proper procedures and performance standards in receiving, checking, and marking merchandise, as well as on adequate safeguards and physical conditions which will contribute to the prevention of pilferage.
4. Increased risks in merchandise handling—receiving, marking, warehousing, delivery, transfer, etc.—brought about by shortcuts originating from expense-reduction pressures.
5. Inadequate control of employee packages; failure to insist on central exit points for employees, with spot-checking of packages.
6. Use of clerk-wrap without adequate checks.
7. Unnecessary shortage risks in customer fitting rooms due to poor location and multiple exits, or lack of control of garments taken into these rooms.
8. Lack of control in handling and procedures governing customer re-

turns and the returned goods room.
9. Improper handling of returns to vendors.
10. Weaknesses in workroom procedures which increase possibilities for pilferage and/or damage to merchandise.
11. Inadequate safeguards in stock-keeping procedures and areas.
12. Lack of control over unused address labels to prevent fraudulent movement of merchandise out of store.
13. Improper handling of oversold merchandise and back order procedures, resulting in duplicated orders.
14. Inadequate protection of premises before and after store hours.
15. Failure to insist on a high degree of clarity and accuracy by sales people in recording names and addresses on sales checks, particularly charge-takes.
16. Failure to obtain correct handling of employee discount transactions by sales people.[24]

Preventative Methods

Based on an analysis of the problem areas encountered in operations a number of specific preventative measures can be developed. These include the following:
1. A comprehensive employee package control, spot checking.
2. Thorough check on all irregularities to their ultimate resolution.
3. Confidential police check of employees.
4. Thorough and reliable applicant investigation.
5. Coordinated action with police in matters of mutual security interests.
6. Cooperation with other local security departments.
7. Instructive talks on security with new employees.
8. Prosecution where warranted.
9. Firm-wide understanding of company rules and regulations.
10. Follow-up on overs and shorts to correct innocent errors.
11. Spot checking of markups and markdowns.
12. Circulation of test letters on credit returns and follow-ups.
13. Collecting security information from employees.
14. Inspection to observe employee operations.
15. Surprise examination of truck, warehouse, receiving and shipping room merchandise as well as merchandise en route to floor from stock.
16. After hours examination of premises for signs and symptoms not discernible during working period.
17. Designed casual meetings with former employees for informative purposes.
18. Development of loyal sources of information.
19. Storage protection of portable, valuable, or popular merchandise.
20. Department meetings for security discussions.
21. Surveillances and concentrated patrol in a suspected area.
22. Use of informants or plants or police facilities.
23. Interviews with customers.
24. Neighborhood checks and inquiries.[25]

Loss Prevention Function

Every security office responsible for a loss prevention program should be involved in the following activities:
1. Isolating exposure to loss.
2. Developing the means to minimize such exposure.
3. Auditing existing loss preventive procedures.
4. Conducting an internal loss prevention training and public relations program.
5. Providing watchful patrol and emergency action.
6. Investigating suspicious or questionable occurrences.
7. Maintaining an awareness of newly developed devices and procedures.[26]

Embezzlement

Embezzlement occurs when a person takes money or property which came legally into his possession as a responsibility to his position, and while in possession of this property or money appropriates the same to his use or the use of another than the lawful owner.[27] A number of positive theories are available in regard to the embezzler which range from D. R. Cressey's theory that embezzlement "occurs when a pressing nonsharable problem concurs with opportunities to embezzle without detection and rationalization of the conversion not a subsequent justification but as a cause preceding the conversion."[28] Another theory presented by Surety officials is that anyone will embezzle when the temptation is great and concealment is easy.

A number of findings are available concerning embezzlement. They include the fact that embezzlement is very frequent, costly in economic loss, and widespread. Embezzlers are found in all phases of society, and very few embezzlers are actually brought into the formal criminal justice system. They are rather sanctioned through private means.

The Embezzler

The embezzler is usually a trusted employee of the organization and has the appearance of being an important person in the community. He normally enjoys the respect of his fellow workers and friends and is considered to be a "respectable" person in his community. The Surety Association of America lists six major reasons why people with good reputations embezzle. They are: (1) gambling, (2) extravagant living standards, (3) unusual family expenses, (4) undesirable associates, (5) inadequate income, and (6) resentment or revenge.

Methods of Embezzlement

S. J. Curtis in his book, *Modern Retail Security*, lists the following more common methods used by employees to embezzle money:

1. The dishonest employee issues a check in payment of bills for a ficticious supplier. He has the check sent to a prearranged address. Later, he cashes the check through a dummy company or by fake endorsements.
2. He invoices goods above established prices and gets a cash kickback from the resource.
3. He raises the amount on checks, invoices, or vouchers after they have been officially approved.
4. He issues and cashes checks for returned purchases not actually returned.
5. He pockets the proceeds of cash sales and does not record the transaction.
6. He pockets collections made on presumably uncollectible accounts (bad debts).
7. He pockets small amounts from incoming payments on accounts and applies subsequent remittances on other items to cover the shortages (similar to bank check "kiting").
8. He forges checks and destroys them when returned by the bank. He then conceals these transactions by forced footings in the cash books or by raising the amounts of legitimate checks.
9. Overcharges customers and pockets the difference.
10. He pads the payrolls as to the rates, time worked, or number of employees.
11. He fails to record return purchases and steals an equal amount of cash.
12. He steals checks made payable to cash.
13. He pays creditors' invoices twice and appropriates the second check.
14. He increases the amounts of creditors' debts and pockets the excess or splits it with the creditor.
15. He pockets unclaimed wages.
16. He pockets portions of collections made from customers and offsets them on the books by improper credits for allowances or discounts.

Internal Controls of Embezzlement

A number of internal control mechanisms can be established to reduce the possibility of internal loss by embezzlement. These controls revolve around accounting procedures and specific procedural mechanisms for insuring adequate checks and controls of all financial transactions within the organization.

Business and Industrial Espionage

Business and industrial espionage can be described as all efforts designed or developed to gain all information possible about a particular organization. It could include products, new products, operating procedures, manufacturing process, or related functional data. Many millions of dollars are being spent by business and industry in the United States and worldwide to obtain information about competitors and their operations. While it is difficult to ascertain an exact amount of the losses sustained by business and industry due to competitive intelligence gathering procedures and operations, it is safe to assume that billions of dollars are lost annually to organizations through activities termed industrial or business espionage. There are three motivational factors behind business or industrial espionage: competition, individual reward, and political advantage. Much of the material gathered in an industrial or business espionage context is for use within the business or industry to gain competitive advantage by the firm collecting the information over the particular organization from which the information is obtained. In other cases, information of a specific type is collected and sold to the highest bidder by individuals seeking personal gain. These individuals would sell information back to the company from which it was stolen if they provided the most lucrative arrangement to the thief. Lastly, political advantage is often sought when agents of foreign governments obtain information about new industrial processes or business operations so that their country would not be subjected to long research and development periods for a new product.

Sources of Information

All information to be collected through the use of espionage techniques can be divided into two classes: open information and closed information. Over 90 percent of all the information required to be collected for espionage purposes can be obtained through open sources. Open sources include periodicals, trade journals, presentations of employees before learned societies, and so forth. The obtaining of closed information requires the use of covert means. The covert means that are available are (1) clandestine listening devices, (2) bribery, (3) fraud, (4) theft, (5) misrepresentation, and (6) illegal entry.

In the collection of information, either from open or closed sources, a wide range of key areas in an organization are particularly vulnerable to penetration by those seeking company secrets. These sources include the following:

MANAGEMENT NEGLIGENCE. This normally includes failure on the part of management to establish policies and procedures which preclude the disclosure of, or dissemination of information about new developments to individuals in the organization who do not need to know the information to successfully carry on their activities. Similarly, information can be disclosed by management officials who are prone to discuss company business freely.

ACCIDENTAL DISCLOSURES. The loss of information by accidental disclosure either through loose conversation or lost or poor control. Poor information control systems can be quite costly. While accidental losses are difficult to control, it can be kept to a minimum through appropriate selection procedures for personnel and control systems for material. Likewise, press releases must be reviewed so that information is not inadvertently sent out which may be helpful to competitors.

SALES PERSONNEL. In attempting to promote sales for the organization, salesmen very often present facts about product development not generally for public con-

Kinds of Information	Industries Most Interested	Companies Most Interested
Pricing	Retail and wholesale trade Construction Manufacturing consumer products	250-1,000 employees
Promotional Strategy	Retail and wholesale trade Banking Communications	Under 250 employees
Research and Development	Engineering Manufacturing industrial products	Over 10,000 employees 1,001-10,000 employees
Sales Statistics	Transportation Retail and wholesale trade Communications	1,001-10,000 employees
Manufacturing Processes	Engineering Manufacturing industrial products	Over 10,000 employees
Cost Data	Construction Engineering	1,001-10,000 employees
Expansion plans	Banking Transportation	Over 10,000 employees

Figure 17-2. Kind of information about competitors that executives feel important to their companies and their industry. (Courtesy of Edward E. Furash: Industrial espionage. *Harvard Business Review,* November-December 1959.)

sumption. Studies have documented that a considerable portion of the information about competitors' new products and processes are obtained from their salesmen or suppliers.[29]

SUPPLIERS, PURCHASING DEPARTMENT AND CONSULTANTS. Information about the supplier requirements of the organization can greatly assist a competitor in determining what types of new products or procedures are being utilized by the organization. They may be obtained from vendors as well as from the purchasing department. Consultants to the organization likewise may do considerable damage since they can be exposed to a wide variety of sensitive material within the organization.

STAFF PERSONNEL. This group includes

Kinds of Information	Industries Most Interested	Companies Most Interested
Competitive bids	Construction Engineering	Under 250 employees
Produce styling	Manufacturing consumer goods	1,001-10,000 employees
Financing	Banking Transportation	Under 250 employees
Patents and infringements	Manufacturing industrial products	Over 10,000 employees
Executive compensation	Banking Consulting Communications	Under 250 employees

Figure 17-3. Kind of information about competitors that executives feel important to their companies and their industry. (Courtesy of Edward E. Furash: Industrial espionage. *Harvard Business Review,* November-December 1959.)

people who have no direct access to classified or company secrets. However, the nature of their work, either janitorial or secretarial, allows these individuals to have access throughout the organization; where as other staff individuals, technicians, etc., may be limited to certain rooms, buildings, or sections of the facility. Thus, this individual is a common target for competitive intelligence purposes. In addition to these specific sources, another source which is of critical importance is that of the transient employee. The upward mobility of a person in an organizational structure is a management fact of life. A person may work in one organization for a few years and then move to an organization, very likely within the same industry or business group, by way of advancement. He takes with him the knowledge of one organization's activities to another; quite often to the detriment to his former employer.

Business and industrial managers utilize material obtained from open and closed sources in the organizational decision making process. The amount of emphasis placed upon this type of information varies considerably within industries and businesses. It does, however, play a role in the decisions made by all businesses and industries.

Control

Unquestionably, there is little possibility of protecting all of a company's activities from disclosure. It is, however, essential for the protection of important information that a fully integrated utilization of the processes of physical, personnel, and information security be applied in a manner most applicable to the needs of the organization. These may be viewed in two categories; technical and nontechnical security. Technical security would include various types of alarms, anti-intrusion and surveillance equipment to provide physical security. Nontechnical security can be provided through employee indoctrination, security education programs, and background investigations as well as a constant surveillance by security personnel of visitors, speeches made by employees, classified materials, limiting access to important documents, limiting reproduction of materials, control and destruction of waste materials. The use of employee secrecy agreements or employment agreements are also most beneficial and desirable to reduce losses of trade secrets. Most court decisions and litigation concerning the theft of trade secrets places a requirement on the organization having such materials that (1) employees be told that information is to be protected, (2) provide safeguards to protect it, (3) assist employees in protecting the information. If these conditions are not met, it is difficult to prove that information was secret and that proper safeguards were provided to prevent any unauthorized disclosure. The employee secrecy agreement serves the purpose of notifying the employee that confidential information will be provided to him during his employment and that it may not be disclosed to persons without a "need-to-know" either during or after a period of employment.

AREA CONTROL

The purpose of area control is to ensure that only authorized personnel and vehicles are admitted to a protected area. The extent of the controls required depends upon the criticality and vulnerability of the facility and its products as well as upon the degree of security considered necessary to protect the facility from damage, or the interruption of its activities.

In providing for area control a number of specific functions must be performed by the security organization. These procedures include the physical, information, and personnel security processes. They include personnel identification and control, lock and key administration, traffic and parking control, and protective lighting.

Personnel Identification and Control

A positive personnel identification and control system must be established and maintained in order to achieve required compartmentalization, preclude unauthorized entry, and facilitate authorized entry to restricted areas at personnel control

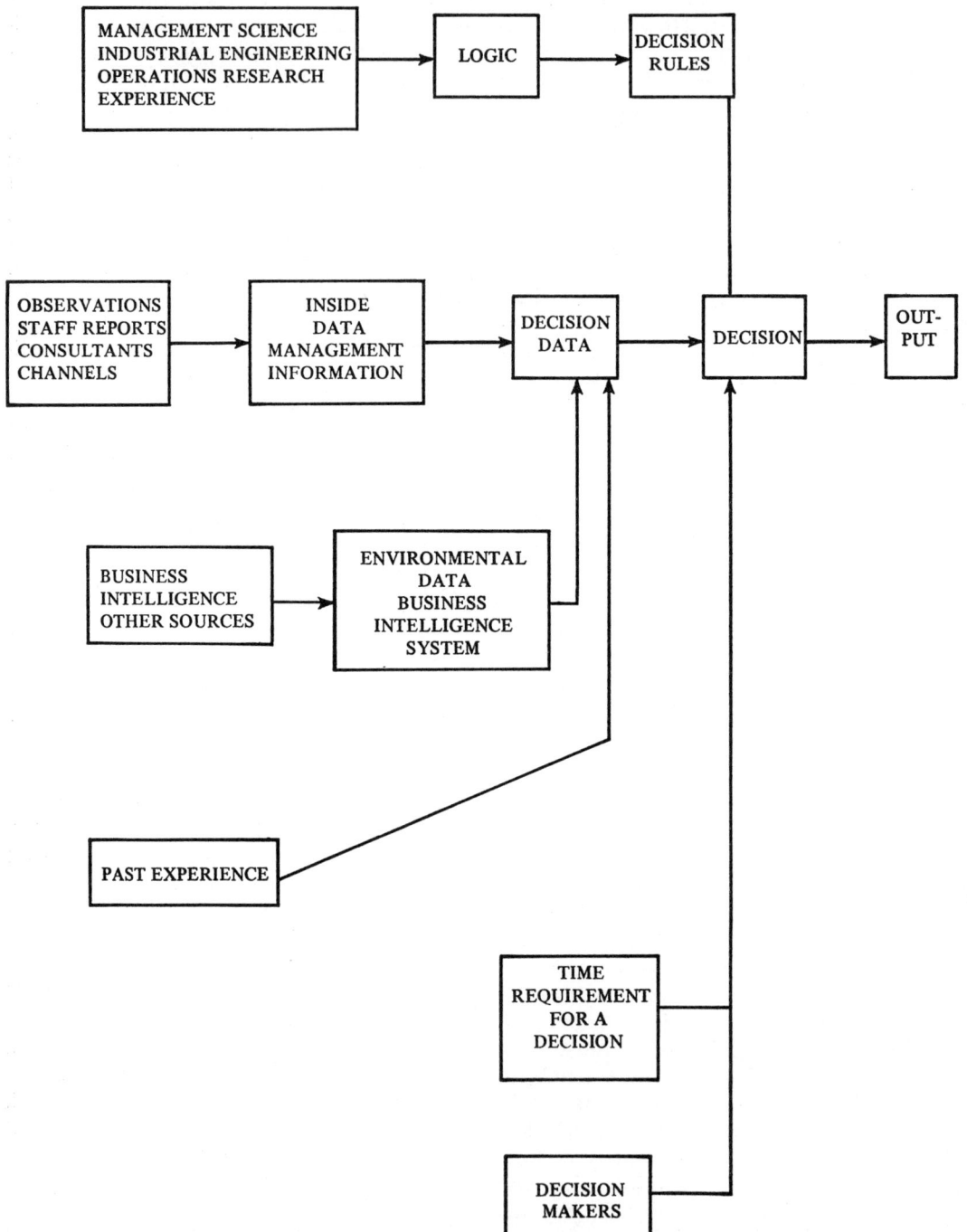

Figure 17-4. The role of intelligence in the decision-making process. (Courtesy of Dow Jones-Irwin, Inc., 1966, p. 6.)

points. Access lists, personal recognition, security identification cards and badges, badge exchange procedures, and personnel escorts are elements which contribute to the effectiveness of identification and control systems. The best control is provided when systems incorporate all of the following elements:
1. Initially determining who has a valid requirement to be in an area.
2. Limiting access to those persons who have a right and need to be there.
3. Establish procedures for positive identification of persons within, and of persons authorized access into areas.
4. Issue special passes or badges to personnel authorized access to restricted areas.
5. Use access lists.
6. Use identification codes.

Lock and Key Administration

Individuals responsible for the security of an organization should insist that adequate locks be provided for access control for the organization. Similarly, insistence should be made that proper controls be exerted over keys for the locks and equipment. A proper key control system is easy to plan, install, and maintain. Such a system should provide management with current information pertaining to keys issued. It should also pinpoint areas which have been compromised because of lost or stolen keys.

A key control system is composed of two major parts. The first of these is the key cabinet which holds an original key for each lock in the system. Attached to each key is a coded tag showing where the lock has been installed. The second part is a card file. It contains references to the code numbers and a listing of persons to whom the keys have been issued. Keys to areas in which security requirements are more stringent should be kept in a cabinet and issued only to those so designated for a specified length of time. In such cases, a separate card file is initiated to show the date and time of issue of the key. If a required degree of key control and administration is being maintained, regular inspection should be made part of that system. Grand master, submaster, and master keys should be inspected regularly but more often than keys normally used in the system.

Unquestionably, if the facility is large enough to support a security office the control of locking systems should be located there. Similarly, if the facility is large enough to warrant the hiring of a full-time lock maintenance man, a qualified locksmith should be hired and administered by the security department.

Protecting Master Keys

The grand master and master key to the system must be afforded additional security precautions. Loss, theft, or compromise of keys in these categories usually reduces the security of the key system. Grand master keys, with few exceptions should never be issued. They should be kept to the lowest possible number and issued only to those who have an absolute necessity to utilize them. They should be kept in a vault or safe and should be given the same protection as the most sensitive material in an organization. Master keys should be, likewise, extremely limited in issue and should be kept in a securely locked place when not in use.

Locking Devices

A lock is a device designed to prevent its being opened except by means of a particular key or by means of manipulation of parts according to a predetermined combination of numbers or letters. Locks can be divided into two general categories depending on the means of opening: key locks and combination locks.

An analysis of a locking device is of little value without a comparable analysis of the technique of its installation. This extends, as well, to include the construction of an entire lock assembly, since some locks that are comparatively highly resistant to picking manipulation may be passed quite easily without resorting to these time-consuming processes.

In picking, instruments are introduced into the keyway of a key-operated lock and the action of the key is manually duplicated. Bypassing is the method whereby the actual locking device is left untouched—the intruder circumvents the devices. Manipulation is a term used in connection with combination safelocks and combination padlocks. It is the art of opening a combination lock by using the human senses of sight, sound, and touch without tools or prior knowledge of the combination setting. The relative security factor of a locking device is measured in time; it is the degree of delay a lock offers against neutralization by a knowledgeable person. The purpose of a lock or locking device is to delay an unauthorized individual access to an area or specific container. All locks and locking devices are merely delaying devices and should never be considered to be more than that.

In order to plan for security measures properly, the capabilities and limitations of the locking devices employed must be known.

WARDED LOCKS. The relative security of a warded lock is zero. Warded padlocks may be identified by a combination of three points: the case, the opening of the shackle, and the keyway. Generally, the case is constructed of light weight metal or laminated iron plates. As a rule that shackle is ejected from the case directly vertical, or straight out, whereas another type lock has a shackle which hinges open in an arc. The keyway is one of two types: it will case, or it will rotate in the locked position without having any effect on the opening of the lock.

DISC TUMBLER LOCKS. The disc tumbler-type lock was designed primarily for the automobile industry and is of American origin. However, the device satisfied its need so well that its use was expanded to other fields such as padlocks, desk locks, and cabinet locks. This was due also to its low cost of manufacture and its mass production capabilities.

The disc tumbler lock is mass produced to sell at a moderate cost and it is made of easy wearing materials, such as brass. The life expectancy of a disc tumbler lock is short. Because of its mass production, the disc tumbler lock does not offer the security it might under different production techniques.

Disc tumbler locks, also referred to as *wafer* tumbler locks, offer more security than warded devices. It is a device which will give more privacy than security.

The relative security of a disc tumbler lock is approximately three minutes.

PIN TUMBLER LOCKS. The pin tumbler lock is of American origin, invented by Lirus Yale, Sr. in 1844 and improved upon by his son in 1865. The pin tumbler lock is a fairly secure device for household and ordinary uses. It has great latitude of use in its master keying arrangement, and its basic security can be increased by the use of mushroom-type pins.

The relative security of a pin tumbler lock is approximately ten minutes, although this time may be lengthened or shortened by the use of mushroom pins or master pins, respectively.

LEVER LOCKS. The relative security of a lever tumbler lock is more difficult to define than that of other locks. This is due, principally, to the wide range of the mechanism.

While the *average* pin tumbler lock offers more security than the *average* lever lock, the best lever lock is more secure than the best pin tumbler lock. An example of the best type of lever lock is the type used for safe deposit boxes. In this complex lever lock, picking is seldom attempted since such a device presents too great a time delay and is not readily adaptable to other installations such as doors or desks, because of the size and cost. It can be said that on a dollar for dollar basis more security will be obtained from the pin tumbler lock.

Simple lever locks, such as those found

on cabinets, desks, chests, and lockers, offer very little security. This is because of the small number and inferior type of levers. Lever door locks offer *fair* to *good* security.

Security in lever locks may be gauged by the number and types of levers in the lock. The more levers, the greater the security. To increase the security of a lever lock, levers with false gates and serrated faces should be used.

The relative security of a lever lock ranges from *very poor* to *excellent*.

For identifying purposes the locks are broken down into three categories:

1. *Lever padlocks*. Lever padlocks are generally identified by a shackle which hinges to open and a keyway which will rotate in the locked position without having any effect on the opening of the lock.
2. *Cabinet, chest, or desk locks*. These locks are generally identified in one of two ways:
   a. The keyway will rotate freely without having any effect on the opening of the lock, or
   b. The keyway will rotate from approximately 25 to 45 degrees in one direction and back in the same path, all without opening the lock.
3. *Lever door locks*. Lever door locks are generally identified by dual escutcheon plates, one for the doorknob and one for the keyway.

COMBINATION LOCKS. Combination locks usually provide far greater security than any type of key lock. Combination locks are much more costly to buy, maintain, and require longer times to open. Combination locks provide a security feature in that only a limited number of persons can have access to the combination required to open the lock. The combination lock has a wide variety of uses. These include bank vaults, safes and filing cabinets used for storing classified information, lockers where personal belongings or valuable records or equipment are kept, and vaults for jewelry, money or valuable articles.

COMBINATION PADLOCK. Combination padlocks usually have two riding discs with a fixed, unchangeable combination, i.e. the combination that is on the lock when it comes from the factory cannot be changed and must remain the same for the life of the lock. Therefore, the security of the device is limited in that the combination cannot be changed when there is a change of personnel. Some devices of more modern design have three riding discs and a changeable combination. Generally, such devices do not offer more relative security then their routine counterparts, but serve as convenience items in that the combination can be changed upon a turnover of personnel. Of primary consideration in determining the security of a device such as this is to determine the degree of accessibility to the internal working parts of the lock.

All other factors being equal, the more riding discs there are in a combination lock, the more secure the device. In some combination padlocks, the lock may be opened with a key as well as the proper combination. This lock was designed for industrial and academic institutions.

Virtually all combination padlocks have serial numbers stamped somewhere on the exterior of the case. Quite a few of these numbers have been incorporated into reference books and cross-references to the factory set combination. Any such serial number should be obliterated, since in some cases, it serves the same purpose as the combination.

The relative security of combination padlocks ranges from practically zero to approximately forty minutes.

Selecting the Lock

When a new locking system is being considered, it should be understood that no lock, regardless of the make or the type, will offer absolute security. Therefore, the longer the period of delay, the better the lock. A lock with close tolerances provides better security.

Traffic Management and Security

Traffic planning, control, and enforcement are traditionally vested on the municipal level in the local law enforcement agency; likewise, these functions are normally placed in a security department.[30]

In developing programs for internal traffic management and security, a number of specific factors must be considered. The first and most important consideration is that of the planning and operation of facility access points (access and egress) and parking. Closely associated with the first point is coordination of facility traffic needs with local law enforcement agencies. Environmental and security management factors must likewise be included in a meaningful traffic program.

Access Points

Traffic congestion is a major factor to take into consideration in design and utilization of an entrance and exit point. Large industrial plants employing many people on a multishift basis, particularly those situated in semiurban, suburban, and rural environments, have found that, of necessity, appropriate managerial attention must be given to the problem of large-scale employee movements.[31]

Traffic engineers should be consulted to better develop a control system which may provide or improve a safer environment for traffic on the public streets and for vehicles moving in and out of parking areas. Driving lanes, signal systems, variable speeds, and timing are the factors to be controlled or managed. Security personnel should be directly involved in traffic discussion as one of many groups consulted. Likewise, fencing, movement inside the parking area, lighting, guard patrols, and intrusion detection devices, although of a technical nature, should directly involve security personnel and management in the planning and administration.

Environmental factors or hazards from weather, terrain, and man-made errors can often be modified or eliminated if close liaison work is done in the planning and construction stages of the entrance-exit ramp systems. Snow melting systems, water drainage and other preventive designs could be instituted and installed before final construction, thus helping both the highway traffic and security personnel. This assistance would be provided through less loss of life and individual vehicle damage. This would be particularly evident in times of inclement weather where power failures or other unexpected situations arise. The guard or traffic officer can be of great assistance in controlling traffic movement. Darkness brings special hazards to the individual and the parked car in most parking areas.

Parking

Company parking lot facilities are rarely designed to accommodate the fullest possible first shift and office groups, plus a second shift. As a plant population grows, the pressure on parking facilities increases to saturation; therefore, additional real estate must be sought and devoted to parking.[32]

Before any permanent plans can be made, a thorough survey should be conducted to uncover possible parking areas. Major areas to consider would include old railroad right-of-ways in the facility area, clearing of slum areas adjacent to the plant, roof parking, bus services, and the advocating of car pools.

After agreeing on a parking area, the next step involves the actual "layout" of the area. This involves getting the greatest possible number of cars in the lot. Three areas which are also a major concern is the space between cars, assignment of spaces, and especially relevant to management, low cost construction.

Vehicle Control

Vehicular control and security are normally the functions of security officers. In the parking area, the standard "policing" instrument of control and prevention can be used. Automatic control gates, fencing, and anti-intrusion devices are often used controlling traffic and theft. If the organization has the manpower, facilities, and finances, vehicle registration of permanent work personnel can be used constructively. Written and published regulations are a necessity in any vehicle control program.

It is imperative that *all* personnel be directed to comprehend and utilize the regulations concerning traffic. An exception to this is the visitor vehicle. This can easily be remedied by appropriately designed and designated parking spaces. The easiest method for control of illegal parking is periodic physical vehicle inspections. If any irregularities occur, proper remedial action can take place.

Specific security in parking areas may be gained by using perimeter barriers, fences, controlled entrances, perimeter lighting, armed or unarmed guard personnel and in some cases, dogs. Auto thefts and loss control can often be deterred by the simple personal precaution of locking vehicles.

The lack of good security systems increases the losses per capita in each parking area. A well designed and controlled area obviously has a better probability ratio of less crime than an area poorly constructed with little or no security.

Protective Lighting

The function of a protective lighting system is to provide uninterrupted light during darkness and periods of low visibility. Light has value as a deterrent, but is primarily for illumination. Normally protective lighting requires less candlepower than working light, except at identification and inspection points. Each area of a plant presents its particular problem based upon the physical layout, terrain, atmospheric conditions, and the protective requirements. Data is available from the manufacturers of lighting equipment which will assist in designing a lighting system.

Power Sources

Usually, the primary power source at a facility is a local public utility. As facility control seldom extends beyond the perimeter, the interest of the security officer begins where the power feeder lines enter. An alternate source of power should be available where the primary power source is subject to frequent interruptions or failures. Stand-by gasoline-driven generators that start automatically upon the failure of outside power will ensure continuous light, but may be inadequate for operations. Battery-powered portable lights and/or generator-powered or battery-powered stationary lights should be available at special control points for the use of guards in case of a complete power failure.

Uses of Protective Lighting

Protective lighting is used to increase the effectiveness of security officers by increasing their visual range during the hours of darkness or in locations where natural lighting does not reach or is insufficient to provide the desired visibility. Outdoor protective lighting has considerable value as a deterrent to thieves and vandals and may make the job of the saboteur more difficult, but the most important function of lighting is to help see quickly and accurately at night. The seeing tasks confronting security officers are seldom casual, more often they are critical. Seeing tasks are varied, and in each situation careful study is needed to provide the best visibility that is practicable for such duties as identification of badges and people at gates, inspection of vehicles, stopping attempts at illegal entry, detection of intruders both outside and inside buildings and other structures, and inspection of usual or suspicious circumstances. Another function of lighting is to protect the security officer by keeping him in comparative darkness. This means planning a lighting layout in relation to guard patrol paths, the direction in which the guard moves, and all possible approaches of the intruder. In general, sources of glare are pointed toward the intruder and away from the officer.

Protective lighting obviously performs many useful functions. There are, however, specific uses and activities which re-

quire protective lighting. Requirements can be enumerated as follows:
1. Brightness.
2. Contrast.
3. Silhouette seeing.
4. Shadow elimination.
5. Boundary lighting, glare projection method.
6. Boundary lighting, near adjoining property.
7. Boundary lighting, buildings close to fences.
8. Boundary lighting, floodlights.
9. Active entrances.
10. Locked gates.
11. Guard towers.
12. Vital locations.
13. Inactive areas.
14. Building surroundings.
15. Shipping areas.
16. Waterfronts, piers, and docks.
17. Parking lots.
18. Interior protective lighting.
19. Emergency lighting.

Occupational Safety and Health Act

The Occupational Safety and Health Act[33] of 1970 established employee protection standards which apply to five million workplaces and sixty million workers has become a major responsibility of protective services programs. There are sixteen technical Subparts in the Act which can be divided into seven categories. These categories[34] include:
1. Workplace Standards
2. Machines and Equipment Standards
3. Materials Standards
4. Employee Standards
5. Power Source Standards
6. Process Standards
7. Administrative Regulations

Workplace Standards

There are certain basic safety and health standards which apply strictly to the workplace, a building, or other work location. These include safety of floors or other working surfaces, protection of floor and wall openings, access and exit requirements, sanitation, and fire and emergency protection.

Machines and Equipment Standards

When machines and equipment are added to the workplace, new elements of risk come into play. Standards are included for risks involving machine guarding, operational techniques, special safety devices, inspection and maintenance, mounting, anchoring, grounding, and other protection.

Materials Standards

Materials which are utilized, processed, or applied on the job add to the hazards. There are standards covering materials which yield dangerous or toxic fumes or mists, ignitable and/or explosive dusts, and other atmospheric contaminants. There are standards for safe storage and handling of compressed gases and flammable and combustible liquids as well as more stable materials used in production processes.

Employee Standards

When the employee is added to the workplace, other variables and related technical standards become important. What medical and first aid services are required? What personal protective equipment and devices must be provided? Are licenses or other accreditation documents required? What about special training or educational requirements?

Power Source Standards

The power source utilized creates additional hazards. Electrical, pneumatic, hydraulic, steam, explosive actuated, and other sources of power have standards.

Process Standards

Some standards cover a special process or a special industry. Welding, cutting and brazing, spray finishing, abrasive blasting, and utilization of dip tanks are hazardous processes. Special standards exist for textile and bakery operations, sawmills, pulpwood logging, and agriculture.

Administrative Standards

In addition to the safety and health standards in Part 1910, there are administrative requirements in Parts 1903, 1904, 1905, 1911, and 1912 for all employers whatever the size of the work establishment. Every employer must display an OSHA poster stating the rights and obligations of employees and employers; keep injury, illness, and exposure records; report fatalities and multiple hospital injury cases (5 or more); and post an annual summary of injuries and illnesses.

In administering the Act, the Labor Department's Occupational Safety and Health Administration (OSHA) issues standards—rules for safe and/or healthful working conditions, tools, equipment, facilities, and processes—and conducts inspections to assure they are followed.[35]

Standards

Most of OSHA's standards are not new. Business has operated under them for years as national consensus standards—those agreed upon by members of such groups as the American National Standards Institute and the National Fire Protection Association—or as established Federal standards under other laws, such as the Walsh-Healey Act.

In developing new standards, OSHA invites full participation by employers and employees and gives their views full consideration.

Enforcement

When OSHA compliance officers discover hazards in inspected establishments, employers may be issued citations listing alleged violations, and penalties and abatement periods may be proposed.

The employer may contest these before the Occupational Safety and Health Review Commission. The three-member Commission—appointed by the President—presumes the employer to be free of violations and puts the burden of proof on OSHA.

State Plans

The Act is not an attempt to increase Federal authority. To the contrary, the Act encourages States to develop plans for their own job safety and health programs.

OSHA grants States up to 50 percent of the cost of operating an OSHA-approved plan that meets the basic test of being "at least as effective as" OSHA's program. All states currently have such plans.

Coverage

The Act and the standards issued by OSHA apply to every employer with one or more employees, except those covered by other Federal legislation such as the Atomic Energy Act and the Coal Mine Safety Act.

Under the Act, employers have the general duty of providing employees employment and a place of employment free from recognized hazards to safety and health, and the specific duty of complying with OSHA standards.

Employees must comply with standards and with job safety and health rules and regulations that apply to their own conduct.

The protective services program in many organizations have been given the responsibility for insuring compliance with OSHA Standards. The combined Safety-Security function often involves safety as well as security inspections and services. Since the goal for protection is the prevention of losses to the organization, reducing losses due to accident or job-related illness is integral to the safety-security function.

Fire Protection

The protection of facilities from fire is one of the most important responsibilities of many security forces. Most major disasters stem from industrial or business fires. Fire hazards must be eliminated and if fires do occur, they must be caught in their early stages and be extinguished promptly. This is necessary to safeguard the organization and its facilities for continued operations.

The unsafe handling of flammable liquids produces the highest incidence of loss of life and injury due to fires which are caused by these accidents. There are additionally a multitude of causes for fires which include lack of ventilation, improper procedures in processing, and poor maintenance of equipment. However, in almost every case where a flash fire or an explosion occurs, they can be traced and directly attributed to failure of individuals to properly perform their activities.

In order to properly maintain a fire prevention program three types of activities must take place as part of a total fire protection program. These include automatic sprinklers, physical surveillance of critical fire prone areas and general facilities, and thorough inspections of fire fighting equipment and general work areas.

Security personnel generally have little responsibility for the placement of sprinkler systems in the facility. They should, however, inspect and assist in the maintenance of these systems. The responsibility of security personnel for surveillance of facilities is coupled with the responsibility to perform inspections of general operating facilities and of fire fighting equipment. These inspections range from regular tours by security officers to check for the presence of fires or safety hazards in the facility, to daily, weekly, monthly, semiannual, and annual checks of facilities and equipment.[36]

Security personnel should have a thorough understanding of the various types of fires and the various classifications of fire extinguishers to control these fires. These include:

Class A fires are those which include the burning of wood, paper, textiles, and other carbonaceous materials. Extinguishment of this type of fire is by cooling and quenching. Extinguishers containing water, soda and acid, foam, and the special solution of an alkali metal salt which is in the loaded-stream extinguisher should be used on this type of fire.

Class B fires are those caused by flammable liquids. Extinguishers to be used on this type of fire are foam, and loaded-stream, carbon dioxide, dry-chemical, and vaporizing liquid.

Class C fires are those which start in live electrical equipment. The extinguishers that should be used for such fires are carbon dioxide, dry-chemical, and vaporizing liquid, because they are nonconductors of electricity.

One of the additional fire prevention responsibilities of a security department is the organization and training of individual employees in fire prevention and fire fighting activities. A security plan for any facility should always include the organization and training of employees to fight fires. It should further include evacuation plans and emergency operational plans.

Sprinkler Systems

A sprinkler system is little more than a series of pipes or tubing which convey water throughout a protected facility to extinguish fires. The pipes may be always filled with water as in a "wet" system or filled with a gas or empty "dry" depending upon the type of facility and protective need.

Water is discharged from the pipes through sprinkler "heads" which are spaced at regular intervals throughout the protected areas. The heads are held closed by a fusible link consisting of a soft piece of metal which melts when surrounding temperature reaches a specified level. When the link melts, the head is opened and the water allowed to flow, thus extinguishing the source of the heat.

Water is supplied to the pipes and sprinkler heads through "risers" which are large pipes through which water rises to the level of the ceiling pipes and heads. The rises normally contain gauges and alarms which indicate that water is flowing when a sprinkler is "set off."

Water is supplied to the risers and the sprinklers from a "main." The "main" connects risers to the central water supply. They are equipped with a PIV (Post Indicator Valve) which controls the flow of water. The PIV's are open at all times to insure that adequate water is available for the sprinklers. If a sprinkler should fuse and open, the water supply can be stopped by closing the PIV. This would turn off water to the entire system served by the PIV.[37]

The type of system will determine the amount of visual and physical inspections that must be made of the system. Many sprinkler systems are connected to alarm systems which monitor water flow, temperature, rate of heat rise, ionization, pressure drop, etc.

If automatic alarms are used, prompt response is required by security personnel to determine the extent of damage or the exact nature of the problem.

Computer Security

The protection of data processing (EDP) facilities and the functions they perform for business, government, and industry is one of the most difficult and complex that the protective services function must undertake. The program of protection involves five components:

- Physical site
- Equipment
- Software
- Operations
- Personnel

As with any system, the interaction of these components is a proper sequence and under the correct conditions is essential for effective and efficient operations. The application of appropriate equipment and control techniques is key to protective program success.[38]

The protection of data processing facilities and operations is further complicated by the nature of the equipment and operations. Information processing often takes place in remote locations with a number of facilities preparing data for entry into the system or having direct access to information or equipment use from multiple location. The opportunities for fraud, embezzlement, manipulation of information and assets as well as physical damage to equipment and operations are plentiful.[39]

Physical Site

The physical location of the computer facilities is as critical as the protective measures taken to prevent losses from it. The selection of a site for the computer facility should, while not reducing efficien-

EQUIPMENT MOST COMMONLY USED TO IMPROVE EDP SECURITY

Type of Equipment	Mission
Badge Control Systems	Control access to and within the data center.
Fire Detection and Fighting Systems	Detect heat and smoke, signal alarms, initiate firefighting system.
Power Backup Systems	Provide reserve power in cases of extended power outage. Computer system remains down until power system is started.
Uninterruptible Power System	Reserve power with capability for instantaneous cut-over. Computers, air conditioning, and other equipment continue uninterrupted.
Magnetic Detector—Access Control	Detect and confine within a set of special doors anyone carrying magnetic materials either into or from the data center.
Intrusion Alarms, Closed—Circuit Television Monitors	Used in conjunction with central guard systems.
Fire Doors—One-Way Emergency Doors	Used to convert side doors, stairway doors, etc., to prevent entrance but allow emergency egress.
Paper Shredders	
Equipment Covers	Plastic sheeting used in emergency situations to protect equipment cabinets.

Figure 17-5. Courtesy Bradnt R. Allen, Computer security. *Data Management,* February 1972, p. 27.

cy of operation be in a location where hazards are kept to a minimum. Areas which are prone to crime, fires, flooding, power failures, poor communications, high-traffic areas, or other factors necessary to operations such as postal, food, or medical facilities should be avoided.

Computer facilities require a controlled environment both from an equipment and a personnel perspective. The physical security measures for the facility provide the protection for the designed environment and often are the features which define and maintain it. The monitoring of environmental systems, power failure alarms, water flow alarms, intrusion detection and access control systems all assist in environmental maintenance. They also protect the facility from losses.

A wide variety of equipment is used to provide protection for computer facilities. Figure 17-5 provides a partial listing of the more common types used from EDP protection. It can be noted that much of this equipment is used in other application and is applied to the specialized needs of the computer center.

Equipment

The protection of the EDP hardware is accomplished by the control devices and monitoring systems used for the entire facility. In addition, the unauthorized use of equipment, malicious destruction, sabotage, and vandalism must be avoided or reduced. Water damage, heat, magnetism, radiation, and fire are also major problems.

In systems which are located in multiple facilities, the telecommunications systems which connect them must be protected. Network security, as well as terminal procedures and security, must be maintained. Controls are also necessary for data input, processing, and output to insure that only authorized work is performed on the machines. Millions of dollars have been lost through fraud by programmers or machine operators who manipulated computers for their personal gain. Likewise, there have been many instances of employees who "went into business" processing data using their employers computers for other clients. Their employers absorbed their operating costs and salaries while they received the profits. System controls, procedural control, and access controls are all necessary to protect equipment.

Software

The development of information for use within the computer requires that data be converted in "machine useable" form, programs be prepared to direct the equipment's processing of it; and instructions be provided for the machines on what to do with the data once the processing is completed. The raw data used by the machines, the processed data and data being stored must be protected. Raw data is often retained for "back-up" in cases of emergency or for audit purposes. Processed data are often the operating records used by the organization in its day-to-day activities as is the data in storage.

The storage, duplication, and dispersal of software information is vital to organizational operations and survival. Computers only do what a program directs using the data provided. If either the data or the programs are defective or absent, the computer will not function properly. The protection of software during all phases of operations is a major concern for the protective program developer or manager.

Operations can not be significantly retarded solely for security. Prudent protective measures for insuring continuity of operations is, however, essential. The balancing of these two considerations in program development and operations will result in effective yet carefully accepted risks being assumed by the organization.

Operations

Computer systems often operate on a twenty-four-hour a day basis. Their operations are subject to strict procedural controls which include:

- Job Submission
- Dispatching
- Job Processing

- Output Disposition
- Operators Activity Log
- Tape and Disk Library
- Storage and Waste
- Downtime and Maintenance
- Off-Shift
- Emergency

Each set of procedures attempts to prescribe the precise manner in which each activity is to be performed. Safeguards are or should be "built into" each of these controls to insure that equipment and operations are carried out in the best interests of the organization.

Personnel

Personnel controls require that access to the computer facility be limited to only those persons having an absolute need to enter the area. Likewise, the use of equipment is restricted to those responsible for machine operations. The separation of responsibilities within the computer center likewise reduces the opportunity for any one person to manipulate the system without raising suspicion. Personnel develop the programs, prepare data for input, and manage the operations of the system. Personnel are likewise the greatest risk present in the computer operation. The degree of screening or surveillance of their activities relate directly to the criticality of the computer operation within the organization. The higher the criticality, the greater the risk personnel pose and the higher the security requires for them and the computer operation.

Insurance

Insurance is something that almost all types of business and industry utilize as a form of protection. Insurance requirements often specify the type of protection required before a policy will be issued or to qualify for a premium discount. For many organizations, security is equated to insurance. Insurance is used to "cover losses" through theft, fires, vandalism, etc. Only those minimal protective features or devices absolutely required to obtain coverage or maintain it are utilized. Such attitudes are common and have added to the increases in crime and insurance losses; both of which have caused insurance premium increases.

The concept behind insurance coverage is the "pooling of risk." Groups of people and businesses insure against major losses by pooling funds in a common reimbursement program. The "rate of risk" of various members of the group determines the amount of money each is required to contribute to the common protection fund. When losses increase and payment to cover them increases, the contributions or "premiums" by all members in the group must increase.

The "risk" of each of the individual members of the group contributes to the probability of losses. Anything which can contribute to reduced losses or risk of loss is a positive contribution to the financial welfare of the entire group. Risk reduction programs, the increased use of protective measures, and improved loss prevention management all contribute to such reductions. Dependence on insurance as a security measure might appear to be an attractive and low-cost means of protection while, in reality, it unnecessarily transfers the true cost of loss to others in the group.

Insurance coverage should be placed in its proper perspective in a total loss prevention program. Insurance will provide replacement of inventory or property lost, stolen, damaged or destroyed; it will not prevent the loss from happening. Security measures, loss prevention, and risk management programs of which insurance is a component will, however, materially aid in reducing loss. Security measures provide the "before the incident" protection, but if they are unsuccessful in preventing the loss, the insurance program will compensate the organization.

High-loss businesses or areas have experienced difficulty in obtaining adequate coverage or in some cases any protection at all. The Federal Crime Insurance Program was created to insure that all businesses that want insurance will have an opportunity to buy it should conventional sources feel that the risk is too high.

Specific types of protection in the form of alarm systems or guard services are often required by insurance companies before the issuance of policies of coverage. The loss experiences of various types of businesses dictate the protection required and the limits for the policy. The standards established by insurance companies in many cases have become the maximums that businesses will employ even though the requirements are listed as *minimum requirements* by the insurance companies. Thus, *minimums become maximums* for the purpose of protection and the saving of a few dollars.

It can be argued that a business must be cost-conscious and not spend more than absolutely necessary for anything, including security. While this is true, the question about cost is not fully answered by insurance. Most insurance policies cover *direct losses*. Indirect losses such as lost sales, lowered employee morale, customer loss, etc., are not covered by insurance. They can, however, be prevented by security programs. The old adage of "an ounce of prevention is worth a pound of cure" still applies to the trade-offs that must be made between insurance and prevention programs. Both are necessary, they supplement each other, but security prevents while insurance only reimburses for loss.

Employee Bonding

Some types of losses can be covered by insurance as a normal and necessary portion of the security program. Employee fidelity bonding is a prime example of such protection. Through fidelity bonding, an employer can protect his operations against the actions of a dishonest employee(s) who is covered either by an individual fidelity bond or through a "blanket bond" covering all employees.

A fidelity bond requires that the individual employee be investigated by the bonding company and follow prescribed procedures in handling the business of the employer. If in carrying out the assigned duties the employee defrauds, steals, or damages the business of the employer, the bonding company indemnifies the employer for the amount of his protection policy.

Bonding should be a normal, protective feature of any security or insurance program. It should be utilized for any employee who occupies a position of trust or who could damage the company financially by his actions.

FOOTNOTES

1. Cressey, Donald R.: *Other People's Money*. Glencoe, Ill., The Free Press, 1953, p. 5.
2. Curtis: S. J.: *Modern Retail Security*. Springfield, Thomas, 1960, p. 167.
3. Clinard, M. B.: *Black Market,* 1952, p. 310.
4. Sutherland, Edwin H.: *White Collar Crime*. New York, Hold, 1949, p. 234.
5. Reckless, Walter C.: *Crime Problem,* 2nd ed. New York, Appleton-Century-Crofts, 1955, p. 223.
6. Lane, R.: Why businessmen violate the law. *Journal of Criminal Law, Criminology and Police Science, 163:*163, 1953.
7. Newman, Donald J.: *White Collar Crime, Law and Contemporary Problems,* No. 23, Autumn, 1958, p. 749.
8. Taft, Donald R.: *Criminology,* 3rd ed. Macmillan, 1956, p. 339.
9. Vold, George B.: *Theoretical Criminology*. New York, Oxford University Press, 1958, p. 257.
10. Clinard: *Black Market,* p. 261.
11. Newman: *White Collar Crime, Law and Contemporary Problems,* p. 751.
12. Fuller, R. C.: Morals and the criminal law. *Journal of Criminal Law, Criminology and Police Science, 32:*624, 1942.
13. Lane: Why businessmen violate the law, p. 165.
14. Newman, p. 752.
15. Douglas, Paul H.: *Ethics in Government,* 1952.
16. Taft: *Criminology,* p. 757.
17. Congressional Record, February 6, 1967, H986.
18. See Appendix B in Chapter 5 for details of common elements of white collar crime.
19. Jaspan, Norman: Employee theft—barometer of mismanagement. *Security World, 2:* no. 7, 30, October 1965.
20. Edwards, Loren E.: *Shoplifting and Shrink-*

age Protection. Springfield, Thomas, 1958, p. 3.
21. Pratt, Lester A.: *Embezzlement Controls for Business Enterprises.* Baltimore, Fidelity and Deporte Co., 1966, p. 8-9.
22. Astor, Saul D.: More shortages—the discount dilemma. *Security World,* 2: no. 1, January 1965.
23. Illinois Retail Merchants Association. *The Problem of Store Protection.* 1961, p. 22-24.
24. National Retail Merchants Association. *Stock Shortages: Their Causes and Prevention.* New York, 1959, p. 63-66.
25. Florida State Retailers Association Special Report, p. 3-4.
26. Astor, Saul D.: *Testing for Profit Leaks.* Loss Prevention Security—International Security Conference Unit 3. Security World Publishing Company, 1968, p. 7.
27. Curtis, J.: *Modern Retail Security.* Springfield, Thomas, 1960, p. 962.
28. Cressey: *Other People's Money.*
29. *Harvard Business Review:* November-December, 1959, p. 10.
30. Davis, James: Traffic management, a case history. *Industrial Security,* June 1966, p. 20.
31. Davis: Traffic management, a case history.
32. Davis, p. 22.
33. Williams-Steiger Occupational Safety and Health Act of 1970 (84 Stat. 1593, 1600; 29 U.S.C. 655, 657) and in Secretary of Labor's Order No. 12-71 (36 FR 8754), Part 1910 of Title 29 of the Code of Federal Regulations.
34. Guide of Applying Safety and Health Standards, U.S. Department of Labor, Occupational Safety and Health Administration, Washington, D.C., October 1972, #2072, p. 1.
35. Fact Sheet on the Occupational Safety and Health Act of 1970, OSHA 2220, U.S. Department of Labor, Occupational Safety and Health Administration, Washington, D.C., December 1974.
36. See Appendix 17A for a Fire Equipment Inspection format.
37. A Standard Operating Procedure for use of an Industrial Sprinkler (Wet) system is presented in Appendix 17B.
38. Appendix 17C provides a series of EDP Audit Guidelines for Security.
39. Appendix 17D presents a case study of EDP Protective System Development.

BOOKS
WHITE COLLAR CRIME

Barnes, Harry E., and Teeters, Negley K.: *New Horizons in Criminology.* Englewood Cliffs, Prentice-Hall, 1959.
Beveridge, W. I. B.: *The Ethics of Research.* New York, Vintage Books, 1950.
Bloch, Herbert A., and Geis, Gilbert: *Man, Crime, and Society: The Forms of Criminal Behavior.* New York, Random House, Inc., 1962.
Bonger, William Adrian: *Criminality and Economic Conditions.* Trans. by Henry P. Horton. Boston, Little, Brown and Co., 1916.
Braham, Vernon C., and Kutash, Samuel B.: *Encyclopedia of Criminology.* New York, Philosophical Library.
Bromberg, Walter: *Crime and the Mind: A Psychiatric Analysis of Crime and Punishment.* New York, Macmillan, 1965.
Cardwell, Harvey: *The Principles of Audit Surveillance.* Princeton, Van Nostrand, 1960.
Cavan, Ruth Shonle: *Criminology,* 3rd ed. New York, Thomas Crowell Co., 1962.
Clinard, Marshall B.: *Sociology of Deviant Behavior.* 3rd ed. New York, Holt, Rinehart, and Winston, 1968.
Cohen, Albert: *The Sutherland Papers.* Bloomington, Indiana, 1956.
Cole, William E., and Miller, Charles H.: *Social Problems: A Sociological Interpretation.* New York, McKay, 1965.
Fuller, John G.: *The Gentlemen Conspirators: The Story of Price-Fixers in the Electrical Industry.* New York, Grove Press, 1962.
Geis, Gilbert: *The White Collar Criminal: The Offender in Business and the Professions.* New York, Atherton, 1967.
Gentry, Curt: *The Vulnerable Americans.* Garden City, New York, Doubleday, 1966.
Gibney, Frank: *The Operations.* New York, Harper, 1960.
Hartung, Frank E.: *Crime Law and Society.* Detroit, Wayne, 1965.
Jaspan, Norman, and Black, Hillel: *The Thief in the White Collar.* Philadelphia, Lippincott, 1960.
Jastrow, Joseph, Ph.D.: *Piloting Your Life.* New York, Greenberg Publisher, 1930.
Lever, Harry, and Young, Joseph: *Wartime Racketeering.* New York, Putnam's, 1945.
MacDonald, John M.: *Psychiatry and the Criminal.* Springfield, Thomas, 1958.

Mannheim, H.: *Comparative Criminology.* Boston, Houghton-Mifflin, 1965.

Sutherland, E. H.: *White Collar Crime.* New York, Dryden Press, 1949.

Tompkins, Dorothy Campbell: *White Collar Crime—A Bibliography.* University of California, Berkeley, 1967.

Weinberg, Arthur and Lila: *The Muckrakers.* New York, Simon and Schuster, 1961.

PUBLICATIONS OF THE GOVERNMENT, LEARNED SOCIETIES, AND OTHER ORGANIZATIONS

Podalsky, E.: The swindler: A fascinating sociopath. *Pakistan Medical Journal,* October 1957, p. 5-8.

PERIODICALS

Ahern, James F.: Casualty fraud: A billion-dollar industry. *Security Register, 1:* no. 3, 15-17, 1974.

Allen, Francis A.: Criminal justice, legal values and the rehabilitative idea. *J Criminal Law, 50:*226, 232, September/October 1959.

American Bar Association. Report of committee on mercenary crime. *J Criminal Law, 23:*94-100, May/June 1932.

Aubert, Wilhelm: White collar crime and social structure. *Amer J Social, 58:*263-271, November 1952.

Bacon, Seldon D.: Review of Sutherland, white collar crime. *American Sociological Review, 15:*309-310, April 1950.

Baker, Russell: The big brain invasion goes on and on. *The New York Times,* September 10, 1964.

Battelle, Phyllis: Invisible advertising. *New York Journal American,* September 17, 1957.

Bauer, Bertrand N.: Truth in lending: College business students' opinions of Caveat Emptor, fraud and deception. *American Business Law Journal, 4:*156-161, Fall, 1966.

Bayley, David H.: The effects of corruption in a developing nation. *Western Political Science Quarterly, 19:*719-732, December 1966.

Berkovitch, Israel: Fire, smoke and dangerous gases. *Security Gazette,* October 1974, p. 372.

Bowen, William: Who owns what's in your head? *Fortune,* July 1964, p. 175.

Burnstein, Harvey: Not so petty larceny: millions of dollars lost to pilferage, embezzlement, malicious mischief and other breaches of industrial security. *Harvard Business Review, 37:*72-78, May/June 1959.

Caldwell, Ribert C.: A reexamination of the concept of white collar crime. *Federal Probation, 22:*30-36, March 1958.

Chicago crime commission: Why honest people steal. *J Criminal Law, Criminology, Police Sci, 38:*94-103, July/August 1947.

Clinard, Marshall B.: Sociologists and American criminology. *J Criminal Law, Criminology Police Sci, 41:*549-577, 1951.

Cressey, Donald R.: Application and verification of the differential association theory. *J Criminal Law, Criminology, Police Sci, 43:*43-52, May/June, 1952.

———: The criminal violation of financial trust. *American Sociol Rev,* 1950.

———: Foreword. In Sutherland, Edwin H. (ed.), *White Collar Crime.* New York, Holt, 1961, p. iii-xii.

———: The respectable criminal: Why some of our best friends are crooks. *Transaction, 2:*12-15, March/April 1965.

Curran, Barbara A.: *Trends in Consumer Credit Legislation.* Chicago, University of Chicago Press, 1965.

Curtis, Sargent J.: The problem of crime and kids. *Industrial Security,* April 1962.

England, Ralph W., Jr.: The independent offender. *Current History, 52:*334-340+, June 1967.

Finklestein, Louis: The businessman's moral failure. *Fortune, 58:*116-117ff, September 1958.

Florita, Dr. Giorgio: Enquiry into the cause of crime. *J Criminal Law, Criminology Police Sci, 44:* no. 1, 1953.

Frank, Stanley B.: Beware of home-repair racketeers. *Saturday Evening Post,* 17ff, July 21, 1956.

Fuller, Richard C.: Morals and the criminal law. *Journal of Criminal Law and Criminology, 32:*624-630, March/April 1942.

Gambling menace, The. *Security Management Plant and Property Protection,* April 10, 1973, p. 1-3.

Geis, Gilbert: Toward a delineation of white collar offenses. *Sociological Inquiry,* Spring, 1962, p. 160-171.

Gibbons, Don C., and Garrity, Donald L.: Definition and analysis of certain criminal types. *J Criminal Law, Criminology, Police Sci,* March 1962, p. 27-35.

Gibney, Frank: The crooks in the white collars. *Life,* October 14, 1957, p. 162-165.

Hadden, Tom: The origins and the development of conspiracy to defraud. *American Journal of Legal History, 11:*25-40, January 1967.

Hartung, Frank E.: The white collar thief. *Security World,* March 1966, p. 30.

Hazard, John N.: Soviet socialism and embezzlement. *Washington Law Review, 26:*301-320, November 1951.

Hazard, Leland: Are big businessmen crooks? *The Atlantic, 208:*57-61, November 1961.

Hewitt, William H.: Combating the white collar criminal. *Law and Order,* February 1963.

Insalata, S. John: Deceptive business practices: Criminals in cuff links. *Vital Speeches of the Day, 29:*473-475, May 15, 1963.

Jaspan, Norman: Employee theft: barometer of mismanagement. *Security World, 2:* no. 7, October 1965.

———: How can you curtail employee dishonesty? *American Business,* November 1959.

———: Internal control of fraud and embezzlement. *National Public Accountant, 7:* 6-7, May 1962.

———: Internal theft and fraud . . . A "human relations" problem. *Drug and Cosmetic Industry.*

———: Thieves in white collars. *Sales Management, 82:*44, February 20, 1959.

Kessler, Friedrich: The protection of the consumer under modern sales law. *Yale Law Journal, 74:*262-284, December 1964.

Keysor, Charles W.: Beware of genteel crooks. *Commerce Magazine, 52:*20ff, April 1955.

Kossack, Nathaniel E.: Scram: The planned bankruptcy racket. *New York Certified Public Accountant, 35:*417-423, June 1965.

Kostelanetz: The defense of the white collar accused: a symposium. *Amer Criminal Law Quart, 3:*124-145, Spring, 1965.

Lane, Robert E.: Why businessmen violate the law. *J Criminal Law, Criminology Police Sci, 44:*151-165, July/August 1953.

Myers, Robert S.: The rise and fall of fee-splitting. *Bulletin of the American College of Surgeons, 40:*507-509, 523, November/December 1955.

Myers, J. H.: Predicting credit risk with a numerical scoring system. *Journal Applied Psychology, 47:*348-352, 1963.

Newman, Donald J.: Public attitudes toward a form of white collar crime. *Social Problems, 4:*228-232, January 1957.

———: White collar crime. *Law and Contemporary Problems, 23:*735-753, Autumn, 1958.

Quinney, Richard: The study of white collar crime: Toward a reorientation in theory and research. *Journal of Criminal Law, Criminology, and Police Science, 55:*208-214, June 1964.

Rising, Nelson: Contours of conflict: Protection of the defaulting consumer. *UCLA Law Review, 13:*348-365, January 1966.

Sherwin, Robert: White collar crime, conventional crime and Merton's deviant behavior theory. *Wisconsin Sociol, 2:*7-10, Spring, 1963.

Shulman, Harry M.: Cultural aspect of criminal responsibility. *J Criminal Law, Criminology, Police Sci, 43:*323-327, September/October 1952.

Smith, Richard Austin: The incredible electrical conspiracy. *Fortune, 63:*132-137, April 1961; and *63:*161-164, May 1961.

Smith, Ralph A.: The incredible electrical conspiracy. *Fortune, 64:*132-137, 170, 172, 175, 179, 180, April 1961; May 1961, p. 161-164, 210, 212, 217, 218, 221, 222.

Sorenson, Robert C.: Review of Sutherland, white collar crime. *Journal of Criminal Law, Criminology and Police Science, 41:* 80-82, May/June 1950.

Tavel, Emilie: White collar crime probed. *Christian Science Monitor,* March 4, 1964.

Tyler, Harold R., Jr., Harris, B., and Kostelanetz: The defense of the white collar accused: A symposium. *Amer Criminal Law Quart, 3:*124-145, Spring, 1965.

White collar criminality. *American Sociological Review,* February 1940.

Whitman, Howard: Why some doctors should be in jail. *Collier's, 132:*23-27, October 30, 1953.

UNPUBLISHED MATERIAL

Bisset, William: The economic sanction: An alternative method for the deterrence of white collar crime in America. (B Paper), Michigan State Univ., 1974.

Campbell, Richard N.: White collar crime. Masters Paper for Michigan State Univ., 1965.

BOOKS
INDUSTRIAL ESPIONAGE

Alden, Lurton, et al.: *Competitive Intelligence: Information Espionage, and Decision Making.* Boston, Mass., C. I. Associates, 1959.

Bergier, Jacques: *L'Espionnage industriel.* Paris, Hachette, 1969.

Beriot, Louis: *L'invasion; de l'espionnage economique a' ului de la vie privee.* Paris, S. Michel, 1971.

Bursk, E. C., and Fenn, D. H.: *Planning the Future Strategy of Your Business.* New York, McGraw-Hill Book Co., Inc., 1956.

Baker, Anthony G.: *Competitive Espionage.* Beverly Shores, Ind., Industrial Research, Inc., *IV:* no. 4, April 1962.

Calkins, Clinch: *Spy Overhead—The Story of Industrial Espionage.* New York, Harcourt, Brace and Company, 1937.

Coman, E. T.: *Sources of Business Information.* Englewood Cliffs, N.J., Prentice-Hall, Inc., 1949.

Dauten, Carl A.: *Business Fluctuations and Forecasting.* Cincinnati, Ohio, South-Western Publishing Co., Inc., 1954.

Greene, Jay R., and Sisson, Roger L.: *Dynamic Management Decision Games.* New York, John Wiley & Sons, Inc., 1959.

Greene, Richard M., Jr.: *Business Intelligence and Espionage.* Homewood, Ill., Dow Jones-Irwin, 1966.

Greenlaw, P. S.: *Business Simulation.* Englewood Cliffs, N.J., Prentice-Hall, Inc., 1962.

Hamilton, Peter: *Espionage and Subversion in an Industrial Society.* London, Hutchinson, 1969.

Hanebuth, Klaus: *Das Auskunftsrecht in europaischea Wertschaftsrecht Rechtsgreendlagen and Handhabung.* Baden—Baden Nomos Verlagsgesellschafts, 1967.

Hichson, Philip: *Industrial Counter Espionage.* London, Spectator Publications, 1968.

Hodges, Luther H.: *The Business Conscience.* Englewood Cliffs, N.J., Prentice-Hall, Inc., 1963.

Hug, Theodor: Der Wirtschaftliche. *Nachrichtendienst im schweizerischen Recht.* Zurich, Juris—Verlag, 1961.

Markham, Jesse William: *Competition in the Rayon Industry.* Cambridge, Mass., Harvard University Press, 1952.

Payne, Ronald: *Private Spies.* London, 1967.

Spy Overhead, The Story of Industrial Espionage. New York, Harcourt Brace & Co., 1937.

Turner, A.: *The Law of Trade Secrets.* London, Sweet and Maxwell, 1962.

Wade, Worth: *Industrial Espionage and Misuse of Trade Secrets.* Ardmore, Advance House, 1965.

PUBLICATIONS OF THE GOVERNMENT, LEARNED SOCIETIES, AND OTHER ORGANIZATIONS

American Management Association, R and D Division. Trade secrets—a management overview. *Management Bulletin No. 64,* 1965.

Australia. *Report of Royal Commission on Espionage.* Sydney, Commonwealth of Australia. A. H. Pettifier, 1955.

Bigelow, C. G.: Bibliography on project planning and control by network analysis; 1959-1961. *Operations Research,* September 1962, p. 728-731.

Bishop, J., and Ryan, C. T.: Use marketing information for making major management decisions. *ISA-J,* May 1962, p. 41-45.

Blum, Richard H.: Surveillance and espionage in a free society. A report by the Planning Group on Intelligence and Security for the Policy Council of a Democratic National Committee. New York, Praeger Publishers, 1972.

Bonner, Stanley Z.: Reconciling short-range profit prospects with long-range goals. Section 2. *NAA Bulletin,* July 1959, p. 37-46.

Competitive Intelligence: Information, Espionage and Decision-Making: A Special Report for Businessmen. Watertown, Mass, C.I. Associates, 1959.

———. Harvard University Graduate School of Business Administration, Boston, 1959.

Cottam, Richard: *Competitive Interference and Twentieth-Century Diplomacy.* Pittsburgh, Penn., University of Pittsburgh Press, 1967.

Desk Book on Organized Crime. Available from the Chamber of Commerce of the United States, 1615 H Street, N.W., Washington, D.C. 20006.

Ellis, R.: *Trade Secrets.* New York, Baker, Voorhis and Company, 1953.

Engberg, Edward: *The Spy in the Corporate Structure.* Published by the World Publishing Company, 2231 West 110th Street, Cleveland, Ohio 44102.

Feidel, Arthur H.: What the general practitioner should know about trade secrets and employee agreements. Ronald L. Tanitch. Philadelphia Joint Committee on Continuing Legal Education of the Law Institute and the American Bar Association, 1973.

Gray, L. Patrick, III: Domestic intelligence. *The Mosler Security Letter,* December 14, 1972.

Greene, Richard: *High Cost of a Blabbermouth.* Chicago, Enterprise Publications, 1962.

Machlup, Fritz: *The Production and Distribution of Knowledge in the United States.* Princeton, Princeton University Press, 1962.

Milgrin, Roger M.: *Trade Secrets.* New York, M. Bender, 1969.

National Industrial Conference Board Division of Personnel Administration. Employee Patent and Secrecy Agreement. New York National Conference Board, 1965.

Someone's Stealing From You. (film 16mm, Store Management Group. New York, Nat. Retail Merchants Assn., 1967.

PERIODICALS

Allen, Eugene V.: The value of internal intelligence. *Industrial Security,* August 1971.

American Cyanamid. Industrial spies hurt U.S. industry. *Journal American,* July 11, 1962.

American Management Association R and D Division. Trade secrets, a management overview. *Management Bulletin,* No. 64, 1965.

Anonymous and Staff Written. Espionage and subversion in an industrial society. *Security World, V:* no. 10, November 1968.

Aries, others are named in lifting trade secrets. *Oil Paint and Drug Report,* August 14, 1961, p. 5.

Baen, Jack M.: Protecting the company's marketing secrets, Part I. *Security World, 5:* no. 6, June 1968.

Baram, Michael S.: Trade secrets: What price loyalty. *Harvard Business Review,* November/December 1968, p. 66.

Bartenstein, Fred, Jr.: Research espionage: A threat to our national security. *Food and Drug Cosmetic Law Journal,* 17:813-827, December 1962.

———: Industrial espionage—a national and international problem. *Industrial Security, 6:* no. 4, October 1962.

———: Research espionage. *Food and Drug Cosmetic Law J, 17:* 1962.

———: Theft of research data: a national menace. *Research Management,* July 1963.

Bate, Frank L.: The protection of company knowledge from theft—legal remedies. *Research Management, VII:* no. 4, 1964.

Beauvois, J. J.: International intelligence for the international enterprise. *California Management Review,* Winter, 1961, p. 39-46.

Belda, Bertrand J.: Income and expense planning. *Journal of Accountancy,* November 1959, p. 46-50.

Bienvenu, Leonard P.: The industrial employee faces the Soviet spy. *Industrial Security,* June 1965.

Bill of safeguard scientific secrets of industrial research. *Chemical and Engineering News,* April 6, 1964.

Bills would make trade secret theft a crime. *Time,* March 26, 1965.

Bingham, H. L.: Six types of firms and the management data they need. *Business,* 1957, p. 84-91.

Blake, Harlan M.: Employee agreements not to compete. *Harvard Law Review, 73:* no. 4, 625-691, February 1960.

Blake, Robert M.: Employee dishonesty can ruin you if accounts lack proper coverage. *Credit and Financial Management,* October 1960.

Block, Victor: When a secret is better than a patent. *Popular Mechanics,* November 1965, p. 110.

Bloom, Murray Teigh: How to protect those "valuable papers." *Reader's Digest,* December 1965.

Boltwood, Parke: Confessions of an auto spy. *SAGA,* November 1965, p. 20.

Brean, H.: Everybody is dishonest. *Life, 45:*70, November 24, 1958.

Brennan, Charles D.: Subversion and security. *Industrial Security,* April 1968.

Brenton, Myron: They spy for industry. *Popular Mechanics,* May 1962.

Brown, Robert L.: Marketing espionage. *Sales Management,* December 4, 1964, p. 23.

Buge, Edward W.: How to collect and use business intelligence data. *The Office,* July 1964, p. 69.

Burgess, R. W.: Meaning of the new business census. *Dun's Review and Modern Industry,* June 1956, p. 41-42.

Burstein, Harvey: Not so petty larceny. *Harvard Business Review,* May 8, 1959, p. 72.

Business espionage and mobile research. *Security World,* April 1966, p. 23.

Business espionage, Part I. A professional spy tells—"How I steal company secrets; Part II, How your company can thwart a spy. *Business Management,* October 1965, p. 57.

Business forecasting: Different systems the forecasters are using. *Business Week,* September 1955, p. 90-92.

Business espionage, Part I. *Business Management,* October 1965.

Business spies seen as growing peril. *The Blade,* Toledo, Ohio, February 13, 1964.

Business spies still busy. *Industrial Management,* June, 1969, p. 58-59.

Business spies cost industry $2 million yearly. San Francisco, *Commercial News,* October 14, 1965.

Business spy. *Wall Street Journal,* March 3, 1959.

Can a competitor "palm off" your product as his own. *Business and the Law,* April 1965.

Capon, Frank S.: Essentials of corporate planning. *Controller,* May 1960, p. 218, 220, 222.

Carley, William M.: Crackdown on trade secrets thefts seen as laws are enacted, civil suits filed. *Wall Street Journal,* August 29, 1968.

———: The secret stealers. *Wall Street Journal,* October 5, 1962.

Cassady, Ralph: The intelligence function and business competition. *California Management Review,* Spring 1964.

Cassels, Louis, and Randall, R. L.: Six steps to better planning. *Nation's Business,* August 1961.

Chemical Brain Control. *Science News Letter,* 85:214, April 4, 1964.

Christensen, N. C.: What about spies in business. *Personnel,* 1963.

Clapham, J. C.: Need for research in planning mechanization. *Engineering Journal,* August 1961, p. 47-49.

Coe, B. P.: Capacity stretch. *Industrial and Engineering Chemical,* March 1962, p. 47-51.

Cohen, Sidney: How the retailer uses census tracts. *Journal of the American Statistical Association,* September 1956, p. 506.

Coleman, J. J., and Abrams, I. J.: Mathematical model for operational readiness. *Operations Research,* January 1962, p. 126-138.

Coleman, L. A., and Cole, C. B.: The effect of shifting employees on trade secrets. *Business Lawyer,* January 1959.

———: How much of your knowledge is yours? Law of proprietary rights. *Hydrocarbon Process and Petroleum Refiner,* June 1961, p. 208-212.

Collection of data from firms. *Nature,* September 1962, p. 500-501.

Colm, Gerhard: How good are long-range projections of GNP for business planning. *California Management Review,* Winter, 1959.

Competitive intelligence: Information, espionage and decision making. *Industrial Security,* 1959.

Coppock, Joseph D.: *Economics of the Business Firm.* New York, McGraw-Hill Book Co., Inc., 1959.

Cornetta, Anna: Firms improve ethics. *Democrat and Chronicle,* May 25, 1965.

———: Spies hurt U.S. industry. Nation's business suffers in world market. *Democrat and Chronicle,* May 24, 1965.

———: Spies in business; industry seeks curb on secret stealing. *Democrat and Chronicle,* May 23, 1965.

Correa, Mathias F.: Protection of trade secrets. *Business Lawyer,* January 1963.

Corrigan, William H.: Protecting intellectual property. *Industrial Security,* April 1966, p. 4.

Court says Goodrich can't bar competitor from hiring former key space suit man. *Wall Street Journal,* January 25, 1963.

Court supports Universal Match in suit against Vendo and three former employees. *Wall Street Journal,* October 21, 1963.

Coville, D. M.: How to measure the effectiveness of manufacturing planning. *Tool and Manufacturing Engineers,* August 1962, p. 62-64.

Cox, T.: I've got a secret, or have I? *Management Methods,* February 1959, p. 43-45.

Crackdown on in-plant pilferage. *Business Week,* 64: no. 27, July 6, 1957.

Cyanamid says Italians bought pirated secrets. *Oil, Paint and Drug Report,* July 2, 1962, p. 3.

Daniel, P. J.: Application of operations research for site planning facilities support. *Aerospace Engineering,* June 1961.

Daniel, James: Spies invade big business. Evening and Sunday Bulletin. *Reader's Digest,* January 1962, p. 93.

Delmage, Sherman H.: The eavesdropper's target—business and industry. *Industrial Security,* June 1966.

Denzel, James L.: Security—the right arm of scientific research. *FBI Law Enforcement Bulletin,* September 1965, p. 8.

Diebold, John: Automation: Its impact on business and labor. *NRA Planning Pamphlet,* May 1959, p. 64.

Dishonest practices: How management can prevent them. *Textile Industries,* February 1959, p. 100-103.

Do you know the law on trade secrets? *Management Methods,* May 1959.

Document control: Plugging paper leaks at DuPont. *Occupational Hazards,* September 1971, p. 34.

Does your employer own your knowledge? *Chemical Engineering,* June 13, 1960.

Donovan, Robert: Trade secrets. *Security World,* 4: no. 4, April 1967.

Drawing a bead on drug pirates. *Chemical Week,* August 4, 1962.

Dyment, Robert: Continuing safety program is the answer in Jamestown, N.Y. *Law and Order,* 2: February 1962.

Eggert, F. John: Are your engineering drawings protected from pirating? *Plant Engineerings*, May 1966.

Engineering Joint Council. Does your employer own your knowledge? *Chemical Engineering*, June 13, 1960.

——. DuPont sues Von Kohorn on trade secrets. *Chemical and Engineering News*, December 1961.

Espionage and subversion in an industrial society. *Security World, 5:* no. 10, November 1968.

Espionage in business and industry. *Wall Street Journal*, March 3, 1959.

Farmer, Richard: Firm secrets and their protection. *Industrial Security*, January 1965.

FBI smashes plot to sell secret formulas to DuPont. *Corning Leader*, March 12, 1966.

Federal grand jury has indicated three former employees of American Cyanamid Lederle Laboratory Division. *Chemical and Engineering News*, November 1962.

Ferreting out the spies in business. *Business Week*, April 2, 1966.

Former employee stole Cyanamid's secrets New York judge rules. *Wall Street Journal*, January 10, 1964.

Fremed, R. F.: Does your employer own your knowledge? *Chemical Engineering*, July 28, 1958.

Friedlander, Mort: Profile of a plant thief. *Personnel Journal*.

Furash, Edward E.: Problems in review—industrial espionage. *Harvard Business Review*, November/December 1959.

Fuchs, Gerald J., and Thompson, G. Clark: Management of new product development. *Business Record*, October 1960, p. 36-39.

——: Sources of economic intelligence. *Business Record*, September 1960, p. 27-33.

Gearhart, Donald C.: Controlling security costs in a research and development organization —panelist. *Industrial Security*, October 1960.

Gehman, Richard: Executing spies. *Cosmopolitan*, February 1958, p. 70-75.

Georgia legislature levels its guns at pirates of industrial processes. *Cyanamid News*, May 10, 1964.

Gerstel, Stephen: Drug spy collects interesting facts. *UPI Release*, Washington, D.C., April 13, 1966.

Goddard, R. J.: Industrial espionage—an overview. *Industrial Security*, p. 4-12.

Golin, Milton: Beware the secret stealers. *Mechanix Illustrated*, February 1964.

Gomolak, Louis S.: How to track down your own private eye. *College and University Business, 42:*91, April 1967.

Gookin, R. Burt: Profit planning and control at Heinz. *American Business*, September 1959.

Gordon, Robert Aaron: Business fluctuations. *Harper's*, 1952, p. 127-152, 449-483.

Granzeier, Frank J.: The competitive advantage. *Industrial Security*, November 1963.

——: Guarding against industrial espionage. *Management Rev*, January 1964.

Gray, A. W.: Can your cherished trade secrets be pirated legally? *American Machinist/Metalworking Manufacturing*, February 1960, p. 106-06.

——: Don't lose a patent by premature use. *Chemical Engineer*, January 1962, p. 117-118.

——: Novelty and patentability. *Machine Design*, August 1962, p. 79-80.

——: Protection of trade secrets. *Audio*, April 1962, p. 23.

——: Trade secrets; what the courts have to say. *Product Engineering*, October 1962, p. 70.

——: What the courts say about trade secrets. *Petroleum Refiner*, February 1957, p. 226.

——: When a salesman switches jobs. *Purchasing*, August 27, 1962, p. 77-78.

Guilfoyle, Joseph M.: How busy investigator ferrets out corporate secrets for his clients. *Wall Street Journal*, March 1962.

Harper, M.: Business detectives would stalk information; new profession to aid management proposed. *Petroleum Engineer*, September 1960, p. A20d-A20.

Harris, Ray M.: Trade secrets as they affect the government. *The Business Lawyer*, April 1963.

——: Trade secrets as they a affect the government. *The Business Lawyer*, July 1963.

Hawthorne, R.: Patents and progress. Ed. *Space-Aeronautics*, May 1960, p. 17.

Hooper, Ken: How firms spy on rivals. *The Times Union*, Jacksonville, Florida, April 22, 1965.

Hoover, J. Edgar: American-Soviet espionage target. *Industrial Security*, no. 1, April 1964.

——: FBI investigation of fraud. *J Accountancy, 120:*34-39, July 1965.

——: Embezzlement, their causes and consequences. *N.Y. Certified Public Accountants, 33:*540-548, August 1963.

———: Why Reds make friends with businessmen. *Nations Business,* May 1962, p. 78-80.

———: The U.S. businessman faces the Soviet spy. *Harvard Business Review,* January/February 1964, p. 140-161.

Hopkinson, Tom M.: New battleground—consumer interest. *Harvard Business Review,* 42:97-104, October 1964.

How I steal company secrets, Part I. *Business Management,* October 1965.

How Russia spies: The new game. *Newsweek,* October 11, 1971, p. 31.

How secure can you keep your design secrets? *Product Engineering,* July 4, 1966, p. 87-95.

How to protect yourself against business spies. *Management Methods,* June 1960, p. 58-62.

How paper shredders protect business data. *The Office,* September 1971, p. 58.

How your company can thwart a spy, Part II. *Business Management,* October 1965.

How to guard against dishonesty. *Management Review,* August 1964.

How to keep employees honest. *Administration Management,* August 1963.

How your company can thwart a spy, Part II. *Business Management,* October 1965.

Hoyt, Douglas: The computer as a target for the industrial spy. *Assets Protection,* Spring, 1975, p. 41-48.

Hutchinson, H.: Security in a research organization. *Security World, 5:* no. 11.

Industrial espionage: Operation air bubble. *Newsweek,* April 19, 1965, p. 79.

Industrial espionage: A modern menace. *Occupational Hazards,* October 1966.

Industrial spying goes big league. *Business Week,* October 6, 1962, p. 65-66.

Industrial spying essential to market research. *Advertising Age,* November 1964.

Intelligence function and business competition, The. *California Management Review,* Spring, 1964.

Jeter, R. G.: The space suit case—refitted. *The Business Lawyer,* July 1963.

———: Trade secrets—the space suit case. *The Business Lawyer,* April 1963.

Johnson, Don S.: Office executives have long been fair game for fraud. *The Office,* December 1968, p. 41-45.

Jones, R. V.: Scientific intelligence. *Journal of the Royal United Service Institution, XCII:* 352-369, August 1947.

Jones, Stacy V.: Loss of "secrets" vexes companies. *New York Times,* August 2, 1964.

Kandlik, Ed.: Spies in your plant. *Pittsburgh Press,* January 6, 1960.

Kast, R., and Rosenzeig, James: Keeping an eye on the opposition: Hercules Powder's Information Center. *Chemical Week,* 1957, p. 78-80.

Kemp, L. M., and White, C. M.: Basic sources of business information. *Special Libraries,* April 1958, p. 160-163.

Kitshoff, A. B.: Industrial espionage. *Security World,* April 1969.

Kleiler, F. M.: Keys to rich uncle's treasure: Federal agencies have priceless data. *American Business* March 1958, p. 13-15.

Klein: The technical trade secret quadrangle: A survey. *N.Y.U. Law Rev, 55:*437, 1960.

Kobler, John: Who stole the formula. *Saturday Evening Post,* June 29, 1963.

Kurnow, Ernest, Ottman, Frederick R., and Glasser, Gerald: *Statistics for Business Decisions.* Homewood, Illinois, Richard D. Irwin, Inc., 1959.

Labine, R. A.: Truth about industrial spying. *Chemical Engineering,* February 1960, p. 121-126.

LaForge, Charles P.: The leak in the dike: the petty larceny of ideas. *Security World, 2:* no. 3, May 1965.

Laing, Philip P.: Hidden crime. *Police, 4:* no. 1, September/October 1959.

Lambert, William: The broad scope of trade secrets. *Amer Perfumer Cosmetics, 78:* July 1963.

Lapp, R. E.: Scientists say secrecy hides fallout threat. *Industrial Security,* August 9, 1958.

Larke, A. G.: Your company secrets: are they safe? *Dun's Review and Modern Industry,* August 1958.

Lawson, Herbert G.: Thefts of U.S. firm's secrets pose questions on pricing and patents. *Wall Street Journal,* July 8, 1963.

Lawsuits unlimited. *Chemical Week,* November 25, 1961, p. 23.

Legal hurdle for job-hoppers. *Business Week,* June 1, 1963.

Levin, M.: Industrial security of intellectual property. *32:* no. 6, June 1965.

Liebhofsky, Douglas S.: Industry secrets and the skilled employee. *N.Y.U. Law Rev,* April 1963.

Lightner, Max W.: How much is research worth to a company? *Metals Rev,* October 1965.

Lipman, M.: Are your secrets safe? *Home Appliance Builder,* June 1964.

Lockheed-California knows how to keep a secret. *Occupational Hazards,* May 1967.

Lockheed points the way to plant security. *Occupational Hazards,* January 1962.

Loomis, Roland L.: The problems of security in research and development. *Industrial Security, 9:* no. 4, August 1965.

Luhn, H. P.: A business intelligence system. *IBM Journal of Research and Development,* October 1958, p. 314-319.

MacCutheon, Richard H.: Patent or secrecy. *Tool Engineer,* January 1957.

Make your salesmen intelligence agents. *Sales Management,* October 21, 1960.

Malcolm, D. G., Roseboom, J. H., Clark, C. E., and Fazar, W.: Application of a technique for R & D program evaluation. *Operations Research,* September/October 1959.

Manley, Marian C.: *Business Information—How to Find and Use It.* New York, Harper Bros., 1955, 265pp.

Marketing intelligence systems—a dew line for marketing men. *Business Management,* January 1966.

Marvin, Phillip: Can trade secrets be protected? *Machine Design, 36:* no. 12, 112, July 8, 1965.

Mati Hari in industry? Firm says rival hired girl spy. *Pittsburgh Press,* March 11, 1959.

McAloney, S. H.: Ford instrument tells how to publicize a secret. *Industrial Marketing,* May 1958.

McClain, J. A., Jr.: Injunction relief against employees using confidential information. *Kentucky Law J, 23.*

McCutcheon, J. W.: Employee approximation of trade secrets. *Soap and Chemical Specialties,* July 1955, p. 83.

McTierman, Charles E.: Employees and trade secrets. *J Patent Office Society, 16:* no. 12, December 1959.

Merck. Kobler, John: Who stole the formula? *Saturday Evening Post,* June 29, 1963, p. 84.

———: Amprolium coccidiostat is the subject of a French lawsuit involving Merck and Dr. Robert S. Aries. *Chemical and Engineering News,* February 26, 1962.

———: Hearing of first Merck-Aries suit in Paris postponed. *Chemical and Engineering News,* July 1962.

———: Two more suits filed in Merck-Aries battle. *Chemical and Engineering News,* July 30, 1962.

———: Second drug firm sues pirate ring. *Washington News,* **June 22, 1962.**

———: Expands security program to block theft of company secrets. *Manager's Letter,* April 20, 1963.

Mergott, Winston: Protecting your business against dishonesty. *Management Aids for Small Manufacturers,* December 1957.

Metzdorff, Howard A.: Module concept of intelligence gathering. *Police Chief,* February 1975, p. 40-44.

Milgrim, R.: *Business Organizations—Trade Secrets,* Rev. Ed. Albany, N.Y., Matthew Bender & Co., Inc., 1972.

———: Trends in trade secret litigation. *Assets Protection,* Spring, 1975, p. 9-12.

Miller, A. Ross: What industry can do to prevent espionage and sabotage. *Industrial Security,* October 1962.

Minnesota Mining sues two former employees. *Wall Street Journal,* February 13, 1963.

Monat, Paul: Americans talk too much. Spy in the U.S. *Industrial Security,* Harper & Bros.

Morris, Sid: My vendors the spies. *The Office,* December 1968, p. 46-49.

National Association of Cost Accountants: Protection of vital corporate records. New York, Nat. Assn. of Cost Accountants, January 1954.

National Industrial Conference Board. Employee patent and secrecy agreements. *Studies in Personnel Policy,* no. 199, 1965.

Neal, H. R.: How Detroit guards auto secrets. *Iron Age,* November 1957, p. 98.

Need help with market research? Your biggest answer man, **Uncle Sam.** *American Business,* February 1959, p. 17-19.

Newcomb, Robinson: Plan your part in the boom. *Nation's Business,* July 1959, p. 38-39.

Olsen, Askel G.: How to plan research for profit, Part I: Objectives and costs. *Food Engineering,* November 1959, p. 39-41.

———: How to plan research for profit, Part II: Managing the team. *Food Engineering,* December 1959, p. 99-100.

O'Meara, John Roger: How smaller companies protect their trade secrets. *The Conference Board,* 1971.

One Thousand-one Embezzlers. A study in defacations in business. Baltimore, United States Fidelity and Guaranty Co, 1950. 51pp.

Packard, Vance: The walls do have ears. *New York Times Magazine,* September 20, 1964, p. 23, 114-116.

Patents decline—secrets increase. *Materials and Methods, 44:* 1956.

Patterson, Morehead: How to prepare for future company growth. *Iron Age,* October 1959, p. 75-77.

Penrose, Edith Tilton: *The Theory of the Growth of the Firm.* New York, John Wiley & Sons, Inc., 1959.

Perham, John: The great game of corporate espionage. *Dun's,* October 1970, p. 33.

Piracy—a rising worry for U.S. business. *U.S. News and World Report,* September 3, 1962.

Popper, Herbert: How safe are your company's secrets? *Chemical Engineering,* May 23, 1966, p. 157.

Powers, M. C.: Piracy—20th century style: formulas, patents, trademarks, or what have you. *NAM News,* March 13, 1964.

Preston, Lee E.: A longer look at the sixties. *California Management Review,* Fall, 1960.

Protecting your company's business secrets. *Research Institute of America.* Report of the Month. Section II of two sections. New York, August 23, 1963.

Protection of intellectual property against industrial espionage and theft. *The New York Times,* March 16, 1965.

Queeney, Jack: Computer spies: New worry for business. *Chicago's American,* January 16, 1969.

Raphael, M.: Guides to business information sources. *Business Literature,* May 1957, p. 142-143.

Reber, Jim: The essence of industrial espionage. *Security Industry and Product News,* July/August 1975, p. 24, 29, 30.

Re-suit for Lummus Company vs. Phillip Witt and Robert Calogne. *Chemical Engineering,* August 8, 1960.

Richardson, William W.: The accountant's opportunities as a profit planner. *NAA Bulletin,* July 1960, p. 3-10.

———: Significance of company forward planning. *NAA Bulletin,* January 1960.

Rooting out data that distort business forecasting. *Business Week,* March 19, 1960, p. 190-192.

Rossman, Martin: Agencies get hush-hush on ideas they intend to blare. *Los Angeles Times,* February 7, 1972.

Safeguarding trade secrets. *Dun's Review and Modern Industry,* March 1960.

Scher, V. A.: Protecting your new ideas. *Textile World,* June 1960, p. 38-39.

Schrader, E. W.: Engineers and trade secrets. *Design News,* July 1965.

Secrecy and invention agreements. William F. Weigel. *Research Management, VII:* no. 4, 1964.

Secrets stealers: Firms seek to curb job hoppers who take their confidential plans. *Wall Street Journal,* October 5, 1962.

Security management: From perimeter guarding to keeping corporate secrets. *Occupational Hazards,* September 1971, p. 31.

Sheeran, S. R.: Intelligence on rivals comes from all over. *Advertising Age,* March 22, 1965.

Strawn, R. B.: Protecting the buyer of record protection. Boston, National Fire Protection Assn., October 1961.

Sims, E. Ralph, Jr.: Industrial technical intelligence: Tool for long-term planning and prevention of technological surprise. *Advanced Management,* July 1959, p. 18-20.

Skolasky, George E.: Foreign countries stealing U.S. products cause loss of revenue. *Corning Leader,* July 12, 1962.

Smith, Richard Austin: Business Espionage. *Fortune,* May 1956, p. 118.

———: Business espionage: Long-range planning for management. *Harper's,* 1958, p. 358-372.

Sokolsky, George E.: Foreign countries stealing U.S. products loss of revenue. *Corning Leader,* Corning, N.Y., July 12, 1962.

Sperry Rand wins suits against former employees in trade secret piracy. *Wall Street Journal,* October 19, 1964.

Spying. *Esquire Magazine,* May 1966.

Spying for Profit. *Time,* July 1959, p. 59-60.

Stedman: Inventions and public policy. *Law and Contemporary Problems,* 1947.

Stessin, Lawrence: I spy becomes big business. *The New York Times Magazine,* November 28, 1965.

Stevens, Warren C.: The office spy system. *Modern Office Procedures,* August 1965.

Stewart, N.: When to tell your secrets; manager's need to know must be weighed against risks of carelessness. *Nation's Business,* June 1962, p. 62-64.

Stolen stocks and bonds for sale. *Changing Times, The Kiplinger Magazine,* February 1971, p. 28.

Suit by Frochim against Merck (France) for alleged patent infringement. *Chemical and Engineering News,* January 14, 1963.

Taking dead aim at industrial Mata Haris. *Factory,* March 1961, p. 164-165.

Talking too much can hurt you; three-year story of a patent fight. *Business Week,* January 28, 1956, p. 116.

Taylor, Robert L.: Secrecy and the dissemination of scientific information. *Industrial Security,* June 1971.
Thefts of U.S. firm's secrets pose questions on pricing and patents. *Wall Street Journal,* July 8, 1963.
Theft of drug secrets. *Washington Daily News,* June 21, 1962.
Time-industrial and business section: Bills would make trade secrets theft a crime. *Time,* March 26, 1965.
Tomassan, Robert: Chemist admits sale of secrets. *New York Times,* July 9, 1964.
Trademarks: identity crisis. *Newsweek,* July 3, 1972, p. 63. United States v. Mayfield, 65 Cr 143 (E.D.N.Y. 1965).
Uncovering your competitor's costs. *Chemical Engineering,* November 26, 1962.
Wackenhut, George R.: Business espionage. *Industrial Security,* February 1966, p. 4.
———: Is your department letting out secrets? *Chemical Purchasing,* May 1965, p. 10.
———: Legal remedies and recovery of stolen patent rights. *Industrial Security,* December 1964.
———: *Patents for Technical Personnel.* Ardmore, Advance House, 1956.
———: *The Corporate Patent Department.* Ardmore, Advance House, 1963.
Walsh, John H.: The business of espionage. *Industrial Security,* April 1971.
Weigel, William F.: Secrecy and invention agreements. *Research Management, VII:* no. 4, 1964.
What is a trade secret? *Machine Design,* March 1960, p. 105-107.
Williams, James H.: Trade secrets—too easy to steal. *Security World,* November 1964.
Yeager, P. B.: Company secrets have reasonable protection. *Nation's Business,* October 1959, p. 14.

UNPUBLISHED MATERIAL

Ingram, Dennis J.: Employee mobility and loss of trade secrets. Michigan State Univ. Unpublished graduate B Paper, 1965.
Otis, J. L.: Psychological espionage. Unpublished address to American Psychological Association, 1957.

BOOKS
COMPUTER

Brown, William F., Ph.D.: *Computer and Software Security,* ed. Greenlee, M. Blake, and Jacobson, Robert V. New York, AMR International, Inc., 1971.
Burck, G.: *The Computer Age.* New York, Harper & Row, 1965.
Farr, M. A. L., Chadwick, B., and Wong, K. K.: *Security for Computer Systems.* London, NCC Publications, 1972.
Hamilton, Peter: *Computer Security.* London, Associated Business Programs.
Hoyt, Douglas B.: *Computer Security Handbook.* Chairman, Computer Security Research Group. New York, 1973.
Hemphill, Charles F., Jr., and Hemphill, John M.: *Security Procedures for Computer Systems.* Homewood, Ill., Dow Jones-Irwin Co., 1973.
Krauss, Leonard I.: Safe, *Security Audit and Field Evaluation for Computer Facilities and Information Systems.* Amacom, 1972.
Leibholz, Stephen W., and Wilson, Louis D.: *Users' Guide to Computer Crime.* Radnor, Pa., Chilton Book Co., 1974.
Martin, James: *Security, Accuracy, and Privacy in Computer Systems.* Englewood Cliffs, N.J., Prentice-Hall, Inc., 1973.
Momboisse, R. M.: *Industrial Security for Strikes, Riots and Disasters.* Springfield, Thomas, 1968.
Parker, Donn B., Nycum, Susan, and Oura, S. Stephen: *Computer Abuse.* Stanford Research Institute, Calif., 1973.
U.S. Congress. House. Subcommittee of the Committee on Government Operations. The Computer and Invasion of Privacy. Hearing, 89th Cong., 2nd Sess., July 26, 27, and 28, 1966. Washington, Government Printing Office, 1966.
Van Tassel, Dennis: *Computer Security Management.* Englewood Cliffs, N.J., Prentice-Hall, 1972.
Rule, James: *Private Lives and Public Surveillance: Social Control in the Computer Age.* New York, Schocken Books.
Stessin, L., and Wit, I.: *The Disloyal Employee.* New York, Man & Manager, Inc., 1967.

PUBLICATIONS OF THE GOVERNMENT, LEARNED SOCIETIES, AND OTHER ORGANIZATIONS

Approach to reality in EDP audit. The NCR Company—Consultant and CPA Liaison, Dayton, Ohio, 1969.
Auditing Bank EDP Systems. Bank Administration Institute, 1968.
Automated data processing state governments.

Chicago, Public Administration Service, 1965.

Automation in government, 1963. Washington, D.C., American Society for Public Administration, 1963.

Baran, Paul: On distributed communications: IX Security, secrecy and tamper-free considerations. Rand Corporation, Santa Monica, Calif.

Berg, Philip J.: The plane facts about data center accidents. May 1970.

Bingham, H. W.: Security techniques for EDP of multi-level classified information. U.S. Government Research and Development Report, December 1965.

Brown, R.: Electronics at the point of sale. Ovum Ltd., London.

Brown, W. F.: Computer and software security. AMR International, New York, 1971.

Carroll, J. M., and McLelland, P. M.: Fast "infinite-key" privacy transformation for resource-sharing systems. Proceedings, Fall Joint Computer Conference, 1970, p. 223-230.

Checklist for evaluation of data processing systems. NCUMA Convention, 1968.

Cheek, R. C.: Fail-safe power and environmental facilities for a large computer installation. Proceedings, Fall Joint Computer Conference, 1968, p. 51-56.

Chubb-Mosler and Taylor. A guide to data center security.

Computer and cryptography. University of California, Berkeley, October 14, 1970.

Computer based management for information and control. New York American Management Association, 1963.

————: Computer crime. Proceedings, Fall Joint Computer Conference, 1970.

Condon, Richard A.: Information retrieval: a Parkinsonian approach. FMA, Inc.

————: The considerations of data security in a computer environment. IBM Publication.

————: Cryptographic techniques for computers. Proceedings, Spring Joint Computer Conference, 1969.

Electronic computer data processing equipment. National Fire Protection Association, Boston, 1972.

Freedman, Samuel B.: Vital records protection through the use of microfilms. Bell and Howell Co., December 1963.

Garrison, W. A., and Ramamoorthy, C. V.: Privacy and security in data banks. U.S. Government Research and Development Report, 1970.

Glaser, E. L.: A brief description of privacy measures in the multics operating system. Proceedings, Spring Joint Computer Conference, 1967, p. 303-304.

Hawkins, David H.: How to safeguard your software! *Computer Decisions,* June 1972.

Humphrey, S. M.: Impact of computer developments. Proceedings of the Association of Computing Machinery, 1959, p. 17.

Hunter, H. E., and Conway, J.: Lawrence (MA)—use of AVCO data analysis and prediction techniques (ADAPT)—final report. 1972.

IBM. *The Considerations of Data Security in a Computer Environment.*

IBM. *The Considerations of Physical Security in a Computer Environment.*

Individual liberties the administration of criminal justice. Temp. State Comm. Constit. Conv., Albany, N.Y., March 16, 1967.

INFO: Information network & file organization. Dade County Computer Center, Miami, Fla., 1967.

Internal Audit of EDP Systems Institute of Internal Auditors, 170 Broadway, New York, N.Y.

Jeffery, Seymour: Executive guide to computer security. U.S. Dept. of Commerce National Bureau of Standards and Association for Computing Machinery, 16 pages.

Krauss, Leonard I.: *Security Audit and Field Evaluation for Computer Facilities and Information Systems.* Amacom, U.S.A., 1972.

McGuire, E. P.: Target for terrorists. *Conference Board Record,* August 1971.

Molho, Lee M.: Hardware aspects of secure computing. Proceedings of the Spring Joint Computer Conference, 1970.

Myers, Charles A.: Some implications of computers for management. Industrial Relations Research Association, December 28, 1966.

————: The nature of computer related crime. Sanford Research Institute, Menlo Park, Calif., May 1972.

Niblett, B.: Computer security and the law. *Computerguard Black File,* no. 1, May 1972.

Oppenheimer, G., and Clancy, K. P.: Considerations for software protection and recovery from hardware failures in a multi-access, multi-programming, single processor system. Proceedings, Fall Joint Computer Conference, 1968, p. 29-37.

Peat, Marwick, Mitchell & Co. Protecting management's computer investment. Accounting and Auditing Information Letter, 1969.

Peters, B.: Security considerations in a multiprogrammed computer system. Proceedings, Spring Joint Computer Conference, 1967, p. 283-286.

Petersen, H. E., and Turn, R.: Security of computerised information systems. U.S. Government Research and Development Report, July 1970.

Proceedings of American Society for Industrial Security; computer security panel. Washington, D.C., September 18, 1969.

Protection of essential records—the methods and equipment. Southern California Chapter of American Records Management Association, May 6, 1963.

Rowe, B. C.: Privacy, computers and you. Proceedings, NCCL Workshop on the Data Bank Society, November 1970.

———: Security controls in the ADEPT-50 time-sharing system. Proceedings, Fall Joint Computer Conference, 1969, p. 119-133.

———: System implications of information privacy. Proceedings, Spring, Joint Computer Conference, 1967, p. 291-300.

Ware, W. L. H.: Security and privacy in computer systems. Proceedings, Spring Joint Computer Conference, 1967, p. 279-282.

Weiss, Harold: Reducing the risk of destruction. *Proceedings*, 1969.

PERIODICALS

Accounting system uses "lock and key" to prevent payment default, copying. *Computerworld*, May 20, 1970.

Adams, Donald, and Mullarkey, John F.: A survey of audit software. *The Journal of Accountancy*, September 1972.

ADAPSO speaks out on privacy. *Computer World, 9:* no. 44, 12, November 4, 1970.

Adelson, A.: Crooked operators to embezzle money from companies. *Wall Street Journal*, April 5, 1968.

Adelson, A.: Embezzlement by computer. *Security World*, September 1965.

Adelson, A.: Whir, blink-jackpot. *The Wall Street Journal*, April 5, 1968.

Aftermath of Sir George Williams University computer center destruction in February 1969. *Computer World, 4:* no. 17, 1, 6-7, April 29, 1970.

Agency collects bills previously paid. *Computer World, 5:* no. 9, March 3, 1971.

Alert program spots credit ring. *Computer World, 4:* no. 49, 1, December 9, 1970.

Allen, Brandt: Danger ahead! Safeguard your computer. *Harvard Business Review, 46:*97-101, November-December 1968.

Allen, Brandt R.: Computer security. *Data Management*, January 1972 and February 1972.

Allen, B. R.: Computer fraud. *Financial Executive*, May 1971.

All's well that ends well. *Computer World, 4:* no. 5, 4, December 16, 1970.

American Express sued for $25,000. *Computer World, 4:* no. 50, 4, December 16, 1970.

Anderson, A. F.: Records protection in the age of EDP. *The Office,* June 1966.

Antitrust suit charges rearrangement of data. *Computer World, 5:* no. 2, 4, March 24, 1971.

Anti-war protestors erase 1,000 DOW tapes. *Computer World, 3:* no. 48, 1, December 3, 1969.

American society for industrial security: DOD industrial security clearance program. *Industrial Security*, December 1969, p. 9-16.

Amir, M.: Computer embezzlement—prevention and control. *The Computer Bulletin,* November 1971, p. 387-400.

Analyze computer use for proper protection. *Business Insurance,* July 1971, p. 34.

Are your EDP operations insured? *Modern Office Procedures,* May 1966.

Army data-bank curtailed, but legislation lacking. *Computer World, 5:* no. 5, 6, February 1971.

Astor, S. D.: An investigator talks of embezzlement and robbery. *The Office,* September 1971.

Auditing EDP . . . by computer. *Administrative Management,* December 1970.

The auditor encounters electronic data processing. *Price Waterhouse and Co.,* International Business Machines Corp., 1968.

Auditor must be involved in DP, ACM speaker says. *Computer World, 5:* no. 8, 6, February 24, 1971.

Babcock, J. D.: A brief description of privacy measures in the time-sharing system. Proceedings, Spring Joint Computer Conference, 1967.

Bachi, R., and Baron, R.: Confidentiality problems related to data-banks. *IAG Quarterly Journal, 3:* no. 1, 43-68.

Background information provided on data

banks. *Computer World, 4:* no. 52, 10A, December 30, 1970.

Baen, J. M.: Protecting the company's marketing secrets. *Security World,* June 1968.

Baird, Lindsay L., Jr.: An analytical approach to identifying computer vulnerability. *Security Management,* May 1974, p. 6-11.

Ball, Ian: America's computer snoopers. *Daily Telegraph,* April 24, 1971, p. 12.

Baker, H. R., Leach, P. B., Singleterry, C. R., and Zisman, W. A.: Cleaning by surface displacement of water and oils. *Industrial and Engineering Chemistry,* June 1967.

Banks spending for computer security in the "Wild West." *Information Week, 10:* no. 41, October 12, 1970.

Banzhaf, John F., III: When your computer needs a lawyer. *Communications of the ACM,* August 1968.

Bates, A.: Privacy—a useful concept. *Social Forces,* 1964.

Bates, Robert E.: Auditing the advanced computer systems. *Management Accounting,* June 1970.

Bates, W. S.: Security of computer based information systems. *Datamation, 16:* no. 5, 60-65, May 1970.

Behrens, Carl: Computer and security. *Science News,* June 3, 1967.

Bergamini, D.: Government by computers. *The Reporter, XXV:* no. 3, 21-28, August 17, 1961.

Beardsley, Charles W.: Is your computer insecure? *Spectrum,* January 1972, p. 67.

Bigelow, Robert P.: Some legal aspects of commercial remote access computer services. *Datamation,* August 1969.

Berson, T. A.: Sleuthing your data center. *Computer Decisions, 3:* no. 6, 6, June 1971.

Best data sabotage plan wins. *Computer World, 4:* no. 41, October 14, 1970.

Bevans, M. J.: Internal ID systems aid security. *Administrative Management,* April 1969.

Bigelow, R. P.: Automation and the law. *Boston Bar Journal, VI:*31-33, September 1962.

Bigelow, R. P.: Contract caveats. *Computer World,* September 15, 1970.

———: Legal aspects of proprietary software. *Datamation,* October 1968.

———: Legal and security issues posed by computer utilities. *Harvard Business Review, 45:* no. 5, September-October 1967.

Binns, James: Why man to man defense for EDP audit control? *Data Management,* October 1969.

Bill would limit credit card practices. *Computer World, 5:*2, February 3, 1971.

Bisco, R. L.: Social science data archives a review of developments. *American Political Science Review, LX:*93-109, March 1966.

Blackouts inevitable. *Computer World, 5:* no. 8, 10, February 24, 1971.

Blumenthal, F.: Do you love your computer? Keep it warm. *Parade,* April 4, 1971.

Boehm, G. A. W.: The next generation of computers. *Fortune, LX:*132-135, March 1959.

Bomb demolishes army computer complex. *Computer World, 4:* no. 35, 1-4, September 2, 1970.

Boni, Gregory M.: Impact of electronic data processing on auditing. *The Journal of Accountancy,* September 1963.

Bootleg bribe buys computer time. *Computer World,* September 30, 1970.

Boyer, Jerome T.: Fire protection of computers. *Industrial Security,* October 1971, p. 27.

Braithewaite, Chris: Computers destroy personal privacy. *Canadian Security Gazette,* October 1970.

Brandon, Dick: Does your contract really protect you? *Computer Decisions,* December 1971.

Brictson, R. C.: Computers and privacy implications of a management tool. *SDC,* March 14, 1968.

Bride, Edward J.: Audit trails lost in computerization. *Computer World,* April 29, 1970.

———: Malpractice charged to CDC data center. *Computer World,* July 7, 1971.

———: NAS warns of despair in privacy invasion fight. *Computer World,* October 25, 1972.

———: DP center invaded. *Computer World,* July 15, 1970.

———: Firms offer card, key systems for data security. *Computer World,* August 26, 1970.

———: First program patent issued under new rule. *Computer World,* June 24, 1970.

Briggs, P. L.: Students demolish computer center. *Computer World,* February 26, 1969.

Brown, Patrick: Computers—controls and fraud prevention. *Canadian Security Gazette Magazine,* October 1971.

Browne, Peter S.: Computer security, a survey—1972. *Data Processor,* August 1972.

Browne, P. S.: Computer security—a risk management approach. *Security Register, 1:* no. 1, 22-25, 61, 63, November 1973.

Buckley, J. L.: Computers, automation, and security. *Law and Order,* March 1965.

———: The future of computers in security and law enforcement. (Pts. 1 and 2) *Law* and *Order,* August 1965, p. 36-38; September 1965, p. 48-50, 52.

Buckley, J. L.: Central index file—Europe. *Law and Order,* February 1964, p. 50-53.

Burleigh, William P., Jr.: Retrieval systems and micro-records. American Records Management Association.

Burns takes security seriously. *Computer World,* 5: no. 2, 14, January 13, 1971.

Burt, K. H.: Computer center security: Protecting the Achilles heel. *The Magazine of Bank Administration,* April 1970.

Businesses not security conscious. *Computer World,* 5: no. 19, 1, May 12, 1971.

Calculated computer errors manipulate three banks' security, million lost. *Computer World,* 4: no. 12, March 23, 1970.

California Earthquake. *Computers and Automation,* 20: no. 5, 33, May 1971.

Campbell, A., and Woods, A.: Computers and freedom. *Law and Computer Technology,* June 1969.

Canada builds debtor data bank. *Computer World,* 4: no. 52, 1, December 30, 1970.

Canadian data banks concern grows. *Computer World,* 4: no. 50, 11, December 16, 1970.

Cantor, Lon: Electronic intrusion alarms. *Electronics World,* 80: no. 3 and 4, September and October 1968.

Carley, W. M.: Computer companies are hauled into court by flurry of lawsuits. *Wall Street Journal,* November 30, 1970.

Carlson, P.: Bank protects its memory. *Banking,* April 1971.

Carmichael, Dr. D. R.: Fraud in EDP systems. *The Internal Auditor,* May-June 1969.

Carney, P. L.: Police say Mafia's DP use impedes crime prevention. *Computer World,* December 2, 1970.

Carr, Peter F.: Most DP centers lax in arranging backup facilities. *Computer World,* July 15, 1970.

———: NBS says DP center noise may cause loss of hearing. *Computer World,* July 1, 1970.

———: Limiting access to centers called a major problem. *Computer World,* June 24, 1970, p. 2.

———: Poor security leaves DP facilities ripe for sabotage. *Computer World,* June 17, 1970, p. 1.

Chandler, Donald B.: Solving the "invisible record" problem. *Price Waterhouse Review,* June 1970.

Chesson, F. W.: Computers and cryptology. *Datamation,* January 1973.

Christiansen, Theodore W.: Data processing systems controls. *Journal of System Management,* June 1969.

Chu, A. L. C.: Computer security, the corporate Achilles heel. *Business Automation,* 18: no. 3, 32, February 1971.

———: The need to know—the right to privacy. *Business Automation,* 18: no. 8, 30, June 1971.

Chu, Albert L. C.: The need to know . . . the right to privacy. *Business Automation,* June 1, 1971, p. 24.

City DPERS seek crisis funds. *Computer World,* 5: no. 14, 1, April 7.

Clive de Paula, C.: Problems of auditing data: The external auditor and computers. *The Computer Journal,* no. 3, 1960.

Company security practices. *Conference Board Record,* October 1967.

Computers and auditing: a conference report. *Datamation,* July 15, 1970.

Computer bomb damage studied. *Computer World,* September 9, 1970.

Computer "capers" herald new crime wave of embezzlement. *The National Underwriter,* August 20, 1971, p. 1.

Computer center occupied for bargaining position. *Computer World,* April 27, 1970, p. 1.

Computer companies are hauled into court by flurry of lawsuits. *Wall Street Journal,* CLXXVI: no. 107, 1, November 30, 1970.

Computer communication security. *Industrial Security,* June 1970, p. 24.

Computer con men are on the rise. *Charlotte Observer,* August 16, 1970, p. 1.

Computer designs tamperproof computer. *Data Management,* September 1969, p. 55.

Computer file privacy in Great Britain. *The Office,* September 1971, p. 41-44.

Computer fire detection systems. Manual of Prytronics, Inc., Cedar Knolls, N.J., 1970.

Computers: How do you keep the secrets secret? *Los Angeles Times,* November 23, 1971, p. 1.

Computers in LA Feb. 9, nothing earthshaking. *Datamation,* 17: no. 6, 56, March 15, 1971.

Computer insurance. *The Accountant,* April 6, 1968.

Computer plays big role in defrauding welfare unit. *Computer World,* 3: no. 14, 7, April 9, 1969.

Computer plays big role in defrauding welfare unit. *Computer World,* 4: no. 40, October 7, 1970.

Computer power in small packages. *Electrical World,* 173: no. 1, 51, January 5, 1970.

Computers require special protection. Newark, N.J., *Evening News,* May 19, 1970, p. 1.

Computer security: Backup and recovery methods. *EDP Analyzer,* January 1972.

Computer security. *Data Processing Digest, 16:* no. 10, 35, October 1970.

Computer security is sensitive area. *Industry Week, 167:*13, October 5, 1970.

Computers show resiliency after earthquake. *Computer World, 5:* no. 7, February 17, 1971.

Computer security. *Industrial Security,* December 1969.

Computer systems security has received a boost by the introduction of a new key reader assembly. *Security Systems Digest,* August 5, 1970, p. 4.

Computers takes rap in securities swindle. *Datamation,* August 1968, p. 111.

Computer tape: Cleanliness counts. *Administrative Management,* May 1969, p. 32.

Congress asked to probe computer billing systems. *Computer World,* October 29, 1969.

Congdon, Frank: Operation madcap—method for automatic document control and processing—panelist. *Industrial Security,* October 1962, p. 37.

Considerations of data security in a computer environment, The, *IBM,* 1969, p. 36.

Considerations of physical security in a computer environment. *IBM,* 1972, p. 37.

Conway, R. W., Maxwell, W. L., and Morgan, H. L.: On the implementation of security measures in information systems. *Comm. ACM, 15:* no. 4, 211-220, April 1972.

Cook, A. D.: EDP defends against disaster. *Electronic News,* December 29, 1969, p. 33.

Cooperation urged for national data bank. *Computer World, 4:* no. 42, October 21, 1970.

Costlier protection hits campus centers. *Computer World, 4:* no. 31, August 5, 1970.

Coultas, F. W.: A compact high-speed digital data link. *Security Gazette,* May 1975, p. 164-168.

Crime information data storage system tested. *Security World, 2:* no. 1, January 1965.

Cross, Richard F.: Tighter security for computers. *Industrial Security,* August 1971.

———: Cryptographic techniques for computers: Substitution methods. *Information and Storage Retrieval,* June 1970.

———: Information security in a computer environment. *Computers and Automation,* July 1969.

———: A contingency plan for catastrophe. *Datamation, 17:* no. 13, 30, July 1, 1971.

Daenzer, B. J.: Fact-finding techniques in risk analysis. *American Management Association,* New York, 1970.

Dahl, A., and Patrick, R. L.: Voting systems. *Datamation, 16:* no. 5, 81-82, May 1970.

Dansinger, Sheldon J.: Proprietary protection of computer programs. *Computers and Automation,* February 1968.

———: Dansinger cites ways to ward off computer embezzlement problems. *Computer World,* June 19, 1968, p. 4.

———: Danger of total corporate amnesia. *Financial Executive,* June 1969.

———: Embezzling primer. *Computers and Automation,* 1967.

Darby, Edwin: Fighting computer thieves. *Chicago Sun-Times,* June 24, 1970, p. 74.

Data bank hearings open to call for control agency. *Computer World, 5:* no. 9, 1, March 3, 1971.

Data processing errors and omissions insurance. *Banking,* April 1971, p. 38, 76.

Data processing may receive scrutiny at FTC hearing on credit card billing. *Computer World, 4:* no. 42, October 21, 1970.

Data processing security. *Security Systems Digest,* July 8, 1970, p. 5.

Data security in the CDB. *EDP Analyzer,* May 1970.

Data security in the CBD (i.e. corporate database). *EDP Analyzer,* May 1970.

Davidson, T. A.: Computer information privacy. *The Office,* August 1969, p. 10-12.

Davis, A. G.: Security of the computer center. *Industrial Security, 15:* no. 2, 20, April 1971.

Davis, F.: What do we mean by "right to privacy." *South Dakota Law Review, IV:*1-24, Spring, 1959.

Davis, Gordon B.: Auditing and EDP. *American Institute of Certified Public Accountants,* 1968.

Davis, Morton S.: Service bureaus need to improve data security. *Computer World,* August 26, 1970.

DeLair, W. E.: Security responsibilities of a time-sharing service company. *Transdata Corporation,* October 25, 1969.

Dennis, J. B.: A position paper on computing and communications. *Comm. ACM, 11:* no. 5, 370-377, May 1968.

Dennis, Robert L.: Security in the computer environment. *System Development Corporation,* 1968.

Detroit's canvassers axe punch card vote. *Computer World, 4:* no. 47, November 25, 1970.

Devitt, R. G.: Cut expenses by taking care of

your tape. *Computer Decisions,* October 1970, p. 42.

Diamond, T. D., and Krallinger, J. C.: Controls and audit trails for real-time systems. *Internal Auditor,* November-December 1968.

er World, September 18, 1968.

Donovan, Robert D.: An automated document declassification system. *Journal of the National Classification Management Society,* 1966, p. 138-143.

DP center dig out in hurricane's wake. *Computer World,* August 19, 1970.

DP centers feel the brunt of hurricane's fury. *Computer World,* 4: no. 32, 1, August 12, 1970.

DP center "invaded." *Computer World,* 4: no. 28, 1, 4, July 15, 1970.

DP figures in bank loss of $128,000. *Computer World,* 5: no. 5, 1, February 3, 1971.

DP centers find new fire extinguishing agent system. *Computer World,* 5: no. 10, 6, March 10, 1971.

DP fraud—mum's the word. *Computer World,* 5: no. 2, 6, March 24, 1971.

Duggan, Michael A.: Software protection. *Computer World,* June 1969, p. 1.

———: Software protection. *Datamation,* June 1969.

Dutages rated a leading cause. *Computer World,* 5: no. 19, 2, May 12, 1971.

Dwyer, John N.: Auditing computer operations. *Banks Monthly,* May 15, 1969.

Electric brownouts add unforeseen risks for company using computer. *Business Insurance,* September 13, 1971, p. 16.

Electronic security in the computer room. *Banking,* 62:86, May 1970.

Employees accused of illegal computer use. *Datamation,* December 1967, p. 78.

Employee charged in program theft. *Computer World,* 5: no. 12, 1, March 10, 1971.

1970 environment and security supplement. *Various Computer World,* 4: no. 34, August 26, 1970.

Ernst, M. L.: Management, the computer, and society. *Computers and Automation,* September 1971, p. 8-14.

———: What else will computers do to us. *Wall Street Journal,* October 21, 1970.

Errors and omissions coverage vital for those who process other's data. *Business Insurance,* July 5, 1971, p. 38.

Explosion damages computer center, injures three at University of Kansas. *Lowell Sun,* no date.

Fair credit bill would protect against false billing. *Computer World,* 4: no. 32, August 12, 1970.

Farmer, J., Springer, C., and Strumwasser, M. J.: Cheating the vote counting systems. *Datamation,* 16: no. 5, 76-80, May 1970.

Fast circuits may be more prone to failure from everyday shock. *Computer World,* 5: no. 3, 1, January 20, 1971.

Fatal blazes in modern office skyscrapers stir charges of unsafe building practices. *Wall Street Journal,* December 8, 1970, p. 40.

Fazar, W.: Federal information communities the systems approach. American Political Science Association, New York City, September 6-10, 1966.

FBI accuses youth of tapping T/S service, copying data files. *Computer World,* 4: no. 30, 1, July 29, 1970.

FBI's Hoover criticizes CQ's NCIC coverage. *Computer World,* 4: no. 43, 10, October 28, 1970.

FBI to add "rap" file. *Computer World,* 4: no. 51, December 23, 1970.

FBI tracks wandering Wang. *Business Automation,* April 1969.

Federal employee receives $27,054 courtesy of "computer assisted error." *Computer World,* 4: no. 41, October 14, 1970.

Felsman, Robert A., Chrisman, Thomas L., Hope, Henry W., Holder, John E., and Medlock, V. Bryan, Jr.: Computer program protection. *Texas Bar Journal,* January 1971, p. 33, 40, 53-61.

Firebombs damage a computer center. *The Office,* August 1970, p. 42-43.

Fire defenses for computer rooms. *Occupational Hazards,* December 1968.

Fire hazards in new buildings. *The Office,* October, 1970.

Firms offer card, key systems for data security. *Computer World,* August 26, 1970.

Firms sue in mailing list theft. *Computer World,* 4: no. 27, 1-2, July 8, 1970.

First program patent issued under new rule. *Computer World,* June 24, 1970.

Fitzpatrick, Robert J.: The influence of EDP on internal control. *The Controller,* March 1961, p. 123.

Fortifying your business security. *The Office,* August 1969.

Foster, C. C.: Data banks: A position paper. *Computers and Automation,* March 1971, p. 28

Fowler, G.: Offices in city to improve security. *New York Times,* October 17, 1971.

Freed, R. N.: *Materials and Cases on Comput-*

ers and Law, 3rd ed. Boston, Widett & Kruger, 1971.

Fraud in EDP systems. *N.Y. Certified Public Accountant,* January 1970.

———: Computer fraud—a management trap. *Business Horizons,* June 1969.

Friedman, T. D.: The authorization problem in shared files. *IBM Systems Journal,* no. 4, p. 259-280, 1970.

Frost, H. M.: Fidelity and dishonesty—exposures and coverages. *The Growth of Corporate Insurance Management,* Bulletin 58, American Management Association, New York, 1964.

Gallati, R. R. J.: Criminal justice systems and the right to privacy. *Public Automation,* July 1967.

Garland, Robert F.: Computer programs—control and security. *Management Accounting,* December 1966.

Gellman, H. S.: Using the computer to steal. *Computers and Automation, 20:* no. 4, 16, April 1971.

Gellman, Harvey S.: Using the computer to steal. *Computers and Automation,* April 1971, p. 16-18.

Gibson, Willeam S.: Protection of EDP, equipment and control in its operation from the security standpoint. *The Interpreter,* June 1971.

Gill, W. A.: Federal-state-local relationships in data processing. *New York University,* April 2, 1966.

Godbout, William: Computer theft by computer. *Security World,* May 1971, p. 22-24.

Goodman, John V.: Auditing magnetic tape systems. *The Computer Journal,* July 1964.

Government offices lose things too. *The Office,* August 1970.

Gower, R. G.: Computer fire protection. *Data processing,* March-April 1972, p. 140-141.

Graham, Robert M.: Protection in an information processing utility. *Communications ACM,* May 1968.

Grant, C. B. S.: Will students wreck your computer center? *Data Processing,* May 1969, p. 62-63.

Great game of corporate espionage, the. *Dun's,* October 1970.

Greenberg, David: Lawsuits against computer firms. *Administrative Management,* April 1971, p. 59-60.

Greenberg, Harold: Privacy and security. *Data Systems News,* August 1969, p. 8-9.

Grenier, E. J., Jr.: Computers and privacy—a proposal for self-regulation. *Proc. of ACM,* October 1969, p. 231-269.

Ground data corporation says it is now marketing two new communications systems scramblers. *Security Systems Digest,* July 22, 1970, p. 7.

Gruenberger, Fred: Program testing and validating. *Datamation,* July 1968.

GSA tightens office building security. *The Office, 73:* no. 2, 32, February 1971.

Guard that computer. *Nation's Business,* April 1971, p. 84-86.

Gudie, A. H.: Are you ready for EDP? *The Electrical Distributor,* June 1971, p. 24.

Guise, R. F., Jr.: File security. *Data Systems News, 10:* no. 11, November 1969.

Gunton, A.: Recovery procedures for direct access commercial systems. *Computer Journal, 13:* no. 2, 123-126, May 1970.

Hallinan, Arthur: Internal audit of a computer disaster plan. *The Internal Auditor,* November-December 1970.

Halsbury, Earl of: Lord Halsbury speaks on computer privacy. *Computers and Automation, 19:* no. 7, 42-43, July 1970.

Halting the electronic hijacker. *Management Review,* November 1968, p. 45-50.

Halting the electronic hijacker. *Modern Office Procedures,* September 1968.

Hanlon, Joseph: Anti-war protestors erase 1,000 Dow tapes. *Computer World,* December 3, 1969, p. 1.

———: Ten students convicted in 1969 computer center burning. *Computer World,* April 29, 1970, p. 1.

———: FTC warns: Illegal bills draw fines. *Computer World,* July 7, 1971.

Harris, Richard D.: EDP Systems Audits. *Data Management,* Sept. 1971.

Harrison, William L.: Program Testing. *Data Management,* December 1969.

Has the mafia permeated the computer community? *Computer World,* August 18, 1968, September 11, 1968.

Hecksher, A.: The invasion of privacy—the reshaping of privacy. *American Scholar,* 1959.

Heeschen, Paul E.: Auditing data processing administration activities, operational auditing applied to EDP. *The Internal Auditor,* November-December 1970.

Hemphill, C. F., Jr.: Preventing damage to EDP systems. *Administrative Management,* April 1969.

Highlights of a security plan devised by experts. *Occupational Hazards,* March 1969.

Hiles, Richard A.: Paper Shredders. *Modern Office Procedures,* February 1963.

Hill, O. A., Jr.: The role of the auditor with

respect to internal control and fraud. *The Internal Auditor*, May-June 1968.

Hines, Harold H., Jr.: Letter to the Editor. *Harvard Business Review*, May-June 1969.

Hirsch, P.: The world's biggest data bank. *Datamation, 16:* no. 5, 66-73, May 1970.

Hirschfield, R. A.: Security in on-line systems—a primer for management. *Computers and Automation, 20:* no. 9, September 1971.

Hoffman, Lance J.: Computers and privacy: A survey. *Computing Surveys*, June 1969.

———: The formulary model for access control and privacy in computer systems. *SLAC Report No. 117*, May 1970.

———: Security and Privacy in Computer Systems. Los Angeles, Melville Publishing Co., 1973.

———, and Miller, W. F.: Getting a personal dossier from a statistical data bank. *Datamation*, May 1970.

Holmes, F. W.: Software security. American Management Association, Briefing Session #6373-60, April 15, 1970.

Horton, F.: Privacy safeguards urged. *EDP Weekly, 11:* no. 23, 3, September 21, 1970.

Housing computers. *Administrative Management*, February 1971, p. 34-35.

How bad guys thwart computers. *The Office*, September 1970, p. 36-39.

Howes, Paul R.: EDP security: Is your guard up? *Management Review*, July 1971, p. 29-32.

———: EDP security: Is your guard up? *Price Waterhouse Review*, Spring, 1971.

How I steal company secrets. *Business Management*, October 1965.

How paper shredders protect business data. *The Office*, September 1971, p. 58.

How safe are your business secrets? *Business Management*, March 1968.

How to find out what your competitors are up to. *Management Methods*, July 1960, p. 50-52.

How to make sure nobody knows your business. *Modern Office Procedures*, July 1970.

How to murder a computer. *Chicago Seed*, 1970.

How to protect against the million dollar racket. *Modern Office Procedures*, March 1968.

How vulnerable is the computer system. *ADP, The Diebold Group, Inc., 15:* no. 5, March 8, 1971.

How your company can thwart a spy. *Business Management*, October 1965.

How secure are your computers? *Industry Week*, September 22, 1975, p. 40-43.

Huggins, P.: Employee charged in program theft. *Computer World*, March 10, 1971.

———: Programmer thankful for "bug" during computer center bombing. *Computer World*, May 27, 1970.

———: Computer plays big role in defrauding welfare unit. *Computer World*, October 7, 1970.

———: Rebuilt Fresno state DP center follows tight security. *Computer World*, July 8, 1970.

Hurtado, Corydon D.: General audit techniques in data processing. *Data Management*, October 1970.

Hutt, A. E.: Backup and recovery in a real-time system. *Real Time, Infotech State of the Art*, Report 3, 1971.

IBM puts Volkswagon back on the road three days after a total-loss fire. *Wall Street Journal*, April 21, 1971. (Advertisement)

Individual responsibility. *Data Systems News*, February 1969.

Information security in a computer environment. *Computers and Automation*, July 1969.

In-office ID systems aid company security. *Administrative Management*, April 1969, p. 24-28.

Inside Eastern's data center. *Business Week*, February 5, 1972, p. 60-61.

Insuring EDP departments against disaster. *Supervisory Management*, September 1966, p. 44-46.

Introduction to CODE. *Economatics*, Pasadena, Calif., February 13, 1969.

Invasion of privacy, the. *Computers and Automation, 19:* no. 4, 6-29, April 1970.

Invasion of privacy. *University of Pittsburgh Law Review, XIX:*98-111, Fall, 1957.

Is system "theft" scandal or rumor? *Computer World*, May 29, 1968.

Immel, A. R.: Sabotage, accidents and fraud cause woes for computer centers. *Wall Street Journal*, March 22, 1971.

Jackson, W. A.: Fire protection systems. *Data Processing*, March-April 1969.

Jacobs, Morton C.: Patent protection of computer programs. *Communications of the ACM*, October 1964.

Jacobson, R. V.: Providing data security. *Automation*, June 1970, p. 85-90.

———: Planning for backup facilities. *Computer Services, 2:* no. 3, 22-29, May/June 1970.

———: Providing security protection for computer files. *Best's Review*, May 1970, p. 42-44.

———: Providing security protection for computer files. *Best's Review, Property Liability Insurance Edition*, June 1970.

Jacobsen, Robert V.: Cornerstones for computer security. *Security Register*, January-February 1974, p. 32-35.

James, I. T.: Fraud and the computer. *Data Systems*, July 1969.

Jasper, David P.: A discussion of checkpoint/restart. *Software Age*, October 1969.

John, Richard C., and Nissen, Thomas J.: Evaluating internal control in EDP audits. *The Journal of Accountancy*, February 1970.

Johnson, D.: Control and prevention of thefts of proprietary information. *Industrial Security*, February 1968.

Johnson, Carl B.: Protection primer for EDP records. *Banking*, December 1969, p. 85-86.

———: Six types of firms and the data each needs. *American Business*, May 1959, p. 7-9.

Journal warns of dishonest "computer-operators." *Computer World*, 2: no. 16, April 17, 1968.

Jury gives damages against IBM in user warranty case. *Computer World*, April 10, 1968.

Just plain grabbing is becoming old hat to securities thieves. *Wall Street Journal*, October 26, 1970.

Kahn, David: Modern cryptology. *Scientific American*, 215: no. 1, 38-46, July 1966.

Karst, K. L.: Legal control over the accuracy and accessibility of stored personal data. *Law and Contemporary Problems*, 31:342-376, Spring, 1966.

Kaufman, Felix: Data systems that cross company boundaries. *Harvard Business Review*, 44:141-145, 148-152, 1966.

Keebler, Jim: Speedy computer people. *Automation*, May 1971, p. 39.

———: Keeping confidential information confidential. *Systems Management*, February 1969, p. 14-15.

Kessler, J. N.: New electronics joins national war on crime. *Electronic Design*, February 18, 1971.

Keys, Edward G.: The auditors role in new systems development. *The Internal Auditor*, September-October 1971.

King, D. B.: Electronic surveillance and constitutional rights—some recent developments and observations. *George Washington Law Review*, XXX:240-268, October 1964.

Kingsbury, Arthur: Guidelines for security education; introduction to computer security. *Security Management*, 18: no. 2, 47, May 1974.

Kjeldaas, P. M.: Security in software. *IAG Quarterly Journal*, 3:25-26.

Kleinschrod, W. A.: What to do about bomb scares. *Administrative Management*, August 1970.

Koefod, Curtis F.: The handling and storage of computer tape. *Data Processing*, July 1969, p. 20-23.

Krauss, L. I.: Computer-based management information systems. *American Management Association*, New York, 1970.

Lachter, Lewis E.: Preventing business-secret espionage. *Administrative Management*, December 1965.

L. A. justice systems safeguards called adequate. *Computer World*, February 27, 1974, p. 32.

Lampson, B. W.: Dynamic protection structures. Proceedings, Fall Joint Computer Conference, 1969, p. 27-38.

Lange, D.: Employees called biggest risk at centers. *Computer World*, June 23, 1971.

Lang, William, Jr.: Backup files are a must. *Administrative Management*, October 1971, p. 55.

———: Get it in writing. *Administrative Management*, August 1971, p. 12.

Lauren, R.: Reliability of data banks. *Datamation*, 16: no. 5, 88-89, May 1970.

Laver, Murray: User's influence on computer systems design. *Datamation*, October 1969, p. 107-113.

Lawlor, Reed C.: Copyright aspects of computer usage. *Communications of the ACM*, October 1964.

———: Lawyer's warnings: Let customer beware in computer contracting. *Computer World*, January 13, 1971.

Leaky center may lose vendor support. *Computer World*, 4: no. 40, October 7, 1970.

Lear, J.: Whither personal privacy. *Saturday Review*, July 23, 1966.

Leavitt, Don: Cipher/1 designed for assurance of total file privacy. *Computer World*, June 10, 1970.

Lefer, H.: How to shield your office against crime. *Modern Office Procedures*, 15: no. 4, 21-29, April 1970.

Legal aspects of commercial remote access computer services, some. *Datamation*, August 1969.

Legal protection of computer programs. *Harvard Business Review*, March-April 1965.

Let customer beware in computer contract. *Computer World,* 5: no. 2, 1, January 13, 1971.

Levine, R. A.: How to protect your EDP records. *New York Certified Public Accountant,* May 1969, p. 353-356.

Lewis, William F.: Auditing concepts and on-line computer systems. *The Arthur Young Journal,* Winter/Spring 1971.

———: Auditing on-line computer systems. *The Journal of Accountancy,* October 1971.

Light plane lights ADR's fire. *Datamation,* January 1970.

Linde, R. R., Weissman, C., and Fox, C. E.: The ADEPT-50 time-sharing system. Proceedings, Fall Joint Computer Conference, 1969, p. 39-50.

Limiting access to centers called a major problem. *Computer World,* 4: no. 25, June 24, 1970.

Lobel, Jerome: Auditing in the new systems environment. *The Journal of Accountancy,* September 1971.

Looking at fire hazards. *Fire Journal,* May 1970.

Lutter, Frederick H.: Keeping the computer secure. *Administrative Management,* October 1970, p. 10-14.

———: Protecting the database. *Administrative Management,* 31:10, November 1970.

MacDonald M. B., Jr., and Brown, J. K.: Company security practices. *The Conference Board Record,* October 1967, p. 40-47.

Machines get a charge from nylon panties. *Computer World,* 5: no. 8, 4, February 24, 1970.

Machine says no, police don't go: Conflicting addresses cited. *Computer World,* January 13, 1971.

Mackail, J. J.: Planning and organizing for a centralized document control system. *Industrial Security,* January 1962.

Macy, J. W., Jr.: Automated government. *Saturday Review,* August 23, 1966, p. 21.

Madists invade IBM French office. *Computer World,* 5: no. 8, 4, February 24, 1971.

Man and the computer. American Association for Advancement of Science, 1962.

Mandell, M.: Computer scare talk. *New York Times,* May 9, 1971, p. F3.

"Manipulation" of Penn Central computers cited in boxcar theft. *Computer World,* 5: no. 13, 1, March 31, 1971.

Manufacturer has special responsibility for security safeguard, says FCC's Lee. *Computer World,* 4: no. 50, 7, December 16, 1970.

Manufacturer safeguards for data called inadequate. *Computer World,* 4: no. 45, 3, November 11, 1970.

Manufacturers seen responsible for providing adequate data security. *Computer Decisions,* 3: no. 5, 6, May 1971.

Martin, B. A.: Guidelines for contracting for computer related services. *Computers and Automation,* April 1970.

McCarthy, J.: Information. *Scientific American,* CCXV: no. 3, 65-72, September 1966.

McLaughlin, R. A.: Privacy is not a personal thing—maybe Orwell was right. *Datamation,* April 1969.

Melloan, George: Automation backlash speedy office machines pour out enough paper to bury their uses. *Wall Street Journal,* February 28, 1966.

Meldman, J. A.: Centralized information systems and the legal right to privacy. *Marquette Law Review,* 52: no. 3, Fall, 1969.

Menard hearing to test data bank policy. *Computer World,* September 1970.

Menkus, B.: Retention of data . . . for the long term. *Datamation,* September 15, 1971.

Merritt, M.: DP figures in bank loss of $128,000. *Computer World,* February 3, 1971.

Milestone near in program theft case. *Computer World,* July 21, 1971.

Miller, R. I.: Computers and the law of privacy. *Datamation,* 14: no. 9, 49-55, September 1968.

Miller, A. R.: The national data center and personal privacy. *The Atlantic,* 220: no. 5, November 1967.

———, and Miller, W. F.: Getting a personal dossier from a statistical data bank. *Datamation,* May 1970.

Mintz, Harold K.: Safeguarding computer information. *Computers and Automation,* September 1969, p. 10.

———: Safeguarding computer information. *Software Age,* May 1970.

Michael, D. N.: Speculations on relation of computer to individual freedom and the right to privacy. *George Washington Law Review,* XXX:270-286, October 1964.

Moore, William C.: Riot plan worked. *The Office,* August 1970.

More work needed to solve problem of data security. *Computer World,* May 27, 1970, p. 6.

Morton, Thomas J.: Bomb demolishes army

computer complex. *Computer World,* September 2, 1970.

———: DP centers dig out in hurricane's wake. *Computer World,* August 19, 1970.

———: DP centers feel the brunt of hurricane's fury. *Computer World,* August 12, 1970.

———: FBI accuses youth of tapping T/S service, copying data files. *Computer World,* July 29, 1970.

———: Firms sue in mailing list theft. *Computer World,* July 8, 1970.

———: Psychologist views "insecurity" at DP centers. *Computer World,* July 22, 1970.

———: Prevention of public access "key" to DP center security. *Computer World,* June 9, 1971, p. 2.

Morran, J. Richard: How does your bank stack up in insurance against EDP losses? *Banking,* April 1970, p. 36-37.

Mossberg, W.: Junk dealer makes a tidy profit selling GM its own microfilm. *Wall Street Journal,* April 29, 1971.

Most DP centers lax in arranging backup facilities. *Computer World,* July 15, 1970, p. 4.

Most U.S. companies called lax in security of data processing centers. *Business Insurance,* June 22, 1970, p. 18.

Motorist gets stung by "small bugs." *Computer World, 5:* no. 2, 6, January 13, 1971.

Mroz, Gene P.: Computer "bug" control. *Journal of Data Management,* January 1970.

Nelson, F. B.: Campus computers—target for militants and almost anyone else. *Datamation, 16:* no. 13, 37-38, October 15, 1970.

Neuman, E. W., and Riley, R.: Protecting the computer in a process environment. *Control Engineering,* September 1970.

———: Danger ahead! Safeguard your computer. *Harvard Business Review,* November-December 1968.

Neville, Haig: Letter to the Editor, *Harvard Business Review,* May-June 1969.

———: Insurance for data processing. *Datamation,* July 1966.

New computer systems are developed to solve many problems at once. *Wall Street Journal,* March 25, 1966.

New threats and new defenses. *Banking,* August 1970.

New range of computer room storage and forms processing equipment. *Mech Acct Mangmt (GB), 4:* no. 5, 59, May 15, 1969.

Nielsen, George E.: Computerized access controls. *Industrial Security,* October 1968, p. 32-36.

Nigra, A. L.: Auditing acquisitions of data processing equipment. *The Internal Auditor,* January-February 1968.

Norman, J. L.: Reducing telephone network errors. *Datamation,* October 1, 1971, p. 24-31.

No sticky buns in the computer room—a guide to magnetic tape handling. *Computer Management, 5:* no. 4, 58-60, April 15, 1970.

Not just army snooping came up at Ervin hearings—EDP examined, too. *Datamation, 7:* no. 8, 41, April 15, 1971.

Numbers racket used data cards. *Computer World,* June 18, 1969.

O'Brien, T.: ADP systems get action from DOD. *Industrial Security,* June 1970.

———: Office crime. *Administrative Management,* November 1971.

Ortiz, J. V.: Constant-Power system for computers. *Elect Constr Maintenance, 69:* no. 1, 96-97, January 1970.

Ottenberg, Miriam: Electronic tax fraud investigated at IRS. *The Evening Star,* Washington, D.C., June 24, 1970.

Outages rated a leading cause. *Computer World, 5:* no. 19, 2, May 12, 1971.

Packard, V.: Don't tell it to the computers. *New York Times Magazine,* January 8, 1967.

Pacific telephone sued for erroneous billing. *Computer World,* September 30, 1970.

Palmer, R. R., and Duma, W. J.: Auditing with computers. *Banker's Monthly Magazine,* January 15, 1969.

Parker, D. B.: Ethics in the arts and sciences of information processing. *Control Data Corp.,* U.S.A.

Passport office maintains inclusive data bank. *Computer World, 5:* no. 8, 3, February 24, 1971.

Pauley, Charles: Audit responsibilities in the design of computerized systems. *The Internal Auditor,* July-August 1969.

Penn, S.: Computers will bring problems along with their many benefits. *Wall Street Journal,* December 20, 1966.

Perham, J.: The computer: A target. *Dun's Review,* January 1971, p. 34.

———: The computer: A target. *Management Service Bulletin, 7:*37, September 1970.

Personal protection urged. *Data Processing,* April 1971.

Plan for an unwanted reward. *Business Automation,* February 1967.

Planning for your new computer. *Computer Decisions, 2:* no. 12, December 1970.

Plot thickens in plotting program theft. *Datamation, 7:* no. 8, 47, April 15, 1971.

Plugging the leaks in computer security. *Harvard Business Review*, September-October 1969.

Plug-to-plug combustible. *Computer World, 4:* no. 40, October 7, 1970.

Police say Mafia's DP use impedes crime prevention. *Computer World, 4:* no. 48, 1, December 2, 1970.

Poor security leaves DP facilities ripe for sabotage. *Computer World, 4:* no. 24, 1, June 17, 1970.

Porter, Thomas W.: Evaluating internal controls in EDP systems. *Journal of Accountancy*, August 1964, p. 34-40.

Pratt, L. A.: Loss exposure hazards under bank automation. *Burroughs Clearing House, 55:* no. 1, 18, October 1970.

Precautions preclude misuse of student data. *Computer World, 4:* no. 9, 1, March 4, 1970.

Presnick, Walter: Protecting your computer's security. *Data Systems News*, February 1970.

———: Preventing damage to EDP systems. *Administrative Management*, April 1969.

Prevention of public access "key" to DP center security. *Computer World, 5:* no. 23, 2, June 9, 1971.

Price, D. G., and Mulvihill, D. E.: The present and future use of computers in state government. *Public Administration Review, XXV:* no. 2, 142-150, June 1965.

Privacy and the computer—an issue becomes a crusade. *Data Systems News, 11:* nos. 8-9, August/September 1970.

Privacy and security. *Data Systems*, August 1969, p. 9.

Privacy commission chairman suggests licensing plan. *Computer World, 4:* no. 45, November 11, 1970.

Privacy DP digest. *Data Processing Digest, 16:* no. 10, 34, October 1970.

Privacy—debate will rage and confuse the issues, but a national data center will become reality. *Business Automation*, January 1970.

Privacy. *Law and Contemporary Problems, XXXI:* no. 2, Spring, 1966.

Privacy thing, the. *Business Automation, 18:* no. 7, 80, May 1, 1971.

Problems and potential solution in computer control. *Industrial Security*, April 1969, p. 34-50.

Problems of liability for the EDP service industry. *Computers and Automation*, September 1970.

Programmer thankful for "bug" during computer center bombing. *Computer World, 4:* no. 21, 1-4, May 27, 1970.

Program plagiarism alleged in U.K. case. *Datamation*, June 1968, p. 91.

Prosser, W.: Privacy. *California Law Review, LXVIII:* no. 3, 383-423, August 1960.

Protecting your computer—ask what it does and what you need to back up. *Business Insurance*, July 5, 1971, p. 31-32.

Protection against fire. *Data Processing in New Zealand*, March 1970.

Protecting EDP systems from Fifth Column attacks. *Computer Decisions*, June 1972.

Protection management (newsletter). Man & Manager, Inc., New York.

Protecting your computer's security. *Data Systems News*, February 1970.

Providing security protection for computer files. *Best's Review* (Life Edition), May 1970.

Providing the right environment. *Elect Rev (GB), 185:* no. 22, 796-800, November 28, 1969.

Psychologist views "insecurity" at DP centers. *Computer World, 4:* no. 29, 1, July 22, 1970.

Pushbutton locks for computer room security. *The Office*, March 1971, p. 161-163.

Queeny, Jack: Computer spies: New worry for business. *Chicago's American*, January 16, 1969.

"Questions" extend data bank hearing. *Computer World, 5:* no. 11, 1, March 17, 1971.

Quinn, D. E.: The clandestine listening threat. *Industrial Security*, October 1968.

Radar wipes out IRS tape, consultant cites poor ground. *Computer World, 4:* no. 52, 1, December 30, 1970.

Radical rumblings heeded, centers increase security. *Computer World, 4:* no. 41, October 14, 1970.

Real DP crime may blossom. *Computer World, 4:* no. 47, November 25, 1970.

Rebuilt Fresno state DP center follows tight security. *Computer World*, July 8, 1970.

Reider, H. R.: Maintaining the security of computer records. *Burrough's Clearing House*, February 1971, p. 28.

Reservations center prefers "wet look." *Computer World, 4:* no. 41, October 14, 1970.

Revere, R.: The invasion of privacy—technology and the claims of community. *American Scholar*, 1958, p. 416.

"Revolutionary-force" bombs IBM office. *Computer World, 4:* no. 11, 1-4, March 18, 1970.

Reynolds, J. H.: Computer misuse: A look at vulnerable areas. *Best's Review,* May 1971, p. 71.

Right of privacy and medical computing. *Datamation,* April 1970.

Robertson, J.: Major drive to protect personal privacy due. *Electronic News,* February 12, 1970.

Robsham, E. P.: Computer protects 1,020-unit building. *Security World, 10:* no. 10, 16-17, 19-20, November 1973.

Rowan, T. C.: Computer technology and social change. American Orthopsychiatric Association, 1965.

Ruebhausen, O. M., and Brim, D. G., Jr.: Privacy and behavioral research. *Columbia Law Review, LXV*:1184-1211, November 1965.

Ruling sought for software patents. *Computer World, 5:* no. 15, 2, April 14, 1971.

Rush, H. M., and Brown, J. K.: The drug problem in business. *Conference Board Record,* March 1971.

Sabotage, accidents and fraud cause woes for computer centers. *Wall Street Journal, CLXXVII:* no. 55, March 22, 1971.

Sabotage course shows action may have bad effect on society. *Computer World, 4:* no. 47, 1, November 25, 1970.

Safeguarding time—sharing privacy—an all-out war on data snooping. *Electronics,* April 17, 1967.

Safe source says some safes are safer. *Computer World, 4:* no. 42, October 1970.

Scandinavia's first data theft occurs at service bureau. *Computer World,* November 18, 1970.

Scaletta, Phillip J.: The legal ramifications of the computer age. *Data Management,* November 1970, p. 10-14.

Schiedermayer, P. L.: The many aspects of computer security. *The Police Chief,* July 1970, p. 20.

Schroeder, M. D., and Saltzer, J. H.: A hardware architecture for implementing protection rings. *Comm ACM, 15:* no. 3, 157-170, March 1972.

Science, technology and the law. *Saturday Review, LI:* no. 31, 39-52, August 3, 1968.

Scoms, Louis, Jr.: Environmental factors: how vulnerable are you? *Data Security,* Illinois.

———: Security in the computer complex. *Computers and Automation,* November 1970.

Scotese, Peter G.: What top management expects of EDP. *Business Automation,* February 1, 1971, p. 48.

Secure communications. *The Communications User,* January-February 1971.

Security and defenses for the computer room. *Occupational Hazards,* December 1968.

Security breach leads to police data theft. *Computer World, 5:* no. 6, February 10, 1961.

Security cut damage from DP center blast. *Computer World, 4:* no. 51, December 23, 1970.

Security in the computer complex. *Computers and Automation,* November 1970.

Security in a computer environment. *Computers and Automation,* July 1969, p. 24-26.

Security in data processing. *Data Processor,* February 1973.

Security defenses for the computer room. *Management Review,* May 1969, p. 67-68.

Security dictated bank's choice of terminals. *Computer World,* September 18, 1968.

Security for business. *The Office, 74:* no. 3, September 1971.

Security letter. *Security Letter, Inc., 11:* nos. 12 and 24, 1972.

Security men thrive on the wages of fear. *Business Week,* no. 2129, 112-114, June 20, 1970.

Security of the computer center. *EDP Analyzer,* December 1971.

Security products survey. *The Office,* August 1970, p. 44-55.

Security supplement. *Computer World,* August 26, 1970.

1971 security supplement. *Computer World, 5:* no. 26, June 30, 1971.

Service bureau head gets $85,000 in bank suit. *Computer World, 4:* no. 45, 12, November 11, 1970.

Shannon, C. E.: Communication theory of secrecy systems. *Bell System Technical Journal,* October 1949.

Sheffield, R. J.: EDP audit techniques. *The Internal Auditor,* November-December 1969.

Shils, E.: Privacy and power. American Political Science Association, New York City, September 6-10.

———: Privacy—its constitution and vicissitudes. *Law and Contemporary Problems, 31*:281, Spring, 1966.

Skatrud, R. D.: Cryptographic techniques in data processing. *Computer Services,* July/August 1970, p. 13.

Skyscraper pitted against bomb terrorists. *Occupational Hazards,* March 1971.

Slom, S. H.: Some misled computers rebel against firms and refuse to bill customers for services. *Wall Street Journal,* June 22, 1971.

Smith, T. J.: Internal auditing of controls for data processing departments. *The Internal Auditor,* May-June 1968.

Smith, T. J.: Internal controls for data processing. *Computers and Automation,* November 1969.

Some quick tips for surviving brownouts. *Factory,* May 1971, p. 26.

Some tips on computer security. *Industry Week,* August 2, 1970, p. 22.

Sprague, R. E.: The invasion of privacy and a national information utility for individuals. *Computers and Automation, 19:* no. 1, 48-49, January 1970.

State bans punched card noting as city sues vendor, even weather a problem. *Computer World, 4:* no. 52, 3, December 30, 1970.

Sticking up a computer. *Innovation,* no. 7, 1969.

Stolle, C. D.: Computer-based audits. *Management Adviser,* May-June 1971.

Stone, J.: Man and machine in the search for justice. *Stanford Law Review, XVI:*515-560, May 1964.

Stoppages beset Dartmouth T/S. *Computer World, 5:* no. 5, 1, February 3, 1971.

Storer, N.: Large-scale data collections and the protection of privacy. Social Science Research Council, 1967.

Stubbs, James: Filing and finding computer tapes. *Administrative Management,* May 1969, p. 56.

Students demolish computer center. *Computer World, 3:* no. 8, 1-4, February 26, 1969.

Students protest Lads, occupy center. *Computer World, 5:* no .8, 4, February 24, 1971.

Suit hinges on programs. *Computer World, 4:* no. 50, December 16, 1970.

Summers, Garth E.: Providing reliable power for computer systems. *Plant Engineering,* January 7, 1971.

System development for regional, state, and local government. *System Dev Corp Magazine, VII:* no. 10, 1-27, October 1965.

Tagen, W. G.: Educating the internal auditor in EDP. *The Internal Auditor,* January-February 1970.

Tang, D. T., and Chien, T. R.: Coding for error control. *IBM Systems Journal,* no. 1, 48-84, 1969.

Tape library management. *Computer Management, 5:* no. 5, 16-20, May 15, 1970.

Tape recertification vital to NASA savings. *Computer World, 5:* no 12 March 10 1971.

Tamperproof computer. *Datamation* September 1969.

Tassel, J.: Information security in a computer environment. *Computers and Automation,* July 1969.

Taylor, A.: Directors fortunes being risked by DP departments. *Computer World, 4:* no. 52, December 23, 1970.

———: The questions arise who will control DP problems. *Computer World, 4:* no. 45, November 11, 1970.

Taylor, Robert L., and Feingold, Robert S.: Computer data protection. *Industrial Security,* August 1970, p. 20-29.

Technician seized, accused of picking computer's brain. *Los Angeles Times,* March 3, 1971, p. 8.

Telephones used in program theft. *Business Automation,* April 1971, p. 7.

Ten-point guide offers EDP security; privacy. *Business Insurance,* July 5, 1971, p. 31-32.

Term theft of computer data sophisticated crime. *Business Insurance,* June 9, 1969, p. 7.

Technology of computer destruction, the. *Seed Magazine* (Underground).

Thief inside, the. *The Office,* August 1970.

Thief outside, the. *The Office, 72:*35-38, August 1970.

Thomason, Francis J.: Management controls in EDP. *The Internal Auditor,* March-April 1969.

———: The week the computers stopped. *Datamation,* April 1967.

Thompson, T. R.: Problems of auditing data: Internal audit. *The Computer Journal,* no. 3, 1960.

Thorne, Jack F.: The audit of real time systems. *Data Management,* May 1970, p. 14.

———: Internal control of real-time systems. *Data Management,* June 1970, p. 34-37.

Titus, J. P.: Security and privacy. *Communications of the ACM, 10:*379-380, June 1967.

Tomeski, E. A., and Wescott, R. W.: The clarification, unification and integration of information storage and retrieval. *New York Management Dynamics,* 1961.

Trade-off consideration in security system design. University of California, Berkeley, October 14, 1970.

TWA burroughs trade suits on contract. *Computer World,* October 28, 1970.

Twenty students take over DP center, promise they don't plan any damage. *Computer World, 4:* no. 47, November 25, 1970.

Twigg, Terry: Need to keep digital data secure. *Electronic Design,* November 9, 1972, p. 68.

Two arrested in threat to destroy DP center. *Computer World, 4:* no. 32, 1, August 12, 1970.

$290,000 awarded in libel damages to an insurance broker suing retail credit company. *Computers and Automation,* May 1971, p. 32.

U.S. court awards user $480,811 in damages against IBM's SBC. *Computer World,* April 16, 1969.

U.S. marshall releases federal fugitive because of incomplete data in computer. *Computer World, 5:* no. 3, 2, January 20, 1971.

User-vendor contract problems can be avoided. *Computer World,* March 31, 1971.

User wins over $1 million in contract suit against WU. *Computer World,* April 7, 1971.

U.S. passport office maintains inclusive data bank. *Computer World, 5:* no. 8, 3, February 24, 1971.

Van Tussel, Dennis: Advanced cryptographic techniques for computers. *Communications of the ACM,* December 1969.

Verba, J.: Protecting your EDP investment. *Management Services, 7:*37, September 1970.

Violence by rebels threatens center. *Computer World, 4:* no. 3, 40, October 7, 1970.

Voltage unit solves firm's DP troubles. *Computer World, 5:* no. 2, 24, January 13, 1971.

Vanishing trail, the. *Bell Telephone Magazine,* July-August, 1968.

Wackenhut, G. R.: Business is the target of bombings and bomb hoaxes. *The Office,* September 1971.

Walsh, Timothy J., and Healy, Richard J.: Data processing. *Protection of Assets Manual, 1:* 1974.

Warren, S. D., and Brandeis, L. D.: The right to privacy. *Harvard Law Review, IV:* no. 5, 193-220, February 1890.

Washington airport shuttle crippled by driver strike. *Computer World, 4:* no. 31, 1, August 5, 1970.

Wasserman, Joseph J.: Auditing the computer. *Management Review,* October 1968.

———: Control in an EDP environment. *The Internal Auditor,* September-October 1970.

———: Plugging the leaks in computer security. *Harvard Business Review,* September-October 1969.

———: The vanishing trail. *Bell Telephone Magazine,* July-August 1968.

———: Protecting your computer's security. *Data Systems,* February 1970, p. 17.

Waterson, L.: Data banks can protect privacy. *Banking, 60:* January 1968.

Weapons and ferromagnetic objects detected by a new magnetic searcher. *Computers and Automation,* January 1971, p. 51.

Wearstler, Earl W.: A computer center is for safety, not for show. *Banking,* April 1971, p. 70-72.

Weeks, J. K.: Comparative law of privacy. *Clev Mar Law Review, XXII:*484-502, September 1963.

Weissman, C.: Trade-off considerations in security system design. *Data Management,* April 1972, p. 14-19.

Wessel, Milton R.: Computer services and the law. *Business Automation,* November 1970.

———: Legal protection of computer programs. *Harvard Business Review,* March-April 1965, p. 97-105.

Westin, A.: Balancing the conflicting demands of privacy, disclosure, and surveillance. *Columbia Law Review, LXVI:* no. 7, 1205-1253, November 1966.

———: New laws will protect your privacy. *Think,* May/June 1969.

———: Science, privacy, and freedom—issues and proposals for the 1970's. *Columbia Law Review, LXVI:* no. 6, 1004-1008, June 1966.

Wessler, J., Myers, E., and Garner, W. D.: Physical security—facts and fancies. *Datamation, 17:* no. 13, 34, July 1, 1971.

Whelan, Thomas: Software security. *American Management Association,* November 17-19, 1969.

Whisenand, P. M., and Medak, G. M.: Security, justice, and the computer. *Datamation, 17:* no. 12, 24, June 15, 1971.

Who watches the watchers. *Data Systems News,* December 1970.

Why the public dislikes computers. *Computers and Automation, 20:* no. 5, 7, May 1971

Wilson, T.: Air-conditioning in the computer room. *Data Processing, 11:* no. 2, 167-168, March 1969.

Wirecutters, acid used on computer. *Computer World, 3:* no. 14, 7, April 9, 1969.

"No great feat to wiretap" says Canadian computer professor. *Computer World, 4:* no. 47, November 25, 1970.

Worley, A. R.: Practical aspects of data communication. *Datamation,* October 1969, p. 60-66.

Wolsey, Dr. Marvin M.: EDP systems controls. *Data Management,* September 1971.

Wright, G. M.: Data center protection. *Canadian Security Gazette,* May 1971.
Yippies, convene, discuss methods of DP sabotage. *Computer World, 5:* no. 15, 2, April 14, 1971.
Your computer: How secure? *Chemical Engineering,* November 2, 1970.
Youth convicted of computer fire. *Los Angeles Times,* February 10, 1971, p. 16.
Youth indicted in data file copying. *Computer World, 4:* no. 45, 3, November 11, 1970.

UNPUBLISHED DATA

Gray, Ronald: Computer and privacy. Unpublished graduate thesis, Michigan State Univ., 1969.

BOOKS
EMBEZZLEMENT
HISTORICAL/UNIQUE DOCUMENTS

Arnon, Hermann: *Die untueue.* Tubingen, Druck von H. Laupp, Jr., 1894.
Auchincloss, Louis: *The Embezzler.* Boston, Houghton-Mifflin Co., 1966.
Arrogave, Jaime: *El peculado; Manual para jueces, functionarios deinstruccion.* Colombia, 1951.
Bauer, Birgil: *Die Unterschlagung: zur historischen, krimologischen und strafredktlichen.* Franfurt am Main, 1970.
Bernacchi, Osuaido C.: *Defraudaciones y estafas.* Buenos Aires, 1937.
Brugge, W.: *Het fraudeprobleem.* Belgium, 1965.
Busch, Sorensen A.: *Bedragericr og underslack.* Denmark, 1968.
Cain, James Mallahand: *The Embezzler.* New York, Avon Book Co., 1944.
Dearnlev, Irdine H.: *Fraud & Embezzlement, Including the Use of Office Machinery, Statistics, etc. as Preventive Measures.* London, Sir I. Pitman & Sons Ltd., 1933.
Employee Theft, 1001 Embezzlers. Baltimore, Fidelity and Deposit Company, 1937.
Ferreira Lopes, Ciceroed: *Peculato, Molda falsa a falsificacao d documentos (1923).* Rio de Janeiro, 1933.
40 Thieves. Baltimore, Fidelity and Deposit Company, 1969.
Dezeure, Roland: *Algemene practische rechtsveryameling.* Brussels, 1968.
Diaz Vasconcelos, Luis Antonio: *El delito financiero.* Guatemala, 1942.
Dopler, Jacob: *Der ungetreue Rechnung.* Beamble, 1684.
Haacke, Harry H., and San Souce, William B.: *How to Reduce Embezzlement Losses.* New York, Royal-Globe Insurance Companies.
Heimer, Einer Henick: *Forskingringsbrottet.* Sweden, 1926.
Husnt, Mahmud Na;tb: *Embezzlement.* Egypt, 1972.
Keller, Albert Edward: *Embezzlement & Internal Control.* WA Warner-Arms Publishing Co., 1946.
Leglisse Cordero, Salvador: *Delitos equiparados al abuso de confianya.* Mexico, 1965.
Nuvoolone, Pietro: *L'infedelta' pakrimoniale nel diritto penale.* Milano, 1941.
Politoff Lifschitz, Sergio: *El delito de apropiacion indebiden.* Santiago, Chile, 1957.
Pesumil Aragunde, Manuel: *Embezzlement.* Dominican Republic, 1943.
Silva, Francisco de Oliveira: *Apropriacao indebita, estelionato e "habeas corpus."* Rio de Janeiro, 1956.
Smith, Paul Ignatius: *Ind. Intelligence and Espionage.* London, Business Books, 1970.
Tejera Y Garcia, Diego Vicente: *La Malversacion de Caudales publicos.* Cuba, 1924.
Tourinmo, Demetrio Cyriaco Ferreira: *Do peculato.* Salvador, Brasil, 1954.
Wales, Stephen H.: *Embezzlement and Its Control.* Richmond, Ind., Igelman Printers and Publishers, 1965.
Weiser, May el: *Neve strafrechtliche bestimmungen.* Austria, 1932.

PUBLICATIONS OF THE GOVERNMENT, LEARNED SOCIETIES, AND OTHER ORGANIZATIONS

Halper, Stanley D.: Embezzlement—detection and control. Address before National Association of Credit Men, S. D. Leidesdorf & Co., 1968.
U.S. Congress, House Committee of the Judiciary. Embezzlement of Indian tribal organization property. Washington, Government Printing Office, June 1956.
U.S. Fidelity & Guaranty Co. 1001 embezzlers; a study in business. Baltimore, U.S. Fidelity & Guaranty Co., 1937.
Thomas, Alfred A.: The temptations of employees who handle money; what can the employer do to protect himself and them? Dayton, Ohio, 1905.

PERIODICALS

Ashdown, William: The psychology of embezzlement. *The Bankers Magazine,* April 1926.
Astor, Saul D.: An investigator talks of embezzlement and robbery. *The Office,* September 1971, p. 55-57.

Barrett, A. R.: The era of fraud and embezzlement. *The Arean*, 1895, p. 197.

Barnett, Chris: Security systems: An ounce of detection is worth a pound of cash receipts. *Mainliner*, December 1973, p. 25-27.

Conner, George A.: Embezzlement: The crime of honest people. *Credit and Financial Management*, 63:12, October 1961.

Cressey, Donald R.: Embezzlement: Robbery by trust. *Security World*, May 1965.

Criminal violation of financial trust, the *American Sociological Review*, December 1950.

Embezzlement controls, part I. *Security World*, 3: no. 4, April 1966.

Embezzlement controls, part II. *Security World*, 3: no. 5, May 1966.

Embezzlement controls, part III. *Security World*, 3: no. 6, June 1966.

Embezzlers, the trusted thieves. *Fortune*, November 1957.

Haacke, Harry H.: Embezzlement losses. *Security World*, 5: no. 6, June 1968.

How to Reduce Embezzlement Losses. New York, Royal-Glove Insurance Company.

Internal control of fraud and embezzlement. *Nat Public Accountant*, 7: May 1962.

Keysor, Charles W.: Do you have an embezzler on your payroll? *Commerce Magazine*, 51: 19-20, November 1954.

Kostelanetz, B.: The auditor meets the thief. *N.Y. Cert. Pub. Accountants 458*, July 1951.

Moran, Christopher J.: Preventing embezzlement. Small Business Administration, Small Marketer's Aids, No. 151. Washington, Government Printing Office, January 1973.

Pennington, W. J.: Embezzling: Cases and cautions: what can be done to stop the thief in the white collar. *J Accountancy*, 118:47-51, July 1964.

Pratt, Lester A.: *Embezzlement Controls and Other Safeguards for Banks.* Baltimore, Fidelity and Deposit Company, 1958.

———: *Embezzlement Controls for Business Enterprises.* Baltimore, Fidelity and Deposit Company, 1952.

———: Embezzlement controls. *Security World*, April 1966, p. 10-13.

Redfern, E. K.: How weak accounting systems encourage employee embezzlement: Four case histories. *Journal of Accountancy*, 92: 82-86, July 1951.

Sans Soucie, William B.: Embezzlement losses. *Security World*, 5: no. 6, June 1968.

Seavey, Warren A.: Embezzlement by agent of two principles: Contribution? *Harvard Law Review*, 64:431-436, January 1951.

Sederberg, Arelo: Bank embezzlements soar; robbery count rises, too. *Los Angeles Times*, Business and Finance Section. November 9, 1965.

Skivira, Gregory: Watch who's watching electronic accounts. *Detroit Free Press*, September 28, 1975.

Taylor, Roy C.: Methods of embezzlement—and protective measures. *NACA Bulletin* (National Association of Cost Accountants), 34:747-754, February 1953.

———: Methods of embezzlement—and protection measures. *Nat Assn Cost Accountants Bull*, 34:747-754, February 1953.

UNPUBLISHED MATERIAL

Elliott, Raymond W.: Computer embezzlement prevention and control. Unpublished presentation at the International Security Conference. Security World Publishing Co., Inc., February 1971.

Klotz, W. H.: Figures aren't always what they seem. Address to the New York Chapter of the Institute of Internal Affairs (mimeographed).

Pratt, Lester A.: Embezzlement controls for business enterprises. Private Printing, March 1966.

Redden, Elizabeth A.: Embezzlement: A study of one kind of criminal behavior with prediction tables based on fidelity insurance records. Ph.D. dissertation, Univ. of Chicago, 1939.

BOOKS

PILFERAGE

Cadmus, Bradford, and Child, Arthur: *Internal Controls Against Fraud and Waste.* New York, Prentice-Hall, 1953.

Cooper, Herbert Noel: *Fifteen Years a Store Detective.* London, Pallas Publ. Co., Ltd., 1940.

Curtis, Bob: *Security Control: Internal Theft.* New York, Chain Store Age Books, 1973, 361 pp.

Rudnitsky, Charles P., and Wolff, Leslie M.: *How to Stop Pilferage in Business and Industry.* New York, Pilot Books, 1961.

Stevenson, Thomas M.: *How to Be a Store Detective.* A Handbook for the American Retail Industry & Store Detective, Student or Professional. Chicago, Grayce Publ. Co., 1963.

Wenchell, Prentice: *I Was a House Detective.* New York, Dutton, 1954.

PUBLICATIONS OF THE GOVERNMENT, LEARNED SOCIETIES, AND OTHER ORGANIZATIONS

American Management Association. Controls and coverage against dishonesty loss. New York, Insurance No. 85, 1950.

An inside job. Store Management Series. New York, Nat. Retail Merchants Assn. Filmstrip.

Astor, Saul D.: Preventing retail theft. Washington, Government Printing Office, February 1966.

———, and Gugas, Chris: *Loss Prevention Security*. New York, Security World Publishing Company.

Atkins, John Leslie: Industrial espionage and trade secrets: A bibliography. Coventry, Cadig. Leaison Centre, 1972.

Bank Administration Institute, Audit Commission. A study of internal fraud in banks. Parkridge, Ill., Bank Administration Institute, 1972.

Bernstein, Mark: Preventing losses due to employee pilferage. p. 2. President, Willmark Service System, Inc., New York.

Bock, Robert H.: *An Analysis of the Long-Range Planning Process in a Select Group of Business Firms*. Lafayette, Ind., Purdue University Press, 1960.

Case Studies in Internal Control, No. 1. New York, The Textile Company, American Institute of Accountants, 1950. 60 pp.

———, No. 2. New York. The Machine Manufacturing Company. American Institute of Accountants, 1950. 38 pp.

Crime Loss Control. American Mutual Liability Insurance Company.

Crime Loss Prevention. Chicago, Continental National American Institute.

Estes, G. B.: Loss control. Speech given at Center for Police Training, Indiana University, 1960.

Galvin, Raymond T.: The cost of crime. *Michigan State Univ. Bureau Economic Res*, 7:1-2, November 1965.

Greedy hands. Store Management Group. New York, Nat. Retail Merchants, Assn., 1962. (Filmstrip).

Gross, Samual S.: Internal Auditing Methods. Address presented at Top Management Business Security Seminar, Michigan State, April 1963.

How Much Honesty Insurance. New York, Surety Assn of America, 1961.

Industrial Security III. Theft control procedures. Studies in Business Policy, No. 70. National Conference Board, Inc., New York, p. 7.

Industrial Security III. Theft control procedures. National Industrial Conference Board, Inc., New York, 1954.

Internal Theft: Investigation and Control—An Anthology. Los Angeles, Security World Publishing Co.

National Retail Merchants Association. *National Retail Merchants Association Stock Shortages . . . Their Causes and Prevention*. New York, Nat. Retail Merchants' Assn., 1959.

Plant pilferage. Highway Safety Foundation, Mansfield, Ohio. (Filmstrip)

Shortage Control Manual. Detroit, The J. L. Hudson Co.

Stock Shortages—Their Causes and Prevention. New York, Nat'l Retail Merchants Assn.

Store Protection. Boston, Retail Trade Board. Greater Boston Chamber of Commerce, 1954.

Suggested standards for industrial safeguards. Special Bulletin No. 7. Washington, U.S. Department of Labor, 1942.

Supplementary comments concerning employee honesty. New York, Norman Jaspen Associates.

U.S. General Accounting Office. Examination of fraudulent transactions relating to the accounts of military dispursing officers. Report to the Congress of the U.S. by the Comptroller General of the U.S. WA, 1961.

U.S. Navy Safety Precautions. Washington, Government Printing Office, 1953. 501 pp.

PERIODICALS

American Institute of Accountants. Case studies in internal control. No. 2. New York, The Textile Company, 1950.

Anderson, William D.: Establishing effective retail loss prevention attitudes. *Security Management*, May 1974, p. 45-46.

Arkin, Joseph: The $2 billion theft. *Security World*, February 1966, p. 29.

———: How to prove a "fidelity loss." *Security World*, 5: no. 4, April 1968.

Astor, Saul D.: More shortages—discount dilemma. *Security World*, January 1965.

———: An analysis of employee dishonesty. *Department Store Economist*, 29: no. 8, August 1966.

———: Combating collusive employee theft. *Retail Control*, May 1964.

———: Operation audits to teach security. *Security World*, 2: no. 5, July/August 1965.

———: Astor's laws of loss prevention. *Security World,* June 1969.

———: Shoplifting: Far greater than we know? *Security World,* December 1969.

———: Undercover investigation: A view from the grass roots. *Security World,* September 1969.

———: Internal crime, Part I. *Security World,* September 1970.

———: Internal crime, Part II. *Security World,* October 1970.

———: Security or loss prevention? Part I. *Security World, V:* no. 1, January 1968.

———: Security or loss prevention? Part II. *Security World, V:* no. 2, February 1968.

———: Testing for profit leaks. Part I. *Security World, V:* no. 7, July/August 1968.

Atherton, Raymond M.: Theft and pilferage as personal and industrial problem. *Industrial Security,* 1963.

Bachman, William E.: Pilferage, payola, protection. *Office Executive, 26:*26-27, September 1961.

Bernstein, Joseph E.: Preventing pilferage losses. *Retail Control,* March 1956.

Berton, Lee: Robbing the boss. *Wall Street Journal,* August 19, 1964.

Bleecker, Philip: Controlling internal theft. *Chain Store Age,* 1960.

Blunt, I.: *Insurance Activities of the Silk Association of American, Inc.* New York, Am. Management Assn., 1932.

———: *Controls and Coverage Against Dishonesty Loss.* Insurance Series. New York, Am. Management Assn., 1950.

Bollard, R. L.: Industrial pilferage—unplanned profit-sharing. *Mill and Factory,* November 1963.

———: Safeguards against employee dishonesty. *Credit and Financial Management, 6:* no. 11, October 1964.

Buckley, J. L.: Shaping employee attitudes toward security. *Law and Order, 2:* no. 6, June 1963.

———: Your company's security program—how does it rate? *Office Executive,* September 1961.

———: Business uses the lie detector. *Business Week,* June 18, 1960, p. 98-106.

Burnstein, Harvey: Stock shortage control: The personnel aspect. *Management Rev.,* November 1960.

Calvert, Arthur: How closed-circuit TV cuts pilferage losses for chain store. *Chain Store Age,* September 1959.

Can you sue a former employee for having delayed a company's progress? *The Businessman and the Law Issue,* December 1, 1965.

Carlson, Gus R.: A financial menace to industry. *Security World,* January 1971.

Case study. A shortage prevention program for lessees. *Security World,* October 1969.

Conner, George A.: Open your eyes to theft. *Office Executive, 29:* no. 4, 21-24, April 1954.

Controlling shortages and improving protection. Nat. Retail Dry Goods Assn., 1953.

Crime is cancerous. *The Office,* August 1969, p. 118.

Curtis, Sargent J.: Dishonesty, the sinister cancer. *Industrial Security,* April 1963.

Davis, John R.: Packaging can curb pilferage. *Police, 7:* no. 4, March/April 1963.

———: The combating of theft from within . . . what is the answer? *Police, 9:* November/December 1964.

Deskbook on organized crime. Organized crime and American business, Part 1. *Security World,* March 1970.

———: Organized crime and American business, Part 2. *Security World,* April 1970.

Edd, J. A.: Employee theft. *Canadian Security Gazette,* April 1971.

Elam, Ralph C.: Meaningful identification for equipment in local schools. *Security World,* October 1974, p. 63.

Factory Management and Maintenance. Petty pilferage—not-so-petty-problem. *CXII:* no. 9, September 1964.

———. Stop the thief in your plant. *CXII:* no. 9, September 1954.

Falk, R. Neal: Employee thefts are on the increase at a formidable rate. *National Council on Crime and Delinquency, 45:*5, November/December 1966.

Flanagan, J. J.: The war against pilferage. *Security World,* April 1969.

Foster, Reginald: Keeping the Christmas spirit. *Security and Protection,* December 1974, p. 4-7.

Fowler, Frederick: Preventing crime in works. *Security Gazette,* September 1961.

Froemming, Roger G.: Why internal control. *Auditgram, 36:* no. 11, November 1960.

———: Security: Is yours as good as you think? *Systems Department,* 1967.

———: Why internal control. *Auditgram, 36:* no. 11, November 1960.

Gardner, John: Package policies. *Security Gazette,* March 1975, p. 93.

Garner, Louis E.: Business counterspies. *Fortune,* June 1956, p. 18, 20.

Gay, William O.: The craft of the pilferer. *Police Journal,* 7:226.

Goldstein, Morton: Reduction of shortages in retain stores. *Industrial Security,* October 1971.

Greenest are ripest for plucking, The. *The Office,* September 1971, p. 49-51.

Greyhound Food Management Inc. I am here to help you with receivables. Systems Department, No. 2. Detroit, Mich., Greyhound Inc., 1967.

Hardware retailer: Internal theft in rural community. *Security World,* April 1970.

Harwell, Edward M.: Checkout management. *Security World Magazine, II:* no. 8, November 1965.

Hansen, Paul: The dishonest employee. *Industrial Security,* December 1964.

Hass, Robert C.: Inventory shrinkage control. *Security World, I:* no. 1, July 1964.

Hemphill, Charles F., Jr.: If Christmas means "temporary." . . . *Security World,* December 1974, p. 27+.

———: Limiting loss potential from employee theft. *Security World,* February 1975, p. 28-29.

———: Tightening security at the will-call counter. *Security World,* May 1975, p. 23+.

Hoffman, E. E.: *Billion Dollar Check Racket.* Tarzana, Calif., Renay, 1964.

How can the package help to discourage pilferage in self-service selling? *Modern Packaging,* 32:61-62, August 1959.

How to stop stealing in your plant. *Factory Management and Maintenance,* September 1954, p. 84.

How to guard against dishonesty. *Management Review,* August 1964.

How to plug holes in company security. *Business Management,* July 1962.

How to stop plant thievery. *Occupational Hazards,* February 1964.

How to stop plant thievery. *Industrial Security,* February 1964.

Jaspan, Norman: Stopping employee theft before it starts. *Management Review,* January 1960, p. 20.

———: Employee dishonesty and indifference. Presentation at National Association of Accountants, New York. Norman Jaspan Associates, Febuary 19, 1968.

———: The hidden conspiracy against profits. Presentation at the American Management Association, New York. Norman Jaspan Associates, 1967.

———: *Thieves on the Payroll.* New York, Investigations, Inc., 1968.

King, John A.: Three rules to help you stop thieves in your plant. *Factory,* August 1959.

King, Paul A.: You can crack down on employee thefts. *Factory,* April 1961.

Klein Schrod, Walter A.: Crisis in office crime. *Administrative Management.* November 1971, p. 24-27.

Laird, Donald A.: Psychology and the crooked employee. *Management Review, 12:* no. 4, April 1950.

Lefer, Henry: How to preserve your business lifeblood. *Modern Office Procedures,* April 1971, p. 21-27.

Leighton, Alexander H.: Preventive psychiatry in industry. *Management News,* January 28, 1949.

Liability awards pose another "loss" problem. *Protect Newsletter,* September 1973.

Lowell, Leonard S.: Employee theft: We create our own problems. *Security World, 2:* no. 3, May 1965.

McDowell, Carl W.: Marine underwriter's view of containerization security. *Security World,* July/August 1971.

Meiggs, Walter B.: *Principles of Auditing.* Homewood, Ill., Irwin, 1964.

Melville, N. M.: Prevention of pilferage. *Security World,* June 1969.

Mooney, Bernard J.: Employee dishonesty vs. profits. *Industrial Security,* August 1965, p. 26.

Natale, John: Countering internal theft. *Security World, 6:* no. 1, January 1969.

National Association of College Stores. Pilferage and how to cope with it. *Publisher's Weekly,* May 16, 1966.

National Retail Dry Goods Assoc. Controlling shortages and improving protection. A report prepared by the National Retail Dry Goods Assoc., 1953.

Penland, Jack D.: Auditors view of theft. *Industrial Security, 10:* no. 4, August 1966.

Petty pilferage—not so petty problem. *Factory Management and Maintenance,* September 1964.

Pratt, Lester A.: The detection and prevention of employee dishonesty. *Industrial Security,* October 1962.

Quinn, Davis: The magnitude of losses from dishonesty. *Industrial Security,* December 1964.

RH-prevention medicine for employee dishonesty. *Management Review,* November 1960.

Riemer, Svend H.: Embezzlement: pathological basis. *Journal of Criminal Law, 32*:411-423, November/December 1941.

Rise of dishonesty. *U.S. News World Rep.,* November 1951.

Ross, I.: Thievery in the plant. *Fortune,* October 1961; November 1961.

Schaefer, Herman J.: The importance of internal audits. *Industrial Security,* October 1962.

Schwartz, Dr. Howard: Good supervision: Your best internal security. *Security World,* April 1975, p. 48-49.

Scruton, Robert A.: Inventory shrinkage control. *Security World Magazine, II:* no. 4, June 1965.

Security or loss prevention? Part I. *Security World, 5:* no. 1, January 1968.

Shortage control is a symptom. *Parents,* January 1952.

Silberford, E. J.: Theft of drug cultures. *New York Herald Tribune,* October 30, 1962.

Skowronek, David: How to slam the door on plant thieves. *Business Management,* March 1962.

Smith, H. Wayne: Packing can curb pilferage. *Police, 7:*44-46, March/April 1963.

Smith, Jon L.: Investigative casebook: Cash register theft. *Security World,* January 1971.

Stock, warehouse and enroute controls. *Industrial Security,* October 1962, p. 77.

Stop stealing in your plant. *Factory Management and Maintenance, 112:* no. 9, September, 1954.

Stop the thief in your plant. *Factory Management and Maintenance,* September 1964.

Stores Mutual Protection Association. *Manual for Store Security.* New York, 1954.

Sutherland, Erwin H.: Crime and business. *Ann Amer Acad Political Social Sci, 217:* 1941.

———: White collar criminality. *Amer Sociol Rev, 5:*1-12, February 1940.

Taylor, R. C.: Preventive medicine for employee dishonesty. *The Management Review,* November 1960, p. 20-28.

Taylor, Roy C.: Put the lid on larceny! *Supervisory Management, 6:* no. 2, February 1961.

Testing for profit leaks. *Security World, 5:* no. 7, July/August 1968.

Thayer, Frank, and Bowen, James: How papers can prevent cash and inventory pilferage. *Industrial Security,* February 25, 1956.

Theft and capture: At stake, $2000,000. *Occupational Hazards,* August 1969, p. 27.

Theft and pilferage as personal and industrial problems. *Industrial Security,* October 1963, p. 40.

Thefts—internal. *Industrial Security,* October 1960, p. 18.

To catch a thief . . . wrap security in a tight package. *Buildings,* May 1972.

Tocchio, O. J.: Employee theft: It can be curtailed. *Police, 7:* no. 1, 53, 1962.

———: Shoplifting—the scourge of mercantile establishments. *Police, 6:* no. 5, 8, 1962.

Walsh, John H.: Pilferage in industry. *Security World,* July/August 1966, p. 17.

Walsh, Timothy: Guidelines to handling thefts in industry. *Industrial Security,* April 1962.

Webster, George C.: *Reducing Stock Shrinkage in Small Firms.* Small Business Administration, 1959.

Why let pilfering drain your company. *Fleet Owner,* April 1960.

Wilkins, George: Loss prevention—make it work. *Canadian Security Gazette,* Summer, 1971.

Wirth, Charles J., and Richards, George: Internal control. *Auditgram, 38:* no. 4, April 1962.

UNPUBLISHED MATERIAL

Astor, Saul D.: Implementation of procedures: Methods and controls. Address presented at Top Management Business Security Seminal, Michigan State, April 17, 1963.

Colling, Russell: An evaluation of certain psychological deterrents to employee pilferage and their apparent effectiveness. Unpublished Michigan State thesis. Fall, 1966.

Conser, James Andrew: The private adjudication of dishonest employees in selected retail establishments in Arlington County, Virginia. Thesis, Michigan State Univ., 1974.

Freeman, Eugene M.: A study of the employee theft problem in selected manufacturing enterprises and proposals for control of employee dishonesty. Michigan State Univ. thesis. Fall, 1963.

Horning, Donald Neal Michael: Blue collar theft: A study of pilfering by industrial workers. Indiana University, Ph.D., Vol. 25-04. DAI., p. 2650, 1963.

Leek, Everett P.: A study of hourly rated industrial employee attitudes toward theft from industrial plants in Saginaw, Michigan. Unpublished Michigan State Univ. thesis. Fall, 1965.

Myers, D.: Control of employee dishonesty.

Unpublished graduate B-paper, Michigan State Univ., 1968.
Ozenne, Tim Oliver: The economics of theft and security choice. Michigan University Microfilms, Inc., 1972, 99 pp.
Ripberger, Raymond C.: Theft from selected Indiana businessmen in 1971. Organized crime implications. Theses, Indiana University, April 1974.
Robin, Gerald: Employee as offender. A sociological analysis of occupational crime. Ph.D. dissertation, Univ. of Pennsylvania.
Steeno, David Lawrence: Creating a secure environment for loss prevention in retail discount stores. B Paper, Michigan State Univ., 1973.
Westin, Paul R.: Administration of a security program designed to detect and prevent internal losses. Unpublished graduate B paper, Michigan State Univ.
White, Russell E.: Relationships of employee morale to theft. Address presented at Top Management Business Security Seminar, Michigan State Univ., April 18, 1963.

BOOKS
OSHA

Blake, Roland P.: *Industrial Safety,* 2nd ed. New York, Prentice-Hall, 1953.
Factory Mutual Engineering Division. *Handbook of Industrial Loss Prevention.* New York, McGraw, 1959.
Dickie, A. L.: *Production with Safety.* New York, McGraw-Hill, 1947.
Handbook of Industrial Safety Standards. Baltimore, United States Fidelity and Guaranty Company, 1954, 315 p.
Heinrich, H. W.: *Industrial Accident Prevention,* 3rd ed. New York, McGraw-Hill, 1953.
———: *Industrial Accident Prevention—A Scientific Approach.* New York, McGraw-Hill, 1950, 470 pp.
National Safety Council. *Accident Prevention Manual—For Industrial Operations.* Chicago, Nat. Safety Council, 1955.

PERIODICALS

Bennett, Leonard E.: Entertaining with safety. *Law & Order, 8:* no. 9, 1960.
Bleicken, Gerhard D.: Emergency planning—a national responsibility. John Hancock Mutual Life Insurance Co. *Industrial Safety.*
Breeding, C. A.: Safety and security in aerospace industries—panelist. *Industrial Security, 46:* October 1963.
Chapman, Paul B.: Accident prevention—retail department stores—panelist. *Industrial Security, 35:* October 1960.
DeKay, R. C.: Safety at the sidewalk. *Law & Order, 8:* no. 11, 1960.
Hamilton, Margaret: The industrial safety officer. *Security Gazette,* May 1971.
Handley, Thomas: Safety—panic and its control. *Industrial Security,* December 1964.
Healy, Richard J.: Safety and security: Industrial buildings. Washington, *Building Research Institute,* March/April 1967, p. 15.
Hemeon, W. C. L.: *Plant and Process Ventilations.* New York, Industrial Pr, 1955.
Hewins, Gilbert M.: Security new hat. *Industrial Security, 6:* Por Bio., July 1963.
Humphreys, William Y.: The plant emergency control center. *Industrial Security, 36:* Por Bio., July 1960.
Jeppesen, C. L.: Safety and security in aerospace industries—panelist. *Industrial Security, 46:* Bio, October 1963.
Johnson, Howard O.: Snow safety measures. *Police Chief, 19:* no. 3, March 1962.
Kingsburg, Arthur: Guidelines for security education; an introduction to occupational safety. *Security Management, 17:* no. 5, 24, November 1973.
Kralovec, Dr. Dalibor W.: Safety—panic and its control—panelist. *Industrial Security, 67:* Por Bio., December 1964.
Kreidel, Francis A.: Cooperation and compatibility of the safety program with industrial security—panelist. *Industrial Security, 22:* Por Bio., December 1964.
Kutcher, R. D., and Lang, E. J.: Plant security a problem? Chain belt solved theirs. *Plant Engineering,* April 1963.
McCants, L. C., Col. USAF: Safety and security in aerospace industries—panelist. *Industrial Security, 44:* Bio., October 1963.
Montgomery, G. E.: Safety symposium, seminar workshop. *Industrial Security,* October 1961.
National Safety Council. Accidents in the meat packing industry. Chicago, National Safety Council, 1953.
———. Accident facts. Chicago, National Safety Council, 1964.
National Safety News. Published monthly by National Safety Council, 425 N. Michigan Ave., Chicago, Ill., 60611.
Ohio Legislative Service Commission. Industrial safety enforcement in Ohio. Report #36. Columbus State House, 1959.
Olsen, Olaf: Loss prevention by internal control. *Industrial Security,* December 1964.
Peterson, D. C., and Weaver, D. A.: Criteria to

niche safety. *Industrial Security,* August 1966.

Proetz, William F.: Safety devices demonstrated during university seminar. *Law and Order, 8:* no. 7, 1960.

Rockwell, Dr. Thomas: Safety—panic and its control—panelist. *Industrial Security, 67:* Por Bio., December 1964.

Role of the security department in safety. *Security Education Briefs,* OAK Security Inc., *2:* no. 11.

Safety equipment puts police emphasis on personal service. *College and University Business, 48:*99, April 1970.

Sampson, Arthur F.: Life safety systems for high-rise structures. *Fire Journal,* July 1971, p. 8-10.

Security and fire issue. *Occupational Hazards,* June 1967.

Security and fire protection. *Occupational Hazards,* January 1966.

Simpson, George H.: The plant welfare program in emergency. *Industrial Security, 54:* Por Bio., July 1960.

Stevens, Richard E.: Security versus life safety from fire. *Security World,* April 1971.

Stickney, C. W.: Techniques of fire investigation—panelist. *Industrial Security, 11:* October 1960.

Stone, James B.: Industrial manual and association for emergencies. *Industrial Security,* July 1960.

Strobl, Walter M.: Plant fire protection. *Security Management,* March 1974, p. 16-22.

Supervisors Safety Manual. Chicago, Nat. Safety Council, 1956, 354 pp.

Tarrington, Arthur E.: The security/safety merger. *Security World,* January 1975, p. 36-39.

———: The security/safety merger, Part II. *Security World,* March 1975, p. 28+.

Taylor, James D.: Plant police services in an emergency. *Industrial Security,* July 1960.

U.S. Department of Labor, Bureau of Labor Standards. *Safety Subjects.* Washington, Government Printing Office, 1953.

Weaver, D. A.: Criteria to niche safety. *Industrial Security, 4:*29, Bio., 1966.

Welton, Harry: Planned subversion is aimed at industry. *Security Gazette,* March 1963.

Wheland, R. L.: Emergency headquarters plan —panelist. *Industrial Security,* 30, October 1960.

White, Will W., Brig. Gen. USAF (Ret.): Emergency and civil defense planning—panelist. *Industrial Security, 27:* Por Bio., October 1962.

White, Stanhope: The health and safety at work, etc., Act. 1974. *Top Security,* July 1975, p. 120-121.

Wilcox, Warren J.: Lockheed-Georgia is ready for its next fire: are you? *Industrial Security,* December 1968.

Yandell, Kenneth E.: The emergency company headquarters. *Industrial Security, 103:* Por Bio., July 1960.

Zucker, John P.: Guide to plant fire protection. *Occupational Hazards,* 1964-1965.

BOOKS

FIRE

Bond, Horatio: *NFPA Inspectional Manuel.* Boston, National Fire Protection Assoc., 1950.

Crosby-Fiske-Forster: *N.F.P.A. Handbook of Fire Protection,* 10th ed. Boston, National Fire Prevention Association, 1948.

Moulton, Robert S.: *Handbook of Fire Protection.* Boston, Nat. Fire Protection Assn., 1948.

National Fire Protection Association. *Fire Protection Handbook.* National Fire Protection Association, 1969.

Stecher, G. E., and Lendall, H. N.: *Fire Prevention and Protection Fundamentals.* Philadelphia, Spectator, 1953.

Tyron, George H.: *Fire Protection Handbook.* Boston, National Fire Protection Association, 1962.

Underwriters' Laboratories. *Fire Protection Equipment List.* Northbrook, Ill., Underwriters' Laboratories.

Wels, Byron: *Fire and Theft Security Systems.* Blue Ridge Summit, Pa., Tab Books, 1971.

PUBLICATIONS OF THE GOVERNMENT, LEARNED SOCIETIES, AND OTHER ORGANIZATIONS

Emrath, P. C.: *Explosives and Incendiaries Used in War.* Lexington, Univ. of Kentucky, 1941.

Fire Extinguishing Equipment. Bureau of Facilities. Washington, Government Printing Office, 1970.

Douglas and Thompson: *Some Studies in Chemical Fire Hazards.* Stillwater, Oklahoma, Oklahoma Agriculture and Mechanical College, November 1949, 16 pp.

Fire Department of the City of Los Angeles with Office of Civil Defense, City of Los Angeles. Plant protection. January 1958.

Industrial Protection Training Series. Lafayette, Indiana. Public Safety Institute, Purdue University, 1941.

Lewis-Yarnell: *Pathological Firesetting.* New York, Nervous and Mental Disease Monographs, 1951, 437 pp.

National Board of Fire Underwriters. *First Aid Fire Appliances.* New York, Nat. Board Fire Underwriters, 1950.

———. *Private Fire Brigades.* New York, Nat. Board Fire Underwriters, 1949.

National Fire Protection Association. *Protection Against Fire Exposure of Openings in Fire Resistive Walls.* Boston, No. 80A.

———: *Protection of Library Collections.* Boston, N.F.P.A., 1970.

National Fire Protection Company. *Fire Protection and Suppression for Oil Storage Tanks.* Feltham, England, Nat. Fire Protection Co., Ltd.

National Fire Protection Association. *Protection of Museum Collections.* Boston, N.F.P.A., 1969.

———. *Protection of Records.* Boston, N.F.P.A., 1970.

Office of the State Fire Marshall. *Excerpts From Law Relating to the Office of State Fire Marshall.* Sacramento, State Printing Office, 1961.

Stecher, G. E., and Lendall, H. N.: *Fire Prevention and Protection Fundamentals.* Philadelphia, Spectator, 1953, 744 pp.

PERIODICALS

Ahern, John J.: Techniques of fire investigation—panel moderator. *Industrial Security, 11:* October 1960.

Amoraso, Louis J.: Where is your company in a fire emergency? *Security World,* March 1975, p. 20+.

An after-fire program. *Security World, 2:* no. 5, July/August 1965.

Anonymous and Staff Written: Dormitory fire kills nine. *Fire Journal, V:* no. 1, January 1968.

———: 25 die in department store fire. *Fire Journal, V:* no. 2, February 1968.

———: Fires to learn by. *Fire Journal, V:* no. 6, June 1968; *V:* no. 7, July/August 1968.

———: Fraternity house fire. *Fire Journal, V:* no. 11, December 1968.

Are you prepared for a strike? *Occupational Hazards,* April 1966.

Armistead, J. C.: DuPont's Louisville works explosion. *World Security, III:* no. 2, February 1966.

Auck, S. E.: How Underwriters Laboratory tests fire extinguishers. *Security World, 3:* no. 5, May 1966.

Baaden, Walter J., Jr.: Security from fire. *Industrial Security,* April 1970.

Badin, Fred: The infernal problem. *Industrial Security,* June 1965.

Barlay, Stephen: Is Britain burning? *Top Security,* June 1975, p. 66-68.

Beck, Sanford E.: What a corporate security director must know about insurance. *Security Management,* May 1974, p. 25-26.

Best, Richard H.: Central index fire—Europe—panel moderator. *Industrial Security, 56:* Bio., October 1963.

Betz, G. M.: Design fire-resistant materials into your building plan. *Plant Engineering, 17:* 133, January 1, 1963.

Blast and fallout protection for an industrial plant. *Plant Engineering,* December 1961.

Bond, Horatio, and Nolting, Orin S.: *Municipal Fire Administration.* Chicago, Int. City Manager's Assn., 1950.

———: General management responsibility for effects of fire on operations. *Industrial Security,* April 1967.

———: *Industrial Fire Brigades.* Boston, Nat. Fire Protection Assn., 1954.

———: *NFPA Inspection Manual.* Boston, Nat. Fire Protection Assn., 1950.

———: *Research on Fire.* Boston, Nat. Fire Protection Assn., 1957.

———: Fire extinguisher service and service fraud. *Security World,* February 1970.

Darling, Don: $11 million for missing fire door. *Security World,* July/August 1975, p. 14-15.

Davis, T. L.: *The Chemistry of Powder and Explosives.* New York, Wiley, 1941.

Doe, Everett: Supermarket fire experience. *Security World, 3:* no. 4, April 1966.

———: Fire-loading in retail stores. *Security World,* February 1969.

Don't let your plant go up in smoke. *Factory,* February 1963.

Dougherty, Frank L.: Protection of "know-how" in the chemical field—panelist. *Industrial Security, 23:* October 1960.

Coffey, E. P.: Sabotage through fire. *Fire Engineering, 94:*123, 140, 1941.

Corrosive gas—leak detection suspected of causing fires and explosions. *Industrial Fire and Security Report,* January 10, 1975, p. 6.

Cohen, Jerry, and Murphy, William S.: *Burn, Baby, Burn.* New York, E. P. Dutton and Company, 1966.

Eastham, Gerald: Reasonable fire safety. *Security Gazette,* May 1975, p. 171-172.

Factory Mutual Record. Fighting fires of the fire-lighter. *Security World,* November 1969.

———. Prepare your plant for flood and fire. *Security World,* February 1970.

Federal government organized centralized fire prevention program. *Industrial Fire & Security Report,* January 10, 1975, p. 1-2.

Fire Department and the City of Los Angeles with Office of Civil Defense, City of Los Angeles. *Plant Protection,* January 1958.

Fire dangers in high buildings. *Security and Protection,* September 1974, p. 40.

Fire Journal. Fires to learn by, Part I. *Security World,* February 1970.

———. Fires to learn by, Part II. *Security World,* May 1970.

Fire hazards in new buildings. *The Office,* October 1970, p. 32.

Fire is a full-time menace. *The Office,* October 1970, p. 32.

Fireman and Electrical Equipment—A Guide to Self Protection. Ann Arbor, Mich., University of Michigan Press, Bulletin No. 280, 1952, 52 pp.

Fire protection by cooperation. *Security Management,* March 1974, p. 16.

Harvey, Bruce: Fire hazards in libraries. *Library Security,* January 1975, p. 1, 6, 7.

Hayes, Fred V.: Fire and the high rise structure. *Security Management,* September 1974, p. 13-15.

Johnson, J. E.: Engineering early warning fire detection. Privately printed. Cedar Knolls, N.J., 1969.

Jones, Charles L.: *Safety in Lacquer Plants.* Wilmington, Hercules Powder Company.

Keir, Are: An effective fire fighting program. *Industrial Security,* April 1971.

Kingsbury, Arthur: Guidelines for security education: Introduction to fire prevention. *Security Management, 18:* no. 1, 25, March 1974.

Kissel, Gary: Fire safety in Las Vegas. *Security World, 1:* no. 2, September 1964.

Klinger, Keith E.: Fire security is vital. *Security World, 1:* no. 1, July/August 1964.

———: Industrial fire protection. *Security World, 1:* no. 1, July 1964.

Major property-loss fires of 1970. *Fire Journal,* May 1971, p. 28-39.

Meagher, Walter: Supermarket fire security planning. *Security World, 1:* no. 3, November 1964.

Miller, Homer M.: Fire planning for supermarkets. *Security World,* February 1971.

National Board of Fire Underwriters. *Private Fire Brigades,* New York, N.B.F.U., 1949.

———. *First Aid Fire Appliances.* New York, N.B.F.U., 1950.

———. *Employee Organization for Fire Safety.* Boston, N.B.F.U., 1951.

———. *Care and Maintenance of Sprinkler Systems.* New York, N.B.F.U., 1954.

———. *Fire Prevention Code.* New York, N.B.F.U., 1956.

———. *Burglary Resistant Safes.* No. 687, Chicago, N.B.F.U.

———. *Fire Resistance Classification of Record Protection Equipment.* No. 72, Chicago, N.B.F.U., July 1952.

———. *Fire Resistance Classification of Vault and Fire Storage Room Doors.* No. 155 and No. 669. Chicago, N.B.F.U., December 1941.

———. *Security File Containers.* No. 505, Chicago, N.B.F.U.

———. *Installation of Fire Doors and Windows.* Boston, No. 80, 1962.

———. *Standards for the Installation of Sprinkler Systems.* No. 13, Boston, N.B.F.U. 1963.

———. *Standards for Air Conditioning Systems.* No. 90A, Boston, N.B.F.U., 1963.

———. *Standard for Storage and Handling of Cellulose Nitrate Motion Picture Film.* No. 40, Boston, N.B.F.U.

———. *Plant Protection Handbook.* Studies in Business Policy. No. 64, New York, N.B.F.U., 1953.

Parr, Edward M.: Some basics of sprinkler protection. *Security World,* February 1969.

Peek, Edward: Don't give fire a chance to spread. *Security Gazette,* March 1975, p. 94.

———: Fire hazards of that extra bit of warmth. *Security Gazette,* November 1974, p. 419.

———: Smoking—more than just a health hazard. *Security Gazette,* October 1974, p. 383.

Prepare your plant for flood and fire. *Security World Magazine,* author unknown. February 1970.

Ritz, Richard E.: A high-rise fire-resistive office building with automatic sprinklers installed throughout. *Fire Journal,* September 1969.

Ryan, T. A.: Safeguarding records after the fire. *Security World, 2:* no. 1, January/February 1965.

Savage, Donald T.: Fire prevention and control—panel moderator. *Industrial Security, 70:* Por Bio., December 1964.

Schrire, T.: *Emergencies—Casualty Organization and Treatment.* Springfield, Thomas, 1962.

Segal, Louis: Flameproofing: facts and fallacies. *Security World, 4:* no. 5, May 1967.

Shaw, Paul: Halon 1301. *Security Register,* November 1973, p. 33-37.

Shuger, Dr. Leroy W.: New defense against fire. *Industrial Security,* July 1959.

Status quo: fire and safety in security, The. *Security World,* February 1970.

Steinmetz, Dr. Richard: Techniques of fire investigation—panelist. *Industrial Security, 12:* October 1960.

Three hundred twenty-five die in department store fire. *Security World, 5:* no. 2, February 1968.

Titus, Charles M.: Pre-riot retail fire training: A sequel. *Security World,* October 1966, p. 35.

Transue, J. M.: Fire is your business. *Industrial Security,* January 1958.

Tuvo, Richard L.: Dry chemical fire extinguishers. *Security World, 3:* no. 7, July/August 1966.

Wallace, Donald B.: The industrial fire brigade in a civil defense emergency. *Industrial Security,* July 1962.

Watrous, Laurence D.: Plymouth meeting mall fire. A fire to learn by. *Security World,* February 1971.

BOOKS
ARSON

Battle, Brendan P., and Weston, Paul B.: *Arson.* New York, Arco Publishing Company.

PUBLICATIONS OF THE GOVERNMENT, LEARNED SOCIETIES, AND OTHER ORGANIZATIONS

Steinmetz, R. C.: Arson in times of war. Boston, National Fire Protection Assn., 1940.

Straeter-Crawford: *Techniques of Arson Investigation.* Los Angeles, R. L. Straeter, 1955.

———: *Techniques of Arson Investigation.* Los Angeles, R. L. Straeter, 1965.

PERIODICALS

Aids to the detection of explosives. *Security Gazette,* February 1975, p. 48+.

Arson: Its terrifying new dimension. *Occupational Hazards,* June 1970, p. 39.

Arson, window breakage push damage costs up. *Nation's Schools,* September 1969, p. 84.

Barrett: Rope nets for the handling of bombs and suspicious packages. *J Criminal Law, Criminology, Police Sci,* May-June 1941.

Bennett, Glenn D.: Laboratory aids for the arson investigator. *Industrial Security,* April 1964.

Boise, Robert J.: Industrial security and the arsonist. *Industrial Security,* July 1961.

Buzby, W. J.: Arson and the security officer. *Security World,* June 1971.

Conroy, Louis N.: Emergency, riot and disaster control plan for retailers. *Industrial Security,* August 1966.

Dierst, Glenn V.: Arson prevention and control —panelist. *Industrial Security,* December 1964, p. 23.

Donovan, Robert: Strike defense. *Security World, 4:* no. 2, February 1967.

Electronics: New hope for vandalism control. *Nation's Schools, 81:* no. 4, April 1968.

Faulstich, W. L.: Anatomy of a demonstration, Part III. *Security World,* October 1966, p. 26-31.

Five motives for arson. *Occupational Hazards,* June 1970, p. 43.

French, Harvey M.: Current arson problems. *Security World,* February 1971.

———: Investigate that fire! *Security World,* February 1969.

Gardner, John: Arson in schools. *Security Gazette,* February 1975, p. 56.

Hamilton, Peter: The urban guerrilla—a new challenge to security. *Security Gazette,* January 1971.

If strife hits the plants. *Business Week,* June 28, 1969.

Jones, Adrian H.: Internal defense against insurgency; six cases. Washington Center for Research and Social Systems, American University, 1966.

Lavau, George: Political pressures by interest groups in France. *Interest Groups on Four Continents,* 1958, p. 60-95.

L'Hote, John D.: Detroit fights theft and arson. *American School and University, 42:* no. 11, July 1970.

National Petroleum Council. Security principles for the petroleum and gas industries. Washington, D.C., May 1955.

Peek, Edward: Protecting business premises against arson. *Security Gazette,* August 1974, p. 309.

Post, Richard S.: Investigation of arson for security officers. *Security Education Briefs,* OAK Security Inc., *1:* no. 8.

Pros and cons of paying ransom, The. *Assets Protection,* Summer, 1975, p. 29-32.

Thomas, Robert: Militant community organizations funding. *Security World,* July/August 1969.

Vandiver, James V.: Extortion investigation. *Assets Protection,* Summer, 1975, p. 20-28.

Walsh, Timothy J., and Healy, Richard J.: Kidnapping, extortion, and terrorism. *Protection of Assets Manual, 2:* California, The Merritt Company, 1974.

Wilson, Harry W.: Million dollar school arson. *Security World,* March 1975, p. 62-63.

Wilson, W.: Corporate vulnerability to crime. *Conference Board Record,* August 1969.

Wilson, Ralph: Will an office thief strike tonight? *Modern Office Procedures,* June 1964, p. 19.

APPENDIX 17A

INSPECTION OF FIRE PROTECTION EQUIPMENT: XYZ CORPORATION

The following schedule of inspection of fire protection equipment performed by XYZ security department provides a composite picture of the inspectional and surveillance activities of a security department.

Daily Inspections

Trucks:
1. If engine is started, it is to run until water temperature reaches 180 degrees before being shut off.
2. Check ammeter operation for generator charging.
3. Check windshield wipers.
4. Check vacuum boosters on brakes.
5. Check engine oil level.
6. Check water level in radiator.
7. Check fuel by reading gauge.
8. Check all lights.
9. Check horn and siren.
10. Check foot brake.
11. Check hand brake.
12. Check tires (visual).
13. Check booster tank, water level.
14. Check radio.
15. Check pump primer oil.
16. Report all malfunctions to supervisor.

Equipment:
1. Check equipment on truck against truck checklist to determine that all equipment is on the truck and in proper place.
2. Check each piece of equipment to see if it is in proper working condition and check all seals to see that they are not broken.
3. Check position of valves on mounted equipment.
4. Check working parts of equipment to see that moving parts are free and easily moved (except cock valves on dry chemical units).
5. Check equipment for accessibility and removal from truck.
6. Report all malfunctions and deficiencies to supervisor.

Extinguishers:
1. Check all extinguishers daily to see that they are accessible, noting if any of the extinguishers are damaged.
2. Check the seals to see if any have been broken. If any seals are broken, weigh the CO_2 extinguisher. If the weight of the extinguisher shows a decrease of more than 10 percent of the total weight of the extinguisher, the extinguisher must be brought into the station and refilled. If the weight is within the range of full to 10 percent less, it is permissible to reseal the extinguisher and leave it in service. If the extinguisher is a dry chemical type and the seal is broken, the level of the dry chemical must be checked. The cartridge must be removed and inspected to see if it has been punc-

tured. If the cartridge is not punctured and the powder is at the designated level, the extinguisher may be assembled and sealed. If either the powder is low or the cartridge is punctured, the extinguisher must be brought to the fire station and properly recharged.
3. Report all extinguishers in need of service to supervisor.

Weekly Inspections

Trucks:
1. Check tires for proper pressure with tire pressure gauge.
2. Check batteries using hydrometer. If readings are 1200 or below, connect charger to the batteries.
3. Flush the booster tanks.
4. Report all deficiencies to supervisor.

Weekend Inspection Schedule:
1. Check sprinkler systems using inspector's test valve.
2. Check fire alarms by operating the alarm, using key to open box.
3. Check sectional controls valves by actually moving valve to see that valve is in either the open or closed position, whichever it is supposed to be.
4. Sprinkler control valves to be checked by same procedure as sectional control valve.
5. Inner control valves to be checked by same procedure as sectional control valves.
6. Check hydrants by visual inspection. Note if caps are in place, hand tight, if wrench is on hydrant, and that hydrant is not obstructed.
7. Report all deficiencies to supervisor.

Monthly Inspections

Building Inspections:
1. Each building is to be inspected monthly for fire hazards.
2. Buildings will be inspected with attention given to proper storage pertaining to sprinkler systems, allowing proper clearances for efficient sprinkler operations.
3. Wiring will be checked for temporary hook-ups and dangerous use of extension cords.
4. Poor housekeeping is a contributing factor in causing fires. Check for poor housekeeping practices and ways in which they can be eliminated.
5. When flammable liquids are used or stored in a building, check to see that proper provisions have been made for this type operation.
6. Check storage and operations to see that first equipment is not blocked.
7. Check for violations of smoking rules.
8. Report all deficiencies to supervisor.

Operational Checks on Truck Equipment:
1. Dry chemical units will be activated, using as little dry chemical as possible.
2. Foam units will be actuated, condition and quality of foam will be noted. Foam tanks will be cleaned and contaminated foam removed.
3. Report any unusual equipment condition to supervisor.

Hose Rebedded:
1. Bedded hose will be removed from trucks and rebedded with sections coupled together, hand tight. Fold hose in different places so folds will not be in the hose for a period longer than thirty days.
2. Observe condition of gaskets, butts, and hose at this time.

Assigned Location Extinguishers:
1. Assigned location extinguishers will be inspected monthly.
2. The inspection will consist of general visual condition.
3. Check hose to see if it is defective or obstructed.
4. Check to see that the seals are not broken and for signs of tampering.
5. If the seal is broken, the extinguisher is to be taken apart and checked (except CO_2 extinguishers, which will be weighed).
6. If all parts are in proper condition, extinguisher may be assembled and sealed.
7. Report all unserviceable equipment to supervisor.

Dry Chemical Extinguishers:
1. Dry chemical extinguishers are to be checked monthly by dumping dry chemical and observing the condition of the powder. If the powder is not lumpy from moisture, its continued use is permissible.
2. The CO_2 cartridge will be removed and checked for proper weight.
3. The hose will be tested for obstructions by blowing air through it.
4. Gaskets will be checked for condition and proper seating.
5. If there are no deficiencies, the extinguisher can be reassembled and marked as shown in the inspection book.

Semiannual Inspections

Pumper:
1. The CO_2 unit on the pumper will be disassembled and cylinders weighed. If the weight is within 10 percent of weight indicated on cylinder heads, they may be marked "OK" and the system reassembled.

Fire Hose Test:
1. All fire hose will be tested with water pressure every six months.
2. To test, connect hose to pumper and bleed off the air in the hose. Hose may be tested in any lengths as long as all air is bled from hose.
3. The $2\frac{1}{2}$ inch double jacket hose will be tested at 200 lbs. pressure for a period of 3 minutes.
4. The $2\frac{1}{2}$ inch single jacket and the $1\frac{1}{2}$ inch hose will be tested at 150 lbs. pressure for a period of 3 minutes.
5. During the test, observe the hose for pinhole leaks and for couplings pulled from the hose.
6. After the test, break *all* hose at couplings, drain, roll into doughnuts and assemble at pumper for bedding. Hose will be coupled hand tight when bedding.
7. Only dry hose will be reloaded. Wet hose will be hung in tower and replaced with dry hose.
8. Record test on hose record cards in file.

Buildings:
1. A fire inspection survey has been completed on all buildings. Results of the building surveys are on file in the Fire Protection Office.
2. Reinspections will be made on these buildings every six months.
3. The buildings will be checked against the original inspection. Any differences or changes noted in construction or of contents will be noted in a supplemental report and affixed to the original inspection.
4. Construction changes are to be noted on original drawings.

Fire Hydrants:
Fire hydrants will be flushed every six months.

Control Valves:
All sprinkler control valves, sectional control valves, and county connection valves will be closed and opened, once every six months as a preventive maintenance procedure.

Annual Inspections

Fixed CO_2 Systems:
1. All fixed CO_2 systems will be dismantled and cylinders weighed.
2. A check will be made of all moving parts of the system and if they operate satisfactorily, the system will then be reassembled.
3. If any deficiencies are found, they are to be corrected before the system is reassembled.
4. Report all unserviceable equipment to supervisor.

Extinguisher Recharge:
1. All foam and soda acid extinguishers will be recharged the first month of each year.
2. Extinguishers will be taken apart and contents dumped, hoses and gaskets will be checked.
3. The extinguishers will be cleaned and polished.
4. New charges will be prepared and extinguishers charged.
5. Extinguishers will then be sealed and condition registered in extinguisher record book.

Dry Chemical Extinguishers:
1. To check dry chemical extinguishers, disassemble and dump powder, checking for lumps due to moisture.
2. Check hose for obstruction and weigh the CO_2 cartridge.
3. If all parts are in good condition, reassemble extinguisher.
4. If any part is defective, replace it and then reassemble the extinguisher.
5. Record work in record book.

Weigh CO_2 Extinguishers:
1. Weigh all CO_2 extinguishers.
2. Weight must be within 10 percent of the weight stamped on head of cylinder.
3. Check hose and horns for breaks and obstructions.
4. Replace any part that is defective.
5. Record condition and work done in extinguisher record book.

APPENDIX 17B

TYPICAL INSTRUCTIONS FOR USE OF INDUSTRIAL (WET) SPRINKLER SYSTEMS

I. Overhead sprinklers are always filled with water. In the event of accidental discharge, jam red tapered stick found on all fire extinguishers in the plants into round yoke to cut off water spray. *You will get wet.* Ladders are found throughout the plants.
II. Sprinklers are supplied water as follows:
  A. Plant No. 1
   1. Six Systems
     a. There are eight *red* painted numbered post indicator risers on outside of building.
     b. *Under no conditions touch risers No. 7 and No. 8.* They shut down the whole system and have the numbers painted on them.
     c. Each riser has an indicator window which should read "OPEN."
     d. There is a wrench on each riser, sealed and wired down.
     e. There is a drain valve inside the building near an adjustment to the outside riser.
     f. The valve handles or knob is painted green and is round.
     g. There is a pressure gauge next to it.
   2. To *shut off* a discharging line
     a. Break the Seal and shut off the water at P.I.V. with wrench. All wrenches are interchangeable.
     b. Go inside and open drain valve behind door marked with red and white metal sign. This will allow all water in that system to drain out.
   3. To *locate* system covering discharging lines
     a. Alarm rings at Enunciator panel located on wall over doorway to Personnel from Production area.
     b. Number of affected system will show on Enunciator Panel.
     c. Location: No. 1 and No. 2—P.I.V. west of Cafeteria outside door. Drain—Inside on wall of kitchen storeroom. No. 3—at N.W. corner of bldg. *P.I.V. of risers No. 7 and No. 8 are also nearby. Do not touch.* Drain valve is inside of wall in Blue Print Room. No. 4 is adjacent to Personnel Entrance, the drain valve is on wall behind cloak room beside Personnel Entrance. No. 5 and No. 6 P.I.V.

are just East of Accounting Entrance Door. The Drain valves are behind panel on wall at S.W. corner of the Accounting Office.
B. Plant No. 2
 1. There are *5 riser systems* in Plant No. 2 and they are one of *two types.*
   a. *Post Indicator Valve*—a red vertical post with a window indicating OPEN or CLOSED and a wired and sealed wrench.
   b. Rising Stem—wheel to open or close valve. When stem is all the way out, it is open and vice versa.
   c. There is a drain valve for each system with a Pressure Gauge. These are located nearby the risers.
 2. To *shut off* discharging system
   a. Break seal and close valve at riser, shutting off water.
   b. Drain system by opening drain valve.
 3. To *locate* system covering discharging line
   a. Systems all have an alarm bell in the area of the valve and pressure gauge and it will sound when line starts discharging.
   b. There is no enunciator panel. You will have to follow the sound and check the pressure gauges.
   c. Riser System No. 1—Rising stem wheel valve controls outside next to Raw Material Receiving door on east side. Alarm bell is outside. Drain valve is red painted with sign inside north of recciving door.
   d. System No. 2—P.I.V. outside N.W. corner of Interplant Shipping Room.

   Drain valve inside same room, the alarm is also in the N.W. corner.
   e. System No. 3—Rising stem wheel controlled valve—*Inside*—behind wall panel in Punch Press Inspection room. Drain valve is about 6′ up on the riser, behind wall. It can also be reached thru the Cafeteria on North wall, open wood panel door, high on wall near Drinking Fountain in center.
   f. System No. 4—Inside in floor under metal cover in Die Vault, S.W. corner. Has a T-bar valve control handle. Drain valve on other side of wall, red painted, in punch press area.
   g. System No. 5—Rising stem wheel handle control valve outside in firehose shed against north wall. Drain valve inside on north wall at center, under Sprinkler Sign. The Alarm Bell is outside.

APPENDIX 17C

AUDITING THE SECURITY SYSTEM*

Evaluating the effectiveness of an installed security system is essential to its success. Performing audits in-house is particularly advantageous because it permits continuous monitoring.

Internal security audits are highly recommended. They must be performed by people knowledgeable in data processing

---

* Reprinted by permission from "The Considerations of Physical Security in a Computer Environment" 1st. Ed. © 1972 by International Business Machines Corporation, pp. 26-29.

and auditing principles. The auditors should not assume responsibility for the development of a security control system, but should evaluate the established controls and procedures. Also, they should not be responsible for enforcing control; responsibility for enforcement rests with local management. The objectives of the internal security audit should be to determine the adequacy of security controls.

A full internal security audit should be performed at a frequency appropriate to the size of the organization. More specialized or selective audits can then be scheduled periodically to review the strengths and weaknesses identified in the full-scale audit.

It is advisable to have the audit conducted by a department other than that which is normally responsible for security. In other words, it would not be logical to have the computer center or the security staff perform an audit of their own functions.

The checklists that follow are general guides to procedures and points that should be examined in detail and should be modified to meet the needs of a specific user.

SAMPLE CHECKLIST 1—
GENERAL AUDIT GUIDELINES

A. Adoption of Overall EDP Security Procedures

1. Are responsibilities for security firmly fixed and clearly understood by all those having such responsibilities?
2. Are there established security standards, procedures, and guidelines?
3. Is compliance with the security standards, procedures, and guidelines readily auditable?
4. Are variances from security procedures recorded, and are corrective measures taken?
5. Are there periodic independent security audits scheduled, and are these schedules followed?

B. Internal Controls

1. Are existing controls adequate to protect incoming data from time of receipt, through conversion, to entry into the system?
2. Are existing controls adequate to protect system output through delivery to the end user?
3. Are controls established for exceptions to the normal handling of input/output data and documents?
4. Are errors and exceptions brought to the attention of the people responsible for correcting them?
5. Are adequate measures provided to restrict access to the computer room facilities?
6. Are the storage areas that are being used to store sensitive data files, operating procedures, and documentation sufficiently secure?
7. Is there compliance with adequate procedures for the control of operator coverage, operator console logging, and operator access to sensitive tapes, disks, programs, and documentation?
8. Are adequate library controls established for sensitive tape and disk volumes? This would include records of usage, copies at different locations, an adequate retention plan, an inventory, and a list of authorized users.
9. Are authorization procedures established for overtime computer usage and for use of programs, tapes, disks, and documentation?
10. Is proper separation of duties established for controlling sensitive information, in monitoring established controls, maintaining copies of tapes, disks, etc., at a separate location, and checking and recording receipt and distribution of input/output?
11. Are the security procedures established for blanking or purging intermediate storage (including scratch tapes and disks) after computer processing applied to information sufficiently sensitive to warrant this additional step?
12. Are control and review procedures

established for monitoring program changes and patches?
13. Is there an adequate internal awareness program on security?
14. Are adequate security practices applied to all application development activities?
15. Are teleprocessing security needs properly defined and satisfied?

C. Vital Records

Is there an effective vital records program?

SAMPLE CHECKLIST 2—MAGNETIC TAPE (AND DISK) CONTROLS

A. Internal Controls

1. Is the responsibility for the administration of the tape library properly defined and applied?
2. Is the tape library adequately controlled during all shifts?
3. Are all tapes containing unusually sensitive data properly segregated and controlled?
4. Is a separate access log maintained for these tapes?
5. Is the tape library adequately protected against fire?
6. Are files backed up to permit reconstruction?
7. Review the method of tape shipments to other locations.
   a. Are tapes sufficiently protected against damage, loss, or interception?
   b. If sent on a nonreturn basis, is interplant billing rendered?
   c. Are tapes sent directly to the corresponding tape library?
8. Review the master tape replacement operation.
   a. Is the old master retained pending run verification?
   b. Is the old master monitored by the tape librarian to prevent misuse or premature scratching?
   c. Is the entire replacement operation adequately controlled?
   d. Is the "son-father-grandfather" backup theory used?
9. Are procedures adequate for ordering and controlling new tapes?
10. Is there a listing that cross-references job number and tape location in the tape library?
11. Are library personnel job responsibilities clearly defined?
12. Does the tape library manager receive reports and take corrective action for:
    a. Tapes on loan an excessive length of time
    b. Tapes not located in periodic inventories
    c. Tapes authorized for release or return but which cannot be located
    d. Tapes for which a responsible person is not identified (usually because of changes in personnel or job responsibilities)

B. Processing Timeliness and Accuracy

1. Are job control sheets consistent, and do they accurately record reel numbers?
2. Are tape reels clearly identified as to job number, reel number, storage location, etc.?
3. Examine tapes in all programming departments, and any other departments, which use a large number of tapes. Test-check the tape library's records to determine that these tapes have been properly signed out—especially tapes that have been on loan for a long time.

C. Exceptions—Validation

1. Check the "programmers' hold area" and programming department for any:
   a. Unauthorized possession of tapes, especially master or confidential tapes
   b. Stockpiling or nonuse of scratch and testing tapes, resulting in a shortage of tapes for use by the tape library
2. Determine if any tapes have been restricted for one individual's or de-

partment's use for an extended length of time.
3. Is there follow-up for release of dormant restricted tapes?
4. Is there follow-up for return of confidential tapes to locked cabinet?

SAMPLE CHECKLIST 3—OVERALL DP SECURITY

A. Internal Controls

1. Are predetermined totals or item counts maintained within the DP operation and compared with keypunch, unit record, or computer output before being sent to the customers?
These controls should apply to accounts payable and ledger distribution input, as well as any job where input data is used in more than one program. The person maintaining controls should not be one who is processing the data, such as unit record operator.
2. Determine if the management of using departments approves the programs before they become operational.
3. Are all jobs or revisions supported by a written request?
4. Is approval and justification required before development of new programs within an established system?
5. Can changes be made to programs without the using departments' knowledge and approval?
6. Is the procedure for updating program documentation effective?
7. Key entry: are all important data fields verified?
8. Are limit checks included in appropriate programs? For example, dollar limitations may be programmed for payroll checks.

B. Processing Timeliness and Accuracy

1. Is program documentation adequate? Does documentation include:
    a. Systems flowchart showing portion of the system the program represents and equipment used (keypunch, unit record, computer, etc.)?
    b. Program flowchart—logic of program instructions?
    c. Narrative description—what is the objective of the program? What data is the user receiving?
    d. Program listing (printout of the source deck)?
    e. Sample of output (listings, cards, etc.)?
    f. Record layouts (data included in tape and punched card fields, etc.)?
2. Are operator instructions clear and complete? Do operator instructions for running each job include:
    a. Identification of all machine components used?
    b. Identification of all input/output forms?
    c. Explanation of purpose of run?
    d. Detailed setup (tape reels, cards, disk to be used, etc.) and end-of-run instructions?
    e. Identification of all manual switch settings?
    f. Identification of all possible programmed halts and prescribed restart instructions?
3. Are data control personnel provided with schedules listing dates on which programs will be run, due-in and due-out time, customer-provided input data, and distribution of output? Is the workflow monitored?
4. Is the backlog of jobs reasonable? Review for excessive delays.
5. Are target dates established for "hands-on" time rerun (because of operator or programmer error) realistic?
6. Are all applicable items sent to vital records and updated as required?

C. Exceptions—Validation

1. Has responsibility been established for following up all input errors to ensure that they are properly corrected and returned for processing?
2. Are the exceptions (or significant events) logged by machine operators reviewed by management, and is action taken?

3. Are reasons determined and corrective action taken for rerun hours (machine—operator—input—program)?
4. Are all physical inventory variances resolved?

D. Equipment Accountability and Billing

1. Are physical inventories taken periodically and reconciled to equipment rental invoices? The auditor should perform a physical inventory and reconcile to latest equipment rental invoices.
2. Is equipment rental being billed correctly?
3. Are samples of submitted billable time authorizations compared with billing?

E. Program Patching

1. Are patches adequately documented?
2. Is program (source) listing annotated with patch number, date, and explanation?
3. Are operator instructions revised to include steps to be taken if program failure occurs because of incorrect patches?
4. Is there excessive patching?

F. Control of Computer Usage

1. Are tests, assemblies, and "hands-on" time (computer operation by the programmer) adequately controlled?
   a. Are programmers required to obtain written permission from their department manager for all "hands-on" time?
   b. Is management able to determine whether programmers are making excessive tests and assemblies because of poor programming techniques?
   c. Are targets established for reasonable "hands-on" time rerun because of operator or programmer error?
2. Are equipment utilization reports generated to aid in planning and justifying continued use of equipment?
3. Are there reports identifying equipment used by classifications—that is, productive time, test and assembly time, rerun time, no scheduled activity time, and programmer "hands-on" time?
4. Are there adequate procedures for distributing costs to using departments?

G. Operations

1. Do operators have access to program flowcharts, source decks, program listings, etc.? (These are not necessary to operators' duties and should be maintained outside the computer room to prevent changes to programs or operation by computer operators.)
2. Is an operating log maintained to record any significant events and action taken by the operator? (Proper recording would indicate whether operators were following instructions for halts in programs, etc.)
3. Trace the flow of operational data through the computer and/or machine room; is adequate control maintained over the input/output data?
4. Is there adequate separation of responsibilities for security?
5. Do service requests include pertinent data, estimated run time, and programming hours required?
6. Has management approval of all requests been secured?
7. Are rerun hours (machine—operator—input—program) controlled?

APPENDIX 17D

A CASE STUDY—IBM'S ADVANCED ADMINISTRATIVE SYSTEM*

The Advanced Administrative System (AAS) is the administrative system for the Data Processing Division of IBM. It handles accounts receivable, order entry, scheduling, commission accounting, customer master records, asset and billing control, and payroll for the branch offices of the Division.

The system is oriented toward administrative rather than management information and, therefore, functions more as a data collection system than as an inquiry system. (It feeds the Advanced Information System, which is the Division's management inquiry system.)

AAS serves approximately 6,000 administrative people in about 300 locations. Communication is through display terminals, over voice-grade lines, and through strategically placed line concentrators. These concentrators are connected to the central location via high-speed lines. The supporting hardware complement is changed with changes in load and current technology.

Communication between the user and the system takes place in a conversational mode; the user supplies a minimum of data, which the system validates with feedback. The system provides supportive default options for details that the user does not know; for example, System: ENTER CUSTOMER NUMBER, IF UNKNOWN DEPRESS SHIFT-AND-ENTER; if the user enters a customer number, the system validates it by feeding back the name and address; if the user defaults, the system requests limited data describing the customer, feeds back customers meeting the description, and asks the user to pick one by line number.

By letting many locations access the central file, AAS does away with the need to maintain copies of the file locally, thus saving money but also making business easier for everyone using the same data. However, improved access to concentrated data has hazards associated with it.

SECURITY PHILOSOPHY

Security is considered to be a tradeoff against utility; it competes with other objectives for resources. Alternatives are evaluated in terms of "acceptable business risks." The measures of adequacy of security that are used in AAS are:

1. Can the same degree of security be maintained in the AAS environment as before?
2. Considering the economics, have we done what a prudent management would do?

By improving access, by concentrating data and assets, and by eliminating paper files which could be used for backup, AAS complicates the task of achieving security. To the degree that it does, new tools are provided. Among these new tools is the system itself, since it does a good job of ensuring the uniform application of standards, policies, and procedures.

The resources, organization, procedures, and enforcement techniques that AAS uses to try to achieve security serve as an example of how protective security measures can be implemented. It should not be inferred that these measures are complete or ideal; they evolved through trial and error and are subject to continual evaluation, testing, refinement, and adjustment. Likewise, it is not suggested that all these measures are appropriate for all computer users. AAS is a relatively large and complex system, but probably contains some security features that would be of interest to other organizations, large or small.

---

* Reprinted by permission from "The Considerations of Physical Security in a Computer Environment," 1st ed. 1972 by International Business Machines Corporation, pp. 30-36.

PHYSICAL SECURITY CONSIDERATIONS

Building

One of the most significant resources provided for the security of AAS is the building in which it is located. The building contains 67,500 square feet of space on three floors and is windowless. It is constructed of steel and both precast and cast-in-place concrete.

Access to the building is restricted by having only five entrances. Normal access to the computer room is restricted to those people whose jobs require it on a regular basis. This would include operations and control personnel, but would exclude all but a few programmers who require a special testing environment. Access is controlled by badge-actuated locks.

Access on an exception basis is by job need also, but is controlled by personnel in the scheduling room. A manager participating in a test or conducting a guest might be permitted exceptional access, for example. This is controlled by a log and a hand-pass that must be surrendered on exiting.

Note that normal work, such as tests and batch jobs, enters the computer room through the scheduling room. Jobs that are hand-carried by authorized programmers may run only on limited-resource test machines. All machines, even test machines, are run by operations personnel in accordance with standing or special instructions from their own management.

Power and Communications

Because of the building's location, exposure to noise on the electrical power lines and to brownouts and blackouts is high. To protect against this hazard, AAS has installed a power source that supplies noise-free power at a constant voltage.

The standby power supply supports a circuit called the essential power circuit. This circuit drives the realtime system; batch, test, and offline systems are on another circuit, which is driven by a diesel generator.

There is a similar exposure to a failure on the part of the communications common carrier. However, performance to date indicates that the communications system fails more softly than the power system, partly because service is modular. There is an exposure to a total loss of communications for an extended period, but the alternatives are expensive enough to make it an acceptable business risk at this time.

While AAS functions in realtime mode, it does not have to respond to a realtime stimulus. If a record is not processed this minute, hour, or day, it will still be there. This differs from an airline reservation system, for example, where the customer may go to another airline. Therefore, while delay is costly, it is not catastrophic.

Tape Library

The physical AAS data base is a significant resource, and some special assets are employed for its protection. Among these are the tape library, the vault, and a remote underground facility.

Each floor of the machine building is served by a tape library. Each has a capacity of 27,500 reels, expandable to 37,500. The library is protected by two-hour rated fire walls and doors. The walls extend from ceiling to subfloor. The air-conditioning ducts serving the libraries are equipped with fire dampers.

The libraries are protected by independent Halon* flooding systems which are activated by heat detectors. The heat detectors are also wired to annunciator panels in the security office and facilities control (see "Fire" below).

The fire doors open into a work area which is separated from the machine room. Individual tapes can be passed to the machine room over a normally open counter. Groups of tapes for large jobs can be taken into the machine room on a cart via a normally closed door.

Tapes are pulled by the librarian in re-

---

* Registered trademark of E. I. duPont de Nemours & Co.

sponse to "buck slips," production work orders signed by operations management, or requests from an operator. All tapes are logged out by the librarian. He also retains the authorization for releasing the tape.

The vault and the underground facility are used to store the backup files for the program library and the realtime data base. The vault protects magnetic tape from temperatures of 1500°F for up to four hours. The underground installation is a large, commercially operated, secure facility in a remote location.

In the machine room or the library, it is possible to restore the data base and the program library to its current status. In the vault, it is possible to restore to end-of-processing on the previous night. In the underground facility, restoration is to end-of-processing on each of the two previous Fridays. Each of the backups requires 100-120 tapes. The underground facility has two iterations to permit backup in the event of damage in transit.

Fire

A concentration of data and hardware such as in AAS justifies concern for fire. AAS depends upon early warning for fire protection.

Heat and smoke detectors are installed in both the floor and ceiling plenums (floor is supply air and ceiling is return air). They are set to ignore cigarette smoke, but will trip on pipes, cigars, or smoke generated by power tools.

The detectors are connected to annunciator panels in facilities control and in the security office. When a signal is received, it is confirmed at both panels, and trained maintenance personnel are dispatched to the indicated area. The indications are to room, floor, and floor or ceiling. A true fire would usually trip a ceiling detector because air is being drawn floor-to-ceiling. The ceiling detectors have indicator lights to localize the alarm within the room.

Communication is maintained between the maintenance personnel and the security personnel via radio transceiver. Maintenance personnel will attempt to control the fire using portable CO_2 fire extinguishers. Single switches near each door can be used to kill electrical power to the whole room. As a last resort, fire hoses located just outside the machine rooms are long enough to reach any point in the room.

The security personnel will call the fire department, and will order partial evacuation or total evacuation by using the public address system or the fire evacuation alarm, as appropriate.

Note that the only thing that takes place automatically is warning. All steps beyond that are planned, but taken only as appropriate. This philosophy is used for two reasons. First, as discussed elsewhere in this manual, automatic response systems often are hazards in themselves. Second, in a complex as large as AAS, it is difficult to apply automatic responses selectively. For instance, automatic shutdown of the air movement system cannot be done selectively, since the floor and ceiling constitute a continuous plenum. Because the most likely kind of fire will be localized, and a larger one will probably require time to get under way, there should be time to make a decision. An automatic CO_2 system exemplifies both considerations. If used over the whole floor, it could disrupt operations and endanger personnel. On the other hand, the only way to localize a fire effectively is by a hand-carried extinguisher.

The concentration of computing power in one location makes it important to keep the system running. Response to potential hazards must be measured and judicious. Thus, there is no automatic response. However, if AAS were not a three-shifts-per-day, seven-days a week operation, automatic response systems would be more necessary.

FUNCTIONAL ORGANIZATION OF AAS

Security is provided for in AAS by including it in the mission of all organizational entities, by establishing missions in such a way that sensitive duties are separated into two or more organizations, and by ensuring that these organizations check on each other.

Obviously, security is not the overriding concern, and it need not be. Security is generally compatible with other management objectives, and the organization that includes this objective is generally similar to the one that would result if security were not included.

Basically, AAS customers are the branch offices and the administrative functional areas (that is, sales order control, asset and billing control, product scheduling, and internal asset management). AAS develops programs and procedures to meet the needs and specifications of the functional areas. It provides processing services to both the functional areas and the branch offices. Branch office communication is via a visual display, and services are controlled by the realtime security system. Services to the functional areas may be via a visual display or in batch (defined as requiring access to the AAS data base) or offline modes. Batch and offline services are provided and controlled by AAS Systems Integration and Control (SIC). Thus, the development groups write programs for the functional groups, and the programs are run at the request of the functional groups but under control of SIC. In other words, three groups are involved, each receiving services from and controlling the others.

The organization of SIC illustrates how it controls the development and functional areas; it contains the following groups:

1. Data base management is responsible for the physical data base (logical content of the data base is the responsibility of the functional area), allocates space, decides when to reorganize or reconstruct, and authorizes all batch runs.
2. Application management is responsible for the program libraries, controls all changes, and decides what subloads should be concatenated and in what order.
3. Production control is responsible for scheduling the hardware resources; all production jobs are submitted to operations through this group.
4. Testing control is responsible for providing test resources and controlling testing, participates with the development group in planning and evaluating tests, and approves of all changes to the realtime library.
5. Operations is responsible for running all jobs on all systems; it is currently organized into two groups: the first, data base operations, works under the direction of production control and is responsible for running realtime and batch; the second, test system operations, works under the direction of testing control and runs all jobs not requiring access to the AAS data base.

MAINTAINING CONTROL IN AAS

The need for consistency in a system like AAS is apparent. To achieve consistency, several interfaces have been established—for the programmer, operations, and users. These do not hamper performance; rather, they eliminate duplication of effort and facilitate documentation, thus making programmers even more productive.

Programmer Interfaces

The first of these interfaces is the Data Definition System. This system assists the programmer in defining and documenting records in accordance with previously defined terms. It ensures that he conforms to AAS documentation standards, uses standard labels, and is aware if he duplicates data already in the system.

The second major interface is provided by the AAS macros. This set of high-level macros is designed specifically for this installation; all application programs are written using these macros so as to make field attributes and registers transparent to the programmer. They provide standard interfaces to other subsystems (for example, data management, teleprocessing, and security) and produce only reentrant code.

They permit a programmer to write code for this complex system with relatively little experience and training.

While the macros are providing all these services to the programmer, they also control him. Note that he cannot violate register conventions, write nonreentrant code, or bypass security, teleprocessing, or data management because the macros do not contain facilities for any of these alternatives. If a programmer wants to do something that is beyond the scope of these macros, he may request a new macro from another group that is responsible for standards. This concept of exercising control at the programming language level is very powerful and flexible.

The third major interface is the Data Management System. This system functions as a file clerk for the application programmer. It files and retrieves records in the data base, maintains indexes, maintains a complete audit trail, and provides for reconstruction. The application programmer communicates to data management across the channel-to-channel adapter via AAS macros. However, these macros and the interface ensure that he cannot bypass the audit trail or otherwise violate standards.

The AAS macros also provide an interface between the application program and the security system. Security identifies the user to the application program, tells it what the user wishes to do, and verifies that he is both authorized by line management and capable, according to the AAS standards of training, to do it. This interface relieves the application programmer of those responsibilities and ensures that they are carried out (that is, there is no way for the user to get to the application program without going through security).

All of these can be viewed as program-to-program interfaces, or the division of sensitive programs among multiple programmers.

The AAS test system is another interface that the programmer maintains. The testing control group provides two testing resources, permitting four different kinds of tests. The first testing resource is called the test driver. In unit mode, the test driver permits a single program module to run in a low-risk, isolated environment. All of its interfaces to other systems, programs, and the data base are simulated and monitored for the programmer. The other mode under the test driver is called string test, in which several modules can be tested together. Their interfaces are real; all others are simulated.

The second testing resource is called the model test system. In this system, a module to be tested runs with the other programs and subsystems with which it will run in realtime, but a model of the data base is used. The first mode of test under this system is called paper test; in this mode, primary input (user terminal input) is entered on cards, and output is on paper. The second mode is called tube test; primary input is from the display terminal, and output is on the display. Printouts of the work area are available on demand.

The programmer uses these systems through the testing control group, which is responsible for the availability and control of the test system. The group helps the programmer select appropriate test modes for his test and assists in designing the test plan; it participates with the programmer in the actual test and helps him determine when the test is adequate. The testing control group provides management with visibility into the testing process, so that management is in a position to note variances and take corrective action. (Some development groups carry this process a step farther. One programmer codes a module, while another prepares the test data from the same specification. They work together on the actual testing.)

The programmer does not test directly on the machine. In fact, he does not even interface directly to operations, but rather through the testing control group. The control group represents a service that makes it unnecessary for the programmer to go near the machine.

Another service that the programmer uses is the library maintenance system. This system keeps track of all source and object versions of every program module in the system. The programmer need not maintain card decks; he can recompile simply by submitting change cards. Each object module and its source can be referenced by a unique historical job number. The library maintenance system is under control of the testing group. It relieves the programmer of the burden of maintaining card decks. It provides management with a history of every module, telling who is responsible for it and what it looked like at any time in the past. It also provides a backup in case an object module is lost or damaged. Thus, it performs as an efficient audit trail.

The program is finally promoted to the realtime run library. This is the most sensitive interface that the programmer maintains, since his program exercises control over the real data base.

Because promotion is such a sensitive function, eight different people or groups participate in it. The programmer initiates the process by filling in a "promotion sheet." He identifies each module by name and job number. He states the purpose of the change and prerequisites, if any. His management signs the sheet signifying that the change is approved and that whatever tests were deemed appropriate were made to ensure that the program will do what it is specified to do and nothing more.

Next, the sheet is signed by the testing control group signifying that the module has been tested in accordance with its standards. With those approvals, the sheet is forwarded to the application control group, which uses it as its authority to move the program to the live library. However, the move is accomplished through operations, and operations requires authorization from application control to run the job which performs the move.

The application control and online control groups make the decision to exercise the new program in the realtime system. Online control is concerned with the risk to system performance (measured against response-time and mean time-to-interrupt criteria) associated with the changes. The group attempts to minimize this risk by spreading changes over time, and by maintaining a fallback strategy and capability. It will stop changes altogether if performance degrades dramatically or is poor enough to adversely affect the ability of the user to do his job.

After the program is finally exercised in the real system, it is evaluated by the user and the functional group. This evaluation constitutes the final check that everyone else has done his job.

The programmer, test control, operations, application control, online control, the functional area, and finally the user all participate in getting the program from design to final acceptance test.

Normally, management does not allow programmers to exercise their own programs in the realtime system. This is to protect them from the risks associated with their ability to exercise blind options to cause the program to do something beyond its intended scope. For example, the job description of the individual who wrote the SIGN-ON program provides that he will never use the realtime system without management supervision. (See "User/Terminal Interface" for how others are excluded from using certain programs.)

Operations Interface

There are two operations groups in AAS: data base operations is responsible for operating all systems that use the AAS data base; testing and support operations is responsible for operating the test systems and support system. Production control is data base operations' customer. While testing and support operations performs some services for production control, its principal customer is testing control.

The other interfaces maintained by operations are the operating system (via the card reader) and the tape library. Viewed in this way, the control of operations is not significantly different from what it would be in a batch environment.

Input and output are controlled through the production control shop. The console log and the production work orders (PWO's) constitute the audit trail. For every job run, operations should have a work order to justify it. Thus, production control must be involved, and since it cannot initiate PWO's, yet another group is brought in (typically, PWO's are originated by the functional area).

The key concept to security in the computer room is that of "control through service." A simple and effective way to exercise control is by providing a service, and computer operations is such a service. It makes it unnecessary for a user to run the system because a trained, experienced operator is always available. Furthermore, management has the control required to ensure that the system is used efficiently and as intended. This can be viewed, of course, as separation of sensitive duties by specialization.

User/Terminal Interface

Responsibility within the AAS security system is divided as follows.

AAS USER. The user is responsible for maintaining security in his use of the system. He should protect his security code from disclosure or compromise and report any such events to his manager. He should protect against observation by other parties of sensitive data displayed or printed in the course of his use of the system.

SECURITY OFFICER. This man is responsible for the secure use of the AAS system in his location. Normally, he is cleared for all transactions required to carry out the mission of his location. Within his location he may add new users to the system, and he may delegate any of his authority to them. In cooperation with the security administrator, he may delegate authority to other locations, change security codes, assign users to another location, and create additional security officers. He is responsible for detecting variances from standards in use of the system and for taking appropriate corrective action.

SECURITY ADMINISTRATOR. This individual is responsible for the day-to-day administration of the system. He functions as staff for all levels of management, delegates authority over the system in accordance with standards or directives provided by responsible management, and is responsible for noting variances from standards and recommending corrective action.

APPLICATION SERVICES. The responsibility for the development of the security system rests with the application services group of the systems and programming design department. This responsibility includes the establishment of objectives and the development of all programs, procedures, and standards required to meet those objectives.

APPLICATION DEVELOPMENT GROUPS. Authority is delegated and controlled in terms of logical transactions. The development groups are responsible for designing transactions in accordance with security standards. These standards require that transactions be limited in scope, be predictable, and provide for separation of sensitive duties. Transactions must be limited in scope so that the security officer has maximum flexibility in matching the scope of a person's job with his functional control over the system. Transactions must be predictable so that the security officer understands what he has delegated.

AAS Security System

The components of the AAS security system are the identification, authorization, and audit trail subsystems.

IDENTIFICATION SUBSYSTEM. An AAS user is identified by his IBM employee serial number. To indicate to the system that he is the person identified by a given number, the user must also enter a four-letter security code. This code is unique to the user and is known only to him and the system. He must enter the code once for each transaction that he wishes to perform.

The code is generated automatically by the system. It is changed monthly (to protect against an undetected compromise) or upon demand (in the event of a detected compromise).

AUTHORIZATION SUBSYSTEM. A user must be authorized by his management to perform a transaction. The system maintains a list of the transactions for which each user is authorized. Management is provided with a realtime transaction—security authorization—for creating, adding to, or deleting from this list.

The rule used by a manager to delegate authority is: "a user should be able to do anything within the system that is required for him to do his job; that, and nothing more." Note that this is a line management decision; only the user's manager can make it. It is for this reason that in a system which has thousands of users a realtime capability is provided in AAS. In a smaller system it would be possible to delegate the authorization function to a single staff person and still remain responsive.

For some transactions a user, in addition to being authorized by his own management, must meet an AAS prerequisite; this is normally the completion of a computer-assisted instruction (CAI) course which covers the transaction. When a user meets the prerequisite, it is recorded in his capability profile. This is usually handled automatically by the system; however, it can also be entered by the security administrator.

When a user indicates that he wishes to perform a transaction, the system verifies that he is authorized to do so by his management. It then verifies whether there is or is not a prerequisite for that transaction and, is there is, determines if the user has met it. Only if the user is authorized and, when appropriate, capable, will he be permitted to process the transaction.

AUDIT TRAIL SUBSYSTEM. The audit trail in AAS is provided by the data file journal. This file contains one entry for each change to the data base, and one for each transaction. The entry for a change to the data base points to the terminal, the transaction, and the transaction record. The transaction record points to the user. Thus, every change to the data base can be traced back to the person responsible.

In addition to the data file journal, a record is written of all security variances (security variances log), and of all changes to the user profiles. The log is sorted each month by location, and is distributed to local management. This management determines whether and what corrective action is needed, and when it is to be taken.

To permit more timely corrective action, the terminal is "locked" (made unresponsive) if two variances are detected before a valid transaction is completed. The terminal can be unlocked only by the security officer in that location, or by someone he has authorized to do so. At unlock time the system feeds back, through the terminal, everything it knows about the variances. This permits local management to determine whether a violation has occurred, and what corrective action is required. Note that most variances are keying errors, not violations. The manager on the spot is the only person in a position to determine which variances are violations.

Chapter 18

EXTERNAL PROBLEMS

THE PROTECTION of an organization involves the security for physical facilities, operations, personnel, and visitors. Threats to the continuity of operations, personnel, and facilities can be controlled when initiated from inside the organization more easily than those from external sources. The external environment is also controllable but to a lesser degree. External threats can be directed against any aspect of operations or personnel without warning. The protective program of the organization must, therefore, determine the risks present and develop control and prevention programs to reduce or eliminate them.

ARSON

Willful or malicious burning, with or without the intent to defraud, constitutes a criminal act which is high on the security priority list. Arson can be initiated either for gain or other motivations including revenge, intimidation, spite, etc. It may also be used to conceal the occurrence of other crimes such as burglary or embezzlement. It can also be the result of pathological fire setters (pyromania) or vandalism.

The security department prevents arson by denying access to the facility to anyone who could set a fire. Since incendiary devices with time delays can be used, the officers must follow adequate inspectional procedures to discover potentially dangerous situations before they result in damage to the facility. Fire, smoke, heat, and rate of rise detectors all assist in locating fires at an early stage. The alert security officer should, however, discover the ingredients of the potential arson before they ignite and are detected by the sensors.

Inspections conducted by the security officer should include those areas which could indicate the possibility of arson conditions being present. Such conditions include:

- Indications of forcible entry
- Unusually large or small inventories
- Irregularities with electrical, gas, or sprinkler systems
- Unusual or unfamiliar odors
- Unusually cluttered or unsafe condition

In addition to noting these prefire conditions, the officer should make careful note of the type of smoke, flame, or odors present when he discovered the fire. The characteristics of various fires indicate the materials used to cause them.

- White smoke—Vegetable compounds (hay, phosphorus)
- Black smoke—Petroleum products (gasoline, oil, grease)
- Yellow or Brownish-yellow—Sulfur, nitric acid, hydrochloric acid, smokeless gun powder

The presence of smoke indicating the presence of materials not normally found in the areas could indicate the possibility of arson and would assist the arson investigator in conducting the investigation.

BOMB THREATS

The threat of destruction from bombing has become an increasingly serious problem for protective programs. The use of bombing by terrorists, criminals, and psychopaths to gain political, financial, or personal satisfaction has reached epidemic proportions throughout the world. Extortion by bombing or threats of bombing; indescriminate terror by bombing high-traffic public and private facilities; and selective bombings for political propaganda all have a similar effect: loss of property, lives, and disruptions of normal human activity. Any or all of these effects are desired by the bomber(s). The responsibility of the security program is to prevent the bomb attack; locate and neutralize the bomb, and apprehend the attacker.

Depending on the motivation of the attacker, capabilities, associations, and visibility, the attack might not be prevented through normal intelligence collection techniques. The defensive systems established, however, should be able to prevent any attack, or at least, minimize the effects of it. Physical protection, site hardening, and internal controls including searching, identification system, and access control all attempt to lower risk. Likewise, police operation to arrest known terrorists and extremists or to monitor their activities are often used to control their ability to perpetrate an attack.

The protection of individual facilities either governmental or proprietary is, however, accomplished by appropriate security measures being developed and implemented.[1] The application of physical and personnel security processes will insure that the risk is lowered and the threat minimized.

BURGLARY

The unlawful entering of a building to commit a felony or theft is one of the most common problems of security programs. There were over three million burglaries in 1974 with a total loss of $1.2 billion. Residential losses amounted to $758 million with nonresidential losses being $423 million. The average loss per burglary was $391.

The security department has the role of both attempting to prevent the occurrence of burglary and investigating any which occur. The application of physical security measures and the use of alert and well-trained security officers in conjunction with well-developed procedures and controls will eliminate much of the risk of burglary. Since burglary is a crime of opportunity, the effective security program lowers the risk of this criminal attack and reduces the opportunity of the criminal.

Cooperative programs with other businesses in the area of the facility to be protected as well as active cooperation with public law enforcement agencies will strengthen individual protective measures. Likewise, the application of site hardening and intrusion detection equipment will significantly increase the protection levels for the facility.

CIVIL DISTURBANCES

Civil disorders, riots, and violent confrontations between groups have posed serious problems for the continuity of business and governmental operations. Protection from losses, continuity of functions, and preventing injury or loss of lives requires adequate preplanning.

Sound security programs have the basic components for facility and employee protection in the event of these conditions. The coordination and aid of other police, fire, or protective agencies is, however, essential in any major confrontation.[2] Internal procedures and controls to reduce damage or injury should be made as well as providing for emergency shut-down of the facility.

CREDIT CARDS

The fraudulent use of credit cards is one of the major threats facing the business community. Millions of dollars annually are stolen from business and banking institutions through the illegal credit card transaction. Stolen and lost credit cards are used by individuals and organized groups to acquire cash, merchandise, and services on a worldwide scale. Losses are due to improper security and identification procedures by firms who accept the cards or by processing procedures by credit card companies. The liability to holders of credit cards has been limited to $50, placing the responsibility for protection of card company assets directly on its internal protective system.

Verification of persons using the card at the point of sale is critical to security of operations. Likewise, insuring that sufficient credit is available to insure payment of the merchandise or service being purchased is essential. Such information must be present in a convenient manner from the card company for sellers to use the necessary verification systems.

EXECUTIVE PROTECTION

The protection of dignitaries and corporate executives, their families, or employees has become an increasingly serious problem. Loss of life, injury, holding for ransom or political advantage are extremely serious outcomes because of ineffective or nonexistent security measures. The damage, embarrassment, loss of privacy to the corporation or government far outweigh the costs associated with sound protective programs.

Executive protection involves understanding the possible threats directed against the executive, his family, home, activities, while traveling or working or at leisure. Involved as well is understanding the sources, be they criminal or political. Effective protection requires that a thorough knowledge of the person to be protected and a commitment on his part to participate in a protective program. Unless there is a willingness to participate and a corresponding change in life and working style, protection cannot be more than minimal and haphazard.

Control, coordination, and communications are essential for the execution of any security operation for protection of dignitaries. Planning is the major consideration in the security of dignitaries. Good protection is an intangible quality. Protection is of the highest quality and is most successful when nothing is allowed to disturb the peace and security of the person protected.

There are three basic considerations in the security of dignitaries. First, protection is a buffer established around him which will prevent an attack or absorb the shock of an attack. Since absolute security is seldom possible, these security officers should always operate in such a manner that attempt attack will have the least possible chance of success. To achieve this protection, it should be established "in-depth" through a system of barriers blocking all logical approaches to the dignitary. This way, an assailant is forced to disclose his intentions before he gets too close to the dignitary. Second, protection must be thoroughly planned in advance. Three objectives must be kept in mind during the planning stages: (1) The dignitary must be protected from all hazards regardless of whether they are caused by design, negligence, or accidents; (2) the protection must not unnecessarily interfere with freedom of action expected of and by the dignitary, and (3) the protection must be surprise-proof and flexible enough to respond to any emergency.

Third, effective intelligence on potential threats is essential. Without current information about threats or the activities of potential sources of threat, the best security program is at a disadvantage.

Countermeasures

A considerable amount of protective technology exists for use in executive protection. Planning for protection must include the most appropriate use of available means of protection including:
1. Alarm systems and sensory devices
2. Visual surveillance equipment
3. Bodyguards
4. Defensive Tactics—training
    a. while working
    b. while traveling
    c. while at home or at play
    d. if held hostage
5. Site-hardening equipment

As in any security program, the extent to which executive protective measures are to be employed will be determined by the risk or risk potential of the person to be protected.

EXTORTION

The use of threats to obtain money or other valuable considerations has been a problem for business and industry and government for a considerable period of time. While it is not preventable, its effectiveness can be minimized with the application of comprehensive planning protective programs.

ROBBERY

The forcible taking of money or valuables is as old as man himself. The 441,290 robberies in the United States in 1974 indicated the magnitude of the current problem. Weapons are most often used in the commission of this violent crime with $142 million being lost in 1974.

Security precautions and planning can reduce the risk of robbery but, as with any security program, cannot guarantee against it. The design, procedures, and training of personnel directly affect the possibility of robbery.

The occurrence of a robbery indicates that the robber perceived a target as being poorly protected and thereby vulnerable to his particular level of criminal sophistication. The amount of risk or level of criminal sophistication possessed by the robber increases as the level of protection is raised. Therefore, if high-value items or large amounts of cash are present, unless protected by procedures or physical measures, they are vulnerable to even criminals with low levels of sophistication. Thus, to insure that risk is raised for the criminal and that only the most determined criminals will be able to successfully attack a facility, high security is required.

TERRORISM

Terroristic acts are committed for political, criminal, or psychological reasons. They are extremely difficult to prevent but can be made more difficult for even the determined terrorist by the application of sound security principles. Terrorists attacks require the selection of a target, information collection and surveillance of it, planning for the operation to insure access, surprise, and successful completion.

Terroristic acts may either be discriminate by being focused against specific targets or persons for definite purposes. They may also be indiscriminate to cause general unrest, suspicion, or antagonism. Protection against the former is much easier than the latter. When targets are identifiable, appropriate security measures can be taken. When indiscriminate terror is initiated, security measures are almost always ineffectual unless they affect the movement and living patterns of large numbers of people. Since showing the ineffectualness of government to provide protection or private protective measures is often the aim of indiscriminate terror, the desired result is obtained in the effort to prevent the acts.

Terrorist groups are found in all parts of the world and operate to advance specific ideological, political or criminal goals. Adequate protection from them and the prevention of damage from their actions requires effective intelligence collection and effective security measures.

LOSS CONTROL

Shoplifting

Shoplifting is a form of theft known as larceny. The two major elements of shoplifting are (1) the taking of merchandise, and (2) the carrying away of merchandise with the intention of theft. Appendix 5A summarizes the shoplifting laws in all fifty states of the United States and details specifically a number of them. In some states, shoplifting is considered a felony; in others, a misdemeanor. The trend appears to be making shoplifting a felony offense. There are five major points to remember when attempting to prove the offense of shoplifting:

1. The suspect had the intention to steal the merchandise.
2. The proprietor must be able to identify the stolen articles.
3. The individual in question must still have in his possession the stolen articles.
4. The merchandise must be identified as property of the store.
5. The person must have been seen taking the merchandise.

A wide variety of individuals are engaged in shoplifting. The major categories of shoplifters are as follows:

PROFESSIONALS. These individuals are

very much like all "professional criminals" except that their vocation is that of shoplifting. Professional shoplifters are in the numerical minority when considering the total number of those engaged in shoplifting. The major concern of these individuals is the great amount and value of the merchandise that they steal. Many of these individuals will steal to order, and more particularly, they steal for quality and not a quantity of merchandise.

DRUG ADDICTS. These individuals are very difficult to work with since they steal in order to support some type of narcotic habit.

JUVENILES. There are two subcategories of juvenile shoplifters:
1. Lower class juveniles—These individuals either steal for need or the desire for a certain object. In many cases, their particular subculture condones and encourages this type of activity.
2. Upper or middle class juveniles—These individuals steal for a wide variety of reasons, none associated with need. They steal for initiation into a gang, for excitement, or possibly out of sheer boredom with their way of life.

DISORDERED PERSONALITY. Similarly there are two subcategories of the *disordered personality* shoplifter.
1. Psychogeneric—These are persons with personality disorders.
2. Sociogeneric—These are individuals with the inability to interact within groups or have some type of social disorder.

AMATEURS. This is the largest single group of shoplifters. Most amateurs steal for noneconomic reasons and usually steal items for their own use. These individuals usually do not have a sustained contact with a criminal subculture and many are motivated by "spur of the moment" or "I thought it was a good idea at the time" type of rationale. Many more women than men are placed into this amateur category. A normal housewife-type is often considered a subcategory of the amateur shoplifters.

Practical Approaches to Prevent Shoplifting

It is unquestionably more beneficial to prevent shoplifting from taking place than to make apprehensions of those committing the act. Proper and appropriate preventative measures eliminate the opportunity to steal and likewise reduce the risk of customer embarrassment and legal entanglements for the overly judicious use of security measures. Prevention measures involve the interaction of the use of physical equipment and personnel to provide a total preventative approach to shoplifting. Physical measures include display of material, television cameras, one-way mirrors, and announcements on public address systems. Personnel resources include trained security officers and store operating personnel.

Shoplifting Countermeasures

Specific countermeasures can be taken against shoplifting. They include the following:
1. Education programs for employees—Programs should be developed which acquaint employees with the techniques of the shoplifter in all categories; what to look for and how to deal with one if an incident is discovered.
2. Poster and envelope stuffers—Posters should be developed which portray the damage done by shoplifters. Educational materials should also be developed which are placed in employee pay envelopes to continually make them aware of the problem of shoplifting.
3. A state of curiosity should be developed in sales personnel. Sales personnel should be mentally questioning the presence and activities of individuals within their area of responsibility.
4. Display arrangements—Merchandise should be presented to the public in such a manner as is consistent with the security needs of a particular type of item.

5. Shrinkage meetings—Regular meetings should be held to discuss the particular problems of loss associated with store operations. Specific problem areas should be pinpointed and additional efforts should be made to reduce losses in these areas.
6. Training for rush seasons—Special sales personnel should be given training in regard to procedure for handling seasonal sales rushes. Losses are considerably higher during rush seasons; therefore, all care should be taken to insure that all old as well as special employees are made aware of the particular dangers involved.
7. Job training—Each sales person in the store should thoroughly know the merchandise they handle and the proper procedures for handling it.
8. Trained security staff—It is essential that a well trained security staff be available to supplement the activities of general store operating personnel in the protection of property from shoplifters.
9. Mirrors and one-way viewing windows—These devices are useful in many applications. They increase the coverage of the store by operating personnel as well as security officers.
10. List of stolen merchandise—A list of stolen merchandise should be developed and prominently posted at each cash register. This would insure that employees know what type of merchandise is being taken and would further provide a safeguard against this stolen merchandise being returned to the store for cash refunds.
11. Electronic surveillance equipment—Electronic surveillance equipment such as closed circuit TV cameras can be used to great advantage to provide surveillance of critical areas against shoplifting losses.
12. Physical and internal security surveys—Surveys of security needs should be conducted regularly to insure that security measures are keeping pace with security needs.
13. Procedural audits—Periodic audit of procedures should be conducted to insure that loopholes have not developed which result in losses in profit.
14. Written policy—Written policies in regard to shoplifting should be developed and disseminated to all employees to insure that uniformity exists in the handling and processing of shoplifters.
15. Liaison—Close liaisons should be developed and maintained with all local law enforcement agencies. This will ensure that shoplifters and information about shoplifting losses are made available to all those with a need to know.

SOCIAL PROBLEMS

A variety of social problems which employees bring to the job can adversely affect organizational activities. The most common problems are alcohol, drugs, and gambling. The first two affect not only the employees ability to be productive but may lead to individual theft or fraud to provide money to pay for his habit. Gambling, likewise, may cause employees to steal to participate or pay off debts. Gambling, however, contributes to low morale and losses to productivity when allowed to continue uncontrolled in the organization.

Alcohol and drug counseling programs are utilized by government, business, and industry to assist employees overcome their involvements with them. The security department has a responsibility to assist in the identification of problem employees to prevent damage to the organization and the individual. The sources of drugs should be identified and removed from the facility by appropriate investigative and enforcement actions. Likewise, gambling should not be tolerated on any scale by employees while at work.

TRANSPORTATION

Transportation security involves a wide variety of protective programs to insure the safe and protective movements of cargo from one point to another in a product distribution network.[3] Transportation security includes trucks, rail, bus, water, and air transport methods. It requires the development of maintenance of protective programs to insure the safe arrival of cargo from its point of origin to its final point of destination. It provides protection against theft, burglary, hijacking, pilferage, sabotage or vandalism of vehicles or cargo in transit.

APPENDIX 18A

OPERATIONAL POLICY FOR BOMB THREATS*

PROCEDURE FOR HANDLING BOMB THREATS

1. Keep the caller on the line as long as possible. Ask the caller to repeat the message. Record every word spoken by the person making the call.
2. If the caller does not indicate the location of the bomb or the time of possible detonation, the person receiving the call should ask the caller to provide this information.
3. Inform the caller that the building or area is or may be occupied and the detonation of a bomb could result in death or serious injury to innocent people.
4. Pay particular attention for any strange or peculiar background noises such as: motors running, background music and the type music and any other noises which might give even a remote clue as to the place from which the call is being made.
5. Listen closely to the voice (male—female), voice quality, accents, and speech impediments. Immediately after the caller hangs up, the person receiving the call should report this information to the mall manager.
6. The security manager should report this information immediately to the local police department, fire department, FBI and other agencies as deemed necessary. The sequence of notification should be determined after consultation with such local agencies.
7. In order to obtain an Army Bomb Disposal Unit, the local police must first find and identify the bomb and then notify the proper Army Disposal Unit for the particular plant involved according to locality.

PROCEDURE FOR EVACUATION IN BOMB THREAT SITUATIONS

The decision whether to evacuate or not to evacuate is the responsibility of the building manager. If a determination to evacuate is made, the following procedure is recommended for evacuation.

1. The signal for evacuating the building in the event of a bomb threat should be similar or the same as that used for evacuation in the event of a fire, if such a system exists. The use of a different signal for a bomb threat may tend to create unnecessary excitement and confusion during the process of evacuation. It may be necessary to walk through the areas with portable loud speakers to inform the employees and customers of the evacuation. Emergency shut-down procedures already in existence should be utilized if evacuation is determined necessary. All electricity, gas and fuel lines should be cut off at the main switch or valve.

---

* Courtesy Post and Associates, Security Consultants.

2. Priority of evacuation would be determined by suspected location of the bomb, i.e. whether in a building, outside, etc. It is recommended to evacuate the floor levels above the danger area in order to remove those personnel from the extremes of danger as quickly as possible. Training in this type evacuation should be available from police or fire units within the immediate area.
3. When the police and fire departments arrive at the mall, the contents, operation, floor plan, etc. will be strange to them unless they have been through the area at some previous date. Thus, it is extremely important that the evacuation unit be thoroughly trained and thoroughly familiar with the areas being evacuated.
4. If the area or building is evacuated, controls must be established immediately to prevent unauthorized access to the building. If proper coordination has been effected with the local police and other agencies, they may assist in establishing controls to prevent reentry into the area or building until the danger has passed.
5. Remove personnel to a safe distance from the building to protect them against debris and other flying objects in the event there is an explosion.
6. Preemergency plans should include a temporary relocation in the event an explosion materializes and the area or building is rendered untenable for a considerable period of time.

PROCEDURE FOR SEARCH IN BOMB THREAT SITUATIONS

The evacuation unit or a separate unit should be trained in bomb search techniques, but not in the techniques of neutralizing, removing or otherwise having contact with the device. The search unit should be thoroughly familiar with area, floor plan of the mall, etc. To be proficient in searching a building, they must be thoroughly familiar with all walkways, hallways, restrooms, locker rooms, false ceiling areas and every conceivable location in the building where an explosive or incendiary device might be concealed. Guidelines for the search are listed below.

1. During the period of search, a rapid two-way communication system is a must. The existing telephone system would probably be most efficient. The use of radios during the search can be dangerous because the radio beam could cause a premature detonation of an electric initiator (blasting cap).
2. During the search particular attention should be given to such areas as elevator shafts, all ceiling areas, restrooms, locker rooms, access doors and crawl spaces in restrooms and other areas which are used as a means of immediate access to plumbing fixtures, electrical fixtures, etc., utility and other closet areas, areas under stairwells, boiler or furnace rooms, flammable storage areas, main switches and valves, e.g. electric, gas, and fuel, indoor trash receptacles, record storage areas, mail rooms, ceiling lights with easily removable panels and fire hose racks.

Although this list is not totally complete, it does put emphasis on the areas where a time delayed explosive or incendiary device might be concealed. Each area in the building should review the possible areas of concealment and have a listing for the search units use.

3. *If a strange or suspicious object is encountered it should not be touched.* Its location and a description as can best be provided should be reported to the security manager **immediately.**

External Problems

4. If the danger zone is identified as located, the area should be blocked off or barricaded with a clear zone of three hundred feet until the object has been removed or disarmed or danger has otherwise passed.
5. During the search medical personnel should be alerted to stand by in the event of an accident involving an explosion of the device.
6. Preemergency plans should include a temporary relocation in the event an explosion materializes and the area or building is rendered untenable for a considerable period of time.

APPENDIX 18B

STANDARDS FOR CARGO SECURITY*

PHYSICAL SECURITY STANDARDS

All cargo handling and storage facilities should provide a physical barrier against unauthorized access to cargo. Usually this will require a covered structure with walls, and apertures which can be securely closed and locked. In addition, fencing may be needed:

1. As supplementary protection to prevent unauthorized persons and vehicles from entering cargo storage and handling areas.
2. As sole protection for open storage of bulk cargo or large articles which do not require covered storage because they cannot be easily pilfered or removed without mechanical handling equipment or which have their own inherent security (containers).

Buildings

General Standard

All buildings used to house cargo and associated support buildings should be constructed of materials which resist unlawful entry. The integrity of the structure must be maintained by periodic inspection and repair. Security protection should be provided for all doors and windows.

Recommended Specifications

1. Equip all exterior doors and windows with locks.
2. Protect all windows through which entry can be made from ground level by safety glass, wire mesh, or bars.
3. Similarly safeguard all glassed-in areas where shipping documents are processed.
4. Construct all delivery and receiving doors of steel or other material that will prevent or deter unlawful entry and keep them closed and locked when not in use.
5. Where fencing is impractical or guards insufficient, equip the building with an intrusion detection or alarm system.
6. Inspectors must insure particularly that there are no avenues for surreptitious entry through floors, roofs, or adjacent buildings.

---

* Treasury Department, Bureau of Customs, T.D. 72-66, March 1972, 2nd printing, Washington, D.C.

Fencing

General Standard

Where fencing is required, it should enclose an area around cargo storage structures, support buildings, and exterior stored cargo sufficient to provide maneuvering space for pick-up and delivery vehicles and to prevent use of buildings or cargo to surmount the fence. The fence line must be inspected regularly for integrity and any damage promptly repaired.

Recommended Specifications

1. Install chain link type fencing with at least six gauge, two-inch mesh and at least eight feet high (not including a barbed wire extension). If the level on which the fence is constructed is lower than the area outside the fence line, increase the height of the fence to provide an effective eight-foot fence at all points.
2. Top the fence with a two-foot barbed wire extension, consisting of three strands of barbed wire, properly spaced and at a 45° angle to the vertical.
3. Place fence posts on the inside of the fence and secure them in a cement foundation at least two feet deep.
4. Ensure that objects or persons cannot pass beneath the fencing by providing:
   a. Cement aprons not less than six inches thick, or
   b. Frame piping, or
   c. U-shaped stakes driven approximately two feet into the ground.
5. Avoid any condition which compromises the fence line. Prohibit the placing of containers, dunnage, cargo, vehicles, or any other item that may facilitate unlawful entry adjacent to the fence line.
6. Where necessary, install bumpers or fence guards to prevent damage by vehicles.

Gates

General Standard

The number of gates in fences should be the minimum necessary for access. All fence gates should be at least as substantial as the fence. Gates through which vehicles or personnel enter or exit should be manned or under observation by management or security personnel.

Recommended Specifications

1. Equip gates with a deadlocking bolt or a substantially equivalent lock which does not require use of a chain. All hardware connecting the lock to the gate should be strong enough to withstand constant use and attempts to defeat the locking device.
2. Construct swing-type gates so that they may be secured to the ground when closed.
3. Separate gates for personnel and vehicle traffic are desirable.

Gate Houses

General Standard

Operators of facilities handling a substantial volume of cargo should maintain a manned gate house at all vehicle entrances and exits during business hours.

Recommended Specifications

1. Set the gate house back from the gate so that vehicles can be stopped and examined on terminal property.
2. Equip the gate house with a telephone or other communication system.
3. Clear the area around the gate house of any encumbrances that restrict the guard's line of vision.
4. Post prominently on the exterior of all gate houses signs advising drivers and visitors of the conditions of entry. Include in conditions of entry a notice that all vehicles and personnel entering the area are subject to search.

PARKING

General Standard

Private passenger vehicles should be prohibited from parking in cargo areas or immediately adjacent to cargo storage buildings. Access to employee parking areas should be subject to security controls.

Recommended Specifications

1. Locate parking areas outside of fenced operational areas, or at least a substantial distance from cargo handling and storage areas or buildings and support buildings.
2. Require employees exiting to the parking area from the cargo area to pass through an area under the supervision of management or security personnel. Require employees desiring to return to their private vehicles during hours of employment to notify management and/or security personnel.
3. Allow parking in employee parking areas by permit only. Maintain a record of each issued permit, listing the vehicles registration number, model, color, and year. The permit should consist of a numbered decal, tag, sticker, or sign placed in a uniform location on the vehicle.
4. Issue to vendors and other visitors temporary parking permits which allow parking in a designated area under security controls.

Lighting

General Standard

Adequate lighting should be provided for the following areas:
1. Entrances, exits and around gate houses
2. Cargo areas, including container, trailer, aircraft and rail-car holding areas
3. Along fence lines and stringpieces
4. Parking areas

Recommended Specifications

1. The Society of Illuminating Engineers recommends the following light intensities measured at ground level:
   a. Vehicle and pedestrian areas   2.0 foot candles
   b. Vital structures and other sensitive areas   2.0 foot candles
   c. Unattended outdoor parking areas   1.0 foot candle
2. Illuminate all vehicle and pedestrian gates, perimeter fence lines, and other outer areas with mercury vapor, sodium vapor, power quartz lamps or sub-

stantially similar high intensity lighting, employing a minimum of 400 watts per fixture. Locate lights thirty feet above ground level and properly spaced to provide the appropriate light intensity for the area to be illuminated.
3. Establish a system of planned maintenance.
4. Protect lighting subject to vandalism by wire screening or other substantially equivalent means.

Locks, Locking Devices, and Key Control

General Standard

Locks or locking devices used on buildings, gates and equipment should be so constructed as to provide positive protection against unauthorized entry. The issuance of all locks and keys should be controlled by management or security personnel.

Recommended Specifications

1. Use only locks having (a) multiple pin tumblers, (b) dead-locking bolts, (c) interchangeable cores, and (d) serial numbers.
2. To facilitate detection of unauthorized locks, use only locks of standard manufacture displaying the owner's company name.
3. Number all keys and obtain a signature from the recipient when issued. Maintain a control file for all keys. Restrict the distribution of master keys to persons whose responsibilities require them to have one.
4. Safeguard all unissued or duplicate keys.
5. Remove and secure keys from cargo handling equipment and vehicles when not in actual use.

High-Risk Cargo

General Standard

Adequate space capable of being locked, sealed, or otherwise secured for storage of high-value cargo and packages which have been broken prior to or during the course of unloading must be provided at each cargo handling building. When such cargo must be transported a substantial distance from the point of unloading to the special security area, vehicles capable of being locked or otherwise secured must be used.*

Recommended Specifications

1. Construct special security rooms, cribs, or vaults so as to resist forcible entry on all sides and from underneath and overhead.
2. Locate such special security areas, where possible, so that management and/or security personnel may keep them under continuous observation. Otherwise, install an alarm system or provide for inspection at frequent intervals.
3. Release merchandise from such an area only in the presence of authorized supervisors and/or security personnel.
4. Log all movements of merchandise in or out of a special security area, showing date, time, condition of cargo upon receipt, name of truckman, and company making pick-up and registration number of equipment used.

---

* The standards are required by Customs Regulation (19 C.F.R. 4.30).

PROCEDURAL SECURITY STANDARDS
Personnel Screening
General Standard

Operators of cargo handling facilities should conduct employment screening of prospective employees.*

Recommended Specifications

1. Require all personnel, including maintenance and clerical personnel, who will have access to cargo areas to submit a detailed employment application which contains a photograph of the applicant and lists his residences and prior employment for the preceding ten years.
2. Screen all such employment applicants for:
    a. verification of address and prior employment,
    b. credit record, and
    c. if possible, criminal record.

Security Personnel
General Standard

Operators of cargo handling facilities should employ a Security Officer or assign a particular officer of the firm to be responsible for security. All operators handling a substantial volume of international cargo should provide guards to protect the cargo.

Recommended Specifications

1. Employ the number of guards required to provide adequate security for the size of each facility and the volume of cargo handled. Alarm systems, closed circuit television and other security devices may reduce the number of guards needed.
2. Train all company employee guard forces or insure that contract guard forces are trained in:
    a. Methods of patrolling terminals and warehouses.
    b. Use of firearms and other equipment that may be furnished.
    c. Report writing, log and record keeping.
    d. Identification of security problems and specific trouble areas.
3. Equip guard forces with uniforms which are complete, distinctive and authoritative in appearance.
4. Provide firearms, vehicles, communications systems, and other equipment deemed necessary for the successful performance of the guard function.
5. Insist on physical fitness as a prime consideration in selecting a guard force. Require guards to undergo self-defense training similar to that of police agencies. Require a physical examination at least once a year.
6. Furnish each guard a manual covering operating procedures and standards of conduct, and a clear statement of what management expects of him.

---

* Customs regulations already require international carriers, proprietors of bonded warehouses, and customhouse brokers to submit employee lists upon request from the District Director of Customs. Such lists must contain the name, address, social security number, and date and place of birth of each employee and be kept up to date (Customs Regulations, 19, C.F.R. 4.30 (m), 19.3 and 111.28).

Communications

General Standard

Adequate and reliable communications between elements of the terminal security force and from the security force to local police should be provided.

Recommended Specifications

1. Provide security personnel with a telephone at fixed posts or two-way radio, intercom or other type of equipment providing voice communication capability within the company.
2. Arrange assured means (telephone, radio, or special alarm line) for summoning assistance from local police forces.

Identification System

General Standard

All operators of facilities handling a substantial volume of cargo should employ an identification card system to identify personnel authorized to enter cargo and document processing areas.

Recommended Specifications

1. Include on the ID card: (a) a physical description or, preferably, a color photograph of the holder, (b) name and address, (c) social security number, (d) date of birth, (e) employer's Customs license number, if any, (f) signature of holder, and (g) reasonable expiration date.
2. Laminate all cards to prevent alterations and assign each card a control number.
3. Recover ID cards from terminated employees.
4. Require each employee to display his ID card to gain access to the facility, to cargo areas within the facility, and to areas where shipping documents are processed. Preferably, the ID card should be displayed so that it is visible at all times that the employee is within the facility.

Independent Contractors

General Standard

The background and corporate structure of independent contractors providing janitorial service, refuse disposal, or other services should be verified. Access by independent contractors to the facility should be under security controls.

Recommended Specifications

1. Periodically examine independent contractor vehicles which are parked in or near cargo areas.
2. Permit independent contractor employees to enter only those areas necessary for their particular work; permit them access to cargo areas and areas where shipping documents are located only under the supervision of security and/or management personnel.
3. Require independent contractors to display identification similar to that required by the facility for its own employees.

Cargo Quantity Control

General Standard

Cargo should be tallied at time of delivery to the consignee or his agent. In the event of any discrepancies at time of delivery, a U.S. Customs Form 5931 or a duplicate copy of the amended cargo manifest must be completed and submitted to Customs by the carrier or his agent.*

Recommended Specifications

1. To facilitate accurate delivery of cargo, terminal operators should maintain and continuously up-date a location chart or list of all cargo received.
2. Segregate imported cargo, cargo for export, and domestic cargo.
3. Carriers should arrange procedures with each terminal operator to insure that all overages and shortages are reported to Customs.

Delivery Procedures

General Standard

Gate passes should be issued to truckmen and other onward carriers to control and identify those authorized to enter the facility. Verification of the identity and authority of the carrier requesting delivery of cargo should be made prior to the cargo's release.

Recommended Specifications

1. Require truckmen to submit proper personal identification (such as a driver's license or Customs I.D. card) and a vehicle registration certificate before being issued a gate pass and being permitted to enter the facility; require them to surrender the gate pass before leaving the facility.
2. Seal containers and trailers and note the seal number on the gate pass before delivery is effected. Verify the seal number when the gate pass is surrendered at the gate.
3. Require the company name of all onward carriers to be clearly shown on all equipment. Do not accept temporary placards or cardboard signs as proper identification of equipment. Require carriers using leased equipment to submit the lease agreement for inspection and note the leasing company's name on the delivery order.
4. Release cargo only to the carrier specified in the delivery order unless a release authorizing delivery to another carrier, signed by the original carrier, is presented and verified. Accept only original copies of the delivery or pick-up orders.
5. Personnel processing prelodged delivery or pick-up orders should verify the identity of the truckman and the trucking company before releasing the pick-up order. Limit access to areas where such documentation is processed or held to authorized personnel and rigorously safeguard all shipping documents from theft or unauthorized observation.
6. Conduct delivery and receiving operations at separate docks or doors, if feasible.
7. Tally salvage and accumulated unclaimed cargo at the time of delivery and

---

* All international carriers are required by Customs Regulations to make discrepancy reports (19 C.F.R. 4.12 (a), 6.7 (h), 15.8, 18.2 (b), 18.6 (b), (c), 123.9).

have management representatives and/or security personnel verify that only properly released items are included. If a terminal has truck scales, weigh the vehicle used to remove bulk salvage cargo (bales and drums) when empty and loaded.

Containerized Shipments and Seals

General Standard

All containers, trailers, rail cars and air cargo lockers entering or leaving a facility should be sealed. Mounted and high-value containerized shipments should receive special security attention.

Recommended Specifications

1. Inspect seals whenever a sealed containerized shipment enters or leaves a facility. If the seals are not intact or there is evidence of tampering or the seal numbers are incorrect, notify security and/or management personnel and tally the cargo.
2. Seal unsealed containerized shipments at the point of entry to the facility and note the seal number on the shipping documents. Seal all containerized shipments leaving the facility and note the seal number on the shipping documents.
3. Release seals to as few persons as possible. Require all persons handling seals to maintain strict control of the seals assigned and to store them in a secure place.
4. Maintain a seal distribution log which indicates to whom seals have been released.
5. Where possible, secure containers by butting or "marrying" their door ends against each other. However, do not butt them against a perimeter fence or building wall if that will compromise the protection provided by the fence or wall. In stacking containers, place those containing high value merchandise on top.
6. Locate high-value merchandise in mounted containers or trailers in a special security holding area where it can be observed by management and/or security personnel.
7. When containers are mounted on frames, secure the fifth-wheel by a pin-lock which meets the minimum standards for locks and is constructed to withstand normal abuse from equipment. Hold designated management and/or security personnel responsible for storage and control of pin-locks.
8. Restrict access to special security holding areas and permit the release of containers or trailers from such areas only in the presence of management representatives and/or security personnel.
9. Log movements of containers in or out of a special security holding area, showing: date, time, seal number, name of truckman and company making pick-up, and registration number of equipment used.

Security Education

General Standards

Management should institute a security awareness program for all personnel.

Recommended Specifications

1. Conduct a program of periodic security seminars for all employees involved

in cargo handling and documentation processing, stressing the importance of:
   a. maintaining legible and accurate cargo tallies;
   b. processing only legible documents;
   c. writing only in ink or ball point pen;
   d. completing all information required by shipping documents; and
   e. obtaining clearly written signatures.
   f. safeguarding the confidentialty of shipping and entry documents, and
   g. maintaining good cargo security generally.
2. Include in the security awareness program posters, stickers, payroll stuffers, monetary incentives, and properly worded reward signs. (Appropriate signs can be obtained from the Bureau of Customs field offices.)

FOOTNOTES

1. Appendix 18A presents a sample Bomb Threat Policy for a proprietary facility.
2. See Appendix 14H for a Comprehensive Plan for Protection in the Event of Civil Disorder.
3. Appendix 18B presents the Cargo Security Standards Development by the U.S. Treasury Department, Bureau of Customs.

BOOKS

CARGO

Bray, Samuel E., and Hurley, Robert: *Freight Security Manual.* Los Angeles, Security World Publishing Co., Inc., 1969.

PUBLICATIONS OF THE GOVERNMENT, LEARNED SOCIETIES, AND OTHER ORGANIZATIONS

Cargo security handbook for shippers and receivers. *Dot,* 1972, p. 38.
Cargo theft and organized crime. Department of Justice, LEAA, October 1972.
Physical security guidelines for the prevention of loss and theft of cargo in the transportation system, draft for coordination. *Dot,* 1971, p. 237.
Port security study for Wilmington, Delaware. Westinghouse Justice Institute, 1973.
U.S. Department of Transportation. Physical security guidelines for the prevention of loss and theft of cargo in the transportation system, 1971.
———. Guidelines for the physical security of cargo, Office of Secretary, May 1972.
U.S. Department of the Treasury, Bureau of Customs. Standards for cargo security, 1972.

PERIODICALS

CARGO

Cargo, the $1 billion theft. *Occupational Hazards,* April 1969, p. 77.
Cargo pilferage and the police. *Police Journal,* 1929, p. 132-149.
Cargo theft. *Industrial Security,* August 1970, p. 4.
Car insurance. *Security and Protection,* October 1974, p. 17-18.
Cooke, Max: Looking ahead: Freight theft prevention. *Security World,* September 1971.
Guidelines for the physical security of cargo. *Dot,* 1972.
Harris, Harvey T., Jr.: Air cargo security. *Industrial Security,* August 1965, p. 4.
Hill, Y. J.: Transportation security—literature survey. *Dot,* 1972, p. 187.
Judge, J. F.: Goods in transit—target for thieves. *Security Management, 17:* no. 4, 7-10, 13, 15-17, 29, 53, September 1973.
McFarlane, Harry L.: J. C. Penney saves profit by preventing cargo loss. *Industrial Security,* August 1970, p. 12.
Sanders, James D.: Senate committee hears cargo theft data. *Industrial Security,* February 1971.
Smith, Gordon C.: Stock, warehouse and enroute controls—panel moderator. *Industrial Security,* October 1962, p. 77.
Smith, Arthur T.: Port security. *Industrial Security,* April 1965.
Sweat, Milton: Freight security. *Security World Magazine,* November and December 1969.
Trucking security: From dock to delivery. *Occupational Hazards,* November 1966.
Trucking security manual—recommended draft. *National Assn. of Transportation Security,* p. 30.

Tyska, Louis A.: Confusion, conspiracy and the common denominator in cargo theft. *Security Management,* January 1975, p. 8-10.

Venning, Colin: Goods in transit . . . what losses tell us. *Canadian Security Gazette,* October 1970.

Ward, Daniel A.: A transportation security prospectus. *Security Management,* September 1974, p. 8-11.

Webster, Cecil D.: Cleaning house in the transportation industry. *Security World,* May 1970.

UNPUBLISHED MATERIAL

Harris, Harvey T.: International cargo thefts and losses: an identification of variables. Unpublished thesis, Michigan State Univ., Fall, 1966.

BOOKS

EXTORTION PROTECTION

Nelms, Henning, LL.B.: *The Defense Book.* Arlington, Police Science Series, Inc., 1974.

PUBLICATIONS OF THE GOVERNMENT, LEARNED SOCIETIES, AND OTHER ORGANIZATIONS

Executive Protection Handbook. Florida, Burns International Investigation Bureau, 1973.

Lubman, Stanley: Mao and mediation: Politics and dispute resolution in Communist China. *California Law Review, 55:*1284, 1967.

PERIODICALS

Adkins, Elmer H., Jr.: Protection of American industrial dignitaries and facilities overseas. *Security Management,* July 1974, p. 14+.

Anonymous or Staff Written: Riot loss prevention: One corporate program. *Security World, 5:* no. 3, March 1968.

Bradford, Ralph, Lt.: Technical casebook: Extortion. *Security World, 2:* no. 8, November 1965.

Cote, Robert: Terrorist activities in Montreal. *Security World,* November 1969.

Directory frauds—how to avoid being duped. *Security Gazette,* May 1975, p. 173-174.

Drive to halt terror bombings. *U.S. News and World Report,* March 15, 1971, p. 17-19.

Dudley, W. E., and Glavni Vrag: A quick look at the opposition. *Industrial Security,* December 1968.

Extortion. *Security World, 2:* no. 8, November 1965.

Fitzpatrick, T. K.: The semantics of terror. *Security Register, 1:* no. 4, 21-23, 1974.

Fraudulent telephone calls. *Training Key,* No. 193. Maryland, Professional Standards Division, International Association of Chiefs of Police, 1973.

High schools: Next target for unrest. *U.S. News and World Report, 67:* no. 12, September 22, 1969.

Mahoney, Harry T.: After a terrorist attack; business as usual. *Security Management,* March 1975, p. 16-19.

McDavitt, Victor: If attack should come. *Industrial Security,* January 1962.

McGuire, Patrick E.: Targets for terrorists. *The Conference Board Record,* August 1971.

Militants. *Time,* August 30, 1971, p. 27.

UNPUBLISHED MATERIAL

Lorwin, Val R.: All colors but red, interest groups and political parties in Belgium. Unpublished paper, Center for Advanced Study in the Behavioral Sciences, 1962.

Mei, Ko-Wang: The theory and practice of modern guerrilla warfare. Masters paper, Michigan State Univ., 1965.

BOOKS

ROBBERY

Arendt, H.: *On Violence.* New York, Harcourt, Brace, World, 1969.

American Friends Service Committee. *Struggle for Justice: A Report on Crime and Punishment in America.* New York, Hill & Wang, 1971.

Acton, H. B.: *The Philosophy of Punishment.* New York, Macmillan, 1969.

Bay, C.: *The Structure of Freedom.* Stanford, Calif., Stanford University Press, 1970.

Bell, G., Randall, E., and Roeder, J. E. R.: *Urban Environments and Human Behavior.* Stroudsburg, Pa., Dowden, Hutchinson, and Ross, 1973.

Chambliss, W. J.: *Crime and the Legal Process.* New York, McGraw-Hill, 1969.

Clinard, M. B., and Quinney, R.: *Criminal Behavior Systems: A Typology.* New York, Holt, Rinehart & Winston, 1967.

Cole, R. B.: *The Application of Security Sys-*

tems and Hardware. Springfield, Thomas, 1970.
Committee for Economic Development. *Reducing Crime and Assuring Justice*. New York, Author, 1972.
Conklin, J. E.: *Robbery and the Criminal Justice System*. Philadelphia, Lippincott, 1972.
Dressler, D. (Ed.): *Readings in Criminology and Penology*. New York, Columbia University Press, 1964.
Erickson, R. J., Crow, W. J., Zurcher, L. A., and Connett, A. V.: *Paroled but Not Free*. New York, Behavioral Publications, 1973.
Fawcett, J. (Ed.): *Dynamics of Violence*. Chicago, American Medical Association, 1971.
Gibson, W. B.: *The Fine Art of Robbery*. New York, Grosset & Dunlap, 1966.
Hacker, A.: Getting used to mugging. *The New York Review of Books*, April 19, 1973, p. 9-14.
Hanewicz, W. B., and Shields, R. O.: Environmental security: The exploration of an idea. In Post, R. (Ed.), *Determining Security Needs*. Madison, Wisc., Oak Security Publications, 1973, p. 75-94.
Healy, R. J.: *Design for Security*. New York, Wiley & Sons, 1968.
Hemphill, C. F.: *Security for Business and Industry*. Homewood, Ill., Dow Jones-Irwin, 1971.
Hughes, M. M. (Ed.): *Successful Retail Security*. Los Angeles, Calif., Security World Publishing.
Hunt, M.: *The Mugging*. New York, Atheneum, 1972.
Irwin, J.: *The Felon*. Englewood Cliffs, N.J., Prentice-Hall, 1970.
Jackson, B.: *Outside the Law: A Thief's Primer*. New Brunswick, N.J., Transaction Books, Rutgers University, 1972.
Jeffrey, C. R.: *Crime Prevention through Environmental Design*. Beverly Hills, Sage Publications, 1971.
Mandelbaum, A. J.: *Fundamentals of Protective Systems*. Springfield, Thomas, 1973.
May, R.: *Power and Innocence*. New York, Norton, 1972.
McCormick, Mona: *Robbery Prevention*. La Jolla, Western Behavioral Sciences, 1974.
McLennan, B. N. (Ed.): *Crime in Urban Society*. New York, Dunellen, 1970.
Menninger, K.: *The Crime of Punishment*. New York, Viking Press, 1966.
Morris, N., and Hawkins, G.: *The Honest Politician's Guide to Crime Control*. Chicago, University of Chicago Press, 1970.

Oliver, E., and Wilson, J.: *Practical Security in Commerce and Industry*. 2nd ed. Essex, England, Gower, 1972.
Oughton, Frederick: *The Big Steal: The Realities of Robbery*. London, Neville Spearman, 1963.
Palmer, S.: *The Prevention of Crime*. New York, Behavioral Publications, 1973.
Pursuit, D. G., et al.: *Police Programs for Preventing Crime and Delinquency*. Springfield, Thomas, 1972.
Reckless, W. C.: *The Crime Problem*. 5th ed. New York, Appleton-Century-Crofts, 1973.
Sutherland, E. H.: *The Professional Thief*. Chicago, University of Chicago Press, 1937.
Tappan, P. W.: *Crime, Justice and Correction*. New York, McGraw-Hill, 1960.
Toch, H.: The convict as researcher. In Maple, T., and Matteson, D. W. (Eds.), *Aggression, Hostility, and Violence*. New York, Holt, Rinehart, Winston, 1973.
———: *Violent Men*. Chicago, Aldine, 1969.
von Hentig, H.: *The Criminal and His Victim*. 1948. Reprint. Hamden, Conn., Archon Books, 1967.
Wolfgang, M. E., and Ferracuti, F.: *The Subculture of Violence*. New York, Tavistock, 1967.
Woods, A.: *Crime Prevention*. 1918. Reprint. New York, Arno, 1971.
Zimring, F. E., and Hawkins, G. J.: *Deterrence: The Legal Threat in Crime Control*. Chicago, University of Chicago Press, 1973.

PUBLICATIONS OF THE GOVERNMENT, LEARNED SOCIETIES, AND OTHER ORGANIZATIONS

Acietuno, Thomas, and Matchett, Michael: Street robbery victims in Oakland. *The Prevention and Control of Robbery*, Vol. 1, University of California, Davis, The Center on Administration of Criminal Justice, April 1973.
Chaiken, Jan J., Lawless, Michael W., and Stevenson, Keith A.: *The Impact of Police Activity on Crime: Robberies on the New York City Subway System*. New York City, The Rand Corporation, 1974.
Conklin, John E.: *Robbery and the Criminal Justice System*. New York City, The Rand Corporation, 1974.
Curtis, S. J.: *Preventing Burglary and Robbery Loss*. Small Business Administration, Small

Marketers Aids No. 134. Washington, Government Printing Office, May 1968.

Einstandter, Therner J.: The social organization of armed robbery. *Social Problems*, Summer, 1969, p. 17.

Feeny, Floyd, and Weir, Adrianne (Eds.): The response of the police and other agencies to robbery. *The Prevention and Control of Robbery*, Vol. 4, University of California, Davis, The Center on Administration of Criminal Justice, April 1973.

Gunn, Lawrence: Commercial robbery in a medium-sized city: Columbus, Georgia. Washington, D.C., the Mitre Corp., November 1973.

Holcomb, R. L.: Armed robbery. *Iowa City, Inst. of Pub. Affairs*, Univ. of Iowa, 1949.

Kay, Julius: About the unknown robber. *Justice of Peace and Local Government Review, 129:* March 20, 1965.

Lorenz, R. W.: Armed robbery prevention. *Abstracts on Criminology and Penology*.

Los Angeles Police Department, Training Division. *Preliminary Investigation of Robbery*. Los Angeles, Calif., Los Angeles Police Department Training Division.

National Institute of Law Enforcement and Criminal Justice. The Crime of Robbery in the United States. Washington, Government Printing Office, 1971.

Roebuck, J. R., and Cadwallader, M. L.: The Negro armed robber as a criminal type: The construction and application of a typology. *Pacific Sociological Review, 4:* Spring, 1961.

Sagalyn, A.: *The Crime of Robbery in the United States*. U.S. Department of Justice, Law Enforcement Assistance Administration, National Institute of Law Enforcement and Criminal Justice. Washington, Government Printing Office, January 1971.

Scarr, H. A.: *Patterns of Burglary*. U.S. Department of Justice, Law Enforcement Assistance Administration, National Institute of Law Enforcement and Criminal Justice. Washington, Government Printing Office, February 1972.

Stepichev, S., and Moiseenko, G.: Dangerous recidivism in the light of the study of recidivists sentenced for armed or unarmed robbery. *Sovetakaia Isutisiia, 1:* 1968.

U.S. Department of Justice. *The Crime of Robbery in the United States*. National Institute of Law Enforcement and Criminal Justice, LEAA, ICR, January 1971.

University of California, Davis, The Center on Administration of Criminal Justice. *The Prevention and Control of Robbery*. Davis, Author, April 1973; Summary, February 1974.

Van Court, Charles: The history and concept of robbery. *The Prevention and Control of Robbery*, Vol. 5, University of California, Davis, The Center on Administration of Criminal Justice, April 1973.

Ward, Richard H., Ward, Thomas J., and Feeley, Fayne: *Police Robbery Control Manual*. Washington, Government Printing Office, April 1975.

Wilcox, Susan: The geography of robbery. *The Prevention and Control of Robbery*, Vol. 3, University of California, Davis, The Center on Administration of Criminal Justice, April 1973.

PERIODICALS

Andrews, J. A.: Robbery. *Criminal Law Review*, October 1966.

Armed robbery. *Security World*, January 1967, p. 37-38.

Bottoms, A. M.: Police tactics against robbery. *NILECJ*, 1971.

Clerks and Robbers. Store Management Series, New York, Nat. Retail Merchants Assn. (Filmstrip).

DeBawn, Everett: The heist: The theory and practice of armed robbery. *Harpers Magazine, 20:* February 1950.

Le Mouel, F.: Direct and indirect methods of preventing holdups, II. *Abstracts on Criminology and Penology*.

Lessen your chances of being robbed. *Security World*, April 1970, p. 37.

Lorenz, R. W.: Armed robbery prevention. *Security World, 7(1):*12-15, 1970.

Normandeau, Andre: Robbery in Philadelphia and London. *British Journal of Criminology, 9:* no. 1, 1969.

Sykes, G. M.: The social organization of armed robbery. *Social Problems, 17(1):*64-83, 1969.

Titus, Charles M.: Armed robbery—prepare for it before it happens! *Security World Magazine, 1:* no. 1, July 1964.

Traini, R.: You're never too old to be robbed. *Security Gazette*, 1971.

UNPUBLISHED MATERIAL

Carey, James T.: Armed robbery: A career study. Werner Julius Einstadter. School of Criminal Justice, University of California, Berkeley.

Einstadter, Werner J.: Armed robbery: A ca-

reer study in perspective. Doctor of Criminology theses, University of California, Berkeley, 1966.
Graham, George W.: One hundred burglars and one hundred robbers. Master's thesis, University of Texas, 1941.
Newman, O.: Trends and patterns in crimes of robbery. Unpublished doctoral dissertation, University of Pennsylvania, 1968b.
Parsons, Robert L.: Some advantages of cooperation between retail management and law enforcement officials in the prevention and control of robberies and burglaries of retail establishments. Unpublished Graduate B Paper, Michigan State Univ., 1968.
Syvrud, G.: The victim of robbery. Dissertation. Washington State University, 1967. Ann Arbor, Mich., University Microfilms, Inc., 1967.

BOOKS
BURGLARY

Arnold, R. Y.: *The Burglars are Coming.* California, Arnold Publishing, 1972.
Bennett, George E.: *Fraud—Its Control Through Accounts.* New York, Century, 1930.
Blum, Richard H.: *Deceivers and Deceived.* Springfield, Thomas, 1972.
Heiner, W., and Henier, J. E. (eds.): *A Burglar's Life; or The Stirring Adventures of the Great English Burglar Mark Jeffrey.* London, Angus, 1968.
Thinet, Louis: *Histories de Voleurs.* Paris, A. Fayard el, c. 1929.

PUBLICATIONS OF THE GOVERNMENT, LEARNED SOCIETIES, AND OTHER ORGANIZATIONS

Guha Roy, Nihar Ranjan: War against crime: A mine of information on modern security from maximum protection against pilferage, burglary, and juvenile delinquency reinforced with most up-to-date and latest data prodigious proportions from England, USA and West Germany. Calcutta, Security Publishers, 1966.
Loopholes in business security. Winter Park, Florida, Florida State Retailers Assn.
Scarr, H.: *Patterns of Burglary.* 2nd ed. Report prepared under NILECJ Grant No. 72-0002-6, June 1973.

PERIODICALS

Bowers, John: Big city thieves. *Harper's Magazine,* 234:50-54, February 1967.
Bradford, Ralph, Lt.: . . . Reference charts for handwriting, check and check protector comparison. *Security World,* 1: no. 2, Septemper 1964.
———: . . . Don't invite forgery. *Security World,* 1: no. 3, November 1964.
———: Technical course: Introduction to the Bradford system of check classification. *Security World,* 1: no. 3, November 1964.
———: . . . Check protector identifications. *Security World,* 2: no. 7, October 1965.
Burglary—a white paper. *Security World,* November 1974, p. 11+.
Clavi, Francois de: Historie generale des lamons. *F.D.C. Lyonnois.* Paris, A. Coulon, 1639.
Check protector classification and identification. *Security World,* 2: no. 7, October 1965.
Combat crime—burglary. *Crime Prevention News,* June 1975, p. 13.
Eckelmann, Clarence F.: Theft of electric energy. *Security Management,* January 1975, p. 18-21.
Friedman, Albert B.: The scatological rites of burglars. *Western Folklore,* 27:171-179, July 1968.
Gately, Glenn S.: An autopsy of a fraud. *Industrial Security,* December 1971.
Girard, Paul J.: Burglary trends and protection. *Journal of Criminal Law, Criminology, and Police Science,* 50:511-518, 1960.
Grout, Edward Harold: *Burglary Risks in Relation to Society, Law, and Insurance.* London, Sir I, Pitman and Sons, 1927.
Haveman, E.: History of a burglar. *Life,* 43: 147-151, October 7, 1957.
Holcomb, R. L.: Protection against burglary. Iowa City, *Inst. of Pub. Affairs,* Univ. of Iowa, 1953.
Introduction to the Bradford system of check classification. *Security World,* 1: no. 3, November 1964.
Jones, O. A.: Forgeries, frauds and worthless checks as they apply to retail stores—panelist. *Industrial Security,* October 1960, p. 36.
Law abiding safe-crackers, The. *Evening Record,* January 15, 1960.
Paholke, Sgt. Arthur R.: Burglary: The scene, investigation, prosecution. *Security Register,* November 1973, p. 53+.
Post, Richard S.: Burglary prevention and investigation for security officers. *Security Education Briefs,* OAK Security Inc., 1: no. 9.
Shabecoff, Philip: Thievery rising in U.S. industry. *New York Times,* June 16, 1963.
Theft: Is this what it's coming to? *Special Re-*

port, National Office Products Association, February/March, 1971.

Twentieth century pirates. *Newark Star Ledger,* July 26, 1962.

Two cases of theft and capture. *Occupational Hazards,* June 1966.

Two billion dollar theft. *Security World,* February 1966.

The $2 million touch. *Occupational Hazards,* April 1966.

Valkis, A. G.: Theft in department stores. Fourth Int. Criminological Congress, 1960.

Walsh, Timothy J., and Healy, Richard J.: Theft and fraud. *Protection of Assets Manual,* Vol. 1. California, The Merrit Co., 1974.

Weygand, Leroy C.: A new American liberty—the right to steal. *Industrial Security,* August 1971.

Wheeler, Keith: Brotherly boom in burglaries. *Life Magazine,* August 6, 1965, p. 71.

Willmer, M. A. P.: But what about the crimesmen? *Security Gazette,* August 1971.

Wilson, John: The changing pattern of commercial and industrial security. *Police Review,* April 1972, p. 501.

UNPUBLISHED MATERIAL

Reynolds, Morgan Owen: Crimes for profit: Economics of theft. Ph.D. thesis, Dept. of Economics, University of Wisconsin, 1971.

BOOKS

CREDIT CARDS

Maurer, D. W.: *The Big Con: The Story of the Confidence Man and the Confidence Game.* New York, Bobbs-Merrill, 1940.

Melville, H.: *The Confidence Man.* New York, Hendricks House, 1954.

Sternitsky, Julius L.: *Forgery and Fictitious Checks.* Springfield, Thomas, 1955.

PUBLICATIONS OF THE GOVERNMENT, LEARNED SOCIETIES, AND OTHER ORGANIZATIONS

Scharnack, John J.: Fraud. Chicago, Ill., Manufacturers Costs Assn., March 14, 1962. (Mimeographed)

PERIODICALS

CREDIT CARDS

Bad Checks. Records and Identification, Police Department, Long Beach, Calif., Dreis Investigation Bureau.

Barbash, J. T.: Compensation and the crime of pigeon dropping. *Journal of Clinical Psychology,* 8:92-94, 1952.

Barlay, Stephen: The perfect crime. *Top Security,* September 1975, p. 18-182.

Bray, Samuel: Credit card fraud can be controlled. *Security World,* December 1969.

California counterfeit twenty. *Security World,* 2: no. 4, June 1965.

Counterfeit currency. *Security World,* 1: no. 2, November 1964.

Countering the safe-breaker. *Security Gazette,* October 1961.

Decker, Briane A.: Colorado Bank Americard combats fraud in use of credit cards. *Industrial Security,* February 1968, p. 8.

Dodge, Robert L.: Credit card fraud investigations. *Security World,* July/August 1971.

Don't invite forgery. *Security World,* 1: no. 3, November 1964.

Ellson, D. G.: A report of research on detection of deception. Dept. of Psychology, Univ. of Indiana, 1952.

Fraud. *Bulletin No. 239,* Illinois Manufacturers Costs Assn., February 1962.

Gee, Harold F.: Broad form crime insurance primer. Chicago, Continental Natl. Amer. Institute, April 1963.

How to spot a phoney bill. *International Security Review,* September 18, 1972, p. 1, 13.

Keane, Timothy J.: Credit card fraud. *Security Management,* January 1974, p. 17-21.

Keating, James H.: Paper bandits. *Security World,* 1: no. 2, September-October 1964.

Kellett, William: Check recovery: A cooperative solution. *Security World,* 2: no. 2, March-April 1965.

Loomis, Roland L.: Plant protection reporting by data cards. *Industrial Security,* October 1966.

Marsh, Quinton N.: Check fraud and kiting. *Security Management,* May 1974, p. 13-18.

Mitchell, Ewan: The case of the worthless cheque. *Security and Protection,* April 1975, p. 24.

Schur, E. M.: Sociological analysis of confidence swindling. *Journal of Criminal Law, Criminology and Political Science,* September 1948, p. 296-304.

Skolnick, Jerome: The Berkeley scheme: Neighbourhood police. *The Nation,* March 22, 1971.

Sturgis, John C.: Policing the world-wide charge cards. *Security World,* July/August 1969.

The plastic card in 1984. Interview. *Security Management,* January 1974, p. 8-11.

Tocchio, O. J.: Counterfeiting: Another merchant's dilemma. *Police, 8:* no. 2, 44-47, November-December 1963.

Trade dollars for worthless checks. *American Stores,* 1953.

Walker, Don H.: The wide wide world of plastic cards. *Security Management,* January 1974, p. 6-7.

――――: The production and protection of credit cards. *Security Management,* January 1974, p. 12-16.

Williams, H. E.: Plastic crime. *Industrial Security,* October 1971.

BOOKS

SHOPLIFTING

Almer, C. H.: *The Fight Against Shoplifting in Sweden and W. Germany.* Boras, Sweden, Ab Boras Tryckservice, 1971.

Anonymous and Staff Written: Shoplifting in 1860, From Mayhew's London. 5: no. 1, January 1968.

Bath, Kenneth Charles: *An Introduction to Retail Security.* Broadway, Store Detectives Ltd., 1967.

Cameron, Mary O.: *Department Store Shoplifting.* Ann Arbor, Univ. of Michigan.

――――: *The Booster and the Snitch.* New York, Free Press, 1964.

Curtis, B.: *Security Control External Theft.* New York, Chain Store Publishing Corp., 1971, p. 382.

Curtis, S. J.: *Modern Retail Security.* Springfield, Thomas, 1960.

Edwards, L. E.: *Shoplifting and Shrinkage Protection for Stores.* Springfield, Thomas, 1962.

Frin, J. D., Sherman, E., Maskell, J. S., and Arthur, M.: *Selected Cases on the Law of Shoplifting.* Springfield, Thomas, 1975.

Gibbens, T. C. N., and Prince, Joyce: *Shoplifting.* London, The Institute for the Study and Treatment of Delinquency, 1962.

Hughes, Mary Margaret (Ed.): *Successful Retail Security.* Los Angeles, Security World Publishing Co., Inc., 1974.

Kaufmann, Arthur C.: *Combating Shoplifting.* New York, National Retail Merchants Association, 1974.

Leininger, Sheryl (Ed.): *Internal Theft: Investigation and Control.* Los Angeles, Security World Publishing Co., Inc., 1975.

Nader, L. R.: *Protecting Your Business Against Employee Thefts, Shoplifters, and Other Hazards.* Pilot Books, 1971.

Nystrom, Paul H.: *Retail Store Operations.* New York, Ronald, 1937.

Owens, Ruth M.: *The Human Element of Security.* December 1969.

Rogers, Keith M.: *Detection and Prevention of Business Losses.* New York, Arco, 1962.

Wright, K. G.: *The Shopkeeper's Security Manual.* Gateshead, Northumberland Press Limited, 1971.

PUBLICATIONS OF THE GOVERNMENT, LEARNED SOCIETIES, AND OTHER ORGANIZATIONS

SHOPLIFTING

Almer, C. H.: *The Fight Against Shoplifting in Sweden and W. Germany.* AB Boras Trychservice, Sweden, 1971.

Douglas, Johnson: Business prevention studies. McCann Erickson (draft of paper).

Hughes, M. M.: *Successful Retail Security— An Anthology.* Los Angeles, Security World Publishing Co., 1974.

Inbau, Fred E.: *Manual for Store Protection.* The Retail Special Service Association, 1951, p. 37.

Morris, Howard Britton: *Elements of Successful Plant Protection.* Louisville, Industrial Training Associates, 1953.

National Industrial Conference Board. *Industrial Security, Part III, Theft Control Procedures.* New York, Nat. Indust. Conf. Board, 1954.

Small Business Administration. *Preventing Retail Theft.* Washington, D.C., 1966.

Retail Anthology. Volumes 1, 2, and 3. Los Angeles, Security World Publishing Co., 1967.

Rogers, Keith M.: *Coping with Shoplifters.* Los Angeles, Security World Publishing Co., 1965.

Sears, Roebuck & Co. *Retail Security Manual.* Chicago, Sears, Roebuck & Co., 1963.

Self-service techniques, pilferage control. *Publisher's Weekly, 175:* no. 30, May 1959.

Shoplifter, The. Highway Safety Foundation, Mansfield, Ohio. (Film)

Store Managers National Retail Merchants Association. *They Shall Not Pass.* New York.

Wright, Kenneth: *The Shopkeeper's Security Manual.* London, Tom Stacey, Ltd.

PERIODICALS

SHOPLIFTING

Angelino, Henry: Shoplifting: A critical review. *Security Management, 1-5*:17-22, Spring, 1953.

———: Riot precautions for retail merchants, Detroit's Retail Merchant Association. *Security World, 5:* no. 7, July/August 1968.

Anybody might be a shoplifter. *Bluebook*, March 1954.

Arieff, Alex J., and Bowie, Carol G.: Some psychiatric aspects of shoplifting. *J Criminal Psychol, 8:* January 1947.

Astor, S. D.: Mastering the impossible situation in loss prevention. *Security World, 10:* no. 2, 32-33, 35-36, February 1973.

———: Shoplifting survey. *Security World*, March 1971.

Bennet, H. M.: Shoplifting in midtown. *Criminal Law Review*, 1968, p. 413.

Bowen, Croswell: How do you know a shoplifter? *Pageant*, May 1956.

Buckner, Al: Designing security into retail facilities. *Security World, 4:* no. 8, September 1967.

Business Week. Shoplifting: The pinch that hurts. June 27, 1970.

Buttars, Clair: Shipping theft. *Security World, 5:* no. 3, March 1968.

———: Trucking security. *Security World, 3:* no. 1, December/January 1966.

Campus shoplifting from Wall Street Journal. *Security World, 5:* no. 5, May 1968.

Carroll, William E.: An executive look at security. *Industrial Security, 11:* no. 3, August 1967.

Carter, Carl L.: Help enforce loss prevention. *Security Management*, 1960.

Clinic on shoplifting. Chicago Retail Merchants Association, 1954.

Cole, Richard B.: Retail loss prevention; an educational program. *Security World, 5:* no. 6, June 1968.

Curtis, S. J.: Juvenile shoplifters. *Industrial Security*, January 1961.

———: Security for the smaller store. *Police, 6:* nos. 3 and 4, January/April, 1962.

———: Retail/security organization. *Industrial Security*, January 1962, p. 14.

Debo, Charles: The key of shoplifting control. *Security World*, July 1964, p. 44.

DeGan, William H.: Theft and pilferage in the retail industry. *Industrial Security*, April 1964, p. 35.

DeSantis, Amelia: Anybody might be a shoplifter. *Bluebook*, March 1954.

Dickins, B. M.: Shops, shoplifting and law enforcement. *Criminal Law Review*, 1969, p. 464.

Dornfeld, G. R.: The shoplifter. *FBI Law Enforcement Bulletin, 36:* no. 12, 2, 1967.

Doyle, James J.: Retail protection. *Industrial Security*, April 1960, p. 12.

Edward, Loren F.: The challenge of store security. *Police, 3:* no. 4, March-April 1959.

Elmes, Frank: Police and commercial security. *Police: 44:*55, 1971.

Fiegel, Al: Telling it like it is to teenagers: An anti-shoplifting program. *Security World*, July/August 1970.

Gardner, John: Theft loss totals show crime prevention pays. *Security Gazette*, December 1974, p. 402.

Gibbens, Joyce: Shoplifting. London, The Institute for the Study and Treatment of Delinquency, 1962.

Glaser, N. T.: Shopping services. *Security World, 3:* no. 7, July/August 1966.

Griffen, R.: Shoplifting apprehensions: A statistical report. *Security World, 6:* no. 10, 33, 1969.

———: The shoplifter; shadow or substance? *Security World, 4:* no. 9, 1967.

———: Shoplifting; a moral dilemma. *Security World, 4:* no. 4, January 1967.

———: Shoplifting; a statistical report. *Security World, 2:* no. 5, July/August 1965.

———: Shoplifting facts from figures. *Security World, 3:* no. 11, December 1966.

———: Shoplifting in supermarkets. *Police Chief, 31:* July 1964.

———: Behavioral patterns in shoplifting. *Security World*, September 1971.

———: 10,212 shoplifting apprehensions. *Security World*, November 1969.

Hodgson, Tom: *You Can Do Something About Shoplifting*. Minnesota Retail Federation, Inc., Minneapolis, Minn.

Hoffman, G. H.: Department store security; a working programme. *Security World, 3:* no. 8 and 10, 17, 1966.

Hole, Robert R.: Shoplifting apprehensions can be made to stick, Part II. *Security World*, February 1972.

Hoover, J. Edgar: Women with sticky fingers. *Auditgram, 38:* no. 1, January 1962.

Bowen, Croswell: How do you know a shoplifter? *Pageant*, May 1956.

Howard, John P.: A community anti-shoplifting program. *Security World*, February 1966, p. 12.

How to stop shoplifters—a guide for employees. Florida State Retailers Assn.

———: A community anti-shoplifting program. *Security World, 2:* no. 2, December 1966.

Inbaue, F. E.: Protection and recapture of

merchandise from shoplifters. *Illinois Law Review, 46:* no. 6, 887, 1952.

———: Manual for store protection. Chicago, Retail Special Services Associations, 1951.

Jepson, N.: Shoplifting: A comment. *Criminal Law Review,* 1968, p. 430.

Juvenile shoplifters. *Industrial Security,* January 1961, p. 14.

Kunz, Armand D.: Criminal law—robbery—corpse as victims. *Wayne Law Review,* Spring, 1962, p. 8.

Landis, John C.: First time shoplifter: A judge's view. *Security World, 1:* no. 1, July 1964.

Lanning, J. J.: How to curb shoplifting. *Bulletin of the Society of Professional Investigators,* February 1969.

Laubach, A. C.: Protecting company assets. *Industrial Security, 10:* no. 1, February 1966.

———: Lie detector tests are a blight on retailing. *Retail Clerk Advocate,* October, 1960.

Lubach, A. C.: Protecting company assets. *Industrial Security,* February 1966, p. 24.

Lawder, Lee E.: Shoplifting—a growing occupation. *Law and Order, 12:* July 1964.

Legal aspects of shoplifting. *STLJ, 3:* Summer-Fall, 1958.

Lodge, J. S.: Anti-shoplifting devices—how effective are they? *Security Gazette,* December 1969.

Loss prevention check list. New York, National Retail Merchants Association.

McIntyre, Fred T.: Duties of a department store security director. *Industrial Security,* October 1967, p. 8.

Marks, D. A.: Retail store security in Iceland. *Top Security,* September 1975, p. 204-206.

Mapes, G.: Campus shoplifting. *Security World, 5:* no. 5, 29, 1968.

Maynard, F.: The housewives crime—and what makes them do it. *Good Housekeeping,* October 1967.

Merrick, B.: Shoplifting, a microcosm. *The Criminologist, 5:* no. 18, 68, 1970.

Mewer, Wesley O.: Shoplifting—big business. *Industrial Security,* January 1963.

Moore, P.: Clear objectives are a prerequisite of retail security efficiency. *Security Gazette,* February 1975, p. 42-43.

———: Is prevention preached whilst detection is practised? *Security Gazette,* December 1974, p. 444-446.

———: Shoplifters and the guilt of the innocent. *Security Gazette,* June 1975, p. 208-209.

One in ten shoppers is a shoplifter. *New York Times Magazine,* March 10, 1970.

Ordway, John A.: Successful: Court treatment of shoplifters. *J Criminal Law, Criminology, Police Science, 53:* no. 3, September-November 1962.

Osborough, W. Nial: Immunity for the English shoplifter. *Security World, 2:* no. 7, October 1965.

Paine, David: Retail security. *Canadian Police Chief,* April 1975.

Pollack, Jack H.: The war on shoplifters. *This Week,* October 1949.

Problem of store protection. Chicago, Illinois Retail Merchants Assn, April 12, 1961.

Protection and recapture of merchandise from shoplifters, The. *Illinois Law Rev, 46:* no. 6, 1952.

Ptacek, Bernarr: Who's watching the store? *Industrial Security,* June 1967.

Pugh, John: Shop thefts. *Security and Protection,* March 1975, p. 9-11.

Rayen, Alan G.: The next decade in retail security. *Industrial Security,* February 1971.

Recommended procedure when cashing checks for strangers. *L.A. Police, Form, 12:* no. 29, Rev. Los Angeles.

Retail Trade Board. Store protection. Boston, Greater Boston Chamber of Commerce, 1954.

Reynolds, E. Stanley: Plugging the profit drain. *Industrial Security,* February 1970.

Robin, Gerald D.: The American customer: Shopper or shoplifter. *Police, 8:* no. 3, January-February 1964.

Rouke, Fabian L.: Psychology of the retail criminal (business criminal, pilferer, neurotic theft). *Police, 4:*28-32, March-April 1960.

———: Shoplifting: Its symbolic motivation. *NPPA J, 3:* January 1957.

Save teenage futures—stop juvenile shoplifting. New York, National Retail Merchants Assn.

Security of retail shops and private dwellings. *Security Gazette, 12:* no. 4, 136, 1970.

Shadowed shoppers. *Security Gazette, 12:* no. 8, 306, 1970.

Shoplifting: A moral dilemma. *Security World,* January 1967, p. 14-15.

Shoplifting in 1860, from Mayhew's London. *Security World, 5:* January 1968.

Shoplifting and the law of arrest. *Yale Law Journal, 62:* no. 5.

Shoplifting—prevention and control. *Store Managers Guide,* no. 99, 1961.

Shoplifting racket tricks of the trade. *Amer Stores Inc.,* 1953.

Smith, C. Gorgon: Burglary of retail stores—panelist. *Industrial Security,* October 1965, p. 42.

Snatch thief is in for a big surprise, The. *Security and Protection,* April 1975, p. 12-14.

Some vital points of shopfront protection. *Security Gazette,* December 1971.

Spencer, Klaw: Shoplifting. *Harper's Bazaar,* September 1950.

Sterling, Stewart: Stop that shoplifter! *Saturday Evening Post,* October 22, 1949.

Stevenson, M. W.: Burglary and shoplifting as related to retail business—panelist. *Industrial Security,* October 1960, p. 35.

Strom, Maggie: An ounce of prevention. . . . *Security World,* December 1974, p. 40.

Titus, Charles M.: Shoplifting security: One company's problem, Part II. *Security World, 2:* no. 8, November 1965.

Tocchio, O. J.: Shoplifting—the source of mercantile establishment. *Police, 6:* no. 5, May-June 1962.

———: Shoplifting—the source of mercantile establishment. *Police, 7:* no. 1, September-October 1962.

———: Shoplifting—the source of mercantile establishment. *Police, 7:* no. 2, November-December 1962.

Verrill, Addison, H.: Reducing shoplifting losses. *Small Marketers Aid No. 129,* Washington, Government Printing Office, September 1967.

View of shoplifting in the affluent society. *Commercial Service System,* 1964.

Vincent, Barbara: Electronics application pioneers for retailing. *Security World,* September 1974, p. 34-35.

———: Retail security to become even more important. *Security and Protection,* May 1975, p. 8-9.

Waltz, J. R.: Shoplifting and the law of arrest: The merchants' dilemma. *Yale Law Journal, 62:* no. 5, 188, 1953.

Wendell, Roy: Shoplifting a growing menace. *Kiwanis Magazine,* February 1953.

Why shoplifters get caught. *Coronet,* August 1950.

Willmer, M. A.: Shoplifting—some lessons from the post. *Security Gazette,* July 1975, p. 246.

Wisher, Curtis: Teenage shoplifting: Who, where, when, how? *Security World,* November 1968, p. 16-20.

UNPUBLISHED MATERIAL

SHOPLIFTING

Reynolds, Morgan: The economics of theft. Paper presented to the Western Economic Association Conference, August 1971.

Schlager, Robert W.: Direct employee theft in a retail department store. Unpublished Graduate B-Paper, Michigan State Univ., 1967.

Steffensmeir, Darrell John: Respectability and deviance: An observational study of reactions to shoplifting. Ph.D. thesis, Department of Sociology, University of Iowa, 1972.

Chapter 19

ISSUES IN SECURITY

SECURITY AND LOSS prevention are dynamic and in a constant state of change. The past several years have brought dramatic changes into the operations and services provided by those engaged in the protective services. The legal basis for operations, the training required for licensing, the type and quality of hardware, and many other factors have changed. The prognosis is for more change in the future. The problems of yesterday have been responded to within the limitations of the then available technology and managerial know-how. The basic issues inherent in the provision of protective services have not, on the other hand, been resolved.

The issues present in the protective service can be categorized as:

BASIC:[1]
    Environmental design and behavioral modification and control
    Public and private police relationships and protective policy development

FUNCTIONAL:[2]
    Management decision-making
    Hardware and systems

The issues which affect the practitioners are in an almost daily state of flux. Consequently any listing of specific problems and issues will always be out of date. Bearing this limitation in mind, the compilation of issues and problem areas in the appendices in need of research are presented. These categories allow for the introduction of issues, questions, and problems on a timely yet systematic basis. Issues which are resolved either permanently or temporarily can be removed but may be reinstated when the need arises.

PUBLIC AND PRIVATE PROTECTIVE SERVICES RELATIONSHIPS

Public law enforcement agencies have assumed a position of being the most visibly present and responsible agency for preventing crime and delinquency in United States cities. Public police are viewed as the primary mechanism through which government provides public protective services to the communities and provide necessary protective functions. It has long been recognized in municipal police circles that the ability of the department to provide protective services is limited; based on the amount of available resources for support of police activities. The amount of manpower, equipment, and the number of calls for services that can be answered, relate very directly to the following factors:

1. the revenue for protective services provided to the policy agency,
2. the ability of the police executives to organize and to respond to problems appropriately,
3. the training level of personnel delivering protective services, and
4. support by the community for police operations.

These factors and the assumptions that the public police are primary means of protection of community have not sufficiently taken into consideration the condition in existence in many major metropolitan communities of the existence of a significantly large group of persons involved in private and protective services. Recent studies[3] have indicated that, in most major metropolitan areas, the ratio of public police to private security personnel is 1:1.

Other estimates indicate that the ratio will increase to a 60:40 in favor of the private sector by 1977. The situation is similar in Canada with a ratio of 2:1 and in some cases 3:1 ratio between private and public protective services in large municipalities. A similar condition exists in England with 3:1 and in some cases 4:1 ratio between private and public agencies.

Figure 19-1 indicates that the public agencies in a community only provide a

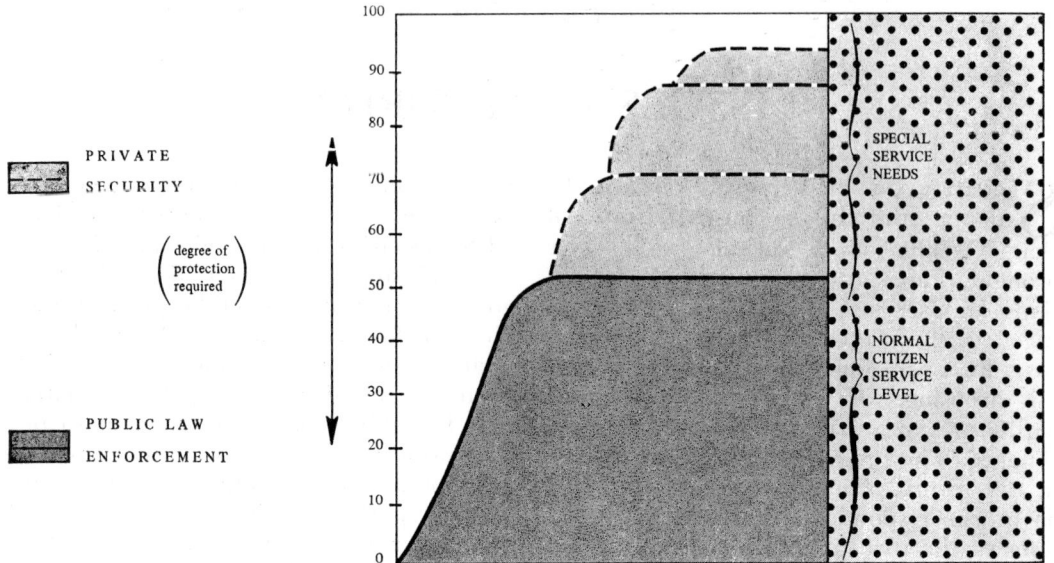

Figure 19-1. Protection provided individuals in society.

level of protection for something less than the total amount of protective services required in a given community. Individual needs which are present over and above the basic services provided by the public agencies can only be provided through other than public means or through public means supported at individual expense for services rendered.[4] In such cases, the question that can be legitimately asked is, "What relationship do these private agencies have to the overall level of performance of public agencies in that community?" What interrelationships exist between the public and private sectors? These questions are particularly significant if one looks at the existence and use of private central alarm services or connect direct police alarms to protect the high-risk buildings and activities within a community, or the use of large private police forces or security firms to patrol and enforce criminal and civil regulations in a large segment of the community.

What effect on crime deterrence is present with the increase in private security personnel?

Obviously, if a business or industrial plant is paying money over and above their tax support for municipal police services, they expect to get and probably perceive some additional value from the private enforcement or protective services they purchase. If their perceptions are correct, what effect does the presence of these privately paid supplementary security forces have on the overall crime rate (positively or negatively) in a community? If crime displacement theory is correct, the question should be raised of the transference of risk from the people who have financial means to pay for additional protection over and above the level provided by police to those less able to pay the supplement in the community. Thus, the risk of crime for the low-income, inner-city dweller is increased as opposed to the inner-city business or industrial firm which can provide for additional protection.

It is unknown at this time what relationships exist between public and private enforcement and security agencies in communities which have large amounts of private sector personnel operating. What are the overt and quasi-official relationships between public and private agencies which assist both in completing their respective assignments? Public policy for the com-

munity, and effective private agencies actions for their clients should dictate some action. Answers to these questions are of particular relevance to the extent that the following factors seem to be operative in society in general:
1. the ability of municipalities to continue increasing the revenue levels for protective services is decreasing,
2. the called-for services within municipal police agencies are increasing;
3. if our assumptions about citizen service levels (See Fig. 19-1) are correct, this can only result in the lowering of the ability of municipal protective public agencies to continue providing an overall level service comparable to what has been provided in past years.

The development of a community protective service network requires that an analysis be conducted of all factors within a community which in some way contribute to the overall protection or perceive protection of the members of that community. Thus, with a large percentage of the population (although admittedly the most financially able segment of society) providing supplementary protection, the question must be raised of the incorporation of the prevalent practice of providing private protective services into public policy arena. The admission on the one hand, of police inability to provide a uniformily acceptable level of service to everybody is currently recognized. On the other hand, there is tacit admission that does not permit the incorporation of the operational reality into planning for reorganized services series or alternative approaches to providing protection within a community. It is, for example, not possible to either delegate or subcontract various preventive or protective services to nonpolice agencies (either public or private) unless public agencies admit that these agencies do perform a meaningful function. While this has not occurred in fact, it does occur in practice in many municipalities. To make such an assumption for a public police agency does, however, require it to make an admission that it cannot adequately cope with the protection of all property and persons within a community. It thus requires a sharing of responsibility which police agencies have not been willing to make up to this time.

The question of the reorganization of municipal protective services and possibly the development of some alternative models of community protection based on a shared responsibility between the public and the private sector or between police or other types of public protective agencies is by no means a unique suggestion. A form of this model was the operational plan for protection which existed prior to the Metropolitan Police (London) Organization and the model that persists in this country today. This division has, however, not been recognized as a part of a formal criminal justice system in this country.

The problem, however, remains that very little is known about the effectiveness of the private sector of protective services in reducing crime, disorder, or preventing loss for a client. Many indicators are present that seem to indicate client satisfaction with services rendered. The growth in dollar sales volume, and the growth of privately funded protective services has been indicated to be between 10 and 15 percent per year since 1968. Thus, almost a 100 percent growth in the private protective field has occurred in the last seven years.

Very few businesses can afford to pay for an overhead item, which security has often been considered to be, without gaining some perceived or tangible benefits from it. This, however, is not an acceptable kind of empirical data upon which public policy issues can be based. If the goals and standards of private security are to be clearly defined and delineated, they must be based on empirical data about the effect their use in a given community has for the overall level of protection.

Basic to any discussion of issues in protective services is the role of all facets of public and private organizations. The roles of neither the public police nor security agencies have not been clearly defined, while considerable role ambiguity exists for the entire private sector and its activities. These uncertainties have been transmitted to the public and users of these services who are also uncertain as to what level and quality of service they should expect or are entitled to receive. Even if roles were to be crystalized, there is little evidence that the situation would be alleviated since new alignments in any service function bring new problems of expectations and quality of service. In such a situation, there is also no evidence that the new arrangements for service would be any better than the current fragmented relationship where the users were not sure who was to perform a given task.

Barriers to Cooperation

The Private Security Advisory Council (PSAC) indicated that there were nine major areas which reduced the possibility of cooperation between the public police and private security. These areas are:
1. Lack of mutual respect
2. Corruption
3. Lack of cooperation
4. Lack of two-way data flow
5. Competition from police "moonlighting"
6. Restrictions by the police
7. Police have inadequate knowledge of private security functions
8. Private security lacks a unified professional voice
9. Police/private security need better standards of professionalism

Both the Rand Corporation study and the National Advisory Commission on Criminal Justice Goals and Standards Report—Police listed problem areas which affect the relations between public and private protection agencies.

These problems include:
1. Excessive use of force by private agencies
2. Illegal arrests by security officers
3. Impersonation of police officers by security personnel
4. Invasion of privacy by security officers
5. Abuse of authority
6. Illegal and inappropriate access to police records
7. Nonreporting of crimes
8. Dishonest and poor business practices
9. Poor response to alarms
10. Private system of criminal justice
11. High false alarm rate
12. Accountability less than public police
13. More discretion for arrest, search and seizure than police.

The basic issue facing protective services planners and decision-makers is the development of an effective delivery system to meet the diverse and often contradictory needs of society and private interests. Most crime control strategies produce various side effects among which are prevention, displacement, and escalation. While prevention is most desirable, displacement does occur as does the increasing of related criminal activity as the prevented or controlled crimes become more difficult. The quality of life of the users of a protective program should not be diminished by the development of a secure working or living environment.

In the United States compliance and enforcement must be obtained by dissuasion from criminal behavior rather than authoritarian control and repressive measures. A very fine distinction, however, exists between compliance and a controlled environment. Personal freedoms have historically been reduced by increased security measures. The social costs of security cannot be outweighed by the short-term benefit of reducing visible manifestations of crime or insecurity.

There is little question about the scope of private security activities in the United States. The Private Security Task Force Report clearly indicates the direction this function will move in the future. Unlike some police experts who view private security as a "necessary evil" in which "profit is all," or some security experts who feel that "statistics is all" to the police, the Task Force indicates that a meaningful, complimentary, and mutually supportive relationship will assist not only police and security agencies but the users of these services have a safe and secure community. The major issue is, however, what form these working relationships will take.

FOOTNOTES

1. Appendix 19A presents a listing of the major areas of need for basic research in the protective services.
2. Appendix 19B presents a listing of functional problem areas.
3. Rand Report, *Nature and Extent of Private Police in the U.S.,* 1971.
4. The Experience with Public Housing Police in Boston Through a Municipal Security Force. In Dallas, for example, specialized police such as airport and port authorities provide individual protective needs. Consequently, there is some indication that in every community where private police, either publicly or privately funded, exist these agencies perform some type of crime prevention, crime deterrence function in addition to that provided by the public agencies.

BOOKS

Abrahamsen, David: *The Psychology of Crime.* New York, Columbia University Press, 1960.
———: *Crime and the Human Mind.* New York, Columbia, 1944.
Adorno, T. W., Frenkel-Brunswick, E., and Levinson, D. J.: *The Authoritarian Personality.* New York, Harper, 1950.
Alexander, Christopher: *The City as a Mechanism for Sustaining Human Contact, in Environment for Man.* Bloomington, Indiana University Press, 1967.
Arendt, Hannah: *The Human Condition.* New York, Anchor ed., 1959.
Ashby, W. R.: *Design for a Brain.* New York, John Wiley & Sons, 1952.
Asratyan, E. A.: *Pavlov, His Life and Work.* Moscow, Foreign Languages Publishing House, 1953.
Babkin: *B. P. Pavlov, A Biography.* London, V. Gollancz, 1951.
Bachrach, Arthur (Ed.): *Experimental Foundations of Clinical Psychology.* New York, Basic Books, 1962.
Banton, Michael: *The Policeman in the Community.* New York, Basic Books, 1964.
Bauer, R. A., Inkeles, A., and Kluckhohn, C.: *How the Soviet System Works.* New York, Vantage Press, 1956.
Bay, Christian: *The Structure of Freedom.* Stanford University Press, 1970.
Beck, F., and Godin, W.: *Russian Purge and the Extraction of Confession.* New York, Viking Press, Inc., 1950.
Berger, Monroe (Ed.): *Freedom and Control in Modern Society.* New York, 1964.
Berle, Adolph A.: *The 20th Century Capitalist Revolution.* New York, Harcourt, 1954.
Beynon, Ray Mane: *The Importance of Feeling Inferior.* New York, Harper & Brothers.
Bishop, Cecil: *Women and Crime.* London, Chatto and Windus, 1931.
Blake, Jusith, and Kinglsey, Davis: *Handbook of Modern Sociology.* Chicago, Rand McNally and Co., 1964.
Blau, P. M.: *The Dynamics of Bureaucracy.* Chicago, 1955.
———, and Scott, W. R.: *Formal Organizations.* San Francisco, 1962.
Blum, R. H., et al: *Drugs II: Society and Drugs.* San Francisco, Jossey-Bass, 1969.
Booth, Ernest: *Stealing Through Life.* New York, Alfred A. Knopf, 1929.
Brannon, W. T.: *Yellow Kid Weil. The Autobiography of America's Master Swindler.* Chicago, Ziff-Davis, 1948.
Brasol, Boris: *The Element of Crime.* New York, Oxford University Press.
Breeze, Burtis Burr: *Psychology.* New York, Charles Scribner's Sons, 1917.
Bromberg, W.: *The Mind of Man.* London, Hamish Hamilton, 1937.
Burke, Shifty (pseud.): *Peterman: Memoirs of a Safe-Breaker.* London, Barker, 1966.
Camnion, B., and Stearns, M.: *Crooks Are Human Too.* Englewood Cliffs, N.J., Prentice-Hall, 1957.
Chalmers, David M.: *Hooded Americanism: The History of the Klu Klux Klan.* Chicago, Ill., Quadrangle Books, 1968.
Chappell, D., and Wilson, P. R.: *The Police and the Public in Australia and New Zealand.* St. Lucia, Queens Land, Australia.
Drug abuse in industry. Philadelphia, Pa., *The Journal of Industrial Medicine and Surgery,* 1970.
Dudycha, G. J.: *Psychology for Law Enforcement Officers.* Springfield, Thomas, 1955.
Elliot, Mabel A.: *Crime in a Modern Society.* New York, Harper, 1952.
Felley, V. L.: *American Law Enforcement.* Boston, Holbrook Press, 1973.

Frey, Elmer L.: *The Tax Dodger.* New York, Greenberg Publishers, 1958.
Germann, A. C., Day, and Gallati, R.: *Introduction to Law Enforcement.* Springfield, Thomas, 1967.
———: *Theft, Law and Society.* Indianapolis, Bobbs, 1952.
Hayward, Arthur L.: *Lives of the Most Remarkable Criminals.* London, George Routledge and Sons, Ltd., 1920.
Hibbert, Christopher: *The Roots of Evil: A Social History of Crime and Punishment.* Boston, Little, 1963.
Hickey, Hiel: *The Gentleman Was a Thief.* New York, Holt, Rinehart, and Winston, 1961.
Irwin, John: *The Felon.* New Jersey, Printice-Hall, Inc., 1970.
Karn, H. W., and Gilmer, B.: *Reading in Industrial and Business Psychology.* New York, McGraw-Hill, 1952.
Karpman, Benjamin: *The Sex Offender and His Offense.* New York, Julian, 1954.
Kenney, John P., and Williams, John B.: *Police Operations.* Springfield, Thomas.
Lane, R., and Sears, D.: *Public Opinion.* New Jersey, Printice-Hall, 1964.
Lawshe, C. H.: *Psychology of Industrial Relations.* New York, McGraw, 1953.
Lee, Dorothy: *Freedom and Culture.* Englewood Cliffs, N.J., 1959.
Leonhard, W.: *Child of the Revolution.* Chicago, Henry Regnery, 1958.
Maier, Norman R. F.: *Psychology in Industry,* 2nd ed. New York, Houghton Mifflin, 1955.
Mannheim, H.: *Social Aspects of Crime in England Between the Wars.* London, Allen and Unwin.
McClintock, F. H., and Gibson, Evelyn: *Robbery in London.* New York, Macmillan and Co., 1961.
Packard, Vance: *The Naked Society.* New York, McKay, 1954.
Packer, H.: *The Limits of the Criminal Sanction.* Stanford, Stanford University Press, 1968.
Pearl, Arthur and Riessman, Frank: *New Careers for the Poor.* New York, Free Press, 1965, p. 88.
Pollack, Otto: *Criminality of Women.* Philadelphia, 1950.
Reckless, Walter C.: *The Crime Problem.* New York, Appleton, 1961.
Reid, Ed., and Demaris, Ovid: *The Green Felt Jungle.* New York, Pocket Books, 1964.
Rein, Martin: *Social Policy: Issues of Choice and Change.* New York, Random House, 1970. p. 236-237.
Reinhardt, Guenther: *Crime Without Punishment.* New York, Hermitage House, 1952.
———: *Sex Perversions and Sex Crimes.* Springfield, Thomas, 1957.
Report: A Panel Discussion on Drugs in Industry. New York, Burns Security Institute, 1975.
Roebuck, Julian B.: *Criminal Typology.* Springfield, Thomas, 1967.
Rose, Arnold M.: *Sociology: The Study of Human Relations.* New York, Knoft, 1965.
Scarne, J.: *Scarne's Complete Guide to Gambling.* New York, Simon & Schuster, 1961.
Schwartz, L. B., and Goldstein, S. R.: *Law Enforcement Handbook for Police.* West Publishing Co., 1970.
Sellin, Thorsten, and Savitz, Leonard: *A Bibliographical Manual of the Student of Criminology,* 1: no. 3.
Simmel, G.: *The Sociology of George Simmel.* New York, Free Press, 1964.
Smith, Alexander: *A History of the Lives and Robberies of the Most Notorious Highwaymen.* London, George Routledge and Sons, Ltd., 1926.
Williams, David: *Not in the Public Interest.* London, Hutchinson, 1965.

PUBLICATIONS OF THE GOVERNMENT, LEARNED SOCIETIES, AND OTHER ORGANIZATIONS

Altman, Stanley M.: The design of automatic surveillance systems for urban freeways. Institute of Brooklyn, Ph.D., 1967.
Belson, William A.: The extent of stealing by London boys and some of its origins. London, The Survey Research Center, London School of Economics and Political Science, 1968.
Biderman, Albert D., Johnson, Louise A., McIntyre, Jennie, and Weir, Adrienne: Report on a pilot study in the District of Columbia on victimization and attitudes toward law enforcement. The President's Commission on Law Enforcement and Administration of Justice. Field Surveys I, Washington, D.C.
Blum, R., and Gomber, E.: President's Commission on Law Enforcement and Administration of Justice—Police Field Procedures Report. Police Procedures Advisory Group.

———: Drugs and violence. *National Commission on the Causes and Prevention of Violence Staff Study Series, 13:* Washington, Government Printing Office, 1970.

Boggs, Sarah L.: The ecology of crime occurrence in St. Louis: A reconceptualization of crime rates and patterns. Dissertation, Washington University, 1964.

Caldwell, Morris G.: Personality trends in the youthful male offender. *Journal of Criminal Law, Criminology and Police Science, 49:*405-416, 1959.

Caper—Crime Analysis—Project Evaluation—*NILECJ,* 1972.

Chicago Crime Commission—Action Summary—1971, Chicago. Chicago Crime Commission, 1971.

Chicago Police Department. Operations Research Task Force. Allocation of Resources in the Chicago Police Force, November 1969.

Cutler, Stephen, and Reiss, Albert C., Jr.: Crimes against public and quasi-public organizations in Boston, Chicago, and Washington, D.C. A report to the President's Commission on Law Enforcement and Administration of Justice. *Business and Organization Survey Report,* October 9, 1966.

District of Columbia—Comprehensive Plan and Action Programs for Law Enforcement and Criminal Justice, April 1970. District of Columbia, Offices of Criminal Justice and Annual.

Dobrogaev, S. M.: *Speech Reflexes.* New York, National Committee for a Free Europe, 1953.

Dodge, D. C., and McMahon, J. S.: New York, State Crime Control Projects in the Fourth Dept.—The First Year. New York State Crime Control Planning Board.

Drug abuse as a business problem. New York Chamber of Commerce, 1970.

Elliott, J. F., and Sardino, T. J.: Experimental evaluation of the crime control team organization concept. *Police,* May/June 1970.

———: Detection and interception capability of one and two man patrol units. GE Electronics Laboratory, 1969.

Ennis, Philip N.: Criminal victimization in the United States: A report of a national survey. The President's Commission on Law Enforcement and Administration of Justice.

Evaluation of crime control programs. National Institute of Law Enforcement and Criminal Justice, LEAA, April 1972.

Fabian, Robert: *The Anatomy of Crime.* London, Pelham, 1970.

Falk, Gerhard J.: The influence of season on the crime rate. *Journal of Criminal Law, Criminology, and Police Science, 54:*456-469, December 1963.

Federal Bureau of Investigation. Crime in the United States. Uniform Crime Reports. Washington, Government Printing Office, 1975.

Giffin, K.: Recent research on inter-personal trust. Paper presented to the Annual Convention of the Speech Association of America, Los Angeles, December 1967.

Gray, B. M.: *Crime Specific Planning—An Overview.* William and Mary College, Metro Criminal Justice Center, 1973.

Greenhalgh, W. F.: A town's rate of serious crime against property and its association with some broad social factors. London, Home Office, Scientific Advisor's Branch, February 1964.

Illinois Law Enforcement Commission Annual Report, 1969. Illinois Law Enforcement Commission, 1970.

Information sharing: The hidden challenge of criminal justice. New York State Identification and Intelligence System, 1965.

International Association of Chiefs of Police Training Bulletin, Aggressive Patrol, No. 21. Published by the Field Division of I.A.C.P., Washington, 1965.

Kansas City-South Patrol Division-Proactive Patrol Deployment Project. Kansas City Police Dept.

Kansas City-Special Operations Division-Apprehension Orientated Patrol Deployment Project. Kansas City Police Dept.

Kaufmann, Ulrich George: *How to Outsmart Criminals.* Brooklin P. Gauss' Sons, 1967.

Knopf, T. A.: Youth Patrols: An experiment in community participation. Waltham, Mass., Brandeis University Lemberg Center for the Study of Violence, 1969.

Knots, Terry Ann: Youth patrols: An experiment in community participation. Brandeis University Lemberg Center for the Study of Violence, 1969.

Law Enforcement Conferences on Campus. Eastern Kentucky Univ. LEAA, 1st Grant.

Louisiana Crime Control Goals—New Directions 6907. Louisiana Comm. on Law Enforcement & Administration of Criminal Justice, 1969.

MacLeod, R. B.: The phenomenological approach to social psychology. *Person Perception and Interpersonal Behavior,* Stanford University Press, 1958.

Management Guide on Alcoholism and Other Behavioral Problems. Chicago, Kemper Insurance Company.

Milwaukee County Safe Streets Council—Recommended Law Enforcement Plan. Ernst & Ernst, 1970.

Napote, J.: Interpol versus the underworld of narcotics. *UNESCO Courier,* May 1968, p. 24-28.

OMAMA-Douglas County Metropolitan Criminal Justice Center—Ninety Day Report. NILECJ, 1972.

Piven, Herman, and Alcabes, Abraham: *The Crisis of Qualified Manpower for Criminal Justice,* Vol. 1. Office of Juvenile Delinquency and Youth Development, 1969.

Police manpower and equipment resources study. *Ontario Police,* 2:30, 1972.

Police—Report of the National Advisory Commission on Criminal Justice Standards and Goals, 1973. National Advisory Commission on Criminal Justice Standards and Goals.

Powell, M. D., and Murray, D.: *Regional Criminal Justice Planning—Manuel for local officials, Part 1—Regional Criminal Justice Planning, and local officials.* National Association of Counties Research Foundation, 1971.

President's Commission on Law Enforcement and Administration of Justice. The challenge of crime in a free society. Washington, Government Printing Office, 1967.

———: Criminal victimization in the United States: A report of a national survey. Field Survey II by Phillip H. Ennis, Washington, Government Printing Office, 1967.

Proceedings of the First National Conference on Crime Control. Washingotn, March 28-29, 1967.

Reiss, A. J., Jr.: Field Surveys III: Studies in crime and law enforcement in major metropolitan areas. Washington, Government Printing Office, 1967.

Slott, I., and Sprecher, W. M.: *Cost Effectiveness and Criminal Justice.* NILECJ, New York, 1972.

Smith, D. C., and Salerno, R. F.: *Use of Strategies in Organized Crime Control.* Baltimore, William and Wilkins, 1970.

South Africa, Commission of Inquiry into Matters Relating to the Security of the State, Pretoria, Statsdrakker Publishing, 1971.

Stewart, H. Wayne: *Drug Abuse in Industry.* Miami, Halos and Associates, Inc., 1970.

Training police as specialists in family crisis intervention. National Institute of Law Enforcement and Criminal Justice, Washington, Government Printing Office, 1970.

Urban Renewal Administration. Design objectives in urban renewal. Technical Guide #16. Washington, Government Printing Office, 1965.

U.S. Department of Defense. Rules for the avoidance of organization conflicts of interest. Washington, Government Printing Office, 1963.

Use of polygraphs as "lie detectors" by the federal government. Hearings before a subcommittee of the House Committee on Government Operations, 88th Cong., 2nd Sess., 1964.

Weelder, R.; *Progress and Revolution. A Study of the Issues of Our Age.* New York, International Universities Press, 1967.

PERIODICALS

Alexander, Donald G.: Is marijuana really dangerous? *Nation's Cities,* December 1969.

Alexander, Franz: Observations on organizational factors affecting creativity. *The Creative Organization,* Chicago, 1965.

Allport, Gordon: The American soldier. *Journal of Abnormal and Social Psychology,* LXV: 1959.

Barthol, R. P.: Learn how to make decisions. *Petroleum Engineer,* October 1960, p. 10-12.

Barton, John: Pseudo psychos. *American Mercury,* October 1956, p. 5-12.

Bauer, R.: Some trends in sources of alienation form the Soviet system. *Public Opinion Quarterly,* 19:275-291, 1955.

Beattle, R. H., and Kenney, J. P.: Aggressive crimes. *The Annals of the American Academy of Political and Social Science,* 1966.

Blum, R., and Osterloh, W.: The polygraph examination as a means for detecting truth and falsehood in stories presented by police informers. *Journal Criminal Law, Criminology and Police Science, 58:* no. 1, 133-137, 1968.

Boggs, Sara: Urban crime patterns. *American Sociological Review,* December 1965.

Bologna, G. J.: Drug abuse is everybody's business. *Industrial Security,* August 1971.

Bottom, Norman R., Jr., Ph.D.: Mind frozen in motion . . . graphology. *Security Management,* November 1974, p. 22-26.

Bowman, C. C.: Loneliness and social change. *American Journal of Psychiatry,* 1955, p. 194.

Bradford, Ralph, Lt.: . . . Handwriting comparison for the security officer. *Security World, 1:* no. 1, July 1964.

———: Technical courses: Elements of handwriting comparison, Part I. *Security World, 2:* no. 1, January 1965.

———: Technical courses: Elements of handwriting comparison, Part II. *Security World, 2:* no. 2, March 1965.

———: Technical courses: Elements of handwriting comparison, Part III. *Security World, 2:* no. 3, May 1965.

———: . . . Once too often. *Security World, 2:* no. 2, March 1965.

Brain monitor operates inside astronauts' helmet. *Science News Letter,* August 31, 1963.

Bromberg, W.: Liar in delinquency and crime. *Nervous Child, 1:* no. 4, 351-357, 1941.

———, and Keiser, S.: Psychology of a swindler. *American Journal of Psychiatry, 4:* no. 2, 1441-1458, 1938.

Brooks, S. H., and Whitman, I. R.: Transportation. *Data Processing Yearbook,* 1962-63, p. 177.

Brown, Brendan: Rising crime rate and education. *Catholic Educational Review, 62:* October 1964.

Browning, Norma Lee: Shady ladies. *The Chicago Tribune,* August 22, 1948.

Burnham, David: Bronx police aim at indoor crime. *The New York Times,* December 25, 1969.

———: Detectives here make arrests in less than 6% of robberies. *The New York Times,* December 20, 1970.

———: Fear of muggers looms large in public concern over crime. *The New York Times,* December 29, 1968.

———: Study discloses wide disparity in crime and police efficiency. *The New York Times,* February 1972.

Callahan, Richard A., and Sarmanian, Jack: Narcotics—a trip to nowhere. *Industrial Security,* April 1971.

Carroll, Joseph F., Lt. Gen.: Security role in engineering total peace. *Industrial Security, 1:* no. 31, July 1957.

Chapin, Barbara: Shh . . . the need for do-nothing quietude. *Recreation,* November 1964, p. 437.

Clark, John P., and Wenninger, Eugene P.: Socio-economic class and area as correlates of illegal behavior among juveniles. *American Sociological Review, 27:* no. 6, 826-834, December 1962.

Clark, Tom C.: Justice: The American heritage—your responsibility. *Industrial Security, 12:*24, October 1968.

Clarke, A. C.: The world of the communications satellite. *Astronautics and Aeronautics,* February 1964, p. 46-47.

Cockrell, Hulon D.: The truth about operation abolition. *Industrial Security,* April 1962, p. 12.

Cohen, Albert K.: An evaluation of Gault by a sociologist. Symposium on Juvenile Problems: In Re Gault, *Indiana Law Journal.*

Coleman, Lee: What Is American? A study of alleged American traits. *Social Forces,* 1941, p. 492.

Cordrey, J., and Pence, G. K.: Analysis of team policing in Dayton, Ohio. *Police Chief, 32:* no. 8, 44-45, August 1972.

Criminal statistics 1974. *Top Security,* September 1975, p. 186.

Dash, Samuel: Cracks in the foundation of criminal justice. *Illinois Law Rev, 46:* 1951.

Davies, M.: Offense behavior and the classification of offenders. *British Journal of Criminology, 9:*39-50, 1969.

Davis, James A.: Traffic management—a case history. *Industrial History,* June 1966, p. 20.

Decade of military operations research in perspective; symposium. *Operations Research,* November 1960, p. 798-860.

Dentler, Robert A., and Monroe, Lawrence J.: Social correlates of early adolescent theft. *American Sociological Review, 26:* no. 5, October 1961.

Derrick, C. D.: Interrogation by hypnosis. *The Police Chief,* March 1959, p. 26-29.

Dession, George: Psychiatry and the conditioning of criminal justice. *Yale Law Journal Review, XLVII:*319-340, January 1938.

Drug abuse: Don't do it yourself. *Occupational Hazards,* July 1971, p. 46.

FBI is on trail of drug pirates. *Cleveland Press News,* June 21, 1962.

Ferracuti, Franco, Hernandes, Rosita Perez, and Wolfgang, Marvin E.: A study of police errors in crime classification. *Journal of Criminal Law, Criminology, and Police Science, 53:*113-119, March 1962.

Fooner, Michael: Victim-induced criminality. *Science, 153:*1080-1083, 1966.

———: Some problems in evaluation of proposals for victim compensation. *International Criminal Police Review, 22:*66-71, 1967.

Foster, Reginald: Alcoholism and industry. *Security and Protection,* November 1974, p. 4-6.

Foster, Willard O., Jr.: The invisible alcoholic. *Industrial Security, 2:*9, December 1967.

Gay, William O.: Communication and crime. *Police Journal, 46:*109.

Gardner, John: Acts of God and days of grace. *Security Gazette,* September 1974, p. 342.

Ger Stacker, C. A.: Living with confrontation. *Security World,* September 1970.

Giffin, K.: The contribution of studies of source credibility to a theory of interpersonal trust in the communication process. *Psychological Bulletin,* No. 68, 104-120, 1967.

Gottschalk, L. A.: The use of drugs in interrogation. *The Manipulation of Human Behavior,* 1961, p. 96-141.

Greenberg, Frank: Violence today. *Security World,* April 1970, p. 16-20.

Gregory, G. H., and Warnock, William J.: The military drug abuse program: Time for change. *Industrial Security,* December 1971.

Guard and the worker, is antagonism inevitable? *Occupational Hazards,* July 1971, p. 44.

Harari, H., and McDavid, J. W.: Situation influence on moral justice: A study of "finking." *Journal Personality and Social Psychology, 11:* no. 3, 240-245, 1960.

Hartlep, Felix: How to keep from getting murdered. *American Weekly,* Hearst Publishing Co., Inc., New York, 1955.

Hazard, Geoffrey C.: The politicalization of youth crime. *Security World,* December 1960.

Heading off an energy crisis. *Nation's Business,* July 1971, p. 26.

Heiss, J. S.: The DYAD views and newcomer: A study of perception. *Human Relations, 16:* no. 3, 241-248, 1963.

Hoffeld, D. R.: Effect of incentive upon information use in a choice situation. *Psychology Reports, 13:* no. 2, 547-550, 1963.

Irving, J. F.: State planning agencies and crime control. *Police Law Quarterly, 1:* no. 4, 12-28, July 1972.

Jeffery, Ray C.: Criminal behavior and learning theory. *Journal of Criminal Law, Criminology, and Police Science, 56:* no. 3.

Karpman, B.: Lying. *Journal of Criminal Law, Criminology and Political Science, 49:* 135-137, 1949.

Kiesler, C., and Kiesler, Sara: The role of forewarning in persuasive communications. *Journal Abnormal Social Psychology, 68:* no. 5, 547-549, 1964.

Lee, Walter R.: Organized crime. *Canadian Security Gazette,* May 1971.

Lottier, Stuart: A tension theory of criminal behavior. *Amer Sociol Rev, 7:*840-848, December 1942.

Lucas, Ferris E.: Wanted: Change of attitude. *Amer County Gov,* February 1966.

Manis, M., and Blake, J. B.: Interpretation of persuasive messages as a function of prior immunization. *Journal Abnormal and Social Psychology, 66:* no. 2, 225-230, 1963.

Mendelson, Wallace: The neo-behavioral approach to the judicial process: A critique. *American Political Science Review, 57:*593-603, 1963.

Morgan, John J. B.: *The Psychology of Abnormal People.* New York, Longmans, Green and Company, 1928.

Mullins, C. J., and Force, R. C.: Rater accuracy as a generalized ability. *Journal Applied Psychology, 46:*191-193, 1962.

Network TV "riot policy." *Security World,* October 1970.

Neville, Haig G.: Letters to the editor. *Harvard Business Review,* May/June 1969, p. 40-42.

Newman, C. L.: War on crime. *Federal Probation, 20:* no. 4, 35-38, December 1966.

Nichols, John F., and Bannon, James D.: STRESS zero visibility policing. *Police Chief, 39:* no. 6, 1972.

Obstacles to redress for crime victims. *Security Gazette,* September 1974, p. 340.

Ploscowe, Morris: Crime in a competitive society. *Annals of American Academy of Political and Social Science, 217:*105-111, September 1941.

Polygraph's enemies, The. *Law and Order,* November 1964, p. 46.

Polygraphing politicians. *International Security Review,* September 18, 1972, p. 2.

Post, Richard S.: "Contemporary Protective Services," *Security Register,* Vol. 1, Jan.-Feb. 1974.

———: "Creating Corporate Freedom," *Industrial Security,* March 1972.

———: "Creating a Secure Environment," *Industrial Security,* December 1970.

Prostano, Emanuel T., and Piccirillo, Martin L.: Law enforcement: A selective bibliography. *Libraries Unlimited,* 1974.

Quinney, R.: Crime in political perspective.

American Behavioral Scientist, December 1964, p. 19-22.

Rawson, Ralph W.: Labor strife—a national dilemma. *Industrial Security*, August 1965, p. 28.

Reid, John E., and Inbau, Fred E.: Truth and deception, the polygraph (lie detector). *Technique*, Baltimore, 1966.

Reiss, Albert J., Jr., and Rhodes, Albert Lewis: Delinquency and social class structure. *Amer Socio Rev, 26:* no. 5, October 1961.

Rim, Y.: Machiavellianism and decisions involving risk. *British Journal of Social and Clinical Psychology, 5:* no. 1, 30-36, 1966.

Roclefs, H. M.: The tension of citizenship: Private man and public duty. New York, 1957.

Roebuck, J. B.: The Negro numbers man as a criminal type: The construction and application of a typology. *Journal of Criminal Law, Criminology and Political Science, 54:*48-60, 1963.

———, and Johnson, R.: The "short con" man. *Crime and Delinquency, 10:*236-248, 1964.

Roper, Elmo: How powerful are the persuaders? *Saturday Review, 40:* October 5, 1957.

———: Discrimination in industry. *Industry and Labor Relations Review, 5:*595, 1952.

Rud, F.: The social psychopathology of schizophrenic states. *Journal of Clinical and Experimental Psychopathology, 12:* 1951.

Ruebbhausen, O. M., and Erin, O. G., Jr.: Privacy and behavioral research. *American Psychologist, 21:,* and *Columbia Law Review, 65:*1184, 1965.

Rytten, J. E.: How to tap a telephone and recording and the police profession. *Law and Order*, January, March and May 1955.

Schurr, Robert: There is a way to cure addicts. *Industrial Security*, October 1971.

Schwenk, Edmund H.: Our readers inform us: German judicial precedent on the polygraph. *Security World, 2:* no. 5, September 1965.

Schwitzgebel, Ralph: *Street Corner Research.* Cambridge, Harvard University Press, 1964.

Smith, Burke M.: The polygraph. *Scientific American, 216:* January 5, 1967.

Smith, Philip A.: The impact of computers on psychological research. *SDC*, October 1950, p. 119.

Spence, K. W., and Farber, I. E.: Conditioning and extinction as a function of anxiety. *Journal of Experimental Psychology, 45:* 1953.

Stricker, L. J.: The true deceiver. *Psychological Bulletin, 63:* no. 1, 13-29, 1967.

Talese, Gay: Most hidden persuasion. *The New York Times Magazine*, January 12, 1958, p. 59.

Technical course: Elements of handwriting comparison, Part I. *Security World, 2:* no. 1, January 1965.

Technical course: Elements of handwriting comparison, Part II. *Security World, 2:* no. 2, March 1965.

Technical course: Elements of handwriting comparison, Part III. *Security World, 2:* no. 3, May 1965.

Thomas, John J.: The state of the art—1970. *Police Chief, 37:* no. 8, 63, August 1970.

Tosun, Oztekin: Economic crimes in Turkey. *Istanbul Universitesi Hakuk Fkaultesi Mecmuasi, 26:* nos. 1-4, 3-15, 1961.

Tulsa panel probes professionalism conduct. *Chemical Engineering*, December 12, 1950.

Who is the criminal? *Amer Sociol Rev, 12:*96-102, February 1947.

Why boys steal. *Atlantic*, October 1951.

Williams, G.: The correlation of allegiance and protection. *Cambridge Law Journal, 1:*54-76, 1948.

Wittles, David G.: Why cops turn crooked. *Saturday Evening Post*, April 23, 1949.

Zelkind, Joseph G.: Is your office part of the drug scene? *Administrative Management*, October 1970, p. 40.

UNPUBLISHED MATERIAL

Cerovsky, Stephen M.: Drug abuse in industry. Michigan State Masters paper, 1970.

Feavel, John S., Knuth, Louis S., and Lind, Lloyd H.: Probation versus incarceration for felony burglars. Master's thesis, Univ. of Wisconsin, 1968.

Hanewicz, Wayne: A critique of some assumptions apparently underlying social scientific analysis with particular reference to the field of criminal justice. Unpublished Michigan State thesis, Fall, 1968.

Kaplan, Barbara: Victimology: Analysis, evaluation and potential. Unpublished Ph.D. dissertation, John Jay, College of Criminal Justice, 1970.

APPENDIX 19A

BASIC PROBLEM AREAS

I. *Basic Research*
   A. What constitutes risk-taking behavior? How can it be measured? What are the effects of training, sensory devices, guard task performance, working environment, and physical condition of the facility on the vulnerability for loss?
   B. How can unsecure acts and/or conditions be measured? What is meant by secure or unsecure? Are these related to consequences or to the antecedent conditions?
   C. Can a systematic classification be made of unsecure acts that are likely to be repeated?
   D. To what extent do sudden physical or combined stress conditions affect the probability of experiencing a loss?

Consumer Protection Areas

1. Development of cost effectiveness studies for alarm/security systems
2. Effectiveness of security hardware (such as alarm systems)
3. Studies for alternative use of security systems, hardware or personnel

Education

1. Development of educational programs for security personnel
2. Development of academic programs for the administration of public and private security programs versus police science or law enforcement administration programs

II. *Environmental Design and Behavioral Control Research*
   A. What is the influence of supervision on individual worker security awareness? What kinds of supervisory characteristics are generally most effective?
   B. In what ways can communication through the use of posters, visual aids, exhibits, and signs be improved to increase employee motivation? What are the criteria for effective communication as applied to loss prevention?
   C. How can we design security equipment to complement the physical and psychological abilities of workers? How can we use, to better advantage, the principles of matching equipment design to physical abilities?
   D. How effective are inspections in maintaining secure conditions? Are there other methods which are better or more effective in determining unsecure conditions than either loss reports or inspections?

Architectural Areas

1. Development of architectural standards to:
   a. Design to reduce incidence of crime
   b. Design to provide a secure environment for conduct of business and activities
   c. Create an environment conducive to productivity and insure security
2. Development of new designs for high-rise residential apartment build-

ings to reduce incidence of crime and promote a psychologically pleasing environment

AREAS OF SECURITY WHICH REQUIRE RESEARCH AND INVESTIGATION

Legal/Social Areas

1. Legal aspects of retail and commercial security programs
2. Private justice systems (how private firms and security organizations adjudicate cases)
3. Licensing and regulation of private security services
4. Relationships of private security to public protective agencies in the community
5. Development of alternative models of public and private protection in a community
6. Reallocation of traditional police tasks to private security services
7. Invasion of privacy by private security as well as public law enforcement agencies
8. Invasion of privacy and relationship to overall social interaction
9. The social implications of current background investigative policies and procedures by both public and private security agencies
10. Additional studies on national data center (implications for individual freedom and privacy)
11. Development of computer simulation models of protection using various security components such as private police, public police, insurance, electronic hardware, internal control procedures, etc.
12. Psychological studies of the impact of security measures on employee morale and their effectiveness and efficiency
13. Development of a system for collecting information relative to currently nonreported criminal activities of an industrial or proprietary nature
14. Development of alternate models of public security as applicable to specific communities

III. *Management Decision-Making Research*

   A. What constitutes a reasonable loss prevention budget? What formulations should be used to arrive at this decision?
   B. How should management measure its security performance? Are present cost formulations valid estimates of loss costs?
   C. How dependable or significant are near losses in the evaluation of security status or loss potential of plants and/or areas?
   D. Among establishments of different sizes and types, how are the security programs best organized and administered? What are the distinguishing characteristics of "successful" programs?

Administration and Management

1. Development of standards for dishonesty and disloyalty in employee activities with respect to company property or trade secrets
2. Development of adequate means for evaluating security officer performance

IV. Systems Research

Areas of Security Where Technical Investigation Is Required

1. Development of test standards and specifications for:
   a. intrusion devices (sensors)
   b. communication links
   c. system control equipment
   d. alarm reporting systems
2. Develop procedures to check the effectiveness of installed alarm systems and specifications for certification.
3. Investigate trade-offs required to provide minimum false alarm rates for maximum system sensitivity and effectiveness.
4. Investigate and classify environmental noise and spurious responses. Use this information to update system designs to reject these signals.
5. Develop criteria to aid in deciding which type of sensor or which mix of sensors will be the most effective in a specific environment.
6. Develop criteria to aid in deciding which type of communications link and alarm reporting technique is best for a given application or level of security.

Engineering

1. Development of alarm hardware specifications for design application
2. Development of certification standards for intrusion detection systems
3. Development of low-cost noncompromisable security systems
4. Development of low-cost home security systems
5. Development of false alarm "free" security systems

Additional Investigation Areas

1. Radiation patterns within metropolitan areas as related to reliable radio communication.
2. Computerized monitoring and dispatch of police and guard vehicles.
3. Determination of judicial and penal costs associated with:
   a. crimes against property
   b. crimes against persons
4. Development of characterization parameters to classify organizations and/or communities according to nature of security risks and features of optimum security systems.
   a. dollar inventory
   b. attractiveness of inventory
      (1) personal use
      (2) sale for gain
      (3) portability
   c. degree of public access
   d. character of criminal interest
      (1) shoplifters
      (2) major organization
      (3) local groups
5. Cost-effectiveness of redundancy techniques
6. Develop modelling schemes for evaluating system effectiveness.

APPENDIX 19B

FUNCTIONAL PROBLEM AREAS*

ASSAULTS:
  Can an assaulted worker hold management responsible for failure to furnish guards?
  Can employees refuse night work for fear of assaults?
  Can a worker who is mugged sue the company for having its business in a high-crime area?
BACKGROUND INVESTIGATION:
  When should you check an employee's past life before promoting him?
  Can an employee be fired for preemployment misconduct?
  Can Uncle Sam delay a contractor while investigating its employees' backgrounds?
  May a job applicant be rejected on the basis of a confidential report?
BLACKLISTS:
  Is blacklisting a crime?
  What rights does an ex-employee have against a boss who threatens to blacklist him?
BOMB SCARES:
  Are workers entitled to be paid if sent home because of a bomb threat?
  Must the company notify its workers if it receives a bomb threat?
  Are repeated bomb threats an excuse for failure to meet production schedules?
  Can the company be sued by an injured worker if it ignores a bomb threat?
BONDING:
  Can an employee refuse to be bonded?
  What must you do to get your money back when a bonded employee steals?
  Can the company fire a worker if his bond is cancelled?
  Does the employee fidelity bond cover top-echelon dishonesty?
  What are the dangers in covering up employee defalcations?
BOOKKEEPERS & ACCOUNTANTS:
  Can you recover from the bank on checks raised by your bookkeeper?
  Is your accountant responsible for failing to detect employee embezzlement?
  Must an accountant tell IRS all he knows about you?
BRIBERY & KICKBACKS:
  If your purchasing agent was bribed, can you refuse to pay for the goods?
  Are kickbacks tax deductible?
  Is it a crime to bribe a plant worker for trade secrets?
  Can management close its eyes to the way its sales chief gets business?
  Can you make your employee return bribes he took?
  Is it a crime to accept a kickback?
BUGGING—WIRE-TAPS—PHONE INTERCEPTS:
  Can the security force plant mikes to ascertain the source of threats?
  Can a phone operator be fired for eavesdropping?
  Can you use a tracer gadget as evidence that an employee made anonymous phone calls?

---

* Reprinted from *Protection Management,* a twice-a-month newsletter published by Man & Manager Publications, 799 Broadway, New York, N.Y. 10003. Used with permission.

Can you instruct one of your employees to listen in on a phone conversation?
Can the switchboard operator monitor phone calls on the company's lines?
Can you use a secret tape recording to prove an ex-employee stole your process?
Can a tape recorder be used to get evidence against an employee?

BURGLARY & BURGLAR ALARMS:
Must the insurance company pay for burglary from an unlocked safe?
Does burglary insurance cover you for "mysterious disappearance"?
Can a guard sue the company if he is shot by a burglar?
Can your theft insurance be invalidated because your burglar alarm didn't work?
Must the insurance company pay if you forgot to turn on your burglar alarm?

CIVIL RIGHTS PROTESTS (See RIOTS):

CONFESSIONS:
If an employee gives a statement to the police, can his company use it against him?
Can a worker take "the fifth" when questioned by management?
Must a worker be advised of his rights before being questioned by management?
Is a confession obtained "behind closed doors" any good?

CRANK CALLS:
Are "voice prints" admissible to prove the source of crank calls?
Is it the company's business if an employee annoys a superior at home?

CREDIT CARDS:
If someone finds your credit card and goes on a spending spree, are you responsible?
How serious an offense is misuse of a company credit card?
How should management handle company credit cards held by departing employees?

CRIMINAL CHARGES:
Can you fire a worker who's been arrested and charged with a serious crime?
Must the company rehire an ex-employee who's on probation for a crime?
Can you drop the charges if an employee makes restitution?
Can you safely tell the employees a co-worker is charged with a crime?

CRIMINAL RECORDS:
What are the dangers in hiring a man with a criminal record?
Can you fire a worker for not revealing his police record during his hiring interview?
Must the company retain a worker who's awaiting a jail sentence?
When can you not fire an employee who's convicted of a crime?
Must you reinstate a veteran who does not have an honorable discharge?
Can you turn down a job applicant because he has a record of arrests?

DESKS & LOCKERS:
Is management liable for personal property stolen from company lockers?
Can the company make a new rule barring "early birds" from the locker room?
Can a worker's locker be searched without his/her okay?

DETECTIVES:
When can you legally use a detective in your business?
When can you hire private detectives to masquerade as employees?

DISCRIMINATION:
Can a job applicant be rejected as a security risk because he's a homosexual?
Is an alien permitted to work on a Government contract?

Can you ask a job applicant whether he/she ever changed his/her name?
Can you refuse to hire a civil rights activist?

DISHONESTY:
How much proof do you need to justify a discharge for dishonesty?
How promptly must you act on complaints of employee dishonesty?
Is an employee who accepts part of the loot as guilty as the thief?
Does possession of stolen property "prove" an employee guilty of theft?
Can an office temporary agency be held responsible for furnishing a dishonest employee?

DISLOYALTY:
Can your employee become one of your suppliers without disclosing it?
Can an employee who's secretly planning to leave urge others to do the same?
Can the company refuse to pay an employee because of his disloyal conduct?
How far can employees go in organizing a competing company behind your back?
Can management demand a "loyalty oath" from employees?
Can a worker be fired for testifying against his company in a lawsuit?

DRUG ABUSE:
Is it discrimination to refuse a job to an ex-drug addict?
What tests can you use to determine if an employee is a drug addict?
Can you require employees to attend lectures on drug abuse?
Can employees be frisked for drugs?

EMBEZZLEMENT–FORGERY:
Does the employee fidelity bond cover top-echelon dishonesty?
If an officer embezzles company income, must the corporation pay a tax on it?
How promptly must you examine your bank statements?
Can you demand the right to examine your employees' bank accounts?
What steps should you take to reduce the possibility of check forgery?

EMPLOYEE IDENTIFICATION—FINGERPRINTING:
Can the company require employees to wear name tags?
Can a worker be disciplined for refusing to sign in?
Can you insist that all job applicants be fingerprinted?

EMPLOYEE RAIDING:
Can you hire away an expert and pick his brains about your competitor's operations?
Can a departing executive urge his staff to come with him?
Can you bar a competior from hiring away your skilled workers?
When can you hire away an executive who is under contract to another company?

ENTRAPMENT:
Can management "go along" with a fraud scheme to catch the culprit?
Is it proper to "test" an employee's honesty?
Can you hire a private detective to pose as a worker to uncover employee pilferage?

EXECUTIVES–SUPERVISORS:
Can a director be barred from the plant premises?
If a company "big-wig" embezzles money, must fellow officers make it good?
How responsible is a supervisor for undetected pilferage?
Can the president call a stockholders' meeting to reveal top-echelon dishonesty?
Can an executive be sued for disclosing the salaries his company pays?

Can an executive be held personally liable for failure to institute an adequate security program?

FALSE ARREST:
Is a store liable for false arrest if its management violated company rules for handling shoplifters?
Can an employee sue for false arrest if questioned by management behind closed doors?
How long can you detain a suspect without being charged with false arrest?

FILES—DOCUMENTS—PLANS—CUSTOMER LISTS:
What personal files can an employee take with him when he leaves his company?
Can correspondence removed from your files by an employee be used against you?
Is a customer list always "confidential"?
Is a departing employee entitled to take away the contents of his desk?

FIREARMS:
Can you suspend a worker who carries a gun for which he has a permit?
How should you handle a pistol-packing employee?
Should management disarm its plant guards?
Is the company answerable for unnecessary use of firearms?
Does management have an obligation to institute a weapons safety program for its guards?

FRISKING:
Can an employee be fired if he refuses to empty his pockets?
Can the company inspect an employee's purse?
Can the company install an inspectoscope without the union's okay?
Can you adopt tough gate inspection rules to reduce pilferage without consulting with the union?

GAMBLING:
Can you adopt a rule barring all forms of gambling on company property?
Can management be prosecuted for permitting gambling on company property?

GOVERNMENT CONTRACTS—SECURITY RISKS:
Can you fire a worker who cannot get a security clearance?
Can a worker be denied a security clearance without the right to confront his accusers?
Can the company be sued for libel for calling an employee a security risk?
Can management require a loyalty oath before giving an employee clearance?

GUARDS—MISCONDUCT:
Can a guard with many years' service be fired for a first offense of sleeping on duty?
Does laxity in company rules bar firing a guard for misconduct?
Can a guard demand damages for the "humiliation" of being fired?

HANDWRITING EXPERTS:
Can management fire a worker on the testimony of a handwriting expert?
Can a suspect be required to furnish a sample of his handwriting?
How can a handwriting sample determine if an employee is a drug addict?

HIJACKING:
If goods are hijacked, who foots the loss: Trucker, shipper or consignee?
How promptly must you file a claim for a hijacking loss?
Can a trucker limit its liability for a hijacking loss?

HIRING PRACTICES—JOB APPLICATIONS:
Is it a crime for an employee to misrepresent his education?
Can a worker be fired for omitting his prison record from his application?
Can a customer sue you for carelessness in hiring an employee?

INDUSTRIAL ESPIONAGE—COMPETITORS:
When can you plant a spy in your competitor's business?
Is it a crime to bribe a worker for copies of his company's secret formulas?
Can you stop outsiders from taking aerial photos of your plant?
Can a competior copy your secret machines?
Are the executives liable if their company pirates a competitor's trade secrets?
Must an employee tell you what he saw in a competitor's plant?

INFORMERS—UNDERCOVER AGENTS:
Is discharge too severe for the employee who refuses to be a "squealer"?
Is an anonymous tip enough to hang a theft charge on an employee?
Does a worker have the right to confront his accusers?
Can you plant an agent in your warehouse to help reduce thefts?
Can you fire an employee on reports of an industrial spy?
Can you ask an employee to "keep an eye" on coworkers?

INSURANCE:
Can an insurance company refuse to pay because your employee was too easily frightened?
Can you sue your broker for failing to insure you?
How promptly must you notify the insurance company about a claim?
Can you still collect even though you violated the term of the policy?
What are the dangers of lying to the insurance company about the loss?

INVENTORY RECORDS:
Can the insurance company refuse to pay because your inventory records were stolen?
Can an outside accountant be held liable for failing to verify inventory records?
What steps should be taken to verify the accuracy of inventory records?

INVESTIGATIONS:
Is it dangerous to question customers at random to curb shoplifting?
How should management handle the questioning of fellow workers about an employee's honesty?
Can workers be disciplined for refusing to cooperate in an investigation?
Can a worker refuse to testify against coworkers charged with theft?
If an employee witnesses a crime, does he have a "duty" to report it?

LIBEL & SLANDER:
Can you get into trouble if you tell others you fired a worker for dishonesty?
Are you free to tell the government what you think about an employee?
How safe is it to speak frankly to employees?
Can the boss be sued for "telling off" employees he just fired?
Can an employee sue for slander for receiving a poor reference?
What should you say to a credit inquiry about a former employee?

LIE DETECTORS:
Has an employee a right to refuse a lie detector test?
Does a suspected worker have the right to demand a lie detector test?

LOAN SHARKS—RACKETEERING:
Can you discharge a "perfect worker" for being a loan shark on the side?

What should you do if racketeers try to "muscle in" on your business?
Can you fire an executive because he has mobster connections?
Should management discipline a worker who is in trouble with a loan shark?

MOONLIGHTING:
Can a worker be fired if he moonlights for a competitor?
Can you make a blanket rule prohibiting employees from working for competitors in any capacity?

OFF-DUTY BEHAVIOR:
Can you control the off-job behavior of an employee?
Can a worker be fired for the company he keeps?
Can you enforce a rule prohibiting off-plant drinking?

PARKING LOTS:
Are you liable if someone is assaulted on the company parking lot?
Is the company responsible for the theft of goods from employees' cars?
Do you have the right to remove unauthorized vehicles from the company parking lot?

PHOTOGRAPHING:
Can you demand photos from job applicants?
Can you photograph workers as part of a safety study?
Can you photograph workers distributing union literature?
Can you require workers to wear ID tags with their photographs?

PLANT LIGHTING:
Can the neighbors force you to stop illuminating your plant at night?
Can you sue the electric company if your plant is burglarized during a power failure?
Can a worker sue the company if he is mugged in a dark passageway?
Does management have an obligation to install an emergency power supply?

PLANT RULES:
Can you make a new rule barring "early birds" from the locker room?
Can you adopt strict gate inspection rules without consulting with the union?

POLICE—FBI:
Can you ask the FBI if a job applicant has a criminal record?
Will the FBI help businessmen to improve plant security?
Can you be held liable for delay in summoning the police?

PROSECUTION:
After accusing an employee of a crime, can the company withdraw the charge?
Can a worker be fired for theft although the evidence couldn't convict him?
If a fired worker is acquitted, must the company rehire him?

RIOTS—PROTESTS—DEMONSTRATIONS:
Are employees entitled to extra pay if their departure is delayed by a race riot?
Can you sue the city if your factory is burned by a mob?
Whom can you sue if your property is looted during a riot?
Is a riot an "act of God"?
Can you bar demonstrations outside your plant for fear of violence?
Can a worker take time off to participate in a demonstration?
Can you refuse to hire an applicant who was arrested in a civil rights protest?

SABOTAGE:
Can you discipline a group of workers for refusing to tell who was sabotaging production?
What evidence is sufficient to prove a worker guilty of sabotage?
Can a worker be fired for sabotage on circumstantial evidence only?

Is a plant guard's eye-witness testimony sufficient proof of sabotage?
Can you transfer an entire work crew to counter suspected sabotage?
SEARCHES:
Can you break into a worker's home to reclaim your stolen property?
Can employee's lockers be searched for missing tools?
SECURITY VIOLATIONS & CARELESSNESS:
If the help fails to lock up, can you still collect on your burglary insurance?
Can a negligent worker void your insurance coverage?
Is an employee right in quitting because his/her boss is careless about money?
Can a worker be disciplined for failing to report a security violation?
Can you deny promotion to a worker who is indifferent to security requirements?
SENTRY DOGS:
Are you liable if your guard dog attacks an innocent passerby?
Does the kennel which sells you a sentry dog warrant its efficiency?
STRIKES—PICKETING:
Can you order plant guards to photograph pickets?
What must you do to protect non-strikers from being assaulted?
SURVEILLANCE:
Can management single out an employee for "close observation"?
How discreet must you be in placing a suspect under surveillance?
TAX DEDUCTIONS:
Can you take a tax deduction on a theft from the company?
Can you get a tax deduction if an article is "lost"?
If an executive embezzles company income, must the corporation pay a tax on it?
THEFT & PILFERAGE:
Is discharge too severe for the worker who talked about stealing—but didn't do it?
What penalty is proper for a fifteen-year man who steals $5 worth of materials?
If an employee and supervisor are caught stealing, can the management man get a lesser penalty than the worker?
Can you fire a worker for stealing even though he does not walk out with the company property?
Can a worker be fired for stealing candy from a vending machine?
Must a worker who is caught stealing be paid his wages?
Is an employee a thief for using company property without permission?
TOOLS:
Is the company liable for the theft of employee tools?
Can employees be required to pay for lost tools?
Is discovery of tools in a worker's locker enough proof for discharge for theft?
TRADE SECRETS—PATENTS—INVENTIONS:
Can you block an ex-employee from revealing trade secrets without a written agreement?
Can employees who leave your company copy your "know-how"?
Can an executive be forever barred from disclosing a trade secret?
Can you stop a former employee from using a business idea developed in your office?
Are salaries a company pays its executives a trade secret?
Can an employee reveal company data if he was not told they were trade secrets?

Can an inventor who walks off the job be barred from working elsewhere?
Can you hire a technician to duplicate competitors' products?
UNIONS:
Can you fire a supervisor who would not spy on a union?
Is a union official entitled to enter your plant any time he wants to?
Can you fire a union steward who refused to allow his briefcase to be inspected?
Must the union be notified before an employee is fired for theft?
Can management deny union officers access to certain parts of its plant?
Can the company refuse to let a union examine its books?
VANDALISM:
Can you take a tax deduction if your property is damaged by vandals?
Does your insurance policy cover you for damage caused by vandals?
VIOLENCE—FIGHTING:
Can you use force to eject an abusive employee?
Can management discipline workers for off-premises fighting?
Can a worker collect compensation if injured during strike violence?

INDEX

A

Access points, 802
Act respecting official secrets, 601
Air transportation security, 192
Airport, security of, 403
Alarms, 250
   burglar, 453
   central station, 388
   components, 504
   definitions and terms, 550
   procedures, 504
Alcoholism, 233
American Society for Industrial Security, 20
Area control, 797
Armored car, 314
Army Criminal Investigation Command, 186
Army Intelligence Agency, 186
Army Security Agency (ASA), 185
Arrest, 109, 134
Arson, 865
   bibliography, 844, 847
Assault, 113, 281
Assistant Secretary of Defense, 183
Atomic Energy Act, 85
Atomic Energy Commission (AEC), 85, 188
Attorney General, organization list, 660
Auditing the security system, 852

B

Background investigation, 639
Bad checks, 232
Barriers
   animal, 503
   energy, 503
   human, 502
   natural, 502
   structural, 502
   systems, 501
   time delays on criminal attack, 477
Becker, T. M., 61
Bomb threats, 865
   operational policy, 871
Border patrol, 177
Braun, Michael, 133
Building, exteriors, 250
Bureau of Alcohol, Tobacco and Firearms, 181
Burglary, 233, 247, 250, 452, 866
   bibliography, 885
   checklist, 273

Business
   cost of crime, 230-231
   espionage, 795
   management for crime prevention, 245
   problems, 234
   resources, 246
   security, 220
   security bibliography, 308
Business and Defense Services Administration, 198
Business Terminal, security of, 403

C

Canadian security, structure, 201
Campus Security, 397
   data collection handbook, 425
   university security and safety guide (Syracuse), 441
Cargo security, 873
   bibliography, 881
Cash, theft, 262
Categories of security, 10
Censorship, office of, 199
Central Intelligence Agency (CIA), 172
Central Intelligence Group (CIG), 170
Central station, 313, 388
Checks
   cashing procedure, 259
   fraudulent, 258
Checklist, burglary, 273
Choice rationale, 739
Citizen and community action, 454
Civil Aeronautics Board (CAB), 192
Civil aviation security, 192
Civil disturbances, 866
Civil Service Commission, 189
Civil service war regulation, 77
Classification of public and private police, 62
Classified materials, distribution, 589
Clearance, 633
Clearance of foreign nationals, 638
Coast Guard, U.S., 192
Code of ethics, 339
Collective security programs, 457
College and university programs, 771
Cognizant security officers, 345
Commercial protective services, 311
Commercial security
   bibliography, 320
   competition, 316

growth and development, 315
training and education, 317, 769
Commission on Government Security, 198
Committee on Un-American Activities, 78
Committee to Combat Terrorism, 175
Committees to Coordinate Internal Security, 80
Community crime prevention, 451
Community patrols, 455
Community volunteers, 455
Company rules and policies, 690
Comptroller of the Currency Report of Crime, 423
Comptroller of the Currency Report on Security Devices, 420
Computer security, 807
  bibliography, 821
  case study—IBM's advanced administrative system, 857
Confidential, 588, 645
Confidential status of employee loyalty records, 80
Consolidated Federal Law Enforcement Training Center, 182
Constabuli, 29
Consultants, 314
Consumer credit reporting, 652
Contemporary security definitions, 13
Cost of crime prevention, 245
Cost of crimes, 230
Counterfeiting, 232
Courier services, 314
Courts, security of, 398, 447
Courts of Star Chamber, 30
Credit cards, 866
  bibliography, 886
Credit Reporting Agencies, 652
Crime, concealment, 241
Crime control, procedures, 247
Crime prevention
  community, 451
  cost, 245
  history, 459
  management, 245
  police, 455
Crime related problems, 233
Crimes, types, 232
Criminal Justice Information System, 594
Criminal opportunity, 451, 452
Curfew, 30
Customs security, enforcement responsibilities, **204**
Customs Service, U.S., 180

D

Defamation, 113
Defense Contract Administration, 343
Defense Contract Administration Services Region, 345

Defense Industrial Security
  adjudication division, 344
  administrative division, 344
  central index file division, 344
  clearance office, 344
  processing division, 344
  international programs division, 344
Defense Intelligence Agency (DIA), 184
Defense Investigative Service (DIS), 184
Defense Supply Agency Contract Administration Services, 185
Defensible space, 453
Definitions, security, 10, 11, 13, 14
Department of Agriculture, 189
Department of Air Force, 187
Department of Army, 185
Department of Defense, 184
  colleges, 183
  information security programs, 621
Department of Housing and Urban Development (HUD), 190
Department of Interior, 191
Department of Justice, 175
Department of State, 174
Department of Transportation, 191
Department of the Treasury, 179
Derogatory information, 632
Development of private sector of the criminal justice system, 47
Divisions of security, 15
Doors, 249
Drug abuse and alcoholism, 233
Drug Enforcement Administration (DEA), 178

E

Educational program development, 771
  bibliography, 775
Electronic protection equipment, 314
Embezzlement, 793
  bibliography, 837
  methods, 794
Emergency preparedness questionnaire, 377
Emergency services, 491
Emotional distress, 115
Employee screening, **629**
Employee theft, 261, 789
Employment, scope of, 115
Energy Research and Development, U.S., 188
Enforcement, 480
Environmental design, 453
Environmental Protection Agency, 193
Equipment, companies, 508
Espionage, business, 795
  industrial, 795
  statutes, 88

Index

Ethical considerations, 268
Ethics, 268, 319
Executive protection, 867
External loss, 227
External problems, 865
External security, 739
  bibliography, 881
Extortion, 867
  bibliography, 882

F

Facility security program, 198
False imprisonment, 117, 279
Federal Aviation Act, 92
Federal Aviation Agency (FAA), 191
Federal Bureau of Investigation (FBI), 176
  organizational chart, 177
Federal Communications Commission (FCC), 193
Federal Deposit Insurance Corporation (FDIC), 193
Federal Fire Council, 198
Federal Home Loan Bank Board, 194
Federal Maritime Commission, 194
Federal Power Commission, 194
Federal Protective Service, 401
Federal Reserve System, 195
Federal Trade Commission, 195
Fidelity bonds, 225
Fire protection
  bibliography, 844, 847
  inspection of equipment, 848
  sprinkler systems, 851
Force, use of, 118
Framework of security, 16
Functional problem areas, 905

G

General Service Administration (GSA), 195
Generic security functions, 471, 475
  bibliography, 495
Glossary, 550
Government programs, 341
  bibliography, 206
  development, 341
  organization, 341
  security programs, 170
Government security, history, 170
Governmental security systems, 167

H

Hammurabi Codes, 28
Hatch Act, 77
Historical development
  bibliography, 36
  governmental security, 170

History of crime prevention, 459
History of security, 27
  American development, 32
  ancient period, 27
  Anglo-Saxon, 28
  modern period, 31
  Norman period, 29
  present, 35
  ward and watch, 30
  Westminster period, 29
Hospital, health care and nursing home security, 398
Hudson, F. W., 459

I

Immigration and Naturalization Service, 177
Individual protection program, 449
Individual protection security, bibliography, 467
Industrial and Commercial Security Association of South Africa, Ltd., 21
Industrial defense, 168, 372
Industrial Emergency Plan Outline Against Civil Disorders, 383
Industrial espionage, 795
  bibliography, 813
Industrial Police and Security Association (England), 21
Industrial security, 168
  bibliography, 366
  programs, 94, 340
Information security, 478, 577
  accountability, 586
  bibliography, 597
  handling, 585
  initial marking, 585
  overclassification, 585
  proprietory applications, 594
  responsibility for safeguarding, 586
  restrictions, 586
  safeguarding, 587
  sources, 795
  transmittance, 585
Inspections, 484-485
  staff, 742
  types, 742
Institutional Protective Services, 394
  bank, 396
  principles, 394
  types, 396
Institutional security, bibliography, 404
Insurance, 222, 809
Issues, functional problem areas, 902
Inter-agency classification review, 199
Internal loss, 224
Internal organization for security, 737

Internal problems, 787
Internal Revenue Service (IRS), 181
Internal security, 738
  bibliography, 811
Internal Security Act, 81
International Security Associations, 20
Intelligence and the decision making process, 798
Interrogation, 138
Interstate Commerce Commission (ICC), 195, 196
Inventory shortages, 232
Investigations, 483
Investigative processes, 265
Issues in security, 891
  basic problem areas, 902
  bibliography, 895

J

Joint Chiefs of Staff, 183
Justice of the Peace, 30

L

Law
  arrest, 109
  assault, 113
  battery, 281
  bibliography, 156
  defamation, 113
  emotional distress, 115
  false imprisonment, 117
  malicious prosecution, 119
  physical evidence, 117
  polygraph, 121
  privacy, 120
  search, 124
  shoplifting, 125, 301, 302
  special police, 127
  tort, 278
  warnings (Miranda Rule), 128
Law Enforcement Assistance Administration (LEAA), 178
Lee, David J., 133
Legal authority, 51
Legal basis, 76, 96, 98, 99, 103
  commercial protective services, 98
  governmental security programs, 76
  private protective services, 96
Legal framework, 50
Legal guidelines for security officers, 109
Legal powers and limitations for private police forces, 133
Liability, legal basis, 103
  tort, 278
Library, security of, 400

Licensing and regulations, 319
Lighting, 298, 454, 503, 803
Local government, 30
Lock and key administration, 799
Locks, 248
Locksmiths, 315
Loss, categories, 223
  control, 868
  direct, 220
  estimate, 235
  external, 227
  high conditions, 223
  internal prevention, 224
  store manager talk, 275
Loss prevention in business, 221
Loyalty, 632
Loyalty review board, 79, 199

M

Magna Carta, 29
Malicious prosecution, 119
Marine Corps, U.S., 187
Marshalls, U.S., 179
McPherson, Marlys, 47
Military installations, 80
Military personnel security program, 84
Minimum security devices and procedures for Federal Reserve banks and state member banks, 409, 416
Minimum security measures by governmental agencies, 84
Museum, security of, 400

N

NASA, 346
NASA security program, 93
National Advisory Commission on Criminal Justice Goals and Standards, 894
National Agency Check, 639
National Association of Food Chains, 236
National intelligence structure, 171
National Labor Relations Board, 196
National Park Service, 191
National Science Foundation, 196
National Security Act, 79
National Security Agency (NSA), 188
National Security Council, 170, 171
National Security Resources Board, 199
Naval Intelligence, U.S., 186
Navy Cryptology, 187
Need to know, 631
Negligence, 104
Nuclear Regulatory Commission, 189
Number of persons in security, 35

Index

O

Occupational Safety and Health Act (OSHA), 804
Office of Civil Aviation Security, 192
Office of Industrial Security Defense Supply Agency, 342
Office of Law Enforcement (OLE), 179
Office of Special Investigations (OSI), 187
Office of Strategic Services (OSS), 170
Office security, 233
Order-maintenance, 19
Organization for security, 735
    bibliography, 743
Organization response to security, 736
Organized crime, 233
Organized Crime and Racketeering (OCRS), 178
OSHA, bibliography, 833

P

Park, security of, 400
Parking, 802
Personnel security, 479, 628
    bibliography, 646
    governmental, 631
    investigation, 639
    proprietary, 628
Physical barriers, 27
Physical evidence, 117
Physical protection, 533
Physical security, 478, 499
    barriers, 502
    bibliography, 510
    definitions and terms, 564
    survey, 500
    surveys, governmental, 540
    systems, 509
    systems components, 499
Pilferage, 788, 790
    bibliography, 838
Planning security
    bibliography, 673
    checklist, 679
    cycle, 670
    decision making, 665
    disaster planning, 672
    emergency management plan, 696, 700
    emergency planning program, 692
    framework of security policy development, 667
    guide to industrial civil defense planning, 724
    industrial defense plan against civil disturbance and sabotage, 702
    industrial sabotage, 722
    mechanics of protective services planning, 680
    organizational factors affecting protective service planning, 669
    policy development, 665
    process, 665, 666, 670
    security policy, guidelines, 666
    steps, 668
    survey, 681
Police and public, 464
Police crime prevention, 455
Political scientists and security, 47
Polygraph, 121
Postal Service, U.S., 197
Power to terminate employment, 78
Praetorian Guard, 28
Presidents' Foreign Intelligence Advisory Board, 199
Privacy, 120
Privacy Act of 1974, 607
Private police, in society
    regulation of, 67
    relationships, 56
    structure of system, 52
Private protective services, 8
Private Security Advisory Council (PSAC), 894
Private security programs, 458
Private Security Task Force to the National Advisory Committee on Criminal Justice Standards and Goals, 337
Probable cause, 284
Processes of security, 479
Program effectiveness, measuring, 228
Proprietary industrial program of security, 397
Proprietary security manual, 745
Proprietary security systems, 169
Prosecution, malicious, 281
Protection, individuals, 27, 449, 892
Protection of military installations, 80
Protective agencies, 5
Protective choices, 7
Protective functions, 471, 473
Protective Services, 5, 6, 76, 492
    choices, 450
    commercial, 311
    organization, 474
    processes, 475
    relationships, 891
    resources, 474
    systems, 76
    technology, 473, 475
Public and private police, classification, 62
Public building, security of, 401
Public law, 93, 579, 607
Public utilities, security of, 401

R

Rail terminal, security of, 403
Records protection, 596

Regulatory requirements for security officers, 130-132
Research areas, architectural, 902
  consumer protection areas, 902
  education, 902
  legal, 903
  management decision making, 903
  systems research, 904
Rifas, Richard A., 109
Risk management, 222
Robbery, 233, 251, 252, 868
  bibliography, 882
Royal Canadian Mounted Police, 202

S

Sabotage statutes, 88
Safeguarding information, criteria, 578
Safes, 249
Scott, Thomas, 47
Search, 124, 135
Secret, 588, 645
Secret Service, 180
Securities and Exchange Commission, U.S., 197
Security
  access points, 802
  area control, 797
  areas, 501
  associations, 20, 21, 22
  auditing, 852
  basic problem areas, 902, 905
  business, 220
  business bibliography, 308
  Canadian, 201
  categories, 10
  collective program, 457
  commercial, 311
  commercial, bibliography, 320
  competition, 316
  components of, 497
  computer, 807, 857
  divisions, 15
  enforcement, 480
  ethics, 319
  external, bibliography, 881
  firms, 311-313
  framework, 16
  functional activities, 736
  generic functions, 471
  generic functions, bibliography, 495
  growth and development, 315
  implementing, 223
  individual protection bibliography, 467
  industrial programs, bibliography, 366
  information, 478
  information, bibliography, 597
  inspections, 484, 485
  institutional, bibliography, 404
  internal, bibliography, 811
  introductory bibliography, 25
  investigations, 483
  issues, 891
  issues, bibliography, 895
  legal basis, 76, 96, 98, 99, 103
  legal, bibliography, 156
  licensing and regulations, 319
  office, 233
  organization, 735
  organization, bibliography, 743
  personnel, 479, 628
  personnel, bibliography, 646
  physical, 478
  physical, bibliography, 510
  physical design, 228
  planning and development, 665
  planning, bibliography, 673
  private programs, 458
  problem areas, 785
  procedures, 691
  processes, 479
  programs, 165
  role of, 396
  survey, 681
  training, 317, 318
  training and education, 768, 771-773, 778, 779
  training and education, bibliography, 775
Security and police, similarity, 472
Security classification of foreign countries, 578
Security control of air traffic, 92
Security definitions, 10, 13, 14
  descriptive, 13
  functional, 12
  historical, 11
  management, 12
  normative, 12
  psychological, 11
  sociological, 11
  structural, 12
Security education, industry and government, 772
Security for government employment, 83
Security officer forty hour basic course, 778
Security orientation and guidance for cleared employees, 761
Security protection, types, 478
Security services, 36
Security survey flow chart, 505
Security systems, governmental, 167
  industrial defense, 168
Securus, 10
Selective Service Systems, 197
Sensor-intrusion detection guide, 505

Shire-reive, 29
Shoplifters, apprehension, 277
    types, 255
Shoplifting, 125, 254, 868
    bibliography, 887
    combating, 257
    laws, summary, 301
    methods, 256
Shrink control, 276
Similarity of systems, 169
Situational security, 500
Slander, 280
Small business administration, 197
Small business, theft, 264
Social problems, 870
Software, 808
Special police, 127
    legal basis, 99
Sports facilities, security of, 402
State War Navy Air-Coordinating Committee (SWNACC), 170
Statutes, detention, 284
    espionage, 88
    evaluation, 290
    presumption of intent, 282, 283
    probable cause, 284, 286, 287
    sabotage, 88
Statutes affecting strikebreaking, 45
Statutes of treason in 1352, 30
Statutes of 1285, 29
Statutory encouragement of merchant efforts to apprehend shoplifters, 277
Strategic Services Unit (SSU), 170
Strikebreakers and guards in industrial disputes, use of, 37
Subversive activities control board, 82
Suitability, 633
Summary of state licensing, 130-132

T

Terrorism, 868
Top secret, 587, 644

Tort law, 97
Total loss control, 16
Theft, cash, 262
    curbing, 263
    employee, 261
    merchandise, 262
Trade associations and conferences, training, 772
Traffic management, 802
Training, 317, 318
    bibliography, 775
    literature and materials, 772
    programs, 769
    program development, 773
    proprietary schools, 772
    standards, 771
    suggested curriculum, 778, 779
    trade associations and conferences, 772
Transportation, 870
Transportation terminals, security of, 403

U

U.N. and other international organizations, 82
U.S. Intelligence Board (USIB), 172
Urban cohorts, 28

V

Vandalism, 233
Vehicle control, 802
Vessels and harbors, safeguarding, 82
Veterans Administration (V.A.), 198
Vigiles, 28

W

Warnings (Miranda Rule), 128
What is security, 9
White collar crime, 221, 787
    bibliography, 811
    categories, 243
    elements, 238
    intent, 239
Windows, 249
Working definitions of security, 14

DATE DUE